The Phototrophic Prokaryotes

The Phototrophic Prokaryotes

Edited by

Günter A. Peschek
Wolfgang Löffelhardt
Georg Schmetterer

University of Vienna
Vienna, Austria

Kluwer Academic / Plenum Publishers
New York, Boston, Dordrecht, London, Moscow

Library of Congress Cataloging in Publication Data

The phototrophic prokaryotes / edited by Günter A. Peschek, Wolfgang Löffelhardt, Georg Schmetterer.
    p.   cm.
    Includes bibliographical references and index.
    ISBN 0-306-45923-X
    1. Photosynthetic bacteria. 2. Cyanobacteria. I. Peschek, Günter A. II. Löffelhardt, W. III. Schmetterer, Georg.
QR88.5.P53   1998   98-35189
579.3'8—DC21   CIP

Proceedings of the Ninth International Symposium on Phototrophic Prokaryotes,
held September 6 – 12, 1997, in Vienna, Austria

ISBN 0-306-45923-X

© 1999 Kluwer Academic / Plenum Publishers, New York
233 Spring Street, New York, N.Y. 10013

10 9 8 7 6 5 4 3 2 1

A C.I.P. record for this book is available from the Library of Congress.

Printed in the United States of America

PREFACE

This book contains the Proceedings of the Ninth International Symposium on Phototrophic Prokaryotes (IXth ISPP) which was held in Vienna, Austria, from September 6–12, 1997. In 1973, the far-sighted efforts of Gerhart Drews, Roger Y. Stanier, and Norbert Pfennig launched the first ISPP as a joint forum for scientific discussion of all aspects of research on phototrophic prokaryotes, both anoxygenic and oxygenic (cyanobacteria or blue-green algae). The ISPP International Scientific Committee, a rather loose community of scientists (see page vii for a list of current members), has since been responsible for the organization of the ISPPs on a triennial basis, without any dedicated financial support from a society, membership fees, or similar. The success of the ISPPs is therefore due to the continued enthusiasm of the scientific community for such a forum, as exemplified by the rise in attendance from 79 (1st ISPP) to more than 400 active scientists (plus dependants) from all five continents in Vienna. Comparing the Book of Abstracts of 1973 with that of 1997 reflects the tremendous change that has occurred in the last 25 years in scientific research on phototrophic prokaryotes. Two noteworthy examples of such change are the significant shift in emphasis from anoxyphototrophs to cyanobacteria and the advent of molecular genetics, culminating at the 9th ISPP in the first complete genomic DNA sequence of a phototrophic prokaryote, viz. the cyanobacterium *Synechocystis* sp. PCC6803.

In contrast to all previous ISPPs (see list on page ix), the Vienna meeting is the first one for which a Proceedings volume is issued due to a proposal by Plenum Press and the interest expressed by many participants. The editors would be pleased to see that this becomes a regular institution at future ISPPs. In order to commemorate the 25th anniversary of the ISPPs, this book starts with a chapter on "Historical Perspectives" by three eminent scientists who were already active at the time of the 1st ISPP. The other scientific papers are grouped into chapters 2–7 representing the major research areas in the field. Although the integration of 94 articles into this Proceedings volume was time-consuming and laborious the editors wish to express their hope that the book will contribute to the dissemination, both broadening and deepending, of the interest and eagerness, notably among young colleagues, in the scientific research on this huge group of microorganisms. After all, the biosphere—including human beings—does owe its present shape to them, the phototrophic prokaryotes. Last but not least, we want to express our gratitude to Plenum Press for cooperation and help during the whole process of producing this book.

The Editors

ISPP INTERNATIONAL SCIENTIFIC COMMITTEE

Structure and Development

C. Dow (UK)
R. Feick (Germany)
R. C. Fuller (USA)
J. Golden (USA)
K. J. Hellingwerf (NL)
K. I. Takamiya (J)

Genetics and Molecular Biology

D. A. Bryant (USA)
F. Daldal (USA)
J. Elhai (USA)
N. Tandeau de Marsac (F)
R. Tabita (USA)
J. Willison (F)

Experimental and Physiological Ecology and Taxonomy

P. Caumette (F)
Y. Cohen (IL)
M. Herdman (F)
H. W. Paerl (USA)
L. Stal (NL)
S. Ventura (I)

Metabolism and Bioenergetics

C. J. Howe (UK)
J. Meeks (USA)
J. Ormerod (N) (Chair)
G. A. Peschek (A)
H. G. Trüper (Germany)
D. Zannoni (I)

THE INTERNATIONAL SYMPOSIA ON PHOTOTROPHIC PROKARYOTES (ISPPs)

1st ISPP, September 1973, Freiburg (Germany); Drews, Pfennig & Stanier
2nd ISPP, September 1976, Dundee (UK); Codd & Stewart
3rd ISPP, September 1979, Oxford (UK); Carr & Nichols
4th ISPP, September 1982, Bombannes-Bordeaux (F); Cohen-Bazire
5th ISPP, September 1985, Grindelwald (CH), Zuber et al.
6th ISPP, August 1988, Noordwijkerhout (NL), Mur et al.
7th ISPP, July 1991, Amherst (USA), Fuller et al.
8th ISPP, September 1994, Urbino (I), Zannoni et al.
9th ISPP, September 1997, Wien (A), Peschek et al.
10th ISPP, September 2000, Barcelona (E), Guerrero et al.

CONTENTS

Part 3: Electron Transport and Bioenergetics

Part 4: Genome Analysis and Molecular Biology

Part 7: Phylogeny, Taxonomy, and Evolution

RESEARCH ON PURPLE BACTERIA DURING TWENTY-FOUR YEARS OF INTERNATIONAL SYMPOSIA ON PHOTOSYNTHETIC PROKARYOTES

Gerhart Drews

Institut für Biologie 2, Mikrobiologie
Albert-Ludwigs-Universität
D-79104 Freiburg, Germany

PHOTOSYNTHESIS IN THE 19TH CENTURY

Photosynthesis was discovered by its metabolic aspect, the CO_2 assimilation and the O_2 production of green plants as a light-dependent process catalyzed by chlorophyll (J. Priestley 1772, J. Ingenhousz 1779, N. T. de Saussure, 1804, H. Dutrochet 1837). Justus Liebig demonstrated 1840 that plants can grow on pure inorganic nutrients (HCO_3^-, NH_4^+, SO_4^{2+} and trace elements). The law of maintenance of energy and the unit of heat (calorie) were discovered by J. Robert Mayer (1845), J. P. Joule (1818–1889) and Hermann L. Helmholtz (1821–1894), at the middle of the century, but the energetic aspect of photosynthesis, i.e. the transformation of light energy into chemical energy, was recognized many decades after demonstration of CO_2 fixation. Wilhelm Engelmann wrote 1888 "The purple schizomycetes (bacteria) step in the group of organisms which assimilate like green plants. The bacteriopurpurin (bacteriochlorophyll) is a true chromophyll, so far the actual absorbed light energy is transformed in potential chemical energy".

RESEARCH ON PHOTOSYNTHESIS UNTIL THE EARLY SEVENTIES

The morphology and some physiological aspects of several species of purple bacteria, green bacteria and cyanobacteria (schizophyceae) have been described in the second halve of the 19th century. But the anoxygenic and oxygenic type of photosynthesis and the main physiological and biochemical properties of these organisms were detected not before the first decades of this century. In the fifties and sixties the research and the progress

The Phototrophic Prokaryotes, edited by Peschek *et al.*
Kluwer Academic / Plenum Publishers, New York, 1999.

in that field increased exponentially. This should be demonstrated by few examples mainly taken from research on purple bacteria.

The chemical properties of chlorophyll were described 1913 by Willstätter and Stoll. C.B. van Niel developed 1929 the concept of photosynthesis as a light-induced oxido-reduction process. S. Ruben, M. Randall, M.D. Kamen and J.L. Hyde proved 1941 that the oxygen generated during photosynthesis of plants comes from water and not from CO_2 as assumed during the last century until the period of Warburg.

E.L. Smith isolated 1938 chlorophyll-protein complexes. The photosynthetic membranes, the thylakoids, were discovered 1949 by A. Frey-Wyssling and K. Mühlethaler using electron microscopical techniques; E. Steinmann described 1952 the structure of chloroplasts. The intracytoplasmic membranes of cyanobacteria and purple bacteria were shown 1955 by W. Niklowitz and G. Drews using the same technique. These and other studies opened the way for structural investigations.

In 1948 T. Förster proposed the theory of exciton transfer between light-harvesting pigments. In 1952 L.N.M. Duysens detected the primary photochemical reaction, the bleaching of P890, by flash-induced absorption spectroscopy. The analysis of the intermediates and the kinetics of the charge separation in reaction centers was improved 1955 by introduction of the repetitive light pulse spectroscopy by H.T. Witt. The transduction of light energy in chemical energy by photophosphorylation was discovered 1954 with bacterial chromatophores by A.W. Frenkel and 1959 with plant preparations by D.I. Arnon. The pioneering biophysical studies initiated together with biochemical analyses the research on electron transport.

G. Cohen-Bazire, W.R. Sistrom and R.Y. Stanier observed 1957 the coordinated synthesis of bacteriochlorophyll and protein in purple bacteria after lowering of oxygen tension. The accompanied formation of intracytoplasmic membranes by invagination from the cytoplasmic membrane was described 1962 by P. Giesbrecht and G. Drews. These physiological and morphogenetic studies were the begin of research on cell differentiation of photosynthetic bacteria.

In 1960 R. Hill and F. Bendall proposed the Z-scheme of oxygenic photosynthesis. The concept of the chemiosmotic theory, which explains the formation of the proton motive force and its utilization for ATP production, was proposed in 1961 by P. Mitchell. R.K. Clayton claimed 1963 the different functions of light-harvesting and reaction center bacteriochlorophyll; however, it was known before that the mass of chlorophyll did not participate in the CO_2 assimilation of plants. With these and other work the different partial processes of photosynthesis (light absorption and exciton migration, charge separation, electron transport and $NADP^+$ reduction, formation of the electrochemical proton gradient across the membrane and photophosphorylation) have been acquainted.

The intramembrane particles, the functional units of the photosynthetic apparatus, were visualized by freeze-fracture electron microscopy 1966 in intracytoplasmic membranes of *Rhodopseudomonas viridis* by P. Giesbrecht and G. Drews. In the same year subchromatophore particles were solubilized by detergent treatment and isolated by A.F. Garcia, L.P. Vernon and H.I. Mollenhauer from *Chromatium*. Photochemically active reaction centers were isolated from purple bacteria 1969 by D.W. Reed and R.K. Clayton. In the same year J.D. Mc Elroy, G. Feher and D.C. Mauzerall described the formation of a bacteriochlorophyll radical cation during photooxidation of the primary donor P870 in reaction centers.

J. Aagaard and W.R. Sistrom showed 1972 that the reaction center and the light-harvesting complex I (B870) form a photosynthetic unit of constant size in purple bacteria, while the amount of B800–850 relative to the reaction center was variable. The proteinous subunit structure of the bacterial reaction center of *Rhodobacter sphaeroides* was analyzed 1974 by M.Y. Okamura, L.A. Steiner and G. Feher. Bacteriopheophytin was discovered

1975 as the primary photochemical intermediate state in bacterial reaction centers of type II by P.L. Dutton et al. and J. Fajer et al.

This enormous progress in about 1–2 decades of which only few data could be mentioned, required a platform for communication and exchange of results. This led to the organization of the first symposium on bacterial photosynthesis in Yellow Springs, Ohio, organized in 1963 by H. Gest, A. San Pietro and L.P. Vernon. The first international congress on photosynthesis followed 1968 in Freudenstadt, Germany and the first international Symposium on photosynthetic prokaryotes in Freiburg 1973.

THE PROGRESS DURING TWENTY FOUR YEARS OF ISPP

During the period between the first and the ninth ISPP a large progress in research on photosynthesis was achieved. This should be illustrated by few examples of research on the structure, function and development of the photosynthetic apparatus of anoxygenic photosynthetic bacteria:

Reaction Center and Light-Harvesting Complexes

Reaction center and light-harvesting complexes of several bacteria were purified, the stoichiometry of pigments, proteins and cofactors was analyzed and the primary structure of proteins determined. Computer search, topographic studies and spectroscopic analyses predicted structure models describing the topography of polypeptides with membrane-spanning α-helical regions and conserved pigment-binding domains keeping the pigments in a defined localization. The discovery of chlorosomes by electron microscopy, the detection of substructures by freeze-fracture technique and the development of a chlorosome model initiated studies on the migration of excitation energy and on the molecular organization of chlorosomes and pigment–pigment interactions therein. The coronation of the work on pigment-protein complexes was the determination of the atomic structure of two reaction center and two light-harvesting complexes and the confirmation of the model for the reaction center-light-harvesting B870 core complex by high resolution electron microscopy. These structures seem to be organized in all purple bacteria by the same principles.

The progress in techniques of recombinant molecular genetics paved the way to determine the photosynthetic genes and the amino acid sequence of photosynthetic proteins and made it possible to clone, mutate and express genes, in order to study the function of single amino acids and conserved sequences. The big progress in different techniques of spectroscopy allowed to record electron transport and exciton transfer in the femto second time scale and to determine the intermediates of charge separation in reaction centers and led to a detailed understanding of the kinetics and efficiency in formation of the electric field and the redox potential difference across the membrane. We learned that the polypeptides are not only the scaffold for the pigments and other cofactors but that they are involved in the processes of electron-, proton- and exciton transfer. New techniques, e.g. time-resolved, Fourier transformed I.R spectroscopy will support the studies on the role of proteins in these kinetic processes.

From the point of evolution it was a great progress of comparative biochemistry to find out that in all photosynthetic organisms only two types of reaction centers are present. Organisms with oxygenic photosynthesis, these are the green plants, algae and cyanobacteria, contain two photosystems working in sequence, the type II, the quinone-photosystem II with the water-splitting system and the type I, the iron sulfur type photosystem I, reducing NADP$^+$. Type II reaction center without the water splitting system is the only re-action center of all purple bacteria and *Chloroflexus*. The type I reaction center is present in green sulfur bacteria and Heliobacteria. From the great homology of amino acid se-quences of all reaction center proteins and the principal similar organization and function of cofactors it was predicted that all reaction centers originate from one ancient type.

The evolution of the light-harvesting system resulted in much more types in adapta-tion to different environments. It is interesting that photosynthetic prokaryotes—in spite of their strong homology in reaction center organization - do not belong to one taxonomic group but are spread over numerous branches of the evolutionary tree.

Assembly of the Pigment-Proteins in the Membrane and Regulation of the Biosynthesis

At the first ISPP it was known that oxygen partial pressure and light intensity effect the rate of bacteriochlorophyll synthesis and the protein pattern of the membrane during the morphogenesis. But the interaction between specific proteins and cofactors was un-known. Today we know the structure genes of those polypeptides which are coordinately synthesized together with the pigments and found out how they bind and interact with pig-ments and redox carriers. Several steps and mechanisms of transcriptional and post tran-scriptional processes in regulation of the formation of the photosynthetic apparatus have been elucidated. We learned that a complex regulatory network is responsible for the fine-tuned biosynthetic processes. The knowledge of the assembly process, i.e. the targeting of polypeptides to the membrane, their insertion, folding and binding to the pigments and co-factors is in its infancy. There are numerous helper polypeptides, of which we know only few, which support this process. I believe it is an exciting period to study the morphogene-sis of the bacterial photosynthetic apparatus during adaptation to the environment in its molecular events. The purple bacteria which are facultatively active in photosynthesis are excellent model systems to study cell differentiation.

These few examples showed the progress in a small field of research on anoxygenic prokaryotic photosynthesis during the 24 years of ISPP. The major outcome of these sym-posia was the possibility to have a platform for discussion of all aspects of physiology of photosynthetic prokaryotes, which is not or only in restricted form available on the inter-national congress of photosynthesis or other specialized meetings. It is very clear that the number and types of meetings, where special topics of photosynthetic prokaryotes are dis-cussed, will increase in the future but I hope that the ISPP will continue to use one lan-guage which is understandable for all participants to fulfill its major function to be a forum for the discussion of general ideas and overlapping problems.

REFERENCES

Overview reviews with historical aspects:

Bulloch, W. (1938) The history of bacteriology, Oxford University Press, London
Huzisige, H. and Ke, B. (1993) Photosynth. Res. 38, 185–209
Nordensskiöld, E. (1926) Geschichte der Biologie, Gustav Fischer Vlg., Jena

Original paper:

Engelmann, T.W. (1888) Bot. Ztg. 46, 661–669, 677–689, 693–701, 709–720

Special references mentioned in the text can be requested from the author.

FREIBURG TO VIENNA—LOOKING AT CYANOBACTERIA OVER THE TWENTY-FIVE YEARS

Noel G. Carr

Department of Biological Sciences
University of Warwick
Coventry, CV4 7AL, United Kingdom

In following an invitation to present an overview of the last twenty-five years of cyanobacteriological research it will come as no surprise that my first response will be some kind of general disclaimer with respect to synoptic balance or indeed total accuracy of what follows. The best that one can hope for is a personal, hopefully dispassionate, version of some of the developments and constraints that have happened in our own field during what has surely been a second golden age of biology, coming more or less a century after the first. I will advance an argument that, separate from the revolution of molecular biology that has pervaded virtually the whole of biology, there have been two themes of particular importance which have changed the way in which we look at cyanobacteria and indeed many other bacteria. Firstly there has been a tighter integration between the understanding of biochemistry and ecology. The role of the organism in its natural environment and the ways in which it interacts with other microbes increasingly forms the background in which laboratory experiments are designed. This has consequences with regard to the use and value of "type species" and this will be alluded to further. More importantly, there has been the emergence of the need to understand the evolutionary significance of our knowledge of bacteria. This has arisen directly from the triumph of molecular phylogeny, which has provided for the first time a coherent measure of relatedness. Dobzhansky's well known dictum about understanding biology must now be applied to microorganisms and is perhaps particularly appropriate to cyanobacteria.

As a generalization one could say that when these Symposia started there was much greater confidence in what particular groups of organisms could or could not do in metabolic terms. Possession of particular light harvesting structures, capacity for anaerobic nitrogen-fixation, respiratory use of organic molecules, employment of particular electron donors in photosynthesis—all these and many others firmly defined the identity of the species within the photosynthetic prokaryote group. Nowadays allocation of such metabo-

The Phototrophic Prokaryotes, edited by Peschek *et al.*
Kluwer Academic / Plenum Publishers, New York, 1999.

lic properties is much less secure and many examples in the literature illustrate this. The most striking being the discovery, by Shilo and his group, of the cyanobacterium *Oscillatoria limnetica* during their study of the Solar Lake near Eilat (1). This was the first description of an organism that could shift between oxygen-evolving and anoxic photosynthesis according to environmental change and was later shown, additionally, to have a degree of fermentative capacity. This was surely the ultimate expression of the concept of the unity of the photosynthetic process that was proposed by van Neil in the 1930s. All this has made the naming of organisms and the value of "type species" more uncertain. Furthermore the development of a phylogeny based on 16S rRNA sequence has challenged profoundly ideas of photosynthetic bacterial phylogeny, although it is of interest that this is much less the case with the cyanobacteria, which remain a coherent phylogenetic group..

Twenty five years ago the term "blue green algae" was still prevalent and its use, however historic, disguised the most important evolutionary connection of these organisms, i.e. to the chloroplast. "Cyanobacteria" came to be adopted fairly rapidly, and usually without too much rancour, soon after the arguments for its use were presented by Stanier. The recognition of the endosymbiotic origin of chloroplasts, long supposed by the more perceptive of cell biologists, soon came to be, to use Doolittle's nice phrase, "what all right-thinking people now believe". This had come about by the early application of molecular biology to understanding evolutionary development. The acceptance of the endosymbiotic origin of eukaryotic cells is a story in itself and has been extensively reviewed. More recently, emphasis has moved to understanding the origin of the eukaryotic cell itself (see 2). The conclusions were largely driven by considerations of a cyanobacterial origin of chloroplasts. The establishment of the endosymbiotic theory must be the major advance in cell biology of our time. The recognition of the unitary nature of oxygen-evolving photosynthesis led to the increasing use of cyanobacteria in photosynthetic studies (evident from elsewhere in this Symposia) and to the rather clear understanding that we now have of the process in terms of the chemi-osmotic hypothesis. It is worth remembering that this conceptual advance largely associated with Mitchell and his ideas was, like the endosymbiotic origin of cell organelles, bitterly resisted for many years by much of the scientific community. Much of the enjoyment of science is the observation of the destruction of cherished beliefs and it is therefore worthwhile occasionally to ponder on which of our present accepted views will in course of time prove to be quite wrong.

I will mention two other examples of assumptions whose overturn during the past 25 years have proved rather surprising, at least to me. One is mainly of ecological significance and the other concerned with physiology. The discovery about twenty years ago that a significant proportion of the primary production of the open oceans was driven by a group of hitherto unrecognised unicellular cyanobacteria was of great moment to biological oceanography and extended the global role of these organisms by a vast scale (see 3).The original observation and isolation of phycoerythrin containing cyanobacteria were from the North Atlantic but subsequent isolates from around the world have been shown to be rather similar and phylogenetically grouped away from other unicellular cyanobacteria. The importance of these unicellular cyanobacteria was not only in adding to our knowledge of the engines of primary production but also in emphasising the role of the "microbial loop". Because the cyanobacteria are less open to predation (due to their size) there is recycling and mineralising of a significant proportion of oceanic productivity at the microbial level rather than it entering the accepted food chain.

The second change which has certainly surprised me, has been a direct result of the application of molecular genetics. Mutations, until recently, were produced by a chemical or physical agent and were defined by their phenotype. For many years the analysis of mutated strains has been a major tool, especially in the understanding of metabolic pathways and their control. An essential part of this process was the isolation of the mutant from the wild type, and this was usually achieved by some form of "elective" culture which preferentially favoured the growth of the mutant. Fine, everybody knows this. More recently alteration in genetic capacity has been generated by the precise elimination of a specific gene by a process of gene insertion or deletion. Some "mutations" thus produced would be expected to be without apparent phenotype, just as it was always recognised that many chemically induced mutations were silent. But the deletion of certain specific genes with well characterized products has produced surprising results in that the disruption of a known metabolic process has limited physiological effects. Two laboratories have recently reported the deletion of the *zwf* gene which codes for glucose-6-phosphate dehydrogenase, a key enzyme in the oxidative pentose phosphate pathway of glucose dissimilation and accepted for many years to be central in cyanobacterial carbon dissimilation. In a *Nostoc* sp. the result of *zwf* deletion was limited to preventing growth only in the absence of fixed nitrogen (4) and in the unicellular *Synechococcus* sp. the only phenotypic change was the loss of viability following prolonged incubation in the dark (5). There was no reduction in dark respiration and it is evident that glucose dissimilation was achieved by another, possibly the glycolytic, route. It would appear that metabolism is not the tightly integrated process of thousands of enzymes working in perfect balance as indicated by those metabolic maps on our office walls but is something much looser and more flexible. As increasing, precise alterations in genetic composition are achieved we will begin to see how what proportion of genetic information is necessary for growth in existing environmental conditions and how tight is the metabolic adjustment that has evolved to those conditions.

The 16S and 18S rRNA sequence based unrooted tree that relates Bacteria and the Archaean to the Eukaryotes is a frequently used image and its casual use belies its profound importance. As one who has observed from the sidelines the development of the sequence data that has allowed the contruction of this visual representation of phylogenetic relationships, I would suggest the unrooted tree is an appropriate icon for modern biology. The ways in which the ideas of microbial evolution prevalent when our Symposia began (6) have radically changed with the recognition of the Archaean (2) and have been far reaching. Some long established bacteria groups such as the pseudomonads and purple photosynthetic bacteria have been shown to have no coherent phylogenetic unity. The cyanobacteria however emerge intact forming a distinct lineage, together (as discussed in this Symposium) with the recently discovered *Prochlorococcus* that lack phycobilin based light harvesting antennae—so much for our old friends the "blue-green algae"! What is there to learn from the fact that the cyanobacteria have evolved as a rather tight phylogenetic group, especially in view of their apparently ancient evolutionary origin? Recently, Castenholz has discussed this in some detail and formulated some interesting suggestions centred around aspects of their cell biology and ecological niche (7). Another aspect to be considered may lie in the way in which the majority of cyanobacteria that are used in laboratory work have been brought into culture. The elective culture technique that was devised in the 19th century has been extensively employed and for cyanobacteria means light, autotrophic nutrition and often the absence of fixed nitrogen. The elective procedure itself defines the properties of the organisms that are isolated from a particular environment. Recently the passive technique of amplification of the 16S rRNA gene and comparison of the derived sequences has been used to recognise organisms without their isolation. Results confirm that there is a significant proportion of uncultivated bacteria in natural environments some of which do not reside in any of the so far established phylogenic groups. We know that in some natural environments only a very small proportion of microorganisms can be brought into culture, the open oceans are a good example. Whereas the 16S amplification/sequence determination strategy will tell us where organisms from such an enviroment lie within any defined lineage, such as that of the cyanobacteria, it does not tell us anything of their physiological properties. There is no reason to exclude the possibility that organisms may exist which are phylogenetically located within the cyanobacteria group but may have properties which are quite different from those that we associate with the cyanobacteria that are in culture.

There is no doubt that the organisms that are brought into culture, and on which our knowledge is very largely based, are a less than perfect representative sample of those that are in in the environment. The degree to which this is true varies widely, with some pathogenic bacteria there is probably minimal deviation between the natural populations and laboratory cultures whilst the opposite may be the case with organisms from other, perhaps particularly the oligotrophic, environments. This assertion is supported by the limited success, alluded to above, in bringing into defined axenic culture organisms, including cyanobacteria, from many well studied areas. Why is this so? Speculation is easy and is based on information that has differing degrees of security. I should like to consider briefly two approaches which have been previously discussed in more detail (8).

It is well known that bacteria sometimes act in a consortium with other species that together achieve a particular metabolic degradation or geochemical change; the cyanobacterial mat communities being an example. One could extend this idea to argue that, for example, some/many of the open ocean cyanobacteria exist in a relationship with one or more of the other microbes present involving a tightly inter-dependant exchange of metabolic products which makes it difficult to separate an individual partner. The demonstration, many years ago by Van Baalen (9), that several of his isolated marine cyanobacteria had a requirement for vitamin B12 may be an illustration of this. Whether such partnerships would require physical contact is open to question, the physics of flow around very small entities is complex and can lead to acute micro-gradients of nutrients and of products. Relatively stable microenvironments could be formed between two or more bacteria without their physical adhesion. If such a situation were established, if would certainly make the isolation of one partner-organism more difficult but, nevertheless, need not frustrate an experienced microbiologist. Things become rather more difficult if messages rather than metabolites are what are being exchanged between the partners. There is now recognised to be widespread distribution of signalling systems among gram-negative heterotrophic bacteria which are based on a series of acyl homoserine lactones (10). The concentration of these quite low molecular weight molecules allow quorum sensing within a species, that is to say the ability of each member of a population to assess the number its its own population is are present within a niche and to respond physiologically to that information. A range of attributes, such as luminosity and toxin production, are only expressed when a certain population density of the bacteria in question is present. Can we think of this as unicellular organisms acting as multicellular entities? If the quorum sensing response was to extend to the control of growth itself then the isolation of an axenic culture of such a bacterium by any form of dilution or elective culture would be difficult indeed. One recalls that most of us must have noticed that for effective growth to be guaranteed, a cyanobacterial inoculum is usually fairly substantial. So far quorum sensing appears to be species specific and the process has not yet been found in cyanobacteria, but there would seem no reason why either of these limitations should hold absolutely. If this were to be the case we would be some way towards an explanation of the difficulties experienced in cyanobacterial isolation and axenic culture.

Exponential growth is considered to be a characteristic feature of bacterial division although there are examples from a wide range of organisms where this has been shown not to be the case. There are several reasons why the cyanobacterial examples of non-exponential growth are of interest. The study of the cell cycle in microorganisms has been made possible by modern techniques (see this Symposium) and there remains the continuing interest in the important question as to how cells are selected for differentiation into specialised types such as the heterocyst. In the year prior to the Freiburg meeting, Mitchison and Wilcox showed that in an *Anabaena* strain the vegetative cells always divided unequally, and it was from the smaller daughter cell that any heterocyst arose (11). Unequal division resulted in the smaller daughter taking longer to reach its division than did the larger.This imposed irregularity in the division cycle through the filament which was important in heterocyst spacing. It would also impose deviation from exponential growth. Even more striking was the description of the massive multicellular divisions in the pleurocapsalean cyanobacteria in which several hundred daughter cells were produced at division with considerable decrease in cell size (12). The question that I would raise is this: to what extent is exponential growth found in natural populations of cyanobacteria and does the mode of division contribute to the difficulty in isolation from such populations? Very careful studies with chemostat cultures of the versatile heterotroph *Aerobacter* showed that the proportion of viable cells resulting from division was reduced with decreasing division time; in effect exponential growth was demonstrated only around maximum growth rate (13, see 8). Unfortunately such an experiment cannot be repeated with a cyanobacterium because of the problem of accurate viable counting. If departure from exponential growth is associated with a slower overall growth rate it may be that it is a general feature of oligotrophic environments and that there is selective advantage in constraining division rate. This would result from lack of viability in a proportion of cells produced at division or by unequal cell division resulting in the smaller daughter taking longer (perhaps much longer) to divide. In the *Aerobacter* experiments the proportion of the population which was non-viable was fully reversible when the growth conditions returned to maximal. In contrast, at least in certain *Anabaena* species, unequal division appears to be programmed into the division cycle. The production of daughter cells at division that have markedly different life spans would facilitate maintenance of a stable population in low nutrient conditions. Clearly many natural populations of cyanobacteria have very slow growth rates and it may be that in the process of elective culture, organisms have been preferentially isolated that are atypical of the natural population in that they have escaped (by mutation) programmed contraints on exponential growth. If these conjectures were to be true they would account for some of the difficulties in isolation and, perhaps more significantly, raise questions regarding the application of information obtained with laboratory cultures to the understanding of natural populations.

REFERENCES

1. Cohen Y.,Jorgensen B.B., Padan E. & Shilo M. (1975) Nature, Lond. 257,489–492
2. Doolittle,W.F. (1996) In "Evolution of Microbial Life" eds D.M.Roberts et al Cambridge University Press, pp.1–21
3. Waterbury,J.B.,Watson,S.W.,Valois,F.W. & Franks D.G. (1986) Can. Bull Fish. Aquat.Sci. 214,71–120
4. Summers,M.C.,Wallis,J.G.,Campbell,E.L. & Meeks, J.C. (1995) J. Bact. 177,6184–6194
5. Scanlan,D.J.,Sundaram,S. Newman,J.,Mann,N.H. & Carr,N.G. (1995) J.Bact. 177,2250–2553
6. Stanier, R.Y. (1970) In "Organization and Control in Prokaryotic and Eukaryotic Cells" eds H.P.Charles & B.C.J.G. Knight Cambridge University Press, pp.1–38

7. Castenholz, R.W. (1992) J.Phycol. 28,737–745

8. Carr, N.G. (1995) In "Molecular Ecology of Aquatic Microbes" ed I. Joint,Springer, pp 391–402

9. Van Baalen, C. (1961) Science N.Y. 133,1922–1923

10. Swift,S.,Bainton, N.J. & Winson,M.K. (1994) Trends in Microbiology 2,93–98

11. Mitchison,G.J. & Wilcox, .M. (1972) Nature Lond. 239,110–111

12. Waterbury, J.B. & Stanier, R.Y.(1978) Microbiol. Rev. 42,2–44

13. Tempest D.W., Herbert,D. Phipps, P.J. (1967) In "Microbial Physiology and Continuous Culture" eds E.O.Powell et al H.M.S.O. London pp 240–254

3

BIOENERGETIC AND METABOLIC PROCESS PATTERNS IN ANOXYPHOTOTROPHS

Howard Gest

Photosynthetic Bacteria Group
Department of Biology
Indiana University
Bloomington, Indiana 47405

During the past decade, genetic and molecular biological approaches have given us a much better understanding of how photopigment synthesis is regulated in photosynthetic bacteria, but there still is a large gap in our knowledge of the mechanisms that integrate photophosphorylation and reducing power generation with the overall biosynthesis of cell material.

Cyclic photophosphorylation in purple bacteria was first demonstrated in 1954[1], but the mechanism of generating net reducing power for biosynthesis in anoxyphototrophs has been, and still is, a controversial issue. When the ISPP meetings began in 1973, a number of investigators believed that in some circumstances, the bacteria use a non-cyclic electron pathway similar to that occurring in oxygenic photosynthesis. But the evidence, from in vitro experiments with membrane fragments, was weak and not at all convincing[2]. Those who wanted to believe in a non-cyclic electron flow totally ignored physiological biochemistry from which it was clear that the redox budget and biosynthetic pattern depended strongly on the nature of the accessory electron donor always required for anaerobic photosynthetic growth. This was already evident from experiments of Muller[3] in 1933 with purple sulfur bacteria. There is no doubt that with the various carbon sources of Table 1, the metabolic traffic patterns of carbon and electrons are very different.

The integration of bioenergetics and cell biosynthesis is a problem of immense complexity. Fig. 1 gives an overall view of the key processes involved. Regulation of synthesis of Bchl-membrane complexes obviously does not occur in isolation of other aspects of cell metabolism, and there is no question that production of Bchl and reaction centers must be integrated with the synthesis of major cell components, namely, proteins, nucleic acids, lipids and carbohydrates. It is important to remember that the regulatory processes are *not* perfect--they clearly are designed to aid growing cells in adjusting to chemical and physical changes in the environment. In a sense, the controls operate as buffer systems—but as buffer systems of limited capacity.

The Phototrophic Prokaryotes, edited by Peschek *et al.*
Kluwer Academic / Plenum Publishers, New York, 1999.

Table 1. CO_2 changes during photoheterotrophic growth of purple sulfur bacteria on organic carbon sources.[3] The table lists mean values from a number of trials with each carbon source

Substrate	ΔCO_2 (moles/mole substrate used)
acetate	+ 0.17
lactate	+ 0.29
succinate	+ 0.70
malate	+ 1.22
butyrate	- 0.74

By this I mean that even the most sophisticated controls are usually compensatory and only *help* adjust photopigment biosynthetic rates to an optimal level for the prevailing circumstances even though the overall rate of cell growth may be greatly decreased. This is the familiar case when light intensity is substantially reduced. Bchl-membrane production, which is energy-expensive, increases significantly, but the cell doubling time is greatly prolonged. It is evident that there must be chemical signals that serve as connecting links between the bioenergetic machinery and the large assemblage of biosynthetic enzymes responsible for cell growth. The most logical candidates for a signaling system are

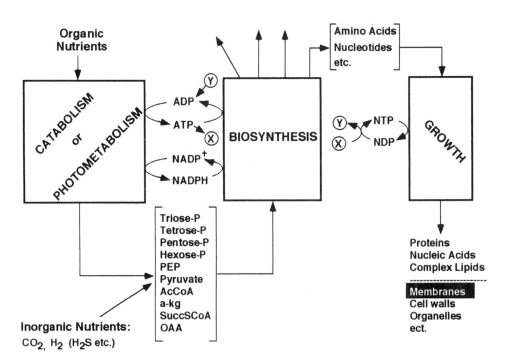

Figure 1. Metabolic process patterns: connections between bioenergetics and biosynthesis. Modified from Atkinson.[4]

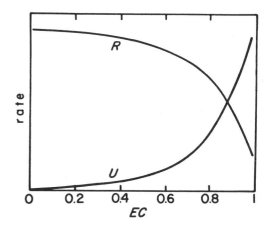

Figure 2. Generalized representation of enzyme activity responses to Energy Charge (EC). R, enzymes involved in ATP regeneration; U, biosynthetic enzymes that use ATP. After Atkinson.[4]

the adenylate nucleotides and certain electron carriers. This was recognized in 1968 by Daniel Atkinson[4], who proposed the Energy Charge Concept of metabolic regulation (see also discussion in refs. 2,5,6). He developed the concept primarily in respect to the adenylate nucleotides and started with the premise that the logic of cell metabolism requires two kinds of response characteristics for a number of critical enzymes or enzyme sequences.

As indicated in Fig. 2, sequences in which ATP is regenerated should respond to an increase in the relative concentration of ATP as shown by the R curve. On the other hand, enzymes or sequences that use ATP should respond as shown by the U curve. Atkinson reasoned that these opposing effects would constitute an effective control system, provided that the steep portions of the curves occur in the same region of what can be called "adenylate energy."

Atkinson defined adenylate energy quantitatively as a function of the three adenylate nucleotides by the equation:

$$\text{Energy Charge} = [ATP] + 0.5[ADP]/[ATP] + [ADP] + [AMP]$$

In essence, the energy charge (EC) reflects the fact that the energy available in the adenylate pool is proportional to the mole fraction of ATP plus 1/2 the mole fraction of ADP. The energy charge value varies from 0, when only AMP is present, to 1.0 when all the adenylate is in the form of ATP. Atkinson and his colleagues measured the activities of a number of enzymes in vitro at different values of EC, with the total size of the nucleotide pool kept constant. And they confirmed that they behaved as either R or U enzymes as would be predicted from their functions in cell metabolism. Moreover, the steep portions of the experimental R and U curves were in the same region of EC, at a value somewhere between 0.8 and 0.9.

What is the situation in living cells? During exponential growth of *E. coli*, the EC is about 0.8, and if it falls below 0.5, the cells die rapidly. It is important to emphasize that enzyme activity cannot be modulated directly by EC as such. The EC is an abstract calculated value that relates the relative concentrations of the adenylate nucleotides. If we say that the EC "affects" the velocity of an enzymatic reaction or sequence, we mean that the enzyme must respond to a change in concentration of one of the nucleotides, or to two of the nucleotides—or to all 3, and that the EC specifies the concentrations.

In other words, EC is an abstract quantity that reflects the energy level of the cell. It can be used as a characteristic that permits us to construct and compare performance curves of enzyme systems in response to the relative concentrations of the three adenylate nucleotides. Of course, the relevant question for photosynthetic cells is: can the EC concept help us interpret the regulation of photosynthetic metabolism? Or, put in another way, does the concept suggest experimental ways of studying the integration of photosynthetic bioenergetics with biosynthesis and cell growth? A review of some research done in my laboratory some years ago is pertinent.[2,5]

We assumed that the Bchl-membrane biosynthetic machinery should behave as an R-type system, that is, as a system whose function is the regeneration of ATP. Since R-type systems show diminished activity at high EC, the synthesis of Bchl-membrane complexes might be expected to be minimal at high light intensity, a fact established a long time ago. Growth rate is, of course, maximal at high light intensity. In contrast to the Bchl complexes, the ribosome content of bacteria is, within limits, directly proportional to growth rate. In other words, at high light intensity, growth rate of purple bacteria and cellular ribosome content are maximal, and Bchl-membrane content is minimal. ATP is being produced by photophosphorylation and is simultaneously used for biosynthesis. The balance is reflected by the nucleotide pools which specify the EC.

One of the most remarkable aspects of the nucleotide pool in bacteria is that ATP is present in essentially catalytic quantities. It has been calculated that an *E. coli* cell contains only about 1,000,000 molecules of ATP; enough for only a half second of biosynthetic activity. The situation in photosynthetic bacteria is undoubtedly essentially the same. So, the instantaneous pool size of ATP does not give a meaningful indication of the high flux of ATP turnover during active growth. This is another way of saying that there is a very tight coupling between production and utilization for biosynthesis. Although the EC in growing cells is well poised overall, there is a physiological range of variation that is influenced by a number of factors.

Integration of energy conversion and biosynthesis in photosynthetic cells presents a special problem in that the photochemical machinery is orders of magnitude faster than any other kind of energy conversion—which brings up the question of how photosynthetic energy conversion was coordinated with slower enzymatic processes in the first place, in the first photosynthetic cells (see hypothesis in refs. 7 and 8). The evolutionary problem was how to stabilize photochemistry and connect it with biochemistry. However it happened, the fact that light comes from outside of the cell means that cell growth is at the mercy of a completely external non-chemical energy source, and this fact has significant physiological consequences. And it also means that it provides us with an easy way of experimentally manipulating the kinetics of energy delivery to the biosynthetic machinery.

We decided to explore the dynamics of the coupling between bioenergetics and biosynthesis by looking at the consequences of exposing photosynthetically growing cultures of *R. capsulatus* to different regimens of intermittent illumination.[9] Fig. 3 shows the growth kinetics of *R. capsulatus* in continuous saturating light and in two regimens of intermittent light. In alternating light and dark periods of only 0.1 second, the biomass doubling time slows down from 130 minutes to 175 minutes and when the intervals are 30 seconds, the growth rate is severely depressed to a doubling time of about 8 hours. The cultures receive the same amount of light per hour (i.e., 30 minutes of saturating illumination), but the efficiency of energetic coupling is greatly affected by the intermittency. The reduction of efficiency with the 30 second periods is such that the cells behave as though they are growing in dim light of about 1/10th the saturating intensity.

The effects of intermittent illumination on some important cell components are indicated in Table 2. It is evident that even very short repeated interruptions in photophosphorylation cause a syndrome of energy stress, apparently caused by derangement of the

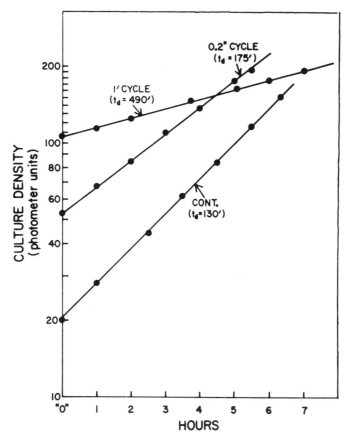

Figure 3. Growth kinetics of *R. capsulatus* in continuous light and in two intermittent light regimens.[9] With intermittent illumination, light and dark periods were always equal; the total time elapsed for one light interval and a succeeding dark interval is designated as "cycle length."

integration between energy conversion and biosynthesis. Comparison of the values for the 30 second and 1.2 second cycles is of special interest. Cells growing in the 1.2 second cycle clearly utilize the incident light energy for biosynthesis with considerably greater efficiency. Moreover, with energy flow pulsed at these frequencies, the usual relationship between growth rate and RNA content seen with continuous energy supply is obviously

Table 2. Effects of intermittent illumination on *R. capsulatus* growth and synthesis of Bchl, protein, and RNA[9]

Regimen	t_d (min)	Bchl*	protein*	RNA*
continuous light	132	0.66	71.3	17.2
intermitt: 1.2 sec	255	1.67	69.4	10.7
intermitt: 30 sec	501	1.19	60.7	11.2

*% of dry weight

violated. Assuming the same ribosome content in both "types" of cells, the results suggest, not surprisingly, that growth rate is a function of both ribosome content and the "average" chemical energy flux.

Another way of experimentally disturbing the integration between energy conversion and biosynthesis is to influence photophosphorylation rate by addition of inorganic arsenate to growing cells. Arsenate substitutes for orthophosphate in a number of biochemical reactions, leading to the formation of unstable esters. Photosynthetic growth rate of R. capsulatus is sverely inhibited when arsenate is present equimolar to orthophosphate in the culture medium.[10] Studies on an arsenate-resistant mutant indicated that its resistance can be explained by a significantly increased photophosphorylation rate of the energy-converting machinery, which can compensate for the decrease in ATP regeneration rate caused by esterification of arsenate in place of phosphate.[10,11] The high photophosphorylation capacity of the mutant presumably enables cells growing in the presence of arsenate to maintain the energy charge within the range necessary for orderly biosynthesis. Repeated subculture of the mutant under certain conditions yields cell populations that have strikingly elevated contents of reaction centers.[11] Membranes from such cells may well be useful experimental systems for various studies.

The most prominent example of metabolic imbalance in the physiology of purple bacteria is the phenomenon of light-dependent production of molecular hydrogen during photoheterotrophic growth.[2] When the bacteria grow on organic acids such as malate or lactate with certain amino acids as nitrogen sources, the overall electron flow pattern is markedly different than when the nitrogen supply is an ammonium salt. For example, with glutamate as nitrogen source, H_2 is evolved continuously during growth, and in large quantity. In contrast, with an ammonium salt, hydrogen is not produced. This striking difference in disposition of electrons from organic substrates is due to the nitrogenase complex in cells grown with glutamate, and its absence in cells growing with an ammonium salt.[12]

Fig. 4 depicts the flow of carbon and electrons in cell producing H_2 during photoheterotrophic growth. The double-bulbed device on the right represents the nitrogenase complex. In the absence of ammonia, the nitrogenase is derepressed, and when molecular nitrogen is absent, the complex functions as a hydrogen-evolving catalyst. With glutamate as the nitrogen source, the supplies of ATP from photophosphorylation and electrons from organic substrates are in excess relative to the biosynthetic machinery, Under these conditions, protons are reduced yielding molecular hydrogen, which is shown here as being discarded by a "hydrogen relief valve." If molecular nitrogen is added, hydrogen evolution stops, because ATP and the electron supply are used for production of ammonia, which is rapidly consumed for the production of amino acids and other nitrogenous compounds.

The "hydrogen relief valve" is construed as a control device that permits "energy idling" when this is required by the balance between generation of ATP and reducing power, on the one hand, and overall biosynthetic rate, on the other.[2,13,14] The energy-dependent formation of H_2 presumably helps maintain energy charge within an acceptable physiological range. The hydrogen valve seems to have a large number of possible settings that will accomodate a considerable range of metabolic traffic conditions. The metabolic traffic, especially of adenylate nucleotide flux and electron flow, can vary greatly depending on the light intensity, the temperature, the oxidation/reduction level of the organic carbon source, and the nature of the nitrogen source. A good example reflecting the influence of the chemical composition of amino acid nitrogen sources is given by the data of Table 3.

The "hydrogen relief valve" offers an excellent system for analyzing EC controls of metabolism, but this has not yet been significantly attacked using the powerful tools of genetics and molecular biology.

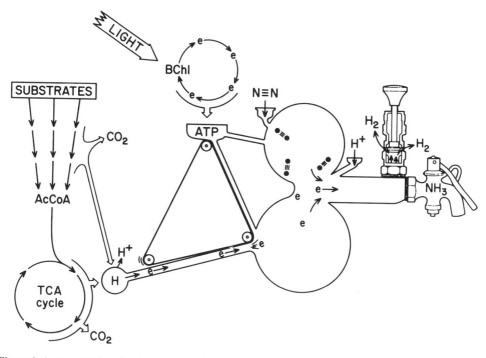

Figure 4. A representation of metabolic conversions occurring during "photoproduction of H_2," catalyzed by nitrogenase. If N_2 is present, H_2 is not produced and reducing power is used instead for reduction of N_2 to ammonia.

I summarize some of the major points discussed above by quoting from a review by C. J. Knowles[6] on "Microbial metabolic regulation by adenine nucleotide pools":"It is not surprising that growing microorganisms closely control their adenine nucleotide content and the proportion of the total pool present as ATP, ADP, and AMP. Nor is it surprising that they have evolved sensitive mechanisms for protecting the adenine nucleotide pools against fluctuations in nutritional status and other environmental factors. Adenylate energy

Table 3. Photoproduction of H_2 and growth rates of *R. capsulatus* in media containing 30 mM DL-lactate and various amino acids (7 mM) as nitrogen sources[13]

N source	t_d (h)	$H_2\uparrow$ (μl/h/ml culture)
Valine	11.0	76
Isoleucine	6.4	68
Leucine	5.6	99
Alanine	3.2	130
Glutamate	3.0	130
Aspartate	2.9	100
Glutamine	2.4	110

charge is a convenient parameter for expressing the state of the adenine nucleotide pool; together with a knowledge of the adenylate kinase equilibrium position, it tells us the exact adenine nucleotide composition of the cell. Moreover, studies of the effect of adenine nucleotides on the activities of isolated enzymes can be assessed using different applied energy charge values as well as ATP to ADP or ATP to AMP ratios, since they relate the in vitro conditions to the adenylate kinase-catalyzed equilibrium of the nucleotides within the cell."

There are, of course, limitations of the EC concept in understanding the dynamics of metabolic control—but it was not offered in the first place as a comprehensive "global" analysis. In fact, it is just a beginning in trying to understand an astonishingly complex network of processes that enables cells to maintain homeostasis.

In my opinion, we tend to think too simplistically about individual regulatory effects and we underestimate the liklihood of multiple interlocking controls. In this connection, it seems reasonable and very likely that EC regulation is interlocked with what could be called Redox Charge Control.[2] The latter was implicated a long time ago by the fact that optimal photophosphorylation by membrane fragments from purple bacteria is dependent on maintenance of a suitable redox potential.[15] More recently, the notion of "molecular redox signalling" has been proposed by J.F. Allen et al.[16] to explain coupling of expression of photosynthesis genes to electron transfer.

The history of biochemical and molecular biological research over the past few decades should teach us that soon after we think we have reached a new and seemingly satisfactory level in understanding the workings of cells, new discoveries reveal still more unanticipated complexity and we are back at the drawing board again. There is a recent instructive example.Using mutant strains of *Rhodospirillum rubrum* and *Rhodobacter sphaeroides*, Joshi and Tabita[17] have obtained evidence for a molecular link between the reductive pentose CO_2 fixation pathway and nitrogen fixation. The mutants are defective in RubisCO and although CO_2 is no longer able to serve as an electron sink, the organisms can dispose of reducing power using the alternative of reducing molecular nitrogen to ammonia. But in contrast to wild type cells, such mutants produce copious quantities of hydrogen even in the presence of ammonia. In other words, the mutants circumvent powerful control mechanisms in order to remove excess reducing power. Joshi and Tabita further suggest that a two component signal transduction system integrates the expression of genes required for the three processes of photosynthesis, nitrogen fixation, and carbon dioxide fixation.

The anoxyphototrophs continue to surprise us with their remarkable properties, and we still have much to learn.

REFERENCES

1. Frenkel, A.W. (1954) J. Am. Chem. Soc. 76, 5568–5569.
2. Gest, H. (1972) in: Adv. Microbial Physiol. (Rose, A.H. and Tempest, D.W., Eds.), vol. 7, pp. 243–282. Academic Press, London and New York.
3. Muller, F.M. (1933) Doctoral Dissertation. Julius Springer, Berlin.
4. Atkinson, D.E. (1977) Cellular Energy Metabolism and its Regulation. Academic Press, New York.
5. Gest, H. (1973) in: Genetics of Industrial Microorganisms/Bacteria. (Vanek, Z., Hostalek, Z., and Cudlin, J., Eds.) pp. 131–143, Academia, Prague.
6. Knowles, C.J. (1977) in: Microbial Energetics (Haddock, B.A. and Hamilton W.A., Eds.) pp. 241–283. Cambridge Univ. Press, Cambridge.
7. Gest, H. (1980) FEMS Microbiol. Lett. 7, 73–77.

8. Gest, H. (1983) in: The Phototrophic Bacteria: Anaerobic Life in the Light (Ormerod, J.G., Ed.) pp. 215–235. Blackwell Scientific, Oxford.

9. Sojka, G.A. and Gest, H. (1968) Proc. Natl. Acad. Sci. U.S.A. 61, 1486–1493.

10. Zilinsky, J.W., Sojka, G.A., and Gest, H. (1971) Biochem. Biophys. Res. Comm. 42, 955–961.

11. Lien, S., San Pietro, A., and Gest, H. (1971) Proc. Natl. Acad. Sci. U.S.A. 68, 1912–1915.

12. Gest, H. (1994) Photosyn. Res. 40, 129–146.

13. Hillmer, P. and Gest, H. (1977) J. Bacteriol. 129, 724–731.

14. Hillmer, P. And Gest, H. (1977) J. Bacteriol. 129, 732–739.

15. Bose, S.K. and Gest, H. (1963) Proc. Natl. Acad. Sci. U.S.A. 49, 337–345.

16. Allen, J.F., Alexciev, K., and Håkansson, G. (1995) Current Biology 5, 869–872.

17. Joshi, H.M. and Tabita, F.R. (1996) Proc. Natl. Acad. Sci. U.S.A. 93, 14515–14520.

4

STRUCTURAL AND FUNCTIONAL ANALYSES OF CYANOBACTERIAL PHOTOSYSTEM I

The Directionality of Electron Transfer

Fan Yang,[1] Gaozhong Shen,[2] Wendy M. Schluchter,[2] Boris Zybailov,[2] Alexander Ganago,[2] John H. Golbeck,[2] and Donald A. Bryant[2]

[1]Department of Chemistry
University of Nebraska
Lincoln, Nebraska 68588
[2]Department of Biochemistry and Molecular Biology
Pennsylvania State University
University Park, Pennsylvania 16802

1. INTRODUCTION

The photosystem I (PS I) reaction center is a multisubunit, photo-oxidoreductase that catalyzes the oxidation of reduced plastocyanin (or cytochrome c_6) in the thylakoid lumen and the reduction of soluble [2Fe-2S] ferredoxin (or flavodoxin) in the stroma. A major advance in our understanding of the reaction center has resulted from the crystallization and X-ray structural analysis of PS I trimers from the thermophilic cyanobacterium *Synechococcus elongatus* by the groups of Witt and Saenger [1]. A detailed structural model of the PS I complex has recently been described at a nominal resolution of 4 Å [2, 3].

Cyanobacterial PS I forms discoidal trimers with a diameter of approximately 210 Å and a maximum thickness of 90 Å [1, 2]. Each PS I monomer is a prolate ellipsoid with a maximum length of 130 Å and a maximum width of 105 Å. Each monomer is composed of single copies of 11 polypeptides, named according to the corresponding genes (see Table 1). At 4 Å resolution, a total of 43 α-helical segments have been defined in the structure. A pseudo-twofold rotation axis C_2 (AB) occurs essentially parallel to the membrane plane normal, and this two-fold symmetry relates 15 α-helices to their pseudo-symmetric partners. These symmetrically displaced helices are presumed to be derived from the sequence related PsaA and PsaB polypeptides, the two largest subunits of the PS I complex (Table 1). Of these 15 helices, 11 are transmembrane, 1 occurs on the lumenal surface, and

The Phototrophic Prokaryotes, edited by Peschek *et al.*
Kluwer Academic / Plenum Publishers, New York, 1999.

Table 1. Components of the *Synechococcus* sp. PCC 7002 Photosystem I complex

Protein	Residues[a]	Mass (Da)[a]	Transmembrane α-helices	Cofactors[b]	Function/Properties
PsaA[c]	739	81,684	11	~100 Chl *a* ~10 β-carotene	Antenna Chl binding; electron transport and cofactor binding;
PsaB[c]	733	81,384	11	2 Phylloquinones 1 [4Fe-4S]	Charge stabilization F_X intermediate acceptor
PsaC[d]	81	8,814	0	2 [4Fe-4S]	F_A and F_B terminal acceptors
PsaD[d]	142	15,650	0	None	PsaC binding/orientation; ferredoxin docking
PsaE[d]	70	7,667	0	None	Cyclic electron transport; ferredoxin docking; A_1 shielding
PsaF[e]	169	18,321	2	3-5 Chl *a*	State transitions; coupling to phycobilisomes; A_1 shielding
PsaI[f]	38	3,957	1	Chl *a*?	Stabilization of PsaL binding; Chl *a* binding?
PsaJ[f]	37	4,231	1	Chl *a*?	State transitions; coupling to phycobilisomes
PsaK[g]	87	8,684	2	Chl *a*	Stabilization of trimers; Chl *a* binding
PsaL[f]	152	15,617	2	Chl *a*	Trimer formation; Chl *a* binding; state transition kinetics
PsaM[f]	31	3,304	1	None	Stabilization of acceptor side proteins; trimer formation

[a]Deduced from gene sequence; includes N-terminal methionine.
[b]Per reaction center monomer.
[c]Processing status unknown.
[d]N-terminal methionine removed in mature protein.
[e]27 amino acids at N-terminus removed in mature protein; actual molecular mass is 15,413 Da.
[f]N-terminal methionine not removed.
[g]8 amino acids at N-terminus removed in mature protein; actual molecular mass is 7,898 Da.

3 occur on the stromal surface of the PsaA/B polypeptides. Most of the electron transport cofactors (see Figure 1 and below), including P700, the chlorophyll acceptor A_0, the phylloquinone acceptor A_1, and the [4Fe-4S] center F_X, are bound by the PsaA/B heterodimer. In addition to the critical electron transport cofactors, PsaA and PsaB are also responsible for binding most of the antenna chlorophylls and ~10–15 β-carotene molecules. Of an anticipated ~100 chlorophyll molecules per P700 in the PS I monomer, 89 have been located in the structure; 83 of these are considered to represent antenna chlorophylls that are not involved in the electron transport chain.

Three water-soluble subunits, PsaC, PsaD, and PsaE form a ridge that projects approximately 30 Å above the membrane plane of the complex on the stromal surface. These polypeptides are easily removed from the complex by treatment with chaotropic agents [4]. The 8.9 kDa PsaC polypeptide binds the terminal electron transfer acceptors, the two [4Fe-4S] clusters denoted as centers F_A and F_B [5]. The PsaD polypeptide is required for stable binding and orientation of the PsaC protein on the PS I complex [6], and PsaD probably also plays a role in the docking of soluble acceptors ferredoxin and flavodoxin to the complex [7]. Three α-helices are assigned to the proteins of the stromal ridge: 2 short helices are found in the PsaC protein, and one helix is tentatively assigned to PsaD. The PsaE polypeptide has been shown to play roles in cyclic electron transport and docking of ferredoxin [8, 9]. The structures of the PsaE proteins of *Synechococcus* sp. PCC 7002 [10] and *Nostoc* sp. PCC 8009 (K. McLaughlin, C. Falzone, G. Shen, D. Bryant, and J. Lecomte, unpublished results) have been determined by multi-dimensional NMR methods.

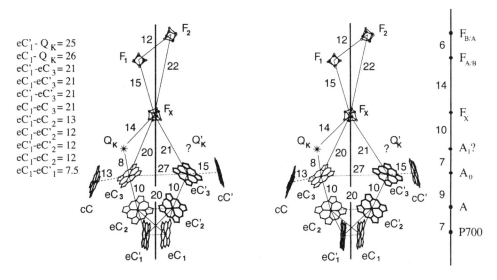

$eC'_1 - Q_K = 25$
$eC_1 - Q_K = 26$
$eC'_1 - eC'_3 = 21$
$eC_1 - eC'_3 = 21$
$eC'_1 - eC'_3 = 21$
$eC_1 - eC_3 = 21$
$eC'_1 - eC_2 = 13$
$eC_1 - eC'_2 = 12$
$eC'_1 - eC'_2 = 12$
$eC_1 - eC_2 = 12$
$eC_1 - eC'_1 = 7.5$

Figure 1. Stereo diagram depicting the cofactors of the electron transfer system as well as the connecting chlorophylls (cC and cC'), which have been proposed to couple the antenna chlorophylls to the electron transport cofactors. This diagram is taken from Schubert et al. [3]. Center-to-center distances (in Å) between neighboring cofactors are indicated. The axis C_2 (AB), parallel to the c-axis and defined to pass through the F_X [4Fe-4S] cluster, relates the pseudosymmetric cofactor branches of the electron transfer system. This axis deviates by up to ~1 Å from the midpoints of the cofactor pairs. In this view eC_1 is to the right of eC_1'. Cofactors with no "prime (')" symbol" are most closely associated with the non-primed α-helices of PsaA/PsaB (**m, n, o**), and those indicated with a "prime (')" symbol" are most closely associated with the "primed" α-helices of PsaA/PsaB (**m', n', o'**). According to the results of the study described in this work, the active branch for electron transport is proposed to be $eC_1 \rightarrow eC_2' \rightarrow eC_3' \rightarrow Q_K \rightarrow F_X$.

The structures of the two proteins are quite similar and reveal a compact protein composed of 5 anti-parallel strands of β-sheet in a 2+3 barrel configuration.

Each PS I monomer additionally contains six membrane-intrinsic proteins that contain one or more transmembrane α-helices. In total, nine transmembrane α-helices and one lumenal-surface helix are assigned to these six proteins (Table 1). The functional roles of these low-molecular-mass proteins have remained uncertain, but PS I assembly and linear electron transport do not appear to be severely impaired in mutants lacking any of these subunits. The X-ray structure model suggests that several of these low-molecular-mass polypeptides, including PsaI, PsaL, PsaF and PsaK, may bind chlorophylls. The PsaL subunit has been suggested to contain two transmembrane α-helices, and trimeric PS I complexes can not be isolated from *psaL* null mutants [11, 12]. This result strongly suggests that the PsaL subunit lies near the C_3 symmetry axis of the trimer in the so-called trimerization domain. The PsaI polypeptide appears to stabilize the binding of the PsaL polypeptide in the trimerization domain [13], and only a very low yield of PS I trimers could be isolated from a *psaI* deletion mutant of *Synechococcus* sp. PCC 7002 [12].

The *psaF* and *psaJ* genes typically form a dicistronic operon, and the products of these two genes are assigned to helices that lie at the outermost edge of the monomeric units of the PS I trimer [14]. In higher plants it has been shown that PsaF can be cross-linked to plastocyanin [15], and similarly, in cyanobacteria, this polypeptide has been cross-linked to cytochrome c_6 [16]. However, mutants lacking PsaF do not appear to be seriously impaired in electron transport [17, 18], and mutants lacking PsaJ have been shown to have little if any phenotype as well [18, 19]. In *Synechococcus* sp. PCC 7002 the fluorescence emission spectra of PS I complexes isolated from mutants lacking PsaF are blue-shifted approximately 3–4 nm relative to those of the wild-type, consistent with the idea that PsaF is a chlorophyll-binding protein. Mutants lacking either PsaF, PsaJ or both, exhibit state-associated fluorescence changes that are greater in amplitude than those of the wild-type strain. Low-temperature fluorescence emission spectra of state-adapted cells suggest that the fluorescence emission characteristics of cells in State 1 are similar to those of the wild-type strain, but that the fluorescence emission from PS I in particular is lower in mutant cells dark-adapted to State 2. Taken together, these results imply that energy transfer from phycobilisomes to PS I complexes is less efficient in mutant cells lacking PsaF or PsaJ (manuscript in preparation).

The remaining two polypeptides, PsaK and PsaM have been suggested to occur in the periphery of the PS I monomers and possibly interact at the boundaries between pairs of monomers to stabilize trimers [3]. Mutants lacking PsaK have fluorescence properties similar to those described for the *psaF* and *psaJ* mutants above, although there is no shift in the fluorescence emission maximum of PS I complexes isolated from the *psaK* mutant (unpublished results). However, a state 2 to state 1 transition produces a larger fluorescence amplitude increase than in the wild type, suggesting some impairment of energy transfer to PS I in State 2 in the *psaK* mutant strain. Mutants lacking *psaM* are much more sensitive to high light intensity than the wild-type and are impaired in cyclic electron transport. These defects are apparently due to a destabilization of the stromal side proteins PsaC, PsaD, and PsaE, and degradation products of PsaC and PsaE have been detected immunologically in thylakoids of the *psaM* mutant grown at or subjected to high light intensity (manuscript in preparation).

The components of the electron transfer chain of PS I are arranged along the pseudo-C_2 (AB) symmetry axis of the monomers (see Figure 1) and are separated from the antenna chlorophylls by a palisade of 10 α-helices [2, 3]. The first pair of Chl a molecules, eC_1 and eC_1', of the electron transfer chain form the symmetric dimer pair parallel to the pseudo-C_2 (AB) axis. Two additional pseudo-symmetric pairs of chlorophylls, eC_2 / eC_2' and eC_3 / eC_3', form a bifurcating electron transfer chain; one of the latter pair is presumed to represent the primary acceptor, A_0, while one of the former pair of chlorophylls is presumed to represent a bridging chlorophyll analogous to the voyeur chlorophyll of the reaction centers of purple bacteria [20, 21]. Two phylloquinones, one of which is known spectroscopically as the secondary electron acceptor A_1 [22], also are found in the PS I complex, but because the head group of napthoquinones are similar in size to the side chains of aromatic amino acids, the locations of these quinones are not yet unequivocally established. In the most recent model for the PS I reaction center, however, a position for one (Q_K) of the anticipated pair of quinones (Q_K / Q_K') was suggested [3] (see Figure 1). The bifurcating electron transport chain converges again at the interpolypeptide F_X [4Fe-4S] center.

Important questions which cannot be addressed by X-ray crystallography are the following: does electron transfer occur through only one branch of the pseudo-symmetric pairs of chlorophylls, as occurs in the bacterial reaction center, or can electron transfer occur through both branches. Moreover, if electron transfer only occurs through one of the two branches, which branch is the active branch? We have sought to answer these questions through a combination of targeted mutagenesis and spectroscopic characterization of the resulting reaction centers. As noted above, the PS I reaction center contains a number of non-symmetry related polypeptides, and we reasoned that selective removal of some of these polypeptides by mutation might lead to reaction centers with altered electron transport properties that might shed new insight into the pathway of electron transport through the complex.

2. RESULTS AND DISCUSSION

Using the detergents Triton X-100 or β-dodecylmaltoside to solubilize thylakoid membranes of *Synechococcus* sp. PCC 7002, PS I complexes were prepared from the wild-type strain of *Synechococcus* sp. PCC 7002 and from mutant strains in which the following genes had been deleted or insertionally inactivated: *psaK, psaL, psaM, psaJ, psaE, psaF, psaE psaF*. The polypeptide compositions of the resulting PS I complexes was verified by SDS-PAGE and immunoblotting (data not shown). When isolated with β-dodecylmaltoside, the PS I complexes isolated from the mutant strains were only depleted of the targeted polypeptide(s). However, as noted previously [7, 12], special precautions are required for the preparation of PS I complexes with Triton X-100, especially for mutants lacking PsaE or PsaF. Prolonged exposure of PS I complexes from the wild-type strain to high concentrations of Triton X-100 led to the depletion of PsaE, PsaF and PsaJ. Complexes prepared from mutants lacking PsaF were observed to be completely depleted of PsaJ and largely depleted of PsaE when isolated with Triton X-100 but not β-dodecylmaltoside. Complexes prepared from the *psaL* mutant were also depleted of PsaI. Finally, complexes prepared from mutants lacking PsaE were similarly depleted of PsaF and PsaJ when isolated with Triton X-100 but not β-dodecylmaltoside. By reducing the exposure time and the chlorophyll:detergent ratio, however, it was possible to minimize the loss of these three polypeptides.

Table 2 shows the rates of flavodoxin photoreduction for various PS I preparations prepared with the two detergents. Complexes prepared from the wild-type with β-dodecylmaltoside gave the highest rates of flavodoxin photoreduction. In general, complexes prepared with Triton X-100 had slightly lower rates of flavodoxin reduction for any given mutant strain. Complexes lacking PsaE and PsaF (+PsaJ) had the lowest rates of flavodoxin reduction; nevertheless, the rate of electron throughput for these complexes was only about two-fold lower than for the wild-type complexes. Some of this effect might be due to altered interactions of flavodoxin with PsaC that could occur when PsaE is missing from the acceptor side [7, 23]. However, these data clearly show that PS I complexes lacking a variety of low-molecular-mass polypeptides retain their capacity for efficient electron transport.

The phylloquinone content of PS I complexes, isolated with Triton X-100 from the wild-type strain and from the *psaE psaF* mutant, were analyzed by solvent extraction followed by high-performance liquid chromatography. The PS I preparation from the *psaE psaF* mutant with Triton X-100 was selected for analysis since the rate of flavodoxin reduction catalyzed by this complex was the lowest of all the complexes tested. The calculated quinone content of the wild-type complex was 2.5 phylloquinones per 100 Chl, while the calculated quinone content of PS I complexes from the *psaE psaF* mutant was 2.6 phylloquinones per 100 Chl. These data, in combination with the observation that the various PS I complexes efficiently carried out the photoreduction of flavodoxin (Table 2), indicate that phylloquinone(s) are not extracted from the PS I complexes of the various mutants during isolation. Therefore, the lower rate of flavodoxin reduction in the *psaE psaF* mutant can not be attributed to the loss of A_1 function.

Isolated PS I complexes were reduced with dithionite at pH 10 and subjected to photoaccumulation conditions by illuminating the samples at 205 K for 40 minutes [24]. The

Table 2. Rates of flavodoxin reduction for PS I complexes isolated from various strains of *Synechococcus* sp. PCC 7002[a]

Strain	Detergent[b]	Rate[c]	%[d]
Wild-type	DM	3000	100[e]
Wild-type	TX-100	2860	95
psaK	DM	2210	74
psaK	TX-100	1990	65
psaJ	DM	2390	80
psaJ	TX-100	2240	75
psaE	DM	2510	84
psaE	TX-100	2200	73
psaF	DM	1800	60
psaF	TX-100	1590	53
psaE psaF	DM	1710	57
psaE psaF	TX-100	1610	54

[a]Rates of flavodoxin photoreduction were measured in a 1.0 ml volume using 15 μM flavodoxin and PS I complexes at 5 μg Chl ml^{-1} in 50 mM Tricine buffer, pH 8.0, 50 mM NaCl, 15 μM cytochrome c_6, 6 mM sodium ascorbate and 0.05% β-dodecylmaltoside. The measurement was made by monitoring the rate of change in absorption of flavodoxin at 467 nm.
[b]DM, β-dodecylmaltoside; TX-100, Triton X-100.
[c]Rate is expressed as micromoles flavodoxin reduced h^{-1} mg Chl^{-1}.
[d]Percentage of rate for wild-type PS I complex prepared with DM.
[e]The 100% value corresponds to an electron through-put rate of 76 s^{-1}.

photoaccumulated samples were then analyzed by electron paramagnetic resonance (EPR) spectroscopy for the presence of the radicals A_1^- and A_0^-. Figure 2 shows a comparison of the X-band (9.4 GHz) and Q-band (34.1 GHz) spectra of A_1^-, A_0^- and P700$^+$ in PS I complexes from the wild-type strain. The radicals A_1^-, A_0^- and P700$^+$ can be distinguished on the basis of three parameters: g-value, linewidth and presence of hyperfine splittings. P700$^+$ has a relatively narrow, featureless derivative lineshape with a g-value of 2.0025 and linewidths of 7.7 G at X-band and 8.5 G at Q-band. The linewidth is nearly field-inde-

Figure 2. Comparison of the EPR spectra of the P700$^+$, A_0^- and A_1^- radical species of PS I complexes of *Synechococcus* sp. PCC 7002 at X-band (9.4 GHz; top) and Q-band (34.1 GHz; bottom) microwave frequencies. The P700$^+$ spectrum was generated by illumination at 100 K with a weak reductant, sodium ascorbate, of an intact PS I complex containing F_A, F_B, F_X and A_1. The A_0^- spectrum was generated by photoaccumulation at 205 K with a strong reductant, dithionite, of a core complex which was depleted of F_A, F_B, F_X and A_1 [25]. The A_1^- spectrum was generated by photoaccumulation at 205 K with a strong reductant, dithionite, of an intact complex containing F_A, F_B, F_X and A_1. X-band EPR spectra were recorded with a Bruker ESP-300E spectrometer equipped with an ESR 900 Oxford Instruments liquid helium cryostat and an Oxford ITC4 temperature controller. All measurements were performed at 100 K. The magnetic field was swept from 3359.749 to 3399.749 Gauss. The microwave frequency was measured with an HP 5340A frequency counter. All spectra were recorded at a microwave power of 63 μW and a modulation amplitude of 2 G. A final sample concentration of 0.4 mg chlorophyll ml^{-1} was used for all measurements. Q-band EPR spectra were measured with a Bruker ESP-300E Q-band spectrometer equipped with an ER 5106QT-low temperature Q-band resonator. All measurements were performed at 100 K using an ER 4118CV liquid nitrogen Dewar and a Eurotherm temperature controller. The magnetic field was measured using an ER 035M NMR Gaussmeter and the microwave frequency was measured using a Hewlett-Packard 5352B 40 GHz frequency counter. All spectra were recorded at a microwave power of 54 μW and a modulation amplitude of 1 G or 7 G. The sample concentration was 1.5 mg chlorophyll ml^{-1} for all measurements. PS I complexes were reduced with sodium hydrosulfite (dithionite) at pH 10 and incubated in the dark for 30 min prior to photoaccumulation, which was carried out at 205 K for 40 minutes.

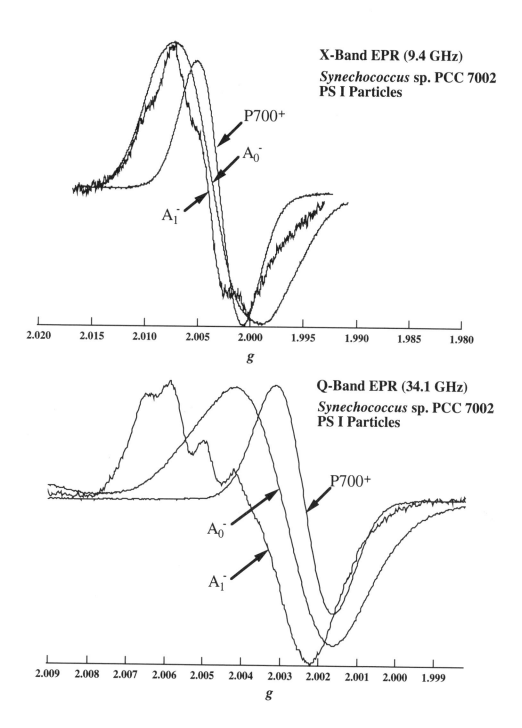

pendent because the spectrum is dominated by hyperfine couplings at both X-band and Q-band. The A_0^- spectrum is broader but also exhibits a featureless, derivative lineshape with a g-value of 2.0034 and linewidths of 14.4 G at X-band and 15.4 G at Q-band. Although the hyperfine couplings similarly dominate the spectrum at both microwave frequencies, the monomeric nature of A_0 results in a broader linewidth when compared with P700, a chlorophyll dimer. The spectral tail which extends into lowfield is probably due to a small amount of A_1 remaining in this sample (the modulation amplitude of 4 G in the A_0 spectrum would not have allowed the hyperfine couplings from A_1 to be resolved). At X-band, A_1^- exhibits a structured, derivative lineshape with superimposed hyperfine couplings derived primarily from the 2-methyl group of phylloquinone [26]; the g-value of the zero cross-over point is 2.0043 and the linewidth is about 9.5 G. However, at Q-band, the field-dependent g-anisotropy becomes the dominant feature in the spectrum, and the field-independent hyperfine couplings derived from the 2-methyl group of phylloquinone are now well-resolved. At this frequency, the spectrum of A_1^- approaches rhombicity, showing g-values for the low-field and high-field peak of 2.0061 and 2.0022 and an overall linewidth of 24 G. In summary, although the spectra of A_1^-, A_0^- and P700$^+$ are not well resolved at X-band, the three radicals are unambiguously distinguishable at Q-band. The data in Figure 2 demonstrate that if changes in the identity, the intensity, or the properties of these radicals do occur, these changes would be readily detectable only at the higher microwave frequency.

Figure 3 shows the results obtained when PS I complexes from the wild-type strain or various mutant strains were isolated with β-dodecylmaltoside, reduced with dithionite at pH 10, subjected to photoaccumulation, and analyzed by Q-band EPR spectroscopy. The spectra of all of the PS I complexes are similar and show that the A_1^- phylloquinone radical accumulates as expected under these conditions. However, dramatically different results are obtained with some but not all of the PS I complexes prepared with Triton X-100. As shown in Figure 4, only the A_1^- radical accumulates after 40 minutes of illumination in complexes prepared with Triton X-100 from the wild-type strain and from mutants lacking PsaK, PsaL/PsaI, PsaM, and PsaE. However, in mutants lacking PsaF/PsaJ or PsaE/PsaF/PsaJ, the A_1^- radical is partially (*psaF* mutant) or completely (*psaE psaF* mutant) replaced by a spectrum with a g-value of 2.0032 and a linewidth of 16.2 G, which are very similar to the values for the A_0^- radical. When PS I complexes are illuminated for much shorter time periods (5 to 15 seconds), both the A_1^- and A_0^- signals were found (data not shown). As discussed above, complexes from these two mutants are also depleted of PsaJ due to a presumed destabilization of PsaJ binding to the complex in the presence of Triton X-100, to polarity effects of the interposon fragment on the downstream *psaJ* gene, or because of both reasons. However, as shown in Figure 4, the removal of PsaJ alone does not affect the photoaccumulation of the A_1^- radical.

These results show that removal of the PsaF and PsaE polypeptides at the periphery of one edge of the PS I complex, along with the use of Triton X-100 as the detergent, results in the appearance of the A_0^- radical during the normal photoaccumulation protocol. The observations that the A_1^- radical is transiently detected in these complexes after a short periods of illumination, in combination with the analytical results indicating the retention of two phylloquinones per P700, strongly implies that under photoaccumulation conditions A_1 becomes doubly reduced [24]. The resulting inability of A_1 to function as an electron donor or acceptor leads to the accumulation of the A_0^- species. Removal of polypeptides PsaK, PsaI, PsaL, or PsaM near the trimerization domain or at the interfaces between trimers has no effect on the photoaccumulation of the A_1^- phylloquinone signal. Thus, it appears that the loss of the PsaF and PsaE polypeptides in the presence of Triton

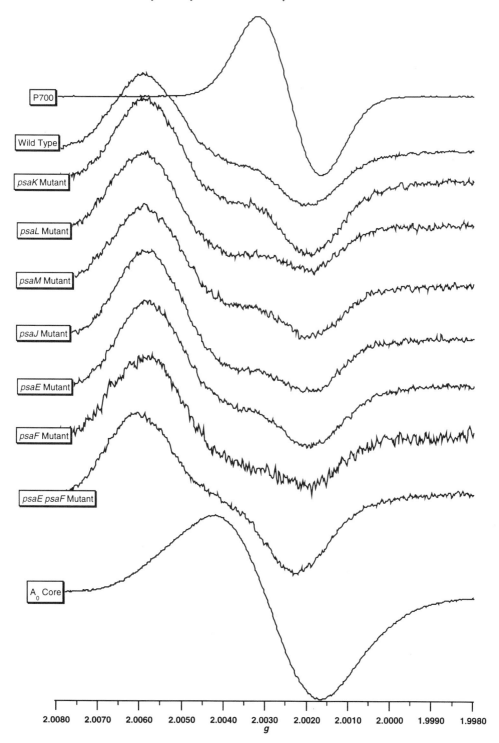

Figure 3. Q-band EPR studies of PS I complexes isolated with β-dodecylmaltoside from various *Synechococcus* sp. PCC 7002 strains. Conditions for photoaccumulation and for EPR measurements were identical to those described in the legend for Figure 2. The spectrum of P700$^+$ and A$_0^-$ are shown at the top and bottom of the figure for comparison.

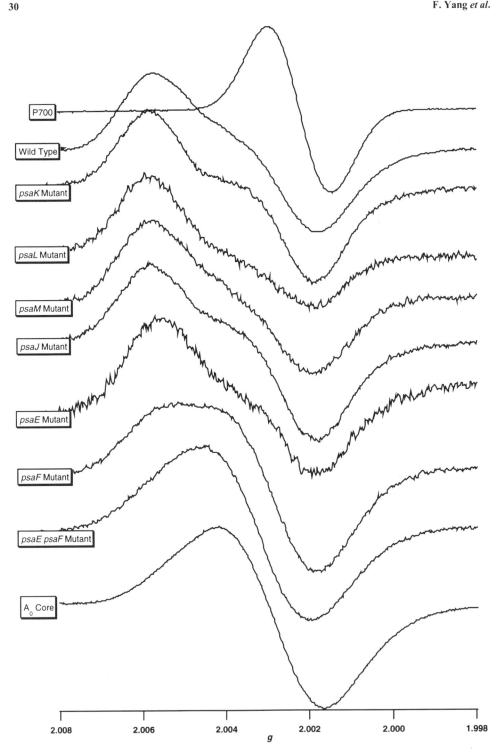

Figure 4. Q-band EPR studies of PS I complexes isolated with Triton X-100 from various *Synechococcus* sp. PCC 7002 strains. Conditions for photoaccumulation and for EPR measurements were identical to those described in the legend for Figure 2. The spectrum of P700$^+$ and A$_0^-$ are shown at the top and bottom of the figure for comparison.

X-100 leads to a modification of the environment surrounding A_1. The double reduction of A_1 would likely be an energetically-demanding event, particularly in the hydrophobic binding pocket which is proposed to bind the phylloquinone. However, if the single reduction were followed by a subsequent protonation, the second electron reduction would be expected to occur at a relatively modest reduction potential, in accordance with the finding that the redox potential of the semiphylloquinone/phylloquinol couple is higher than that of the phylloquinone/semiphylloquinone couple [27]. The second reduction step might also be followed by protonation, resulting in a fully-reduced, phyllohydroquinone species. In the intact complex, protonation is prevented by the sequestered binding region of the phylloquinone in the hydrophobic interior of the reaction center. Although the specific effect of Triton X-100 in this process is not known, some possibilities can be suggested. In the absence of PsaF and PsaE, it is possible that Triton X-100 modifies the polypeptide environment of the A_1 phylloquinone by changing the packing of the α-helices in this region of the complex. Alternatively, it is known that Triton X-100 can extract chlorophylls from higher plant PS I complexes, and it is possible that in the absence of PsaE and PsaF the removal of some chlorophylls occurs which in turn modifies the environment of A_1. Both mechanisms could result in the opening of a water channel that could facilitate the protonation of the singly- and doubly-reduced phylloquinone species.

In the most recent model for PS I [3], a well-defined pocket of electron density immediately adjacent to the short connection between α-helices m' and n' [3] was observed and tentatively assigned as phylloquinone Q_K. This region is nearly perfectly conserved between PsaA and PsaB and is partially conserved in the PshA and PscA reaction center polypeptides of heliobacteria and green sulfur bacteria, respectively. This electron density lies in close proximity to the other components of the electron transport chain and lies between the eC_3 chlorophyll and the F_X center. Moreover, the distance (25 Å ± 1Å) and angle (26° ± 2°) of this electron density relative to P700 are in close agreement with the experimentally determined values of 25.4 Å ± 0.3 Å and 26° ± 5° [28, 29; R. Bittl, personal communication]. It is presumed that a second phylloquinone might be bound near the short connection between α-helices m and n, no clear region of electron density that could be assigned to the Q_K' quinone is observed in this symmetry-related position. However, it is likely that this uncertainty will be removed when the crystal structure is available at higher resolution.

The model for PS I based on the 4 Å data set depicts m and m' as transmembrane helices and n and n' as stromal surface helices. The n helix extends over a region occupied by membrane-spanning *primed* helices, and the n' helix extends over a region occupied by membrane-spanning *non-primed* helices. The intersection of the m and n helices contains a group of conserved aromatic residues which have some similarity to a portion the Q_B binding site of bacterial reaction centers. The n helix is equivalent to the 12-residue helix which connects transmembrane helices D and E, and which also forms a portion the Q_B binding site of the L subunit of the bacterial reaction center. This region contains a number of amino acid residues that have been implicated in protonation of Q_B [30]. In the region most distal from the C_3 threefold symmetry axis of the trimer, the n helix abuts the region occupied by the membrane spanning PsaF/PsaJ polypeptides, whereas the region of the n' helix most distant from the pseudo-twofold axis C_2 (AB) abuts the region occupied by the membrane-spanning PsaL/PsaI and (to a lesser degree) PsaK polypeptides. Since the removal of PsaF (and to an even greater degree PsaE + PsaF) results in the photoaccumulation of A_0^- (and by inference the double reduction of A_1), we infer that the quinone which is active in electron transport is bound near the intersection of the *non-primed* m and n helices. Finally, Schubert et al. [3] have suggested that the non-primed helices correspond to PsaA and the primed helices correspond to PsaB. Hence, our results suggest that the A_1 phylloquinone is bound by the *m* and *n* helices of PsaA.

As discussed above, it is not known whether electron transfer occurs through only one branch of the pseudo-symmetric pairs of chlorophylls or whether electron transfer can occur through both branches. Our results indicate that electron transfer is not only unidirectional through one branch, but that the active electron transfer pathway can be identified. If electron transfer were to occur bidirectionally, double reduction of that phylloquinone which is located on the pathway near the PsaE/PsaF/PsaJ proteins would still allow electron transfer to occur via the phylloquinone on the pathway near the PsaL/PsaI proteins. Based on these results, we therefore propose that electron transfer is unidirectional through that phylloquinone which is located on the non-primed *m* and *n* helices associated with the PsaA polypeptide in PS I. By analogy to the pathway of electron transport in bacterial reaction centers, we suggest that the active pathway for electron transport is as follows: $eC_1 \rightarrow eC_2' \rightarrow eC_3' \rightarrow Q_K \rightarrow F_X$. If this suggestion is correct, it is interesting to note that P700 and Q_K would be most closely associated with PsaA while the accessory and A_0 chlorophylls would be most closely associated with PsaB. The F_X [4Fe-4S] center is formed with two cysteine ligands provide by each subunit. The model suggested from these studies should be further testable by appropriate site-directed mutagenesis studies.

ACKNOWLEDGMENTS

The authors would like to thank W.-D. Schubert and Drs. H. T. Witt , and W. Saenger for communication of results prior to publication. J.H.G. thanks Drs. Art Van der Est, Frazer Macmillan, R. Bittl and Dietmar Stehlik for stimulating discussions on the EPR characterization of A_1 and communication of results prior to publication. This work was supported by National Science Foundation Grants MCB-9205756 and MCB-9723661 to J. H. G. and MCB-9206851 and MCB-9723469 to D. A. B.

REFERENCES

1. Krauß, N., Hinrichs, W., Witt, I., Fromme, P., Pritzkow, W., Dauter, Z., Betzel, C., Wilson, K. S., Witt, H. T. and Saenger, W. (1993). Nature 361, 326–331.
2. Krauß, N., Schubert, W.-D., Klukas, O., Fromme, P., Witt, H. T. and Saenger, W. (1996) Nature Struct. Biol. 3, 965–973.
3. Schubert, W.-D., Klukas, O., Krauß, N., Saenger, W., Fromme, P. and Witt, H. T. (1997) J. Mol. Biol. 272, 741–769.
4. Li, N., Warren, P. V., Golbeck, J. H., Frank, G., Zuber, H. and Bryant, D. A. (1991) Biochim. Biophys. Acta 1059, 215–225.
5. Zhao, J., Li, N., Warren, P. V., Golbeck, J. H. and Bryant, D. A. (1992) Biochemistry 31, 5093–5099.
6. Li, N., Zhao, J., Warren, P. V., Warden, J. T., Bryant, D. A. and Golbeck, J. H. (1991) Biochemistry 30, 7863–7872.
7. Mühlenhoff, U., Zhao, J. and Bryant, D. A. (1996) Eur. J. Biochem. 235, 324–331.
8. Yu, L., Zhao, J., Mühlenhoff, U., Bryant, D. A. and Golbeck, J. H. (1993) Plant Physiol. 103, 171–180.
9. Rousseau, F., Sétif, P., and Lagoutte, B. (1993) EMBO J. 12, 1755–1765.
10. Falzone, C.J., Kao, Y.-H., Zhao, J. Bryant, D. A. and Lecomte, J. T. J. (1994) Biochemistry 33, 6052–6062.
11. Chitnis, V. P., Xu, Q., Yu, L., Golbeck, J. H., Nakamoto, H., Xie, D.-L., and Chitnis, P. R. (1993) J. Biol. Chem. 268, 11678–11684.
12. Schluchter, W. M., Shen, G., Zhao, J. and Bryant, D. A. (1996) Photochem. Photobiol. 64, 53–66.

13. Xu, Q., Yu, L., Hoppe, D., Chitnis, V. P., Odom, W. R., Guikema, J. A. and Chitnis, P. R. (1995) J. Biol. Chem. 270, 16243–16250.
14. Kruip, J., Chitnis, P. R., Lagoutte, B., Rögner, M., and Boekema, E. J. (1997) J. Biol. Chem. 272, 17061–17069.
15. Wynn, R. M. and Malkin, R. (1988) Biochemistry 27, 5863–5869
16. Wynn, R. M., Omaha, J. and Malkin, R. (1989) Biochemistry 28, 5554–5560.
17. Chitnis, P. R., Purvis, P. and Nelson, N. (1991) J. Biol. Chem. 266, 20146–20151.
18. Xu, Q., Yu, L., Chitnis, V. P., and Chitnis, P. R. (1994) J. Biol. Chem. 269, 3204–3211.
19. Xu, Q., Odom, W. R., Guikema, J. A., Chitnis, V. P., and Chitnis, P. R. (1994) Plant Mol. Biol. 26, 291–302.
20. Deisenhofer, J. Epp, O., Miki, K., Huber, R. and Michel, H. (1985) Nature 318, 618–624.
21. Ermler, U., Fritsch, G., Buchannan, S. K. and Michel, H. (1994) Structure 2, 925–936.
22. Golbeck, J. H. and Bryant, D. A. (1991) in Current Topics in Bioenergetics, Vol. 16 (Lee, C. P., ed.), pp. 83–177, Academic Press, New York.
23. Mühlenhoff, U., Kruip, J., Bryant, D. A., Rögner, M., Sétif, P., and Boekema, E. (1996) EMBO J. 15, 488–497.
24. Heathcote, P., Moënne-Loccoz, P., Rigby, S. E. J. and Evans, M. C. W. (1996) Biochemistry 35, 6644–6650.
25. Warren, P. V., Golbeck, J. H. and Warden, J. T. (1993) Biochemistry 32, 849–857.
26. Barry, B. A., Bender, C. J., McIntosh, L., Ferguson-Miller, S. and Babcock, G. T. (1988) Isr. J. Chem. 28, 129–132.
27. Bottin, H. and Sétif, P. (1991) Biochim. Biophys. Acta 1057, 331–336.
28. Bittl, R., Zech, S. G. and Lubitz, W. (1997) Biochemistry 36, 12001–12004.
29. Dzuba, S. A., Hara, H., Kawamori, A., Iwaki, M., Itoh, S. and Tsvetkov, Y. D. (1997) Chem. Phys. Lett. 264, 238–244.
30. Okamura, M. Y. and Feher, G. (1995) in Anoxygenic Photosynthetic Bacteria (Blankenship, R. E., Madigan, M. T., and Bauer, C. E. eds.), pp. 577–594, Kluwer, Dordrecht.

5

STUDIES ON STRUCTURE AND MECHANISM OF PHOTOSYNTHETIC WATER OXIDATION

G. Renger

Max-Volmer-Institut für Biophysikalische Chemie und Biochemie
Str. des 17. Juni 135, 10623 Berlin, Germany

1. INTRODUCTION

The key evolutionary step in the exploitation of solar radiation by photosynthetic water cleavage was the "invention" of a molecular device that enables the light-driven oxidation of two water molecules to molecular oxygen. Two indispensable prerequisites are required to perform water oxidation: (a) the generation of sufficiently oxidising redox equivalents and (b) the cooperation of four oxidising redox equivalents. This goal was achieved about 2–3 billion years ago at the level of prokaryotic photosynthesising organisms. The result was a multimeric pigment protein complex referred to as Photosystem II (PS II) that is anisotropically incorporated in the thylakoid membrane and acts as a water-plastoquinone oxidoreductase (for a recent review see ref. [1]). The overall reaction of PS II leading to formation of membrane bound plastoquinol (PQH_2) and release of molecular oxygen into the atmosphere comprises three reaction sequences: i) photooxidation of a special Chla component (symbolised by P680) and subsequent stabilisation of the primary charge separation by rapid electron transfer from $Pheo^{-\bullet}$ to a specially bound plastoquinone-9 molecule (Q_A) (for a review see ref. [2]); ii) cooperation of four strongly oxidising holes via a sequence of four univalent redox steps at a manganese containing unit, the water oxidising complex (WOC) that leads to molecular oxygen and four protons (for reviews see refs. [3–5]), and iii) cooperation of two reducing equivalents (electrons) via a sequence of two univalent redox steps at a plastoquinone-9 molecule (Q_B), transiently bound into a protein pocket (Q_B-site). This reaction leads to PQH_2 formation (for reviews see refs. [6,7]).

The functional and structural organisation of reaction sequences (i) and (iii) exhibits striking similarities to the corresponding processes that take place in the reaction centers of anoxygenic purple bacteria. Accordingly, the functional redox groups performing these reactions are assumed to be incorporated into a heterodimer consisting of polypeptides D1 and D2 with similar structural arrangements as those of purple bacteria that are bound to a heterodimer composed of the L- and M-subunit [8]. This idea is highly supported by latest values for center to center distances gathered from measurements with magnetic resonance methods. The distance between (B)Pheo and the non-heme iron center was inferred

The Phototrophic Prokaryotes, edited by Peschek *et al.*
Kluwer Academic / Plenum Publishers, New York, 1999.

to be very similar in both types [9] and more importantly also the distance between P680 and Q_A with 27.4 ($\pm$0.3) Å [10] is close to that between P870 and Q_A in *Rhodobacter sphaeroides* with 28.3 Å [11]. Likewise the mode of Q_A binding via hydrogen bridges exhibits characteristic similarities [12]. As a complementary functional analogon, the kinetics of the elementary reactions leading to the radical pair $P680^{+\cdot}Q_A^{-\cdot}$ in PS II are very similar to those of the corresponding processes in purple bacteria (for a discussion see [13] and references therein).

On the other hand there exist marked differences with respect to the properties of the photochemically active pigment P680. Apart from this peculiarity of process (i), the unique feature of PS II is the reaction sequence (ii). Accordingly, the following considerations will be entirely restricted to these two problems with special emphasis on the water oxidising complex.

2. PROPERTIES OF P680

The cation radical $P680^{+\cdot}$ is one of the most oxidising species in biological systems. The reduction potential of $P680^{+\cdot}$ has been estimated to be about +1.1 V [14, 15]. A comparison with the photoactive Chl a in Photosystem I (PS I) referred to as P700 readily shows that the replacement of bacteriochlorophyll (BChl) in purple bacteria by Chl a in PS II cannot be the clue for achieving the goal because the reduction potential of $P700^{+\cdot}$ with +0.5 V (for a review see refs. [16,17] has almost the same value as that of the special pair in purple bacteria (e.g. P870 in *Rb sphaeroides*, see [18]). Furthermore, the reduction potential of monomeric Chl $a^{+\cdot}$ in solution has been found to be only +0.85 V and this value decreases upon dimerization [19]. Based on these findings it is clear that the protein environment around P680 including the possibility of binding positively charged cofactors is of key functional relevance. Recently, the number of hydrogen bonds to the BChl moieties of the special pair in *Rb. sphaeroides* was found to cause profound shifts of the reduction potential. Values as high as +0.765 V were achieved when a maximum number of hydrogen bonds to carbonyl groups was established by site directed mutagenesis [20]. This tool for enhancing the reduction potential, however, does not appear to be the key mechanism for P680 because in Chl a the 2-acetyl-group of BChl is replaced by an ethylene group. Therefore, other possibilites have to be considered such as the presence of a positive charge (e.g. cation cofactor). At present, the underlying mechanism of the protein matrix for making the Chl $a^{+\cdot}$ radical a highly oxidising species (E $\approx$ +1.1 V) is not yet resolved.

The exceptionally high reduction potential of $P680^{+\cdot}$ has additional implications for array and/or properties of other Chl a molecules bound to the protein matrix (see [4, 21]). In order to establish a selective electron transfer from the redox active tyrosine Y_Z (vide infra) to $P680^{+\cdot}$ and to avoid dissipative side reactions the other Chl a-molecules have to either be sufficiently far apart from P680 or also to undergo a drastic shift of the reduction potential. Both alternatives require serious structural constraints that have no corresponding counterpart in other reaction centers. It is a challenge for future research unravelling this problem.

3. PHOTOSYNTHETIC WATER OXIDATION

3.1. Functional Pattern

Based on measurements of the pattern of flash induced oxygen yield in dark adapted samples (whole algal cells, isolated thylakoids) the process of water oxidation was shown to take place via a redox cycle of four univalent redox steps (for review see ref. [22]) that can be summarised by the scheme depicted in Fig. 1. The oxidation pathway is energetically driven by the light induced formation of P680$^{+\cdot}$ that extracts stepwise the electrons from the WOC with tyrosine 161 of polypeptide D1 (Y_Z) as redox mediator (see [3] and references therein). The mechanistic role of Y_Z (electron transfer component or hydrogen abstractor from bound substrate water) is a matter of current debate and will be discussed later in connection with the mechanism of water oxidation. An inspection of Fig. 1 reveals that the understanding of water oxidation requires detailed knowledge on the electronic configuration and the nuclear geometry of the functional center of the WOC in its different redox states S_i. In order to address this point the following description provides experimental results and theoretical consideration on the structure of the WOC, the temperature dependence and kinetic H/D isotope exchange effects of individual redox steps and mechanistic implications.

3.2. Structure of the WOC

In contrast to reaction centers of purple bacteria (see [11] and references therein) and Photosystem I of cyanobacteria [23,24] a corresponding X-ray crystallographic structure or structural model is lacking for PS II so far. Based on the similarities between PS II and purple bacteria rather detailed models have been presented (see [25] and references therein). However, it has to be emphasised that the arrangement of the essential cofactors and Chl a molecules within the D1/D2 heterodimer are mutually exclusive among different proposals

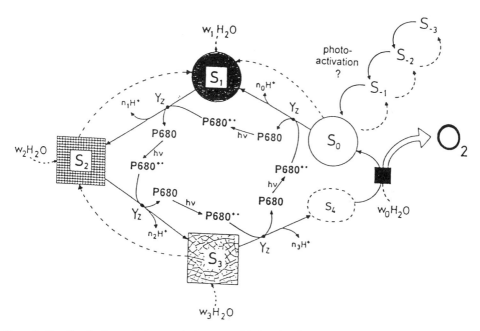

Figure 1. Functional scheme of photosynthetic water oxidation via a four step univalent redox sequence energetically driven by photooxidised P680$^{+\cdot}$ with tyrosine Y_Z as redox active intermediate. The "superreduced" redox states S_{-1}, S_{-2} and S_{-3} can be populated by using hydrophilic reductants like NH_2OH and NH_2NH_2 (for a recent study see [57] and references therein). It is not yet clarified if these states are formed as intermediates in the process of photoactivation. For further details see text.

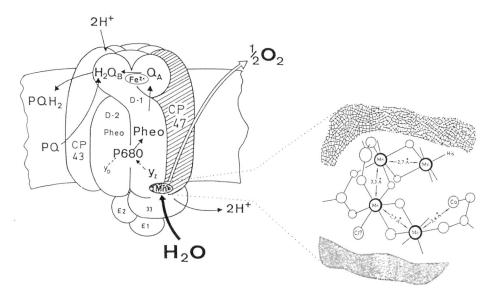

Figure 2. Schematic representation of the PS II complex (left side) and of the tetranuclear manganese cluster and part of its first coordination sphere (modified after ref. [27]) embedded into an asymmetric protein matrix. For further details, see text.

(see e.g. ref. [25] versus [26]) and almost no information on the WOC and P680 properties can be gathered from this type of modelling. Therefore, it appears more appropriate to present only a cartoon of the overal structural organisation of PS II as is shown on the left hand side of Fig. 2. It has been known for a long time that the functionally competent WOC contains four manganese (for a review see ref. [3]). Invaluable structural information on the array of these manganese and the first coordination sphere could be gathered from extended X-ray absorption fine structure (EXAFS) analysis. A widely used model emerging from this data [27] is shown in Fig. 2, right side, but it has to be stressed that this is not the only possible manganese arrangement [28] and therefore at present provides nothing more than a very useful working model. EXAFS data revealed that the first coordination sphere of the manganese is composed of (C)N, O-atoms whereas S-containing ligands can be excluded [27]. Apart from the pure μ-oxo bridges, in general three different types of ligands have to be considered: α) substrate water molecules in different protonation states (α-type ligands), β) aminoacid residues of the surrounding protein matrix (β-type ligands) and γ) other possible inorganic cofactors like Ca^{2+}, chloride or bicarbonate (γ-type ligands).

The vast majority of the experimental data support the idea that substrate water binds to the WOC in redox state S_1 (S_0) and that it is coordinated to manganese (for a recent discussion see ref. [29]). The redox state of this water remains virtually unaffected up to the redox state S_3 as will be outlined in the subsequent section 3.3.4.

With regard to β-type ligands the EXAFS data reveal that carboxylic aminoacid residues (Asp, Glu) and histidine are the most likely candidates. A systematic screening was performed for polypeptides D1 and D2 in the cyanobacterium Synechocystis sp. PCC6803. Asp-170, Glu-189, His-332, Glu-333, His-337 and Ala-344 (C-terminus) of polypeptide D1 ([30,31] and references therein) and Glu-70 of polypeptide D2 [32] were found to be essential for the assembly of a functionally competent and stable WOC. So far none of these residues was unambiguously identified as ligand of the first coordination sphere of manganese.

Regardless of lacking direct proof, it is most likely that the manganese cluster is bound to the D1/D2 heterodimer which houses the other functional groups of PS II, especially P680 and Y_Z. This idea, however, does not exclude the possibility that other subunits are essential constituents either by providing direct ligands or by stabilising the manganese cluster. The latter function is indispensable because the WOC is thermodynamically a non-equilibrium system (for discussion see ref. [29]) that requires a light driven multistep mechanism of low quantum yield referred to as photoactivation (for details see refs. [33–35]) for its assembly. It is known that the extrinsic 33 kDa protein (Psb-O protein) is essential for the stability of the manganese cluster (for a recent review see ref. [36]). Another most interesting component of PS II is the Chl-a binding protein CP47 that is present in all oxygen evolving organisms [32]. It is symbolised by a hatched area in Fig. 2. Hydrophobicity analyses led to the conclusion that CP47 forms six transmembrane helices (see refs. [37,38] and Fig. 3]). A second Chl a-binding protein of similar structure is CP43.

One functional role of proteins CP47 and CP43 is to form the core antenna. In a recent study the folding pattern within the thylakoid membrane of both proteins was proposed to resemble that of transmembrane helices I to VI of the proteins PsaA and PsaB. Sequence homology of >60% have been found for transmembrane helices I and IV of CP43/CP47 and of PsaB/PsaA [39]. The latter proteins build up the heterodimer of PS I which not only provides the matrix of the functional groups for light induced charge separation, but also binds the pigments of the core antenna. It was therefore suggested that a motif of 22 transmembrane helices is characteristic for both PS I and PS II. In the former case the large proteins PsaA and PsaB each contain eleven of them while in PS II the system is split into two parts with the five transmembrane helix part confined to each D1 and D2, respectively, and the six transmembrane helix part to CP47 and CP43, respectively [24]. Within the framework of this proposal the intrinsic transmembrane regions of CP47 and CP43 are nothing more than a counterpart to the core antenna of PS I. A closer inspection of the folding pattern of CP47 depicted in Fig. 3 (CP43 exhibits a similar feature, see ref. [40]), however, reveals a striking structural feature i.e. the presence of an exceptionally large loop (designated as loop E) connecting transmembrane helices V and VI, that is lacking in PsaA and PsaB. This loop with its predicted 191 aminoacid residues is of a size comparable with that of some extrinsic regulatory subunits of the WOC (vide supra). Different lines of evidence suggest a close interaction with the extrinsic manganese stabilising PsbO protein. All data available so far are in line with the idea that the N-terminus of the PsbO protein interacts with a domain of loop E of CP47 between residues 364 and 440 (for review see [36] and references therein) as indicated in Fig. 3. Based on these findings CP47 appears to be not simply a core antenna but probably also involved in establishing a functionally competent WOC. Taking into account the difference to PsaA/PsaB of PS I, an analysis of the possible role of loop E of CP47 seems to be a most promising approach. To address this problem three to eight aminoacid residues were deleted by genetic engineering from conserved and charged regions. A screening of 12 mutants revealed that in some of them drastic effects were observed, i.e. the PS II complex could not be assembled in a stable form and photoautotrophic growth was suppressed (for details see ref. [41]). This finding most convincingly shows that loop E is an indispensable structural element for PS II but does not provide information on a specific role for the WOC. For this reason those mutants are much more interesting that retain PS II activity and simultaneously exhibit modified properties of the WOC. Interestingly, these mutations that are marked in black in Fig. 3 are located in the putative region of interaction with the PsbO protein. A thorough analysis of the properties [42,43] indicates that neither the lifetimes of redox states S_2 and S_3 nor the peak temperatures of the thermoluminescence bands (B and Q-band upshift of 3–7°C compared to WT) are markedly changed in mutants

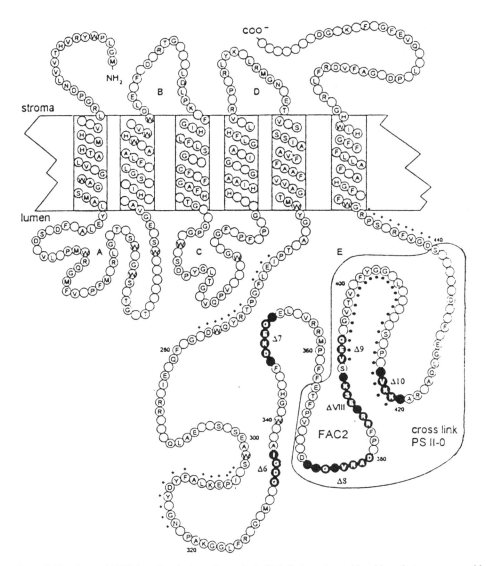

Figure 3. Topology of CP47 based on hydropathy analysis [37]. Only amino acid residues that are conserved between *Synechocystis* sp. PCC6803 and spinach (*Spinacea oleracea*) are explicitly given. The enclosed area describes the region containing sites of cross-linking with the N-terminus of the PS II-O protein. The regions indicated by solid residues are domains that can be deleted without loss of autotrophic growth and oxygen evolution, and upon deletion of regions indicated with crosses mutants grow only photoheterotropically and are not able to assemble a stable PS II complex. Mutants with strongly restricted photoautotrophic growth and drastically reduced PS II content are not shown (for details see ref. [41]). The following abbreviations were used to symbolise the mutants (see also ref. [41]): Δ6, Δ(G333–I336); Δ7, Δ(K347-R352); Δ8, Δ(A373-D380); Δ VIII, Δ(R384-V392); Δ9, Δ(R392-Q394) and Δ10, Δ(D416-F240).

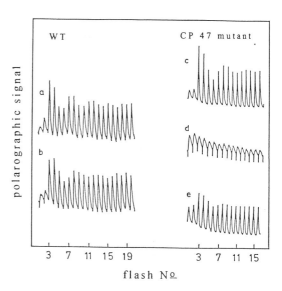

Figure 4. Polarographic signals detected with a Joliot type electrode in cells of WT (traces *a* and *b*) and mutant Δ8 (traces *c*, *d* and *e*) that were preilluminated for 5 min and subsequently dark-adapted for 5 min (traces *a* and *c*) or 24 h (traces *b* and *d*). Trace *e* shows a pattern of a Δ8 mutant after 24 h dark incubation and subsequent illumination with 90 single turnover flashes followed by 5 min dark relaxation before measurements. (For details see refs. [42,43]).

Δ6-Δ10. On the other hand pronounced effects were observed for the stability of the WOC during dark incubation. Likewise, mutant Δ8 also exhibits a lower binding affinity to the PsbO protein. The effects on the characteristic period four oscillation of flash induced oxygen evolution [22] are illustrated in Fig. 4 for mutant Δ8. After 5 min illumination followed by a short dark time (5 min) the oscillation patterns are virtually the same in WT (trace *a*) and mutant Δ8 (trace *c*). However, after thorough dark incubation (24 h) the WOC of the mutant attains an inactive state (the signals of trace *d* are artifacts as shown for the mutants deprived of PS II, see [42]) whereas the WT is only slightly affected. Illumination of the dark adapted inactive mutant Δ8 restores the oscillation pattern as shown by trace *e* in Fig. 4. However, 90 pre-illumination flashes are obviously not sufficient for a full recovery thus indicating that the restoration is a low quantum yield process. Furthermore, the number of flashes required for achieving pattern *c* was smaller by a factor of almost 10 in the presence of Ca^{2+} [43]. These properties are reminiscent of the process for photoactivation of the WOC (vide supra). Therefore, the WOC in mutant Δ8 and similar mutants is inferred to be unstable towards reductive displacement of the manganese cluster and needs to be reassembled via the process of photoactivation (for further details see ref. [43]). The loop E of CP47 is highly conserved (see Fig. 3). It is therefore most likely that this structural element is not only important for a stable WOC in cyanobacteria but of general relevance for all oxygen evolving organisms. Recent studies with CP47 mutants which are additionally deprived of the extrinsic 33kDa proteins by PsbO gene inactivation clearly show the the effects observed in Δ8 and similar mutants are specific for CP47 and not owing to a modified interaction of loop E with the PsbO protein. [Seeliges, Engels, Pistorius, Vemaas and Renger, manuscript in prep.]

With regard to evolutionary development it appears attractive to speculate that the gene encoding a polypeptide of similar size and sequence as loop E was inserted to the psbB (and psbC) gene precursors.

Apart from the indispensable role of proteins and cofactors like Ca^{2+} another class of compounds has to be taken into consideration: lipids. It is known that cardiolipin is an essential constituent of cytochrome c oxidase that catalyses the reverse reaction of water oxi-

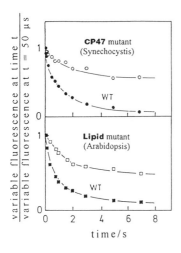

Figure 5. Laser flash induced changes of the normalised variable fluorescence quantum yield at room temperature as a function of time in dark adapted DCMU treated cells of *Synechocystis* sp. PCC6803 wild type (solid circles) and CP47 mutant Δ8 (O) in the upper part [42] and of leaves from *Arabidopsis thaliana* wild type (■) and *dgd1* mutant in the bottom part (for further details see ref. [45]).

dation in the respiratory chain (for a review see ref. [44]). Systematic studies for the WOC are lacking so far. In a recent attempt the properties of an *Arabidopsis thaliana* mutant were analysed that contains a drastically reduced level of digalactosyldiacylglycerol (DGDG). Interestingly, it was found that this mutant exhibits a qualitatively similar modification of the transient flash induced fluorescence change [45] as the CP47 mutant Δ8 of *Synechocystis* PCC6803 when leaves or cells are treated with DCMU. The data are shown in Fig. 5. This finding and additional measurements with much higher time resolution in untreated leaves from the DGDG-mutant led to the conclusion that DGDG affects the reaction pattern of the WOC [45]. The exact role of lipids for the WOC remains to be clarified in future studies.

With regard to manganese coordination within the protein matrix, another point of key relevance has to be taken into account: the protein environment provides not only structural determinants but, most importantly, will play a key functional role for several reasons: i) an asymmetric protein surrounding can give rise to a heterogeneity of the reaction properties of each manganese (for detailed discussions see refs. [4, 46]), ii) the chemical nature of the aminoacid residues and their geometrical array determine the redox properties of manganese and the exchange rate of α- and/or γ-type ligands, and iii) dynamic structural changes are of central relevance for key steps of the reaction pathway from water to molecular oxygen: substrate water uptake, formation of the essential O-O bond, protolytic reactions and O_2 release. Structure-function relations of this type will be discussed in the following sections.

3.3. Mechanism of Water Oxidation in the WOC

The use of different spectroscopic techniques led to the conclusion that the redox transitions $S_0 \rightarrow S_1$ and $S_1 \rightarrow S_2$ within the WOC are metal centered reactions with Mn(II) $\rightarrow$ Mn(III) and Mn(III) $\rightarrow$ Mn(IV) oxidation steps, respectively. On the other hand, $S_2 \rightarrow S_3$ is probably a ligand centered process that might comprise substrate oxygen (for recent reviews see [29,47] and references therein). Further information on the nature of the redox steps leading to water oxidation can be gathered from analyses of their temperature dependencies within the framework of the Marcus theory (see [48] and references therein) and by measuring kinetic H/D isotope exchange effects.

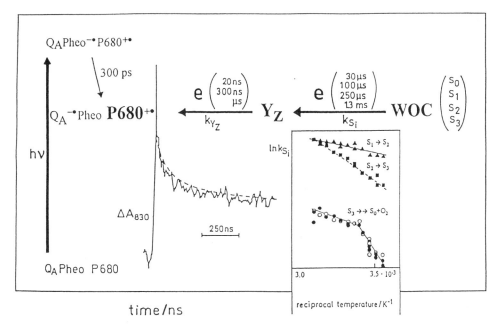

Figure 6. Absorption changes at 830 nm (left side) as a function of time and reciprocal half lifetimes as a function of reciprocal temperature (right side) in PS II fragments from *Synechococcus vulcanus* Copeland. The signal on the left side represents a trace monitored at 33°C, the dashed curve symbolises the data at 0°C. The fast decay of the 830 nm absorption change due to Pheo⁻˙ oxidation by Q_A is not resolved and symbolised by a spike (for further details see [49] and references therein).

3.3.1. Temperature Dependence of the Redox Reactions in Thermophilic Cyanobacteria and Higher Plants. Fig. 6 shows the kinetics of P680⁺˙ oxidation by Y_Z (trace at left side) and the Arrhenius-plot of the temperature dependence of redox transitions in the WOC (insert at the right-hand side, bottom) of the thermophilic cyanobacterium *Synechococcus vulcanus* Copeland. The trace on the left side reveals that the electron transport from Y_Z to P680⁺˙ is only slightly dependent on temperature [49]. More detailed studies in PS II membrane fragments from spinach led to activation energies of about 10 kJ/mol in samples with an intact WOC [50] and of about 27 kJ/mol after destruction of the WOC [75].

The redox transitions of the WOC as a function of temperature were analysed by measuring flash induced UV absorption changes. An Arrhenius plot of the data obtained for *Synechococcus vulcanus* Copeland is shown in the insert of Fig. 6 and the activation energies calculated are compiled in Table 1 together with the values for the corresponding reactions in PS II membrane fragments and PS II core complexes from spinach. An inspection of these data reveals two striking features: a) the activation energies increase with progressing oxidation state S_i of the WOC up to formation of S_3, b) the electron abstraction from the WOC in S_3 exhibits a marked change of E_A at a characteristic temperature ϑ_c, c) the E_A-values are similar in PS II preparations from a thermophilic cyanobacterium and from higher plants.

The latter finding suggests that the reaction coordinates of the WOC remained practically invariant to the evolutionary development of all oxygen evolving organisms from thermophilic cyanobacteria to higher plants. This conclusion implies that the basic functions were already optimised at early stages after its invention. A further consequence emerges from the data of Table 1. As the PsbO protein (33 kDa) is the only extrinsic

Table 1. Activation energies $E_{A,i+1,i}$ of the reactions leading to water oxidation in PS II preparations

Reaction	$E_{A,i+1,i}$ (kJ/mol)		
	Cyanobacteria (*S. vulcanus* Copeland)[a]	Plants (spinach)	
		Membrane fragments[b]	Core[c]
$Y_Z^{OX}S_0 \rightarrow Y_ZS_1$		5	
$Y_Z^{OX}S_1 \rightarrow Y_ZS_2$	9.6	12.0	14.8
$Y_Z^{OX}S_2 \rightarrow Y_ZS_3$	26.8	36.0	35.0
$Y_Z^{OX}S_3 \rightarrow\rightarrow Y_ZS_0 + O_2$			
$\vartheta > \vartheta_c$	15.5	20.0	21.0
$\vartheta < \vartheta_c$	59.4	46.0	67.0
Temperature ϑ_c	+16°C	+6°C	+11°C
Extrinsic proteins	PsbO	PsbO	PsbO
	PsbV	PsbP	
	PsbU	PsbQ	

[a]Koike, H., Hanssum, B., Inoue, Y. and Renger, G. (1987) Biochim. Biophys. Acta 893: 524–533
[b]Renger, G. and Hanssum, B. (1992) FEBS Lett. 299: 28–32
[c]Karge, M., Irrgang, K.-D. and Renger, G. (1997) Biochemistry 36, 8904–8913

subunit that is present in all three types of preparation, the other two are very likely without any effect on the reaction coordinates of the WOC. This idea is supported by the finding that removal of the proteins PsbP (23 kDa) and PsbQ (18 kDa) does not modify the structure of the S_2 multiline EPR signal [51]. Likewise, the replacement of cytochrome C550 (PsbV) and the PsbU protein in cyanobacteria (see ref. [52]) by the PsbP and PsbQ proteins in plants does not directly affect the reaction pattern of the WOC. Therefore, this evolutionary change of the PS II polypeptide composition is probably related to regulatory function(s) that still remain(s) to be unraveled (see Ref. 76).

3.3.2. Analysis of the Redox Reactions within the Framework of the Marcus Theory. The Marcus theory of nonadiabatic electron tunneling cannot be applied to all biological redox reactions in a simple straightforward manner and therefore the results obtained have to be considered with care [48,53]. The experimental data reported in this study were obtained at temperatures above 250 K. In this case the low frequency modes of the proteins are thermally equilibrated and the classical approach of Marcus [54,55] can be used for the rate constant k_{ET}:

$$k_{ET} = \frac{4\pi^2}{h}|V_{ET}|^2 (2\pi\lambda_{ET}k_B T)^{-\frac{1}{2}} \exp\left[-\frac{(\Delta G_{ET}^0 + \lambda_{ET})^2}{4\lambda_{ET} \cdot k_B T}\right] \tag{1}$$

where V_{ET} is the matrix element of electronic coupling between the redox groups, λ_{ET} is the reorganisation energy that reflects all nuclear modes which are coupled with the electron transfer step, ΔG_{ET}° is the standard Gibbs energy of the process, and k_B and h are the Boltzmann and Planck constants, respectively.

The value of V_{ET} depends on the overlapping of the wavefunctions and therefore exhibits an exponential dependence on the van der Waals distance r_{DA} between the redox groups D (donor) and A (acceptor)

$$V_{ET}(r_{DA}) = V_{ET}(O)e^{-\beta r_{DA}} \qquad (2)$$

where β is the distance decay coefficient that characterises the *average* intercalating medium (protein) (for a more detailed discussion see refs. [53,55,56]).

A comparison with the simple Arrhenius equation for the temperature dependence of rate constants

$$k = A \cdot \exp(-E_A / RT) \qquad (3)$$

where A and E_A are the preexponential factor and the activation energy, respectively, and R is the gas constant, leads to the following expression for calculating a physically realistic value λ_{ET} (for details see ref. [58, 75]):

$$\lambda_{ET} = 2(2E_A + RT - \Delta G^{\circ}_{ET})\left\{1 + \sqrt{1 - (\Delta G^{\circ}_{ET})^2 / 2(2E_A + RT - \Delta G^{\circ}_{ET})}\right\} \qquad (4)$$

Likewise, based on an empirical distance-rate constant relationship [59] the following expression can be derived for the van der Waals distances (for details see ref. [58, 75]):

$$r_{DA} = 25 - 1.67 \log k_{ET} - 0.21 E_A \qquad (5)$$

where r_{DA} is in units of Angström and E_A in kJ/mol.

Using values of ΔG° [60] and E_A ([50, 75] and Christen and Renger unpublished) for the electron transfer from Y_Z to $P680^{+\cdot}$ leads to λ_{ET}-values of 0.5 eV and 1.6 eV in samples with a functionally competent and destroyed WOC, respectively. The marked increase of λ_{ET} can be explained by a significantly larger number of water molecules in the environment of Y_Z caused by WOC destruction [29, 61]. For the van der Waals distance r_{DA} between Y_Z and P680 values of 9 ± 2 Å are obtained in both sample types [58,62], i.e. r_{DA} is almost independent of the WOC integrity.

A corresponding analysis for redox transitions in the WOC induced by Y_Z^{OX} gives rise to values of $\lambda_{ET}(Y_Z^{OX}S_0 \rightarrow Y_ZS_1) \approx \lambda_{ET}(Y_Z^{OX}S_1 \rightarrow Y_ZS_2) \approx 0.7$ eV, $\lambda_{ET}(Y_Z^{OX}S_2 \rightarrow Y_ZS_3) \approx$ 16 eV, and $\lambda_{ET}(Y_Z^{OX}S_3 \rightarrow\rightarrow Y_ZS_0 + O_2) \approx 1.1$ eV ($\vartheta > \vartheta_c$) [57,62] and 2.5 eV ($\vartheta < \vartheta_c$). Reorganisation energies of 0.7 eV are typical for biological electron transfer reactions in a less hydrophilic protein environment [55]. Based on the idea that the oxidation of manganese by Y_Z in redox states S_0 and S_1 is determined by the electron transfer step, and applying of Eqn. (5) leads to a van der Waals distance between Y_Z and $(Mn)_4$ of ≥ 15 Å [58, 62, 75]. This rather large distance has most important mechanistic implications as will be briefly outlined in section 3.3.3. Another very interesting point is the markedly higher λ_{ET} value calculated for S_2 oxidation to S_3. It is concluded that this feature reflects an electron transfer that is gated by a structural change (for a general discussion of gated electron transfer reactions see ref. [48]). The idea of significant structural changes coupled with S_3 formation has gained strong support by the drastically different reactivity of S_2 and S_3 towards exogenous reductants (NH_2NH_2, NH_2OH) (for details see [64]) and by recent EXAFS measurements

Table 2. Kinetic H/D isotope exchange effects in PS II
membrane fragments and PS II core from spinach

Reaction	k(H)/k(D)	
	Membrane fragments[a,b]	Core[b]
$Y_Z^{OX} \rightarrow Y_Z S_2$	1.3 ± 0.2	1.6 ± 0.3
$Y_Z^{OX} \rightarrow Y_Z S_3$	1.3 ± 0.2	2.3 ± 0.4
$Y_Z^{OX} S_3 \rightarrow\rightarrow Y_Z S_0 + O_2$	1.4 ± 0.3	1.5 ± 0.2

[a]Renger, G., Bittner, T. and Messinger, J. (1994) Biochem. Soc. Trans. 22, 318–322.
[b]Karge, M., Irrgang, K.-D. and Renger, G. (1997) Biochemistry 36, 8904–8913.

that indicate a marked lengthening of the manganese distance in one of the μ-oxo bridged dimers [65].

As for redox transition $S_2 \rightarrow S_3$ also the oxidation of S_3 by Y_Z that eventually leads to formation of molecular oxygen is assumed to be a gated reaction.

3.3.3. Kinetic H/D Isotope Exchange Effects. It has recently been proposed that $P680^{+\bullet}$ leads to the formation of a neutral $Y_Z^\bullet$ radical which acts as abstractor for hydrogen atoms from H_2O (or OH^-) coordinated to manganese [61,66,67]. Within the framework of this model the oxidation of the manganese in the WOC during the reactions $S_0 \rightarrow S_1$ and $S_1 \rightarrow S_2$ has to be considered as a catalytically gated electron transfer that is triggered by the break of a covalent OH bond. This mode of reaction is expected to give rise to significant kinetic isotope effects. Recent data for the intermolecular electron transfer from tryptophan-tryptophylquinone of methylamindehydrogenase to Cu-containing amicyanin show that a switch from a "pure" electron transfer step to one that is gated by proton transfer gives rise to a marked increase of the kinetic H/D isotope exchange effect from 1.5 to more than 10.0 [68]. Accordingly, the rate constants of electron transfer from Y_Z to $P680^{+\bullet}$ and from the WOC to Y_Z^{OX} were measured in PS II membrane fragments and core complexes from spinach. In samples with intact WOC virtually no effect, i.e. k(H)/k(D) < 1.05, is observed for the dominating ns kinetics of $P680^{+\bullet}$ reduction by Y_Z [69] while after destruction of the WOC the reaction is dominated by μs kinetics with an increased k(H)/k(D) ratio of 2.7 [70]. The latter effect is fully consistent with the idea that the environment of Y_Z becomes enriched with H_2O molecules after removal of the WOC (see section 3.3.2).

The results obtained for electron abstraction from the WOC by Y_Z are compiled in Table 2. Comparatively small values are observed that can be explained by secondary isotope exchange effects (for further discussion see refs. [29,68]). These data are not in favour with the idea of a hydrogen abstractor model. This conclusion is supported by the analysis presented in section 3.3.2 and other experimental evidence on the nature of the protolytic reactions [5]. Of special mechanistic relevance is the distance between Y_Z and the manganese cluster. A large distance of the order of 15 Å excludes the hydrogen abstractor model. However, it has to be emphasised that this value is currently a matter of controverse discussion (for further details see [29, 58] and references therein).

3.3.4. Proposed Model for Water Oxidation. The pivotal step of water oxidation to dioxygen is the formation of the oxygen–oxygen bond. Two decades ago this event had been postulated to occur at the substrate redox level of a peroxidic type species that is binuclearly complexed to manganese [71]. This state was assumed to be preformed in redox state S_3 symbolised by:

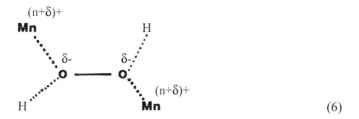

$$(6)$$

where δ denotes a charge distribution parameter between manganese and substrate ligand [72]. As a consequence of this idea and taking into account the structural change of the WOC upon S_3 formation this redox state has been assumed to comprise a redox isomerism of the type:

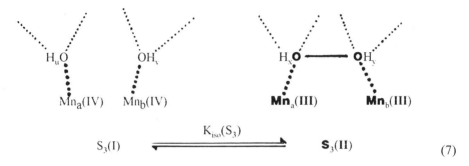

$$(7)$$

where Mn_a and Mn_b are two manganese centers that attain the mixed-valence state $Mn(III)Mn(IV)$ in S_2 (for details see [47] and references cited therein), u, v, x, y are the numbers of protons covalently bound to the oxygen atom of each substrate molecule, heavy dotted lines symbolise coordination of substrate oxygen atoms to manganese, thin dotted lines hydrogen bridges to the protein matrix and $k_{iso}(S_3)$ the equilibrium constant. The idea of a redox isomerism equilibrium is in line with $H_2{}^{18}O/H_2{}^{16}O$ isotope exchange data and with XANES experiments as has been outlined elsewhere [62].

It has to be emphasised that the structure of the WOC in S_3 is inferred to be markedly different from that of S_2. In order to avoid misinterpretation, this phenomenon has to be clearly distinguished from the question as to what extent there exists a conformational difference of the WOC between redox isomers $S_3(I)$ and $S_3(II)$.

An essential point of the redox isomerism concept is the proposal that generation of $Y_Z{}^{ox}$ by P680$^{+\bullet}$ leads to a drastic and "instantaneous" increase of $K_{iso}(S_3)$ followed by oxidant induced reduction of the functional manganese and formation of complexed dioxygen [73]. In this way $S_3(II)$ can be considered as the "entatic" state [74] of the WOC for water oxidation to molecular oxygen (see also [29, 58, 75]).

With respect to the "two manganese template" the "dimer of dimers model" depicted in Fig. 2 (right side) offers two alternative possibilities: i) one "dimer" is redox active during the redox cycle from S_0 through S_3 and binds the substrate water molecules (see [29] and references therein) or ii) the manganese at the open end of each dimer that are not connected via a μ-oxo (carboxylato) bridge and have a distance of about 5.5 Å constitute the "functional unit" for water oxidation [61,65]. The second alternative appears to be very attractive from a mechanistic point of view but depends on the cis-configuration of the "dimer of dimers" model. At present no unambiguous decision can be made.

3.4. Concluding Remarks

The present report provides a short summary of our current stage of knowledge on the functional and structural organisation of photosynthetic water oxidation. As the result a model has been proposed that is summarised in Fig. 7. The key mechanistic point is a postulated redox isomerism in oxidation state S_3 of the WOC that gives rise to preformation of an O–O bond at the substrate redox level of a peroxide. The oxidation of Y_Z to Y_Z^{OX} by $P680^{+\cdot}$ is assumed to cause an "instantaneous" large increase of $K_{iso}(S_3)$ followed by oxidant induced reduction of manganese under formation of molecular oxygen. Furthermore, this mechanism implies that the reactions take place at a binuclear manganese template (in Fig. 7 one dimer is assumed to establish this unit, but this problem is not yet resolved, *vide supra*). The implications of this model have to be tested in future studies.

ACKNOWLEDGMENTS

The author gratefully acknowledges the contributions of all coworkers and colleagues and their papers quoted in this report and the financial support by Deutsche Forschungsgemeinschaft and Fonds der Chemischen Industrie. The author would also like to thank Dr. P. Fromme for very stimulating discussions on evolutionary aspects, S. Hohm-Veit for drawing the figures and A. Menke for competent typing of the manuscript.

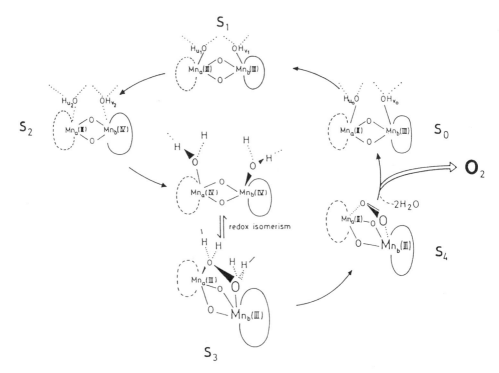

Figure 7. Model of photosynthetic water oxidation at a catalytic binuclear manganese template. The ligands other than μ-oxo bridges and coordinated substrate are summarised by ovals that are dashed and full-lined in order to symbolise the different microenvironment of manganese Mn_a and Mn_b. For further details see text.

REFERENCES

1. Renger, G. (1997) in: Treatise on Bioelectrochemistry, Vol. 2: Bioenergetics (Gräber, P. and Milazzo, G., eds.) pp. 310–358. Birkhäuser Verlag, Basel
2. Renger, G. (1992) in: Topics in Photosynthesis, The Photosystems: Structure, Function and Molecular Biology (Barber, J., ed.), pp. 45–99, Elsevier, Amsterdam
3. Debus, R.J. (1992) Biochim. Biophys. Acta **1102**, 269–352
4. Renger, G. (1993) Photosynth. Res. **38**, 229–247
5. Haumann and Junge (1996) in: Oxygenic Photosynthesis: The Light Reactions (Ort, D.R. and Yocum, C.F., eds), pp. 156–192, Kluwer Academic Publishers, Dordrecht
6. Crofts, A.R. and Wraight, C.A. (1983) Biochim. Biophys. Acta **726**, 149–185
7. Lavergne and Briantais (1996) in: Advances in Photosynthesis Vol. 4 Oxygenic Photosynthesis: The Light Reactions (Ort, D.R., Yocum, C.F., eds), pp 265–287 Kluwer Academic Publishers, Dordrecht
8. Michel, H. and Deisenhofer, J. (1988) Biochemistry **27**, 1–7
9. Deligiannakis, Y. and Rutherford, A.W. (1996) Biochemistry **35**, 11239–11246
10. Zech, S.G., Kurreck, J., Eckert, H.-J., Renger, G., Lubitz, W. and Bittl, R. (1997) FEBS Lett. **414**, 454–456
11. Ermler, U., Fritzsch, S.K., Buchanan, S.K. and Michel, H. (1994) Structure **2**, 925–936
12. MacMillan, F., Lendzian, F., Renger, G. and Lubitz, W. (1995) Biochemistry **34**, 8144–8156
13. Renger, G., Eckert, H.-J., Bergmann, A., Bernarding, J., Liu, B., Napiwotzki, A., Reifarth, F. and Eichler, H.J. (1995) Austr. J. Plant Physiol. **22**, 167–181
14. Jursinic, P., and Govindjee (1977) Photochem. Photobiol **26**, 617–628
15. Klimov, V.V., Allakhverdiev S.I., Demeter, S. and Krasnovsky A.A. (1979) Dokl. Akad. Nauk. SSR **249**, 227–230
16. Golbeck, J.H. (1987) Biochim. Biophys. Acta **895**, 167–204
17. Brettel, K. (1997) Biochim. Biophys. Acta **1318**, 322–373
18. Williams, J.C. Alden, R.G. Murchison, H.A., Peloquin, J.M. Woodbury, N.W. and Allen, J.P. (1992) Biochemistry **31**, 11029–11037
19. Watanabe, T. and Kobayashi, M. (1991) in: Chlorophylls (Scheer, H. ed), pp. 287–315, CRC Press, Boca Raton
20. Liu, K., Murchison, H.A., Nagarajan, V., Allen, J.P. and Williams, J.C. (1994) Proc. Natl. Acad. Sci. USA **91**, 10265–10269
21. van Gorkom, H.J. and Schelvis, J.P.M. (1993) Photosynth. Res. **38**, 297–301
22. Joliot, P. and Kok, B. (1975) in: Bioenergetics of Photosynthesis (Govindjee, ed.), pp. 387–412, Academic Press, New York
23. Krauß, N., Schubert, W.-D., Klukas, O., Fromme, P., Witt, H.T. and Saenger, W. (1996) Nature Struct. Biol. **3**, 965–973
24. Schubert, W.-D., Klukas, O., Krauß, N., Saenger, W., Fromme, P. and Witt, H.T. (1997) J. Mol. Biol. **272**, 741–769
25. Svensson, B., Etchebest, C., Tuffery, P., van Kan, P., Saith, J. and Styring, S. (1996) Biochemistry **36**, 14486–14502
26. Mulkidjanian, A.Y., Cherepanov, D.A., Haumann, M. and Junge, W. (1996) Biochemistry **35**, 3093–3107
27. de Rose, V.J., Mukerji, I., Latimer, M.J., Yachandra, V.K., Sauer, K. and Klein, M.P. (1994) J. Am. Chem. Soc. **116**, 5239–5249
28. Yachandra, V.K., de Rose, V.J., Latimer, M.J., Mukerji, I., Sauer, K. and Klein, M.P. (1993) Science **260**, 675–679
29. Renger, G. (1997) Physiol. Plant. **100**, 828–841
30. Chu, H.A., Hguyen, A.P. and Debus, R.J. (1995a) Biochemistry **34**, 5839–5858
31. Chu, H.A., Hguyen, A.P. and Debus, R.J. (1995b) Biochemistry **34**, 5859–5882
32. Vermaas, W.F.M., Styring, S., Schröder, W. and Andersson, B. (1993) Photosynth. Res.
33. Tamura, N., Inoue, Y. and Cheniae, G. (1989) Biochim. Biophys. Acta **976**, 173–181
34. Chen, C., Kazimir, J. and Cheniae, G.M. (1995) Biochemistry **34**, 13511–13526
35. Zaltsman, L., Ananyev, G.M. Bruntrager, E. and Dismukes, G.C. (1997) Biochemistry **36**, 8914–8922
36. Seidler, A. (1996) Biochim. Biophys. Acta **1277**, 35–60
37. Vermaas, W.F.J. Williams, J.G.K. and Arntzen, C.J. (1987) Plant Mol. Biol. **8**, 317–326
38. Bricker, T.M. (1990) Photosynth. Res. **24**, 1–13
39. Fromme, P., Witt, H.T., Schubert, W.-D., Klukas, O., Saenger, W. and Krauß, N. (1996) Biochim. Biophys. Acta **1275**, 76–83
40. Sayre, R.T. and Wrobel-Boerner, E.A. (1994) Photosynth. Res. **40**, 11–19

41. Haag, E., Eaton-Rye, J., Renger, G. and Vermaas, W.F.J. (1993) Biochemistry **32**, 4444–4454
42. Gleiter, H.M., Haag, E., Shen, J.R., Eaton-Rye, J.J., Inoue, Y. Vermaas, W.F.J. and Renger, G. (1994) Biochemistry **33**, 12063–12071
43. Gleiter, H.M., Haag, E., Shen, J.R., Eaton-Rye, J.J., Seeliger, A.G., Inoue, Y., Vermaas, W.F.J. and Renger, G. (1995) Biochemistry **34**, 6847–6856
44. Babcock, G.T. and Wikström, M.K.F. (1992) Nature **356**, 301–306
45. Reifarth, F., Christen, G., Seeliger, A.G., Dörmann, P., Benning, C. and Renger, G. (1997) Biochemistry **36**, 11769–11776
46. Renger, G. (1987) Photosynthetica **21**, 203–224
47. Roelofs, T.A., Liang, W., Lattimer, M.J., Cinco, R.M., Andrews, J.C., Sauer, K., Yachandra, V.K. and Klein, M.P. (1996) Proc. Natl. Acad. Sci. USA **93**, 3335–3340
48. Davidson, V.L. (1996) Biochemistry **35**, 14035–14039
49. Renger, G. (1988) Chemica Scripta **28A**, 105–109
50. Eckert, H.-J. and Renger, G. (1988) FEBS Lett. **236**, 425–431
51. Fiege, R., Zweygart, W., Bittl, R., Adir, N., Renger, G. and Lubitz, W. (1996)Photosynth. Res. **48**, 227–237
52. Shen, J.-R., Vermaas, W.J.F. and Inoue, Y. (1995) J. Biol. Chem. **270**, 6901–6907
53. Moser, C.C., Page, C.C. Chen, X. and Dutton, P.L. (1997) J. Biol. Inorg. Chem. **2**, 393–398
54. Marcus, R.A. and Sutin, N. (1985) Biochim. Biophys. Acta **811**, 265–322
55. Gray, H.B. and Winkler, J.R. (1996) Annu. Rev. Biochem. **65**, 537–561
56. Lopez-Castillo, J.-M., Mouhim, A.F., Van Binh-Otten, E.N. and Jay-Gerin, J.-P. (1997) J. Am. Chem. Soc. **119**, 1978–1980
57. Messinger, J., Seaton, G.R., Wydrzynski, T., Wacker, U. and Renger, G. (1997) Biochemistry **36**, 6862–6873
58. Renger, G. (1998) in: Concepts in Photobiology: Photosynthesis and Photomorphogenesis (Singhal, G.S, Renger, G., Sopory, S.K., Irrgang, K.-D. and Govindjee, eds), pp. 280–315, Narosa Publishing Co., Delhi and Kluwer Academic Publishers, Dordrecht,
59. Moser, C.C. Page, C.C., Farid, R. and Dutton, P.L. (1995) J.Bioenergetics Biomembrane. **27**, 263–274
60. Rappaport, F., Porter, G., Barber, J., Klug, D., and Lavergne, J. (1995) in: Photosynthesis: from Light to Biosphere, Vol. II (Mathis, P. ed), pp. 345–348, Kluwer Academic Publishers, Dordrecht
61. Tommos, C. and Babcock, G.T. (1997) Acc. Chem. Res. (in press)
62. Karge, M., Irrgang, K.-D. and Renger, G. (1997) Biochemistry **36**, 8904–8913
63. Renger, G. and Hanssum, B. (1992) FEBS Letters **299**, 28–32
64. Messinger, J., Wacker, U. and Renger, G. (1991) Biochemistry **30**, 7852–7862
65. Yachandra, V.K., Sauer, K., and Klein, M.P. (1996) Chem. Rev. **96**, 2927–2950
66. Babcock, G.T. (1995) in: Photosynthesis: From Light to Biosphere, Vol. II (Mathis, P., ed), pp. 209–215, Kluwer Academic Publishers, Dordrecht
67. Britt, R.D. (1996) in: Oxygenic Photosynthesis: The Light Reactions (Ort, D.R. and Yocum, C.F., eds), pp. 137–164, Kluwer Academic Publishers, Dordrecht
68. Bishop, G.R. and Davidson, V.L. (1995) Biochemistry **34**, 12082–12086
69. Karge, M., Irrgang, K.-D., Sellin, S., Feinäugle, R., Liu, B., Eckert, H.-J., Eichler, H.J. and Renger, G. (1996) FEBS Lett. **378**, 140–144
70. Christen, G., Karge, M. Eckert, H.-J. and Renger, G. (1997) Photosynthetica **33**, 529–539
71. Renger, G. (1977) FEBS Lett. **81**, 223–228
72. Renger, G. (1978) in: Photosynthetic Oxygen Evolution (Metzner, H., ed), pp. 229–248 Academic Press, London
73. Renger, G., Bittner, T. and Messinger, J. (1994) Biochem. Soc. Trans. **22**, 318–322
74. Williams, R.J.P. (1995) Eur. J. Biochem. **234**, 363–381
75. Renger, G., Christen, G., Korge, M. Eckert, H.-J. and Ingang, K.-D. (1998) J. BioInorg. Chem. (in press)
76. Enami, I., Kikuchi, S., Fukuda, T. Ohta, H. and Shen, J.-R. (1998) Biochemistry **37**, 2787–2793.

6

DEVELOPMENT OF A *psb*A-less/*psb*D-less STRAIN OF *Synechocystis* sp. PCC 6803 FOR SIMULTANEOUS MUTAGENESIS OF THE D1 AND D2 PROTEINS OF PHOTOSYSTEM II

Svetlana Ermakova-Gerdes and Wim Vermaas

Department of Plant Biology and Center for the Study of Early Events in
    Photosynthesis
Arizona State University
Box 871601, Tempe, Arizona 85287-1601

1. INTRODUCTION

The cyanobacterium *Synechocystis* sp. PCC 6803 has several features that make it an excellent genetic system for the study of fundamental processes such as photosynthesis or respiration, as well as for applied research [1,2]. A major advantage of *Synechocystis* 6803 is that its entire genome has been sequenced recently [3]. This has opened up numerous possibilities to modify genes and entire biochemical pathways by introducing, deleting or altering any part of the *Synechocystis* 6803 genome. An appealing application of genetic engineering in this organism is the simultaneous introduction of mutations into multiple genes, but the extent to which multiple deletions or mutations can be combined in one strain has been limited by the number of currently available selectable markers. So far, seven antibiotic-resistance cassettes have been used successfully in *Synechocystis* 6803. These cassettes confer resistance to erythromycin (Er) [4], chloramphenicol (Cm) [5], gentamycin (Gm) [6], kanamycin (Km) [7], spectinomycin (Sp) [8], streptomycin (Str) [9], and zeocin (Zeo) [10]. In addition, a tetracycline (Tet) resistance cassette has been reported to be useful in this respect [11], but this cassette has proven to be rather unreliable in our hands for routine use. However, even this sizable number of useful cassettes may not be sufficient for genetic studies of complicated biochemical and physiological pathways, where many different genes need to be deleted and possibly later replaced with their mutagenized copies.

One area of emphasis in our group is to investigate which parts of the protein environment are central towards determining the properties of the photosynthetic apparatus, particularly photosystem II (PS II). Even though the two reaction center proteins of photo-

The Phototrophic Prokaryotes, edited by Peschek *et al.*
Kluwer Academic / Plenum Publishers, New York, 1999.

system II, D1 and D2, share a common ancestry, the sequences of these two proteins have diverged sufficiently that electrons flow through only one of two possible electron transfer branches in the reaction center. To study the reasons why electron transfer almost exclusively utilizes only one of the transfer pathways, sequence exchanges between homologous regions of D1 and D2 would be advantageous. Similar experiments have been carried out on reaction centers from purple bacteria [12,13], which are homologous to photosystem II at the acceptor side. We will be particularly interested in studying the results of sequence exchanges at the donor side of photosystem II, as the electron transport pathways at this side of the reaction center are not homologous to those in purple bacteria, and very prominent functional asymmetries exist for example with respect to the properties of the symmetrically arranged redox-active Tyr residues, Y_Z (Tyr161 of D1) and Y_D (Tyr160 of D2).

In order to perform sequence exchanges between D1 and D2, it is necessary to use a host *Synechocystis* 6803 strain lacking all the copies of the *psb*A gene coding for D1 and of *psb*D coding for D2. There are three *psb*A genes and two *psb*D genes in the *Synechocystis* 6803 genome. Moreover, we would like to be able to reintroduce mutated gene copies and to have the option to also introduce a deletion of the photosystem I reaction center genes as resulting mutants are more amenable to *in vivo* photosystem II analysis. This would necessitate the use of eight antibiotics resistance markers. To alleviate the shortage of resistance markers, we have applied a "marker recycling" technique to construct a *psb*A⁻/*psb*D⁻/photosystem I (PS I)-less strain of *Synechocystis* 6803, and to use this strain for simultaneous mutagenesis of the D1 and D2 proteins. This marker recycling technique utilizes *sac*B, which codes for a levan sucrase and expression of which is lethal to the cells if sucrose is added [14, 15]. A marker can be recycled simply by co-introduction of an antibiotic-resistance marker together with *sac*B into *Synechocystis* 6803, followed by segregation of the antibiotic-resistant, sucrose-sensitive mutant, which in turn is followed by transformation with a marker-less deletion construct and selection for sucrose resistance. The resulting strain carries the desired gene deletion and yet does not have a selectable marker at the site of the deletion.

In this paper the creation of a *Synechocystis* 6803 mutant lacking all five D1 and D2 genes as well as PS I is described. As one of the D2 genes is linked to that of the chlorophyll-binding CP43 protein of photosystem II, the CP43 gene is deleted as well in this strain. Use of this strain to introduce sequence exchanges between small homologous domains in the D1 and D2 proteins will be the subject of a future publication.

2. MATERIALS AND METHODS

Synechocystis 6803 strains were cultivated in BG-11 medium supplemented with 5 mM glucose (10 mM for PS I-less strains). Solid medium was supplemented with 1.5% agar, 0.3% $Na_2S_2O_3$ and 10 mM TES/NaOH buffer, pH 8.2. For PS I-less strains $NaNO_3$ in BG-11 was partially substituted with NH_4NO_3 (the final concentration of ammonia was 4.5 mM). PS I-containing strains were grown at a light intensity of 50 µmol photons $m^{-2}\cdot s^{-1}$; PS I-less strains were propagated at dim light (5 µmol photons $m^{-2}\cdot s^{-1}$). When necessary, BG-11 was supplemented with antibiotics in the following concentrations: erythromycin, 0.5–15 µg/ml; chloramphenicol, 10 µg/ml; gentamycin, 5–10 µg/ml; kanamycin, 15–50 µg/ml; spectinomycin, 15 µg/ml; and zeocin, 2–10 µg/ml. Upon initial selection the lowest antibiotic concentration was used, then concentrations were gradually increased during segregation of transformants. When noted, sucrose was added to the final concentration of 1–5%. *Synechocystis* 6803 transformation was performed essentially as previously described [16].

Fluorescence emission spectra were determined using a Fluorolog 2 instrument (SPEX Industries Inc., NJ). Whole cells were resuspended in 25 mM HEPES/NaOH, pH 7.0 in 60% (v/v) glycerol (for measurements upon excitation at 440 nm, exciting chlorophyll) or in the absence of glycerol (for measurements upon excitation at 560 nm, exciting mostly phycobiliproteins). Spectra were recorded at 77 K.

Chlorophyll *a* concentrations were determined in a UV-160 spectrophotometer according to [17]. Chlorophyll was extracted from cells with 100% methanol.

Isolation of *Synechocystis* 6803 chromosomal DNA for use as template in PCR amplification of different loci was performed as follows. A loopful of cells from a plate was resuspended in TE, washed, resuspended in 200 µl TE again, 200 µl of a 1:1 mix of TE : glass beads and 300 µl of phenol were added, and the cells were broken with mini Beadbeater for 30 seconds. The mixtures were centrifuged for 5 minutes and aqueous phase was transferred to a new eppendorf tube. This phase was extracted with phenol/chloroform 1:1 mix and chloroform. DNA was precipitated with 0.5 volume of 7.5 M NH_4OAc and 1.5 volumes of ethanol, and the DNA pellet was washed with 70% ethanol, dried and resuspended in 20 µl TE.

3. RESULTS AND DISCUSSION

Several different *Synechocystis* 6803 strains lacking either the *psb*A or *psb*D genes (in the PS I-containing as well as in the PS I-less genetic backgrounds) have been previously constructed and employed for mutagenesis of D1 or D2 by different groups [18–23]. However, for simultaneous mutagenesis of the D1 and D2 it was necessary to develop a strain lacking all the copies of both genes. Our strategy was to start with the *psb*DIC$^-$/*psb*DII$^-$ strain [23] and to introduce sequentially deletions into the three copies of *psb*A.

3.1. Cloning of the *psb*A Genes and Their Flanking Regions

Synechocystis 6803 contains three *psb*A copies per genome [18], one of which (*psb*AI) presumably is a pseudogene as no D1 can be detected in mutants lacking the other two *psb*A gene copies [24]. At the time this project was started the genomic *Synechocystis* 6803 sequence was not yet available, and genomic DNA of the *Synechocystis* 6803 strain with all three copies of *psb*A interrupted by different antibiotic-resistance markers [18] was used for cloning of *psb*AI (interrupted by a Km-resistance cassette) and *psb*AIII (interrupted by a Cm-resistance cassette). This approach was chosen (1) to avoid difficulties regarding cloning functional *psb*A genes that appear to be toxic for *E. coli* and (2) to take advantage of the convenient antibiotics-resistance selection of the desired clones. Genomic DNA of the triple-interruption mutant was digested with *Xba* I and *Nhe* I and ligated into the *Xba* I site of the pUC118 polylinker to produce a genomic library of the *psb*A$^-$ *Synechocystis* 6803 strain [18]. Km-resistant *E. coli* transformants contained *psb*AI interrupted by a Km-resistance cassette in a 9 kb fragment of *Synechocystis* 6803 DNA. A 4.3 kb *Hpa* I/*Eco*R I fragment containing the interrupted *psb*AI with 0.8 and 1.2 kb of upstream and downstream regions, respectively, was subcloned into the *Sph* I/*Eco*R I sites of pUC118 (Figure 1), resulting in plasmid pAI-Km. The *psb*AIII gene with a Cm-resistance cassette inserted into the gene was recovered within a 16 kb *Synechocystis* 6803 DNA fragment and a 4.0 kb *Acc* I/*Pac* I fragment carrying the interrupted *psb*AIII was subcloned into the *Acc* I/*Bam*H I sites of the pUC118 polylinker (Figure 1), resulting in the pAIII-Cm plasmid.

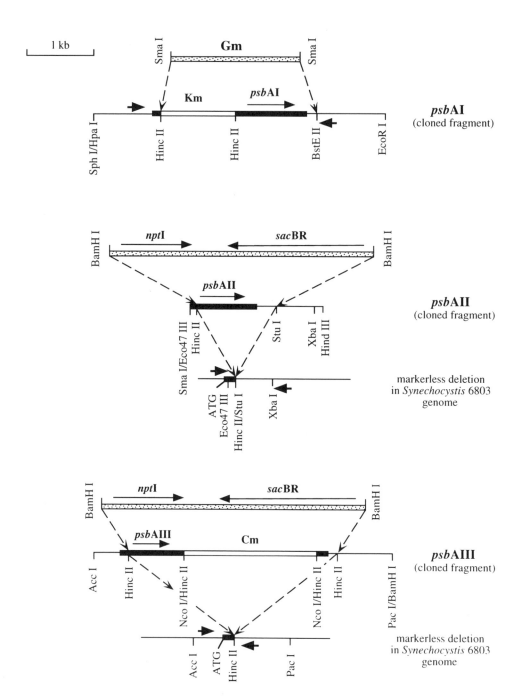

Figure 1. Scheme of cloning and deleting the *psb*A genes from the *Synechocystis* 6803 *psb*DIC⁻/*psb*DII⁻ strain as described in the text. The positions of the primers used in PCR amplification of the *psb*A regions from the wild type and the constructed *psb*DIC⁻/*psb*DII⁻/*psb*AI⁻/*psb*AII⁻/*psb*AIII⁻ strain are indicated.

The *psb*AII gene (excluding the first 10 codons) and a 0.9 kb sequence immediately downstream of this gene was amplified by PCR from wild-type *Synechocystis* 6803 genomic DNA. After cutting the PCR product with *Eco*47 III and *Hin*d III (which cut 27 and 10 bp from the ends, respectively), the resulting 1.94 kb DNA fragment was cloned into the *Sma* I/*Hin*d III restriction sites of the pUC19 polylinker (plasmid pAII).

3.2. Deletion of the *psb*A Genes

In order to delete *psb*AI from the *Synechocystis* 6803 genome, the plasmid pΔAI-Gm was constructed from pAI-Km by replacing a 2.3 kb *Hin*c II/*Bst*E II fragment including most of *psb*AI (except for the 5′ 122 bp) and a 0.2 kb region immediately downstream of the gene by the 1.9 kb *Sma* I/*Sma* I fragment from pHP45Gm^r (a gift of Dr. R. Debus) that leads to gentamycin resistance [6] (Figure 1). The pΔAI-Gm plasmid was used to transform the Sp^r/Str^r/Cm^r *psb*DIC$^-$/*psb*DII$^-$ strain of *Synechocystis* 6803 lacking both copies of the D2 gene as well as the CP43 gene [23].

The resulting mutant lacks both *psb*D copies and *psb*AI, but retains *psb*AII and *psb*AIII. However, four out of seven available antibiotic-resistance markers already have been introduced into this strain, and two more markers (Km^r and Er^r) needed to be reserved for application in selectable vectors designed for mutagenesis and consecutive reintroduction of *psb*DI and *psb*AII into the *psb*DIC$^-$/*psb*DII$^-$/*psb*AI$^-$/*psb*AII$^-$/*psb*AIII$^-$ *Synechocystis* 6803 strain. In addition, the last available marker (Zeo^r) was reserved for deletion of the *psa*AB operon following the deletion of all copies of *psb*A and *psb*D.

Hence, to delete the remaining *psb*AII and *psb*AIII genes, we employed counterselection against the conditionally lethal *Bacillus* sp. *sac*B gene [14]. This gene encodes levan sucrase, production of which is lethal to many gram-negative bacteria in the presence of 5% sucrose, presumably because of the production of fructose derivatives. DNA constructs containing *sac*B have been used in a variety of bacteria to select for allelic exchange, to entrap insertion sequences, to isolate double recombinants, etc. (for instance, see [25] and references therein). We wished to employ this gene for yet a different task, "marker recycling". Another cyanobacterium, *Anabaena* sp., already had been shown to be susceptible to sucrose when bearing the *sac*B gene [15], and therefore it was not very surprising that the presence of *sac*B in the *Synechocystis* 6803 genome was lethal on sucrose-containing media (data not shown). We did not observe a measurable effect of *sac*B on the growth rate of *Synechocystis* 6803 in the absence of sucrose (data not shown).

In order to delete *psb*AIII, a 2.9 kb *Hin*c II/*Hin*c II pAIII-Cm fragment that contained 85% of *psb*AIII (except for the 5′ 174 bp) together with 97 bp of downstream region and the Cm^r gene (Figure 1) were replaced with the 3.9 kb *Bam*H I/*Bam*H I DNA fragment of the pRL250 plasmid [15] containing the *npt*I-*sac*BR cassette. The resultant plasmid pΔAIII-SacKm was used to transform the *psb*DIC$^-$/*psb*DII$^-$/*psb*AI$^-$ strain of *Synechocystis* 6803 (note that upon deletion of *psb*AI the Km-resistance marker originally inserted in *psb*AI had been removed and replaced by a Gm-resistance cassette). Transformants were propagated, and selection for resistance to increasing Km concentrations (15–50 µg/ml) for several culture transfers led to deletion of *psb*AIII from all genome copies of the *Synechocystis* 6803 transformant. After complete segregation was confirmed by PCR analysis (data not shown), the *psb*DIC$^-$/*psb*DII$^-$/*psb*AI$^-$/*psb*AIII$^-$/*sac*BKm^r strain was transformed with the markerless *psb*AIII deletion plasmid pDAIII containing the same deletion as pDAIII-SacKm but without any marker inserted. Exposure of the transformation mixture to sucrose four days after the transformation event led to several sucrose-resistant colonies, which were expected to either have the *npt*I-*sac*B cassette deleted or to lack active *sac*B because of other reasons (for example, spontaneous mutation). To distinguish between these two possibilities, sucrose-resistant colonies were plated in parallel on glucose-containing BG-11 with and without Km. Colonies that were sucrose-resistant and Km-sensitive had most likely lost both genes (*sac*B and Km^r) due to recombination with pΔAIII, whereas colonies that were resistant to both sucrose and kanamycin probably had *sac*B inactivated by mechanisms other than transformation. In different experiments the yield of sucrose-resistant, Km-sensitive colonies was 60–90%. The presence of a markerless *psb*AIII deletion in the colonies that were selected was confirmed by PCR (Figure 2).

A similar procedure was used to delete *psb*AII from the *psb*DIC⁻/*psb*DII⁻/*psb*AI⁻ /*psb*AIII⁻ strain of *Synechocystis* 6803. The deletion introduced in *psb*AII comprised a 1.2 kb *Hin*c II/*Stu* I fragment that contained the 3' 85% of *psb*AII and a 283 bp region directly downstream of *psb*AII (Figure 1).

To make sure that all the three deletions introduced had been segregated to homozygosity, PCR was performed using chromosomal DNA of the constructed *psb*DIC⁻/*psb*DII⁻ /*psb*AI⁻/ *psb*AII⁻/*psb*AIII⁻ *Synechocystis* 6803 strain as template and appropriate sets of primers to amplify the three copies of *psb*A independently. This strain did not produce any PCR products of the sizes expected for one of the wild-type *psb*A genes, but gave rise to PCR fragments of sizes corresponding to deletion versions of these genes (Figure 2).

To verify that the strain did not have inadvertently introduced mutations that affected the phenotype of the organism, the *psb*DIC⁻/*psb*DII⁻/*psb*AI⁻/*psb*AII⁻/*psb*AIII⁻ *Synechocystis* 6803 strain was transformed with wild-type *psb*DIC and *psb*AII. The resulting transformant is photoautotrophic and is indistinguishable from wild-type in its properties (data not shown), indicating that properties observed in the mutant and its derivatives are due solely to the deletions that have been introduced.

3.3. Deletion of Photosystem I from the *psb*DIC⁻/*psb*DII⁻/*psb*AI⁻/ *psb*AII⁻/*psb*AIII⁻ Strain

PS I-less strains of *Synechocystis* 6803 [26, 27] have several advantages for photosystem II characterization, including (1) an increased variable fluorescence yield [28], (2) decreased levels of EPR signals not associated with photosystem II thus facilitating Signal II EPR characterization in thylakoids and intact cells, and (3) an improved photosystem II accumulation in cells of selected mutants with donor side modifications (P. Manna, R. LoBrutto, and W. Vermaas, submitted). In order to facilitate analysis of mutants with altered D1 and D2 genes introduced into the *psb*DIC⁻/*psb*DII⁻/*psb*AI⁻/ *psb*AII⁻/*psb*AIII⁻ *Synecho-*

1 2 3 4 5 6 7 8 9

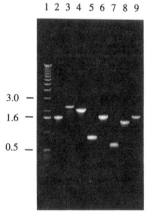

3.0 —

1.6 —

0.5 —

Figure 2. Lanes 2 - 7: amplification by PCR of the *psb*A genes using total chromosomal DNA of the wild-type (lanes 2, 4, 6) and *psb*DIC⁻/*psb*DII⁻ /*psb*AI⁻/*psb*AII⁻/*psb*AIII⁻ strain (lanes 3, 5, 7) as templates. Lane 1: 1 kb DNA ladder. Lanes 2 and 3: amplification of the *psb*AI region. Lanes 4 and 5: amplification of the *psb*AII region. Lanes 6 and 7: amplification of the *psb*AIII region. The positions of the primers used in these PCR experiments are indicated in Figure 1. Estimated sizes of PCR products for the wild-type and the mutant strains are: 1.6 and 2.3 kb (*psb*AI amplification), 2.0 and 0.8 kb (*psb*AII amplification), 1.54 and 0.53 kb (*psb*AIII amplification). Lanes 8 - 9: amplification by PCR of the *psa*AB region. The template was chromosomal DNA of the wild-type *Synechocystis* 6803 (lane 8) or chromosomal DNA of the constructed *psb*DIC⁻/*psb*DII⁻/*psb*AI⁻/*psb*AII⁻/*psb*AIII⁻/*psa*AB⁻ strain (lane 9). Estimated sizes of PCR products in lanes 8 and 9 are: 1.42 kb and 1.6 kb.

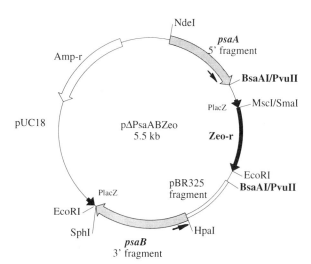

Figure 3. Map of the plasmid pΔPsaABZeo, used to delete *psa*AB from the *psb*DIC⁻/*psb*DII⁻/*psb*AI⁻/*psb*AII⁻/*psb*AIII⁻*Synechocystis* 6803 strain. The positions of the primers used for PCR amplification of the *psa*AB region (see Figure 2) are indicated.

cystis 6803 strain, we wished to delete the *psa*AB gene cluster from the this strain. To do so, the Cm-resistance cassette in the pBWV *psa*AB deletion construct [29] was replaced with the *ble* gene, conferring resistance to zeocin [10]. Plasmid pBWV was digested with *Bsa* I to delete a 1.5 kb fragment containing most of the Cm-resistance cassette and then ligated with the 0.9 kb *Pvu* II/*Pvu* II fragment from pZeo1, a plasmid that contained *ble* under the control of the *lac*Z promoter (C. Howitt, personal communication), to yield the plasmid pΔPsaABZeo (Figure 3). Plasmid pΔPsaABZeo was used to transform the *psb*DIC⁻/*psb*DII⁻/*psb*AI⁻/ *psb*AII⁻/*psb*AIII⁻ strain. Transformants were selected on BG-11 plates at low light intensity (5 μmol photons m⁻²·s⁻¹) and supplemented with 2 μg/ml zeocin, 40 μM IPTG and 10 mM glucose. After several restreaks on increasing concentrations of zeocin (up to 10 μg/ml), some colonies could be discerned that were much bluer than the parental strain; this change in color is associated with the loss of PS I, with which most chlorophyll in the cell usually is associated. The bluish colonies were picked, restreaked, and grown in liquid. Indeed, as shown in Figure 4A, the 720 nm emission maximum associated with PS I has disappeared in these strains, indicating that PS I has been lost. In addition, PCR analysis of the *psa*AB locus from these organisms showed the absence of the wild-type gene, indicating full segregation (Figure 2).

3.4. Characterization of the *Synechocystis* 6803 Strain Lacking PS I and the D1, D2, and CP43 Photosystem II Proteins

77 K fluorescence emission spectra obtained upon excitation of whole cells at 440 nm are presented in Figure 4A. As expected, the spectrum of the constructed PS I-less/*psb*DIC⁻/*psb*DII⁻/*psb*AI⁻/*psb*AII⁻/*psb*AIII⁻ strain lacks the 720 nm emission maximum associated with photosystem I chlorophyll. The fluorescence peak at 695 nm originating from photosystem II chlorophyll is absent as well, since functional PS II centers do not assemble in the absence of D1 or D2 [21,30]. However, a broad emission band at 682 nm is present in the spectrum. In cyanobacteria, allophycocyanin-B is known to contribute to fluorescence emission in 680–685 nm region. However, phycobilin contribution to the 682 nm peak obtained upon excitation at 440 nm can be excluded, since upon direct excitation

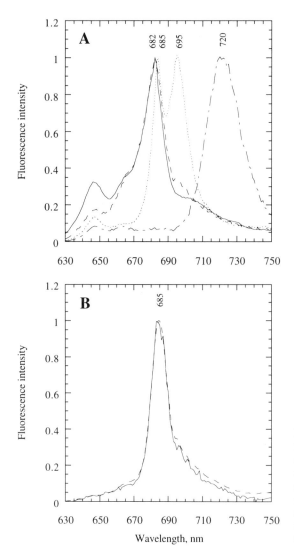

Figure 4. 77. K fluorescence emission spectra of whole cells of the psbDIC$^-$/psbDII$^-$/psbAI$^-$/psbAII$^-$/psbAIII$^-$/ (— - —), psaAB$^-$/psbDIC$^-$/psbDII$^-$/psbAI$^-$/psbAII$^-$/psbAIII$^-$ (——), psaAB$^-$/psbDIC$^-$/psbDII$^-$ (– –) and psaAB$^-$ (······) strains of *Synechocystis* 6803. A: excitation at 440 nm; B: excitation at 560 nm. Each spectrum was normalized to 1.0 at its maximum.

of phycobiliproteins at 560 nm an intensive emission maximum was obtained at 685 nm (Figure 4B). We believe that broad 682 nm emission peak originates from chlorophyll associated with proteins other than the PS II or PS I proteins. Several different *Synechocystis* 6803 strains lacking photosystem I and photosystem II complexes have been constructed before [21,22,31]. Although they are genetically different (they carry deletions of different photosystem II core proteins), in all of them major chlorophyll-binding proteins (PsaAB, D1, D2, CP43, CP47) are either present in drastically reduced amounts or absent from the thylakoid membranes. However, in all of these strains significant amounts of chlorophyll *a* can still be detected [21,22,31]. Interestingly, the amounts and the 77 K fluorescence emission characteristics (upon excitation at 440 nm) of this residual chlorophyll differ slightly in the strains lacking various PS II genes (Table 1). Chlorophyll content in the constructed PS I-less/psbDIC$^-$/psbDII$^-$/psbAI$^-$/psbAII$^-$/psbAIII$^-$ was found to be several times lower than in the parental PS I-less/psbDIC$^-$/psbDII$^-$ strain. The positions of the 77 K fluorescence emission maximum in both strains were identical (Table 1 and Figure 4).

Table 1. The chlorophyll *a* content and fluorescence emission maximum in PS I-less and in different PS I-less/PS II-less *Synechocystis* 6803 strains

Strain genotype	µg Chlorophyll per ml OD$_{730}$	Emission maximum[a] (nm)
*psa*AB⁻	0.634 ± 0.014	685, 695
*psa*AB⁻/*psb*B⁻/*psb*C⁻/*apc*E⁻ [b]	0.12 ± 0.02[b]	678[b]
*psa*AB⁻/*psb*DIC⁻/*psb*DII⁻	0.17 ± 0.03	682
*psa*AB⁻/*psb*DIC⁻/*psb*DII⁻/*psb*AI⁻/*psb*AII⁻/*psb*AIII⁻	0.045 ± 0.013	682

[a]Fluorescence emission spectra were determined at 77 K, excitation wavelength was 440 nm.
[b]Data from [31].

To our knowledge, a simultaneous deletion of six loci that corresponds to eight genes and five proteins sets a new "record". More importantly, it shows that it is technically feasible to delete a large number of genes simultaneously, in excess as compared to the number of available antibiotics resistance markers. This clearly has considerable promise toward pathway engineering in *Synechocystis* 6803.

ACKNOWLEDGMENTS

We thank Dr. Peter Wolk (Michigan State University) for supplying us with pRL250 containing the *npt* I-*sac*BR cassette, Dr. Rick Debus (University of California, Riverside) for the pHP45Gm^r plasmid containing the gentamycin-resistance cassette, and Dr. Crispin Howitt for the zeocin-resistance construct pZeo1. This research is supported by a grant from the National Institutes of Health (GM-51556).

REFERENCES

1. Vermaas, W.F.J. (1996) J. Appl. Phycol. 8, 263–273.
2. Vermaas, W.F.J. (1997) Meth. Enzymol., in press.
3. Kaneko, T., Sato, S., Kotani, H., Tanaka, A., Asamizu, E., Nakamura, Y., Miyajima, N., Hirosawa, M., Sugiura, M., Sasamoto, S., Kimura, T., Hosouchi, T., Matsuno, A., Muraki, A., Nakazaki, N., Naruo, K., Okumura, S., Shimpo, S., Takeuchi, C., Wada, T., Watanabe, A., Yamada, M., Yasuda, M. and Tabata, S. (1996) DNA Research 3, 109–136.
4. Elhai, J. and Wolk, C.P. (1988) Gene 68, 119–138.
5. Chang, A.C.Y. and Cohen, S. (1978) J. Bact. 134, 1141–1156.
6. Yin, J.C., Krebs, M.P. and Reznikoff, W.S. (1988) J. Mol. Biol. 199, 35–45.
7. Oka, A., Sugisake, H. and Takanami, M. (1981) J. Mol. Biol. 147, 217–226.
8. Prentki, P. and Krisch, H.M. (1984) Gene 29, 303–313.
9. Bagdasarian, M., Bagdasarian, M.M., Lurz, R., Nordheim, A., Frey, A. and Timmis, K.N. (1982) in: Bacterial Drug Resistance (Mitsuhashi, S. ed.) pp. 183–197, Japan Scientific Society Press, Tokyo.
10. Drocourt, D., Calmels, T., Reynes, J.P., Baron, M. and Tiraby, G. (1990) Nucl. Acids Res. 18, 4009.
11. Nixon, P.J., Trost, J.T. and Diner, B.A. (1992) Biochemistry 31, 10859–10871.
12. Lin, S., Xiao, W., Eastman, J.E., Murchison, H.A., Taguchi, A.K.W. and Woodbury, N.W. (1996) Biochemistry 35, 3187–3196.
13. Taguchi, A.K.W., Eastman, J.E., Gallo, Jr., D.M., Sheagley, E., Xiao, W. and Woodbury, N.W. (1996) Biochemistry 35, 3175–3186.
14. Ried, J.L. and Collmer, A. (1987) Gene 57, 239–246.
15. Cai, Y.P. and Wolk, C. P. (1990) J. Bact. 172, 3138–3145.
16. Vermaas, W.F.J., Williams, J.G.K. and Arntzen C.J. (1987) Plant Mol. Biol. 8, 317–326.
17. Lichtenthaler, H.K. (1987) Methods Enzymol. 148, 350–382.

18. Jansson, C., Debus, R.J., Osiewacz, H.D., Gurevitz, M. and McIntosh L. (1987) Plant Physiol. 85, 1021–1025.
19. Nixon, P.J., Metz, J.G., Rögner, M. and Diner, B.A. (1990) in: Current Research in Photosynthesis (Baltscheffsky, M., ed.) vol. I, pp. 471–474, Kluwer Academic Publishers, Dordrecht.
20. Debus, R.J., Nguyen, A.P. and Conway, A.B. (1990) in: Current Research in Photosynthesis (Baltscheffsky, M. ed.) vol. I, pp. 829–832, Kluwer Academic Publishers, Dordrecht.
21. Smart, L.B., Bowlby, N.R., Anderson, S.L., Sithole, I. and McIntosh L. (1994) Plant. Physiol. 104, 349–354.
22. Ermakova-Gerdes, S., Shestakov, S. and Vermaas W. (1995) in: Photosynthesis: from light to Biosphere (Mathis, P. ed) vol. I, pp. 483–486, Kluwer Academic Publishers, Dordrecht.
23. Vermaas, W.F.J., Charité, J. and Eggers, B. (1990) in: Current Research in Photosynthesis (Baltscheffsky, M., ed.) vol. I, pp. 231–238, Kluwer Academic Publishers, Dordrecht.
24. Mohamed, A. and Jansson, C. (1989) Plant Mol. Biol. 13, 693–700.
25. Lawes, M. and Maloy, S. (1995) J. Bacteriol. 177, 1383–1387.
26. Smart, L.B., Anderson, S.L. and McIntosh, L. (1991) EMBO J. 10, 3289–3296.
27. Shen, G., Boussiba, S. and Vermaas, W.F.J. (1993) Plant Cell 5, 1853–1863.
28. Vermaas, W.F.J., Shen, G. and Styring, S. (1994) FEBS Lett. 337, 103–108.
29. Boussiba, S. and Vermaas, W.F.J. (1992) in: Research in Photosynthesis (Murata, N. ed.) vol. III, pp. 429–432, Kluwer Academic Publishers, Dordrecht.
30. Vermaas, W.F.J., Ikeuchi, M. and Inoue, Y. (1988) Photosynth. Res. 17, 97–113.
31. Shen, G. and Vermaas, W.F.J. (1994) J. Biol. Chem. 269, 13904–13910.

7

THE MOLECULAR MECHANISMS CONTROLLING COMPLEMENTARY CHROMATIC ADAPTATION

David M. Kehoe and Arthur R. Grossman

Department of Plant Biology
Carnegie Institution of Washington
Stanford, California 94305

1. INTRODUCTION

The phycobilisome (PBS) is a macromolecular complex that can constitute up to 40% of the soluble cellular protein. These light harvesting structures, found in eukaryotic red algae, glaucocystophytes, and the prokaryotic cyanobacteria, absorb light energy in the region of the spectrum from 540 nm to 660 nm. This energy is efficiently transferred from the PBS to the photosynthetic reaction centers as a result of the close association of these complexes with the outer surface of the thylakoid membranes (Porter, 1978; Searle, 1978). Two classes of proteins make up the PBS: the phycobiliproteins, which are pigmented, and the linker polypeptides, which are nonpigmented (Tandeau de Marsac, 1977). Four major species of phycobiliproteins are found in these organisms. Allophycocyanin (AP, A_{max} approximately 650 nm) is found in the core substructure, the part of the PBS that is in direct association with the photosynthetic reaction center and phycocyanin (PC, A_{max} approximately 620 nm), phycoerythrin (PE, A_{max} approximately 565 nm), and phycoerythrocyanin (PEC, A_{max} approximately 540 nm) are present in rods that emanate from the core. Two subunit forms (α and β) of each of these phycobiliproteins exist and are assembled as heterodimers into higher order structures; these aggregates are the building blocks of the PBS. The linker polypeptides allow the assembly of these building blocks into larger, more stable structures and also assist in the efficient transfer of light energy through the PBS and into the chlorophyll-protein complexes within the thylakoid membranes (Glazer, 1982; Glazer, 1985).

During the process of complementary chromatic adaptation (CCA) in the cyanobacterium *Fremyella diplosiphon*, synthesis of the components that comprise the rods of the PBS is altered, resulting in the creation of a new form of PBS that is capable of more efficiently harvesting the most prevalent wavelengths of light in the environment. This shift

The Phototrophic Prokaryotes, edited by Peschek *et al.*
Kluwer Academic / Plenum Publishers, New York, 1999.

in the structure of newly synthesized PBS during CCA is controlled primarily at the level of transcription of the genes encoding the rod phycobiliproteins and linker polypeptides (Oelmüller, 1988a; Oelmüller, 1988b).

Several operons encode components of the PBS. The α and β subunits of AP, encoded by the *apcA1B1* genes, are part of an operon that also contains the genes for the core linker polypeptide, *apcC*, and the anchor protein, *apcE* (Houmard, 1988). None of the genes encoding core components are differentially regulated during CCA. In *F. diplosiphon*, there is one gene set encoding the α and β subunits of PE, and two encoding PC α and β subunits that are important during CCA. The α and β subunits of PE are encoded by the *cpeBA* operon (Mazel, 1986). Transcriptional activity from this operon is high when cells are grown in green light (GL) and barely detectable when cells are grown in red light (RL). The two forms of PC that are important during CCA are called constitutive PC (PC_c) and inducible PC (PC_i), and these are encoded by the *cpcB1A1* and the *cpcB2A2* operons, respectively. As their names imply, *cpcB1A1* is constitutively expressed in both RL and GL, whereas *cpcB2A2* is transcriptionally active in RL and is inactive in GL. Thus the abundance of the mRNAs from the *cpeBA*, *cpcB1A1* and *cpcB2A2* operons reflects the polypeptide composition within the rods of the PBS under any given light condition. In addition, the genes encoding the linker polypeptides associated with the phycobiliproteins are also transcriptionally regulated by RL and GL (Federspiel, 1990; Federspiel, 1992).

Our approach to understanding the signal transduction pathway(s) that controls CCA has been primarily through the generation, isolation, and complementation of mutants in this pathway. Such mutants can be easily identified by their color phenotypes in RL and GL. Wild type *F. diplosiphon* cells grown in RL are blue-green in color as a result of the accumulation of PC_i in the rods of their PBS. However, if the same cells are grown in GL, they will be red due to the incorporation of PE into their PBS rods. Several groups (Cobley, 1983; Tandeau de Marsac, 1983; Bruns, 1989) have isolated numerous mutants in CCA. A number of these (Bruns, 1989; Chiang, 1992; Kehoe, 1996; Kehoe, 1997; Casey, 1997) have been extensively studied, and in many cases the genetic lesions responsible for the mutant phenotypes have been identified. These mutant classes, named after their color phenotypes, include the red (**FdR**), black (**FdBk**), blue (**FdB**), and green (**FdG**) mutants (Table 1). **FdR** strains appear red under all conditions of illumination. In these strains, the expression of PE has become constitutive and PC_i is never expressed; thus it behaves as if it is always in GL. The **FdB** strains are bluer than wild-type cells in RL due to the overexpression of the *cpcB2A2* operon and require more GL to suppress PC_i synthesis than do

Table 1. Summary of the relative phycobiliprotein content of wild type and several *F. diplosiphon* CCA mutants under two light conditions[a]

	Abundance of phycobiliproteins in CCA mutants			
	Green light		Red light	
	PE	PCi	PE	PCi
Wild type	++++	–	+/–	++++
Mutants				
Red (**FdR**)	++++	–	++++	–
Black (**FdBk**)	++	++	++	++
Blue (**FdB**)	++++	++++	+/–	+++++
Green (**FdG**)	–	–	–	++++

[a]The (+) symbol indicates that the protein is synthesized and increasing numbers of (+) symbols denote higher levels of accumulation. The (–) indicates that the phycobiliprotein is not present at detectable levels.

wild-type cells (Casey, 1997). The **FdG** strains exhibit normal PC_i expression, but the PE genes are inactive in both RL and GL. In the **FdBk** mutants there are moderate levels of both PE and PC_i, however, these levels remain the same in RL and GL (Kehoe, 1996).

The first mutant to be complemented was an **FdR** strain (Chiang, 1992). The complementing gene, *rcaC*, encoded a member of the response regulator class of proteins that are part of two component regulatory systems (Parkinson, 1992; Appleby, 1996). Such pathways typically utilize a sensor polypeptide with histidine kinase activity to detect changes in environmental conditions. For CCA, the environmental cue is light quality; therefore the sensor is likely to be a photoreceptor. After activation, the sensor undergoes autophosphorylation at a histidine residue. The phosphoryl group is then transferred from the histidine to an aspartate (D) residue within a conserved receiver domain of the second protein within the system, called the response regulator. Response regulators often contain output domains, such as DNA binding domains, whose activity is regulated by the phosphorylation state of the conserved D residue.

RcaC has been found to be an unusual response regulator in a number of ways (Chiang, 1992; Kehoe, 1996). It is approximately twice as large as most response regulators, and has two (instead of one) conserved D receiver domains; one is at the amino terminus and the other at the carboxy terminus. Furthermore, near the middle of the polypeptide is a H2 block found in components of some complex two component systems called four step phosphorelays (Ishige, 1994; Appleby, 1996), Finally, contiguous with the amino terminal receiver domain is a DNA binding motif. Thus RcaC may act as a transcription factor. Using site directed mutagenesis (SDM), we have conducted a preliminary analyses of the roles of the two conserved D residues during CCA. Our results indicated that the amino terminal D (D51) is the primary site of CCA regulation within RcaC and that this residue is phosphorylated when the cells are grown in RL, and not phosphorylated in GL (Kehoe, 1995).

Below we will briefly summarize the recent progress we have made in our understanding of the CCA signal transduction pathway, including the role of RcaC in this process as well as the identification of several new components of the pathway. Finally, we will provide a model that proposes an initial ordering of the CCA pathway components that have been thus far identified.

2. RESULTS AND DISCUSSION

We were interested in isolating the primary sensor and other components controlling the phosphorylation of RcaC. Our SDM studies of the conserved D residues within RcaC suggested that CCA mutants that were incapable of efficiently phosphorylating the D51 residue of RcaC could not develop a blue-green phenotype when grown in RL. Thus, in order to obtain new mutants in the CCA pathway that were unable to properly phosphorylate RcaC, we mutagenized wild type cells using nitrosoguanidine and grew these cells in RL, selecting mutants that failed to develop the normal blue-green phenotype. We biased our selection of mutants away from additional *rcaC*-**FdR** mutants by transforming the wild type cells with an autonomously replicating plasmid containing a wild type copy of *rcaC* prior to our mutagenesis.

The screen for novel CCA mutants has allowed us to isolate two interesting new phenotypic classes of CCA mutants. The first class is a group of **FdR** mutants that are not complemented by *rcaC*, and the second class is a type of CCA mutant that we had not previously isolated in our laboratory, termed the black (**FdBk**) mutants. Neither of these two mutant classes show any response to changes in light quality at the level of *cpcB2A2* or *cpeBA* transcript accumulation. CCA appears to operate through a single pathway at least at one step (RcaC) prior to the point at which *cpcB2A2* and *cpeBA* are controlled by separate elements. These new mutant classes apparently also contain lesions in components that act within the early part of the pathway, prior to its bifurcation.

The **FdBk** mutants exhibited a phenotype that was particularly interesting; both PE and PC_i levels were intermediate between the RL and GL conditions. Because many sensor kinases also act as the primary phosphatase of their partner response regulator, mutations in sensors often lead to intermediate constitutive phenotypes due to low level phosphorylation of the response regulator by other cellular phosphate sources (Jacobs, 1992; Lukat, 1992; Parkinson, 1992; Wanner, 1992). Thus the **FdBk** strains exhibited a possible sensor mutant phenotype. We succeeded in complementing the **FdBk** mutants with a wild type *F. diplosiphon* genomic DNA library; this resulted in the isolation of a 4.4 kilobase (kb) DNA fragment that contained two complete open reading frames (ORFs) (Kehoe, 1996) (Figure 1).

The first ORF (*rcaE*) encodes a protein, designated RcaE, that has a molecular mass of 74 kilodaltons (kDa). It contains a novel combination of domains. The carboxy-terminal region has similarity to the output domains of histidine kinase sensors, and within the amino-terminal half of the protein, a region of approximately 140 amino acids is similar to the chromophore attachment domain of the phytochromes, which are photoreceptors that have been shown to control a wide range of responses in vascular plants (Kendrick, 1994). This is exciting because it represents the first identification of a gene in a prokaryote with considerable sequence similarity to the phytochromes, a class of proteins that have been previously noted to share sequence similarities in their carboxy-terminal halves with the histidine kinase domains of bacterial sensor kinases (Schneider-Poetsch, 1991). Furthermore, numerous ORFs within the completely sequenced genome of *Synechocystis* spp. PCC6803 have been identified that show sequence similarity to both *rcaE* and the phytochromes (Kaneko, 1996). Because *Synechocystis* spp. PCC6803 is not capable of CCA, it appears likely that phytochrome-like molecules such as RcaE control a range of photoresponses in prokaryotic organisms. We expect that the facile genetic manipulation of prokaryotic organisms will allow us and others to contribute to an understanding of phytochrome-regulated gene expression in higher plants.

The second ORF (*rcaF*), initiated 12 basepairs (bp) downstream of *rcaE*, encodes the protein RcaF. RcaF is predicted to be 124 amino acids and is similar in size and sequence to a number of small response regulators (see Parkinson, 1992) that do not contain

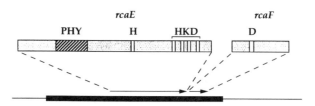

Figure 1. The two complete open reading frames of the 4.4 kb fragment complementing the **FdBk** mutants. *rcaE* contains a region encoding the chromophore attachment domain of higher plant phytochromes (PHY, dark stripes), a histidine phosphorylation site found in sensor proteins of two component regulatory systems (H) and the N, G1, F, and G2 blocks that comprise the histidine kinase domain (HKD) of sensor proteins. *rcaF* contains a region encoding an aspartate phosphorylation domain (D) of response regulator proteins of two component regulatory systems.

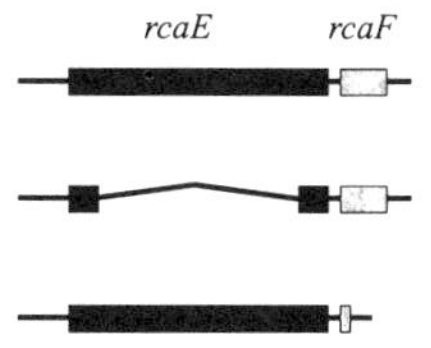

Fragment used in transformation	Complementation Result		
	rcaE-FdBk	rcaE-FdR	rcaF-FdR
rcaE rcaF ▬▬▬▬▭▬	+	+	+
▬▄‿‿‿‿▄▭▬	−	−	+
▬▬▬▬▬▭▬	+	+	−

Figure 2. Results of transformation of the **FdBk** and new **FdR** mutants with various subclones of the region of *F. diplosiphon* genomic DNA containing *rcaE* and *rcaF*. The fragments shown were cloned into an autonomously replicating plasmid and electroporated into these mutants. Transformants were analyzed for their ability (+) or inability (−) to undergo CCA when grown in RL and GL.

identifiable output domains. A similar molecule, *spo0F*, which is part of the sporulation signaling pathway in *Bacillus subtilis*, serves to transfer phosphoryl groups between two histidine residues in a complex type of two component regulatory system called a four step phosphorelay (Appleby, 1996).

We also successfully complemented the new **FdR** mutants, which are phenotypically similar to the previously identified *rcaC*-**FdR** mutants (Kehoe, 1997; Bruns, 1989; Chiang, 1992). Surprisingly, these mutants were complemented by the same 4.4 kb region of the *F. diplosiphon* genome that complemented the **FdBk** mutants (Figure 1).

In order to identify which of these genes were responsible for the complementation of the **FdBk** and new **FdR** mutants, subclones of the 4.4 kb genomic DNA fragment were made that contained either *rcaE* or *rcaF*. These were transformed into the **FdBk** and **FdR** strains and analyzed for their ability to complement the mutant phenotypes (Figure 2). These studies clearly demonstrated that the **FdBk** strains examined are lacking functional RcaE, and that two new classes of **FdR** stains existed, those lacking normal RcaE, and those deficient in RcaF (Kehoe, 1997). We have designated these as the *rcaE*-**FdR** and the *rcaF*-**FdR** strains, respectively.

Since *rcaE* alone is sufficient for complementing both a **FdBk** (Kehoe, 1996) and a **FdR** mutant (Kehoe, 1997), different lesions in *rcaE* must be able to generate different pigment phenotypes. In addition, the **FdR** phenotype results from lesions in at least three separate genes (*rcaC*, *rcaE*, and *rcaF*). We determined the molecular basis for the above observations by isolating and sequencing the region of the genome containing *rcaE* and *rcaF* from several **FdBk**, *rcaE*-**FdR**, and *rcaF*-**FdR** mutants. We found that all of these strains had DNA inserts, ranging from 1.0–1.6 kb, in this region (Figure 3). The insertions were probably the consequence of transposition events that were triggered during the chemical mutagenesis procedure (Chiang, 1992); there is some evidence for stress-induced mobilization of transposable elements in *F. diplosiphon* (Bruns, 1989). The *rcaE*-**FdBk** mutants contained inserts located within 200 bp of the putative translation start site of *rcaE*; these mutants are likely to be null for *rcaE*. DNA insertions were also present within *rcaE*-**FdR** mutants. However, these insertions were positioned between the conserved histidine residue that is typically phosphorylated in sensors and the four conserved motifs critical for histidine kinase activity; this strain probably makes a truncated form of RcaE. It is notable that similarly truncated forms of sensors from other organisms have been shown to have lost their ability to act as a kinase but retain their phosphatase activity (Atkinson, 1993; Cavicchioli, 1995). The *rcaF*-**FdR** mutants analyzed contained DNA insertions located approximately 200 bp downstream of the *rcaF* translation initiation codon; in this strain functional RcaF appears to not be produced.

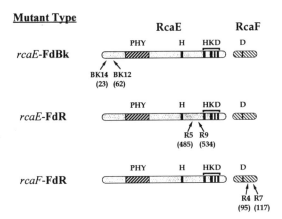

Figure 3. The location of insertions within the proteins encoded by *rcaE* and *rcaF* in three classes of CCA mutants. For *rcaE*-**FdBk** mutants, insertions (arrows) occurred at locations corresponding to the amino terminal end of RcaE. In the *rcaE*-**FdR** mutants, the insertions were located between the regions encoding the conserved histidine residue and the histidine kinase domain. The *rcaF*-**FdR** mutants contained insertions within the region corresponding to the central and carboxy terminal portions of RcaF. Numbers in parentheses refer to the amino acid number corresponding the nucleotide where the insertion occurred. Other symbols are as described in Figure 1.

We have now identified three components of the signal transduction circuitry controlling CCA. To begin to understand the relationships among these elements, we have individually expressed *rcaC*, *rcaE*, and *rcaF* extrachromosomally, either in wild type or in each of the mutant backgrounds, and examined the resulting phenotypes (Kehoe, 1997). When wild type *F. diplosiphon* is transformed with a wild type copy of *rcaC* and grown in cool white fluorescent light, it becomes more blue; PC levels increase and PE levels decline compared to cells transformed with the vector only. It appears that the additional copies of *rcaC* in these cells affects the downstream components of the CCA pathway to some extent. When additional copies of *rcaC* are introduced into the *rcaE*-**FdR** and *rcaF*-**FdR** mutants, or when *rcaF* is introduced into the *rcaE*-**FdBk** mutant, there is also an increase in the PC:PE ratio. This increase is not observed when *rcaF* is introduced into the *rcaC*-**FdR** mutant, however, suggesting that RcaF acts upstream of RcaC. Taken together with the sequence similarity of RcaE to both sensor kinases and plant phytochromes, these results support the proposal that this portion of the signal transduction pathway is linear, and that RcaE senses light quality and transmits a signal to RcaF, which is then passed to RcaC. We do not yet know if additional components of this pathway exist between these elements.

The following model presents a useful framework through which the wild type, *rcaE*-**FdBk**, *rcaE*-**FdR**, *rcaF*-**FdR**, and *rcaC*-**FdR** phenotypes can be better understood (Figure 4). A unique feature of this pathway is the central region of RcaC, which has been proposed to contain a novel motif, called an "H2" motif, based upon sequence similarity to other such domains (Ishige, 1994; Appleby, 1996; Kehoe, 1997). H2 domains have been suggested to be signature domains for a complex type of two component pathway called a four step phosphorelay, in which autophosphorylation occurs at a conserved histidine residue (H1) contained within a "conventional" sensor kinase domain (Ishige, 1994; Appleby, 1996). The phosphoryl group is transferred to a D residue (D1), and from there to a histidine within an "H2" motif, then finally to a second D residue (D2). Because RcaE contains an "H1" resi-

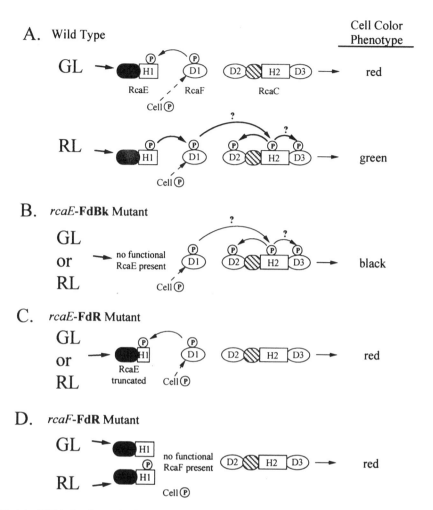

Figure 4. Model of CCA signal transduction in *F. diplosiphon* wild type and CCA mutants. **A.** The three signal transduction components known thus far are the putative sensor RcaE and response regulators RcaF and RcaC. The rate and direction of phosphoryl group flow between the components is indicated by the thickness and direction of the arrows, respectively. The domains of these components that may be phosphorylated are H1 of RcaE, D1 of RcaF, and H2, D2, and D3 of RcaC. In addition, RcaE contains a putative light sensing domain (black oval) and RcaC contains a consensus DNA binding motif (hatched oval). Phosphorylation is stimulated in red light while green light prevents phosphoryl group flow through the pathway, possibly by promoting phosphatase activity of RcaE. This prevents phosphorylation of RcaC (probably at H2 and then D2) and results in red colored cells. RcaF may also be able to undergo phosphorylation via transfer from other cellular phosphate sources (Cell P). **B.** In *rcaE* **FdBk** mutants, no functional RcaE is present. Cellular phosphate sources partially phosphorylate the pool of RcaF in the cell at the D1 residue, resulting in constitutive, intermediate levels of RcaC phosphorylation which in turn results in intermediate levels of accumulation of PE and PC in both red and green light. **C.** *rcaE* **FdR** mutants are proposed to contain a truncated form of RcaE that possesses constitutive phosphatase activity and no kinase activity. In these mutants, phosphoryl groups transferred to D1 of RcaF by other cellular phosphate sources would be removed by the mutant form of RcaE, keeping RcaC in a dephosphorylated state and the cells constitutively red. **D.** In the *rcaF* **FdR** mutants, no functional RcaF is expected to be present. This would block the flow of phosphoryl groups to RcaC and result in a constitutively red phenotype.

due, and RcaC contains an "H2" motif, and our extrachromosal expression studies have indicated that the flow of phosphoryl groups through this pathway (when the sensor is acting as a kinase) is from RcaE to RcaF to RcaC, we propose that phosphoryl groups are being transferred from the H1 of RcaE to the H2 of RcaC via the D1 contained within RcaF. It is not yet clear if the phosphoryl groups then move to both the amino- and the carboxy-terminal D residues within RcaC, or to D51 only. However, our SDM studies of D51 in RcaC suggest that this forward flow of phosphoryl groups to this residue is occurring under RL (Figure 4A). We also propose that in GL, RcaE actively dephosphorylates the CCA pathway. The phenotype of the *rcaE*-**FdBk** mutants supports this, because in cells where RcaE is apparently absent, PE and PC_i are expressed constitutively, at intermediate levels (Figure 4B). As noted previously, similar phenotypes have been noted for sensor mutations in other two component systems; the basal level of phosphorylation is supplied by other cellular phosphate sources. It appears that RcaF is the sole entrance point for donation of phosphoryl groups from donors other than RcaE, because *rcaF* mutants are red, suggesting that RcaC cannot be effectively phosphorylated by sources other than RcaF. The truncated form of RcaE that appears to be made in the *rcaE*-**FdR** mutants is proposed to lack kinase activity but retain phosphatase activity (Figure 4C), based upon similar sensor mutations in other systems (Atkinson, 1993; Cavicchioli, 1995). In these strains, the pathway would be maintained in a constantly dephosphorylated state, resulting in a constitutive red phenotype. Finally, in *rcaF*-**FdR** mutants, the lack of RcaF blocks the flow of phosphates through the system, which results in a constitutive red phenotype (Figure 4D).

Many questions remain concerning the signal transduction pathway controlling CCA. For example, it is still unclear if the CCA phosphorelay story ends with RcaC. We do not know if RcaC binds to the promoters of the phycobiliprotein genes directly and alters their activity or if additional phosphorelay components are required. But with our basic knowledge, we and others are poised to make rapid advances in our understanding of this process. The information gained from the study of the CCA signal transduction system will not only reveal more about the ways in which prokaryotic organisms sense and respond to their light environment, but will also help to understand both the origins and mechanisms of action of the phytochromes.

REFERENCES

Appleby JL, Parkinson JS, and Bourret RB (1996) The multi-step phosphorelay: Not necessarily a road less traveled. Cell 86: 845–848

Atkinson M and Ninfa AJ (1993) Mutational analysis of the bacterial signal-transducing protein kinase/phosphatase nitrogen regulator II (NR_{II} or NtrB). J Bacteriol 175: 7016–7023

Bruns B, Briggs WR, and Grossman AR (1989) Molecular characterization of phycobilisome regulatory mutants in *Fremyella diplosiphon*. J Bacteriol 171: 901–908

Casey ES, Kehoe DM, and Grossman AR (1997) Suppression of mutants aberrant in light intensity responses of complementary chromatic adaptation. J Bacteriol 179: 4599–4606

Cavicchioli R, Schroder I, Constanti M, and Gunsalus RP (1995) The NarX and NarQ sensor-transmitter proteins of *Escherichia coli* each require two conserved histidines for nitrate-dependent signal transduction to NarL. J Bacteriol 177: 2416–2424

Chiang GG, Schaefer MR, and Grossman AR (1992) Complementation of a red-light indifferent cyanobacterial mutant. Proc Natl Acad Sci USA 89: 9415–9419

Cobley JG and Miranda RD (1983) Mutations affecting chromatic adaptation in the cyanobacterium *Fremyella diplosiphon*. J Bacteriol 153: 1486–1492

Federspiel NA and Grossman AR (1990) Characterization of the light-regulated operon encoding the phycoerythrin-associated linker proteins from the cyanobacterium *Fremyella diplosiphon*. J Bacteriol 172: 4072–4081

Federspiel NA and Scott L (1992) Characterization of a light-regulated gene encoding a new phycoerythrin-associated linker protein from the cyanobacterium *Fremyella diplosiphon*. J Bacteriol 179: 5994–5998

Glazer AN (1982) Phycobilisomes: structure and dynamics. Annual Review of Microbiology 36: 173–198

Glazer AN (1985) Light harvesting by phycobilisomes. Annual Review of Biophysics and Biophysical Chemistry 14: 47–77

Houmard J, Capuano V, Cousin T, and Tandeau de Marsac N (1988) Genes encoding core components of the phycobilisome in the cyanobacterium *Calothrix* sp. strain 7601. Occurrence of a multigene family. J Bacteriol 170: 5512–5521

Ishige K, Nagasawa S, Tokishita S-I, and Mizuno T (1994) A novel device of bacterial signal transducers. EMBO J 13: 5195–5202

Jacobs JD, Ludwig JR, Hildebrand M, Kukel A, Feng T-Y, Ord RW, and Volani B (1992) Characterization of two circular plasmids from the marine diatom *Cylindrothece fusiformis*: Plasmids hybridize to chloroplast and nuclear DNA. Mol Gen Genet 233: 302–310

Kaneko T, Sato S, Kotani H, Tanaka A, Asamizu E, Nakamura Y, Miyajima N, Hirosawa M, Sugiura M, Sasamoto S, Kimura T, Hosouchi T, Matsuno A, Muraki A, Nakazaki N, Naruo K, Okumura S, Shimpo S, Takeuchi C, Wada T, Watanabe A, Yamada M, Yasuda M and Tabata S (1996) Sequence analysis of the genome of the unicellular Cyanobacterium *Synechocystis* sp. strain PCC6803. II. Sequence determination of the entire genome and assignment of potential protein-coding regions. DNA Res. 3: 109–136

Kehoe DM and Grossman AR. The use of site directed mutagenesis in the analysis of complementary chromatic adaptation. In Proceedings from the Xth International Photosynthesis Congress: Photosynthesis: from light to biosphere. 1995. Kluwer Academic Publishers

Kehoe DM and Grossman AR (1996) Similarity of a chromatic adaptation sensor to phytochrome and ethylene receptors. Science 273: 1409–1412

Kehoe DM and Grossman AR (1997) New classes of mutants in complementary chromatic adaptation provide evidence for a novel four-step phosphorelay system. J Bacteriol 179: 3914–3921

Kendrick RE and Kronenberg GHM (1994) Photomorphogenesis in Plants. 2nd ed., Dordrecht: Kluwer Academic Publishers.

Lukat GS, McCleary WR, Stock AM, and Stock J, B. (1992) Phosphorylation of bacterial response regulator proteins by low molecular weight phospho-donors. Proc Natl Acad Sci USA 89: 718–722

Mazel D, Guglielmi G, Houmard H, Sidler W, Bryant DA, and Tandeau de Marsac N (1986) Green light induces transcription of the phycoerythrin operon in the cyanobacterium *Calothrix* 7601. Nucleic Acids Res 14: 8279–8290

Oelmüller R, Conley PB, Federspiel N, Briggs WR, and Grossman AR (1988a) Changes in accumulation and synthesis of transcripts encoding phycobilisome components during acclimation of *Fremyella diplosiphon* to different light qualities. Plant Physiol 88: 1077–1083

Oelmüller R, Grossman AR, and Briggs WR (1988b) Photoreversibility of the effect of red and green light pulses on the accumulation in darkness of mRNAs coding for phycocyanin and phycoerythrin in *Fremyella diplosiphon*. Plant Physiol 88: 1084–1091

Parkinson JS and Kofoid EC (1992) Communication modules in bacterial signaling proteins. Annu Rev Genet 26: 71–112

Porter G, Tredwell CJ, Searle GFW, and Barber J (1978) Picosecond time-resolved energy transfer in *Porphyridium cruentum*. Part I. In the intact alga. Biochim Biophys Acta 501: 232–245

Schneider-Poetsch HAW, Braun B, Marx S, and Schaumburg A (1991) Phytochromes and bacterial sensor proteins are related by structural and functional homologies. FEBS Lett 281: 245–249

Searle GFW, Barber J, Porter G, and Tredwell CJ (1978) Picosecond time-resolved energy transfer in *Porphyridium cruentum*. Part II. In the isolated light-harvesting complex (phycobilisomes). Biochim Biophys Acta 501: 246–256

Tandeau de Marsac N (1977) Occurrence and nature of chromatic adaptation in cyanobacteria. J Bacteriol 130: 82–91

Tandeau de Marsac N (1983) Phycobilisomes and complementary adaptation in cyanobacteria. Bulletin de L'Institut Pasteur 81: 201–254

Wanner BL and Wilmes-Riesenberg MR (1992) Involvement of phosphootransacetylase, acetate kinase, and acetyl phosphate synthesis in the control of the phosphate regulation in an *Escherichia coli*. J Bacteriol 174: 2124–2130

8

PHOSPHORYLATION OF β-PHYCOCYANIN IN
Synechocystis sp. PCC 6803

Nicholas H. Mann and Julie Newman

Department of Biological Sciences
University of Warwick
Coventry CV4 7AL, United Kingdom

1. INTRODUCTION

Regulation of protein function and activity by the monoester phosphorylation of specific hydroxyl-containing amino acid residues in proteins has long been recognized as an important feature of metabolic control in eukaryotes and more recently in prokaryotes. In the specific case of the cyanobacteria there is considerable evidence for the occurrence and metabolic significance of monoester protein phosphorylation [for reviews see 1,2], but only in the case of the P_{II} protein and its signalling role in nitrogen status has the process been well characterized [see 3,4]. The harvesting of light is the central process underpinning the photoautotrophic lifestyle of cyanobacteria and there is much evidence for the involvement of protein phosphorylation in this process [for review see 5], though much of the specific details remain to be elucidated. The phycobilisome represents the primary antenna for PSII in cyanobacteria and there is preliminary evidence that a major component of the phycobilisome rod structure, the β-subunit of phycocyanin, is subject to monoester phosphorylation [6,7]. This work is concerned with firmly establishing whether β-phycocyanin is subject to phosphorylation, and if so, what the physiological significance of the process is. The approach was to identify the phosphorylated amino acid residue in β-phycocyanin and then to carry out site-directed mutagenesis of the β-phycocyanin (*cpcB*) gene.

2. MATERIALS AND METHODS

2.1. Bacterial Strains and Growth Conditions

Stock cultures of *Synechocystis* sp. PCC 6803 and the mutant 4R and its derivatives were routinely maintained and grown as described by Bloye *et al.* [8]. *E. coli* strains were maintained on LB agar and grown in LB broth supplemented, as required, with the appropriate antibiotics.

The Phototrophic Prokaryotes, edited by Peschek *et al.*
Kluwer Academic / Plenum Publishers, New York, 1999.

2.2. Preparation of Cell Extracts

11 cyanobacterial cultures were harvested and the cells resuspended in TES buffer (20 mM TES-NaOH pH 7.5) prior to disruption in a French pressure cell. The extract was subjected to centrifugation at $200,000 \times g$ for 1 h and the supernatant collected and stored at 4°C.

2.3. Site-Directed Mutagenesis and Other Recombinant DNA Techniques

A 1.12 kb *Kpn*I fragment from plasmid pPC338 [9] containing the *cpcB* gene and the 5' half of *cpcA* was cloned into *Kpn*I site of pAlter (Promega, Altered sites II in vitro mutagenesis system) to produce pPC001WT. *In vitro* mutagenesis was performed as recommended by the manufacturer using two mutagenic oligonucleotides; PCB-Ala-5' CGGTAATGCTGCCGCTATCGT (21-mer) and PCB-Asp-5' ATCACCGGTAATGCT-GATGCTATCGTTTCCAAC (33-mer). Other recombinant DNA techniques were carried out as described by Maniatis *et al.* [10].

2.4. Conjugations

The cells were harvested from an exponentially growing culture of 4R and resuspended in fresh BG11 medium to a concentration of 2.5×10^8 cells ml^{-1}. 1 ml of an overnight culture of *E. coli* S17.1 transformed with derivatives of pDSK519, was mated with 1 ml of 4R by filtration through a sterile 0.45 µm filter (25 mm diameter, cellulose acetate, Whatman). The filter was placed on a BG11 agar plate without antibiotic and incubated overnight in the light. The cells on the filter were resuspended in 1 ml BG11 medium, concentrated by centrifugation to 0.3 ml, and 100 µl aliquots were spread onto three BG11 agar plates containing kanamycin (10 µg ml^{-1}). Colonies appearing after approximately 10 days were patched out onto fresh plates.

2.5. Protein Kinase Assay and SDS-PAGE

In order to allow for the marked variation in the phycocyanin content of the cell extracts, the amount of protein added to the protein kinase assays was calculated to give constant allophycocyanin (112 µg). Protein kinase assays were carried out in a total reaction volume of 40 µl at 30°C for 10 min. Reactions were started by the addition of 2 µl MgCl$_2$ (100 mM), 2 µl ATP (10 µM) and 5 µCi (185 kBq) of [γ-^{32}P] ATP (specific activity 9.25 MBq $\times$ 10^{12} mmol^{-1}). Reactions were terminated by the addition of an appropriate volume of 4x sample buffer and heating at 100°C for 3 min. The entire sample was loaded onto a gel for analysis by SDS-PAGE. Discontinuous SDS-PAGE was performed using 12.5% (w/v) acrylamide slab gels. The gels were stained for protein with Coomassie Brilliant Blue R250, before drying and autoradiography at −70°C with a Dupont Cronex Lightning Plus intensifying screen.

3. RESULTS AND DISCUSSION

Previous work [7] had established that the 18 kDa polypeptide observed to be phosphorylated at the expense of ATP in cell-free extracts of *Synechocystis* sp. PCC 6803 could be cleaved with staphylococcal V8 protease to produce a 9 kDa fragment which yielded an amino terminal sequence (VFDVFTRVVSQADA) suggesting that it was derived from the amino terminus of β-phycocyanin (MFDVFTRVVSQADA) [11]. Hydrolysis of the phosphorylated protein and subsequent thin layer chromatography identified the phosphorylated residue(s) as a serine. An alignment of phycobiliprotein sequences [12] reveals that there are three serine residues which are absolutely conserved in the amino terminal portions of the β-phycocyanins so far characterized. The serine at position 108 is precluded from being the phosphorylated residue by virtue of the fact that the V8 fragment was only 9 kDa which corresponds to roughly 82 amino acid residues. The conserved serine at position 21 is unlikely to be the phosphorylated residue since it is also conserved in α-phycocyanin which has not been observed to be phosphorylated. This leaves the serine at position 50 as the likely target for phosphorylation. Consequently, it was decided to carry out site-directed mutagenesis of the *cpcB* gene such that serine 50 was converted to a non-phosphorylatable, but similarly sized alanine, or to a negatively charged aspartate, thereby mimicking phosphoserine. The 4R mutant of *Synechocystis* sp. PCC 6803 provides a suitable cellular environment in which to test the effect of these mutations. The 4R mutant which was characterized by Plank et al. [9] is olive green in colour and lacks detectable levels of PC. The 4R phycobilisomes are core structures that lack all rod components, including both PC subunits and the rod-associated linker proteins. Comparison of the *cpcB* genes in 4R and the wild type reveals a C insertion at the 73rd nucleotide position and the resulting frameshift causes premature termination yielding a 32 amino acid peptide of which the first 23 residues correspond to β-PC. Transformation of the 4R mutant to the PC⁺ phenotype was obtained with plasmids (bluescript) carrying the *cpc* operon [9]. The absence of CpcA in the 4R mutant is presumed to result from rapid degradation following its failure to be assembled into the αβ monomer.

The *cpc* operon of Synechocystis sp. PCC 6803 commences with the *cpcB* and *cpcA* genes encoding the β- and α-subunits of phycocyanin respectively which are followed by the *cpcH*, *cpcI* and *cpcD* genes. A 1.12 kb *Kpn*I fragment from pPC338 [9] carrying *cpcB* and approximately half of the *cpcA* gene was cloned into the *Kpn*I site of pALTER for site-directed mutagenesis. Two mutagenic oligonucleotides (PCB-Ala and PCB-Asp) were used respectively to induce a 1 bp mutation in which the serine codon was altered to alanine and a 3 bp mutation yielding an aspartate codon. The mutations were verified by subcloning a *Sst*I/*Xba*I fragment into M13mp18 and sequencing the region containing the mutation using a −40 primer.

In order to test for complementation of the 4R mutation, 1.2 kb *Kpn*I/*Bam*HI fragments containing 230 bp upstream of the *cpcB* gene, the 5′ half of *cpcA* and carrying a short region of vector polylinker and encoding the mutated and wild type *cpcB* genes, respectively, were sub-cloned from pALTER into pDSK519, a broad-host-range plasmid, capable of replication in cyanobacteria. The plasmid constructs were introduced into *E. coli* S17–1, transferred to the 4R mutant by conjugation and selection was made for kanamycin resistance. The exconjugants retained the typical olive-green colouration of 4R indicating that there was no complementation of the 4R mutation. Since even the wild type type genes failed to complement, it was thought that this might reflect a necessity for the *cpcB* and *cpcA* genes to be translated from the same transcript as proximal expression might be required for assembly of the αβ-monomer and consequent stability of the translation products. After 2–3 weeks incubation rare blue-green colonies were observed on the plates, but these were assumed to have arisen by recombination between the chromosome and the plasmid yielding an unmutated *cpc* operon. A series of constructs were made using the wild type genes to establish conditions in which complementation would occur. A plasmid (pPC003) was constructed which contained the same amount of DNA upstream of *cpcB*, but extending through the complete *cpcA* gene to the *Xho*I site near the beginning of *cpcH*. This also failed to complement the 4R mutation and so it was concluded that the *cpcBA* genes were not being efficiently transcribed and that important elements upstream of *cpcB* might be missing. Consequently a plasmid (pPC006) was constructed with approximately 2 kb of the region upstream from *cpcB* and extending through to the *Xho*I site in *cpcH*. This plasmid yielded blue-green exconjugants as did similar constructs carrying the two mutant *cpcB* alleles indicating that the β-phycocyanin with the serine-50 altered to alanine or aspartate could be assembled into the αβ-monomer, the bilins could be attached to the apo-proteins, and the consequent steps in phycobilisome assembly could take place. However, it was noticed during streaking on BG11 plates that 4R carrying the plasmid with the asp-50 allele segregated variants which grew more rapidly, again this might be explained by recombination between the plasmid and the chromosome yielding an unmutated chromosomal *cpc* operon. Protein kinase assays were carried out on cell free extracts from 4R and strains of 4R carrying the complementing wild type and mutant *cpcB* genes (Fig. 1). There are abundant amounts of α- and β-phycocyanin in the wild type *Synechocystis* sp. PCC 6803 and in the 4R mutant carrying the complementing plasmids, but not in the 4R mutant itself. There is an abundant phosphoprotein of 18 kDa in both the wild type and in the 4R strain carrying the wild type complementing plasmid. There is no detectable phosphorylation of this band in the 4R mutant or in the strain carrying the alanine-50 allele, and barely detectable amount in the strain carrying the aspartate-50 allele.

Thus it can be concluded that serine-50 is the site of phosphorylation in β-phycocyanin. The residual phosphorylation of β-phycocyanin in the aspartate-50 strain is assumed to arise from the accumulation of cells in the culture which contain an intact *cpc* operon which has arisen by recombination between the plasmid and the chromosome. We are presently attempting to overcome this problem with merodiploids by constructing haploid strains carrying the various *cpcB* alleles. There are additional alterations in the protein

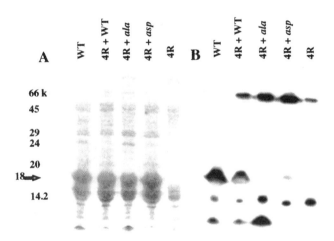

Figure 1. Protein kinase activities. Protein kinase assays were performed on cell-free extracts of *Synechocystis* sp. PCC 6803 wild type and the 4R mutant and of strains of 4R carrying complementing plasmids carrying the unmutated *cpcB* gene and the mutated forms of the gene in which serine-50 is replaced by either an alanine residue or an aspartate. The amount of protein in each assay was calculated to contain constant (112 µg) allophycocyanin. The assay mixtures were subsequently analysed by SDS-PAGE and autoradiography. A, SDS-PAGE gel stained with Coomassie Blue; B, autoradiograph. The positions and sizes of standard proteins are indicated at the left hand side of the SDS-PAGE gel as is the position of the 18 kDa polypeptide.

phosphorylation pattern between the wild type, 4R, and it derivatives. A band of approximately 70 kDa is strongly phosphorylated in all strains except the wild type and a band of less than 10 kDa is particularly strongly phosphorylated in the 4R-asp strain. The identity of these additional phosphoproteins is unknown. We are currently conducting experiments to examine the physiological effects of the alanine-50 and aspartate-50 mutations.

ACKNOWLEDGMENTS

We would like to thank Dr Lamont K. Anderson both for the 4R mutant and plasmids pDSK519 and pPC338 as well as for helpful discussion concerning the project.

REFERENCES

1. Mann, N.H. (1994) Microbiology 140, 3207–3215.
2. Zhang, C-C. (1996) Molec. Microbiol. 20, 9–15.
3. Forchhammer, K. and Tandeau de Marsac, N. (1995a) J. Bacteriol. 177, 2033–2040.
4. Forchhammer K and Tandeau de Marsac N (1995b) J. Bacteriol. 177, 5812–5817.
5. Allen, J.F. (1992) Biochim. Biophys. Acta 1098, 275–335.
6. Harrison, M.A. (1990) Molecular mechanisms of adaptation in the photosynthetic apparatus. PhD thesis, University of Leeds.
7. Tsai, J-W. (1996) Protein phosphorylation in cyanobacterial light-harvesting complexes. PhD thesis, University of Warwick.
8. Bloye, S.A., Silman, N.J., Mann, N.H. and Carr, N.G. (1992). Plant Physiol. 99, 601–606.
9. Plank, T., Toole, C. and Anderson, L.K. (1995) J. Bacteriol. 177, 6798–6803.
10. Maniatis, T., Fritsch, E.F. and Sambrook, J. (1982) Molecular cloning : a laboratory manual. Cold Spring Harbor Laboratory, Cold Spring Harbor, N.Y.
11. Kaneko T, Sato S, Kotani H, Tanaka A, Asamizu E, Nakamura Y, Miyajima N, Hirosawa M, Sugiura M, Sasamoto S, Kimura T, Hosouchi T, Matsuno A, Muraki A, Nakazaki N, Naruo K, Okumura S, Shimpo S, Takeuchi C, Wada T, Watanabe A, Yamada M, Yasuda M. and Tabata S (1996) DNA Res 3: 109–136
12. Apt, K.E., Collier, J.L. and Grossman A.R. (1995) J. Mol. Biol. 248, 79–96.

9

RpbA FUNCTIONS IN TRANSCRIPTIONAL CONTROL OF THE CONSTITUTIVE PHYCOCYANIN GENE SET IN *CALOTHRIX* SP. PCC 7601

Katherine Kahn, Roxanne P. Nieder, and Michael R. Schaefer

School of Biological Sciences
University of Missouri-Kansas City
Kansas City, Missouri 64110

1. INTRODUCTION

Cyanobacteria harvest light energy for photosynthesis with phycobilisomes (PBS). These macromolecular antenna complexes consist of two structural domains: a core which associates with the photosynthetic membrane, and a series of rods that radiate from the core [1]. Both domains are composed of different chromophoric phycobiliproteins and nonchromophoric linker polypeptides. The major cyanobacterial phycobiliproteins are allophycocyanin (AP), phycocyanin (PC), and phycoerythrin (PE), each consisting of an α and β subunit. AP is localized to the core domain, whereas PC and PE are localized to the rods. The linker polypeptides specifically associate with each phycobiliprotein to maintain PBS structure and facilitate efficient energy transfer within the antenna complex.

The cyanobacterium *Calothrix* sp. strain PCC 7601 (hereafter referred to as *Calothrix*) responds to changes in spectral light quality by altering the phycobiliprotein composition of the PBS rods [2–4]. This acclimation response, termed complementary chromatic adaptation (CCA), provides for optimal harvesting of the available light. Green light promotes the synthesis of rods composed of three distal hexamers of PE linked to the core by an invariant hexamer containing a constitutively expressed form of PC (designated PC_1). Conversely, red light promotes the synthesis of rods composed of two distal hexamers containing an inducible form of PC (designated PC_2) linked to the core by the invariant PC_1 hexamer. The change in rod phycobiliprotein composition during CCA is mediated primarily through differential expression of gene sets encoding the PE and PC_2 apoproteins along with their associate linker polypeptides [3].

The isolation, characterization, and complementation of different classes of PBS regulatory mutants have provided significant insight into the sensory and regulatory

The Phototrophic Prokaryotes, edited by Peschek *et al.*
Kluwer Academic / Plenum Publishers, New York, 1999.

mechanisms controlling PE and PC_2 synthesis during CCA in *Calothrix*. Several recent reports support the involvement of a complex linear phosphorelay in the CCA signaling mechanism [5–8]. In contrast, very little is known about the regulatory mechanism(s) controlling the synthesis of the other PBS components in this strain. We have developed a Tn*5469*-based system for identifying new regulatory genes in *Calothrix*. By screening for transposition of Tn*5469*, we have identified a number of PBS regulatory mutants characterized by an extra genomic copy of the transposon. One such mutant, designated FdBM1, exhibits elevated and constitutive synthesis of PC in any illumination [9]. The phenotype of strain FdBM1 is due to Tn*5469*-inactivation of the *rpbA* gene, which predicts a protein that shares structural similarities with characterized bacterial and phage DNA-binding repressor proteins [9]. Here, we provide evidence that the RpbA protein functions as a repressor in the control of transcription from the *cpcB1A1* gene set.

2. MATERIALS AND METHODS

2.1. Strains and Growth Conditions

Strain Fd33 is a short filament mutant of *Fremyella diplosiphon* UTEX 481 (= *Calothrix* sp. strain PCC 7601) which exhibits wild-type CCA [10]. Primary pigment mutant strain FdBM1 was isolated as a spontaneous derivative of strain Fd33 [9]. Cells were grown in liquid or on solid BG-11 medium [11] as previously described [12,13].

2.2. Methods

Transformation and complementation experiments with strains Fd33 and FdBM1 were performed as described by Chiang et al. [13]. Isolation of total RNA was performed as described by Kahn and Schaefer [9]. DNA manipulations, transformation of *E. coli*, plasmid minipreparations, RNA hybridizations, amplification of DNA by the polymerase chain reaction (PCR), and DNA mobility shift analysis were performed using established procedures [14,15].

3. RESULTS AND DISCUSSION

3.1. Characterization of Mutant Strain FdBM1

Calothrix is characterized by three gene sets (*cpcB1A1*, *cpcB2A2*, and *cpcB3A3*) encoding α^{PC} and β^{PC} apoproteins [16]. The *cpcB1A1* gene set (encodes PC_1) is constitutively expressed at low levels under all growth conditions that support phycobilisome biosynthesis; PC_1 and its associate linker polypeptides comprise the invariant core-proximal phycobiliprotein hexamer of each rod, which is required for phycobilisome function. The *cpcB2A2* gene set (encodes PC_2) functions in CCA and is expressed under red illumination. The *cpcB3A3* gene set (encodes PC_3) is expressed only when cells are cultured in a medium lacking sulfur [17]. With the exception of the cysteine residues required for pigment attachment, the α^{PC} and β^{PC} apoproteins encoded by *cpcB3A3* lack sulfur-containing amino acids, which may provide the cells an adaptive advantage during growth under sulfur-limiting conditions. An earlier RNA hybridization analysis showed that in comparison to the parental strain Fd33, the level of transcripts from *cpcB1A1*, but not *cpcB2A2* or *cpcB3A3*, was significantly elevated in strain FdBM1 [9].

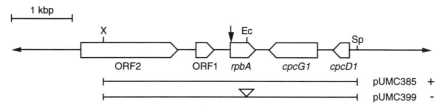

Figure 1. Physical map of the *rpbA* locus for strain Fd33. Open boxes denote size and orientation of *rpbA* and putative flanking genes or open reading frames (ORFs) as determined by sequence analysis. Vertical arrow shows site of Tn*5469* insertion in mutant strain FdBM1. Labeled horizontal bars below map indicate fragments harbored in shuttle vector pPL2.7 for complementation experiments. Symbols to the right of the horizontal bars indicate complementation (+) or noncomplementation (−) of strain FdBM1 by the respective constructs. Plasmid pUMC399 contains an amber codon (designated by small triangle) within the *rpbA* coding region. Restriction sites are shown for enzymes used in cloning experiments. Ec, *Eco*RV; Sp, *Spe*I; X, *Xba*I.

3.1.1. Complementation Analysis. A direct role for *rpbA* in the FdBM1 phenotype was demonstrated by complementation. For this study, two different *rpbA* constructs were introduced into cells of strain FdBM1 and assayed for their ability to complement the mutant strain. Plasmid pUMC385, which carries the intact *rpbA* gene on a 4.1-kbp genomic subclone (Fig. 1), was capable of restoring wild-type pigmentation to strain FdBM1. Plasmid pUMC399 is identical to complementing plasmid pUMC385 with the exception of an amber codon inserted in-frame into the *rpbA* coding region at a unique *Eco*RV site (Fig. 1). Introduction of pUMC399 into FdBM1 cells yielded transformants which were phenotypically indistinguishable from strain FdBM1. These data correlate *rpbA* activity with the phenotype of mutant strain FdBM1.

3.1.2. cpcB1A1 Transcription Analysis. To characterize the FdBM1 phenotype and complementation of strain FdBM1 at the level of PC_1 expression, transcription from *cpcB1A1* was examined by RNA hybridization analysis. Total RNA isolated from cells of strains Fd33/pPL2.7, FdBM1/pUMC399, and FdBM1/pUMC385 cultured in white light was hybridized to a DNA probe specific for *cpcB1A1*. In comparison to the Fd33/pPL2.7 transformant control, the level of the *cpcB1A1* transcript in the noncomplemented FdBM1/pUMC399 transformant was elevated by a factor of 1.7 (Fig. 2, compare lanes 1 and 2). Under the same conditions, the level of the *cpcB1A1* transcript in the complemented FdBM1/pUMC385 transformant was reduced to 55% of that measured for the FdBM1/pUMC399 transformant and 93% of that measured for the Fd33/pPL2.7 transformant (Fig. 2, compare lanes 1, 2 and 3). Thus, the *cpcB1A1* transcript pool in complemented strain FdBM1/pUMC385 was essentially restored to the wild-type level. Because complementation was achieved in trans, the 7% decrease in *cpcB1A1* transcripts in strain FdBM1/pUMC385, relative to that for the Fd33/pPL2.7 transformant, may reflect elevated expression of *rpbA* from the multiple copies of complementing plasmid pUMC385.

3.2. Characterization of RpbA

The *rpbA* gene, which complements mutant strain FdBM1, predicts a polypeptide of 139 amino acids with a molecular mass of 15.7 kDa (Fig. 3). A BLAST search of the GenBank database with the RpbA sequence produced no significant matches. However, this analysis revealed that RpbA contains regions resembling the helix-turn-helix (HTH) motif which is involved in DNA recognition by a number of characterized bacterial and phage

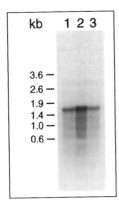

Figure 2. Analysis of *cpcB1A1* transcript levels in the parental, mutant, and complemented mutant strains. Total RNA (5 μg per lane) from strain Fd33/pPL2.7 (lane 1), FdBM1/pPUMC399 (lane 2), and FdBM1/pUMC385 (lane 3) cultured in white light was subjected to RNA blot hybridization against a DNA probe specific for transcripts from *cpcB1A1*. The position of RNA molecular standards is shown to the left of lane 1.

transcription regulator proteins. RpbA contains two HTH motif-like regions: one corresponding to residues 30–49 (designated HTH-1) and the other corresponding to residues 59–78 (designated HTH-2) of the native protein (Fig. 3).

3.2.1. Expression of RpbA in E. coli. To begin a functional characterization of RpbA, the *rpbA* coding region was amplified by PCR and the product ligated into *E. coli* expression vector pET22b. The resulting plasmid, designated pUMC460, provides for inducible expression of a modified RpbA form, designated RpbA-His, which contains six carboxyl-terminal histidine residues (Fig. 3). For the DNA mobility shift analysis presented below, the RpbA-His polypeptide was expressed in *E. coli* strain BL21.

3.2.2. DNA Mobility Shift Analysis. Collectively, the RNA hybridization and RpbA sequence analyses suggested that RpbA may function as a DNA-binding repressor in controlling transcription from *cpcB1A1*. To examine this possibility, a soluble protein extract containing RpbA-His was assayed for binding to a DNA fragment containing the *cpcB1A1* promoter region. An end-labeled 230-bp *Vsp*I-*Bst*98I fragment spanning the characterized [18] *cpcB1A1* transcription start site was used as the binding probe in this study (Fig. 4).

Binding by RpbA-His to the *cpcB1A1* promoter region was examined using a DNA gel mobility shift assay. The 230-bp *cpcB1A1* promoter probe was incubated with a soluble protein extract from *E. coli* strain BL21/pET22b or BL21/pUMC460 and assayed for the formation of a DNA-protein complex. No complexes were detected following incubation of the probe with the BL21/pET22b extract which lacks RpbA-His (Fig. 5, compare lanes 1 and 2). In contrast, a single, slower migrating complex was detected following incubation of the probe with the BL21/pUMC460 extract containing RpbA-His (Fig. 5, lane

```
MPARLQIKAEDCSPLGRFILQYLEEQGISMNRLADLSGVPQPRLRGACFKGTCPTPETLR    60
                             HTH-1

KLARVMGKHHLELYTLAYEARIEKLPEDADDTSLDILMRDLFETARELRLAVPKVRPSKA   120
 HTH-2

KIRKALLELGFSEENDECA(LEHHHHHH)                                  139
```

Figure 3. Predicted amino acid sequence of RpbA. The numbering for the amino acids is shown to the right of the sequence. Residues corresponding to the HTH motif-like regions HTH-1 and HTH-2 are underlined. Residues in parentheses following the native RpbA sequence correspond to the carboxyl-terminal sequence of the 147-residue RpbA-His polypeptide (see below) used in the DNA mobility shift analysis.

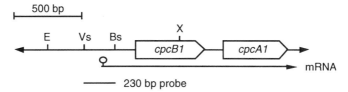

Figure 4. Physical map of the *cpcB1A1* locus in *Calothrix*. Open boxes denote size and orientation of *cpcB1A1* gene set as determined by sequence analysis. Arrow below map represents the 1.5-kb *cpcB1A1* transcript detected by RNA hybridization analysis. Open circle above arrow denotes putative attenuator structure described for the *cpcB1A1* transcript [18]. Labeled bar below map indicates fragment used in DNA mobility shift assays. Restriction sites are shown for enzymes used in generating the DNA fragments. E, *Eco*RI; Bs, *Bst*98I; Vs, *Vsp*I; X, *Xba*I.

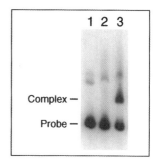

Figure 5. DNA gel mobility shift assay. The 230-bp *cpcB1A1* promoter probe (275 fmol) was electrophoresed on a native 4% acrylamide gel alone (lane 1) or following incubation with a soluble protein extract (2 μg) from *E. coli* strain BL21/pET22b (lane 2) or BL21/pUMC460 (lane 3).

3). These data suggest that RpbA recognizes and binds the *cpcB1A1* promoter region and support the hypothesis that RpbA functions as a repressor in the control of constitutive transcription from *cpcB1A1*. Competition gel mobility shift assays to confirm this hypothesis are in progress.

ACKNOWLEDGMENTS

We thank J. Salmasina for excellent technical assistance. This research was supported by a grant from the National Science Foundation. K. Kahn was supported by a UMKC Chancellor's Interdisciplinary Ph.D. Fellowship.

REFERENCES

1. Sidler, W.A. (1994) in: The Molecular Biology of Cyanobacteria (Bryant, D.A., ed.), pp. 139–216 Kluwer Academic Publishers, Dordrecht.
2. Bogorad, L. (1975) Annual Review of Plant Physiology 26, 369–401.
3. Grossman, A.R. (1990) Plant Cell and Environment 13, 651–666.
4. Tandeau de Marsac, N. (1983) Bulletin de L'Institut Pasteur 81, 201–254.
5. Kehoe, D.M. and Grossman, A.R. (1997) Journal of Bacteriology 179, 3914–3921.
6. Kehoe, D. and Grossman, A.R. (1996) Science 273, 1409–1412.
7. Chiang, G.G., Schaefer, M.R. and Grossman, A.R. (1992) Proceedings of the National Academy of Sciences USA 89, 9415–9419.
8. Casey, E.S., Kehoe, D.M. and Grossman, A.R. (1997) Journal of Bacteriology 179, 4599–4606.

9. Kahn, K. and Schaefer, M.R. Submitted for publication.
10. Cobley, J.G., Zerweck, E., Reyes, R., Mody, A., Seludo-Unson, J.R., Jaeger, H., Weerasuriya, S. and Navankasattusas, S. (1993) Plasmid 30, 90–105.
11. Allen, M.M. (1968) Journal of Phycology 4, 1–4.
12. Bruns, B.U., Briggs, W.R. and Grossman, A.R. (1989) Journal of Bacteriology 171, 901–908.
13. Chiang, G.G., Schaefer, M.R. and Grossman, A.R. (1992) Plant Physiology and Biochemistry 30, 315–325.
14. Ausubel, F.M., Brent, R., Klingston, R.E., Moore, D.D., Seidman, J.G., Smith, J.A. and Struhl, K. (1989) Current Protocols in Molecular Biology. John Wiley and Sons, New York.
15. Sambrook, J., Fritsch, E.F. and Maniatis, T. (1989) Molecular Cloning. Cold Spring Harbor Laboratory, Cold Spring Harbor, N.Y.
16. Grossman, A.R., Schaefer, M.R., Chiang, G.G. and Collier, J.L. (1993) Microbiological Reviews 57, 725–749.
17. Mazel, D. and Marliere, P. (1989) Nature 341, 245–248.
18. Mazel, D., Houmard, J. and Tandeau de Marsac, N. (1988) Molecular and General Genetics 211, 296–304.

10

BIOSYNTHESIS OF PHYCOBILIPROTEINS IN CYANOBACTERIA

Wendy M. Schluchter and Alexander N. Glazer

Department of Molecular and Cell Biology
229 Stanley Hall
University of California
Berkeley, California 94720-3206

1. INTRODUCTION

The characteristic blue-green and red colors of cyanobacteria and red algae reflect the presence of their water-soluble, light-harvesting antenna complexes called phycobilisomes. Phycobiliproteins are the major components of these complexes and can represent 40–50% of the total protein of cyanobacterial cells cultured under low light.[1] Each phycobiliprotein consists of two different polypeptide chains, α (~17 kDa) and β (~18 kDa), each of which carries at least one (and up to 3) covalently attached phycobilins.[1,2] Phycobiliproteins are frequently isolated as $(\alpha\beta)_3$ or as $(\alpha\beta)_6$ hexameric complexes. The assembly of the phycobiliproteins into phycobilisomes is mediated by their interaction with specific linker polypeptides which also modulate the spectroscopic properties of the phycobiliproteins.[3]

A great deal is known about the structure and assembly of the phycobiliproteins. There are three major classes of these proteins which share similar tertiary and quaternary structures: phycoerythrins (PE: λ_{max}~565 nm), phycocyanins (PC: λ_{max}~620 nm), and allophycocyanins (AP: λ_{max}~650 nm).[4–7] Allophycocyanin is the major building block of the core of the phycobilisome, and radiating out from the core are rods composed of phycocyanin-linker polypeptide complexes. In some cyanobacteria phycoerythrin forms the distal portion of the rods.[8] The distinctive absorption spectra of the different phycobiliproteins result from the differences in the numbers and types of bilin chromophores that are attached to their polypeptides through thioether bonds to specific cysteinyl residues. There are four isomeric bilins known in cyanobacteria, differing only in the arrangement of their double bonds: phycocyanobilin (PCB), phycobiliviolin (PXB), phycoerythrobilin (PEB), and phycourobilin (PUB; see Figure 1).[9] Understanding of the biosynthesis of the bilin prosthetic groups and of the ways in which particular bilins are specifically attached to specific cysteinyl residues in the phycobiliproteins is very incomplete. The research presented here reviews the current state of knowledge about these two processes.

The Phototrophic Prokaryotes, edited by Peschek *et al.*
Kluwer Academic / Plenum Publishers, New York, 1999.

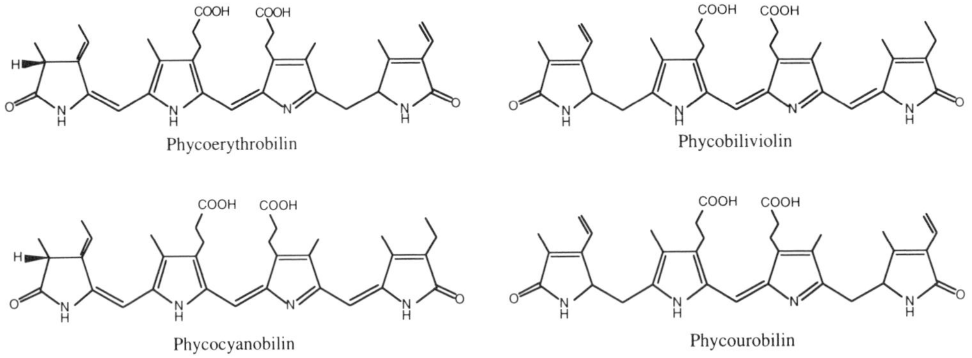

Figure 1. Structures of the four phycobilins present in cyanobacteria.

2. BIOSYNTHESIS OF PHYCOBILINS

Heme oxygenase cleaves heme, the precursor of all phycobilins, at the IXα position to produce biliverdin IXα.[10] In eukaryotic photosynthetic organisms containing phytochrome, phytochromobilin synthase converts biliverdin IXα to phytochromobilin (PΦB),[11,12] the covalently attached photoreversible chromophore of phytochrome.[13–15] The conversion of biliverdin Ixα to phycobilins has been studied in the red alga *Cyanidium caldarium*, the green alga *Mesotaenium caldariorum*, and the cyanobacteria *Synechocystis* species PCC 6701 and PCC 6803. The results of these studies are described below.

2.1. Phycobilin Biosynthesis in *Cyanidium caldarium*

The first biochemical characterization of the phycobilin biosynthetic pathway was performed in the thermoacidophilic red alga *Cyanidium caldarium* by Beale and coworkers.[16–23] All of the experiments were performed with partially purified *Cyanidium* extracts. Heme oxygenase was demonstrated to be the first enzyme in this pathway.[16,18,22,23] Heme oxygenase, which is nuclear-encoded and synthesized on cytoplasmic ribosomes, appears to be induced by light or by heme precursors such as δ-aminolevulinic acid, and protoporphyrin IX and repressed by D-glucose.[24–26] This enzyme requires ferredoxin as reductant.[18,19,22] It was subsequently shown that addition of iron chelators or of acid after incubation of heme with the enzyme enhanced the yield of biliverdin IXα demonstrating that the product of this reaction was not covalently bound to heme oxygenase, and it was also shown that a second reductant such as ascorbate was required for heme oxygenase activity.[23]

Initially, Beale and co-workers showed that incubation of biliverdin Ixα in extracts of *Cyanidium* produced the 3(Z) and 3(E) isomers of PCB.[17] On the basis of product accumulation rates, they proposed that the 3(Z) form preceded the appearance of the 3(E) form. They were unable to isolate any other intermediates until they began fractionating the extract to separate each enzyme. They were able to isolate several other phycobilin products by means of these partially purified extracts. Biliverdin IXα was converted into 15,16-dihydrobiliverdin IXα by reduction of the C15-C16 double bond by an enzyme which used NADPH and ferredoxin as reductants (see Figure 2).[21] Beale and Cornejo pro-

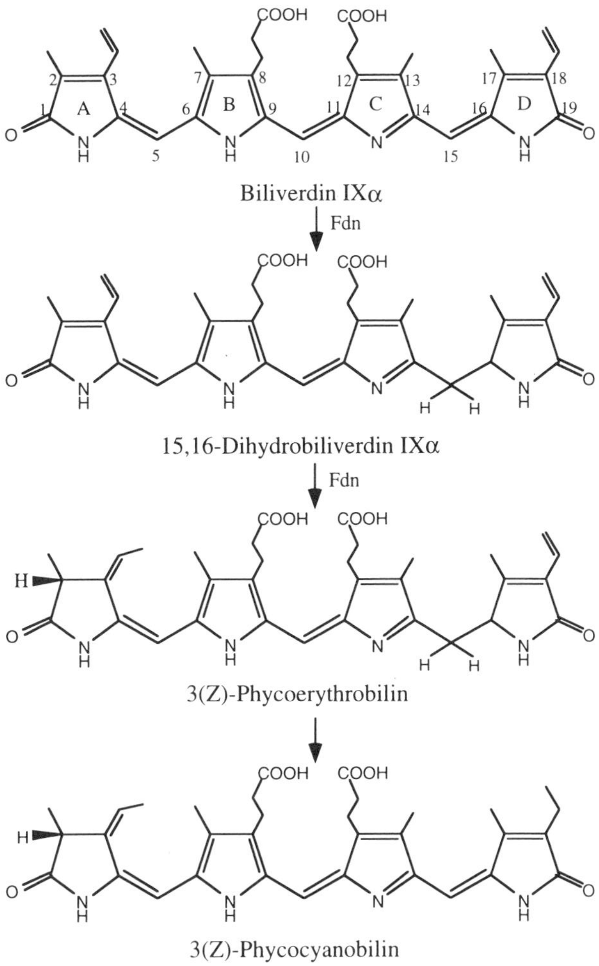

Biliverdin IXα

↓ Fdn

15,16-Dihydrobiliverdin IXα

↓ Fdn

3(Z)-Phycoerythrobilin

↓

3(Z)-Phycocyanobilin

Figure 2. Proposed pathway of phyco-bilin biosynthesis in the thermoacido-philic red alga *Cyanidium caldarium*. The first two steps are ferredoxin-de-pendent reactions.[19-21] Iso-merizations from the 3(Z) to the 3(E) form of PEB and PCB are also believed to be en-zyme-mediated.

posed that heme oxygenase and this biliverdin-reducing enzyme may be associated with each other because both enzymes require ferredoxin and ferredoxin-NADP reductase (which generates reduced ferredoxin when NADPH is supplied).[19] Beale and Cornejo then characterized several phycobilins produced after addition of biliverdin IXα to a *Cyani-dium* extract.[20] They detected both 3(E) PCB, the phycobilin present in phycocyanin, and 3(Z) PCB. If they removed the low molecular weight constituents of this extract, only 3(Z) PCB was produced. Addition of glutathione to the initial extract restored the ability of this extract to produce 3(E) PCB.[20] They found that one fraction produced 3(Z) phycoerythro-bilin (PEB) while another fraction was able to use this as a substrate and produce 3(Z) PCB.[20] They subsequently showed that after addition of 15,16-dihydrobiliverdin, one frac-tion could produce both 3(Z) PEB and 3(Z) PCB.[21] Because *Cyanidium* does not contain phycoerythrin, Cornejo and Beale concluded that PEB must be on the biosynthetic path-way to PCB.[21] The authors proposed a biosynthetic pathway (Figure 2) from biliverdin to 3(E) PCB which included all of the phycobilins that were detected and identified in their *Cyanidium* extracts. The first two steps are 2-electron reductions which require ferredoxin, followed by an isomerization of 3(Z) PEB to 3(Z) PCB. The second reduction step of the double bond at the C2-C3 position on the A ring of the linear tetrapyrrole would yield a product with a vinyl group at C3.[21] However, studies with chemically synthesized bilins have shown that a vinyl group at position 3 isomerizes spontaneously to form a (Z)-ethylidene group.[27] Production of this isomer by synthetic bilin reactions as well as by enzyme-mediated reactions seems to indicate that the (Z)-ethylidene isomer is a stable product of the vinyl isomerization.[21] Studies on the biosynthesis of phytochromobilin which also undergoes reduction at the C2-C3 double bond indicate the 3(Z) ethylidene isomer is the preferred product.[11,12,28]

It is quite surprising that PEB is a precursor to PCB in an organism which does not produce phycoerythrin. Since 3(E) PEB has been shown to serve as a substrate for bilin attachment to apo-phycocyanin *in vitro* both enzymically[29] and non-enzymically[30], phycobilin biosynthetic reactions must either be spatially separated from apo-phycobiliproteins or be processive to prevent accumulation of intermediates such as PEB in sufficient quantities to be accidentally attached to apo-phycocyanin.

While Beale and co-workers have been quite successful in characterizing the enzyme activities, thus far, none of these biosynthetic enzymes have been purified in sufficient quantities to obtain any amino acid sequences. Therefore, no progress has been reported in cloning any of the genes in *Cyanidium* which encode these important enzymes

2.2. Phycobilin Biosynthesis in *Mesotaenium caldariorum*

Mesotaenium is a unicellular green alga which does not produce phycobiliproteins, but does produce a phytochrome with a blue-shifted P_r–P_{fr} difference spectrum under dark-adapted conditions.[28,31] While investigating the cause for the blue-shift in the difference spectrum, Lagarias and co-workers discovered that the spectrum of this phytochrome closely resembled that of oat phytochrome reconstituted with PCB.[28] Therefore, they investigated whether this green alga could in fact synthesize PCB. When biliverdin was added to *Mesotaenium* soluble protein extracts from dark-adapted chloroplasts, 3(Z) phytochromobilin (PΦB) and 3(Z) PCB were produced in a ferredoxin-dependent manner. Production of the 3(E) isomers of both PCB and PΦB was observed, but it was unclear whether the formation of these compounds was enzyme-catalyzed. Wu et al. did careful time-course assays and established precursor/product relationships among all products detected.[28] They concluded that the biosynthetic pathway from biliverdin IXα to 3(Z) PCB proceeds *via* 3(Z) PΦB as shown in Figure 3. They also demonstrated that 3(E) PCB formed an adduct with recombinant algal phytochrome that was indistinguishable from the blue-shifted native phytochrome that was purified from these dark-adapted cells. This is the first report of the presence of PCB in a non-phycobiliprotein-containing organism and presents a more direct route for the synthesis of PCB than the pathway described for *Cyanidium*. The authors failed to detect 15,16-dihydrobiliverdin or PEB as intermediates in PCB synthesis in *Mesotaenium*. They also noted in passing that in their examination of the PCB biosynthetic pathway in *Cyanidium*, they detected 3(Z) PΦB as an intermediate.[28] This raises a question as to whether there are two different pathways for the synthesis of PCB in *Cyanidium* and perhaps in cyanobacteria as well. The report documenting PCB as the functional chromophore on a native phytochrome invites speculation about what chromophore is present on the cyanobacterial phytochrome-like protein, PlpA, shown to be required for growth of *Synechocystis* sp. PCC 6803 under blue light conditions[32] as well as on the cyanobacterial complementary chromatic adaptation sensor, RcaE, discovered recently in *Fremyella diplosiphon*.[33]

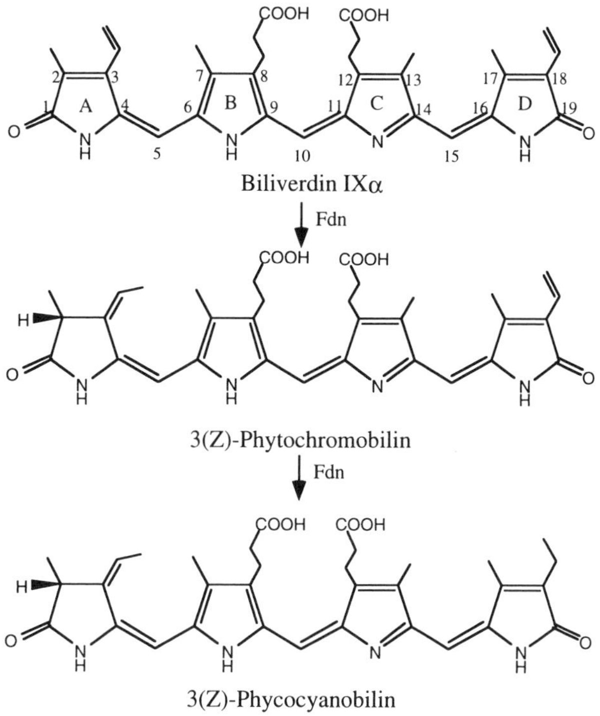

Biliverdin IXα

Fdn

3(Z)-Phytochromobilin

Fdn

3(Z)-Phycocyanobilin

Figure 3. Pathway from biliverdin IXα to phycocyanobilin observed in dark-adapted chloroplasts from the green alga *Mesotaenium caldariorum*. Both reductions are ferredoxin dependent. Production of the 3(E) isomers of phytochromobilin and phycocyanobilin were observed, but it is unclear whether production of these isomers is enzyme-mediated.[28]

2.3. Phycobilin Biosynthesis in Cyanobacteria

Progress has been reported recently in the study of cyanobacterial phycobilin biosynthesis. Cornejo and Beale have had some success in detecting enzyme activities in whole cell extracts of *Synechocystis* sp. PCC 6701 and *Synechocystis* sp. PCC 6803.[34] They have succeeded in characterizing heme oxygenase activity in fractionated extracts of *Synechocystis* sp. PCC 6701. By using mesoheme as a substrate, it is possible to study the activity of this enzyme in crude extracts because mesobiliverdin cannot be further metabolized into other phycobilins. Cyanobacterial heme oxygenase, like its red algal counterpart, also requires two reductants: ferredoxin and another reductant such as Trolox, a soluble vitamin E analog, or ascorbate, but it is less sensitive to inhibition by protoporphyrin IX than the red algal enzyme.[34]

These researchers were also successful in detecting 3(Z) PEB and 3(Z) PCB after addition of biliverdin to unfractionated extracts in *Synechocystis* sp. PCC 6701, but did not detect any 3(E) isomers. This organism contains both phycoerythrin and phycocyanin which may account for the accumulation of both of these phycobilins. Only 3(Z) PCB was detected in *Synechocystis* sp. PCC 6803, an organism which contains only PCB attached to its phycobiliproteins (Table 1).

The fact that no 15,16-dihydrobiliverdin IXα was detected in their assay system and that PEB was only detected in *Synechocystis* sp. PCC 6701, an organism which attaches this bilin to its phycobiliproteins, does not allow one to choose between the potential candidate pathways of phycobilin biosynthesis in cyanobacteria. Therefore, it is still an open question as to what pathway exists in cyanobacteria, whether cyanobacteria can produce

Table 1. PCB-bearing polypeptides in *Synechocystis* sp. PCC 6803, the genes which encode them, and the number of bilins on each polypeptide

Gene	Phycobiliprotein	Phycobilins
apcA	α^{AP}	1 PCB
apcB	β^{AP}	1 PCB
apcD	α^{APB}	1 PCB
apcE	L_{CM}	1 PCB
apcF	$\beta^{18.5}$	1 PCB
cpcA	α^{PC}	1 PCB
cpcB	β^{PC}	2 PCB

PΦB, and whether there is only one pathway for the synthesis of phycobilins. The accumulation of only the 3(Z) isomers raises the question as to whether the true precursor phycobilin in the reaction leading to the attachment of phycobilins to apo-phycobiliproteins is the 3(Z) isomer and not the 3(E) isomer. The 3(E) isomer is the major product of bilin cleavage from phycobiliproteins by methanolysis.[17,35,36] It will be necessary to fractionate cyanobacterial extracts in order to isolate each biosynthetic enzyme and establish the substrate and product. Discovery of the genes which encode these enzymes will also be absolutely necessary to the conclusive establishment of the biosynthetic pathway of phycobilins in cyanobacteria.

2.3.1. Putative Phycobilin Biosynthetic Genes in Synechocystis sp. PCC 6803. Now that a complete *Synechocystis* sp. PCC 6803 genome sequence is available,[37] one can search for the phycobilin biosynthetic genes based on homologies to enzymes that catalyze similar reactions in other organisms. For example, in a search of this genome for homologs of heme oxygenases, two ORFs (sll1184 and sll1875) show 46–57% identity with heme oxygenase from *Porphyra purpurea*.[37] The presence of two heme oxygenases in cyanobacteria raises many interesting questions. Do all cyanobacteria contain two enzymes? Does one function in phycobilin biosynthesis and the other in heme catabolism? Does one function as a stress response enzyme?

Since no phycobilin biosynthetic gene beyond heme oxygenase had been identified, we looked for homologs of known enzymes from other systems which catalyze reactions generally similar to those expected to be catalyzed by the cyanobacterial phycobilin biosynthetic enzymes. In both of the biosynthetic pathways described, two reduction steps require reduced ferredoxin as an electron donor. However, there is no conserved domain in proteins which has been shown to bind ferredoxin; the interaction of ferredoxin (which is an acidic protein) with several ferredoxin-dependent enzymes has been shown to be through positively charged residues on these enzymes.[38,39]

Animals possess biliverdin reductase, an enzyme in heme catabolism, which reduces the C10-C11 double bond of biliverdin IXα with electrons from NADPH, to produce bilirubin IXα. This reaction is very similar to that required to produce 15,16- dihydrobiliverdin from biliverdin. Therefore, we searched the *Synechocystis* sp. PCC 6803 genome for homologs of biliverdin reductase sequences from rat and human. An ORF of 328 amino acids (slr1784)[37] showed 20–21% identity to both sequences.[40] We cloned this gene, named it *bvdR*, and overexpressed it in *E. coli* both in its native form and as a His-tagged fusion protein. Both forms of the enzyme reduced biliverdin using NADPH as the sole reductant to produce bilirubin IXα.[40] This is the first report of bilirubin formation in a bacterium. The K_m for biliverdin was 1.3 µM while the pH optimum for this reaction was at 5.8. Interestingly, rat BvdR was shown to catalyze the reduction of phycobilin substrates at the C10 position.[41] Indeed overexpression of rat BvdR in *Arabidopsis,* caused the plants to display aberrant photomorphogenesis, presumably because the enzyme destroys PΦB.[42] Therefore, we determined the affinities of the cyanobacterial BvdR for biliverdin and PCB. The K_m for PCB was at least 10-fold higher than that for biliverdin[40], suggesting that this substrate specificity is sufficient to prevent interference by BvdR with phycobiliprotein biosynthesis *in vivo.*

We expected that since this enzyme produced bilirubin, it catalyzed a step in the heme catabolic pathway in cyanobacteria and that a mutant, if it displayed any phenotype, would show increased phycobiliprotein synthesis. To our surprise, a segregated *bvdR* interposon mutant produces no phycocyanin and only 85% of the normal amount of allophycocyanin-containing phycobilisome cores.[40] Interestingly, pseudorevertants which can produce 30–50% of normal levels of phycocyanin arise at high frequency. We are confident that the phenotype displayed by this *bvdR* mutant is due solely to the absence of *bvdR* for several reasons. The *bvdR* gene appears to be co-transcribed with an upstream regulatory ORF resembling a regulatory component of a sensory transduction system. The transformants which segregated were created using a construction in which the direction of transcription of the interposon cartridge was the same as that of the *reg-bvdR* operon.[40] The levels of transcription of this upstream regulatory gene were very similar in the wild-type, *bvdR* mutant, and *bvdR* pseudorevertant (W. M. Schluchter and A. N.Glazer, unpublished results). We have also overexpressed the His-tagged *Synechocystis* sp. PCC 6803 *bvdR* construct in *Anabaena* sp. PCC 7120. These cells appear very yellow-green in color presumably due to excess bilirubin production, and synthesize varying amounts of phycobiliprotein (Y. A. Cai, W. M. Schluchter, and A. N. Glazer, unpublished results), hinting that the level of this enzyme or of bilirubin IXα may play some sort of regulatory role.

How can one explain the selective blockage of phycocyanin synthesis in *bvdR* mutants? Biliverdin IXα is a branch point between phycobilin biosynthesis and heme catabolism (see Figure 4). Control of the flux of biliverdin IXα between the catabolic reaction leading to bilirubin and the biosynthetic pathways leading to the phycobilins may provide one means of regulating the biosynthesis of phycobiliproteins. Could it be that bilirubin directly or indirectly modulates the flux of biliverdin into the biosynthetic

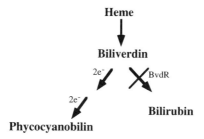

Figure 4. Diagram of the relationship of the phycobilin biosynthetic pathway and a heme catabolic pathway. Biliverdin is at the branch point of these two pathways. Two two-electron reductions are required in order to transform biliverdin IXα to phycocyanobilin. Biliverdin reductase (BvdR) catalyzes the NADPH-dependent reduction of biliverdin to produce bilirubin IXα, presumed to be a heme catabolic reaction. Blockage of this reaction in *bvdR* mutants affects the flux into the phycobilin biosynthetic pathway as demonstrated by absence of phycocyanin.[40]

pathway? This could happen by the control of the transcription of phycocyanin structural genes. We have examined the transcription of the $cpcBAC_1C_2D$ operon in *Synechocystis* sp. PCC 6803 wild-type cells, *bvdR* mutant, and *bvdR* pseudorevertants. The amounts of the major *cpcBA* transcript appear to be within a factor of two of one another and, therefore, do not correspond to the amounts of phycocyanin present in these organisms (W. M. Schluchter and A. N.Glazer, unpublished results). It could be that high levels of biliverdin or low levels of bilirubin serve as repressors of the transcription of phycobilin biosynthetic genes or, either directly or indirectly, of the activity of the enzymes which they encode.

Another possibility is that bilirubin plays a significant a role in scavenging reactive oxygen species in cyanobacteria as it is believed to do in humans.[43,44] Perhaps the *bvdR* mutant is compensating for its inability to deal with reactive oxygen species generated by reduced ferredoxin by lowering its photosynthetic rate by decreasing its absorption cross-section for visible light. Under these circumstances, one might expect that the pseudorevertants would have higher levels of superoxide dismutase as a way to offset the absence of bilirubin. However, one would expect that if this were true that the *bvdR* mutant would be extremely sensitive to growth under high light which does not seem to be the case (W. M. Schluchter and A. N. Glazer, unpublished results).

Another possibility that might explain the phenotype of this mutant is that bilirubin is a biosynthetic intermediate of a phycobilin which is a chromophore for a light receptor that regulates transcription of phycocyanin biosynthetic genes. In any case, examination of the pseudorevertants to determine which DNA mutations present in these cells can partially complement the *bvdR* mutant will help to determine the role of biliverdin reductase in normal phycobiliprotein biosynthesis.

2.3.2. Other Putative Phycobilin Biosynthetic Genes. There are several other genes which show similarity to *bvdR* in the *Synechocystis* sp. PCC 6803 genome. We have cloned two of these and are in the process of generating interposon mutants and of overexpressing these genes in *E. coli* in order to determine if they have any phycobilin reductase activity.

3. PHYCOBILIN ATTACHMENT TO APO-PHYCOBILIPROTEINS

There are three different mechanisms by which phycobilins might attach to apo-phycobiliproteins. The first and most straight-forward way would be autocatalysis. The addition might take place spontaneously during the folding of each apo-phycobiliprotein. The spontaneous addition of PΦB to apo-phytochrome is one example of this process.[15] However, when PCB or PEB are added to apo-phycocyanin or apo-phycoerythrin subunits *in vitro*, unnatural adducts predominate. These products are more oxidized bilins with an extra double bond at C2-C3 in ring A.[29,30,45–48] There is no discrimination between PCB and PEB addition to apo-phycobiliproteins *in vitro* that would account for site selectivity *in vivo*. This tends to argue against the autocatalysis hypothesis.

Another scenario would envisage a small set of lyases, each of which recognizes multiple phycobilin attachment sites, in analogy with other enzymes involved in post-translational modification of proteins, such as glycosylases and phosphatases. One line of evidence against this hypothesis is that the stereochemistry of PCB at the three attachment sites in C-phycocyanin is different with the α-84 and β-82 bilins being attached at C3' of the bilin in the *R* configuration and the β-153 bilin in the *S*.[7]

A third possibility is that there is one enzyme which recognizes each particular phycobilin addition site and assures that the correct bilin is attached at the appropriate site. The evidence in favor of this bilin attachment process was the discovery that the *Synechococcus* sp. PCC 7002 *cpcE* and *cpcF* genes encoded a heterodimeric phycocyanin α subunit PCB lyase and that mutation in either of these genes affected only PCB addition to the α subunit.[29,47,49,50] Other *cpcE* and/or *cpcF* mutants have been characterized in *Synechococcus* sp. PCC 7942,[51] *Anabaena* sp. PCC 7120 (W.M. Schluchter and A. N. Glazer, manuscript in preparation), and in *Calothrix* sp. PCC 7601.[52] In all of these cases, the mutants produce significantly reduced amounts of phycocyanin. Pseudorevertants of the *Synechococcus* sp. PCC 7002 *cpcEF* mutants arose at high frequency, and when the phycocyanin produced by one of these mutants was examined, it was found to have an amino acid substitution, Tyr-Cys, at position α-129.[50] This mutation presumably allowed a different lyase to recognize this site and catalyze the addition of PCB to the α-subunit of phycocyanin. Jung et al. showed that a mutation in one or both of the *pecEF* genes of *Anabaena* sp. PCC 7120, genes which show a high degree of similarity with *cpcE* and *cpcF*, affected the level of the phycoerythrocyanin holo-α-subunit.[53] The phycoerythrocyanin α-subunit purified from a *pecEF* mutant was found to contain a PCB-adduct instead of the PXB chromophore which is normally present. These results suggest that PecEF form a heterodimeric phycoerythrocyanin α subunit PXB lyase, and that in the absence of PecEF, another lyase can recognize this site and add PCB.

There are several other candidate genes for lyases found in operons together with phycobiliprotein structural genes. These ORFs share limited similarity with CpcEF and PecEF but contain a highly conserved region of 10 amino acids dubbed the "E-Z motif".[54] The *mpeUV*, *cpeYZ*, and *rpcEF* genes are clustered together with the structural genes for R-phycoerythrin I, R-phycoerythrin II and R-phycocyanin II in the marine unicellular cyanobacterium *Synechococcus* sp. WH8020.[54] These genes have been inferred to encode lyases based upon their similarity in sequence to other known lyase genes, their position close to phycobiliprotein structural genes, and their size (all known lyases are 200–300 amino acids in length). Recently, Schaefer and colleagues isolated a *Fremyella diplosiphon* secondary pigment mutant from an *rcaC* mutant strain which produced much less phycoerythrin.[55] They found that a transposon had inserted into the *cpeY* gene which shows similarity to other lyase genes such as *Synechococcus* sp. PCC 7002 *cpcE*.[55] This mutant produced 46% of the amount of phycoerythrin produced by the parent *rcaC* mutant. Therefore, the authors suggested that this gene was involved in the biosynthesis of phycoerythrin due to its similarity to other lyases and the phenotype of the mutant.

3.1. Putative Lyases in *Synechocystis* sp. PCC 6803

Synechocystis sp. PCC 6803 contains 8 different PCB attachment sites on its phycobiliproteins (see Table 1) which would imply that there may be as many as 16 genes involved in bilin attachment (if all lyases are assumed to be heterodimeric proteins). However, there are only five open reading frames in the *Synechocystis* sp. PCC 6803 genome which show some similarity to known lyases.[37] The *cpcE* (slr1878) and *cpcF* (sll1051) genes are obvious homologs to *cpcE* and *cpcF* of *Synechococcus* sp. PCC 7002. These genes are not contiguous and therefore are not co-transcribed. The remaining three ORFs are less conserved overall, but the central regions of these proteins (amino acids 30–130) do contain regions that are conserved, including most of the E-Z motif. We have cloned four of these five genes and have made interposon mutants to determine the phenotype to see if loss of any of these genes affects phycobiliprotein synthesis. We have not yet named these genes because the data are insufficient to assign a specific function for each of the encoded proteins.

3.1.1. CpcE (slr1878). This 272-residue ORF is 61% similar to *Synechococcus* sp. PCC 7002 CpcE. The *cpcE* gene was cloned and inactivated by the insertion of an interposon cartridge conferring resistance to chloramphenicol. After several passages, colonies that appeared yellow-green in color were found to contain no copies of the wild-type *cpcE* gene. The mutant produced phycobilisomes with about 5–10% of the wild type amount of phycocyanin but normal allophycocyanin cores (W. M. Schluchter and A. N. Glazer, unpublished observations). This phenotype is similar to that of the *cpcE* mutant from *Synechococcus* sp. PCC 7002. The latter gene encodes one of the polypeptides of the heterodimeric phycocyanin α subunit PCB lyase.[49,50] However, the *Synechocystis* sp. PCC 6803 *cpcE* mutant does not accumulate any excess apo-α-PC or holo-β-PC, and pseudorevertants which are capable of synthesizing more phycocyanin have not been observed.

3.1.2. slr1098. Slr1098 is the ORF in the *Synechocystis* sp. PCC 6803 genome most closely similar to CpcE (slr1878). This ORF is 252 residues long and is overall ~46% similar to CpcE (slr1878). We inactivated this gene by insertion of an interposon which confers resistance to chloramphenicol, transformed *Synechocystis* sp. PCC 6803 cells with two constructs with different orientations of the interposon cartridge, and selected for segregation. Transformants generated with both constructs formed yellowish-green colonies that were shown to have no remaining copies of the wild-type slr1098 gene. We purified phycobilisomes from both mutant and wild-type cells grown under high light conditions $(100–120 \ \mu E \ m^{-2} \ s^{-1})$ and found that the mutant produced phycobilisomes that were significantly smaller than those found in wild-type cells as judged from their sedimentation rates in sucrose density gradients. When the phycobiliproteins present in these phycobilisomes were separated by C4 reverse phase HPLC, no apo-phycobiliproteins were observed. When we compared the ratio of the amount of α^{PC} to β^{AP} and compared this to the ratio obtained for wild-type phycobilisomes, we found that the mutant produced only ~30% of the phycocyanin that was found in wild-type phycobilisomes (W. M. Schluchter and A. N. Glazer, unpublished results). The L_{CM} and α^{AP} and β^{AP} proteins were chromophorylated as judged by their HPLC profiles and absorption spectra and by zinc-enhanced fluorescence of the polypeptides after separation by SDS-PAGE. The fluorescence emission peaks of phycobilisomes isolated from two different isolates of the mutant were 665 and 667 nm compared to the emission maximum of 663 nm obtained for wild-type phycobilisomes. It is possible that this red shift in the fluorescence emission peak is due to an abnormal phycobilin adduct in the terminal energy acceptor in the core of the phycobilisome. However, from previous analyses of nonenzymic addition of PCB to apo-phycobiliproteins, the product of this type of addition, mesobiliverdin, displays a 20-nm red shift in its fluorescence emission peak relative to PCB.[45,47] More careful analysis of this mutant will be required to determine if all of the core polypeptides contain normal PCB chromophores. It is quite clear that this mutant does not produce normal amounts of phycocyanin, similar to the *cpeYZ* mutant phenotype observed in *Fremyella diplosiphon*.[55] The phenotype of this mutant along with the homology to the CpcE protein suggest that slr1098 is a lyase. Assignment of the substrate specificity of this lyase will require *in vitro* addition analyses.

3.1.3. slr1687. Slr1687 is approximately 30% similar to *Synechocystis* sp. PCC 6803 CpcE and is 233 residues long. Interposon mutants were generated (interposons were in both orientations) and all transformants examined had fully segregated indicating that this gene is not required for growth. However, there is little difference between the mutant and the wild-type when phycobiliprotein levels are compared in cells grown under the same conditions. Phycobilisomes purified from this mutant have the same absorption maximum (623 nm), and appear to have the same ratio of PC/AP as judged by SDS-PAGE. The fluorescence emission maxima for phycobilisomes purified from the wild-type and Δslr1687 mutant are at 662 nm and 663 nm, respectively (W. M. Schluchter and A. N. Glazer, unpublished results). We do know that this gene is transcribed on a transcript of 1.8 kb in wild-type exponentially growing cells. Several experiments remain to be completed for this mutant. The absolute ratio of PC:AP and of PCB:chl must be quantitated for this mutant, and growth rates need to be determined for this mutant as well. Therefore, we cannot positively say that this ORF encodes a lyase.

   3.1.4. sll1663. Sll1663 is 220 residues long and is only ~20% similar to *Synechocystis* sp. PCC 6803 CpcE, but is 30% similar to slr1687. Cloning and insertional inactivation of this gene using two different constructions were successful, and all transformants tested had segregated. Initially, colonies appeared yellow-green, but over time cells appeared more blue. When their phycobilisomes were purified, they had normal absorption and fluorescence spectra. Initial analyses of the phycobiliproteins purified from this mutant by SDS-PAGE followed by Zn^{2+}-enhanced fluorescence showed no differences from wild-type phycobiliproteins (W. M. Schluchter and A. N. Glazer, unpublished results). Careful quantitation of phycobiliprotein levels will be required, but these initial analyses do not indicate a phenotype as yet.

   We plan to generate double and triple mutants in these putative lyases. If one of these ORFs can substitute for another, we may not discover the phenotype of these mutants until we examine one in which at least two of these genes are inactivated. We also plan to produce these proteins in *E. coli* and test for their ability to add PCB to apo-phycobiliproteins and to transfer PCB from holo-phycobiliproteins to apo-phycobiliproteins.[47]

4. CONCLUDING REMARKS

   It is obvious that much more work needs to be done to establish unequivocally the phycobilin biosynthetic pathway beyond biliverdin in cyanobacteria. To our knowledge, no well-characterized phycobilin-minus mutant has yet been described. Is this because there is a phycobilin-bearing photoreceptor that is absolutely required for growth? If we cannot identify phycobilin biosynthetic genes based upon similarities to genes such as biliverdin reductase, then a biochemical approach may offer the most promising method to identifying these genes.

   There are several candidates for lyases in *Synechocystis* sp. PCC 6803, but many experiments remain to be done. Clearly, if there is indeed a lyase for every site, then most of these lyases do not share significant sequence similarity with known lyases. The phenotype of Δslr1098 suggests that this protein is a lyase that is involved in adding PCB to phycocyanin. However, its substrate is not obvious and *in vitro* biochemical assays will be required to identify it. It will be necessary to generate double and triple mutants as well as to characterize these putative lyases biochemically *in vitro* to determine if they play any role in phycobilin biosynthesis. These experiments are currently underway.

ACKNOWLEDGMENTS

This research was supported by the National Institutes of Health (Grant GM28994), the W. M. Keck Foundation, and by the Lucille P. Markey Charitable Trust. W.M.S. was supported by a National Research Service Award (GM16935).

REFERENCES

1. Glazer, A. N. (1989) J. Biol. Chem. 264: 1–4.
2. Ong, L. J., and Glazer, A. N. (1991) J. Biol. Chem. 266, 9515–9527
3. Glazer, A. N. (1984) Biochim. Biophys. Acta 768, 29–51
4. Duerring, M., Schmidt, G.B. and Huber, R., (1991) J. Molec. Biol. 217, 577–592
5. Ficner, R., Lobeck, K., Schmidt, G. and Huber, R., (1992) J. Molec. Biol. 228, 935–950
6. Brejc, K., Ficner, R., Huber, R. and Steinbacher, S., 1995J. Mol. Biol. 249: 424–440
7. Schirmer, T., Huber, R., Schneider, M., Bode, W., Miller, M., and Hackert, M. L. (1986) J. Mol. Biol. 188: 651–676
8. Glazer, A.N. (1994) Advances in Molecular and Cell Biology 10, 119–149
9. Glazer, A. N. (1988) Meth. Enzymol. 167, 291–303
10. Schmid, R., and McDonagh, A. F. (1978) in The Metabolic Basis of Inherited Disease (Stanbury, J.B. and Wyngaarden, J.B., eds) pp. 1221–1257, McGraw Hill, New York
11. Terry, M. J., and Lagarias, J. C. (1991) J. Biol. Chem. 266, 22215–22221
12. Terry, M. J., and McDowell, M. T., and Lagarias, J. C. (1995) J. Biol. Chem. 270, 11111–11118
13. Lagarias, J. C. and Rapoport, L. H. (1980) J. Am. Chem. Soc. 102, 4821–4828
14. Furuya, M. (1993) Annu. Rev. Plant Physiol. Plant Mol. Biol. 44, 617–645
15. Quail, P. H., Boyla, M. T., Parks, B. M., Short, T. W., Xu, Y., and Wagner, D. (1995) Science 268, 675–680
16. Beale, S. I., and Cornejo, J. (1983) Arch. Biochem. Biophys. 227, 279–286
17. Beale, S. I., and Cornejo, J. (1984) Plant Physiol. 76, 7–15
18. Beale, S. I., and Cornejo, J. (1984) Arch. Biochem. Biophys. 235, 371–384
19. Beale, S. I., and Cornejo, J. (1991) J. Biol. Chem. 266, 22328–22332
20. Beale, S. I., and Cornejo, J. (1991) J. Biol. Chem. 266, 22333–22340
21. Beale, S. I., and Cornejo, J. (1991) J. Biol. Chem. 266, 22341–22345
22. Rhie, G., and Beale, S. I. (1992) J. Biol. Chem. 267, 16088–16093
23. Rhie, G., and Beale, S. I. (1995) Arch. Biochem. Biophys. 320, 182–194
24. Troxler, R. F., Lin, S., and Offner, G. D. (1989) J . Biol . Chem. 264, 20596–20601
25. Lin, S., Offner, G. D., and Troxler, R. F. (1990) Plant Physiol. 93, 772–777
26. Rhie, G., and Beale, S. I. (1994) J. Biol. Chem. 269, 9620–9626
27. Gossauer, A., Nydegger, F., Benedikt, E., and Köst, H.-P. (1989) Helv. Chim. Acta. 72, 518–529
28. Wu, S.-H., McDowell, M. T., and Lagarias, J. C. (1997) J. Biol. Chem. In press
29. Fairchild, C. D., and Glazer, A. N. (1994) J. Biol.Chem. 269, 8686–8694.
30. Arciero, D. M., Dallas, J. L., and Glazer, A. N. (1988) J. Biol. Chem. 263, 18358–18363
31. Kidd, D. G., and Lagarias, J. C. (1990) J. Biol. Chem. 265, 7029–7033
32. Wilde, A., Churin, Y., Schubert, H., and Börner, T. (1997) FEBS Lett. 406, 89–92
33. Kehoe, D., and Grossman, A. (1996) Science 273, 1409–1412
34. Cornejo, J., and Beale, S. I. (1997) Photosyn. Res. 51, 223–230
35. Cole, W. J., Chapman, D. J., and Siegelman, W. H. (1967) J. Am. Chem. Soc. 89, 3643–3645
36. Rudiger, W., Brandlmeier, T., Blos, I., Gossauer, A., Weller, J.-P. (1980) Z. Naturforsch. 35c, 763–769
37. Kaneko, T., Sato, S., Kotani, H., Tanaka, A., Asamizu, E., Nakamura, Y., Miyajima, N., Hirosawa, M., Sugiura, M., Sasamoto, S., Kimura, T., Hosouchi, T., Matsuno, A., Muraki, A., Nakazaki, N., Naruo, K., Okumura, S., Shimpo, S., Takeuchi, C., Wada, T., Watanabe, A., Yamada, M., Yasuda, M., and Tabata, S. (1996) DNA Res. 3, 109–136
38. Knaff, D. B., and Hirosawa, M. (1991) Biochim. Biophys. Acta 1056, 92–125
39. Karplus, P. A., and Bruns, C. M. (1994) J. Bioenerg. Biomemb. 26, 89–99
40. Schluchter, W. M., and Glazer, A. N. (1997) J. Biol. Chem. 272, 13562–13569
41. Terry, M. J., Maines, M. D., and Lagarias, J. C. (1993) J . Biol. Chem. 268, 26099–26106
42. Lagarias, D. M., Crepeau, M. W., Maines, M. D., and Lagarias, J. C. (1997) Plant Cell 9, 675–688

43. Stocker, R., Yamamoto, Y., McDonagh, A. F., Glazer, A. N., and Ames, B. N. (1987) Science 235, 1043–1046
44. Dennery, P.A., McDonagh, A.F., Spitz, D.R., Rodgers, P.A., and Stevenson, D.K. (1995) Free Radical Biol. Med. 19, 395–404
45. Arciero, D. M., Bryant, D. A., and Glazer, A. N. (1988) J. Biol. Chem. 263, 18343–18349.
46. Arciero, D. M., Dallas, J. L., and Glazer, A. N. (1988) J. Biol. Chem. 263, 18350–18357.
47. Fairchild, C. D., Zhao, J., Zhou, J. Colson, S. E., Bryant, D. A., and Glazer, A. N. (1992) Proc. Nat. Acad. Sci. 89, 7017–7021
48. Fairchild, C. D. and Glazer, A. N. (1994) J. Biol. Chem. 269, 28988–28996
49. Zhou, J., Gasparich, G. E., Stirewalt, V. L., deLorimier, R., and Bryant, D. A. (1992) J. Biol. Chem. 267, 16138–16145
50. Swanson, R. V., Zhou, J., Leary, J. A., Williams, T., de Lorimier, R., Bryant, D. A., and Glazer, A. N. (1992). J. Biol. Chem. 267, 16146–16154
51. Bhalerao, R. P., Kind, L. K., and Gustafsson, P. (1994) Plant Mol. Biol. 26, 313–326
52. Tandeau de Marsac, N., Mazel, D., Capuano, V., Damerval, T., Houmard, J. (1990) In Molecular Biology of Membrane-Bound Complexes in Phototropic Bacteria, (Drews, D. and Dawes, E. A., eds) pp143–153, Plenum, NY
53. Jung, L. J., Chan, C. F., and Glazer, A. N. (1995) J. Biol. Chem. 270, 12877–12884
54. Wilbanks, S. M., and Glazer, A. N. (1993) J. Biol. Chem. 268, 1226–1235
55. Kahn, K., Mazel, D., Houmard, J., Tandeau de Marsac, N., and Schaefer, M. R. (1997) J. Bacteriol. 179, 998–1006

11

THE STRUCTURE AND FUNCTION OF THE LH2 ANTENNA COMPLEX FROM *RHODOPSEUDOMONAS ACIDOPHILA* STRAIN 10050

R. J. Cogdell,[1] S. M. Prince,[2] A. A. Freer,[2] T. D. Howard,[1] N. W. Isaacs,[2] A. M. Hawthornthwaite-Lawless,[3] M. Z. Papiz,[3] and G. McDermott[4]

[1]Division of Biochemistry and Molecular Biology
[2]Department of Chemistry
University of Glasgow
Glasgow G12 8QQ, United Kingdom
[3]CCLRC Daresbury Laboratory
Daresbury, Warrington, WA4 4DD, United Kingdom
[4]Lawrence Berkeley Labs
University of California at Berkeley
Berkeley, California 94720-1460

1. INTRODUCTION

Since the last conference on Photosynthetic Prokaryotes the structure of the LH2 antenna complex from the purple non-sulphur photosynthetic bacterium *Rhodopseudomonas acidophila* strain 10050 has been determined (1). This integral membrane pigment-protein complex consists of a circular array of heterodimers, each comprising and α- and β-apoprotein which non-covalently bind three molecules of Bchl*a* and, most probably, two molecules of the carotenoid rhodopin glucoside (Figure 1). The pigments are enclosed by two concentric rings of transmembrane α-helices. The inner ring, radius 13Å, is composed of α-apoproteins and the outer ring, radius 34Å, is formed from the β-apoproteins. The long axes of these transmembrane α-helices are approximately normal to the presumed membrane plane. The structure is 'capped' top and bottom by the N- and C-termini of the apoproteins which fold over and interact with each other at the hydrophilic surfaces of the complex. The Bchl*a* molecules form two spectrally distinct sets. Eighteen Bchl*a*'s, with their central Mg^{++} liganded to the protein via conserved histidine residues (α his 31 and β his 30), form a strongly coupled array with the plane of their bacteriochlorin rings normal to the membrane plane (parallel to the long axes of the transmembrane α-helices). These

The Phototrophic Prokaryotes, edited by Peschek *et al.*
Kluwer Academic / Plenum Publishers, New York, 1999.

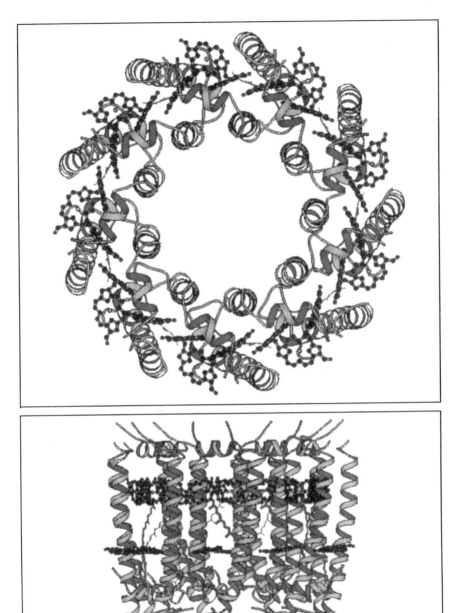

Figure 1. The overall structure of the LH2 antenna complex from *Rps. acidophila* strain 10050. **Top:** A = schematic view of the LH2 nonameric complex viewed from the periplasmic side of the membrane. **Bottom:** The view from within the membrane The phytol chains have been removed from the Bchl's for clarity.

Bchl*a* molecules form the so called B850 Bchls. Nine monomeric Bchl*a* molecules, separated by a centre to centre distance of 21.2Å, lie with their bacteriochlorin rings parallel to the membrane place between the β-apoproteins. Originally it was suggested (1) that the central Mg^{++} stem of these Bchl*a*'s was ligated to the N-formyl group on α-met 1. However as described below this tentative conclusion is now called into question. These 9 Bchl*a*s form the so called B800 Bchl*a*'s.

One carotenoid molecule, per αβ apoprotein pair, is also well resolved. This carotenoid has its sugar head group located in a hydrophilic pocket on the cytoplasmic side of the structure, where the partially disordered glucoside group is seen making a number of contacts with charged residues on adjacent helices. The conjugated region of the carotenoid then passes close to the edge of the B800 bacteriochlorin ring (closest distance 3.4Å) and traverses the length of the LH2 complex passing over the face of the α-bound B850 bacteriochlorin (closest distance 3.6Å). This carotenoid is in an all-*trans* configuration and forms a 'lazy S-shape' (2). In the first description of this structure (1), a region of electron density located on the other side of the α-bound B850 Bchl*a* was suggested to represent a bound molecule of β-octylglucoside (the detergent in which the complex was crystallised). However spectroscopic and stoichiometric analysis of the carotenoid content of this complex had suggested that there should be a second carotenoid present per αβ apoprotein pair (3). More recently we have crystallised the LH2 complex from *Rps. acidophila* in the presence of the detergent lauryldimethylamine-*N*-oxide (LDAO): in this case the antenna complex has never seen β-octylglucoside. Analysis of these crystals (S. Prince and K. McKluskey, unpublished data) has suggested that this electron density represents part (about 1/3rd) of the second carotenoid molecule. In this case the glucoside group is on the periplasmic side of the complex and the carotenoid chain then passes over the outer face of the α-bound B850 bacteriochlorin [see (2) for a fuller description of this]. The carotenoids therefore 'sandwich' the α-bound B850 bacteriochlorin. Interestingly there is further support for this location of the second carotenoid from the 2D projection map of a homologous LH2 complex from *Rhodovulum sulphidophilum* (4) where extra density, in comparison to the *Rps. acidophila* structure, could represent the remainder of the second carotenoid. Our extrapolations of the remaining 2/3rds of the carotenoid place it in a rather peripheral groove in the structure assuming that it adopts an all-*trans* configuration and passes close to the edge of the other side of B800 bacteriochlorins ring.

2. RECENT IMPROVEMENTS IN THE STRUCTURE

The X-ray diffraction pattern recorded at room temperature shows anisotropy of resolution and pronounced diffuse scattering distributions. Since these features are very likely to be temperature dependant we attempted to establish a protocol for cryocooling. As a result of trial and error a mini-dialysis procedure was developed using sucrose as the cryoprotectant. Then following rapid freezing it was possible to collect a ~95% complete native data set with an improved resolution of 2.0Å. This data is currently being fully refined so the crystallographic data presented in Table 1 is only preliminary. In general there are no large differences between the structure as presented at 2.5Å (1) to that now resolved to 2.0Å.

One region where there is, however, a clear difference due to the improved resolution is in the region of the fifth ligand to the central Mg^{++} atom of the B800 Bchl*a*. Comparison of the electron densities in this area (Figure 2) show that at 2.0Å resolution the density of the extension of the N-terminal methionine residue is clear bifurcated. This suggests that our original assignment of this density as that of an N-formyl group is probably incorrect. Further work is needed to determine exactly the chemical identity of this ligand. Elsewhere the improved resolution has largely confirmed the structural model presented at 2.5Å resolution (1, 5) and shows that there are no significant structural changes on any cooling (none bigger than the co-ordinate errors).

Table 1. Refinement of LH2 at 2.0Å and 120K[a]

Resolution limits	32.0–2.0Å
No. independant non-hydrogen atoms	3146
No. reflections	50188
[R]Cryst	19.7%
[R]Free	23.1%
RMS deviation from target geometry[b]	0.039–0.47Å
Mean B-factor	46.34
RMS deviation on PC superimposition	0.235Å (PC1–PC2)
	0.237Å (PC1–PC2)
Estimated rms accuracy[c]	0.26A

[a]Refinement is not yet complete.
[b]Bonded and non-bonded distance restraints.
[c]From the CCP4 (CCP4, 1994) program SIGMAA (15)f.
Unit cells: a = b = 117.1, c = 298.4Å (hexagonal index), Space group R32

3. ENERGY TRANSFER REACTIONS TAKING PLACE WITH LH2

There has been a flurry of studies designed to investigate the photochemical properties of LH2 in recent years. This is undoubtedly due to a combination of the availability of detailed structural information and the well resolved spectra of the different pigment groups in LH2 (see 5–7 for some recent detailed reviews of this). The basic photochemical reaction taking place can be summarised as follows. If the B800 Bchl*as* are excited with a fs excitation pulse then a very rapid energy transfer from excited B800* to B850 is seen. this takes 0.7ps at room temp, ~1.8ps at 77K and ~2.4ps at 4K [see (6) for a recent review]. Moreover from the time dependence of the anisotropy of the decay of the B800* signal it has been determined that there is also rather fast B800 → B800 energy transfer (the anisotropy decays with a time constant of ~0.35ps). Once the electronic excitation energy is in the B850 manifold it has a variety of fates. If the LH2 complex is in the photosynthetic membrane then there is a fast energy transfer to LH1, which takes about 3.3ps at room temperature and ~5ps at 77K (6,7). If on the other hand the LH2 is pure in detergent solution, with no coupled LH1 complexes, then the B850* excited state lasts for about ~1ns. Then it can decay by a variety of routes such as fluorescence, intersystem crossing to produce Bchl triplets (which then sensitize the production of carotenoid triplets) and other non-radiative decay processes. If the decay of the anisotropy of the B850* excited state is monitored it can be seen that at room temperature the deamyterase is ~130fs. This then implies ultrafast energy transfer within the B850 ring. These reactions are summarised in Figure 3.

At this point we need to ask what are the possible energy transfer mechanisms which could account for these properties? The simple and classical Förster (weak interaction) resonance energy transfer mechanism seems to account well for the rates of singlet-singlet energy transfer between the well separated B800 Bchl molecules (6). However that is then as far as the current agreement extends. It is not yet clear what the best way to consider the spectroscopic properties of the B850 Bchls is. If the basic structure of the complex is considered and then used as a base to calculate the spectroscopic properties of B850, then the absorption and CD properties can only be accounted for by assuming that the B850 ring acts as a 'super molecule' (8,9). In other words that the 18 Bchl molecules form a strongly exciton coupled system. Interpretation of some of the fs and ps energy transfer experiments carried out on LH2, however, has led to the different conclusion that the excited state, B850*,

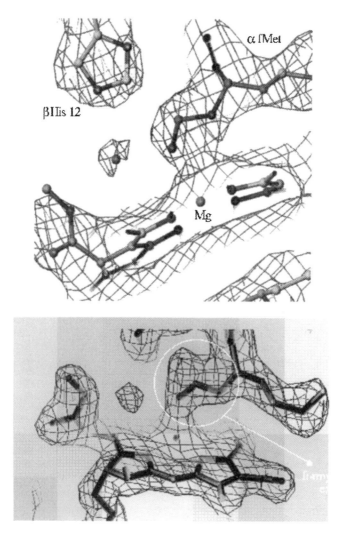

Figure 2. A comparison of the electron density map in the region of the B800 Bchl in the 2.5 and 2.0Å resolution structures. **Top:** The 2.5Å data. **Bottom:** The 2.0Å data. Notice how the electron density extension of α Met is now clearly bifurcated in the 2.0Å resolution data. This suggests that the N-formyl group that has been modelled is a wrong assignment.

is more localised onto 2–4 Bchl molecules [for example see (10–13)]. Calculation of the exciton states of the B850 ring, assuming complete delocalisation, predict an upper exciton state which lies very close in energy to the B800 absorption band. Some people therefore assume that B800 → B850 singlet-singlet energy transfer proceeds via this higher energy exciton state, which explains the weak dependence of the rate of B800 → B850 energy transfer on the spectral overlap between the main B800 and B850 bands [for example (13)]. Others have used the Förster formalism (14). Once the electronic excitation has reached the B850 ring then these two different views on the 'nature' of the B850 Bchl manifold lead to two separate interpretations to explain the ultrafast decay of the anisotropy of B850* (6). From the delocalised viewpoint this decay represents relaxation among the various excited

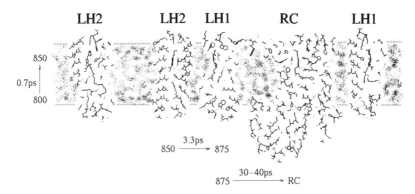

Figure 3. A section through a model of the purple bacterial photosynthetic unit. A section through model presented in [16] viewed from within the membrane. This view highlights the similarity in depth at which the B850 Bchl's from LH2, the B875 Bchl's from LH1 and the special pair 'Bchl's in the reaction centre are positioned in the membrane. The typical energy transfer times between the different groups of Bchl's are also shown. A colour version of this is available at the following website: http://www.cham.gla.ac.uk/protein/LH2/1h2.html.

states, whereas if the excited state is more localised i.e. on a pair of Bchl's then this decay could be explained by the excited state 'hopping' between dimers. Further experimental data is required to unequivocally solve this problem.

ACKNOWLEDGMENTS

This work was supported by grants from the BBSRC, the Human Frontiers of Science and the Gatsby Foundation.

REFERENCES

1. McDermott, G., Prince, S.M., Freer, A.A., Hawthornthwaite-Lawless, A.M., Papiz, M.Z., Cogdell, R.J. and Isaacs, N.W. (1995) Nature **374**, 517–521.
2. Freer, A.A., Prince, S.M., Sauer, K., Papiz, M.Z., Hawthornthwaite-Lawless, A.M., McDermott, G., Cogdell, R.J. and Isaacs, N.W. (1996) Structure **4**, 449–462.
3. Gardiner, A.T., Cogdell, R.J. and Takaichi, S. (1993) Photosyn. Res. **38**, 139–168.
4. Montoya, G., Cyrklaff, M. and Sinning, I. (1995) J. Mol. Biol. **250**, 1–10.
5. Prince, S.M., Papiz, M.Z., Freer, A.A., McDermott, G., Hawthornthwaite-Lawless, A.M., Cogdell, R.J. and Isaacs, N.W. (1997) J. Mol. Biol. **248**, 412–423.
6. Pullerits, T. and Sundström, V. (1996) Acc. Chem. Res. **29**, 381–389.
7. Sundström, V. and van Grondelle, R. (1995) In 'Anoxygenic Photosynthetic Bacteria' (Blankenship, R.E., Madigan, M.T. and Bauer, C.E. eds.) pp 349–372, Kluwer Academic Publishers, The Netherlands.
8. Sauer, K., Cogdell, R.J., Prince, S.M., Freer, A.A., Isaacs, N.W. and Scheer, H. (1996) Photochem. Photobiol. **64**, 564–576.
9. Alden, R.G., Johnson, E., Nagurajan, V., Parson, W.W., Law, C.T. and Cogdell, R.J. (1997) J. Phys. Chem. **101**, 4667–4680.
10. Pullerits, T., Chachisvilis, M. and Sundström, V. (1996) J. Phys. Chem. **100**, 10787–10792.
11. Jimenez, R., Dikshit, S.N., Bradforth, S.E. and Fleming, G.R. (1996) J. Phys. Chem. **100**, 6825–6834.
12. Monshouwer, R., Abrahamsson, M., van Mourik, F. and van Grondelle, R. (1997) J. Phys. Chem. B. in press.

13. Reddy, M.S., Wu, H.-M., Jankowiak, R., Picorel, R., Cogdell, R.J. and Small, G.J. (1996) Photosyn. Res. **248**, 177–289.

14. Hess, S., Visscher, K.J., Pullerits, T., Sundström, V., Fowler, G.J.S. and Hunter, C.N. (1994) Biochemistry **33**, 8300–8305.

15. Read, R.J. (1986) In 'Advances in Protein Chemistry' (Ainfinan, J.B., Anson, M.L., Edsall, J.T. and Richards, R.M., eds.) Vol. **23**, 283–478, Academic Press, New York and London.

16. Papiz, M.Z., Prince, S.M., Hawthornthwaite-Lawless, A.M., McDermott, G., Freer, A.A., Isaacs, N.W. and Cogdell, R.J. (1996) TIPS **1**, 198–206.

12

INTRAPROTEIN ELECTRON TRANSFER

Illustrations from the Photosynthetic Reaction Center

C. C. Moser, C. C. Page, X. Chen, and P. L. Dutton

Johnson Research Foundation
Department of Biochemistry and Biophysics
University of Pennsylvania
Philadelphia, Pennsylvania 19104

1. INTRODUCTION

Photosynthetic reaction centers have provided the best system in which to study natural intraprotein electron transfer. With an array of colorful redox centers conveniently activated by light, the reaction center provides access to a network of reactions over the variety of distances and free energies that has made it possible to understand the basics of design of natural electron transfer systems [1–3]. Indeed, it first became clear that tunneling is central to biological electron transfer after the observation that the photochemistry of the reaction center operates even at liquid helium temperatures [4].

The tunneling nature of electron transfer in the reaction center (RC) leads us to consider non-adiabatic electron transfer theory which is useful for tunneling reactions when the donor and acceptor are clearly separate from one another and no bonds are made or broken during electron transfer. Central to this theory is a simple relation for the electron transfer rate called Fermi's Golden Rule: [5]

$$Ket=(2\pi/\hbar)\ V(r)^2\ FC \tag{1}$$

This rule separates the terms which are dependent on the tunneling electron itself, from the term which is dependent on nuclear rearrangement. The former is included in $V(r)$, the electronic coupling between the donor and the acceptor, reflecting the very small and distance dependent overlap of the electronic wavefunctions of the electron on the donor and the electron on the acceptor. The latter is included in FC, the Franck-Condon weighted density of states.

The Phototrophic Prokaryotes, edited by Peschek *et al.*
Kluwer Academic / Plenum Publishers, New York, 1999.

2. ELECTRONIC TUNNELING TERMS

An intuitive understanding of the first, electronic term is rather simple. An electron placed on a donor, such as a bacteriopheophytin, senses the surrounding protein as an insulator; that is, the electron does not have enough classical energy to move away from the donor and penetrate the insulating barrier potential on all sides. Yet a quantum description informs us that the electron can exist with some small probability inside this barrier, the probability vanishing quickly as the barrier grows in height or the distance away from the donor becomes large. In this sense the electron can tunnel through the barrier.

The most extreme barrier is provided by a vacuum, where the barrier is approximated by the vacuum ionization energy of a chlorin, about 8 eV [6]. Electron tunneling between an idealized donor and acceptor separated by vacuum would then be expected to fall off exponentially with distance, with an exponential coefficient, usually symbolized as β, of about 2.8Å^{-1}. On the other hand, if the donor and acceptor are linked directly by a rigid covalent bridge, then the barrier height is much reduced; it requires less energy to place an electron in the vicinity of positively charged nuclei, an energy roughly corresponding to promoting the electron to the orbitals of these intervening atoms. Experimentally, it is found that β for covalently bridged systems is about 0.9Å^{-1} [7], suggesting an effective barrier height of about 0.8 eV. This barrier is larger than the 0.5 eV ($\beta=0.7\text{Å}^{-1}$) of our first estimate [1]. In a structurally heterogeneous protein medium the barrier will usually be between these two extremes. We will address below the intriguing question of whether natural selection has drawn from the heterogeneity to evolve protein structure to assist productive long range electron transfer and hinder physiologically unproductive electron transfer.

In view of the extraordinary complexity of protein electron transfer systems from a quantum mechanical computational view, and in view of the rather severe experimental uncertainties surrounding the measurement of the parameters of electron transfer theory, we prefer to use models of electron tunneling which are simple, practical, unambiguous and empirically based. Such models do not purport to capture all the physical subtleties of quantum mechanics and as such are not "real", but they do allow ready prediction of electron transfer rates and an appreciation of the engineering of natural systems, as well as a means to guide the engineering of systems of our own design [8,9].

Our first practical concern is the definition of distance. We have found the edge-to-edge distances to be the most useful, where the cofactor includes all the ring atoms of non-metallic redox centers such as flavins and includes the oxygens attached to the rings of quinones and aromatic radicals. Redox centers with metals include the immediate liganding atoms, while hemes and chlorins include the conjugated porphyrin macrocycle as well. Non-conjugated substituents on the edge of the porphyrin ring are not included.

Our second concern is to address the barrier between the donor and acceptor. We have found success with a simple model that examines the packing of the medium atoms in the region between the donor and acceptor. Thus we sample the region between donor and acceptor, declaring the region within the united atom van der Waals radius of any atom to contribute a low barrier, typical of covalently bridged systems ($\beta=0.9\text{Å}^{-1}$), and the region outside this radius to contribute a high vacuum-like barrier ($\beta=2.8\text{Å}^{-1}$). Typical protein packing densities around 74% [10] lead to an effective β of about 1.4Å^{-1}. Indeed a β of this value is consistent with experiments with the photosynthetic reaction center. Figure 1 shows the dependence of the log of the free energy optimized electron transfer rate for various physiologically productive and unproductive electron transfer reactions in the reaction centers of *Rb. sphaeroides* and *Rp. viridis* (filled symbols). These points tend to cluster about a line with a slope of $\beta=1.4\text{Å}^{-1}$ and an intercept at van der Waals contact of 10^{13} s^{-1}.

At least part of the distribution about the average line appears to reflect the natural structural inhomogeneities in the protein; our calculated rates based on these packing densities are shown as open symbols in Figure 1. For example, the 14Å electron tunneling from Q_A to Q_B, which are separated by an unusually well packed medium of bridging histidines more or less directly connecting the quinones to the central non-heme iron and hence to each other, should be ~50 times faster than average, comparable to the fast phase that has been reported for this reaction [11].

Under conditions in which the detailed packing of a particular protein is unknown, but a fairly good estimate of the distance between donor and acceptor is possible, then we would expect a range of rates due to random packing variations with a standard deviation

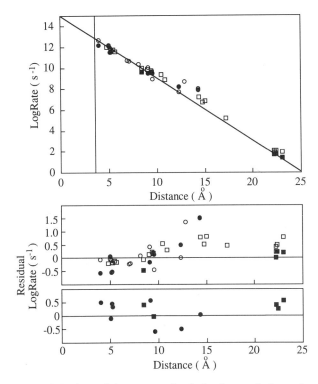

Figure 1. The free energy optimized rate of electron tunneling in the photosynthetic reaction centers of Rp. viridis and Rb. sphaeroides (filled symbols) falls off approximately exponentially with edge-to-edge distance. The average slope (in natural log units) is $\beta=\sim1.4\text{Å}^{-1}$ (solid line). There is no tendency for physiologically productive charge separations (filled circles) to be faster than average, nor for physiologically unproductive electron transfers (■) to be slower than average; thus the structure of the protein medium does not appear to have been naturally selected to modulate electron transfer rates. Optimum rates for theoretical electron transfers between many pairs of redox centers in reaction centers can be calculated by taking into account the packing inhomogeneities between these redox centers, as described in the text (O and □). The middle panel shows the residual deviations of the experimental and calculated rates from the line corresponding to an average protein $\beta=1.4\text{Å}^{-1}$. The bottom panel shows the residual deviations of the experimental optimal rates from a rate calculation that reflects the heterogeneous packing of the protein medium between the redox cofactors. The standard deviations of these calculated rates from the average distance dependence is less than half a log unit. Q_A to Q_B electron transfer takes place in a medium that is unusually well packed in both *Rb. sphaeroides*, where the free energy optimized rate has been experimentally estimated (● at 14.3Å) and in *Rp. viridis*, where the optimized rate has not been experimentally estimated (O at 12.8Å).

of about a half a log unit. This sort of packing induced rate variance is comparable to a ~1Å change in the edge-to-edge distance. Clearly even relatively small changes in distance can have more profound effects on rates than typical variations in protein packing.

The intriguing question arises: given that variations in protein packing can modulate electron transfer rates, is there any evidence that billions of years of evolution has led to the design of electron transfer proteins such that the physiologically productive reactions are accelerated by a denser packing of the intervening medium, while the harmful unproductive reactions are hindered by the design of light packing and better insulation in the intervening medium? The answer seems to be no. Figure 1 illustrates that for the RC there is no systematic tendency for productive reactions to have a smaller beta than unproductive charge recombination reactions. Nor is there any evidence for any such packing trend in any other natural electron transfer system that we have examined. Even the RC electron transfer at the dense end of the packing distribution, Q_A to Q_B, does not appear to take advantage of the relatively fast tunneling rate; while part of the RC population exhibits a fast phase, much of the population seems to be limited by some sort of adiabatic reaction, possibly involving proton binding. It also appears that Q_B can assume a number of different geometries in different RC crystals. As we will discuss below, natural selection seems to focus on other factors besides protein packing and β to guide electrons in productive directions.

3. NUCLEAR REARRANGEMENT TERMS

Another means to modulate electron transfer rates is by changes in the FC or nuclear rearrangement term of Fermi's Golden Rule. A very successful means of describing this term was provided by Marcus [12,13]. In this view, the equilibrium nuclear geometry of the reactant (i.e. the electron on the donor) is distorted along an assumed simple harmonic oscillator potential, until it increasingly resembles the equilibrium nuclear geometry of the product, (which has the electron on the acceptor). The amount of energy required to distort the geometry of the reactant to that of the product but without transferring the electron, is called the "reorganization energy," symbolized as λ. In this model, an electron transfer of modest free energy becomes faster as the driving force of the reaction is increased. The rate reaches a maximum or optimum when the free energy released matches the reorganization energy. As the driving force is increased further, the reaction is overpowered, and the rate decreases, defining the Marcus "inverted region." The overall response of the tunneling rate to free energy is a Gaussian, with a maximum at the reorganization energy, and a standard deviation (σ) of

$$\sigma = \tilde{A}(2 \lambda k_B T) \tag{2}$$

where k_B is the Boltzmann constant and T is the absolute temperature. This leads to a temperature dependent electron transfer rate which in this classical model is expected to be dramatic for transfers far from the reorganization energy.

Obtaining a sufficiently detailed free energy dependence of a tunneling reaction to define the free energy optimized rate, and hence the reorganization energy, demands a great deal of the experimentalist. It is usually insufficient simply to obtain the temperature dependence of the reaction at a single free energy and extrapolate a reorganization energy, partly because the temperature dependence observed may be that of a bond making-bond breaking adiabatic reaction that does not reflect the tunneling rate, and partly because the free energy dependence may not be Marcus like. For example, an extensive exploration of electron transfers to and from quinone in the RC in which the Q_A has been extracted and replaced with exotic quinones of different free energies and the temperature varied from liquid helium to room temperature [14,15], has shown that at least some electron transfers are coupled to high frequency vibrations and that the standard deviation of the free energy dependence is larger than that expected by Marcus theory. Thus we prefer to use a FC expression that recognizes the quantization of the nuclear reorganizations that accompany electron tunneling, such as the semi-classical expression of Hopfield [16], in which the free energy dependence is also a Gaussian, but the standard deviation is now

$$\sigma = \tilde{A}(\lambda \, \hbar \, \omega \, \coth(\hbar \, \omega/2k_B T)) \tag{3}$$

which reduces to the Marcus expression at the high temperature limit, but becomes temperature independent at low temperatures. At least for some photosynthetic reactions, the characteristic frequency coupled to electron transfer ($\hbar \, \omega$) is about 70 meV. At room temperature, this leads to a Gaussian free energy vs. rate distribution that is only modestly broader than the Marcus expression.

The relatively robust photosynthetic RC protein has tolerated chemical modification of cofactors and mutagenic changes that have better defined the free energy dependence of many of these reactions.

4. EMPIRICAL TUNNELING EXPRESSION AND ENGINEERING

We can summarize the typical electron tunneling rate for intraprotein electron transfer with this expression:

$$\log_{10}Ket \, (s^{-1}) = 15 - (0.6 \text{Å}^{-1})R - (3.1 eV^{-1}) \, (\Delta G - \lambda)^2/\lambda \tag{4}$$

where R is the edge-to-edge distance in units of Å, ΔG is the free energy of the reaction and λ is the reorganization energy, both in eV. This simple tunneling expression does not consider protein structure heterogeneity, and assumes an average β value of 1.4Å^{-1}. Nevertheless, it is quite useful for estimating tunneling rates with a standard deviation of about half a log unit. When experimental rates are significantly slower, we look for slow adiabatic rate limiting events, such as conformational changes and proton motion.

Natural systems will include not only exothermic, but also endothermic reactions. It may be a surprise to some that it is possible to tunnel in an uphill endothermic reaction. A nice example is provided by the nuclear tunneling in Tutton salts [17]. Thus we expect that there is good reason to extend the quantized Gaussian free energy relationship of the above expressions into the endothermic region, leading to non-Boltzmann tunneling equilibria. Nevertheless, until it becomes clear that such non-Boltzmann equilibria are effective in biological electron transfer systems, we will use a less controversial approximation and assume that tunneling reactions lead to Boltzmann equilibria. Because the quantized expression does not differ greatly from the classical expression at room temperature, the error in such an estimate should not be unreasonably large. An endothermic tunneling rate can be calculated by applying the free energy dependent Boltzmann equilibrium constant to the corresponding exothermic rate estimated as above.

Figure 2 illustrates the use of these expressions in the case of the cytochrome c hemes of *Rp. viridis*. Despite the considerable endothermic and exothermic swings in free

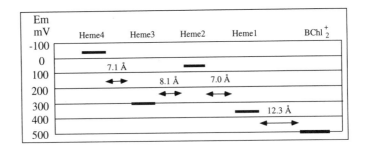

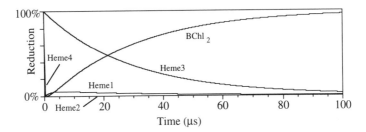

Figure 2. Electron transfer from the most distant heme of *viridis* (heme 4) to the bacteriochlorophyll dimer (BChl$_2$) can take place over a distance of 70Å in tens of microsecond by tunneling, because the hemes form a typical redox chain with edge-to-edge distances of 7 to 13Å, even though midpoint potentials lead to a combination of exothermic and endothermic reactions. A theoretical simulation of the concentration of each intermediate reduced state, starting with reduced heme 4 and using the tunneling expressions in the text, (assuming a reorganization energy of 1 eV) is shown at bottom.

energy from center to center, it is clear that a truly long distance electron transfer over 70Å can take place in microseconds (consistent with measured rates [18]) via a chain of redox centers, each redox center separated from the next by 7 to 13Å. Chains of this form, including unfavorable steps, appear to be common in biological systems. Indeed, it appears that the overall structure of electron transfer systems is to guide electrons along chains of redox centers separated by gaps of similar magnitude, in which rapid non-adiabatic electron tunneling can take place, to connect local adiabatic redox sites at which either protons are coupled to electron transfer, or redox energy is coupled to transmembrane pH gradients and electric fields.

REFERENCES

1. Moser, C.C., Keske, J.M., Warncke, K., Farid, R.S. and Dutton, P.L. (1992) Nature 355, 796–802.
2. Moser, C.C. and Dutton, P.L. (1992) Biochim. Biophys. Acta 1101, 171–6.
3. Moser, C.C., Page, C.C., Chen, X. and Dutton, P.L. (1997) J. Bio-inorg. Chem. 2, 393–398.
4. Devault, D. and Chance, B. (1966) Biophysical J. 6, 825–847.
5. Devault, D. (1980) Quart. Rev. Biophys. 13, 387–564.
6. Meot-Ner, M., Green, J.H. and Adler, A.D. (1973) Ann. N. Y. Acad; Sci. 206, 641–648.
7. Smalley, J.F., Feldberg, S.W., Chidsey, C.E.D., Linford, M.R., Newton, M.D. and Liu, Y.-P. (1995) J. Phys. Chem. 99, 13141–13149.
8. Robertson, D.E. et al. (1994) Nature 368, 425–432.
9. Gibney, B.R., Mulholland, S.E., Rabanal, F. and Dutton, P.L. (1996) Proc. Natl. Acad. Sci. U. S. A. 93, 15041–15046.

10. Levitt, M., Gerstein, M., Huang, E., Subbiah, S. and Tsai, J. (1997) Annu. Rev. Biochem. 66, 549–579.
11. Li, J., Gilroy, D. and Gunner, M.R. (1996) Biophys. J. 70, SUAM2.
12. Marcus, R.A. and Sutin, N. (1985) Biochim. Biophys. Acta 811, 265–322.
13. Marcus, R.A. (1956) J. Chem. Phys. 24, 966–978.
14. Gunner, M.R. and Dutton, P.L. (1989) J. Ame. Chem. Soc. 111, 3400–3412.
15. Gunner, M.R., Robertson, D.E. and Dutton, P.L. (1986) J. Phys. Chem. 90, 3783–3795.
16. Hopfield, J.J. (1974) Proc.e Natl. Acad. Sci. U.S.A. 71, 3640–3644.
17. Trapani, A.P. and Strauss, H.L. (1989) J. Amer. Chem. Soc. 111, 910–917.
18. Shopes, R.J., Holten, D., Levine, L. and Wraight, C.A. (1987) Photosynth. Res. 12, 165–180.

13

STRUCTURAL AND FUNCTIONAL ANALYSIS OF THE ORF1696/PucC FAMILY OF LIGHT-HARVESTING COMPLEX ASSEMBLY PROTEINS

C. Y. Young and J. T. Beatty

Department of Microbiology and Immunology
University of British Columbia
300-6174 University Blvd.
Vancouver, BC, V6T 1Z3

1. ABSTRACT

The amino acid sequences of *Rhodobacter capsulatus* ORF1696 and PucC proteins were analyzed in alignments and by comparison of topology models. An ORF1696::PucC fusion protein was assessed for function in LHI and LHII assembly. We also evaluated sequence alignments of *R. capsulatus* ORF1696 and PucC proteins with homologues from other purple non sulfur and oxygenic photosynthetic bacteria. We propose that a function of tetrapyrrole delivery to nascent light-harvesting complexes is shared by this family of ORF1696/PucC homologous proteins.

2. INTRODUCTION

There are two types of light-harvesting (LH) complex synthesized by the purple, non-sulfur bacterium *Rhodobacter capsulatus*. These are the LHI complex, which is directly associated with the reaction center (RC), and the LHII complex. Both LHI and LHII are located within the intracytoplasmic membrane (ICM), and function to absorb and transmit light energy to the RC complex, which drives the process of light energy transduction to chemical energy [1].

The LHI complex is thought to be composed of oligomers of a subunit that consists of: one molecule of each of the α and β polypeptides, which span the ICM due to a central hydrophobic α-helical region; two interacting ("dimer") bacteriochlorophyll (bchl) molecules (B870); and two carotenoid molecules [1]. *In vivo*, the formation of the LHI com-

The Phototrophic Prokaryotes, edited by Peschek *et al.*
Kluwer Academic / Plenum Publishers, New York, 1999.

plex requires the presence of bchl and has been proposed to be dependent on the electrostatic attraction of oppositely charged amino acids present in the cytoplasmically located N-terminal domains of the α and β [2]. It is likely that the LHI complex consists of about 16 LHI subunits (αβ: bchl$_2$: crt$_2$) arranged in a ring with a central hole occupied by the RC, by analogy with the *Rhodospirillum rubrum* LHI complex [3]. The *in vitro* reconstitution of the LHI complex has been achieved by combination of purified LHI α and β polypeptides with bchl in detergent emulsions, although the yield of *R. capsulatus* LHI is lower than that from *Rhodobacter sphaeroides* and *Rsp. rubrum* [4].

The *R. capsulatus* LHII complex is thought to similarly consist of oligomers of a subunit consisting of membrane-spanning α and β polypeptides, a protein of unknown function (γ), three bchl molecules (a B800 monomer and a B850 dimer) and two carotenoid molecules [1]. The determination of the three dimensional structure of *Rhodopseudomonas acidophila* and *Rhodospirillum molischianum* LHII complex crystals revealed oligomers of nine or eight subunits (respectively) arranged in a ring [5,6]. A low resolution structural determination of the *R. capsulatus* LHII complex suggests it to have a similar organization [7].

The steady-state levels of the *R. capsulatus* LHI and LHII complexes are greatly reduced in *ORF1696* and *pucC* mutants, respectively. A transposon disruption of the *pucC* gene led to a loss of the LHII complex [8], and other experiments demonstrated that *trans*-complementation of a *pucC* deletion strain restored the level of the LHII complex nearly to that of the wild type [9]. Disruptions of the *ORF1696* gene with a Km^r cartridge resulted in a 60 to 70% decrease in the level of the LHI complex [10,11] and complementation of *ORF1696* disruptions in *trans* increased the LHI level [11]. The ORF1696 has been shown to act as a catalyst of LHI complex assembly [11].

In this paper we compare and analyze amino acid sequence alignments of the *R. capsulatus* ORF1696 and PucC proteins and membrane topology models which have been proposed for ORF1696 and PucC [11,12], as well as an alignment with homologues of other purple non sulfur bacteria, with regard to possible structures and functions of these proteins. The results of a domain switching experiment involving the ORF1696 and PucC proteins of *R. capsulatus* are presented, and an alignment that includes homologues from oxygenic phototrophic bacteria is discussed.

3. MATERIALS AND METHODS

3.1. Bacterial Strains and Growth Conditions

Rhodobacter capsulatus strains ΔLHII and ΔStu were described [9,11], as were semiaerobic growth conditions [9] and spectral measurements [11]. Analyses of spectra was performed using Grams 386 software (Galactic Industries).

3.2. Membrane Topology Models

The derivation of the PucC membrane topology model was described [12]. The ORF1696 membrane topology model was derived by analysis of the ORF1696 primary amino acid sequence with the TopPred II 1.1 software program [13] using the GES, GvH1, and KD hydropathy algorithms to arrive at a consensus model. The consensus model was tested experimentally by construction and evaluation of translationally in-frame fusions to the *E. coli pho'A* and *lac'Z* alleles at sites along the length of the *ORF1696* gene sequence [11], similarly to the approach used to generate the PucC topology model [12].

3.3. Sequence Alignments

Amino acid sequences of ORF1696 and PucC homologues were acquired from the Genbank sequence database, except for the *Prochlorococcus marinus* sequence which was kindly provided by W. Hess (personal communication). Alignment and presentation of amino acid sequences was done using the Geneworks 2.45 software program (Intelligenetics, Inc.).

3.4. Construction of the *ORF1696'::puc'C* Fusion Vector pFUS1

Plasmid pB25–1 [11], which contains a gene fusion between the 5' 607 nucleotides of the *ORF1696* gene fused translationally in-frame with the *E. coli phoA* allele in pUC19::phoA [14], were linearized at the fusion joint by digestion with *Kpn* I, and was treated with T4 DNA polymerase to generate blunt ends. The linearized pB25-1 plasmid was digested with *Eco*R I and a 3.4 kb vector DNA fragment including the sequences encoding the 5' segment of *ORF1696* was separated from smaller DNA fragments by agarose gel electrophoresis and purified.

Plasmid pUC::*pucC*(+) [15], which contains the entire *pucC* gene subcloned into pUC13, was linearized by digestion with *Nar* I, and then treated with the Klenow fragment of DNA polymerase I to produce blunt ends. The linearized, blunt-ended pUC::*pucC*(+) plasmid DNA was digested with *Eco*R I to generate a 1 kb DNA fragment that included *pucC* sequences from nucleotide 623 to 1386, which was separated from vector DNA by gel electrophoresis, purified, and ligated with the 3.4 kb DNA fragment derived from pB25–1 to yield plasmid pUC19::FUS. The DNA sequence of the *ORF1696'::puc'C* fusion joint was determined to ensure that the fusion was translationally in-frame and consisted of the predicted codons.

Plasmid pUC19::FUS was digested with *Hinc* II, ligated with an *Eco*R I linker (5'-CCGGAATTCCGG-3') to introduce an *Eco*R I restriction site 5' to the *ORF1696* sequences, and then was digested with *Eco*R I. The resultant 1.8 kb *ORF1696'::puc'C* DNA fragment was separated from vector DNA by gel electrophoresis, purified, and ligated with the broad host range plasmid pRR5C [11], which had been previously digested with *Eco*R I. The resultant broad host range *ORF1696::pucC* fusion vector pFUS1 was used in *trans*-complementation experiments as described below.

4. RESULTS AND DISCUSSION

4.1. Comparison of Primary Sequences of Purple Bacterial ORF1696/PucC Homologues

The *R. capsulatus* ORF1696 protein was shown to catalyze LHI assembly, and the PucC protein is thought to function similarly in that both are required for optimal levels of the respective LH complexes [11,12]. On the basis of *in vitro* studies [4] the assembly of the LHI complex may consist of two stages: 1) association of six molecules (the α protein, the β protein, 2 bchl and 2 carotenoids) to form the subunit (B820); 2) oligomerization of subunits to form the ring-shaped holocomplex (B870). It is possible that in the absence of catalysis the first stage would limit the rate of the overall process, since it would depend on random collision of six molecules of four different structures. Once formed, the subunits are identical and so there are fewer degrees of freedom restricting their associa-

```
Rc orf1696    M---ILSRRM IGSLAMTWLP FADAASETLP LRQLLRLSLF QVSVGMAQVL    47
Rc pucC       MGYRAFALKN LARHAPKYLP FADVASEEVP LSRLLRLSLF QITVGMTLTL    50

              LLGTLNRVMI LELGVPALVV AAMISIPVLV APFRAILGHR SDTYRSALGW    97
              LAGTLNRVMI VELAVPASLV SVMLAMPMLF APFRTLIGFK SDTHKSALGL   100

              KRVPYLWFGS LWQMGGLALM PFSLILLSG- DQTM-GPAWA GEAFAGVAFL   145
              RRAPWIWKGT IYQFGGFAIM PFALLVLSGF GESVDAPRWI GMSAAALAFL   150

              MAGVGMHMTQ TAGLALAADR ATEETRPQVV ALLYVMFLIG MGISAVIVGW   195
              LVGAGVHIVQ TAGLALATDL VAEEDQPKVV GLMYVMLLFG MVISALVYGA   200

              LLRDFDQITL IRVVQGCGAM TLVLNVIALW KQEVM-RPMT KAEREAPRQS   244
              LLADYTPGRL IQVIQGTALA SVVLNMAAMW KQEAVSRDRA RQMETAEHPT   250

              FREAWGLLAA ETGALRLLAT VMVGTLAFSM QDVLLEPYGG QVLGLKVGQT   294
              FKEAFGLLMG RPGMLALLTV IALGTFGFGM ADVLLEPYGG QALHLTVGET   300

              TWLTAGWAFG ALVGFIWSAR RLSQGAVAHR VAARGLLVGI VAFTAVLFSP   344
              TKLTALFALG TLAGFGTASR VLGNGARPMR WSA-GCTDRV PGFVAIIMSS   349

              LFGSK-V-LF FASAMGIGLG SGMFGIATLT VAMMVVVRGA SGIALGAWGA   392
              LISQDGIWLF LAGTFAVGLG IGLFGHATLT ATMRTAPADR IGLALGAWGA   399

              AQATAAGLAV FIGGATRDLV AHAAAAGYLG SLHSPALGYT VVYVTEIGLL   442
              VQATAAGLGV ALAGVVRDGL V-ALPGTFGS GVVGP---YN TVFAIEALIL   445

              FITLAVLGPL VRPGSLFPKK PEAGEARIGL AEFPT                  477
              IVAIAFAVPL LKRG------ -----GR--- -----                  461
```

Figure 1. Alignment of *R. capsulatus* ORF1696 and PucC primary sequences. Identical residues are shaded.

tion. Thus, we hypothesize that a function of ORF1696 (and PucC) in LH assembly is to catalyze subunit formation, which implies ORF1696/PucC interactions with LH pigments and proteins.

As the first step in our structure-function analysis of ORF1696 and PucC we examined their primary sequences. An alignment of the *R. capsulatus* 477 amino acid ORF1696 and the 461 amino acid PucC protein sequences is shown in Fig. 1, which yields 215 identical amino acids (45 to 47% identity), more-or-less evenly distributed.

When homologues from three other purple non-sulfur bacteria are included in an alignment (Fig. 2), the number of identical residues drops to 138 (29 to 30% identity), with the distribution of these amino acids skewed to the N-terminal halves.

The sequence Gly/Ala-X-X-X-His seems to be involved in bchl binding and is present in most purple bacterial LH α and β polypeptides [16,17]. A Gly-X-X-X-His motif is also present at the binding sites for the accessory bchl molecules in *R. capsulatus* RC L and M polypeptides, and *R. sphaeroides* and *Rhodopseudomonas viridis* homologues in which the α-helical nature of this sequence was shown by X-ray crystallography [18].

The *R. capsulatus* ORF1696 His-152 is conserved among all ORF1696 and PucC homologues from purple bacteria (Fig. 2) as part of a Gly-X-X-X-His motif. However, substitution of the His-152 residue with either Asn or Phe by site-directed mutagenesis indicated this His residue is not required for the function of the ORF1696 protein [11]. Therefore, if a His residue coordinates bchl in the ORF1696-catalyzed assembly of LHI, it must be one of the other four, less conserved His residues, for example His-86, His-323,

```
Rs pucC      MSRIA----E HLVRIGPRFL PFADAASDQL PLRKLLRLSL FQVAVGMAIV    46
Rf pucC      MNRLSKMAVN RIATVGPRFL PFAEAASEDL PLSRLLRLSM FQVSVGMAMV    50
Rc.pucC      MGYRA-FALK NLARHAPKYL PFADVASEEV PLSRLLRLSL FQITVGMTLT    49
Rc orf1696   MI-LSRRMIG S-LAMT--WL PFADAASETL PLRQLLRLSL FQVSVGMAQV    46
Rr G115      MRGLNASLAR RWLSVAPRFL PFADAATKEL PLGRLLRLSL FQVTVGMAGV    50

             LLVGTLNRVM IVELKVPASV VGIMISLPLL FAPFRALIGF KSDTHVSALG    96
             LLVGTLNRVM IVELEVPASI VGIMISLPLL FAPFRALIGF KSDTHKSALG   100
             LLAGTLNRVM IVELAVPASV VSVMLAMPML FAPFRTLIGF KSDTHKSALG    99
             LLLGTLNRVM ILELGVPALV VAAMISIPVL VAPFRAILGH RSDTYRSALG    96
             LLTGTLNRVM IVELGVPTWL VAVMVALPIL FAPFRVLIGF RSDTHRSVLG   100

             WRRVPWIYRG TLALWGGFAI MPFALIVLGG QGYAEGQPFW LGVSSAALAF   146
             WRRVPYIWKG TLLQWGGFAI MPFALIVLSG QESAAGAPEW IGILSAAVSF   150
             LRRAPWIWKG TIYQFGGFAI MPFALLVLSG FGESVDAPRW IGMSAAALAF   149
             WKRVPYLWFG SLWQMGGLAL MPFSLILLSG -DQTMG-PAW AGEAFAGVAF   144
             WRRVPYIWMG TLLQFGGFAV MPFALFVLAG -DTAAPLPPV VGELCAGVAF   149

             LMVGGGVHTI QTVGLALATD LAPREDQPKV VGLMYVVLLI SMIFASIGFG   196
             LLVGAGVHTV QTVGLALATD LAPREDQPNV VGLMYVMLLV GMIVSALLFG   200
             LLVGAGVHIV QTAGLALATD LVAEEDQPKV VGLMYVMLLF GMVISALVYG   199
             LMAGVGMHMT QTAGLALAAD RATEETRPQV VALLYVMFLI GMGISAVIVG   194
             LLVGAGIHTT QTAGLALATD LAPEASRPRV VALLYVMLLI GMTVSSFGLG   199

             WLLDPYYDAQ LIKVISGVAV AVFFLNMIAL WKMEPRNRAF T---VKPEKE   243
             MWLEDFYHAK LIKVIQGAAV ATMVFNVIAL WKMEARDRVR ARQRLEGDPE   250
             ALLADYTPGR LIQVIQGTAL ASVVLNMAAM WKQEAVSRDR ARQ-METAEH   248
             WLLRDFDQIT LIRVVQGCGA MTLVLNVIAL WKQEVM-RPM TKA-EREAPR   242
             ALLEDFSPLR LIQVVQGAAA LTLVLNLVAL WKQEAR-QP- ALT-RPDAPR   246

             PEFGDHWREF ISRENALHGL IVIGLGTLGF GMADVILEPY GGEVLSMTVA   293
             PSFREAWGLF TRGPNARRLL WVIGLGTLGF GLSDVLLEPF GGQVLDMSVA   300
             PTFKEAFGLL MGRPGMLALL TVIALGTFGF GMADVLLEPY GGQALHLTVG   298
             QSFREAWGLL AAETGALRLL ATVMVGTLAF SMQDVLLEPY GGQVLGLKVG   292
             PSFSQRWGAF STRGRPARLL CVVGLGTAGF TMQDILLEPY GGEILHLSVG   296

             ETTRLTATFA GGGLVGFWLA SWVLGRGFDP LRMAFLGAAA GLPGFFAIMG   343
             ATTKLTAAVA GGTLVGFAWA SRVLSRGYDP MAMAGWGAVV GLPAFAIITF   350
             ETTKLTALFA LGTLAGFGTA SRVLGNGARP MRWSA-GCTD RVPGFVAIIM   347
             QTTWLTAGWA FGALVGFIWS ARRLSQGAVA HRVAARGLLV GIVAFTAVLF   342
             ATTMLTAMMA TGTLVGFALA ARALGRGAEP HRLAAYGLLA GIAAFCAVIF   346

             AT--EMTNVW VFLL------ --------G TLVVGFGGGL FSHGTLTATM   376
             SA--TLQSEP VFVF------ --------G TMMAGFGAGL FSHGTLTATM   383
             SSLISQDGIW LFLA------ --------G TFAVGLGIGL FGHATLTATM   382
             SPLF---GSK VLFFASAMGI GLGSGMFGIA TLTVAM---- ----------   375
             SGPL---GSP VLFRAGSLLI GFGSGLFSVG TLTAAM---- ----------   379

             RLAPKEQVGL ALGAWGAVQA TAAGVAIAGA GVLRDILQAM --PDLSGYGP   424
             RSAPKAQVGL ALGAWGAVQA TSAGAGIALG GVFRDAVRIR --RSAIS-SA   430
             RTAPADRIGL ALGAWGAVQA TAAGLGVALA GVVRDGLVAL --PGTFGSGV   430
             MVVVRGASGI ALGAWGAAQA TAAGLAVFIG GATRDLVAHA AAAGYLG-SL   424
             ALADETVSGM ALGAWGAVQA TATGAAVALG GGLRDGVSSL AAHGLLGEAL   429

             GAP---YVAV FALEAGFLFL TMIVILPLLR SAL------- ---AARRL----- 459
             RRP---Y--S LSTRSNWRCW S--------R PSS------- ---SATGF----- 455
             VGP---YNTV FAIEALILIV A----IAFAV PLL------- ---KRGGR----- 461
             HSPALGYTVV YVTEIGLLFI TLAVLGPLVR PGSLFPKKPE AGEARIGLAEFPT 477
             TTAHTGYGFV YLVEVVLLFT TLAIIGPLVR TAG--HRASQ SSEGRFGLAEFPG 480
```

Figure 2. Alignment of purple bacterial ORF1696 and PucC homologues. *Rs, Rhodobacter sphaeroides*; *Rf, Rhodovulum sulfidophilum*; *Rc, Rhodobacter capsulatus*; *Rr, Rhodospirillum rubrum*. Identical residues are shaded.

His-414, and His-425. His-86 is not highly conserved , but it is part of a Ala-X-X-X-His motif in *R. capsulatus*. His-323 is a conserved residue in ORF1696 and in the *Rsp. rubrum* G115 protein as part of a Gly-X-X-X-His motif, but is not present at this position in the PucC homologues. His-414 and His-425 appear to be less likely to bind bchl based on their Asp-X-X-X-His and Leu-X-X-X-His sequences, respectively. However, it could be argued that bchl binding in a transient manner (for delivery to nascent LH complexes) would not include magnesium coordination by a His residue, but instead would involve relatively weak interactions with the macrocycle.

Although the primary sequence alignments (Figs. 1 and 2) show significant amino acid sequence similarity between *R. capsulatus* ORF1696, PucC and purple bacterial homologues, there is a limited amount of information in these one-dimensional data, and so as the second step in our analysis we compared the proposed membrane topologies of *R. capsulatus* ORF1696 and PucC proteins to see if these two-dimensional structural representations could offer insight into structure-function relationships.

4.2. Topology Models of *R. capsulatus* ORF1696 and PucC Proteins

The membrane topology models of the *R. capsulatus* ORF1696 and PucC proteins (Fig. 3) consist of 12 transmembrane segments, seven cytoplasmic domains enriched for the positively charged amino acids Arg and Lys, and six periplasmic domains.

Although the models are superficially similar, the sizes of the proposed domains P3, C4, P5, C5, P6 and the C-termini differ more than other aqueous segments. As noted above, one function of ORF1696 and PucC could be in the delivery of bchl to nascent LH complexes. If binding of bchl by ORF1696 and PucC occurs, the amino acid sequences and topologies of the binding sites would be expected to be similar in order to provide a similar bchl-binding site conformation. The N-terminal halves of purple bacterial homologues are more conserved than the C-terminal halves (Fig. 2), and the differences in sizes of the P1, C1, P2, C2, C3 and P4 topological segments are less than those of C-terminal segments (Fig. 3). If these similarities relate to a function in bchl binding, it would implicate the N-terminal or central regions of these proteins.

In addition to bchl delivery, an alternative function of these proteins might be in the translocation and juxtaposition of LH α and β polypeptides in the ICM. Tests of the PucC topology model with most fusions of truncated *pucC* alleles to *phoA* and *lacZ* were consistent with the model. However, results of the enzyme fusion analyses in the last periplasmic loop were not as clear as in other regions of the protein [12]. Apparent anomalies were also observed for enzyme fusions in the C-terminal region of ORF1696 [11]. This suggests that the C-terminal domain of both of these proteins is somewhat flexible in its topology, and perhaps this is related to a function that is important for LH complex assembly. Furthermore, results obtained with one ORF1696 mutant (ΔNae) suggested that the C-terminal 13 amino acids of ORF1696 are important for function [11].

In the topology models, the P6 domain consists primarily of hydrophobic amino acids with the only charged amino acids, Arg-409 and Asp-410 of ORF1696 (Arg-416 and Asp-417 in PucC), located at the beginning of this periplasmic loop (Fig. 3). The C5 domain also contains several hydrophobic amino acids just before the penultimate transmembrane segment (#11), as does the segment of P6 immediately preceding the last transmembrane segment (#12). Perhaps the relatively hydrophobic character of these regions of ORF1696 and PucC allows for some movement, such that some of the amino acids in the C5 and P6 domains migrate into and out of the membrane. If so, a mobile, conformationally variable region such as this could be envisioned to participate in the

ORF1696

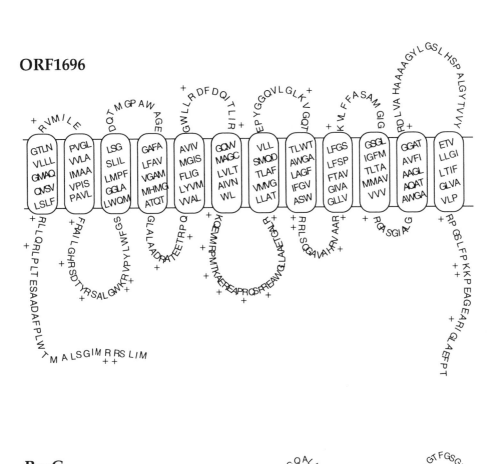

PucC

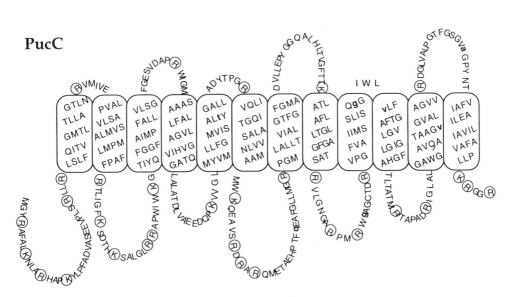

Figure 3. Topological models of *R. capsulatus* ORF1696 and PucC proteins. Positively charged residues are indicated by either + symbols (ORF1696) or by encircling (PucC).

translocation of the LHI α and β polypeptides, pulling them through the membrane to assist proper juxtaposition of transmembrane α-helices of these proteins. It would be interesting to evaluate if there is any charge complementarity between these regions and the N- and C-termini of the LH proteins.

The specificity of the ORF1696 and PucC proteins for the assembly of the LHI and LHII complexes, respectively, must be a function of differences in structure. The greatest differences between the two models is in the number and content of amino acids in the proposed three extramembranous domains P3, P5 and the C-terminus. The P3 domain of PucC is predicted to consist of seven amino acid residues, one of which is positively charged (Arg-209) and one of which is negatively charged (Asp-204), whereas P3 of ORF1696 is predicted to have fourteen amino acids of which two are positively charged (Arg-198 and Arg-207) and two of which are negatively charged (Asp-199 and Asp-201). Differences between the two models are greater in the P5 domain, which consists of three amino acids in PucC and twelve amino acids in ORF1696, including one positively charged residue (Lys-349). The proposed C-terminal cytoplasmic domains of both proteins are enriched for positively charged residues (Lys-457, Arg-458, and Arg-461 for PucC and Arg-454, Lys-461, Lys-462, and Arg-468 for ORF1696), but in PucC this domain contains only five amino acids whereas there are twenty-three amino acids comprising this domain of ORF1696. Furthermore, there are three negatively charged residues within the ORF1696 C-terminal domain (Glu-463, Glu-466, and Glu-473). The C-terminus and P5 domains of ORF1696 could be more bulky, globular domains, compared to the same regions of PucC, because of the greater number of amino acids in these regions of ORF1696. These differences might contribute to the specificity of these proteins for the LHI or LHII α and/or β polypeptides.

The two hypothetical functions described above (bchl delivery, and LH α and β protein juxtaposition) are not mutually exclusive. The relative similarity in primary amino acid sequence and topology of the N-terminal halves of the proteins, in contrast to the greater dissimilarities found nearer the C-termini (Figs. 2 and 3), could suggest that two functional regions exist in these proteins. Thus, the more conserved N-terminal half could be important for a similar transient binding of bchl molecules that are transferred to LH α and β polypeptides during membrane insertion, whereas the less similar C-terminal region could bestow specificity for either the LHI or LHII polypeptides. A weakness of these thoughts is that they do not explicitly account for the potentially different delivery of B870/B850 *versus* B800 bchl.

These speculations on ORF1696 and PucC function assume that these proteins give rise to the subunit forms of LH complex precursors. The pigment-protein subunits would then oligomerize to form the mature LHI rings, formed by protein-assisted mechanisms or occurring spontaneously, as has been observed *in vitro* for LHI [4]. However, if these two stages of assembly exist *in vivo*, it is possible that ORF1696, PucC and homologues catalyze both processes.

4.3. *Trans*-Complementation of *ORF1696* and *pucC* Mutants with an *ORF1696'::puc'C* Fusion

As noted above, the ORF1696 and PucC proteins are thought to have analogous roles in LH complex assembly, although their specificities differ in that mutant ORF1696 proteins reduce LHI complex assembly whereas mutant PucC proteins result in the complete loss of the LHII complex [9,11]. We investigated the specificity of *R. capsulatus* ORF1696 N-terminal and PucC C-terminal segments by constructing a translationally in-

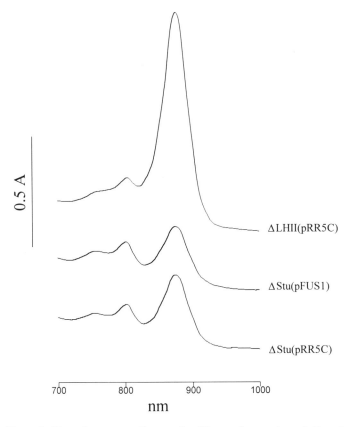

Figure 4. Absorption spectra of intact cells of *R. capsulatus* strains as indicated.

frame fusion of a 5′ segment of the *ORF1696* gene with a 3′ segment of the *pucC* gene. The chimaeric ORF1696::PucC fusion protein had the deduced amino acid sequence ...FDQIT::GRILQ.... at the fusion joint located within the P3 domain (Fig. 3), and thus consisted of amino acids 1 to 192 of ORF1696 and 209 to 461 of PucC for a total of 444 amino acid residues. In theory, the fusion protein would be expected to have the same membrane topology as ORF1696 and PucC.

Plasmid pFUS1 was introduced into the *ORF1696* mutant ΔStu, to see if the fusion protein would enhance the level of the LHI complex. As shown in Fig. 4, spectral analysis of semiaerobically grown whole cells of ΔStu(pFUS1) compared to the controls did not reveal a significant increase in the 875 nm (LHI) peak. This result contrasts with the complementation of ΔStu with the *ORF1696* gene [11]. Thus, the ORF1696::PucC fusion protein was not able to complement the effects of the ΔStu *ORF1696* chromosomal mutation; negative results were also obtained in pFUS1 complementation experiments on a *pucC* deletion strain (data not shown).

The ORF1696 and PucC sequence alignments and topology models discussed above suggest the presence of two distinct regions which might be significant for the function of these proteins. However, it appears that if discrete N- and C-terminal regions exist they are not interchangeable, based on the experiments with the chimaeric protein, and thus the entire wild type protein in each case seems to be required for function. Gene disruption studies of the *ORF1696* and *pucC* genes showed that disruptions relatively close to the C-terminus have deleterious effects on LH complex levels in *R. capsulatus* [11,15]. For example, a Km[r] cartridge insertion into the *ORF1696* gene, which resulted in the replacement of the last 13 amino acids at the C-terminus of the ORF1696 protein, caused a 50% reduction in the LHI complex [11]. Although ORF1696 and PucC exhibit a high degree of sequence identity their functions in maintaining LHI and LHII complex levels, respectively, could result from interactions of domains of the proteins distant from one another in the primary sequence. Thus, the results of the domain swapping experiment could be explained by a conformation of the ORF1696::PucC fusion protein that was not conducive to assembly of either of these LH complexes.

4.4. Amino Acid Sequence Alignments of Purple Non-Sulfur ORF1696 and PucC Proteins and Oxygenic Bacterial Homologues

Homologues of *ORF1696* have been found in the purple non-sulfur bacteria *R. viridis* [19]), *Rsp. rubrum* [20] and *R. sphaeroides* [10,21], and are similarly located 5' to the *puhA* gene of these species. Furthermore, *pucC* gene homologues are present in bacteria that contain the LHII complex [22,23], and homologues have been discovered in the cyanobacterium *Synechocystis* PCC6803 [24], and *Prochlorococcus marinus* (W. Hess, personal communication). An amino acid sequence alignment of all of the known ORF1696 and PucC homologues whose complete sequences have been deduced from the nucleotide sequences of genes is shown in Fig. 5.

The most significant cluster of identical amino acids is located around domain P4 of the *R. capsulatus* topology models, with a smaller cluster of identical residues located in transmembrane segment #11 (Fig. 3).

Since cyanobacteria and *Prochlorococcus* do not contain purple bacterial LH complexes, the properties of ORF1696 homologues in these organisms could provide a hint about a general function of this class of proteins. There are great differences in the amino acid sequences of purple bacterial and cyanobacterial LH complexes [17, 25]. In contrast, the chlorophyll *a* (chl) and bchl molecules differ only in that chl has a vinyl group at pyrrole ring I whereas bchl has an acetyl group at this position, and pyrrole ring II of chl is unsaturated whereas the same ring of bchl is saturated [26]. We speculate that the primary amino acid sequence conservation in ORF1696-like proteins is due to a common function in tetrapyrrole translocation for LH complex assembly in these otherwise diverse photosynthetic organisms, and that the P4 domain (Fig. 3) is important for this function.

5. SUMMARY

Comparisons of the *R. capsulatus* ORF1696 and PucC primary sequences and membrane topology models indicate that they are structurally quite similar, yet the differences in amino acid sequences and structures must account for their differing specificities for the LHI and LHII complexes. The results obtained with the ORF1696::PucC fusion protein indicated that the function of both proteins was lost, perhaps because the specificities of the wild type ORF1696 and PucC proteins for the two different LH complexes is due to long-range interactions between different regions of the proteins. Similarities in the topological organization of the proteins in the N-terminal half and the centrally located amino acid sequence conservation may indicate a common conformation in these regions which might

```
Spc         MTVSESMAGS PLPKLPLVTM FRLGLFQMGL GIMSLLTLGV LNRVLIDELA VLPWVAATAI AMYQFVSPFK VWCGQLSDSQ RLWGYHRTGY VWLGALGFTI   100
Pmar        MKKAQHL--- ---K-PL-NL LRLSLFQGCL GCLAVIFAGM LNRIMITELA FPAILVGGGL AFEQLVAPSR            PIKGRKRTPY IYLGSAAFCF    92
Rc orf1696  MI-LSRRMIG S-LA-MT--W LPFADAASET LPLRQLLRLS LFQVSVGMAQ VLLLGTLNRV MILELGVPAL V-VAAMISIP VLVAPFRAIL GHRSDTYRSA    94
Rr G115     MRGLNASLAR RWLS-VAPRF LPFADAATKE LPLGRLLRLS LFQVTVGMAG MIVELGVPTW L-VAVMVALP ILFAPFRVLI GFRSDTHRSV    98
Rs pucC     MSRIA---E  HLVR-IGPRF LPFADAASDQ LPLRKLLRLS LFQVAVGMAI MIVELKVPAS V-VGIMISLP LLFAPFRALI GFKSDTHVSA    94
Rf pucC     MNRLSKMAVN RIAt-VGPRF LPFAEAASED LPLSRLLRLS MFQVSVGMAM MIVELEVPAS I-VGIMISLP LLFAPFRALI GFKSDTHKSA    98
Rc pucC     MGYRA-FALK NLAR-HAPKY LPFADVASEE VPLSRLLRLS LFQITVGMTL TLLAGTLNRV MIVELAVPAS L-VSVMLAMP MLFAPFRTLI GFKSDTHKSA    97

Spc         LSFIALQVVW QLGLSLQNNG WGALTIFWSI VLGAVFAAYG VTLSLSSTPF DNRSKLVGIV WSMLMVGIVV GAIVSSRLLN TPEICGPALL   200
Pmar        LAVLSIPIIF LTEKALAQGS FAAISA-SVI CLCSLFALYG LAISMSTTPY LALVIDLTDE KERPKAVGI1 WCMLTIGIIV GAIAIS--IT TKSLDG---   185
Rc orf1696  LGWKRVPYLW FGSL-WQMGG LALMPF-SLI LLSG-DQTMG -PAWAGEAFA GVAFLMAGVG MHMTQTAGLA LAADRATEET RPQVVA-LLY VMFLIGMGIS   189
Rr G115     LGWRRVPYIW MGTL-LQFGG FAVMPF-ALF VLAG-DTAAP LPPVVGELCA GVAFLLVGAG IHTTQTAGLA LATDLAPEAS RPRVVA-LLY VMLLIGMTVS   194
Rs pucC     LGWRRVPWIY RGTL-ALWGG FAIMPF-ALI VLGGQGYAEG QPFWLGVSSA ALAFLMVGGG VHTIQTVGLA LATDLAPRED QPKVVG-LMY VVLLISMIFA   191
Rf pucC     LGWRRVPYIW KGTL-LQWGG FAIMPF-ALI VLSGQESAAG APEWIGILSA AVSFLLVGAG VHTVQTVGLA LATDLAPRED QPNVVG-LMY VMLLVGMIVS   195
Rc pucC     LGLRRAPWIW KGTI-YQFGG FAIMPF-ALL VLSGFGESVD APRWIGMSAA ALAFLLVGAG VHIVQTAGLA LATDLVAEED QPKVVG-LMY VMLLFGMVIS   194

Spc         NADALLVKKT VDIAQLQRGI NPVFIIMPAI VVFLAWLATV GVEKKYSRFG DRSGGREDEI TLGQALKVLT ASRQTAIFFG FLLLLTLSLF MQDAVLEPYG   300
Pmar        ITDPALLQPT LQQFML-RVS TIIFII--SI IS--CW---- GIEPKSKSLT KGSNKHRQEI GLKSAWSLIR SSKQIFIFFA FLIFYTLGLF LQDPILESFG   276
Rc orf1696  AVIVGWLLRD FDQITLIRVV QGCGAMTLVL NVIALW---- KQEVM-RPMT KA-ERREAPRQ SFREAWGLLA AETGALRLLA TVMVGTLAFS MQDVLLEPYG   283
Rr G115     SFGLGALLED FSPLRLIQVV QGAAALTLVL NLVALW---- KQEAR-QP-A LT-RPDAPRP TRGRPARLLC VVGLGTAGFT MQDILLEPYG   287
Rs pucC     SIGFGWLLDP YYDAQLIKVI SGVAVAVFFL NMIALW---- KMEPRNRAFT --VKPEKEP  EFGDHWREFI SRENALHGLI VIGLGTLGFG MADVLLEPYG   284
Rf pucC     ALLFGMWLED FYHAKLIKVI QGAAVATMVF NVIALW---- KMEARDRVRA RQRLEGDPEP SFREAWGLFT RGPNARRLLW VIGLGTLGFG LSDVLLEPFG   291
Rc pucC     ALVYGALLAD YTPGRLIQVI QGTALASVVL NMAAMW---- KQEAVSRDRA RQ-METAEHP TFKEAFGLLM GRPGMLALLT VIALGTFGFG MADVLLEPYG   289
```

Figure 5. Alignment of anoxygenic and oxygenic bacterial ORF1696, PucC and homologous proteins. *Spc*, *Synechocystis* PCC6803; *Pmar*, *Prochlorococcus marinus*; other abbreviations as in Fig. 2. Identical residues are shaded.

```
GEVFNLCISE TTQLNAFFGM GTLLGIGSTG FFVVPRLGKQ RTTSLGCALA ALCFSLLILA G--FQQNVTL LKSGLLFFGL ASGMITAGAT SLMLDLTAVE  398
GEVFNLPISQ TTLLNAFWGI GTLIGLLIGG LLIIPSIGKF SAAKLGCWLI AISLGLLVIS G--ALENSNL LFLVLFIFGV AAGIATNSAL SLMLDLTLPE  374
GQVLGLKVGQ TTWLTAGWAF GALVGFIWSA RRLSQGAVAH RVAARGLLVG PLFG--SKVL FFASAMGIGL GSGMFGIATL TVAMMVVVRG              381
GEILHLSVGA TTMLTAMMAT GTLVGFALAA RALGRGAEPH RLAAYGLLAG IAAFCAVIFS GPLG--SPVL FRAGSLLIGF GSGLFSVGTL TAAMALADET  385
GEVLSMTVAE TTRLTATFAG GGLVGFWLAS WVLGRGFDPL RMAFLGAAAG T--EMTNVWV FLLGTLVVGF GGGLFSHGTL TATMRLAPKE              382
GQVLDMSVAA TTKLTAAVAG GTLVGFAWAS RVLSRGYDPM AMAGWGAVVG A--TLQSEPV FVFGTMMAGF GAGLFSHGTL TATMRSAPKA              389
GQALHLTVGE TTKLTALFAL GTLAGFGTAS RVLGNGARPM RWSA-GCTDR VPGFVAIIMS SLISQDGIWL FLAGTFAVGL GIGLFGHATL TATMRTAPAD  388


TAGTFIGAWG LAQSISRGLA TVAGGTVLNI GKAL---FAN --AVLA---- YGLVFALQAL GLILSIFLLN KVNVREF--Q DNAKTAIATV MAGDLDG     484
VAGTFVGVWG LAQALSRAMG KLIGGGLLDL GRII---GGN DNSLFA---- FSFVFSIEII IIIISIFILN KVSISKF--K NETSAKMSEI LMSDLD-     461
ASGIALGAWG AAQATAAGLA VFIGGATRDL VAHAAAAGYL G-SLHSPALG YTVVVTEIG  LLFITLAVLG PLVRPGSLFP KKPEAGEARI GLAEFPT     477
VSGMALGAWG AVQATATGAA VALGGGLRDG VSSLAAHGLL GEALTTAHTG YGFVYLVEVV LLFTTLAIIG PLVRTAG--H RASQSSEGRF GLAEFPG     480
QVGLALGAWG AVQATAAGVA IAGAGVLRDI LQAM--PDLS GYGPGAP--- YVAVFALEAG FLFLTMIVIL PLLRSAL--- -------AAR RL-----     459
QVGLALGAWG AVQATSAGAG IALGGVFRDA VRIR--RSAI S-SARRP--- Y--SLSTRSN WRCWS----- ---RPSS--- -------SAT GF-----     455
RIGLALGAWG AVQATAAGLG VALAGVVRDG LVAL--PGTF GSGVVGP--- YNTVFAIEAL ILIVA----I AFAVPLL--- -------KRG GR-----     461
```

Figure 5. (*continued*)

mediate the transient binding of bchl. There are a number of His residues present in both proteins and in ORF1696/PucC homologues which conceivably could participate in bchl binding. The C-terminal regions of both proteins seem to have flexible topologies and may participate in LH α and β polypeptide translocation.

The alignment of all ORF1696 and PucC homologues, including those from the phylogenetically distant and physiologically dissimilar *Synechocystis* and *Prochlorococcus* species revealed that the region of most significant conservation extends from the periplasmic end of transmembrane segment #7, through P4 and into transmembrane segment # 8 (compare Figs. 3 and 5). Thus, if this family of proteins shares a common function in tetrapyrrole delivery to LH complexes, it seems that an important component is P4 and flanking residues, and that tetrapyrrole binding occurs through interactions other than between a His and the magnesium of a bchl or chl molecule.

REFERENCES

1. Drews, G. and Golecki, J.R. (1995) Structure, molecular organization, and biosynthesis of membranes of purple bacteria in: Anoxygenic photosynthetic bacteria, pp. 231–257 (Blankenship, R.E., Madigan, M.T. and Bauer, C.E., Eds.) Kluwer Academic Publishers, Dordrecht, The Netherlands.
2. Drews, G. (1996) Formation of the light-harvesting complex (B870) of anoxygenic phototrophic purple bacteria. Arch. Microbiol. 166, 151–159.
3. Walz, T. and Ghosh, R. (1997) Two-dimensional crystallization of the light-harvesting I-reaction centre photounit from *Rhodospirillum rubrum*. J. Mol. Biol. 265, 107–111.
4. Loach, P.A. and Parkes-Loach, P.S. (1995) Structure-function relationships in core light-harvesting complexes (LHI) as determined by characterization of the structural subunit and by reconstitution experiments in: Anoxygenic photosynthetic bacteria (Blankenship, R.E., Madigan, M.T. and Bauer, C.E., Eds.) Kluwer Academic Publishers, Dordrecht, The Netherlands.
5. McDermott, G., Prince, S.M., Freer, A.A., Hawthornthwaite-Lawless, A.M., Papiz, M.Z., Cogdell, R.J. and Isaacs, N.W. (1995) Crystal structure of an integral membrane light-harvesting complex from photosynthetic bacteria. Nature (London) 374, 517–521.
6. Koepke, J., Hu, X.C., Muenke, C., Schulten, K. and Michel, H. (1996) The crystal structure of the light-harvesting complex II (B800–850) from Rhodospirillum molischianum. Structure 4, 581–597.
7. Oling, F., Boekma, E.J., deZarate, I.O., Visschers, R., vanGrondelle, R., Keegstra, W., Brisson, A. and Picorel, R. (1996) Two-dimensional crystals of LH2 light-harvesting complexes from *Ectothiorhodospira* sp. and *Rhodobacter capsulatus* investigated by electron microscopy. Biochim. Biophys. Acta 1273, 44–50.
8. Tichy, H.V., Oberlé, B., Stiehle, H., Schiltz, E. and Drews, G. (1989) Genes downstream from *pucB* and *pucA* are essential for formation of the B800–850 complex of *Rhodobacter capsulatus*. J. Bacteriol. 171, 4914–4922.
9. LeBlanc, H.N. and Beatty, J.T. (1993) *Rhodobacter capsulatus puc* operon: promoter location, transcript sizes and effects of deletions on photosynthetic growth. J. Gen. Microbiol. 139, 101–109.
10. Bauer, C.E., Buggy, J., Yang, Z. and Marrs, B.L. (1991) The superoperonal organization of genes for pigment biosynthesis and reaction center proteins is a conserved feature in *R. capsulatus*: analysis of overlapping *bchB* and *puhA* transcripts. Mol. Gen. Genet. 228, 438–444.
11. Young, C. (1997) The role of the *Rhodobacter capsulatus* integral membrane protein ORF1696 in light-harvesting I complex assembly. Ph. D. thesis, University of British Columbia, Vancouver.
12. LeBlanc, H.N. and Beatty, J.T. (1996) Topological analysis of the *Rhodobacter capsulatus* PucC protein and effects of C-terminal deletions on light-harvesting complex II. J. Bacteriol. 178, 4801–4806.
13. Claros, M.G. and Heinje, G.v. (1995) TopPred II, version 1.1: Prediction of transmembrane segments in integral membrane proteins, and the putative topologies. CABIOS 10, 685–686.
14. Bingle, W., Kurtz, H.D. and Smit, J. (1993) An "all-purpose" cellulase reporter for gene fusion studies and applications to the paracrystalline surface (S)-layer protein of *Caulobacter crescentus*. Can. J. Microbiol. 39, 70–80.
15. LeBlanc, H. (1995) Directed mutagenesis and gene fusion analysis of the *Rhodobacter capsulatus puc* operon. Ph. D. thesis, University of British Columbia, Vancouver.
16. Bylina, E.J., Robles, S.J. and Youvan, D.C. (1988) Directed mutations affecting the putative bacteriochlorophyll-binding sites in the light-harvesting I antenna of *Rhodobacter capsulatus*. Israel J. Chem. 28, 73–78.

17. Zuber, H. and Cogdell, R.J. (1995) Structure and organization of purple bacterial antenna complexes in: Anoxygenic photosynthetic bacteria (Blankenship, R.E., Madigan, M.T. and Bauer, C.E., Eds.) Kluwer Academic Publishers, Dordrecht, The Netherlands.

18. Lancaster, C.R.D., Ermler, U. and Michel, H. (1995) The structures of photosynthetic reaction centers from purple bacteria as revealed by X-ray crystallography in: Anoxygenic photosynthetic bacteria (Blankenship, R.E., Madigan, M.T. and Bauer, C.E., Eds.) Kluwer Academic Publishers, Dordrecht, The Netherlands.

19. Wiessner, C. (1990) Molekularbiologishe analyse der gene des phototosynthetischen apparates von *Rhodopseudomonas viridis*. Ph. D. thesis, Wolfgang Goethe-Universität, Frankfurt.

20. Bérard, J. and Gingras, G. (1991) The *puh* structural gene coding for the H subunit of the *Rhodospirillum rubrum* photoreaction center. Biochem. Cell. Biol. 69, 122–131.

21. Donohue, T.J., Hoger, J.H. and Kaplan, S. (1986) Cloning and expression of the *Rhodobacter sphaeroides* reaction center H gene. J. Bacteriol. 168, 953–961.

22. Gibson, L.C.D., McGlynn, P., Chaudhri, M. and Hunter, C.N. (1992) A putative coproporphyrinogen III oxidase in *Rhodobacter sphaeroides*. II. Analysis of a region of the genome endoding *hemF* and the *puc* operon. Molec. Microbiol. 6, 3171–3186.

23. Hagemann, G.E., Katsiou, E., Forkl, H., Steindorf, A.C.J. and Tadros, M.H. (1997) Expression of the *puc* operon from *Rhodovulum sulfidophilum*. Biochim. Biophys. Acta 1351, 341–358.

24. Kaneko, T., Sato, S., Kotani, H., Tanaka, A., Asamizu, E., Nakamura, Y., Miyajima, N., Hirosawa, M., Sugiura, M., Sasamoto, S., Kimura, T., Hosouchi, T., Matsuno, A., Muraki, A., Nakazaki, N., Naruo, K., Okumura, S., Shimpo, S., Takeuchi, C., Wada, T., Watanabe, A., Yamada, M., Yasuda, M., Tabata, S. (1996) Sequence analysis of the genome of the unicellular cyanobacterium *Synechocystis* sp. strain PCC6803. II. Sequence determination of the entire genome and assignment of potential protein-coding regions. DNA Research 3, 109–136.

25. Laroche, J., Vanderstaay, G.W.M., Partensky, F., Ducret, A., Aebersold, R., Li, R., Golden, S.S., Hiller, R.G., Wrench, P.M., Larkum, A.W.D., Green, B.R. (1996) Independent evolution of the prochlorophyllite and green plant chlorophyll A/B light-harvesting proteins. Proc. Natl. Acad. Sci. USA 93, 15244–15248.

26. Lawlor, D.W. (1993) Photosynthesis: molecular, physiological and environmental processes. Longman Scientific and Technical, Burnt Mill.

14

TRANSCRIPTIONAL REGULATION OF *puf* AND *puc* OPERON EXPRESSION IN *RHODOBACTER CAPSULATUS* BY THE DNA BINDING PROTEIN RegA

S. Katharina Hemschemeier, Michaela Kirndörfer, Ulrike Ebel, and Gabriele Klug

Institute for Microbiology and Molecular Biology
Frankfurter Str. 107
D-35392 Giessen, Germany

1. INTRODUCTION

The formation of photosynthetic complexes in the facultative phototrophic bacterium *Rhodobacter capsulatus* is regulated by oxygen. Oxygen affects the expression of photosynthesis genes at multiple levels by acting on rates of transcription, on the stability of mRNA, and on posttranslational steps. Some years ago the two component system RegB/RegA was shown to be involved in transmission of the oxygen signal. RegB can undergo autophosphorylation and consequently functions as a phosphor donor for RegA (Sganga and Bauer, 1992; Mosley et al., 1994). RegA was suggested to be an intermediate in signal transduction and to transfer the phosphate group to an unknown regulatory protein that regulates transcription of photosynthesis genes by binding to DNA sequences in promoter regions (Sganga and Bauer, 1992).

By in vitro DNA binding studies using wild type or His-tagged RegA and two RegA mutants we show that RegA specifically binds to DNA sequences upstream of the *puf* and *puc* operons encoding reaction center and light harvesting proteins (Fig. 1). RegA can be phosphorylated in vitro using acetyl phosphate as small phosphor donor, but phosphorylation seems not to be essential for DNA binding as shown by gel shift assays.

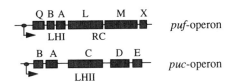

Figure 1. Operons coding for light- harvesting (LH) and reaction center (RC) proteins of the *Rhodobacter capsulatus* photosynthetic apparatus.

The Phototrophic Prokaryotes, edited by Peschek *et al.*
Kluwer Academic / Plenum Publishers, New York, 1999.

2. RESULTS AND DISCUSSION

2.1. Isolation of RegA and Phosphorylation in Vitro

Native and His-tagged RegA were both purified and compared in in vitro assays with regard to their properties. Both protein preparations behaved virtually identical in gel shift and footprint experiments, but in contrast to the His-tagged RegA native RegA could only poorly be phosphorylated. This might be due to the fact that the protocol for the isolation of native RegA involved a denaturation/renaturation step.

RegA, like many other response regulators, autophosphorylates in the presence of acetyl phosphate (Fig. 2). Phosphorylation is magnesium-dependent and requires at least 5 mM $MgCl_2$. In buffers containing 25 mM EDTA, phosphorylation is completely abolished. We also tested various acetyl phosphate concentrations (10–60 mM). The affinity of RegA to this artificial phospho donor is rather low, and RegA has a K_m of about 30 mM for acetyl phosphate. After phosphorylation of RegA with radioactive acetyl phosphate followed by a cold chase, we observed no decrease of the radioactive RegA-phosphate over the time period of one hour. This indicates that RegA has only a weak intrinsic phosphatase activity and in vivo inactivation may occur mainly by RegB.

2.2. Gel Shift Assays and Footprints

Native and His-tagged RegA were both tested in gel shift assays for their DNA binding properties. The *puf* upstream region (168 bp DNA fragment) includes a region of dyad symmetry which is supposed to be required for transcriptional regulation, the *puc* fragment comprises 300 bp of the upstream *puc* regulatory region. Both fragments were equally shifted by RegA and RegA phosphate (Fig. 4). No differences in the affinity or amount of protein bound to the DNA fragment could be observed after phosphorylation, the K_d was approximately 70 nM for both. This result was somewhat surprising since in most two component systems, the affinity for regulatory sequences is dramatically enhanced after phosphorylation.

In order to characterize the binding site for RegA and RegA-phosphate, we performed footprints with both *puf* and *puc* promoter fragments. Several RegA protected regions were identified (Fig. 3), but a clear consensus sequence for RegA binding could not be defined. RegA binds close to DNA motifs which contain large stretches of C-residues. RegA and RegA-phosphate seem to protect similar sites in the *puc* promoter region, but the protection pattern of phosphorylated and unphosphorylated RegA is clearly different for the *puf* promoter fragment. The region of dyad symmetry is only bound by RegA-phosphate. We assume that RegA-phosphate and a postulated negative repressor, PPBP (Taremi and Marrs, 1990; Klug 1991) compete for binding in this region. Induction of *puf* and *puc* transcription may require both a release of the repressor PPBP and binding of the activa-

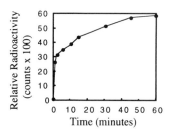

Figure 2. Phosphorylation of His-tagged RegA in the presence of acetyl phosphate. His-RegA was incubated at 37°C with 40 mM acetyl phosphate in 50 mM Tris, pH 7.6, 10 mM KCl, and 5 mM $MgCl_2$. At time points as indicated, 6 μl samples were removed and added to 2 μl SDS-gel loading buffer (Maniatis, 1982). The samples were separated in an SDS-gel and the gel was dried. Radioactivity associated with the RegA-protein was visualized and quantified (relative radioactivity × 100) using a phosphorimager screen.

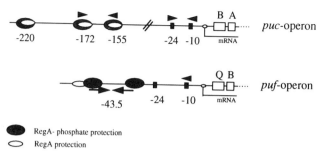

Figure 3. RegA and RegA-phosphate protection in the *puf* and *puc* promoter. Large arrows indicate the region with dyad symmetry, small arrows indicate ACATT and TGTAA repeats which might be involved in oxygen regulation.

tor, RegA. Our in vitro data suggest that binding of RegA to its target DNA does not necessarily require phosphorylation. On the other hand, phosphorylation is essential for efficient transcriptional activation since expression of these genes is impaired in RegB-mutants (data not shown).

2.3. Construction of Mutant RegA Proteins

Sganga and Bauer (1992) postulated that despite its homologies to reponse regulators, RegA is not involved in the DNA binding itself but rather acts as an intermediate in the signal transmission cascade. However, according to a computer compilation (Dodd and Evan 1990) RegA contains a region with a high potential to form a helix-turn-helix motif. To investigate whether this sequence motif is required for DNA binding, we constructed two mutant forms of RegA.

Plasmid SP72::regA* codes for a RegA protein lacking the last 13 C-terminal amino acids which are part of the helix-turn-helix motif. Plasmid SP72::regAPro encodes a RegA with a leucine167-proline exchange which is supposed to disturb the proper folding of the DNA binding motif. Neither one of these two RegA mutants was able to bind to *puf*- or *puc* specific DNA fragments indicating that the helix-turn-helix motif is indeed required for efficient DNA binding (Fig. 4). We therefore conclude that the reponse regulator RegA directly interacts with the *puf* and *puc* promoter region and is not merely an intermediate in this signalling cascade.

Figure 4. Gel shift with native and mutant RegA. A–E: puc promoter fragment and native RegA. A: without competitor DNA, B–E: 1x, 5x, 10x, and 50x excess of unspecific competitor DNA, respectively. F: RegA* lacking the C-terminus, G: mutant RegA[Pro], H: control without RegA. Lanes F–H contain no unspecific competitor DNA.

2.4. Truncated RegA: Properties of the Activator Domain

Most response regulators have two clearly seperated protein domains: the receiver domain where phosphorylation occurs, and the activator or DNA binding domain. Da Re et al. (1994) proposed an interesting model for the activation of response regulators: the DNA binding domain is normally occluded by the receiver domain. Phosphorylation of the receiver induces a conformational change, thereby releasing the occlusion of the DNA binding domain by the receiver. In this case, both phosphorylated response regulator and the C-terminal domain alone should have a much higher affinity to their DNA target. In order to test this for RegA, we isolated truncated RegA protein which contains only the C-terminal domain. This mutant RegA (RegA Activator) was tested in gel shift and in vivo experiments. RegA Activator shifted both *puf* and *puc* promoter fragments with the same affinity as full length and phosphorylated RegA. However, after addition of only 5 fold unspecific DNA as competitor binding of the RegA Activator was reduced. These data suggest that for RegA, phosphorylation may not be required to release repression by the receiver domain and to enhance DNA binding. This is also supported by the finding that no conformational change or dimerization of RegA-phosphate could be detected in native gels. Phosphorylation may however be required for efficient transcription initiation. This is also supported by in vivo studies: the truncated RegA Activator is able to induce *puf* and *puc* transcription, but the expression is about 50% reduced as compared to transcriptional induction by wild type RegA.

REFERENCES

Da Re, S., Bertagnoli, S., Fourment J., Reyat, J.-M., Kahn, D. (1994). Nucl. Acids -Res. 22:1555–1561
Dodd, I.B., Egan, J.B. (1990). Nucleic Acids Res. 18: 5019–5026
Klug, G. (1991). Mol. Gen. Genet. 226:167–176
Mosley, C.S., Suzuki, J.Y., Bauer, C.E. (1994). J. Bacteriol. 176:7566–7573
Sambrook,G., Fritsch, E.F., Maniatis, T. (1989). Second ed., CSH Laboratory Press, New York.
Sganga, M.W.,Bauer, C.E. (1992). Cell 68:945–954
Taremi, S.S., Marrs, B.L. (1990). In: Hauska, G., Thauer, R. (eds.) The molecular basis of bacterial metabolism, pp 146–151. Springer, Berlin, Heidelberg, New York

15

MUTATION ANALYSIS AND REGULATION OF PpsR

The *Rhodobacter sphaeroides* Repressor of Photopigment and Light Harvesting Complex-II Expression

Mark Gomelsky, Hye-Joo Lee, and Samuel Kaplan

Department of Microbiology and Molecular Genetics
University of Texas Medical School
Houston, Texas 77030

1. INTRODUCTION

In the presence of oxygen, *Rhodobacter sphaeroides* derives energy from aerobic respiration. When oxygen tension decreases below certain threshold levels, it develops intracytoplasmic membrane vesicles which house the photosystem (PS) comprised of the reaction center and light harvesting (LH) complexes. When light becomes available under anaerobic conditions, the bacterium uses PS for capturing photons and ultimately converting light energy into biochemical energy. In the absence of light, the PS is produced gratuitously. This suggests that oxygen is a major environmental factor affecting PS development in *R. sphaeroides*. Under anaerobic conditions, light intensity affects the cellular abundance and proportion of photocomplexes [1].

Development of the PS depends to a great extent on expression of PS genes, i.e. the genes encoding structural proteins and assembly factors for the LH and reaction center complexes, as well as the genes encoding enzymes for bacteriochlorophyll (Bchl) and carotenoid (Crt) biosynthesis. Under high oxygen tension, expression of most of these genes is suppressed to low basal levels. Several pathways of oxygen dependent regulation of PS gene expression have been identified over the last few years [2]. One pathway involves transcriptional repression under aerobic conditions and is mediated by the PpsR protein [3] (also known as CrtJ in a related bacterium *Rhodobacter capsulatus* [4]). The C-terminal portion of PpsR contains a helix-turn-helix (HTH) motif which enables this repressor to bind DNA [5]. The binding site for PpsR is a palindrome TGT-N_{12}-ACA, where N is a nucleotide. This sequence is present upstream of the *puc* operon encoding protein components of the LH II complex as well as upstream of a number of *bch* and *crt* genes and operons involved in Bchl and Crt biosynthesis. Therefore, one can infer that the role of the

The Phototrophic Prokaryotes, edited by Peschek *et al.*
Kluwer Academic / Plenum Publishers, New York, 1999.

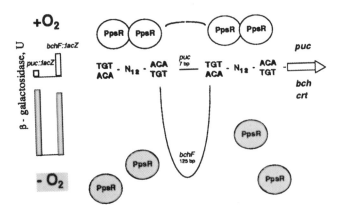

Figure 1. Model of PpsR mediated repression of PS genes.

PpsR repressor is to ensure the coordinate production of Bchl, Crt and proteins for the LH II complex (Fig. 1). Further, by controlling the amount of available Bchl, PpsR also affects the abundance of the reaction center and LH I complexes. Hence, PpsR represents a crucial regulatory factor governing development of PS by regulating PS gene expression. The PpsR regulon is solely comprised of PS genes. This juxtaposes PpsR to the other oxygen regulators of PS gene expression from *R. sphaeroides*, e.g. FnrL and the Prr two-component system, which have a substantially broader range of targets [2].

In this report we address questions of how PpsR mediated repression works, i.e., how the extent of repression is controlled, what signal PpsR senses and responds to, what regulatory cascades PpsR is involved in, and what structural features of PpsR are important for repression.

2. RESULTS AND DISCUSSION

2.1. *ppsR* Gene Expression

The extent of PpsR mediated repression depends on oxygen tension. In order to test whether derepression in the absence of oxygen is brought about by lowered expression of *ppsR*, or by a posttranslational modification(s) of PpsR, we assessed *ppsR::lacZ* expression under various oxygen tensions. We found that expression of a *ppsR::lacZ* fusion is largely independent of oxygen tension (Fig. 2). There was some increase in *ppsR::lacZ* expression under anaerobic as compared to aerobic conditions which implies that slightly more repressor is produced under anaerobic conditions and therefore PS gene expression should be lower, not higher as experimentally observed. Hence, changes in *ppsR* gene expression do not contribute to PS gene derepression under anaerobic conditions. We therefore speculate that posttranslational modification(s) of PpsR is a more plausible mechanism of regulating the extent of PS gene repression [6].

2.2. PpsR as a Redox Sensitive Transcriptional Repressor

2.2.1. PpsR and Oxygen Sensing. Because oxygen tension is the environmental factor which determines the extent of PpsR mediated repression, we wanted to know whether or not PpsR senses oxygen directly. We tested this hypothesis by monitoring the pheno-

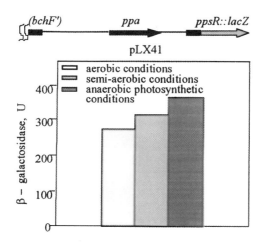

Figure 2. *ppsR* gene expression. Reproduced from [6] with modifications.

type of the wild type containing extra copies of the *ppsR* gene in *trans*, in strain 2.4.1(pPNs). The latter strain showed drastically decreased pigmentation compared to wild type when grown under low oxygen tension and it was unable to grow photosynthetically despite prolonged incubation. PS gene expression in this strain was also substantially lower when compared to wild type [5] and expression did not increase in response to oxygen removal. Spectral analysis of 2.4.1(pPNs) grown under fully anaerobic dark conditions revealed only traces of photocomplexes (Fig. 3). Therefore, the absence of oxygen alone is insufficient to relieve repression when PpsR is present in excess. These observations suggest (although do not prove) that PpsR is not inactivated by removal of oxygen and hence it does not sense oxygen directly; instead, it is likely that it requires other cellular factors to communicate "the absence of oxygen".

2.2.2. PpsR as a Light Intensity Dependent Regulator of PS Gene Expression under Anaerobic Conditions. We found that PpsR can act as a repressor under fully anaerobic conditions. This is best demonstrated by comparing the spectra of photosynthetically grown cells of the wild type and the PpsR null mutant, PPS1 (Fig. 4). When grown anaerobically under high light intensity, the latter strain produced approximately 4.5 times more LH II complex compared to the wild type strain. Hence, when light is abundant and the LH II complex is not required for growth, PpsR suppresses excessive production of the LH II complex in the wild type, while the PpsR null mutant produces maximum amounts of the LH II complex. Under low light intensity, when the LH II complex becomes critical for cell growth, the difference in the amount of this complex between the two strains diminishes indicating that wild type PpsR no longer represses LH II complex expression. These observations indicate that under fully anaerobic conditions, PpsR functions as a light intensity dependent regulator of PS gene expression [6].

Whether PpsR is capable of direct interactions with light is unclear. However, no biochemical evidence of such interactions exists for PpsR (unpublished data) or its *R. capsulatus* homolog, CrtJ [7]. We therefore propose that PpsR alone does not sense primary environmental stimuli, oxygen or light. Instead, it responds to an intrinsic downstream signal which reflects changes in both oxygen tension and light intensity. We envision PpsR as being able to either sense cellular redox poise directly or by interacting with a redox messenger.

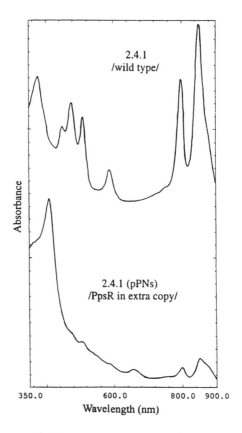

Figure 3. Spectra of cells grown under anaerobic-dark (+DMSO) conditions.

2.3. AppA as a Redox Dependent Modulator of PpsR Repressor Activity

We have identified one of the components of the putative signal transduction pathway communicating the cellular redox state to PpsR, namely AppA [8]. Initial biochemical characterization of AppA reveals that it contains a flavin cofactor and an iron-sulfur cluster and therefore is likely to be involved in a redox reaction(s) (unpublished data). Below we summarize the existing evidence linking AppA and PpsR in one regulatory cascade.

1. When cellular levels of AppA are increased, e.g., by placing the *appA* gene in extra copy, the genes belonging to the PpsR regulon are derepressed even under aerobic conditions. When AppA is absent, e.g., in an AppA null mutant, these same genes are repressed despite removal of oxygen resulting in the impairment of the AppA null mutant in transition from aerobic to anaerobic photosynthetic growth [8]. In fact, the phenotype of the AppA null mutant is similar to the phenotype of wild type moderately overexpressing the PpsR repressor.

2. AppA acts independently of other known transcriptional regulators of PS gene expression, e.g., FnrL and the Prr two-component system. Further, the impairment imposed by an *appA* null mutation cannot be overcome by activation of either the Prr system or FnrL [6]. In contrast, the AppA null phenotype depends on the presence of a functional PpsR. The impairment imposed by an *appA* null

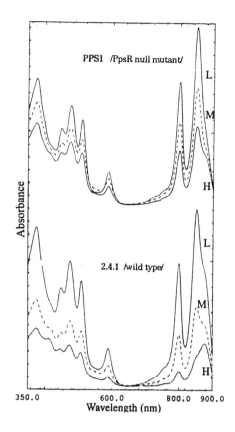

Figure 4. Spectra of cells grown photosynthetically at various light intensities, L, low; M, medium (dashed line), H, high. Reproduced from [6] with modifications.

mutation is overcome by inactivating *ppsR*, i.e., the AppA PpsR double mutant appears to be identical to the PpsR null mutant [6].

3. Introduction into the AppA null mutant of extra copies of the PpsR binding sites, TGT-N$_{12}$-ACA, partially suppresses the impairment of this mutant, presumably by titrating available PpsR repressor. This confirms that removal of PpsR is sufficient to overcome the AppA null phenotype (unpublished data).

4. Majority of suppressors of the AppA null mutant which regain the ability to transit from aerobic to anaerobic photosynthetic growth were found to be localized to the *ppsR* gene (see 2.4) [6].

5. Finally, there is evidence that AppA antagonizes PpsR mediated repression when both *appA* and *ppsR* are expressed in the heterologous host, *Paracoccus denitrificans* [6].

These observations strongly suggests that AppA acts upstream of PpsR in the same regulatory cascade. We believe that AppA modulates PpsR repressor activity by communicating a redox signal to PpsR. It is unclear whether AppA interacts with PpsR directly or through other mediators.

2.4. Isolation and Analysis of PpsR Mutations

Based on the selection for photosynthesis competent suppressors of the *appA* null mutation we isolated and analyzed over twenty mutations in the *ppsR* gene ([6] and un-

Table 1. Characterization of the PpsR mutations (PpsR*)

PpsR* in *trans* (Mutation)	Repression level achieved by introducing PpsR* into *P. denitrificans* (Miller U)		Competition with wild type PpsR (Miller U)
	puc::lacZ	*bchF::lacZ*	*bchF::lacZ + ppsR*
Wild type PpsR (None)	15	18	17
PPS52 (L27-P)	122	46	ND
PPS3 (L107-P)	19	17	ND
PPS18 (C251-R)	1202	185	30
PPS40 (C424-A)	2278	774	35
P[R] (Δ HTH motif)	2160	706	52
PPSΔ40 (Δ40 N-terminal residues)	1157	230	23
None (Vector)	2133	717	21

published data). The phenotypes of these supressors and repressor activities of the corresponding mutant PpsR proteins (PpsR*) vary significantly. Analysis of the selected mutations is presented below (Table 1).

2.4.1. Two Cys Residues Conserved in Both PpsR and CrtJ Are Crucial for Repressor Activity. Mutations which resulted in (a nearly) inactive PpsR, were identified as amino acid substitutions in two Cys residues present in PpsR, i.e. Cys(251)-Arg and Cys(424)-Ala. It is worth noting that these are the only Cys residues which are also conserved in CrtJ [3]. The Cys(251)-Arg mutation was found in three independent isolates. Further, two mutations in the immediate vicinity of Cys(251) were also identified supporting the importance of Cys(251). To assess the degree of repressor activity of the PpsR* proteins, we expressed these in *P. denitrificans* along with *lacZ* fusions to PpsR target genes. The repression imposed by the PPS18 mutant containing the Cys(251)-Arg substitution was only a fraction of the level of the wild type PpsR (Table 1).

The second mutation, Cys(424)-Ala, rendered PpsR* completely inactive. Cys(424) is positioned within the HTH motif of PpsR, therefore its substitution could potentially interfere with the secondary structure of this domain and therefore directly impair DNA binding. Because computer stimulated predictions revealed no change in the secondary structure as the result of this mutation, it is more likely that the nature of this residue, i.e., Cys, is important.

We believe that the highly reactive and redox sensitive thiol groups of the above discussed Cys residues are involved in keeping PpsR in a redox state appropriate for DNA binding. The nature of such involvement, e.g., metal or other ligand binding, oxidation or formation of disulfide bonds, is currently unknown.

2.4.2. Role of the PpsR N-Terminus. Several point mutations were grouped in the N-terminus of PpsR. Surprisingly, when expressed in *P. denitrificans*, some of the PpsR* forms corresponding to these mutations, e.g. PPS3, showed repressor activities equal to that of the wild type (Table 1). How could these alterations suppress the impairment of the *R. sphaeroides* AppA null mutant? We propose that the N-terminus is important for stabilization of the PpsR-DNA complex, e.g., through oligomerization of PpsR (see 2.4.3.). An impairment in PpsR* oligomerization could be sufficient for moderate derepression of the PS genes in *R. sphaeroides*, yet such impairment might be masked in *P. denitrificans* where PpsR* is present in excess due to multiple *ppsR** gene copies. The importance of the amino terminus is further stressed by the phenotype associated with the deletion derivative of PpsR, PPSΔ40. This mutant lacking as few as 40 amino acids at the N-terminus retained very little repressor activity (Table 1) and was impaired in oligomerization (see 2.4.3.).

2.4.3. Genetic Evidence for PpsR Oligomerization. The PpsR binding site, TGT-N$_{12}$-ACA, possesses dyad symmetry implying that PpsR binds as a dimer. Furthermore, in many instances, the PpsR binding sites are positioned in tandem. This provides the possibility of a cooperative interaction between PpsR dimers. To assess the importance and structural determinants for oligomerization, we introduced various PpsR* along with the wild type PpsR protein into *P. denitrificans* in order to allow formation of heterologous oligomers. Heterologous oligomers were anticipated to be less efficient in DNA binding when compared to wild type homogenous oligomers resulting in increased expression of a reporter gene.

The PpsR* deletion derivative which lacks the HTH motif, P[R], brought about a 2.5-fold derepression when introduced into *P. denitrificans* together with the wild type PpsR repressor (Table 1). This observation confirms the existence of protein-protein interactions, i.e. oligomerization. In contrast, introduction of PPSΔ40 had no effect on repression levels (Table 1), supporting our assumption that the N-terminal domain of PpsR could be involved in oligomerization.

3. CONCLUSIONS

- PpsR is involved in both oxygen- and light intensity dependent regulation of PS gene expression.
- Extent of PpsR mediated repression is controlled primarily by posttranslational modification(s) of PpsR.
- PpsR is unlikely to directly sense the environmental stimuli, instead it responds to changes in the cellular redox poise.
- Cysteine residues, conserved in both PpsR and CrtJ, are critical for repressor activity.
- PpsR acts as a dimer/multimer and the N-terminal domain could be important for oligomerization.
- A flavoprotein, AppA is involved in redox dependent modulation of PpsR repressor activity.

ACKNOWLEDGMENTS

We are grateful to Joy L. Marshall for isolation and spectral analysis of some of the AppA PpsR double mutants and to Adrian E. Simmons for assistance in using software. This work was supported by NIH grant GM15590 to S.K.

REFERENCES

1. Cohen-Bazire, G., Sistrom, W.R. and Stanier, R.Y. (1956) J. Cell. Comp. Physiol. 49, 25–68.
2. Zeilstra-Ryalls, J., Gomelsky, M., Eraso, J. M., Yeliseev, A. A., O'Gara, J. P. and Kaplan, S. (1997) Submitted.

3. Penfold, R. J. and Pemberton, J. M. (1994) J. Bacteriol. 176, 2869–2876.
4. Ponnampalam, S. N., Buggy, J. J. and Bauer, C. E. (1995) J. Bacteriol. 177, 2990–2997.
5. Gomelsky, M. and Kaplan, S. (1995) J. Bacteriol. 177, 1634–1637.
6. Gomelsky, M. and Kaplan, S. (1997) J. Bacteriol. 179, 128–134.
7. Ponnampalam, S. N. and Bauer, C. E. (1997) J. Biol. Chem. 272, 18391–18396.
8. Gomelsky, M. and Kaplan, S. (1995) J. Bacteriol. 177, 4609–4618.

16

A PUTATIVE TRANSLATIONAL REGULATOR OF PHOTOSYNTHESIS GENE EXPRESSION

Isolation and Characterisation of the *pif*C Gene Encoding Translation Initiation Factor 3, Which Is Required for Normal Photosynthetic Complex Formation in *Rhodobacter sphaeroides* NCIB 8253

Slobodan Babic,[1] C. Neil Hunter,[2] Nina Rakhlin,[3] Robert W. Simons,[3] and Mary K. Phillips-Jones[1]

[1]Department of Microbiology
University of Leeds
Leeds, LS2 9JT, United Kingdom
[2]Robert Hill Institute for Photosynthesis
Department of Molecular Biology and Biotechnology
University of Sheffield
Sheffield, S10 2UH, United Kingdom
[3]Department of Microbiology and Molecular Genetics
University of California
Los Angeles, California

1. INTRODUCTION

Regulation of expression of photosynthesis genes in purple bacteria is clearly complex and appears to be exerted at a number of different levels in the cell. This is not surprising, given the complex arrangement of the numerous genes involved, the importance of the stoichiometry of the final products incorporated into the membrane and the necessity of adjusting the levels of some of these products in response to changing external influences such as light and oxygen (Fig. 1). Much of the environmental regulation appears to be exerted at the level of photosynthesis gene transcription (for recent reviews see [1,2]; for a review of light regulators see [3]). Other studies have shown the importance of post-transcriptional mRNA degradatative processes and post-translational effects in determining final levels of membrane-bound complexes [4,5]. However, the process of transla-

The Phototrophic Prokaryotes, edited by Peschek *et al.*
Kluwer Academic / Plenum Publishers, New York, 1999.

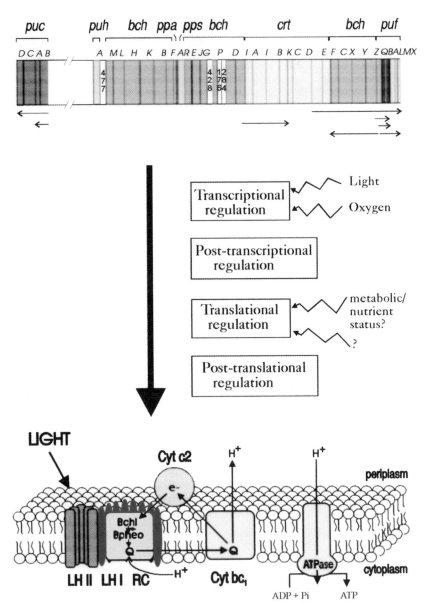

Figure 1. Possible and known levels at which photosynthesis gene expression may be regulated in *Rhodobacter sphaeroides*. The photosynthesis gene cluster is shown (top)[3], together with some of the gene products assembled in the membrane (bottom). Boxes show genetic levels at which regulation can/may occur. Redrawn and reproduced with permission [3].

tion has received relatively little attention, (though one recent study has now revealed that translation has an important specific role in controlling rates of *puf* gene expression [6]). This general lack of attention is surprising since, if gene expression in other prokaryotes is anything to go by, then one might expect translation to be an appropriate level at which to coordinate the synthesis of photosynthetic components with other metabolic processes within the cell, particularly protein synthesis.

Amongst the many factors required for translation is translation initiation factor 3 (IF3). This protein has been characterised in *Escherichia coli* [7], *Bacillus stearothermophilus* [8] and *Myxococcus xanthus* [9,10] and has been identified in a number of enteric bacteria [11]. It is known to have a number of important roles during protein synthesis including proofreading the AUG codon-initiator tRNA^{fMet} complex on the 30S ribosomal subunit during translation initiation [12,13], dissociation of the 70S complex into 50S and 30S subunits to ensure a plentiful supply of 30S subunits for new rounds of translation initiation [14], and it is also able to discriminate atypical start codons, repressing translation initiation from those codons [15,16]. It is largely the first two functions which are thought to make IF3 an essential protein in the cell. Recently, two studies showed that IF3 possesses an additional role; it appears to be involved in the selective expression of different genes or groups of genes in *E. coli* and in *M. xanthus*. In *E. coli*, IF3 appeared to repress expression of the *rec*J gene (thus affecting the recombination capacity of the cell) [17], whilst in *M. xanthus*, IF3 (known as Dsg in this organism) is required for fruiting body development and sporulation [10]. These effects, particularly those involving developmental genes, raised the possibility that IF3 may exert a selective effect on the expression of photosynthesis genes in photosynthetic bacteria. We have undertaken a study of the IF3 homologue in *R. sphaeroides* and present evidence that it appears to be required for normal photosynthetic complex formation in *Rhodobacter sphaeroides* under semi aerobic conditions [18].

2. ISOLATION OF THE *R. sphaeroides* IF3 PROTEIN AND THE ENCODING GENE

The protein was first identified during studies of another distinct protein, PORP, which behaves as a repressor of *puc* gene transcription in this organism [19]. It was partially purified by chromatography using S-Sepharose [18,19], and purified from PORP and other proteins present by band excision from SDS-polyacrylamide gels followed by elution of the protein using diffusion and electroelution methods. The purified protein was then digested with cyanogen bromide and the fragments run out on gels to enable N-terminal sequencing of each of the 5 peptide bands obtained. Using the N-terminal sequence data obtained, degenerate PCR primers were synthesised in order to obtain fragments of the encoding gene. One of the gene fragments obtained was then used to isolate a full-length clone from a pSUP202 gene library. Fig. 2a shows a subclone possessing a 4.4 kb *Bam*HI fragment carrying the IF3 gene, which has been designated *pifC* (photosynthesis-affecting initiation factor). The sequence of the gene is shown in Fig. 2b; note the possession of an unusual start codon (AUA) which is typical of all IF3 genes characterised to date.

3. FEATURES OF THE PifC PROTEIN SEQUENCE

Fig. 3 shows a comparison of the PifC sequence with the *E. coli* IF3 and *M. xanthus* Dsg sequences. PifC shows high sequence homology with both IF3 sequences; it shares 60% identity (79% similarity) with the *E. coli* IF3 sequence and 46% identity (66% similarity) with the *M. xanthus* Dsg sequence. The *E. coli* and *M. xanthus* proteins themselves share 49% identity (67% similarity). PifC also possesses all the conserved residues which have so far been attributed direct roles in IF3 function, including Tyr-110 and Lys-113 (corresponding to Tyr-107 and Lys-110 in *E. coli* implicated in ribosome binding ability of IF3), Phe-73 and Tyr-78 (corresponding to Tyr-70 and Tyr-75 in *E.coli* IF3 implicated in

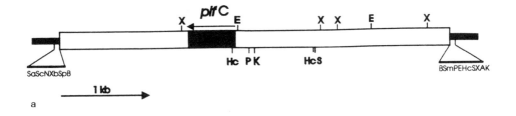

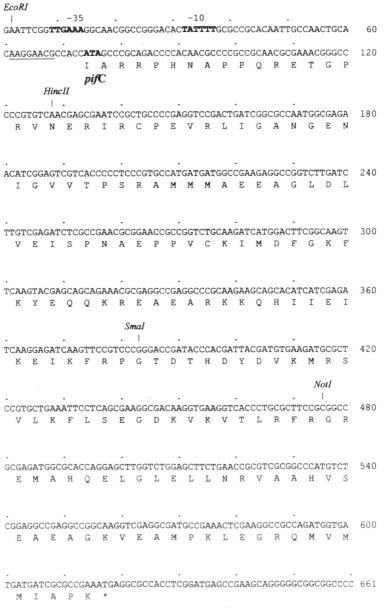

EcoRI
| . -35 . . -10 . .
GAATTCGG**TTGAAA**GGCAACGGCCGGGACAC**TATTTT**GCGCCGCACAATTGCCAACTGCA   60

CAAGGAACGCCACC**ATA**GCCCGCAGACCCCACAACGCCCCGCCGCAACGCGAAACGGGCC   120
        I A R R R P H N A P P Q R E T G P

*pif*C

HincII
| .
CCCGTGTCAACGAGCGAATCCGCTGCCCCGAGGTCCGACTGATCGGCGCCAATGGCGAGA   180
 R V N E R I R C P E V R L I G A N G E N

. . . . . .
ACATCGGAGTCGTCACCCCCTCCCGTGCCATGATGATGGCCGAAGAGGCCGGTCTTGATC   240
 I G V V T P S R A M M M A E E A G L D L

. . . . . .
TTGTCGAGATCTCGCCGAACGCGGAACCGCCGGTCTGCAAGATCATGGACTTCGGCAAGT   300
 V E I S P N A E P P V C K I M D F G K F

. . . . . .
TCAAGTACGAGCAGCAGAAACGCGAGGCCGAGGCCCGCAAGAAGCAGCACATCATCGAGA   360
 K Y E Q Q K R E A E A R K K Q H I I E I

SmaI
| .
TCAAGGAGATCAAGTTCCGTCCCGGGACCGATACCCACGATTACGATGTGAAGATGCGCT   420
 K E I K F R P G T D T H D Y D V K M R S

                                                       NotI
. . . . . |
CCGTGCTGAAATTCCTCAGCGAAGGCGACAAGGTGAAGGTCACCCTGCGCTTCCGCGGCC   480
 V L K F L S E G D K V K V T L R F R G R

. . . . . .
GCGAGATGGCGCACCAGGAGCTTGGTCTGGAGCTTCTGAACCGCGTCGCGGCCCATGTCT   540
 E M A H Q E L G L E L L N R V A A H V S

. . . . . .
CGGAGGCCGAGGCCGGCAAGGTCGAGGCGATGCCGAAACTCGAAGGCCGCCAGATGGTGA   600
 E A E A G K V E A M P K L E G R Q M V M

. . . . . .
b   TGATGATCGCGCCGAAATGAGGCGCCACCTCGGATGAGCCGAAGCAGGGGGCGGCGGCCCC   661
   M I A P K *

Figure 2. The *R.sphaeroides* *pif*C gene region (a) restriction map of a 4.4 kb *Bam*HI *pif*C fragment cloned in pBluescript-SK and (b) the nucleotide sequence of the *pif*C gene. Features shown in bold include a putative promoter consensus sequence of the *E. coli* σ[70]-type (base positions 9 and 32), and the ATA initiation codon (base 75–78). The putative ribosome binding site is underlined starting at position 62. Restriction enzyme sites are abbreviated: E, *Eco*RI; B, *Bam*HI; Hc, *Hinc*II; X, *Xho*I; S, *Sal*I; Sm, *Sma*I; P, *Pst*I; A, *Apa*I; K, *Kpn*I. Reproduced with permission [18].

```
                                                                      •
EcInfC:   MKGGKRVQTARPNRINGEIRAQEVRLTGLEGEQLGIVSLREALEKAEEAGVDLVEISPNAEP       62
                *  |*   ** ** ***** *   ** ||*|*|   *  ***** |****|********* *
RsPifC:   MARRPHNAPPQRETGPRVNERIRCPEVRLIGANGENIGVVTPSRAMMAEEAGLDLVEISPNAEP       65
               *      * ** *** ***** *  ***||*   *  *|| ** ***** *|*
MxDsg :   MIREQRSSRGGSRDQRTNRRIRAREVRVVGSDGSQLGVMPLEAALDRARTEGLDLVEISPMASP       64

                                                 •
EcInfC:   PVCRIMDYGKFLYEKSKSSKEQKKKQKVIQVKEIKFRPGTDEGDYQVKLRSLIRFLEEGDKAKIT     127
            ***|***|*** *   *  | *** *   |**|||*** ** *****|||||*** *|*
RsPifC:   PVCKIMDFGKFKYEQQKREAEARKQHIIEIKEIKFRPGTDHDYDVKMRSVLKFLSEGDKVKVT      130
            ***|***|*** |  |  ||** *   |-|**|** ** *||*|  * |*|-|-  * **
MxDsg:    PVCKIMDYGKFKYEEKKKASEAKRAQVTVLLKEVKLRPKTEEHDYEFKVRNTRRFIEDGNKAKVV     129

                                        •
EcInfC:   LRFRGREMAHQQIGMEVLNRVKDDLQEL-AV-VESFPTKIEGRQMIMVLAPKKKQ              180
          *****|****||* *|**** || *  *|  *  |* *|**|* | |***
RsPifC:   LRFRGREMAHQELGLELLNRVAAHVSEAEAGKVEAMP-KLEGRQVMMIAPK                  181
          | *****|| |  |  ** |- |   **  |  *  |*|||**
MxDsg:    IQFRGREITHREQGTAILDDVAKDLKDV-AV-VEQMP-RMEGRLMFMILAPTPKVAQKA....       185
```

Figure 3. Comparison of the *R. sphaeroides* PifC sequence (RsPifC) with the sequences of *E. coli* IF3 (EcIF3) and *M. xanthus* Dsg (MxDsg) proteins. Identical residues are shown using * whereas similar residues are marked by |. Residues shown in bold in the *E. coli* IF3 sequence are those which have been assigned specific roles associated with IF3 function or which are thought to be essential for full IF3 function; those shown in bold in the other two sequences are those which share identity or similarity with those functionally-important residues. Underlined residues the RsPifC sequence are those also identified during the N-terminal sequencing of peptide fragments. Reproduced with permission [18].

proofreading and Tyr-75 also implicated in start codon discrimination), Val-18 and Val-57 (corresponding to hydrophobic Ile-15 and Val-54 required for contacting the C-terminal helix of the N-terminal domain in IF3) and finally, residues 79–92 (which align with the basic linker of IF3 which separate the two domains of the protein).

In view of the high level of overall similarity of the PifC sequence with the *E. coli* and *M. xanthus* IF3 sequences, together with the presence of an atypical AUA start codon for the *pif*C gene, we conclude that PifC is most likely to be the IF3 homologue of *R.sphaeroides*. Further evidence for this is provided by our finding that PifC, in common with *E. coli* IF3 (encoded by the *inf*C gene), possesses a discriminatory function towards atypical start codons. This was shown by complementation studies using an *E. coli* *inf*C362 strain which lacks this discriminatory function. In these studies *inf*C362 mutant strains, which carry a chromosomal *lac*Z reporter gene fused to the 5′-end of an *inf*C gene carrying atypical (CUG) or typical (AUG) start codons, showed that the presence of the *pif*C gene resulted in a partial complementation of IF3 discriminatory function [16,18].

4. PARTIAL INSERTIONAL INACTIVATION OF *pif*C GENE AND THE EFFECT ON PHOTOSYNTHESIS GENE EXPRESSION

A disruption of the *pif*C gene was made by inserting a kanamycin resistance cassette at the *Hinc*II site located at base position 54 of the *pif*C gene (Fig. 2). Following introduction of the resulting construct into *R.sphaeroides* wildtype, a number of transconjugants (in which double cross over events at the chromosome were subsequently demonstrated), were obtained. In semi-aerobic liquid culture, these mutants were visibly less pigmented (though this was less obvious on solid plates). Fig. 4 shows the levels of spectral complexes in mutant and wild type cells grown semi-aerobically or anaerobically in the light. Mutant cells grown semi aerobically possessed reduced overall levels (Fig. 4a); total bacteriochlorophyll levels were also measured and were 66% of wild type [18]. When mutant and wild type strains were grown anaerobically in the light (photosynthetically), there was no obvious difference in complex levels observed (Fig. 4b).

Although the effects of the *pif*C knock-out may appear relatively small, it is now clear that the *pif*C mutant still produces a PifC protein and this protein is likely to retain partial IF3 function. This was demonstrated by complementation studies using the *E. coli inf*C362 strain described above. The plasmid used to construct the *R. sphaeroides* knock-out mutant, which possesses a kanamycin resistance cassette inserted in *pif*C, was still able to complement the *inf*C362 mutation as shown by our *lac*Z reporter measurements, though not to the same level observed with the wild type *pif*C gene (see [18] for full details of how this was carried out). In other words, the gene has only been partially knocked out. Presumably the effect of the mutation might be expected to be even greater if PifC activity could be lowered still further. It is interesting to note that total knock out of *inf*C in *E.coli* results in a non viable phenotype. It is tempting to speculate that this will also be the case for a *pif*C total knock out.

Recently, we produced a polyclonal PifC antibody using pure overexpressed His-tagged PifC protein [18]. The antibody was used in Western blotting experiments of extracts of wild type and mutant cells. Fig. 5 shows that a PifC protein is indeed still produced in the *pif*C mutant as well as in the wild type strain, that it is produced in almost the same amounts and that it is almost full length. There are two possible explanations which account for all the observations noted above and for the observed partial activity in the mutant construct. Firstly, there may be more than one copy of this presumably essential gene, and so far just one gene has been insertionally inactivated (though our Southern

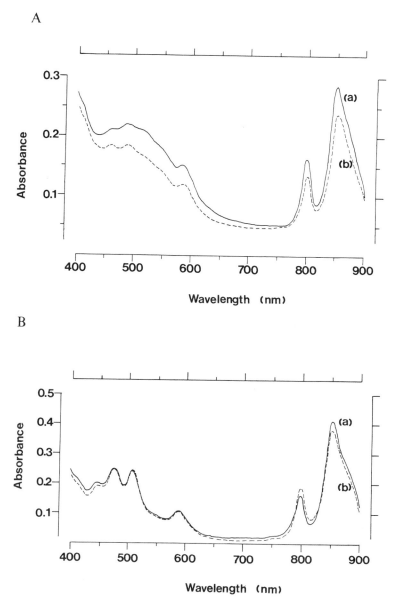

Figure 4. Absorbance spectra of *R. sphaeroides* whole cells. Cells were grown (A) semi-aerobically in the dark at 34°C or (B) anaerobically in the light at 30°C. (a) room temperature spectra of wild type cells; (b) room temperature spectra of the *pifC* mutant cells. All cultures were grown as described in [18]. Reproduced with permission [18].

data to date suggest the presence of only one gene). Secondly, it is known that the use of the Tn903 kanamycin resistance cassette can lead to inframe fusions with a small section of the 5′ end of a gene present in the cassette, leading to expression of a fusion protein. This possibility could lead to a partially active PifC protein being produced, since the insertion site for the cassette was at base position 54 in the gene (removing only 17 residues from the N-terminus of *pifC*).

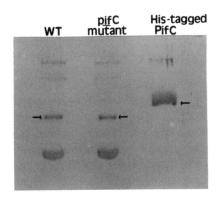

Figure 5. Western blot of wild type and *pif*C mutant cell extracts of *R. sphaeroides* using a polyclonal PifC antibody. Cells were grown semi-aerobically at 34°C [18]. Wild type (25 μg total protein), *pif*C mutant (29 μg total protein) extracts or pure His-tagged PifC protein (3.75 μg) are shown. The signals specific to PifC are shown by the arrows; their masses were shown to be: wild type PifC, 23 kDa; mutant PifC, 21.5 kDa and aberrant His$_6$-tagged PifC protein, 26.6 kDa. Reproduced with permission [18].

The mechanism(s) by which PifC might exert its selective effect on photosynthetic complex levels in semi aerobic cells of *R. sphaeroides* is currently unknown. In the case of the selective effect of *M. xanthus* Dsg towards sporulation and fruiting body development genes, the protein possesses an extra C-terminal portion (which is not present in either *E. coli* IF3 or PifC) which has been shown to be required for the selective effect towards these genes [9,10]. In the case of the selective effect of IF3 towards *rec*J in *E. coli*, it is currently believed that the effect is exerted through the discriminatory function of IF3 towards atypical start codons though *rec*J itself possesses a typical GUG start codon [17]. None of the photosynthesis genes characterised to date possess atypical start codons either; thus, it would appear that the mechanism underlying the PifC selective effect may be similar to that observed for IF3 towards *rec*J. Thus, in common with *rec*J selective expression in *E. coli*, the precise mechanism by which a discriminatory selective function might operate is not yet clear.

Finally, studies with the partial knock out mutant described above have indicated that PifC exerts an effect on complex levels in semi aerobic cells but not significantly in anaerobic/light-grown cells, suggesting some environmental regulatory effect. However, since our mutant is only partial, it is difficult to draw any firm conclusions about this aspect; further studies in this area are required to determine whether light or anaerobiosis (or both) play a role or whether the selective effect of PifC occurs as a response to less direct intracellular signals.

REFERENCES

1. Bauer, C.E. (1995) Chapter 58 in: Anoxygenic Photosynthetic Bacteria. Ed. Blankenship, R.E., Madigan, M.L.T. & Bauer, C.E. pp. 1221–1234. Kluwer Academic Publishers, Dordrecht, The Netherlands
2. Bauer, C.E. & Bird, T.H. (1996) Cell 85, 5–8
3. Phillips-Jones, M.K. In: Microbial Responses to Light and Time. SGM Symposium Series No 56. Cambridge University Press, Cambridge, UK (*in press*).
4. Klug, G. (1993) Mol. Microbiol. 9, 1–7
5. Klug, G. (1995) Chapter 59 in: Anoxygenic Photosynthetic Bacteria. Ed. Blankenship, R.E., Madigan, M.L.T. & Bauer, C.E. pp. 1235–1244. Kluwer Academic Publishers, Dordrecht, The Netherlands.
6. Gong, L. & Kaplan, S. (1996) Microbiology (UK) 142, 2057–2069
7. Noller, H.F. (1991) Annu. Rev. Biochem 60, 191–227
8. Pon, C.L., Brombach, M., Thamm, S. & Gualerzi, C.O. (1989) Mol. Gen. Genet. 218, 355–357
9. Cheng, Y.L., Kalman, L.V. & Kaiser, D. (1994) J. Bacteriol. 176, 1427–1433
10. Kalman, L.V., Cheng, Y.L. & Kaiser, D. (1994) J. Bacteriol. 176, 1434–1442
11. Liveris, D., Schwartz, J.J., Geertman, R. & Schwartz, I. (1993) FEMS Microbiol. Lett. 112, 211–216

12. Gualerzi, C.O. 7 Pon, C.L. (1990) Biochemistry 29, 5881–5889
13. Hartz, D., McPheeters, D.S. & Gold, L. (1990) In: The ribosome: structure, function and evolution. Ed. Hill, W.E., Dahlberg, A., Garrett, R.A., Moore, P.B., Schlessinger, D. & Warner, J.R. pp. 275–280. American Society for Microbiology.
14. Godefroy-Colburn, T., Wolfe, A.D., Dondon, J., Grunberg-Manago, M., Dessen, P. & Pantalini, D. (1975) J. Mol. Biol. 94, 461–478
15. Sacerdot, C., Chiaruttini, C., Engst, K., Graffe, M., Milet, M., Mathy, N., Dondon, J. & Springer, M. (1996) Mol. Microbiol. 21, 331–346
16. Sussman, J.K., Simons, E.L. & Simons, R.W. (1996) Mol. Microbiol. 21, 347–360
17. Haggerty, T.J. & Lovett, S.T. (1993) J. Bacteriol. 175, 6118–6125
18. Babic, S., Hunter, C.N., Rakhlin, N., Simons, R.W. & Phillips-Jones, M.K. Eur. J. Biochem. (*in press*).
19. McGlynn, P. & Hunter, C.N. (1992) J. Biol. Chem. 267, 11098–11103

A MUTATION THAT AFFECTS ISOPRENOID BIOSYNTHESIS RESULTS IN ALTERED EXPRESSION OF PHOTOSYNTHESIS GENES AND SYNTHESIS OF THE PHOTOSYNTHETIC APPARATUS IN *RHODOBACTER CAPSULATUS*

David Nickens, Joseph J. Buggy, and Carl E. Bauer*

Department of Biology
Indiana University
Bloomington, Indiana 47405

1. INTRODUCTION

Observations of the repressing effects of high light intensity on synthesis of the purple bacterial photosystem were first reported in 1957 by Cohen-Bazire *et al.* [1]. Their study demonstrated that shifting a growing culture from low to high light intensity resulted in an abrupt decrease in synthesis of the bacterial photosystem. Subsequent physiological studies demonstrated that photosynthetic bacteria respond to alterations in light intensity by adjusting the amounts of reaction centers, light harvesting-I, and light harvesting-II complexes located in an intracytoplasmic membrane[2-4]. Studies of the regulation of photopigment synthesis have supported earlier observations concerning light and oxygen control of biosynthesis of reaction centers, light harvesting-I, and light harvesting-II complexes[5-7].

Lien and Gest[8] observed that the amount of light harvesting-II complex in *R. capsulatus* was regulated to a larger extent then that of the reaction center:light harvesting-I complexes which appeared to remain in a fixed 1:1 stoichiometry[8-10]. They proposed that synthesis of light harvesting-II is dependent upon the "energy state" of the cell[11] such that, as light becomes limiting, the level of light harvesting-II increases to funnel more light energy to the reaction center complexes. This effectively maintains ATP production and growth rate under light limiting conditions. When discussing the effect of light intensity

---

* Corresponding author: Dr. Carl E. Bauer, Department of Biology, Indiana University, Jordan Hall, Bloomington, Indiana 47405. Phone: (812)855-6595; Fax: (812)855-6705; E-mail; cbauer@bio.indiana.edu.

The Phototrophic Prokaryotes, edited by Peschek *et al.*
Kluwer Academic / Plenum Publishers, New York, 1999.

on photosystem synthesis the photosynthetic apparatus (PA) is often used to refer to the entire unit involved in light capture surrounding a reaction center. The surface area of each PA increases as light intensity is reduced increasing the amount of light harvesting-II complex relative to the reaction center:light harvesting-I core.[10,12,13] Cells respond to diminishing light intensity by increasing the number of PAs per cell and increasing the light harvesting-II/light harvesting-I:reaction center ratio[9,10]. High-light grown cells compensate for a reduced number of reaction centers and for reduced size of each PA, by increasing the photophosphorylation rate per reaction center under saturating light conditions[10].

The mechanism of regulating synthesis of the PA by light is quite complex and involves transcriptional as well as posttranscriptional processes. High light intensity has been shown to cause a reduction in the steady state levels of many mRNAs involved in synthesis of the bacterial photosystem.[14–16] Changes in steady state levels of mRNA can be a result of changes in mRNA transcription or stability, but the molecular mechanism by which light intensity regulates photosynthetic gene expression has remained elusive. A DNA binding protein, HvrA, has been shown to act as a light controlled regulator of the *puf* and *puh* operons in *R. capsulatus* that code for structural polypeptides of the light harvesting-I and reaction center complexes[17]. Transcription from the *puc* promoter, which codes for structural polypeptides of the light harvesting-II complex as well as the bacteriochlorophyll and carotenoid promoters are unaffected by mutation of HvrA[17,18]. Transcription from the *puc* promoter is, however, highly regulated by light indicating that additional uncharacterized transcription factors are involved in light regulation of light harvesting II proteins.

A previous study has established that ORF176 (*idi*) from the photosynthetic gene cluster of *R. capsulatus,* codes for the enzyme isopentenyl diphosphate isomerase[19]. This enzyme is involved in early stages of isoprenoid metabolism and is thought to be required for synthesis of such compounds as geranylgeranyl diphosphate which is needed for carotenoid and bacteriochlorophyll biosynthesis. However, interposon mutagenesis at the 95th codon of *idi*, referred to as *hvr*C, had no effect on photosynthetic growth rate under high-light intensities indicating that a redundant pathway for isoprenoid biosynthesis must also occur in this species[20]. In this report, we demonstrate that disruption of *idi* results in a severe impairment in dim-light photosynthetic growth capability. This appears to be the consequence of the inability of *idi* disrupted cells to increase biosynthesis of bacteriochlorophyll or other photopigments upon a reduction of light intensity presumably as a result of limiting levels of isoprenoid precursors. Curiously, gene fusions of *lacZ* to bacteriochlorophyll *(bch)*, carotenoid *(crt), puf, puh* and *puc* operon promoter regions all exhibited significantly elevated levels of transcription activity in DB176 under dim-light growth conditions. We speculate that increased dim-light transcription from photopigment genes is caused by a drop in the energy status of the cells due to the inability to increase photopigment levels as light intensity falls. This implies that a system exists that measures the energy status (or energy charge) of the cell generating a signal that subsequently regulates photosynthesis gene expression.

2. MATERIALS AND METHODS

2.1. Bacterial Strains and Cultures

R. capsulatus strains SB1003 (wild-type) and DB176[20] and were routinely grown at 34°C in peptone-yeast extract (PY) broth or agar medium[21]. Spectinomycin was used at a final concentration of 10 µg/ml for maintenance of plasmids. Rifampicin was used as a counter selection for transconjugants at a concentration of 100 µg/ml. Aerobic/dark growth was achieved by growing 20 ml cultures in 250 ml Erlenmeyer flasks with shaking (300 rpm). Anaerobic conditions were achieved by growing cultures in completely filled screw cap tubes of 17-ml capacity. Photosynthetic cultures were grown anaerobically under high-light conditions (high light=7000 lux) or low-light conditions (low light=200 lux) using banks of incandescent lumiline (Sylvania 30W) lamps. Cultures of *R. capsulatus* were harvested at a cell density of ca. 1.5×10^8 to prevent oxygen depletion in aerobic cultures or self-shading in photosynthetic cultures. *E. coli* cultures were routinely grown aerobically at 37°C in Luria broth (LB) or agar medium[22]. Spectinomycin was added at 100 µg/ml, while ampicillin was added at 200 µg/ml for maintenance of plasmids.

2.2. Growth Yield vs. Light Intensity Measurements and Absorption Spectra

Growth curves were prepared with the wild type strain SB1003 and the mutant *hvr*C under photosynthetic growth conditions as described above. Replicate cultures were grown at light intensities of 200, 600, 1000, 2000, 4000, or 7000 lux and cell density was plotted against light intensity. Culture densities were measured with a Klett-Summerson photometer fitted with a no. 66 filter. Absorption spectra of photosynthetic pigments in cell free extracts prepared by sonication were determined as previously described[23]. Extracts were prepared with cells grown photosynthetically in PY medium at 80, 250, and 6000 lux.

2.3. DNA Manipulations and Promoter Probe Analysis

Standard recombinant DNA techniques were performed using *E. coli* strain DH5α as the host[22]. Restriction enzymes and T4 DNA ligase were purchased from New England BioLabs, Inc., Beverly, Mass. and were used according to the manufacturers recommendations. The promoter probe vector pZM400, which was constructed specifically for use in *R. capsulatus* was used to make transcriptional fusions of the *puc*URS to a promoterless *lacZ* gene[24]. Mobilization of plasmid pDN13S (*puclacZ* fusion[24]) from *E. coli* DH5α to *R. capsulatus* strains was accomplished by triparental mating with *E. coli* strain HB101 harboring pRK2013, a helper plasmid for conjugation of plasmids with RK2 origins of replication[24,26]. Transconjugants were purified by repeated streaking of plates of RCV medium[21] with 10 µg/ml spectinomycin. Construction of plasmids encoding translational fusions of the *bchC*, *pucB*, *puf*, and *puh* promoter regions to *lacZ* were previously reported by Buggy et al. [18].

2.4. β-Galactosidase Assays

Cultures used for β-galactosidase assays were grown aerobically in the dark, or photosynthetically under high light or low light conditions as described above. Cells were harvested at the culture densities described above and cell-free extracts were prepared and assayed for β-galactosidase activity as previously described[27]. The protein concentration of each extract was determined by Bradford's assay[28]. Final results are reported as the amount of *o*-nitrophenyl-β-galactoside hydrolyzed per minute per milligram of total protein (µmoles/min/mg protein). Reported β-galactosidase values all had standard deviations of ≤ 6.2%.

3. RESULTS

3.1. *idi* Is Essential for Normal Light-Regulated Photopigment Biosynthesis and Photosynthetic Growth Capabilities

Shown in Figure 1 are the results of spectral analysis of *R. capsulatus* cells grown under high light (5,000 lux) and dim-light (500 lux) growth conditions. As shown in panel A, wild type cells exhibit a characteristic 1.5-fold reduction in photopigment levels as

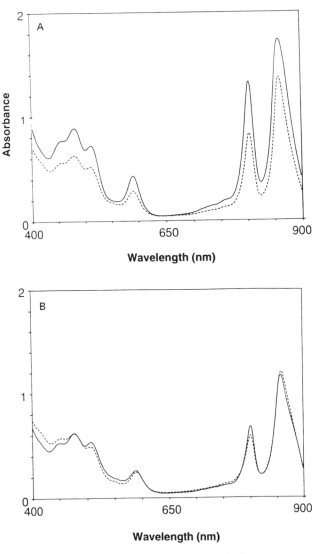

Figure 1. *In vivo* absorbance spectrum of SB1003 (w.t.) and DB176 (*idi*). Extracts were prepared and analyzed as described in the Materials and Methods section. Panel A shows the absorbance spectra of cell free extracts from SB1003 grown under high-light (5000 lux, dotted line) and low light (500 lux, solid line) photosynthetic growth conditions. Panel B represents the same experiment performed with cell free extracts from DB176 grown under identical conditions.

light intensity is increased. In comparison, DB176 cells, which contains a Km^R insertion in the 95th codon of *ipi*, exhibits constitutively low amounts of photopigment biosynthesis (panel B).

The effect of an *idi* disruption on photosynthetic growth capabilities is shown in Figure 2. Under high light intensity (6,000 lux) DB176 show only a slight impairment in growth. However, there is a much more pronounced reduction when that cells are grown under moderate (250 lux) and low (80 lux) light intensity (panels B and C, respectively).

A plot of relative growth rates of wild type versus DB176 cells under varying light intensities is shown in Figure 3. Wild type *R. capsulatus* grows with similar doubling times over the range of 1000 to 7000 lux with a steady decline in growth rate as light intensity falls below 1000 lux. 1000 lux thus appears to represent a point where wild type cells can no longer increase levels of photopigments necessary to maintain a yield of ATP needed to maintain an optimum growth rate. The growth rate of DB176 during high light conditions (7000 lux) was virtually identical to that observed with SB1003, but as light intensity decreases DB176 is unable to maintain its optimum growth rate. Figure 3 graphs the doubling times of DB176 and SB1003 over a range of 80 to 7000 lux showing the growth yield per unit of light. The area between the SB1003 curve and the DB176 curve represents the difference in growth yield between SB1003 and DB176 at each light intensity.

3.2. Photosynthesis Gene Expression Is Constitutively High in an *idi* Mutant

We also analyzed the affect of an *idi* disruption on photosynthesis gene expression using a variety of reporter plasmids that we previously constructed. As shown in the bar graphs in Figure 4A, B and C, using translational fusions we observe significantly elevated expression of the *puf* and *puh* operons that encode the light harvesting-I and reaction center structural proteins as well as with *bchC* which codes for an enzyme involved in bacteriochlorophyll biosynthesis. One interesting observation is that even though expression of the *puf*, *puh* and *bchC* operons are elevated they still retain a measurable amount of regulation in response to alterations in light intensity. In contrast, a translational fusion that measures *puc* expression (*pucB::lacZ*) exhibits both elevated and deregulated expression in regards to light intensity (Fig. 4D). The deregulation of *puc* expression appears to be caused by a post-transcriptional effect since a *puc* transcriptional expression vector *pucURS lacZ*, which fuses the *puc*URS to *lacZ* 57 bp upstream of the *pucB* translational start site, exhibits elevated as well as light-regulated control in DB176 (Fig. 4E).

4. DISCUSSION

It was reported that *idi* encodes a novel truncated form of isopentenyl diphosphate isomerase (IPPase) which catalyzes an early step in isoprenoid biosynthesis by converting isopentenyl diphosphate (IPP) to dimethylallyl diphosphate (DMAPP)[19]. IPP and DMAPP function as precursors for synthesis of geranylgeranyl diphosphate which is utilized as a substrate for synthesis of bacteriochlorophyll[19,29]. Additional isoprenoid products such as carotenoids, ubiquinones, dolichols and hopanoids require IPP and DMAPP as precursers so it is surprising that disruption of *idi* is not lethal. Southern blot analysis has indicated that a second copy of *idi* may exist, which would account for the nonessentiality of this gene[19]. The ability of DB176 to exhibit normal anaerobic high-light and aerobic dark

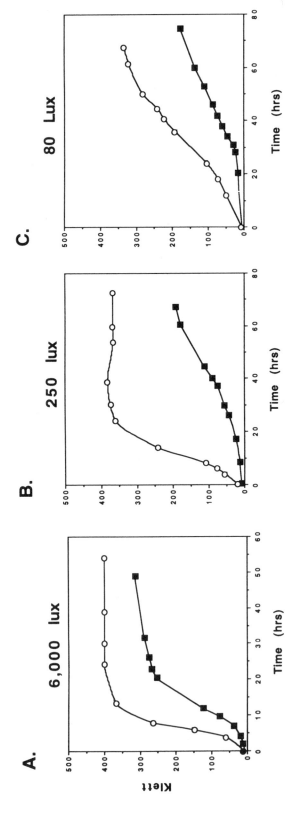

Figure 2. Growth curves of SB1003 (O) and DB176 (■) grown photosynthetically at 6000 lux (Panel A), 250 lux (Panel B), 80 lux (Panel C).

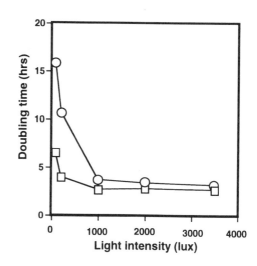

Figure 3. Doubling times of SB1003 (□) and DB176 (O) grown photosynthetically at 80 lux, 200 lux, 1000 lux, 2000 lux, and 3500 lux.

growth rates demonstrates that all components of the photosystem and respiratory chain are functioning in DB176. This result also supports the conclusion that a second IPPase must be present in this species[19]. The phenotype of the *idi* disrupted strain, which is a reduced capability to synthesize high levels of bacteriochlorophyll and/or carotenoids, coupled with the genes placement in the photosynthesis gene cluster, suggests that *idi* may function as an "accessory" enzyme that provides additional IPPase activity during conditions of high level photopigment biosynthesis. Failure to increase photopigment synthesis as light intensity decreases explains why DB176 is unable to maintain wild type low-light growth rates.

Since DB176 synthesizes constitutively low amounts of photopigments, it was surprising that expression of all photosystem genes tested exhibit a 2 to 3-fold elevation of expression. Given that *idi* codes for an enzyme in isoprenoid metabolism, it is apparent that the observed effects on transcription is an indirect consequence of limiting isoprenoid biosynthesis. Another notable observation is that DB176 cells are defective in high-light repression of *puc* operon expression when assayed using translational expression vectors, but still capable of exhibiting high-light repression of *puc* operon expression when assayed using transcriptional fusions. These observations indicate that the *puc* operon is differentially regulated from other photosynthesis genes in regards to high-light repression and further supports the conclusion that a posttranscriptional mechanism is involved in regulation of light harvesting-II complex formation[30,31].

One possible explanation for the observed transcription effects is that the cell has a mechanism for controlling photosynthesis gene expression in response to the energy charge of the cells. As light energy becomes limiting the resulting decrease in energy charge signals the cell to increase production of the photosynthetic unit by increasing transcription of photosynthesis genes. Since DB176 is unable to elevate photopigment synthesis these cells would exhibit a more severe reduction in energy charge than would wild type cells which are capable of altering photopigment levels. Consequently expression would be higher in DB176 than in wild type cells. A more detailed understanding of this process will have to await the isolation and characterization of additional transcription factors that are involved in controlling photosynthetic gene expression in response to alterations in light intensity.

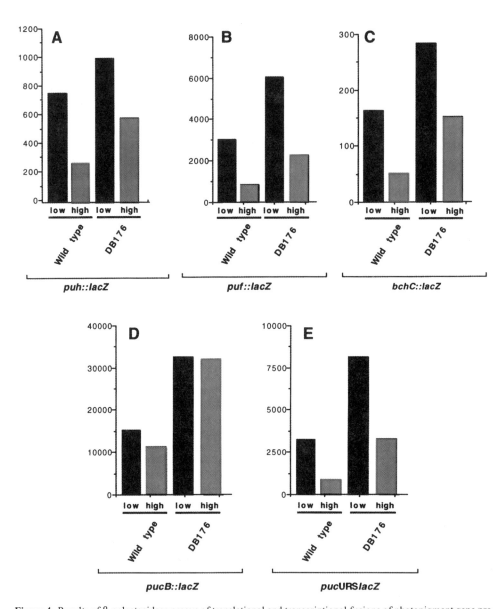

Figure 4. Results of β-galactosidase assays of translational and transcriptional fusions of photopigment gene promoter regions to *lacZ*. SB1003 (w.t.) and DB176 (*idi*) were grown photosynthetically at 7000 lux (high-light) or 200 lux (low-light). A) *puh::lacZ* translational fusion, B) *puf::lacZ* translational fusion, C) *bchC::lacZ* translational fusion, D) *pucB::lacZ* translational fusion, and E) *puc*URS:*lacZ* transcriptional fusion.

ACKNOWLEDGMENTS

This study was supported by National Institutes of Health grants GM 539040 and GM 00618 to CEB.

REFERENCES

1. Cohen-Bazire, G., Sistrom, W. R. and Stanier R. Y. (1957) J. Cell. Comp. Physiol. 49, 25–68.
2. Aagard, J. and Sistrom, W. R. (1972) Photochem. Photobiol. 15, 209–225.
3. Drews, G. (1985) Microbiol. Rev. 49:59–70.
4. Oberlé, B., Tichy, H. V. and Drews, G. (1990) in: Molecular Biology of Membrane-Bound Complexes in Phototrophic Bacteria. (Drews, G. and Dawes E. A. eds.), pp. 77–84. Plenum Press, New York.
5. Sganga, M. W. and Bauer, C. E. (1992) Cell 68, 945–954.
6. Bauer, C. E., Buggy, J. J., Yang, Z. and Marrs, B. L. (1993) Trends in Genet. 9, 56–60.
7. Bauer, C. E. (1995), in: Anoxygenic Photosynthetic Bacteria. (Blankenship, R. E., Madigan, M. T. and Bauer C. E. eds.). pp. 1221–1234, Kluwer Academic Publishers, Dordrecht, Boston, London.
8. Lien, S., Gest, H. and San Pietro, A. (1973) Bioenergetics 4, 423–434.
9. Reidl, H., Golecki, J. R. and Drews, G. (1983) Biophys. Acta. 725, 455–463.
10. Reidl, H., Golecki, J. R. and Drews, G. (1985) Biochem. Biophys. Acta. 808, 328–333.
11. Zilinsky, J. W., Sojka, G. A. and Gest, H. (1971) Biochem. Biophys. Res. Commun. 42, 955–1915.
12. Drews, G. (1986) Microbial. Rev. 49, 59–70.
13. Zuber, H. and Cogdell, R. J. (1995) in: Anoxygenic Photosynthetic Bacteria. (Blankenship, R. E., Madigan, M. T. and Bauer C. E., eds.). pp. 315–348. Kluwer Academic Publishers, Dordrecht, Boston, London.
14. Clark, W. G., Davidson, E. and Marrs, B. L. (1984) J. Bacteriol. 157, 945–948.
15. Klug, G., Kaufmann, N. and Drews, G. (1985) Proc. Natl. Acad. Sci. U.S.A. 82, 6485–6489.
16. Zhu, Y. S. and Hearst, J. E. (1986) Proc. Natl. Acad. Sci. U.S.A. 83, 7613–7617.
17. Kuadio, J. L. (1997) Ph.D Dissertation. Indiana Unviversity.
18. Buggy, J. J., Sganga, M. W. and Bauer, C. E. (1994) J. Bacteriol. 176, 6936–6943.
19. Hahn, F. M., Baker, J. A. and Poulter, C. D. (1996) J. Bacteriol. 178, 619–624.
20. Bollivar, D. W., Suzuki, J. Y., Beatty, J. T., Dobrowolski, J. M. and Bauer C. E. (1994) J. Mol. Biol. 237, 622–640.
21. Weaver, P. F., Wall, J. D. and Gest H. (1975) Arch. Microbiol. 105, 207–216.
22. Sambrook, J., Fritsch, E. F. and Maniatis, T. (1989) Cold Spring Harbor Laboratory, Cold Spring Harbor, N.Y.
23. Sojka, G. A., Freeze, H. H. and Gest H., (1970) Arch. Biochem. Biophys 136, 57–580.
24. Ma, D., Cook, D. N., O'Brien, D. A., and Hearst J. E. (1993) J. Bacteriol. 175, 2037–2045.
25. Nickens D. G.. (1997) Ph.D Dissertation. Indiana Unviversity.
26. Ditta , G., Schmidhauser, T., Yakobsen, E., Lu, P., Liang, X. Y., Finlay, D. R., Guiney, D. and Helinsky, D. R. (1985) Plasmid 13, 149–153.
27. Young, D. A., Bauer, C. E., Williams, J. C., and Marrs, B. L. (1989) Mol. Gen. Genet. 218, 1–12.
28. Bradford, M. M. (1976) Anal. Biochem. 72, 248–254.
29. Bollivar, D. W., Wang, S., Allen, J. P. and Bauer, C. E. (1994) Biochemistry. 33, 12763–12768.
30. Zucconi, A. P. and Beatty, J. T. (1988) J. Bacteriol. 170, 877–882.
31. Yurkova, N. and Beatty, J. T. (1996) FEMS Microbiol. Lett. 145, 221–225.

18

THE POLYPEPTIDES *puc*D AND *puc*E STABILIZE THE LHII (B800–850) LIGHT-HARVESTING COMPLEX OF *RHODOBACTER CAPSULATUS* 37B4 AND SUPPORT AN EFFECTIVE ASSEMBLY

Friedemann Weber,* Christiane Kortlüke, and Gerhart Drews

Institut für Biologie 2, Mikrobiologie
Albert-Ludwigs-Universität
Schaenzlestr. 1, D-79104 Freiburg, Germany

1. INTRODUCTION

The *puc* operon of *Rhodobacter capsulatus* consists of the genes *pucBACDE*. The genes *pucB* and *pucA* encode the pigment-binding polypeptides LHIIβ and LHIIα and *pucC* encodes a regulatory protein, essential for the expression of the LHII complex. The function of the gene product PucD is unknown; *pucE* encodes the γ polypeptide of the LHII peripheral light-harvesting complex. The LHII complex is formed in variable amounts mainly under the control of the light intensity (1–3). PucC is a membrane-bound protein which spans the membrane presumably 12 times. The C terminal transmembrane segments are important for the function (4). Mutation of *pucC* by transposon insertion or deletion results in a strong reduction of *puc* mRNA (2,5) and a complete inhibition of the formation of the LHII complex (1–4). The gene products of *pucDE* seem to stabilize the LHII complex (2) and their absence inhibits growth (3). For the investigation of the *puc* operon the mutant strain NK3 of *Rb. capsulatus* was used up to now. This strain contains a Tn5 insertion in *pucC* (1,2,5) which did not suppress completely the expression of the *puc* operon but has a polar effect on transcription. Therefore, strains were constructed which have chromosomal deletions in *pucBACDE, pucDE* and *pucD,* respectively. The expression of *puc* genes in these and plasmid-reconstituted strains was investigated.

* Present address: Institut für Medizinische Mikrobiologie und Hygiene, Abt. Virologie, D-79008 Freiburg.

The Phototrophic Prokaryotes, edited by Peschek *et al.*
Kluwer Academic / Plenum Publishers, New York, 1999.

2. MATERIALS AND METHODS

2.1. Bacterial Strains and Growth Conditions

Rb. capsulatus 37b4 (DSM 983), the interposon mutants Δ*pucBACDE,* Δ*pucDE* and Δ*pucD* derived from 37b4 and the *Escherichia coli* strains DH5α (6) and S17–1 (7) and the plasmids pLEP4.5 (2), pSUP202 (7), pG3PN1.5 (Tichy; containing the 1.5 kb *PstI-NruI*-fragment of the *puc* operon), pG7AE′ (Kortlüke; containing the 3099 bp *ApaI-EcoRI* fragment from the *PstI-EcoRI* fragment of the *puc* operon) were used. The growth conditions were described elsewhere (1, 2).

2.2. Construction of the Plasmids with *puc* Deletions

From the plasmid pG3PN1.5 the 1053 bp restriction fragment *EheI-ClaI* was isolated and cloned into the *SmaI-ClaI* digested vector pGEM®-7Zf(+) (Promega), resulting in pG7EC1.0. Recombinant clones contain the 5′ upstream region of the *puc* operon and a small part of *pucB* (B′ Fig. 1). The next step was the isolation of the Ω-streptomycin-spectinomycin resistance cassette from pHP45Ω after digestion with *Hind*III and cloning into pG7EC1.0. The *ScaI*-[*Bam*HI]-fragment of pG7EC1.Ω containing the *puc* upstream region, *B′* and the Ω-interposon was ligated with the *XmnI-ScaI*-fragment of pG7AE′ (3′ flanking sequence of the *puc* operon and a small part of *pucE* (′E)) resulting in pG7Δ BACDE::Ω (Fig. 1a). The *Eco*RI

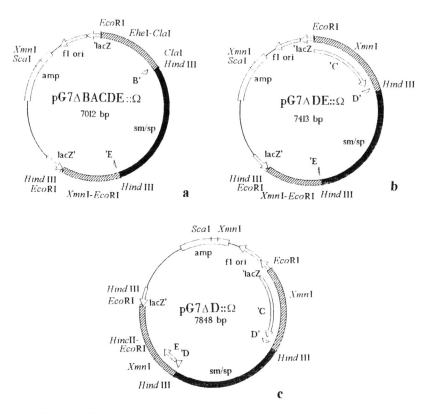

Figure 1. The plasmids pG7ΔBACDE::Ω (a), pG7ΔDE::Ω (b) and pG7ΔD::Ω (c).

fragment from this plasmid containing the Ω cassette with the flanking 5' and 3' elements of the *puc* operon but no *puc* genes were cloned into the *Eco*RI site of pSUP202 and transformed into *E. coli* S17–1. By using the same strategy the plasmids pG7ΔDE::Ω and pG7ΔD::Ω were constructed (Fig. 1b,c). The 5' flanking region was derived from the 1443 bp *Xmn*I fragment of pG7AE' and the 3' flanking region from the *Xmn*I-*Sca*I-fragment of pG7AE'. The 3' flanking sequence for the *pucD* deletion was the *Hinc*II-*Sca*I-fragment o pG7AE'.

The correct sequence of the cloned DNA was tested by restriction analysis and resistance expression of the plasmids in *E. coli*.

3. RESULTS AND DISCUSSION

3.1. Construction and Isolation of the Deletion Mutants

The hybrid plasmids pSUPΔBACDE::Ω, pSUPΔDE::Ω and pSUPΔD::Ω (see Materials and Methods and compare with Fig. 1) were transferred from *E. coli* S17–1 to *Rb. capsulatus* by conjugation. Since pSUP202 derivatives cannot be established as plasmids in *Rb. capsulatus*, cells which have the interposon integrated in the chromosome can be selected by the interposon encoded spectinomycin resistance. Homogenote recombinants, generated by double crossover, contained only the interposon but not the plasmid encoded tetracyclin resistance. To differentiate between homogenote and heterogenote recombination, the transconjugants were tested for tetracyclin resistance. Between 15 and 35% of the recombinants were homogenotes. In the homogenote recombinants the absence of vector sequences and the presence of *puc* sequences of defined size was proved by Southern hybridization (not shown).

3.2. Characterization of the Deletion Mutants

The wild type strain of *Rb. capsulatus* and the three deletion mutants were grown semiaerobically in the dark or anaerobically in the light, harvested in the late logarithmic growth phase and the intracytoplasmic membrane fraction was isolated (9). The absorption spectra of the membrane fractions showed a complete absence of the LHII complex in the Δ*pucBACDE* mutant (Fig. 2a,b) and the lack of LHIIα, β and γ polypeptides in the tricine SDS polyacrylamide gels (10) (Fig. 3). In addition, the bacteriochlorophyll concentration was lower than in the wild type strain (Tab. 1).

The absorption spectra of mutant strains Δ*pucDE* and Δ*pucD* indicated that the LHII complex was formed but in lower amounts. In both cases the ratio of the peak heights 800/855 nm was lower in the mutant than in the wild-type strain. The shoulder at 870 nm indicated an increase of the B870 (LHI, core complex)/B800–850 ratio in the mutant strains compared with 37b4 (Fig. 2c,d).

As expected, the γ-polypeptide of the LHII complex was absent from the protein patterns of the membrane. The α and β polypeptides of the mutants, however, were present but in lower concentrations as in 37b4 (Fig. 3, lane 3 and 4). The PucD protein could not be detected on the polyacrylamide gels. The bacteriochlorophyll concentration of membranes of ΔDE and ΔD mutants was about 60% of that of the strain 37b4 (Tab. 1).

3.3. Transconjugants of the Δ*pucBACDE* Mutant

Plasmids with partially deleted *puc* genes (2) were transconjugated to the Δ*pucBACDE*::Ω mutant. The absorption spectra of membrane fractions from these tran-

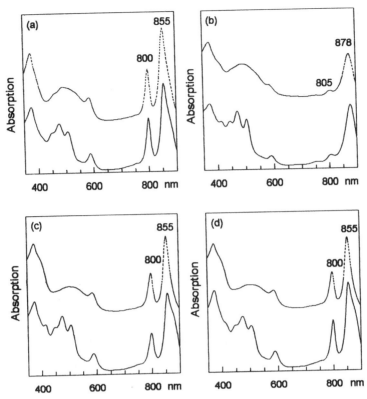

Figure 2. Absorption spectra of membranes from *Rb. capsulatus* strains. The cells were grown anaerobically in low light (——) or semiaerobically in the dark (------). **a.** 37b4, wild-type, **b.** Δ*pucBACDE*, **c.** Δ*pucDE*, **d.** Δ*pucD*. The absorption spectra of membranes from semiaerobically grown cells are transferred 0.3 absorption units to the top.

sconjugants showed a partial or complete reconstitution of the wild type character. Interestingly, the pLΔC (containing *pucBADE*) transconjugant was not only LHII-negative but was also reduced in the formation of the reaction center-core complex (not shown). It was striking that a reconstitution with the plasmids pLΔE (*pucBACD*) and pLΔDE (*pucBAC*) resulted in a lowering of the 800 nm peak. In contrast, the absorption spectrum of mem-

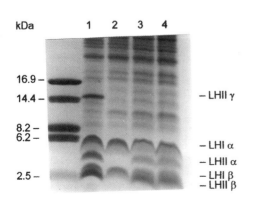

Figure 3. Protein pattern of purified intracytoplasmic membranes of *Rb. capsulatus*. Equal amounts of protein were applied to Tricine SDS polyacrylamide gel electrophoresis. Molecular mass standards in kDa. Lane 1, 37b4; 2, Δ*pucBACDE*; 3, Δ*pucDE*; 4, Δ*pucD*.

Table 1. The pigment content of purified intracytoplasmic
membranes of *Rhodobacter capsulatus*

Strain	BChl (nmol/mg protein)	Crt (nmol/mg protein)
37b4	42.32	22.57
Δ*pucBACDE*	22.95	20.25
Δ*pucDE*	24.85	18.32
Δ*pucD*	26.14	20.6

BChl, bacteriochlorophyll; Crt, carotenoid

branes from the transconjugant pLΔD (*pucBACE*) showed no significant differences to the wild type spectrum.

The results with the *puc* deletion mutants confirmed the results obtained with the mutant NK3 which has a Tn5 insertion in the *pucC* gene (2,5). It is concluded that PucC is essential for the assembly of the LHII complex and has a strong effect on the transcription of the *puc* operon (2). PucE supports the assembly of the LHII complex and its stability (2). The isolated LHII complex without the γ-polypeptide has the same spectral properties (11) but is unstable. In comparison with the wild-type complex, especially the 800 nm peak but also the 850 nm peak were lowered very fast (2). It is interesting that a γ-polypeptide has not been detected in LHII complexes of other species (12) and not found in the atomic structures of two crystallized LHII complexes (13,14). The γ-polypeptide is part of the LHII complex of *Rb. capsulatus;* it might be involved in the assembly process. PucD has not been identified on SDS polyacrylamide gels (C. Kortlüke, unpublished).

The Δ*pucBACDE*::Ω deletion strain has been used to construct a Δ*puf* Δ*puc* double deletion mutant for heterogenous expression of photosynthetic genes (15).

ACKNOWLEDGMENTS

The work was supported by grants of the Deutsche Forschungsgemeinschaft and the Fonds der Chemischen Industrie.

REFERENCES

1. Tichy, H.V., Oberlé, B., Stiehle, H., Schiltz, E. and Drews, G. (1989) J. Bacteriol. 171, 4914–4922
2. Tichy, H.V., Albien, K.U., Gad'om,N. and Drews, G. (1991) EMBO Journal 10, 2949–2955
3. LeBlanc, H.N. and Beatty, J.T. (1993) J. Gen. Microbiol. 139,101–109
4. LeBlanc, H.N. and Beatty, J.T. (1996) J. Bacteriol. 178, 4801–4806
5. Oberlé, B., Tichy, H.V., Hornberg, U. and Drews, G. (1990) in: Molecular Biology of Membrane-Bound Complexes in Phototrophic Bacteria, Drews, G. and Dawes, E.A., eds., pp.77–84, Plenum Press, New York
6. Hanahan, D.C. (1985) In: DNA Cloning, Vol.I, Glover, D.M., ed., pp 109–135, IRL Press, Oxford
7. Simon, R., Priefer, U. and Pühler, A. (1983) Bio/Technology 1, 37–45
8. Prentki, P. and Krisch, H.M. (1984) Gene 29, 303–313
9. Klug, G., Kaufmann, N. and Drews, G. (1985) Proc. Natl. Acad. Sci. USA 82,6485–6489
10. Schägger, H. and Jagow, G. (1987) Anal. Biochem. 166, 368–379
11. Tadros, M.H., Garcia, A.F., Drews, G., Gad'on, N. and Skatchkov, M.P. (1990) Biochim. Biophys. Acta 1019, 245–249
12. Zuber, H. and Cogdell, R. (1995) in: Anoxygenic Photosynthetic Bacteria, Blankenship, R.E., Madigan, M.T. and Bauer, C.E., eds., pp. 315–348, Kluwer Academic Publ., Dordrecht

13. McDermott, G., Prince, S.M., Freer, A.A., Hawthornthwaite-Lawless, A., Papitz, M., Cogdell, R. and Isaacs, N. (1995) Nature 374, 517–521
14. Koepke, J., Hu, X., Muenke, C., Schulten, K. and Michel, H. (1996) Structure 4, 581–597
15. Kortlüke, C., Breese, K., Gad'on, N., Labahn, A. and Drews, G. (1997) J. Bacteriol. 179, in press.

19

GENES ENCODING LIGHT-HARVESTING, REACTION CENTER, AND CYTOCHROME BIOGENESIS PROTEINS IN *CHROMATIUM VINOSUM*

Gary E. Corson, Kenji V. P. Nagashima,[2] Katsumi Matsuura,[2] Yumiko Sakuragi,[2] Ruwanthi Wettasinghe,[3] Hong Qin,[1] Randy Allen,[3,4] Yie Lane Chen,[1] and David B. Knaff[1,3]

[1]Department of Chemistry and Biochemistry
Texas Tech University
Lubbock, Texas 79409-1061
[2]Department of Biology
Tokyo Metropolitan University
[3]Institute for Biotechnology
Texas Tech University
Lubbock, Texas 79409-1061
[4]Departments of Biological Sciences and Plant and Soil Sciences
Texas Tech University
Lubbock, Texas 79409-3131

1. INTRODUCTION

Experiments demonstrating the low-temperature photoxidation of a membrane associated, *c*-type cytochrome in the photosynthetic purple sulfur bacterium *Chromatium vinosum* provided some of the first evidence that the reaction centers of certain photosynthetic bacteria might contain a cytochrome subunit (1), an hypothesis supported by an early partial characterization of a purified *Chr. vinosum* reaction center preparation (2). Spectral measurements at defined oxidation-reduction potentials and heme orientation studies of the *Chr. vinosum* reaction center *in situ* (3–5), indicated that the four hemes of the cytochrome subunit can be grouped into one set of two high-potential hemes (with approximate E_m values of +360 and +330 mV) and one set of low-potential hemes (with approximate E_m values of +30 and −10 mV). A comparison (4,5) of the redox and spectroscopic properties of the *Chr. vinosum* reaction center hemes with those of the corresponding *Rhodopseudomonas viridis* reaction center, for which a 2.3 Å resolution tertiary

The Phototrophic Prokaryotes, edited by Peschek *et al.*
Kluwer Academic / Plenum Publishers, New York, 1999.

structure is available (6), suggested that in *Chr. vinosum* the four hemes were arranged in a pattern in which high and low potential hemes alternate, with the highest potential heme being in closest proximity to P870, the bacteriochlorophyll *a* dimer that serves as the primary electron donor.

Amino acid sequences of the tetraheme cytochrome subunit (encoded by the *pufC* gene) of the reaction center are known for several photosynthetic purple bacteria, including *Rps. viridis* (7), *Rubrivivax gelatinosus* (formerly called *Rhodocyclus gelatinosus*) (8), *Rhodospirillum molischianum* (9) and the thermophilic *Chromatium tepidum* (10). As the only sequence currently available for this reaction center subunit from a purple sulfur bacterium is that for the *Chr. tepidum* protein, additional sequences are of interest to help identify structurally important portions of the protein, such as those that may be involved in the docking of the soluble electron donors. Novel docking sites may be present in *Chr. vinosum*, where cytochrome c_8 (11,12) and/or HiPIP (13) is/are likely to be the immediate donor(s) to the reaction center, instead of cytochrome c_2, which serves as the electron donor to a reaction center heme in *Rps. viridis* (14) and *Rvi. gelatinosus* (15). Additional sequences for the *pufC* gene product and for the L and M reaction center subunits are also of interest for phylogenetic studies (5,8,9). As the *Chr. vinosum* reaction center has been so extensively studied from a biophysical point of view, the sequences of genes coding for reaction center subunits from this bacterium are of particular interest. Sequences of the genes encoding *Chr. vinosum* antenna proteins should also be of considerable interest for phylogenetic studies. We have obtained complete sequences for the following *Chr. vinosum* genes: *pufL* and *pufM*, coding for the reaction center L and M subunits; *pufC* ; and *pufB* and *A*, located immediately upstream from *pufL* and coding for antenna proteins. We have also obtained a sequence for most of a second *Chr. vinosum pufB* gene, located downstream from *pufC*, a gene arrangement that is unique to *Chr. vinosum*, and for the 3′ portion of the *bchZ* gene, located upstream from the *puf* operon and coding for an enzyme of the bacteriochlorophyll biosynthesis pathway. As part of our attempt to understand the biogenesis of the cytochrome *c*-containing subunit of the *Chr. vinosum* reaction center, we have also sequenced completely one gene and a portion of another gene that exhibit considerable sequence homolgy to the *helX* and *ccl1* genes that have been shown to be involved in cytochrome *c* biogenesis in other bacteria (16).

2. RESULTS

Both strands of a 3.2 kb *Eco*RI fragment of *Chr. vinosum* DNA, which hybridized with a probe prepared from a portion of the *Rps. viridis pufC* gene, were completely sequenced. Comparison of the sequence of this 3.2 kb fragment to the sequences of *puf* operons from purple non-sulfur bacteria, indicated that the *Chr. vinosum* DNA fragment contained, in the following order, the 3′ end of *pufA*, the complete *pufL,M* and *C* genes and the 5′ end of a second *pufB*-like gene. PCR amplification of *Chr. vinosum* DNA, using primers designed against well-conserved regions in *bchZ*, *pufL*, *pufM* and *pufC*, produced a 1.3 kb DNA fragment when *bchZ* and *pufL* primers were used, a 1.5 kb fragment when *pufL* and *pufM* fragments were used and a 1.1 kb fragment when *pufM* and *pufC* primers were used. These lengths correspond well to those expected from the arrangement of genes in other *puf* operons. Sequencing of the 1.3 kb PCR fragment showed that it contained the 3′ end of the *bchZ* gene, the complete *pufB* and *pufA* genes and the 5′ end of the *pufL* gene. Sequencing of the 1.5 kb PCR fragment showed that it contained the rest of *pufL* and the 5′ end of *pufM*. The 1.1 kb PCR fragment was shown to contain the remaining part of *pufM* and most of *pufC*. For those portions of these PCR products that over-lapped with the 3.2 kb genomic *Chr. vinosum* DNA fragment, identical sequences were obtained from the two types of samples, confirming the accuracy of the sequences. The order of these genes in *Chr. vinosum* is identical to that found for *Rps. viridis* and *Rsp. molis-chianum*, and there is no evidence for any additional open reading frames between *pufB* and *pufA* or between *bchZ* and *pufB*, of the type found in *Rvi. gelatinosus* (8).

PCR amplification, using one primer based on a known sequence near the 3' end of the 3.2 kb *Eco*RI *Chr. vinosum* DNA fragment and a second primer based on a conserved sequence near the 3' end of *pufB* genes produced a PCR product that was used to complete the sequencing of almost all of the second, downstream *pufB* gene. With the exception of *Rvi gelatinosus*, where the *crtD* and *crtC* genes encoding enzymes of the spirilloxanthin and hy-droxyspheroidene biosynthesis pathways are located downstream from *pufC* (17), *Chr. vinosum* is the only currently-known example of a bacterium in which a gene associated with the photosynthetic light-harvesting apparatus is found immediately downstream from *pufC*.

A comparison of the sequences for the *Chr. vinosum pufL* and *pufM* gene products to those of other photosynthetic bacteria showed that residues involved in binding bacterio-chlorophyll, bacteriopheophytin, non-heme iron and quinone at the Q_A and Q_B sites are highly conserved. The tyrosine that is located between the primary donor and its reaction center cytochrome reductant in *Rps. viridis* (L162, Ref. 6,18) is conserved in the *Chr. vinosum* L subunit, as are amino acid residues such as Glu L212 and Ser L223 (*Rhodobacter sphaeroides* numbering) that have been implicated in proton transfer from the solvent to the Q_B site (19). However, in *Chr. vinosum*, the residue that corresponds to Asp L210 in *Rba. sphaeroides* (another residue implicated in protonation of Q_B) is a glutamate. The *Chr. vinosum* L subunit contains a small amino acid insert between the likely positions of the first and second membrane-spanning helices, as was previously found to be the case for the *Chr. tepidum* L subunit (10).

A comparison of the *Chr. vinosum pufC* sequence to that of the *Rps. viridis* reaction center cytochrome subunit indicates that all four of the C–CH heme-binding domains are conserved, as are the one histidine and three methionines that serve as axial heme ligands. As was first demonstrated for the *Rps. viridis pufC* gene product (7,20), and subsequently found likely to be the case for other reaction center cytochrome subunits (8,10), the *Chr. vinosum* cytochrome subunit has a N-terminal transit sequence with a cleavage cite for a signal peptidase II. In the case of *Chr. vinosum*, this cleavage motif consists of the amino acids VLLGC at positions −4 through 1. It appears likely that the conserved cysteine residue becomes the N-terminus of the mature cytochrome subunit, where it is covalently modified by fatty acid attachment (20).

Translation of the *Chr. vinosum pufA* gene gave an amino acid sequence that was similar to, but not identical to that found previously by direct amino acid sequencing of the isolated antenna protein (21). A comparison of these sequences suggests that there may be some post-translation processing near the N-terminus of the protein. This may also be the case for the product of the upstream *Chr. vinosum pufB* gene. Aside from the N-termi-nal region, the amino acid sequence produced from translating the base sequence of the upstream *pufB* gene differs at only a single position from that obtained by direct amino acid sequencing, raising the possibility of an error in one of the sequences. Translation of the uniquely located downstream *Chr. vinosum pufB* gene gives a sequence different from all of the published sequences obtained by direct sequencing of purified *Chr. vinosum* an-tenna proteins. All three of the *Chr. vinosum* antenna protein sequences contain the highly conserved histidine residue that serves as an axial ligand to magnesium in binding bacte-riochlorophyll.

Considerable progress has been made recently in identifying genes that code for proteins involved in cytochrome *c* biogenesis in the purple non-sulfur bacterium *Rhodobacter capsulatus* (16), but virtually nothing is known about the process in purple sulfur bacteria. To investigate this question, a 2.1 kb PCR product, amplified using *Chr. vinosum* DNA as a template, was completely sequenced. The 1.44 kb at the 5′ end of the PCR product contains an ORF that appears to be about 73% of a *ccl1*-like gene, including the stop codon. At the amino acid level, the product of this *Chr. vinosum* gene is 48% identical to the *Rb. capsulatus ccl1* gene product, 53% identical to the *E. coli ccmF* gene product and 47% identical to the *Bradyrhizobium japonicum cycK* gene product. After 10 bases a complete 498 bp ORF that appears to code for the *Chr. vinosum helX* gene begins. The 165 amino acid *Chr. vinosum helX* gene product is 46%, 43% and 41% identical to the *E. coli ccmG*, *B. japonicum tlpB* and *Rb. capsulatus helX* gene products, respectively. In *Rba. capsulatus* and *B. japonicum*, the *ccl1* (*cycK*) gene is followed by the *ccl2* (*cycL*) gene, not by *helX* (16). However, a gene arrangement similar to that we have found for the *Chr. vinosum* genes has been reported for *E. coli* (16).

REFERENCES

1. Kihara, T. and Dutton, P.L. (1970) Biochim. Biophys. Acta 205, 196–204.
2. Lin, L. and Thornber, J.P. (1975) Photochem. Photobiol. 22, 37–40.
3. Alegria, G. and Dutton, P.L. (1990) Biophys. J. 57, W-Pos 607.
4. Nitschke, W., Jubault-Bregler, M. and Rutherford, A.W. (1993) Biochemistry 32, 8871–8879.
5. Nitschke, W. and Dracheva, S. (1995) in : Anoxygenic Photosynthetic Bacteria (Blankenship, R.E., Madigan, M.T. and Bauer, C.E., eds.) pp. 775–805, Kluwer, Dordrecht.
6. Deisenhofer, J., Epp, O., Sinning, I. and Michel, H. (1995) J. Mol. Biol. 246, 429–457.
7. Weyer, K.A., Lottspeich, F., Gruenberg, H., Lang. F., Oesterhelt, D. and Michel, H. (1987) EMBO J. 8, 2197–2202.
8. Nagashima, K.V.P., Matsuura, K., Ohyama, S. and Shimada, K. (1994) J. Biol. Chem. 269, 2477–2484.
9. Nagashima, K.V.P., Matsuura, K. and Shimada, K. (1997) Photosyn. Res. 50, 61–70.
10. Fathir, I., Tanaka, K., Yoza, K., Kojima, A., Kobayashi, M., Wang, Z.-Y., Lottspeich, F. and Nozawa, T. (1997) Photosyn. Res. 51, 71–82.
11. Kerfeld, C.A., Yeates, T.O. and Knaff, D.B. (1966) Biochemistry 35, 7812–7818.
12. Samyn, B., De Smet, L., Van Driessche, G., Meyer, T.E., Bartsch, R.G., Cusanovich, M.A. and Van Beeumen, J.J. (1996) Eur. J. Biochem. 236, 689–696.
13. Schoepp, B., Parot, P., Menin, L., Gaillard, J., Richaud, P. and Verméglio, A. (1995) Biochemistry 34, 117–11742.
14. Knaff, D.B., Willie, A., Long, J.E.,Kriauciunas, A., Durham, B. and Millett, F. (1991) Biochemistry 30, 1303–1310.
15. Matsuura, K., Fukushima, A., Shimada, K. and Satoh, T. (1988) FEBS Lett. 237, 21–25.
16. Kranz, R.G. and Beckman, D.L. (1995) in: Anoxygenic Photosynthetic Bacteria (Blankenship, R.E., Madigan, M.T. and Bauer, C.E., eds.) pp. 709–723, Kluwer, Dordrecht.
17. Ouchane, S., Picaud, M., Vernotte, C., Reiss-Husson, F. and Astier, C. (1997) J. Biol. Chem. 272, 1670–1676.
18. Dohse, B., Mathis, P., Wachtveitl, J., Laussermair, E., Iwata, S., Michel, H. and Oesterhelt, D. (1995) Biochemistry 34, 11335–11343.
19. Okamura, M.Y. and Feher, G.(1995) in : Anoxygenic Photosynthetic Bacteria (Blankenship, R.E., Madigan, M.T. and Bauer, C.E., eds.) pp. 577–594, Kluwer, Dordrecht.
20. Weyer, K.A., Schäfer, W., Lottspeich, F. and Michel, H. (1987) Biochemistry 26, 2909–2914.
21. Brunisholz, R.A. and Zuber, H. (1992) J. Photochem. Photobiol. B 15, 113–140.

THE HOMODIMERIC REACTION CENTER OF *CHLOROBIUM*

Christine Hager-Braun, Rainer Zimmermann, and Günter Hauska

Lehrstuhl für Zellbiologie und Pflanzenphysiologie
Universität Regensburg
93053 Regensburg, Germany

1. INTRODUCTION

Chlorobiaceae, the green sulfur bacteria are strict anaerobes, adapted to photosynthesis in dim light. The special organization of their photosynthetic apparatus as depicted schematically in Fig. 1. Light is efficiently captured in huge antennae, the chlorosomes[1], which contain thousands of bacteriochlorophylls *c* or *d* in tubular stacks, and are located on the inner surface of the cell membrane. The excitation energy is transferred from the BChl *c* or *d* stacks absorbing maximally around 750 nm, via BChl *a* in chlorosomes and in the FMO-protein[1,2] absorbing around 800 nm, to the P840-RC in the membrane. Energy transfer from the chlorosomes[3], as well as through BChl *a*[4] is quenched under oxidizing conditions—a redox control which protects the organism from damage by oxygen. Chlorosomes are elsewhere found only in the *Chloroflexaceae*[1].

Phototrophy is scattered throughout the eubacterial phyla[5], as is the occurrence of the RC-types, which are distinguished by the terminal electron acceptors[6]. Type 1, with FeS-clusters as terminal acceptors (FeS-type) occurs in PS 1 of cyanobacteria (and chloroplasts), and in *Chlorobiaceae* and *Heliobacteria*. Type 2, with quinones as terminal acceptors (Q-type) is found in PS 2, in the purple bacteria *Rhodospirillaceae* and *Chromatiaceae* and in the green *Chloroflexaceae*.

A remarkable trait of *Chlorobiaceae*, only recently recognized and shared with the gram-positive *Heliobacteria*, is the homodimeric structure of the RC. The core is built by two identical membrane proteins, in contrast to all other RCs, which have a heterodimeric core structure. The evidence comes from peptide analysis and Southern blotting[7,8]. A symmetric core is also indicated by measurements of the spin density distribution in the two bacteriochlorophylls of P[+][9,10] as well as by resonance Raman studies on P[+][11] and the primary acceptor Ao[12]. A homodimeric structure is pertinent to the question why the symmetry has been broken in evolution. Why are oxygen tolerant RCs pseudosymmetric, with

The Phototrophic Prokaryotes, edited by Peschek *et al.*
Kluwer Academic / Plenum Publishers, New York, 1999.

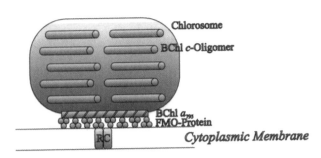

Cytoplasm

Figure 1. Organization of the *Chlorobium* photosynthetic apparatus. FMO-protein stands for "Fenna, Mathews, Olson"-protein[2], RC for reaction center, and BChl for bacteriochlorophyll.

two branches of electron transfer components from the primary donor to the terminal acceptor, and why is one of them preferred?[10,13]

Several reviews on the *Chlorobium* RC have been published[14–17]. In this article we are summarizing our recent progress[38,51]. A brief description of the electron transport system in *Chlorobium* is followed by the isolation of the P840-RC in two forms, a functionally competent complex and a core-subcomplex. After describing their properties we will compare the amino acid sequences of the two central subunits PscA and -B from *C.limicola* f.sp. *thiosulfatophilum* and *C.tepidum*. We will then present the expression of the subunit PscB in *E.coli* and the reconstitution of its FeS-centers. Finally the nature of the electron acceptor A1 will be discussed. It is controversial to what extent it resembles A1 in PS1 of cyanobacteria and chloroplasts.

2. THE ELECTRON TRANSPORT SYSTEM OF *CHLOROBIUM*

Electron transfer in green sulfur bacteria is thought to proceed from excited P840 via Ao and A1 to the FeS-centers X, A and B, and through ferredoxin and ferredoxin-NAD$^+$ oxidoreductase (FNR) to NAD$^+$ (Fig. 2). FNR eluded isolation from *Chlorobium* so far, however (D. Bryant, personal communication).

Oxidized P840 is rereduced by sulfide, either in a short pathway via flavocytochrome c[18], or by sulfide-quinone oxidoreductase[19], involving the menaquinone pool and the cytochrome bc-complex[20]. Some *Chlorobium* species also utilize thiosulfate as electron donor. P840, Ao, A1 and the FeS-center X are bound by the large core-protein PscA, while the FeS-centers A and B reside on the protein PscB, as concluded from the primary structures.[7] This cascade closely resembles the electron acceptor sequence in PS1[14–16]. There are subtle differences, however. First of all, the core is homodimeric, built by two identical subunits PscA,[7] unlike the heterodimeric core of PS1 with PsaA and PsaB. Secondly, the chlorophylls in P and Ao are different in *Chlorobium*. P840 is a special pair of BChl a, while Ao is a Chl a-molecule,[21] a feature welcome to biophysicists for specific LASER excitation, for instance in resonance Raman spectroscopy.[12] Another difference regards the relative redox potentials of FeS-centers A and B.[23]

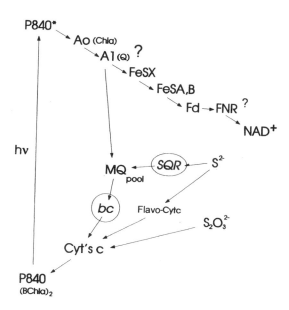

Figure 2. The electron transport system of *Chlorobium*. The components are positioned according to their redox potentials. (B)Chl stands for (bacterio)chlorophyll, Ao and A1 for the primary and secondary electron acceptor, respectively, Fd for ferredoxin, FNR for ferredoxin-NAD$^+$ oxidoreductase, SQR for sulfide-quinone oxidoreductase, Q for quinone, MQ for menaquinone, bc for cytochrome bc-complex, and cyt for cytochrome. Two open questions, the nature of A1 and the participation of FNR are marked.

The nature of A1, which is firmly bound phylloquinone in PS1[23,24], is still an open question. *Chlorobium* membranes contain menaquinone and two derivatives of it, but none of these is retained in photochemically active RCs[25]. Menaquinone may function as an A1-like acceptor in membranes but we conclude that it is not an obligatory intermediate between Ao and FeS-X. It is only losely bound and may exchange with the quinone pool, like Q_B in purple bacteria. However, contradicting results exist (see below).

3. ISOLATION AND CHARACTERIZATION

Our first protocol for isolating the P840-RC was designed for *Chlorobium limicola* using octyl glucoside as detergent[26]. Later we switched to *C. tepidum*, a moderate thermophile[27], and to a new procedure using Triton X-100. It yields a functionally more stable form of the RC in better yields, and can also be applied to *C. limicola*[28].

3.1. Isolation Procedure and Composition

Two advantages of our procedure may be worth mentioning. Removal of oxygen from the buffers by addition of FMN and illumination with white light[29] allowed for working outside an anaerobic tent. Secondly, the RC is solubilized from the total membrane fraction by Triton X-100, leaving the chlorosomes behind. This increases the yield substantially over chlorosome depleted membranes as the starting material. The solubilization is followed by ion exchange chromatography on DEAE-cellulose and sucrose density gradient (SDG) centrifugation. Three pigmented bands are observed in the SDG, free pigments on the top, mainly residual BChl *c* partially decomposed to bacteriophaeophytin *c*, and two forms of the P840-RC, a brownish-green fraction at about 18% sucrose, and a bluish-green fraction at about 25%. Figure 3 also shows the absorption spectra and polypeptide patterns of the two RC-bands.

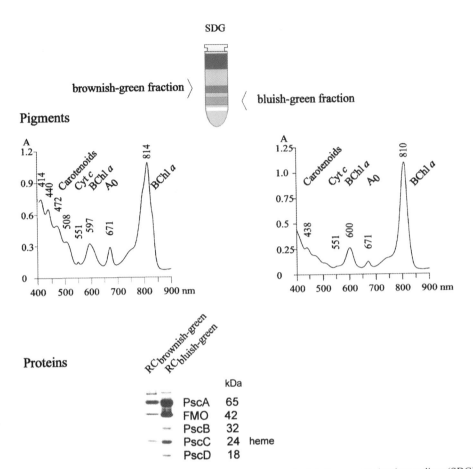

Figure 3. Isolated forms of the P840-reaction center. The upper part depicts the sucrose density gradient (SDG), the middle part shows the spectra, and the bottom part the polypeptide composition of the two RC-forms. PscA to -D denote the RC-subunits identified so far.[65] Other abbreviations are given in the former legends.

From the total of solubilized BChl *a* we obtain around 15% in the RC-fractions on SDG, about 3% in the upper, brownish-green and 12% in the lower, bluish-green band. The distribution between the two bands varies somewhat. Since the upper band is enriched in RC about three times (s. below), these values correspond to yields of 9 and 12% RC in the two fractions. The average protein/BChl *a*-ratios on weight basis were 14 in the brownish-green and 8 in the bluish-green fraction.

Comparable P840-RC preparations are available in several groups now, from *C. tepidum*[30–32] and from *C. vibrioforme*[33]. A photoactive RC-core preparation of remarkably simple structure has recently been achieved from *Prosthecochloris aestuarii*[34].

3.1.1. Polypeptide Composition. At the bottom of Fig. 3 the SDS-PAGE polypeptide patterns of the two RC-fractions with the apparent MW-values is shown. The bluish-green fraction consists of five polypeptides, PscA-D and FMO. PscA represents the core polypeptide of 83 kDa which binds P840, Ao, A1 and FeS-X and resembles the PS1-polypeptides PsaA and -B[7,35]. Like them it migrates as a diffuse band anomalously fast at

65 kDa, what is typical for membrane spanning, hydrophobic proteins. PscB is a 23 kDa-protein and carries two FeS-centers. It resembles the 8 kDa-subunit PsaC from PS1, but with a positively charged N-terminal extension[7], which causes its slowed down migration at 32 kDa[36]. The heme-active band PscC represents cytochrome c-551, while PscD does not carry a redox center. The average densitometric subunit stoichiometry from Coomassie stained gels is about 2:0.5:1.5:1 for PscA:B:C:D. The actual stoichiometry may well be 2:1:2:1, if cytochrome photobleaching (see below) and a poor staining of the positively charged PscB are taken into account. The band above PscA represents the not fully un-folded, partially still pigmented dimer of PscA. The composition of the brownish-green band is reduced to PscA and PscC, with some residual FMO, and is similar to our previous isolate with octyl glucoside, which additionally contained the FeS-subunit PscB[26].

The amount of FMO varies in both fractions. In the bluish-green the FMO protein is present in variable excess, and in two states, bound to the RC and as FMO trimers[1,2], which band at only slightly higher sucrose density. The contamination by FMO rises up to 6 trimers/RC with increasing salt concentration in the eluates from the DEAE-cellulose column, and can be removed by further purification of the bluish-green fraction, by gel fil-tration on Superdex 200, by chromatography on OH-apatite, or by native electrophoresis[37]. The removal of FMO resulted in a shift of the absorption peak from 810 up to 813 nm. Be-tween one to two trimer equivalents remained bound to the RC (3 to 6 FMO/RC). On the other hand, FMO could be totally removed from the brownish green fraction indicating that FMO is not bound to this RC form.

Triton X-100 could be exchanged for the detergents Thesit (Boehringer), dodecyl mal-toside and Hecameg (BioRad) during sucrose density gradient centrifugations of the eluate from the DEAE cellulose column. With all three the band pattern with the two forms of the RC and their polypeptide composition was similar to Fig. 3, but more smearing of the bands was observed with dodecyl maltoside and Hecameg. In all cases the RC-fraction at higher sucrose density was cw-photoactive to a similar extent found with Triton X-100 (Fig. 4).

Treatment of the bluish-green fraction with the chaotropic agents urea, NaBr, NaI, NaSCN and $NaClO_4$ specifically removes the smallest subunit PscD, only at elevated con-centrations also the cytochrome PscC dissociates.[37]

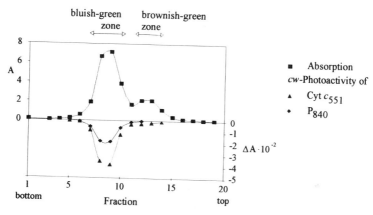

Figure 4. Distribution of photoactivity in the sucrose density gradient. The absorption of the Bchl a-peak at 810–814 nm (Fig. 3) is scaled on the left. Photoactivity, scaled on the right, is given as the amplitude of bleaching in continuous light (cw), at 610 nm for P840 and 551 nm for cytochrome c, both with 540 nm as reference. It was measured in a dual wavelength spectrophotometer (Aminco DW 2 UV-Vis).

Proteolytic digestion with thermolysin of chlorosome depleted membrane vesicles from *C. tepidum*[20] resulted in the loss of PscD in W-blots, followed by a loss of PscB, the cytochrome *c* PscC being resistant[37]. This demonstrates that the vesicles are inside-out, and that PscD and PscB sit on the cytoplasmic surface, like the arrangement of PsaD and -C in PS1[38].

The brownish-green RC-core fraction was analyzed by single particle averaging in scanning transmission electron microscopy[39]. The particle had a molecular mass of 240 kDa, with a length of 14 nm and a width of 8 nm and was similar in dimensions to the cyanobacterial PS1 core complex[40]. The averages revealed a pseudo two-fold symmetry axis in further support of a homodimeric core structure.

3.1.2. Pigment Composition. In the middle of Fig. 3 the optical spectra of the two RC-forms are shown. The RC-core is characterized by asymmetric Qx (597 nm) and Qy (814 nm) absorption peaks, an absorption ratio of 835/814 nm > 0.5, and higher amounts of Chl *a*-671 and cytochrome *c*-551 per BChl *a*. The bluish-green fraction, due to its FMO content, has more symmetric peaks at 600 and 810 nm, and a 835/810 ratio of only 0.2, similar to chlorosome depleted membranes.

Table 1 summarizes average pigment compositions determined.

More recent experiments showed that BChl *a* is extracted from FMO and the RC-core at rates different enough to allow the distinction between BChl *a* bound to FMO and RC. Moreover, the extinction coefficient for BChl *a* at 810 nm in FMO is significantly higher than the one at 814 nm in the RC (Griesbeck et al., in preparation). Thus the values in Table 1 need some correction. Nevertheless, about 2 FMO trimers with 42 BChl *a* molecules[1,2] are present in the average bluish-green RC fraction, only slightly less than in membranes. This corresponds to about 6 FMO polypeptides per RC, which falls into the range reported for densitograms above. The brownish-green fraction is contaminated by less than one FMO protein. BChl *c* and Chl *a*-671 cannot by clearly distinguished in methanolic aceton extracts, and have been determined by HPLC (asterisks in Table 1). Clearly there is more Chl *a*-671 present than the 2 molecules Ao.

Table 1. Pigment composition of membranes and P840-RC preparations from *Chlorobium tepidum*

	Membranes	RC	RC-core
BChl *c*	20–70	none[b]	none[b]
BChl *a*	80	70	25
Chl *a*-671	6[b]	5	5
Carotenoids	30	8	8
Heme *c*	6	2	2
MQ-7	30	< 0.2	< 0.1
CQ	40	< 0.2	< 0.1
Polar Q	5	< 0.2	< 0.1

[a]Average values for the molar ratios of pigments per RC are given for chlorosome depleted membranes[20], the bluish-green RC fraction and the brownish-green RC-core. The amount of RC present was estimated by the absorbance ratio of 835 to 810 nm[28]. The content of pigments was determined in extracts with methanolic acetone (2:7), using the mM extinction coefficients of 86 at 669 nm for BChl *c*[41], 75 at 770 for BChl *a*[42], 82 at 660 nm for Chl *a*-671[43], and 150 at 508 nm for the carotenoids[44]. The cytochrome *c* content was measured by redox spectrometry using a differential mM extinction coefficient of 20.
[b]N.-U. Frigaard/Odense, unpublished.

3.2. Photoactivity

Fig. 4 shows the distribution of P840 and cytochrome c-551 cw-photobleaching in the sucrose density gradient, together with bacteriochlorophyll absorption. Photobleaching of P840 is confined to the bluish-green zone, while a slow residual bleaching of the cytochrome is observed also in the brownish-green RC-core. The ratio of BChl a to photobleachable P840 is about 80 in the bluish-green fraction, if a mM extinction coefficient of 30 is taken for the 610–540 nm change[42]. This compares well to the value of 70 BChl per total RC in Table 1, demonstrating that nearly all the RCs present are photoactive.

Using a differential mM extinction coefficient of 20 for the cytochrome change, about 3 cytochromes c/RC are photooxidized in the bluish-green fraction of Fig. 4. The values range from 1.5 to 3 in different preparations, which corresponds to the average of 2 hemes c/RC found with redox spectrometry (Table 1). In the brownish-green fraction, which also contains 2 hemes c/RC, the extent of photooxidizable cytochrome c is much lower, probably because the slow rate is largely competed out by rereduction. In membranes, on the other hand, cytochrome c is photooxidized very efficiently. Essentially all the heme c present, up to six per RC (Table 1), is photooxidized. The most rapid component is not cytochrome c-551, a c-553 being advocated as the immediate electron donor to P840[+].[45]

The cw-photobleaching of the bluish-green fraction is rather stable. The extents of P840[+] formation after storage for a week at RT, on ice or in frozen state at −20°C, were 80, 94 and 94%, respectively. Even after 2 months the corresponding values were 10, 50 and 67%.[37]

Photobleaching of P840 was rapidly lost during proteolysis, which parallels the disappearance of PscD in W-blots. Also treatments with chaotropic agents (s. above) lead to a parallel loss of PscD and photoactivity, which could not be regained after removing the agents by dialysis.

The functional competence of the bluish-green fraction was also tested by EPR[46]. Photoreduction of two FeS-centers with similarities to FeS-A and -B of PS1 could be clearly demonstrated. The reduction of a third center resembling FeS-X was less clear, but the features were similar to the signal in membranes, which was substantiated by its spatial orientation. Photoreduced FeS centers had EPR-spectra of similar shape in membranes and isolated RCs. Only their redox potentials, which had been reported to be significantly lower than the corresponding ones in PS1[47], were shifted to somewhat higher values. Furthermore, comparison of the spin polarized P840 triplet signal shows that over 90% of electron transfer proceeds beyond Ao in membranes as well as in isolated RCs. Both findings demonstrate that the bluish-green RC fraction is functionally fully competent, which is additionally supported by transmembrane charge separation in P840-RC proteoliposomes[25]. NADP-reduction with ferredoxin and ferredoxin-NADP reductase, both from spinach, has been demonstrated in a similar preparation from *C. vibrioforme*[33].

4. GENES

All five genes for the polypeptides in the bluish-green P840-RC have been cloned and sequenced for *C. limicola* and partially for other species as discussed before.[28] The genes *pscA* for the core polypeptide and *pscB* for the FeS-A/B protein form a transcription unit[7], while *pscC*[48], *pscD*[28] and *fmo*[49] are found in isolated loci.

We now obtained the PCR sequence of the transcription unit *pscAB* also for *C. tepidum*[50]. For *pscB* we also sequenced a genomic clone. This gene is 84% identical while *pscA* is 90% identical to the nucleotide sequence of *C. limicola*[7]. The deduced primary structure of PscA was 95% identical to the one for *C. limicola*. However, one striking difference was found. Residues 285 to 294 of *C. tepidum* read AIGYINIALG followed by CI instead of HLRHQHRAWV followed by I without the C in *C.limicola*. This difference has been verified in 4 independent PCR clones. Thus only 19 histidines per PscA are found in *C. tepidum* compared to 21 in *C. limicola* (at position 589 *C. tepidum* carries a H instead of a Q). The two core proteins of PS1 PsaA and PsaB contain about the double number of histidines, 42 and 39, which results in a higher pigmentation by chlorophylls relative to PscA.

Sequence alignment of the type 1 RC-core polypeptides suggests that the 11 transmembrane helices in PsaA/B of PS1 are conserved in *Chlorobium* PscA, as well as in PshA of *Heliobacillus*[8]. Particularly conserved is the C-terminal half, from putative transmembrane helix VII to XI which forms the central part and binds P, Ao, A1 and FeS-X. Crystallography of PS1 revealed that this part resembles the 5 helix folding of the type 2 RC-core in purple bacteria[51,52]. Since from electron crystallography this fold just emerges also for PS2[53], it seems to be common to all RCs. The N-terminal parts in the type 1-RCs with the first six transmembrane spans function as antennae.[51,52] In analogy to PsaA/B of PS1 P840 is considered to be bound to the conserved H622 in helix X, Ao to conserevd H376 in helix VI and FeS-cluster Fx to conserved C527 and C536 in PscA.

The homodimeric structure of the RC-core in *Chlorobium* and *Heliobacillus* is based on two observations. No second related gene is found in DNA digests, and all the proteolytic peptides obtained, 7 for PscA and 13 for PshA, are represented by the genes *psc*A or *psh*A[7,8]. That the core also in *Chlorobium* and *Heliobacillus* is dimeric, not monomeric, follows from its analogy to the PsaA/B heterodimer of PS1. PscA carries only 2 of the 4 required cysteines to bind the 4Fe4S-cluster Fx. Recently, the dimeric structure of the *Chlorobium* RC-core has been directly observed by electron microscopy[39].

The double FeS-center PscB protein in *C. tepidum* shows 90% identity and 99% similarity to the one of *C. limicola*. The differences are largely confined to the N-terminal, positively charged, P- and A-rich extension. The C-terminal, highly conserved part resembles PsaC of PS1 and hosts the FeS-center binding peptides with 4 cysteines in each one. The folding of this part is such that the first three and the last of the eight cysteines bind FeS-B, the rest FeS-A[54]. The exchange of the two positively charged residues KR between the sixth and seventh cysteine in PsaC for the neutral residues SA in PscB has been advocated to explain the drop in redox potential of FeS-A in *Chlorobium* compared to PS1[7]. This was recently substantiated by targeted mutation of PsaC in *Chlamydomonas reinhardtii*.[55]

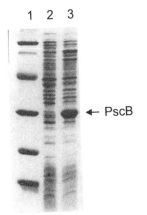

Figure 5. The PscB expression protein of *Chlorobium tepidum* in SDS-PAGE. Lane 1 contains the protein standards corresponding to 95, 68, 45, 32, 21 and 14 kDa, lane 2 and 3 show total *E. coli*-extracts, before and after induction with IPTG. The genomic *Nru*I-fragment in pUC19 and containing the *psc*B-gene was excised with *Pvu*II/*Sal*I and was ligated into the expression vector pTrc99A[66] cut with *Sma*I/*Sal*I. The construct was transformed into *E.coli* BL12.

5. HETEROLOGOUS EXPRESSION

Attempts to express PscA in *E. coli*, in full length or in several truncated forms were without success so far. Expression of PscB was achieved, on the other hand.[50] Fig. 5 shows total *E. coli* extracts before and after induction.

The expression PscB was almost exclusively found in inclusion bodies from where it was solubilized with 7 M urea. During this isolation the PscB protein at 32 kDa was partially (about 30%) degraded to a fragment migrating at 20 kDa, although the protease inhibitors PMSF, EDTA and p-amino benzamidine had been added. The fragment has been cleaved after a tyrosine, starting with the peptide VKPKAVPPP and losing about two thirds of the N-terminal extension. Since this left the FeS-center binding region intact the mixture of the 32 and 20 kDa apoproteins were used for reconstituting the FeS-centers[56]. The EPR spectra at pH 10, before and after addition of excess dithionite are shown in Fig. 6.

The spectrum in presence of dithionite is similar to the one of PsaC dissociated from PS1[56], with signals much broader than in bound form, not allowing the resolution into centers A and B. Redox titration is in progress. Our next goal is to rebind the expressed and reconstituted PscB to the brownish-green P840-RC core fraction in the presence of PscD, and to test whether narrowing with resolution of the EPR spectrum into the two centers[46] is reobtained, as was achieved by rebinding PsaC to PS1.[56]

6. THE ELECTRON ACCEPTOR A1

The RC of PS1 contains two bound phylloquinones which are functioning as A1, the electron acceptor between Ao and FeS-X[24,52,57] (see ref. 58 for an extensive recent review).

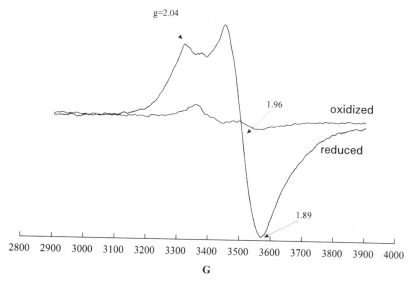

Figure 6. EPR-spectrum of the reconstituded PscB expression protein. The FeS-centers were reconstituted into PscB after solubilization from inclusion bodies with urea, dilution into a mixture containing mercaptoethanol, ferric chloride and sulfide[58] and were measured with EPR after concentration, at high pH before and after addition of dithionite. Details will be published elsewhere. EPR conditions: 15 K, microwave frequency 9.456 GHz, microwave power 20 mW, modulation amplitude 10 G at 100 kHz.

By analogy MK-7 was taken to fullfill this function in *Chlorobium*[59]. To our surprise though none of the three quinones occurring in membranes could be detected in the P840-RC by chemical analysis, neither in the brownish-green RC core nor in the functionally competent bluish-green fraction[25] (Table 1). In a similar preparation, but using dodecyl maltoside instead of Triton X-100, from *C.vibrioforme*, as well as from *C. tepidum* up to two MK-7/RC have been detected by HPLC (N-U Frigaard, H-V Scheller/Odense, personal communication). However, the lack of MK-7 in our preparations has been substantiated recently by N-U Frigaard. Moreover, in the preparation from *C. vibrioforme* a photoreduced radical with properties of a menasemiquinone was observed by EPR in Q-band, but was absent in our bluish-green fraction from *C. tepidum* (JH Golbeck, personal communication). On the other hand, in time resolved EPR with our bluish-green fraction the change in the polarization pattern of the radical pair assigned to P840$^+$FeS-X$^-$ at two different microwave frequencies suggests an intermediate acceptor between Ao and FeS-X[60]. Since none of the three quinones known for *Chlorobium* is present, the chemical nature of this acceptor remains to be determined.

We conclude that menaquinone functions as an A1-like electron acceptor in *Chlorobium* but is not essential for electron transport to the FeS-centers. Rather it may divert electrons into cyclic electron transport, being more losely bound than phylloquinone in PS1 and exchanging with the quinone pool like Q_B in type 2-RCs. A similar situation may exist in *Heliobacteria*, where extraction of menaquinone from membranes did not affect electron transfer to the terminal acceptors,[61] although two menaquinone molecules remained in a photoactive RC preparation which consisted of the single polypeptide PshA[62].

Less tight binding of A1 in *Chlorobium* and *Heliobacillus* is indicated by sequence comparison of the core polypeptides, because the two aromatic residues YW suggested to be involved in A1-binding in PS1[63] on the basis of a pulsed EPR analysis of PS1 crystals[57], are replaced by SR in the two organisms.

7. CONCLUSION

Fig. 8 summarizes our notion on electron transport in *Chlorobium* in a topographical model, which corresponds to Fig. 2 and thus in addition to the P840-RC includes sulfide-

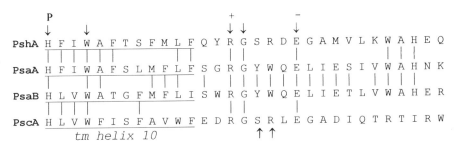

Figure 7. Alignment of the putative A1-binding peptides in the core proteins of FeS-type reaction centers. The region shown comprises the C-terminal half of transmembrane helix X, also denoted *m*, and parallel helix *n* of PS1[51,52]. On the top the heliobacterial PshA[8], at the bottom the chlorobial sequence PscA[7], and in the middle the two core proteins PsaA/PsaB of PS1 from maize[67] are shown. Identical amino acids in neighbouring sequences are indicated by bars, totally conserved ones additionally by arrows from the top. The histidine binding the special pair is marked by P, and the two aromatic residues in PscA/-B binding A1 by arrows from below.

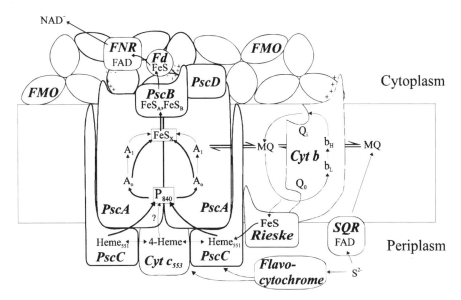

Figure 8. Topography of electron transport in the *Chlorobium* membrane. Qi and Qo mark the sites of quinol oxidation and quinone reduction in the cytochrome bc-complex, and b_L and b_H stand for low - and high potential heme *b*, respectively. Other abbreviations are given in the former legends.

quinone oxidoreductase (SQR), the menaquinone pool, the cytochrome bc-complex, as well as ferredoxin and the postulated ferredoxin-NAD$^+$ oxidoreductase (FNR).

We suggest that electron transport from sulfide proceeds via SQR[19], the menaquinone pool and the cytochrome bc-complex[20] to the homodimeric P840-RC, which may possess two equally used electron transfer branches. The immediate electron donor to the RC may be cytochrome *c*-553[45], but the involvement of cytochrome *c*-551 in appreciable rates of NAD photoreduction by isolated RCs lacking cytochrome *c*-553 has been documented[33]. The key conclusion depicted in Fig. 8 is that A1 is exchanging rapidly with the menaquinone pool for cyclic electron flow, but is not required for electron transport to NAD via the FeS-centers.

It is intriguing to point out in this context that also in PS1 the exchange of A1 with unbound phylloquinone has been measured by ESP EPR[24,64]. It takes hours to days at 4°C in the dark, but only minutes in the light or at 37°C in the dark. Thus the difference of A1 in *Chlorobium* and PS1 may be a quantitative rather than a qualitative one.

REFERENCES

1. Blankenship RE, Olson JM and Miller M (1995) in: Anoxygenic Photosynthetic Bacteria (Blankenship RE, Madigan MT and Bauer CE eds.) pp 399–435, Kluwer Acad.Publ.
2. Olson JM (1980) Biochim.Biophys.Acta 594: 33–51
3. Wang J, Brune DC and Blankenship RE (1990) Biochim.Biophys.Acta 1015: 457–463
4. Zhou W, LoBrutto R, Lin S and Blankenship RE (1994) Photosynth.Res. 41: 89–96
5. Woese CR (1987) Microbiol.Reviews 51: 221–271
6. Blankenship RE (1992) Photosynth.Res. 33: 91–111

7. Büttner M, Xie D-L, Nelson H, Pinther W, Hauska G and Nelson N (1992) Proc.Natl.Acad.Sci.USA 83: 8135–8139
8. Liebl U, Mockenstorm-Wilson M, Trost JT, Bruce DC, Blankenship RE and Vermaas W (1993) Proc.Natl.Acad.Sci. USA 7124–7128
9. Rigby SEJ, Thapar R, Evans MCW and Heathcote P (1994) FEBS Lett. 330: 24–28
10. Huber M (1997) Photosynth.Res. 52: 1–26
11. Feiler U, Albouy D, Robert B and Mattioli TA (1995) Biochemistry 34: 11099–11105
12. Feiler U, Albouy D, Porcet C, Mattioli TA, Lutz M and Robert B (1994) Biochemistry 33: 7594–7599
13. Plato M, Möbius K, Michel-Beyerle ME, Bixon M and Jortner J (1988) J.Am.Chem.Soc. 110: 7279–7285
14. Feiler U and Hauska G (1995) in: Anoxygenic Photosynthetic Bacteria (Blankenship RE, Madigan MT and Bauer CE eds.) pp 665–685, Kluwer Acad.Publ.
15. Nitschke W and Lockau U (1993) Physiologia Plantarum 88: 372–381
16. Golbeck JH (1994) Proc.Natl.Acad.Sci.USA 90: 1642–1646
17. Sakurai H, Kusumoto N and Inoue K (1996) Photochem.Photobiol. 64: 5–13
18. Kusai K and Yamanaka T (1973) Biochim.Biophys.Acta 325: 304–314
19. Shahak Y, Arieli A, Padan E and Hauska G (1992) FEBS Lett. 299: 127–130
20. Klughammer C, Hager C, Padan E, Schütz M, Schreiber U, Shahak Y and Hauska G (1995) Photosynth.Res. 43: 27–34
21. Nujis AM, Vasmel H, Joppe HLP, Duysens LNM and Amesz J (1985) Biochim.Biophys.Acta 807: 24–34
22. Nitschke W, Feiler U and Rutherford AW (1990) Biochemistry 29: 3834–3842
23. Schoeder H-U and Lockau W (1986) FEBS Lett. 199: 23–27
24. Ostafin AE and Weber S (1997) Biochim.Biophys.Acta 1320: 195–207
25. Frankenberg N, Hager-Braun C, Feiler U, Fuhrmann M, Rogl H, Schneebauer N, Nelson N and Hauska G (1996) Photochem.Photobiol. 64: 14–19
26. Hurt EC and Hauska G (1984) FEBS Lett. 168: 149–154
27. Wahlund TM, Woese CR, Castenholz RW and Madigan MT (1991) Arch.Microbiol. 156: 81–90
28. Hager-Braun C, Xie D-L, Jarosch U, Herold E, Büttner M, Zimmermann R, Deutzmann R, Hauska G and Nelson N (1995) Biochemistry 34: 9617–9624
29. Nelson N, Nelson H and Racker E (1972) Photochem.Photobiol. 16: 481–489
30. Kusumoto N, Inoue K and Sakurai H (1995) Photosynth.Res. 43: 107–112
31. Oh-Oka H, Kamei S and Matsubara H (1995) FEBS Lett. 365: 30–34
32. Oh-Oka H, Kakutani S, Kamei S, Matsubara H, Iwaki M and Itoh S (1995) Biochemistry 34: 13091–13097
33. Kjaer B and Scheller H-V (1996) Photosynth.Res. 47: 33–39
34. Francke C, Permentier HP, Franken EM, Neerken S and Amesz J (1997) Biochemistry, in press
35. Chitnis PR, Xu Q, Chitnis VP and Nechushtai R (1995) Photosynth.Res. 44: 23–40
36. Illinger N, Xie D-L, Hauska G and Nelson N (1994) Photosynth.Res. 38: 111–114
37. Hager-Braun C (1997) thesis, university of Regensburg/ Germany
38. Kruip J, Chitnis PR, Lagoutte B, Rügner M and Boekema EJ (1997) J.Biol.Chem. 272: 17061–17069
39. Tsiotis G, Hager-Braun C, Woplensinger B, Engel A and Hauska G (1997) Biochim.Biophys.Acta, in press
40. Kruip J, Bald E, Boekema E and Rögner M (1994) Photosynth.Res. 40: 279–286
41. Stanier RY and Smith JH (1960) Biochim.Biophys.Acta 41: 478–484
42. Olson JM, Giddings H and Shaw EK (1980) Biochim.Biophys.Acta 449: 197–208
43. McKinney G (1941) J.Biol.Chem. 140: 315–321
44. Young AJ (1993) in: Carotenoids in Photosynthesis (Young AJ and Britton G eds.) pp 16–71, Chapman & Hall, London
45. Albouy D, Joliot P, Robert B and Nitschke W (1997) Eur.J.Biochem. 249: 630–636
46. Hager-Braun C, Jarosch U, Hauska G, Nitschke W and Riedel A (1997) Photosynth.Res. 51: 127–136
47. Nitschke W, Feiler U and Rutherford AW (1990) Biochemistry 29: 3834–3842
48. Okkels JS, Kjaer B, Hansson O, Swendsen I, Lindberg-Moeller B and Scheller H-V (1992) J.Biol.Chem. 267: 21139–21145
49. Dracheva S, Williams J-A and Blankenship RE (1992) in: Research in Photosynthesis (Murata N ed.) Vol I, pp 53–56, Kluwer Acad.Publ.
50. Zimmermann R (1997) thesis, university of Regensburg/ Germany
51. Krauß N, Schubert W-D, Klukas O, Fromme P, Witt HT and Saenger W (1996) Nature Struct.Biol. 3: 965–973
52. Schubert W-D, Klukas O, Krauß N, Saenger W, Fromme P and Witt HT (1997) J.Mol.Biol. 272: 741–769
53. Rhee KH, Morris EP, Zheleva D, Hankamer B, Kühlbrandt W and Barber J (1997) Nature 389: 522–526
54. Mehari T, Qiao F, Scott MP, Nellis DF, Zhao J, Bryant DA and Golbeck JH (1995) J.Biol.Chem. 270: 28108–28117

55. Fischer N, Setif P and Rochaix J-D (1997) Biochemistry 36: 93–102
56. Mehari T, Parrett KG, Warren PV and Golbeck JH (1991) Biochim.Biophys.Acta 1056: 139–14857) Bittl R, Zech SG, Fromme P, Witt HT and Lubitz W (1997) Biochemistry 36: 12001–12004
58. Brettel K (1997) Biochim.Biophys.Acta 1318: 322–373
59. Hauska G (1988) Trends.Biochem.Sci. 13: 415–416
60. van der Est A, Hager-Braun C, Leibl W, Hauska G and Stehlik D (1998) in preparation
61. Kleinherenbrink FAM, Ikegami I, Hiraishi A, Otte SCM and Amesz J (1993) Biochim.Biophys.Acta 1142: 69–73
62. Trost JT and Blankenship RE (1989) Biochemistry 28: 9898–9904
63. van der Est A, Prisner T, Bittl R, Fromme P, Lubitz W, Möbius K and Stehlik D (1997) J.Phys.Chem. B 101: 1437–1443
64. Rustandi RR, Snyder SW, Biggins J, Norris JR and Thurnauer MC (1992) Biochim.Biophys.Acta 1101:311–320
65. Bryant DA (1994) Photosynth.Res. 41: 27–28
66. Amann E, Ochs B and Abel K-J (1988) Gene 69: 301–315
67. Fish LE, Kück U and Bogorad L (1985) J.Biol.Chem. 260: 1413–1421

21

A NOVEL PHOTOSYSTEM I-TYPE REACTION CENTER FROM AN OXYGENIC PROKARYOTE *ACARYOCHLORIS MARINA* THAT CONTAINS CHLOROPHYLL *D*

Shigeru Itoh,[1] Masayo Iwaki,[1] Qiang Hu,[2] Hideaki Miyashita,[2] and Shigetoh Miyachi[3]

[1]National Institute of Basic Biology
Okazaki 444, Japan
[2]Marine Biotech. Inst.
Kamaishi, Lab.
Kamaishi, Iwate 026, Japan
[3]Marine Biotech. Inst.
Tokyo 113, Japan

1. INTRODUCTION

Chlorophyll *a* that absorbs red light, supports the oxygenic photosynthesis of plants and cyanobacteria by its photochemical reactions. Neither the other pigments such as chlorophylls b, c, phycobilins and carotenoids that just harvest light of different colors and transfer energy to chlorophyll *a*, nor bacteriochlorophylls that are used in anoxygenic bacterial photosynthesis, have thus far been shown to replace the role of chlorophyll *a* in the photosystems (PS) I and II reaction center complexes of oxygenic photosynthesis. Biosynthesis of chlorophyll *a*, therefore, has been assumed to be preceded by the establishment of oxygenic photosynthesis. We report that the role of chlorophyll *a* in PS I reaction center complex can be replaced by another pigment, chlorophyll *d* that can absorb far-red light.

Miyashita et al. [1] recently isolated a new type of oxygenic photosynthetic prokaryote, named *Acaryochloris marina*, from a species of colonial Acidians. Cells of *A. marina* grow photoautotrophically, and contain chlorophylls *d* and *a* in a molecular ratio of 100:2 with a low amount of phycobilins. Chlorophyll *d* was originally found in red algae as a minor pigment, and its role in photosynthesis has not been clear [2]. *A. marina* cells exhibited normal oxygen evolution under the illumination suggesting the involvement of this pigment in photosynthesis [3].

The Phototrophic Prokaryotes, edited by Peschek *et al.*
Kluwer Academic / Plenum Publishers, New York, 1999.

We purified PS I reaction center from *A. marina* cells that contains chlorophyll *d* as the primary electron donor that was named P740 according to its red peak in difference spectrum [4]. The finding indicates a fully active PS I photochemistry without chlorophyll *a* in oxygenic photosynthesis.

2. MATERIALS AND METHODS

Acaryochloris marina was batch-cultured at 28°C in a mineral medium at an average light intensity of 20 μmol m^{-2} s^{-1}. Cells were broken by treatment with sonic oscillation and PSI complex was isolated from thylakoid membranes solubilized at 4°C for 1 h with 0.8% b-dodecyl maltoside/20 mM Bis-Tris (pH 7.0)/10% glycerol/10 mM NaCl/5 mM $CaCl_2$/2 mM EDTA-Na_2, and followed by a linear density gradient (10–30%) of sucrose centrifugation for 16 h at 40,000 rpm. Among the three pigment-containing fractions the lower green band contained PS I complex. The PS I complex was further purified by an anion exchange chromatography.

Measurements of absorption changes were done as described elsewhere with a split beam spectrophotometer [5].

3. RESULTS AND DISCUSSION

Room temperature absorption spectrum of the purified PS I RC complex resembled those of intact cells or thylakoid membranes [3]. It showed a red maximum at 708 nm with a minor shoulder at round 680 nm and a Soret maximum at 460 nm. The peaks at 708 and 460 can be ascribed to chlorophyll *d* as expected from the high content of chlorophyll *d*. The spectrum differs significantly from the ever-reported spectra of PS I reaction center complexes that gives peaks at 680 and 430 nm due to chlorophyll *a*. The absorption maxima of the new PSI complex are red-shifted by 30 nm compared with those of ever-known chlorophyll *a*-type PSI preparations [3].

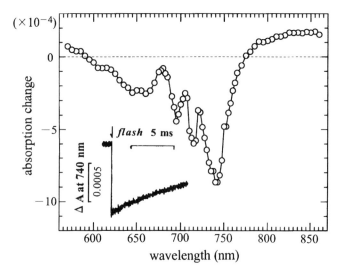

Figure 1. Laser flash-induced difference absorption spectrum at 15°C of PS I preparation of *Acaryochloris marina*. Insert shows the time course of absorption change after the flash excitation.

When *A. marina* PS I RC was excited by a 10 ns laser flash light, a rapid absorption change was induced followed by a decay time of 30 ms at room temperature, as shown in Fig. 1. Point-to point difference spectrum of the change induced by the flash gave a complex difference spectrum with negative peaks at 740, 715, 695 and 650 nm, and positive ones at 720, 705 and 680 nm. The spectrum significantly differs from that of P700 which is a dimer of chlorophyll *a* and gives a peak around 700 nm. The bleaches at 740 nm in the present PSI preparation indicates its origin as the bleach of chlorophyll *d* since chloro-

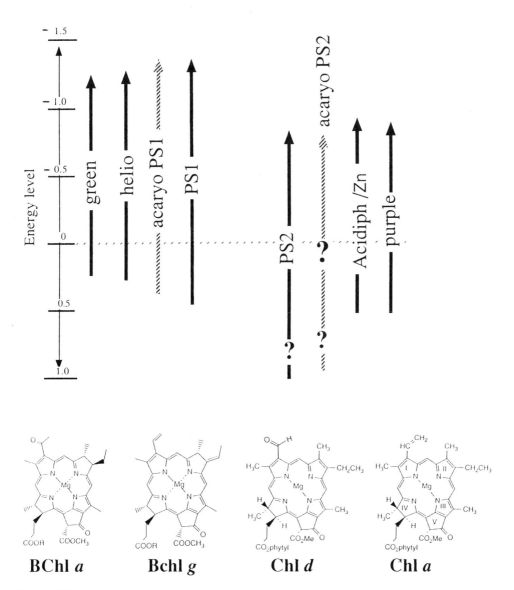

Figure 2. Schematic representation of the redox levels of the donor and acceptor sides of reaction centers of *Acaryochrolis marina* (acaryo), green sulfur bacteria (green), heliobacteria (helio), purple photosynthetic bacteria (purple), Zn-bacteriochlorophyll a-containing *Acidiphilium rubrum* (Acidiph/Zn) and plant/cyanobacterial (no indications) PS I and PS II. Molecular structures of bacteriochlorophylls (BChl) and chlorophylls (Chl) are also shown.

phyll *a* has no absorption around these wavelengths. The shifts at 670–710 nm can be attributed to the electrochromic shifts of chlorophylls *d* or *a* close to the donor [4].

The results above indicates that the chlorophyll *d* function as the electron donor chlorophyll in *A. marina* PS I reaction center. We name this donor chlorophyll *d* to be P740 after its red most peak wavelength. P740 seems to be embedded in polypeptides that composition and amino acid sequences are highly homologous to those in the ever-known chl *a*-type PS I reaction centers. The extent of light-induced change suggests one P740 per around 100 chlorophyll *d*. The purified RC contained about 70 chlorophyll *d* molecules per chlorophyll *a* on HPLC analysis, and, therefore, the amount of chlorophyll *a* seems rather too small to explain P740 as a dimer of chlorophyll *a* supporting the conclusion above.

We compared the redox levels of the donor and acceptor sides of newly found PS I of *A. marina* with those in various oxygenic and anoxygenic photosynthetic organisms (Fig. 2). It is seen that PS I/green sulfur bacterial type reaction centers give similar reducing power and the differences in the energies of absorbed light appear to be compensated by the variation of the redox levels in the oxidizing side. Similar rule also seems to exist for the PS II/ purple bacterial type reaction centers, although they have larger variations in their oxidizing sides. The reaction center pigment in PS II of *A. marina*, which has not been determined yet, is quite interesting with this respect. It is rather surprising this rule also seems to apply to the reaction center of an acidic bacterium, *Acidiphilium rubrum* that was recently shown to use Zn instead of Mg as a central metal of bacteriochlorophyll *a* and to perform normal photosynthesis according to our study [6]. Natural photosynthetic reaction centers, thus, seem to be designed to produce constant reducing powers, either to reduce ferredoxins or to reduce quinone, even with different light energies.

ACKNOWLEDGMENTS

We thank Yoshio Sasaki for N-terminal amino acid sequence analysis and Hiroshi Urata for assistance with the preparation of the figures. This work was performed as a part of the Industrial Science and Technology Frontier Program supported by New Energy and Industrial Technology Development Organization.

REFERENCES

1. Miyashita, H., Ikemomot, H., Kurano, N., Adachi, K., Chihara, M. and Miyachi, S. (1996) Nature 383, 402.
2. Manning, W. M. and Strain, H.H. (1943) J. Biol. Chem. 151, 1–19.
3. Miyashita, H., Adachi, K., Kurano, N., Ikemoto, H., Chihara, M. and Miyachi, S. (1997) Plant Cell Phisol., 38, 274–281.
4. Hu, Q., Miyashita, H., Iwasaki, I., Kurano, N., Miyachi, S., Kobayashi, M., Iwaki, M. and Itoh, S. (1997) submitted
5. Iwaki, M. and Itoh, S. (1991) Biochemistry, 30, 5347–5352.
6. Wakao, N., Yokoi, N., Isoyama, N., Hiraishi, A., Shimada, K., Kobayashi, M., Kise, H., Iwaki, M., Itoh, S., Takaichi, S. and Sakurai, Y. (1996) Plant Cell Physiol., 37, 889–893

22

SENSING OF GREEN LIGHT IN COMPLEMENTARY CHROMATIC ADAPTATION OF THE CYANOBACTERIUM *CALOTHRIX SP.*

Involvement of a Rhodopsin?

Hans C. P. Matthijs,[1] Jeroen H. Geerdink,[1] Hans Balke,[1] Andrea Haker,[1,2] Hendrik Schubert,[1,*] Luuc R. Mur,[1] and Klaas J. Hellingwerf[2]

[1]Laboratory for Microbiology (ARISE/MB)
University of Amsterdam
Nieuwe Achtergracht 127
1018 WS Amsterdam, The Netherlands
[2]E. C. Slater Institute
University of Amsterdam
Amsterdam, The Netherlands

1. INTRODUCTION

Light harvesting is essential for phototrophic organisms. The reaction centers of PS2 and PS1 in cyanobacteria contain chl a,[†] a ubiquitous pigment that absorbs light in two spectral domains. In the blue, absorption is maximal around 436 nm and in the red a maximum is found at about 682 nm. Additional light harvesting capacity is present in cyanobacteria. The phycobilisome antennae are situated on the stroma-exposed outside of the thylakoid membranes. The phycobilisomes enlarge the optical cross section of the organisms into the orange (around 628 nm) via the light harvesting function of the constitutively present blue chromophore phycocyanin. Additional light harvesting may also be induced in (selected) cyanobacteria after culture in orange light. Green light is largely left unabsorbed by most oxygenic phototrophic organisms. However, some cyanobacteria and also the eukaryotic red algae are able to harvest green light by means of another phycobilisome bound chromophore: the reddishly coloured phycoerythrin. The adaptation to different colours of light through changes in the phycobilisome rod structure polypeptide composition and matching complementation

* Present address: Fakultät Biologie, Universität Rostock, D-18055, Rostock
† Abbreviations: CCA, complementary chromatic adaptation; chl a, chlorophyll a; PC, phycocyanin; PE, phycoerythrin.

The Phototrophic Prokaryotes, edited by Peschek *et al.*
Kluwer Academic / Plenum Publishers, New York, 1999.

with the pigments phycoerythrin and/or phycocyanin has been studied in great detail in as far as the structure and function of the phycobilisome is concerned [1,2]. Evidence for the involvement of a two component signal transduction system has been revealed [3]. Evidence has been presented that the effector may function as a kinase [4]. Questions on the mechanism(s) involved in the sensing of the light quality (the sensor) have been posed [5]. Light reception and regulation of gene transcription has been related to the photosensor phytochrome [6,7]. A role for phytochrome has been advocated from complementation analysis of a "black" mutant in the cyanobacterium *Fremyella* [6]. Some recent observations, in the non CCA adapting cyanobacterium *Synechocystis* in which the accumulation of *phy* transcripts was studied indicated a role in light flux (light intensity) sensing [8]. Wilde et al. of the same team [9] reported that in addition to the *phy* gene histidine kinase motives of two component bacterial signal transduction may be involved in light-sensing. This might also apply to light quality sensing in the blue as was found from a deletion mutagenesis approach [9]. It might be concluded that phytochrome plays a broad role in light sensing, not only in quantitative light flux sensing, but also in a qualitative sense [7–9]. In this the phytochrome sensor likely plays a role in the red (and possibly blue) domain of the spectrum. Sensor systems in biology that act in the green comprise retinal linked to opsins, the so called rhodopsins [10–13]. In *Fremyella* and *Calothrix* retinal is present in minute amounts [14]. The research in this contribution reports on our continued efforts to disclose a possible role for a rhodopsin type sensor in the regulation of CCA in *Calothrix* [10,11,14].

2. MATERIALS AND METHODS

2.1. Strains and Culture

Two different isolates of the cyanobacterium *Calothrix* sp. have been used in our studies. One strain, referred to as *Fremyella diplosiphon* originated from the Pasteur Culture Collection in Paris and was long time ago (around 1980) transfered to us after intermediate storage in the collection of cyanobacteria at the University of Utrecht. The other strain, refered to as *Calothrix* sp. PCC 7601 was recently acquired from the Pasteur culture collection. Both strains were grown at 28°C in standard BG11 medium. Monochromatic light was obtained through coloured transparent plastic foil (Lee filters, Andover, England, #105 Orange for orange light conditions and a combination of #353 Lighter blue and #010 medium yellow for green light conditions, in some cases #353 was replaced by #172, lagoon blue to reduce the light transmission in the red domain above 680 nm. Light was provided to continuous cultures at a flux of 40 μmoles$\cdot$m$^{-2}\cdot$s^{-1} in culture. More recently we have started to apply light emitting diodes in culture [15]. These devices allow very narrow chromatic definition. In the green domain we used model L-53 SGC (565 nm) and L-934 IT (625 nm) in the orange domain. Apart from the fine spectral properties the use of LEDS as light source for culture allows these lamps to be switched on/off very swiftly in the subsecond time range. The total daily photon may be tuned electronically to arrive in all ratio's from the orange or green lamps without changing at large the geometry of the light supply as would be the case while using ordinary lamps. The LEDS thus promise an advantage in signal transduction studies which we are currently pursuing.

2.2. Use of Inhibitors and Reconstitution

Retinal biosynthesis was inhibited by nicotine addition (100 μM, final concentration). This inhibitor acts in the biosynthetic pathway of carotenoids, it prevents the cyclisation step in the transformation of lycopene into β-carotene [16]. The latter compound is the direct precusor of retinal. Addition of retinal to inhibited cells in order to test reconstitution was done with retinal (1 μM, final). In some experiments additional DTE (1 mM, final) has been used. β-carotene was disolved in n-hexane and 10 fold diluted in ethanol before addition to cultures (0.1% hexane, 1% ethanol and 1 mM β-carotene, final concentrations). In all experiments batch cultures of cells were inoculated in 25 ml of BG11 in 100 ml erlenmeyer flasks. After 48 h. additions of nicotine and reconstitution agents were made, after 1 h of incubation to the light conditions were switched to arrange for CCA (from orange to green or white, or from green to orange) as indicated in the results section. The experiment was continued for another 48 to 144 h, after which time growth in control cells was complete. In the presence of nicotine bleaching occurred after 48 to 72 h, any CCA in the presence of additions was therefore recorded during the first three days prior to cell death. Changes in the ratio of phycoerythrin to phycocyanin and of either pigment relative to protein was used as a measure for CCA ability.

2.3. Retinal Binding Studies

Fractionation of *Calothrix* cells from continuous cultures involved French Press cell breakage, treatment with Triton X 100 at a chl to detergent ratio of 25 [14] and (sucrose gradient) centrifugation. Membranes or subfractions thereoff with retinal (or radiolabeled ^{3}H retinal) were studied by double or dual wavelength spectroscopy to reveal a spectral change on an Aminco DW2000 spectrophotometer or from radio-autography of SDS-PAGE separated proteins (Geerdink et al., in prep.). Hydroxylamine washes were applied as in [13].

3. RESULTS AND DISCUSSION

3.1. Complementary Chromatic Adaptation, Inhibition by Nicotine

In view of the possible involvement of a rhodopsin in the regulation of CCA, we tested the effects of inhibition of its biosynthesis. In *Fremyella* a fast arrest of PE synthesis (in response to a transfer from orange to green (or white) light was observed in the presence of nicotine (Fig. 1).

An instantaneous inhibition in *Calothrix* PCC 7601 could not be revealed even with 100 μM nicotine, the inhibitor concentration that effected a complete arrest of PE synthesis in *Fremyella*. In *Calothrix* PCC 7601 PE synthesis continued for some 24 h after the nicotine addition. After this transient period, overall biosynthesis started to relapse, cells gradually became bleached and at 28°C a full growth arrest was observed after 3 days, similar for both cultures.

Nicotine showed inhibition of β-carotene synthesis in both *Calothrix* strains used in this work. HPLC analysis of acetone extracts showed the accumulation of lycopene in all cultures with nicotine (not shown). Our assumption is that the β-carotene pool in *Calothrix* PCC 7601 available for retinal synthesis may be slightly larger or better recruitable than in *Fremyella*. This would enable *Calothrix* to sustain a sufficiently large enough pool of functionally active rhodopsin as long as the cells would be able to perform biosynthesis. The effects of nicotine on the cell physiology and growth in time became also evident from a kinetic study of the changes in the carotenoid composition. The chl to β-carotene ratio rapidly rose during the first 2 or 3 days after nicotine addition, afterwards the ratio remains constant (Fig. 2). Differences in the kinetics were observed in response to differences in culture conditions (Fig. 2, top) and strains (not shown).

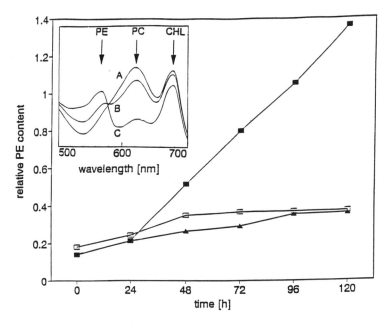

Figure 1. Effect of nicotine addition on CCA (from orange to green light), measured as the relative presence of PE. (▲), green light with 50 μM nicotine; (□), white light with 50 μM nicotine; (■), green light control. The insert shows the spectral region of interest (72 h following the light switch or addition of nicotine. A, orange light control; B, white light plus nicotine; C, green light control.

Effects in time of nicotine were also visible in studies of CCA. The data presented in Fig. 3 show that the full inhibition of CCA in *Fremyella* could be reversed in a number of cases by the addition of low concentrations of retinal to the cultures. This reconstitution was possible without but better in the presence of dithiothreitol, a reductant used in similar studies with archaebacteria to prevent oxidation of the aldehyde function [12,17]. Each experiment included in Table 1 comprised 8 independent groups of 3 erlenmeyer flasks. We assume that in cases in which the inhibition of CCA could not be reversed by retinal, cell bleaching occured before phycoerythrin synthesis could be detected from absorbance spectra.

The obvious experiment to use β-carotene for reconstitution has been tried, we were not able to homogenously distribute this hydrophobic substance in water, lack of transfer into the cells may explain why CCA, after inhibition by nicotine could not become reconstituted with β-carotene in *Fremyella*. It remains an open question whether the inhibition of CCA by nicotine in Fremyella is really a consequence of depletion of retinal and lack of functional rhodopsin. Attempts were therefore made to identify the potential retinal binding opsin protein that could be involved in CCA. For reasons of future exchange or comparison of results, all work below was continued with *Calothrix* PCC 7601 only.

3.2. Properties of *Calothrix* Membranes in Retinal Binding

Samples with the best binding properties (amplitude of the 540 nm peak, relative to protein) in spectroscopy as well as on gels were those taken from cultures the day following the change from orange light to green light.

In the experiments in which we combined isolated membrane preparations with retinal, it was on the one hand realised that absence of retinal binding might be the conse-

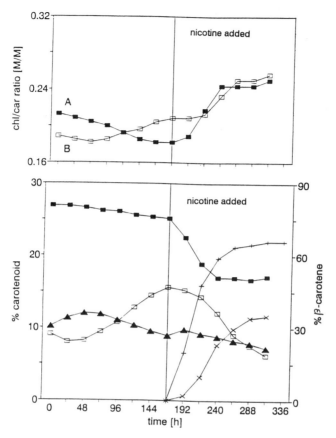

Figure 2. Changes in the relative pigment content of acetone extractable pigments in *Fremyella*, effect of nicotine addition. (*top*) ratio of chl a to β-carotene. A, cells in green light; B, cells in orange light. (*bottom*), details on carotenoid changes in culture A. (■), β-carotene; (+), lycopene; (□), xanthophyll; (▲), echinenone The effect of nicotine was fast and lasted in the growth conditions used (20°C) some 2 to 4 days. At 28°C this time time span was reduced to 1 to 2 days (data not shown). Afterwards cells stop all biosynthesis.

Figure 3. A comparison of visible light spectra of Fremyella, reconstitution of CCA through retinal and DTE addition in nicotine inhibited samples. Cells were pregrown in orange light, and except for (■), orange light control and (+), orange light control with 100 μM nicotine present, the cultures were transfered to green light at the start of the experiment, 2 days before the spectra were obtained. Additions were all made at the time that the light was switched from orange to green. (●), green light with nicotine; (×), green light with nicotine and retinal (1 μM, final); (▲), green light with nicotine, with retinal and additionally added DTE (1 mM) final; (◆), green light control.

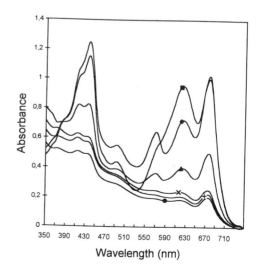

Table 1. Reconstitution tests of CCA in Fremyella diplosiphon, counteracting the inhibition by nicotine[a]

Pre culture light	Test light	Additions			CCA (number of positive responses)	
		Nicotine (µM)	Retinal (µM)	DTE (mM)	Relative PE content	Relative growth rate
Orange	Orange	—	—	—	3.6	100
Orange	White	—	—	—	100	96
"	"	100	—	—	8 (3)	38
"	"	100	0.5	—	14 (8)	39
"	"	100	1.0	—	16 (13)	34
"	"	100	10.0	—	12 (11)	
"	"	100	50.0	—	7 (16)	22
"	"	100	0.5	1	14 (15)	42
"	"	100	1.0	1	92 (11)	45
"	"	100	10	1	63 (9)	36
"	"	100	50	1	58 (3)	32
"	"	100	—	1	23 (4)	n.d.

[a]Erlenmeyer flasks of 100 ml with 25 ml of culture were pre grown in orange light before transfer to white light during incubation with additions as listed. The additions were made 1 h prior to the light switch. Spectra of the contents of the erlenmeyer flasks were obtained on an Aminco DW 2000 spectrophotometer. The PE presence was judged from the area underneath the spectra after normalisation on chl a content. Calculation was done with conversion factors from [18]. Only a limited number of CCA positive samples (that is when PE presence is above the orange light control) was found, as indicated in brackets, each experimental group had 3 erlenmeyers, the experiment was repeated 8 times, hence n=24. The numbers given for PE presence relative to the maximally adapting control had a large standard error of the mean, which amounted to 35% (not shown).

quence of a lack of free binding sites with endogenous retinal being bound to the opsin at all times. A large aspecific signal on the other hand would indicate that other proteins might provide a less characteristic but effective binding pocket for retinal. To anticipate we compared membranes with a different growth history, even from organisms without CCA ability. In difference spectroscopy an elegant method was designed which allowed determination of spectral differences of about one thousand's of an absorbance unit

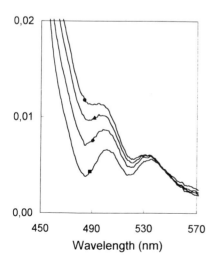

Figure 4. The kinetics of the development of spectral changes after addition of retinal (10 µM) to total membranes (10 µg chl·ml^{-1}). Spectra were taken between a reference blank to which only ethanol was added, and a measuring cuvette to which retinal in ethanol was added. Spectra were recorded after (■) 1 min, (♦), 2 min, (▲), 5 min and (●), 10 min. The signal around 510 nm (quite large in this experiment varied in size relative to 540 nm one, or v.v.).

against a background of 1 or 2 absorbance units. In double beam spectroscopy actual spectra that way were obtained. We noticed that a peak at 540 nm emerged quicly after mixing of the sample with retinal (Fig. 4). This peak overlaps surprisingly well with a peak in the long time known action spectrum of CCA in the green, which is also maximal at 540 nm [5]. Another spectral change at 490 to 510 nm was also related to the addition of retinal. The known action of the 540 nm signaling absorbance peak is to bring about the green light response i.e. the induction of phycoerythrin synthesis. After mixing of retinal and membranes, a broad and strong absorbance change without specific peaks below 500 nm gradually started to develop. In dual wavelength spectroscopy, (one beam with alternating measuring wavelength, i.e. 540 nm and reference at 600 nm through a single cuvette) the very fast growth of the 540 nm peak was estimated to be nearly fully established in less than 30 seconds after mixing (data not shown). Interestingly, tests with membranes from *Prochlorothrix hollandica*, *Planktothrix*, and *Microcystis* did not reveal the fast spectral change at 540 nm (not shown, but see [14]). Tests with hydroxylamine, which has been documented to free opsins from retinal in tests with other organisms [17], yielded only a disappointing decrease of the binding efficiency.

The spectroscopic observations were combined with protein separation and autoradiography studies. Binding of the tritiated retinal (which we synthesized ourselves, Geerdink et al., manuscript in prep.) appeared to be prominent to two polypeptides only. The binding between retinal and the putative opsin(s), in order to become visible on the autorads needed the transformation of the retinal-(lysine?) Schiff's base bond into a covalent bond by means of the strong reductant sodiumcyanogen borohydride (data not shown). The staining intensity of the two bands that would light up on the autoradiographs was rather low, exposure was routinely extended to 30 or more days (with amplifier, Geerdink et al., in prep.).

A correlation between the size of the spectral change and the blackening on the X-ray film was observed. Membrane fractionation with detergents and subsequent separation on sucrose gradients showed presence of the retinal binding capacity at a narrowly defined

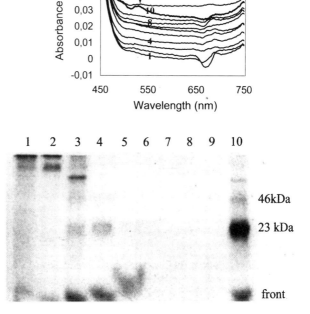

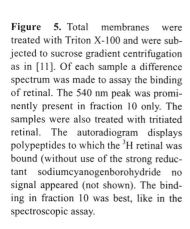

Figure 5. Total membranes were treated with Triton X-100 and were subjected to sucrose gradient centrifugation as in [11]. Of each sample a difference spectrum was made to assay the binding of retinal. The 540 nm peak was prominently present in fraction 10 only. The samples were also treated with tritiated retinal. The autoradiogram displays polypeptides to which the ^{3}H retinal was bound (without use of the strong reductant sodiumcyanogenborohydride no signal appeared (not shown). The binding in fraction 10 was best, like in the spectroscopic assay.

sucrose density. In this fraction (denoted as 10 in Fig. 5) typical cytoplasmic membrane carotenoids had accumulated and hardly any chlorophyll was present (not shown). The samples that were fractionated on a sucrose gradient were analysed. It is obvious from Fig. 5 that the sample with the largest spectral change at 540 nm also yielded the strongest signals on the X-ray film. Bands of two different molecular masses (21 to 23 kD and around 46 kD bands were routinely found labeled. However, hardly any protein was present in these bands (not shown).

Analysis of a FPLC separated fraction with appreciable ^{3}H counts, yielded only very faintly staining bands (silver stain) with masses of approximately 21 to 23 kD and of about 46 kD. These mass numbers were nicely in accordance with the earlier results from the gel data following sucrose gradient fractionation (cf. Fig. 5). Following determination of the amino terminal part of the putative opsin protein, inactivation mutagenesis or deletion of the gene that encodes the opsin would potentially help to learn more about the regulation of CCA and light quality signaling in cyanobacteria.

ACKNOWLEDGMENTS

We are grateful to Dr. W Hoff and Prof. J. Spudich, University of Texas, Houston, TX, USA for providing an initial supply of radiolabeled retinal, to Dr. H. Overkleeft, Organic Chemistry, University of Amsterdam for help in design of a pathway for the synthesis of the radiolabeled product ourselves, to U. van Hess for the design of the LED equiped culture set up, to Dr. W.R. Briggs (Carnegie Inst., Stanford, CA, USA), the Inst. Pasteur team from Paris and Prof. S. Golden (TX, USA) for helpful suggestions and inspiring discussions and to our undergraduate students, W. Spijker, A. Sletterink and M. van Abbe.

REFERENCES

1. Sidler, W. (1994) in The Molecular Biology of Cyanobacteria Bryant, D.A. ed., Kluwer Acad. Publ. pp. 139–216
2. Grossman, A.R., Scheafer, M.R., Chiang, G.G. and Collier, J.L.(1993) Microbiol. Rev. 57, 725–749.
3. Sobczyk, A., Schyns, G., Tandeau de Marsac, N., and Houmard, J. (1993) EMBO J. 12, 997–1004
4. Allen, J.F. and Matthijs, H.C.P. (1997) TIPS 2, 41–43
5. Haury, J.F. and Bogorad, L. (1977) Plant Physiol. 60, 835–839.
6. Chiang, G.G., Schaefer, M.R. and Grossman, A.R. (1992) Proc. Natl. Acad. Sci. USA 89, 9415–9419
7. Kehoe, D.M. and Grossman, A.R. (1996)Science 273, 1409–1412
8. Hughes, J, Lamparter, T., Mittmann, F, Hartmann, E., Gaertner, W, Wilde, A. and Boerner, T. (1997) Nature 386, 663 (1997).
9. Wilde, A., Churin, Y., Schubert, H. and Boerner, T. (1997) FEBS Lett. 406, 89–92
10. Hellingwerf, K.J., Crielaard, W., Hoff, W.D., Matthijs, H.C.P., Mur, L.R. and van Rotterdam, B.J.(1994) Antonie van Leeuwenhoek 65, 331–347
11. Hoff, W.D., Matthijs, H.C.P., Schubert, H., Crielaard, W. and Hellingwerf, K.J. (1994) Biophys. Chem. 56, 193–199
12. Beckmann, M. and Hegemann, P. (1991) Biochemistry 30, 3692–3697.
13. Barsanti, L., Passarelli, V., Lenzi, P, Walne, P.L. and Gualteri, P. (1993) Vision Res. 33, 2043–2050
14. Geerdink, J.H., Haker, A., Matthijs, H.C.P., Hoff, W.D., Hellingwerf, K.J. and Luuc, R. Mur (1995) in Mathis, P. ed., Photosynthesis: from Light to Biosphere, Vol. 1, Kluwer Acad. Publ., 303–306
15. Matthijs, H.C.P., Balke, H., van Hes, U., Kroon, B.M.A., Mur, L.R. and Binot, R.A. (1996) Biotechnol. and Bioengineer. 50, 98–107
16. Howes, C.D. and Batra, P.P. (1970) Biochim. Biophys. Acta 222, 174–179
17. Gualtieri, P, Pelosi, P., Passarelli, V. and Barsanti, L. (1992) Biochim. Biophys. Acta 1117, 55–59.
18. Tandeau de Marsac, N.and Houmard, J. (1988) Methods in Enzymol. 167, 318–328.

23

CHARACTERIZATION OF A *SYNECHOCYSTIS* PHYTOCHROME

Thomas Börner,[1] Thomas Hübschmann,[1] Annegret Wilde,[1] Yuri Churin,[1] Jon Hughes,[2] and Tilman Lamparter[2]

[1]Institut für Biologie
Humboldt-Universität
Chausseestr. 117, 10115 Berlin, Germany
[2]Institut für Pflanzenphysiologie und Mikrobiologie
Freie Universität
Königin-Luise-Str. 12-16, D-14195 Berlin, Germany

1. INTRODUCTION

Phytochromes are photoreceptors common to lower and higher plants. Plant phytochrome apoproteins autocatalytically attach phytochromobilin, a linear tetrapyrrole chromophore [1]. They regulate a wide variety of photomorphogenetic responses of plants to light [2]. The total sequence of the *Synechocystis* sp. PCC 6803 chromosome revealed the existence of a putative phytochrome gene (*phy*, ORF slr0473) [3]. The *phy* gene potentially encodes a protein (PhyS) that exhibits homology to phytochromes throughout its length, in particular in the N-terminal region including the chromophore binding site [4]. Moreover, the putative *Synechocystis* phytochrome shows distinct homology to sensory histidine kinases. Sensory histidine kinases and the so-called response regulators form the "two component systems" of signal transduction in bacteria [5]. Two component systems have been discovered more recently also in eukaryotes including components involved in the ethylene-induced signal transduction in *Arabidopsis* [6]. The *Synechocystis* chromosome encodes a number of putative response regulators and histidine kinases [3]. Several of these kinases contain stretches of amino acids with limited similarity to the N-terminal domain of phytochromes and to domains of the ethylene receptor protein of *Arabidopsis* (summarized in Table 1). We named these putative histidine kinases "phytochrome like proteins" (abbreviated: *plp*) [7]. We disrupted the gene of one of these *plp*'s (*plpA*, encoded by ORF sll1124 [3]) by directed insertional mutagenesis. The obtained mutant was found to be unable to grow under blue light [7].

In the following, evidence is reported indicating that PhyS is a functioning cyanobacterial phytochrome.

The Phototrophic Prokaryotes, edited by Peschek *et al.*
Kluwer Academic / Plenum Publishers, New York, 1999.

Table 1. Open reading frames in the genome of *Synechocystis* with similarity to genes encoding a phytochrome (PhyE) and a putative ethylene receptor (ETR1) of *Arabidopsis*[a]

	phy1	phy2	phy3	R2L	T2L	His-kinase
slr0473 (PhyS)	+++	++	+++			yes
sll1124 (PlpA)	+		++	+	++	yes
sll0821 (PlpC)	+++		++	++	++	no
sll1473 (PlpD)	++		++		++	no
slr1212 (PlpB)	++			+	+	yes
slr1393 (PlpE)	++			++	+	yes
slr1969	+					yes

[a]Phy1, phy2 and phy3 represent subdomains of the chromophore attachment region of PhyE; R2L and T2L designate amino acid motifs that are conserved in the putative ethylene receptors of plants including ETR1 [12]. Plus signs indicate the degree of similarity with "+++" - high similarity and "+" - low similarity.

2. RESULTS AND DISCUSSION

2.1. Autoassembly, Spectral Characteristics, and Phosphorylation

We expressed the *phy* gene in *E. coli*. The isolated recombinant protein was found to autoassemble with phycocyanobilin (PCB) likely to be the authentic chromophore within cyanobacterial cells [8] as well as with the phytochromobilin as known from plant phytochromes [4]. After autoassembly the sample was divided and each portion irradiated with far-red or red light. Two thermostable forms of PhyS were observed (Fig. 1). Photoconversion is a characteristic feature of phytochromes. Initial photoconversion kinetics were monitored spectrophotometrically at 658 and 730 nm during irradiation. Light intensity and PhyS concentration was adjusted to get first order kinetics during the 1st minute.

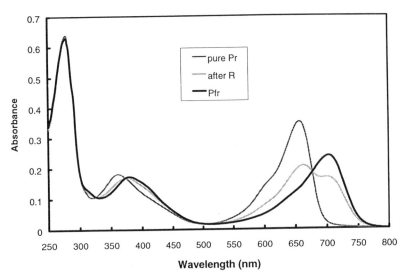

Figure 1. Spectra of PhyS after assembly with PCB in the dark (pure Pr) and after photoconversion with saturating red light (657 nm). The Pfr spectrum is calculated from the other spectra using the data from photoconversion kinetic measurements.

chromophor	-	+	+
irradiation (720 nm)	-	-	+

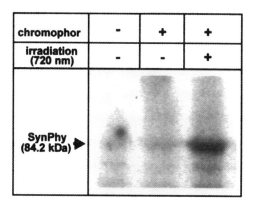

SynPhy (84.2 kDa) ▶

Figure 2. *In vitro* autophosphorylation of recombinant PhyS protein. Recombinant PhyS protein with attached (chromophor +) and without phytochromobilin (chromophor −) was irradiated by day light (irradiation −) or far red light (irradiation +) for 5 min. After incubation with [γ-^{32}P]ATP the protein was analyzed by SDS-PAGE and autoradiography.

These data were used to calculate $[Pfr]_{max, 658 nm}$ to be 0.68, allowing the Pfr spectrum to be derived from the raw absorbance data [8]. Spectra are shown in Fig. 1. The spectral characteristics of PhyS were remarkably similar to those of plant phytochromes [4,8]. Attachment of phycocyanobilin as chromophore led to a blue-shift of the absorbance maxima and isosbestic point (658, 702, and 677 nm, respectively), also known from plant phytochromes [e.g. 9]. PhyS with phytochromobilin as chromophore showed spectra comparable to those obtained with oat phytochrome A [4].

The first step of signal transduction by sensory histidine kinases is autophosphorylation in dependence upon an environmental stimulus. Thus, if PhyS functions as a phytochrome, one would expect a red- and/or far red-dependent autophosphorylation. Therefore, recombinant PhyS protein with phytochromobilin as chromophore was irradiated with day light and far red light, respectively, and assayed *in vitro* for autophosphorylattion [10]. Our data indicate a higher rate of autophosphorylation of recombinant PhyS protein after irradiation with far red light versus irradiation with day light (Fig. 2).

2.2. Inactivation of the *phy* Gene

As a first step in the process of elucidating the function of PhyS we inactivated the *phy* gene by insertional mutagenesis. An intragenic BalI fragment was replaced by a chloramphenicol resistance gene cartridge from pACYC184. Chloramphenicol-resistant colonies were selected essentially as described [11]. Southern hybridization demonstrated that the transformed cells contained only the inactivated version of the *phy* gene (Fig. 3A). Northern hybridization did not reveal a *phy* transcript in wild-type and mutant cells. Obviously, transcription of this gene is very weak. Therefore, we used an RNase protection assay to detect the transcript. For this purpose, an *in vitro* labelled *phy*-antisense probe was hybridized with total RNA from wild-type and mutant cells and subjected to RNase degradation. The probe used was complementary to a 3'-region of the *phy* gene downstream the *cat* cassette insertion. Whereas in wild-type cells a *phy* mRNA was detectable, no such messenger was found in the mutant (Fig. 3B). Thus, Southern hybridization and RNase protection assay indicate that the chloramphenicol-resistant mutant cells lacked PhyS (although, according to the construct used for mutagenesis, a truncated version of the protein might still be synthesized). The *phy*⁻ cells grow well under white, red, green or blue light. This finding was surprising in view of our data on another mutant lacking one of the phytochrome-like proteins: insertional mutagenesis of the *plpA* gene led to cells unable to grow under blue light [7].

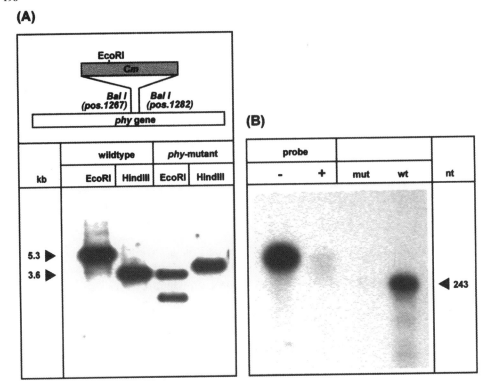

Figure 3. Insertional inactivation of the *phy* gene and RNase protection analysis of *phy* gene transcription in wildtype and *phy*-mutant cells of *Synechocystis* 6803. **(A)** Total DNA of wildtype and *phy*-mutant cells was digested with *Eco*RI and *Hind*III, separated by gel electrophoresis, transferred onto nylon membrane and probed with a *phy* gene DNA fragment. The sizes of wildtype DNA fragments are indicated at the left in kilobases. The strategy of insertional inactivation is shown in the scheme; Cm, chloramphenicol resistance. **(B)** Total RNA of wildtype (wt) and *phy*-mutant (mut) cells was hybridized with an *in vitro* labelled *phy* gene antisense probe and subjected to RNase degradation. Lane (−) shows the probe itself without RNase treatment, lane (+) shows the probe after RNase degradation. The size of the protected fragment is given at the right in nucleotides.

Taken together, our results support the idea that PhyS represents a cyanobacterial phytochrome which functions as a light-sensing histidine kinase. The next step in signal transduction should be the transfer of the phosphate group from PhyS to a response regulator protein [5]. Such a regulator protein is encoded by ORF slr0474 which is located next to the *phy* gene on the chromosome. We generated also a mutant version of ORF slr0474. Currently, we are investigating the mutants for a phenotype differing from the wild type in order to gain information about the function of PhyS and the putative response regulator in light-dependent signal transduction.

ACKNOWLEDGMENTS

Part of this work was funded by the Deutsche Forschungsgemeinschaft (T.B., J.H., T.L.). We appreciate the contribution of the team at Kazuka Institute in making the *Synechocystis* genome data available to the scientific community. We are grateful to Kurt Schaffner, Silvia Braslavski and Wolfgang Gärtner (MPI, Mühlheim) for generous support and valuable discussions, to Ingo Lindner for providing phycocyanobilin, to Franz

Mittmann for initial cloning of *phy* in *E. coli*, and to Tabea Börner and Sabine Buchert for excellent technical help.

REFERENCES

1. Rüdiger, W., Thümmler, F. (1994) in: Photomorphogenesis in Plants (R.E. Kendrick, G.H.M. Kronenberg, eds), 2nd Edition, pp. 51–70, Kluwer Academic Publ., Dordrecht.
2. Koornneef M., Kendrick, R.E. (1994) in: Photomorphogenesis in Plants (R.E. Kendrick, G.H.M. Kronenberg, eds), 2nd Edition, pp. 601–630, Kluwer Academic Publ., Dordrecht.
3. Kaneko, H., Sato, S., Kotani, H., Tanaka, A., Asamizu, E., Nakamura, Y., Miyajima, N., Hirosawa, M., Sugiura, M., Sasamoto, S., Kimura, T., Hosochi, T., Matsuno, A., Muraki, A., Nakazaki, N., Naruo, K., Okumura, S., Shimpo, S., Takeuchi, C., Wada, T., Watanabe, A., Yamada, M., Yasuda, M. and Tabata, S. (1996) DNA Res. 3, 109–136.
4. Lamparter, T., Mittmann, F., Gärtner, W., Börner, T., Hughes, J. (1997) Proc. Natl. Acad. Sci. USA, in press.
5. Hoch, J.A., Silway, T.J. (eds.) (1995) Two Component Signal Transduction. ASM Press, Washington D.C.
6. Hua, J., Chang, C., Sun, Q., Meyerowitz, E.M. (1995) Science 269, 1712–1714.
7. Wilde, A., Churin, Y., Schubert, H., Börner, T. (1997) FEBS Lett. 406, 89–92.
8. Hughes, J., Lamparter, T., Mittmann, F., Hartmann, E., Gärtner, W., Wilde, A., Börner, T. (1997) Nature 386, 663.
9. Kunkel, T., Neuhaus, G., Batschauer, A., Chua, N.-H., Schäfer, E. (1996) Plant J. 10, 625–636.
10. Bloye, S.A., Silman, N.J., Mann, N.H., Carr, N.G. (1992) Plant Physiol 99, 601–606.
11. Wilde, A., Härtel, H., Hübschmann, T., Hoffmann, P., Shestakov, S.V., Börner, T. (1995) Plant Cell 7, 649–658.
12. Kehoe, D.M., Grossman, A.R. (1996) Science 273, 1409–1411.

24

PHOTOSYNTHESIS AND RESPIRATION OF CYANOBACTERIA

Bioenergetic Significance and Molecular Interactions

Günter A. Peschek

Biophysical Chemistry Group
Institute of Physical Chemistry
University of Vienna
Währingerstraße 42, A-1090 Wien, Austria

INTRODUCTION

Cyanobacteria (blue-green algae) are oxygenic phototrophic bacteria carrying out water-splitting, O_2-releasing (viz., plant-type) photosynthesis and water-forming, O_2-reducing respiration in one and the same "noncompartmentalized" prokaryotic cell [1]. Two types of morphologically more or less separate membrane systems occur in cyanobacteria, the chlorophyll-containing intracytoplasmic or thylakoid membranes (ICM, "photosynthetic lamellae") and the chlorophyll-free cytoplasmic or plasma membranes (CM). Both ICM and CM partly contain identical electron transport components forming a dual function photosynthetic-respiratory assembly in ICM but, as necessitated by the absence of chlorophyll, a purely respiratory chain in CM. Both ICM and CM surround individual and osmotically autonomous cellular compartments, viz. the intrathylakoid space and the cytosolic space, respectively. (The periplasmic space, characteristic of Gram-negative bacteria and sandwiched together with other cell wall constituents such as peptidoglycan between CM and the outer membrane, is in osmotic equilibrium with the external medium of the cell and, therefore, is not counted as intracellular compartment.).

The advent of primordial cyanobacteria in a previously anoxic, O_2-free, biosphere about 3.2 billion years ago [2] marked the decisive turning point in evolution which has made possible the ensuing development of all "higher" forms of life, from primitive eukaryotes to *Homo sapiens*. Clearly, as the primordial cyanobacteria were the first to liberate bulk amounts of ("toxic"! See Ref. 3) oxygen gas into the anoxic biosphere they also must have been the first to be affected by it. Thus it seems likely that cyanobacteria were (among) the first to elaborate a mechanism for aerobic respiration essentially by modifying and adapting pre-existing photosynthetic electron transport systems ("conversion

The Phototrophic Prokaryotes, edited by Peschek *et al.*
Kluwer Academic / Plenum Publishers, New York, 1999.

hypothesis", see Refs. 4,5). Thereby the fundamental chemiosmotic (membrane) principle of energy conversion could be retained. A recent phylogenetic tree-based claim that aerobic respiration evolved from denitrification (e.g. NO reduction; see Refs. 6,7), representing a repetition of a still older idea put forward by Egami et al. [8] faces the problem that the stability (life time) of such oxidized nitrogen species is negligible in the presence of sufficient reducing agents (sulfide, ferrous iron, H_2, CO, NH_3, etc.) as are believed to have been present in the primeval atmosphere and biosphere, and as can be verified by Urey-Miller type experiments (e.g. [9]). On the other hand, some primitive type of anaerobic respiration ("pre-respiration", see Ref. 5) using electron acceptors readily available also in the absence of O_2, such as fumarate, might well have preceded oxygenic photosynthesis and thus, a fortiori, aerobic respiration.

The present article will discuss common features and separate traits of oxygenic photosynthesis and aerobic respiration in cyanobacteria. Emphasis will rest on membrane-bound (electron transport) components and on the comparison of such components with those of chloroplasts and mitochondria. Functionally, our results on cyanobacteria could be reconciled with a so-called "generalized endosymbiont hypothesis" according to which, eventually, both chloroplasts and mitochondria of eukaryotes originated from endocyanelles. This assumption gains new interest in view of the recent discovery of vestiges of oxygenic photosynthetic organelles ("thylakosomes") in several protists such as *Psalteriomonas lanterna* and representatives of the parasitic genus *Apicomplexa* [10,11].

THE STEADY-STATE OF OUR TERRESTRIAL BIOSPHERE AND ATMOSPHERE

Eq. 1[*] gives the chemical stability criterion of our biosphere in terms of the indispensable balance between oxygenic photosynthesis and aerobic respiration. In the endergonic direction (carbohydrate i.e. biomass synthesis from CO_2 and H_2O) eq. 1 is driven by light. Other self-sustaining biological systems such as sulfureta or chemosynthetic primary producers surrounding the so-called "black smokers" are without any significant quantitative impact on terrestrial life.

$$6\ CO_2 + 6\ H_2O = C_6H_{12}O_6 + 6\ O_2; \quad \Delta G'_o = 2821.5\ kJ \tag{1}$$

An estimated 100 billion tons of carbon (in the form of CO_2) per year is converted into biomass by plant-type photosynthesis (primary biomass production) and the equivalent amount of O_2 is thereby released from water according to eq. 1. Recent estimates assign between 20 and 30% of this world-wide primary productivity to cyanobacteria, in particular to small unicellular *Synechococcus* species [12] and likewise unicellular planctonic *Prochlorophytes* [13] which, though not especially concentrated anywhere in the euphotic zone of the oceans are nevertheless extremely widespread in *all* oceans. Thus apart from the evolutionary privilege of having been the first to deal with oxygen gas, also ecologically the cyanobacteria claim the privilege of ranking highest among photosynthetic primary producers.

Fig. 1 shows exchange rates between, and pool sizes of, oxygen reservoirs in the atmosphere. It can be calculated that the mean residence time (turnover time) of oxygen is 6000 years in O_2, 500 years in CO_2, but 20 million years in H_2O [4].

---

[*] Mechanistically, conforming to the well-known fundamental Van Niel equation of autotrophic CO_2 fixation, another 6 H_2O would have to be added to both sides of the equation.

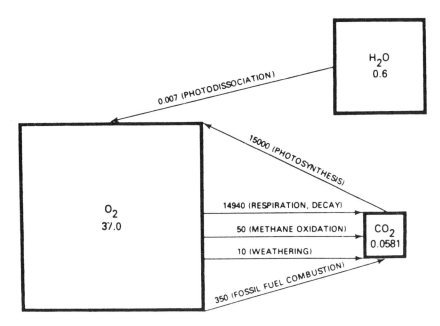

RESERVOIRS ARE IN EMOLES O_2 AND RATES ARE IN EMOLES O_2/MYR

Figure 1. Oxygen exchange pattern in the atmosphere; pool sizes in squares. One Emole (erdamol) = 10^{18} moles. One Myr = 10^6 years. Exchange rates given in Emoles O_2/Myr. Pool sizes are given in Emoles. After Ref. 14.

EVOLUTION OF THE O_2 CONTENT OF THE ATMOSPHERE

Fig. 2 shows the increase in the O_2 concentration of the atmosphere during the last four billion years. Before the oxygen gas entered the atmosphere 1.2 billion years of chemical reduction by the huge amount of reductants (sulfide, ferrous iron) in the water bodies, where the cyanobacteria produced it, had elapsed. However, 0.4 billion years thereafter already the first eukaryotic cells appeared (as evidenced by the microfossil record [15]). Note that there is no primary anaerobe among extant eukaryotes, anaerobic protists or worms reflecting the drop of respiration by adaptation to peculiar anaerobic habitats.

With the advent of eukaryotes a violent increase in the atmosphere's oxygen budget ensued. The Pasteur point was reached 0.6 billion years later marking the time when the global energetic efficiency (ATP production) of respiration exceeded that of fermentation (substrate level phosphorylation) due to increasing availability of oxygen. Finally, at the Berkner-Marshall point an atmospheric O_2 concentration (approx. 2%, vol/vol) was reached when a substantial ozone layer started to build up in the upper atmosphere, thus protecting the surface of the earth from deleterious UV and other ionizing radiation from the outer space, opening up the era of terrestrial life just at the beginning of the paleozoic period of earth's history. During this time the initially almost unlimited proliferation of land plants led to a transient overshoot in the atmosphere's O_2 concentration (cf. the massive paleozoic vegetation that gave rise to our present-day fossil fuel reserves) which, through a likewise massive development of animals finally approached the contemporary level of 21% (vol/vol). Table 1 summarizes the crucial stages in the evolution of our universe and our earth as viewed against the background of the creation myth (Genesis) given

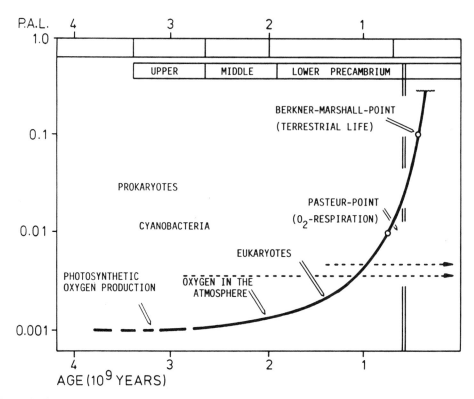

Figure 2. The oxygen content of the earth's atmosphere during the previous four billion years. P.A.L., present atmospheric level (fractions of 21% v/v). The diagram has been compiled from [4,14,16,17].

Table 1. Summary of the crucial steps in the evolution of the universe and of our earth. Left-hand side: Original sentence (Mose 1,1) introducing the creation myth described in the Holy Bible

בְּרֵאשִׁת

	Origin or appearance of	Years ago $(\times 10^9)$
בָּרָא אֱלֹהִים		
אֵת הַשָּׁמַיִם	Universe	18
וְאֵת הָאָרֶץ:	Sun	10
וְהָאָרֶץ הָיְתָה	Earth (solid crust)	4.6
	Life (prokaryotes)	3.5
תֹהוּ וָבֹהוּ	CYANOBACTERIA (O_2!)	3.2
וְחֹשֶׁךְ עַל־פְּנֵי	Atmospheric O_2	2
	Eukaryotes	1.6
תְהוֹם	Terrestrial biosphere (O_3-shield)	0.4
וְרוּחַ אֱלֹהִים	Homo erectus	0.0007–4
	Homo sapiens	0.00005
מְרַחֶפֶת		

עַל־פְּנֵי הַמָּיִם:

by Mose 1,1 who successfully competes with sophisticated astrophysics for the most comprehensible explanation of how the universe might have come into existence.

The oxygen gas released into the biosphere by the primordial cyanobacteria in ever increasing amounts must have threatened all of the still anaerobic organisms including the cyanobacteria themselves which, of course, were closest to the oxygen they produced. Thus long before any specialization towards respiratory metabolism the pre-existing membrane-bound photosynthetic electron flow chain adopted respiratory functions through integration of a dehydrogenase on the low-potential side and an oxidase (O_2-reducing) on the high-potential side. Fig. 3 shows the basic principle of such dual-function photosynthetic-respiratory assembly as we still know it from extant cyanobacteria [18,19]. In view of the functional similarity between respiratory and photosynthetic electron-transport components such transformation cannot have been too difficult.

With the advent of eukaryotes (most probably through some primary endosymbiotic event independent relics of which might still be the extant forms of endosymbionts such as *Cyanophora paradoxa* [20]) a specialization of a cell's bioenergetic processes occurred: Photosynthesis was compartmentalized within the chloroplast (presumably derived from an endosymbiotic cyanobacterium [21]) and respiration was compartmentalized within the mitochondrion (derived from an endosymbiotic protobacterium such as *Paracoccus denitrificans*, or from an endosymbiotic cyanobacterium as well?). Fig. 4 compares bioenergetic membrane functions in a cyanobacterium, a mitochondrion, and a chloroplast. The striking functional analogy of the membranes, e.g. with respect to sidedness and polarity of H^+ translocation, electron transfer into and out of the membranes, ATP synthesis, etc. is evident.

CYANOBACTERIA VIEWED AS "FREE-LIVING CHLOROMITOCHONDRIA"—THE FUNDAMENTAL BIOENERGETIC DUALITY

As we have seen, bioenergetically, the balance between oxygenic photosynthesis and aerobic respiration (Eq. 1) keeps our world going, so to speak. The only organisms that accommodate both processes in a single (prokaryotic!) cell are the cyanobacteria. They should, therefore, display both chloroplast and mitochondrial bioenergetic features at the same time in the same cell. Indeed, during the last decade kinetic, immunological, spectroscopic and many other types of experiments on cyanobacteria, most of them performed in my laboratory, have led to the conclusion that cyanobacteria may contain two separate, yet functionally identical membrane-bound electron transport (and other bioenergetic) components one of which bears

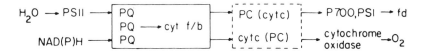

Figure 3. Dual-function photosynthetic-respiratory electron transport in cyanobacterial thylakoid membranes (ICM). Membrane-bound common components are boxed by solid lines, soluble common components are boxed by broken lines. PSII, PSI, photosystems II and I; PQ, plastoquinone; cyt, cytochrome; PC, plastocyanin; fd, ferredoxin; P700, chemically active chlorophyll a species in PSI.

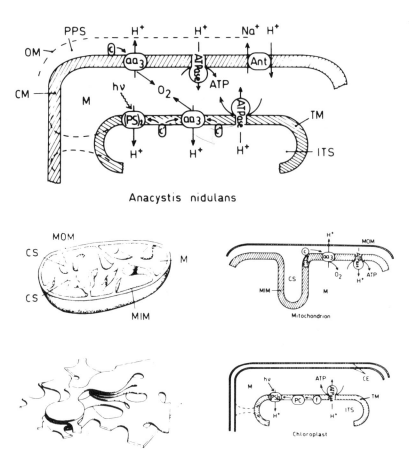

Figure 4. Comparison of bioenergetic membrane functions in a cyanobacterium, a mitochondrion and a chloroplast. Left-hand side: hypothetical stereographic views. CM, plasma or cytoplasmic membrane; ICM (TM), intracytoplasmic or thylakoid membrane; PPS, periplasmic space; OM, outer membrane; MOM, mitochondrial outer membrane; MIM, mitochondrial inner membrane; ITS, intrathylakoid space; M, matrix; Ant, Na^+/H^+ antiporter; aa_3, cytochrome oxidase.

close relationship to mitochondrial protagonists while the other bears close relationship to those described in chloroplasts (Fig. 5). The cyanobacterial species with most conspicuous features of such duality is *Synechocystis* PCC6803 [22,23]. Less pronounced but anyhow experimentally clearly distinguishable is the duality in *Anacystis nidulans* [23] and in *Anabaena* PCC7120 [24]. Further research will have to show to what extent the duality can be generalized among cyanobacteria, or to which species it might apply.

In the long run, as every protein should be coded by its own gene molecular genetics will have the last word. On the other hand, cyanobacterial genetics is still in its infancy and it is only now that the group of M. Sugiura in Japan is tentatively presenting a complete genetic map of *Synechocystis* (Sugiura, personal communication). At any rate, the rather boring and monotonous job of gene sequencing can only show if, and how, a given protein might possibly be coded for in a given cell; it will never give any clue, however, to the integral functioning of proteins in a cell, but it is this functioning on which life depends.

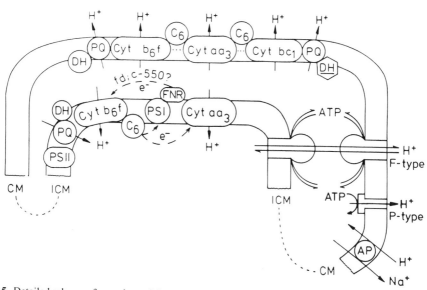

Figure 5. Detailed scheme of cyanobacterial membrane functions and sidedness showing the dual-function photosynthetic/respiratory electron transport chain in the ICM and the dual-structural purely respiratory chain in the CM. FNR, fd-NADP reductase; c_6, soluble cytochrome c-554 (may functionally replace plastocyanin in certain species); DH, NADH dehydrogenases, NDH1 boxed, NDH2 circled. For other abbreviations see Fig. 3. See the text for discussion.

To sum up the bioenergetic duality as it applies in *Synechocystis* and, to roughly the same extent also in *Anacystis*, the best investigated species so far, reflects the following situation: Typical components of oxygenic photosynthesis are a high-potential photosystem II and a low-potential photosystem I, plastoquinone, NADPH as the functional reductant in the Calvin cycle and, though not a bioenergetic component, proper, ß-carotene. Typical components of mitochondrial respiration are the H^+-pumping, multi-subunit NADH dehydrogenase [22,24,25] and the aa_3-type cytochrome-c oxidase [19,26–28]. A bacterial trait which the latter might have retained is the ability of a single COI polypeptide to combine either with heme A or heme O according to ambient oxygen concentrations [29–31]. Like other bacterial oxidases the cyanobacterial enzyme contains much less than the mitochondrial 13 subunits [32,33], but in addition to subunits I, II and III which are characteristic of most bacterial oxidases, too [34], and a small ctaFx (ORF4) gene product which is totally unrelated to other small bacterial subunits IV, the cta operon of *Synechocystis* contains a ctaF gene which may code for a large (187 aa, one or two transmembrane helices) "mitochondria-like" subunit IV which, in addition, may confer the ATP-/ADP-regulated activity pattern (see Ref. 35, 36) on the cyanobacterial enzyme [37,38].

The mitochondria-like NADH dehydrogenase was discovered, again in the form of a few subunits only while the mitochondrial complex I comprises up to 42 subunits [39], in chloroplasts, too, where its genuine function is totally mysterious but it might be a relic from days when chloroplasts were still cyanobacteria, so to speak, according to the endosymbiont hypothesis [21]. In cyanobacteria the problem with this enzyme would be that by far the most efficient, if not only, substrate dehydrogenation pathway is the oxidative pentose phosphate cycle and that this invariably produces NADPH instead of NADH [see

Refs. 18, 19]. According to an early, yet fairly convincing experiment [40], therefore, NADPH is still regarded the natural electron donor to the cyanobacterial respiratory chain in vivo. *Synechocystis* (and, to a lesser extent, also *Anacystis* and *Anabaena*) membranes, particularly CM [22,41] reduced cytochrome-c and artificial quinones with NADH in a reaction markedly inhibited by rotenone and piericidin A and gave specific immunoblotting with antibodies raised against certain subunits of chloroplast, mitochondrial and cyanobacterial NADH dehydrogenases [23]. On the other hand there is no doubt that isolated and highly purified membranes from cyanobacteria do oxidize both NADH and NADPH, specific rates with the former usually even being significantly higher than with the latter [27,42]. Unfortunately, also the search for an efficient NADPH/NAD$^+$ transhydrogenase in cyanobacteria has been negative so far. Thus the question of the "true" electron donor to the respiratory chain of cyanobacteria in vivo must be left open.

The joint presence of a chloroplast-like (b_6f-type) and a mitochondria-like (bc_1-type) cytochrome-c_6 (or plastocyanin) reductase in *Synechocystis* (perhaps less conclusively also in *Anacystis* and *Anabaena*), was inferred from the fact that NADH-cyt-c reduction was markedly sensitive to antimycin A, particularly when the reaction was catalyzed by CM [22,41]. Immunoblotting of the cyanobacterial membrane proteins with monospecific and monoclonal antibodies raised against beef heart mitochondrial or *P. denitrificans* cytochrome c_1 gave clearcut positive results [22,23].

It seems therefore that at least certain cyanobacteria (such as *Synechocystis* and *Anacystis*) contain both chloroplast-like and mitochondria-like NADH dehydrogenase and cytochrome-c reductase (bc-complex) and that the concentration of the "mitochondrial" complexes is at least higher in CM than in ICM.

A final piece of bioenergetic duality is evident from the observation that cyanobacterial CM may contain both F-type and P-type ATPases while ICM contains the bioenergetically most relevant F-type (ATP synthase) only. This conclusion was inferred from Jagendorf-type (base-acid shift) experiments [19], functional studies on isolated membranes [43] and immunogold labeling [23,43]. The latter approach has also confirmed the presence of aa_3-type cytochrome oxidase (in both CM and ICM), of cytochrome c_1 and NDH1 subunits (mostly in CM), cytochromes f and subunit IV (of the b_6f complex), the Rieske protein, the Mn-stabilizing protein of PSII in both CM and ICM (but mostly ICM), and the soluble cytochrome c_6 in both intrathylakoid and periplasmic spaces (see Refs. 44–48).

ACKNOWLEDGMENTS

Work in the author's laboratory has been generously supported by the Austrian Ministry for Science and Research, the Austrian Research Community, the Austrian Science Foundation and the Kulturamt der Stadt Wien. Thanks are due to Miss Alexandra Messner, Miss Irene Steininger, Mr. Thomas Zehetbauer and, in particular, to Mr. Otto Kuntner for devoted and most skilful technical assistance during the many years.

REFERENCES

1. Stanier, R.Y. and Cohen-Bazire, G. (1977) Annu.Rev.Microbiol. 31, 225–274.
2. Barghoorn, E.S. and Schopf, J.W. (1966) Science 150, 758–763.
3. Morris, J.G. (1975) Adv.microbial Physiol. 12, 169–246.
4. Broda, E. (1975) The Evolution of the Bioenergetic Processes, Pergamon Press, Oxford.
5. Broda, E. and Peschek, G.A. (1979) J.theor.Biol. 81, 201–212.

6. Castresana, J. and Saraste, M. (1995) TIBS 20:443–448.
7. Castresana, J., Lübben, M. and Saraste, M. (1995) J.Mol.Biol: 250:202–210.
8. Egami, F. (1974) Origins of Life 5: 405–411.
9. Zohner, A. and Broda, E. (1979) Origins of Life 9:291–298.
10. Hackstein, J.H.P., Schubert, H., v.d. Berg, M., Brul, S., Derksen, J.W.M. and Matthijs, H.C.P. (1993/94) Endocytobiosis & Cell Res. 10, 261 (Abstract 261).
11. Hackstein, J.H.P., Mackenstedt, U., Mehlhorn, H., Meijerink, J.P.P., Schubert, H. and Leunissen, J.A.M. (1995) Parasitol.Res. 81:207–216.
12. Waterbury, J.B., Watson, S.W., Guillard, R.R.L. and Brand, L.E. (1979) Nature 277:293–294.
13. Chisholm, S.W., Olson, R.J., Zettler, E.R., Goericke, R., Waterbury, J.B., Welschmeyer, N.A. (1988) Nature 340:340–343.
14. Gilbert, D.L. (ed., 1981) Oxygen and Living Processes: An Interdisciplinary Approach. Springer Verlag, New York, Inc.
15. Knoll, A.H. (1992) Science 256, 622–627.
16. Rutten, G.M. (1966) Paleogeogr. Paleoclimatol.Paleoecol. 2, 47–57.
17. Schidlowski, M. (1971) Geol.Rundschau 60, 1351–1384.
18. Peschek, G.A. (1984) Subcell.Biochem. 10, 85–191.
19. Peschek, G.A. (1987) in The Cyanobacteria (Fay,P. and Van Baalen, C., eds.), 119- 161, Elsevier, Biomedical Division, Amsterdam - New York - Oxford.
20. Wasmann, C.C., Loeffelhardt, W. and Bohnert, H.J. (1987) in P. Fay and C. van Baalen (eds.): The Cyanobacteria, Elsevier, Amsterdam, pp. 303–324.
21. Gray, M.W: and Doolottle, W.F. (1982) Microbiol. Rev. 46:1–42.
22. Dzelzkalns, V.A., Obinger, C., Regelsberger, G., Niederhauser, H., Kamensek, M., Peschek, G.A. and Bogorad, L. (1994) Plant Physiol. 106, 1435–1442.
23. Dworsky, A., Mayer, B., Regelsberger, G., Fromwald, S. and Peschek, G.A. (1995) Bioelectrochem.Bioenerg. 38, 35–43.
24. Howitt, C.A., Smith, G.D. and Day, D.A. (1993) Biochim. Biophys. Acta 1141:313- 320.
25. Berger, S., Ellersiek, U. and Steinmüller, K. (1991) FEBS Lett. 286:129–132.
26. Peschek, G.A., Wastyn, M., Trnka, M., Molitor, V., Fry, I.V. and Packer, L. (1989) Biochemistry 28, 3057–3063.
27. Wastyn, M., Achatz, A., Molitor, V. and Peschek, G.A. (1988) Biochim.Biophys.Acta 935, 217–224.
28. Molitor, V. and Peschek, G.A. (1986) FEBS Lett. 195, 145–150.
29. Peschek, G.A., Wastyn, M., Fromwald, S. and Mayer, B. (1995) FEBS Lett. 371, 89- 93.
30. Peschek, G.A., Alge, D., Fromwald, S. and Mayer, B. (1995) J.Biol.Chem. 270, 27937–27941.
31. Auer, G., Mayer, B., Wastyn, M., Fromwald, S., Eghbalzad, K., Alge, D. and Peschek, G.A. (1995) Biochem.Mol.Biol.International 37, 1173–1185.
32. Ludwig, B. (1987) FEMS Microbiol.Rev. 46, 41–56.
33. Kuhn-Nentwich, L. and Kadenbach, B. (1984) FEBS Lett. 172:189–192.
34. Saraste, M. (1990) Q.Rev.Biophys. 23, 331–366.
35. Hüther, F.-J. and Kadenbach, B. (1986) FEBS Lett. 207, 89–94.
36. Hüther, F.-J. and Kadenbach, B. (1987) Biochem.Biophys.Res.Commun. 147, 1268- 1275.
37. Zimmermann, U. (1990) M.Sc. Thesis, University of Vienna, Austria.
38. Wastyn, M. and Peschek, G.A. (1988) EBEC Short Reports, Vol.5, p.94.
39. Weiss, H., Friedrich, T., Hofhaus, G. and Preis, D. (1991) Eur.J.Biochem. 197:563- 576.
40. Biggins, J. (1969) J.Bacteriol. 99:570–575.
41. Kraushaar, H., Hager, S., Wastyn, M. and Peschek, G.A. (1990) FEBS Lett. 273, 227- 231.
42. Peschek, G.A., Wastyn, M., Molitor, V., Kraushaar, H., Obinger, C. and Matthijs, H.C.P. (1989) in Highlights Modern Biochem., Vol.1 (Kotyk, A., Skoda,J., Paces, V. and Kosta, V., eds.), 893–902, VSP, Zeist, The Netherlands.
43. Neisser, A., Fromwald, S., Schmatzberger, A. and Peschek, G.A. (1994) Biochem.Biophys.Res.Commun. 200, 884–892.
44. Sherman, D.M., Troyan, T.A. and Sherman, L.A. (1994) Plant Physiol. 106, 251–262.
45. Peschek, G.A., Obinger, C., Fromwald, S. and Bergman, B. (1994) FEMS Microbiol.Letters 124, 431–438.
46. Peschek, G.A., Obinger, C., Sherman, D.M. and Sherman, L.A. (1994) Biochim.Biophys.Acta 1187, 369–372.
47. Bergman, B., Siddiqui, P.J.A., Carpenter, E.J. and Peschek, G.A. (1993) Appl.Environ.Microbiol. 59, 3239–3244.
48. Serrano, A., Giménez, P., Scherer, S. & Böger, P. (1990) Arch.Microbiol. 154, 614- 618.

25

THE COMPLEX I-HOMOLOGOUS NAD(P)H-PLASTOQUINONE-OXIDOREDUCTASE OF *SYNECHOCYSTIS* SP. PCC6803

Stefan Kösling,[1] Jens Appel,[2] Rüdiger Schulz,[2] and Klaus Steinmüller[1]

[1]Institut für Entwicklungs- und Molekularbiologie der Pflanzen
Heinrich-Heine-Universität
Universitätsstrasse 1, D-40225 Düsseldorf
[2]Fachbereich Biologie/Botanik
Philipps-Universität
Karl-von-Frisch Strasse, D-35032 Marburg

1. INTRODUCTION

Cyanobacteria are unique among photosynthetic prokaryotes in three different aspects. 1) they contain two photosystems, 2) their photosystem II is able to use water as the electron donor and to evolve oxygen, and 3) they carry a complete respiratory chain not only on the cytoplasmic membrane but also on the intracytoplasmic (thylakoid) membrane. A recent review by Schmetterer [1] describes the multitude of electron carriers present in the cytoplasmic and the thylakoid membranes of cyanobacteria. While the components of the photosynthetic electron transport chain which resides solely on the thylakoid membrane have been intensively studied, the organization of the respiratory chain has been less well characterized. By immunoblot studies it has been shown, that the respiratory chain in cyanobacteria contains a complex I-homologous NAD(P)H-dehydrogenase [2], a cytochrome b_6/f complex [3] and a cytochrome c-oxidase [4].

2. THE COMPLEX I-HOMOLOGOUS NAD(P)H-DEHYDROGENASE OF *SYNECHOCYSTIS* SP. PCC6803

2.1. Subunits of the Hydrogenase May Form the NAD(P)H-Oxidizing Part of the Enzyme

The NADH-ubiquinone-oxidoreductase is the first enzyme of the respiratory chain in mitochondria and eubacteria (complex I, [5]). The structurally most simple complex I

The Phototrophic Prokaryotes, edited by Peschek *et al.*
Kluwer Academic / Plenum Publishers, New York, 1999.

so far characterized, is that of *E. coli* [6]. It consists of 14 different subunits (NUO-A-N), which are all encoded in a single operon [7]. The enzyme can be fragmented into three subcomplexes: 1) a peripheral NADH-oxidizing subcomplex with three proteins (NUO-E, -F and -G), 2) a connecting fragment with four proteins, and 3) a membrane fragment, which contains the remaining seven subunits.

In the genome of the unicellular cyanobacterium *Synechocystis* sp. PCC6803 eleven genes homologous to genes encoding complex I of *E. coli* have been found (*ndhA-F*, [8–11]). The corresponding NDH proteins are homologous to the four subunits of the connecting and the seven subunits of the membrane fragment of *E. coli* complex I. However, no polypeptides corresponding to the three subunits of the NADH-oxidizing part of the enzyme have been described so far.

The recently published sequence of the complete genome of *Synechocystis* sp. PCC6803 [12] provides the opportunity to look for the missing subunits by a computer homology search. In fact, no reading frames, whose products can be unambiguously addressed as subunits of the complex I-homologous NAD(P)H-dehydrogenase, were found. There are, however, three genes with homology to the missing subunits, but they were recently identified as genes encoding subunits of a NAD(P)$^+$-reducing hydrogenase (HOX-E, -F and -U, [13]). The homology between subunits of complex I of mitochondria and the hydrogenase of *Alcaligenes eutrophus* has been first recognized by Pilkington et al. [14]. In Fig. 1, we compare the relationships between subunits of bacterial and mitochondrial complex I, and the hydrogenase enzymes of *A. eutrophus* and *Synechocystis* sp. PCC6803.

Because of the lack of "genuine" *ndh* genes homologous to *nuoE*, *-F* and *-G* of *E. coli* complex I, it is possible that the NAD(P)H-oxidizing part of the cyanobacterial complex I is formed by the subunits HOX-E, -F and -U of the hydrogenase, as has been proposed by Appel and Schulz for *Synechocystis* sp. PCC6803 [13] and Schmitz and Bothe for *Anabaena variabilis* [15].

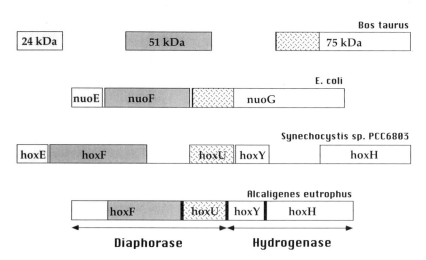

Figure 1. Homology between subunits of complex I from mitochondria and *E. coli* and the hydrogenase enzymes of *Synechocystis* sp. PCC6803 and *A. eutrophus*, indicated by different shadings. The 24, 51 and 75 kDa subunits of mitochondrial complex I are encoded in the nucleus. The *nuo* genes of *E. coli* are part of an operon which contains all 14 genes of complex I. The *hox* genes of *Synechocystis* sp. PCC6803 and *A. eutrophus* are also organized in operons.

2.2. The *ndhD* and *ndhF* Gene "Family" of *Synechocystis* sp. PCC6803

The complete sequence analysis of *Synechocystis* sp. PCC6803 has revealed another unexpected feature of the genes encoding subunits of the cyanobacterial complex I-homologous NAD(P)H-dehydrogenase. Nine out of eleven *ndh* genes (*ndhA, -B, -C, -E, -G, -H, -I, -J, -K*) are present as single copies, but five homologous reading frames were classified as *ndhD* genes and four ones as *ndhF* genes [12]. Table 1 lists the similarities between the deduced amino acid sequences of the different *ndhD* genes and those for the already well described *psbA* gene family. While all PSII-A proteins are very similar (A2 and A3 even identical) to each other, and also share a high similarity with the plastidial protein of tobacco, the NDH-D proteins differ substantially. NDH-D1 and -D2 have a similarity of about 55%, but their similarity to NDH-D3–5 is considerably lower (22–36%).

The presence of multiple genes for *ndhD* and *ndhF* in *Synechocystis* sp. PCC6803 may indicate that different forms of the NDH-complex exist. We have therefore analysed, if the expression of the genes *ndhD1–3* is differentially regulated during auto- and mixotrophic growth. Transcripts for *ndhD1* are found in cells grown in the presence of glucose, while in autotrophically grown cells the RNA is barely detectable. In contrast, *ndhD2* is expressed under both growth conditions. No transcripts for *ndhD3* were found. When mixotrophically grown cells are placed in the dark for one day, the mRNA steady state concentration for *ndhD1* increases, while it slightly decreases for *ndhD2* (Fig. 2).

In a further experiment, we analysed, if the different *ndhD* copies can substitute each other functionally. The genes *ndhD1–3* were inactivated by inserting a kanamycin-resistance marker into the reading frame of *ndhD1* and a chloramphenicol-resistance marker into *ndhD2* and *ndhD3*. Mutants were selected and tested by Southern blot analysis for complete segregation.

The mutants were then characterized by measuring their chlorophyll fluorescence induction curves with the pulse amplitude-modulated fluorometer. It has been shown by Mi et al., [16] for *Synechocystis* sp. PCC6803, that after switching-off actinic light an increase in chlorophyll fluorescence is observed, which indicates that the plastoquinone pool becomes reduced in the dark. A mutant of *Synechocystis* sp. PCC6803, in which *ndhB* is inactivated, lacks this dark-rise of fluorescence. Moreover, the dark-rise can be abol-

Table 1. Comparison of the deduced amino acid sequences of the three *psbA* and the five *ndhD* genes of *Synechocystis* sp. PCC6803[a]

	A1	A2	A3	Tobacco		
psbA1 (slr1181)	100	84.4	84.4	79.3		
psbA2 (slr1311)		100	100	85.0		
psbA3 (sll1867)			100	85.0		

	D1	D2	D3	D4	D5	Tobacco
ndhD1 (slr0331)	100	54.6	35.9	33.9	21.6	52.1
ndhD2 (slr1291)		100	34.1	32.3	23.6	42.2
ndhD3 (sll0027)			100	53.2	22.6	31.2
ndhD4 (sll1733)				100	21.8	31.9
ndhD5 (slr2007)					100	24.0

[a]The designations of the reading frames according to Kaneko et al. [12] are given in parenthesis. The designations *ndhD1* to *ndhD5* are tentatively.

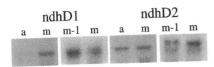

Figure 2. Northern blot analysis of *ndhD1* and *-D2*. The RNA was isolated from autotrophic (a) and mixotrophic (m) cells or mixotrophic cultures, that had been incubated in darkness for one day (m-1). The blots were hybridized with specific gene probes for both genes. 5 μg of RNA were loaded in each lane.

ished by rotenone, an inhibitor of complex I. This indicates, that the dark-rise of chlorophyll fluorescence is caused by the NAD(P)H-dehydrogenase.

Fig. 3 shows the amplitudes of the dark-rise in wild type and *ndhD1–3* mutant cells of *Synechocystis* sp. PCC6803. After illumination with actinic light of an intensity of 320 μmol/m^2·s in the wild type a rapid transient increase in fluorescence is observed, which is even higher as the level of steady state fluorescence. After about 2 min in the dark, the level of fluorescence has dropped down to the level of minimal fluorescence Fo. The *ndhD1* mutant shows a transient dark-rise, albeit with a much smaller amplitude, and also the return to the Fo-level within 2 min. In contrast, mutants in which *ndhD2* and *-D3* have been inactivated, display a slower increase of the dark-rise and the fluorescence yield remains above Fo.

As mentioned before, in the *ndhB* mutant, the dark-rise of chlorophyll fluorescence is abolished completely [16]. Since *ndhB* is a single copy gene in *Synechocystis* sp. PCC6803, it is likely that no functional enzyme is synthesized in this mutant. All *ndhD* mutants however, display an increase in fluorescence in the dark. This indicates, that the function of a deleted *ndhD* copy can be substituted by another gene copy, even though large differences in the amino acid sequences between the different polypeptides exist (see Table 1). In the *ndhD1* mutant, a less active but still adjustable form of the enzyme may be formed, while the *ndhD2* and *-D3* mutants probably contain enzymes, whose activities can not be regulated any longer.

Presently, it is not clear, which *ndhD* copies substitute in the different mutants. The results of the northern blots show, that at least *ndhD1* and *-D2* are expressed at a signifi-

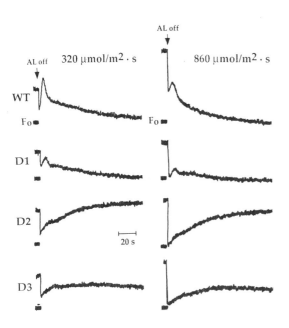

Figure 3. Chlorophyll fluorescence of wild type and mutants in which *ndhD1, -D2* and *-D3* have been inactivated. The cells were illuminated with actinic light of 320 or 860 μmol/m^2·s for 4 min, then the light was switched off and the chlorophyll fluorescence was recorded.

cant level. Therefore it is likely, that different forms of the NDH-complex exist in one cell—each one with a different function in respiration, photosynthetic electron transport and hydrogen metabolism.

ACKNOWLEDGMENTS

This work was supported by a grant of the Deutsche Forschungsgemeinschaft to K. St. (SFB 189) and to R. S. (Schu 1053/1-1 and 1-2).

REFERENCES

1. Schmetterer, G. (1994) in: The Molecular Biology of Cyanobacteria (Bryant, D.A., Ed.), pp. 409–435, Kluwer Academic Publ., Dordrecht.
2. Berger, S., Ellersiek, U. and Steinmüller, K. (1991) FEBS Lett. 286, 129–132.
3. Kraushaar, H., Hager, S., Wastyn, M. and Peschek, G.A. (1990) FEBS Lett. 273, 227–231.
4. Molitor, V., Trnka, M. and Peschek G.A. (1987) Curr. Microbiol. 14, 263–268.
5. Friedrich, T., Steinmüller, K. and Weiss, H. (1995) FEBS Lett. 367, 107–111.
6. Leif, H., Sled, V.D., Ohnishi, T., Weiss, H. and Friedrich, T. (1995) Eur. J. Biochem. 230, 538–548.
7. Weidner, U., Geier, S., Ptock, A., Friedrich, T., Leif, H. and Weiss, H. (1993) J. Mol. Biol. 233, 109–122.
8. Steinmüller, K., Ley, A.C., Steinmetz, A.A., Sayre, R.T. and Bogorad, L. (1989) Mol. Gen. Genet. 216, 60–69.
9. Ogawa, T. (1991) Proc. Natl. Acad. Sci. USA 88, 4275–4279.
10. Ellersiek, U. and Steinmüller, K. (1992) Plant Mol. Biol. 20, 1097–1110.
11. Dzelzkalns, V.A., Obinger, C., Regelsberger, G., Niederhauser, H., Kamensek, M., Peschek, G.A. and Bogorad, L. (1994) Plant Physiol. 106, 1435–1442.
12. Kaneko, T., Sato, S., Kotani, H., Tanaka, A., Asamizu, E., Nakamura, Y., Miyajima, N., Hirosawa, M., Sugiura, M., Sasamoto, S., Kimura, T., Hosouchi, T., Matsuno, A., Muraki, A., Nakazaki, N., Naruo, K., Okumura, S., Shimpo, S., Takeuchi, C., Wada, T., Watanabe, A., Yamada, M., Yasuda, M. and Tabata, S. (1996) DNA Res 3, 109–136.
13. Appel, J. and Schulz, R. (1996) Biochim. Biophys. Acta 1298, 141–147
14. Pilkington, S.J., Skehel, M., Gennis, R.B., and Walker J.E. (1990) Biochemistry 30, 2166–2175
15. Schmitz, O. and Bothe, H. (1996) Naturwiss. 83, 525–527
16. Mi, H., Endo, T., Ogawa, T. and Asada, K. (1995) Plant Cell Physiol. 36, 661–668.

26

SULFIDE-DEPENDENT ANOXYGENIC PHOTOSYNTHESIS IN PROKARYOTES

Sulfide-Quinone Reductase (SQR), the Initial Step

Y. Shahak,[1] M. Schütz,[2] M. Bronstein,[3] C. Griesbeck,[2] G. Hauska,[2] and E. Padan[3]

[1]Institute of Horticulture
The Volcani Center
Bet-Dagan 50250, Israel
[2]Lehrstuhl für Zellbiologie und Pflanzenphysiologie
Universität Regensburg
93040 Regensburg, Germany
[3]Division of Microbial and Molecular Ecology
The Hebrew University
Jerusalem 91904, Israel

1. INTRODUCTION

Anoxygenic photosynthesis with sulfide serving as the electron donor is a property unique to prokaryotes. Most photosynthetic bacteria can grow photoautotrophically using sulfide (as well as few other inorganic sulfur compounds) as electron donors for CO_2 fixation (see ref. 1 for recent review). Cyanobacteria are exceptional in the world of phototrophic prokaryotes. With the closely related prochlorophytes, they are the only eubacteria that can perform plant-type oxygenic photosynthesis, using two photosystems in series and water as the electron donor. However, some species of cyanobacteria can facultatively shift to anoxygenic, sulfide-dependent photosynthesis in which only PS I is involved[2–5]. This unique capacity to shift between anoxygenic and oxygenic photosynthesis was discovered in various strains of cyanobacteria, evolutionarily distant from each other[3]. It was considered to represent a primitive relic of the evolution of photosynthesis[6]. Of these strains, the filamentous cyanobacterium *Oscillatoria limnetica* has been studied most extensively . *O. limnetica* shifts to anoxygenic photosynthesis 2–3 hours after incubation in the presence of sulfide and light, in an inducible process specific to sulfide[7,8]. Depending on the growth conditions, the induced cells perform several sulfide-dependent reactions: CO_2 fixation,[3,7–9] H_2 evolution[10] or N_2 fixation.[11]

The Phototrophic Prokaryotes, edited by Peschek *et al.*
Kluwer Academic / Plenum Publishers, New York, 1999.

217

In spite of its wide use, the initial step in the sulfide oxidation pathway is still a matter of debate. In several species, the sulfide oxidizing enzyme was proposed to be flavocytochrome c. This enzyme, which can catalyze electron transfer from sulfide to a variety of small c-type cytochromes[12] and is composed of haemoprotein and flavoprotein subunits, has recently been crystallized[13]. However, since flavocytochrome *c* has not been found in several sulfide oxidizing bacteria (e.g. *Rhodobacter capsulatus*), its role in the major pathway of sulfide oxidation has been questioned[1,14]. Alternatively, the transfer of electrons from sulfide directly into the quinone pool was proposed, based on the inhibition by quinone-analogs, as well as energetic considerations[15,16].

We have taken advantage of the unique properties of the cyanobacterial system to track the inducible enzyme which enables the photosynthetic sulfide oxidation. Our approach led to the discovery of sulfide-quinone reductase (SQR; E.C.1.5.5.´.), an enzyme which catalyzes sulfide-quinone oxido-reduction, and found widely-spread amongst green-sulfur bacteria, purple "non-sulfur" bacteria, and most recently purple-sulfur and green "non-sulfur" bacteria. It was also found in non-photosynthetic prokaryotes (chemoautotrophs), and even a eukaryotic system. In this presentation we review our research of the last decade, and update with the recent results.

2. THE DISCOVERY OF SQR IN *O. LIMNETICA*

2.1. Cell-Free System

The development of a cell-free system which maintains the capacity to photosynthetically oxidize sulfide[17] led to the discovery of the initial step of sulfide-dependent electron transport in *O. limnetica* . This capacity was found to reside in the thylakoids isolated from cells which had been induced by sulfide. The induced thylakoids can catalyze sulfide-dependent photoreduction of either NADP or methyl viologen. These light reactions had high affinities to sulfide (apparent Km between 20–40 μM), were coupled to proton-pumping, sensitive to all inhibitors of the cytochrome *b6f* complex, and were not lost upon washing the thylakoids[17]. Non-induced thylakoids could also photo-oxidize sulfide. However in this case the reaction had a low affinity to sulfide (apparent Km in the mM range) and was insensitive to the specific inhibitors[17,18]. It was, therefore suggested that the adaptation of *O. limnetica* to anoxygenic photosynthesis involves the induction of a thylakoid factor which enables sulfide photo-oxidation via the cytochrome *b6f* complex[17].

2.2. What Is the Immediate Electron Acceptor?

Based on the above results, two potential acceptors were anticipated: the quinone pool or the cytochrome *b6f* complex. Since the reduction of the endogenous native quinone cannot be easily determined, we tested the first alternative by the application of external, short-chain quinones. Indeed, we found that induced *O. limnetica* thylakoids can catalyze sulfide-dependent reduction of externally added quinones in the dark. The reaction could be detected by colorimetric determination of sulfide oxidation. Alternatively, assay conditions were developed to continuously monitor quinone reduction, using dual-wavelength spectrophotometry at the UV range[19]. This dark reaction resembled NADP photoreduction in its inducibility and high affinity to sulfide (apparent Km of about 2 μM), while it differed from the light reaction in its differential sensitivity to quinone-analog inhibitors (e.g. insensitivity to DBMIB). The affinity to plastoquinone-1 was about 30 μM, and Vmax 80 μmol PQ-1 reduced/mg chlorophyll/h. The induction of sulfide-quinone oxidoreduction activity was prevented if chloramphenicol was present during the sulfide-induction period. We therefore suggested that the inducible enzyme which enables sulfide oxidation is an electron carrier which transfers electrons from sulfide to the PQ pool, and named it SQR: sulfide-quinone reductase[19].

2.3. Purification of *O. limnetica* SQR

In addition to providing the definition of the inducible enzyme, the direct activity assay developed for SQR further enabled us to isolate the enzyme, as unlike sulfide-dependent NADP photoreduction, the quinone reduction did not require additional membranal components. SQR was solubilized from induced *O. limnetica* thylakoids by mild detergent treatment (dodecyl-maltoside plus sodium cholate), and purified by ammonium sulfate fractionation followed by three-step HPLC on gel-filtration and hydrophobic columns. Dodecyl-maltoside was present throughout the purification to stabilize the enzyme. SDS-PAGE of the most active fractions indicated a single band of 57 kDa, while the native size, as estimated by gel filtration, ranged between 67–80 kDa. This was taken to suggest that the active soluble enzyme is composed of a single polypeptide, surrounded by 20–46 detergent molecules[20,21].

The conclusion that the purified protein is indeed SQR was based on the following evidences:

1. Western blots of thylakoids as well as solubilized preparations, probed with antibodies raised against the denatured polypeptide, indicated a major band of 57 kDa in the samples prepared exclusively from induced cells. In addition, [^{35}S]-methionine present during sulfide-induction of *O. limnetica* cells yielded the labeling of one major membranal protein of the same apparent size on denaturing gels[21]. It is worth mentioning that in addition to the membranal SQR, soluble proteins of low molecular weight (11.5 and 12.5 kDa on SDS-PAGE) were also heavily labeled during the induction. They were located at the periplasmic soluble fraction, and speculated to be related to the exclusion of elemental sulfur out of the cells[22].

2. The purified enzyme resembled the membrane-bound SQR in its Vmax (100–150 μmol PQ-1 reduced/mg protein/h) and apparent Km for sulfide (about 8 μM) and PQ-1 (32 μM), as well as in its specificity to the quinone substrate[21].

3. A detailed comparative study of quinone analog inhibitors fully supported the identity of the purified enzyme being the SQR. The differential sensitivity pattern of both the isolated and membrane bound enzymes were the same, with Aurachin C being an exceptionally efficient inhibitor, while both DNP-INT and DBMIB had no effect[23]. KCN also inhibits SQR activity, providing it is mixed with the enzyme prior to sulfide addition[21] (Table 1).

2.4. What Is the Prosthetic Group?

Absorption and fluorescence spectra of the purified enzyme were typical of flavoproteins. The absorption and fluorescence-excitation spectra showed peaks at 280 nm and two broad peaks around 370 and 460 nm. The fluorescence emission of SQR peaked at 527 nm. The prominent excitation peak at 280 nm may reflect energy transfer between an aromatic residue and the flavin moiety, indicating close proximity between them. Both the absorption and fluorescence intensities were specifically quenched by sulfide[21]. However,

Table 1. Comparison of the SQR enzyme of three prokaryotes[a]

Property	*O. limnetica*	*Chlorobium*	*Rb. capsulatus*
Inducibility	Yes	No (constitutive)	Yes
Km (H$_2$S, μM)	8	ND	2
Km (Q, μM)	32 (PQ-1)	<20 (PQ-1)	2 (decUQ)
Vmax (μmol/mgChl/h)	120	50	140
Best Q-substrate	PQ	ND	UQ
Best Q-inhibitor (I50, nM)	Aura (7)	Stigma (5)	Aura (230)
KCN inhibition[b] (I50, μM)	15	10	100
Prosthetic group	FAD[c]	ND	FAD[b]
Solubilization	DM+cholate	ND	Thesit (or cholate or NaBr)
Size (SDS-PAGE,KDa)	57	ND	55

Abbreviations: ND, not determined; Q, quinone; Aura, Aurachin C; Stigma, Stigmatellin. I50: inhibitor concentration required for 50% inhibition of SQR activity.
[a]Data from references [19–21,23,29–33].
[b]KCN added before sulfide.
[c]Based on spectral and sequence analyses.

the inhibitor KCN, which is known to affect the absorption spectrum of flavocytochrome c[24] did not affect the spectra of SQR[21], suggesting that in this case its target site might reside at the sulfide-interacting site, rather than the flavin binding site (Fig. 1). The spectral results were further supported by sequence analyses. The N-terminus (29 residues) of the purified enzyme was sequenced and found to contain the first 8 (out of 11) fingerprint residues typical of the ADP-binding site characterizing many NAD(P)/FAD containing proteins[21,25,26]. This was an important step ahead, since in addition to providing this valuable information, it later established the close homology with the bacterial SQR (see below).

3. SQR IN CHLOROBIUM AND RHODOBACTER CAPSULATUS MEMBRANES

3.1. Membranal Activity

Photosynthetic bacteria, unlike cyanobacteria, contain only one reaction center. The reaction center of most green bacteria resembles the higher-plant photosystem I (PSI)[27], while that of purple bacteria resembles PS II[28]. We applied the SQR activity assay to representatives of the two bacterial types: the green-sulfur bacterium *Cb. limicola* f.*thiosulfatophilum* and the purple "non-sulfur" bacterium *Rb. capsulatus*. Photosynthetic membranes prepared from both species could catalyze sulfide-quinone oxido-reduction[29,30]. The occurrence in such diverse species was taken to indicate the universal importance of SQR in sulfide-dependent anoxygenic photosynthesis in prokaryotes. Table 1 compares the main properties of the three systems. The bacterial enzymes resemble the cyanobacterial SQR in their size, high affinity to sulfide (in the μM range) and KCN-sensitivity. However, they differ in their quinone-substrate specificity, and individual selective sensitivity to quinone-analog inhibitors[23], a result consistent with the occurrence of different natural quinone acceptors in these species (PQ in *Oscillatoria*, MQ or *Chlorobium*-quinone in *Chlorobium*, and UQ in *Rhodobacter*). It might therefore be expected that each SQR has a slightly modified quinone-binding site, to fit its native substrate. The three systems also differ in their regulation of the enzyme synthesis, being sulfide-inducible in *Oscillatoria* while constitutive in *Chlorobium* (SQR activity occurs in membranes of *Cb. limicola* grown on sulfide as well as thiosulfate[29]). In *Rb. capsulatus* bio-

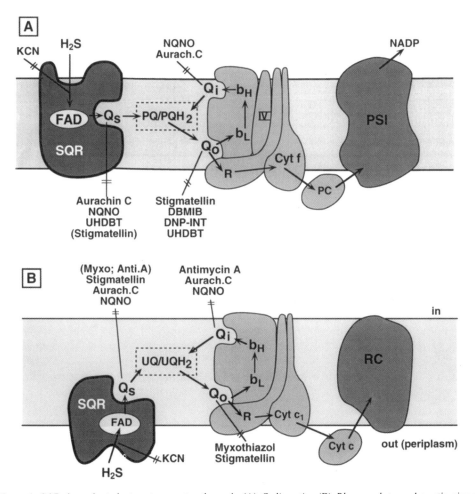

Figure 1. SQR-dependent electron transport pathway in (A) *O. limnetica* (B) *Rb. capsulatus*: schematic views. The different localization and orientation of SQR in the two systems represent two hypothetical alternative views which await further experimental support. The sites of inhibition of the SQR and the cytochrome complex by different quinone analogs are illustrated. Parentheses indicate lower potency. The proposed site of inhibition of KCN is based on its lack of effect on the spectroscopic properties of the bound-FAD, and its competition with sulfide. Abbreviations: Cyt, cytochrome; Qs, quinone reduction site on SQR; Qo and Qi, quinone oxidation and reduction sites, respectively, on the cytochrome complex; b_L and b_H, low- and high-potential cytochrome *b*-haems; R, Rieske protein; PC, plastocyanin; PSI, photosystem I; RC, reaction center.

chemical and reporter-gene studies suggest that both sulfide and oxygen concentrations interplay in the induction (Schütz et al., unpublished data).

3.2. The Electron Transport Pathway

Sulfide-dependent electron transport pathway was studied in two bacterial systems by LED-array spectrometry[30,31]. The reduction by sulfide of exogenous ubiquinone was compared to the reduction of cytochromes *b* and *c* in chromatophores of *Rb. capsulatus*. From titrations with sulfide, values for Vmax of 300 and 10 (µmoles reduced/mg bacteriochlorophyll a/h), and Km of 5 and 3 µM were estimated, for decyl-ubiquinone- and cytochrome *c*-reduction, respectively. Both reactions were sensitive to KCN. Effects of inhibitors interfering with quinone binding sites have suggested that the electron transport from sulfide in *Rb. capsulatus* employs the cytochrome *bc1*-complex via the ubiquinone pool[30]. Studies of chlorosome-free membranes of *Chlorobium tepidum* supported these results, and further demonstrated sulfide-dependent oxidant-induced reduction of cytochrome *b*[31], indicating the operation of a Q cycle. It was therefore concluded that the main pathway of sulfide oxidation by prokaryotes proceeds via the quinone pool and the cytochrome *bc*-complex[30,31]:

Sulfide → SQR → Q-pool → Cytochrome b6f/bc1 complex → reaction center

4. CLONING THE *RB. CAPSULATUS* SQR GENE

4.1. Purification of the *Rb. capsulatus* Enzyme

Compared with the cyanobacterial enzyme, SQR of *Rb. capsulatus* seems to be less firmly bound to the membrane, as indicated by the ease of its release. The solubilization of SQR from membranes of *Rb. capsulatus* could be achieved by treatments with 0.5% cholate, or 2M NaBr[20,23]. Still better yield and stability of the isolated enzyme were obtained by the use of the non-ionic detergent Thesit[32,33]. The enzyme was purified by sequential steps of DEAE-column chromatography, sucrose density gradient and Mono-Q HPLC. Like *O. limetica*, the bacterial enzyme consists of a single polypeptide with an apparent size of 55 kDa on SDS-PAGE. It also has similar absorption and fluorescence spectra, indicative of the flavin content[33]. The apparent K_m of the purified SQR for sulfide, as well as decyl-UQ was about 2 µM (Table 1). These K_m values are similar to those obtained for the reduction of exogenous decyl-UQ (5 µM) or cytochromes (3 µM) in chromatophores of *Rb. capsulatus*[30] as well as for the sulfide dependent growth of this bacterium[34]. The affinity for both substrates is higher than that of the cyanobacterial enzyme (Table 1).

4.2. Cloning

The *sqr* gene of *Rb. capsulatus* was cloned by PCR amplified probes obtained with oligonucleotides deduced from the peptide sequences of the N-terminus and additional four partial peptides of the purified SQR. Elements identified in the promoter region: a putative ribosome binding site (GGAGG) which fits the translation initiation sequence of many *Rb. capsulatus* genes[35]; the sequence TTGACA is identical to the −35 region of the *E. coli*-like σ70 promoter consensus sequence[35]. A sequence encoding a transcript of a hairpin-like structure, containing the stop codon occurs at the end of the sqr-gene (33). The full sequence published in ref. (33) was later found to contain a few sequencing mistakes. The data bank sequence (GenBank/EBI access number x97478) was updated with the correction in September 1997.

The SQR polypeptide deduced from the corrected sequence consists of 427 amino acids with a MW of 47 kDa. The theoretical isoelectric point is 7.8 with a net charge of +1. Structure prediction analysis gives 36% α-helices and 23% β-sheets. Hydropathy analysis did not indicate any membrane spanning or anchoring α-helix.

The first amino acid of the mature SQR, which was identified by the peptide sequencing, is alanine. Thus, either the preceding methionine or an amino terminal leader peptide of 16 amino acid residues including the other methionine is cleaved off[33].

The N-terminus of the SQR from *Rb. capsulatus* is highly homologous to that of *O. limnetica*[21,32,33]: 48% (14/29) identity and 72% (21/29) similarity. The bacterial SQR, like the cyano-enzyme, also contains the expected fingerprint residues of the ADP-binding site[21,33]. The *Rb. capsulatus* SQR shows a score of 10 out of 11 fingerprint residues within the N-terminal 36 amino acids. Secondary structure analysis for both SQR-enzymes predicts a βαβ-fold, consistent with the one proposed for the ADP-part of NAD(P)/FAD binding domains, with the exception that the loop is positioned between the first β-sheet and the α-helix[33]. Unlike the NAD(P)-binding domains, the FAD-binding domains of many flavoenzymes, including SQR, appear to be located close to the amino terminus[25,26].

Two additional segments of the bacterial SQR are suggested to be involved in FAD-binding, since they contain motives involved in the binding of the FAD in two enzymes of known 3D structure, glutathione reductase[35,37] and lipoamide dehydrogenase[38], and other pyridine nucleotide oxidoreductases. Taken together, the sequence similarities suggest that the flavin of the SQR is a FAD which is bound to the protein moiety as in pyridine nucleotide-disulfide oxidoreductases.

Except for the three regions involved in FAD binding, no other homologies were detected for SQR in the EMBL Database.

Recently, we have also obtained a mutant of *Rb. capsulatus* by insertional inactivation of the gene *via* homologous recombination. The mutant cannot grow photoautotrophically in a sulfide dependent way, indicating the essential role of SQR under these conditiones (Schütz, unpublished data).

4.3. Expression

The SQR was expressed under the control of T7 promoter in an active form in *E. coli* BL21(DE3). After induction, a dominant band around 55 kDa (on SDS-PAGE), which reacted with purified antibodies against the *Rb. capsulatus* SQR, appeared only in extracts obtained from cells containing the *sqr*-gene. These extracts were active in reducing decyl-UQ with sulfide, indicating that the FAD, and if present, additional cofactors, are incorporated in the correct way by *E. coli*. A substantial portion of the expressed SQR obtained from *E. coli* (30–50%) was found in the soluble fraction. This may be taken to support the localization of SQR at the periphery of the *Rb. capsulatus* membrane.

5. OCCURRENCE OF SQR IN ADDITIONAL SYSTEMS

During the last year we have broadened our study to many sulfide-oxidizing biological systems, including non-photosynthetic ones. Table 2 summarizes the state of our present knowledge in each system. So far, we succeeded in detecting SQR activity in all sulfide-oxidizing systems tested. The unpublished results are described below. All activities were found to be sensitive to quinone-analog inhibitors, and diminished by boiling the membrane samples.

5.1. Cyanobacteria

Aphanothece halophytica, a unicellular cyanobacterium, is capable of anoxygenic sulfide-dependent photosynthesis, although it cannot live on sulfide[3]. We succeeded in detecting SQR activity in *A. halophytica* 7418 few hours after the addition of 1 mM sulfide under anaerobic conditions in the light. Surprisingly, very high activity could be assayed

Table 2. SQR abundance amongst sulfide-oxidizing systems[a]

Biological system	Species	State of SQR resolution
Photosynthetic prokaryotes		
Cyanobacteria		
Filamentous	*Oscillatoria limnetica*	isolated & purified[21]
Unicellular	*Aphanothece halophytica*	partially purified
Purple "non-sulfur"	*Rhodobacter capsulatus*	purified, cloned, expressed[33]
Purple "sulfur"	*Chromatium vinosum*	membranal activity[b]
Green "sulfur"	*Chlorobium limicola*	membranal activity[29]
	Chlorobium tepidum	membranal activity[31]
Green "nonsulfur"	*Chloroflexus aurantiacus*	membrane crude extract
Non-photosynthetic prokaryotes		
Chemoautotrophs	*Paracoccus denitrificans*	membranal activity
	Aquifex aeolicus	membranal activity
Eukaryotes		
Mitochondria of lugworm	*Arenicola marina*	membranal activity

[a]Except for the data cited by reference numbers, all other information relates to recent, unpublished results.
[b]Our measurements and also H. Trüper, Bonn, personal communication.

in intact cells. On a chlorophyll basis, the rates obtained for *A. halophytica* (180 µmol PQ-1 reduced/mg Chl/hr) were about 50% higher than previously obtained with *O. limnetica* thylakoids. Attempts to further increase the accessibility of the substrates (sulfide and PQ-1) by either permeabilization of the cells by toluene, or lysosyme treatments, or by sonication did not improve the detected rate (M. Bronstein, unpublished data). This might indicate that in *Aphanothece* SQR is facing the periplasm, or that this system is more permeable to the substrates, as compared with *O. limnetica*. The *A. halophytica* enzyme has been solublized according to the *O. limnetica* protocol[21]. The partially purified active fractions show a major band of 55 kDa on SDS-PAGE (M. Bronstein, unpublished data).

5.2. Purple-Sulfur Bacteria

The Chromatiaceae are easily distinguished by the presence of microscopically observable globules of elemental sulfur formed intracellularly when they are grown on sulfide or thiosulfate[1]. Membranes of *Chromatium vinosum* (DSM 180) cells (a gift of Dr. J. Schwenn) were isolated and assayed according to[33]. The obtained rates (70 µmol decyl-UQ reduced/mg Bchl/h; 5.4 µmol decyl-UQ reduced/mg protein/h) were in the same range as previously determined for *Rb. capsulatus*[30]. Myxothiazol (40 µM) and Aurachin C (12 µM) caused 75% and 90% inhibition, respectively (Schütz, unpublished data).

5.3. Green "Non-Sulfur" Bacteria

The family Chloroflexaceae, or green gliding (green "non-sulfur") bacteria, belongs to a phylum which branched off early in the evolutionary history of Bacteria[39]. The only species of that family that has been obtained in pure culture and characterized biochemically is *Chloroflexus aurantiacus*, a thermophile capable of slow photoautotrophic growth while oxidizing sulfide to elemental sulfur[1]. Membranes prepared from *Chloroflexus aurantiacus* (J-10; a gift of Dr. G. Fuchs) according to[33], exhibited rather low SQR activity with decyl-UQ as the acceptor (0.03 µmol decyl-UQ reduced/mg protein/h). The apparent K_m values were 3 µM for decyl-UQ, and 30 µM for sulfide (Schütz, unpublished data).

5.4. Non-Photosynthetic Systems

Sulfide is also oxidized by various non-photosynthetic organisms including ciliates, marine worms, invertebrates, different symbiotic systems and chemolithotrophic bacteria.[40-43] It is tempting to suggest that SQR-like enzymes operate in these systems as well.

5.4.1. Chemoautotrophic Bacteria. The abundant soil bacterium *Paracoccus denitrificans* (GB17) is closely related to *Rb. capsulatus* within the alpha-subdivision of Proteobacteria. The oxidative metabolism of inorganic sulfur compounds in this system has recently been reviewed[44]. We have detected SQR activity in membranes (a gift of Prof. C. Friedrich) of *P. denitrificans* with rates somewhat higher (18 μmol decyl-UQ reduced/mg protein/h) than in the systems mentioned above. The apparent K_m-values for decyl-UQ and sulfide were 3 μM and 26 μM, respectively. The pI50 values for inhibitors were 4.65. (Myxothiazol), 4.70 (Stigmatellin) and 4.82 (Antimycin A). KCN did not inhibit decyl-UQ reduction. Further spectroscopic studies of membranes demonstrated sulfide-dependent cytochrome b_{560} and c_{553} reduction, as well as oxidant-induced reduction of cytochrome *b*. In the presence of oxygen, the cytochromes reoxidized rapidly. The reoxidation was sensitive to KCN (Schütz and Klughammer, in preparation). This was indicative of the operation of a Q-cycle mechanism and transfer of electrons from sulfide to the terminal cytochrome oxidase via the cytochrome *bc*-complex.

Aquifex aeolicus is a microaerobic chemolithoautotroph which oxidizes sulfide to sulfate. It is a thermophile found in marine hydrothermal vents, with optimal growth temperature around 83°C (R. Huber, personal communication). Membranes of *A. aeolicus* (a gift of Dr. R. Huber) were prepared and assayed as for *Rb. capsulatus*[33]. They exhibited substantial SQR rates even at room temperature (about 80 μmol decyl-UQ reduced/mg protein/h). The apparent K_m values for decyl-UQ and sulfide were 2 μM and 10 μM, respectively. The pI50 values for inhibitors were 5.35 (Myxothiazol), 4.75 (Stigmatellin), 4.63 (Antimycin A), and 1.6 (KCN; M. Schütz, unpublished data). *Aquifex* is the most deeply rooting eubacterial trait, closest to Archaebacteria[45]. The occurrence of SQR in this system indicates that SQR is a very early enzyme in evolution.

5.4.2. Lugworm. This is the first eukaryote system in which SQR activity has been detected so far. The lugworm *Arenicola marina* is able to survive rather high levels of sulfide, which it oxidizes to thiosulfate. Sulfide oxidation is localized in the mitochondria of the body wall musculature. Inhibitor studies indicate that electrons from sulfide oxidation enter the respiratory chain either at the level of UQ, or complex III[43], in a reaction which is coupled to ATP synthesis[46]. Mitochondria were freshly isolated[43] from *Arenicola* worms collected from the intertidal flats near Zierikzee (Holland), prior to SQR assay. Qualitative measurements revealed sulfide-dependent decyl-UQ reduction activity (6 μmol decyl-UQ reduced/mg protein/h) sensitive to heat treatment and some quinone analog inhibitors (Klein, Schütz, Hauska and Grieshaber, unpublished data).

6. OPEN QUESTIONS

It is well established by now that SQR is a wide-spread enzyme amongst sulfide oxidizing organisms. However, there are still many unknowns about the structure, function and regulation of SQR:

1. The localization of the enzyme: how far is it embedded in the membrane? Does it have an anchor counter-part? Is the sulfide oxidation site facing the cytoplasmic or rather the periplasmic side of the photosynthetic membrane? Fig. 1 illustrates the supposed electron transfer pathway in SQR from sulfide *via* the sulfide-oxidizing site and the FAD cofactor to the quinone acceptor at the quinone binding site (Qs), and further on along the electron transport chain of *O. limnetica* (Fig. 1A) and *Rb. capsulatus* (Fig. 1B). The target points of the inhibitors are also illustrated in this scheme, with the less efficient inhibitors marked in parentheses. The illustrated integral *vs* peripheral localization of the enzymes is based on the apparent hydrophobic properties of the cyanobacterial enzyme, compared with the relative ease of extraction of SQR from both *Rb. capsulatus* membranes and the *E. coli* expression systems. However, the localization of the sulfide oxidizing site on the cytoplasmic *vs* periplasmic side is still an open question.

2. What is the major role of SQR: bioenergetics or rather detoxification?

3. How is the gene regulated? What is the mechanism of sulfide-induction?

These and other questions should be the target of future studies.

ACKNOWLEDGMENTS

This research is being supported by the Deutsche Forschungsgemeinschaft (DFG), and the Israel Science Foundation (ISF) administered by The Israel Academy of Science and Humanities. We are grateful to Prof. J. Schwenn for *Chromatium vinosum* cells, to Prof. G. Fuchs for *Chloroflexus aurantiacus* cell, Prof. C. Friedrich for *Paracoccus denitrificans* membranes, Dr. R. Huber for *Aquifex aeolicus* cells and Prof. M. K. Grieshaber and co-workers for the preparation of *Arenicola marina* mitochondria and co-operative measurements. Thanks are also due to the Moshe Shilo Minerva Center for Marine Biogeochemistry (Israel).

REFERENCES

1. Brune, D.C. (1995) Sulfur compounds as photosynthetic electron donors. In: Anoxygenic Photosynthetic Bacteria (Blankenship, R.E., Madigan, M.T. and Bauer, C.E., eds.) pp. 847–870, Kluwer Academic Publishers, The Netherlands.
2. Padan, E. (1979) Facultative anoxygenic photosynthesis in cyanobacteria. Annu. Rev. Plant Physiol. 30, 27–40.
3. Garlick, S., Oren, A. and Padan, E. (1977) Occurrence of facultative anoxygenic photosynthesis among filamentous unicellular cyanobacteria. J. Bacteriol. 129, 623–629.
4. Belkin, S., Shahak, Y. and Padan, E. (1988) Anoxygenic photosynthetic electron transport. In: Methods in Enzymology. (Packer, L. and Glazer, A. N., eds.) Vol. 167, pp. 380–386, Academic Press, San Diego.
5. Stal, J.L. (1995) Physiological ecology of cyanobacteria in microbial mats and other communities. New Phytol. 131, 1–32.
6. Padan, E. (1989) Combined molecular and physiological approach to anoxygenic photosynthesis of cyanobacteria. In: Microbial Mats, Physiological Ecology of Benthic Microbial Communities (Cohen, Y. and Rosenberg, E., eds.) pp. 277–282.
7. Cohen, Y., Padan, E and Shilo, M. (1975) Facultative anoxygenic photosynthesis in the cyanobacterium *Oscillatoria limnetica*. J. Bacteriol. 123, 855–861.
8. Oren, A. and Padan, E. (1978) Induction of anaerobic photoautotrophic growth in the cyanobacterium *Oscillatoria limnetica*. J. Bacteriol. 133, 558–563.

9. Cohen, Y., Jorgensen, B.B., Padan, E and Shilo, M. (1975) Sulfide-dependent anoxygenic photosynthesis in the cyanobacterium *Oscillatoria limnetica* Nature (London) 257, 489–492.

10. Belkin, S. and Padan, E. (1978) Sulfide-dependent hydrogen evolution in the cyanobacterium *Oscillatoria limnetica*. FEBS Lett. 94, 291–294.

11. Belkin, S., Arieli, B. and Padan E. (1982) Sulfide-dependent electron transport in *Oscillatoria limnetica*. Isr. J. Botany 31, 199–200.

12. Knaff, D.B. and Kampf, C. (1987) In: New Comprehensive Biochemistry (Ametz, J. ed.) Vol. 15, pp. 199–211, Elsevier, Amsterdam.

13. Chen, Z., Koh M., Van Dreissche G., Van Beeumen J.J., Bartsch R.G., Meyer T.E., Cusanovich M.A. and Mathews F.S. (1994) The structure of flavocytochrome c sulfide dehydrogenase from a purple phototrophic bacterium. Science 266, 430–432.

14. Steinmetz, M.A., Trüper H.G. and Fischer, U. (1983) Cytochrome *c*-555 and iron-sulfur proteins of the non-thiosulfate-utilizing green sulfur bacterium *Chlorobium vibrioforme*. Arch. Microbiol. 135, 186–190.

15. Brune, D.C., and Trüper, H. G. (1986) Noncyclic electron transport in chromatophores of photolithotrophically grown *Rhodobacter sulfidophilus*. Arch. Microbiol. 145, 295–301.

16. Trumpower B.L. (1990) Cytochrome *bc1* complexes of microorganisms. Microbiol. Rev. 54, 101–129.

17. Shahak, Y., Arieli, B., Binder, B. and Padan, E. (1987) Sulfide-dependent photosynthetic electron flow coupled to proton translocation in thylakoids of the cyanobacterium *Oscillatoria limnetica.* Arch. Biochem. Biophys. 259, 605–615.

18. Slooten, L., de Smet, M. and Sybesma, C. (1989) Sulfide-dependent electron transport in thylakoids from the cyanobacterium *Oscillatoria limnetica*. Biochim. Biophys. Acta 973, 272–280.

19. Arieli, B., Padan, E. and Shahak, Y. (1991) Sulfide induced sulfide-quinone reductase activity in thylakoids of *Oscillatoria limnetica*. J. Biol. Chem. 266, 104–111.

20. Shahak, Y., Arieli, B., Hauska, G., Herrmann, I. and Padan, E. (1992) Isolation of sulfide-quinone reductase (SQR) from prokaryotes. Phyton 32, 133–137.

21. Arieli, B., Shahak, Y., Taglicht, D., Hauska, G. and Padan, E. (1994) Purification and characterization of sulfide-quinone reductase (SQR), a novel enzyme driving anoxygenic photosynthesis in *Oscillatoria limnetica*. J. Biol. Chem., 269, 5705–5711.

22. Arieli, B, Binder, B., Shahak, Y. and Padan, E. (1989) Sulfide induces the synthesis of a periplasmic protein in the cyanobacterium *Oscillatoria limnetica*. J. Bacteriol. 171: 699–702.

23. Shahak, Y., Hauska, G., Herrmann, I., Arieli, B., Taglicht, D. and Padan, E. (1992) Sulfide-quinone reductase (SQR) drives anoxygenic photosynthesis in prokaryotes. In: Research in Photosynthes is (Murata, N., ed.), Vol. II, pp 483–486 Kluwer Academic Publishers, The Netherlands.

24. Yamanaka, T. and Kusai, A. (1976) In Flavins and Flavoproteins (Singer, T. P., ed) pp. 292–301, Elsevier, Amsterdam.

25. Wierenga, R.K., Terpstra, P., and Hol, W.G.J. (1986) Prediction of the occurrence of the ADP-binding βαβ-fold in proteins, using an amino acid sequence fingerprint. J. Mol. Biol. 187, 101–107.

26. Eggink, G., Engel, H., Vriend, G., Terpstra, P. and Witholt, B. (1990) Ruberdoxin reductase of *Pseudomonas oleovorans*. Structural relationship to other flavoprotein oxidoreductases based on one NAD and two FAD fingerprints. J. Mol. Biol. 212, 135–142.

27. Feiler, U. and Hauska, G. (1995) The reaction center from green sulfur bacteria. In: Anoxygenic Photosynthetic Bacteria (Blankenship, R.E., Madigan, M.T. and Bauer, C.E., eds.) pp. 665–685, Kluwer Academic Publishers, Dordrecht, Holland.

28. Buchanan, S.K., Fritzsch, G., Ermler, U. and Michel, H. (1993) New crystal form of the photosynthetic reaction center from *Rhodobacter spheroides* of improved diffraction quality. J. Mol. Biol. 230, 1311–1314.

29. Shahak, Y., Arieli, B., Padan, E. and Hauska, G. (1992) Sulfide-quinone reductase (SQR) activity in *Chlorobium*. FEBS Lett. 299, 127–130.

30. Shahak, Y., Klughammer, C., Schreiber, U., Padan, E., Herrmann, I. and Hauska, G. (1994) Sulfide-quinone and sulfide-cytochrome reduction in *Rhodobacter capsulatus*. Photosynt. Res. 39, 175–181.

31. Klughammer, C., Hager, C., Padan, E., Schütz, M., Schreiber, U., Shahak, Y. and Hauska, G. (1995) Reduction of cytochromes with menaquinol and sulfide in membranes from green sulfur bacteria. Photosynt. Res. 43, 27–34.

32. Schütz, M., Shahak, Y., Padan, E. and Hauska, G. (1996) Purification and characterization of the sulfide-quinone reductase (SQR) of *Rhodobacter capsulatus* DSM 155. Proc. Xth International Congress of Photosynthesis, Montpellier, France (Mathis, P., ed.) Vol II, pp 673–676, Kluwer Academic Publishers.

33. Schütz, M., Shahak, Y., Padan, E. and Hauska, G. (1997) Sulfide-quinone reductase from *Rhodobacter capsulatus*: purification, cloning and expression. J.Biol.Chem., 272, 9890–9894.

34. Hansen, T., and Van Gemerden, H. (1972) Sulfide utilization by purple nonsulfur bacteria. Arch. Microbiol. 86, 49–56.

35. Alberti, M., Burke, D.H. and Hearst, J.E. (1995) Structure and sequence of the photosynthesis gene cluster. In: Anoxygenic Photosynthetic Bacteria (Blankenship, R.E., Madigan, M.T., and Bauer, C.E., eds) pp. 1083–1106, Kluwer Academic Publishers, Dordrecht, The Netherlands.

36. Karplus, P.A. and Schulz, G.E. (1987) Refined Structure of Glutathion Reductase at 1,54A Resolution. J. Mol. Biol. 195, 701–729.

37. Krauth-Siegel, R.L., Blatterspiel, R., Saleh, M., Schiltz, E., Schirmer, R.H. and Untucht-Grau, R. (1982) Glutathion Reductase from Human Erythrocytes. Eur. J. Biochem. 121, 259–267.

38. Schierbeck, A.J., Swarte, M.B.A., Dijkstra, B.W., Vriend, G., Read, R.J., Hol, W.G.J., Drenth, J. and Betzel, C. (1989) X-ray Structure of Lipoamide Dehydrogenase from *Azotobacter vinelandii* determined by a Combination of Molecular and Isomorphous Replacement Techniques. J. Mol. Biol. 206, 365–379.

39. Woese CR (1987) Bacterial evolution. Microbiol. Reviews 51: 221–271.

40. Fenchel, T.M. and Riedl, R.J. (1970) The sulfide system. A new biotic community underneath the oxidised layer of marine sand bottom. Mar. Biol. 7, 255–268.

41. Fenchel, T. and Bernar, C. (1995) Mats of colourless sulphur bacteria. I. Major microbial processes. Mar. Ecol. Prog. Ser. 128, 161–170.

42. Schiemer, F., Novak, R. and Ott, J. (1990) Metabolic studies on thiobiotic free-living nematodes and their symbiotic microorganisms. Mar. Biol. 106, 129–137.

43. Völkel, S. and Grieshaber, M.K. (1996) Mitochondrial sulphide oxidation in the lugworm *Arenicola marina*. Evidence for alternative electron pathways. Eur. J. Biochem. 235: 231–377.

44. Kelly, D.P., Shergill, J.K., Lu, W-P. and Wood, A.P. (1997) Oxidative metabolism of inorganic sulfur compounds by bacteria. Antonie van Leeuwenhock 71, 95–107.

45. Burggraf, S., Olsen, G.J., Stetter, K.O. and Woese, C.R. (1992) A phylogenetic analysis of *Aquifex pyrophilus*. Syst. Appl. Microb. 15, 352–356

46. Völkel, S. and Grieshaber, M.K. (1997) Sulphide oxidation and oxidative phosphorylation in the mitochondria of the lugworm *Arenicola marina*. J. Exp. Biol. 200, 83–92.

27

MECHANISTIC ASPECTS OF THE Q_0-SITE OF THE bc_1-COMPLEX AS REVEALED BY MUTAGENESIS STUDIES, AND THE CRYSTALLOGRAPHIC STRUCTURE

A. R. Crofts,[1] Blanca Barquera,[1] R. B. Gennis,[1] R. Kuras,[1] Mariana Guergova-Kuras,[1] and E. A. Berry[2]

[1]Center for Biophysics and Computational Biology
University of Illinois at Urbana-Champaign
[2]Lawrence Berkeley National Laboratory
University of California at Berkeley

INTRODUCTION

Ubiquinol:cytochrome c oxidoreductase (bc_1-complex) is a central component of the electron transfer system in almost all organisms, occurring ubiquitously in respiratory and photosynthetic chains of mitochondria and bacteria, and (as the b_6f-complex) in the photosynthetic chain of oxygenic photosynthesis (see refs. 1–3 for recent reviews). A modified Q-cycle (4–9) accounts well for a large body of experimental data from studies in which the function has been explored. The mechanism involves two catalytic sites for oxidation or reduction of the quinone couple, and a third site for electron transfer to cytochrome c. The function of the two quinone reactive sites has been explored biochemically, and characterized using kinetic spectroscopy, and their activities differentiated through use of inhibitors that act specifically at one site or the other (1–13). The quinol oxidizing site (Q_0-site) catalyzes a unique reaction in which the electrons from QH_2 are passed to separate electron transport chains in the complex through a bifurcated reaction. Further information on the occupancy of the Q_0-site has been obtained though study of the interaction of the occupant with the 2Fe2S center of the reduced ISP, observed through changes in the EPR spectrum of this center (33–40). In addition to the effects observed with stigmatellin and UHDBT (10–13,33), a well-defined band at $g_x = 1.800$ has been observed when quinone occupies the pocket. The dependence of these spectral changes on ambient redox potential, and on the extent of extraction of quinone, has lead Ding (34,35) and colleagues to suggest a double occupancy of the site by weakly (Q_{ow}) and strongly (Q_{os}) binding quinone species.

The Phototrophic Prokaryotes, edited by Peschek *et al.*
Kluwer Academic / Plenum Publishers, New York, 1999.

Although the modified Q-cycle mechanism is well supported, there remain several problems in understanding the mechanism of catalysis at the molecular level. A more detailed understanding of the structure-function relationship at the catalytic sites requires a structural context, and this has recently been provided by X-ray crystallographic studies (15–23). In this paper, we review the information on mechanism obtained from studies of mutations that effect the Q_o-site, and consider this in the context of the detailed structural information from crystallographic studies of the chicken heart mitochondrial bc_1-complex (20–22).

METHODS

The crystallographic determination of the structures of the complexes is discussed at length elsewhere (22). Molecular engineering protocols used were similar to those previously reported (24–26), and will be presented in detail elsewhere (Barquera et al., in preparation). Biophysical characterization of mutant strains was performed using procedures previously described (5,6,9,11).

RESULTS AND DISCUSSION

Structure of the Catalytic Subunits

The catalytic subunits, cytochrome b, cytochrome c_1 and the iron sulfur protein, have been well conserved across the bacterial/eukaryote evolutionary divide, allowing us to interpret data from experiments with *Rb. sphaeroides* in the context of the mitochondrial structures. The X-ray crystallographic work has confirmed the general features anticipated in earlier work (1,29,30), and provided a wealth of additional detail (22,23). The location of the quinone reactive sites in the avian structure (22) has been determined using inhibitors shown through functional studies to be specific for each site. Crystals have been grown in the presence of three Q_o-site inhibitors, stigmatellin, myxothiazol, and UHDBT, and the Q_i-site inhibitor antimycin. The position of the inhibitors has been determined from the electron density difference from the native crystals. The structure of the protein has not yet been determined independently for most of the inhibitor containing crystals, but, apart from changes in a few key residues (to be described in context), the cytochrome b subunit does not appear to undergo any major changes in structure on binding inhibitors.

In contrast, the ISP changes its position dramatically on binding stigmatellin, showing a rotational displacement in which the 2Fe2S center moves by ~16 Å from a position close to cyt c_1 in the absence, to a docking interface on cyt b in the presence of inhibitor. Theoretical considerations suggest that the structure would not be competent in all electron transfer steps in either of the positions found, and we have suggested that the ISP must move between these positions during turn over. Figure 1 summarizes the mechanism we have proposed, in which the movement of the catalytic domain of the ISP between reaction interfaces on cytochrome b and cytochrome c_1 subunits is required for electron transfer from quinol to cyt c_1 (31).

The Q_o-Site

An extensive literature on mutations that affect inhibitor binding or kinetic function at the Q_o-site has been summarized recently by Brasseur et al. (27); our own recent work on

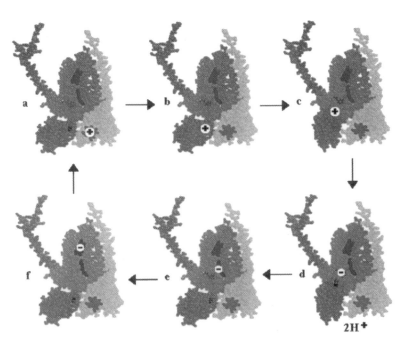

Figure 1. Scheme showing the reactions of quinol oxidation at the Q_o-site. The complex is represented by the three catalytic subunits (cyt b - red; cyt c_1 - green; ISP - blue), shown as transparent spacefilling models, revealing the metal centers. The transfer of an oxidizing equivalent to cyt c_1 is shown by the ringed "+" (a), which is transferred to the ISP (in ISP_R position) (b); the oxidation of QH_2 (a–c) is indicated by the appearance of an electron (ringed "–") at the Q_o-site, indicating formation of a semiquinone (d) (in c and d, ISP in ISP_O position), and its transfer to the b-cytochrome chain (e, f), leaving quinone (e), which leaves the site (f). Note that the electron transfer from QH_2 to the 2Fe2S center to form the semiquinone anion, releases $2H^+$ to the aqueous phase (d), balancing the charges in the transition (c–d).

site-directed mutations in the ef-loop, including the -PEWY- span, will be presented in detail elsewhere (28, and Barquera et al., in preparation). This work has identified spans in the sequence likely to contribute structurally; the C-terminal end of the transmembrane helix C, the N-terminal section of the cd loop, including the amphipathic cd-helix, the ef loop, including a coil from the end of the helix E, the -PEWY- turn, the small ef helix connecting the -PEWY- turn to transmembrane helix, and the N-terminal end of helix F. In the structure, these spans define two features: i) a quinol binding pocket, occupying a volume in which a different electron density is seen in crystals containing myxothiazol or stigmatellin, and ii) a concave surface, which is a docking interface for the ISP. The transmembrane helices contributing to the site flare out to provide the volume, and to form the sides of the pocket, the -PEWY- loop acts as a spacer, and the amphipathic helices provide the bottom. They also, on their other faces, provide most of the ISP binding surface, with the ef-loop providing additional residues. From examination of the placing of these domains in the structure it is apparent that mutations in either can interfere with the catalytic mechanism.

Differential effects on binding of the two classes of inhibitor acting at the site have been noted, with strains resistant to stigmatellin, but not myxothiazol or mucidin, and vice versa, and strains resistant to both inhibitors (1,27,28). Binding studies have shown that occupation is mutually exclusive (3,10,12,13,32–35), so that only one occupant (quinone, stigmatellin, myxothiazol, or UHDBT) can be in the site, and the crystallographic studies confirm these earlier results. Mutations have several different effects on the changes in EPR spectrum of the Rieske center at $g_x = 1.800$, which have been attributed to interaction with quinone at the Q_o-site. However, for any particular mutation, the effects on signals attributed to Q_{ow} and Q_{os} have been the same.

The Q_o-site is shown in greater detail in Fig. 2, in which residues known to effect inhibitor binding or catalysis are shown as stick models. The locations of stigmatellin and myxothiazol are modeled on the basis of the electron density differences observed in separate crystals containing these inhibitors, compared to the native structure.

Kinetics in the High Potential Chain

Myxothiazol binds at the Q_o-site of cyt b, and displaces quinol, preventing the delivery of electrons to the 2Fe2S center of the ISP. This simplifies the electron transfer chain so that the reactions of the high potential components can be examined separately. In *Rb. sphaeroides*, oxidizing equivalents from the primary donor (P^+/P) of photochemical reaction center (RC) are delivered to cytochrome c_1 by cytochrome c_2, with a $t_{1/2}$; of about 150 μs. The electron transfer from the Rieske center to cyt c_1 cannot be resolved, because it is not rate determining. However, kinetic arguments suggest that the rate constant must be in the range 10^5 s^{-1} (5,7,9). In the absence of myxothiazol, the bifurcated reaction occurs, resulting in cytochrome b reduction. A ~200 μs lag observed before onset of reduction of cyt b_H is contributed by several processes: the delivery of oxidizing equivalents to cyt c_1 (~150 μs), the oxidation of the 2Fe2S center (<10 μs), and the movement of the ISP to its reaction interface with QH_2. The reactions of the ISP with its partners (QH_2 bound at the Q_o-site of cyt b, and the heme of cyt c_1) (Fig. 1) require a movement of ~16 Å between two reaction domains on these separate subunits.

Kinetics of Quinol Oxidation

In the oxidation of quinol, the kinetics of electron transfer to the high potential chain and the cytochrome b chain are closely matched. The system behaves as if the oxidation involved a concerted, two-electron reaction with simultaneous delivery of electrons to two separate chains. However, it is generally accepted that this behavior reflects a sequential reaction in which the intermediate semiquinone is consumed rapidly, and never attains a significant concentration.

$$Q_oH_2.(2Fe2S^+.c_1{}^+).(b_L{}^+.b_H{}^+.Q_i) = Q_o{}^{\cdot-}.(2Fe2S.c_1)^+.(b_L{}^+.b_H{}^+.Q_i) + 2H_p{}^+ \quad (\sim 600 \text{ μs})$$

$$Q_o{}^{\cdot-}.(2Fe2S.c_1)^+.(b_L{}^+.b_H{}^+.Q_i) = Q_o.(2Fe2S.c_1)^+.(b_L.b_H{}^+.Q_i) \quad (< 60 \text{ μs})$$

Overall:

$$Q_oH_2.(2Fe2S^+.c_1{}^+).(b_L{}^+.b_H{}^+.Q_i) = Q_o.(2Fe2S.c_1)^+.(b_L.b_H{}^+.Q_i) + 2H_p{}^+$$

Because the reduction of cyt b_L by $Q_o{}^{\cdot-}$, and of cyt b_H by cyt b_L are not rate limiting (34,36), the overall reaction can be assayed by measuring the rate of reduction of cyt b_H in the presence of antimycin; quinol oxidation is the rate limiting step under conditions of saturation with substrate.

Crofts and Wang (9) developed a model that accounted well for the kinetic features of quinol oxidation; although this provided an excellent fit to the kinetics over a wide

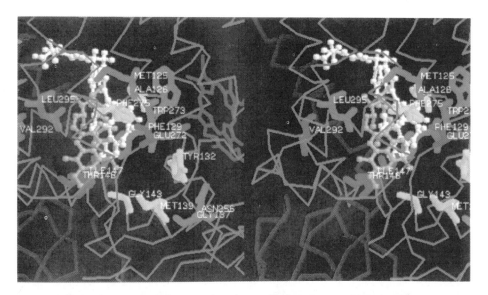

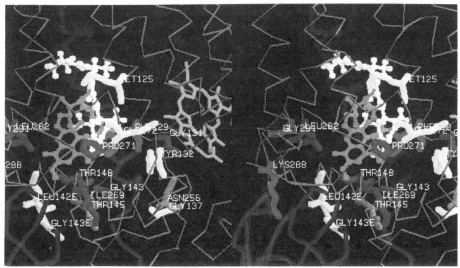

Figure 2. Residues at which mutation leads to modification of function. **A.** Inhibitor resistance sites. Note that in many cases, several mutations have been generated at a site, with differential effects on inhibitor binding, depending on the nature of the change. The following color-coding is therefore somewhat simplistic. Myxothiazol resistant sites are shown in yellow, stigmatellin resistance sites are shown in red. Residues at which modification leads to resistance to both inhibitors are shown in green. Stigmatellin and myxothiazol are shown as ball and stick models, colored cyan and white respectively, occupying the volumes in which they are found in separate crystals. The heme of cyt b_L is shown as a magenta wireframe model. The iron sulfur protein is shown in the stigmatellin-induced configuration, as a blue backbone model. **B.** Residues at which mutation produces a modified $g_x = 1.800$ band. Yellow, normal occupancy but no rate; red, empty, but moderate rate; magenta, empty, and little or no rate; violet, partially occupied, slow rate, hyper sensitive to myxothiazol. Heme is orange. Inhibitors as ball and stick models; stigmatellin is green, myxothiazol is white. The iron sulfur protein (chain E in Berry's structure) is shown as a blue backbone model, with Leu-142 and Gly-143 colored gray and brown. Changes at these residues also modify the kinetics of the Q_o-site, and the $g_x = 1.800$ signal.

range of conditions, one aspect of the mechanism of quinol oxidation was not explicitly addressed. It has been a long standing paradox that the bifurcated reaction at the Q_o-site delivered electrons stoichiometrically to the two separate chains, despite the strong thermodynamic potential favoring delivery of both electrons to the high potential chain (see Brandt (37) for a recent discussion).

Double-Occupancy Model

Earlier work had shown an effect of the redox state of the quinone pool on the EPR spectrum of the 2Fe2S center of the complex (33,38,39). More recently, Ding et al. (34,35,40), using wild-type and mutant strains from *Rb. capsulatus*, have investigated the effects of extraction of ubiquinone, and concluded that different spectral changes can be detected at different local concentrations. They suggested that these reflect two different bound quinones at the Q_o-site, called Q_{os} and Q_{ow} for strongly and weakly binding species, occupying separate domains, at both of which Q and QH_2 bind with equal affinity (i.e. with no change in E_m with respect to the pool). They have extended these observations to mutant strains, and kinetic studies of turnover and inhibitor binding, and have discussed possible mechanisms and atomic details of ligation. Brandt (43) has recently expanded on the double occupancy model to suggest an explanation for the observation that electrons do not "leak" rapidly from the strongly reducing semiquinone at the Q_o-site into the high potential chain. The extensive work on the g_x signals of the 2Fe2S center has contributed a useful set of data on interactions of quinone and inhibitor species with the ISP, and on occupancy of the Q_o-site. Nevertheless, there are some problems with the interpretation offered (28).

Effects of Q_o-Site Occupancy on EPR Spectra of the Iron Sulfur Protein

In summary, with the Q-pool oxidized and ISP reduced, a sharp band at $g_x = 1.80$ is seen. When the pool is reduced, or Q extracted, or myxothiazol bound at the Q_o-site, the g_x signal shifts from 1.80 to ~1.765, and becomes smaller (34,35,39,40). When the pool is extracted so as to leave ~2 Q/RC (1 as Q_A, plus ~2Q/bc_1-complex in the preparations used) an intermediate signal at $g_x = 1.783$ is seen. Several possible explanations for these effects on the EPR spectrum can be offered:

i. Ding et al. (35,40) suggested that the $g_x=1.783$ signal was due to the Q_{os} species, and that this would likely be the one forming a strong interaction with the ISP. From the structure, this would translate to Q_{os} occupying the distal end of the pocket (where stigmatellin binds). Q_{ow} was suggested to be an exchangeable species, through which coupling to the quinone pool could occur. In the context of the structure, Q_{ow} would be expected to bind at the proximal (myxothiazol) end, nearer heme b_L. Since the $g_x=1.800$ signal is ascribed to Q_{ow}, the interaction with the 2Fe2S center indicated by the signal would have to be indirect, and mediated by Q_{os}.

ii. The $g_x = 1.800$ signal is due to ISP_{red} interacting directly with Q at the Q_o-site, and is lost when the site is either empty, occupied by inhibitor, or by QH_2. We believe this explanation requires the fewest *ad hoc* hypotheses. However, if this is the case, then under conditions in which the quinone pool is oxidized before flash-activation, the reduced ISP would not be in position for its reaction with cytochrome c_1. It would be necessary to postulate that the diffusion time between cyt b and cyt c_1 docking interfaces is rapid compared to the <10 μs oxidation time of 2Fe2S observed, and that the binding at the cyt b interface under these conditions (Q oxidized, ISP reduced) must not be so great as to prevent the ISP leaving within this time. We discuss these parameters in greater detail elsewhere. (31).

In deciding between these explanations, the following points should also be considered:

i. None of the structures (22,23) show a density in the uninhibited site expected for a tightly bound quinone species. In the Berry structures, a weak density in the unoccupied site is observed, which may represent a quinone species, either weakly bound with low occupancy in a fixed position, or mobile in the site. In the Xia et al. structure, the authors reported that no loss of density from the pocket was observed on binding inhibitors at the Q_o-site, although a loss of density was observed at the Q_i-site on binding antimycin, and this was attributed to displacement of quinone (23). Since their crystals contained sub-stoichiometric levels of quinone (~0.6 Q/complex), these results would suggest that the Q_i-site binds the quinone more tightly than the Q_o-site.

ii. In the structures from both groups, the electron densities of inhibitors in the stigmatellin and myxothiazol crystals occupy overlapping volumes, suggesting, in line with competition studies, that these two inhibitors are mutually exclusive occupants of the site (see Fig. 2). Although at the current structural resolution, we cannot exclude the possibility that two ubiquinones could occupy the same site, it seems probable that only one quinone can occupy this volume.

iii. The intermediate line-shape at $g_x = 1.783$, seen when ~2 Q/bc_1-complex are present, is attributed to Q_{os}, a species that binds 30-fold more tightly than Q_{ow} (34,35). Paradoxically, the other properties of these two species are similar. This paradox could be resolved if the line shape change reflected only one species. In attributing the 1.783 signal, Dutton and colleagues assumed a tight binding species. A difficulty with this assumption comes from the heterogeneity in distribution of redox components likely in a population of chromatophores (cf. 42). Extraction of chromatophores to the point giving rise to the Q_{os} signal would leave the distribution of quinone random, so that some chromatophores would have an excess of quinone over the 2 per bc_1-complex, and we would expect to see a sharp peak at 1.800 due to Q_{ow}. In another fraction, the quinone would be below this ratio, leaving some bc_1-complex without a quinone, with a shallow peak at 1.765. At the very least, we would expect in the overall population a peak representing the convolution of these two spectra, at a position intermediate between them, as previously seen in pentane extracted mitochondrial complex III by de Vries et al. (39).

iv. The environment of the 2Fe2S center, and the protein interactions of the Rieske subunit, will change substantially as the ISP moves. In addition, the interaction between the spins of the reduced 2Fe2S center, and the oxidized cyt b_L (52) might also be expected to change during catalysis, since the distance between them would change by ~8 Å. Some perturbation of the spectrum of the 2Fe2S center might be attributed to these changes.

We suggest a simpler interpretation of the $g_x = 1.800$ signal and the changes seen on reduction, extraction, or in mutant strains.

a. The $g_x=1.800$ signal is associated with a complex formed between quinone bound in the Q_o-site at the distal end, and the reduced ISP docked firmly at the interface with cytochrome b.

b. The signal is lost whenever this complex cannot form: - on reduction of Q, on binding of inhibitors, on extraction of quinone, in strains where mutation prevents binding at the distal end of the pocket, and in strains where mutation prevents docking of the ISP.

c. Only one quinone species occupies the site. The spectroscopic effects ($g_x = 1.783$) attributed to Q_{os} arise from the statistical distribution of quinone in the heterogeneous chromatophore population, with additional contributions from changes in environment of the ISP due to changes in its interactions accompanying extraction of the last occupant, to leave a vacant site.

As an alternative to c), a modified double-occupancy model might be suggested in which Q_{ow} occupies the distal site, and Q_{os} the proximal site, and the latter serves as an electron shuttle to heme b_L on formation of the semiquinone on the former.

The Paradox of the Bifurcated Reaction

In the presence of antimycin, the Q_o-site is free to turn over, and under steady state conditions in which quinol is available in the pool, the b-cytochrome chain becomes reduced, and the ISP is oxidized by the high potential acceptor. Under these circumstances, de Vries was able to detect a small amount of semiquinone at the Q_o-site. Since antimycin is an effective inhibitor, and net electron transfer is blocked under these conditions, the semiquinone must not pass its second electron to the ISP. How is this thermodynamically favored reaction prevented? Several factors might be involved:

i. a set of forces in the binding domain which disfavor the binding of the oxidized ISP, or

ii. displacement of the semiquinone by ligand exchange to a position deeper in the pocket, and closer to its reaction partner, the cyt b_L heme, or

iii. a conformational change in which movement of one or more residues "insulates" the quinone binding pocket from the ISP domain.

These are not mutually exclusive; indeed some contribution from each is likely to occur, since the liganding interaction between Q_o-site occupant and the ISP will contribute to the force favoring occupancy, and any movement of the occupant of the site will likely involve some rearrangement of side chains.

We favor the second of these possibilities as the main factor. The possibility of such a movement is supported by the deeper position in the pocket of myxothiazol compared to stigmatellin (Fig. 2). The recent evidence for a substantial movement of Q_B in its binding pocket when it is reduced to semiquinone in the bacterial reaction center (41), also shows the possibility of such movement. Stowell et al. suggest that the activation barrier which gives the high temperature dependence of electron transfer from $Q_A^- Q_B$ to $Q_A Q_B^-$, might reflect this movement, and such considerations could obviously be applied to the reaction at the Q_o-site. In support of the third possibilitiy, the position of the head of stigmatellin (as currently modeled) overlaps the volume occupied by the ring of a tyrosine side chain (Tyr-279) in the uninhibited complex, and preliminary analysis suggests that the sidechain rotates from this position on binding of inhibitor (Berry, E.A., unpublished). In the absence of inhibitor, the tyrosine sits at the opening between the Q-binding pocket and the

ISP interface. It seems likely that any occupant of the pocket that interacts with the ISP can only do so if the tyrosine is displaced. We suggest that the tyrosine might act as a trapdoor controlling access to the occupant of the binding pocket. Stigmatellin also occupies some common volume with the position of the ring of Pro-271 in the uninhibited structure, which also appears to be involved in a small conformational change on occupation of the distal end of the pocket.

The Distribution of Residues at Which Mutation Results in Modified Function: Distinct Domains and Modes of Action

The extensive set of data provided by Ding et al. (34,35,40) on occupancy of the site in different mutant strains, as reflected in the $g_x = 1.800$ signal of the 2Fe2S center, and our own work in collaboration with these authors (28), provide important clues about the mechanism. At the present resolution of the structure, we can distinguish several different classes of mutational change, which are color-coded in Fig. 2A (inhibitor resistance sites) and B (residues where change leads to a modified $g_x = 1.800$ signal).

i. *Residue changes at the interface with the ISP.* Changes at Lys-288, Leu-282, Thr-145, Gly-143, Ile-269, which can be seen to project into the ISP interfacial surface in the Fig. 2B, eliminate the $g_x=1.800$ signal, which we have attributed to interaction between quinone at the Q_o-site and the reduced ISP. In some cases this is associated with a loss of activity, but some of these strains show substantial turnover. It seems possible that the mutations that eliminate or reduce the g_x signal interfere with the ISP binding, and prevent formation of the complex between Q and ISP_{red} that gives rise to this signal. The steric effects that give rise to loss of the EPR band might also be expected to interfere with the access of the ISP_{ox} to QH_2 at the site. The variable affects on turnover number would then be explained by the decrease in the residence time of the ISP at the docking interface, and therefore the probability of formation of the reaction complex.

ii. *Residue changes at the proximal end of the Q_o-binding pocket.* Residue changes (yellow in Fig. 2B above), which lead to loss of activity without loss of the $g_x = 1.800$ signal, cluster at the proximal end of the pocket, closer to the heme of cyt b_L, where myxothiazol binds. Many of the myxothiazol resistance strains also have residue changes at this end of the pocket (yellow and green in Fig. 2A). It seems reasonable to conclude that these mutations do not interfere with the binding of either Q at the distal end, or the ISP. However, they do prevent oxidation of QH_2. The inhibition of oxidation in these mutants must reflect some effect relating to the occupancy of the proximal end of the pocket. This could either be a second quinone species involved in electron transfer to the b_L heme (in a modified double-occupancy model), or a need for movement of the semiquinone to the proximal position to bring it closer to cyt b_L to allow rapid electron transfer.

iii. *Residue changes at the distal end of the Q_o-binding pocket.* Consistent with the model developed above, changes to residues impinging on the stigmatellin binding domain lead to loss of the $g_x = 1.800$ signal (red and magenta in Fig. 2B), and variable effects on electron transfer rate and inhibitor binding. In general, as might be expected, many of the mutational changes giving resistance to stigmatellin are found in this set (red and green residues in Fig. 2A).

CONCLUSIONS

We suggest as an alternative to the double-occupancy model, the following mechanism for turnover of the Q_o-site of the complex:

a. Oxidizing equivalents are transferred to the site through cytochrome c_1 and the ISP. The mechanism involves a movement of the ISP between reaction domains on cytochrome c_1 and cytochrome b (Fig.1).

b. The reaction proceeds from a complex between the oxidized ISP and quinol. The reaction complex involves QH_2 bound at the distal end of the pocket, and the oxidized ISP docked tightly at the interface on cytochrome b. Binding of QH_2 and formation of the complex requires a displacement of Tyr-279, to allow access of the ISP to the quinol. This binding likely also leads to release of 1 H^+ through an interaction at the site that changes the effective pK of the QH_2/QH^- .H^+ dissociation reaction.

c. Electron transfer from QH_2 to the ISP leads to formation of semiquinone.

d. The semiquinone moves in the pocket to the proximal end, near heme b_L.

e. The Tyr-279 trap-door flips back to its "closed" position, insulating the semiquinone from further reaction with the ISP.

f. The semiquinone passes its electron to the cyt b_L heme, and the quinone exits the site.

g. At some point between c) and f), the second proton is released. This might be required for formation of the semiquinone, or its movement in the pocket, or the transfer of the second electron. We suggest that the pH dependence of the activation barrier (37) reflects this second deprotonation, and that the activation barrier is in one of these steps.

h. Meanwhile, formation of the semiquinone liberates the reduced ISP so that it can deliver an electron to cytochrome c_1 by the tethered diffusion mechanism we have previously suggested.

ACKNOWLEDGMENTS

We are grateful to Professor Thomas Link for kindly providing us with the coordinates of the ISP soluble fragment before these were publicly available. We acknowledge with gratitude the support for this research provided by NIH grants GM 35438 (to ARC) and DK 44842 (to EAB), and by the Office of Health and Environmental Research, US Department of Energy, under contract DE-AC03-76SF00098 (EAB). The work was partially done at SSRL which is operated by the Department of Energy, Division of Chemical/Material Sciences. The SSRL Biotechnology Program is supported by the National Institutes of Health Biomedical Resource Technology Program, Division of Research Resources.

REFERENCES

1. Gennis, R.B., Barquera, B., Hacker, B., van Doren, S.R., Arnaud, S., Crofts, A.R., Davidson, E., Gray, K.A. and Daldal, F. (1993) J. Bioenerg. Biomembr. 25, 195–210.
2. Brandt, U. and Trumpower, B.L. (1994) CRC Crit. Rev. Biochem. 29, 165–197.
3. Schägger, H., Brandt, U., Gencic, S. and von Jagow, G. (1995) Methods Enzymol. 260, 82–96.

4. Mitchell, P. (1976) J. Theor. Biol. 62, 327–367
5. Crofts, A. R., Meinhardt, S. W., Jones, K. R. and Snozzi, M. (1983) Biochim. Biophys. Acta, 723, 202–218
6. Glaser, E.G. and Crofts, A.R. (1984). Biochim. Biophys. Acta, 766, 322–333.
7. Crofts, A. R. (1985). In: The Enzymes of Biological Membranes, (Martonosi, A.N., ed.), Vol. 4, pp. 347–382, Plenum Publ. Corp., New York.
8. Robertson, D.E. and Dutton, P.L. (1988) Biochim. Biophys. Acta 935, 273–291.
9. Crofts, A.R. and Wang, Z. (1989) Photosynth. Res. 22, 69–87
10. Bowyer, J. R., Dutton, P. L., Prince, R. C., & Crofts, A. R. (1980) Biochim. Biophys. Acta 592, 445–460.
11. Meinhardt, S.W. and Crofts, A.R. (1982). FEBS Lett., 149, 217–222.
12. Link, T.A., Haase, U., Brandt, U. and von Jagow, G. (1993) J. Bioenerg. Biomemb. 25, 221–232
13. von Jagow, G. and Link, T.A. (1986) Methods Enzymol. 126, 253–271
14. Crofts, A. R. and Wraight, C. A. (1983). Biochim. Biophys. Acta, 726, 149–186.
15. Iwata, S., Saynovits, M., Link, T.A. and Michel, H. (1996) Structure 1996, 4: 567–579.
16. Xia, D., Yu, C.-A., Deisenhofer, J., Xia, J.-Z. and Yu, L. (1996) Abstracts. Biophys. Soc. Meeting, Baltimore, Feb. 1996.
17. Xia, D., Kim, H., Deisenhofer, J., Yu, C.-A., Xia, J.-Z. and Yu, L. (1996) Abstracts, #SO355. XVII Meeting, Intntl. Union Crystallography, Seattle.
18. Kim, H., Xia, D., Deisenhofer, J., Yu, C.-A., Kachurin, A. and Yu, L. (1996) Abstracts, #SO356. XVII Meeting, Intntl. Union Crystallography, Seattle.
19. Yu, C.-A., Xia, J.-Z., Kachurin, A.M., Yu, L., Xia, D., Kim, H. and Deisenhofer, J. (1996) Biochim. Biophys. Acta 1275, 47–53.
20. Berry, E.A., Zhang, Z., Huang, L.-S., Chi, Y.-I., and Kim, S.-H. (1997) Abstracts, Biophy. Soc. Ann. Meeting, New Orleans, March 1997, Tu-PM-G4
21. Berry, E.A., Shulmeister, V.M., Huang, L.S. and Kim, S. H. (1995) Acta Crystallographica 51, 235–239.
22. Zhang, Z., Huang, L., Chi, Y.-I., Kim, K.K., Crofts, A.R., Berry, E.A., and Kim, S.-H. (1997) Manuscript in preparation.
23. Xia, D., Yu, C.-A., Kim, H., Xia, J.-Z., Kachurin, A.M., Zhang, L., Yu, L., Deisenhofer, J. (1997) Science, 277, 60–66
24. Yun, C.-H., Beci, R., Crofts, A.R., Kaplan, S. and Gennis, R.B. (1990) Europ. J. Biochem. 194, 399–411
25. Yun, C.-H., Crofts, A.R. and Gennis, R.B. (1991) Biochemistry, 30, 6747–6754
26. Konishi, K., Van Doren, S.R., Kramer, D.M., Crofts, A.R., and Gennis, R.B. (1991) J. Biol. Chem. 266, 14270–14276
27. Brasseur, G., Sami Saribas, A. and Daldal, F. (1996) Biochim. Biophys. Acta 1275, 61–69.
28. Crofts, A.R., Barquera, B., Bechmann, G., Guergova, M, Salcedo-Hernandez, R., Hacker, B., Hong, S. and Gennis, R.B. (1995) In Photosynthesis: from light to biosphere. (Mathis, P., ed.), Vol. II, pp. 493–500. Kluwer Academic Publ., Dordrecht.
29. Crofts, A.R., Hacker, B., Barquera, B., Yun, C.-H. and Gennis, R. (1992) Biochim. Biophys. Acta 1101, 162–165
30. Degli Esposti, M., De Vries, S., Crimi, M., Ghelli, A., Patarnello, T., and Meyer, A. (1993) Biochim. Biophys. Acta 1143, 240–271
31. Berry, E.A., Zhang, Z., Huang, L.-S., Kuras, R. , Guergova-Kuras, M. and Crofts, A.R. (1997) Abstract, Meeting on "Reaction centers of photosynthetic purple bacteria: structure, spectroscopy, dynamics", Cadarache, June 1997
32. Meinhardt, S.W. and Crofts, A.R. (1982). FEBS Lett. 149, 223–227.
33. Matsuura, K., Bowyer, J.R., Ohnishi, T., & Dutton, P.L. (1983) J. Biol. Chem. 258, 1571–1579.
34. Ding, H., Robertson, D.E., Daldal, F. and Dutton, P.L. (1992) Biochemistry 31, 3144–3158.
35. Ding, H., Moser, C.C., Robertson, D.E., Tokito, M.K., Daldal, F. and Dutton, P.L. (1995) Biochemistry 34, 15979–15996.
36. Glaser, E.G., Meinhardt, S.W. and Crofts, A.R. (1984) FEBS Lett. 178, 336–342
37. Brandt, U. (1996) Biochim. Biophys. Acta 1275, 41–46.
38. Siedow, J.N., Power, S., De La Rosa, F.F. and Palmer, G. (1978) J. Biol. Chem. 253, 2392–2399
39. De Vries, S., Albracht, S.P.J., Berden, J. and Slater, E.C. (1982) Biochim. Biophys. Acta 681, 41–53
40. Ding, H., Daldal, F. and Dutton, P.L. (1995) Biochemistry 34, 15997–16003
41. Stowell, M. H. B., McPhillips, T. M., Rees, D. C., Soltis, S. M., Abresch, E. and Feher, G. (1997) Science, 276, 812–815
42. Crofts, A.R., Guergova-Kuras, M., Hong, S. and Bechmann, G. (1997) Photosynth. Research, in press.

28

CORRELATION BETWEEN CYTOCHROME bc_1 STRUCTURE AND FUNCTION

Spectroscopic and Kinetic Observations on Q_O Site Occupancy and Dynamics

R. Eryl Sharp,[1] Aimee Palmitessa,[1] Brian R. Gibney,[1] Fevzi Daldal,[2] Christopher C. Moser,[1] and P. Leslie Dutton[1]

[1]Johnson Research Foundation
Department of Biochemistry and Biophysics
[2]Department of Biology
University of Pennsylvania
Philadelphia, Pennsylvania 19104

1. INTRODUCTION

In energy-transducing membranes, the free energy derived from redox reactions is coupled to charge separation and proton gradient formation by a sequence of membrane-spanning, multisubunit complexes (1,2). A key obligatory component within these systems from both photosynthetic and respiratory organisms is ubihydroquinone-cytochrome c oxidoreductase (cyt bc_1 complex). During respiration, this complex derives electrons from NADH or succinate via the ubiquinone pool (Q_{pool}) as ubihydroquinone (QH_2), subsequently reducing ferri cytochrome c generated by cyt c oxidase and molecular oxygen. In photosynthetic prokaryotes, the cyt bc_1 complex participates in a light-induced cyclic electron-transfer system with the reaction center protein which delivers QH_2 to the Q_{pool} and generates ferri cyt c, which is re-reduced by cyt bc_1. The common theme of cyt bc_1 function is the utilization of the 190 (mitochondria), or 250 mV (photosynthetic bacteria) redox potential difference between the Q_{pool} and cyt c to drive the separation of charge and vectorial pumping of protons across the supporting membrane to generate a protonmotive force, utilized to drive ATP synthesis (3,4).

For all organisms, the key redox functional components of the cyt bc_1 complex through which electron transfer is mediated, reside in three distinct subunits: one containing a high-potential [2Fe-2S] cluster, one a cyt c_1 and another cytochrome subunit with low- and high-potential b-type hemes (cyt b_L and b_H, respectively) (5,6). Integral to the

The Phototrophic Prokaryotes, edited by Peschek *et al.*
Kluwer Academic / Plenum Publishers, New York, 1999.

electron-transfer activity and the postulated loci of proton binding and release are two discrete sites for ubiquinone redox catalysis, designated Q_o and Q_i, termed as such on the basis of their proton output and input functions, respectively. These two sites are situated at opposite sides of the membrane and form the basis for the Q-cycle mechanism, which is the most commonly accepted description of the cyt bc_1 protonmotive function (7,8).

In photosynthetic prokaryotes, the Q_o site is located at the periplasmic face of the cytoplasmic membrane (9,10), where it catalyzes the two-electron oxidation of QH_2 to Q, the primary event of energy conversion in the cyt bc_1 complex. This oxidation involves the cooperation of two one-electron redox centers, the [2Fe-2S] cluster and cyt b_L (10,11), which flank the Q_o site. In order for one complete turnover of the cyt bc_1 complex to occur, it is clear that two QH_2 molecules are required to be oxidized at the Q_o site (12). However, at present little is known, in any species regarding the number of ubiquinone occupants, dynamics and mechanism of this process. In the conventional model, the Q_o site is envisioned as binding one QH_2 at a time and performing two separate, serial oxidations.

Ding *et al* have proposed an alternate working model, based on Q_o occupancy and kinetic analysis of wild-type and mutant cyt bc_1 complexes that is entirely consistent with the basic precedents of the Q-cycle hypothesis (3,4). These studies indicated that the Q_o site is able to accommodate two ubiquinone molecules within two distinct binding domains, one of which was determined to have a high affinity and the other a lower affinity for Q/QH_2 and were accordingly designated the Q_{os} (strong) and Q_{ow} (weak) domains, respectively. On the basis of this a plausible model for one complete turnover of the cyt bc_1 complex was presented which did not require exchange of Q/QH_2 with the Q_{pool}. The other ubiquinone binding site, the Q_i site, is located at the cytoplasmic face of the membrane near cyt b_H (10), where it catalyzes the net reduction of one bound Q to QH_2 in two discrete one-electron reduction steps, via a stable ubisemiquinone intermediate (13–16).

Recently, the crystal structure of the cyt bc_1 complex isolated from bovine heart mitochondria has been reported at a resolution of 0.29 nm (17). The complex crystallizes as a dimer and is believed to function as such *in vivo* (18). Each isolated monomer is composed of eleven distinct subunits (19), consisting of 2165 amino acid residues, with a total mass of 248 kDa. In the described structure, about 86% of the total residues present are visible in the electron density map (17): these include all of the cyt b subunit, both b_L and b_H hemes and four accessory subunits; core 1, core 2, subunit VI and subunit VII (which are not present in prokaryotic cyt bc_1 complexes). However, in the case of the cyt c_1 and [2Fe-2S] subunits, only the electron density due to the transmembrane anchor regions and redox active metal centers could be assigned with any degree of confidence, implying that the extramembrane domains of these subunits are highly mobile in the crystal (17). Indeed, no intercomplex crystal contacts occur between the domains of the cyt c_1 and [2Fe-2S] subunits that extend into the intermembrance space (equivalent to the periplasmic space in prokaryotes). It is proposed that the observed disorder maybe a prerequisite for domain mobility as a requirement for cyt bc_1 function (17). Additionally, the binding sites of two specific cyt bc_1 complex inhibitors could be identified in co-crystals of inhibitors and the cyt bc_1 complex. Myxothiazol binds specifically to the Q_o site, where it inhibits QH_2 oxidation (20) and consistent with this, the electron density for this inhibitor was located in a pocket within the cyt b subunit situated between the b_L heme and [2Fe-2S] cluster. Antimycin *A* specifically binds to the Q_i site where it inhibits Q reduction (21) and the electron density for this inhibitor was located near the cyt b_H heme.

The advent of the cyt bc_1 crystal structure raises some important structure/function issues that need to be addressed. Firstly, the distance between the [2Fe-2S] cluster and the cyt c_1 heme iron is 3.1 nm, which as the structure stands, is longer than that expected for the observed electron-transfer rate constant, unless a major conformational change occurs during catalysis (4,17). Secondly, in the difference density map of the antimycin: cyt bc_1 complex and the native crystal, a region of negative electron density close to the antimycin, was attributed to bound ubiquinone (17). However, no ubiquinone molecules could be located in the electron density map for the Q_o site. The lack of bound ubiquinone in this structure is not surprising, as the purified complex prior to crystallization has only one mole of ubiquinone present per two moles of cyt b subunit (17) and presumably was located at the Q_i site. Loss of ubiquinone upon solubilisation and subsequent isolation of the cyt bc_1 complex from the membrane is not uncommon and has been observed for the complex from *R. capsulatus* (22). Unfortunately, the lack of any visible electron density for ubiquinone in the Q_o site precludes any evidence for binding of one or two ubiquinones, although the size of the pocket between the b_L heme and [2Fe-2S] cluster is conceivably large enough to accommodate two ubiquinone molecules (17).

The electron paramagnetic resonance (EPR) positional lineshape of the reduced [2Fe-2S] cluster is highly sensitive to the degree and nature of the Q_o site occupants (Q/QH$_2$ and inhibitors) (3,11). Previously, we have noted that a 1% addition of absolute ethanol to preparations of *R. capsulatus* chromatophores dramatically alters the [2Fe-2S] EPR lineshape, implying that ethanol effectively *uncouples* the [2Fe-2S] cluster from the Q_o site. Preparation of the solubilized, isolated bovine cyt bc_1 complex for crystallization involved the use of buffers containing high concentrations of glycerol (20% *v/v*) (17), which we considered may have a similar effect upon the Q_o site occupancy as ethanol. In this report we describe a parallel spectroscopic and kinetic investigation of the effect of ethanol and glycerol upon the cyt bc_1 complex in *R. capsulatus* chromatophores and tentatively propose a model for cyt bc_1 function at the level of the Q_o site, which is in accord with our previous proposals, the reported crystal structure and the evidence put forth in this report.

2. EXPERIMENTAL PROCEDURES

2.1. Bacterial Growth, Harvesting, and Chromatophore Preparation

This was performed as previously described (3,4). For most of the experiments involving glycerol, the chromatophores were prepared as above, but in buffer containing 20% glycerol by volume. This was necessary because glycerol does not readily diffuse across the chromatophore membranes when added exogenously to the samples. The concentration of cyt b and cyt c and bacteriochlorophyll in the preparations was determined as previously described (23).

2.2. EPR Spectroscopy and Sample Preparation

EPR measurements were performed on a Bruker ESP300E spectrometer. Temperature control was maintained by an Oxford ESR model 900 continous flow cryostat interfaced with an Oxford model ITC4 temperature controller. Frequency was measured by a Hewlett-Packard model 5350B frequency counter. Typical operating procedures for [2Fe-2S]$^+$ detection were: sample temperature, 20 K; microwave frequency, 9.474 GHz; microwave power, 2 mW; modulation frequency, 100 kHz; modulation amplitude, 19.8 G and time constant, 164 ms. Sample concentrations were 10 μM reaction center. Samples for

EPR spectroscopy were poised anaerobically at redox potentials < 230 mV, in order to maintain the [2Fe-2S] cluster in the reduced, paramagentic state (E_{m7} = 280 mV). Typically, for monitoring the characteristics of Q_O site occupancy when the Q_{pool} is completely oxidised (E_{m7} = 90 mV), chromatophore prerparations were poised at 200 mV in the presence of 50 μM diaminodurene (DAD) and *N*-methyldibenzopyrazine (PMS) as mediators, analysis of the Q_O site in the presence of inhibitors was performed in the presence of ascorbate as a reductant (E_{m7} < 50 mV). As previously noted and relevent to this study, all redox mediators added to poised EPR samples of chromatophore preparations were dissolved in DMSO which has no effect upon the [2Fe-2S] cluster EPR lineshape (3). EPR investigation of the Q_i site occupancy were performed under optimal conditions for ubisemiquinone formation, at pH 9.0, with samples poised at 50 mV in the presence of 1,2-napthoquinone as a mediator. The antimycin-sensitive semiquinone was assayed from the isotropic g = 2.005 resonance recorded as above, but at 20 mW power and a temperature of 143 K.

2.3. Flash Photolysis of Chromatophores

Flash-activated turnover of the cyt bc_1 complex was performed on a Biomedical Johnson Foundation Type (University of Pennsylavnia) single-wavelength spectrophotometer fitted with an anaerobic redox cuvette. Single, (full width at half-height, 8 μs) pulses of actinic light were delivered to the cuvette perpendicular to the measuring beam. The reaction center concentration was determined from the flash-activated absorbance change of the bacteriochlorophyll dimer oxidation (forming $[BChl]_2^+$) at 602–540 nm in chromatophores poised at 380 mV, so that the cyt c_2 was oxidised prior to activation (24). When cyt c_2 is oxidised, the flash generated $[BChl]_2^+$ is stable for milliseconds and is readily measured. Kinetics of cyt b reduction were monitored at 560–570 nm, cyt c_1+c_2 oxidation and re-reduction were measured at 550–540 nm (24). Data was collected and averaged on a LeCroy binary oscilloscope. The kinetic data were fit to single or double exponential kinetics, as was appropriate using a non-linear iterative procedure on the software package IGOR (Wavemetrics). All experiments were performed at room temperature in 50 mM MOPS, 100 mM KCl buffer, pH 7.0 at a reaction center $[BChl]_2$ concentration of 0.2 μM. The samples were maintained under anaerobic conditions by continually flushing the cuvette with water-saturated argon. The ambient redox potential was reported by a platinum measuring electrode and a Ag/AgCl reference electrode, relative to the SHE and the potential adjusted by addition of either sodium dithionite or potassium ferricyanide as reductant and oxidant, respectively. The system was maintained at redox equilibrium by the presence of the mediators DAD, 2-hydroxy-1,4-napthoquinone, duroquinone at 10 μM and PMS, *N*-ethyldibenzopyrazine ethosulfate (PES) at 2 μM concentration.

3. RESULTS

3.1. EPR Spectral Properties of the [2Fe-2S] Cluster under Varying Q_O Site Occupancy

The left hand panel of Figure 1 illustrates the EPR spectra of the cyt bc_1 complex [2Fe-2S] cluster in wild-type *R. capsulatus* chromatophores under various solvent conditions. Spectrum (a) was aquired from chromatophores poised at 200 mV, at which poten-

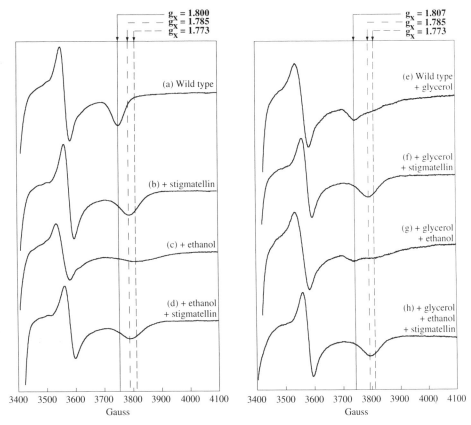

Figure 1. EPR spectra of reduced, paramagnetic [2Fe-2S] cluster in *R. capsulatus* chromatophores containing the cyt bc_1 complex, suspended at a reaction center concentration of 10 μM (0.6:1 ratio of cyt bc_1 to reaction centre). All samples were prepared at a redox poise of 200 mV, except for the experiments with stigmatellin, where the samples were reduced by the addition of a limiting amount of ascorbate (the EPR spectrum of the [2Fe-2S] cluster in the presence of stigmatellin is independent of redox poise at $E_h < 230$ mV). Each spectrum is an addition of five succesive scans. Spectrum (a), chromatophores suspended in buffer solution (50 mM MOPS, 100 mM KCl, pH 7.0); (b), + 10 μM stigmatellin; (c), + 170 mM ethanol; (d), 170 mM ethanol + 10 μM stigmatellin; (e), chromatophores prepared in buffer solution containing 20% glycerol (*v/v*); (f), 20% glycerol + 10 μM stigmatellin; (g), 20% glycerol + 170 mM ethanol; (h), 20% glycerol + 170 mM ethanol + 10 μM stigmatellin.

tial the Q_{pool} is completely oxidised and exhibits a prominent g_x signal at 1.800; this is characterisitic of cyt bc_1 complex with the Q_o site fully occupied by Q (3,4). Spectrum (b) shows the effect of addition of a slight molar excess relative to the cyt bc_1 concentration of the high affinity Q_o site specific inhibitor, stigmatellin, which abolishes the g_x signal at 1.800 and replaces it with the characteristic stigmatellin induced signal at g_x = 1.785. As mentioned in the figure legend, this effect is independent of the redox poise of the Q_{pool} and also of the occupancy of the Q_o site (3,4). The addition of 1% (*v/v*) absolute ethanol (170 mM) to the chromatophore suspension poised at 200 mV results in a very broad g_x signal centered around 1.773 (spectrum c). The course of the ethanol titration (data not shown) gave g_x signal lineshapes which gradually lost intensity at 1.800 and shifted upfield to 1.773. Further addition of ethanol to 340 mM did not result in any change in the lineshape of the EPR signal from spectrum c, illustrating saturation at 170 mM ethanol. Spectrum d shows that when stigmatellin is added to chromatophores suspended in buffer plus 170 mM ethanol, the EPR spectrum displays the characteristic stigmatellin g_x signal at 1.785. The right hand panel of Figure 1 shows the effect of 20% glycerol upon the [2Fe-2S] EPR signal. Spectrum (e) was acquired under identical conditions to (a), except for the presence of glycerol. This spectrum is more complicated than that obtained with ethanol (c) and appears to be composed of at least two components: a lower intensity signal compared to that obtained in the absence of glycerol (a), due to ubiquinone at $g_x = 1.807$, (this is a reproducible small downfield shift in resonance from the value of 1.800 obtained in the absence of glycerol); and a shallow, broad resonance centered around $g_x = 1.773$. Spectrum (f) again shows the characteristic stigmatellin signal when this inhibitor was added to the 20% glycerol suspended chromatophores. Spectrum (g) was acquired after the addition of 170 mM ethanol and again is more complicated than that of (c), where the g_x signal at 1.807 is further diminished in intensity and the signal at 1.773 becomes more apparent. A possible reason for the lack of a complete shift in the g_x signal to 1.773 in the spectrum (g) is that ethanol is highly soluble in glycerol and the effective ethanol concentration in the Q_o site is maybe less than the saturating 170 mM. Upon addition of stigmatellin to the glycerol/ethanol sample, the $g_x = 1.785$ signal is recovered (h).

Figure 2 shows the effect of adding glycerol exogenously to chromatophores prepared in the standard fashion (top spectrum) compared to chromatophores prepared in the presence of 20% glycerol (bottom spectrum). The EPR spectra are collected under identical conditions to that described for Figure 1 (a & d). One can see that the two are qualitatively similar, however when adding glycerol exogenously, sufficient time needs to be allowed for equilibration to occur (24 hours), as the chromatophore membranes are relatively impermeable to glycerol and the Q_o site is located on the inside of the vesicles.

3.2. Effect of Glycerol upon the Q_i Site Ubiquinone Occupancy

Unlike the Q_o site, the ubiquinone occupancy of the Q_i site can be directly determined from the EPR signature of the stable ubisemiquinone radical ($Q^{\bullet -}$), which is maximally formed at a potential of 50 mV, pH 9.0. The EPR spectrum of $Q^{\bullet -}$ is centered at $g_{iso} = 2.005$ and the intensity of this signal is greatly diminished by the addition of antimycin. The relative amount of $Q^{\bullet -}$ in the Q_i site can be ascertained from the difference EPR spectra between wild-type minus antimycin inhibited cyt bc_1. EPR spectra of chromatophore samples which were identical, but for the absence and presence of 20% glycerol revealed that there was no difference in the amplitude of the signal, implying that addition of glycerol does not perturb the Q_i site occupancy (data not shown).

3.3. Kinetic Analysis of Cyt bc_1 Flash Induced Turnover

Single and multiple flash induced cytochrome bc_1 turnover kinetics were monitored in wild-type *R. capsulatus* chromatophores as described in the experimental procedures. The chromatophores were investigated under three different buffer conditions: buffer alone (50 mM MOPS, 100 mM KCl, pH 7.0); plus 1% ethanol; plus 20% glycerol. Each experiment was performed in triplicate on newly prepared samples to ensure reproducibility. No significant differences were observed for ubihydroquinone oxidation at the Q_o site in the presence of both glycerol and ethanol, compared to that in buffer alone (data not shown).

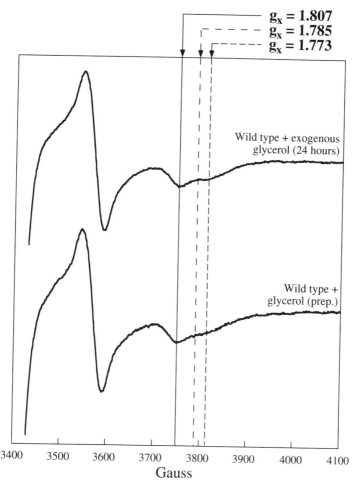

Figure 2. EPR spectra of the $[2Fe-2S]^+$ cluster in *R. capsulatus* chromatophores containing the cyt bc_1 complex. Experimental conditions as for Figure 1, with the chromatophore suspension poised at 200 mV. The upper spectrum was aquired after chromatophores that had been prepared in buffer solution were exposed to exogenously added 20% glycerol for 24 hours. The lower spectrum is from chromatophores prepared in buffer containing 20% glycerol.

4. DISCUSSION

4.1. Differential Effect of Alcohols on the Q_O and Q_i Site Occupancy and Cytochrome bc_1 Function

The data presented in Figure 1 (spectrum c) confirms the previously reported observation that addition of 1% ethanol to cytochrome bc_1 chromatophores with the Q_{pool} oxidised completely abolishes the prominent EPR signal at $g_x = 1.800$ and replaces it with a much broader feature centered at $g_x = 1.773$. This $[2Fe-2S]$ EPR signature induced by ethanol is very similar to that observed for one or two ubihydroquinones bound in the Q_O site, which have a $g_x = 1.777$ (3,4). It seems likely that in both cases (either the presence of ethanol or

QH_2), the hydroxyl functions on these molecules are influencing the [2Fe-2S] EPR lineshape in a similar fashion by hydrogen bonding to the histidine ligands of the cluster (3).

The effect of glycerol is more complicated than that of ethanol, since under the experimental conditions explored in this report, an intermediate effect is observed. The g_x signal at 1.800 is significantly decreased in intensity and slightly shifted to lower field, but is not completely abolished. Addition of ethanol to this sample decreases the intensity at $g_x = 1.800$ further and renders the $g_x = 1.773$ feature more obvious. Clearly, glycerol and ethanol have similar effects upon the [2Fe-2S] EPR signature, but to differing extents. In chromatophores which are completely or partially extracted of ubiquinone (cyt bc_1 Q_O site occupancy ≤ 1 Q), the EPR spectra of the ethanol and glycerol samples are identical under the same experimental conditions as described for the unextracted material. In both cases, the g_x signal is shifted from 1.765 (indicative of a ubiquinone free Q_O site) downfield to 1.773 (data not shown). Of note is the fact that these spectra are also the same as that obtained for addition of ethanol to unextracted chromatophores. This implies that the end-point effect of ethanol and glycerol on the [2Fe-2S] EPR spectra are the same, but that the glycerol affect is not manifested fully unless the Q_O site is at least partially occupied with ubiquinone. Our interpretation of this data is that ethanol at 170 mM effectively competes for the [2Fe-2S] cluster histidines and *uncouples* the sensitivity of the [2Fe-2S] cluster to the ubiquinones in the Q_O site. However, from the kinetic measurements it is clear that ethanol does not cause the ubiquinone to dissociate, since cyt bc_1 turnover is unaffected by its presence. Glycerol, evidently is a less effective competitor for [2Fe-2S] cluster than ethanol by at least a factor of 10-fold, since at a concentration of 2.7 M (20% v/v) it only partly alleviates the ubiquinone *coupling* to the [2Fe-2S] cluster.

Addition of the specific, tight binding Q_O site inhibitor, stigmatellin, to cyt bc_1 chromatophores under any of the experimental conditions investigated, always yielded the characteristic $g_x = 1.785$ signal, due to stigmatellin binding at the Q_O site and completely displacing the ubiquinone. We propose that under these circumstances, the [2Fe-2S] cluster is tightly locked in position at the Q_O site by direct interaction of the cluster histidine ligands with stigmatellin (*coupled*).

4.2. Influence of the EPR and Kinetic Experiments Upon the Cyt bc_1 Crystal Structure Interpretation

As described in the introduction, the reported crystal structure model of bovine cyt bc_1 complex appears devoid of ubiquinone at the Q_O site (17). Purification and subsequent crystallization of the complex were performed in the presence of 20% glycerol. We have shown in this study that prokaryotic cyt bc_1 complexes embeded within the chromatophore membrane and therefore experiencing a similar environment to that *in vivo*, are also affected by the presence of glycerol, specifically at the level of the Q_O site. The EPR spectral signature of the [2Fe-2S] cluster indicates that glycerol, like ethanol may interact via its hydroxyl functions with the histidine ligands of the cluster. In chromatophores, the presence of glycerol does not appear to induce dissociation of ubiquinone from the Q_O site, but it does tend to *uncouple* the [2Fe-2S] cluster from the Q_O site. The lack of observable ubiquinone in crystallized cyt bc_1 maybe due to the fact that glycerol induces dissociation of the [2Fe-2S] subunit from the Q_O site, thereby weakening ubiquinone binding.

4.3. A Proximal-Distal Model for Cytochrome bc_1 Function at the Q_O Site

Another feature of the reported cyt bc_1 structural model which has important functional ramifications is that parts of the cyt c_1 and the [2Fe-2S] subunit as a whole are highly disordered, only the positions of the metal redox centers could be deduced with certainty. Also, the 3.1 nm distance between the iron of the cyt c_1 heme and the [2Fe-2S] cluster is too large to support the anticipated electron transfer rate constant during turn-over, without some major conformational change occuring (4,17). This mechanistic co-nundrum and a likely solution to the distance problem is bourne out by the work of E. A. Berry and colleagues on the crystal structure of cyt bc_1 complexes from a number of dif-ferent mammalian sources (personal communication). They find two distinct structural forms: in co-crystals of stigmatellin and cyt bc_1, the [2Fe-2S] subunit is positioned close to the Q_O site and the distance between the cluster and cyt c_1 heme is comparable to that of the reported bovine structure (17); in the absence of Q_O site inhibitors, however, the [2Fe-2S] subunit position has shifted about 1.6 nm closer to cyt c_1, where the electron transfer distance is within physiological range (24). On the basis of this data, a parallel paper in these proceedings reports a proximal:distal type model for the movement of the [2Fe-2S] subunit between the Q_O site and cyt c_1 during turnover (25).

The EPR data discussed in this report can be incorporated into the reported model (24) and extend its applicabitily to cyt bc_1 function during turnover. Figure 3 schemati-cally illustrates the basic tenets of this model. Box A describes the state that cyt bc_1 exists in when the Q_O site is fully occupied by Q, the [2Fe-2S] cluster is *coupled* to the Q_O site ubiquinones and is located in the proximal equilibrium position (bold arrow upwards). Progressing from left to right in Figure 3; (B) illustrates an intermediate effect observed with subsaturating alcohol concentrations, where two equipopulated EPR signals are ob-served and the [2Fe-2S] subunit position is presumed to be split between the proximal and

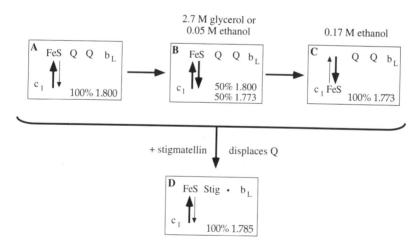

Figure 3. Schematic summary of EPR data and the structure: function correlation. Each box represents the cyt bc_1 complex, showing the characteristic EPR signal g_x component induced by the Q_O site environment of the [2Fe-2S] cluster. FeS and b_L represent the [2Fe-2S] cluster and the low potential heme, that flank the proximal and distal boundaries of the Q_O site and c_1 the cyt c_1 subunit, respectively. The arrows indicate the equilibrium that is pre-sumed to exist between the locality of the [2Fe-2S] subunit, either proximal and *coupled* to the Q_O site - Box A (bold arrow directed upwards, under the given conditions the fraction of the equilibrium population represented by the lighter arrow is insignificant), or *uncoupled* and distal to the Q_O site - Box C (bold arrow directed downwards). Box B represents an intermediate situation when the [2Fe-2S] cluster is partially *uncoupled* and the [2Fe-2S] subunit equilibrium population is equally distributed between the proximal and distal positions (bold arrows in both directions).

distal positions (bold arrows in both directions). The situation in box (C) occurs with saturating alcohol concentrations (170 mM ethanol); the [2Fe-2S] cluster is effectively *uncoupled*, shifting the equilibrium position of the [2Fe-2S] subunit distal to the Q_O site. Stigmatellin, as befits a specific, tight binding inhibitor always induces the same [2Fe-2S] EPR signal regardless of the Q_O site population, strongly *coupling* the [2Fe-2S] cluster to the Q_O site and shifting the [2Fe-2S] subunit equilibrium to the proximal position (D).

ACKNOWLEDGMENTS

This work was supported by the Public Health Service Grants GM 27309 to P. L. D. and GM 38237 to F. D. R.E.S. and B.R.G. gratefully acknowledge receipt of postdoctoral fellowships from the North Atlantic Treaty Organisation (NATO) and the NIH, respectively. We also thank Dr. E. A. Berry for making available preliminary structural data of the chicken cytochrome bc_1 complex.

REFERENCES

1. Brandt, U. & Trumpower, B. L. (1994) *Crit. Rev. Biochem. Mol. Biol.* 29, 165–197.
2. Gray, K. A. & Daldal, F. (1995) in *Anoxygenic Photosynthetic Bacteria* (Blankenship, Madigan & Bauer, Eds) 725–745, Kluwer Academic Publishers, Dordrecht, Netherlands.
3. Ding, H., Robertson, D. E., Daldal, F. & Dutton, P. L. (1992) *Biochemistry* 31, 3144–3158.
4. Ding, H., Moser, C. C., Robertson, D. E., Tokito, M. K., Daldal, F. & Dutton, P. L. (1995) *Biochemistry* 34, 15979–15996.
5. Dutton, P. L., Wilson, D. P. & Lee, C. P. (1970) *Biochemistry* 9, 5077–5082.
6. Dutton, P. L. & Jackson, J. B. (1972) *Eur. J. Biochem.* 30, 495–510.]
7. Mitchell, P. (1975) *FEBS Lett.* 56, 1–6.
8. Mitchell, P. (1976) *J. Theor. Biol.* 62, 327–367.
9. Glaser, E. G. & Crofts, A. R. (1984) *Biochem. Biophys.Acta* 766, 322–333.
10. Robertson, D. E. & Dutton, P. L. (1988) *Biochem. Biophys Acta* 935, 273–291.
11. Robertson, D. E., Daldal, F. & Dutton, P. L. (1990) *Biochemistry* 29, 11249–11260.
12. Lawford, H. G. & Garland, P. B. (1983) *Biochem. J.* 130, 1029–1044.
13. Ohnishi, T. & Trumpower, B. L. (1980) *J. Biol. Chem.* 255, 3278–3284.
14. De Vries, S., Albracht, S. P. J., Berden, J. A., Marres, C. A. M. & Slater, E. C. (1983) *Biochem. Biophys. Acta* 723, 91–103.
15. Robertson, D. E., Prince, A. C., Bowyer, J. R., Matsurua, K., Dutton, P. L. & Ohnishi, T. (1984) *J. Biol. Chem.* 259, 1758–1763.
16. Salerno, J. C., Osgood, M., Liu, Y., Tayla, H. & Scholes, C. P. (1990) *Biochemistry* 29, 6987–6993.
17. Xia, D., Yu, C-A., Kim, H., Xia, J-Z, Kachurin, A. M., Zhang, L.,Yu, L & Deisenhofer, J. (1997) *Science* 277, 60–66.
18. Akiba, T., Toyoshima, C., Matsunaga, T., Kawamoto, M., Kubota, T., Fukuyama, K., Namba, K. & Matusbara, H. (1996) *Nature Struc. Biol.* 3, 553–561.
19. Gonzolez-Halphen, D., Lindorf, M. A. & Capaldi, R. A. (1988) *Biochemistry* 27, 7021–7031.
20. von Jagow, G. & Engle W. D. (1981) *FEBS Lett.* 136, 19–24.
21. Robertson, D. E., Giangiacomo, K. M., de Vries, S., Moser, C. C. & Dutton, P. L. (1984) *FEBS Lett.* 178, 343–350.
22. Robertson, D. E., Ding, H., Chelminski, P. R., Slaughter, C., Hsu, J., Moomaw, C., Tokito, M. K., Daldal, F. & Dutton, P. L. (1993) *Biochemistry* 32, 1310–1317.
23. Gray, K. A., Dutton, P. L. & Daldal, F. (1994) *Biochemistry* 33, 723–733.
24. Moser, C. C., Keske, J. M., Warncke, K., Farid, R. S. & Dutton, P. L. (1992) *Nature* 355, 796–802.
25. Crofts, A. R., Barquera, B., Gennis, R. B., Kuras, R., Guergova-Kuras, M. & Berry, E. A. (these proceedings).

29

IDENTIFICATION OF PLASTOQUINONE-CYTOCHROME/b₆f REDUCTASE PATHWAYS IN DIRECT OR INDIRECT PHOTOSYSTEM 1 DRIVEN CYCLIC ELECTRON FLOW IN *SYNECHOCYSTIS* PCC 6803

Robert Jeanjean,[1] Jasper J. van Thor,[2] Michel Havaux,[3] Françoise Joset,[1] and Hans C. P. Matthijs[1,2]

[1]Unité de Métabolisme Energétique, LCB-CNRS
31 Chemin Joseph Aiguier
13402 Marseille Cedex 20, France
[2]Laboratorium voor Microbiologie (ARISE/MB and E.C. Slater Institute)
Universiteit van Amsterdam
Nieuwe Achtergracht 127, 1018 WS Amsterdam, The Netherlands
[3]CEA/Cadarache
Département d'Ecophysiologie Végétale et de Microbiologie
13108 Saint-Paul-lez-Durance, France

1. INTRODUCTION[*]

In cyanobacteria and chloroplasts, concerted action of PS2 and PS1 in linear oxygenic photosynthetic electron transfer renders NADPH and ATP according to the Z-scheme. Maintenance of a stable phosphate potential is secured by PS1 cyclic photophosphorylation (1,2). In regular photoautotrophic growth, the latter has a very limited capacity (2–4), and is probably used for finetuning of ATP generation capacity and regulation (5). More pronounced usage of PS1 cyclic photophosphorylation and demonstration of the involvement of other PQ reductase pathways which under standard conditions may remain repressed or be only marginally active, can be envisaged during

---

[*] *Abbreviations:* flvd, flavodoxin; FNR, ferredoxin-NADP⁺ oxidoreductase; M55N, mutant M55 in normal medium; M55S, ibid in medium plus 550 mM NaCl; NEM, N-ethyl maleimide; P700, the reaction center of PS1; PAS, photoacoustic spectrometry; PAM, pulse amplitude modulation fluorimetry; PS1, photosystem 1; PQ, plastoquinone; WTN/S, wildtype in normal/salt added medium.

The Phototrophic Prokaryotes, edited by Peschek *et al.*
Kluwer Academic / Plenum Publishers, New York, 1999.

energetically more demanding growth conditions, such as in the case of cultivation in high salt medium (6–9). Quinone reductase(s), which connect the stroma exposed side of PS1 and the PQ-cytochrome b_6f complex, have not yet been fully disclosed in cyanobacteria (1). A mutant of *Synechocystis* PCC6803 without NDH1 activity (denoted M55, acquired by deletion of subunit 2, NdhB, of the NAD(P)H dehydrogenase complex), showed strongly reduced respiratory and indirect PS1 cyclic electron transfer rates (10–12). To envisage a role for NDH1 in photoautotrophic growth, i.e. to judge the need of its suggested contribution to indirect PS1 cyclic electron transfer, we posed the question whether replacing quinone reductase is necessary to permit growth of M55. This approach thus would enable study of any other direct or indirect quinone reductases in absence of the obscuring major NDH1 catalysed pathway. The existence of a direct PS1-quinone reductase which acts without the intermediate involvement of NAD(P)H as a cofactor, FQR, has been revealed in chloroplasts from inhibition by antimycin (1). This putative FQR pathway has never been unambiguously demonstrated in cyanobacteria. The role of ferredoxin-NADP$^+$ oxidoreductase in chloroplasts in a PS1 cyclic route separate from FQR has been suggested from the distinct inhibitory effects of antimycin and the sulfhydryl-alkylating agent NEM (13). Soluble proteins are involved in electron transfer from PS1 to FNR i.e. ferredoxin and/or flavodoxin (1,3). A docking site for ferredoxin and FNR is provided by PsaE, one of the stromal exposed proteins of the PS1 complex (1,3,4,15,16).

Cultivation of *Synechocystis* PCC 6803 in high salt medium was used as a means to induce maximal cyclic photophosphorylation capacity (8). The work presented renders evidence for three separate, but functionally comparable in parallel operating PQ reductase routes. Of these, route III is inducible and likely passes via flavodoxine (17) and FNR (this work). Routes will be discussed in view of their direct or indirect role in PS1 cyclic electron transfer.

2. MATERIALS AND METHODS

2.1. Strains and Culture

Synechocystis PCC6803 and M55, with a functionally impaired NDH1 complex (10), were cultivated under continuous illumination at a light intensity of approximately 50 μmol m^{-2} s^{-1} at 34°C in modified Allen's mineral medium (8). This medium is called normal medium (denoted by N) throughout the paper. In case of high salt medium the NaCl concentration was raised from 50 to 550 mM. All cultures in high salt medium (denoted by S) were routinely inoculated from stocks freshly pregrown in normal medium. Cultivation in high salt before measurements was for 4 to 5 generations. Agitation and C_i supply were provided by bubbling with 3% CO_2 in air.

2.2. Sample Preparation and Use of Inhibitors

Cells in the exponential phase of growth (corresponding to 6 to 9.10^7 cells/ml, i.e. 6 to 9 μg Chl/ml) were routinely suspended in growth medium at 80 to 100 μg Chl/ml directly before use. The chlorophyll content determination was as in (8). 1000x solutions of inhibitors were freshly prepared. Antimycin (from *Streptomyces*, Sigma) was dissolved in ethanol. Rotenone (Sigma) was prepared in DMSO, and NEM (Sigma) in water. After addition of inhibitors, a standard 10 min incubation period in darkness at 34°C was obeyed to secure proper diffusion into the cells.

2.3. Photoacoustic Spectroscopy, Measurements of P700 Re-Reduction Rates and PS1 Cyclic Photophosphorylation Capacity Assays

Use of the PAS and PAM techniques was as in (8, 14). For phosporylation assays suspensions of *Synechocystis* cells (80 μg Chl/ml) were stirred in the presence of the PS2 inhibitor atrazine (5 μM) and KCN (0.1 mM) to eliminate linear photo- and oxidative phosphorylations to take place. Other inhibitors were present as indicated. Samples were pre-incubated in darkness for 10 min. White light (400 μE.s^{-1}.m^{-2}) was provided by a quartz-iodide halogen lamp. After 5 minutes of illumination, 400 μl aliquots were withdrawn and immediately mixed with 40 μl of an ice cold solution of 50% perchloric acid (0.6 M) in precooled tubes. The collected samples were kept on ice for 5 minutes. The samples were then neutralized with 40 μl KOH (0.54 M) plus 80 μl KHCO$_3$ (0.17 M). Next, 40 μl Tris-acetate buffer (pH 7.0, 50 mM) was added to ensure neutral pH. All concentrations are final. The denatured proteins and precipitated KClO$_4$ were removed by centrifugation (12000xg, 5 min, 4°C). The ATP content in the supernatant of these samples was estimated by the luciferin-luciferase assay (Boehringer-Mannheim).

2.4. SDS-PAGE, Immunoblotting, Northern Analysis and Diaphorase Assay

SDS-PAGE and immunoblotting were done according to standard procedures. A polyclonal antibody raised against the *Anabaena* PCC7119 FNR (a kind gift of Dr. M.L. Peleato, Zaragoza, Spain), was used for immunodecoration, and the staining intensity was estimated with a gel scanner in the refractive mode. Spectrophotometric diaphorase assays at 430 nm was with cells broken in TAPS 20 mM, pH 8.9 directly before use, and with NADPH as electron donor and K-ferricyanide as acceptor [18]. RNA extraction to judge m-RNA abundance was as in [19], petH quantification through Northern analysis and phospho-imaging were done according to standard procedures.

3. RESULTS AND DISCUSSION

3.1. Effects of Exposure to Medium with Salt

Transfer of WT cells of *Synechocystis* or of its derived mutant M55 in their early phase of logarithmic growth from normal medium to high salt medium caused a temporary growth arrest during 4 to 8 h. After this transient period, growth resumed at a slower rate (Table 1). Adaptation to salt gave rise to changes in the rate of dark respiration, as in [6].

Table 1. Rates of exponential growth and whole cell respiration in WT and its derived NDH1 impaired mutant M55[a]

	WTN	WTS	M55N	M55S
Growth rate[b]	5.5 ± 0.5 (3)	8.5 ± 0.5 (3)	6.5 ± 0.5 (3)	10.5 ± 0.5 (3)
Dark respiration[c]	390 ± 40 (14)	780 ± 70 (11)	80 ± 10 (5)	175 ± 15 (6)
ibid + Rotenone (20 μM)	320 (2)	515 (2)	75 (2)	180 (2)

[a]Data represent rates in normal growth medium (50 mM NaCl, WTN, M55N) and in high salt medium (550 mM NaCl, WTS, M55S). Use of rotenone (20 μM), and other methods are presented in Experimental procedures. Data are mean values of separate experiments (± SD).

[b]Doubling time in h during growth.

[c]nmoles oxygen·mg^{-1}·Chl·min^{-1}.

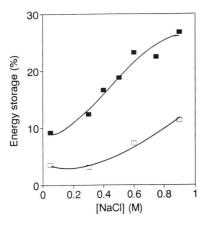

Figure 1. Energy storage in *Synechocystis* PCC6803 (WT) and the derived NDH1 impaired mutant M55 after cultivation in the presence of different salt concentrations. Far-red light was dosed at a constant flux of 8 W.m^{-2}. (■), WT; (□), M55.

The NDH1 inhibitor rotenone (20 µM) provoked stronger inhibiton of respiration in WTS, but respiration in M55S remained insentive to this inhibitor. The NDH2 inhibitor flavone (20–500 µM) exerted no inhibition in WTS nor M55S (not shown).

3.2. Inducible Capacity Changes of PS1 Cyclic Electron Transfer

The increase of respiratory capacity coincided with increased PS1 cyclic electron transfer (Fig. 1). It is suggested from these data that electron transfer pathways which mediate the increased energy generation capacity, other than *via* the NDH1 route in WTS, are likely operating in M55S. Table 2 illustrates the effects of adaptation and inhibitors on energy storage and P700$^+$ reduction. Antimycin inhibited appreciably in M55N and hardly in the other samples. NEM did not inhibit in WTN, but showed substantial inhibition in WTS, M55S and M55N. Combination of both inhibitors resulted in progressive inhibition in M55N, M55S and WTS. Rest activity (from transfer via NDH1) was found in both WT samples, but in M55 any remaining activity was nearly absent. The inhibition in WTN was stronger than would be expected from direct cumulative effects of the separate inhibitors. In photophosporylation assays with WTN and M55N the ATP level showed a transient collapse immediately upon exposure to salt (Fig. 2). This drop was followed by a fast though not maximal recovery, which could be effected in the WT through cyclic photophosphory-

Table 2. PS1 capacity in *Synechocystis* PCC6803 WT and M55 grown in normal medium or in high salt medium^a

Addition	Energy storage efficiency per strain (%)				Halftime of P700 re-reduction per strain (ms)			
	WTN	WTS	M55N	M55S	WTN	WTS	M55N	M55S
None	18	21	7	11	300 ± 50 (15)	180 ± 16 (8)	980 ± 60 (5)	580 ± 30 (5)
Antimycin	17	20	3.5	9	340 ± 60 (9)	196 ± 16 (5)	1620 ± 200 (5)	670 ± 80 (3)
NEM	17	16	4	3	380 ± 70 (12)	360 ± 60 (6)	1130 ± 120 (4)	1110 ± 170 (3)
Both	11	15	<1	<1	590 ± 90 (7)	390 ± 40 (3)	1800 ± 170 (5)	1290 ± 90 (3)

^{a}The effect of electron transfer inhibitors added at a saturating concentration, i.e. antimycin 50 µM or NEM 1 mM, and a combination of both, has been monitored by photoacoustic spectrometry and the P700$^+$ reduction assay. Photochemical energy storage and the reduction of P700$^+$ have been determined with far-red light 8 W m^{-2}. Energy storage values are the average of 3 to 5 experiments (in this technique standard deviations are well below 10% of the mean (not shown). Reduction of P700 data are mean values (± SD).

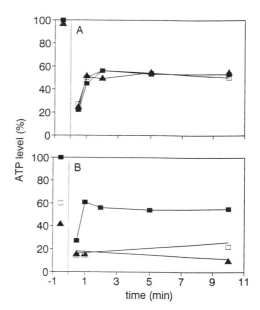

Figure 2. The effect of antimycin on the rate of PS1 cyclic photophosphorylation in *Synechocystis* PCC6803 immediately after raising the NaCl content from 50 mM to 550 mM. **A**, WTN; **B**, M55N. (■), linear plus cyclic photo-phosphorylation; (□), cyclic photophosphorylation only (5 μM atrazine present); (Δ), cyclic photophosphorylation plus antimycin (50 μM).

lation only, but not in M55N. Interestingly, this incapability of M55N directly after exposure to high salt was restored to nearly the control value after prolonged incubation i.e. in M55S (Table 3). This new cyclic photophosphorylation capacity showed inhibitor sensitivity as in Table 2.

3.3. Cumulative Effect of Inhibitors, Indication for in Parallel Operating PQ Reduction Pathways

Different concentrations of the separately added inhibitors antimycin and NEM or combinations of both were tested in P700$^+$ reduction assays. In the presence of NEM, antimycin sensitivity in the WT was already demonstrated at low concentrations (I_{50} = 10 μM in whole cells) (Fig. 3). A similar agonistic effect was shown in the reverse case: antimycin increased the inhibitory potential of NEM; in this antimycin was already effective at 5 μM and more at higher concentrations (not shown). Our current hypothesis of a function of the PsaE protein in a pathway sensitive to antimycin was arrived at via these typi-

Table 3. Steady state ATP synthesis capacity in WT and M55[a]

Activity measured (± addition)	Strain and condition of growth			
	WTN	WTS	M55N	M55S
Total photophosphorylation	330 ± 40 (7)	310 ± 10 (3)	310 ± 30 (8)	300 ± 30 (6)
Cyclic photophosphorylation	320 ± 30 (8)	300 ± 40 (6)	190 ± 20 (10)	280 ± 20 (3)
ibid + Antimycin	330 ± 30 (3)	320 ± 40 (3)	160 ± 30 (5)	230, 270
ibid + NEM	270 ± 30 (4)	210 ± 20 (3)	140, 160	190 ± 20 (4)
ibid + Antimycin + NEM	210 ± 30 (4)	210 ± 20 (3)	70, 70	100, 120

[a]Total photophosphorylation denotes linear and PS1 mediated cyclic photophosphorylation. Dark oxidative phosphorylation was inhibited with 0.1 mM KCN. Prior to illumination cells were depleted of ATP by incubation in darkness. The numbers listed denote ATP in nmoles ATP/mg Chl present after 5 min illumination with strong white light. Data are mean values of separate experiments (number in brackets) ± SD. Details are presented in Methods.

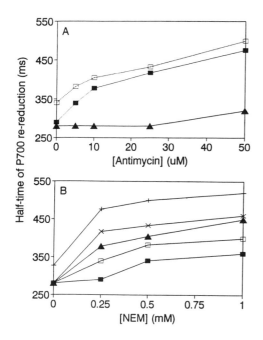

Figure 3. Inhibition of P700$^+$ reduction in *Synechocystis* by antimycin and NEM: concentration dependence and agonistic effects. A, Inhibition by antimycin in the presence of NEM at differerent concentrations: (■), control; (□), 0.25 mM NEM; (Δ) 0.5 mM NEM. B, Inhibition by NEM as a function of antimycin presence, at different concentrations: (■), 0 μM; (●), 5 μM; (Δ), 10 μM; (×), 20 μM; (+), 50 μM.

cal experiments and were supported by observations in a PsaE deletion mutant (Jeanjean *et al.*, in prep.). In all cases of adaptation to high salt studied in this work, the observed increase of PS1 cyclic capacity has coincided with a larger NEM-sensitive contribution to the overall capacity of P700$^+$ reduction and energy storage values.

3.4. A Putative Role for the Amino-Terminal Extension of FNR

A comparison between crude preparations of broken cells from WTN and WTS cultures on Western blots challenged with a polyclonal antibody raised against FNR rendered increased staining density of a 42 kDa protein after growth in high salt medium. Northern blot analysis of total RNAs from WTN and WT cells that were exposed during 8 h to salt stress yielded 4 times higher *petH* transcript levels (Fig. 4).

Similar Western analysis of M55N and M55S also showed increased FNR presence, which coincided with higher diaphorase activity (Table 4). It is our hypothesis that induc-

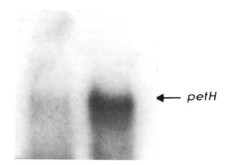

← *petH*

Figure 4. Northern blot of total mRNA after hybridization with the *petH* gene. The same quantity of RNA was loaded in each lane. left, WTN; right, WTS.

Table 4. Diaphorase activity and relative immuno-decoration of FNR in preparations of broken cells of WTN, WTS, M55N and M55S[a]

Measurement	Type of sample			
	WTN	WTS	M55N	M55S
Diaphorase activity (U/mg chl)	0.439	0.845	1.09	1.68
Refractive index	100	140	255	320

[a]The refractive indices of the 42 kDa FNR band were obtained from decorated Western blots, and are expressed as % of the control.

tion of flavodoxin and FNR after exposure to high salt may render a clue towards ATP generation in PS1 cyclic electron transfer in the absence of NDH1. In this, the typical aminoterminal extension of FNR from cyanobacteria [19–21] may support reversible membrane attachment to enhance dehydrogenase activity of the enzyme such as to effectively render PQ reductase activity. Attachment and release could proceed via the very positively

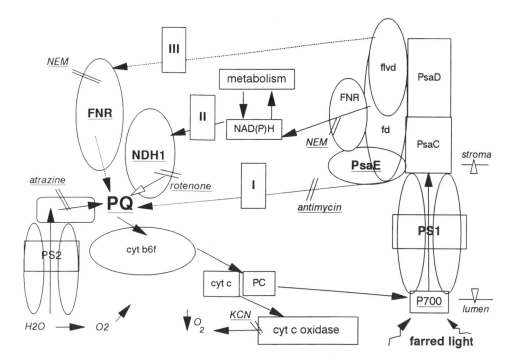

Figure 5. A scheme of plastoquinone/cytochrome b6f reductase routes that catalyze indirect and direct PS1 cyclic electron transfer in the cyanobacterium *Synechocystis* PCC6803. The in parallel operating PSI mediated quinone reductase cycles, indicated by arrows, labeled [I], [II] and [III] respectively, may explain why it has often been observed that singly added inhibitors are due to remain without detectable effect. We define Cycle I (FQR) to be direct and it likely proceeds *via* PsaE. It can be inhibited by low concentrations of antimycin (I_{50} < 10 µM). Cycle II is indirect, it exerts its action *via* NDH1 and is rotenone-sensitive. This route is regarded indirect since it relies on intermediary NAD(P)H present in the cytoplasm. Because of this property Cycle II plays a key role at the intersection of the catabolic and anabolic metabolism, also in terms of net electron flux capacity [10–12]. Cycle III is indicible and is mediated by FNR possibly in a direct way. This preliminary conclusion was derived from experiments with a truncated petH mutant (data not shown, Van Thor et al., in prep.). This route is NEM-sensitive (I_{50}<0.5 mM). Exposure to energy demanding conditions by which the cellular ATP turn-over increases, such as through cultivation in high salt medium, induces flvd (flavodoxin) and FNR, here organized as catalysts in the newly introduced Cycle III.

charged CpcD like domain (the isoelectric point is about 11 [19–21]). The homology between the CpcD like domain and the N-terminal part of FNR has delivered an explanation for the co-isolation of FNR with phycobilisomes [20] The functional meaning of this rather surprising positioning of FNR inside a cyanobacterial cell is currently studied (Van Thor et al., in prep.).

Preliminary results of experiments with *petH* mutated by an aminoterminal truncation, which essentially rendered typical chloroplast length FNR [21] showed a phenotype which lacked the contribution to direct $P700^+$ reduction after farred excitation (manuscript in prep.). We propose impairment of the inducible route III in this mutant (cf. Fig. 5). Interestingly, we observed that wild type activity in $NADP^+$ photoreduction was retained in this mutant (van Thor et al., in prep.).

ACKNOWLEDGMENTS

This work was supported in France by the C.N.R.S., the Ministère de l'Education Nationale and the Commissariat à l'Energie Atomique (CEA), in the Netherlands by the Dutch Society for Life Sciences (SLW), with financial aid of the Dutch Organisation for Scientific Research (NWO). HCPM was recipient of a visitors grant of the Chamber of Commerce of the City of Marseilles. The authors are very much indebted to Dr. T.Ogawa (Nagoya, Japan) for the M55 strain and to Dr. M.L. Peleato (Zaragoza, Spain) for the gift of anti FNR polyclonal antibodies.

REFERENCES

1. Bendall, D.S. and Manasse, R.S. (1995) Biochim. Biophys. Acta **1229**, 23–38
2. Myers, J. (1987) Photosynth. Res. **14**, 55–69
3. Yu, L., Zhao, J., Mühlenhoff, U., Bryant, D.A. and Golbeck, J.H. (1993) Plant Physiol. **103**, 171–180
4. Chitnis, P.R., Xu, Q., Chitnis, V.P. & Nechusthai, R. (1995) Photosynth. Res. **44**, 23–40
5. Herbert, S.K., Martin, R.E. and Fork, D.C. (1995) Photosynth. Res. **46**, 277–285
6. Fry, J.V., Huflejt, M., Erber, W.W.A., Peschek, G.A. and Packer, L. (1986) Arch. Biochem. Biophys. **244**, 686–691
7. Schubert, H. and Hagemann, M. (1990) FEMS Microbiol. Lett., **71**, 169–172
8. Jeanjean, R., Matthijs, H.C.P., Onana, B., Havaux, M. & Joset, F. (1993) Plant Cell Physiol. **34**, 1073–1079
9. Joset, F., Jeanjean, R. and Hagemann, M.(1996) Physiol. Plant., **96**, 738–744
10. Ogawa,T (1991) Proc. Natl. Acad. Sci. USA **88**, 4275–4279
11. Mi, H., Endo,T., Schreiber, U. and Asada, K. (1992) Plant Cell Physiol. **33**, 1099–1105
12. Mi, H., Endo, T., Schreiber, U., Ogawa, T. and Asada, K. (1994) Plant Cell Physiol. **35**, 163–173
13. Shahak, Y., Crowther, D. and Hind, G. (1981) Biochim. Biophys. Acta **636**, 234–243
14. Ravenel, J., Peltier, G and Havaux, M. (1994) Planta **193**, 251–259
15. Lelong, C., Sétif, P., Lagoutte, B. and Bottin, H. (1994) J. Biol. Chem. **269**, 10034–10039
16. Mühlenhoff, U., Kruip, J., Bryant, D.A., Rögner, M., Sétif, P. and Boekema, E. (1996) EMBO J. **15**, 488–497
17. Fulda, S. and Hagemann, M. (1995) J. Plant Physiol. **146**, 520–526
18. Matthijs, H.C.P., Coughlan, S.J. and Hind, G. (1986) J. Biol. Chem. **261**, 12154–12158
19. Van Thor, J.J., Hellingwerf, K.J. and Matthijs, H.C.P. (1997) Plant Mol. Biol., *in the press*
20. Schluchter, W.A. and Bryant, D.A. (1992) Biochemistry **31**, 3092–3102
21. Van Thor, J.J., Matthijs, H.C.P., Hellingwerf, K.J. and Mur, L.R. (1995) in Photosynthesis from light to biosphere, Vol. III, Mathis, P. ed., Kluwer Ac. Publ., pp. 521–524.

30

THE CYTOCHROMES c OF CYANOBACTERIA

C. Kerfeld,[1] K. K. Ho,[2] and D. W. Krogmann[3]

[1]UCLA-DOE Laboratory of Structural Biology
University of California Los Angeles, UCLA-MB/168
Los Angeles, California 90024-1570
[2]School of Biological Sciences
National University of Singapore
Singapore 119260
[3]Biochemistry Department
Purdue University
West Lafayette, Indiana 47906

1. INTRODUCTION

C type cytochromes are a very valuable and long preserved invention in the history of life. The distinguishing characteristic of these molecules is a heme prosthetic group attached to a protein by two covalent thioether bonds. The attachment site in the protein is recognized by its signature amino acid sequence -Cys-X-X-Cys-His-. With the completion of the nucleotide sequence of the genome of *Synechocystis* PCC 6803 by the Kazusa Institute (Internet address http://www.Kazusa.or.jp), one can recognize this binding site in at least five open reading frames and four of these are identified with gene product proteins that have been isolated and partially characterized. These four are cytochrome f, cytochrome c_6, cytochrome M (all of these are easily reduced by ascorbate, so are high potential cytochromes) and low potential cytochrome c_{549} (reduced by dithionite).

Cytochrome f is found in all oxygenic photosynthetic cyanobacteria, algae and higher plants. Cytochrome f is a membrane bound catalyst participating in electron transfer from PSII to PSI catalysing the transfer of electrons from the Rieske iron sulfur protein to cytochrome c_6 or plastocyanin.

Cytochrome c_6 is a water soluble protein found only in cyanobacteria and algae. The presence of this cytochrome in the periplasmic space of cyanobacteria suggests a function in respiration. Cytochrome c_6 is also found in the thylakoids of cyanobacteria and algae where it functions in transferring electrons from cytochrome f to the PSI reaction center. This function is shared with the copper protein plastocyanin. The limited solubility of copper in natural waters may necessitate the use of an alternative to plastocyanin, cytochrome

The Phototrophic Prokaryotes, edited by Peschek *et al.*
Kluwer Academic / Plenum Publishers, New York, 1999.

c_6, for survival in the planktonic photosynthesizers. The ability of terrestrial plants to exchange protons for the abundant copper bound to soil may have freed higher plants from the need for cytochrome c_6.

Cytochrome M is present in cyanobacteria where it is found in very low concentration. It is probably present in algae as well but, as yet, there is no evidence from amino acid or nucleotide sequencing to confirm its identity in these eukaryotes. The function of cytochrome M is unknown, but it may participate in the same electron transfer step as cytochrome c_6 or plastocyanin between cytochrome f and PSI.

Low potential cytochrome c_{549} is also found in cyanobacteria and algae and its persistence in these organisms may relate to exigencies of the planktonic life style. An essential role for the low potential cytochrome c_{549} in O_2 evolution has been established. The cytochrome is expressed constitutively as a membrane bound structural participant essential to O_2 evolution in PSII. The cytochrome appears to play a structural role which does not involve redox activity. The low potential cytochrome c_{549} in PSII is replaced by another protein, lacking heme, in higher plants. There are sporadic reports of this cytochrome in a soluble form which may indicate another function in which redox catalysis is involved.

2. CYTOCHROME f

Cytochrome f catalyses the transfer of electrons from the Rieske iron sulfur protein to cytochrome c_6 or plastocyanin. Cytochrome f contains approximately 288 amino acids and a single heme located about 20 residues in from the N terminus. The isoelectric point of cytochrome f is 4.5 in cyanobacteria and 6 to 9 in algae and higher plants. The redox potentials are in the range of 330 to 395 mV. This cytochrome is unique among the c type cytochromes in using a tyrosine residue at the N terminus as the sixth ligand to the heme iron (Figure 1). The other c cytochromes use as a sixth ligand a methionine (high potential c) or a histidine (low potential c) which are 44 or 52 residues from the heme toward the C terminus. Gray[1] identified a hydrophobic region near the C terminus of cytochrome f which serves as a transmembrane anchor and which is cleaved by a protease that is activated by organic solvent during cytochrome purification from cruciferous plants. This provides a more tractable form of cytochrome f for crystallization. Martinez *et al.* crystallized the large 251 amino acid N terminal fragment of turnip cytochrome f and solved its three dimensional structure.[2] Recently K.K. Ho has found a similar proteolytic activity at work in preparations of cytochrome f from the cyanobacterium *Arthrospira maxima*. Ho purified this protein lacking its membrane anchor and C. Kerfeld crystallized it and solved the structure.

The structure of turnip cytochrome f revealed several features that are novel among c-type cytochromes including a 2 domain structure predominantly composed of beta sheet, the aforementioned amino terminal axial ligand to the heme (Figure 1) and a buried water chain proposed to function in proton translocation. The primary structures of the cyanobacterial cytochromes c are similar to those of higher plants. Figure 1 shows a hypothetical model of cytochrome f from *Nostoc*, based on turnip cytochrome f structure. Morand *et al.*[3] identified by covalent crosslinking of spinach plastocyanin to cytochrome f a site of interaction between the two proteins. Table I shows the amino acid sequences in this region from cyanobacterial cytochromes f. A fragment of spinach cytochrome f (residues (186–198) identified by amino acid sequencing was linked via Lys 186 to Asp 44 in plastocyanin (in a fragment containing residues 31 to 53). The primary structure of cytochrome f from *Nostoc*, the only representative of section IV or V cyanobacteria, shows a remarkable increase in negative charge in this putative docking region (Table 1 and Fig-

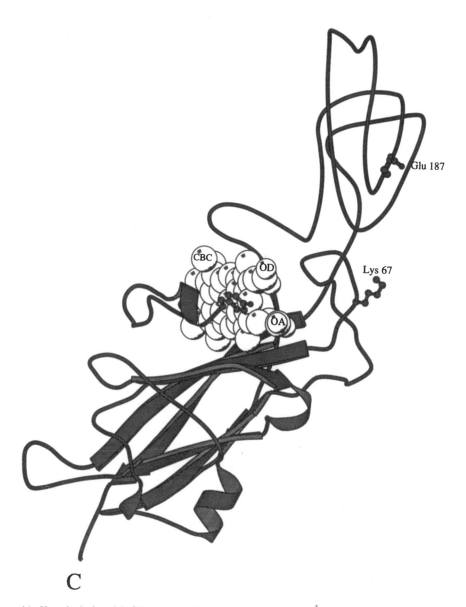

C

Figure 1A. Hypothetical model of *Nostoc* cytochrome f, modeled on the 2.3 Å turnip structure (Protein Data Bank Code 1CTM) using SWISS-MODEL[25]. Solvent-exposed atoms of the heme are labeled (OD and OA are propionate oxygen atoms of the A and D pyrole rings; CBC is the methyl group adjacent to the carbon atom involved in the covalent linkage to Cys 24). The smaller of the two folding domains is at the top of the large domain of the molecule. The tyrosine axial ligand residues in the vicinity of the putative docking site for the soluble electron carrier are shown in ball and stick representation.

ure 1B). *Nostoc* synthesizes a cytochrome c_6 and plastocyanin with a basic isoelectric point (discussed below).

The cytochrome b6f complex can translocate 1–2 protons for each electron it transfers from plastoquinol to plastocyanin. Refinement of the turnip cytochrome f structure to higher resolution revealed the presence of a buried chain of five water molecules, pro-

Figure 1B. Space filling model of the hypothetical *Nostoc* cytochrome f showing charged amino acid side chains (black for Arg and Lys, light gray for Asp and Glu). The orientation is that shown in Figure 1A. Figures 1 and 2 were prepared with MOLSCRIPT[26].

posed to function in proton translocation. This "proton wire" extends from the histidine axial ligand toward Lys 67 (*Nostoc* numbering; Figure 1) which has also been implicated in interaction with plastocyanin.[1]

3. CYTOCHROME c₆

This cytochrome catalyses the transfer of electrons from cytochrome f to P700—a role which it may share with plastocyanin when copper is available. As noted earlier, cy-

Table 1. Electron acceptor docking site on cytochrome f

				186								pI of acceptor
	+	+	−	+								
Spinach	R	K	E	K	G	—	G	Y	E	I	T	pH 3.0
			−									
Synechocystis PCC 6803	A	L	E	A	G	—	G	Y	Q	L	I	
			−									
Synechococcus PCC 7002	V	N	E	A	A	A	G	T	D	I	T	
			−									
Arthrospira maxima	A	L	E	A	G	G	Y	Q	V	V	L	5.2
	+	−	−					+		+		
Phormidium luridum	K	A	E	D	G	S	A	R	V	K	I	
	+	−	−	−	−	−				+		
Nostoc PCC 7906	K	E	E	D	E	D	G	N	V	K	Q	pH 8.8

anobacteria contain cytochrome c_6 in the periplasmic space and substantial amounts of the cytochrome are seen in locations distant from the photosynthetic membranes.[4,5] Zhang et al.[6] found that deletion of the cytochrome c_6 gene from Synechocystis PCC 6803 and deprivation of copper to eliminate plastocyanin synthesis did not affect either photoautotrophic growth or respiration.

Cytochrome c_6 is a small (90–98 amino acid residues), water soluble protein with a single heme bound near the N terminus at residues 14 and 17. The single copy gene which encodes this protein shows a leader sequence which could direct the cytochrome to both the periplasmic space and the interior compartment of the thylakoid membrane. The cytochrome, in parallel with the plastocyanin, shows a striking variation in isoelectric point depending on the cyanobacterial source. The pI is acidic in cytochrome c_6 and plastocyanin isolated from cells of the first three sections of the Stanier classification[7] of cyanobacteria which include the unicellular and the simple filamentous forms. Cells from the fourth and fifth sections produce basic cytochrome c_6 and plastocyanin.[8] This change might impose a complementary change in the proteins which provide docking sites for the reception and donation of electrons. This possibility receives support from an observed variation in amino acid sequence in the proposed electron acceptor docking region of cytochrome f.

The docking site for cytochrome c_6 on the Photosystem I reaction center has yet to be established. A crystal structure for a PSI reaction center of a cyanobacterium is a likely prospect and the tools of molecular biology (e.g. Xu et al.[9]) and of enzyme kinetics (e.g. Hervas et al.[10]) will surely illuminate the interaction of cytochrome c_6 with PSI.

In 1982 Ludwig et al.[11] published a preliminary model of the first structure of a cytochrome c_6 isolated from Anacystis nidulans. The structure confirmed that the overall fold of the molecule was similar to that of other short chain (~90 amino acids) cytochromes in having a predominantly alpha helical fold that encloses most of the heme prosthetic group (Figure 2). More recently, high resolution structures of cytochrome c_6 from two algal sources were described (Kerfeld et al.,[12] Frazao et al.[13]). These models allow more detailed analysis of the protein structure. One noted difference between cytochrome c_6 and the eubacterial cytochromes c_2 is in the degree of solvent exposure of specific atoms of the heme. This may be functionally relevant since electron transfer is thought to be through surface exposed heme atoms. In cytochrome c_6 the propionate oxygen atoms of the pyrrole D ring of the heme (Figure 2) are solvent exposed; in cytochrome c_2 they are buried within the protein interior.

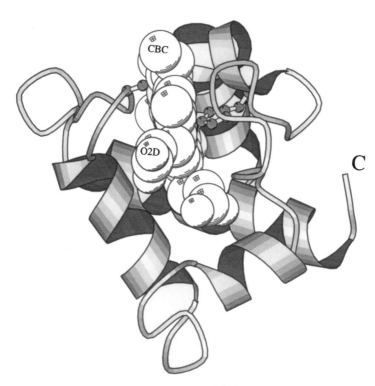

Figure 2A. Ribbon diagram of *Arthrospira maxima* cytochrome c_6 (Kerfeld & Krogmann, manuscript in preparation). The histidine and methionine axial ligands are in ball and stick representation. The heme is shown in space filling representation with the two most solvent exposed atoms shown in white and labeled as in Figure 1.

Another interesting characteristic of cytochrome c_6 noted from the structural studies is their tendency to pack as oligomers within the crystal. The oligomerization interface encompasses the exposed atoms of the heme, resulting in a heme-to-heme distance of about 14 Å. The *A. nidulans* cytochrome c_6 forms a noncrystallographic dimer. A similar crystallographic dimer was observed in one crystal form of *Chlamydomonas reinhardtii* cytochrome c_6, while a second crystal form grown from a different protein preparation formed a crystallographic trimer. A similar trimer was observed in *Monoraphidium braunii* cytochrome c_6 crystals.[13] The oligomerization occludes accessibility to heme atoms that are surface exposed in the monomer and thus has been suggested to have a functional relevance. For example, forming oligomers may be a more efficient means of transferring electrons between an oligomer b6f complex and PSI. Likewise the oligomerization shields surface exposed charge and therefore may be involved in altering the electrostatic surface of the molecule, a feature which may be important in recognition of membrane-bound redox partners.

Purified cytochrome c_6 from *A. maxima* has been fractionated into different mass forms that are the result of different levels of post-translational modification (see below). Two of these isoforms have been crystallized and the structure of one (+16 Da form) is shown in Figure 2. Two of the three molecules in the asymmetric unit form a molecular dimer. A second crystal form of *A. maxima* cytochrome c_6 grown from a protein preparation free of post-translational modification is especially unusual; it promises a very interesting

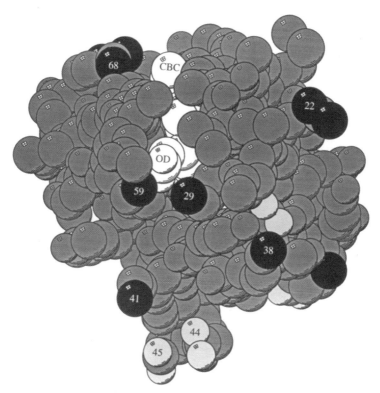

Figure 2B. Space filling model of *A. maxima* cytochrome c_6 showing charged amino acid side chains (black for Arg and Lys, light gray for Asp and Glu). The orientation of the model is that shown in Figure 2A.

oligomeric arrangement within the crystal. The crystal unit cell is enormous (a=b=c=237 Å) and there are approximately 12 molecules in the asymmetric unit, possibly in a complex with tetrahedral symmetry. It is extremely improbable that a naturally monomeric protein would exhibit this crystallization behavior.

Figure 2B shows the charged residues on the surface of *A. maxima* cytochrome c_6 about the exposed atoms of the heme. Despite the acidic pI of the protein, most of the charged amino acids on this surface are basic. Most notable are the two positively charged residues flanking the exposed propionate D oxygen atoms; they are highly conserved in the primary structures of cytochrome c_6, despite a wide variation in overall pI of the molecule in different organisms.

Isoforms of cytochrome c_6 can be separated by ion exchange chromatography and by isoelectric focusing.[8] The underlying reason for this was narrowed to a suggestion of post-translation modification since cytochrome c_6 is known to be encoded by a single copy gene in several species of cyanobacteria. Protein mass spectrometry offered the possibility of defining post-translational modifications. A large sample (to enhance detection of small amounts of isoforms) of reduced *A. maxima* cytochrome c_6, in 5 mM Tris buffer, pH 7.8, was loaded onto a DEAE column equilibrated with the same buffer and eluted with a very gradually increasing concentration of NaCl up to 150 mM. Four bands with identical absorption spectra were eluted. Table 2 shows that each band contains a family of mass peaks detected by electrospray mass spectrometry. The first two peaks are of high abundance and

Table 2. *Arthrospira maxima* cytochrome c_6 bands
resolved on DE 52 (calculated mass = 9,855)

A	B	C	D
—	9,851	9,850	9,850
9,869	9,867 (Δ16)	9,865 (Δ15)	9,866 (Δ16)
9,883 (Δ14)	9,884 (Δ17)	9,883 (Δ18)	9,884 (Δ18)
9,899 (Δ16)	—	9,901 (Δ18)	9,901 (Δ17)
9,918 (Δ19)	—	9,915 (Δ14)	—
9,940 (Δ22)	—	—	9,930
—	9,951	9,950	9,949 (Δ19)
9,968	9,964 (Δ13)	—	—
9,983 (Δ15)	9,982 (Δ18)	—	—
10,003 (Δ20)	10,003 (Δ21)	—	—

the remainder are of progressively lower amounts. The mass differences are from 14 to 20 Daltons. Given an accuracy of 0.1% in these measurements, one suspects methylation (+14) and oxygen insertion (+16) may be possible. One supposes that the different bands are oligomers with different surface charges that cause ion exchange separation. There is clear evidence for the oxidation of Met 26 to both the sulfoxide and the sulfone in cytochrome c_6 of *Chlamydomonas reinhardtii*[13]. The aforementioned two crystal forms, dimer and trimer, differed in the level of post-translational modification. There are two methionines in the cytochrome c_6 of *A. maxima* which could provide five peaks each differing by a mass of 16. When a sample of this cytochrome was used by Jung *et al.*[14] in their development of a high affinity assay for cyanobacterial PSI reduction, they found that the specific activity of the cytochrome was doubled by pretreatment with mercaptoethanol. This improvement might result from the reduction of a methionine sulfoxide back to methionine. We reduced a sample of this cytochrome with an excess of ascorbic acid to fully reduce the iron in the conventional sense. On adding mercaptoethanol or dithionite to this sample, a very small increment of further reduction was detected from the spectrum. *A. maxima* cytochrome c_6 has two methionines. One is immediately adjacent to the fifth iron ligand His in the Cys-X-X-Cys-His-Met 19 heme binding site. The other is Met 63 which is the sixth ligand to the iron. Oxidation of either of these methionines to the sulfoxide might perturb the redox activity or create an inhibitor of normal catalytic activity.

4. CYTOCHROME M

This is a small water soluble cytochrome whose identity was established by Malakhov *et al.*[15] who discovered the gene for it in *Synechocystis* PCC 6803. These authors recognized the heme binding site and slight resemblances in primary structure to both cytochrome c_6 and to mitochondrial respiratory cytochrome c. The open reading frame begins with a leader sequence, as is the case for all four cyanobacterial cytochrome c genes. The mature protein is 76 amino acid residues in length with a heme attachment site at residues 16 to 20. There are methionine residues at positions 55 and 65—either of which might serve as the sixth iron ligand. The predicted pI is 7.3. K.K. Ho has recently purified a cytochrome c from *Synechocystis* PCC 6803 with a mass and purification behavior predicted by the cytochrome M gene. This protein resembles the cytochrome c_{552} from *Anacystis nidulans* described by Holton and Myers in 1963.[16] They found this cytochrome to be present at low abundance—1/22,000 Chl—and this was Ho's experience as well. Omata and Murata[17] found cytochrome c_{552} bound to the thylakoid membrane and released it by sonication. A low molecular weight cytochrome M of neutral charge was found in the green alga *Bryopsis maxima* by Kimura *et al.*[18] and is probably one of the cytochromes recognized by Powls *et al.*[19] in the green alga *Scenedesmus obliquus*. There is no evidence of the presence of this cytochrome in higher plants.

The function of cytochrome M is unknown. Malakhov *et al.*[15] deleted the cytochrome M gene from *Synechocystis* PCC 6803 and no effect was found on the rates of photoautotrophic growth, photosynthetic O_2 evolution or respiration. Metzger *et al.*[20] created a mutant strain of *Synechocystis* PCC 6803 which lacked the genes for both cytochrome M and cytochrome c_6. These mutant cells were deprived of copper to diminish the amount of plastocyanin. In the strain lacking in all three electron carriers, photoautotrophic growth occurred, respiration was unaltered and there was a substantial decrease in the rate of electron transfer from the cytochrome b_6f complex to PSI. Cytochrome M may serve as an electron transfer agent in photosynthesis. Clearly, there is much more to be learned about cytochrome M structure and function. It is likely that cytochrome M will be renamed either by a commission or by experts on cytochrome nomenclature.

5. LOW POTENTIAL CYTOCHROME c_{549}

This cytochrome is a 15 kDa extrinsic protein on the oxygen evolving core complex of PSII where it serves with a non heme protein of 10–12 kD as a structural element necessary in O_2 production.[21] Both of these proteins are found in cyanobacteria and algae, but they have been replaced in higher plants by a 23 kDa and a 17 kDa protein—neither of which contains heme. Nishiyama *et al.*[22] have shown that photosynthetic O_2 evolution is stabilized against heat inactivation by this cytochrome in the thermophile *Synechococcus* PCC 7002. Low potential cytochromes c_{549} are acidic proteins of 135 amino acid residues length with a single heme covalently bound at residues 38 and 41. The sixth ligand to the iron is His 93. Amino acid sequences have been determined for this protein isolated from *Microcystis aeruginosa*, and the gene has been sequenced from *Synechocystis* PCC 6803 and *Synechococcus* PCC 7002.[23] Cytochrome c_{549} has been isolated from *A. maxima*. Crystals that diffract X-rays to 2.3 Å resolution have been grown and the structure determination is in progress (Kerfeld, Yeates and Krogmann, unpublished). Navarro *et al.* have published a careful assessment of the physicochemical properties of the cytochrome isolated from *Synechocystis* PCC 6803.[24] The low redox potential −250 to −314 mV suggests a redox function in hydrogen metabolism or some other low potential electron transfer process.

REFERENCES

1. Gray, JC. 1992 Cytochrome f: structure, function and biosynthesis. *Photosynth. Res.* **34**: 359–374.
2. Martinez, SE, Huang, D, Szczepaniak, A, Cramer, WA, Smith, JL. 1994 Crystal structure of the chloroplast cytochrome f reveals a novel cytochrome fold and unexpected heme ligation. *Structure* **2**: 95–105.
3. Morand, LZ, Frame, MK, Colvert, KK, Johnson, DA, Krogmann, DW, Davis, DJ. 1989 Plastocyanin cytochrome f interaction. *Biochem.* **28**: 8039–8047.
4. Obinger, C, Knepper, J-C, Zimmermann, U, Peschek, GA. 1990 Identification of a periplasmic c-type cytochrome as electron donor to the plasma membrane-bound cytochrome oxidase of the cyanobacterium *Nostoc* MAC. *Biochem. Biophys. Res. Com.* **169**: 492–501.

5. Serrano, A, Gimenez, P, Scherer, S, Böger, P. 1990 Cellular localization of cytochrome c_{553} in the N_2-fixing cyanobacterium *Anabaena variabilis*. *Arch. Microbiol.* **154**: 614–618.

6. Zhang, L, Pakrasi, H, Whitmarsh, J. 1994 Photoautotrophic growth of the cyanobacterium *Synechocystis* PCC 6803 in the absence of cytochrome c_{553} and plastocyanin. *J. Biol. Chem.* **269**: 5036–5042.

7. Rippka, R, Deruelles, J, Waterbury, JB, Herdman, M, Stanier, RY. 1979 Generic assignments, strain histories and properties of pure cultures of cyanobacteria. *J. Gen. Microbiol.* **III**: 1–61.

8. Ho, KK, Krogmann, DW. 1984 Electron donors to P700 in cyanobacteria and algae - an instance of unusual genetic variability. *Biochim. Biophys. Acta* **766**: 310–316.

9. Xu, Q, Chitnis, VP, Chitnis, PR. 1994 Function and organization of photosystem I in a cyanobacterial mutant strain that lacks Psa F and Psa J subunits. *J. Biol. Chem.* **269**: 3205–3211.

10. Hervas, M, Navarro, JA, Diaz, A, De La Rosa, MA. 1996 Comparative thermodynamic analysis by laser flash absorption spectroscopy of photosystem I reduction by plastocyanin and cytochrome c_6 in *Anabaena* PCC 7119, *Synechocystis* PCC 6803 and spinach. *Biochem.* **35**: 2693–2698.

11. Ludwig, ML, Pattridge, KA, Powers, TB, Dickerson, RE, Takano, T. 1982 Structure analysis of ferricytochrome c from the cyanobacterium *Anacystis nidulans*. In *Electron Transport and Oxygen Utilization* (Ho, C ed.) pp. 27–32. Esevier, North Holland.

12. Kerfeld, CA, Anwar, HP, Interrante, R, Merchant, S, Yates, TO. 1995 The structure of chloroplast cytochrome c_6 at 1.9 Å resolution: evidence for functional oligomerization. *J. Mol. Biol.* **250**: 627–647.

13. Frazao, C, Sores, CM, Corrando, MA, Pohl, J, Dauter, Z, Wilson, KS, Hervas, M, Navarro, JA, De la Rosa, MA and Sheldrick, GM. 1995. *Ab initio* determination of the crystal structure of cytochrome c_6 and comparison with plastocyanin. *Structure* **3**: 1539–1169.

14. Jung, YS, Yu, L and Golbeck, JH. 1995 Reconstitution of iron sulfur center F_B results in complete restoration of NADP$^+$ photoreduction in Hg-treated photosystem I complexes from *Synechococcus* sp. PCC 6301. *Photosynth. Res.* **46**: 249–255.

15. Malakhov, MP, Wada, H, Los, DA, Semenenko, VE, Murata, N. 1994 A new type of cytochrome c from *Synechocystis* PCC 6803. *J. Plant. Physiol.* **144**: 259–264.

16. Holton, RW, Myers, J. 1963 Cytochromes of a blue-green alga: extraction of a c type with a strongly negative redox potential. *Science* **142**: 234–235.

17. Omata, T, Murata, N. 1984 Cytochromes and prenyl quinones in preparations of cytoplasmic and thylakoid membranes from the cyanobacterium (blue-green alga) *Anacystis nidulans*. *Biochim. Biophys. Acta.* **766**: 395–402.

18. Kimura, Y, Yamasaki, T, Matsuzaki, E. 1997 Cytochrome components of the green alga *Bryopsis maxima*. *Plant Cell Physiol.* **18**: 317–324.

19. Powls, R, Wong, J, Bishop, NI. 1969 Electron transfer components of wild-type and photosynthetic mutants of *Scenedesmus obliquus* D3. *Biochim. Biophys. Acta* **180**: 490–499.

20. Metzger, S.U., Pakrasi, H, Whitmarsh, J. 1997 Characterization of a mutant strain of *Synechocystis* PCC 6803 that lacks cytochrome c_6 and cytochrome M. Abstracts of this meeting.

21. Shen, J-R, Inoue, Y. 1993 Binding and functional properties of two new extrinsic components, cytochrome c-550 and a 12 kDa protein, in cyanobacterial photosystem II. *Biochem.* **32**: 1825–1832.

22. Nishiyama, Y, Hayashi, H, Watanabe, T, Murata, N. 1994 Photosynthetic oxygen evolution is stabilized by cytochrome c_{550} against heat inactivation in *Synechococcus* sp. PCC 7002. *Plant Physiol.* **105**: 1313–1319.

23. Enami, I, Murayamo, H, Onta, H, Kamo, M, Nakazato, K, Shen, J-R. 1995 Isolation and characterization of a photosystem II complex from the red alga Cyanidium caldarium: association of cytochrome c-550 and a 12 kDA protein with the complex. *Biochim. Biophys. Acta* **1232**: 208–216.

24. Navarro, JA, Hervas, M, De La Corda, B, De La Rosa, M. 1995 Purification and Physicochemical properties of the low potential cytochrome c_{549} from the cyanobacterium *Synechocystis* PCC 6803. *Arch. Biochem. Biophys.* **318**: 46–52.

25. Peitsch, MC. 1996. ProMod and Swiss-Model: Internet based tools for automated comparative protein modelling. *Biochem. Soc. Trans.* **24**: 274–279.

26. Kraulis, PJ. 1991. MOLSCRIPT: A program to produce both detailed and schematic plots of protein structures. *J. Appl. Crystallogr.* **24**: 946–950.

31

CYTOCHROME c_6 FROM *SYNECHOCOCCUS* SP. PCC 7002

Chris Nomura and Donald A. Bryant

Department of Biochemistry and Molecular Biology
Pennsylvania State University
University Park, Pennsylvania 16802

1. INTRODUCTION

Cytochrome c_6 (cyt c_6; formerly cyt c-553) is a small, soluble iron-heme protein from the "c" class of cytochromes that are characterized by their covalently bound heme groups. Examples of c-type cytochromes include the soluble cytochromes of the mitochondria, mobile cyt c-555 in anoxygenic green photosynthetic bacteria, cyt c_2 in purple photosynthetic bacteria, and the membrane-bound cyt c_1 and cyt f associated with the cyt bc_1 and b_6f complexes [1]. In cyanobacteria and eukaryotic algae, cyt c_6 is used as a mobile carrier of electrons between the cyt b_6f complex and P700 of the Photosystem I reaction center. Many of these organisms use either cyt c_6 or plastocyanin (PC) as the mobile carrier of electrons between cytb_6f complex and the P700 reaction center of photosystem I. Whether an organism uses PC or cyt c_6 as an electron donor to PS I is dependent on environmental conditions. When cells are grown in the presence of sufficient copper, cells synthesize and utilize PC for electron transport from cyt b_6f to PS I; however, in the absence of copper, cells synthesize and utilize cyt c_6. Cyt c_6 undergoes a reversible oxidation-reduction reaction similar to that of PC during photosynthetic electron transport. In addition to its role in photosynthetic electron transport, cyt c_6 may also play a role in respiratory electron transport in cyanobacteria. Lockau has presented evidence in *Anabaena variabilis* that both cyt c_6 and PC have dual roles as electron carriers for photosynthesis and respiration [2]. Previous experiments have demonstrated that cyt c_6 depleted membranes of *Nostoc muscorum* could be made competent in transferring electrons to PS I and cyt oxidase by the addition of cyt c_6 [3].

The gene encoding cyt c_6 has been insertionally inactivated by an antibiotic cartridge in *Synechococcus* sp. strain PCC 7942 [4] or deleted in the case of *Synechocystis* sp. strain PCC 6803 [5]. In these interposon insertion/deletion mutants, cells were still able to grow at wild-type rates. It was presumed that other physiological electron donors present within these organisms could replace cyt c_6. In *Synechocystis* sp. strain PCC 6803, the primary

The Phototrophic Prokaryotes, edited by Peschek *et al.*
Kluwer Academic / Plenum Publishers, New York, 1999.

electron donor is PC; cyt c_6 is expressed only under copper depleted conditions. However, since the genes encoding cyt c_6 and PC (petJ and petE respectively) cannot both be deleted within the same strain, it appears that even under copper depleted conditions, Synechocystis sp. PCC 6803 is still able to produce some minimal amount of functional PC [6]. It has also recently been discovered that Synechococcus sp. strain PCC 7942 has a petE gene encoding a functional PC [7]. This mobile electron carrier was most likely responsible for the wild-type phenotype seen in the interposon mutant from [4].

Some cyanobacteria, e.g. Synechococcus sp. PCC 7002, have no detectable plastocyanin, and in these cases it is assumed that cyt c_6 is the sole electron carrier between the cyt b_6f complex and photosystem I [8]. In order to examine further the role of cyt c_6 in Synechococcus sp. strain PCC 7002, the petJ gene encoding cyt c6 was cloned, sequenced and characterized. Unlike the previously mentioned cyanobacterial species, cyt c_6 in Synechococcus sp. strain PCC 7002 appears to be the sole mobile electron carrier in this species, and it is essential for cell viability under various growth conditions. We have currently undertaken a study to determine if genes encoding mobile electron carriers (PC and cyt c_6) from Synechocystis sp. strain PCC 6803 can substitute for the Synechococcus sp. PCC 7002 cyt c_6.

2. RESULTS AND DISCUSSION

2.1. Cloning and Sequence Analysis of Synechococcus sp. Strain PCC 7002 Cytochrome c_6

Initial Southern blot hybridization screens to identify the petJ gene of Synechococcus sp. PCC 7002 using heterologous PCR product probes (petJ genes from Synechocystis sp. PCC 6803 or Synechococcus sp. PCC 7942) at low stringency were unsuccessful; thus, the Synechococcus sp. strain PCC 7002 protein was purified. The N-terminal sequence of the purified cyt c_6 was determined to be ADAAAGAQVFAAN(C)A. Based upon this sequence, the degenerate oligonucleotide: 5'-GCT GA(T/C) GCT GCT GCT GGT GCT CA(A/G) GTT TT(T/C) GCT GCT AA(T/C) TG(T/C) GC-3' was synthesized for use as a forward primer. Based upon a highly conserved region (DVAAYVLDQAEKGW) near the carboxy terminus of the protein, the degenerate oligonucleotide: 5'-CCA CGC TTT TTC TGC TTG GTC GAG TAC GTA GCG TGC NAC (A/G)TC-3' was synthesized for use as the reverse primer for PCR amplification of an internal region of Synechococcus sp. PCC 7002 petJ gene for use as a hybridization probe. Based upon the conserved heme attachment site amino acid sequence (NCAACH), an internal degenerate PCR primer: 5'-AA(T/C) TG(T/C) GC(T/C) GC(T/C) TG(C/T) C-3' was also synthesized. The resultant PCR products were used as probes for Southern analysis of 7002 chromosomal DNA digested with multiple restriction enzymes. A hybridizing HindIII-XbaI fragment of approximately 1.7 kbp was chosen for cloning into pUC19. This fragment was sequenced (Genbank Accession number: AF020306) and shown to contain a single, complete open reading frame of 351 bp that initiates with a TTG codon and predicts a protein of 117 amino acids (see Figure 1A); the petJ gene lies between the partial sequences of two other open reading frames. The gene products of these partial reading frames are most closely related to the hypothetical proteins sll0415 and sll0185 from the Synechocystis sp. PCC 6803 genomic database [9] and would be transcribed in the −3 and −2 reading frames respectively. The 7002 petJ gene is transcribed from the opposite strand from the +3 reading frame. The petJ gene predicts presequence of 24 amino acids that displays many characteristics (a basic N-terminal region

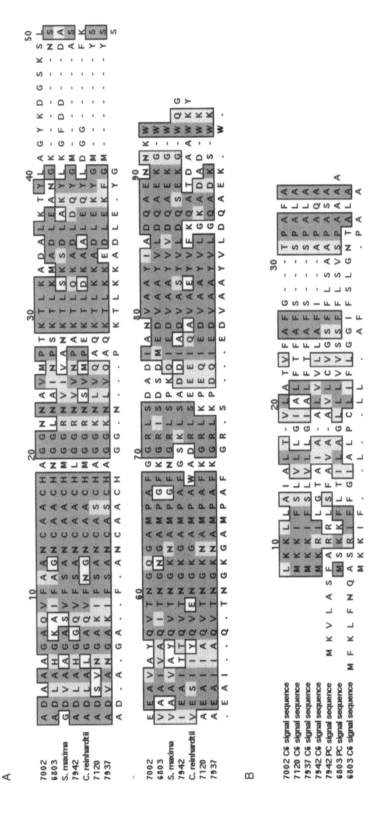

Figure 1. ClustalW multiple protein sequence alignment. **A.** Deduced cyt c_6 amino acid sequences were aligned by the ClustalW multiple protein sequence alignment function in the MacVector program. Dark gray shading indicates identical or majority identical amino acids, the lighter gray shading indicates conserved amino acids, and no shading indicates non-conserved amino acids. Hyphens indicate gaps introduced to maximize the alignment. The consensus cyt c_6 sequence is indicated below. Organisms are indicated as follows: 7002, *Synechococcus* sp. PCC 7002; 6803, *S. maxima, Spirulina maxima; C. reinhardtii, Chlamydomonas reinhardtii;* 7937, *Anabaena* sp. PCC 7937; 7942, *Synechococcus* sp. PCC 7120, *Anabaena* sp. PCC 7120; 6803, *Synechocystis* sp. PCC 6803. **B.** Deduced cyanobacterial signal peptide sequences for cyt c_6 and plastocyanin from various organisms were aligned as described above. Dark gray shading indicates identical or majority identical amino acids, the lighter gray shading indicates conserved amino acids, and no shading indicates non-conserved amino acids. Hyphens indicate gaps introduced to maximize the alignment. Conserved or identical signal sequence amino acids are indicated below. Organisms are as described above. The signal peptide of cyt c_6 from *Synechococcus* sp. PCC 7002 shares many of the traits of typical prokaryotic signal peptide sequences (see text).

with two lysines (n-region), a central hydrophobic region (h-region) and a C-terminal region that follows the '−3, −1 rule' where the amino acids at the −3 and −1 positions are alanines) common for prokaryotic signal peptides [10,11]. A comparison of selected cyanobacterial signal sequences is included in Figure 1B. The mature protein begins at amino acid 25 (Ala), contains 93 amino acids and has a predicted molecular mass of 9.4 kDa and a pI of 4.89. There is a potential transcriptional terminator (-18 kcal mol^{-1}) located between nucleotide positions 970 and 990 about 150 bases downstream from the stop codon of $petJ$. Transcript analyses indicate the $petJ$ gene is transcribed as a monocistronic transcript of approximately 600 nucleotides.

A comparison of mature cyt c_6 proteins generated by ClustalW [12] is shown in Figure 1A. Cyt c_6 from *Synechococcus* sp. PCC 7002 contains 24 amino acids that are conserved in all other sequences for this protein but overall is only 46–54% identical to other cyt c_6 proteins. It is interesting to note that the protein contains an insertion of 7 amino acids starting at glycine 42 through lysine 48. The *Spirulina maxima* cyt c_6 appears to be most closely related to the *Synechococcus* sp. PCC 7002 cyt c_6.

2.2. Attempted Insertional Inactivation of the *petJ* Gene by Interposon Mutagenesis

The *aphII* gene, encoding aminoglycoside 3′-phosphotransferase and conferring kanamycin resistance, was inserted into a unique *Bst*XI site in the coding region of the 7002 *petJ* gene. This construct was used to transform *Synechococcus* sp. PCC 7002 strain PR6000 and kanamycin-resistant transformants were selected for growth in liquid media. Unlike previously described mutagenesis of the *petJ* genes from *Synechocystis* sp. PCC 6803 [5] or *Synechococcus* sp. PCC 7942 [4] for which the functional copy of the gene encoding cyt c_6 was eliminated by complete segregation of alleles, the *petJ* and *petJ::aphII* alleles failed to segregate completely in *Synechococcus* sp. PCC 7002, implying that a wild type copy of the gene is required for cell viability. Since the *petJ* gene is transcribed as a monocistronic transcript, a polarity effect on an adjacent essential gene is unlikely.

2.3. Attempted Functional Substitution and Incomplete Segregation of the *petJ* Gene in *Synechococcus* sp. Strain PCC 7002

Because of the failure of the gene to segregate under the initial conditions tested, constructs were made to determine whether or not functional analogs of cyt c_6 from *Synechocystis* sp. PCC 6803 could substitute functionally for the *Synechococcus* sp. PCC 7002 cyt c_6. The *petE*, *petJ*, and *cytM* genes were PCR amplified from *Synechocystis* sp. PCC 6803 genomic DNA; the 5′ ends of the genes were modified to contain an *Afl*III site overlapping the start codon while the 3′ ends of the PCR products had either a *Bam*HI or *Stu*I site for cloning purposes. The PCR products were isolated and digested with *Afl*III and either *Bam*HI or *Stu*I and inserted into plasmid pSE280 at the *Afl*III-compatible *Nco*I site so that the genes would be under the control of the *E. coli trc* promoter [13]. It has been demonstrated previously that introduced genes could be expressed by the *trc* promoter in cyanobacteria [14]. Constructs containing foreign mobile electron carrier genes were inserted into platform vector pLAT3 or pLAT2. The pLAT3 vector contains the ½ fragment from pHP45½ and a partial polylinker inserted between the coding sequence of the *Synechococcus* sp. PCC 7002 *amtA* gene. The pLAT2 vector contains a marker for erythromycin resistance and a polylinker between *Synechococcus* sp. PCC 7002 *argE3* gene flanking sequences. These sites were previously determined to be neutral sites within the *Synechococcus* sp. PCC 7002 genome. The constructs were used to transform the *petJ* /*petJ::aphII* merodiploid strain. Streptomycin/kanamycin or erythromycin/kanamycin resistant transformants were selected and grown in liquid media with the appropriate antibiotics, and segregation was assayed by Southern hybridization. All of these constructs failed to induce full segregation in the *petJ/petJ::aphII* merodiploid strain.

Attempts to detect a *petE* gene in *Synechococcus* sp. strain PCC 7002 by Southern hybridization using various probes have been unsuccessful. Nevertheless, caution must be exercised when making an argument for the absence of plastocyanin, since similar screening for plastocyanin in *Synechococcus* sp. PCC 7942 failed to reveal the presence of the gene sequence even though it was present [7]. However, the inability to achieve segregation of a *petJ* mutant under various growth conditions implies the absence of plastocyanin (and of any cryptic cytochrome genes that could functionally substitute for cyt c_6) in *Synechococcus* sp. strain PCC 7002, since a similar result is seen with *Synechocystis* sp. strain PCC 6803: if the *petE* gene is deleted, the *petJ* gene cannot be deleted, and vice-versa [6].

Another explanation for the lack of segregation could be that although PC might be present in *Synechococcus* sp. PCC 7002, it is unable to substitute functionally for cyt c_6 under the conditions tested. A similar situation has been observed in the purple non-sulfur bacteria *Rhodobacter sphaeroides* and *R. capsulatus*. In *R. capsulatus*, there are two electron carriers, cyt c_2 and cyt c_y that can functionally substitute for one another between the cyt bc_1 complex and the reaction center. If the gene encoding cyt c_2 is deleted in *R. capsulatus*, the mutant can still grow photosynthetically using cyt c_y. The case for *R. sphaeroides* is quite different although the two organisms are closely related. Deletion of the gene encoding cyt c_2 results in a non-photosynthetic mutant in *R. sphaeroides*, although *R. sphaeroides* also has a gene for cyt c_y. However, if the *R. capsulatus* cyt c_y gene is expressed in the *R. sphaeroides* cyt c_2-less strain, cells again become photoautotrophic [15]. *Synechococcus* sp. PCC 7002 may be capable of producing plastocyanin, but perhaps this PC is unable to substitute for the major mobile electron carrier, cyt c_6.

At the time that this project was begun, the role of cyt c_M was unknown. This question has been recently addressed [6], and it appears that cyt c_M is necessary for respiration but cannot substitute for plastocyanin or cyt c_6 in *Synechocystis* sp. PCC 6803. Thus it is not surprising that the *cytM* gene product could not substitute for cyt c_6. However, it is nonetheless surprising that the *Synechocystis* sp. PCC 6803 cyt c_6 and PC proteins could not substitute for the *Synechococcus* sp. PCC 7002 cyt c_6. These results differ from the functional substitution experiments performed in the purple non-sulfur bacteria [15,16], and suggest that only the native cyt c_6 of *Synechococcus* sp. PCC 7002 can function as a mobile electron carrier in this organism. It is important that we determine whether adequate expression levels are achieved from the platform constructs through Northern analyses and whether the foreign proteins are being produced in *Synechococcus* sp. PCC 7002 by Western analysis with antibodies specific for cyt c_6 and plastocyanin from *Synechocystis* sp. PCC 6803. Preliminary studies show that the *Synechocystis* sp. PCC 6803 *petJ* and *petE* genes are being transcribed from the *trc* promoter in *Synechococcus* sp. PCC 7002.

REFERENCES

1. Dickerson, R.E. (1980) in: The evolution of protein structure and function (Sigman, D.S. and Brazier, M.A.B., eds.), pp. 173–202, Academic Press, New York.
2. Lockau, W. (1981) Arch. Microbiol. 128, 336–340.

3. Stürzl, E., Scherer, S. and Böger, P. (1982) Photosynthesis Res. 3, 191–201.
4. Laudenbach, D.E., Herbert, S.K., McDowell, C., Fork, D.C., Grossman, A.R. and Straus, N.A. (1990) Plant Cell 2, 913–924.
5. Zhang, L., Pakrasi, H.B. and Whitmarsh, J. (1994) J. Biol. Chem. 269, 5036–5042.
6. Manna, P. and Vermaas, W. (1997) Photosynthesis Res., in press.
7. Clarke, A.K. and Campbell, D. (1996) Plant Physiol. 112, 1551–1561.
8. Sandmann, G. and Böger, P. (1981) in: Photosynthesis, vol. II (Akoyunoglou, G., ed.), pp. 623–632, Balaban International Science Services, Philadelphia.
9. Kaneko, T. et al. (1996) DNA Res. 3, 109–136.
10. von Heijne, G. (1986) Nucleic Acids Res. 14, 4683–4690.
11. Gierasch, L.M. (1989) Biochemistry 28, 923–930.
12. Thompson, J.D., Higgins, D.G. and Gibson, T.J. (1994) Nucleic Acids Res. 22, 4673–4680.
13. Brosius, J. (1989) DNA 8, 759–77.
14. Geerts, D., Bovy, A., de Vrieze, G., Borrias, M., and Weisbeek, P. (1995) Microbiology 141, 831–841.
15. Jenney, Jr., F. E., Prince, R.C. and Daldal, F. (1996) Biochim. Biophys. Acta 1273, 159–164.
16. Jenney, Jr., F. E. and Daldal, F. (1993) EMBO J. 12, 1283–1292.

32

MOLECULAR PROPERTIES OF SOLUBLE CYTOCHROME c-552 AND ITS PARTICIPATION IN SULFUR METABOLISM OF *OSCILLATORIA* STRAIN Bo32

Ines Frier, Jörg Rethmeier, and Ulrich Fischer

Universität Bremen
Zentrum für Umweltforschung und Umwelttechnologie (UFT) und
    Fachbereich 2
Abt. Marine Mikrobiologie
Leobener Str., D-28359 Bremen, Germany

1. INTRODUCTION

Many cyanobacteria are not only able to perform an oxygenic photosynthesis with water as electron donor but also an anoxygenic one by using sulfide [1–3]. The last mentioned process also provides the organisms with enough energy to maintain their metabolism. Sulfide penetrates into the cells and can inhibit the electron transport from photosystem II (PS II) to photosystem I (PS I), even at low concentrations; probably by binding at the plastoquinone pool so that oxygen is no longer produced [4,5]. Since the binding of sulfide is reversible, oxygenic photosynthesis immediately starts again when sulfide has disappeared from the environment of the cells [6]. Studies with *Oscillatoria limnetica* have shown that PS II is already inhibited by sulfide concentrations > 0.1 mM and that the organism then performs a PS II independent anoxygenic photosynthesis only with PS I, as known for anoxygenic phototrophic sulfur bacteria [7,8]. Under sulfidic conditions and during a lag-time of 2–3 hrs, a sulfide-quinone reductase (SQR) is synthesized by the organism. This enzyme is necessary to drive anoxygenic photosynthesis by transferring electrons from sulfide to the plastoquinol pool of PS I [9,10]. During cyanobacterial anoxygenic photosynthesis, sulfide is oxidized either to elemental sulfur, thiosulfate, sulfite or polysulfides [see 3]. Electron transfer proteins, such as cytochromes, are known to be involved in sulfide or thiosulfate oxidation by anoxyphotobacteria (for review see [11]). While there exist much information on soluble c-type cytochromes from anoxygenic phototrophic sulfur bacteria, only little is known about such electron transfer proteins from cyanobacteria. Most studies concentrate on cytochrome c_6 (formerly c_{554} or c_{553}) and

The Phototrophic Prokaryotes, edited by Peschek *et al.*
Kluwer Academic / Plenum Publishers, New York, 1999.

the autoxidizable low potential cytochrome c-550 (for review see [12–14]). The *in vivo* function of the last mentioned cytochrome is not established, but concerning its low redox-potential of −260 mV an involvement in sulfur metabolism is assumed [12,14]. Cytochrome c_6 functions as electron-carrier between membrane-bound cytochrome b_6/f complex and P-700 and is functionally equivalent to plastocyanin concerning this electron transfer system [12]. The amount of plastocyanin and cytochrome c_{553} in cyanobacteria is regulated by the copper concentration available [15] or by sulfide and iron, respectively, as shown in this paper for cytochrome c-552 from *Oscillatoria* Bo32.

2. MATERIALS AND METHODS

2.1. Organism and Culture Conditions

Oscillatoria Bo32 was cultivated in artificial seawater medium ASN III using Erlenmeyer flasks of different size [3]. Iron-free culturing was obtained by omitting ferric ammonium citrate. Anaerobic sulfidic growth of strain Bo 32 was performed in gas-tight, sealed 250 ml Erlenmeyer flasks. Inoculated culture media were gased with nitrogen for 15–30 min before sulfide was added (final concentration: 1 mM). All cultures were incubated at a light intensity of 7–8 $\mu E \cdot m^{-2} \cdot s^{-1}$ and at room temperature.

2.2. Cytochrome Isolation

Cells were harvested by centrifugation at 35,000 × *g* for 20 min, washed twice in 0.9 M potassium phosphate buffer, pH 7.0, suspended in the same buffer (1 g cell material/3 ml) and broken up by sonification. Cell particles were removed by centrifugation at 35,000 × *g* (20 min) and 125,000 × *g* (3 hrs). The supernatant of the ultracentrifugation was desalted on Sephadex G-25 and the eluate was chromatographed on a DEAE-52 cellulose column, equilibrated in 5 mM K-phosphate buffer, pH 7.0. Cytochrome c-552 was eluted with the same buffer, but 20 mM instead of 5 mM.

2.3. NADP$^+$-Photoreduction

Sulfide driven NADP$^+$-photoreduction was tested spectrophotometrically and was performed with the thylakoid-membranes and cell-free extracts of normally grown and sulfide adapted cells of *Oscillatoria* strain Bo32 according to the methods described by [16,17].

2.4. Molecular Weight, Isoelectric Point, and Redox Potential

Molecular weight and isoelectric point were determined by SDS-gel electrophoresis and flat bed electrofocussing using the chemicals and equipment from Pharmacia LKB according to the methods given in the suppliers instruction manuals. Redox potential measurement was carried out under anaerobic conditions and continuous stirring in a specially constructed quartz cell with thread and rubber septum to insert needles and microelectrodes and was determined according to [18].

2.5. Analytical Determinations

Inorganic sulfur compounds were determined as described by [19].

3. RESULTS

3.1. Anaerobic Sulfide Oxidation

Sulfide (0.4 mM) was stoichiometrically photooxidized to thiosulfate (0.2 mM) by *Oscillatoria* Bo32 with the formation of sulfite as intermediate sulfur compound. Elemental sulfur or polysulfides were not formed during autotrophic growth with sulfide under anaerobic conditions.

3.2. Spectrophotometrical and Molecular Properties of Cytochrome c-552

The alkaline pyridine absorption spectra of the isolated *Oscillatoria* Bo32 cytochrome exhibit 3 maxima at 552 nm (α-peak), 522 nm (β-peak) and 415 nm (γ-peak) in the reduced state and only one maximum at 408 nm in the oxidized form, typical for c-type cytochromes. Cytochrome c-552 is an acidic hemoprotein, non-autoxidizable and does not react with CO and can therefore be regarded as a typical "low-spin" cytochrome. As confirmed by native and SDS-gel electrophoresis, pure cytochrome c-552 was already obtained after one single chromatographic purification step on DEAE-cellulose (best purity index obtained: $A_{268}/A_{415} = 0.24$ and $A_{268}/A_{552} = 0.59$). The monomeric cytochrome has an M_r of 23,500 (determined by SDS-polyacrylamide gel electrophoresis), an isoelectric point of pH 4.8 and a mid-point oxidation reduction potential at pH 7.0 of +280 mV. All molecular properties are summarized in Table 1 and compared with corresponding data of two other cyanobacterial c_6-type cytochromes.

3.3. Cytochrome Yield

Normally grown cells of *Oscillatoria* Bo32 possess 6.1 nmol cytochrome c-552/g cell material. Cultivated under anaerobic conditions with 1 mM sulfide in the medium, cells contain twice as much of this hemoprotein, while the cytochrome c-552 content was reduced to 2/3 when the cells were grown in the absence of iron. Details are given in Table 2.

3.4. NADP⁺-Photoreduction

Sulfide-driven $NADP^+$-photoreduction was performed with thylakoid-membranes (10 µg *chl a*/ml) and soluble enzyme fractions from regularly grown and sulfide adapted

Table 1. Comparison of molecular properties of soluble cyanobacterial c_6-type cytochromes

Organism	Maxima (nm)		Molecular weight	Isoelectric point (pH)	Redox potential (mV)	Purity index
	Reduced	Oxidized				
Oscillatoria Bo32	552 (α)		23,500	4.8	+280	0.24 (A_{268}/A_{415})
	522 (β)					0.53 (A_{268}/A_{552})
	415 (γ)	408 (γ)				
Spirulina platensis[a]	553.6 (α)		9,500	4.9	+350	1.0 ($A_{280}/A_{553.6}$)
	523 (β)					
	416 (γ)	410 (γ)				
Anacystis nidulans[a]	554 (α)		13,000	3.8	+350	0.84 (A_{275}/A_{554})
	522.5 (β)					
	416.5 (γ)	411 (γ)				

[a]Data taken from [14].

Table 2. Cytochrome c-552 content of normally grown *Oscillatoria* Bo32 cells, in the presence of sulfide (1 mM) and in the absence of iron

Parameter	Culture conditions and yield (nmol)		
	Aerobic	Anaerobic with sulfide	Aerobic without iron
Cyt c-552/g cell material	6.1	11.7	3.6
Cyt c-552/mg chlorophyll *a*	2.1	4.0	1.7

cells of *Oscillatoria* Bo32 in the presence or absence of cell-own cytochrome c-552 (0.025 nmol/ml). The enzyme fraction used in the test system contains ferredoxin and ferredoxin-NADP$^+$-reductase, both necessary for the photoreduction of NADP$^+$. Determination of NADP$^+$-photoreduction was carried out in quartz cells, illuminated with a light intensity of 24.5 μE·m^{-1}·sec^{-1} and wavelengths > 680 nm to ensure an operating of only PS I. The reaction was started upon addition of sulfide (1 mM). Absorbance changes at 340 nm were followed for 1 hour. The results are summarized in Table 3.

4. DISCUSSION

Like many other cyanobacteria, *Oscillatoria* Bo32 is also able to perform an anoxygenic photosynthesis using sulfide as electron donor, which is stoichiometrically photooxidized to thiosulfate [3,20]. During this process, no elemental sulfur was produced, but sulfite appeared as intermediate sulfur compound in the medium. As recently shown by Rabenstein and co-workers [3], no sulfite was formed from sulfide in the presence of DCMU. Obviously, this process seems to be connected to PS II activity. On the other hand, it was demonstrated that in *Anacystis nidulans* both photosystems (I and II) were involved in the oxidation of sulfite [21]. An anaerobic sulfide oxidation to thiosulfate with intermediate sulfite formation by cyanobacteria seems to be more common than generally assumed so far. However, such an intermediate sulfite production during sulfide oxidation is not restricted to oxyphotobacteria, since this has been observed already for the purple non-sulfur bacterium *Rhodovulum sulfidophilum* [22].

An adaptation of cyanobacteria to sulfidic conditions implies that the organism is equipped with all enzymes and electron transport proteins necessary for an efficient use of sulfide as photosynthetic electron donor. Recently, it has been shown that during a

Table 3. Sulfide-driven NADP$^+$-photoreduction with cell-free extracts of normally grown and sulfide adapted cells of *Oscillatoria* Bo32 with or without addition of cell-own cytochrome c-552

Cell extracts and conditions	NADP$^+$-photoreduced (μmol/mg chl *a*/h)
Normally grown cells	
– Cyt. c-552	3
+ Cyt c-552	9
Sulfide adapted cells	
– Cyt c-552	10
+ Cyt c-552	10

lag-time of 2–3 hrs a sulfide-quinone-reductase (SQR) was synthesized, responsible for the utilization of sulfide to perform an effective anoxygenic photosynthesis [9,10]. From anoxyphotobacteria it is known that cytochromes are involved in dissimilatory sulfur metabolism [11]. This may be the case also for soluble cytochrome c-552 from *Oscillatoria* Bo32. In comparison to aerobically grown cells, the organism synthesized twice as much of this hemoprotein under anaerobic sulfidic conditions. Regarding the redox potential of +280 mV, cytochrome c-552 may function as an electron mediator between the plastoquinone pool (~0 mV) and the Fe_2S_2-centre of the "Rieske-Protein" (+290 mV) by accelerating the transport of electrons derived from sulfide oxidation into the transport chain of PS I. It has been reported for different cyanobacteria that cytochromes c-553 can act as possible donor to PS I [see 23]. With the exception of the higher molecular weight, all other molecular properties of *Oscillatoria* Bo32 cytochrome c-552 lie in the order of magnitude, found for other corresponding cyanobacterial soluble c-type cytochromes and may therefore be regarded as a c_6-type cytochrome.

Sulfide-dependent electron transport was studied with thylakoids from *Oscillatoria limnetica* where two different pathways of $NADP^+$-photoreduction, a "specific" and a "non-specific" one, were found [5,16]. The "specific" pathway has a high affinity to sulfide (K_m: 24 μM) and was found only in sulfide-adapted cells, while the "non-specific" pathway occurs in sulfide-adapted cells as well as in normally grown ones and exhibits low affinity to sulfide (K_m: ~1 mM). Addition of isolated cytochrome c-552 to cell-free extracts of normally grown *Oscillatoria* Bo32 cells caused a 3-fold increase of sulfide-driven $NADP^+$-photoreduction (see Table 3). No effect was observed with thylakoids from sulfide-adapted cells. One possible explanation for this behaviour could be that sulfide grown cells contain already enough cytochrome c-552 (twice as much as normally ones, see Table 2) to perform an efficient anaerobic photooxidation of sulfide. Therefore, cytochrome c-552 seems to be necessary for this process, exactly as cytochrome c-553 of *Anabaena variabilis* for $NADP^+$-photoreduction with H_2 [24]. Since sulfide-driven $NADP^+$-photoreduction in *Oscillatoria* Bo32 is not only dependent on the cellular cytochrome c-552 content but also on the sulfide concentration available, it is likely that in this organism the above mentioned "non-specific" pathway of anaerobic sulfide oxidation is realized.

ACKNOWLEDGMENT

This study was supported by the Bundesministerium für Bildung, Wissenschaft, Forschung und Technologie under the project DYSMON (03F0123D).

REFERENCES

1. Garlick, S., Oven, A. and Padan, E. (1977) J. Bacteriol. 129, 623–629
2. Padan, E. (1979) Ann. Rev. Plant Physiol. 30, 27–40
3. Rabenstein, A., Rethmeier, J. and Fischer, U. (1995) Z. Naturforsch. 50c, 769–774
4. Howsley, R. and Pearson, H.W. (1979) FEMS Microbiol. Lett. 6, 287–292
5. Slooten, L., de Smet, M. and Sybesma, C. (1989) Biochim. Biophys. Acta 973, 272–280
6. Padan, E. and Cohen, Y. (1982) in: The Biology of Cyanobacteria (Carr, N.G. and Whitton, B.A. eds) pp. 215–235, Blackwell, Oxford
7. Cohen, Y., Padan, E. and Shilo, M. (1975) J. Bacteriol. 123, 855–861
8. Cohen, Y., JØrgensen, B.B., Revsbech, N.P. and Poplawski, R. (1986) Appl. Environ. Microbiol. 51, 398–407

9. Arieli, B., Padan, E. and Shahak, Y. (1991) J. Biol. Chem. 266, 104–111
10. Arieli, B., Shahak, Y., Taglicht, D., Hauska, G. and Padan, E. (1994) J. Biol. Chem. 269, 5705–5711
11. Fischer, U. (1984) in: Sulfur. Its Significance for Chemistry, for the Geo-, Bio- and Cosmosphere and Technology (Müller, A. and Krebs, B. eds) pp. 383–407, Elsevier, Amsterdam
12. Matsubara, H. and Wada, K. (1988) in: Methods in Enzymology (Packer, L. and Glazer, A.N. eds) Cyanobacteria. vol 167, pp. 387- 410, Academic Press, San Diego
13. Yamaka, T. (1992) The Biochemistry of Bacterial Cytochromes, Japan Scientific Societies Press, Tokyo and Springer, Berlin
14. Monrad, L.Z., Cheng, R.H. and Krogmann, D.W. (1994) in: The Molecular Biology of Cyanobacteria (Bryant, D.A. ed) pp. 381–407, Kluwer, Dordrecht, Boston, London
15. Sandmann, G. and Böger, P. (1980) Plant Sci. Lett. 17, 417–424
16. Sybesma, C. and Slooten, L. (1987) in: Progress in Photosynthesis Research (Biggins, J. ed) Voll II, pp. 12,633–12,636, Nijhoff Publishers, Dordrecht
17. Rethmeier, J. (1991) Diploma thesis, Oldenburg University, Germany
18. Steinmetz, M.A. and Fischer, U. (1982) Arch. Microbiol. 131, 19–26
19. Rethmeier, J., Rabenstein, A., Langer M. and Fischer U. (1997) J. Chromat. 760, 295–302
20. De Wit, R. and van Gemerden, H. (1987) FEMS Microbiol. Ecol. 45, 7–13
21. Peschek, G.A. (1978) Arch. Microbiol. 119, 313–322
22. Neutzling, O., Pfleiderer, C. and Trüper, H.G. (1985) J. Gen. Microbiol. 131, 791–798
23. Schmetterer, G. (1994) in: The Molecular Biology of Cyanobacteria (Bryant, D.A. ed) pp. 409–435, Kluwer, Dordrecht, Boston, London
24. Schrautemeier, B., Böhme, H., and Böger, P. (1985) Biochim. Biophys. Acta 807, 147–154

33

MEMBRANE-ASSOCIATED *C*-TYPE CYTOCHROMES OF *RHODOBACTER CAPSULATUS* INVOLVED IN PHOTOSYNTHESIS AND RESPIRATION

Hannu Myllykallio,[*] Hans-Georg Koch, and Fevzi Daldal

Department of Biology
Plant Science Institute
University of Pennsylvania
Philadelphia, Pennsylvania 19104

1. INTRODUCTION

Purple nonsulfur facultative phototrophs, e.g. *Rhodobacter* species (*Rhodobacter capsulatus* and *Rhodobacter sphaeroides*) provide excellent model systems for studying the structure, function, regulation and biogenesis of cytochrome (cyt) complexes involved in photosynthesis (Ps) and respiration (Res). In these species Ps electron transport pathway is cyclic. It involves the photochemical reaction center (RC) and the ubihydroquinone: cyt *c* oxidoreductase (also called the cyt bc_1 complex) connected to each other by *c*-type cyts and the ubihydroquinone (UQ) pool (Figure 1). On the other hand, the Res pathways are often branched at the UQ pool and contain several terminal oxidases thought to work at different oxygen tensions. *R. capsulatus* has a quinol oxidase (Q_{ox}) and a cyt *c* oxidase (C_{ox}) of cbb_3-type while *R. sphaeroides* also has an additional cyt *c* oxidase of aa_3-type (Figure 1). These electron transport components are involved in production of energy which is essential for cell survival. They are responsible for translocation of protons across the cellular membrane to produce a proton gradient which is then used for ATP synthesis. Because of their crucial importance for life they are often multiplicated with overlapping roles, so that the absence of any one of them, although restrictive under some growth modes, is rarely lethal for the organism. Most of the energy transduction complexes contain *c*-type cyts which are attached to the cytoplasmic membrane with their covalently bound heme groups facing the periplasm. During the electron transport events

---

* Hannu Myllykallio and Hans-Georg Koch contributed equally to these studies.

The Phototrophic Prokaryotes, edited by Peschek *et al.*
Kluwer Academic / Plenum Publishers, New York, 1999.

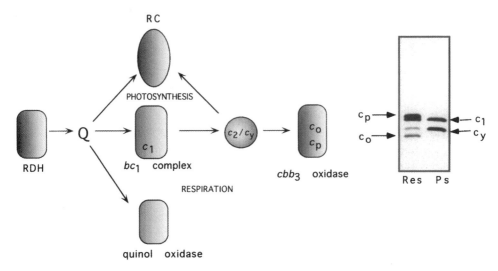

Figure 1. Photosynthetic (Ps) and respiratory (Res) electron transfer pathways of *Rhodobacter capsulatus*. RC, RDH, Q, c_1, c_2, c_y, c_o and c_p correspond to the photochemical reaction center, respiratory dehydrogenases, Q pool, cyt c_1 subunit of the cyt bc_1 complex, the electron carriers cyts c_2 and c_y, and the subunits cyts c_o and c_p of the cyt cbb_3 oxidase, respectively.

these subunits interact closely with other soluble, freely diffusible cyts present in the periplasmic space.

The nature and abundance of the c-type cyts change with cellular energy requirements in function of the growth mode used (Figure 1). In membranes of *R. capsulatus* strain MT1131 grown under Res conditions in enriched, or minimal, media four distinct c-type cyts of approximate molecular mass of 32, 31, 29 and 28 kDa are readily detectable (Figure 1). Of these, the 29 kDa cyt c_y is a functional homologue of the well studied soluble electron carrier cyt c_2, and the 31 kDa cyt c_1 is a subunit of the cyt bc_1 complex which catalyzes oxidation of ubiquinol and reduction of the cyts c_2 and c_y (Jenney and Daldal, 1993). On the other hand, the mono heme cyt c_o (28 kDa) and diheme cyt c_p (32 kDa) are subunits of the cbb_3-type cyt c oxidase (Gray et al., 1994; Koch et al., 1997) which catalyzes oxidation of the cyts c_2 and c_y and conversion of molecular oxygen to water. In membranes of cells grown under Ps conditions only the cyts c_1 and c_y are present (Figure 1), indicating that the cyts c_p and c_o, hence the cyt c oxidase, is regulated by the availability of oxygen. Additional c-type cyts are also detectable in cell membranes of both *R. capsulatus* and *R. sphaeroides*, but often their molecular nature and function are unknown.

Below, our recent studies related to the cyts c_1, c_y, and c_o/c_p are summarized.

2. CYTOCHROME c_1

The very recent resolution of the three dimensional structure of the cyt bc_1 complex from bovine heart mitochondria (Xie et al., 1997; Berry et al., 1997) revealed that cyt c_1 is very distinct in its folding features from cyt f of the cyt b_6f complex which is its chloroplast counterpart (Martinez et al., 1994). Mitochondrial cyt c_1 is unlike the elongated cyt f and has a globular shape common to various cyts c of different sources. It is highly homologous to bacterial cyts c_1 and has a general cyt *c* fold only slightly modified by addition and elimination of specific regions. The heme attachment site of both the cyts c_1 and *f* are of canonical nature (Cys-Xaa-Yaa-Cys-His), and both proteins are anchored to the membrane by a carboxyl-terminal transmembrane helix. However, it is now definite that the sixth ligands of their heme iron atom are different. In the case of cyt c_1 this is a conserved, carboxyl terminally located methionine residue, which has been pinpointed previously by site-directed mutagenesis studies in *R. capsulatus* (Gray et al., 1992). On the other hand, in cyt *f* the amino group of the amino-terminal tyrosine residue, freed up by processing of its signal sequence, serves as the sixth ligand (Martinez et al., 1994). Several other similarities and differences have also been observed between the cyts c_1 and *f* and will be discussed by E. Berry in an upcoming article describing the structure of the cyt bc_1 complex (Berry et al., personal communication).

3. CYTOCHROME c_y

3.1. Structural Properties of cyt c_y

The Ps growth of *R. sphaeroides* is dependent on the presence of the cyt c_2 as a secondary electron carrier (Donohue et al., 1988), while that of *R. capsulatus* continues in its absence (Daldal et al., 1986; Prince et al., 1986). In the latter species, a membrane-bound cyt, termed cyt c_y, was identified as an alternative electron donor to the RC (Jenney and Daldal, 1993; Jenney et al., 1994). *R. capsulatus* cyt c_y has now been purified to homogeneity by epitope tagging, and its properties have been characterized in details (Myllykallio et al., 1997). It has a typical visible spectrum for a *c*-type cyt, and its E_{m7} is approximately 365 mV, which closely resembles that of cyt c_2. Its amino acid sequence (Jenney and Daldal, 1993) together with the recent protein chemical data (Myllykallio et al., 1997) indicate that *R. capsulatus* cyt c_y is composed of three domains (Figure 2). Its amino-terminal domain contains a hydrophobic, unclipped signal sequence which anchors to the cytoplasmic membrane a mono-heme *c*-type cyt domain highly homologous to mitochondrial cyts *c*. The membrane anchor and the cyt *c* domains of cyt c_y are separated by a relatively long, and apparently highly flexible, linker domain of about 70 amino acid residues rich in alanine and proline (Figure 2). The amino-terminal "anchor-linker" domains are sufficient to attach to the cytoplasmic membrane other non naturally membrane-bound cyts. For example, a membrane-attached hybrid cyt, MA-cyt c_2, constructed by using the anchor-linker domains of cyt c_y and the cyt *c* domain of cyt c_2, is able to support both Ps and Res growth as effectively as the soluble cyt c_2 (Table 1). Clearly, *R. capsulatus* cyt c_y is a genuine member of a growing subclass of membrane-bound *c*-type cyts (Bott et al., 1991; Jenney and Daldal, 1993; Turba et al., 1995), and is unique among them by its dual function in both Ps (Jenney and Daldal, 1993) and Res (Hochkoeppler et al., 1995) electron transport. Recently, a homologue of cyt c_y has also been encountered in *R. sphaeroides* (Zeilstra-Ryalls and Kaplan, 1996), but currently its function is unknown.

Earlier experiments have established that *R. capsulatus* Res growth becomes strictly dependent on the cyt bc_1/cyt *c* oxidase branch, hence sensitive to the cyt bc_1 complex inhibitor myxothiazol, when the Q_{ox}-dependent branch is eliminated (Marrs and Gest, 1973; Daldal, 1988) (*e.g.*, pRK415/M6G-G4/S4 (Q_{ox}^- cyt c_2^-) in Table 1). In this background simultaneous inactivation of the cyts c_2 and c_y is impossible under Res growth conditions, strongly implying that these cyts are the only electron carriers capable of connecting the cyt bc_1 complex to the cyt *c* oxidase (Daldal, 1988; Hochkoeppler et al., 1995) (Figure 1).

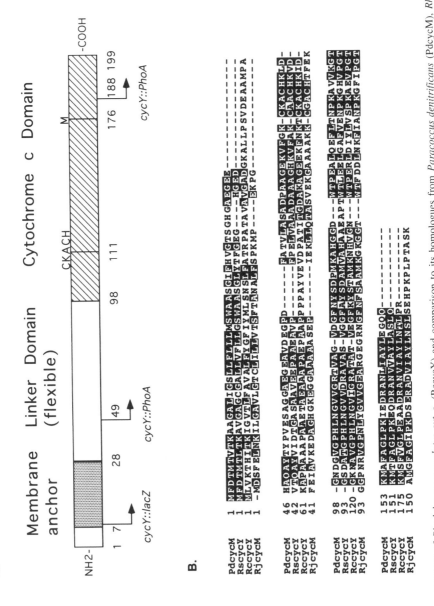

Figure 2. Primary structure of *Rhodobacter capsulatus* cyt c_y (RccycY) and comparison to its homologues from *Paracoccus denitrificans* (PdcycM), *Rhodobacter sphaeroides* (RscycY) and *Bradyrhizobium japonicum* (BjcycM). Identical residues are shown in black boxes.

Table 1. *R. sphaeroides* cyt c_y is able to support electron transfer from the cyt bc_1 complex to the cyt cbb_3 oxidase in *R. capsulatus*

Strain (*R. capsulatus*)	Genotype	Electron carrier(s) (cyts c_2 or c_y)	Terminal oxidase(s)	Ps[c]	Res[c] (dt/min)	Myx[c] (Res)	Nadi
MT1131	*crtD121* Rif[R]	*Rc[1]* cyt c_2 + *Rc* cyt c_y	cyt cbb_3 and Qox	+	+ (142)	+	+
pRK415/M6G-G4/S4[b]	*crtD121* Qox[−] Δ(*cycA*::*Kan*)	*Rc* cyt c_y	cyt cbb_3	+	+ (199)	−	+
pRKSH3/SL3	*crtD121* Qox[−] Δ(*cycA*::*Kan*) Δ(*cycY*::*Spe*)	*Rc* cyt c_2 on pRK415	cyt cbb_3	+	+ (265)	−	+
pHM8/SL3	*crtD121* Qox[−] Δ(*cycA*::*Kan*) Δ(*cycY*::*Spe*)	Membrane-attached *Rc* cyt c_2 on pRK415	cyt cbb_3	+	+ (264)	−	+
pHM13/SL3	*crtD121* Qox[−] Δ(*cycA*::*Kan*) Δ(*cycY*::*Spe*)	*Rs* cyt c_y on pRK415	cyt cbb_3	−	+ (344)	−	+

[a]*Rc* and *Rs* correspond to *R. capsulatus* and *R. sphaeroides*, respectively.

[b]All plasmid containing strains are tetracycline resistant, and SL3 is a derivative of M6G-G4/S4 into which the *cycY*:: *spe* deletion was introduced using GTA in the presence of appropriate plasmids carrying the alternate electron carriers as listed.

[c]Ps, Res and Myx correspond to photosynthetic, respiratory and respiratory growth in the presence of 10 μM myxothiazol in MPYE medium, respectively.

We used this background to probe the role of *R. sphaeroides* cyt c_y in Res electron transport by constructing various *R. capsulatus* strains which lacked the Q_{ox}-dependent pathway, and where the *R. capsulatus* cyt c_y, MA-cyt c_2 or *R. sphaeroides* cyt c_y were the obligatory electron carriers (Table 1). Growth properties of these strains clearly demonstrated that while *R. sphaeroides* cyt c_y is unable to participate in Ps electron transfer, none the less it was capable to support Res growth of *R. capsulatus*. More over, since a cyt c_2^- c_y^+ mutant of *R. sphaeroides* is Ps⁻ (Donohue et al., 1988) and a cyt c_y^- c_2^+ mutant is Ps⁺ (data not shown), it seems unlikely that the *R. sphaeroides* cyt c_y takes part in Ps electron transport even in its own background. Conversely, the *R. capsulatus* cyt c_y is functional not only in its own background, but is also able to complement a cyt c_2^- *R. sphaeroides* strain to a Ps⁺ phenotype (Jenney et al., 1996). In this regard, it is noteworthy that the flexible linker domain of *R. capsulatus* cyt c_y, proposed to be essential for its interactions with its redox partners (Myllykallio et al., 1997), is approximately 25 amino acid residues longer than that of *R. sphaeroides* cyt c_y (Figure 2). The molecular basis for the observed functional differences between the two cyt c_y homologues from these closely related *Rhodobacter* species is under investigation using appropriate constructions.

3.2. Spectroscopic Properties of cyt c_y

Flash-activated transient kinetics obtained using various *R. capsulatus* mutants indicated that two electron transfer pathways, one via the soluble cyt c_2 and one via the membrane-bound cyt c_y, operate independently between the RC and the cyt bc_1 complex in a wild-type strain of *R. capsulatus* (Jenney et al., 1994). These pathways have different kinetics for RC reduction and cyt c oxidation, implying that the cyts c_2 and c_y interact with their redox partners in different ways. Early observations indicated that oxidation of cyt c_y is very fast (*i.e.*, less than 100 μsec) (Prince et al., 1986; Jenney et al., 1994). Our very recent measurements with higher time resolution have confirmed and extended these observations (H. Myllykallio et al., in preparation). In particular, they demonstrated that the RC oxidation by cyt c_y is as rapid (roughly a few μsec for its fastest phase) as that mediated by cyt c_2 docked to the RC in its "proximal" configuration (Overfield and Wraight, 1980; Moser and Dutton, 1988; Tiede and Dutton, 1993) but its diffusion is restricted in comparison to the latter. In addition, they also revealed that cyt c_y functions as a more efficient electron donor to the RC during multiple turnovers of the cyclic electron transfer pathway. Whether this feature is indeed facilitated by a close physical association between the RC, cyt bc_1 complex and cyt c_y forming a "structural supercomplex", as suggested previously by the medium-dependent absence of cyt c_y when the cyt bc_1 complex is genetically eliminated (Jenney et al., 1994), remains unknown.

4. CYTOCHROMES c_o AND c_p

4.1. cyt cbb_3 Oxidase of *R. capsulatus*

R. capsulatus is unique in comparison to other phylogenetically related species, like *R. sphaeroides* and *P. denitrificans*, in that it has no aa_3 type cyt c oxidase (Klemme and Schlegel, 1968; Zannoni, 1995). Its only cyt c oxidase has been purified previously and characterized as being a novel cbb_3-type cyt c oxidase without a Cu_A center (Gray et al., 1994). This enzyme is composed of at least a membrane-integral b-type cyt (CcoN/subunit I) with a low-spin heme b and a high-spin heme b_3/Cu_B binuclear center, and two membrane-anchored *c*-type cyts (CcoO/cyt c_o and CcoP/cyt c_p). It has a unique active site possibly conferring a very high affinity for its substrate, oxygen (Wang et al., 1995). Its structural genes (*ccoNOQP*) have been cloned and sequenced from two different *R. capsulatus* strains (Thoeny-Meyer et al., 1994; Koch et al., 1997), and the existence of a similar cyt *c* oxidase has also been demonstrated in several other bacteria, including *P. denitrificans* (de Gier et al., 1996), *R. sphaeroides* (Garcia-Horsman et al., 1994) and some *Rhizobia*, where their homologues have been named *fixNOQP* (Preisig et al., 1993). The purified cyt cbb_3 oxidase of *B. japonicum* exhibits a high affinity for oxygen, and is required to support respiration under oxygen-limited growth during symbiotic nitrogen fixation (Preisig et al., 1996).

4.2. NADI⁻ Mutants of *R. capsulatus*

The presence of only one-type of cyt *c* oxidase in *R. capsulatus* allows efficient utilization of the NADI (α-*NA*phthol + *DMPD* + O_2 → *I*ndophenol blue + H_2O) reaction (Keilin, 1965) for identification of mutations affecting cyt cbb_3 oxidase activity. Marrs and Gest (1973) have reported the first *R. capsulatus* mutants defective in Res electron transport chain. Of these mutants M5 was unable to catalyze the NADI-reaction and unable to grow by respiration, hence deficient in both terminal oxidases. Two additional NADI⁻ mutants, M7G and M4, which were proficient in both Res and Ps growth (thus defective only in the cyt *c* oxidase) were also reported. Our finding that *R. capsulatus* cyt *c* oxidase is a novel type cyt *c* oxidase (Gray et al., 1994) incited us to study the molecular nature of the defects in these mutants as well as to search for additional NADI⁻ mutants. About 30 000 mutagenized colonies of *R. capsulatus* wild type strain MT1131 were screened after EMS mutagenesis, and 25 independent mutants unable to perform the NADI reaction at a wild-type level but proficient in both Res and Ps growth were retained. These mutants, together with M7G and M4 were analyzed for their membrane-bound *c*-type cyt profiles, the presence of subunit I (CcoN) antigen and their TMPD induced oxygen consumption activities (Table 2). Membranes derived from most of these mutants were devoid of all subunits of cyt cbb_3 oxidase, with the exception of M7G where the subunits CcoN and CcoO/cyt c_o were readily detectable. On the other hand, in some mutants like BK5, all subunits were present at, or close to, wild type level despite their NADI⁻ phenotype. Finally, the NADI⁻ˢˡᵒʷ mutants MR2 and IJ1 had reduced amounts of CcoN/subunit I, CcoO/cyt c_o and CcoP/cyt c_p in addition to their lower cyt c_1 and c_y contents. When oxygen consumption rates of membranes were measured polarographically in the presence of ascorbate and TMPD (Table 2), with the exception of MR2 and IJ1, less than 5% of the wild type activity was detectable in all mutants, confirming their cyt cbb_3 oxidase-negative phenotype. MR2 and IJ1 showed about 20% of the wild type activity, which was consistent with their NADI⁻ˢˡᵒʷ phenotype. Furthermore, reduced amounts of cyt c_1 in both mutants resulted in a reduced cyt bc_1 complex activity (about 20% of that of a wild type strain). However, whether their apparent deficiency in cyt c_y confers them a Ps⁻ phenotype in the absence of the cyt c_2 remains to be seen. In any event, the overall data revealed two different NADI phenotypes (NADI⁻ and NADI⁻ˢˡᵒʷ) and three different subunit profiles, in addition to that of M7G (*i.e.*, presence of CcoN/subunit I and CcoO/cyt c_o in the absence of CcoP/cyt c_p), which was exceptional.

4.3. Identification of the Genes Complementing the NADI⁻ Mutants

Plasmids complementing appropriate NADI⁻ mutants to a NADI⁺ phenotype were isolated using transferable chromosomal libraries and characterized. Sequence analyses of

Table 2. Phenotypic and genetic characterization of NADI⁻ mutants of *R. capsulatus*

Strain	Oxygen uptake-activity (%)[a]	Complementation			CcoN[c]	Cytochrome *c* profile[d]					
		pOX15[b]	pBK1	pMRC		c_p	c_1	c_y	c_o	c_2	
Wild-type	100	N/A[g]	N/A	N/A	+	+	+	+	+	+	
Class I											
M4	1.8	+	−	−	−	−	+	+	−	+	
MR1	1.1	+	−	−	−	−	+	+	−	+	
M7G	2.1	+	−	−	+	−	+	+	+	+	
MG1	0.3	+	−	−	−	−	+	+	−	+	
OH2	1.1	+	−	−	−	−	+	+	−	+	
DB8	0	+	−	−	−	−	+	+	−	+	
GK32	0	+	−	−	−	−	+	+	−	+	
Class II											
BK5	0.7	−	+	−	+	+	+	+	+	+	
BK6	0.9	−	+	−	+	+	+	+	+	+	
SS33	0.6	−	+	−	−	−	+	+	−	+	
SS1[e]	3.9	−	+	−	(−)	(−)	+	+	(−)	+	
SS2[e]	3.9	−	+	−	(−)	(−)	+	+	(−)	+	
Class III											
MR2[f]	17	−	−	+	(+)	(+)	(+)	−	(+)	(+)	
IJ1[f]	19	−	−	+	(+)	(+)	(+)	−	(+)	(+)	

[a]Oxygen uptake activity was measured polarographically using ascorbate and TMBZ as substrates. 100% activity corresponds to 34. 5 μmol O_2/h per mg protein.
[b]In addition to the mutants listed above, OH1, SS24, SS25 and IW2 were also classified as class I mutants since they were complemented by pOX15.
[c]Detected by polyclonal antibodies against CcoN.
[d]Based on TMBZ-stained SDS-PAGE.
[e]Traces of CcoN, CcoP and CcoO were detectable in SS1 and SS2.
[f]Small amounts of CcoN, CcoP and CcoO were detectable in MR2 and IJ1.
[g]Not applicable.

the plasmid pOX15, complementing M4, MR1, M7G, OH2, MG1, GK32 and DB8, revealed that it carried the structural genes of cyt cbb_3 oxidase (*ccoNOQP*) (Figure 3). Among these *ccoNOQP* mutants two different phenotypes were observed (Table 2). The mutation in M7G, which contained CcoN and CcoO/cyt c_o, was identified to be a TGG to TGA single base pair change corresponding to position 267 of CcoP/cyt c_p, and resulting in a carboxyl-terminally truncated version of this gene product. However, in three additional CcoP/cyt c_p mutants none of the subunits of cyt cbb_3 oxidase were detectable. The absence of all subunits of cyt cbb_3 oxidase was also characteristic for the *ccoN* mutant M4 and the *ccoO* mutant MR1 (Table 2). These findings indicated that all subunits are required for the presence of an active cyt cbb_3 oxidase in the membranes.

Finally, upstream of all *ccoNOQP* clusters from various species analyzed thus far, a conserved open reading frame encoding 277 amino acids (ORF277) is present (de Gier et al., 1996; Zeilstra-Ryalls and Kaplan, 1995) (Figure 3). However, since two ORF277 insertion mutations were found to have no effect on either the cyt cbb_3 activity or the ability of *R. capsulatus* to grow under Ps and Res growth conditions, an essential role for this gene product on cyt cbb_3 oxidase activity in this species seems unlikely.

The second class of mutants was complemented by pBKIC, carrying the *ccoGHIS* cluster, located downstream of *ccoNOQP* and presumably encoding a Cu-specific transport system (Figure 3). Functional description of these gene products is currently based on significant similarities between CcoI and copper uptake ATPases (Solioz and Vulpe, 1996), and the identification of cysteine-rich motifs within CcoG, which might bind [4Fe-

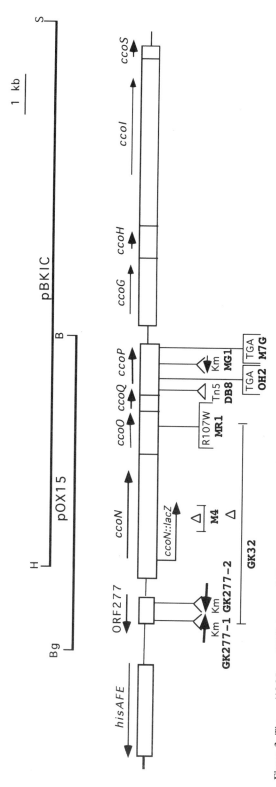

Figure 3. The *ccoNOQP-ccoGHIS* clusters of *Rhodobacter capsulatus*. The nature and location of the mutations found in the NADI⁻ mutants M4, MR1, DB8, OH2, MG1 and M7G are shown. The plasmids pOX15 and pBK1C contain the *ccoNOQP* and *ccoGHIS* clusters, respectively. GK32 is a deletion-insertion mutant lacking the *cbb₃* oxidase activity and *hisAFE* correspond to several histidine biosynthesis genes. See the text for the other genes.

4S] clusters. No recognizable motifs or similarities to known proteins were found for CcoH or CcoS, thus their functions remain unknown. Again, two different phenotypes were observed for this class of mutants complemented by pBKIC. In *ccoI* mutants (SS33, SS1 and SS2) extremely small amounts, if any, of the cyt cbb_3 oxidase subunits were detectable. Interestingly, this phenotype was also observed when a wild type strain was grown on a Cu-free medium, supporting a possible role of CcoI in Cu-transport. On the other hand, the *ccoS* mutants BK5 and BK6 apparently assembled an inactive cyt cbb_3 oxidase since all subunits of this enzyme were present in the membranes. However, attempts to reactivate this cyt cbb_3 oxidase by incubating membranes of this mutant with $CuSO_4$ were unsuccessful.

The mutation(s) in the third class of mutants (MR2 and IJ1) had a more general effect on all membrane bound cyts. They decreased the amount of not only the cyt cbb_3 oxidase but also the cyt bc_1 complex and cyt c_y, and yet the plasmid pMRC, isolated from a chromosomal library, was able to restore simultaneously all these effects in both mutants. Moreover, detection of wild-type level ß-galactosidase activity in merodiploids carrying a *lacZ::ccoN* translational fusion under both Ps and Res growth conditions on both enriched and minimal media, indicated that these mutations had no effect on the expression of *ccoNOQP*, but they rather affected the assembly or biogenesis of the cyt cbb_3 oxidase. Finally, among the 25 NADI⁻ mutants analyzed, five remained not complemented by either pOX15, pBKIC or pMRC, suggesting that they may define additional genes required for the production of an active cyt cbb_3 oxidase in *R. capsulatus*.

5. SUMMARY AND FUTURE OUTLOOK

In summary, the biogenesis of cyt complexes is a very interesting and yet a highly complex biological process which proves to be an extremely fertile ground of research. Our findings have already revealed that the NADI phenotype provides a reliable screen for the isolation of mutants defective in various steps of this process, and that the assembly of *R. capsulatus* cyt cbb_3 oxidase requires, in addition to the *ccoNOQP* and *ccoGHIS* clusters described in this work, at least two additional sets of gene products that are now emerging. Their studies should further contribute to our understanding of how the cyt complexes reach their maturity in bacteria.

ACKNOWLEDGMENTS

Work in this laboratory was supported by grants GM 38237 from NIH and 91ER20052 from DOE.

REFERENCES

Berry, E.A., Zhang, Z., Huang, L.S., Chi, Y-I., and S-H. Kim, 1997. *Biophysical Journal A.* p. 21
Bott, M., Ritz, D., and H. Hennecke. 1991. *J. Bacteriol.* **173:** 6766–6772.
Daldal, F. 1988. *J. Bacteriol.* **170:** 2388–2391
Daldal, F., Cheng, S., Applebaum, J., Davidson, E., and R. C. Prince. 1986. *Proc. Natl. Acad. Sci. U.S.A.* **83:** 2012–2016.
Donohue, T.J., McEwan, A.G., van Doren, S. and A. C. Crofts. 1988. *Biochemistry* **27:** 1918–1925
Garcia-Horsman, A., E. Berry, J.P. Shapleigh, J.O. Alben, and R.B. Gennis. 1994. *Biochemistry* **33:** 3113–3119.

de Gier, J.-W., M. Schepper., W.N.M. Reijnders, S.J. van Dyck, D.J. Slotboom, A. Warne, M. Saraste, K. Kraab, M. Finel, A.H. Stouthamer, R.J.M. van Spanning, and J. van der Oost. 1996. *Mol. Microbiol.* **20**: 1247–1260.

Gray, K. A., E. Davidson and F. Daldal. 1992. *Biochemistry* **31**: 11864–11873.

Gray, K.A., M. Grooms, H. Myllykallio, C. Moomaw, C. Slaughter, and F. Daldal. 1994. *Biochemistry* **33**: 3120–3127.

Hochkoeppler, A., Jenney, F. E., Lang, S. E., Zannoni, D., and F. Daldal. 1995. *J. Bacteriol.* **177**: 608–613.

Jenney, F. E., and F. Daldal. 1993. *EMBO J.* **12**: 1283–1292.

Jenney, F. E., Prince, R. C., and F. Daldal. 1994. *Biochemistry* **33**: 2496–2502.

Jenney, F.E., Prince, R.C., and F. Daldal. 1996. *Biochim. Biophys. Acta* **1273**: 159–164

Keilin, D. 1966. The history of cell respiration and cytochrome. Cambridge University Press, Cambridge.

Klemme, J.-H., and H.G. Schlegel. 1969. *Arch. Mikrobiol.* **68**: 326–354.

Koch, H.-G., Hwang, O., and F. Daldal. 1997. *J. Bacteriol.,* submitted

Marrs, B. L., and H. Gest. 1973. *J. Bacteriol.* **114**: 1045–1051.

Moser, C. C. and P. L. Dutton. 1988. *Biochemistry.* **27**: 2450–2461.

Martinez, S. E., D. Huang, A. Szczepaniak, W. A. Cramer and J. L. Smith. 1994. *Structure* **2**: 95–105

Myllykallio, H., Jenney, F. E., Moomaw, C. R., Slaughter, C. A., & Daldal, F. 1997. *J. Bacteriol.* **179**: 2623–2631.

Overfield, R. E. and C. A. Wraight. 1980. *Biochemistry* **19**: 3322–3327.

Preisig O., D. Anthamatten, and H. Hennecke. 1993. *Proc. Natl. Acad. Sc.* USA **90**: 3309–3313.

Preisig O., R. Zufferey, L. Thöny-Meyer, C. Appleby, and H. Hennecke. 1996. *J. Bacteriol.* **178**: 1532–1538.

Prince, R. C., Davidson, E., Haith, C. E., and F. Daldal. 1986. *Biochemistry* **25**: 5208–5214.

Solioz, M., and Vulpe, C. 1996. *TIBS* **21**: 237–241

Tiede, D. M. and P. L. Dutton. 1993 In *"The Photosynthetic Reaction Center."* Vol. 1, pp. 257–288. Academic Press. New York.

Thöny-Meyer, L, C. Beck, O. Preisig, and H. Hennecke. 1994. *Mol. Micro.* **174**: 705–716.

Turba, A., Jetzek, M., and B. Ludwig. 1995. *Eur. J. Biochem.* **231**: 259–265.

Wang, J., K.A. Gray, F. Daldal, and D.L. Rousseau. 1995. *J. Am. Chem. Soc.* **117**: 9363–9364.

Zannoni, D., R.E. Blankenship, M.T. Madigan and C.E. Bauer (Eds.), 1995. In Anoxygenic Photosynthetic Bacteria. Kluwer Academic Publishers, Dordrecht, Netherlands.

Xie, D., C-A Yu, H. Kim, J-Z. Xia, A. Kachurin, L. Zhang, L. Yu and J. Deisenhofer. 1997. *Science* **277**: 60–66.

Zeilstra-Ryalls, J.H., and S. Kaplan. 1995. *J. Bacteriol.* **177**: 6422–6431.

34

ON THE ROLE OF SOLUBLE REDOX CARRIERS ALTERNATIVE TO CYTOCHROME c_2 AS DONORS TO TETRAHEME-TYPE REACTION CENTERS AND CYTOCHROME OXIDASES

Alejandro Hochkoeppler,[1] Ilaria Principi,[1] Patrizia Bonora,[1] Stefano Ciurli,[2] and Davide Zannoni[1]

[1]Department of Biology
University of Bologna
Via Irnerio 42, 40126 Bologna, Italy
[2]Institute of Agricultural Chemistry
University of Bologna
Viale Berti-Pichat 10, 40127 Bologna, Italy

1. INTRODUCTION

In bacterial phototrophs, the light-induced oxidation of a reaction center (RC) bacteriochlorophyll dimer (P^+) ultimately reduces a quinone pool (Q-pool) which is subsequently oxidized by a cytochrome bc_1 complex. To generate chemical energy from subsequent photons of light, P^+ must be reduced to the ground state. This cyclic electron transport is linked to formation of a proton electrochemical gradient ($\Delta\mu_H^+$) which is used to drive ATP synthesis.

The RC of purple bacteria is the best understood complex from both structural and functional aspects. In its minimal version it is formed by similar 30 kDa core membrane-spanning polypeptides which bind the bacteriochlorophyll (BChl)-special pair and the primary electron acceptor quinones (Qa) [1]. There is considerable similarity in the core structure of the RC complex among different bacterial species [2]. Interestingly, the architecture of the core subunits of cyanobacterial and plant PSII complexes are similar to that of the purple bacterial RC [3]. However several other bacterial species of phototrophs contain an additional 45 kDa tetraheme cytochrome c-subunit which is the immediate electron donor to the special BChl-pair. These types of photosynthetic (PS) apparatuses follow the general model, with a few exceptions, that cytochrome c_2 is missing in species that contain a membrane-bound tetraheme c/RC [4]. From current analysis it appears that tetraheme c-type RCs are more common than those where a soluble electron donor, i.e. cytochrome c_2,

The Phototrophic Prokaryotes, edited by Peschek *et al.*
Kluwer Academic / Plenum Publishers, New York, 1999.

interacts directly with the core subunit. Nonetheless, these latter photosynthetic systems (those present in *Rhodobacter capsulatus*, *Rb. sphaeroides* and *Rhodospirillum rubrum*) are the most well studied and until recently, they have been taken as examples of anoxygenic-type photosynthesis [5].

The facultative phototrophs, *Rhodoferax fermentans* and *Rhodospirillum salinarum* do not express cytochrome c_2 and contain RCs with membrane-bound tetraheme-cytochromes c acting as immediate electron donors to the RC-special pairs [5]. We have recently shown that the high-potential iron-sulfur protein, HiPIP, of *Rf. fermentans* is kinetically competent in reducing one of the c-type hemes (c-556) of the tetraheme-RC [6]. Work in press [7] indicates that a soluble cytochrome c_8 might have, in addition to HiPIP, a role in photosynthesis of *Rf. fermentans* while the role of the two HiPIPs (HiPIP-iso1 and HiPIP-iso2) isolated from *Rs. salinarum* needs to be defined (see however below).

This article will first address some recent and early results on the role of cytochromes and HiPIPs acting as electron carriers alternative to cytochrome c_2 in facultative phototrophs, and then discuss new data on the role of HiPIP in oxidative electron transport of the halophilic phototroph *Rs. salinarum*. In recent years a few other examples of HiPIP involved in respiratory oxidative processes have been reported, i.e. those related with the HiPIP-type cluster core in membranes from the thermohalophilic aerobe, *Rhodothermus marinus* [8] and the HiPIP from aerobically-dark grown cells of *Rf. fermentans* [9].

2. RESULTS AND DISCUSSION

2.1. The Case of *Rb. capsulatus*: Soluble-Cytochrome c_2 and Membrane-Attached Cytochrome c_y, Coexist

As stated in §1, the most detailed information among the photosynthetic systems is available on the interaction between cytochrome c_2 and those RCs lacking a membrane-bound c-tetraheme. Genetic analysis of cytochrome c_2 function is particularly advanced in *Rb. sphaeroides* and *Rb. capsulatus*, also because these bacterial species grow under conditions where cytochrome c_2 is not involved, i.e. aerobic growth [10]. During the last decade, 'gene knock-out' experiments were carried out in *Rb. sphaeroides* and *Rb. capsulatus* under permissive aerobic-growth conditions which yielded mutants lacking only cytochrome c_2 (*cycA* gene). Notably, a *Rb. sphaeroides* c_2^- mutant was unable to grow by photosynthetic growth [11] but a *Rb. capsulatus* c_2^- mutant was PS$^+$ [12]. *Rb. sphaeroides* c_2^- mutants can give rise to PS$^+$ pseudorevertants [13] called *spd* for 'suppressor of the photosynthetic deficiency'. These *spd* pseudorevertants overproduced a 'new soluble cytochrome c_2 named iso-cytochrome c_2 (*cycI* gene) [14]. The mutants lacking iso-cytochrome c_2 are not defective for PS, aerobic or anaerobic dark growth so that the actual function of iso-cytochrome c_2 in wild type *Rb. sphaeroides* remains unknown. Analyses of the *Rb. capsulatus* c_2^- mutant constructed by Daldal and co-workers indicated a membrane bound cytochrome c (initially named c_x) as being responsible for the cytochrome c_2 independent PS growth of *Rb. capsulatus* [15–17]. Recent genetic [18] and biochemical [19] studies identified this component as a novel membrane-attached (MA) cytochrome c, named c_y, encoded by *cycY* gene. Deletion of *cycY*, like that of *cycA*, had no major effect on the PS growth of *Rb. capsulatus* but inactivation of both, yielded a PS$^-$ mutant which could be complemented to PS$^+$ growth by either *cycA* or *cycY*. Estimates of the cellular abundance of cytochrome c_y are not yet available in different *Rb. capsulatus* genetic backgrounds but it has been shown that cytochrome c_y concentration is dependent on nutrient supply [17]. Although cytochrome c_y-like components are more common than initially anticipated as they have been encountered in facultative anaerobes like *Paracoccus denitrificans* and *Bradyrhizobium japonicum* [20,21], their presence in phototrophs lacking soluble cytochrome c₂ (*Rhodoferax, Rhodocyclus, Chromatium* and *Ectothiorodospira* species) has to be demonstrated.

Studies by Daldal and collaborators demonstrated that cytochrome c_y is the prototype of a new class of cytochromes [18,22]. Cytochrome c_y is a two domain protein constituted of a (C)-terminal cytochrome *c* domain that is homologous to many soluble *c*-type cytochromes, and an amino(N)-terminal domain attaching the heme domain to the periplasmic face of the membrane. The N-terminal domain of cytochrome c_y is formed by a proximal 'anchor' and a distal 'linker' subdomains. The anchor is homologous to prokaryotic signal sequences, but it remains uncleaved during translocation, leaving it attached to the membrane. These unusual features of cytochrome c_y raise important issues related to its binding and electron transfer to its redox partners, e.g. antenna complexes, as well as to its regulated expression in response to oxygen and light.

In early studies on the topology of cytochrome c_y, a cytochrome c₂ derivative attached to the membrane (MA-c₂) *via* the N-terminal domain of cytochrome c_y and a 'soluble' derivative of cytochrome c_y, (S-c_y) were constructed (Myllikallio and Daldal, personal communication). These studies demonstrated that the N-terminal domain of cytochrome c_y is capable to anchor a soluble cytochrome c₂ to the membrane. On the other hand cytochrome S-c_y is not produced in adequate amounts to support the photosynthetic growth (PS⁺) of FJ2, a mutant deficient in both cytochromes c₂ and c_y; however, its Ps⁺ revertants were readily observed, but yet, their molecular nature(s) remains to be determined.

Cytochrome c_y is a direct electron donor to the *cbb₃* oxidase of *Rb. capsulatus* [23]. In the absence of the quinol oxidase dependent alternative respiratory pathway, *Rb. capsulatus* mutants lacking either cytochrome c₂ or cytochrome c_y (M6G-G4/S4 or SL1 mutants) are still able to grow by respiration in a manner absolutely dependent on the *bc₁* complex. This finding demonstrated that either one of these two cytochromes is sufficient to support the *bc₁*-dependent respiratory growth. Thus, as in the case of photosynthesis [18], cytochrome c₂ and cytochrome c_y are alternate electron carriers in respiration and only the inactivation of both of them, as in FJ2, makes the respiratory electron flow insensitive to inhibitors of the *bc₁* complex [23].

2.2. Are HiPIPs Electron Donors to Photochemical Reaction Centers? The Case of *Rhodoferax fermentans*

Bacterial species in the genera *Rhodocyclus (Rc), Rhodospirillum (Rs), Rhodoferax (Rf), Chromatium (Ch)* and *Ectothiorhodospira (Ec)* contain RC complexes with a tetra-

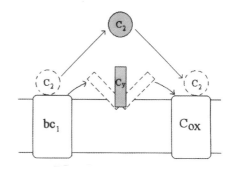

Figure 1. Speculative model of the electron transport between the *bc₁* complex and the cytochrome *c* oxidase in *Rb. capsulatus.* Interrupted and continuous circles symbolize complexed- and free-forms of cytochrome c₂, respectively, while interrupted and continuous rectangles symbolize the oscillating movement of cytochrome c_y distal "linker" allowing electron transfer between the *bc₁* and cytochrome *c* oxidase complexes.

heme cytochrome *c* as direct electron donors to the bacteriochlorophyll dimer [24]; in addition, most species appear to lack soluble proteins related to cytochrome c_2. These observations led to suggest that either a soluble mediator is not required (cytochrome c_y present?) or that proteins other than cytochromes of *c*-type must be considered for this role. A candidate group of electron donors suitable for the role of carriers alternative to cytochrome c_2 is that of the high-potential iron-sulfur proteins, HiPIP. These soluble carriers are relatively abundant among phototrophs [25]; thus they have been extensively investigated as electron transfer-models [4, 26, 27] and their structural/spectroscopic features are well defined [28]. Although the HiPIPs show optimal thermodynamic properties to couple the bc_1 and reaction centers [29], their role in electron transfer has been unambiguously demonstrated only quite recently in whole cells of *Rubrivivax gelatinosus* [30] and in membranes and isolated RCs from *Rhodoferax fermentans* [31, 6] a genus closely related to *Rhodocyclus* [32]. The four hemes of the *Rf. fermentans* RC-subunit have $Em_{7.0}$ of +354 mV (*c*-556), +294 mV (*c*-560), +79 mV (*c*-551) and 0 mV whereas the HiPIP has an $Em_{7.0}$ of +351 mV. We have recently demonstrated that the HiPIP of *Rf. fermentans* is a very efficient electron carrier to the photosynthetic RC [6]. Indeed, the rapid photoxidation of heme *c*-556 belonging to the reaction center is followed, in the presence of HiPIP, by a slower reduction having a second-order rate constant of 4.8×10^7 M^{-1} s^{-1}. The limiting value of k_{obs} at high HiPIP concentration is 95 s^{-1}. The rate constant of a faster phase, determined at 556 and 425 nm, is approximately 3×10^5 s^{-1}. This value is not dependent on HiPIP concentration, indicating that it is related to a first-order process. Whether or not members of the HiPIP family are competent electron donors to reaction centers in other photosynthetic bacteria remains to be demonstrated.

The complete aminoacid sequence of the HiPIP from *Rf. fermentans* has recently been determined [33]. The sequence contains 75 residues, with 11 positive charges, 10 negative charges, and one histidine residue. The molecular mass of the apo-HiPIP is 7849.64 Da with a multiple sequence alignment conservation of Tyr19 and Gly75 (*Chromatium vinosum* numbering) in addition to the four [Fe_4S_4]-bound cysteines. The HiPIP from *Rf. fermentans* is largely similar (57% similarity) to the HiPIP from the β-1 subgroup photosynthetic bacterium *Rubrivivax gelatinosus*.

2.3. Cytochrome c_8: A Possible Electron Donor to the Photochemical Reaction Center of *Rhodoferax fermentans*

The presence of a soluble high-potential cytochrome *c*-551 in extracts of the purple bacterium *Ch. vinosum* has been first reported by Bartsch [34]. Van Grondelle et al. [35] and Knaff et al. [36] reported that cytochrome *c*-551 is able to connect the bc_1 complex to the RC, having an E_m of +240 mV and being periplasmically located [37]. Its quantity is only about 15–20% of the level of HiPIP, cytochrome *c'*, flavo-cytochrome *c* or ferredoxin. In its isolated form, cytochrome *c*-551 is reduced during the flavo-cytochrome *c* dependent oxidation of sulfide [38]. Cytochrome *c*-551 from *Ch. vinosum* has recently been sequenced [39] and found to belong to the cytochrome c_8 class (formerly called *Pseudomonas* cytochrome *c*-551). Whether *Ch. vinosum* cytochrome c_8 actually functions as a mediator of photosynthetic electron transport *in vivo* remains difficult to conclude. We have recently purified from photosynthetic cells of *Rf. fermentans* a soluble cytochrome *c* with $Em_{7.0}$ of +285 mV and Mr of 12 kDa [7]. Based on partial sequence determination of the N-terminus (AEVLAKAGCLACHTXI) the protein has been identified as highly similar to cytochromes of the cytochrome c_8 class. Unfortunately, the low concentration of *Rf. fermentans* cytochrome c_8 has greatly limited our experimental capacity to

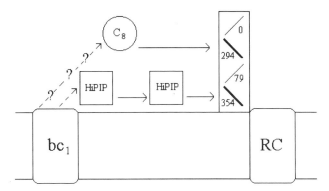

Figure 2. Tentative model of photosynthetic electron flow between the bc_1 complex and the tetraheme c-type reaction center of *Rf. fermentans*. Question marks symbolize the lack of data about the interaction between cytochrome c_8 and the bc_1 complex while the role of HiPIP in photocyclic electron flow is, at present, more than a working hypothesis (see text for details and Refs.). Numbers are the $E_{m7.0}$ of the four hemes depicted in their HLHL (High-Low-High-Low) arrangement.

define its physiological role by *in vitro* assay. Nonetheless, time-resolved spectrophotometric studies show that cytochrome c_8 reduces the tetraheme subunit of the photosynthetic RC, in a fast (sub-ms) and a slow (ms) phase. Competition experiments in the presence of both cytochrome c_8 and HiPIP show that c_8 photooxidation is decreased upon addition of HiPIP. These observations suggest that cytochrome c_8 and HiPIP might play alternative roles in photosynthetic electron transfer of *Rf. fermentans*.

2.4. HiPIP-iso1: A Possible Electron Donor to the Cytochrome Oxidase of aa_3-Type of *Rhodospirillum salinarum*

In a previous work [9] we have shown that membranes from aerobically dark-grown cells of *Rf. fermentans* catalyze significant levels of HiPIP oxido-reductase activities. This finding, along with the one indicating the possible role of a membrane-bound HiPIP in respiration of *Rhodothermus marinus* [8], raises the question whether the role of the HiPIP as carrier in oxidative electron flow is restricted to a few species or it is more widely distributed among those bacteria lacking soluble c-type cytochromes, e.g. halophilic phototrophs [25].

Extracts of cells of the halophilic facultative phototroph *Rs. salinarum* grown aerobically in the dark contain small amounts of cytochrome c' along with two HiPIPs (named HiPIP-iso1 and HiPIP-iso2) and a cytochrome c (c-551) with spectroscopic, thermodynamic, and concentration features identical to those present in light-grown cells [40, 41]. The isolated HiPIP-iso1 shows molar extinction coefficients at 385 nm and 453 nm of 16 and 18 mM^{-1} cm^{-1} in the reduced and oxidized forms, respectively (see spectra in Fig. 3).

The HiPIP-iso1 oxidation kinetics at 490 nm induced by membrane fragments at different HiPIP concentrations, are shown in Fig. 4 (panel A). The results clearly show that the rate constant of the process does not vary on HiPIP concentrations indicating a first-order kinetic. Similarly, the oxidation of reduced horse heart cytochrome c at 550 nm is a first-order process; in this case, however, a strong dependence on cytochrome c concentration is clearly apparent (Fig. 4 B).

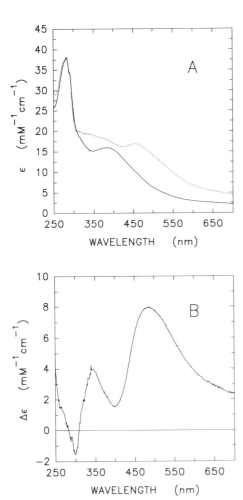

Figure 3. A. Absorption spectrum of the reduced (continuous line) and oxidized (dotted line) HiPIP-iso1; B. Oxidized-minus-reduced difference spectrum. Purification of the HiPIP-iso1 and absorption spectra were performed as in [9].

In Figure 5, the k_{obs} as a function of variable concentrations of either cytochrome c or HiPIP-iso1 are shown. Notably, the k_{obs} of the HiPIP is approx 0.35×10^{-2} s^{-1}, a value which is comparable with the one reported in *Paracoccus denitrificans* for oxidation of reduced cytochrome c-550, the physiological donor to the aa_3-type oxidase (1×10^{-2} s^{-1}) [42]. In both types of bacterial membranes the use of mitochondrial c-type cytochrome instead of their respective physiological donors (HiPIP or c-550) raises the k_{obs} to 3×10^{-2} s^{-1} and 4.6×10^{-2} s^{-1} in *P. denitrificans* and *Rs. salinarum*, respectively (see [42] and Fig. 5).

In Figure 6 the reduction kinetics of both HiPIP and mitochondrial cytochrome c, each used at 24 µM concentration, as catalyzed by membranes from aerobically dark-grown *Rs. salinarum*, are shown. Apparently the HiPIP reduction kinetic is three times faster than that seen with cytochrome c, this value being actually twice as much by considering the relative bc_1 molar concentrations in both samples. Thus, although the mitochondrial c-type cytochrome is more efficient than the HiPIP in donating to the aa_3-type oxidase of *Rs. salinarum* [41], the HiPIP is a better electron acceptor from the bc_1 complex. Further experiments, to be reported elsewhere, show that the HiPIP is indeed a functional component of the respiratory electron transport chain of *Rs. salinarum*.

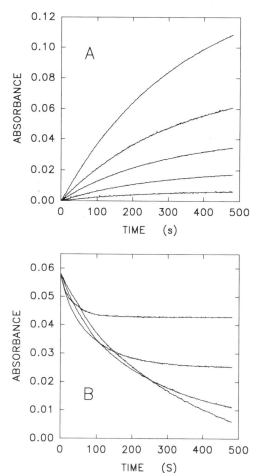

Figure 4. A. Oxidation kinetics of increasing concentrations (1, 3, 6, 12, 24 µM, bottom to the top) of HiPIP-iso1 catalized by membrane fragments of aerobically grown *Rs. salinarum* (0.2 nM cytochromes aa_3) at 490 nm; B. Oxidation kinetics of increasing concentrations (1, 3, 6, 9 µM, top to the bottom) of horse-heart cytochrome *c* (SIGMA Co., USA) catalyzed by membranes from *Rs. salinarum* (0.25 nM cytochromes aa_3). Cell growth and membrane isolation procedure were performed as described by Moschettini et al. [41]; oxidation kinetics were measured at 25°C using a Jasco 7850 spectrophotometer.

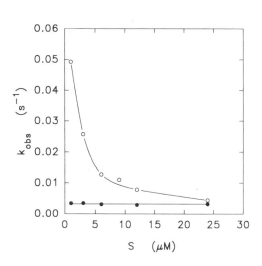

Figure 5. Variation of the k_{obs} as a function of variable concentrations of horse heart cytochrome *c* (empty points) and HiPIP (full points).

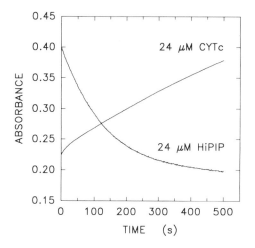

Figure 6. Reduction kinetics of horse heart cytochrome *c* and HiPIP both used at 24 μM concentration as catalyzed by membranes of *Rs. salinarum*. Experiments performed in the presence of 50 μM KCN (see text for further details). In cytochrome *c*- and HiPIP-assays, the concentrations of the bc_1 complexes were 12 nM and 6.7 nM, respectively.

3. CONCLUSIONS

The discovery of two structurally distinct electron carriers operating between the RC, bc_1 complex, and cytochrome *c* oxidase of *Rb. capsulatus* raises the question of how the soluble cytochrome c_2 and the membrane-attached cytochrome c_y interact with their physiological partners. Hopefully, the mutants recently constructed by Daldal and coworkers (Myllikallio and Daldal, personal communication, see §2.1) carrying a soluble derivative of cytochrome c_y (S-c_y) and a membrane-attached cytochrome c_2 (MA-c_2) will help us in clarifying this intriguing aspect. As far as we know, the kinetic data available tend to show that cytochrome c_y is less accessible to the respiratory substrates than soluble cytochrome c_2 [23]; in addition, early results by Jones et al. [16] indicated that soluble cytochrome c_2 is photooxidized more rapidly than cytochrome c_y (formerly named cytochrome c_x). Thus, although it is fairly established that cytochromes c_2 and c_y are equally effective in both photosynthesis and respiration of *Rb. capsulatus*, their actual physiological role is far from being totally clear.

The facultative phototrophs, *Rhodoferax fermentans* and *Rhodospirillum salinarum* do not express cytochrome c_2 and contain RCs with membrane-bound tetraheme-cytochromes *c* acting as immediate electron donors to the RC-special pairs. The high-potential iron-sulfur protein, HiPIP, of *Rf. fermentans* is kinetically competent in reducing one of the *c*-type hemes (*c*-556) of the tetraheme-RC [6]. Work in press [7] indicates that a soluble cytochrome c_8 (Mr of 12 kDa, E $_{m7.0}$ of +285 mV) might have, in addition to HiPIP, a role in photosynthesis of *Rf. fermentans*. From this and previous results, we have recently proposed a working hypothesis [7] in which a whole series of different soluble electron carriers might be involved in photosynthetic electron transfer to the *c*-type tetraheme. Each soluble redox protein would then be able to transfer electrons to an heme group of the tetraheme subunit at appropriate reduction potentials to optimize the photosynthetic growth of the organism in relation with available substrates.

The role in photosynthetic electron transfer of the two HiPIPs isolated from *Rs. salinarum* needs to be defined. At this experimental stage, it can only be concluded that the HiPIP-iso1 is involved in electron transfer to the cytochrome oxidase complex (this study). In addition, data to be reported elsewhere, support a role of the HiPIP-iso1 as electron carrier between the bc_1 complex and the aa_3-type cytochrome oxidase of *Rs. salinarum*. Notably, this is the first report (as far as we know) of a HiPIP oxido-reductase activity in membranes from halophilic photosynthetic bacteria.

ACKNOWLEDGMENTS

D.Z. likes to thank the Organizing Committee of the IX ISPP (6–12 Sept, 1997, Vienna, Austria) for the kind invitation and for financial support (Italia/Austria Scientific Bilaterial Treaty). D.Z. and S.C. are also grateful to Consiglio Nazionale delle Ricerche (CNR) and Ministero Università e Ricerca Scientifica e Tecnologica (MURST) of Italy for funding this research.

REFERENCES

1. Deisenhofer, J., Epp, O., Miki, K., Huber, R. and Michel, H. (1985) Nature 318, 618–624
2. Chang, C.H., El-Kabbani, O., Tiede, D., Norris, J. and Shiffer, M. (1991) Biochemistry 30, 5352–5360
3. Michel, H. and Deisenhofer, J. (1988) Biochemistry 27, 1–7
4. Meyer, T.E., Bartsch, R.G., Cusanovich, M.A. and Tollin, G. (1993) Biochemistry 32, 4719–4726
5. Blankenship, R.E., Madigan, M.T. and Bauer, C.E., eds (1995) Anoxygenic Photosynthetic Bacteria. Advances in Photosynthesis Series, Kluwer Acad.Publ. The Netherlands
6. Hochkoeppler, A., Zannoni, D., Ciurli, S., Meyer, T.E., Cusanovich, M.A. and Tollin, G. (1996) Proc Natl Acad Sci USA 93, 6998–7002
7. Hochkoeppler, A., Ciurli, S., Venturoli, G. and Zannoni, D. (1997) Photosynt Res (in press)
8. Pereira, M.M., Antunes, A.M., Nunes, O.C., da Costa, M.S. and Teixeira, M. (1994) FEBS Letters 352, 327–330
9. Hochkoeppler, A., Kofod, P. and Zannoni, D. (1995) FEBS Letters 375, 197–200
10. Zannoni, D. (1995) in: Anoxygenic Photosynthetic Bacteria (Blankenship, R.E., Madigan, M.T., Bauer, C.E., Eds), pp. 949–971, Kluwer Acad.Publs. The Netherlands
11. Donohue, T.J., Mc Ewan, A.G. and Kaplan, S. (1986) J Bacteriol 168, 962–972
12. Prince, R.C., Davidson, E., Haith, C. and Daldal, F. (1986) Biochemistry 25, 5208–5212
13. Rott, M.A. and Donohue, T.J. (1990) J Bacteriol 172, 1954–1961
14. Fitch, J., Cannac, V.,Meyer, T.E., Cusanovich, M.A., Tollin, G., van Beeumen J., Rott, M.A. and Donohue, T.J. (1989) Arch Biochem Biophys 271, 502–507
15. Prince, R.C. and Daldal, F. (1987) Biochim Biophys Acta 894, 370–378
16. Jones, M.R., Mc Ewan, A.G. and Jackson, J.B. (1990) Biochim Biophys Acta 1019, 59–66
17. Zannoni, D., Venturoli, G. and Daldal, F. (1992) Arch Microbiol 157, 367–374
18. Jenney, F.E. and Daldal, F. (1993) EMBO J 12, 1283–1292
19. Jenney, F.E., Prince, R.E. and Daldal, F. (1994) Biochemistry 33, 2496–2502
20. Bott, N., Ritz D. and Hennecke, H. (1991) J Bacteriol 173, 6766–6772
21. Turba, A., Jetzek, M. and Ludwig, B. (1995) Eur J Biochem 885, 1–7
22. Zannoni, D. and Daldal, F. (1993) Arch Microbiol 160, 413–423
23. Hochkoeppler, A., Jenney, F.E., Lang, S.E., Zannoni, D. and Daldal, F. (1995) J Bacteriol 177, 608–613
24. Meyer, T.E. and Donohue, T.J. (1995) in: Anoxygenic Photosynthetic Bacteria (Blankenship, R.E., Madigan, M.T., Bauer, C.E., Eds), pp. 725–745, Kluwer Acad. Publs. The Netherlands
25. Bartsch, R.G. (1991) Biochim Biophys Acta 1058, 28–30
26. Aprahamian, G. and Feinberg, B.A. (1981) Biochemistry 20, 915–919
27. Jackman, M.P., Lim, M.-C. and Sykes, A.G. (1988) J Chem Soc Dalton Trans 2843–2850
28. Bertini, I., Ciurli, S. and Luchinat, C. (1995) Struct Bonding 83, 1–53
29. Meyer, T.E., Przysiecki, C.T., Watkins, J.A., Bhattacharyya, A., Simondsen, R.P., Cusanovich, M.A. and Tollin, G. (1983) Proc Natl Acad Scie USA 80, 6740–6744
30. Schoepp, B., Parot, P., Menin, L., Gaillard, J., Richaud, P. and Vermeglio, A. (1995) Biochemistry 34, 11736–11742
31. Hochkoeppler, A., Ciurli, S., Venturoli, G. and Zannoni, D. (1995) FEBS Lett 357, 70–74

32. Hirahishi, A., Hoshino, Y. and Satoh, T. (1991) Arch Microbiol 155, 330–336
33. Van Driessche, G., Ciurli, S., Hochkoeppler, A. and Van Beeumen, J.J. (1997) Eur J Biochem 244, 371–377
34. Bartsch, R.G. (1971) Methods in Enzymol 23A, 344–363
35. Van Grondelle, R., Duysens, L.N.M., van der Wel, J. and van der Wal, H.N. (1977) Biochim Biophys Acta 461, 188–201
36. Knaff, D.B., Whetstone, R. and Carr, J.W. (1980) Biochim Biophys Acta 590, 50–58
37. Gray, G.O., Gaul, D.F. and Knaff, D. (1983) Arch Biochem Biophys 222, 78–86
38. Davidson, M.W., Gray, G.O. and Knaff, D. (1985) FEBS Lett 187, 155–159
39. Samyn, B., De Smet, L., van Driessche, G., Meyer, T.E., Bartsch, R.G., Cusanovich, M.A. and van Beeumen, J.J. (1996) Eur J Biochem 236, 689–696
40. Meyer, T.E., Fitch, J.C., Bartsch, R.G., Tollin, G. and Cusanovich, M.A. (1990) Biochim Biophys Acta 1017, 118–124
41. Moschettini, G., Hochkoeppler, A., Monti, B., Benelli, B. and Zannoni, D. (1997) Arch Microbiol (in press)
42. Bolgiano, B., Smith, L. and Davis, H.C. (1988) Biochim Biophys Acta 933, 341–350

35

CHARACTERIZATION OF SOLUBLE FORMS OF THE MEMBRANE BOUND ISO-CYTOCHROME c2 FROM *RHODOBACTER CAPSULATUS*

T. E. Meyer, J. C. Fitch, G. Tollin, and M. A. Cusanovich

Department of Biochemistry
University of Arizona
Tucson, Arizona 85721

Soluble cytochrome c2 functions to couple the membrane bound cytochrome bc1 complex to photosynthetic reaction centers. When the cytochrome c2 gene is inactivated, *Rb. capsulatus* is still capable of photosynthetic growth due to the presence of a constitutively produced membrane bound isozyme of cytochrome c2 known as cytochrome cy (Jenney & Daldal, EMBO J. 12, 1283–1292 (1993)). We have now isolated small amounts of two forms of the membrane bound cytochrome c2 isozyme which are naturally soluble. Forms A and B separate on CM-cellulose chromatography. They have similar amino acid compositions, which are like the C-terminus of the translated isozyme gene and unlike that of cytochrome c2. The N-terminal sequence of form A is: EPAAPPPPAYVEVDPATITG... and of form B is: APPPPAYVEV..., indicating non-specific proteolytic cleavage at more than one site. The absorption spectra and redox potential of the two forms are like those of cytochrome c2. The kinetics of interaction of form B with the photosynthetic reaction center indicate that there is no stable complex formed and that the reaction is entirely second-order. The rate constant for iso-c2 is more than an order of magnitude smaller at low ionic strength and the electostatic effect is one half that of cytochrome c2. Thus, reaction centers show significant specificity for the natural electron donor.

1. INTRODUCTION

Cytochrome c2 is the electron donor to photosynthetic reaction centers in *Rhodobacter sphaeroides*. Thus, when the gene for cytochrome c2 is inactivated by mutagenesis, photosynthetic growth capacity is lost [1]. Spontaneous pseudo-revertants regain the ability to grow photosynthetically due to production of a cytochrome c2 isozyme [2–5]. The level of iso-c2 in wild type cells is only a few percent that of c2, whereas in the pseudo-

The Phototrophic Prokaryotes, edited by Peschek *et al.*
Kluwer Academic / Plenum Publishers, New York, 1999.

revertants, it is about 20–40% the level of wild type c2. The isozyme is not produced at sufficient levels to support photosynthetic growth in wild type cells but presumably functions in glutathione dependent formaldehyde oxidation [6]. The kinetics of oxidation of iso-c2 by photosynthetic reaction centers are similar to those of c2, although the binding is 40-fold weaker [7], consistent with greater specificity for the natural electron donor.

In contrast to the results with *Rb. sphaeroides*, knockout mutants of *Rb. capsulatus* cytochrome c2 remain capable of photosynthetic growth [8]. However, an isozyme of c2, known as cy, was also cloned from *Rb. capsulatus* [9]. Double knockout mutants were then shown to be photosynthetically incompetent, whereas addition of either c2 or iso-c2 genes restored growth [10]. Thus, c2 is the obligate electron donor in both *Rb. capsulatus* and *Rb. sphaeroides* and presumably in other species for which c2 is the dominant soluble cytochrome c. The *Rb. capsulatus* iso-c2 was solubilized with detergent and shown to have absorption spectra and redox potential similar to c2 [11]. We have now purified naturally soluble forms of *Rb. capsulatus* iso-c2 and compare wild type and iso-c2 from *Rb. capsulatus* and *Rb. sphaeroides*.

2. RESULTS AND DISCUSSION

2.1. Protein Isolation

Rb. capsulatus strain MT-G4/S4, ultimately derived from wild type strain B10, has the cytochrome c2 gene inactivated [8]. It was grown aerobically to stationary phase on a standard malate medium in a 14 liter fermenter connected to a titrimeter which continuously maintained the pH at 7 while introducing fresh medium. The cells were harvested after one week by ultrafiltration (Millipore Pellicon Cassette System), they were disrupted in a Ribi cell fractionator, and membranes also removed by ultrafiltration. The extract was concentrated and desalted by ultrafiltration and adsorbed to DEAE-Sephacel from 1 mM Tris-Cl, pH 8, 1 mM mercapto-ethanol, and eluted with 10 mM Tris-Cl, 15 mM NaCl. The protein was desalted and adsorbed to CM-cellulose from 1 mM phosphate, pH 5.5. Two major fractions were resolved by application of a 20–160 mM NaCl gradient in 10 mM phosphate pH 5.5. There was 1.5 times as much form B as form A (in order of elution) and about 5 micromoles in total. Although it did not significantly improve the already pure protein, a reverse ammonium sulfate gradient elution was carried out on Phenyl-Sepharose.

2.2. Protein Properties

The absorption spectra of the two forms of iso-c2 were virtually identical to one another and very similar to those of cytochrome c2. The amino acid compositions of the two forms were also similar to one another. They were different from that of cytochrome c2 [12], but were like that of the translated C-terminus of the iso-c2 gene [9]. The N-terminal sequence of form A was: EPAAPPPPAYVEVDPATITG... and of form B: APPPPAYVEV.... Form A is identical to positions 92–111 and form B to 95–104 of the translated sequence of the iso-c2 gene [9], indicating solubilization via endogenous non-specific proteolysis. The presence of an extra Glu in form A is consistent with its more acidic character and separation from form B on ion exchange chromatography. The redox potential is 360 mV, which is very similar to the 368 mV reported for c2 [13], and to that of detergent solubilized iso-c2 [11].

2.3. Quantity of Soluble Isozyme

Previously, it had been reported that c2 knockout mutants had no significant amount of soluble cytochrome c and double knockouts to which the isozyme gene was added in trans also failed to produce significant soluble protein [9]. However, the quantity of protein we have obtained is about 25% the level of c2 in wild type cells and is the best yield we have obtained. The levels vary from one preparation to another and average closer to 5–10%. It is possible that our relatively long growth conditions contributed to solubilization. Nevertheless, the quantity of soluble isozyme is comparable to the amount of isozyme which supports photosynthetic growth in *Rb. sphaeroides*, which raises the question whether soluble or membrane bound iso-c2 functions in photosynthesis in *Rb. capsulatus* knockout mutants. Previous work suggests that the membrane bound form is functional [14] although this deserves more study.

2.4. Interaction with Reaction Centers

The reaction of *Rb. sphaeroides* c2 with *Rb. sphaeroides* photosynthetic reaction centers is biphasic. A fast first order reaction ($\approx 10^6$ s^{-1}) is presumably due to tightly bound c2 ($K_d \approx 1$ μM) whereas a second-order reaction ($\approx 10^9$ M^{-1} s^{-1} at low ionic strength) is due to unbound protein [15–17]. The *Rb. sphaeroides* iso-c2 was found to be photosynthetically competent and has similar kinetic rate constants as native protein although binding is weaker [7]. It was also reported that *Rb. capsulatus* c2 reacts with *Rb. sphaeroides* reaction centers similarly to *Rb. sphaeroides* c2, although the binding constant is somewhat weaker [16]. Both *Rb. capsulatus* and *Rb. sphaeroides* c2 reactions show strong dependence on ionic strength, indicating a favorable electrostatic interaction due to complementary positive charges localized at the active site of c2 and negatively charged residues on the reaction center.

We were unable to measure a first-order rate constant for the reaction of *Rb. capsulatus* cytochrome c2 with *Rb. sphaeroides* reaction centers, but it is apparent from the decrease in amplitude of the second-order reaction with increasing protein concentration that a stable complex is formed (K_d 1.25 μM) which is as strong as that of *Rb. sphaeroides* c2. We obtained a second-order rate constant for *Rb. capsulatus* c2 of 4.2×10^8 M^{-1} s^{-1} at low ionic strength which is comparable to that reported previously [16]. Another indication that c2 forms a stable complex with reaction centers is a biphasic ionic strength effect in which the rate constant increases with ionic strength before decreasing at higher ionic strengths.

In contrast to the results with c2, the reaction of *Rb. capsulatus* iso-c2 with reaction centers is entirely second-order and no stable complex is formed. Thus, there is no significant change in amplitude of the absorbance transient with protein concentration and the ionic strength effect is simple. The second-order rate constant we obtained for *Rb. capsulatus* iso-c2 is more than an order of magnitude smaller at low ionic strength (3.6×10^7 M^{-1} s^{-1}) than that of c2. The electrostatic effect for iso-c2 was quantified according to Watkins et al [18], resulting in a fitting parameter, V_{ii} (related to the interaction energy), which is one half that of c2, i.e. -4.6 vs. -9.2. All else being equal, this suggests that there are only half as many charged residues at the site of electron transfer in iso-c2 as in c2. The rate constants extrapolated to infinite ionic strength are 1.4×10^6 for iso-c2 vs. 6.5×10^6 M^{-1} s^{-1} for c2. These values are sufficiently alike to suggest that both proteins are interacting at similar sites on the reaction center.

In conclusion, *Rb. capsulatus* iso-c2 is less reactive than c2 by an order of magnitude at low ionic strength, it does not form a stable complex with reaction centers, and it has fewer charged residues at the active site. Taken together, these results indicate that reaction centers are highly specific for c2. Although cells can grow as well with iso-c2 as with c2, it is because the rate limiting step in photosynthesis is at some other point in the pathway.

ACKNOWLEDGMENTS

This work was supported in part by grant GM21277 from the National Institutes of Health. We thank Mel Okamura and Ed Abresch for the gift of photosynthetic reaction centers.

REFERENCES

1. Donohue, T.J., McEwan, A.G., Van Doren, S., Crofts, A.R., & Kaplan, S. (1988) Biochemistry *27*, 1918–1925.
2. Fitch, J.C., Cannac, V., Meyer, T.E., Cusanovich, M.A., Tollin, G., Van Beeumen, J.J., Rott, M.A., & Donohue, T.J. (1989) Arch. Biochem. Biophys. *271*, 502–507.
3. Rott, M.A. & Donohue, T.J. (1990) J. Bacteriol. *172*, 1954–1961.
4. Rott, M.A., Fitch, J., Meyer, T.E., & Donohue, T.J. (1992) Arch. Biochem. Biophys. *292*, 576–582.
5. Rott, M.A., Witthuhn, V.C., Schilke, B.A., Soranno, M., Ali, A., & Donohue, T.J. (1993) J. Bacteriol. *175*, 358–366.
6. Barber, R.D., Rott, M.A., & Donohue, T.J. (1996) J. Bacteriol. *178*, 1386–1393.
7. Witthuhn, V.C. Jr., Gao, J., Hong, S., Halls, S., Rott, M.A., Wraight, C.A., Crofts, A.R., & Donohue, T.J. (1997) Biochemistry *36*, 903–911.
8. Daldal, F., Cheng, S., Applebaum, J., Davidson, E., & Prince, R.C. (1986) Proc. Natl. Acad. Sci. USA *83*, 2012–2016.
9. Jenney, F.E.Jr. & Daldal, F. (1993) EMBO J. *12*, 1283–1292.
10. Jenney, F.E.Jr., Prince, R.C., & Daldal, F. (1994) Biochemistry *33*, 2496–2502.
11. Myllykallio, H., Jenney, F.E. Jr., Moomaw, C.R., Slaughter, C.A., & Daldal, F. ()1997) J. Bacteriol. *179*, 2623–2631.
12. Ambler, R.P., Daniel, M., Hermoso, J., Meyer, T.E., Bartsch, R.G., & Kamen, M.D. (1979) Nature *278*, 659–660.
13. Pettigrew, G.W., Meyer, T.E., Bartsch, R.G., & Kamen, M.D. (1975) Biochim. Biophys. Acta *430*, 197–208.
14. Jenney, F.E. Jr., Prince, R.C., & Daldal, F. (1996) Biochim. Biophys. Acta *1273*, 159–164.
15. Wachtveitl, J., Farchaus, J.W., Mathis, P., & Oesterhelt, D. (1993) Biochemistry *32*, 10894–10904.
16. Tiede, D.M., Vashishta, A., & Gunner, M.R. (1993) Biochem. *32*, 4515–4531.
17. Wang, S., Li, X., Williams, J.C., Allen, J.P., & Mathis, P. (1994) Biochem. *33*, 8306–8312.
18. Watkins, J.A., Cusanovich, M.A., Meyer, T.E., & Tollin, G. (1994) Protein Sci. *3*, 2104–2114.

36

THE 18 kDa CYTOCHROME c_{553} FROM *HELIOBACTERIUM GESTII* IS A LIPOPROTEIN

Ingrid Albert,[1] Hans Grav,[2] and Hartmut Michel[1]

[1]Max-Planck-Institut für Biophysik
Heinrich-Hoffmann-Straße 7
D-60528 Frankfurt am Main
[2]Institute for Nutrition Research
University of Oslo
Oslo, Norway

1. INTRODUCTION

The strictly anaerobic phototrophic heliobacteria are grouped within the Gram-positive line of bacteria on the basis of 16S ribosomal RNA sequences [1,2]. Detailed information is available for the electron transfer chains of Gram-negative bacteria, whereas electron transfer in Gram-positive bacteria is only poorly understood to date. Cytochromes are a rather heterogeneous group of proteins concerned with electron transfer reactions. There is some spectroscopic evidence that a cytochrome c_{553} takes part in the reduction of the oxidized photosynthetic reaction center of heliobacteria [3–7]. In membranes of heliobacteria at least three different heme containing proteins can be distinguished by SDS-PAGE (sodium dodecyl sulphate polyacrylamide gel electrophoresis) analysis [3,8,9]. Redox titrations of membranes by optical and EPR spectroscopy reveal the presence of 5 or 6 membrane associated heme c components [7,10]. All c-type hemes are virtually indistinguishable by their optical spectra. The most abundant heme containing protein in the membranes of *H. gestii* is a c-type cytochrome with an apparent molecular mass of 18 kDa and an α-absorption band at 553 nm. This cytochrome c_{553} is very strongly associated with two proteins of 32 and 42 kDa respectively [11]. In the present study we focus on the biochemical and genetic characterization of this cytochrome c_{553}.

2. MATERIALS AND METHODS

2.1. DNA Isolation and Manipulation

Standard recombinant DNA techniques were used according to [12], if not indicated otherwise. Genomic DNA from *H. gestii* was prepared from 1 day old liquid cultures as de-

The Phototrophic Prokaryotes, edited by Peschek *et al.*
Kluwer Academic / Plenum Publishers, New York, 1999.

scribed by [13]. Degenerate oligonucleotides were derived from internal amino acid sequences of the 18 kDa cytochrome c_{553}. The 250 bp PCR product that was obtained with the
primer P1 (5'-GAG CAG GCD CCV GCN CCN GC-3') and P5 (5'-GCN GGN GGC ATN
GTN GCY TG -3') was used to screen a genomic library of *Eco* RI cut genomic DNA from *H.
gestii* in the Bluescript SK(+) vector. Positive clones obtained from the colony hybridization
were further examined by dot blots and Southern blots. All hybridization techniques were carried out with DIG labeled DNA probes (Boehringer Mannheim, Germany). Primer directed
sequencing was used to determine the entire gene sequence of the 18 kDa cytochrome c_{553}.
Nucleotide sequences were aligned and analyzed using the GCG Analysis software package.

2.2. Protein Purification

H. gestii Chainat strain, ATCC 43375 was obtained from Dr. M.T. Madigan (Southern
Illinois University, Carbondale, Illinois, USA). The cells were grown with continuous illumination (120W) in 1 l bottles in LYE medium (as PYE medium [14] with pyruvate replaced by
1.5 ml/l DL-lactic acid) and with 2.5 mM sodium ascorbate, under strictly anaerobic conditions. After 1–2 days of growth the cells were spun down at 5000×g and stored at –20°C until
used. All subsequent steps were carried out at +4°C with buffers purged with nitrogen for 5
min. Thawed cells were suspended in 20 mM Tris, pH 8.0, 5 mM sodium ascorbate. The cells
were disrupted by passing three times through a French pressure cell (120 000 psi) in the presence of DNase. Crude cell fragments were spun down by centrifugation at 5000×g. Membrane fragments were collected by ultracentrifugation at 100 000×g for one hour. The
membranes were solubilized with 1% Mega 9 and 1% sodium cholate in the above mentioned
buffer. Pefabloc was added to give 0.5 mM. The final BChl concentration was 1 mM BChl*g*
(OD 780=1.00). The solution was stirred for 40 min in the dark on ice and was then directly
loaded onto a sucrose density gradient (20–50%) in 20 mM Tris-HCl, pH 8.0, 5 mM sodium
ascorbate, 0.5% Mega 9, 0.1% cholate. After centrifugation for 14 hours at 42 000×g, the
fraction containing the 18 kDa cytochrome c_{553} was loaded onto a hydroxyapatite column
(Bio-Gel HTP, Bio-Rad) equilibrated with 10 mM sodium phosphate buffer, pH 7.0, 0.3%
Mega 9, 0.1% cholate. The bound proteins were eluted with a linear gradient of 10–100 mM
sodium phosphate buffer. The protein solution was concentrated (centriprep, 30 kDa cutoff,
Amicon), and the sample was loaded onto a TSKgel 3000SW column (Tosohaas) connected
to an HPLC system that was equilibrated with 50 mM potassium phosphate buffer, pH 7.0,
0.3% Mega 9. For protein sequencing and for the determination of the fatty acids the 18 kDa
cytochrome c_{553} was completely separated from the two proteins of 32 and 42 kDa by incubation at 80°C in the presence of 5% SDS for 5 min. The material was then subjected to molecular sieve chromatography on the above mentioned HPLC column equilibrated with 50 mM
Tris, pH 7.0, 0.3% SDS. Internal peptide sequences were obtained after endoproteolytic
cleavage with LysC and AspN. Proteins were sequenced with an Applied Biosystems gas
phase sequencer 477A connecteded to an on-line 120 APTH analyzer.

2.3. Analysis of Fatty Acid Methyl Esters

Fatty acid methyl esters (FAME) were prepared by treating the isolated acetone precipitated cytochrome c_{553} with 1M HCl/methanol for 20 hours at 85°C [15]. The FAME
were extracted into hexane and the solvent was evaporated with a stream of nitrogen. For
the measurement the FAME were redissolved in hexane. FAME mass spectra were recorded on a Shimadzu QP2000 GCMS (Shimadzu Europe, Duisburg, Germany) equipped
with a S.G.E. (Ringwood, Vict., Australia) BP1 nonpolar quartz capillary column (50 m,

0.15 mm i.d.). Electrospray mass spectra of the entire molecule were obtained using a Finnigan Ion Trap 700 mass spectrometer (San Jose, Ca., USA). Both spectrometers were operated in electron impact mode at 70 eV.

3. RESULTS AND DISCUSSION

3.1. The Gene *cyhA* Encoding the 18 kDa Cytochrome c_{553}

In order to identify the gene coding for the 18 kDa cytochrome c_{553} we prepared degenerate PCR primers based on the internal peptide sequences. With one primer combination a 250 bp product was obtained from genomic DNA of *H. gestii* that was cloned and sequenced. The deduced amino acid sequence was identical to that of two internal peptide sequences. Furthermore the same reading frame contained the sequence CITCH with the characteristic heme binding motif CXXCH of c-type cytochromes. The PCR product was used to screen a partial genomic library prepared from *Eco*RI restriction fragments of *H. gestii* genomic DNA. The gene *cyh*A of the 18 kDa cytochrome c_{553} was completely included in a 4.6 kbp genomic DNA fragment. The coding strand and the deduced amino acid sequence are presented in Figure 1. The gene consists of 429 bp and is preceded by a Shine-Dalgarno sequence (AGGAGG) 10 bp upstream of the methionine start codon.The

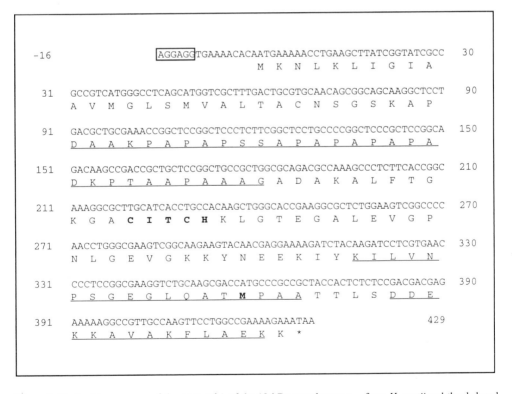

Figure 1. Nucleotide sequence of the gene *cyh*A of the 18 kDa cytochroem c_{553} from *H. gestii* and the deduced amino acid sequence. The amino acid sequences that were known from the internal peptide sequences are underlined. The heme binding motif and the methionine that is the sixth ligand of the heme iron are represented in bold letters. The Shine-Dalgarno sequence is boxed.

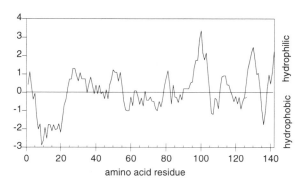

Figure 2. Hydropathy plot of the *cyhA* gene product [18].

primary structure comprises 142 amino acids with the heme binding motif located in the center of the protein. 42 amino acid residues from the heme binding site, in the C-terminal part of the protein, a methionine residue is located that is the only candidate for the sixth axial ligand of the heme iron.

3.2. Determination of the N-Terminal Modifications

In order to obtain the exact molecular mass of the 18 kDa cytochrome c_{553}, the cytochrome was separated from the two other subunits by molecular sieve chromatography in the presence of SDS. The molecular mass of the isolated cytochrome was determined to 13.171 kDa by electrospray ionization (ESI) mass spectrometry. This value is significantly lower than the molecular mass calculated from the amino acid composition with 14.169 kDa. C-terminal protein sequencing confirmed the last five amino acid residues of the translated protein sequence. The discrepancy of the molecular masses can most likely be explained by a proteolytic modification of the N-terminus which was also found to be blocked. This situation is found for some bacterial lipoproteins [16,17]. In these proteins a signal peptide is cleaved off, leaving a cystein at the N-terminus. This cystein is then modified by the formation of a thioether with a glycerol moiety that contains two fatty acid molecules.

The Kyte-Doolittle hydrophobicity plot in Figure 2 demonstrates that the cytochrome primary structure is divided into a hydrophobic part of 18 amino acid residues at the N-terminus and a hydrophilic domain that is harboring the heme. The N-terminal hy-

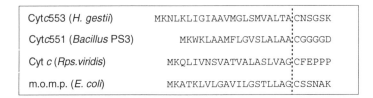

Figure 3. Comparison of the N-terminal amino acid sequence from the 18 kDa cytochrome c_{553} with the N-terminal signal peptid sequences from bacterial lipoproteins: cytochrome c_{551} from *Bacillus* PS3 [17], the tetraheme cytochrome *c* from *Rhodopseudomonas viridis* [19], the "major outer membrane protein" from *E. coli* [20]. The putative signal peptidase cleavage site is indicated as a dotted line.

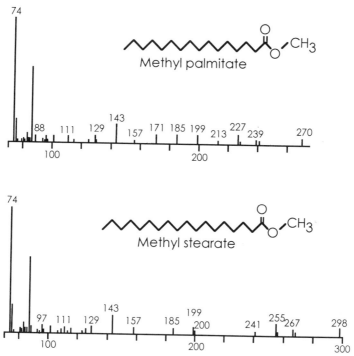

Figure 4. Electron impact mass spectra of the FAME prepared from the isolated 18 kDa cytochrome c_{553}. The upper fragmentation pattern is attributed to methyl palmitate, the lower to methyl stearate.

drophobic stretch is preceded by two positively charged residues which are characteristic for N-terminal signal peptides. Figure 3 shows the comparison of the N-terminal sequence of the gene product of *cyhA* from *H. gestii* with the N-terminal amino acid sequences of bacterial proteins that are known to be processed by a signal peptidase. The stretch of non-polar residues is followed by alanine and cysteine in position −1 and +1, forming a potential cleavage site for the signal peptidase II. Taking into account that the protein is blocked at the N-terminus, this cysteine should also carry an additional modification, for example an acetylation or acylation. In order to confirm this model we had primarily to prove the existence of the covalently bound fatty acid molecules. The fatty acids were transformed into their respective methyl esters (FAME). The fragmentation pattern of the FAME prepared from the isolated cytochrome c_{553} are given in Figure 4. They can clearly be attributed to methyl palmitate and methyl stearate. This result suggests that one palmitate and one stearate molecule are bound to the cytochrome c_{553}.

Assuming that the additional modification of the N-terminus consists of an acylation and that a glycerol residue mediates the junction of the cysteine and the fatty acids, the mass of the amino acid residues +1 to +122, the protoheme, an acetyl residue, a glycerol, one palmitate and one stearate is 13.176 kDa = 11.941 + 0.618 + 0.043 + 0.073 + 0.239 + 0.267 − 0.006. This value is very close to the mass determined by ESI mass spectrometry. A model for the structure of the mature, processed cytochrome c_{553} is given in Figure 5. Comparable models were reported for the tetraheme cytochrome *c* subunit of the photosynthetic reaction center of *Rhodopseudomonas viridis* [19], the major outer membrane protein from *E. coli* [16] and for the cytochrome c_{551} from *Bacillus* PS3 [17].

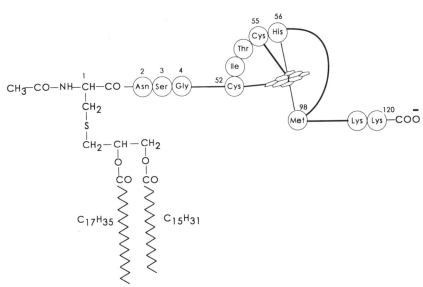

Figure 5. Model for the proposed structure of the mature, processed 18 kDa cytochrome c_{553}. The acetylation of the N-terminus as well as the relative position of the two fatty acid residues on the glycerol component have not been determined.

ACKNOWLEDGMENTS

We wish to thank Dr. J. Ormerod for very helpful discussion. We are grateful to Dr. J. Kellermann for the N-terminal protein sequencing and to Mrs. Pries for performing the mass spectroscopy.

REFERENCES

1. Woese, C.R., Debrunner-Vossbrinck, B.A., Oyaizu, H., Stackebrandt, E. and Ludwig, W. (1985) Science 229, 762–765.
2. Woese, C.R. (1987) Microbiol. Rev. 2, 221–271.
3. Fuller, R.C., Sprague, S.G., Gest, H. and Blankenship, R.E. (1985) FEBS Lett. 182, 345–349.
4. Prince, R. C., Gest, H. and Blankenship, R. E. (1985) Biochim. Biophys. Acta 810, 377–384.
5. Vos, M.H., Klaassen, H.E. and van Gorkom, H.J. (1989) Biochim. Biophys. Acta, 973, 163–169.
6. Kleinherenbrink, F.A.M. and Amesz, L.J. (1993) Biochim. Biophys. Acta 1143, 77- 83.
7. Nitschke, W., Liebl, U., Matsuura,K. and Kramer, D.M. (1995) Biochemistry 34, 11831–11839.
8. Trost, J.T., Brune, D.C. and Blankenship, R.E. (1992) Photosynth. Res. 32, 11–22.
9. Nitschke, W. and Liebl, U. (1992) in Murata, N. (ed.) Research in Photosynthesis Vol. III, 507–510, Kluwer Academic Publishers, Netherlands.
10. Nitschke, W., Schoepp, B., Floss, B., Schricker, A., Rutherford, A.W. and Liebl, U. (1996) Eur. J. Biochem. 242, 695–702.
11. Albert, I. and Michel, H. (1995) in P. Mathis (ed.) Photosynthesis: From Light to Biosphere, Vol. II, 523–526, Kluwer Academic Publishers, Netherlands.
12. Sambrook, J., Fritsch, E.F. and Maniatis, T. (1989) Molecular Cloning: A Laboratory Manual, Cold Spring Harbour, Laboratory Press, USA.
13. Laussermair, E. and Oesterhelt, D. (1992) EMBO J. 11, 777–783.
14. Madigan, M.T. (1992) in Balows, A., Trüper, H.G., Dworkin M. and Schleifer, K.-H. (eds.) The Prokaryotes, 2. ed., 1982–1992, Springer, New York.

15. Aase, B., Jantzen, E., Bryn, K. and Ormerod, J. (1994) Photosynth. Res. 41, 67–74.
16. Hantke, K. and Braun, V. (1973) Eur. J. Biochem. 34, 284–296.
17. Fujiwara, Y., Oka, M., Hamamoto, T. and Sone, N. (1993) Biochim. Biophys. Acta 1141, 213–219.
18. Kyte, J. and Doolittle, R.F. (1982) J. Mol. Biol. 157, 105–132.
19. Weyer, K.A., Schäfer, W., Lottspeich, F. and Michel, H. (1987) Biochemistry 26, 2909–2914.
20. Nakamura, K. and Inouye, M. (1979) Cell 18, 1109–1117.

37

ELECTRON TRANSFER BETWEEN CYTOCHROME *f* AND PLASTOCYANIN IN *PHORMIDIUM LAMINOSUM*

Derek S. Bendall, Michael J. Wagner, Beatrix G. Schlarb,
Tim Robert Söllick,* Marcellus Ubbink,† and Christopher J. Howe

Cambridge Centre for Molecular Recognition and Department of
    Biochemistry
University of Cambridge
Tennis Court Road, Cambridge CB2 1QW, United Kingdom

1. INTRODUCTION

The reaction between cytochrome *f* and plastocyanin is central to the photosynthetic electron transport system of all oxygenic organisms. The overall aim of the work described here is to understand how the rate of such an electron transfer reaction is controlled by the structures of the proteins involved and the chemical environment in which they find themselves *in vivo*. The theory of electron transfer in proteins has been developed mainly with regard to intramolecular electron transfer[1-4]. Study of the bacterial photosynthetic reaction centre and of haemproteins chemically modified with ruthenium complexes to provide a second, photoactivatable redox centre, have led to the conclusion that rate constants are determined on the one hand by the relation between driving force and reorganization energy, and on the other by an electronic factor that decays approximately exponentially with distance. The latter factor may be modified significantly by the structure of the intervening protein and thus by the optimum 'pathway' for the electron.

In the case of redox proteins that interact by diffusion the above principles apply to the transient complex formed between them, and the intrinsic rate of electron transfer is a very sensitive function of the configuration of the complex. Key steps in the reaction are thus the formation of the initial encounter complex and vibrational movements to seek out the optimum configuration for electron transfer[5,6]. There is a conflict of interest here. A high rate of reaction implies a large equilibrium binding consant, K_A $(=k_{on}/k_{off})$, but to

* Present address: Max-Planck-Institut für Züchtungsforschung, Carl-von-Linné Weg 10, D-50829, Germany.
† Leiden Institute of Chemistry, Gorlaeus Laboratories, P.O. Box 9502, 2300 RA Leiden, The Netherlands.

The Phototrophic Prokaryotes, edited by Peschek *et al.*
Kluwer Academic / Plenum Publishers, New York, 1999.

avoid the system getting stuck after reaction has taken place the dissociation rate constant, k_{off}, must also be large. The reaction between cytochrome f and plastocyanin is a classic example of this type of reaction which must involve a dynamic and transient complex.

It seems that different redox systems may have adopted different strategies to solve this fundamental kinetic problem, but as yet there is no single case which is fully understood. In the case of the photooxidation of plastocyanin by photosystem I there is evidence that dissociation is aided by a higher redox potential of plastocyanin in the bound form compared to the free[7]. Evidence is lacking for such an effect in the reduction of plastocyanin by cytochrome f. Kinetic measurements[8] and cross-linking studies[9] have suggested that the initial electrostatic binding is not optimal for electron transfer and that internal rearrangement is necessary. An interpretation of rearrangement is offered below.

Despite this uncertainty, it is clear that electrostatic effects frequently make an important contribution to the kinetics of reaction between redox proteins in solution. With physiological partners there is usually an attractive force resulting from complementary charges on the two proteins. Because the effective charges are usually localized the electrostatic effect causes a preorientation as the proteins approach each other[10,11], as well as an increase in K_A. There may also be another, less obvious, benefit arising from electrostatic effects. This results from the strongly hydrated nature of charged groups, together with the long-range, non-directional nature of coulombic forces. Thus surface charges provide slippery surfaces which discourage close association with exclusion of water and encourage a dynamic interaction with a large number of configurations in fast exchange[5]. The balance between electrostatic and short-range forces is therefore important.

In this paper we will summarize our views on the structural basis for the kinetics of reaction between cytochrome f and plastocyanin, and discuss the similarities and differences of the reaction in higher plants and cyanobacteria. The reaction is studied in solution using the large, soluble haem-containing fragment of cytochrome f from which the membrane anchor has been removed[12,13].

2. THE IMPORTANCE OF CHARGE-CHARGE INTERACTIONS

Higher-plant plastocyanin is a strongly acidic molecule, with acidic groups concentrated on one face of the molecule on either side of Tyr-83[14,15]. Ionic strength effects show that these acidic groups interact with basic groups on cytochrome f, even though the latter is overall weakly acidic[16]. A ridge of basic residues in cytochrome f has been identified as likely to be involved in the interaction, partly from the fact that Asp-44 of spinach plastocyanin can be chemically cross-linked to Lys-187 of turnip cytochrome f[17]. A study of the kinetics of reaction of a series of mutants of spinach plastocyanin with turnip cytochrome f has identified several residues on the acidic face that are involved in binding[18]. In our hands the interaction energy at pH 6.0 depends on the total charge in the acidic patch region and is indifferent to the location of the charges (Figure 1). On the other hand, Lee *et al.*[19] have reported that under different conditions residues in the 'large' patch (42–45) are effective and those in the 'small' patch much less so.

Whilst the pattern of charge in plastocyanin and cytochrome f is highly conserved amongst higher plants, it is markedly different in cyanobacteria. The results in Figure 2 show that the rate of reaction between the cytochrome bf complex and plastocyanin from the moderately thermophilic cyanobacterium, *Phormidium laminosum*, decreases with increasing ionic strength. This is qualitatively similar to the behaviour of the higher-plant proteins, although the effect is less marked. However, the effects of ionic strength on the

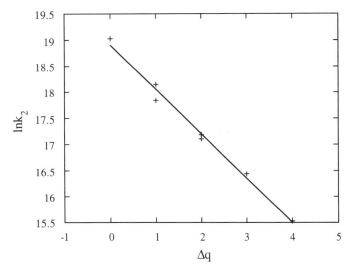

Figure 1. Effect of loss of negative charge on the second order rate constant for reduction of spinach plastocyanin and mutants by turnip cytochrome *f*. Measurements were made with the following mutants: D42N, E43N, E43K, E43Q/D44N, E59K/E60Q, E59K/E60Q/E43N. Replotted from data given by Kannt *et al.*[18]

reactions with the strongly basic cytochrome *c* and the strongly acidic pea plastocyanin suggest that the dominant charge at the reaction site of *P. laminosum* cytochrome *f* is negative, so that the relevant local charge on *P. laminosum* plastocyanin must be positive, which is the inverse of the higher-plant case. These differences are illustrated graphically in Figures 3 and 4, which respectively show pea plastocyanin side by side with *P. laminosum* plastocyanin and turnip cytochrome *f* side by side with *P. laminosum* cytochrome *f*, with acidic and basic residues highlighted. The basic ridge starting with Lys-187 and lying above the exposed haem propionates is clear (Figure 4). In the *P. laminosum* protein the residue corresponding to Lys-187 in turnip is Asp-188. The area corresponding to the basic ridge is predominantly acidic, but the net charge in this region is less negative than it is positive in turnip. In fact the edge of the molecule to the right of the haem, as shown, is also acidic and might provide an attractive site for positive groups in plastocyanin. In pea plastocyanin Tyr-83 is surrounded by acidic residues (Figure 3) whereas the corresponding Tyr-88 of the *P. laminosum* protein has acidic and basic residues in more evenly matched numbers in its immediate vicinity. The net effect is probably weakly basic. The copper ligand, His-87 in pea and His-92 in *P. laminosum*, lies apart from these charged residues in a hydrophobic area.

The role of specific charged residues in *P. laminosum* plastocyanin has been studied in our laboratory by site-directed mutagenesis and expression in *E. coli*. Rates of reduction of various mutants by a recombinant form of *P. laminosum* cytochrome *f* were measured in a stopped flow spectrophotometer. The results were strikingly different from the behaviour of the higher-plant proteins shown in Figure 1. An example of the type of behaviour observed is shown in Figure 5. With the wild-type plastocyanin the observed first order rate constant, k_1, is linearly dependent on plastocyanin concentration, corresponding to a second-order rate constant, k_2, of 4.23×10^7 $M^{-1}s^{-1}$, which is considerably slower than was observed for pea plastocyanin and turnip cytochrome *f*[18]. The behaviour of two mutants is also shown in which two charged sidechains lying adjacent to each other below Tyr-88

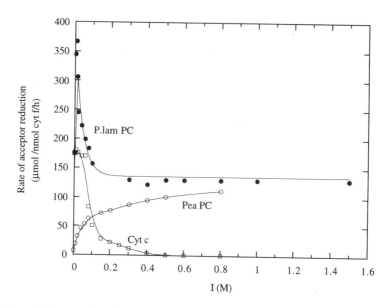

Figure 2. Effect of ionic strength on the rate of reduction of *P. laminosum* plastocyanin, horse heart cytochrome *c* and pea plastocyanin by the cytochrome *bf* complex purified from *P. laminosum*. Reductant: 10 µM plastoquinol-1. Redrawn from[18].

(Figure 3) have had their charge inverted, K53E and E54K. Surprisingly, the behaviour of the two mutants is almost identical, even though the net charge of one is −2 compared to wild type, and of the other +2. In both there is evidence of rate saturation, indicating that there has been a marked drop in the intrinsic rate constant for electron transfer within the complex, compared to wild type. On the other hand, k_2 is slightly increased in both mu-

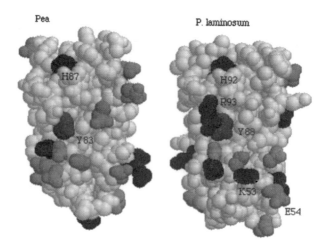

Figure 3. Distribution of acidic and basic residues on the surfaces of plastocyanin from pea and *P. laminosum*. Sidechains of acidic residues are shown in light grey, and of basic residues in dark grey. The structures shown are drawn in Rasmol and were prepared by homology modelling of pea on french bean plastocyanin[15] and of *P. laminosum* on the crystal structure of *Anabaena* plastocyanin (H.C. Freeman, personal communication).

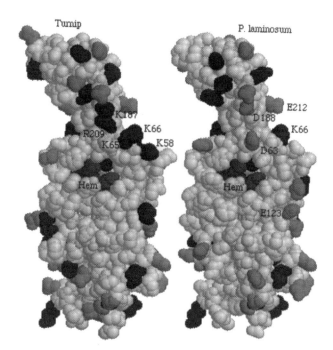

Figure 4. Distribution of acidic and basic residues on the surfaces of cytochrome *f* from turnip and *P. laminosum*. Sidechains of acidic residues are shown in light grey, and of basic residues in dark grey. The structures shown are drawn in Rasmol using the crystal structure of turnip cytochrome *f*[20] and a model of *P. laminosum* cytochrome *f* based on the turnip structure.

tants. These results suggest that the binding constant, K_A, of both mutants is significantly larger than for wild type, but the configuration of the complex is less favourable for electron transfer. Moreover, they indicate that individual charges may have a more specific role than suggested above for spinach plastocyanin, and that different modes of binding are possible.

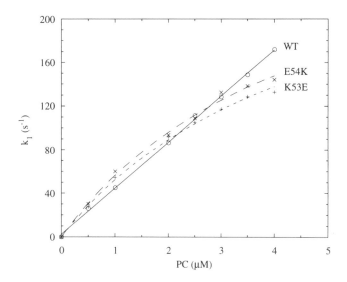

Figure 5. Observed pseudo-first order rate constants for reduction of wild-type and mutant recombinant plastocyanins of *P. laminosum* by homologous cytochrome *f*.

3. THE REACTION SITE OF PLASTOCYANIN

The structure of plastocyanin suggests two most likely pathways for an electron from the molecular surface to the buried copper atom. The most direct is through the copper ligand His-87 (plants) which is located in a hydrophobic region of the surface, sometimes referred to as the 'northern' site. A second pathway involves the surface-exposed Tyr-83 and the ligand Cys-84. Although less direct, coupling through this pathway is strengthened by strong π-bonding between the Cu atom and the Cys-S atom[21]. Studies with small molecules have shown that both routes can be utilised[22-27]. There is now substantial experimental and theoretical evidence that cytochrome *c* binds to plastocyanin at the acidic site and reacts predominantly via Tyr-83[28-30].

For cytochrome *f* as electron donor we have recently determined the kinetic behaviour of mutants of pea (Y83L) and *P.laminosum* (Y88L) plastocyanins in the expectation that there would be a substantial change in rate upon substitution of the smaller leucine sidechain for tyrosine (Figure 6). With both pea and *P. laminosum* plastocyanins the leucine substitution caused a small reduction in rate, but the effect was much smaller than might have been expected if Tyr-83 or -88 formed part of the major route of electron transfer. These results seem to be more consistent with the view that in both systems the electron from cytochrome *f* is transferred to the copper atom via the exposed histidine ligand in the hydrophobic patch. Although in the case of the pea protein this contradicts our earlier work which reached the opposite conclusion[31] it should be noted that a different source of cytochrome *f* was then used, which gave substantially lower rates, even with wild-type protein, than are now obtained with turnip cytochrome *f*. The view that the electron from cytochrome *f* goes mainly via the 'northern' hydrophobic patch, probably involving His-87 or -92, is consistent with work described in the next section.

4. STRUCTURE OF THE COMPLEX OF PLASTOCYANIN AND CYTOCHROME *f*

Calculation of the electrostatic potential field around the two protein molecules has formed the basis of computer modelling of their interaction. Pearson *et al.*[32] used manual docking to propose three forms of the complex. In the first, there would be strong electrostatic interaction but an unfavourable configuration for electron transfer. Rearrangement into either of two other forms would bring His-87 close to the haem and favour hydrophobic interactions. Ullmann *et al.*[33] have used a more sophisticated modelling procedure in which initial docking is brought about by a Monte Carlo method based on electrostatic calculations followed by refinement of the structures of the complex by molecular dynamics simulation with water treated explicitly. This approach led to consideration of six possible structures for the complex. The one most favourable for electron transfer brought the haem group of cytochrome *f* in contact with the northern site of plastocyanin. However, this structure was not the one of lowest energy, nor did it maximize electrostatic interactions.

We have used a novel approach to the problem in which the structure of the complex formed between spinach plastocyanin and turnip cytochrome *f* is modelled by restrained molecular dynamics in which the restraints are provided by an extensive set of experimental data[34]. These were mostly provided by NMR measurements of chemical shift changes when plastocyanin forms a complex with cytochrome *f* in either the reduced or the oxidized form. In the latter case paramagnetic effects are included and allow the calculation of distances

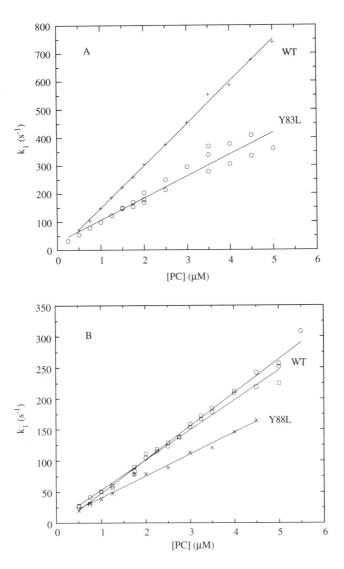

Figure 6. Pseudo-first order rate constants for the reduction of plastocyanin by cytochrome *f*. A, reduction of wild-type and mutant (Y83L) pea plastocyanins by turnip cytochrome *f*. B, reduction of wild-type (two separate experiments) and mutant (Y88L) *P. laminosum* plastocyanins by homologous cytochrome *f*.

and orientations with regard to the magnetic susceptibility tensor of the haem group. Experimental data on electrostatic interactions[18] were also included as restraints. This analysis leads to a single structure for the complex (Figure 7), even when the strength of the coulombic interactions is varied by adjustment of the ionic strength. In this structure the $N^{\varepsilon 2}$ atom of His-87 is in Van der Waals contact with ring atoms of Tyr-1 and Phe-4, thus providing an efficient pathway for electron transfer from the haem iron to the copper atom.

As mentioned above, kinetic measurements[8] and cross-linking experiments[9] have suggested that the initial electrostatic complex is not optimal for electron transfer and that rearrangement of the complex is necessary. Together with our own results and the modelling

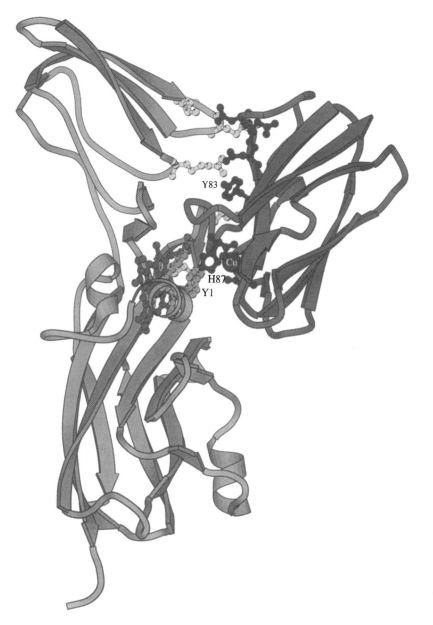

Figure 7. Structure of the complex between spinach plastocyanin and turnip cytochrome *f* determined with paramagnetic NMR and restrained molecular dynamics. Cytochrome *f* is shown in pale grey on the left and plastocyanin in dark grey on the right. Sidechains of some charged residues likely to be involved in stabilizing the structure are shown near the top, and the close interaction between H87 of plastocyanin and Y1 of cytochrome *f* is also shown. The haem ring is shown in medium grey ball-and-stick representation. Drawn with Molscript[35].

already referred to, these experiments suggest a two-step model for complex formation. The first step is the formation of an initial encounter complex under the influence of electrostatic forces. This would be a relatively unspecific, dynamic complex in which many different relative orientations of the proteins are in fast exchange. The function of the charged groups is to provide slippery, attractive surfaces, without exclusion of water from the interface. This is the kind of complex which has been successfully modelled by Brownian dynamics studies[36]. In the second step water is excluded from the interface, allowing a closer approach and the formation of a specific complex under the influence of short-range forces such as hydrogen bonding and the hydrophobic effect. The kinetics would be strongly dependent on the structure of the specific complex and on the equilibrium constant between the two kinds of complex. Charge-charge interactions may play an important role in orientating the complex in a manner which is favourable for electron transfer. Effects of this kind may explain the behaviour of charge mutants of *P. laminosum* plastocyanin (Figure 5). In the non-physiological complex formed between pea plastocyanin and horse heart cytochrome *c* no pseudo-contact shifts were observed[28], suggesting that the equilibrium constant is strongly towards the initial, electrostatic complex. This would be consistent with a low rate constant for electron transfer compared with the reaction with cytochrome *f* as donor.[18,37]

5. EVOLUTION OF SURFACE CHARGES IN PLASTOCYANIN

The most striking difference between the plastocyanins of cyanobacteria and higher plants lies in the pattern of surface charges. The experiments described above emphasise the functional importance of these charges for the reaction with cytochrome *f* in solution. The high degree of conservation of the negative patches either side of Tyr-83 in higher-plant plastocyanins strongly suggests that these charges also have an important function *in vivo*. Cyanobacteria are thought to share a common evolutionary origin with the chloroplasts of green plants. How is it then that the charge patterns of plastocyanin are so different in the two groups of organism? The pattern of an acidic plastocyanin interacting with a basic patch on cytochrome *f* is stable within plants and algae, but seems to have arisen from proteins with an inverse pattern of charge distribution. At first sight this is improbable from an evolutionary point of view.

The analysis of the problem that we give below is based on only small numbers of cyanobacterial sequences, so that the argument is inevitably tentative. Nevertheless, its validity is supported by the fact that the sequences available come from organisms spread widely across the spectrum of cyanobacterial types.

Alignments of the relative parts of the sequences of plastocyanin and cytochrome *f* are shown in Figures 8 and 9 for green plants (including algae) and cyanobacteria. Green plants show complete conservation of the DED motif at positions 42–45, and most show two or three acidic residues within the region 58–61. A Glu residue is frequently to be found at position 45 and Asp is often found at position 51, which is a little farther from Tyr-83 but on the same face of the molecule. In both these regions the cyanobacterial sequences are variable amongst themselves and show little similarity to plant sequences. In *P. laminosum* positions 42 and 43 (poplar numbering) contain Asp residues, but these are immediately followed by a Lys. The region corresponding to 58–61 contains a His and no acidic residue. Figure 3 shows how acidic and basic residues tend to be paired on the face of the protein containing the exposed Tyr (Tyr-88 in *P. laminosum*). A net positive charge is assured by the presence of an Arg residue at position 93 (equivalent to 88 in poplar), which is completely conserved in cyanobacteria but a neutral residue in green plants.

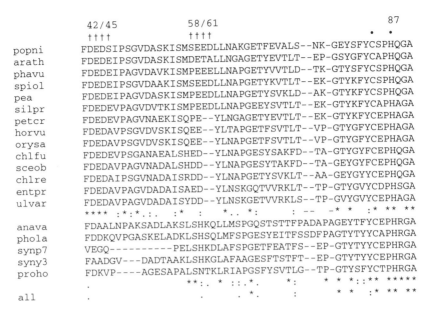

Figure 8. Alignment of partial plastocyanin sequences from green plants and from cyanobacteria. Residues involved in copper binding (●) and cytochrome *f* binding (†) are marked. Alignments were made with ClustalW v.1.7. Sequences were obtained from SWISS-PROT: popni, *Populus nigra*; arath, *Arabidopsis thaliana*; phavu, *Phaseolus vulgaris*; spiol, *Spinacia oleracea*; pea, *Pisum sativum*; silpr, *Silene pratensis*; petcr, *Petroselinum crispum*; horvu, *Hordeum vulgare*; orysa, *Oryza sativa*; chlfu, *Chlorella fusca*; sceob, *Scenedesmus obliquus*; chlre, *Chlamydomonas reinhardtii*; entpr, *Enteromorpha prolifera*; ulvar, *Ulva arasakii*; anava, *Anabaena variabilis*; phola, *Phormidium laminosum*; synp7, *Synechococcus* sp. (strain PCC 7942) (*Anacystis nidulans* R2); syny3, *Synechocystis* sp. (strain PCC 6803); proho, *Prochlorothrix hollandica*.

The alignment of partial cytochrome *f* sequences (Figure 9) shows parts of the sequence that contain basic residues in turnip and spinach which have been identified as forming a positively charged patch interacting with plastocyanin. These are K58, K65, K66, K187 and R209, which are completely conserved amongst green plants except for R209. Only for K66 does the conservation extend to cyanobacteria. In *P. laminosum* the corresponding positions are occupied by neutral residues except for K66 and D188 (*P. laminosum* numbering). The basic ridge of turnip has become an acidic patch comprising residues D63, E187, D188 and E212. A second acidic patch, centred on E123, occurs further down the right hand side of the molecule (Figure 4), but this is not well conserved amongst known cyanobacterial sequences.

The sequence alignments therefore suggest that it may be a general feature of cyanobacteria that the reactive local charges on plastocyanin are positive and on cytochrome *f* negative, the reverse of the eukaryotic case. In an electron transport chain the charge on plastocyanin should also be compatible with the charge on its acceptor. In chloroplasts the acceptor is P700 at the reaction centre of PSI and binding is thought to be strongly influenced by a group of basic residues near the N-terminus of the small subunit, PsaF[38]. An alignment of eukaryotic and cyanobacterial N-terminal sequences of PsaF is shown in Figure 10. Six basic residues are marked that are thought to be involved in binding plastocyanin and which are conserved amongst the small number of known higher-plant sequences. These residues are all thought to be present in one of a pair of amphipathic α-helices[38]. The basic residues are completely absent from the cyanobacterial sequences and

```
               58        65/66                    187                          209
               †         ††                       †                            †
brara   IPYDMQLKQVLANGKKGALN.....KILRK--EKGG----YEITIVDASNERQVIDIIPRG
spiol   IPYDMQLKQVLANGKKGGLN.....KIVRK--EKGG----YEINIADASDRREVVDIIPRG
wheat   IPYDMQLKQVLANGKKGGLN.....KILRK--EKGG----YEISIVDASDGRQVIDIIPPG
pea     IPYDMQVKQVLANGKKGALN.....QIIRK--EKGG----YEITIVDASDGSEVIDIIPPG
marpo   IPYDMQIKQVLANGKKGSLN.....KIFRK--EKGG----YEITIDDISDGHKVVDISAAG
pinth   IPYDMQMKQVLANGKKGALN.....KILRK--EKGG----YEITIDNTSDGGQVVDIVPPG
chlre   LPYDKQVKQVLANGKKGDLN.....AITALSEKKGG----FEVSIEKAN-GEVVVDKIPAG
        .:*** *.********** **      *     :***   :*: * . .  *.* ..*

nossp   IPYDTSAQQVGADGSKVGLN.....KIAKE--EDEDGNVKYQVNIQPES-GDVVVDTVPAG
phola   IPYDHSVQQVQADGSKGPLN.....AIAKA--EDGS----ARVKIRTED-GTTIVDKIPAG
synp2   IPYDHSQQQVLGDGSKGGLN.....NIAVN--EAAG----TDITISTEA-G-EVIDTIPAG
syny3   IPYDLDSQQVLGDGSKGGLN.....EVNAL--EAGG----YQLILTTAD-G-TETVDIPAG
        **** . *** .**** **    :     *       :  :.    *      :***

all     :*** . :** .:*.*   **    :      :       : :          ..*
```

Figure 9. Alignment of partial cytochrome *f* sequences from green plants (selected) and cyanobacteria. Residues implicated in binding plastocyanin in plants are marked †. Alignments were made with ClustalW v.1.7. Sequences were obtained from SWISS-PROT: brara, *Brassica rapa*; spiol, *Spinacia oleracea*; wheat, *Triticum aestivum*; pea, *Pisum sativum*; marpo, *Marchantia polymorpha*; pinth, *Pinus thunbergii*; chlre, *Chlamydomonas reinhardtii*; nossp, *Nostoc* sp. (strain PCC 7906); phola, *Phormidium laminosum*; synp2, *Synechococcus* sp. (strain PCC 7002) (*Agmenellum quadruplicatum*); syny3, *Synechocystis* sp. (strain PCC 6803).

an amphipathic helix cannot be predicted from the sequence. Acidic residues are conserved at two positions, but there is no evidence that they are involved in binding to cytochrome *f*. Thus, charge-charge interactions between plastocyanin and PsaF do not seem to be important in cyanobacteria, consistent with observations that removal of PsaF by detergent[39] or mutation[40,41] does not impair electron transfer.

Plastocyanin in cyanobacteria has a more diverse role than in chloroplasts, because in the former, but not the latter, it acts as a substrate for cytochrome oxidase. Cytochrome oxidase is thought to be an ancient enzyme that may have arisen even earlier than oxy-

```
                    •    •   •     •        •
flatr   -DISGLTPCKESKQFAKREKQSLKKLESSLKLYAPDSAPALAIKATMEKTKRRFDNYGKQ
spiol   -DIAGLTPCKESKQFAKREKQALKKLQASLKLYADDSAPALAIKATMEKTKKRFDNYGKY
horvu   -DIAGLTPCKESKAFAKREKQSVKKLNSSLKKYAPDSAPALAIQATIDKTKRRFENYGKF
odosi   -DIGGLTKCSESPAFEKRLKASVKKLEQRMGKYEAGSPPALALQQQIERTQARFDKYSRS
chlre   -DIAGLTPCSESKAYAKLEKKELKTLEKRLKQYEADSAPAVALKATMERTKARFANYAKA
        **.*** *.** :   *  *   :*.*:   :  *   .*.**:*::  :::*:  ** :*..
              †                              †

anavc   -LGADLTPCAENPAFQALAKN------ARNTTADPQSG-----QK-------RFERYSQA
synen   -DVAGLVPCKDSPAFQKRAAA------AVNTTADPASG-----QK-------RFERYSQA
sync7   DSLSHLTPCSESAAYKQRAKN------FRNTTADPNSG-----QN-------RAAAYSEA
syny3   DDFANLTPCSENPAYLAKSKN------FLNTTNDPNSG-----KI-------RAERYASA
        :  *.**  ::..*:   :      *** ** **   :         *   *:.*

all     . *  *  :. :              *       :        *   *.
```

Figure 10. Alignment of PsaF N-terminal sequences from green plants and cyanobacteria. Positions at which lysines are conserved in higher plants, and occur variously in lower plants but not in cyanobacteria, are marked (●)[38]. Two positions at which conserved acidic residues occur in cyanobacteria are identified (†). Alignments were made with ClustalW v.1.7. Sequences were obtained from SWISS-PROT, except for anavc (EMBL entry AVPSAFJ) and sync7[42]; chlre, *Chlamydomonas reinhardtii*; cyapa, *Cyanophora paradoxa*; flatr, *Flaveria trinervia*; horvu, *Hordeum vulgare*; odosi, *Odontella sinensis*; porpu, *Porphyra purpurea*; spiol, *Spinacea olereacea*; anavc, *Anabaena variabilis* ATCC 29413; synen, *Synechococcus elongatus* Naegeli; sync7, *Synechococcus* PCC 7002; syny3, *Synechocystis* PCC 6803.

genic photosynthesis.[43] A consistent feature seems to be the requirement for an electron donor with positive charges at the interaction site. The alignment of sequences of subunit II (cox2) from green plants and the two known cyanobacterial sequences offers an explanation (Figure 11). Two acidic residues thought to be involved in binding cytochrome c[44,45] are shown by the recent determination of crystal structures[46,47] to be very close to or identical with residues involved in the binding of Cu_A. Thus there are likely to be overriding structural and functional reasons why these residues must be acidic and have remained so throughout evolution.

The conclusion from these considerations is that the main determinant of the positive charge on plastocyanin in cyanobacteria is the need for a negative charge at the binding site on cytochrome oxidase, since we have seen that cytochrome f may adopt either a negative or a positive net local charge without impairment of function, and binding to PSI is not strongly determined by charge. This constraint would have been removed early in the evolution of eukaryotic organisms by the loss of cytochrome oxidase from chloroplasts. Although one might have expected the prevailing cyanobacterial pattern of positive plastocyanin and negative cytochrome f to have been maintained, we have shown above (Figure 5) that this situation is finely balanced, at least in the case of *P. laminosum*. Conversion of a Lys to a Glu (a single transition mutation) causes an increase in the second order rate constant, probably by an increase in binding to cytochrome f. The choice of a negatively charged plastocyanin in chloroplasts would have been a chance event, but one which rapidly became irreversible because of the need for complementary mutations in cytochrome f, and eventually in PsaF as well.

6. CONCLUSIONS

It is now possible to describe the interaction between cytochrome f and plastocyanin in solution so as to provide a structural interpretation of the kinetics. Studies of the effects of ionic strength on kinetics and the behaviour of specific mutants have emphasized the

```
                         †                                              †
                         •                                    • • •          •
pea      LRIIVTPADVPHSWAVPSLGVKCDAVPGRLNQISISVQREGVYYGQCSEICGTNHAF-PIVVEAV
soybn    LRIIVTPADVPHSWAVPSLGVKCDAVPGRLNQISISVQREGVYYGQCSEICGTNHAFTPIVVEAV
dauca    LRIIVTSADVPHSWAVPSSGVKCDAVPGRLNQISISVQREGVYYGQCSEICGTNHAFTPIVVEAV
betvu    IRIIVTSADVLHSWAVPSSGVKCDAVPGRLNQTSILVQREGVYYGQCSEICGTNHAFMPIVVEAV
orysa    LRMIVTPADVLHSWAVPSSGVKCDAVPGRSNLTSISVQREGVYYGQCSEICGTNHAFTPIVVEAV
wheat    LRMIVTPADVLHSWAVPSLGVKCDAVPGRLNLTSILVQREGVYYGQCSEICGTNHAFMPIVVEAV
oenbe    LRLIVTSADVLHSWAVPSLGVKCDAVPGRLNQISMLVQREGVYYGQCSEICGTNHAFMPIVIEAV
vigun    LRVLITSADVLHSWAVPSLGVKCDAVPGRLNQISTFIQREGVYYGQCSEICGTNHAFMPIVVEAV
marpo    LRMIITSADVLHSWAVPSLGVKCDAVPGRLNQTSIFIKREGVYYGQCSELCGTNHGFMPIVVEAV
         :*::*.***.*** ******* *********** *   *   :*:*********.*****.* ***.***

synvu    VQLNLSARDVIHSFWVPQFRLKQDAIPG-VPTTRFKATKVGTYPVVCAELCGGYHGAMRTQVIVH
syny3    VQLNMEAGDVIHAFWIPQLRLKQDVIPGRGSTLVFNASTPGQYPVICAELCGAYHGGMKSVFYAH
         ****:.* ****.**:.**:.*****.***  .*  *:*:. * ***:******.***.*:: ..*
         ::: :  . ** *:: :*.  :* *.:** .      * *  *.*:.** *.  . .*
```

Figure 11. Alignment of partial cox2 sequences from green plants and cyanobacteria. Residues involved in Cu binding (•)[46] and cytochrome c binding (†)[44,45] are marked. Alignments were made with ClustalW v.1.7. Sequences were obtained from SWISS-PROT: betvu, *Beta vulgaris*; pea, *Pisum sativum*; wheat, *Triticum aestivum*; marpo, *Marchantia polymorpha*; cyaca, *Cyanidium caldarium*; synvu, *Synechococcus vulcanus*; syny3, *Synechocystis* PCC 6803).

role of charge-charge interactions. Mutant studies suggest that the pathway of electron transfer involves the 'northern' hydrophobic site which includes the exposed copper ligand, His-87. NMR studies of the effects of binding of plastocyanin to either oxidized or reduced cytochrome *f* on the chemical shifts of individual protons have allowed the use of restrained molecular dynamics to develop a model of the structure of the complex. Binding is proposed to take place in two stages. The first is the formation of an initial encounter complex under the influence of purely electrostatic forces. In this complex several different orientations are in fast exchange with one another. They are also in rapid equilibrium with a second, specific form of the complex in which water is excluded from the interface, allowing a closer approach under the influence of short range forces. This complex provides an efficient pathway of electron transfer via His-87.

The above model was developed with higher-plant proteins, and the appropriate NMR studies have not yet been performed with a cyanobacterial system. Nevertheless, mutant studies with proteins from *Phormidium laminosum* suggest that electron transfer also involves the exposed copper ligand, His-92. The most striking difference between the plant and cyanobacterial systems is that the effective charges of the two proteins tend to be reversed. The evolutionary significance of this difference is discussed in terms of a series of alignments of eukaryotic and cyanobacterial sequences for cytochrome *f*, plastocyanin, PsaF (subunit II of photosystem I) and cox2. It is suggested that the most important constraint on the charge properties of the cyanobacterial proteins is the fact that plastocyanin is an electron donor to cytochrome oxidase as well as photosystem I and that cytochrome oxidase has conserved acidic residues at the donor binding site. Removal of this constraint in chloroplasts allowed the development of negatively charged plastocyanin, but this change rapidly became irreversible.

REFERENCES

1. Marcus, R.A. and Sutin, N. (1985) Biochim. Biophys. Acta 811, 265–322.
2. Gray, H.B. and Winkler, J.R. (1996) Annu. Rev. Biochem. 65, 537–561.
3. Moser, C.C. & Dutton, P.L. (1996) in Protein Electron Transfer (Bendall, D.S., ed.), pp. 1–21, Bios Scientific Publishers, Oxford.
4. Beratan, D.N. & Onuchic, J.N. (1996) in Protein Electron Transfer (Bendall, D.S., ed.), pp. 23–42, Bios Scientific Publishers, Oxford.
5. Bendall, D.S. (1996) in Protein Electron Transfer (Bendall, D.S., ed.), pp. 43–68, Bios Scientific Publishers, Oxford.
6. Davidson, V.L. (1996) Biochemistry 35, 14035–14039.
7. Drepper, F., Hippler, M., Nitschke, W. and Haehnel, W. (1996) Biochemistry 35, 1282–1295.
8. Meyer, T.E., Zhao, Z.G., Cusanovich, M.A. and Tollin, G. (1993) Biochemistry 32, 4552–4559.
9. Qin, L. and Kostic, N.M. (1993) Biochemistry 32, 6073–6080.
10. Margoliash, E. and Bosshard, H.R. (1983) TIBS 8, 316–320.
11. Koppenol, W.H. and Margoliash, E. (1982) J. Biol. Chem. 257, 4426–4437.
12. Gray, J.C. (1992) Photosynthesis Research 34, 359–374.
13. Willey, D.L., Auffret, A.D. and Gray, J.C. (1984) Cell 36, 555–562.
14. Guss, J.M. and Freeman, H.C. (1983) J. Mol. Biol. 169, 521–563.
15. Chazin, W.J. and Wright, P.E. (1988) J. Mol. Biol. 202, 623–636.
16. Niwa, S., Ishikawa, H., Nikai, S. and Takabe, T. (1980) J. Biochem. 88, 1177–1183.
17. Morand, L.Z., Frame, M.K,, Colvert, K.K., Johnson, D.A., Krogmann, D.W. and Davis, D.J. (1989) Biochemistry 28, 8039–8047.
18. Kannt, A., Young, S. and Bendall, D.S. (1996) Biochim. Biophys. Acta 1277, 115–126.
19. Lee, B.H., Hibino, T., Takabe, T. and Weisbeek, P.J. (1995) J. Biochem. (Tokyo) 117, 1209–1217.
20. Martinez, S.E., Huang, D., Szczepaniak, A., Cramer, W.A. and Smith, J.L. (1994) Structure 2, 95–105.
21. Lowery, M.D., Guckert, J.A., Gebhard, M.S. and Solomon, E.I. (1993) J. Am. Chem. Soc. 115, 3012–3013.

22. Armstrong, F.A., Driscoll, P.C., Hill, H.A.O. and Redfield, C. (1986) J. Inorg. Chem. 28, 171–180.
23. Sykes, A.G. (1991) Structure and Bonding 75, 175–224.
24. Sinclair-Day, J.D. and Sykes, A.G. (1986) J. Chem. Soc. Dalton Trans. 2069–2073.
25. Sykes, A.G. (1985) Chem. Soc. Rev. 14, 283–314.
26. Cookson, D.J., Hayes, M.T. and Wright, P.E. (1980) Biochim. Biophys. Acta 591, 162–176.
27. Pladziewicz, J.R. and Brenner, M.S. (1987) Inorg. Chem. 26, 3629–3634.
28. Ubbink, M. and Bendall, D.S. (1997) Biochemistry 36, 6326–6335.
29. Modi, S., He, S., Gray, J.C. and Bendall, D.S. (1992) Biochim. Biophys. Acta 1101, 64–68.
30. Roberts, V.A., Freeman, H.C., Olson, A.J., Tainer, J.A. and Getzoff, E.D. (1991) J. Biol. Chem. 266, 13431–13441.
31. He, S., Modi, S., Bendall, D.S. and Gray, J.C. (1991) EMBO J. 10, 4011–4016.
32. Pearson, D.C.,Jr., Gross, E.L. and David, E.S. (1996) Biophys. J. 71, 64–76.
33. Ullmann, G.M., Knapp, E.-W. and Kostic, N.M. (1997) J. Am. Chem. Soc. 119, 42–52.
34. Ubbink, M., Ejdebäck, M., Karlsson, B.G. & Bendall, D.S. (1997) (Submitted).
35. Kraulis, P.J. (1991) J. Appl. Crystal. 24, 946–950.
36. Northrup, S.H. (1996) in Protein Electron Transfer (Bendall, D.S., ed.), pp. 69–97, Bios Scientific Publishers, Oxford.
37. Modi, S., Nordling, M., Lundberg, L.G., Hansson, Ö. and Bendall, D.S. (1992) Biochim. Biophys. Acta 1102, 85–90.
38. Hippler, M., Reichert, J., Sutter, M., Zak, E., Altschmied, L., Schröer, U., Herrmann, R.G. and Haehnel, W. (1996) EMBO J. 15, 6374–6384.
39. Hatanaka, H., Sonoike, K., Hirano, M. and Katoh, S. (1993) Biochim. Biophys. Acta 1141, 45–51.
40. Chitnis, P.R., Purvis, D. and Nelson, N. (1991) J. Biol. Chem. 266, 20146–20151.
41. Xu, Q., Yu, L., Chitnis, V.P. and Chitnis, P.R. (1994) J. Biol. Chem. 269, 3205–3211.
42. Golbeck, J.H. (1994) in The Molecular Biology of Cyanobacteria (Bryant, D.A., ed.), pp. 319–360, Kluwer Academic Publishers, Dordrecht.
43. Castresana, J., Lübben, M., Saraste, M. and Higgins, D.G. (1994) EMBO J. 13, 2516–2525.
44. Witt, H., Zickermann, V. and Ludwig, B. (1995) Biochim. Biophys. Acta 1230, 74–76.
45. Speno, H., Taheri, M.R., Sieburth, D. and Martin, C.T. (1995) J. Biol. Chem. 270, 25363–25369.
46. Iwata, S., Ostermeier, C., Ludwig, B. and Michel, H. (1995) Nature 376, 660–669.
47. Tsukihara, T., Aoyama, H., Yamashita, E., Tomizaki, T., Yamaguchi, H., Shinzawa-Itoh, K., Nakashima, R., Yaono, R. and Yoshikawa, S. (1995) Science 269, 1069–1074.

38

THE ROLE OF MAGNESIUM AND ITS ASSOCIATED WATER CHANNEL IN ACTIVITY AND REGULATION OF CYTOCHROME *c* OXIDASE

Laurence Florens,[1] Curtis Hoganson,[2] John McCracken,[2] John Fetter,[1] Denise A. Mills,[1] Gerald T. Babcock,[2] and Shelagh Ferguson-Miller[1]

[1]Department of Biochemistry
[2]Department of Chemistry
Michigan State University
East Lansing, Michigan 48824

1. INTRODUCTION

Research on cytochrome *c* oxidase has moved into a new era with the recent resolution of the crystal structures of bacterial [1] and beef heart [2] enzymes, showing their remarkable similarity. The resolution of the crystal structures has specified the spatial organization of the metal centers and defined some possible routes for proton translocation within the molecule. Three distinct pathways for protons are expected in the cytochrome *c* oxidase: two entries for the pumped and substrate protons and an unidirectional exit route. An apparent water channel, which could function as the proton exit pathway, is clearly visible in the beef heart oxidase X-ray structure [2] (Figure 1): it is immediately above the active site and connects it to the exterior of the membrane.

Using *Rhodobacter sphaeroides* cytochrome *c* oxidase (an excellent structural and functional model [3] of the more complex mitochondrial enzyme), our lab demonstrated the presence of a Mg/Mn site [4], which is now confirmed to be in a central location by the X-ray structures (Figure 1). This site is found in all eukaryotic cytochrome *c* oxidases. Although its function is still unknown, different roles have been predicted for the Mn/Mg site. Directly at the interface between the subunits I and II, sharing a ligand with the binuclear Cu_A center (which receives the electrons from cytochrome *c*) and immediately above the heme a_3-Cu_B center (where the oxygen chemistry occurs), the Mn/Mg site could have a structural role, stabilizing this critical interface between the two catalytic subunits. Furthermore, the Mg/Mn center is located adjacent to the apparent water channel. The proton

The Phototrophic Prokaryotes, edited by Peschek *et al.*
Kluwer Academic / Plenum Publishers, New York, 1999.

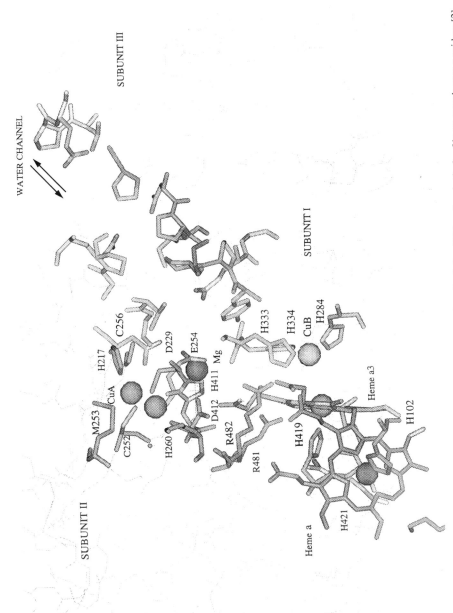

Figure 1. Organization of the metal centers and water channel at the subunit I and II interface in the beef heart cytochrome *c* oxidase [2].

exit pathway is a logical site for mediating control of proton pumping efficiency and of respiration rate by the membrane potential. Control of reversibility of the exit pathway is proposed as a way to control the efficiency of the proton pump without disrupting the oxygen chemistry. If so, the Mg/Mn site could play an important role in these respiratory control processes.

The newly available 3D data provide the basis for an even more incisive mutational approach combined with powerful time-resolved spectroscopies. Mutants of the Mg/Mn ligands [4,5] have been constructed to discern how important an intact Mg/Mn site is for enzyme stability and control of electron and proton transfer. In this study, we have used these mutants to test the role of the Mg/Mn site as a structural element and as a regulator of proton pumping efficiency and respiratory control. In particular, we have examined the thermal stability of the enzymes, the pH dependence and deuterium isotope kinetic effect on the activity. The results are discussed in relation with the extent of structural and functional perturbation that occurs at the Mg/Mn binding site upon mutation.

2. EXPERIMENTAL PROCEDURES

2.1. Protein Production: Construction of Histidine-Tagged, Overproducing Mutant Strains

To overcome protein production limitations, the H411Q and D412A mutations [5] were introduced into a His-tagged, large plasmid to obtain over-producing strains of *Rhodobacter sphaeroides* [6]. The His-tagged cytochrome *c* oxidases showed the same properties as the non His-tagged forms preliminary described [5]. The large amounts of proteins, which were easily and rapidly obtained by Ni^{2+}-NTA affinity chromatography method [7], allowed us to performed an in-depth structural and functional characterization of the Mg/Mn site mutated cytochrome *c* oxidases.

2.2. Metal Content Analysis

The amounts of Mn^{2+} (paramagnetic ion, detectable by EPR in contrast to Mg which is silent) incorporated in the protein has been shown to depend on the [Mg] to [Mn] ratio in the growth medium [4]. The strains carrying the mutated enzymes were grown on different media. The presence or loss of the metal ion was assessed by analyzing the metal content of the mutated enzymes either by EPR (Electron Paramagnetic Resonance) or by ICP (Inductively Coupled Plasma Emission). The structural integrity of the neighboring redox centers, Cu_A and hemes a and a_3, was also investigated by ICP, UV-visible spectroscopy, and EPR.

2.3. Thermal Stability Study

The transitions which occur upon heat denaturation in the optical spectra of cytochrome *c* oxidase were monitored using a Beckman DU 650 spectrophotometer equipped with a High Performance Temperature Controller allowing us to scan a wide range of temperature with an increase rate of 1°C/min. The denaturation temperatures for the heme region and Cu_A site were determined by calculating the first derivative of the absorbance spectra: the T_m was the temperature with the minimum first derivative value.

Temperature inactivation was investigated by incubating WT and mutant cytochrome *c* oxidases at different temperatures for 1min., taking the sample out and placing them in ice. The residual oxygen consumption activity of these different samples was then measured, randomly, at 25°C. Inactivation temperatures ($T_{1/2}$) were the incubation temperature at which is measured half of the initial oxygen consumption activity of the enzyme (at 25°C).

2.4. pH Dependence of the Activity of Purified Cytochrome *c* Oxidases

The oxygen consumption activity of the enzymes was measured polarographically by using a Gilson Model 5/6H oxygraph at 25°C, as described in [3].The reaction medium contained 0.05 % lauryl maltoside, 2.5 mM ascorbate, 1 mM TMPD, and from 5 to 15 nM of oxidase. The reaction was started by adding cytochrome *c* to a final concentration of 0.04 mM. The buffers were prepared to keep a constant ionic strength over the entire pH range and were: 17 mM Potassium Acetate-HCl from pH 4.5 to 5.5, 27 mM Sodium Cacodylate-HCl at pH 5.5, 36 mM Sodium Cacodylate-HCl at pH 6.0, 50 mM Sodium Cacodylate-HCl at pH 6.5, 36 mM Sodium Cacodylate-HCl at pH 7.0, 27 mM Sodium Cacodylate-HCl at pH 7.5, 17 mM Bis-Tris-Propane-HCl (BTP) at pH 6.5, 20 mM BTP at pH 7.0, 36 mM BTP at pH 7.5, 30 mM BTP at pH 8.0, 36 mM BTP at pH 8.5, 50 mM BTP at pH 9.0, 100 mM BTP at pH 9.5, and 50 mM Borate-NaOH pH 9.5 and 10.0.

For the stability to pH experiments, the enzyme samples were incubated for 10 min in the buffers used for the pH dependence study and the remaining activity was measured in 20 mM Bis-Tris-Propane-HCl, pH 7.0.

The turnover numbers were calculated as described in [8] and plotted as a function of the pH checked after the reaction. At the cytochrome *c* concentration and ionic strength used, the measured turnover numbers correspond to the maximal activity of the enzyme.

2.5. Deuterium Isotope Kinetic Effects

The use of a stopped-flow spectrophotometer allowed us to measure the effect of D_2O only on the outside channel: the mixing time being very short, there is minimal leaking of D_2O through the lipid membrane and minimal effect of D_2O on the inside channels. The enzymes were reconstituted in lipid vesicles as described in [3] and the cytochrome *c* oxidation rates were measured as a function of increasing amounts of D_2O on the outside of the membrane. The rate of cytochrome *c* oxidation was determined by following the decrease in absorbance at 550 nm after mixing of reconstituted cytochrome *c* oxidase (0.08 µM final concentration) with reduced cytochrome *c* (2 µM final concentration). The turnover numbers were the rate of cytochrome *c* oxidation multiplied by the cytochrome *c* concentration and divided by the enzyme concentration.

2.6. D$_2$O Access to the Manganese Site

Two sets of experiments, ESE (Electron Spin Echo)-detected EPR and three pulsed ESEEM, were performed on WT oxidase with Mn and H411Q with Mg incorporated into the metal site. A sample (*ca* 30 µM final concentration) was diluted in deutered buffer (50 mM KH_2PO_4-KOH, pH 7.6, 0.1 mM EDTA, 0.5 mM KCl, 0.1% lauryl maltoside) and immediately transferred into EPR tube and frozen in liquid nitrogen. ESE-detected EPR and three pulsed ESEEM spectra were recorded on a home-built pulsed EPR spectrophotometer [9].

3. RESULTS AND DISCUSSION

3.1. Spectral Characterization of Mutants with No Metal (D412A) or Altered Metal Binding (H411Q) at the Mg/Mn Site

Characterization of proteins after site-directed mutagenesis is an essential step for understanding the cause of the altered functional properties. In the case of the cytochrome c oxidase, we have at our disposal a variety of spectroscopic techniques with which to investigate the structural environment of the redox centers of the molecule, and thus obtain insight into the degree of localization of the effect of a mutation.

3.1.1. Magnesium and Manganese Content. It appeared that, unlike D412A and all the Mg/Mn site mutants described so far [4], H411Q still retains the ability to bind Mn or Mg atoms to near stoichiometric levels (0.6 to 0.7 mol/mol of protein, as determined by ICP). The overall lineshape of EPR spectrum arising from the Mn atom in H411Q is similar to the WT (Figure 2), indicating that the Mn is still in a six-coordinate octahedral ligation environment. However, the observed minor differences show that there is some structural change at the Mn binding site, as would be expected if a glutamine nitrogen replaced the histidine 411 nitrogen.

3.1.2. Copper Sites and Hemes. Mutation at the Mg/Mn site, especially a mutation leading to the loss of the positively charged metal, like in the D412A enzyme, is likely to induce some structural and functional changes at the neighboring redox sites: the binuclear Cu_A center and the hemes a and a_3 (Figure 1). Visible and EPR analysis of the binuclear Cu_A site—in spite of a shared ligand with Mg—showed little or no modification

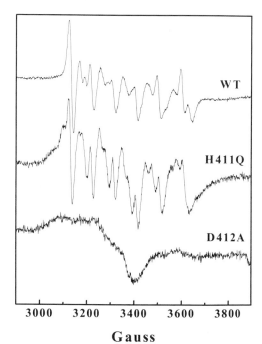

Figure 2. X-band EPR spectra at 110 K of the WT oxidase and Mg/Mn site mutants, isolated from *Rhodobacter sphaeroides* cells grown in high [Mn], low [Mg] conditions, showing (or not) the signal due to the Mn atom. The sample concentrations are *ca* 30 µM, in 50 mM KH_2PO_4-KOH, pH 7.6, 0.1 mM EDTA, 0.5 mM KCl, 0.1% lauryl maltoside. The EPR spectra were recorded as described in [4].

Table 1. Denaturation temperatures (T_m), measured spectrally, of the heme region and of the Cu_A site in the oxidized state. Inactivation temperatures ($T_{1/2}$), measured by activity loss, of WT and mutant cytochrome *c* oxidases

	Spectral denaturation		Activity loss
	T_m (°C) heme region	T_m (°C) Cu_A site	$T_{1/2}$ (°C)
WT-Mg	64.8	64.9	56.4
H411Q-Mg	51.5	52.01	53.9
D412A	52.8	55.0	55.3

from these mutations. Resonance Raman on H411Q, D412N [10] and H411A/Y [10] revealed no modification in the vibrational mode of the hemes a or a_3.

Interestingly, we have in our possession two differently affected kinds of Mg/Mn site mutants as regard the presence or absence of metal. This could account for the differently altered rates of oxygen consumption (Table 2): H441Q is as active nearly as the WT protein at pH 7.5 (retains about 85% activity), while all the other mutants that have lost the metal ion (D412A, H411A/Y [4], D412N [4]) retain less than 50% activity. Since, neither the Cu_A site nor the heme sites show spectral evidence of alteration in these mutants, they cannot be responsible for the observed functional change upon mutation at the Mg/Mn site. However, the heme propionates, the Mg/Mn site, the Cu_B ligands and a number of water molecules create an extended hydrogen bound network [11], that may control pK_a values involved in water or proton movements.

3.2. Testing a Role in Structure/Stability: Studies of Thermostability of Mg/Mn Site Mutants

3.2.1. Thermal Denaturation of Cytochrome c Oxidase Monitored Spectrally. The optical spectra of aa_3 type oxidases show intense absorbance maxima (Soret peak) at 420 nm in the oxidized state. These features are good markers for the integrity of the heme *a* and a_3 local surroundings. The Cu_A site of the cytochrome *c* oxidase displays a broad absorption band at 850 nm in the oxidized state. Since these redox centers are at the interface between subunits I and II, following changes in their spectra is a convenient way to investigate the behavior of this region as a function of denaturing conditions.

3.2.1.1. Wild-Type Oxidase. Increasing the temperature induced a diminution of the absorbance at 420 and 850 nm in the oxidized state. The temperature dependencies of the absorbance are S-shaped which is characteristic of cooperative "all-or-none" melting process. The decrease in absorbance in the Soret region results from the shift of the Soret peak from 420 to 412 nm. This blue-shift of the wavelength indicates that the heme region becomes more exposed to solvent while the heat denaturation occurs. This larger exposure to solvent may result from the loosening of the subunit I/II interface, opening wider the water channel connecting the hemes to the outside. The signal due to the Cu_A at 850 nm disappears as the temperature increases, evidence of the release of the coppers by the heat-denatured molecule. The denaturation temperatures measured for these redox sites are very similar (Table 1), suggesting a close connection between these metal sites located in two different subunits.

3.2.1.2. Mg/Mn Site Mutants. The alteration of the Mg/Mn binding site (H411Q) or the absence of the metal (D412A) induce a large decrease (10 to 12°C) in the thermo-

Table 2. Observed pK_a and maximum turnover numbers of the oxygen consumption activity for purified cytochrome *c* oxidases

	WT	H411Q	D412A
Turnover$_{max}$ (e^-/sec)	1550 ± 40	1380 ± 30	510 ± 10
pK_1	5.1 ± 0.05	5.3 ± 0.05	4.8 ± 0.1
pK_2	8 ± 0.05	8.7 ± 0.05	8.4 ± 0.05

stability of the heme region and Cu_A site in the oxidized state (Table 1). As a rule, single amino acid substitutions alter protein thermostability, but significant changes (10–15°C) are relatively rare. Therefore, we can conclude that the metal at the Mg/Mn site plays a significant role in stabilizing the heme and Cu_A regions.

3.2.2. Temperature Inactivation of the Enzyme Monitored by Activity. The observed inactivation temperatures are very similar for WT and mutants enzymes (Table 1). The inactivation of the enzymes precedes the thermal denaturation of the hemes and Cu_A suggesting this involves another critical, rate-determining feature of the enzyme, such as the spectrally silent Cu_B center.

3.2.3. Conclusions. Although the structural characterization of the H411Q mutant showed the presence of the metal in this protein, the altered ligation of the Mg/Mn seems to affect the stability of the adjacent metal centers to the same extent as the loss of the metal (D412A). In metalloproteins, the protein/metal combination obeys rigid constraints: the metal imposes its own parameters to the protein and participates in establishing the structure of the molecule. Any modification of these constraints should result in a significant effect on the stability of the protein, as observed. The Mg/Mn site appears to play a significant role in the stability of its neighboring redox centers, presumably *via* its involvement as a bridge between subunits and in the hydrogen bond network.

3.3. Testing a Role in Regulation of Rate and Efficiency: Effect of pH on Oxidase Activity

The effect of pH on the activity of the cytochrome *c* oxidase is of particular interest because hydrogen ions participate directly in the catalytic function of the enzyme (oxygen reduction to water and proton translocation). Steady-state oxygen consumption and electron transfer activities are highly pH dependent: turnover rates decrease as the pH increases [12]. This suggests that some proton binding (or release) steps may modulate intramolecular electron transfer [13]. We have proposed that the Mg at the subunits I-II interface may be important for maintaining the pK_a of a residue (or residues) at a suitably low value to release the protons to the more acidic side of the membrane but not take them back up. To determine if this is the case, the pH dependence of the activity of WT and mutant enzymes was examined in the purified state.

3.3.1. pH Dependence of the Activity of Purified Enzymes. The pH dependence of the activity of the beef heart cytochrome *c* oxidase, in the solubilized state [12] and reconstituted form [13] as well, can be described by a model in which at least three sites can take a proton (with pK_a equal to 4.8, 6.5 and 8.0). The pH profiles of the *Rhodobacter sphaeroides* WT enzymes in the purified state showed a bell-shaped curve with a pH opti-

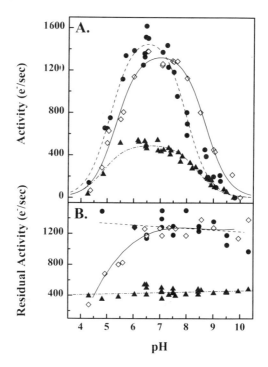

Figure 3. pH dependence of the oxygen consumption activity (A) and stability to pH (B) of purified enzymes. WT (●), H411Q (◇), D412A (▲). Each point is the mean of at least 3 experiments. (---), (—) and (-·-) are the fit to a diprotic model [$v_{obs}=V_{max}/(1+[H^+]/K_1+K_2/[H^+])$] of WT, H411Q and D412A data points, respectively.

mum (Figure 3). The apparent decrease in activity at low pH has been ascribed to the dramatic increase in the K_m for cytochrome c [12]. The resulting pH optimum is obvious in the case of the WT enzyme, while the turnover numbers measured between pH 5.5 and 8.0 are virtually the same for H411Q and D412A. Two apparent pK_a were determined from the pH profiles of these enzymes which evidence an alkaline shift of 0.4 to 0.7 pH unit in case of the Mg/Mn site mutants (Table 2).

The effects of the pH are not due to changes in stability, since preincubation at different pH values (from pH 6.5 to 10.4) did not affect the activity when assayed at pH 7 (Figure 3). In the case of H411Q mutant, the decrease in activity after preincubation at pH lower than 6 (Figure 3) explains the observed slight shift of its acidic pK compared to WT (Table 2).

3.3.2. pH Effect on the Mn Coordination. In an attempt to identify the amino acid residues involved in the pH regulation of the cytochrome c oxidase, we also examined the effect of pH on the EPR spectra of the WT and H411Q enzymes If one of the Mg/Mn ligand undergoes a protonation, some shifts in the EPR lineshape can be expected. However, the EPR spectra of enzymes dialyzed against buffers with pH ranging from 5.5 to 9.5 do not show any significant change compared to the EPR spectrum of the enzymes at pH 7.5. The protonation state of the metal ligands does not appear to change during the pH titration of the cytochrome c oxidase. These preliminary EPR data would suggest that none of the pK_a values are due to direct ligands of the Mg/Mn center.

3.4. Testing the Mg/Mn Involvement in the Water Channel: Deuterium Isotope Effects and D_2O Access

Organized chains of water molecules have been proposed to be part of proton channels: hydrogen bonded chains of water molecules have been described in the crystal structure of the photosynthetic reaction center and invoked in the movement of protons [14] and when the proton conducting capacity of bacteriorhodopsin [15] and ATP-synthetase [16] is lost, FT-IR (Fourier Transformed Infra-Red) studies showed the disappearance of a particular infrared continuum ascribed to hydrogen-bonded chains. For both enzymes, proton pathways have been proposed involving side chains and structural water molecules. A number of water molecules are resolved in the 3D structure of cytochrome c oxidases [11], at the subunit I/II interface where the proton exit channel is predicted. Water molecules that provide three out of six ligands for the Mg metal, form part of this proposed water channel.

3.4.1. Stopped-Flow Analysis of Externally Added D_2O on Activity of Coupled Oxidase. D_2O molecules form stronger hydrogen bonds than H_2O molecules and consequently exchanging H_2O for D_2O is likely to slow down any process involving water movements. The rate of cytochrome c oxidation decreases linearly with the increase in D_2O concentrations and a k_H/k_D ratio can be defined as the rate of cytochrome c oxidation in 100% H_2O to the extrapolated rate in 100% D_2O. D_2O exchange induces a decrease of about 37% of the cytochrome c oxidation rate in the WT enzyme. In the case of Mg/Mn site mutants, the decrease induced by D_2O exchange is 54.2% and 62.4% for H411Q and D412A, respectively.

3.4.2. Water/D_2O Access to the Mg/Mn Site. ESEEM is a technique of choice to examine the magnetic nuclei in the vicinity of the Mn^{2+} atom, especially because it can evidence water molecules bound to transition metals [17]. The ESE-detected EPR spectrum of the WT is a composite of contributions from the Cu_A center and the Mn atom (Figure 4). As a control, an ESE-detected EPR spectrum of a sample containing Mg was taken under identical conditions: the control sample showed only absorption from the Cu(II) and demonstrated that the Mn(II) signal can be studied 300 Gauss up field from the Cu(II) signal (Figure 4).

A three pulse ESEEM measurement was then done on the WT enzyme at 3400 Gauss, where only the Mn contributes to the signal. Figure 5 shows that, during a few minutes mixing time, all the water molecules were very quickly exchanged for D_2O: only deuterons (2.17 MHz) and no protons (13.3 MHz) were detected, indicating that most of the waters coordinated to the metal had been substituted with D_2O.

These ESEEM experiments showed the feasibility of this method to detect the solvent accessibility of the Mn atom; by combining electron spin echo with rapid-mix freeze-quench technology, we will be able to determine the kinetics of the water flow in the outside channel.

3.4.3. Conclusions. In the Mg/Mn site mutants, the larger effect on the rate of cytochrome c oxidation by the presence of D_2O on the outside is consistent with a disturbed water channel, allowing more access of D_2O. One can expect that if the water molecules generated at the binuclear center during the oxygen reduction cannot leave this environment as fast because of the presence of tighter binding D_2O, inhibition might occur. If proton exit is also occurring through this same pathway, it may also be slowed by the stronger bonding of the deuterium ions to critical residues.

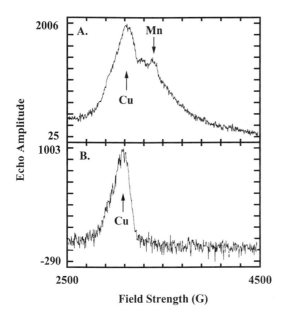

Figure 4. ESE-detected EPR spectra of WT-Mn (A) and H441Q-Mg (B) cytochrome *c* oxidases. Conditions as described in Experimental Procedures.

3.5. Concluding Remarks

Studies on the Mg/Mn site mutants, H411Q and D412A, have shown that the metal site is one of the key players in the structural and functional properties of the proposed exit water channel:

1. The modification of the Mg/Mn site induces a significant decrease in thermostability of the redox centers at the subunit I/II interface, indicating a role in stability.
2. Alteration and/or loss of the metal ion at the Mg site alters the pH dependence of the purified enzyme, indicating regulation of activity at this external center.
3. The Mg/Mn site is rapidly accessible to D_2O and externally added D_2O preferentially inhibits the turnover of mutants in the Mg/Mn site, consistent with a role of this metal site in the structure of the proposed water/proton exit channel.

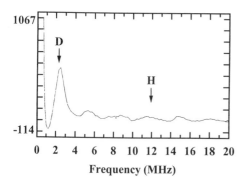

Figure 5. Fourier transform of a three pulse ESEEM collected at 3400 G showing the contribution from deuterium exchanged into the Mn(II) binding site upon mixing.

Defining the location and properties of the proton/water exit route is critical to understanding the mechanism of energy transduction and how efficiency is controlled. These studies support a role for the Mg in the exit process, thus providing a useful tool for further investigation of the kinetics of proton/water movement in this region.

ACKNOWLEDGMENTS

This work was supported by NIH GM 26916 (to S.F.M), NIH 1F32 GM 18205 (to D.A.M.) and was done during the tenure of a Research Fellowship of the American Heart Association, Michigan Affiliate (to L.F.).

REFERENCES

1. Iwata, S., Ostermeier, C., Ludwig, B., & Michel, H. (1995) Nature (London) 376, 660–669.
2. Tsukihara, T., Aoyama, H., Yashimata, E., Tomozaki, T., Yamaguchi, K. Shinzawa-Itoh, K., Nakashima, R., Yaono, R., & Yoshikawa, S. (1995) Science 269, 1069–1074.
3. Hosler, J. P., Fetter, J., Tecklenburg, M. M. J., Espe, M., Lerma, C. & Ferguson-Miller, S., (1992) J. Biol. Chem. 267, 24264–24272.
4. Hosler, J. P., Espe, M. P., Zhen, Y., Babcock, G. T. & Ferguson-Miller, S. (1995) Biochemistry 34, 7586–7592.
5. Fetter, J. (1995) Search for residues critical to proton pumping in cytochrome c oxidase, Ph.D., Michigan State University, East Lansing.
6. Zhen, Y., Follman, K., Qian, J., Howard, T., Nilson, T., Dahn, M., Hamer, A.G., Hosler, J.P., & Ferguson-Miller, S. Overexpression and purification of cytochrome c oxidase from Rhodobacter sphaeroides. Submitted.
7. Mitchell, D.M., & Gennis, R.B. (1995) FEBS Lett. 368, 148–150.
8. Thompson, D.A., & Ferguson-Miller, S. (1983) Biochemistry 22, 3178–3187.
9. McCracken, J., Shin, D.-H., & Dye, J.L. (1992) Appl. Magn. Reson. 3, 305–316.
10. Hosler, J.P., Shapleigh, J.P., Tecklenburg, M.M.J., Thomas, J.W., Kim, Y., Espe, M., Fetter, J., Babcock, G.T., Alben, J.O., Gennis, R.B., Ferguson-Miller, S. (1994) Biochemistry, 33, 1194–1201.
11. Ostermeier, C., Harrenga, A., Ermler, U. & H. Michel (1997) Proc. Natl. Acad. Sci USA 94, 10547–10553.
12. Wilms, J., Van Rijn, J.L.M., & van Gelder, B.F. (1980) Biochim Biophys. Acta 590, 17–23.
13. Thronstrom, P., Soussi, B., Arvidsson, G. & Malmstrom, B.G. (1984) Chem. Scr. 24, 230–237.
14. Baciou, L. & Michel, H. (1995) Biochemistry 34, 7967–7972.
15. Olejnik, J., Brzezinski, B., & Zundel, G. (1992) J. Mol. Struct. 271, 157–173.
16. Bartl, F., Deckers-Hebestreit, G., Altendorf, C., & Zundel, G. (1995) Biophys. J. 68, 104–110.
17. McCracken, J., and Friedenberg, S. (1994) J. Phys. Chem. 98, 467–473.

39

ADENYLATE REGULATION OF THE CYANOBACTERIAL CYTOCHROME c OXIDASE

Daniel Alge, Marnik Wastyn, Christian Mayer, Christian Jungwirth,
Ulrike Zimmermann, Roland Zoder, Susanne Fromwald, and
Günter A. Peschek

Biophysical Chemistry Group
Institute of Physical Chemistry
University of Vienna
A-1090 Vienna, Austria

1. INTRODUCTION

According to common reasoning primordial cyanobacteria were the first to introduce molecular oxygen into a previously near-anoxic biosphere (1–3). Thereby they basically initiated the whole evolutionary succession from aerobic prokaryotes to "primitive" eukaryotes, metaphyta, metazoa and up to the very *Homo sapiens* all of which essentially depend on aerobic respiration and oxidative phosphorylation for energy supply (4–6). Having identified the terminal respiratory oxidase in almost thirty different strains and species of cyanobacteria as a mitochondria-like aa3-type cytochrome c oxidase by means of conventional biochemical and immunological techniques (7–10) we succeeded in cloning and sequencing a cta operon-like genomic structure from *Synechocystis sp. PCC6803* with a very high degree of homology to other bacterial cta operons, particularly as far as the ctaCDE genes are concerned which, in eukaryotes, encode the mitochondrial subunits I-III (11–16). A very similar cta operon was recently published for the taxonomically more uncertain cyanobacterium *Synechococcus vulcanus* (17,18). When the entire genomic sequence of *Synechocystis 6803* was published (Ref. 19; available also on the internet under *http://www.kazusa.or.jp/cyano*) it turned out that this genome contained potential coding regions for an aa3-type cytochrome c oxidase, and an aa3-type and a d-type quinol oxidase. Yet, while Sugiura's cyt-c oxidase displayed >95% deduced amino acid identity to the biochemically characterized cyanobacterial cyt-c oxidase (11–13) and >50% identity to the deduced *S. vulcanus* proteins (17,18) no indications of a functional quinol oxidase were so far detected in any wild-type cyanobacterium (20–23). Therefore, considering the fact that primordial cyanobacteria most probably not only were the first oxygenic photosythesizers but also the first aerobic respirers we have suggested that, of the two known types of aerobic terminal oxidases, viz. cytochrome c and quinol oxidases (24), the cytochrome c oxidase preceded the quinol oxidase in evolution (25).

The Phototrophic Prokaryotes, edited by Peschek *et al.*
Kluwer Academic / Plenum Publishers, New York, 1999.

Extended analysis of the cta operon from our *Synechocystis* (but also from *Anacystis nidulans* (*Synechococcus* sp. PCC6301), results to be presented elsewhere) showed (i) that a small open reading frame ("ORF4" or ctaFb gene) potentially coding for a 44-amino-acid "bacterial subunit-IV" but otherwise unrelated to known subunits-IV from bacteria (see Fig. 4) was inserted after the ctaE gene, (ii) that still another open reading frame putatively coding for 187 amino acids (ctaFm gene) followed the ctaFb gene and (iii) that this ctaFm gene was followed by a pronounced terminator structure (see Figs. 2 and 3). In *S. vulcanus* such terminator structure was found already after the ctaFb gene (18), and in Sugiura's *Synechocystis* 6803 sequence (19) neither ctaFb nor ctaFm genes were detected which, however, might be related to partial losses of gene fragments during DNA isolation (26). Still more interesting, the deduced ctaFm gene product of our *Synechocystis* 6803 (as well as our *Anacystis nidulans*; not to be discussed here), near its N-terminus, showed striking sequence similarities to adenylate-binding proteins such as ATP synthase and adenylate kinase. The transcribed COIVm protein could thus be responsible for the specific regulation of cyanobacterial cytochrome c oxidase activities by ATP (inhibition) and ADP (stimulation). The respiratory situation in cyanobacteria might, therefore, be similar to the recently established allosteric adenylate regulation of mitochondrial cytochrome c oxidase (30,31) whose (nuclear encoded!) subunit IV appears to play a decisive role for energy charge effects (viz. ATP binding) on the enzyme (32). Evolutionarily speaking it is thus conceivable that, on the way from respiring prokaryotes to eukaryotes, cyanobacteria were the first to elaborate an (adenylate-regulated) mitochondria-like subunit IV. Later on in eukaryotic cells the gene coding for this "pre-mitochondrial" subunit IV, i.e. the ctaFm gene, was transferred and integrated into the nuclear genome as was the majority of originally prokaryotic genes in mitochondria, chloroplasts, and even endocyanelles anyway (33). To most cyanobacteria, which are known to be metabolically not very versatile and to respire, in fairly constant environments basically for the generation of maintenance energy during periods of darkness only, it might be physiologically meaningful and economic to slow down respiration (hence, ATP synthesis) when enough ATP is available and to stimulate it when ATP is limiting.

The present article deals with the entire cta operon of *Synechocystis* sp. PCC6803 and deduced proteins (including the detailed structure of ctaFm and deduced protein) and the adenylate regulation of membrane-bound and liposomal cyanobacterial cytochrome c oxidase.

2. MATERIALS AND METHODS

All genetic manipulations performed on genomic DNA from *Synechocystis* 6803 (and also *Anacystis*; not shown here) were described in detail previously (11–13). Nucleotide sequence data reported in these papers were deposited in the EMBL Data Library, Heidelberg, Germany (accession number X53746). For comparison, the complete genomic sequence of the organism (19) is available on the internet under *http://www.kazusa.or.jp/cyano* and the cta sequence data of *Synechococcus vulcanus* can be found in Refs. 17 and 18. Data on the cta deduced protein sequences of *Anacystis nidulans* (*Synechococcus* sp. PCC6301) which display, on an average, 50–60% amino acid identity to *Synechocystis* (including the ctaFm gene) will soon be published elsewhere.

Milligram amounts of cytochrome c oxidase from crude membrane preparations of cyanobacteria (see Concluding Remarks) were isolated and purified by affinity chromatography on immobilized cytochrome c from *Saccharomyces cerevisiae* (34) and antibody raised against membrane-eluted subunits I and II of *Anacystis nidulans* cytochrome c oxi-

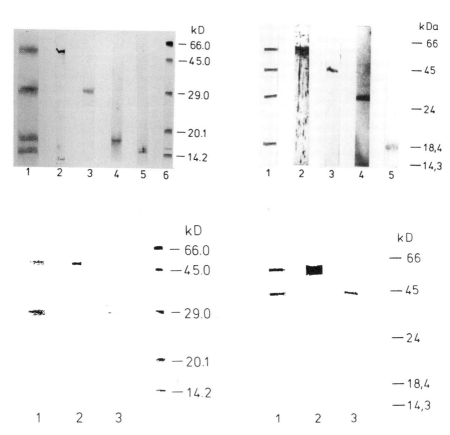

Figure 1. Immunoblotting of isolated and purified cyt-c oxidases from *Anacystis nidulans* (*Synechococcus* 6301) (panels A and C) and *Synechocystis* 6803 (panels B and D) either freshly prepared and purified by affinity chromatography (Refs. 36–38) (panels A and B) or after treatment with 75 mM lauryl maltoside at 4°C over night (panels C and D). Polyclonal antibodies raised against authentic CO polypeptides from *Paracoccus denitrificans* (courtesy of B. Ludwig, Frankfurt, Germany, and A. Azzi, Bern, Switzerland) and rat liver mitochondria (courtesy of B. Kadenbach, Marburg, Germany) gave qualitatively identical results when applied to cyanobacterial subunits I, II and III on lanes 2, 3 and 4, respectively. The unexpected cross reaction with antibody raised against mitochondrial COIV protein (kindly donated by B. Kadenbach, Marburg, Germany) is seen on lane 5. Lanes 1 show Coomassie stained SDS-PAGE patters of freshly prepared (A, B) and 75-mM lauryl maltoside treated (C, D) oxidase preparations from *Anacystis* (A, C) and *Synechocystis* (B, D). Lanes on the right-hand side of each panel show marker proteins (Sigma MW-SDS-70L-Kit).

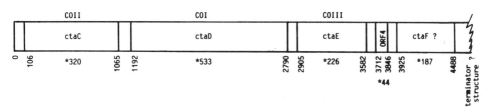

Figure 2. The schematically drawn cta operon of *Synechocystis 6803* (11–13) supplemented by the recently discovered mitochondria-like ctaFm gene and terminator structure (14). Compare the slightly different cta operon of *Synechococcus vulcanus* (18) and the cta gene sequences of Sugiura's *Synechocystis* 6803 (19). Numbers on the Figure refer to the number of coding nucleotides (vertical) and of deduced numbers of amino acids (*horizontal).

```
                                                         ctaF? ──→
                                                                              KpnI
         3860        3880        3900        3920        3940        3960
GGTTGATGGCCCCTGGTATCCAATCTTGCGGGCTTAATCTGACCTTAATCTAATATTGCTCCCAGCTTAAAGGAGAGTTACCCATGCCCGAAAATTGGTACCTCCTGATCATTCTG
                                                                            =====
G *                                                                  M   A   R   K   I   G   T   S   S   I   I   L

         3980        4000        4020        4040        4060        4080
TACGTGGATTCGGTGCGAGCGGATTCCGGAACCGATAACCCCGGCTGGGCGCCAGACCATCTTGATGCGGCCGCCATTAACTCCAGGTCGAGGATGTGCCTCGAAGGAA
 Y   V   D   S   V   R   A   D   S   G   T   D   N   P   G   V   G   A   Q   T   H   L   D   A   A   G   M   E   I   N   S   R   F   E   D   V   P   S   K   E

         4100        4120        4140        4160        4180        4200
CAGAACGATGGGCCCCGTCGAACGGAAAAGGCACGGAGCAACAGAAGCCCCGTGGGCCCAGCTCATTATCAACGAGAAGCAGATTGTGGTACATTATCTACTTCAGTGGGTTCCC
 Q   N   D   G   P   P   S   N   G   K   G   T   E   Q   Q   K   P   P   W   A   Q   L   I   I   N   E   K   Q   I   V   W   Y   I   I   Y   F   Q   W   V   P

                                                 BamHI
         4220        4240        4260        4280        4300        4320
CGCCCCCGTTGATTAACAAACGGGAGTCGTCGTACGCAGCGGAAACCCGTAATCATCTTTTTACTGTTCGCTGTTGGGATCTTCGCGGTGTTGATTATGGTTGATAAAGGGGGCCCGCT
 R   R   P   L   I   N   K   R   E   S   S   Y   A   A   K   P   V   I   I   F   L   L   F   A   V   G   I   F   A   V   L   I   M   V   D   K   G   G   P   A

         4340        4360        4380        4400        4420        4440
ACGTATAGGAAGGAGGTGGGAATTGAAAAACTCCGATGAATGGCTGAGGACCAACAAGCGAACCCTTTGGGGCTACTCCACAGTAACACCAAACCCTGCGAAGAGGAC
 T   Y   R   K   E   W   E   L   K   N   S   D   E   W   L   R   T   K   Q   A   N   P   F   G   G   Y   S   T   A   N   T   K   S   T   K   P   C   E   E   D

         4460        4480
GGGCGACGCTGGTTTGGCGCGTTAGTAGCGGTCCCATCATCAAGGGTGAGAAAAGGGGAAATCGGTTTTTTTACACCGGGTTTCCCCTTTTTCCAA
 G   R   R   W   F   G   A   L   V   T   S   H   H   Q   G   *
```

Figure 3. Part of the *Synechocystis* 6803 cta operon (11–13) between ctaE gene and the terminator structure showing, in particular, the nucleotide and deduced amino acid sequence of the newly discovered, putative ctaFm gene. A very similar genomic organization was recently found in the *Anacystis nidulans* chromosomal DNA (in preparation).

```
1   LLAAIQVIFQLYYFMHMNQKGHEAPALKLYSGVFVAFITVLAFVTIIWW  (110)
                    *  *
2   LLAAVQVAFQLYYFMHMSHKGHEFPAMFIYGGVAVMLLVWAFTTVVWW  (110)

3   MNPSPRERSILVNFFHNGHRKNHQH---------------------HRRQTKVKVGNPPLVIAAKDPRLKG  (50)

4   MKEFLTIMSLLEITTDAEFEQETQ----------------GQTKPCWFIFGLPGGVALVG  (44)

5   ASHHEITDHKHGEMDIRHQQATFAG-------------------------FIKGAT  (31)
```

ORFs 4 (subunits IV?) in the cta operons of Bacillus subtilis (1), bacillus PS3 (2), Synechococcus vulcanus (3), Synechocystis sp. PCC6803 (4) and Paracoccus denitrificans (5).

Figure 4. Sequence alignment of the deduced small proteins putatively encoded by bacterial ctaFb genes ("ORFs 4") of *Bacillus subtilis* (line 1), *Bacillus* PS3 (2), *Synechococcus vulcanus* (3), *Synechocystis* 6803 (4), and *Paracoccus denitrificans* (5). Bracketed numbers on the right-hand side of each sequence give the total numbers of amino acids in the deduced proteins.

Table 1. Effect of intraliposomal ATP, ADP and AMP on initial rates of horse heart ferrocytochrome c oxidation and coupled proton extrusion by proteoliposomal cytochrome c oxidases isolated and purified from *Anacystis* and *Synechocystis* (symbolized by **ANA** and **SYN**, respectively)

Intra-liposomal nucleotide	Cytochrome c oxidation		Proton extrusion		H^+/e^-	Cytochrome c oxidation		Proton extrusion		H^+/e^-
	ANA	SYN	ANA	SYN		ANA-	SYN-	ANA-	SYN-	
None	61	95	55	90	0.9	53	83	46	78	0.9
ATP	48	55	13	16	0.3	50	85	48	80	0.9
ADP	135	215	125	205	0.9	56	90	53	85	0.9
AMP	60	90	50	85	0.9	55	86	48	80	0.9

In parallel, cyt-c oxidases from which subunits IVm (and III) had been removed by detergent treatment (see Fig. 1CD; symbolized by **ANA-** and **SYN-** in the Table) were also used for proteoliposome preparation which was performed according to the exhaustive dialysis method of Kaback (39). Cyanobacterial cyt-c oxidase proteoliposomes were preloaded with initially 5–20 mM MgATP, MgADP and MgAMP present in the dialysis buffer during proteoliposome formation. After preparation the proteoliposomes were freed from externally adhering nucleotides by passage through a Sephadex column (for more details see the text.) Cyt-c oxidation and coupled H^+ extrusion were followed spectrophotometrically and through acidification of the weakly buffered assay suspension using a sensitive pH electrode (48–50), respectively. Rates of cyt-c oxidation and proton extrusion are expressed as nmol cyt c or protons/nmol heme a per min. (see Fig. 10). Our cyanobacterial cyt-c oxidase preparations contained 1.8–2.2 nmol heme a/mg protein. Complete release of the nucleotides through exhaustive sonication of the proteoliposomes after the experiments and quantitation of released nucleotides by the firefly method (after enzymatic conversion of ADP and AMP into ATP) showed that, of the 5–20 mM nucleotides initially added approx. 0.1–0.5 mM had stayed entrapped in the proteoliposomes. Values shown in the Table are rounded means from up to fifty independent cyanobacterial cyt-c proteoliposome preparations assayed in the course of altogether six years of experimentation (also see Concluding Remarks), deviations keeping within about 20% of the corresponding mean.

dase (35–38). When the isolated and purified enzyme was treated with 75 mM lauryl maltoside over night the enzyme completely lost subunits III and IVm as evident from immunoblotting (see Fig. 1CD). The truncated enzyme still catalysed near-normal rates of horse heart ferrocytochrome c oxidation but vesicle-entrapped ATP and ADP now were without any effect on the reaction (Table 1). Cyanobacterial cytochrome c oxidase proteoliposomes were prepared with partially hydrogenated (reduced) soybean azolectin according to the exhaustive dialysis method of Kaback (39). MgATP and MgADP as well as MgAMP and glucose-1 phospate for control were added at initial concentrations of 5–20 mM to cell suspensions and enzyme preparations before French pressure cell extrusion and proteoliposome formation, respectively. The concentration of adenylates recovered within the vesicles or proteoliposomes formed, determined on exhaustively sonicated vesicles by the firefly method, was between 0.1 and 0.5 mM.

3. RESULTS AND DISCUSSION

Fig. 1AB shows immunoblots of isolated and purified cytochrome c oxidases from *Anacystis* and *Synechocystis*, respectively, while Fig. 1CD shows the same enzymes after treatment with 75 mM lauryl maltoside (also see Table 1). As already shown previously, upon SDS-PAGE the cyanobacterial cytochrome c oxidases yielded three immunologically cross reactive bands when (polyclonal) antibodies raised against authentic subunits I, II and III derived from either *Paracoccus denitrificans* or rat liver mitochondrial cytochrome c oxidases were used (8,9,35) and an additional fourth band was seen when antibody raised against subunit IV of the mitochondrial cyt-c oxidase was used (40) although mito-chondrial COIV protein is encoded in the nucleus while cyanobacteria obviously do not possess a nucleus at all. The small bacteria-like COIV protein, provided the ctaFb gene is transcribed (see Refs. 18,41), could not be detected so far in immunoblotting experiments. However, the apparent molecular weight of the large, mitochondria-like subunit IV (15.3 and 17.9 kD, respectively; see Fig. 1AB) could well correspond to a protein of 187 amino acids (see Fig. 2 in case of *Synechocystis*).

Fig. 2 schematically shows the entire *Synechocystis* cta operon comprising alto-gether 4488 base pairs and a terminator structure which, together with the end of the ctaFb gene (ORF 4) and the large mitochondria-like subunit IV as deduced from the ctaFm gene, is shown in more detail in Fig. 3. Note that a similar terminator structure, yet without a preceding mitochondria-like COIV protein was described for *Synechococcus vulcanus* (18) while an identical overall arrangement of the cta operon (including ctaFm) seems to obtain in *Anacystis* (results not shown). Fig. 4 compares the sequences of deduced bacte-rial COIV proteins putatively encoded by the ctaFb genes ("ORFs IV"). While *S. vulcanus* COIVb protein (line 3) bears a faint relationship to corresponding *Bacillus subtilis* (1) and *Bacillus* PS3 (2) proteins (see the seemingly conserved phenylalanine and histidine marked with asterisks) no such similarity is seen with *Paracoccus* (5), *Synechocystis* (4) or *Anacystis* (not shown).

Fig. 5 shows models of the membrane topology of *Synechocystis* COI, COII and COIII proteins (i.e. subunits I-III of the cyt-c oxidase; from top to bottom). Note that, since in cyanobacteria the same cyt-c oxidase is found in both CM and ICM the designa-tion "Out" refers to the periplasmic and thylakoid luminal spaces in case of CM and ICM, respectively, while "In" refers to the cytosolic space for both CM and ICM (Ref. 42). The membraneous models were derived from respective Kyte-Doolittle plots (11–13). They are identical to what was derived from the cta operon of *S. vulcanus* (18). Also the similarity to the corresponding model of the *P. denitrificans* cytochrome c oxidase (15) is conspicu-ous. The first two transmembrane helices of subunit III are missing from the cyanobacte-rial enzyme, but subunit I has the usual twelve transmembrane helices, and subunit II has the typical hair-pin structure. All metal-binding amino acid ligands, viz. the eight con-served histidines of subunit I for heme a and the bimetallic heme a3/CuB center, and the six ligands for binding the bimetallic CuA-CuA center in subunit II, viz. two cysteines, two histidines, a methionine and a glutamic acid, are in the correct positions also in the cyanobacterial CO proteins. Subunit III, with the first two transmembrane helices less than the *Paracoccus* counter-part, holds the highly conserved glutamic acid residue that was previously thought to be involved in proton-pumping (E65, = E100 of the *Paracoccus* numbering (15)) but also in cyanobacteria does not contain metal-binding ligands.

Alignments of the (deduced) amino acid sequences of the putative mitochondria-like subunit IV (ctaFm gene product; see Fig. 3) of *Synechocystis* (line1) and the subunits IV of *S. cerevisiae* (2) and bovine heart (3) mitochondrial cyt-c oxidases are shown in Fig. 6. It is seen that there is a 50% amino acid identity between *Synechocystis* and *Saccharomy-ces* COIVm proteins and even the identity between *Synechocystis* and bovine heart still is 20%. Fig. 7 shows a comparison between Kyte-Doolittle plots of (deduced) COIVm pro-teins of *Synechocystis* and beef heart mitochondria. The high degree of structural similar-ity is evident. Fig. 8 illustrates two alternative membraneous models drawn for the

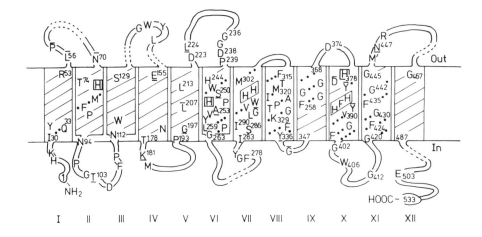

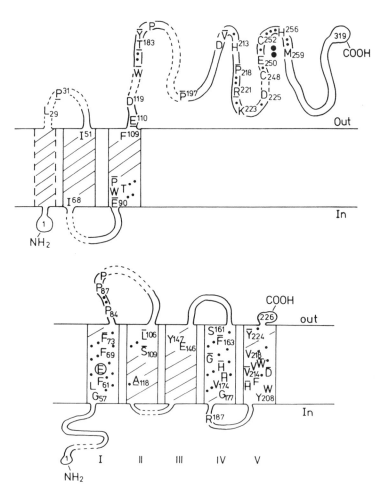

Figure 5. Topographical models of partly transmembraneous subunits I, II and III (from top to bottom) of the *Synechocystis 6803* cytochrome c oxidase as drawn according to the respective hydropathy plots of the ctaCDE-deduced polypeptides, respectively (see Refs. 11–13; also see the text. For comparison with other aa3-type cyt-c oxidases see Ref. 15).

1 putative subunit IV of Synechocystis 6803 cytochrome c oxidase (ctaF gene product)

2 subunit IV of S. cerevisiae cyt c oxidase (Va protein according to Koerner, Hill & Tzagoloff 1985

3 subunit IV of bovine heart cyt c oxidase (Sacher, Steffens & Buse 1979)

```
1 MARKIGTSSI  ILY-V-DSVR  ADSGTDNPGV  GAQTH-L-DA  AGMEINSRFE  DVPSK-EQND  GPPSNG----  KGTEQQK---  ----PPWAQL
2 MLRNTFTRAG  GLSRI-TSVR  F---------  -AQTHALSNA  AVMDLQSRWE  NMPST-EQQD  I-------VS  KLSERQK---  ----LPWAQL
3                         AHGS-----   ----VVKSEDY ALPSYVDRRD  YPLPDVAHVK  NLSASQKALK  EKEKASWSSL
    *  *         ***         ****        *  *  *    *  ***      ***          *            *   **      ****
       +                     +          *  *  *     *  *         *             +         +  ++         + +

1 IINEKQIVWY  IIYFQWVPRR  PLINKRESSY  AAKPVIIFLL  FAVGIFAVLI  MV---DKGGP  -A-TYRKEWE  LKNSDEWLRT  KQANPFGGYS
2 TEPEKQAVWY  ISYGEWGPRR  PVLNKDSSF   IAKGVAAGLL  FSVGLFAVVR  MA----GGQ   DAKTMNKEWQ  LK-SDEYLKS  KNANPWGGYS
3 SIDEKVELYR  LKFKE--SF   AEMNR-STN   EWKTVVGAAM  FFIGFTALLL  IWEKHYVYGP  IPHTFEEEWV  AKQTKRMLDM  KVA-PIQGFS
    ***  **     ****        *          ***  *      *  *  *     *          *  ***  *     **    *     *  *  *
    ++                      +          +  +  +      +  +        + +           +  +       *  +         *  +

1 TANTKSTKPC  EEDGRRWFGA  LVTSHHQG
2 QVQSK
3 AKWDYDKNEW  KK
     *
```

Figure 6. Sequence alignment of the deduced mitochondria-like subunit IV (putative ctaFm gene product) of *Synechocystis 6803* (1) and subunits IV of *Saccharomyces cerevisiae* (2) and bovine heart (3) mitochondrial cytochrome c oxidases. See Refs. 53–55. For discussion see the text. Numbers on the right-hand side of each line give total numbers of amino acids in the respective line. (*) indicates amino acid identities between *S. cerevisiae* and *Synechocystis 6803*. (+) indicates amino acid identities between beef heart and *Synechocystis 6803*.

Synechocystis sp. PCC6803

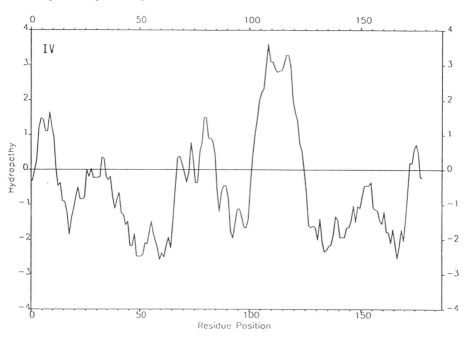

Beef heart

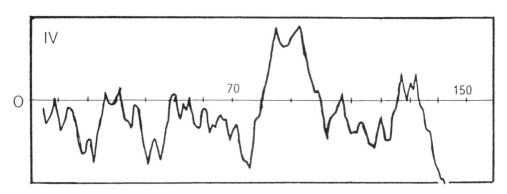

Figure 7. Kyte-Doolittle plots of the deduced mitochondria-like COIV protein of *Synechocystis 6803* (see Fig. 3) and the COIV protein from bovine heart mitochondria (see Fig. 6, line 3).

Synechocystis COIVm protein according to Fig. 7, exhibiting either one or two transmembrane helices. N-terminal domains designated *Walker A, AdKin,* and *Walker B* high-light amino acid stretches with particular sequence similarity to *E. coli* ATPase beta subunit *Walker A* and *B* consensus sequences as well as to part of the adenylate kinase, as seen in more detail from the sequence alignments shown in Fig. 9 (see Refs. 43–46). The adenylate-binding regions of the mitochondria-like subunit IV of the cyanobacterial cytochrome c oxidase could possibly confer allosteric properties with respect to regulation by ATP and

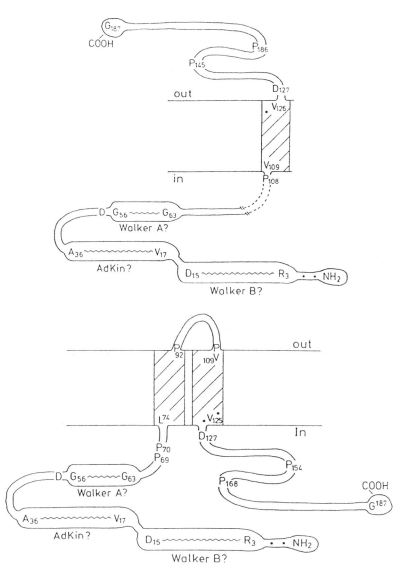

Figure 8. Alternative topographical transmembrane models of the deduced mitochondria-like subunit IV (putative ctaFm gene product) of the *Synechocystis 6803* cyt-c oxidase (see Fig. 3) as drawn with either one (top) or two (bottom) membrane-spanning domains according to the corresponding Kyte Doolittle plot (Fig. 7).

ADP to the enzyme as was previously shown to be the case for membrane-bound and liposomal cyanobacterial cytochrome c oxidase when assayed in the presence of entrapped adenylates (Tables 1 and 2). Results from measurements of both the rates of cyt-c oxidation and of concomitant proton extrusion (47) as paradigmatically shown in Fig. 10 reflected the influence of ATP and ADP on the coupling efficiency (electron/proton ratio) of the proton-pumping cyanobacterial cytochrome c oxidase which, as to cyt-c oxidation, was already documented previously in plasma membranes of intact spheroplasts (see Refs. 48–50; also see the analogous results with *Paracoccus* (51)).

```
1   MARKIGTSSIILY-V-DSVR   DSVRADSGTDNPGVGAQTH-L-DAAGM   SRFEDVPSK-EQNDGPPSNGKGTEQKPPW
2   MLRNTFTRAGGLSRI-TSVR   TSVRF-------AQTHALSNAAVM      SRWENMPST-EQQDI---VSKLSERQKLPW
3   KFRDEGRDVLLFV---DNIY   IYVPADDLTDPAPAVTFAH-L-DATTV   MCPFAKGGKVGLF-GGAGVGKTVNMELIR
4   FERKIGQPTLLLY-V-DAGP   RKVNAEGSVDDVFSQVCTH-L-DTLK   MEEKLKKSKIIFVVGGPGSGKGTQCEKIVQ
     *+++  ++ + +          * +  *   *                     ++   + +  +*++       +
```

```
1   1-18                          15-39                    43-71
    Synechocystis sp. PCC6803 putative COIV protein
2   1-19                          16-32                    36-61
    Saccharomyces cerevisiae COIV protein
3   230-255                       313-336                  137-165
    E. coli ATPase β-subunit
    Walker B consensus    consensus sequence               Walker A consensus
    $RX_nh_4D$            $VXADX_3DX_8HLDA$                 $GX_4GKT$
4   105-122                       171-194                  1-30
    Adenylate kinase
```

Figure 9. Sequence alignment of three (deduced, and putatively) adenylate-binding amino acid stretches of the mitochondria-like COIV protein (putative ctaFm gene product) of *Synechocystis* 6803 (line 1; see Figs. 3 and 8), the *S. cerevisiae* COIV protein (line 2; see Fig. 6 and Refs. 43–45 and 54), the *E. coli* ATPase beta subunit (43,44) showing the so-called Walker A and B and adenylate kinase (AdKin) consensus sequences (line 3; see Ref. 45), and the adenylate-binding sequence of adenylate kinase (line 4; see Ref. 46). (*) indicates amino acid identities between all four polypeptide stretches. (+) indicates amino acid identities between the *Synechocystis* 6803 and either *E. coli* ATPase beta subunits or the adenylate kinase sequence.

Table 2. Effect of intravesicular ATP, ADP and glucose 1-phosohate (for control) on initial rates of horse heart ferrocytochrome c oxidation and coupled proton extrusion by membraneous cyt-c oxidase in CM and ICM vesicles prepared from *Anacystis* and *Synechocystis*, preloaded with nucleotides and G 1-P during French pressure cell extrusion (9,29)

Entrapped compound	Cytochrome c oxidation		Proton extrusion		H^+/e^-	Cytochrome c oxidation		Proton extrusion		H^+/e^-
	ANA CM	SYN CM	ANA CM	SYN CM		ANA ICM	SYN ICM	ANA ICM	SYN ICM	
None	26	3	24	3	0.9	6	38	6	36	0.9
ATP	22	3	7	1	0.3	4	29	2	10	0.3
ADP	36	4	34	4	0.9	8	55	7	50	0.9
G 1-P	25	3	24	3	0.9	6	37	6	36	0.9

After preparation the vesicles were freed from externally adhering nucleotides and G 1-P by passage through a Sephadex column. For details see legend to Table 1. Rates given as nmol cyt-c or protons/mg membrane protein per min. Values shown in the Table are rounded means from >100 independent vesicle preparations of both *Anacystis* and *Synechocystis* CM and ICM vesicles, respectively, prepared and assayed in the course of ten years of experimentation (see Concluding Remarks). Reproducibility of the data obtained always was within approx. 10% of the corresponding mean. It is seen that the proton-pumping cyt-c oxidase of both *Anacystis* and *Synechocystis* shows qualitatively identical behavior towards the nucleotides in both cytoplasmic and thylakoid membranes. At the same time it is obvious that, in *Anacystis* CM contain much more of the enzyme than contain ICM while in *Synechocystis* the opposite situation obtains. (Also see Refs. 8 and 52).

4. CONCLUDING REMARKS

Owing to rather tiny amounts of terminal respiratory oxidase in the membranes of (more or less obligately) phototrophic cyanobacteria, even after growth in salt-stressed conditions which lead to slightly elevated concentrations of respiratory proteins (see Ref. 56), rarely exceeding several nanomol of COX protein per mg membrane protein (22,56,57), phototrophic mass cultures (up to ten 10-L-flasks set up in parallel) of *Anacys-*

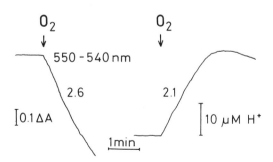

Figure 10. Paradigmatic example of kinetic recorder traces of horse heart cyt-c oxidation (measured by dual wavelength spectophotometry at 550–540 nm) and coupled proton extrusion (measured with a sensitive pH electrode or spectrophotometrically with appropriate pH-indicating dyes). For details see Refs. 47–51. Results from >100 of such measurements on cyanobacterial CM and ICM membrane vesicles and cyt-c oxidase proteoliposome preparations containing entrapped nucleotides, as performed in the course of almost ten years of experimentation (see Concluding Remarks) are shown in Tables 1 and 2.

tis nidulans (Synechococcus sp. PCC6301) and *Synechocystis sp. PCC6803* were continuously run during almost ten years. COX preparations were prepared and pooled from altogether approx. hundred harvests, and stored in concentrated form in 20% (v/v) glycerol at liquid nitrogen temperature. Also owing to the fact that the only reliable and reproducible preparative procedure for the isolation and purification of the COX enzyme from n-octyl glucoside solubilized cyanobacterial membranes still is affinity chromatography whose natural capacity per step of preparation is necessarily limited (36–38), typical yields of COX preparation amounted to a maximum of 50 mg purified COX from 500 g wet weight of cells. This makes it clear that rapid preparation of substantial quantities of terminal respiratory oxidases from cyanobacteria is impossible and that it always takes years until meaningful results on such preparations are obtained. Unfortunately, also the attempts of heterologous expression of cyanobacterial cta genes in *Rhodobacter sphaeroides* did not bring about an improvement of this poor situation (D. Alge, unpublished).

ACKNOWLEDGMENTS

Part of the phototrophic mass cultures of *Anacystis nidulans* were run in the Biochemistry Department of the University of Agriculture, Vienna (courtesy of Dr. C. Obinger and associates). Cloning and sequencing of the cta operon of *Synechocystis* 6803 was partially performed during an EMBO Short Term Fellowship of D.A. in the group of Matti Saraste at the EMBL in Heidelberg. Cloning and sequencing of the cta operon of *Anacystis* is currently being performed, in part, in the Department of Applied Genetics at the University of Agriculture, Vienna (courtesy of Prof. J. Glössl and associates).

REFERENCES

1. Barghoorn, E.S. and Schopf, J.W. (1966) Science 150, 758–763.
2. Broda, E. (1975) The Evolution of the Bioenergetic Processes, Pergamon Press, Oxford.
3. Babcock, G.T. and Wikström, M. (1992) Nature 356, 301–308.
4. Peschek, G.A. (this book, p. 201.)
5. Gilbert, D.L. (ed.) (1981) Oxygen and Living Processes. An Interdisciplinary Approach. Springer Verlag, New York, Inc.
6. Knoll, A.H. (1992) Science 256, 622–627.
7. Peschek, G.A., Schmetterer, G., Lockau, W. and Sleytr, U.B. (1981) Photosynthesis V. Chloroplast Development. (G. Akoyunoglou, ed.), pp. 707–719, Balaban International Science Services, Philadelphia, PA, USA.
8. Peschek, G.A., Wastyn, M., Molitor, V., Trnka, M., Kraushaar, H., Obinger, C. and Matthijs, H.C.P. (1989) in Highlights of Modern Biochemistry (Kotyk, A., Skoda, J., Paces, V. and Kosta, V., eds.), pp. 893–902, VSP International Science Publishers, Zeist, The Netherlands.
9. Peschek, G.A., Wastyn, M., Trnka, M., Molitor, V., Fry, I.V. and Packer, L. (1989) Biochemistry 28, 3057–3063.
10. Peschek, G.A. et al. (unpublished)
11. Alge, D. and Peschek, G.A. (1993) Biochem. Mol. Biol. Intern. 29, 511–525.
12. Alge, D. and Peschek, G.A. (1993) Biochem. Biophys. Res. Commun. 191, 9–17.
13. Alge, D., Schmetterer, G. and Peschek, G.A. (1994) Gene 138, 127–132.
14. Peschek, G.A. (1996) Biochim. Biophys. Acta 1275, 27–32.
15. Saraste, M. (1990) Quart. Rev. Biophys. 23, 331–366.
16. Kadenbach, B., Jarausch, J., Hartmann, R. and Merle, P. (1983) Anal. Biochem. 129, 517–521.
17. Tano, H., Ishizuka, M. and Sone, N. (1991) Biochem. Biophys. Res. Commun. 181, 437–442.
18. Sone, N., Tano, H. and Ishizuka, M. (1993) Biochim. Biophys. Acta 1183, 130–138.

19. Kaneko, T., Sato, S., Kotani, H., Tanaka, A., Asamizu, E., Nakamura, Y., Miyajima, N., Hirosawa, M., Sugiura, M., Sasamoto, S., Kimura, T., Hosouchi, T., Matsuno, A., Yamada, M., Yasuda, M. and Tabata, S. (1996) DNA Research 3, 109–136.
20. Peschek, G.A., Wastyn, M., Fromwald, S. and Mayer, B. (1995) FEBS Lett. 371, 89–93.
21. Peschek, G.A., Alge, D., Fromwald, S. and Mayer, B. (1995) J. Biol. Chem. 270, 27937–27941.
22. Auer, G., Mayer, B., Wastyn, M., Fromwald, S., Eghbalzad, K., Alge, D. and Peschek, G.A. (1995) Biochem. Mol. Biol. Intern. 37, 1173–1185.
23. Fromwald, S., Wastyn, M., Lübben, M. and Peschek, G.A. (this volume)
24. Saraste, M., Holm, L., Lemieux, L., Lübben, M. and van der Oost, J. (1991) Biochem. Soc. Trans. 19, 608–612.
25. Peschek, G.A., Niederhauser, H. and Obinger, C. (1992) EBEC Short Reports, vol. 7, p. 48, Elsevier, Amsterdam.
26. Sugiura, M. (personal communication)
27. Wastyn, M. and Peschek, G.A. (1988) EBEC Short Reports, vol. 5, p. 94, Elsevier, Amsterdam.
28. Peschek, G.A. (1990) in Abstr. 20th FEBS Meeting, p. 128, Budapest, Hungary.
29. Zimmermann, U. (1990) M.Sc. Thesis, University of Vienna, Austria.
30. Hüther, F.-J. and Kadenbach, B. (1986) FEBS Lett. 207, 89–94.
31. Hüther, F.-J. and Kadenbach, B. (1987) Biochem. Biophys. Res. Commun. 147, 1268–1275.
32. Arnold, S. and Kadenbach, B. (1997) Eur. J. Biochem. 249, 350–354.
33. Löffelhardt, W. and Bohnert, H.J. (1994) in The Molecular Biology of Cyanobacteria (D.A. Bryant, ed.) pp. 65–89, Kluwer Academic Publishers, Dordrecht, The Netherlands.
34. Azzi, A. and Gennis, R.B. (1986) Meth. Enzymol. 126, 138–145.
35. Molitor, V., Trnka, M., Erber, W., Steffan, I., Riviere, M.-E., Arrio, B., Springer-Lederer, H. and Peschek, G.A. (1990) Arch. Microbiol. 154, 112–119.
36. Obinger, C. (1991) Doctoral Thesis, University of Vienna, Austria.
37. Niederhauser, H. (1992) Doctoral Thesis, University of Vienna, Austria.
38. Regelsberger, G. (1994) Doctoral Thesis, University of Vienna, Austria.
39. Kaback, H.R. (1971) Meth. Enzymol. 22, 99–120.
40. Dworsky, A., Mayer, B., Regelsberger, G., Fromwald, S. and Peschek, G.A. (1995) Bioelectrochem. Bioenerg. 38, 35–43.
41. Iwata, S., Ostermeier, C., Ludwig, B. and Michel, H. (1995) Nature 376, 660–669.
42. Nitschmann, W.H. and Peschek, G.A. (1986) J. Bacteriol. 168, 1205–1211.
43. Pedersen, P.L. and Amzel, L.M. (1993) J. Biol. Chem. 268, 9937–9940.
44. Thomas, P.J., Garboczi, D.N. and Pedersen, P.L. (1992) J. Biol. Chem. 267, 20331- 20338.
45. Walker, J.E., Saraste, M., Runswick, M.J. and Gay, N. (1982) EMBO J. 1, 945–951.
46. Pai, E.F., Sachsenheimer, W., Schirmer, R.H. and Schulz, G.E. (1977) J. Mol. Biol. 114, 34–45.
47. Wikström, M. and Krab, K. (1979) Biochim. Biophys. Acta 549, 177–222.
48. Peschek, G.A. (1983) J. Bacteriol. 153, 539–542.
49. Peschek, G.A. (1984) Plant Physiol. 75, 968–973.
50. Peschek, G.A., Schmetterer, G., Lauritsch, G., Muchl, R., Kienzl, P.F. and Nitschmann, W.H. (1983) in Photosynthetic Prokaryotes (Papageorgiou, G.C. and Packer, L., eds.) pp. 147–162, Elsevier, Amsterdam.
51. Van Verseveld, H.W. Krab, K. and Stouthamer, A.H. (1981) Biochim. Biophys. Acta 635, 525–534.
52. Peschek, G.A., Obinger, C., Fromwald, S. and Bergman, B. (1994) FEMS Microbiol. Letters 124, 431–438.
53. Kadenbach, B., Kuhn-Nentwig, L. and Büge, U. (1987) Curr. Top. Bioenerg. 15, 113–161.
54. Koerner, T.J., Hill, J. and Tzagoloff, A. (1985) J. Biol. Chem. 260, 9513–9515.
55. Sacher, R., Steffens, G.J. and Buse, G. (1979) Hoppe-Seyler's Z. Physiol. Chem. 360, 1385–1392.
56. Nicholls, P., Obinger, C., Niederhauser, H. and Peschek, G.A. (1992) Biochim. Biophys. Acta 1098, 184–190.
57. Wastyn, M., Achatz, A., Molitor, V. and Peschek, G.A. (1988) Biochim. Biophys. Acta 935, 217–224.

40

EXTENDED HEME PROMISCUITY IN THE CYANOBACTERIAL CYTOCHROME c OXIDASE

Evidence against a Functional d-Type Terminal Oxidase

Susanne Fromwald, Marnik Wastyn, Mathias Lübben,* and
Günter A. Peschek†

Biophysical Chemistry Group
Institute of Physical Chemistry
University of Vienna
A-1090 Vienna, Austria

1. INTRODUCTION

Though there is good evidence that primordial cyanobacteria, having introduced the first substantial quantities of molecular oxygen into an essentially anoxic biosphere by way of their unique oxygenic, plant-type photosynthesis, consequently were the first aerobic respirers as well (1–3), cyanobacterial respiration has not been receiving too much attention so far (4). Only quite recently, after preliminary spectroscopic and kinetic results (5,6), was the terminal respiratory oxidase from *Anacystis nidulans* (*Synechococcus* sp. PCC6301) biochemically, immunologically and genetically characterized as an aa_3-type enzyme (7–9). Immunoscreening of almost thirty different strains and species of cyanobacteria has given evidence that the same type of aa_3-type cytochrome-c oxidase is present in both cytoplasmic and thylakoid membranes (CM and ICM) of all species investigated (10,11). Careful kinetic investigation on the NAD(P)H-oxidizing activity in these membranes clearly demonstrated that the terminal respiratory oxidase in wild-type cyanobacteria is a cytochrome-c oxidase, not a quinol oxidase (12,13), although the addition of external cytochrome c to the isolated membranes was not always necessary to catalyse electron transport between NADH and O_2 (14). Yet, the entire genomic sequence of *Synechocystis* sp. PCC6803 as determined by M. Sugiura and associates (see http://www.kazusa.or.jp/cyano; Ref. 15) has revealed that there

* Present address: Institut für Biophysik, Ruhr-Universität Bochum, Universitätsstraße 150, D-44780 Bochum, Germany
† Corresponding author.

The Phototrophic Prokaryotes, edited by Peschek *et al.*
Kluwer Academic / Plenum Publishers, New York, 1999.

are potential coding regions for another set of ctaC-E proteins (with 30–40% amino acid identity to the established enzyme) and a d-type terminal oxidase (with homology to the *E. coli* quinol oxidase; see Ref. 15). Since, in membranes derived from normally growing laboratory cultures of cyanobacteria during twenty years of thorough spectroscopic investigation we have not (nor has anybody else) discovered the least trace of a d-type cytochrome which, however, due to its peculiar spectral features should be easy to discern (16) we set out growing our cyanobacteria, viz. *Anacystis nidulans*, *Synechocystis* 6803, *Anabaena* 7120 and *Nostoc* 8009, under "nonstandard" laboratory conditions, i.e. with Na-nitrate replaced by ammonium carbonate, Mg-sulfate replaced by thiosulfate, and air/CO_2 flushing replaced by nitrogen/CO_2 flushing. In addition, *Nostoc* and *Anabaena* were also grown under dinitrogen-fixing conditions (17) in otherwise the same modified medium. It has been shown recently that cyanobacteria, phototrophically grown and incubated in low-oxygen conditions ("semi-anaerobically") replaced heme A in subunit I of their cytochrome-c oxidase by heme O, without at the same time altering the character of the enzyme from a cytochrome-c to a quinol oxidase or otherwise affecting electron transport properties of the membranes (12,13,18). Acid-labile hemes extracted into organic solvents had been subjected to the highly sensitive HPLC analysis method recently developed by Lübben et al. (19–21) and the same method was applied in the present investigation which, together with refined spectroscopic analysis showed the presence of small quantities of heme D in membranes (particularly CM) derived from cyanobacteria that had been grown under the "pseudo-reducing" conditions mentioned before. However, this membrane-bound heme D did not give spectral responses which would be expected from a respiratory oxidase, e.g. complex formation with oxygen or carbon monoxide. From our results we conclude that cyanobacteria, under certain strange, viz. "pseudo-reducing" conditions are in fact capable of synthesizing heme D and incorporating it into their usual COI protein (ctaD gene product). Yet, the complex between cytochrome oxidase subunit I and (the rather hydrophilic) heme D is not sufficiently well organized or coordinated to give the usual responses of a heme-type terminal oxidase but rather represents another example of (apparently futile) heme promiscuity (22,23) of the cyanobacterial terminal oxidase.

2. MATERIALS AND METHODS

2.1. Growth and Incubation of the Organisms

Anacystis nidulans (*Synechococcus* 6301), *Synechocystis* 6803, *Anabaena* 7120 and *Nostoc* 8009 (*Nostoc* Mac) were obtained by courtesy of Mme. Rosi Rippka-Herdman, Institut Pasteur, Paris and cultured photoautotrophically in axenic batch cultures (pH 7.8–8.2), medium BG-11 sparged with 1.5% CO_2 in sterile air and illuminated with 15–20 w.m^{-2} warm white fluorescent light at 30–35°C as described (8). Cells were harvested when the cell densities had reached approx. 3 µl packed cell mass (15–35 µg chlorophyll) per ml of culture fluid corresponding to light-limited, linearly growing cells to be designated here as fully aerated or phase-A cells, maximum oxygen concentration in the culture amounting to approx. 350 µM, i.e. > 150% air saturation (12,13,18). Harvesting and resuspending the cells in desired breakage or assay buffers was as described (8). An aliquot of the harvested and resuspended cells was supplemented with 20 µM DCMU while sparging was switched to "technically pure" sterile nitrogen containing approx. 1% oxygen, and continued in the light for 48 hrs prior to harvest ("semi-anaerobic" or phase-B cells). Additionally, the cyanobacteria were also cultivated in modified BG-11 where 0.30 mM

Mg-thiosulfate and 10 mM ammonium carbonate replaced the original Mg-sulfate and Na-nitrate, respectively, and bubbling was with pre-mixed 1.5% CO_2 in N_2 from a high-res-sure gas cylinder, keeping other conditions as given before ("modified semi-anaerobic" or phase-C cells). In both semi-anaerobic cell suspensions the oxygen concentration at the time of final harvest was < 10 μM while the energy charge of these cells stayed at a nor-mal level of close to 0.8 (12,18).

2.2. Isolation and Purification of Cytoplasmic and Thylakoid Membranes

Isolation and purification of cytoplasmic and thylakoid membranes was performed as described in detail previously (7,8). Purified CM and ICM were resuspended in 10 mM K-phosphate buffer pH 7.0.

2.3. Heme Extraction and Identification

Non-covalently bound hemes were extracted from a total of 2.0 ml membrane sus-pensions (30–40 mg protein/ml), each from phase-A, -B, and -C cells of each cyanobacte-rial species employed and divided into 0.05-ml-aliquots for experimental handling, with acetone/HCl (19:1, v/v) followed by ethyl acetate/acetonitrile treatment according to the procedure developed by M. Lübben et al. (19–21). Heme composition was analysed on an ISCO HPLC System 2004i equipped with a Deltapak C18 (13.9 × 150 mm, Zorbax) re-versed-phase HPLC column. Hemes were eluted with acetonitrile/0.5%-trifluoroacetic acid/water gradients (20), and detected spectrophotometrically at 406 nm (UVIS 205 detector, ISCO).

2.4. Spectrophotometry and Polarography

Spectrophotometry and polarography were performed with a Shimadzu UV-3101PC UV-VIS-NIR multi-wavelength spectrophotometer and a YSI Oxygen Monitor, model 53, equipped with a sensitive Clark-type electrode, respectively (see Ref. 18).

2.5. Immunoblotting

Immunoblotting of cyanobacterial membrane proteins with antibodies raised against authentic terminal respiratory oxidases was performed as described (8), using GAR-HRP as the second antibody.

2.6. Standards and Controls

Reproducibility of all quantitative data obtained was within ±15% of the respective means of up to six parallel determinations on membranes isolated from separately harvested batches of cyanobacteria. Standards of hemes B, A and O were prepared by extraction of commercially available bovine cytochrome-c oxidase and reductase (from Sigma) for hemes A and B, respectively, and of membranes from a cytochrome bo_3-overproducing E. coli strain. A standard of E. coli cytochrome bd and a polyclonal antibody against it were kindly donated by R.B. Gennis. Further aliquots of eventually harvested phase-B and -C cells were re-aerated (in the light) for 24 hrs after which time the heme content of the cells had returned to levels of fully aerated cells. No heme D was detected in fully aerated cells grown in ammonium/thiosulfate medium. When the "pseudo-reducing" growth and incubation procedures described before were repeated with *Anabaena* and *Nostoc* after growth under dinitrogen-fixing conditions (*no* combined nitrogen in the medium) the results obtained with membranes derived from these cells were esstentially the same as under nonnitrogen-fixing conditions (not shown). This demonstrates that dinitrogen-fixing conditions do not specifically favor the expression of d-type terminal oxidase in cyanobacteria, contrary to what was recently suggested to be the case in *Azotobacter vinelandii* (27).

3. RESULTS AND DICUSSION

Fig. 1 shows HPLC-heme patterns of extracts from fully aerated phase-A cells (A), semi-anaerobic phase-B cells (B) and ammonium/thiosulfate grown semi-anaerobic phase-C cells. Fig. 1D is the HPL chromatogram of the heme standards. It is seen that, as dis-

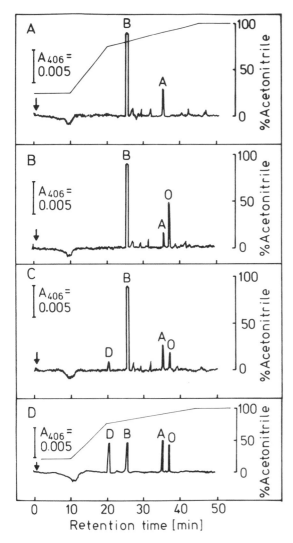

Figure 1. HPLC patterns of acid labile heme groups extracted from cyanobacteria after growth and incubation in fully aerated phase-A conditions (A), in semi-anaerobic phase-B conditions (B; medium BG-11 containing the usual sulfate and nitrate), and in modified semi-anaerobic, viz. "pseudo-reducing" phase-C conditions (C; medium BG-11 containing thiosulfate and ammonium instead of sulfate and nitrate as S- and N-sources, respectively). HPL chromatograms of authentic heme standards are shown in panel D. Extracts from both CM and ICM isolated from *Anacystis*, *Synechocystis*, *Nostoc* and *Anabaena* qualitatively gave the same HPLC pattern in our growth and incubation shift experiments. Also when *Anabaena* and *Nostoc* had been grown under dinitrogen-fixing conditions HPLC patterns of the extracted hemes remained essentially unaltered. For details see the text. Also see Materials and Methods.

cussed previously (12,13,18), growth and incubation under reduced oxygen tension (< 10 μM O_2 in the suspension medium) in the presence of sulfate and nitrate as S- and N-sources, respectively, leads to transient accumulation of heme O, the immediate biosynthetic precursor of heme A (Fig. 1B). Relative levels of hemes A and O strictly and reversibly follow the ambient oxygen concentration available to the cells. Substitution of ammonium and thiosulfate for nitrate and sulfate as N- and S-sources, respectively, in the BG-11 medium alone does not change this heme pattern in fully aerated cells (not shown). However, when the cyanobacteria were grown and incubated semi-anaerobically in the modified, thiosulfate- and ammonium-containing medium the additional appearance of heme D was observed (Fig. 1C). Heme D was never observed in cells growing just under reduced oxygen tension in the presence of sulfate and nitrate (12,13,18). As it is notoriously difficult to exactly quantitate heme contents from such HPLC measurements, different degrees of losses stemming from different degrees of adsorption and occlusion during extraction procedures according to the different lipophilicity of the various hemes (20), the only quantitative statement that can be made from the HPLC pattern of Fig. 1 is that even under the artificial growth conditions imposed here heme D is always synthesized at extremely low amounts only, never exceeding roughly 10% of heme A at most. Clearly, and as expected, the most abundant heme in all four cyanobacterial species examined here was the ubiquitous heme B (12,16). Hemes extracted from both CM and ICM of each of the four species gave a practically identical HPLC pattern in our growth and incubation shift experiments though it must be admitted that, due to a much noisier HPLC background of extracts from chlorophyll-containing (both native and pre-extracted, "de-chlorophyllized"; see Refs. 5,6,12) ICM the rather small heme D peak could be unequivocally identified in CM extracts only.

In spite of the small quantities of heme D separated by HPLC (Fig. 1) it proved possible to chemically convert the HPLC-separated hemes A, O *and* D into their alkaline pyridine ferrohemochromes and to identify them spectroscopically (Fig. 2). Thus the spectral identification of the individual hemes eluting from the HPLC column at 21–22, 26, 35 and 37 min retention time according to increasing lipophilicity confirms the identity with hemes D, B, A and O, respectively, as inferred from the HPLC comparison with authentic standards (Fig. 1). Alkaline pyridine ferrohemochrome spectra of a heme fraction from phase-A cells eluting at 30–36 min retention time from the HPLC column (see Fig. 1) and of a mixture of heme O, A and D standards are shown in Fig. 3A and B, respectively.

Fig. 4 shows artificially reduced minus oxidized difference spectra of 1%-n-octyl-glucoside solubilized CM preparations from phase-C *Anacystis, Synechocystis, Nostoc* or *Anabaena*. (CM preparations from phase-C cells of all four species essentially gave the same spectral pattern.) Dithionite reduced minus 1%-H_2O_2 oxidized preparations revealed the expected α-peaks at 563 nm, 605 nm and 625 nm characteristic of cytochromes o (b), a and d, respectively. Corresponding γ-peaks are seen at 430 nm, 445 nm, and 520 nm (Fig. 4A). While the difference peaks at 563 nm and 605 nm were also conspicuous in membrane samples that had been reduced with NADH and oxidized with atmospheric oxygen, respectively, prior to solubilization (see Refs. 5,6) the 625 nm peak was restricted to artificially reduced vs. oxidized samples (not shown). This corresponds to the observation that also upon CO bubbling (dithionite reduced minus ditionite reduced plus carbon monoxide difference spectra; Fig. 4B) this peak did not respond with the usual shift to shorter wavelength (cf. 563 to 555 nm for cyt o_3 and 605 to 590 nm for cyt a_3. See Refs. 5, 13). This would support the conclusion that the cytochrome d in cyanobacterial membranes does not form spectrally active complexes with either O_2 or CO and thus is not a reactive heme in a terminal respiratory oxidase. Also immunoblotting of phase-C membranes, which do contain heme D

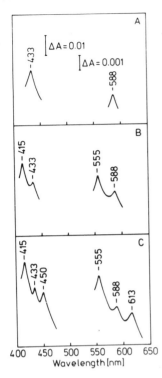

Figure 2. Optical spectra of alkaline pyridine ferrohemochromes prepared from HPLC separated hemes corresponding to membraneous extracts of phase-A (A), phase-B (B) and phase-C (C) cells. (Compare Fig. 1, Materials and Methods, and also the text.)

as judged from HPLC (Fig. 1) and spectroscopic (Fig. 2) evidence, with antibody raised against the *E. coli* bd-oxidase (24,25) did not give the least immunological cross reaction (Fig. 5, lanes 5 and 6) despite 41.3% and 28.6% amino acid identity between the deduced gene products of Sugiura's cydA and cydB genes, respectively, of *Synechocystis* and the authentic proteins from *E. coli* (15). By contrast, antibodies raised against subunits I of *Paracoccus denitrificans* aa$_3$-type (lanes 1 and 2) and *E. coli* bo$_3$-type (lanes 3 and 4) cytochrome-c and quinol oxidases, respectively, did give clear immunological cross reactions (approx. 50 kD) with the cyanobacterial phase-C membrane proteins. Results analogous to those shown here for *Anacystis* membranes were also obtained for both CM and ICM derived from phase-C cells of *Synechocystis*, *Nostoc* and *Anabaena* considering the different apparent molecular weights of their respective COI proteins which had been determined previously and which lie around 40 kD, 45 kD and 50 kD for *Synechocysis*, *Nostoc* and *An-*

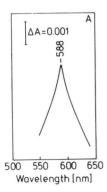

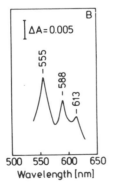

Figure 3. Optical spectra of alkaline pyridine ferrohemochromes prepared from the HPLC fraction eluting from the column at 30–36 min retention time (A) and from a mixture of authentic heme O, A and D standards (B). (Compare Figs. 1 and 2.).

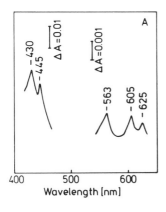
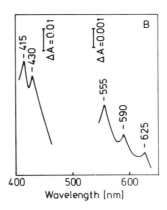

Figure 4. Optical difference spectra of dithionite reduced minus 1%-H_2O_2-oxidized (A) and dithionite reduced plus carbon monoxide minus dithionite reduced (B) CM preparations from phase-C cells after solubilization in 1%-n-octyl glucoside. Only difference peaks corresponding to cytochromes o and a were seen when the membranes, prior to solubilization, had been reduced with NADH and oxidized with atmospheric oxygen (A; see Refs. 5 and 6). Correspondingly, the wavelength of the cyt d peak at 625 nm was not shifted by carbon monoxide either (B).

abaena, respectively (Refs. 10,11). This means that not even highly sensitive immunological techniques were able to localize any cytochrome b.d protein in the cyanobacterial membranes which, on the other hand did give clear-cut evidence for the presence of heme D (Figs. 1 and 2). It seems, therefore, that heme D though it is synthesized in cyanobacteria under very peculiar growth and incubation conditions, and although it may be loosely associated with the conventional COI protein (subunit I of the aa$_3$-type cytochrome-c oxidase; see Refs. 7,12,13) in the sense of "heme promiscuity" (22,23), does not form an integral hemoprotein complex stable enough to represent a fully functional terminal d-type oxidase as in *E. coli* (24,25) or *Azotobacter vinelandii* (26).

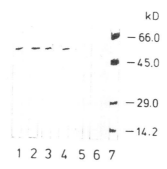

Figure 5. Immunoblots of *Anacystis nidulans* membrane proteins. Immunoblotting was performed according to standard procedures with GAR-HRP as the second antibody (8). 15 and 35 µg SDS solubilized CM and ICM proteins, respectively, were blotted on each lane. Monospecific antibodies raised against subunits I of *Paracoccus denitrificans* aa$_3$-type (lanes 1 and 2), *E. coli* bo$_3$-type (lanes 3 and 4) cyt-c and quinol oxidases, respectively, and against *E. coli* bd-type oxidase (lanes 5 and 6) were probed against CM (lanes 1, 3 and 5) and ICM (lanes 2, 4 and 6) proteins of phase-C cells (see Fig. 1C). Analogous results were obtained with membrane proteins from phase-C *Synechocystis*, *Nostoc* and *Anabaena* and also with *Nostoc* and *Anabaena* phase-C membrane proteins after growth of the organisms under dinitrogen-fixing conditions. For details see the text; also see Materials and Meth-

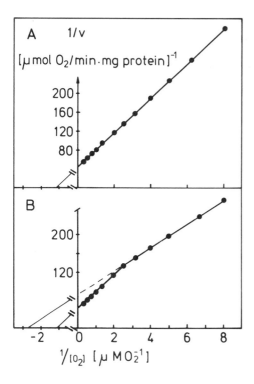

Figure 6. Michaelis-Menten kinetics (Lineweaver-Burk plots) of cyt-c supported oxygen uptake by membrane preparations of fully aerated phase-A cells (A) and modified semi-anaerobic phase-C cells of *Synechocystis*. K_m (O_2) values of approx. 0.85 μM and 0.35 μM are derived for cyt a_3 (fully aerobic; panel A) and cyt o_3 (modified semi-anaerobic, panel B). No third, kinetically distinct reaction with oxygen as would be expected for cyt bd (see Fig. 1C) is evident from the graph. Virtually identical kinetics were measured with membranes isolated from phase-A and phase-C cells of *Anacystis*, *Nostoc* and *Anabaena* (not shown; see Refs. 13 and 18).

Evidence against a functional d-type respiratory oxidase in cyanobacteria may also be concluded from the results shown in Fig. 6 where the kinetics of oxygen uptake supported by phase-C membranes (including the 100% inhibition by as low as 5 μM KCN; not shown) were indistinguishable from the kinetics observed with phase-B membranes in which no heme D can be detected. This result argues against a new and different (viz. d-) type of terminal oxidase which would well be expected to exhibit an affinity towards oxygen at least slightly different from the a_3-type (Fig. 6A) and the o_3-type (Fig. 6B) terminal oxidases present in phase-B membranes.

4. CONCLUDING REMARKS

Cyanobacteria have long been known to be capable of both oxygenic photosynthesis and aerobic respiration whereby a single type of electron transport system present in both CM and ICM serves both processes (Ref. 3 and Fig. 7). All the crucial electron transport components well known from chloroplasts and mitochondria are also present in cyanobacterial membranes, sometimes (immunologically and reactively) more related to their mitochondrial analogues (homologues?), sometimes more related to their thylakoidal analogues/homologues. At any rate the fact that only ICM catalyze photosynthetic electron transport simply depends on the chlorophyll which is contained in ICM only. (Apart from *Gloeobacter violaceus* which does not form thylakoids at all, see Ref. 28). On the other hand, regarding respiration, most respiring bacteria do form branched respiratory chains occasionally with up to three different terminal oxidases in one and the same cell, according to growth conditions. However, all these basically chemoheterotrophic bacteria depend on respiration for sustaining their life processes and it makes sense to them to adjust their terminal

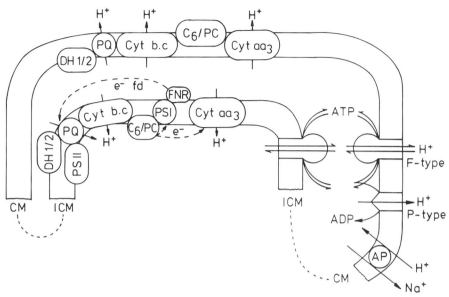

Figure 7. Bioenergetic membrane inventory of a schematically drawn cyanobacterium according to inhibition and immunoblotting studies on isolated and purified membranes and immunogold labeling experiments on intact cells using respective specific antibodies (see Ref. 2). CM, cytoplasmic or plasma membrane; ICM, intracytoplasmic or thylakoid membranes; PSI and PSII, photosystems I and II, respectively; PQ, plastoquinone; DH1/2, NAD(P)H dehydrogenases of either type 1 (proton-pumping, multi-subunit or "mitochondrial") or type 2 (nonproton-pumping, single-subunit or "bacterial"); cyt b.c, cyt-c/PC reductases of either "mitochondrial" type (cyt bc_1 complex) or "chloroplast" type (cyt b_6f complex); cyt aa$_3$, cyt-c oxidase; c_6/PC, either (soluble) c-type cytochrome or (loosely membrane-bound) plastocyanin; FNR, ferredoxin-NADP oxidoreductase; fd, ferredoxin; F-type, proton-translocating reversible ATPase or ATP synthase; P-type, proton-translocating unidirectional ATPase; AP, proton-sodium antiporter. Note that according to experimental evidence obtained thus far cyanobacteria, in terms of reactivity and immunological properties of bioenergetically relevant (membrane) proteins and enzymes, combine *both* mitochondrial *and* chloroplast features (apart from "bacterial features", of course) in a single prokaryotic cell.

oxidase inventory to different partial pressures of ambient oxygen according to the widely varying oxygen affinities of the different terminal oxidases. By contrast, most cyanobacteria are obligate phototrophs which cannot do without light at all and even those few facultative chemoheterotrophs that can grow in darkness do so at a much lower rate than in light so that the physiological significance of chemoheterotrophy in cyanobacteria, apart from recruiting respiratory maintenance energy during dark periods, must be questioned. In terms of the high energy requirement for synthesis and assembly of a large and complicated membrane protein such as an integral terminal oxidase inevitably is, it would certainly not be very wise for cyanobacteria to synthesize a whole array of different terminal oxidases without being able at the same time to make efficient use of them for purposes of growth and proliferation and not only for modest maintenance purposes.

5. NONSTANDARD ABBREVIATIONS

CM, cytoplasmic or plasma membranes; ICM, intracytoplasmic or thylakoid membranes; DCMU, N-(3,4-dichlorophenyl-) N',N'-dimethyl urea; cyt, cytochrome; GAR-HRP, goat anti-rabbit horse radish peroxidase; SDS, sodium docecyl sulfate.

ACKNOWLEDGMENTS

This work was supported by grants from the Austrian Science Foundation and from the austrian Ministery for Science and Research. Particular thanks are due to Dr. R. B. Gennis for the gift of *E. coli* bd-type quinol oxidase as well as of antibody against it. The untiring cooperation of Mr. Otto Kuntner in the experiments is gratefully acknowledged.

REFERENCES

1. Broda, E. and Peschek, G.A. (1979) J. Theor. Biol. 81, 201–212.
2. Peschek, G.A. (1996) Biochem. Soc. Trans. 24, 729–733.
3. Peschek, G.A., this volume, p. 201.
4. Schmetterer, G. (1994) in: The Molecular Biology of Cyanobacteria (D.A. Bryant, ed.), pp. 409–435, Kluwer Academic Publishers, Dordrecht-Boston-London.
5. Peschek, G.A. (1981) Biochim. Biophys. Acta 635, 470–475.
6. Peschek, G.A., Schmetterer, G., Lauritsch, G., Nitschmann, W.H., Kienzl, P.F. and Muchl, R. (1982) Arch. Microbiol. 131, 261–265.
7. Peschek, G.A., Wastyn, M., Trnka, M., Molitor, V., Fry, I.V. and Packer, L. (1989) Biochemistry 28, 3057–3063).
8. Peschek, G.A., Molitor, V., Trnka, M., Wastyn, M. and Erber, W. (1988) Methods. Enzymol. 167, 437–449.
9. Alge, D. and Peschek, G.A. (1993) Biochem. Mol. Biol. Intern. 29, 511–525.
10. Peschek, G.A., Wastyn, M., Molitor, V., Kraushaar, H., Obinger, C. and Matthijs, H.C.P. (1989) in: Highlights in Modern Biochemistry, Vol. 1 (Kotyk, A., Skoda, J., Paces, V. and Kosta, V., eds.), pp. 893–902, VSP Zeist, The Netherlands.
11. Peschek, G.A. et al., unpublished.
12. Peschek, G.A., Wastyn, M., Fromwald, S. and Mayer, B. (1995) FEBS Lett. 371, 89–93.
13. Peschek, G.A., Alge, D., Fromwald, S. and Mayer, B. (1995) J. Biol. Chem. 270, 27937–27941.
14. Flasch, H. (1997) M. Sc. Thesis, University of Vienna, Austria.
15. Kaneko, T., Sato, S., Kotani, H., Tanaka, A., Asamizu, E., Nakamura, Y., Miyajima, N., Hirosawa, M., Sugiura, M., Sasamoto, S., Kimura, T., Hosouchi, T., Matsuno, A., Nakazaki, N., Naruo, K., Okumura, S., Shimpo, S., Takeuchi, C., Wada, T., Watanabe, A., Yamada, M., Yasuda, M. and Tabata, S. (1996) DNA Research 3, 109–136.
16. Lemberg, R. and Barrett, J. (1973) Cytochromes. Academic Press, London and New York.
17. Wastyn, M., Achatz, A., Molitor, V. and Peschek, G.A. (1988) Biochim. Biophys. Acta 935, 217–224.
18. Auer, G., Mayer, B., Wastyn, M., Fromwald, S., Eghbalzad, K., Alge, D. and Peschek, G.A. (1995) Biochem. Mol. Biol. Intern. 37, 1173–1185.
19. Svensson, B., Lübben, M. and Hederstedt, L. (1993) Mol. Microbiol. 10, 193–201.
20. Lübben, M. and Morand, K. (1994) J. Biol. Chem. 269, 21473–21479.
21. Lübben, M. (1995) Biochim. Biophys. Acta 1229, 1–22.
22. Puustinen, A., Morgan, J.E., Verkhovsky, M., Thomas, J.W., Gennis, R.B. and Wikström, M. (1992) Biochemistry 31, 10363–10369.
23. Sone, N. and Fujiwara, Y. (1991) FEBS Lett. 228, 154–158.
24. Anraku, Y. and Gennis, R.B. (1987) Trends Biochem. Sci 12, 262–266.
25. Puustinen, A., Finel, M., Haltia, T., Gennis, R.B. and Wikström, M. (1991) Biochemistry 30, 936–942.
26. Jünemann, S. and Wrigglesworth, J.M. (1995) J. Biol. Chem. 270, 16213–16220.
27. Kelly, M.J.S., Poole, R.K., Yates, M.G. and Kennedy, C. (1990) J. Bacteriol. 172, 6010–6019.
28. Rippka, R., Waterbury, J. and cohen-Bazire, G. (1974) Arch. Mikrobiol. 100, 419–436.

41

CYANOBACTERIAL FERREDOXINS

Herbert Böhme, Carolin Kutzki, and Bernd Masepohl

Botanisches Institut
Universität Bonn
Kirschallee 1, D-53115 Bonn, Germany

1. INTRODUCTION

In higher plants and cyanobacteria light energy is used for oxygenic photosynthetic electron transport. Photosystem II oxidizes water to oxygen and the electrons are transferred from photosystem I to an iron-sulfur protein, called ferredoxin. Cyanobacterial plant-type ferredoxins are negatively charged, small proteins that typically contain a single [2Fe-2S] cluster. They are one-electron carriers with a low redox potential of ≤ 400 mV. Ferredoxins couple a number of reductive biosynthetic pathways to the photosynthetic electron transport chain. These are processes of global importance such as CO_2-fixation, nitrogen assimilation and nitrogen fixation.

In CO_2-fixation, a flavoprotein, the ferredoxin:NADP reductase (FNR), mediates the transfer of electrons from ferredoxin to $NADP^+$ and supplies NADPH for carbohydrate synthesis. Carbohydrate metabolism itself is regulated by the ferredoxin/thioredoxin system. In nitrogen assimilation reduced ferredoxin acts as an electron donor to nitrate reductase (NaR), nitrite reductase (NiR), glutamate synthase (GltS) and—in nitrogen fixing cyanobacteria—also to the nitrogenase system.

We discovered in many nitrogen fixing strains two functionally specialized plant-type ferredoxins: PetF, which participates in electron flow through photosystem I and a second ferredoxin, FdxH, with only ≈51% of sequence identity to PetF, which interacts specifically with nitrogenase reductase, is coinduced with the nitrogenase system and synthesized under nitrogen fixing conditions only[1].

Ferredoxin participates in many different electron transfer processes, so one question concerns the mechanisms that regulate the distribution of these high-energy electrons to different cellular processes. What determines the rate of electron transfer? How accurate is the interaction of ferredoxin with its redox partners? One answer is given by the differential synthesis of participating redox proteins. Another answer can be obtained by site directed mutagenesis.

The Phototrophic Prokaryotes, edited by Peschek *et al.*
Kluwer Academic / Plenum Publishers, New York, 1999.

2. STRUCTURE OF FERREDOXIN

A high resolution 3-D structure of crystallized PetF from *Anabaena* 7120 has been published and recently the structure of recombinant *Anabaena* 7120 FdxH also has been determined[2,3]. In the *Anabaena* 7120 ferredoxin (PetF) both iron atoms of the cluster are tetrahedrally coordinated by 4 cysteinyl-S residues (Cys 41, 46, 49 and 79) and 2 bridging, inorganic sulfur atoms to form a planar ring. This redox center is bound near the surface of the protein with one edge of the iron sulfur cluster more exposed. All known high resolution structures (including that of FdxH) exhibit the same global folding pattern[2,3].

The interaction of ferredoxin with ferredoxin-binding proteins leads to formation of 1:1 complexes, stabilized by electrostatic interactions, to which ferredoxin contributes mainly negative charges[4]. As previously pointed out, acidic residues form two domains of negative surface potential on either side of the iron sulfur cluster and are believed to present the docking sites of the molecule with other redox proteins. Between these domains and just above the iron sulfur cluster is the negative end of the molecular dipole of ferredoxin. This dipole moment helps to steer the protein into proper orientation on the surface of the redox partner[5].

2.1. Overexpression of Ferredoxins and Site Directed Mutagenesis

We have overexpressed in *E. coli* both *petF* and *fdxH* genes resulting in holoproteins indistinguishable from the native proteins[6]. Expression in *E. coli* allowed us to investigate by site-directed mutagenesis the structure-function relationships of the interaction of ferredoxin with its redox partners. Because of the general importance of electrostatic interactions for ferredoxin binding we decided to exchange absolutely conserved acidic residues (from the comparison of 20 cyanobacterial ferredoxin sequences) on *Anabaena* PetF to the corresponding amides i.e. D 22, 23, 28, 67 and 68 by N and E 31, 32, 94 and 95 by Q and subsequently determined the effects on electron transfer involving FNR, NiR, NaR and GltS. All exchanged acidic residues belong to the two domains of negatively charged surface potential surrounding the [2Fe-2S]-cluster.

2.2. Reactions with Ferredoxin:NADP Reductase

When photosystem I reduced the FNR/NADP redox couple, FdxH was half as active as PetF. The most severe inhibition was observed, however, when E94 was replaced by the corresponding amide, in accordance with E94 being a critical residue in interaction with FNR. All other mutants with singly or doubly removed negative charges also had lower activities than wild type (wt) PetF but by far not to that extent. Generally these exchanges also alter the dipole moment of PetF which may be a critical value for the very first steps in the ferredoxin-FNR docking process.

Included in these studies were also PetF mutants in which F65 was replaced by aliphatic or aromatic residues. In NADP-photoreduction F65A and F65I showed no measurable activity. The aromatic residue at position 65 was crucial to rapid electron transfer from reduced ferredoxin to FNR because F65W or F65Y restored wild type properties. Both residues, E94 and F65, are in the direct vicinity of the [2Fe-2S]-cluster and are thought to be a part of the binding domain of PetF[7].

2.3. Reactions with Nitrate- and Nitrite Reductase

Nitrite reductase is an enzyme, which together with nitrate reductase becomes induced upon addition of the substrates nitrate and nitrite. In cyanobacteria both enzymes are reduced by ferredoxin and catalyze the 8e[−] transfer from nitrate to ammonia.

Cyanobacterial nitrite reductase is a monomeric, soluble 52–68 kDa protein with a [4Fe-4S]-cluster and siroheme as prosthetic groups (E_0': −365 mV and −290 mV). Surprisingly, but in accordance with the data on FNR the electron transport activity decreased to large extent when residues E94 and F65 were exchanged. The most dramatic effect was exhibited again by the E94Q mutation: the K_M increased from 2.5 μM (wt) to 8.5 μM and V_{max} decreased by 60% leading to a 80% inhibition of specific electron flow. As with FNR the C-terminal glutamate E94 but not E95 was important for electron transfer to FNR and nitrite reductase. This supports the hypothesis that the ferredoxin-binding sites at FNR and NiR, two non-homologous redox proteins, are structurally related, but not completely identical. The high K_M-value of FdxH (16.5 μM) shows that it is not a ferredoxin optimized for interaction with nitrite reductase.

Nitrate reductase in cyanobacteria is a thylakoid bound protein, has a molecular weight of 76–85 kDa and its activity is ferredoxin dependent. The active center contains a molybdenum cofactor in addition to two [2Fe-2S]-clusters.

With the PetF double mutant E94Q/E95Q electron transfer to NaR was strongly inhibited to 18% of wt PetF. Different to the FNR- and the NiR-assay the corresponding single mutant E94Q did not show the same extent of inhibition (64% of wt PetF). In this case residue E95 seemed to contribute substantially to the ferredoxin-NaR electron transfer. The K_M of 15.5 μM of the wild type was similar to E94Q/E95Q but there was a decrease in V_{max}. The F65A exchange resulted in a reduction of the reaction rate to 36%. F65I still allowed 65% of wt activity. Different to FNR and NiR the optimum electron transfer from ferredoxin to nitrate reductase required a phenylalanine at position 65 because replacement by a different aromatic residue did not restore wild type activity. This could be one explanation why F65 is absolutely conserved among all known cyanobacterial plant-type ferredoxins.

A good adaptation of a protein to its substrate is usually indicated by a low K_M-value. The relatively high K_M of 15.5 μM of the PetF/NaR couple and the lower K_M of PetF/NiR of 2.5 μM may have the physiological meaning to keep the intracellular nitrite concentration at a low level[7].

2.4. Reactions with Glutamate Synthase

The glutamate synthase ($\approx$160 kDa, encoded by the *gltS* gene) with an even lower K_M of $\approx$1 μM is the key enzyme in ammonia assimilation. It catalyzes the transfer of an amino group of glutamine, formed by glutamine synthetase (GlnA), to 2-oxoglutarate in order to produce two molecules of glutamate. In cyanobacteria this reaction is dependent on ferredoxin. The active center contains a [3Fe-4S] cluster and FMN as cofactors (E_0': −190 mV, −180 mV). The *gltS* gene from *Synechocystis* sp. PCC 6803 was expressed in *E. coli* in order to isolate the recombinant holoprotein. The ferredoxin dependent activity was determined by measurements of the glutamate concentration by HPLC. We used recombinant ferredoxins from *Anabaena* 7120 and *Synechocystis* 6803. Neutralization of different negative charges changed wild type activity only marginal. For the double mutant E94Q/E95Q only $\approx$21% (28% with the *Synechocystis* ferredoxin) of the electron transfer activity remained. Heterocyst ferredoxin (FdxH) showed with 9% the lowest rate of electron transfer. Among other findings these data argue against a function of GltS in heterocysts. The other possibility would be that in heterocysts a certain amount of PetF is maintained and reduced vegetative cell ferredoxin is used for glutamate synthesis. The K_M values were almost identical ($\approx$1 μM), while the V_{max} decreased in the E94-mutant (corresponding to E92 in *Synechocystis* ferredoxin). With flavodoxin as electron donor glutamate synthase remained inactive.

2.5. Electron Transfer

The hypothesis that ferredoxin binding proteins, although very different in molecular structure and redox centers, may have an overlapping, but not identical ferredoxin binding domain is in accordance with our studies. Due to an uneven charge distribution, the ferredoxin molecule shows a strong dipole moment, which facilitates the mutual attraction of the corresponding binding proteins at long distances. Electrostatic interactions are not sufficient because the substitution of the aromatic phenylalanine at position 65 severely inhibits electron flow to FNR, NiR and NaR[8]. Long range electrostatic interactions are followed by pairwise, specific ionic (E94) and hydrophobic (F65) interactions over shorter distances at the protein-protein interface. This probably occurs through the release of water molecules from the protein-protein interface, which may establish hydrophobic interactions between non-polar amino acids in the region around the iron sulfur center[9]. These steps allow for electron transfer close to the contact site between both proteins and facilitated release of the oxidized form of ferredoxin to ensure a rapid turnover.

3. HETEROCYST FERREDOXIN

It is notable that FdxH also contains F65 and E94 in identical positions and a negatively charged surface domain close to the Fe-S center. Nevertheless FdxH does not interact very well with nitrate and nitrite reductase ($\approx$20%), as expected, and much worse with GltS (9%). This shows that the charge at position E94 and aromaticity at residue F65 cannot be the only determinants for optimal electron transfer[8].

The heterocyst ferredoxin as "natural variant" of PetF, on the other hand, seems to be optimized for reversed electron flow from NADPH/FNR and specific interaction with nitrogenase. This may be physiologically important because in heterocysts, the compartment where FdxH is synthesized, NADPH is produced by carbon degradation via a highly active oxidative pentose-phosphate pathway. Electrons are transferred from NADPH to FdxH via FNR, which is approximately ten times more abundant in heterocysts as shown at the level of protein content and enzymatic activity as well as by a strongly increased transcription of the *petH* gene[10]. When reduced, FdxH functions as efficient and immediate electron donor to nitrogenase *in vitro*.

Among the structural differences between PetF and FdxH in nitrogen fixing cyanobacteria are conserved, N-terminal lysine residues (K10/K11). NMR data indicate that both lysines are part of a very flexible loop in the heterocyst ferredoxin. Replacing by site directed mutagenesis K10 and K11 of FdxH by E10 and A11, the corresponding amino acids of PetF at this position, we showed that this FdxH mutant protein allowed for only 15% of the nitrogenase activity. From this we imply that lysines 10 and 11, which are some distance away from the iron sulfur center, play an important role in docking ferredoxin to nitrogenase reductase[11].

3.1. Inactivation of the *fdxH* Gene from *Anabaena* 7120

Many physiological and biochemical data support the view that FdxH is the immediate electron donor to nitrogenase. To clarify the role of the heterocyst-specific *Anabaena* ferredoxin in cyanobacterial nitrogen fixation *in vivo*, mutational analysis of the *fdxH* region was carried out. For inactivation of the *fdxH* gene a neomycin interposon was inserted in the *fdxH* coding region (LAK4), whereas BMB92 was deleted for the *fdxH* gene and genes further downstream, including ORF3 and *mop*, and replaced by a neomycin resistance cassette. ORF3, the first gene downstream of *fdxH* shows strong homology to ORF3-products from *Anabaena variabilis* (94% identity) and *Plectonema* 73110 (34% identity). Due to the proximity of *fdxH* and ORF3 genes in *Anabaena* 7120, *Anabaena variabilis* and *Plectonema* 73110, it seems likely that these genes are cotranscribed and, therefore, the ORF3 genes may represent *nif*-associated genes characteristic for filamentous cyanobacteria. The next ORF is oriented in the reverse direction to *fdxH* and ORF3 and has similarities to molybdenum binding proteins, encoded by *mop* genes from *Clostridium pasteurianum*.Western blotting experiments using an specific anti-FdxH antiserum clearly demonstrated that an FdxH-type ferredoxin was no longer synthesized in both mutant strains. Growth rates with nitrate of the mutant strains were similar to the wild-type cultures. The nitrogen fixation phenotype of *Anabaena* wild type and mutant strains was determined by analyzing diazotrophic growth and nitrogenase activities. Diazotrophic growth of LAK4 and BMB92 after 4 days was 40% and 20%, respectively, of wild type *Anabaena*. Maximum nitrogenase activity was inhibited in LAK4 and BMB92 by 50% and 65%, respectively. Both mutant strains, LAK4 and BMB92, showed that FdxH is necessary for maximum nitrogenase activity and optimal diazotrophic growth, but not absolutely essential for nitrogen fixation in cyanobacteria.

Iron limitation causes flavodoxin to be synthesized. *In vitro* it replaces vegetative cell ferredoxin in electron transport from photosystem I to NADP$^+$, but it can not support specific nitrogenase activity, as shown before[12,13]. In *Anabaena* a *nifJ* gene has been described coding for pyruvate:flavodoxin oxidoreductase. In *Klebsiella* this enzyme reduces NifF, a special flavodoxin, serving as a specific electron donor for nitrogenase. NifJ from *Anabaena* was found to be essential for nitrogen fixation under Fe-deficient but not under Fe-replete conditions. So one could not rule out that a different flavodoxin is synthesized in heterocysts and Fe-deficient conditions. To check whether a NifJ/NifF system could "rescue" the growth phenotype of the *fdxH*-mutants under iron limitation, we compared diazotrophic growth and nitrogenase activity of wild type *Anabaena* and the two mutant strains. Doubling times of wild type and the complemented strain decreased generally about 1.5 fold under iron deficiency (1 μM Fe). However, growth and nitrogenase activity in both mutant strains (LAK4 and BMB92) remained low and therefore, FdxH is not replaced by a NifJ/flavodoxin system[14].

4. INACTIVATION OF THE FLAVODOXIN GENE (*isiB*) FROM *SYNECHOCYSTIS* 6803

In our previous *in vitro* assays we also showed that flavodoxin is inactive as electron donor to *Synechocystis* 6803 glutamate synthase (GltS), a key enzyme in nitrogen assimilation[8]. To examine the role of flavodoxin *in vivo* under iron limiting conditions by mutational analysis of the *isiB* gene, we chose the unicellular, non-nitrogen fixing strain *Synechocystis* 6803.

In general, flavodoxins are synthesized in cyanobacteria as functional replacement of ferredoxin when the microorganisms are grown under iron stress induced conditions (*isi*). Iron deficiency also leads to a reduced phycobiliprotein- and chlorophyll *a* content of the cell. By substituting for ferredoxin, flavodoxin synthesis could relieve the organism from its iron demand. The assumption is a programmed compensatory response to iron starvation.

Using the known N-terminal protein sequence[15] and conserved regions of the flavodoxin from other cyanobacteria we were able to clone and sequence the gene and adjacent

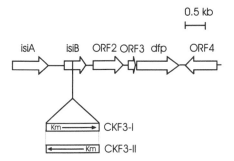

Figure 1. Genetic map of the *Synechocystis* sp. PCC6803 *isiB* gene region. *isiB*, encoding flavodoxin, is downstream of *isiA*, coding for a chlorophyll binding protein; *dfp* encodes a flavoprotein involved in panthotenat metabolism. The *Eco*RI site within the *isiB* gene was used to insert a kanamycin resistance interposon in both orientations resulting in mutant strains CKF3-I and CFK3-II, repectively. Km: kanamycin resistance cassette.

regions. As the whole genomic sequence of *Synechocystis* 6803 is available now on CYANOBASE[16], the following genetic map of the the the *isiB* region can be shown in Fig. 1. Upstream of the *isiB* gene, encoding flavodoxin, the gene *isiA* is situated, which codes for a chlorophyll binding protein; it may replace CP43 under iron limitation. Downstream of the *isiB* gene an open reading frame (ORF2) is located, which has 33% identity to ORF2 of *Anabaena* 7120 and which is found there at a similar position downstream of *isiB*.

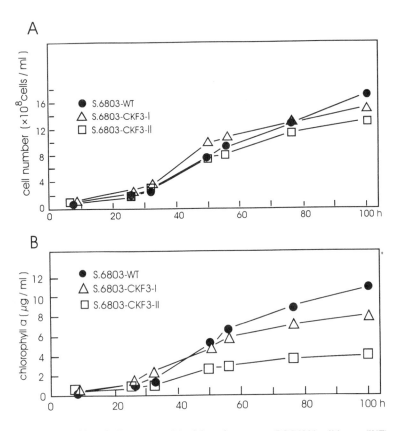

Figure 2. Growth (A) and chlorophyll *a* content (B) of *Synechocystis* sp. PCC6803 wild type (WT) and mutant strains (CKF3-I/II) under iron stress induced conditions (BG11-medium containing 1 μM Fe). The cell number was measured by O.D. values obtained at 730 nm. S.: *Synechocystis*.

By *in vitro* experiments it was shown that flavodoxin can replace ferredoxin in catalyzing electron transfer from photosystem I to FNR leading to photoreduction of NADP⁺, a very important reaction in this organism[17]. We wanted to see whether transcription of flavodoxin is essential for photoautotrophic growth under iron limitation.

For this purpose, two mutants were constructed interrupting the *isiB* gene by a kanamycin resistance cassette. In CKF3-I the kanamycin resistance gene read in the same direction of transcription as *isiB*, whereas in CKF3-II the interposon was inserted in the reverse orientation. Therefore, the mutation in CKF3-II might cause polar effects on the expression of ORF2, assuming that *isiB* and ORF2 are cotranscribed. The *isiB* mutants were completely segregated as demonstrated by Southern analysis. Protein expression analysis by Western blots showed the absence of flavodoxin in both mutants whether the strains were cultivated with a reduced iron content of the medium or not.

Growth curves of the cyanobacterial wild type and mutant cells, measured as cell number (O.D. 730 nm) or chlorophyll *a* content, were quite similar in an iron sufficient medium (30 μM Fe). Under iron-deplete conditions (1 μM Fe), necessary for full induction of flavodoxin, the mutants and wild type *Synechocystis* showed different growth curves (Fig. 2). Compared to wild type cells the chlorophyll *a* content was reduced to 77% in CKF3-I and to 40% in CKF3-II. The cell number, however, remained hardly affected with 89% in CKF3-I and 76% in CKF3-II after 100 h of growth. The mutant CKF3-II, which contained the polar insertion with respect to ORF2 and other downstream genes was much more inhibited than CKF3-I in chlorophyll biosynthesis than in growth as measured by cell number[18]. In addition, the *Synechocystis* mutant CKF3-II was assayed under different Fe-concentrations (8 μM and 3 μM Fe). The results were quite similar to the growth curve obtained with 1 μM Fe.

These results clearly show that flavodoxin (IsiB) of *Synechocystis* 6803 is not essential for photoautotrophic growth under iron limiting conditions. Under these conditions flavodoxin cannot replace all functions of ferredoxin in this cyanobacterium.

REFERENCES

1. Böhme, H. and Haselkorn, R. (1988) Mol. Gen. Genet. 214, 278–285
2. Rypniewsky, W.R., Breiter, D.R., Benning, M.M., Wesenberg, G., Oh, B.-H., Markley, J.L., Rayment, I. and Holden, H.M. (1991) Biochemistry 30, 4126–4131
3. Jacobson, B.L., Chae, Y.K., Markley, J.L., Rayment, I. and Holden, H.M. (1993) Biochemistry 32, 6788–6793
4. Knaff, D.B. and Hirazawa, M. (1991) Biochim. Biophys. Acta, 1056, 93–125
5. De Pascalis, A.R., Jelesarov, I., Ackermann, F., Koppenol, W.H., Hirasawa, M., Knaff, D.B. and Bosshard, H.R. (1993) Protein Sci. 2, 1126–1133
6. Böhme, H. and Haselkorn, R. (1989) Plant Mol. Biol. 12, 667–672
7. Schmitz, S. and Böhme, H. (1995) Biochim. Biophys. Acta 1231, 335–341
8. Schmitz, S., Navarro, F., Kutzki, C.K., Florencio, F.J. and Böhme, H. (1996) Biochim. Biophys. Acta 1277, 135–140
9. Jelesarov, I. and Bosshard, H.R. (1994) Biochemistry 33, 13321–13328
10. Razquin, P., Fillat, M.F., Schmitz, S., Stricker, O., Böhme, H., Gómez-Moreno, C. and Peleato, M.L. (1996) Biochem. J. 316, 157–160
11. Schmitz, S., Schrautemeier, B. and Böhme H. (1993) Mol. Gen. Genet. 240, 455–460
12. Razquin, P., Schmitz, S., Fillat, M.F., Peleato, M.L. and Böhme, H. (1994) J. Bacteriol. 176, 7409–7411
13. Razquin, P., Schmitz, S., Peleato, M.L., Fillat, M.F., Goméz-Moreno, C. and Böhme, H. (1995) Photosynthesis Res 43, 35–40
14. Masepohl, B., Schölisch, K., Görlitz, K., Kutzki, C.K. and Böhme, H. (1997) Mol. Gen. Genet. 253, 770–776

15. Bottin, H., Lagoutte, B. (1992) Biochim. Biophys. Acta 1101, 48–56
16. Kaneko, T., Sato, S., Kotani, H., Tanaka, A., Asamizu, E., Nakamura, Y., Miyajima, N., Hirosawa, M., Sugiura, M., Sasamoto, S., Kimura, T., Hosouchi, T., Matsuno, A., Muraki, A., Nakazaki, N., Naruo, K., Okumura, S., Shimpo, S., Takeuchi, C., Wada, T., Watanabe, A., Yamada, M., Yasuda, M. and Tabata, S. (1996) DNA Res. 3, 109–136
17. Mühlenhoff, U. and Sétif, P. (1996) Biochemistry 35, 1367–1374
18. Kutzki, C.K.and Böhme, H. (1997), unpublished observations

β SUBUNIT THR-159 AND GLU-195 OF THE *RHODOSPIRILLUM RUBRUM* ATP SYNTHASE ARE ESSENTIAL FOR DIVALENT CATION DEPENDENT CATALYSIS

Lubov Nathanson and Zippora Gromet-Elhanan

Department of Biochemistry
The Weizmann Institute of Science
Rehovot 76100, Israel

1. INTRODUCTION

The crystal structures of rat mitochondrial MF_1 at 3.6 Å [1] and bovine MF_1 at 2.8 Å [2] confirm the alternating arrangement of the six large α and β subunits in a closed hexamer. In the high-resolution structure the three catalytic β subunits have different bound nucleotides and display different conformational states. This structure contains also about half of the γ subunit amino-acid residues, composed mainly of the N and C terminal helices, that are embedded within the internal cavity of the $α_3β_3$ hexamer. The asymmetric structure imposed on this hexamer by interaction of the γ-subunit with the three different β-subunits [2], supports the binding change mechanism [3]. This mechanism suggests that ATP synthesis involves transitions between different but interacting catalytic sites via rotation of the γ subunit relative to an $α_3β_3$ subassembly. Full elucidation of the as yet unresolved mechanism of ATP synthesis will depend on identification of all amino-acid residues on the β and γ subunits that participate in catalysis.

The F_1-β subunit of *Rs. rubrum* ($RrF_1β$) provides an especially suitable system for mutational analysis of such residues, based on the $MF_1β$ crystal structure, because: a) Its published amino acid sequence shows the highest, >76% identity with $MF_1β$ [4]; b) A large number of assays have been developed for testing the activity of $RrF_1β$ released from the chromatophore-bound RrF_0F_1 [5–7]; and c) $RrF_1β$ expressed in *E. coli* was found to be as active as the native subunit in all these tests [8,9]. In this study we introduced mutations at the fully conserved positions Thr-159 and Glu-195 of $RrF_1β$ and compared their activities with those of the wild type (WT) subunit.

The Phototrophic Prokaryotes, edited by Peschek *et al.*
Kluwer Academic / Plenum Publishers, New York, 1999.

2. EXPERIMENTAL

E. coli strain LM3115 [10] lacking the whole *unc*-operon were used as a host for re-combinant plasmids under the growth conditions described in [8]. Site directed mutagene-sis at position 159 was carried out as outlined by Nathanson [9]. The procedures for obtaining the mutants at position 195, as well as for isolation of expressed $RrF_1\beta$ subunits from the cytoplasmic fraction of LM3115 cells were described in [11].

The activities of WT and all mutated $RrF_1\beta$ subunits were determined by following their capacity to bind into β-less *Rs. rubrum* chromatophore-bound F_oF_1 ATP synthase and re-store its ATP synthesis and hydrolysis activities. The conditions found optimal for binding of native [5,7] or expressed [8,11] $RrF_1\beta$ into the β-less chromatophores include their incubation for 1 hr at 35°C in a reaction mixture containing: 50 mM Tricine-NaOH (pH 8.0), 25 mM $MgCl_2$, 4 mM ATP, 1 mM DTT, Trace amounts of $RrF_1\alpha$ and saturating concentrations of $RrF_1\beta$. The reconstituted chromatophores were therefore washed twice in 50 mM Tricine-NaOH (pH 8.0) and 10% glycerol to remove the $MgCl_2$ as well as all unbound subunits. Bind-ing of $RrF_1\beta$ into β-less chromatophores was assayed as described in [11]. ATP synthesis and hydrolysis activities of the reconstituted chromatophores were determined according to [7], except that the divalent cation concentrations were varied as indicated.

3. RESULTS AND DISCUSSION

3.1. Activities of $RrF_1\beta$ Mutants at Position 159

$RrF_1\beta$-T159 is the last residue in the glycine rich, p-loop sequence found in α and β subunits of all F_1-ATPases and many other nucleotide binding proteins [12]. The parallel $MF_1\beta O_\gamma$-T163 was identified in the crystal structure as a ligand to Mg^{2+} [2]. In the GTP binding *ras* protein a serine, which replaces the threonine in this sequence, was also found to be a ligand to Mg^{2+} [13]. We have mutated the $RrF_1\beta$-T159 into S as well as A or V and found that $RrF_1\beta$-T159S does bind into β-less chromatophores and restores even higher rates of MgATPase activity (Fig. 1A) as well as Mg-dependent ATP synthesis (Fig. 2A) than those restored by WT $RrF_1\beta$-T159. Chromatophores reconstituted with the other two mutants were, however, completely inactive under all tested $MgCl_2$ concentrations. This

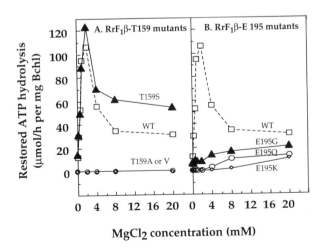

Figure 1. Mg^{2+}-dependent ATP hy-drolysis restored in β-less chromatopho-res reconstituted with $RrF_1\beta$ mutated at positions T159 (A) and E195 (B).

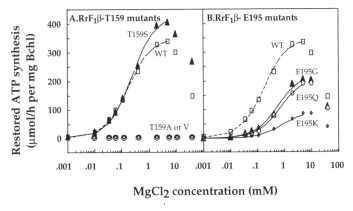

Figure 2. Mg^{2+}-requirement of photophosphorylation restored to β-less chromatophores reconstituted with WT and mutant $RrF_1\beta$ at positions T159 (A) and E195 (B).

absence of restored ATP synthesis or hydrolysis was not due to the inability of the T159A and V mutants to bind into β-less chromatophores, since they were shown to bind to the same extent as the WT and T159S mutant [9].

A very different picture was observed for the restoration of Ca^{2+}-dependent ATP hydrolysis. Here all $RrF_1\beta$-T159 mutants, including T159S, were unable to restore any activity at the whole range of tested $CaCl_2$ concentrations (Fig. 3A). These results reveal an unexpected difference between the ligands for Mg^{2+} and Ca^{2+}, where threonine could not be replaced even by serine.

3.2. Activities of $RrF_1\beta$ Mutants at Position 195

$RrF_1\beta$-E195 is equivalent to $MF_1\beta$-E199, which in the bovine crystal structure [2] is described as pointing into the conical tunnel leading to the catalytic nucleotide binding

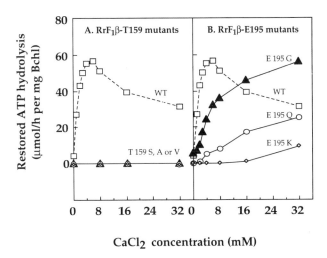

Figure 3. Ca^{2+}-dependent ATP hydrolysis restored in β-less chromatophores reconstituted with $RrF_1\beta$ mutated at positions T159 (A) and E195 (B).

sites. Fig. 1B illustrates that this glutamic acid is involved in F_1 activity, since all $RrF_1\beta$-E195 K, Q, or G mutants did not restore any significant MgATPase activity. They did, however, restore respectively increasing rates of Mg-dependent ATP synthesis (Fig. 2B) and Ca-dependent ATP hydrolysis (Fig. 3B), but required higher divalent cation concentrations than the WT $RrF_1\beta$-E195. This requirement could explain their surprising inability to restore Mg-dependent ATP hydrolysis while restoring ATP synthesis, since MgATPase was found to be much more succeptible to inhibition by excess free Mg^{2+}-ions than the Mg-dependent ATP synthesis [14, and compare Figs. 1 and 2]. The specific inhibition of MgATPase activity is especially important in photosynthetic organisms, where its tight regulation limits the wasteful hydrolysis of low concentrations of ATP in the dark.

4. CONCLUSIONS

In this study we have identified two $RrF_1\beta$ residues, Thr-159 and Glu-195 that are directly involved in the divalent cation dependent RrF_oF_1 catalytic activity. Our work with the T159 mutants revealed a clear difference between the Mg^{2+}- and Ca^{2+} ligands, which could lead to different conformational states of F_1 catalytic sites occupied by either Mg^{2+} or Ca^{2+}. This can explain the complete absence of proton-coupled ATP synthesis in presence of Ca^{2+} [14]. The results obtained with the E195 mutants demonstrate that this residue, which is located at the tunnel leading to the F_1-β catalytic nucleotide binding sites, does also affect their cation binding affinity, thus changing the tight regulation between Mg-dependent ATP synthesis and hydrolysis.

ACKNOWLEDGMENTS

This work was supported by a grant from the Basic Research Foundation administered by the Israel Academy of Sciences and by Humanities, and by the Minerva-Willstätter Center for Research in Photosynthesis, Rehovot.

REFERENCES

1. Bianchet, M., Ysern, X., Hullihen, J., Pedersen, P.L. and Amzel, L.M. (1991) J. Biol. Chem. 266, 197–212.
2. Abrahams, J.P., Leslie, A.G.W., Lutter, R. and Walker, J.E. (1994) Nature 370, 621–628.
3. Boyer, P.D. (1993) Biochim. Biophys. Acta 1140, 215–250.
4. Falk, G., Hampe, A. and Walker, J.E. (1985) Biochem. J. 228, 391–407.
5. Philosoph, S., Binder, A. and Gromet-Elhanan, Z. (1977) J. Biol. Chem. 252, 8747–8752.
6. Khananshvili, D. and Gromet-Elhanan, Z. (1985) Proc. Natl. Acad. Sci. USA 82, 1886–1890.
7. Gromet-Elhanan, Z. and Khananshvili, D. (1986) Methods in Enzymol. 126, 528–538.
8. Nathanson, L. and Gromet-Elhanan, Z. (1995) in: Photosynthesis: from Light to Biospher (Mathis, P., Ed.) Vol. III, pp. 51–54, Kluwer Academic Publishers, Dordrecht.
9. Nathanson, L. (1997) Ph.D. Thesis, The Weizmann Institute of Science.
10. Jensen, P.R. and Michelsen, O. (1992) J. Bacteriol. 174, 7635–7641.
11. Nathanson, L. and Gromet-Elhanan, Z. (1998) J. Biol. Chem. in press.
12. Saraste, M., Sibbald, P.R. and Wittinghofer, A. (1990) Trends Biochem. Sci. 15, 430–434.
13. Tong, L., deVos, A.M., Milburn, M.V. and Kim, S.H. (1991) J. Mol. Biol. 217, 503–516.
14. Gromet-Elhanan, Z. and Weiss, S. (1989) Biochemistry 28, 3645–3650.

43

ENERGY-DEPENDENT REGULATION OF CYANOBACTERIAL AND CHLOROPLAST ATP SYNTHASE

Studies on *Synechocystis* 6803 ATP Synthase with Mutations in Different Subunits

Hendrika S. Van Walraven

Department of Biomolecular Complexity and Dynamics (BCD)
Institute for Molecular Biological Sciences (IMBW)
BioCentrum Amsterdam
Vrije Universiteit
De Boelelaan 1087, 1081 HV Amsterdam, The Netherlands

1. INTRODUCTION

Cyanobacteria are prokaryotes with a high degree of structural and functional similarity to chloroplasts concerning the composition of their photosynthetic membranes (thylakoids [1]). This paper will deal with structure-function relations concerning the activity regulation of the ATP synthase, the terminal enzyme of photosynthetic and respiratory phosphorylation. The overall structure of the ATP synthase is essentially the same among various species. ATP synthesis and hydrolysis are catalysed by the F_1 part (Figure 1) and are linked to the transmembrane transport of protons catalysed by the F_0 part, dissipating or generating a transmembrane proton gradient ($\Delta\bar{\mu}_H^+$), respectively. The physiological role of cyanobacterial and chloroplast ATP synthase is primarily to make ATP; several mechanisms are present to prevent ATP hydrolysis as a waste of energy. In thylakoids from cyanobacteria several enzyme complexes, such as the ATP synthase, are shared for photosynthesis and respiration [2] and this has implications for its regulation.

Whether the ATP synthase performes ATP synthesis or hydrolysis depends on two factors, that have been studied with the use of site-directed mutants from the transformable cyanobacterium *Synechocystis* 6803:

The Phototrophic Prokaryotes, edited by Peschek *et al.*
Kluwer Academic / Plenum Publishers, New York, 1999.

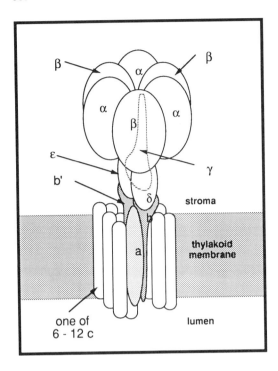

Figure 1. Proposed arrangement of ATP synthase subunits in cyanobacteria and chloroplasts. Subunits of F_1 are α, β, γ, δ and ε and subunits of F_0 are a, b, b' and c. In chloroplasts the latter are known as IV, I, II and III, respectively. Subunits α and β, a and c, and γ are well conserved between chloroplasts and several cyanobacterial strains (see [31]). Recently, it appeared from cross-linking experiments [32] that δ in chloroplast F_1 is not located in the stalk but on the exterior of the α/β crown of the complex.

1. The activation state of the enzyme. In probably all organisms the ATP synthase is latent and requires $\Delta\bar{\mu}_H^+$-dependent conformational changes to become active. In addition to this activation the chloroplast ATP synthase, but not the enzyme from most cyanobacterial strains, can be modulated by its redox state.
2. The value of ΔG. The rate of ATP synthesis or hydrolysis depends on the total driving force of the reactions:

$$\Delta G = \Delta G_p - [H^+/ATP] \cdot \Delta\bar{\mu}_H^+$$

ATP synthesis requires that $\Delta G < 0$ and ATP hydrolysis occurs when $\Delta G > 0$. This equation illustrates the importance of the H^+/ATP ratio.

2. REGULATION OF THE ACTIVITY OF THE ATP SYNTHASES

2.1. Activation of the Enzyme

In Table 1 it is shown that chloroplast ATP synthase in its oxidised form requires a very high $\Delta\bar{\mu}_H^+$ to become active. The result of reduction (thiol-modulation, *in vivo* by light-dependent thioredoxin and *in vitro* by light plus dithiothreitol) is that the enzyme can be activated at lower levels of $\Delta\bar{\mu}_H^+$ [3]. The high value of $\Delta\bar{\mu}_H^+$ to activate the oxidised form makes it very hard for the enzyme to hydrolyse ATP. On the other hand, the reduced form can make ATP at low levels of $\Delta\bar{\mu}_H^+$ when ADP and phosphate are present. There is also a difference in stability between both $\Delta\bar{\mu}_H^+$-activated forms. When the $\Delta\bar{\mu}_H^+$ collapses, e.g. by uncoupling, the active oxidised form inactivates in seconds. The active reduced form inactivates more

Table 1. Summary of analyses of $\Delta\bar\mu_H^+$-dependent activation of thylakoid ATP synthase with the use of acid-base transition-induced uncoupled ATP hydrolysis[a]

Source of thylakoids	State of enzyme	$\Delta\bar\mu_a$ (kJ/mol)	$\Delta\bar\mu_H^+$ onset (kJ/mol)	Reference
Chloroplasts	oxidised	19.3	—	3
	reduced	12.0	—	3
Synechococcus 6716	—	14.2	—	13
Spirulina platensis	oxidised	13.7	—	17
	reduced	8.1	—	17
Synechococcus 6716	—	—	ca. 10	13
Synechocystis 6803 wt	—	—	ca. 10	22
Synechocystis 6803 mutant cys. 1-10	oxidised	—	ca. 10	22
	reduced	—	ca. 6	22

[a] $\Delta\bar\mu_a$, the value of $\Delta\bar\mu_H^+$ at which 50% of the enzyme is in the activated state; $\Delta\bar\mu_H^+$ onset, the value of $\Delta\bar\mu_H^+$ at which activation of the enzyme initiates. wt, wild-type.

slowly and remains stable when ADP is removed. It is suggested that reoxidation provides protection against ATP hydrolysis in the dark. However, reoxidation is very slow. Upon a sudden drop in $\Delta\bar\mu_H^+$ rapid inactivation of the reduced form may take place by binding of ADP and thus a relatively high concentration of ATP in the chloroplast is maintained (see [4]).

Besides this physiological activation mechanism a number of *in vitro* methods has been described, such as treatment with trypsin, heat, methanol, octylglucoside or sulfite ([5] for review). In Fig. 2 the protein sequence of the part of the chloroplast γ-subunit supposedly involved in thiol-modulation and the effects of trypsin, heat and methanol is given. Activation of the ATP synthase by thiol-modulation results from reduction of the disulfide bond between the two cysteines [6]. The structural changes induced in this way can be 'mimicked' by cleavage of trypsin at site I and by heat and methanol, although those treatments do not lead to actual reduction (see e.g. [7]). Sulfite has been shown to compete with phosphate for the phosphate binding site, thereby accelerating ATP hydrolysis and inhibiting ATP synthesis in thylakoid ATP synthase [8, 9].

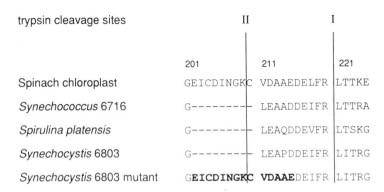

Figure 2. Alignment of the protein region containing the disulfide cysteines of the chloroplast γ-subunit as compared to *Synechococcus* 6716, *Spirulina platensis* and the wild-type and mutant strain cys 1–10 of *Synechocystis* 6803. This *Synechocystis* 6803 mutant contains the chloroplast regulatory segment of 9 amino acids and the following 5 amino acids were adjusted to the spinach sequence [19]. The sites of sequential cleavage in trypsin activation are indicated by arrows I and II.

In the alignment in Fig. 2 it is shown that the supposedly 'regulatory segment' in subunit γ of the chloroplast enzyme (a stretch of 9 amino acids containing the two cysteines) is absent in most cyanobacterial ATP synthases (see also [10,11]). As a consequence cyanobacterial ATP synthase functionally corresponds to the reduced form of the chloroplast enzyme concerning the required $\Delta\bar{\mu}_H^+$ for activation (see Table 1), the low threshold for ATP synthesis activity and stability of the active form [12,13]. In cyanobacteria an oxidised form is probably not required because the thylakoid ATP synthase is also involved in oxidative phosphorylation [2] and has to operate in the dark as well as in the light, in contrast with the chloroplast enzyme. Although the rate of respiration on the thylakoid membranes of cyanobacteria is not very high compared to photophosphorylation (see Scherer et al. [14]) it might be sufficient to prevent ATP hydrolysis. In cells from *Synechococcus* 6716 the dark ATP level is 70% of the level in the light under aerobic conditions, whereas under anaerobic conditions this level drops to 20% [15]. Other *in vitro* activating methods that might involve the cysteines in γ, such as heat and methanol treatment, have no effect on the ATP synthase from *Synechococcus* 6716 [12]. Trypsin still has an activating effect, due to one conserved cleavage site (see Fig. 2). The ATP synthase from *Rhodobacter capsulatus* can be activated in the same way as *Synechococcus* 6716 [16] and also lacks the chloroplast regulatory segment (see [10]).

Interestingly, ATP synthase from the higher developed cyanobacterium *Spirulina platensis* can exist in an oxidised and a reduced form and reacts on methanol [17]. The difference with chloroplasts is that the value of $\Delta\bar{\mu}_H^+$ needed for activation is considerably lower for both forms as given in Table 1. As for chloroplast F_1F_0 the stability of the active oxidised enzyme of *Spirulina platensis* is much lower compared to the active reduced form. The low energy level to activate the reduced enzyme enables this cyanobacterium to synthesise ATP at very low levels of $\Delta\bar{\mu}_H^+$ (see below). From antibody studies (P.A. Austin, personal communication) and sequence data [18] it appears that the cysteines as present in chloroplast γ are absent in *Spirulina* (see Fig. 2). However, the sequence reveals the presence of two non-conserved cysteines in the *Spirulina* γ-subunit which might be involved in redox regulation and in the effect of methanol.

From these studies it appears that the precise site and mechanism of redox regulation are not resolved so far and therefore mutant studies on cyanobacteria constitutively lacking a redox-linked mechanism seem very useful. The strain *Synechocystis* 6803 is suitable for this purpose because it can readily take up DNA. Site-directed mutants have been made of this strain in which a chloroplast-like segment containing the two cysteines is inserted into the γ-subunit [19] (Fig. 2). With these mutants a characterisation of activation requirements compared to the wild-type and chloroplasts has been performed.

After we had overcome various problems in preparing well-coupled thylakoid vesicles from *Synechocystis* 6803 [20] it appeared that the ATP synthase from the wild-type, as for *Synechococcus* 6716, corresponds to the reduced form of the chloroplast F_1F_0 concerning the low energy required for activation and the stability of the active form. Neither heat nor reduction had a stimulating effect on ATP hydrolysis activity in contrast to trypsin [21]. The effect of methanol was relatively small [22]. As shown in Table 1, activation of the mutant ATP synthase in the presence of dithiothreitol initiates at very low $\Delta\bar{\mu}_H^+$. Untreated thylakoids correspond to the wild-type and to reduced chloroplasts concerning $\Delta\bar{\mu}_H^+$ for activation. This indicates the existence of an additional active form of the reduced mutant F_1F_0. The $\Delta\bar{\mu}_H^+$ for activation of the oxidised and reduced mutant F_1F_0 corresponds to those of *Spirulina platensis* but the stability of the active state is different. Both oxidised and reduced mutant F_1F_0 are only stable when ADP is fully removed. When ADP is not removed no ATP hydrolysis occurs at all. As with chloroplast F_1F_0 heat and reduction have an activating effect on ATP hydrolysis activity in the mutant strain. The mutation, however, did not increase the susceptability to methanol [22].

We therefore conclude that the chloroplast regulatory segment in the γ-subunit including the 2 cysteines is indeed involved in redox regulation of $\Delta\bar{\mu}_H^+$-dependent activation, as well as in the effect of heat. The two non-conserved cysteines in γ of *Spirulina platensis* may have a similar function. This corroborates studies with a mutant *Chlamydomonas reinhardtii* strain where the two regulatory cysteines in γ have been mutated. This mutant ATP synthase has lost its redox regulation [23]. However, the factors that determine the diffference in size of $\Delta\bar{\mu}_H^+$ needed for activation and the difference in stability of the active states remains to be determined, as well as the effect of methanol. There is still a fourth cysteine in γ and a cysteine in ε, not present in the studied cyanobacteria, that may play a role.

2.2. Modulation of the H⁺/ATP Ratio

With an active ATP synthase at thermodynamic equilibrium, the phosphate potential (ΔG_p) is proportional to the $\Delta\bar{\mu}_H^+$ according to $\Delta G_p = n \cdot \Delta\bar{\mu}_H^+$, where n is a proportionality factor representing the H⁺/ATP ratio, the number of protons translocated in either direction upon the synthesis or hydrolysis of one molecule of ATP. A number of 3 H⁺/ATP was found for mitochondria and various bacterial strains [24]. However, for chloroplasts and the cyanobacterium *Synechococcus* 6716 the H⁺/ATP ratio was recently corrected to 4 [25]. Fillingame [24] proposed a model for the conformational coupling of proton translocation to catalytical activity, which is related to the stoichiometry of subunits c and β of the protein complex. Such a structure-function relationship may imply the existence of different intrinsic H⁺/ATP ratios between different organisms.

However, our group has not only found that the H⁺/ATP in cyanobacterial vesicles exceeds 3 but also that the H⁺/ATP depends on the environment of the enzyme. The H⁺/ATP can be modulated depending on the protein to lipid ratio of the membrane in reconstituted proteoliposomes [26], the growth temperature and light intensity influencing the fatty acid composition of the photosynthetic membrane [27], and the pH during measurement [28]. Furthermore, *Spirulina platensis* appeared to have a H⁺/ATP of at least 7 in its reduced form [17]. Since this form only requires a very low activation energy (see before) a high H⁺/ATP may enable the latter strain to produce ATP at an extremely low value of $\Delta\bar{\mu}_H^+$. *Spirulina platensis* grows in an extreme environment concerning pH (pH 9–11) just like alkalophilic bacteria which are also able to produce ATP and grow at very low value of $\Delta\bar{\mu}_H^+$ [29].

A molecular mechanism to explain changes in H⁺/ATP has been postulated on the basis of the finding that H⁺/ATP depends on the pH during measurement and that this dependence differs between different ATP synthases [28]. The H⁺/ATP in vesicles from *Synechococcus* 6716 increases when the pH was increased. The same increase in pH led to a decrease in H⁺/ATP in chromatophores from *Rhodospirillum rubrum*. Alignment of the sequences of the a- and c-subunits from the *Synechococcus* 6716 and *Rhodospirillum rubrum* ATP synthases revealed that the distribution of charged residues in the hydrophylic loops near the F_1 side is different between both sources [28]. These differences might affect the H⁺/ATP ratio of the ATP synthase due to a different degree of protonation at a given pH. Mutation of charged amino acids in the a- and c-subunits from the transformable cyanobacterium *Synechocystis* 6803 is a possible way to investigate the molecular mechanism behind the modulation of the H⁺/ATP ratio.

For this purpose membrane vesicles were prepared [20] from a *Synechocystis* 6803 mutant, in which the serine at position 37 in the hydrophilic loop of the c-subunit from the

	27	37	47
Spinach chloroplast	GQGTAAGQAV	**E**GIARQPEAE	GKIRGTLLLS
Synechococcus 6716	GQGNASGQAV	**E**GIARQPEAE	GKIRGTLLLT
Synechocystis 6803	GQGNASGQAV	**S**GIARQPEAE	GKIRGTLLLT
Synechocystis mutant	GQGNASGQAV	**E**GIARQPEAE	GKIRGTLLLT
Rhodospirillum rubrum	GVGNIWANLI	**A**TVGRNPAAK	STVELYGWIG

charge differences with *Rsp. rubrum*:   ↑ ↑   ↑ ↑

Figure 3. Alignment of the amino acid sequences of part of the *c*-subunit from chloroplast and cyanobacterial thylakoid ATP synthase including the mutant *Synechocystis* 6803 strain plc37 and the equivalent part of the *c*-subunit from *Rhodospirillum rubrum*.. In this *Synechocystis* 6803 mutant the serine on position 37 in the hydrophilic loop [30] was replaced by a negatively charged glutamic acid.

wild-type was replaced by a negatively charged glutamic acid residue (strain plc37) (for sequence see [30]). Fig. 3 shows that the *c*-subunit of chloroplasts and the cyanobacterium *Synechococcus* 6716 also contains glutamic acid at this position. In Fig. 4 it is shown that this mutant displays a significantly higher H^+/ATP ratio than the control strain (wild-type with kanamycin resistence). This higher ratio is also observed in chloroplasts and *Synechococcus* 6716 [25]. Furthermore, the pH dependence of the H^+/ATP of strain plc37 resembles that of *Synechococcus* 6716 whereas the control strain resembles *Rhodospirillum rubrum* (see [28]).

These results indicate that thylakoid ATP synthase may exist in different forms with different H^+/ATP ratio's depending on the availability of protonable groups in the hydrophilic loop near the F_1-side of the *c*- and possibly also *a*-subunits. Not only obvious environmental factors such as pH might have an effect on these groups but also growth conditions (temperature, light) that affect membrane protein and lipid composition [27]. Modulation of the H^+/ATP might be an important tool for long-term adaptation and short-time tuning of the efficiency of energy transduction. Cyanobacteria and many other (phototrophic) bacteria are normally exposed to harsh and fluctuating environmental conditions concerning temperature, light, pH and salinity. A large increase in H^+/ATP as observed in *Spirulina platensis* but

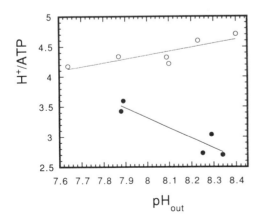

Figure 4. H^+/ATP as a function of external pH of the thylakoid ATP synthases from *Synechocystis* 6803 strain plc37 (m) and the control strain (l, wild-type with kanamycin resistence) determined with active ATP synthase in membrane vesicles from the equilibrium point of ATP synthesis/hydrolysis. $\Delta\tilde{\mu}_H^+$ was applied by acid-base transition.

also smaller changes provide the organisms with the opportunity to continue ATP synthesis in "lower gear" to survive conditions of low $\Delta\bar{\mu}_H{}^+$.

3. CONCLUDING REMARKS

With the help of mutants from *Synechocystis* 6803 two regions of the ATP synthase have been identified that are important for energy-dependent regulation:

1. The presence of the chloroplast regulatory loop in γ containing two cysteines is responsible for some, but not all features of chloroplast redox regulation (thiol-modulation) and for activation by heat but not by methanol.
2. The presence of charged amino acid residues in the hydrophilic loop of the *c*-subunit near the side where F_1 is bound has an effect on the H^+/ATP ratio.

Both $\Delta\bar{\mu}_H{}^+$-dependent activation and H^+/ATP are important regulatory mechanisms of chloroplast and cyanobacterial F_1F_0. The H^+/ATP and $\Delta\bar{\mu}_H{}^+$ for activation (with or without redox regulation) determines under which conditions ATP is made. By such a mechanism an enzyme can be switched down under conditions when ATP synthesis is thermodynamically unfavourable (e.g. oxidised chloroplasts at night). On the other hand, these mechanisms may lead to survival under extreme conditions (low $\Delta\bar{\mu}_H{}^+$ such as with *Spirulina platensis* grown at pH 11).

ACKNOWLEDGMENTS

Ms. M.J.C. Scholts, Ms. Dr. B.E. Krenn, Dr. R.H.A. Bakels, Dr. K. Krab, Dr. J.E. Van Wielink, Prof. Dr. R. Kraayenhof (Vrije Universiteit Amsterdam, The Netherlands) and Prof. R.A. Dilley (Purdue University, West Lafayette, IN, USA) have contributed to this chapter. Prof. Dr. H. Strotmann and co-workers (Universität Düsseldorf, Germany) and Dr. H. Lill (Universität Osnabrück, Germany) are acknowledged for providing various *Synechocystis* 6803 mutant strains and fruitful discussion. This work was supported by the European Union, contract Bio2-CT93-078 and by the Foundation for Chemical Research (SON), with financial aid from the Netherlands Organisation for the Advancement of Research (NWO).

REFERENCES

1. Sherman, L., Bricker, T., Guikema, J. and Pakrasi, H. (1987) in: The Cyanobacteria (Fay, P. and Van Balen, C. eds), pp. 1–53, Elsevier, Amsterdam.
2. Peschek, G.A. (1996) Biochem. Soc. Trans. 24, 729–733.
3. Junesch, U. and Gräber, P. (1987) Biochim. Biophys. Acta 893, 275–288.
4. Gräber, P. (1994) Biochim. Biophys. Acta 1187, 171–176.
5. Strotmann, H. and Bickel-Sandkötter, S. (1984) Annu. Rev. Plant Physiol. 35, 97–129.
6. Nalin, C.M. and McCarty, R.E. (1984) J. Biol. Chem. 259, 7257–7260.
7. Schumann, J., Richter, M.L. and McCarty, R.E. (1985) J. Biol. Chem. 260, 11817–11823.
8. Bakels, R.H.A., Van Walraven, H.S., Van Wielink, J.E., Van der Zwet-de Graaff, I., Krenn, B.E., Krab, K., Berden, J.A. and Kraayenhof, R. (1994) Biochem. Biophys. Res. Commun. 201, 487–492.
9. Bakels, R.H.A., Van Wielink, J.E., Krab, K. and Van Walraven, H.S. (1996) Arch. Biochem. Biophys. 332, 170–174.
10. Werner, S., Schumann, J. and Strotmann, H. (1990) FEBS Lett. 261, 204–208.
11. Van Walraven, H.S., Lutter, R. and Walker, J.E. (1993) Biochem. J. 294, 239–251.

12. Bakels, R.H.A., Van Walraven, H.S., Scholts, M.J.C., Krab, K. and Kraayenhof, R. (1991) Biochim. Biophys. Acta 1058, 225–234.
13. Krab, K., Bakels, R.H.A., Scholts, M.J.C. and Van Walraven, H.S. (1993) Biochim. Biophys. Acta 1141, 197–205.
14. Scherer, S., Almon, H. and Böger, P. (1988) Photosynth. Res. 15, 95–114.
15. Lubberding, H.J. and Schroten, W. (1984) FEMS Microbiol. Lett. 22, 93–96.
16. Turina, P. Rumberg, B., Melandri, B.A. and Gräber, P. (1992) J. Biol. Chem. 267, 11057–11063.
17. Bakels, R.H.A., Van Walraven, H.S., Krab, K. Scholts, M.J.C. and Kraayenhof, R. (1993) Eur. J. Biochem. 213, 957–964.
18. Steinemann, D. and Lill, H. (1995) Biochim. Biophys. Acta 1230, 86–90.
19. Werner-Grüne, S., Gunkel, D., Schumann, J. and Strotmann, H. (1994) Mol. Gen. Genet. 244, 144–150.
20. Scholts, M.J.C., Aardewijn, P. and Van Walraven, H.S. (1996) Photosynth. Res. 47, 301–305.
21. Krenn, B.E., Aardewijn, P., Van Walraven, H.S., Werner-Grüne, S., Strotmann, H. and Kraayenhof, R. (1995) Biochem. Soc. Trans. 23, 757–760.
22. Krenn, B.E., Strotmann, H., Van Walraven, H.S., Scholts, M.J.C. and Kraayenhof, R. (1997) Biochem. J. 322, 841–845.
23. Ross, S.A., Zhang, M.X. and Selman, B.R. (1995) J. Biol. Chem. 270, 9813–9818.
24. Fillingame, R.H. (1992) J. Bioenerg. Biomembr. 24, 485–491.
25. Van Walraven, H.S., Strotmann, H., Schwarz, O. and Rumberg, B. (1996) FEBS Lett. 379, 309–313.
26. Van Walraven, H.S., Scholts, M.J.C., Koppenaal, F., Bakels, R.H.A. and Krab, K. (1990) Biochim. Biophys. Acta 1015, 425–434.
27. Van Walraven, H.S., Hollander, E.E., Scholts, M.J.C. and Kraayenhof, R. (1997) Biochim. Biophys. Acta 1318, 217–224.
28. Krenn, B.E., Van Walraven, H.S., Scholts, M.J.C. and Kraayenhof, R. (1993) Biochem. J. 294, 705–709.
29. Krulwich, T.A. and Guffanti, A.A. (1992) J. Bioenerg. Biomembr. 24, 587–599.
30. Lill, H. and Nelson, N. (1991) Plant Mol. Biol. 17, 641–652.
31. Van Walraven, H.S. and Bakels, R.H.A. (1996) Physiol. Plant. 96, 526–532.
32. Lill, H., Hensel, F., Junge, W. and Engelbrecht, S. (1996) J. Biol. Chem. 271, 32737–32742.

44

THE INORGANIC PYROPHOSPHATE SYNTHASE FROM *RHODOSPIRILLUM RUBRUM* AND ITS GENE

Margareta Baltscheffsky,* Anders Schultz, and Sashi Nadanaciva

Depatment of Biochemistry
Arrhenius Laboratories
Stockholm University
S-106 91 Stockholm, Sweden

1. INTRODUCTION

The inorganic pyrophosphate synthase (PPi synthase, PPase) in *Rhodospirillum rubrum* was first detected in our laboratory as a firmly membrane-bound pyrophosphatase activity, uncoupler stimulated, in the mid-sixties [1]. A few years later H. Baltscheffsky and von Stedingk together with Heldt and Klingenberg [2,3] found that, in the absence of added nucleotides, photophosphorylation yielded inorganic pyrophosphate as end product. Furthermore, this synthesis of PPi is not inhibited, but rather stimulated by the addition of oligomycin. We also found that addition, in the dark, of PPi in the presence of Mg^{++} ions elicited energy requiring redox changes of cytochromes [4]. Simultaneously, Keister reported that it was possible to drive the energy requiring transhydrogenase reaction in *R. rubrum* chromatophores [5] by addition of PPi and Mg^{++}. Johansson established the identity of separate enzyme catalysing the PPi linked reactions in *R. rubrum* chromatophores by finding that antibodies against the ATPase did not impair the PPase [6]. Keister and Minton also showed that it was possible to obtain ATP synthesis in the dark with PPi as the energy source, in *R. rubrum* chromatophores [7]. The capability to transport protons was established by Moyle et al. [8], who found that one proton was translocated across the chromatophore membrane per PPi hydrolysed.

In a reconstitution experiment we succeded to obtain ATP synthesis, driven by hydrolysis of PPi, in azolectin liposomes in which we had incorporated only the F_1F_0 ATPase and the PPi synthase, both enzymes solubilized from *R. rubrum* [9].

In higher plants, a membrane bound K^+ stimulated PPase activity was found by Karlsson [10] in the microsomal fraction from sugar beet roots. Some ten years later Rea

* Corresponding author. Tel: +46 8 162456; Fax: +46 8 153679

The Phototrophic Prokaryotes, edited by Peschek *et al.*
Kluwer Academic / Plenum Publishers, New York, 1999.

and Poole [11] showed that tonoplast membrane vesicles from red beet vacuoles contain two distinct proton pumping enzymes, an ATPase and a PPase. Rea and coworkers were also the first ones to identify and sequence the gene for a vacuolar PPase [12], screening a gene library from *Arabidopsis thaliana* with an antibody against the vacuolar PPase from mung bean (*Vigna radiata*). The same antibody had earlier by us been shown to cross-react with the PPi synthase from *R. rubrum* [13], indicating that there is some similarity between the physiologically unidirectional PPase of plant vacuoles and the bidirectional PPi synthase of *R. rubrum*.

2. MATERIALS AND METHODS

The proton pumping inorganic pyrophosphate synthase was isolated from *R. rubrum* (strain S1) according to Nyrén et al. [14] and purified further after an ethylene glycol gradient by electroelution. Rabbit polyclonal antibody was raised against the electroeluted *R. rubrum* PPase and purified according to the method described by Sambrook et al. [15]. Rat polyclonal antibodies against the vacuolar PPase from *Vigna radiata* were a kind gift by Prof. M. Maeshima. A *R. rubrum* cDNA library constructed in λZAP II was purchased from Clontech Labs Inc. Restriction enzymes and other DNA modification enzymes were obtained from Pharmacia, and Sigma and radioactive nucleotides from Amersham.

For the sequencing we used the dideoxynucleotide chain-termination method [16] using Deaza G/A^{T7} Sequencing kits (Pharmacia).

3. RESULTS

After screening the gene library with purified antibodies against the *R. rubrum* PPi synthase, we found three positive clones, which we termed PP4, PP5 and PP6. These positive plaques were doublechecked with the antibody against the *Vigna radiata* vacuolar PPase, as this has been shown to cross-react with our enzyme [13]. Restriction analysis indicated that the three clones were identical. The *R. rubrum* genome is very GC rich and we decided to sequence PP4 manually in both directions and obtained a sequence with an open reading frame of 1980 bp from a GTG start codon to the TAA termination codon. Eventually we did an automated sequencing of PP5 and PP6 as well, and it turned out that all three clones were indeed identical.

The 1980 bp open reading frame encodes a protein of 660 amino acids with a calculated molecular mass (M_r) of 67,453. The calculated M_r is greater than the apparent M_r (56,000) which we had earlier obtained by SDS-PAGE. Since the membrane-bound *R. rubrum* PPase is known to be an extremely hydrophobic protein [17], it is possible that it moves anomalously on gels giving rise to an underestimated M_r. This is not uncommon with membrane proteins and is ascribed to the binding of nonsaturating amounts of SDS to the protein [18].

Hydropathy plots of the deduced amino acid sequence analysed by the TopPred program [19] supported our earlier observations that the protein is extremely hydrophobic, with 15 transmembrane segments.

4. DISCUSSION

Computer searches of the deduced amino acid sequence of our protein against the Swiss-Prot and NIH databases [20] revealed that the sequence shows partly a striking

```
              130        140        150        160        170        180
R.rub. FGASLISIFARLGGGIFTKCADVGADLVGKVEAGI PEDDPRNPAVIADNVGDNVGDCAGM
       .:.:  ...:.:.:::.:: :::::::::::.: .::::::::::::::::::::::: :::
A.th.  LGGSSMALFGRVGGGIYTKAADVGADLVGKIERNI PEDDPRNPAVIADNVGDNVGDIAGM
       240        250        260        270        280        290

              190        200        210        220        230
R.rub. AADLFETYAVTVVATMVLASIFFAGVPAMTSMMAYPLAIGG----VCILASILGTKFVKL
       ..:::  .::  .   :.,:.:::     :.      . : :::  :..  :!:..  ....: : ..
A.th.  GSDLFGSYAEASCAALVVASISSFGINHDFTAMCYPLLISSMGILVCLI-TLFATDFFEI
       300        310        320        330        340        350
```

Figure 1. Comparison of part of the deduced amino acid sequence of PP4 (*R. rubrum*) with a corresponding part of AVP (*A. thaliana*). The sequence segment includes the predicted loop between transmembrane segments 4 and 5 (AA 135–191) in the *R. rubrum* sequence.

homology with plant vacuolar pyrophosphatases. Our protein has an overall identity of 36–39% with the vacuolar PPases from *Arabidopsis thaliana, Beta vulgaris, Vigna radiata, Oryza sativa, Nicotiana tabacum* and *Hordeum vulgare* [12,21–25]. While the overall identity between our protein and these vacuolar PPases may seem somewhat low, the identity in the major hydrophilic segments connecting the transmembrane segments is high, as high as 81% [26] (Fig. 1). Within one of these hydrophilic segments, between transmembrane segments 4 and 5, is a DhG/AADLVGK motif (DVGADLVGK) found conserved in all sequenced H^+ PPases (Table 1) [27]. This motif is similar to the EX_7KXE motif found in soluble PPases, which, based on the three dimensional structure of the *E. coli* enzyme, has been proposed to participate directly in substrate and Mg^{++} binding [28,29].

Another interesting similarity is the sequence EYYT (in Table 2), located right after our transmembrane segment 8, which is identical to a corresponding sequence in the *Arabidopsis* and other V-PPases, and where the mutation E427Q in the *Arabidopsis* sequence inhibits the H^+ conductance but retains DCCD sensitive PPi hydrolysis, whereas E 427D shows an enhancement of proton pumping [30].

The high percentage of identities in the hydrophilic loops between the transmembrane segments makes it likely that the *R. rubrum* enzyme is rather closely related to the V-PPase family. The great difference is that the *rubrum* enzyme, in contrast to the other members of the family has the ability to synthesize PPi as well as hydrolyze this compound. It is resonable to assume that in the sequence presented in this communication which represents a polypeptide with both a catalytic site for the synthesis as well as the hydrolysis of pyrophosphate and a physiologically bidirectional proton channel we should

Table 1. A putative metal and substrate binding motif [27] in membrane-bound PPases (DhG/AADLVGK)

Organism	Motif	Amino acids in protein
Rhodospirillum rubrum	145 153 D V G A D L V G K "193" "201"	660
Thermotoga maritima	D M A A D L V G K	
Arabidopsis thaliana	257 265 D V G A D L V G K	770

Table 2. Similarities and differences between amino
acid sequences in the region of a glutamate residue
suggested to participate in proton pumping [30]

Organism	Amino acid sequence
Rhodospirillum rubrum	309 L L I W V T E Y Y T G
Thermotoga maritima	"372" L I G F W A E Y Y T S
Arabidopsis thaliana	427 I I G F V T E Y Y T S

be able to find which amino acid(s) is (are) resposible for the bidirectionality, and thus for
the unique capacity to synthesize the "energy-rich" pyrophosphate bond.

ACKNOWLEDGMENTS

This work has been supported by the Carl Tryggers Stiftelse för Vetenskaplig
Forskning and the Magnus Bergvall and Wenner Gren Foundations, which are gratefully
acknowledged. Many thanks are also due to professor Herrick Baltscheffsky for valuable
discussions and unfailing interest.

REFERENCES

1. Baltscheffsky M. (1964) Abstr. 1. Meet. FEBS., London, p. 67..
2. Baltscheffsky H., von Stedingk L.-V., Heldt H.W. and Klingenberg M. (1966) Science 153, 1120–1123.
3. Baltscheffsky, H. and von Stedingk, L.-V. (1966) Biochem. Biophys. Res Commun. 22, 722–728.
4. Baltscheffsky M. (1967) Nature 216, 241–243.
5. Keister, D.L. and Yike, N.J. (1967) Biochem. Biophys. Res. Commun. 24, 519–525.
6. Johansson, B.C. (1975) Doctoral Thesis, University of Stockholm.
7. Keister D.L. and Minton N.J. (1971) Biochem. Biophys. Res. Commun: 42, 932–939
8. Moyle, J., Mitchell, R. and Mitchell, P. (1972) FEBS Lett. 23, 233–236.
9. Nyrén P. and Baltscheffsky M. (1983) FEBS Lett. 155: 125–130
10. Karlsson, J. (1975)Biochim. Biophys. Acta 399, 356–363.
11. Rea, P.A. and Poole, R.J. (1986) Plant. Physiol. 81, 126–129.
12. Sarafian V., Kim Y., Poole R.J. and Rea P.A. (1992) Proc. Natl. Acad. Sci. 89, 1775–1779.
13. Nore, B.F., Sakai-Nore, Y. Maeshima, M., Baltscheffsky, M. and Nyrén, P. (1991) Biochem. Biophys. Res.
    Commun. 181, 962–967
14. Nyrén, P., Nore, B.F. and Strid, Å. (1991) Biochemistry 30, 2883–2887.
15. Sambrook, J., Fritsch, E.F. and Maniatis, T. (1989) Molecular cloning. A laboratory manual, Cold Spring
    Harbor, Laboratory Press.
16. Sanger, F., Nicklen, S. and Coulson, A.R. (1977) Proc. Natl. Acad. Sci. 74, 5463–5467.
17. Shakov, Y.A., Nyrén, P. and Baltscheffsky, M. (1982) FEBS Lett. 146, 177- 180.
18. Maddy, A.H. (1976) J. Theor. Biol. 62: 315–326.
19. Claros, M.G. and von Heijne, G. (1994) Comput. Appl. Biosci. 10, 685–686
20. Altschul, S.F., Gish, W., Miller, W., Myers, E.W. and Lipman, D.J. (1990) J. Mol. Biol. 215, 403–410
21. Kim, Y., Kim, E.J. and Rea, P.A. (1994) Plant Physiol. 106, 375–382.
22. Hung, S.-H., Chiu, S.-J., Lin, L.-Y. and Pan, R.L. (1995) Plant Physiol. 109, 1125–1127.
23. Sakakibara, Y., Kobayashi, H. and Kasamo, K. (1996) Plant Mol. Biol. 31, 1029–1038.
24. Lerchl, J., König, S., Zrenner, R. and Sonnewald, U. (1995) Plant Mol. Biol. 29, 833–840.

25. Tanaka, Y., Chiba, K., Maeda, M. and Maeshima, M. (1993) Biochem. Biophys. Res. Commun. 181, 962–967.
26. Baltscheffsky, M., Nadanaciva, S. and Schultz, A. (1998) (submitted).
27. Baltscheffsky, H., Schultz, A., Persson, B. and Baltscheffsky, M. (1998) (submitted).
28. Cooperman, B.S., Baykov, A.A. and Lahti, R. (1992) Trends Biochem. Sci. 17, 262–266.
29. Rea, P.A., Kim, Y., Sarafian, V., Poole, R.J., Davies, J.M. and Sanders, D. (1992) Trends Biochem. Sci. 17, 348–353.
30. Zhen, R.-G., Kim, E.J. and Rea, P.A. (1997) Proceedings of the First International Meeting on Inorganic Pyrophosphatases. (R.Lahti, ed.) Univ. of Turku, pp. 21–23 (ISBN 951-29-0990-1).

45

PROTON-TRANSLOCATING TRANSHYDROGENASE FROM *RHODOSPIRILLUM RUBRUM*

Kinetic Analysis of the Reaction Catalysed by Recombinant, Nucleotide-Binding Domains Shows that Hydride Transfer Does Not Involve Intermediate Redox Reactions

Jamie D. Venning, Nick N. P. J. Cotton, Philip G. Quirk, Tania Bizouarn, Rachel L. Grimley, Susmita Gupta, and J. Baz Jackson

School of Biochemistry
University of Birmingham
Edgbaston, Birmingham, B15 2TT, United Kingdom

1. INTRODUCTION

Transhydrogenase is found in the cytoplamic membranes of many bacteria. It catalyses the following reaction:

$$NADH + NADP^+ + H^+_{out} \Leftrightarrow NAD^+ + NADPH + H^+_{in}$$

Under physiological conditions the reaction probably proceeds from left to right, in favour of NADPH formation, driven by the protonmotive force generated by either respiratory or photosynthetic electron transport. Under experimental conditions the reaction can be measured in real time using nucleotide analogues. In this report we use acetyl pyridine adenine dinucleotide, $AcPdAD^+$, an NAD^+ analogue, whose reduced form has a characteristic ultra-violet absorbance.

Transhydrogenase has three domains (Fig. 1) [1–3]. Domains I and III protrude from the membrane and possess the nucleotide-binding sites, domain I for NAD^+ and NADH, and domain III for $NADP^+$ and NADPH. Domain II spans the membrane and might comprise 10–12 transmembrane helices. The polypeptide structure of transhydrogenase from

The Phototrophic Prokaryotes, edited by Peschek *et al.*
Kluwer Academic / Plenum Publishers, New York, 1999.

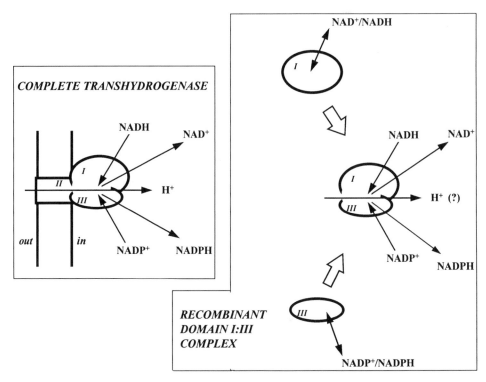

Figure 1. Transhydrogenase and its component domains.

Rhodospirillum rubrum is unique [4], and this opened up a procedure to prepare recombinant forms of the domain I and domain III proteins and express them at high levels in *E. coli* [5–7]. The isolated, purified domain I protein binds NAD^+ and NADH, and the isolated, purified domain III protein binds $NADP^+$ and NADPH [5–7,8]. Subsequently, recombinant domains I and III of transhydrogenase from other species have been expressed and purified [6,9]. A mixture of the domain I and domain III proteins from *R. rubrum* catalyses transhydrogenation (e.g. the reduction of $AcPdAD^+$ by NADPH), even in the complete absence of the membrane-spanning domain II, but the rate of reaction in steady-state is very slow; it is profoundly limited by slow release of product $NADP^+$ from domain III ($k = 0.03$ s^{-1}) [7] -see Fig. 1.

2. RESULTS

Using stopped-flow spectroscopy we can show that, although the steady-state rate of transhydrogenation by the mixture of domains I and III is very slow (see above), the hydride transfer step (from NADPH to $AcPdAD^+$) is extremely fast. Fig. 2 shows that when $AcPdAD^+$ was mixed rapidly with domain I plus domain III plus NADPH, there was a biphasic burst of $AcPdAD^+$ reduction, and of NADPH oxidation. The apparent first order rate constants for the fast and slow phases were approximately 490 s^{-1} and 8.6 s^{-1} at 8°C. This rapid reaction indicates that the apparatus for hydride transfer is located within domains I and III, and that domain II is unlikely to have a role in the redox process itself.

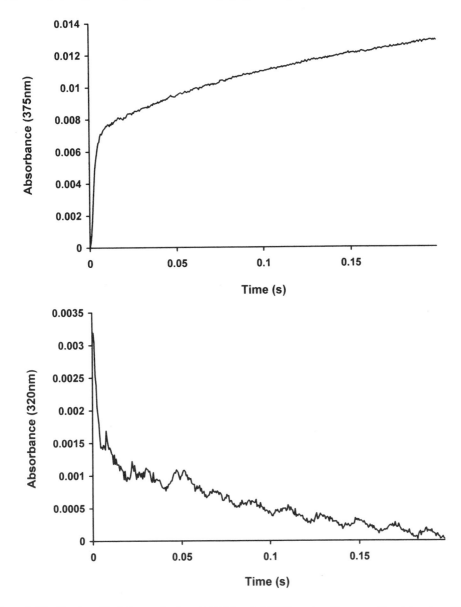

Figure 2. The kinetics of hydride transfer from NADPH to AcPdAD⁺ catalysed by recombinant domains of tran-shydrogenase. Syringe one contained purified recombinant domain I (50 μM), NADPH-loaded domain III (50 μM), and carry-over NADPH from the preincubation (final concentration 70 μM) in 10mM $(NH_4)_2SO_4$, 20 mM Hepes, pH 8.0 (NaOH). Syringe two contained 2 mM AcPdAD⁺, prepared by diluting a 20 mM stock solution of nucleotide in water with 10 mM $(NH_4)_2SO_4$, 20 mM Hepes, pH 8.0 (NaOH). The reaction was initiated by mixing 50 μl from each syringe in the stopped-flow spectrophotometer. Each trace shown in the figure is an average of 10 recordings. Upper panel, AcPdAD⁺ reduction at 375 nm. Lower panel, NADPH oxidation at 320 nm T = 8°C.

Fig. 2 also shows that the kinetics of AcPdAD⁺ reduction were similar to those of NADPH oxidation. This provides firm evidence that the transfer of the hydride proceeds directly between the nucleotides. The possibility of a kinetically undetectable redox intermediate is made unlikely by separate experiments using electrospray mass spectrometry. The molecular masses of the recombinant domains I and III were 40273 and 21469, respectively, which correspond closely with the masses calculated from the amino acid sequence (40276 and 21466, respectively). Thus, there are no covalently bound prosthetic groups or metal centres in the two peripheral domains.

The burst of hydride transfer arises because the rate of $NADP^+$ release from domain III is very slow (see above)—it represents a single turnover of that domain. The reason why the burst is biphasic is still under investigation. The slow phase was replaced by fast phase when the domain I concentration was increased relative to that of domain III (data not shown). A plausible explanation is that the fast component represents hydride transfer within a complex of domains I and III, whereas the slow phase results from slower conformational rearrangements within the proteins.

3. DISCUSSION

The direct hydride transfer indicates that the C4 atoms of the nicotinamide rings of the two nucleotides must be brought into apposition with one another within the domain I:III complex. Currently there are no high resolution structures of transhydrogenase. However, we have previously commented on the sequence similarity between transhydrogenase domain I and alanine dehydrogenase [4], whose x-ray structure has recently been solved by D. Rice and P. Baker (personal communication). The enzyme comprises two subdomains each folded into open, twisted α/β structures which are linked by two helices (a hinge?). The NAD(H)-binding site and the pyruvate/alanine-binding site are located on opposite sides of a cleft between the two subdomains. It is envisaged that, during catalysis, closure of the cleft brings the nucleotide and substrate together to effect hydride transfer. Transhydrogenase domain I is likely to have a similar structure but, as evidenced above, the hydride acceptor for bound NADH is not on the opposite face of the cleft, but within a separate domain (domain III). We must imagine, that during physiological turnover of the protein, domains I and III form an interstitial complex, in which the nicotinamide ring of the $NADP^+$ is presented up to the nicotinamide of NADH bound within a cleft in domain I. How this is achieved now represents a crucial target in the further understanding of this protein.

We have previously pointed out that a key difference between transhydrogenase domain I and alanine dehydrogenase is that the former has an additional sequence of approximately 20 amino acid residues which emanates from the surface of the protein as a mobile loop [9–11]. Stopped-flow experiments reveal that mutations in the loop (e.g. Tyr235Asn) lead to pronounced inhibition (approximately 100-fold) of the hydride transfer step (data not shown). The segmental mobility of the loop residues makes them visible by NMR spectroscopy. NOESY experiments indicate that, upon NADH (or analogue) binding, the loop closes down to make interactions with the adenosine moiety of the nucleotide. The loop can be modelled on to the Rice-Baker structure of alanine dehydrogenase, within 10 Å from the NADH-binding site in the cleft. We propose that, during NADH binding, closure of the loop leads to the proper positioning of the nucleotide to effect hydride transfer to $NADP^+$.

These and other studies [12] suggest that proton translocation by transhydrogenase is not directly coupled to the hydride-transfer step. Our current view is that the H^+-binding and release components of the translocation reaction are coupled to the binding and release of NADP(H) to and from domain III [12,13].

ACKNOWLEDGMENTS

We are grateful to David Rice and Pat Baker of The University of Sheffield for providing us with coordinates of their structure of alanine dehydrogenase prior to publication.

REFERENCES

1. Jackson, J. B. (1991) *Journal of Bioenergetics and Biomembranes* **23**, 715–741.
2. Olausson, T., Fjellstrom, O., Meuller, J., and Rydstrom, J. (1995) *Biochim. Biophys. Acta* **1231**, 1–19.
3. Hatefi, Y., and Yamaguchi, M. (1996) *FASEB J.* **10**, 444–452.
4. Cunningham, I. J., Williams, R., Palmer, T., Thomas, C. M., and Jackson, J. B. (1992) *Biochim. Biophys. Acta* **1100**, 332–338.
5. Diggle, C., Hutton, M., Jones, G. R., Thomas, C. M., and Jackson, J. B. (1995) *Eur. J. Biochem.* **228**, 719–726.
6. Yamaguchi, M., and Hatefi, Y. (1995) *J. Biol. Chem.* **270**, 28165–28168.
7. Diggle, C., Bizouarn, T., Cotton, N. P. J., and Jackson, J. B. (1996) *Eur. J. Biochem.* **241**, 162–170.
8. Bizouarn, T., Diggle, C., and Jackson, J. B. (1996) *Eur. J. Biochem.* **239**, 737–741.
9. Diggle, C., Cotton, N. P. J., Grimley, R. L., Quirk, P. G., Thomas, C. M., and Jackson, J. B. (1995) *Eur. J. Biochem.* **232**, 315–326.
10. Bizouarn, T., Diggle, C., Quirk, P. G., Grimley, R. L., Cotton, N. P. J., Thomas, C. M., and Jackson, J. B. (1996) *J. Biol. Chem.* **271**, 10103–10108.
11. Diggle, C., Quirk, P. G., Bizouarn, T., Grimley, R. L., Cotton, N. P. J., Thomas, C. M., and Jackson, J. B. (1996) *J. Biol. Chem.* **271**, 10109–10115.
12. Hutton, M. N., Day, J. M., and Jackson, J. B. (1994) *Eur. J. Biochem.* **219**, 1041–1051.
13. Bizouarn, T., Grimley, R. L., Cotton, N. P. J., Stilwell, S., Hutton, M., and Jackson, J. B. (1995) *Biochim. Biophys. Acta* **1229**, 49–58.

46

CHARACTERISATION OF REDOX PROTEINS PRESENT IN THE SOLUBLE COMPLEMENT OF *HELIOBACILLUS MOBILIS*

Anna Schricker, Wolfgang Nitschke, and Ursula E. Liebl

BIP/CNRS, 31, chemin Joseph Aiguier
13402 Marseille Cedex 20, France

1. INTRODUCTION

Light-induced electron transport in Heliobacteria has been shown to implicate a RC I-type photosystem and a cytochrome *bc* complex.

Both membrane-integral enzymes, i.e. the photosynthetic reaction centre (Trost and Blankenship 1989, van de Meent *et al.* 1990, Liebl *et al.* 1993) and the cytochrome *bc* complex (Nitschke and Liebl 1992, Liebl, Schoepp, Schricker and Nitschke, unpublished), have been studied to some detail. By contrast, data with respect to soluble redox components are scarce (Lee *et al.* 1995). We therefore examined the supernatant obtained after cell breakage for the presence of redox proteins such as cytochromes, iron sulphur proteins, flavo- or copper proteins.

2. MATERIALS AND METHODS

H. mobilis was grown as described in Nitschke *et al.* (1995). Cells were harvested after 1 day of growth. Approximately 100 g (dry weight) of cell paste were broken in a French pressure cell (3 passages). Membranes and soluble supernatant were separated by two sequential ultracentrifugation steps (2 h 330000 g, 12 h 220000 g). The brown supernatant was subjected to ammonium sulphate fractionation in 5 steps of 20% each. The individual fractions were further characterised by optical and EPR spectroscopy and SDS-PAGE (stained by TMBZ and CBB).

Electrophoresis was performed using a BioRad Mini Protean II system. EPR spectra were taken on a Bruker ESP 300 E spectrometer fitted with an Oxford Instruments Helium cryostat. Optical spectra were recorded using a Cary 5 spectrophotometer.

The Phototrophic Prokaryotes, edited by Peschek *et al.*
Kluwer Academic / Plenum Publishers, New York, 1999.

3. RESULTS

3.1. Rubredoxin

Rubredoxin was detected in the supernatant after precipitation at 100% saturation with ammonium sulphate as evidenced by the presence of the characteristic optical spectrum (peak at 491nm appearing in the redox difference spectra ascorbate reduced minus dithionite reduced). The EPR spectra further support the presence of rubredoxin, although the EPR signal would not be sufficiently specific to identify the rubredoxin from EPR alone. This protein has already been purified and characterised in detail by Lee *et al.* (1995). Our observations, i.e. spectral parameters and redox potential are in agreement with the characteristics determined by Lee *et al.*

3.2. Ferredoxins

Two different ferredoxins could be detected. A more detailed description of the purification procedure and characterisation will be published elsewhere. In the following we will only summarize the essential data.

3.2.1. 8Fe8S-Ferredoxin. Evidence for an 8Fe8S-ferredoxin was obtained by EPR-spectroscopy in the pellet of 60–80% saturation with ammonium sulphate. This finding was consolidated by the optical redox difference spectra showing a broad peak with a maximum at 415nm (spectrum of oxidized sample minus spectrum of reduced sample). An EPR-redox titration yielded an average midpoint potential of the two clusters of −516 mV. At low degrees of reduction, the spectrum of a non-interacting species was observed whereas the fully reduced sample yielded the typical spectrum of paramagnetically interacting 2-cluster 8Fe8S-ferredoxins (Mathews *et al.* 1974).

3.2.2. Single-Cluster Protein, Possibly a 2Fe2S-Ferredoxin. In the supernatant after precipitation at 100% saturation with ammonium sulphate an EPR-signal was detected that can be ascribed to a protein containing a single 2Fe2S- or (less probably according to saturation studies) 4Fe4S-cluster. The midpoint potential of this protein was found to be −518 mV. N-terminal sequencing was performed (to be published elsewhere).

3.3. Cytochromes

It has been reported previously that heliobacteria probably do not contain soluble cytochromes. Nevertheless, in the strongly concentrated samples obtained after ammonium sulphate precipitation, at least five heme-staining bands can be detected (Fig. 1 lanes 2–6). A comparison of the heme proteins present in membrane fragments (Fig. 1, lane 1) to those present in the crude extract after centrifugation (Fig. 1, lane 2) shows that these cytochromes actually correspond to those observed in membranes. This demonstrates that at least part of the membrane-bound cytochromes can be detached from the membrane and can thus be found in the soluble complement. The smaller cytochromes (band III and V, to a lesser degree band IV) seem to become detached more readily than the two larger ones therefore being to some extent enriched in the soluble complement.

The apparent molecular weights of the five heme-staining proteins were determined from their migration on SDS-PAGE with 55 kDa (Fig. 1, band I), 28 kDa (band II), 19 kDa (band III), 17 kDa (band IV) and 11 kDa (bandV).

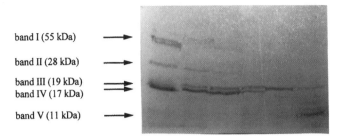

band I (55 kDa)

band II (28 kDa)

band III (19 kDa)
band IV (17 kDa)

band V (11 kDa)

Figure 1. Gel stained by TMBZ for heme bands showing *H. mobilis* membranes (lane 1); supernatant (lane 2); ammonium sulphate fractions of the supernatant: 20–40% (lane 3); 40–60% (lane 4); 60–80% (lane 5); 80–100% (lane 6).

The 11 kDa protein appearing in the last step of precipitation (pellet of 80–100% saturation with ammonium sulphate) may correspond to a degradation product. Otherwise it may be regarded as a candidate for a soluble cytochrome in *H. mobilis*. The 19 kDa cytochrome has until now not been detected, since in membranes it is far less abundant than the 17 kDa cytochrome resulting in only one large heme-staining band on SDS-PAGE of membrane particles. Obviously it is detached more easily than the 17 kDa protein thus appearing in the supernatant.

From characterization of the membrane components three cytochromes have been reported with apparent molecular masses of 50, 30 and 20 kDa (Trost and Blankenship 1990, Nitschke *et al.* 1996) most probably corresponding to the cytochromes of 55, 28 and 17 kDa reported in this work.

The redox potentials of the three cytochromes characterized from membranes have been determined to −60 mV for the 50 kDa protein, 120 mV for the 30 kDa protein and 160 mV for the 20 kDa protein (Nitschke *et al.* 1996). Since the amount of cytochrome obtained in the supernatant was insufficient for performing redox titrations only the criterion of reducibility by ascorbate or dithionite could be used to estimate limits for E_m-values. The obtained ratio of ascorbate-reduced to dithionite-reduced cytochrome for each precipitation step is in line with the relative intensities of the different heme-staining bands on the gels and their presumed identification with the known cytochromes. The 28 kDa and the 17 kDa cytochrome should be reduced by ascorbate whereas the 55 kDa cytochrome would be expected to get only substantially reduced by dithionite. Taking into account the respective ratio of ascorbate-reduced to dithionite-reduced cytochrome found in the different samples leads to the conclusion that the 19 kDa cytochrome must have a rather low E_m being reduced predominantly by dithionite. In the pellet of 60–80% saturation with ammonium sulphate there is twice as much dithionite-reduced than ascorbate-reduced cytochrome while the dominant heme-stain corresponds to the 19 kDa cytochrome with only traces of the 28 and the 17 kDa present (Fig. 1, lane 5). In the pellet of 40–60% saturation with ammonium sulphate bands III and IV corresponding to the 19 and the 17 kDa protein are equally abundant (Fig. 1, lane 4). The ratio ascorbate- to dithionite-reduced cytochrome in this fraction is 1:1, which is again in line with the assumption that the 19 kDa cytochrome has a negative redox potential.

The fact that the different cytochromes precipitate at different saturation steps of the ammonium sulphate fractionation, e.g. the 19 kDa protein appearing mostly in the later steps, the 17 kDa, the 28 kDa and the 55 kDa in the earlier steps can be useful in further purification and characterisation, especially of the 19 kDa cytochrome. The higher molecular weight, the apparently different redox potential and the different behaviour with respect to ammonium sulphate precipitation strongly indicate that the 19 kDa cytochrome differs from the 17 kDa heme protein. It may therefore represent a hitherto overlooked heliobacterial cytochrome.

3.4. Flavo- and Copper Proteins

Flavo- or copper proteins are not present in quantities sufficient to allow detection by optical or EPR-spectroscopy in the crudely fractionated material.

REFERENCES

Lee, W.-Y., Brune, D.C., LoBrutto, R. and Blankenship, R.E. (1995) Arch. Biochem. Biophys. 318, 80–88.

Liebl, U., Mockensturm-Wilson, M., Trost, J.T., Brune, D., Blankenship, R.E. and Vermaas, W.F.J. (1993) Proc. Natl. Acad. Sci. USA 90, 7124–7128.

Liebl, U., Rutherford, A. and Nitschke, W. (1990) FEBS Lett. 261, 427–430.

Liebl, U. (1993) PhD. Thesis, University of Regensburg, FRG.

Mathews, R., Charlton, S., Sands, R.H. and Palmer, G. (1974) J. Biol. Chem. 249, 4326–4328.

van de Meent, E.J., Kleinherenbrink, F.A.M. and Amesz, J. (1990) Biochim. Biophys. Acta 1015, 223–230.

Nitschke, W. and Liebl, U. (1992) In: Murata, N. (ed.) Research in Photosynthesis, vol.3, 507–510. Kluwer Academic Publishers, Dordrecht.

Nitschke, W., Liebl, U., Matsuura, K. and Kramer, D.M. (1995) Biochemistry 34, 11831–11839.

Nitschke, W., Schoepp, B., Floss, B., Schricker, A., Rutherford, A.W. and Liebl, U. (1996) Eur. J. Biochem. 242, 695–702.

Trost J.T. and Blankenship R.E. (1989) Biochemistry. 28, 9898–9904.

Trost J.T. and Blankenship R.E. (1990) In: Baltscheffsky, M. (ed.) Current Research in Photosynthesis, vol. 2, 703–706. Kluwer Academic Publishers, Dordrecht.

47

INCREASED RESPIRATORY ACTIVITY IN LIGHT IN SALT STRESSED *SYNECHOCYSTIS* PCC 6803

Potentiometric Method to Measure Respiratory Electron Transport in Intact Cells Both in Light and in Dark

Ioan Ardelean,[1] Sorin Tunaru,[2] Maria Luiza Flonta,[2] Gabriela Teodosiu,[1] Mădălin Enache,[1] Lucia Dumitru,[1] and G. Zarnea[3]

[1]Institute of Biology
Spl. Independenþei 296
Bucharest 79651, POB 56-53, Romania
[2]Faculty of Biology
Membrane Biophysics Group
Spl. Independenþei 91-95
Bucharest 76201, Romania
[3]Faculty of Biology
Aleea Portocalilor nr. 8
Bucharest, Romania

1. INTRODUCTION

In *Synechocystis* PCC 6803 the energetic changes associated with salt adaptation (0.5 M NaCl) include an increase in the respiratory activity (RA) both in dark (1–16) and in light (3) and an increased energy storage at the level of Photosystem I (PSI) (6,8,17–19). Inspite of the development of our knowledge concerning the role of RA during salt adaptation both *in vivo* (1–16,20–22) and *in vitro* (8–14,22) few major questions still remain, one of them being the occurrence and significance of RA in light (RAL), as originally put forward by Peschek's group (9). The aim of this paper is to develop a new method to check respiratory electron transport in dark and in light in whole cells of *Synechocystis* PCC 6803, by potentiometric monitoring of the oxidation of artificial (2,6 dichlorophenolindophenol $DCPIP_{red}$) and physiological (cyt. c_{red}) electron donors, based on the usefulness of electrochemical method (mediated amperometry) to monitor in intact cyanobacteria RA in dark (1,20,23–25) and photosynthesis (23–25).

The Phototrophic Prokaryotes, edited by Peschek *et al.*
Kluwer Academic / Plenum Publishers, New York, 1999.

2. MATERIALS AND METHODS

2.1. Cyanobacterial Strain and Growth

Synechocystis PCC 6803 was grown in BG_{11}+10 mM HEPES-NaOH buffer (pH = 8.0) at 30°C and 4000 lux with or without 0.5 M NaCl as previously shown (1). For determination only exponentially growing cells were used ($OD_{730\ nm}$ = 0.18–0.40).

2.2. Spheroplast Preparation

Intact spheroplasts were prepared by lysozyme digestion (26). After incubation for 3h at 37°C, spheroplasts were observed by an optical microscope and their concentrated supernatant checked spectrophotometrically for phycocyanin and ferredoxin (26).

2.3. Oxygen Measurements

Oxygen consumption or evolution (under saturated light) were determined as previously described (1) and expressed in nmol O_2 $(\mu gChla)^{-1}$ hr^{-1}. In some oxymetric experiments, artificial electron donors DCPIP (10^{-4} M) (Merck) and N,N,N´,N´-tetramethyl-p-phenylendiamine (TMPD; Sigma) (10^{-4} M) were reduced with excess ascorbate (7×10^{-3} M) (Sigma) so as to maintain the dye in the reduced form during the experiment. Respiratory activity in light was assayed in the presence of DCMU (50×10^{-6} M) (21) without the use of potassium cyanide (KCN). Hydroxylamine (HA) 10 mM was used as artificial electron donor for Photosystem II (PSII) to allow the occurrence of photosynethetic electron transport (PET) without oxygen evolution (27).

2.4. Electrochemical Measurements

Potentiometric measurements were done using platinum and saturated calomel as working and reference electrodes, respectively connected to a digital voltmeter (Radelkis Type OP-208). The electrode was calibrated with saturated quinhidrone solution (potential = 286 mV at pH 6.5, 25°C) (28).

The culture, either intact cells or spheroplasts, was incubated in light or in dark for 10 minutes before the start of potentiometric measurements by adding electron donors (DCPIP 30 μM or cyt. *c* 10 μM, final concentration) and ascorbate. The measurements were done for 15–20 minutes in light and in dark, respectively, so as by plotting the potential versus time to obtain a sigmoid curve. The potential changes occurring during biological oxidation of chemically reduced DCPIP or cyt. *c* were converted to electron donors concentrations following the Nerst equation (29). Enzymatic activity was calculated from the linear part of the curve and expressed in nmol electron donor x $(\mu g\ Chla)^{-1}$ hr^{-1}.

All the experiments were carried out at 30°C and either saturating light intensity or in dark. When cytochrome *c* (Merck) was used as physiological electron donor to cytochrome *c* oxydase, spheroplasts were used instead of intact cells.

In order to compare the amount of electrons used to reduce molecular oxygen (4 electrons/ molecule) and to oxydise DCPIP (2 electrons/molecule) or cyt. *c* (1 electron/molecule) we multiply by 4 the intensity of respiration (expressed as nmol O_2 $(\mu g\ Chla)^{-1}$ hr^{-1}) and by 2 and 1 the intensity of biological oxydation of chemically reduced electron donors (expressed as nmol electron donor $\times$ $(\mu g\ Chla)^{-1}$ hr^{-1}) for DCPIP and cyt. *c* respectively, to obtain nF $\times$ $(\mu g\ Chla)^{-1}$ hr^{-1} (1 Faraday= 6.02×10^{23} electrons).

All the results are obtained from at least five independent experiments and the standard deviation ranges within 10% of the corresponding mean.

3. RESULTS AND DISCUSSION

3.1. Respiratory Activity in Light (DCMU and Hydroxylamine)

In Table 1 the results concerning oxygen consumption in dark and in light in cells treated with DCMU are presented. RAL is higher in salt stressed cells (SSC) than in control.

However, in the presence of DCMU, PQ pool is no longer reduced by PET, a situation that affects respiratory activity (30–32). Thus, it is expected that the measured oxygen consumption in light would be different than the real RAL under physiological conditions (both photosystems in function) (3).

In order to overcome this problem, we measured RA in the presence of HA (10mM), HA being at this concentration an artificial electron donor for PSII (27). As one can see (Table 2), RA is higher in SSC than in control, but the results are difficult to discuss because HA (at 1 mM or 10 mM) is an inhibitor of RA in dark (Table 1), and probably also in light. Thus we expect that the values obtained in the presence of 10 mM HA are lower than the real RAL. This interpretation is sustained by comparing RA value determined in the presence of DCMU and HA (1 mM and 10 mM).

HA at a concentration of 1 mM is an inhibitor of PSII as well as DCMU; however at this concentration HA inhibits also RA in dark in SSC but not in control. Increased sensitivity to 1mM HA could be determined by alternative (soluble) electron carriers active in electron transport during salt stress, as is the case with other inhibitors (8,17).

At 10 mM HA inhibits RAL in control and furthermore in SSC; thus, it is expected that RAL will be also inhibited. By comparing these results (Table 2) with these concerning RAL determined in the presence of DCMU (Table 1) the inhibitory effect of HA (10 mM) on RA either in dark or in light is clearly shown.

This is why inspite of allowing PET to occur, HA (10 mM) cannot be used to measure respiratory activity in light.

From these experiments we concluded that for measuring RAL in intact cells with an oxygen electrode it is necessary to use DCMU in connection with an artificial electron donor for PSII, at a concentration that does not inhibit RA.

Table 1. Respiratory activity in dark (RAD) and in light (RAL) in intact cells of *Synechocystis* PCC 6803

	RAD		RAL	
Growth conditions	A	B	A	B
BG_{11}	78.0	312.0	78.0	312.0
BG_{11} + NaCl	134.0	536.0	134.0	536.0

A = nmol O_2 (µg chl a)$^{-1}$ hr^{-1}; B = nF (µg chl a)$^{-1}$ hr^{-1}.

Table 2. The effect of hydroxylamine (HA) on oxygen consumption nmol O_2 (µg chl a)$^{-1}$ hr^{-1} (–) or production (+) in dark and in light in intact cells of *Synechocystis* PCC 6803

	BG_{11}		BG_{11} + NaCl	
Inhibitor HA	Dark	Light	Dark	Light
Control	−78.0	+256.0	−134.0	+75.0
1 mM	−72.0	−53.0	−87.0	−62.0
10 mM	−31.0	−28.0	−25.0	−31.0

3.2. Chemically Supplemented Respiration

In order to measure the maximal value of oxygen consumption we used artificial electron donors for respiratory electron transport (RET), DCPIP and TMPD (33). In Table 3 the values of RA in dark in SSC and control are presented. Under these conditions, namely chemically supplemented respiration (CSR), because extra electrons are supplied to respiratory chain from chemically reduced (artificial) electron donors.

Up to our best knowledge, this is the first report on CSR in intact whole cells grown under salt stress. CSR is higher than the physiological respiration, result which is sustained in the literature, where higher cytochrome *c* oxidase activity in cell free preparation than in whole cells were shown (6,26,34).

Moreover, the values of CSR are only slightly higher in SSC than in control. These results are sustained by those concerning cyt *c* oxidase activity *in vitro* obtained from salt- and normal-grown cells (6,34).

These results suggest that in intact cells under these conditions, the limiting factor of RA is the amount of electrons available to pass through the respiratory chain.

The experiments concerning CSR enable us to measure the ability of intact cells to take up electrons from chemically reduced electron donors. We further introduce a potentiometric method allowing us to monitor the changes in redox potential during the oxydation of these electron donors by cyanobacteria. Thus, we measured the ability of intact cells to take up electrons from chemically reduced electron donors, both in dark and in light.

3.3. Biological Oxidation of Chemically Reduced-Artificial (DCPIP) and Physiological (Cyt. *c*) Electron Donors

In Table 4 one can see that biological oxidation of chemically reduced electron donors (BOCRED) in dark is higher in control than in SSC both for DCPIP and for cyt. *c*.

Taking into account these results and those concerning oxygen consumption (Tables 1 and 3), we proposed a tentative picture (Fig. 1) to illustrate RAL in intact cells under the op-

Table 3. Chemically supplemented respiration in *Synechocystis* PCC 6803 either by DCPIP (10^{-4} M) or TMPD (10^{-4} M), in the presence of excess ascorbate (70×10^{-3} M)

	BG11		BG11 + NaCl	
Electron donor	A	B	A	B
DCPIP	203.0	812.0	225.0	900.0
TMPD	156.0	624.0	169.0	676.0

A = nmol O_2 (μg chl *a*)$^{-1}$ hr^{-1}; B = nF (μg chl *a*)$^{-1}$ hr^{-1}.

Table 4. Biological oxydation of chemically reduced electron donors by limited amount of ascorbate

	BG11				BG11 + NaCl			
	A		B		A		B	
Electron donor	Dark	Light	Dark	Light	Dark	Light	Dark	Light
DCPIP	215.0	377.0	430.0	754.0	133.0	244.0	266.0	488.0
cyt. *c*	53.0	64.0	53.0	64.0	21.0	46.0	21.0	46.0

A = nmol electron donor (μg chl *a*)$^{-1}$ hr^{-1}; B = nF (μg chl *a*)$^{-1}$ hr^{-1}.

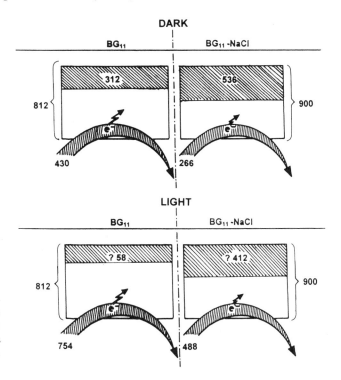

Figure 1. Picture concerning respiratory electron transport in whole cells of *Synechocystis* PCC 6803 both in dark and in light. The values are expressed as nF (μg chl a)$^{-1}$ hr^{-1}. For more explanation see text.

eration of both photosystems. Oxygen consumption values and DCPIP oxydation values are uniformed with respect to the number of electrons involved in these reactions, for a convenient and easier comparison of the results obtained within different experiments.

In dark, BG$_{11}$-grown cells transport during physiological respiration, a quantity of 312 nF (μg chl a)$^{-1}$ (= 78 nmol O$_2$ (μg chl a)$^{-1}$ hr^{-1} while their maximum ability to transport electrons within the respiratory chain is 812 nF (μg chl a)$^{-1}$ hr^{-1} (= 203 nmol O$_2$ (μg chl a)$^{-1}$ hr^{-1}). At the same time BOCRED measured in dark corresponds to 430 nF (μg chl a)$^{-1}$ hr^{-1} (= 2 15 mmol DCPIP (μg chl a)$^{-1}$ hr^{-1}).

In salt stressed cells, the corresponding values are 536 nF (μg chl a)$^{-1}$ hr^{-1}, 900 nF (μg chl a)$^{-1}$ hr^{-1} and 266 nF (μg chl a)$^{-1}$ hr^{-1}, respectively.

These results suggest that BOCRED measures the ability of the respiratory chain to take up electrons. BOCRED is limited by the difference between the CSR and the physiological respiration, namely the resting respiratory activity.

Indeed, in control, this difference measured oxymetrically (812 − 312 = 500 nF (μg chl a)$^{-1}$ hr^{-1}) fits well with the BOCRED determined potentiometrically (430 nF (μg chl a)$^{-1}$ hr^{-1}); almost the same situation is in SSC (compare 900 − 538 = 362 nF (μg chl a)$^{-1}$ hr^{-1} with 266 nF (μg chl a)$^{-1}$ hr^{-1}).

In our opinion, the higher BOCRED in control is due to the fact that in the dark the difference between CSR and physiological respiration is higher than in SSC. Thus, it seems appropriate to claim that BOCRED is a measure of the resting respiratory activity.

Following the above reasoning and assuming that the maximal electron transport capability of the respiratory chain is the same in dark and in light, we can put forward that RAL is less than 58 nF (μg chl a)$^{-1}$ hr^{-1} (812 − 754) and 412 nF (μg chl a)$^{-1}$ hr^{1} (900 − 487) in control and SSC, respectively (Fig. 1).

These values corresponds to the putative values of 14.5 nmol O_2 (μg chl a)$^{-1}$ hr^{-1} and 103 nmol O_2 (μg chl a)$^{-1}$ hr^{-1} in control and SSC, respectively.

Whether or not these are the true values of RAL occurring in intact cells having both photosystems active remains to be further checked.

The potentiometric method introduced here enables us to measure respiratory electron transport in light as well as in dark. Further improvements, including the use of an amperometric device (1,3,20,35) together with "classical methods" for this field could enable the scientists to deeper understand RAL in intact cyanobacteria.

The following directions seem important to better understand the mechanism(s) and the biological significance of the respiratory activity occurring both in light and in dark, as well as the interplay between photosynthesis and respiration in intact cells:

1. The development of electrochemical techniques (both potentiometric and amperometric) to *in vivo* and *in vitro* monitoring the reduction/oxidation of different permeant and impermeant electron acceptors/donors.
2. The use of transgenic cyanobacteria (mainly) with respect to photosynthetic and/or respiratory electron carriers in connection with (specific) inhibitor(s).
3. The growth of (transgenic) cyanobacteria using different carbon sources, light/dark regimes and stresses.

ACKNOWLEDGMENTS

Thanks are due to Professor Günter A. Peschek (Chairman of the Organizing Committee of IXth ISPP) for financial support to I.I.A. for attending the meeting and to Professor J. Schwarz (Hamburg) for helping us in overcoming some technical problems.

REFERENCES

1. Ardelean, I., Ristoiu, V., Flonta, M.L., Zarnea, G. (1995) In Photosynthesis: From Light to Biosphere (Mathis P. ed) vol. IV, 525–528, Kluwer Academic Publishers, The Netherlands
2. Ardelean I (1997) Future prospects In Cyanobacterial nitrogen metabolism and environmental biotechnology . (ed. A.K.Rai) Narosa Publishing House, New Dehli, India pp. 249–272
3. I.Ardelean G. Zarnea (1997). *In* Cyanobacterial Biotechnology (Eds.G. Subramanian. D. Kaushik, G.S. Venkataraman) Publishers M/S Oxford IBH Publishing House, New Dehli, in press
4. Gabbay-Azaria, R., Schoenfeld, M., Tel-Or, S., Messinger, R. and Tel-Or, E. (1992) Arch. Microbiol. 157, 183–190
5. Hagemann M, Erdman N, Wittenburg U (1989) Arch. Hydrobiol. Suppl. 82:425–435
6. Jeanjean R, Matthijs H.C.P.,Onana B., Havaux M., Joset F (1993) Plant Cell Physiol. 34: 1073–1079
7. Jeanjean R,.,Onana B.,Peschek G.A., Joset F(1990) FEMS Microbiol. Lett. 68:125–130
8. Joset F, Jeanjean R, Hagemann M (1996) Physiologia Plantarum 96:738–744
9. Nicholls, P., Obinger, C., Niderhausser, H. and Peschek, G.A. (1992) Biochem. Biophys. Acta 1098, 184–190
10. Nitschmann, W.H.and Peschek, G.A. (1985) Arch. Microbiol. 141, 330–336
11. Peschek G.A. (1996) Biochem. Biophys. Acta 1275:27–32
12. Peschek, G.A., Czerny, T., Schmetterer, G. and Nitschmann, W.H. (1985) Plant Physiol. 79, 278–284
13. Peschek, G.A., Obinger, C., Fromwald, S. and Bergman, B. (1994) FEMS Microbiol. Lett. 124, 431–438
14. Peschek, G.A., Wastyn, M., Trnka, M., Molitor, V., Fry, I.V. and Packer, L.(1989) Biochemistry 28, 3057–3063
15. Rai AK, Abraham G (1993) Bull. Environ. Contam. Toxicol. 51:724–731
16. Scherer S, Ernst , A. Chen T-W, Boger P (1984) Oecologia 62:418–423

17. Jeanjean R, Haveaux M, Ogava T, Joset F, Matthjis HCP (1995) *In* Photosynthesis: From Light to Biosphere (Mathis P. ed) vol. II, pp. 907–910, Kluwer Academic Publishers, The Netherlands
18. Nomura M, Ishitani M, Takabe T, Rai AK, Takabe T (1995) Plant. physiol. 107:703–708
19. Hibino T, Lee BH, Rai AK, Ishikawa H, Kojima H, Tawada M, Shimoyama H, Takabe T (1996) Austral J Plant Physiol 23:321–330)
20. Ardelean, I., Canja, D. and Flonta M.L. (1992) in Research in Photosynthesis (Murata N. ed) pp. 623–626, Kluwer Academic Publishers, The Netherlands
21. Valiente, E.F., Nieva, M., Avendano, M.C. and Maeso, E.S. (1992) Plant Cell Physiol. 33, 307–313
22. Piels D, Gregor W, Schmetterer G. (1997) FEMS- Microbiol. Lett 152:83–88
23. Rawson DM. Wilmer AJ, Turner APF (1987) Toxicity Assessments 2: 325–340.
24. Rawson DM (1988) Int Ind Biotechnol 2: 18–23
25. Rawson DM (1993) *In* Handbook of ecotoxicology (Ed. P Callow) Blackwell Scientific Publications, vol.I, pp. 428–423
26. Peschek G.A, Schmetterer G, Wagesreiter H (1982) Arch. Microbiol. 133:222–224
27. Izawa I (1977) *In* Photosynthesis I (eds. A.Trebst, M. Avron) Springer Verlag, Berlin New York, pp. 266–279
28. Johnson G.N.,Rutherford A.W., Krieger A 91995) Biochem. Biphys. Acta 1229:202–207
29. Jacob H.E (1970) *In* Methods in Microbiology (eds. J.R Norris, D.W. Ribbons) Academic Press London new York, Vol 2, pp.91–125
30. Hirano M, Satoh K, Katoh S (1980) Photosynth. Res. 1: 149–162
31. Mullineaux C.W, Allen J.F (1986) FEBS Lett. 205:155–160
32. Aoki M, Katoh S (1983) Plant&Cell Physiol. 24(B):1379–1386
33. Kimelberg. H.K., Nicholls P (1963) Arch. Biochem. Biophys. 133:327–335
34. G.A. Peschek (1983) J. Bacteriol. 153:539–542.
35. Coleman J.O.D., Hill H.A.O., Walton N.J., Whatley F.R (1983) FEBS Lett. 154 :319–322

48

FROM PLASTID TO CYANOBACTERIAL GENOMES

Masahiro Sugiura

Center for Gene Research
Nagoya University
Nagoya 464-8602, Japan

1. INTRODUCTION

Non-Mendelian inheritance was found based on studies of variegation in higher plants early in this century. Further analysis of variegation revealed that the genetic determinants for these characters were associated with plastids. The demonstration of unique DNA molecules in chloroplasts, half a century later, has led to intensive studies of both the structure and expression of plastid genomes. These studies were accelerated by the development of techniques, like gene cloning and DNA sequencing, in the mid-1970s.

The endosymbiotic theory has proposed that plastids were derived from an ancestral photosynthetic prokaryote related to cyanobacteria [1]. This idea was supported at the molecular level by the observations that nucleotide sequences of plastid 5S rRNAs and 16S rRNA genes are highly similar to those from *Anacystis nidulans* [2,3]. Based on the gene (*rbcL*) sequence encoding the large subunit of Rubisco from higher plant plastids and *A. nidulans*, the divergence time of higher plants and *A. nidulans* was estimated as 2.3–4.6 × 10^8 years [4]. Further analysis of plastid genomes from a wide range of plant species and cyanobacterial genomes provides the fundamental data needed to estimate the origin and evolution of plastids as well as the phylogenetic relationships among photosynthetic organisms.

2. THE PLASTID GENOME

2.1. Overall Structure and Gene Content

As plastid DNA molecules are relatively small and simple, they were selected as one of the first targets of the "genome projects." The entire nucleotide sequences of tobacco and 14 other plastid genomes have been determined to date, disclosing an enormous amount of structural and evolutionary information. Initial efforts on analyzing entire plastid DNA

The Phototrophic Prokaryotes, edited by Peschek *et al.*
Kluwer Academic / Plenum Publishers, New York, 1999.

structures were made for land plants while recent analysis of plastid DNAs from algae, especially red and brown algae, has revealed the presence of many novel genes not found in land plant plastids [5,6, for review].

Figure 1 shows a gene map of the tobacco plastid genome as an example. The genome contains over 100 genes, which can be classified into several groups. The number of genes in the tobacco genome found in 1986 and 1998 is listed in Table 1. We reported 82 different genes and many open reading frames (ORFs) when the entire sequence was published in 1986 [7], and we now (1998) know 105 different genes. The difference is 23. It took over 10 years and more than10 research groups to identify the additional 23 genes: one RNA gene and 22 protein genes. There are still at least 21 ORFs to be identified. This indicates that identification of new genes is an extremely difficult task ever when an entire genome sequence has been determined, and creative experimental approaches are necessary.

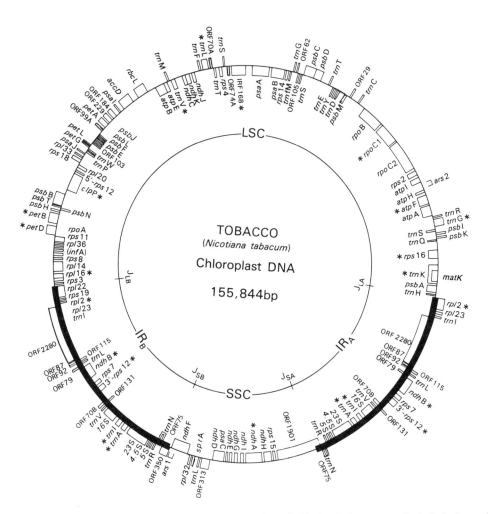

Figure 1. Gene map of the tobacco plastid genome. Genes shown inside the circle are transcribed clockwise, and genes on the outside are transcribed counter-clockwise. Asterisks denote split genes. Major ORFs are included. IRF, intron-containing reading frame; IR, inverted repeat; LSC, large single-copy region; SSC, small single-copy region; J, junctions between IR and LSC/SSC. From the Research Grant Progress Report (1989) with revisions.

Table 1. Numbers of tobacco plastid genes

Components	Genes	
	1986	1998
The photosynthetic apparatus (rubisco, thylakoid membrane)	19	31
The chloroplast genetic system (ribosomal proteins, RNA polymerase)	23	26
NADH dehydrogenase (*ndh*)	6	11
Others (*accD*, *clpP*)	0	2
Ribosomal RNAs	4	4
Transfer RNAs	30	30
Small plastid RNA (*sprA*)	0	1
Number of different genes	82	105
Total number of genes[a]	98	121
ycfs and ORFs (≥ 70 codons)[b]	21	

[a]Sixteen genes are located in the long inverted repeat and hence duplicated.
[b]Nine ORFs are conserved among higher plants and named *ycfs* and the remaining 12 ORFs (≥ 70 codons) are unique to tobacco.

2.2. Modification of Genome Information by RNA Editing

A complete coding information was initially thought to be written in a DNA sequence. However this is not always the case, at least in some plastids, and DNA information is often modified at the level of RNA, by RNA editing as well as RNA splicing. RNA editing is defined as the post-transcriptional modification of pre-RNA to alter its nucleotide sequence through the insertion and deletion of nucleotides, or specific nucleotide substitution, so as to yield functional RNA species. RNA editing in plastids was for the first time reported by Kössel [8]. The genes for ribosomal protein CL2 (*rpl2*) from maize and rice are known to have an ACG codon at the position corresponding to the ATG initiation codon in other plant *rpl2*s analyzed. Analysis of the maize *rpl2* mRNA revealed that the ACG codon is converted to a canonical AUG initiation codon by a C to U base change. RNA editing is not limited to initiation codons but has been observed at internal codons. Most edited codons restore amino acids that are conserved in the corresponding proteins of other plants. Therefore, RNA editing is believed to be functionally significant.

Initially, RNA editing in plastids was reported only in transcripts from angiosperms. Therefore, we examined whether RNA editing occurs in transcripts from a gymnosperm, black pine. We have found 26 C-to-U editing sites in the transcripts from 12 black pine plastid genes. However, a big difference from angiosperms is that two editing sites create a stop codon by CAA to UAA transition [9]. The most striking is the black pine ORF62b (Fig. 2). In the genome, this ORF starts with ATG and ends in TAA, comprising 62 codons. We found RNA editing in the transcript from this region, one editing creates an initiation codon AUG from ACG and the other editing produces a stop codon UAA from CAA, which produce a new reading frame (RF) of 33 codons at the mRNA level. This RF was later found to be a black pine homologue of *petL*. The presence of ATG and in-frame TAA is just by chance, because these triplets are not conserved in other plastid genomes. Therefore, plastid protein structures or protein-coding regions cannot always be predicted from their genome sequences.

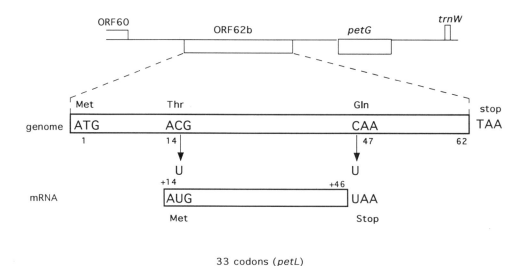

33 codons (*petL*)

Figure 2. Creation of a novel protein-coding region at the RNA level in black pine plastids. The top line represents a portion of the plastid genome. ORF62b is expanded below. Numerals indicate triplet positions starting from ATG. Arrows denote editing sites. The bottom line shows a new reading frame produced by RNA editing [9].

RFs for polypeptides derived from DNA sequences sometimes contain multiple possible initiation codons and it is difficult to assign which is a real initiation codon. We have developed an *in vitro* translation system from tobacco chloroplasts [10]. This system allows us to determine genuine start sites of translation. A notable example is that the initiation codon of tobacco *ndhD* mRNA is the edited AUG from ACG but not the AUG encoded in the genome and located 25 nt upstream from the edited AUG [11].

3. THE GENOME OF CYANOBACTERIA

3.1. Physical Mapping of the Genome

To understand the genome structure, physical maps of the genomes from several representative cyanobacteria have been constructed. The genome of *Anabaena* PCC7120 was cleaved with *Avr*II, *Sal*I and *Pst*I, and the resultant fragments were resolved and ordered by pulsed field gel electrophoresis. The genome was found to be 6.4 Mb in size and circular [12]. Physical maps were then constructed for the genome of *Synechococcus* PCC7002 (2.7 Mb [13]), *Synechocystis* PCC6803 (3.6 Mb [14,15]), and *Anacystis nidulans* 6301 (*Synechococcus* PCC6301, 2.7 Mb [16]). Figure 3 shows physical and gene maps of *Synechococcus* 6301 and *Synechocystis* 6803.

3.2. The Entire Sequence of *Synechocystis* PCC6803 Genome

The complete nucleotide sequence of the genome of *Synechocystis* 6803 was assembled from sequences of partially overlapped cosmid and λ clones and filled gaps by long PCR products [17]. The total length of the genome is 3,573,470 bp. The numbers of structural RNA-coding genes and predicted protein-coding regions (ORFs) are listed in Table 2.

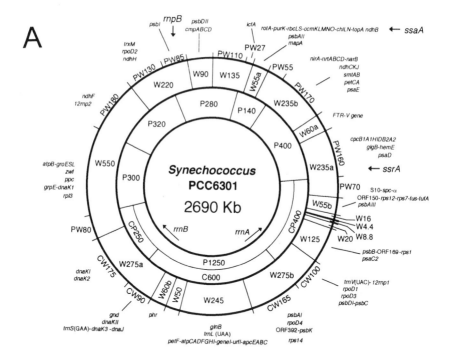

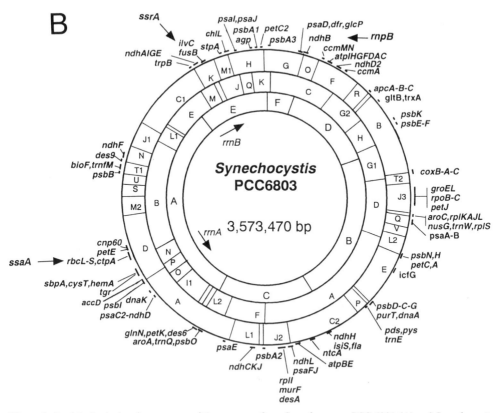

Figure 3. Partial physical and gene maps of the genomes from *Synechococcus* PCC 6301 (A) and *Synechocystis* PCC6803 (B). (A) *Pme*I (P), *Swa*I (W) and I-*Ceu*I (C) fragments. (B) *Asc*I (outer), *Mlu*I (middle) and *Spl*I (inner) fragments. Small stable RNA genes [*rnpB* (M1 RNA), *ssaA* (6Sa RNA) and *ssrA* (10Sa RNA)] are highlighted. Redrawn from Kaneko *et al.* (1996) and Kotani *et al.* (1995).

Table 2. The gene content of Synechocystis sp. PCC 6803

2	rRNA gene clusters (*rrn16–trnI–rrn23–rrn5*)
42	tRNA genes encoding 41 tRNA species
3	small stable RNA genes[a] (M1 RNA, 10Sa RNA, 6Sa RNA)
3,168	predicted protein-coding regions (ORFs)
145 (5%)	reported genes in this species
1259 (40%)	similar to known genes in other organisms
342 (10%)	similar to hypothetical genes
1422 (45%)	unknown

From Kaneko et al. (1996), [a]From Vioque (1992), Watanabe et al. (1997, 1998).

Interestingly, several of the plastid genes are not found in the determined 3.6 Mb sequence. For example, tobacco plastid genes not found in the cyanobaterial sequence are *petL*, *matK*, *ycf1* (ORF 1901) and *ycf2* (ORF 2280) (see Fig. 1). Furthermore, all the plastid rRNA gene clusters so far reported contain a tRNA[Ala] gene in the spacer between *rrn16* and *rrn23*, but this is not the case in this strain. However, *Synechococcus* 6301 has plastid-like *rrn* cluster (*rrn16–trnI–trnA–rrn23–rrn5*).

3.3. Genes Encoding Small Stable RNAs

A number of small stable RNA species other than rRNAs and tRNAs have been found in *E. coli* and several other prokaryotes. These include M1 RNA, 4.5S RNA, 6S RNA, 10Sa RNA, spot 42 RNA, *micF* RNA and *oxyS* RNA. M1 RNA is a part of RNase P which catalyzes the 5′ processing of pre-tRNAs, and 4.5S RNA is a component of the signal recognition particle which is involved in translocation of secretory proteins. 10Sa RNA or tmRNA (*t*RNA-like and *m*RNA-like properties) is involved in *trans*-translation. In plastids, M1-like RNA genes were reported in the cyanelle and a red alga [18], *tscA* RNA in *Chlamydomonas* [19] and *sprA* RNA in tobacco [20].

Computer search for potential small stable (structural or non-coding) RNA genes has not been so successful as in the case of potential protein-coding regions, because of the lack of apparent punctuations such as translation initiation and termination codons to confine DNA regions. Only one stable RNA, M1 RNA, was reported in cyanobacteria [21,22]. Therefore, we have carried out direct isolation and characterization of stable RNAs from *Synechococcus* 6301 because of easier preparation of intact nucleic acids than with *Synechocystis* 6803 and others [23,24]. The total RNA preparation was applied on sucrose gradient centrifugation, and RNA fraction sedimenting between 4S to 16S was labeled at the 5′ or 3′ end and separated by long polyacrylamide gel electrophoreses. Major bands were excised and their partial RNA sequences were determined. We found that a band corresponds to the *E. coli* 10Sa RNA, and that another band shows no similarity to any known RNA sequences and is likely a novel stable RNA species (named as 6Sa RNA based on its size).

To further characterize these two RNA species, we isolated and sequenced the corresponding genes from a genomic library of the 6301 strain. The coding regions were determined by direct RNA sequencing and RNase protection assay. The 6301 10Sa RNA is 394 nt long and shows 55% similarity to that of *E. coli*. A secondary structure of the 6301 10Sa RNA was constructed based on the model proposed for *E. coli* (Fig. 4). It shows a partial tRNA-like structure at the 5′and 3′ ends and a putative tag sequence starting from GCG

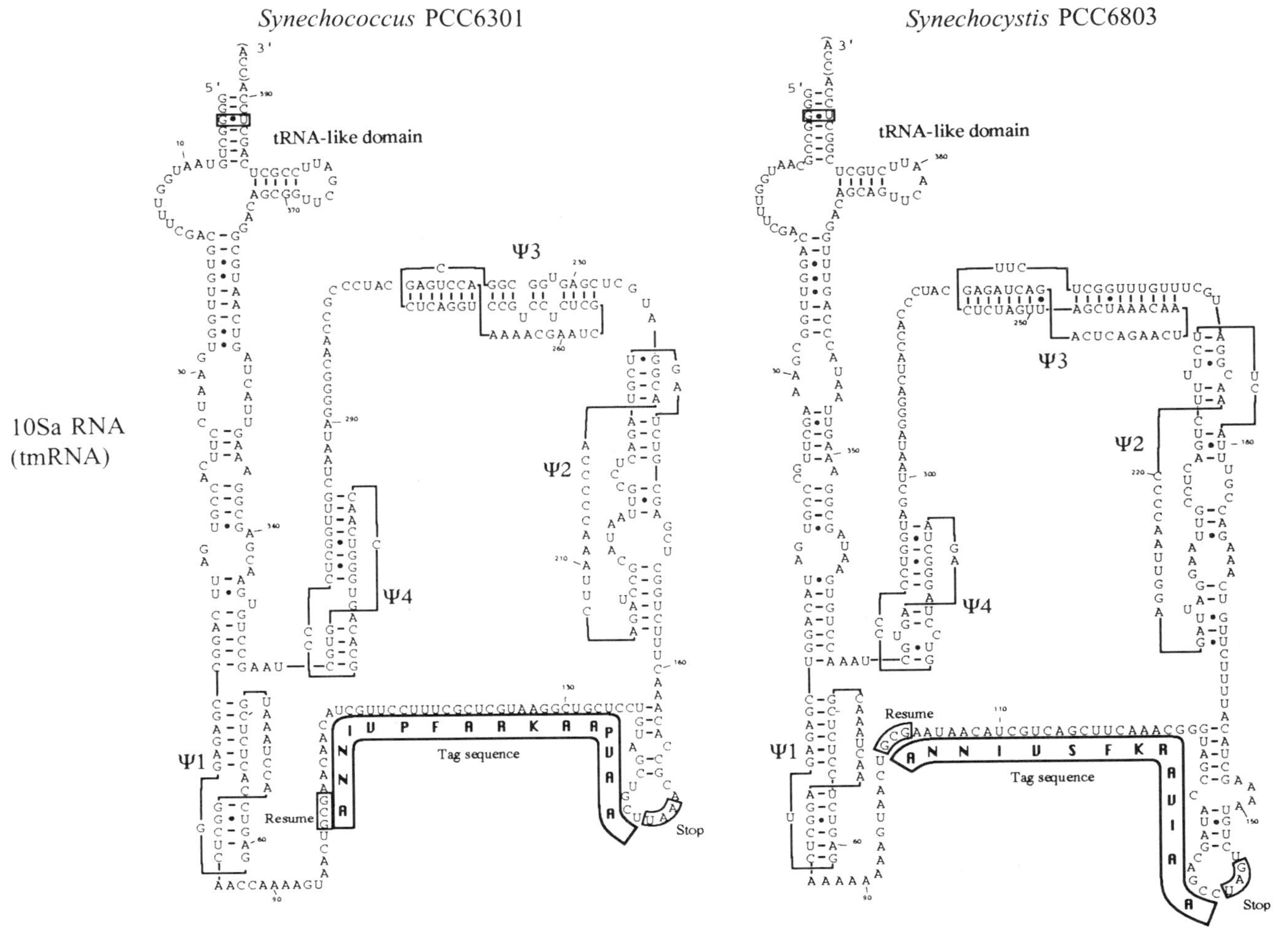

Figure 4. Predicted secondary structures of 10Sa and 6Sa RNAs from *Synechococcus* PCC 6301 and their homologues from *Synechocystis* PCC6803. Redrawn from Watanabe *et al.* (1997, 1998)

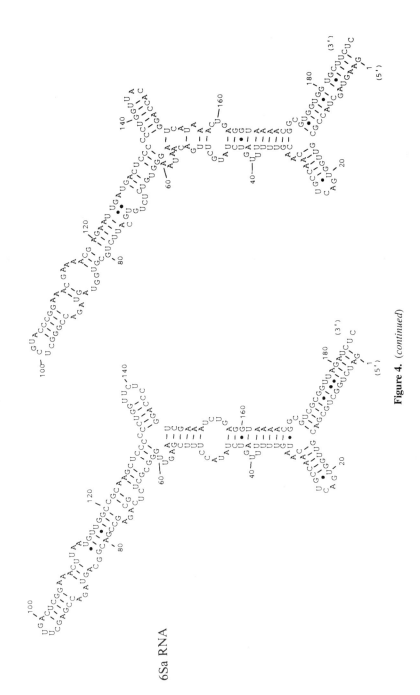

6Sa RNA

Figure 4. (*continued*)

(resume codon) to UAA (stop codon) which is known to be involved in *trans*-translation to rescue stalled ribosomes.

6Sa RNA is 185 nt long and its gene was named *ssaA* (*six S stable RNA A*). Computer-aided prediction of its secondary structure supports its stability (Fig. 4).

Based on the *Synechococcus* sequences, we searched their homologues in the *Synechocystis* 6803 sequence and found the corresponding sequences (Fig. 4). The sequence identity is about 70% among these RNAs.

Recently the number of small stable RNAs or non-mRNA species is extremely increasing in eukaryotic cells, for example, over 100 snoRNAs responsible for rRNA modification have been reported. Unlike *E. coli* or other heterotrophic eubacteria, several eukaryotic type RNA-binding proteins have already been reported in cyanobacteria [25]. We should pay more attention to small stable RNAs and their function in plastids and cyanobacteria.

4. PROBLEMS

A tremendous amount of DNA sequence data is now available from plastids and cyanobacteria, due mainly to successful technological achievement. These data by themselves will not be sufficient to determine biological function but will provide an important basis for appropriate experiments. There are many uncharacterized ORFs in sequenced genomes. One should keep in mind that a portion of these ORFs are real genes encoding polypeptides, and that computers can predict but not prove protein-coding genes. The mere identification of protein-coding regions is not always sufficient. DNA information is sometimes modified at the level of RNA and translation, or even post-translation. Recently novel RNA species which are directly involved in cell function have been reported one after another. Genes encoding such RNAs cannot be predicted from DNA sequences. Experimental approaches are essential to find novel RNA genes.

ACKNOWLEDGMENT

I thank Dr. Mamoru Sugita for critical reading of this article.

REFERENCES

1. Margulis, L.: Origin of Eukaryotic Cells. Yale University Press, New Haven (1970)
2. Dyer, T.A., Bowman C.M.: Nucleotide sequences of chloroplast 5S ribosomal ribonucleic acid in flowering plants. Biochem. J., 183, 595–604 (1979)
3. Tomioka, N. and M. Sugiura: The complete nucleotide sequence of a 16S ribosomal RNA gene from a blue-green alga, *Anacystis nidulans*. Mol. Gen. Genet., 191, 46–50 (1983)
4. Shinozaki, K., C. Yamada, N. Takahata and M. Sugiura: Molecular cloning and sequence analysis of the cyanobacterial gene for the large subunit of ribulose-1, 5-bisphosphate carboxylase/oxygenase. Proc. Natl. Acad. Sci. USA, 80, 4050–4054 (1983)
5. Sugiura, M.: The chloroplast genome. In "10 years in Plant Molecular Biology" (R. Schilperoort and L. Dure, eds), Plant Mol. Biol., 19, 149–168 (1992)
6. Sugiura, M.: The chloroplast genome. In "Essays in Biochemistry" (D. K. Apps and K. F. Tipton, eds), Portland Press, London, pp. 49–57 (1995)
7. Shinozaki, K., M. Ohme, M. Tanaka, T. Wakasugi, N. Hayashida, T. Matsubayashi, N. Zaita, J. Chunwongse, J. Obokata, K. Yamaguchi-Shinozaki, C. Ohto, K. Torazawa, B. Y. Meng, M. Sugita, H. Deno, T.

Kamogashira, K. Yamada, J. Kusuda, F. Takaiwa, A. Kato, N. Tohdoh, H. Shimada and M. Sugiura: The complete nucleotide sequence of the tobacco chloroplast genome: its gene organization and expression. EMBO J., 5, 2043–2049 (1986)

8. Hoch, B., R.M. Maier, K. Appel, G.L. Igloi and H. Kössel: Editing of a chloroplast mRNA by creation of an initiation codon. Nature, 353, 178–180 (1991)

9. Wakasugi, T., T. Hirose, M. Horihata, T. Tsudzuki, H. Kössel and M. Sugiura: Creation of a novel protein-coding region at the RNA level in black pine chloroplasts: The pattern of RNA editing in the gymnosperm chloroplast is different from that in angiosperms. Proc. Natl. Acad. Sci. USA, 93, 8766–8770 (1996)

10. Hirose, T., and M. Sugiura: Cis-acting elements and trans-acting factors for accurate translation of chloroplast psbA mRNAs: development of an in vitro translation system from tobacco chloroplasts. EMBO J., 15, 1687–1695 (1996)

11. Hirose, T., and M. Sugiura: Both RNA editing and RNA cleavage are required for translation of tobacco chloroplast ndhD mRNA: a possible regulatory mechanism for expression of a chloroplast operon consisting of functionally unrelated genes. EMBO J., 6804–6811 (1997)

12. Bancroft, I., C.P. Wolk, and E.V. Oren: Physical and genetic maps of the genome of the heterocyst-forming cyanobacterium Anabaena sp. strain PCC 7120. J. Bacteriol., 171, 5940–5948 (1989)

13. Chen, X., and W.R. Widger: Physical genome map of the unicellular cyanobacterium Synechococcus sp. strain PCC 7002. J. Bacteriol., 175, 5106–5116 (1993)

14. Kotani, H., A. Tanaka, T. Kaneko, S. Sato, M. Sugiura and S. Tabata: Assignment of 82 known genes and gene clusters on the genome of the unicellular cyanobacterium Synechocystis sp. strain PCC6803. DNA Res., 2, 133–142 (1995)

15. Churin, Y.N., I.N. Shalak, T. Börner, and S. V. Shestakov: Physical and genetic map of the chromosome of the unicellular cyanobacterium Synechocystis sp. strain PCC 6803. J. Bacteriol., 177, 3337–3343 (1995)

16. Kaneko, T., T. Matsubayashi, M. Sugita and M. Sugiura: Physical and gene maps of the unicellular cyanobacterium Synechococcus sp. strain PCC6301 genome. Plant Mol. Biol., 31, 193–201 (1996)

17. Kaneko, T., S. Sato, H. Kotani, A. Tanaka, E. Asamizu, Y. Nakamura, N. Miyajima, M. Hirosawa, M. Sugiura, S. Sasamoto, T. Kimura, T. Hosouchi, A. Matsuno, A. Muraki, K. Naruo, S. Okumura, S. Shimpo, C. Takeuchi, T. Wada, A. Watanabe, M. Yamada, M. Yasuda and S. Tabata: Sequence analysis of the genome of the unicellular cyanobacterium synechocystis sp. strain PCC6803. II. sequence determination of the entire genome and assignment of potential protein-coding regions. DNA Res., 3, 109–136 (1996)

18. Löffelhardt, W., H.J. Bohnert, and D.A. Bryant: The cyanelles of Cyanophora paradoxa. Critical Rev. Plant Sci., 16, 393–413 (1997)

19. Goldschmidt-Clermont, M., Y. Choquet, J. Girard-Bascou, F. Michel, M. Schirmer-Rahire, and J.-D. Rochaix: A small chloroplast RNA may be required for trans-splicing in Chlamydomonas reinhardtii. Cell, 65, 135–143 (1991)

20. Vera, A., M. Sugiura: A novel RNA gene in the tobacco plastid genome: its possible role in the maturation of 16S rRNA. EMBO J., 13, 2211–2217 (1994)

21. Vioque, A.: Analysis of the gene encoding the RNA subunit of ribonuclease P from cyanobacteria. Nucleic Acids Res., 20, 6331–6337 (1992)

22. Vioque, A.: The RNase P RNA from cyanobacteria: short tandemly repeated repetitive (STRR) sequences are present within the RNase P RNA gene in heterocyst-forming cyanobacteria. Nucleic Acids Res., 25, 3471–3477 (1997)

23. Watanabe, T., M. Sugiura and M. Sugita: A novel small stable RNA, 6Sa RNA, from the cyanobacterium Synechococcus sp. strain PCC6301. FEBS Lett., 416, 302–306 (1997).

24. Watanabe, T., M. Sugita, M. Sugiura: Identification of 10Sa RNA (tmRNA) homologues from the cyanobacterium Synechococcus sp. strain PCC6301 and related organisms. Biochim. Biophy. Acta, in press (1998)

25. Sugita, M. and M. Sugiura: The existence of eukaryotic ribonucleoprotein consensus sequence-type RNA-binding proteins in a prokaryote, Synechococcus 6301. Nucleic Acids Res., 22, 25–31 (1994)

49

THE *RHODOBACTER CAPSULATUS* GENOME PROJECT

Robert Haselkorn,[1] Michael Fonstein,[1] Yakov Kogan,[1] Vivek Kumar,[1] Natalia Maltsev,[2] A. Jules Milgram,[1] Jan Paces,[1] Vaclav Paces,[3] and Cestmir Vlcek[3]

[1]Department of Molecular Genetics and Cell Biology
University of Chicago
920 East 58 St., Chicago, Illinois 60637
[2]Mathematics and Computer Science
Argonne National Laboratory
Argonne, Illinois
[3]Academy of Sciences of the Czech Republic
Prague, Czech Republic

1. INTRODUCTION

Rhodobacter capsulatus is a widely studied non-sulfur purple bacterium, particularly in the areas of photosynthesis, nitrogen fixation, transport, and utilization of organic substrates. The attraction of *R. capsulatus* is based largely on its convenient system of genetic analysis, the Gene Transfer Agent [1]. With the GTA it is possible to construct deletions of any size with high efficiency. Therefore, it is possible to envision a complete functional analysis of the genome of this bacterium. For that analysis, we need first to determine the entire DNA sequence. With that information as a starting point, we can then prepare chips to study gene expression in the wild type under different conditions and in mutants missing individual regulatory genes. Based on the sequence, the GTA can be used to construct deletions frame by frame or cosmid by cosmid, allowing functions to be assigned to genes or groups of genes. Here, we present a progress report on each of these areas of genome analysis.

2. RESULTS AND DISCUSSION

2.1. Chromosome Architecture

There are two major schools of genome sequence analysis. One is based on a shotgun of the entire chromosome into a sequencing vector, determination of enough

The Phototrophic Prokaryotes, edited by Peschek *et al.*
Kluwer Academic / Plenum Publishers, New York, 1999.

sequences to give many-fold coverage, and then alignment of the sequences using vast and powerful computer programs to assemble contigs. Aside from the expense, this approach is very difficult for genomes larger than 2 Mb and for genomes with substantial numbers of repeated sequence elements. Moreover, since the sequence is not determined until the last fragment is placed, there can be no intermediate conclusions drawn as the sequencing proceeds. The second school, which we favor [2], first creates a library of overlapping fragments of convenient size. These can be cosmids, YACs or BACs. Our characterization of the whole genome began with a long-range restriction map of the chromosomes of several closely related strains of *R. capsulatus* [3]. The physical map for two rarely cutting enzymes was constructed with the aid of lambda and cosmid clones that pointed to junctions between large restriction fragments. Subsequently, from several thousand cosmids in the vector Lorist 6, 192 were selected by numerous hybridizations of large restriction fragments with cosmids, cosmids with cosmids, cosmids with transcripts from their ends, and cosmids with lambda phage clones. The set of 192 cosmids (the encyclopedia) contains all of the 3.7-Mb chromosome and the 134-kb plasmid of *R. capsulatus* strain SB1003. High resolution maps of the entire cosmid set were obtained by *cos*-site labeling and partial restriction, yielding more than 3000 sites for four six-base cutters (Figure 1) [4]. These maps have been used as references for the subsequent sequencing effort. Their accuracy is quite high, the few mistakes being due to closely spaced sites for the same enzyme, leading to loss of the tiny resulting fragment.

We have been interested in bacterial chromosome evolution and so compared the fine-structure restriction maps for two closely-related strains of *R. capsulatus* with our type strain, SB1003. These comparisons produced a remarkable result: in addition to the large-scale inversions and numerous transposable elements described in other systems, the three strains contain chromosomal regions of high restriction site polymorphism interspersed with regions of high conservation (Figure 2). In other words, the chromosome is a mosaic of alternating conserved and polymorphic elements [5]. Sequencing allows us to be quite precise in the description of polymorphism.

2.2. DNA Sequencing

Genome sequencing begins with the overlapping cosmid set. A given cosmid is digested partially with *Sau*3a, which gives surprisingly random cutting for *R. capsulatus* DNA. From a preparative gel, the 4–5 kb fraction is recovered and ligated into a sequencing vector, an M13 derivative for the work in Prague and pGEM for the work in Chicago. A selected number of this set of clones is sequenced from each end. In Prague, the machine is from Pharmacia; in Chicago, an ABI 377. The sequences are assembled into contigs and immediately primer walking is begun. This plan of attack reduces investigator time, redundancy, and eventually cost. In general, we require 80–100 primer walks to complete a cosmid. The great advantage of this procedure is that the sequence is fully determined as we proceed from one cosmid to another, and annotation can be started at once.

Annotation is the rate-limiting step in most large-scale sequencing projects. We have been fortunate to have assistance from the group of Ross Overbeek at Argonne National Laboratory in this effort. They have developed a large number of tools, including ones that take advantage of metabolic reconstruction, in order to make reading frame assignments quite rigorous. A major problem in this field is that a simple blast search may yield a hit on a protein whose function was assigned, in the database, by similarity to something else, and that assignment was weaker still, so that the chain of evidence is in reality not even a

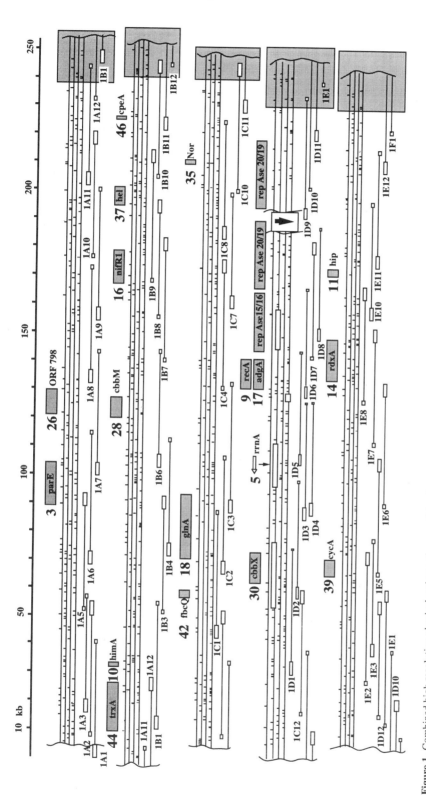

Figure 1. Combined high resolution physical and genetic map of the chromosome of *R. capsulatus* SB1003. Four horizontal lines with vertical ticks for restriction sites represent the physical map of the chromosome, from top to bottom, for *Eco*RV, *Bam*HI, *Hind*III and *Eco*RI. A few areas where the map could not be determined are indicated by empty rectangles. Each cosmid is drawn as a horizontal line, named on the left, e.g. 1A1, 1A2, etc. On each cosmid, the L and R *cos* sites are shown by large and small boxes, respectively. Mapped genes and repeated elements are indicated by boxes above the map. The sizes of these boxes correspond to the minimal hybridizing fragment. The map is read continuously from left to right. Ten to twenty kb marked by the gray boxes at the right end of each 250-kb stretch are repeated at the beginning of the next line, to provide visual continuity. The scale in kb is above the map.

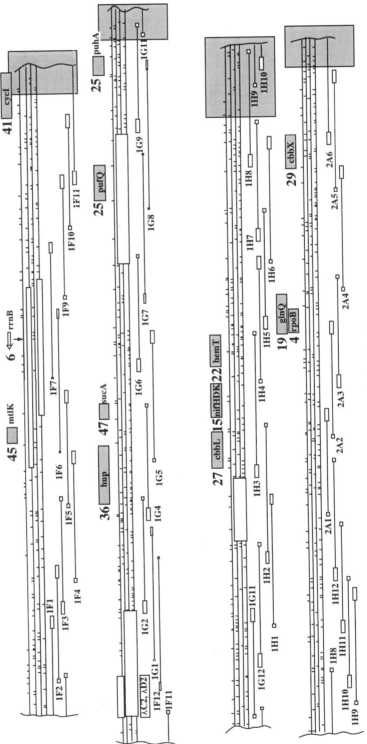

Figure 1. *(continued)*

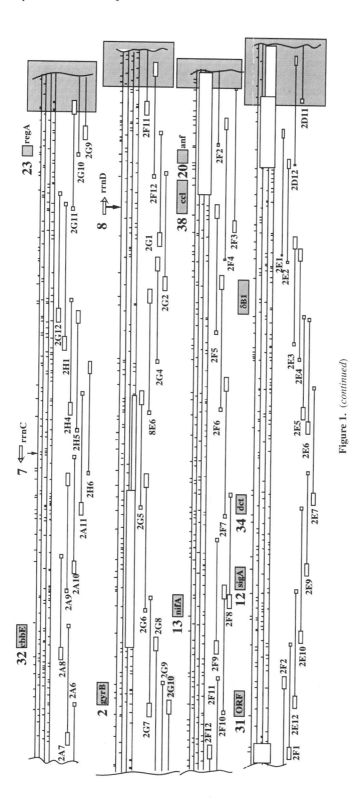

Figure 1. (*continued*)

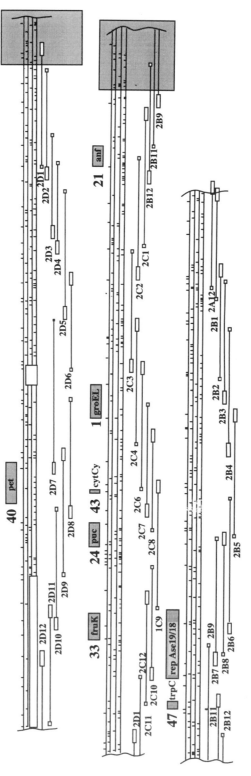

Figure 1. (*continued*)

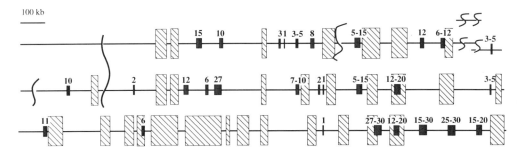

Figure 2. Summary of rearrangements detected in the comparison of the physical maps of three strains of *Rhodobacter capsulatus*: St. Louis, 2.3.1 and SB1003, from top to bottom. The three horizontal lines represent 2.1 Mb from each chromosome. Thick wavy lines indicate the borders of translocations. Gray boxes indicate deletions or insertions whose size appears above the boxes. The hatched rectangles show regions with clustered polymorphism of restriction sites, i.e. fewer than 30% of the sites conserved among all three strains.

thread. With the newer software, assignments are made only if all the searches, including domains and the order of domains, are consistent.

R. capsulatus DNA has 68% GC. This means that there are few stop codons. In addition, codon bias is extreme, so that the algorithm written by C. Halling called CodonUse is usually efficient in predicting the starts and stops of reading frames. Within the first 189 kb, however, there are two regions for which CodonUse fails totally to predict any reading frames at all (Figure 3). For these regions it was necessary to ignore CodonUse and instead to simply consider every frame between starts and stops as possible. Blast searches then allowed assignment of some of these reading frames to proteins of known phages, such as repressors, integrases and tail proteins. Thus, these regions of 30 and 22 kb contain all or parts of phage genomes, whose codons are adapted to life in a very different host [6].

The other feature of this first 189-kb segment worth describing is the set of genes encoding the biosynthetic machinery for cobalamin [6]. Recall that vitamin B_{12} is made by *Salmonella typhimurium* only under anaerobic conditions, while in *Pseudomonas denitrificans* it is made only aerobically [7]. The two pathways contain more than 20 enzymatic steps each, most of which are common. However, in the presence of oxygen, the corrin ring is shrunk early in the pathway and cobalt is inserted early. Anaerobically, ring shrinkage is achieved differently and cobalt is inserted late. Curiously, *R. capsulatus* has genes characteristic of both pathways, so it is possible that B_{12} can be made under either growth condition.

2.3. Applications of the Cosmid Encyclopedia

One hears a great deal these days about expression chips, arrays of oligonucleotides on silicon wafers that permit fluorescently labeled RNA or DNA preparations to be analyzed in one experiment. We have, in fact, been working with the group of A. Mirzabekov at Argonne National Laboratory to develop such chips for the analysis of *Rhodobacter* transcripts. For many purposes, however, it is sufficient to prepare blots of the entire chromosome displayed as restriction fragments on a large gel [4]. 200 lanes show the 560 *Eco*RV fragments in 192 cosmids from the chromosome and six from the large plasmid. This mas-

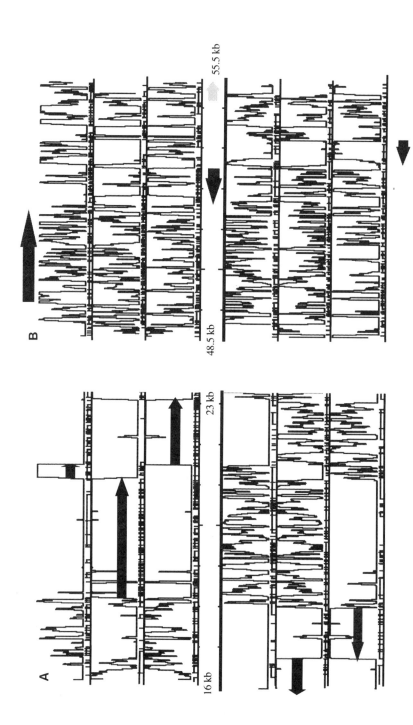

Figure 3. The algorithm CodonUse (C. Halling, Monsanto Co., St. Louis, MO) applied to two regions of sequenced *Rhodobacter* DNA. In (A), the region contains typical *Rhodobacter* DNA, so the program, which computes the probability that a 40-residue window encodes a real protein, yields unambiguous predictions for the location of ORFs. In (B), the DNA probably corresponds to a prophage, making ORF assignments difficult because we do not have a good library of codons known to be preferred by that phage. To the extent that codon use is correlated with expression, we conclude that the phage genes are either a recent invader of the chromosome and/or are not expressed.

ter blot has been used for several applications: mapping genes and gene families [4], global analysis of transcription under different conditions [4], and strain comparisons [5].

2.4. The System of Genetic Analysis

It has been stated often that *R. capsulatus* is the organism of choice for functional analysis of bacterial genomes. This analysis is based on the facile introduction of deletions and substitutions in the chromosome by use of a defective transducing phage called the gene transfer agent, GTA [1]. The GTA has never been found in a lytic form, so phage genes remain uncharacterized and the mechanism by which it packages host DNA is unknown. Nevertheless, there are strains of *R. capsulatus* that produce more GTA particles than other strains, and these "overproducers" are the source of the transducing preparations.

Suppose we wish to attempt the deletion of all the DNA contained in the 35-kb insert of a particular cosmid. First, the cosmid is cut with a restriction enzyme that cuts frequently in *Rhodobacter* DNA but not at all in the vector [8]. The cut ends, which each have a few hundred bases of *Rhodobacter* DNA still attached, are ligated together with a cassette encoding gentamycin resistance. This construct (the deleted cosmid containing Gm^r) is used to transform *E. coli* containing a broad host range self-transmissable plasmid which contains a fragment of the cosmid vector Lorist 6. Homologous recombination in *E. coli* results in construction of chimaeric plasmids (co-integrates) that contain both the deletion construct and the mobilizing vector. This strain is then used to transfer the co-integrate to a GTA-overproducing strain of *Rhodobacter capsulatus* by conjugation. Finally, culture supernatants of the latter strain contain populations of the GTA in which 4.6-kb fragments of all the DNA in the cell are packaged. This size of the package is a little more than one-thousandth that of the chromosome. If packaging is truly random, there ought to be one in a thousand GTA particles with exactly the sequence needed for transduction. In fact, since our target sequence is in a plasmid, there are more copies of the target than of chromosomal DNA so the frequency of good transducing GTA particles is probably higher than one in a thousand.

The last step in deletion strain construction is the addition of the GTA-containing supernatant to a wild-type SB1003 culture, followed by plating on gentamycin [8]. The GTA introduces a 4.6-kb DNA fragment with no replicon. The only reasonable mechanism for stable Gm^r is for the flanking sequences to recombine homologously with their chromosomal counterparts. This introduces the Gm^r cassette into the chromosome in place of the DNA deleted back in the first step of the procedure. We observe two frequency classes of transductants. One class is relatively high, one resistant colony for every thousand transduced cells; the other is found 100 times less frequently. The difference is due to the presence or not of essential genes in the fragment attempting to be deleted. High frequency transductants were found for four of ten cosmids in our first tests; Southern blots confirmed the deletion of the DNA replaced by the cassette [8]. These transductants grew as well as the wild type on RCVB medium. In one case, for which we knew that the DNA deleted from the cosmid included the essential gene *rpoD*, the frequency of transductants was very low. Examination of one of these by Southern hybridization indicated that the transductant had deleted only 2 kb. Apparently, the chromosomal replacement started with a homologous recombination event near the end of the insert, but this surviving transductant had integrated the Gm^r cassette by way of an illegitimate recombination just prior to reaching the *rpoD* gene. In other words, this particular cell and its descendants had both the selectable marker and the retained essential gene.

3. CONCLUSIONS

Rhodobacter capsulatus is an ideal organism for a total sequencing project. We hav demonstrated all of the elements required for such a project: high resolution physical map ping, genome sequencing, genome sequence annotation, determination of global expressic profiles, systematic deletion of cosmid-sized regions of chromosomal DNA to assign func tions to ORFs, and strain comparisons at the level of restriction maps and sequences. Wit suitable funding, all of these aspects of the genome project can be pursued effectively in th near future. It should be obvious that we cannot study the functionality of more than 4,00 ORFs alone. We have made it our practice to supply cosmids upon request and that practic will continue as the sequences become known. The sequence information and some analyt cal tools are available on our web site: (http://capsulapedia.uchicago.edu/capsulapedia capsulapedia/capsulapedia.shtm).

ACKNOWLEDGMENTS

This work was supported by grants from the Dept. of Energy (FG0286ER13546), th NSF (INT 9506881) and the Harris and Frances Block Research Fund of the University o Chicago.

REFERENCES

1. Yen, H. C., Hu, N. T. and Marrs, B. L. (1979) J. Molec. Biol. 131, 157–168.
2. Fonstein, M. and Haselkorn, R. (1995) J. Bacteriol. 177, 3361–3369 .
3. Fonstein, M. and Haselkorn, R. (1993) Proc. Natl. Acad. Sci. USA 90, 2522–2526.
4. Fonstein, M., Koshy, E. G., Nikolskaya, T., Mourachev, P. and Haselkorn, R. (1995) EMBO J. 14 1827–1841.
5. Nikolskaya, T., Fonstein M. and Haselkorn, R. (1995) Proc. Natl. Acad. Sci. USA 92, 10609–10613.
6. Vlcek, C., Paces, V., Maltsev, N., Paces, J., Haselkorn, R. and Fonstein, M. (1997) Proc. Natl . Acad. Sc USA 94, 9384–9388.
7. Roth, J. R., Lawrence, J. G. and Bobik, T. A. (1996) Annu Rev Microbiol 50, 137–181.
8. Kumar, V., Fonstein, M. and Haselkorn, R. (1996) Nature 381, 653–654.

50

CIRCADIAN RHYTHMS OF GENE EXPRESSION IN CYANOBACTERIA

S. S. Golden,[1] M. Ishiura,[2] N. F. Tsinoremas,[1] S. Aoki,[2] S. Kutsuna,[2]
H. Iwasaki,[2] Y. Liu,[3] N. Lebedeva,[1] C. R. Andersson,[1] J. Shelton,[1]
C. H. Johnson,[3] and T. Kondo[2]

[1]Department of Biology
Texas A&M University
College Station, Texas 77843-3258
[2]Department of Biology
Nagoya University
Chikusa, Nagoya 464-01 Japan
[3]Department of Biology
Vanderbilt University
Nashville, Tennessee 37235

1. INTRODUCTION

The cyanobacteria are the simplest organisms known to exhibit circadian rhythms of various metabolic functions as a manifestation of an endogenous biological clock [1]. The unicellular cyanobacterium *Synechococcus* sp. strain PCC 7942 provides an excellent model system with which to analyze the function of the circadian clock and its control over gene expression. *Synechococcus* is readily transformable, and sophisticated tools have been developed for its genetic manipulation [2–4]. Clock control of gene expression is easily demonstrated by using the *Vibrio harveyi luxAB* genes in transcriptional fusions with the promoters of *Synechococcus* genes of interest [5]. The promoter of the *psbAI* gene, one of three genes that encodes the D1 protein of photosystem II, was fused to *luxAB* and this reporter was recombined into the *Synechococcus* genome. The resulting transformants produced light when provided with the long-chain aldehyde substrate of the enzyme (decanal) [5]. Entrainment of the cells to a 12 h light:12 h darkness cycle (LD12:12), followed by incubation in continuous light, resulted in bioluminescence expressed rhythmically, peaking and troughing once per day, with a period of 24 h. The phase of the rhythm could be reset by pulses of darkness, and the period stayed very near to 24 h over a 10°C range of continuous growth temperatures. These features demonstrated that the rhythm of bioluminescence, as a function of *psbAI* expression in *Synechococcus*, conforms to the criteria that have been used to define control by an endogenous circadian oscillator [5].

The Phototrophic Prokaryotes, edited by Peschek *et al.*
Kluwer Academic / Plenum Publishers, New York, 1999.

2. ISOLATION OF MUTANTS

The description of a genuine circadian rhythm in a prokaryote that is genetically ma-
nipulable provided the opportunity to attempt saturation mutagenesis to identify loci that
are central to clock function. A turntable device was designed that would allow hundreds
of colonies on each of 12 agar plates to be sampled periodically for bioluminescence for
days at a time [6]. The turntable and computer software align each plate precisely under a
cooled CCD camera, such that the points of light emanating from colonies can be recognized and identified. Each time a given plate is sampled over a period of days, the spots of
light are recorded, and the data for each colony are tracked. At the end of the screening,
circadian oscillations of bioluminescence from thousands of colonies have been examined
[6]. The wild-type reporter strain that carries the $P_{psbAI::luxAB}$ reporter gene described above
was mutagenized chemically and colonies were screened for alterations in the circadian
rhythm of bioluminescence. Mutant colonies exhibiting periods ranging from 16 h to 60 h,
or apparent arhythmia, were identified [6].

A library of wild-type DNA was created in a vector that will recombine at a neutral
site in the *Synechococcus* genome [7]. When the library was used to transform the mutants,
a number of transformants were identified that had been restored to wild-type rhythmicity,
had more nearly wild-type rhythmicity that the original mutant, or had a more severe phenotype. The transforming DNA was rescued from the transformants of four period mutants,
and the inserts were found to overlap. Restriction mapping and nucleotide sequence analysis
indicated that three contiguous open reading frames, now called *kaiABC*, encode novel
products of unknown function, and lie within the complementing DNA [7]. Additional nucleotide sequence analysis of the *kaiABC* locus from the various period mutants showed that
each carried a missense mutation in one of these three genes. The wild-type *kaiABC* cluster
was able to rescue the phenotypes of all of the altered-period and arhythmic mutants. Results thus far suggest that these three genes are central to clock function in *Synechococcus*.

3. CONTROL OF GENE EXPRESSION BY THE CLOCK

Expression of the *psbAI* gene is clearly under circadian control, as is that of several
other genes whose promoters were fused to *luxAB* [3]. To determine the extent to which
the circadian clock controls gene expression, the reporter genes were integrated randomly
into the *Synechococcus* genome. A vector that contained promoterless *luxAB* genes and a
kanamycin resistance gene was used to clone random fragments of *Synechococcus* DNA
upstream of the reporter genes. Because the vector could not replicate in *Synechococcus*,
kanamycin resistant exconjugants [4] were recovered only when homologous recombination between *Synechococcus* DNA on the plasmid and on the chromosome had occurred,
integrating the plasmid and duplicating the insertion locus [8]. A subset of exconjugants
was bioluminescent as expected—many clones would not have the *luxAB* genes downstream of a promoter—and the magnitude of bioluminescence varied among clones as a
function of the strength of each promoter upstream of the reporter genes [3]. Despite these
variations, all colonies that were bioluminescent showed a circadian oscillation of light
production. However, not all oscillations showed the same timing of gene expression. The
majority of genes (Class 1), like *psbAI*, had a peak of expression that corresponded to late
afternoon, and a symmetric wave form of the bioluminescence profile. A minority of genes
had different shaped wave forms, with peaks at different times of day, including a few
(Class 2) that showed a phase of expression almost opposite that of the Class 1 genes [3].

The apparent global rhythmicity of transcription of the genome, with variation of waveforms and phasing, lead to the following model [3]. A default circadian rhythm of gene expression occurs through a mechanism that is not gene specific, such that in the absence of specific *cis* information, a gene is transcribed with a low-amplitude Class 1 rhythm. Consistent with this hypothesis is the fact that a consensus *E. coli* promoter drives rhythmic gene expression in *Synechococcus* (N.V. Lebedeva and S.S. Golden, unpublished). However, additional layers of control must exist to regulate some genes specifically, such as shifting the phase and increasing the amplitude of the oscillation.

Evidence for the model was obtained when an insertional mutagenesis strategy resulted in a clone that showed low amplitude circadian oscillation from the *psbAI* gene, in which the troughs were elevated without decreasing the peaks of expression [9]. The mutation was an insertion event that decreased or eliminated expression of the *rpoD2* gene, which encodes a group 2 sigma factor. Inactivation of *rpoD2* in other reporter strains, in which different promoters were driving *luxAB* expression, indicated that amplitude of rhythmicity of a subset of genes is dependent on *rpoD2*. One Class 2 gene, *purF*, showed no change in circadian oscillation in an *rpoD2⁻* background, nor did three Class 1 genes other than *psbAI*. However, a fourth Class 1 gene tested, *ndhD*, showed the same low-amplitude phenotype as *psbAI*. These data suggest that rpoD2 is part of an output pathway from the clock that affects circadian expression of a subset of genes [9].

4. CIRCADIAN RHYTHMICITY OF GENE EXPRESSION DURING RAPID GROWTH

The occurrence of circadian rhythms in cyanobacteria, which can have doubling rates more rapid than once per 24 h, provided an opportunity to address the question of whether it is possible for cells to execute a 24 h timing circuit when several generations are required to complete the cycle . Because the growth rate of *Synechococcus* is light-dependent, it is easy to modulate the generation time. In one set of experiments, bioluminescence from the P$_{psbAI::luxAB}$ reporter strain was monitored during microcolony formation, when all cells in the colony were still in a monolayer, and cell division could be directly monitored microscopically. Under conditions in which all cells were dividing with a doubling time of approximately 12 h, bioluminescence increased and oscillation occurred with a 24 h period, consistent with a mathematical model that assumes exponential growth of the culture and oscillation as a sine or cosine function [10].

Another approach confirmed the circadian periodicity in rapidly-dividing *Synechococcus* cells. A wild-type culture was inoculated into a turbidostat, entrained to an LD12:12 cycle, and then kept in continuous exponential-phase growth with a doubling time of approximately 12 h, which corresponds to a 'low' light intensity in which *psbAI* is expressed well [11]. RNA was extracted from samples at regular intervals, and the level of the *psbAI* mRNA was determined by northern blot analysis. Oscillation with a period of approximately 24 h was observed. Another culture was treated similarly, but after 24 h of growth in continuous growth at low-light intensity, the light fluence was increased to stimulate expression of *psbAII* and to accelerate the rate of cell division. An acute induction of *psbAII* mRNA in response to the light shift was observed, followed by 24 h periodicity of oscillation of the *psbAII* message. Under these conditions, the doubling rate was every 5–6 h, such that 4 generations were required to complete one circadian cycle [10]. These results clearly demonstrated that cyanobacteria maintain a circadian timing circuit that is independent of the cell cycle, consistent with the clear rhythm [5] seen in stationary-phase cultures.

5. CLOCK CONTROL OF LIGHT-REGULATED GENE EXPRESSION

The sensitivity of the cells to changes in light intensity was tested throughout the circadian cycle to determine whether this underlying rhythm affects response to environmental signals. A reporter strain in which the high-light-inducible *psbAIII* gene is fused to *luxAB* was grown in a turbidostat, entrained to LD12:12, and then released into constant low-light growth conditions. At intervals, samples were collected, the bioluminescence level was determined, and each sample was exposed to high light for a second measurement. The magnitude of induction of bioluminescence by high light was much greater during a time of 'trough' expression, when *psbAII* expression was at a minimum, than at the peaks of expression. A similar reporter strain which carried an additional mutation in the *kaiAB* locus, which is central to circadian function, was assayed in parallel. In this arhythmic strain, induction was similar at all time points, and *psbAIII* expression did not oscillate. However, the initial dark-to-light transition caused an increase in induction and in overall bioluminescence, suggesting that cells which cannot entrain to the daily cycle and cannot anticipate 'dawn' react acutely to the onset of light. It took approximately 12 h for bioluminescence in this strain to recover to the initial level, which was maintained without rhythmic oscillation during constant low-light growth (J. Shelton and S. S. Golden, unpublished data).

6. CONCLUSIONS

The circadian clock underlies overall metabolic function in *Synechococcus*, having a global effect on transcription. Distinct pathways tie the rhythm of the clock to the control of specific genes. The timing mechanism persists even under growth conditions in which several generations are produced during one 24 h cycle. The sensitivity of cells to environmental stimuli is gated by the clock, such that certain responses are modulated as a function of their timing within the circadian cycle.

REFERENCES

1. Golden, S.S., Ishiura, M., Johnson, C.H. and Kondo, T. (1997) Annu. Rev. Plant Physiol. Plant Mol. Biol. 48, 327–354.
2. Li, R. and Golden, S.S. (1993) Proc. Natl. Acad. Sci. *USA* 90, 11678–11682.
3. Liu, Y., Tsinoremas, N.F., Johnson, C.H., Lebedeva, N.V., Golden, S.S., Ishiura, M. and Kondo, T. (1995) Genes and Development 9, 1469–1478.
4. Tsinoremas, N.F., Kutach, A.K., Strayer, C.A. and Golden, S.S. (1994) J. Bacteriol. 176, 6764–6768.
5. Kondo, T., Strayer, C.A., Kulkarni, R.D., Taylor, W., Ishiura, M., Golden, S.S. and Johnson, C.H. (1993) Proc. Natl. Acad. Sci. *USA* 90, 5672–5676.
6. Kondo, T., Tsinoremas, N.F., Golden, S.S., Johnson, C.H., Kutsuna, S. and Ishiura, M. (1994) Science 266, 1233–1236.
7. Ishiura, M., Aoki, S., Kutsuna, S., Andersson, C.R., Iwasaki, H., Golden, S.S., Johnson, C.J. and Kondo, T. (1997) in preparation.
8. Golden, S.S., Brusslan, J. and Haselkorn, R. (1987) Methods Enzymol. 153, 215–231.
9. Tsinoremas, N.F., Ishiura, M., Kondo, T., Andersson, C.R., Tanaka, K., Takahashi, H., Johnson, C.H. and Golden, S.S. (1996) EMBO 15, 2488–2495.
10. Kondo, T., Mori, T., Lebedeva, N.V., Aoki, S., Ishiura, M. and Golden, S.S. (1997) Science 275, 224–227.
11. Golden, S.S. (1995) J. Bacteriol. 177, 1651–1654.

51

GENOME SEQUENCE SKIMMING OF *RHODOBACTER SPHAEROIDES* 2.4.1[T]

Chromosome II Is a "Run of the Mill" Chromosome

M. Choudhary,[1] C. Mackenzie,[1] G. M. Weinstock,[1,2] and S. Kaplan[1]

[1]Department of Microbiology and Molecular Genetics
[2]Department of Biochemistry and Molecular Biology
University of Texas Medical School
Houston, Texas 77030

1. INTRODUCTION

Rhodobacter sphaeroides 2.4.1[T] is a facultative photoheterotroph belonging to the α-3 subdivision of the *Proteobacteria* [1]. This group of organisms reflects a diverse range of metabolic activities and a complex genome organization. Studies from our laboratory reveal that *R. sphaeroides* 2.4.1[T] possesses two different circular chromosomes, of sizes ~3.0 Mbp (CI), and ~0.9 Mbp (CII) which broke with the long-held dogma regarding our understanding of prokaryotic genome structure and organization [2,3]. Although the concept that the prokaryotic genome can consist of more than a single chromosome has gained universal acceptance, the kinds of genes and their distribution within the second chromosome were previously unknown. Important questions leading to an understanding of the genome complexity of *R. sphaeroides* 2.4.1[T] revolve around the actual genetic content and its organization, present on each chromosome.

In search of the functional role(s) underlying this genomic diversification, a detailed, albeit low resolution, genomic analysis of CII of *R. sphaeroides* 2.4.1[T] was undertaken. We describe here both partial- and fully random sequencing strategies of cosmid inserts containing ~417 kb of unique insert DNA previously mapped and ordered to CII [4]. This generated ~291 kb of unique DNA sequence. Analysis of these sequences suggested that approximately 87% of the total unique sequence was coding. A total of 144 database matches was found representing a wide variety of functions, e.g. amino acid biosynthesis, photosynthesis, nutrient transport, and various regulatory functions. The positions of two previously described ribosomal RNA operons (*rrnB* and *rrnC*) and 3

The Phototrophic Prokaryotes, edited by Peschek *et al.*
Kluwer Academic / Plenum Publishers, New York, 1999.

tRNAs, f-Met, Ala and Ile were confirmed [5]. Two new tRNAs for Met and Val were also identified on this chromosome. We have shown previously that transposon insertions mapping to CII can generate auxotrophic phenotypes [4] suggesting that genes essential for the day to day survival of *R. sphaeroides* are encoded by CII. The diverse range of metabolic functions encoded by CII confirmed the existence of the same major classes of genes as found on single, circular chromosomes of other prokaryotes, i.e., CII appears to be a "run of the mill" chromosome.

2. SEQUENCING STRATEGY

Several bacterial genomes have been now completely sequenced [6–11]. Two main sequencing strategies have been used in these projects, that is, the sequencing of large numbers of random subclones followed by computer assembly (*H. influenzae, M. jannaschii, E. coli,* and *B. subtilis*), or a directed primer walking approach (*M. tuberculosis*). However, both of these methods have high costs because they involve either several fold (~6–10 times) genome coverage, or the requirement for large numbers of primers and other materials to give finished "base perfect" sequence. We have used a low redundancy, semi-random approach to partially sequence CII of *R. sphaeroides*. Our aim was not to generate a base by base account of CII, but rather to generate an overview of the types of genes encoded by CII. The general outline of the sequencing strategy is presented as a flowchart (Fig. 1).

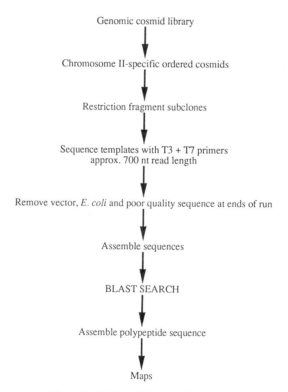

Figure 1. Outline of the sequencing strategy.

2.1. Subcloning Strategies

We evaluated three subcloning strategies. (1) In the first case (cosmid 8536) we sub-cloned restriction fragments which were generated by 4–5 restriction enzymes which cut either rarely (< 3 times) or not at all in the cosmid vector, but relatively frequently (5–10 times) in the cosmid insert. These restriction fragments were subcloned and then sequenced. (2) For cosmid 8603 we generated a DNaseI library of the total cosmid DNA, then screened the subclones by hybridization with the cosmid insert. A random sample of these insert subclones were sequenced. (3) For cosmid 8621 we combined the strategies used for (1) and (2) and screened a cosmid 8621 DNAseI generated library with 8621 *Bam*HI restriction fragments. The *Bam*HI restriction fragments were then subcloned and sequenced. In addition DNaseI subclones which hybridized to two *Bam*HI fragments were considered to be linking subclones and were also sequenced.

Of the three methods described above we finally settled on the first method as it gave an even spread of sequence information which could then be used for mapping gene order. Method (2) proved labor intensive but did not give us the even sequence coverage we had hoped for. In addition it proved more difficult to determine gene order without the advantage of restriction site markers which were present in method (1). Method (3), gave us the best coverage and most information in terms of order. However, these advantages were offset by the considerable amount of hands on time which was required prior to sequencing.

2.2. Automated Sequencing

Templates were sequenced using dye-deoxy terminators purchased from ABI and used as recommended by the manufacturer. Primers annealing to the T3 and T7 sites of pBluescript SK (-) were used for sequencing restriction fragments. Sequencing runs were performed using either an Applied Biosystems Model 377 DNA sequencer as advised by the manufacturer or an ABI 373 (stretch modified) machine.

2.3. Sequence Analysis

Computer analysis was performed using both the Genetics Computer Group (GCG) and Staden software packages. Sequence files were analyzed by using the BLAST server at the National Center of Biotechnology Information (NCBI, Bethesda, MD). Sequences were considered to have given a significant database match when the probability (P) value was $\leq 10^{-3}$ and score ≥ 100. ORF analysis was performed by running the assembled consensus sequences and their complements through the GCG CODONPREFERENCE program using an *R. sphaeroides* codon usage table [12]. All consensus sequences were screened for the presence of tRNAs using the program TRNASCAN [13].

3. RESULTS AND DISCUSSION

3.1. Overview of the Sequence

Subclones were constructed from twenty-three cosmids containing a total of ~417 kb of unique insert DNA. 573 double-stranded DNA templates were sequenced from both ends to yield a total of ~640 kb of edited sequences, about 1.4-fold coverage of the total cosmid insert DNA. The average read length was 559 ± 268 nucleotides with 2.8% ambiguous bases. We have observed that > 700 nucleotides of raw sequence often contained useful data for BLASTX searches, the determination of restriction sites and for fragment assembly. The sequencing summary derived from the data is given in Table 1. A total of 944 sequence fragments containing CII DNA were assembled, and a total of 404 sequence contigs were generated. The average gap length was estimated to be ~500 nucleotides, therefore if required, a large number of gaps could be resolved by primer walking. The total unique DNA sequence was ~291 kb, representing approximately 70% of the total insert coverage. After analysis the sequences could be placed into three categories. Of the total unique sequence, ~131 kb DNA (45%) gave matches to either genes or ORFs (dORFs) in the database. The remaining ~160 kb DNA (55%) of unique sequence did not give database matches. When this latter sequence was analyzed using the CODONPREFERENCE program, 122 kb (42%) was estimated to comprise putative open reading frames (pORFs). These varied from ≥ 50–400 amino acid residues in length. The shortest stretch of sequence which we arbitrarily grouped as a pORF was 50 contiguous amino acids. Due to the fragmentary nature of these data, we have not made any attempt to estimate the number of pORFs. Rather we have presented only the fraction of the sequence which we consider may be coding, but which did not give a database match.

Although not yet completed, the low resolution DNA sequencing of CII confirms the existence of all major groups of genes as applied to the analysis of those genomes sequenced in their entirety. In many respects, and granting the still limited level of analysis, CII is indistinguishable from the classical single, circular bacterial chromosome encoding a wide range of biological functions.

3.1.1. Sequence Composition and Codon Usage. The %G+C composition of the *R. sphaeroides* genome was determined to be 66.0%. The %G+C content was found to be 67.3% and 65.7% for CI and CII respectively.

Table 1. Summary of CII sequencing project

Size of CII	910 kb
Number of cosmids sequenced	23
Total unique insert DNA	417 kb
Double-stranded templates	573
Average edited read length (% ambiguous bases)	559 ± 268 (2.8%)
Total DNA sequenced	640 kb
Total cosmid insert DNA sequence (Includes multiple clones and overlaps)	565 kb
Fold coverage of the total cosmid inserts	1.3x
Total unique DNA sequence	291 kb
Unique sequence coverage	0.7x
Mole G+C%	65.7
DNA resulting in significant database matches	131 kb
DNA resulting in the designation of putative orfs (pORFs) with no database matches	122 kb
DNA resulting in neither an orf nor a database match	38 kb
Number of sequence fragments in random assembly	1145
Number of contigs	404
Average gap length	504 bp
Number of presumptive genes identified	144
Number of ribosomal RNA operons identified	2 (*rrnB* and *rrnC*)
Number of tRNA genes identified	5 (Ala, Ile, f-Met, Met, and Val)

We examined the di- and trinucleotide frequencies of the sequences from both chromosomes and found that the fraction of each di- or trinucleotide on each of the two chromosomes was approximately equivalent.

We also examined the codon usage for genes mapping to CI and CII which had already been placed in Genbank. Here we looked at the relative fraction of redundant codons for Arginine, Leucine and Serine. We found that in general rare codons such as Arg-AGA and Leu-TTA were rare on both chromosomes, whereas relatively frequent codons such as Arg-CGA or Leu-CTG were frequent on both chromosomes. This was also true of amino acids specified by fewer redundant codons such as Lysine and Cysteine.

The virtually identical G+C content, di- and trinucleotide frequencies and occurrence of rare codons existing between CI and CII suggest that CII was not derived exogenously. Therefore we do not have any evidence to support the hypothesis that CII was acquired by horizontal transfer from another organism. This suggests that CI, CII may have arisen either from a single larger replicon or have been evolved *de novo* by the reassortment of existing DNA sequence.

3.1.2. A Variety of Gene Functions is Encoded on CII. Analysis of the coding sequences resulted in the identification of 101 genes of known function and 43 open reading frames of unknown function representing a wide variety of functions [14]. In addition, we have also identified tRNAs for Valine and Methionine and confirmed the position of tRNAs formyl-Methionine, Alanine and Isoleucine and ribosomal RNAs, *rrnB* and *rrnC*.

We don't intend to describe the function of each gene identified on CII but some genes have attracted our attention. The gene tentatively identified as encoding ribosomal protein S1 shows exceedingly high homology at the amino acid sequence level to other S1 proteins. Interestingly, the *rpsA* gene product (30S S1 ribosomal protein) which is a part of the translation apparatus, appears to be present as a single copy on CII when surveyed at a low stringency hybridization condition.

In the case of the tryptophan biosynthesis "operon" the genes are split between CI (*trpCDEG*) and CII (*trpF and trpB*) demonstrating for the first time in bacteria, that genes of a linear biosynthetic pathway can be divided between two chromosomes.

Additionally, several genes seem to be important or even essential when considering the structure of the bacterial cell. For example, the gene *mltB* encodes the enzyme murein hydrolase whose proposed role is as a pacemaker enzyme for murein enlargement or as a hydrolase enzyme during the cell division process involving cell septum formation. The involvement in murein metabolism is significant as it gives both mechanical stability and shape to the bacterium.

Regulatory genes represent ~12.5% of the total genes presently localized on CII. These regulators eg., *aepA, dorX, piaA, amiC,* and *ompR*, seem to mediate various positive and negative transcriptional controls. Also, CII encoded a number of regulatory proteins involving two-component regulatory systems, namely, sensory kinases (*dorS, sasA,* and *baeS*) and response regulators (*dorR* and *copR*).

3.1.3. Unexpected Metabolic Functions. Besides the expected metabolic and biosynthetic functions listed above, CII contained sequences which encoded functions we considered unusual for this photosynthetic organism. That is, we found database matches to *tlyC*, a gene encoding a hemolysin, and *ndvB*, a gene implicated in *Rhizobium* root nodule development in alfalfa. These findings were unanticipated, as *R. sphaeroides* has not been considered to be an animal pathogen nor plant symbiont.

3.1.4. Duplicated Genes and Metabolic Diversity. R. sphaeroides contains a growing number of gene homologues with one member of each pair located on each chromosome e.g., *hemA* and *hemT, rdxA* and *rdxB, cbbPI* and *cbbPII, cbbAI* and *cbbAII, rpoNI* and *rpoNII.* The physiological role(s) of many of these gene homologues have not yet been demonstrated. Recent results [15] involving the *hem*A and *hem*T gene pair of *R. sphaeroides* now reveal that the *hem*T gene is likely to be transcribed by use of the *rpo*E encoded sigma factor and *hem*A by RpoD. Thus, the presence of two genes encoding the same function has added an element of metabolic diversity and perhaps an increased level of adaptability to this organism.

3.2. Comparison to Other Bacterial Genomes

We have compared the relative percentages of functional classes of genes on CII to the chromosomes of *E. coli, Haemophilus influenzae* Rd, *Methanococcus jannaschii, Mycoplasma pneumoniae,* and *M. genitalium.* CII of *R. sphaeroides* 2.4.1^T encodes a diverse set of functions. Some of these functions i.e. cellular processes, biosynthesis of small molecules and intermediary metabolism, are found at approximately the same proportion as in other bacterial chromosomes. However, gene functions classified as involved in macromolecular synthesis (transcription and translation) and cell structure are under-represented, whereas regulatory functions are over-represented when compared to other microbial genomes. Whether this is a true reflection of the content of CII or an artifact of the regions which we have so far examined will require a complete analysis of the chromosome.

Approximately 44% of the total CII coding sequences remain as unassigned pORF DNA which do not give matches to genes in the database. This suggests that even with the large number of genome sequencing projects which have been deposited in the NCBI database there are still many sequences which are unique to CII of this organism and may reflect the metabolic diversities of the members of this genus.

3.3. Sequence Skimming and the Microbial Genomics

We have used a low redundancy, semi-random approach to partially sequence CII of *R. sphaeroides,* the aim being to generate the maximum amount of biological information for the lowest possible cost. We hoped that this approach might allow us to: i) determine if this replicon encodes a specialized set of functions, ii) assess the relationship of this form of genome organization to that routinely observed in other bacteria, iii) provide us with a paradigm with which to compare other bacteria having more than one chromosome and iv) to ascertain if such a genomic organization served as a precursor to the diploid eukaryotes.

From a technical perspective we also wanted to determine how much information could actually be obtained by using a low redundancy sequencing approach. Those few genomes which have been sequenced to "perfection" have provided us with many insights into the bacterial and archaebacterial kingdoms, but perhaps numerous bacterial genomes sequenced to low redundancy could be equally informative both to workers interested in these organisms and from a genetic diversity perspective.

3.4. Completing the Contig Gaps and Functional Study

We found that when restriction enzymes recognizing six base pairs were used to generate template subclones, for the most part, an even distribution of sequence data was obtained from each cosmid. These sequences when used in BLAST searches allowed us to determine which genes may be encoded in a particular cosmid. These fragments of sequence information have allowed us to focus on particular genes of interest which we did not know existed in this organism until the cosmids were partially sequenced. In the case of the DMSO reductase, an initial 700 bp of sequence gave us an indication of the presence of these genes. This has led to the acquisition of a 13 kb region of CII which contains both structural and regulatory genes for the DMSO reductase gene cluster. Functional studies of *dor* genes has directly shown that CII of *R. sphaeroides* encodes an essential gene function, revealing its critical role in cell metabolism under anaerobic, dark growth conditions [16]. Such genes of interest can be readily subcloned either by using the correct combination of restriction fragments, or by PCR. In both approaches the information required for subcloning had been generated by the low redundancy approach.

ACKNOWLEDGMENTS

This work is completely described in October, 1997, issue of Microbiology, 143: 3085–3099, and the research was entirely supported by a grant from the Clayton Foundation for Research.

REFERENCES

1. Woese, C. R., Stackebrandt, E., Weisburg, W. G., Paster, B. J., Madigan, M. T., and others (1984) Syst. Appl. Microbiol. 5, 315–326.
2. Suwanto, A. and Kaplan, S. (1989) J. Bacteriol. 171, 5850–5859.
3. Suwanto, A. and Kaplan, S. (1992) J. Bacteriol. 174, 1135–1145.
4. Choudhary, M., Mackenzie, C., Nereng, K. S., Sodergren, E. S., Weinstock, G. M. and Kaplan, S. (1994) J. Bacteriol. 176, 7694–7702.
5. Dryden, S. C. and Kaplan, S. (1990) Nucl. Acids. Res. 18, 7267–7277.
6. Fleischmann, R. D., Adams, M. D., White, O., Clayton, R. A., Kirkness, E. F., and others (1995) Science 269, 496–512.
7. Fraser, M. Claire, Jeannine, D. G., White, O., Adams, M. D., Clayton, R. A., and others (1995) Science 270, 397–403.
8. Bult, C. J., White, O., Olsen, G. J., Zhou, L., Fleischmann, R. D., and others (1996) Science 273, 1058–1073.
9. Himmelreich, R., Hilbert, H., Plagens, H., Pirkl, E., Li, Bi-Chen, and Herrmann, R. (1996) Nucl. Acids. Res. 24, 4420–4449.
10. Tomb, Jean-F., White, O., Kerlavage, A. R., Clayton, R. A., Sutton, G. G. and others (1997) Nature 388, 539–548.
11. Blattner, Frederick R., Plunkett, G., Bloch, C.A., Perna, N. T., Burland, V. and others (1997) Science 277, 1453–1462.
12. Eraso, J. and Kaplan, S. (unpublished)
13. Fichant, G. A. and Burks, C. (1991) J. Mol. Biol. 220, 659–671.
14. Choudhary, M., Mackenzie, C., Nereng, K. S., Sodergren, E. S., Weinstock, G. M. and Kaplan, S. (1997) Microbiol. 143, 3085–3099.
15. Zielstra-Ryalls, J. and Kaplan, S. (unpublished)
16. Mouncey, N., Choudhary, M. and Kaplan, S. (1997) J. Bacteriol. (in press)

52

TRANSPOSITION OF Tn5 DERIVATIVES IN THE CHROMATICALLY ADAPTING CYANOBACTERIUM, *FREMYELLA DIPLOSIPHON*

John Cobley, Lasika Seneviratne, Lisa Drong, Maya Thounaojam, Jeffrey F. Oda, and Jennifer Carroll

Department of Chemistry
University of San Francisco
2130 Fulton St., San Francisco, California 94117

1. INTRODUCTION

In the cyanobacteria a major fraction of the light used for photosynthesis is initially absorbed by the phycobiliproteins. These are highly pigmented proteins which are assembled together with colorless linker polypeptides into organelles called phycobilisomes. These organelles serve as light-harvesting antennae, and funnel exciton energy into the thylakoid membrane to which they are superficially attached. Within the thylakoid membrane cyanobacterial photosynthesis procedes by a mechanism similar to that in higher plants, viz. two photosystems containing chlorophyll *a* which cooperate to oxidize water.

In some cyanobacterial species the absorption properties of the phycobilisome depend on the color of the light in which the cyanobacterium was grown. Actively growing cultures of such cyanobacteria can dramatically change color when transferred from one color of light to another. To the human eye the color of these cyanobacterial cultures is complementary to the color of the light in which they were grown; cultures from red light appear green whereas cultures from green light appear red. Not surprisingly this adaptive process is known as complementary chromatic adaptation. In dim light chromatic adaptation increases photosynthesis by increasing the absorbancy of the cyanobacterial cell.

Chromatic adaptation has been most extensively studied in the filamentous cyanobacterium, *Fremyella diplosiphon* PCC7601 (for reviews see [1,2]). In this organism green light induces the synthesis of phycoerythrin (PE), a protein with red bilin chromophores, and suppresses the synthesis of phycocyanin (PC) which has blue bilin chromophores. These effects of green and red light are exerted at the level of gene transcription [3–5]. Since there is now much evidence to support the hypothesis that contemporary cyanobac-

The Phototrophic Prokaryotes, edited by Peschek *et al.*
Kluwer Academic / Plenum Publishers, New York, 1999.

teria and plant chloroplasts have a common ancestry, knowledge of cyanobacterial chromatic adaptation could well provide some insight into the mechanisms by which higher plants photoregulate gene expression.

2. HYBRID SHUTTLE PLASMIDS FOR USE WITH *F. DIPLOSIPHON*

Several years ago we established in our laboratory the first system for gene transfer into *F. diplosiphon* [6,7]. The DNA transferred was a hybrid plasmid which we had constructed from a plasmid occurring naturally in *F. diplosiphon* and a plasmid replicating in *E. coli*. The transfer of the hybrid plasmid into *F. diplosiphon* was achieved by triparental mating with *E. coli* essentially as described by Elhai and Wolk [8]. Such a mating involves mixing a cyanobacterial strain (the recipient) with two strains of *E. coli*, one of which carries the conjugal (mobilizing) plasmid, RP4, while the other contains the mobilizable plasmid intended for transfer to the cyanobacterium. Since the *E. coli* part of our hybrid plasmid carried a kanamycin (Km)/neomycin (Nm) resistance gene (originally from Tn903) the transconjugant cyanobacteria could be selected for on neomycin plates. The hybrid plasmid could be isolated from transconjugant *F. diplosiphon* and shuttled back into *E. coli* by transformation.

Using a hybrid shuttle plasmid as vector we have created a mobilizable library of *F. diplosiphon* genomic DNA which when mobilized into *F. diplosiphon* gives rise to merodiploids. Using this approach we have complemented mutations and have cloned genes involved in the regulation of PE expression and in the attachment of phycoerythrobilin to the PE β subunit (unpublished data). We have supplied our hybrid plasmid system to other laboratories where it has proved instrumental in the cloning of novel genes encoding components in the sensory pathway of chromatic adaptation in *F. diplosiphon* [9,10].

3. DNA RESTRICTION IN *F. DIPLOSIPHON*

3.1. Identity of Restriction Barriers in *F. diplosiphon*

Cyanobacterial restriction enzymes can dramatically decrease the frequency at which plasmids can be mobilized into cyanobacteria (see for example [11]). In *F. diplosiphon* two restriction enzymes have been identified [12]; *Fdi*I is an isoschizomer of *Ava*II (GGWCC), and *Fdi*II is an isoschizomer of *Fsp*I (TGCGCA). Initially, to avoid these two possible barriers to gene transfer we constructed hybrid plasmids which totally lacked restriction sites for the enzymes, *Fdi*I and *Fdi*II. This strategy was effective and resulted in mobilization frequencies in some experiments as high as 10^{-2} (fraction of potential recipient cells becoming transconjugant). However, during the course of our work we observed that when either the omega cassette (Sm^{R}Sp^R; [13]), or the *lacZ'* region from pUC 19 were incorporated into a hybrid shuttle plasmid the frequency of mobilization into *F. diplosiphon* decreased at least ten fold [7]. Mobilization frequences obtained with deletion derivatives of these constructs, and computer analysis of their DNA sequences suggested that the decrease in mobilization frequency might be caused by restriction at the palindrome GCATGC (*Sph*I). To test this hypothesis we took the hybrid plasmid, pJCF691 [7], which contains a single copy of GCATGC and removed this palindrome by cutting with *Sph*I, treating with T4 polymerase and religating. The removal of the palindrome resulted in, on average, an 8 to 9 fold increase in mobilization frequency (Table 1).

Table 1. Frequency of plasmid mobilization into *F. diplosiphon* SF33:
Effect of a single copy of the palindrome GCATGC

Plasmid	Number of GCATGC sequences	Frequency of transconjugants (GCATGC unmethylated in donor *E. coli*)		
		Expt. I	Expt. II	Expt. III
pJCF691	1	3.5×10^{-5}	3.5×10^{-5}	7.0×10^{-6}
pJCF69104	0	3.8×10^{-4}	3.2×10^{-4}	2.7×10^{-5}
Average ratio of frequencies (pJCf69101/pJCF691):		8.6		

We also tested to see if methylation of the palindrome GCATGC would protect plasmids from restriction in *F. diplosiphon*. M.*Nla*III methylates the sequence, CATG. When the gene for M.*Nla*III (cloned in pACYC184) was use to methylate DNA in the donor *E. coli,* the plasmid containing one copy of GCATGC (pJCF691) mobilized at close to the same frequency as the plasmid from which the palindrome GCATGC had been deleted (pJCF69104) (Table 2).

To confirm that the effects observed with pJCF691 were specific to the sequence GCATGC and not related to some other aspect of DNA structure we repeated the experiments using a different hybrid plasmid, pJCF626 [7]. Unlike pJCF691 (with its single copy of GCATGC in its omega cassette), pJCF626 has a single copy of GCATGC in a mutliple cloning site (originating from pUC19). The removal of the palindrome from pJCF626 again resulted in, on average, an 8 to 9 fold increase in mobilization frequency (data not shown). When the gene for M.*Nla*III was use to methylate DNA in the donor *E. coli,* pJCF626 mobilized at the same frequency as the isogenic plasmid from which the palindrome GCATGC had been deleted (data not shown). We believe that these results provide clear genetic evidence for the existence in *F. diplosiphon* of a third restriction system, which we are calling *Fdi*III.

3.2. Overcoming the Restriction Barriers in *F. diplosiphon*

Specific methylation of plasmids in donor strains of *E. coli* has proved effective in preventing restriction of plasmids mobilized into the cyanobacterium, *Anabaena* sp. PCC7120 [11]. A generally useful tool for genetic work with *F. diplosiphon* would be a plasmid which replicates in *E. coli* and which encodes enzymes which specifically methylate DNA at restriction sites for *Fdi*I, *Fdi*II and *Fdi*III. We have previously reported [7] the construction of a helper plasmid (pJCF17) which carries the gene for M.*Eco*47II (recognizes GGNCC). The expression of the gene for M.*Eco*47II on pJCF17

Table 2. Frequency of plasmid mobilization into *F. diplosiphon* SF33; M.*Nla*III expression in the donor *E. coli* decreases restriction at the palindrome GCATGC

Plasmid	Number of GCATGC sequences	Frequency of transconjugants (GCATGC methylated in donor *E. coli*)		
		Expt. I	Expt. II	Expt. III
pJCF691	1	9.4×10^{-5}	6.4×10^{-5}	1.8×10^{-5}
pJCF69104	0	2.2×10^{-4}	6.4×10^{-5}	1.7×10^{-5}
Average ratio of frequencies (pJCf69101/pJCF691):		1.4		

is at a level suffient to bring about full protection of pJCF17 from *in vitro* digestion with *Ava*II (the commercially available isochizomer of *Fdi*I). pJCF17 efficiently protects from restriction at *Fdi*I sites when hybrid plasmids are mobilized into *F. diplosiphon*.

To extend the usefulness of pJCF17 we have incorporated into this plasmid a gene which when expressed brings about specific methylation of *Fdi*II sites. This construction was accomplished in two steps. In the first step pJCF17 (Cm^{R}Ap^R) was cut with *Hin*dIII. (The smaller fragment produced is the same fragment as is released from pRL518 cut with *Hin*dIII; see Fig. 1 in [11]). The larger *Hin*dIII fragment was mixed with the 445 bp *Hae*II fragment from pUC19 which encodes *lacZ'* . After treatment with T4 polymerase and T4 ligase a plasmid was isolated (pJCF172) which is Cm^{S}Ap^R and which, when transformed into *E. coli lacZ*ΔM15 and plated in the presence of X-gal, causes colonies to be blue. pJCF172 has two unique restriction enzyme sites (*Hin*dIII and *Sma*I) in the multiple cloning site of *lacZ'*.

The second step involved the use of plasmid 6S (a gift from Marta Meda, New England Biolabs). Plasmid 6S is pBR322 carrying the gene for M.*Mst*I on a *Sau*3AI fragment inserted at the *Bam*HI site. *Mst*I is an isoschizomer of *Fdi*II. The gene for M.*Mst*I was excised from plasmid 6S with *Bam*HI and *Eco*RV, treated with T4 polymerase and ligated into the unique *Sma*I site in pJCF172. The resulting plasmid (pJCF173) when isolated is fully protected from *in vitro* digestion with either *Ava*II or *Fsp*I (the commercially available isochizomers of *Fdi*I and *Fdi*II).

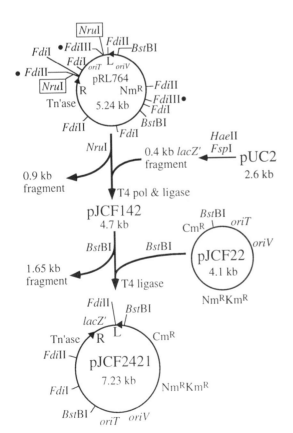

Figure 1. Construction of pJCF2421 which carries Tn5-2421. pUC2 was constructed from pUC7 by digestion with *Eco*RI and religation. pJCF22 is as previously described [7].

None of the hybrid plasmids we had previously constructed contained any *Fdi*II sites. To test if pJCF173 would protect from restriction at *Fdi*II sites in *F. diplosiphon* it was first necessary to create a hybrid plasmid which contained at least one *Fdi*II site. The transposon gamma delta (also called Tn1000) [14] contains a single *Fdi*II site. We therefore used the hybrid plasmid pJCF69101 ($Cm^R Km^R$; [7]) as the target for Tn1000 transposition and screened for clones in which Tn1000 had transposed from the *E.coli* fertility factor, F, into the chloramphenicol resistance gene of pJCF69101. Three plasmids giving the $Cm^S Km^R$ phenotype were obtained and each carried Tn1000 in the predicted location. These three plasmids could be mobilized into *F. diplosiphon* at a 20 to 40 fold higher frequency when pJCF173 was present to methylate both *Fdi*I and *Fdi*II sites as compared to when pJCF17 was present to methylate only *Fdi*I sites.

The general utility of pJCF173 is most convincingly demonstrated by the fact that it allows the mobilization into *F. diplosiphon* of shuttle plasmids possessing several sites for both both *Fdi*I and *Fdi*II. pJCF151 is a derivative of RSF1010 which contains four *Fdi*I sites, three *Fdi*II sites and no sites for *Fdi*III. Mobilization of pJCF151 from *E. coli* containing pJCF17 (protection at *Fdi*I sites but not *Fdi*II sites) did not enable us to detect even a single transconjugant colony of *F. diplosiphon*. From this experiment alone it was not possible to determine whether pJCF151 had been fully restricted at *Fdi*II sites, or whether RSF1010 and its derivatives could not replicate autonomously in *F. diplosiphon*. However, mobilization of pJCF151 from *E. coli* containing pJCF173 (protection at both *Fdi*I and *Fdi*II sites) gave rise to transconjugants at a reasonable frequency ($\geq 10^{-6}$). From such transconjugants pJCF151 can be recovered intact. These experiments serve to illustrate that pJCF173 is a powerful and general tool for the introduction of DNA into *F. diplosiphon*.

4. TRANSPOSON MUTAGENSIS IN *F. DIPLOSIPHON* USING Tn5 DERIVATIVES

Active transposable elements encode the enzymes and DNA sites (usually the element ends) needed for their own transposition [15]. Tn5 [16] has been extensively used in Gram-negative prokaryotes since it is well understood, transposes frequently and transposes randomly. In 1989 Borthakur and Haselkorn [17] showed that Tn5 transposes in the filamentous cyanobacterium, *Anabaena* sp. PCC7120. Recently Tn5 and its derivatives have been successfully used to generate mutants for the analysis of heterocyst differentiation in *Anabaena* sp. PCC7120 and in *Nostoc* sp. ATCC29133 (see for example [18–24]. Of particular utility have been those Tn5 derivatives which have been constructed to contain an origin of DNA replication (*oriV*) coming from an *E. coli* plasmid. When these Tn5 derivatives transpose they place both a selectable antibiotic resistance marker *and* an *oriV* for *E. coli* into the gene which has been insertionally inactived. In a cyanobacterial strain which has undergone mutation by such transposition the transposon and flanking sequences can be easily recovered by isolation of total DNA, restriction digestion, ligation and transformation back into *E. coli*. Thus the transposon not only serves as mutagenic agent but also as a simple means of gene identification. The demonstration of transposition in *F. diplosiphon* would open the door to an analysis of chromatic adaption by transposon mutagenesis.

Four discrete processes must successfully occur if random transposition is to be selected for in a cyanobacterium. First the transposon must enter the cyanobacterial cell with high efficiency. Secondly the transposon must escape degradation by DNA restriction systems. Thirdly the transposon must transpose efficiently and randomly to a cyanobacterial replicon where it will be replicated along with the cyanobacterial genome. Finally a selectable marker gene within the transposon must be expressed to allow for selection of those cells in which transposition has taken place.

A mobilizable suicide plasmid (pRL764) which carries a Tn5 derivative and which gives rise to transposition in *Anabaena* sp. PCC7120 was kindly given to us by P. Wolk and Y. Cai (Michigan State U.). pRL764 contains numerous sites for *Fdi*I, *Fdi*II and *Fdi*III (Fig. 1). Since protection against restriction at *Fdi*III sites by *in vivo* methylation with M.*Nla*III tends to reduce mobilization frequencies (by a mechanism as yet unknown; data not shown) we decided to reconstruct pRL764 to remove its three *Fdi*III sites (black dots in Fig. 1). This was accomplished in two discrete steps as detailed in Fig. 1. The resulting plasmid (pJCF2421; sequence available on request) lacks sites for *Fdi*III. The reconstructed transposon (Tn5-2421) carries within its outside sequences (L and R) the full sequence of plasmid pJCF22 which is already known to mobilize into *F. diplosiphon* at high frequency [7]. Further, it was also known to us that when pJCF22 is integrated into the chromosome of *F. diplosiphon* (by homologous recombination ; our data not shown) its Nm^{R}Km^R marker gene is sufficiently expressed to permit powerful selection of transconjugants with Nm. Thus we were confident that if pJCF2421 (methylated at *Fdi*I and *Fdi*II sites) was mobilized into *F. diplosiphon* the transposon would enter the cyanobacterial frequently, escape restriction and, if transposed into a cyanobacterial replicon, could be powerfully selected for. The only unknown was whether or not transposition would occur at a detectable frequency.

When pJCF2421 was mobilized into *F. diplosiphon* SF220 transconjugants arose which were capable of photoautotrophic growth in the presence of 10 µg/ml Nm. Under these conditions a frequency of 5×10^{-6} Nm^R colonies per potential recipient cell was obtained. Both methylation at *Fdi*I and *Fdi*II sites and removal of *Fdi*III sites from the vector were essential for obtaining Nm^R colonies. The acquisition of Nm^R was clearly a mutagenic process since 1 in 300 of the transconjugants was a color mutant.

5. ANALYSIS OF TWO COLOR MUTANTS GENERATED USING Tn5-2421

Mutant strains *F. diplosiphon* SF2421-1 and SF2421-2 have a wild-type phenotype when grown in red light but produce dramatically less phycoerythrin (PE) when grown in green light. *F. diplosiphon* SF2421-1 when grown in green light contains spectroscopically detectable PE whereas *F. diplosiphon* SF2421-1 does not. Genomic DNA was isolated from these strains and digested with *Bsr*GI (which does not cut in Tn5-2421). After ligation and electroporation into *E. coli* plasmids were recovered which were Cm^{R}Km^R and which contained the single predicted *Bsr*GI site. To identify the genes inactivated by transposition the DNA flanking the transposon in the recovered plasmids was sequenced using primers which hybridize to the transposon ends (to IS50L 5'AGACCAAAACGATCTCAAGAAGAT-CATC3' and to IS50R 5'GGAGGTCACATGGAATATCAGATTCTG3'). In both strains the 9 bp of DNA immediately flanking the transposon at one end is directly repeated in the DNA immediately flanking the other end of the transposon. This 9 bp direct repeat of target DNA is characteristic of Tn5 transposition [16].

In *F. diplosiphon* SF2421-1 the DNA flanking the transposon was identified using the BLAST algorithm [25] to be from the gene *cpeZ* (Fig. 2) and the exact position of transposition determined to be between the C and T of the triplet encoding Pro73. *cpeZ* is

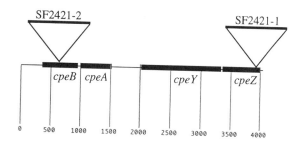

Figure 2. Map of part of the chromosome of *F. diplosiphon* showing the position of insertion of Tn5-2421 in mutant strains *F. diplosiphon* SF2421-1 and SF2421-2.

located downstream of the structural genes for the α and β subunits for PE (*cpeBA*) and is believed is to encode one of the subunits of a heterodimeric lyase which catalyses the attachment of phycoerythrobilin to the PE apoprotein [26]. Mutations in *cpeZ* have not previously been reported. The other subunit of this heterodimeric lyase is believed to be encoded by *cpeY* which is upstream of *cpeZ* (Fig. 2). Mutation in *cpeY* in *F. diplosiphon* causes an apparent decrease in PE to 50% of the amount found in the wild-type [26]. In our strain, mutated in *cpeZ*, there is an apparent decrease in PE to <10% of the amount found in the wild-type (Table 3). Why mutations in *cpeY* and *cpeZ* should result in different levels of apparent PE production is not clear and needs further investigation. Fluorescence emission spectroscopy of intact filaments of *F. diplosiphon* SF2421-1 reveals that on a molar basis the PE in this strain is >10 times more fluorescent than PE in the wild-type. This indicates that the PE present in the mutant strain assembles less efficiently into phycobilisomes than PE in the wild-type.

In *F. diplosiphon* SF2421-2 the DNA flanking the transposon was identified using the BLAST algorithm to be from the gene *cpeB* (Fig. 2, [3]) and the exact position of transposition determined to be immediately downstream of the triplet encoding Glu185. In this strain PE is undetectatable even by fluoresce emission spectroscopy. To our knowledge this is the first instance in *F. diplosiphon* where a mutation has been reported in a gene for a structural component of the phycobilisome.

6. CONCLUSION

Genes have been cloned from *F. diplosiphon* by isolating mutants and complementing the mutations using mobilizable libraries of genomic DNA. This approach has provided considerable insight into the mechanisms by which phycobiliprotein synthesis is regulated ([9,10] and our unpublished data). However, this approach is complex and has

Table 3. Phycobiliprotein ratios in strains of *F. diplosiphon* mutagenized with Tn5-2421 and grown in green light[a]

	Locus	PC/APC	PE/APC
F. diplosiphon SF220	—	1.1	3.0
F. diplosiphon SF2421-1	*cpeZ*[b]	2.5	0.2
F. diplosiphon SF2421-2	*cpeB*[c]	1.0	0.0

[a]Phycobiliprotein ratios are expressed on a molar basis. APC, allophycocyanin; PC, phycocyanin; PE, phycoerythrin.
[b]Between C and T in the triplet encoding Pro73.
[c]Immediately downstream of the triplet encoding Glu185.

limitations. When a mutation has a discernable phenotype this is usually the result of a loss of a function within the cell. To clone the wild-type allele of a mutated gene by complementation requires that the lost function be regained. A gene carried on a hybrid vector is present in the cell at a higher copy number than it would be were it carried in the chromosome. If to regain function requires that the product of a certain gene be present in the cell at a certain concentration then regain of function may not occur with gene being expressed from a high copy number vector. A second limitation of complementation is that dominant mutations cannot be complemented by their wild-type alleles.

We have shown in this paper that by taking a systematic approach it is possible to lower the restriction barriers in *F. diplosiphon* sufficiently to be able to select for transposition of Tn5 derivatives. We have also shown that mutations caused by transposition can be rapidly identified after recovery of the transposon as a plasmid in *E. coli*. Since with transposon mutagenesis the identity of an interrrupted gene can be determined without the necessity of complementation (i.e. the necessity of regaining function) we believe that transposon mutagenesis in *F. diplosiphon* will enable the cloning of a wider range of genes than that afforded by complementation. To date little is known of the mechanisms by which phycobiliprotein synthesis in *F. diplosiphon* is regulated by sulfur [27] or by nitrogen. Transposon mutagenesis with pJCF2421 should facilitate the rapid screening of new mutants for as yet unidentified genes involved in the regulation of phycobiliprotein synthesis, in phycobilisome assembly and in phycobilisome degradation.

ACKNOWLEDGMENTS

This research was supported by awards from Research Corporation and the University of San Francisco Faculty Development Fund. We thank Marta Meda, Rick Morgan and Geoff Wilson of New England Biolabs for donating the genes for the specific DNA methylases. Yuping Cai and Peter Wolk kindly provided pRL764. The DNA primers which hybidize to the ends of Tn5 were a gift from Yuping Cai. Finally, JGC thanks Prof. Frans Smulders (Veterinärmedizinische Universität, Wien) for providing accomodation and kind hospitality in Vienna.

REFERENCES

1. Tandeau de Marsac, N. and Houmard, J. (1993) FEMS Microbiol. Rev. 104, 119–189.
2. Grossman, A.R., Schaefer, M.R., Chiang, G.G. and Collier, J.L. (1993) Microbiol. Rev. 57, 725–749.
3. Mazel, D., Guglielmi, G., Houmard, J., Sidler, W., Bryant, D.A. and Tandeau de Marsac, N. (1986) Nucleic Acids Res. 14, 8279–8290.
4. Houmard, J. (1994) Microbiology 140, 433–441.
5. Sobczyk, A., Bely, A., Tandeau de Marsac, N. and Houmard, J. (1994) Mol. Microbiol. 13, 875–885.
6. Cobley, J.G., Zerweck, E. and Jaeger, H. (1987) Plant Physiol. (Bethesda) 83, 64 (suppl).
7. Cobley, J.G., Zerweck, E., Reyes, R., Mody, A., Seludo-Unson, J.R., Jaeger, H., Weerasuriya, S. and Navankasattusas, S. (1993) Plasmid 30, 90–105.
8. Elhai, J. and Wolk, C. P. (1988) in: Cyanobacteria (Packer, L., Glazer, A., Eds) Methods in Enzymology pp. 747–754, Academic Press, San Diego.
9. Chiang, G.G., Schaefer, M.R. and Grossman, A.R. (1992) Proc. Natl. Acad. Sci. U.S.A. 89, 9415–9419.
10. Kehoe, D.M. and Grossman, A.R. (1996) Science 273, 1409–1412.
11. Elhai, J., Vepritskiy, A., MuroPastor, A.M., Flores, E. and Wolk, C.P. (1997) J. Bacteriol. 179, 1998–2005.
12. van den Hondel, C.A.M., van Leen, R.W., van Arkel, G.A., Duyvestyn, M. and de Waard, A. (1983) FEMS Microbiol. Lett. 16, 7–12.
13. Prentki, P., Binda, A. and Epstein, A. (1991) Gene 103, 17–23.

14. Guyer, M. (1983) in: Recombinant DNA Part C (Wu, R., Grossman, L., Moldave, K., Eds) Methods in Enzymology vol. 101, pp. 363–369, Academic Press, San Diego.
15. Sherrat, D. (1991) Curr. Biol. 1, 192–194.
16. Berg, D. E. (1989) In: Mobile DNA (Berg, D. E., Howe, M. M., editors) pp. 185–210, Am.Soc.Microbiol., Washington.
17. Borthakur, D. and Haselkorn, R. (1989) J. Bacteriol. 171, 5759–5761.
18. Wolk, C.P., Cai, Y.P. and Panoff, J.M. (1991) Proc. Natl. Acad. Sci. U.S.A. 88, 5355–5359.
19. Liang, J., Scappino, L. and Haselkorn, R. (1992) Proc. Natl. Acad. Sci. U.S.A. 89, 5655–5659.
20. Wolk, C.P. (1991) Curr. Opin. Gen. Dev. 1, 336–341.
21. Fernández-Piñas, F., Leganes, F. and Wolk, C.P. (1994) J. Bacteriol. 176, 5277–5283.
22. Maldener, I., Fiedler, G., Ernst, A., Fernández-Piñas, F. and Wolk, C.P. (1994) J. Bacteriol. 176, 7543–7549.
23. Cohen, M.J., Wallis, J.G., Campbell, E.L. and Meeks, J.C. (1994) Microbiology 140, 3233–3240.
24. Khudyakov, I. and Wolk, C.P. (1996) J. Bacteriol. 178, 3572–3577.
25. Altschul, S.F., Gish, W., Miller, W., Myers, E.W. and Lipman, D.J. (1990) J. Mol. Biol. 215,
26. Kahn, K., Mazel, D., Houmard, J., Tandeau de Marsac, N. and Schaefer, M.R. (1997) J. Bacteriol. 179, 998–1006.
27. Mazel, D. and Marliere, P. (1989) Nature 341, 245–248.

53

THE ORGANISATION AND CONTROL OF CELL DIVISION GENES EXPRESSED DURING DIFFERENTIATION IN CYANOBACTERIA

Helen M. Doherty and David G. Adams

Department of Microbiology
University of Leeds
Leeds LS2 9JT, United Kingdom

1. INTRODUCTION

Cell division plays a key role in the differentiation of different cell and filament types in prokaryotes. Differentiation of bacterial endospores in *Bacillus subtilis* depends upon the formation of an asymmetric septum dividing the mother cell from the forespore compartment. In sporulating streptomycetes, colonies develop aerial hyphae with septa, which are distinct from the aseptate vegetative mycelium. In these cases, although the components of the cell division apparatus appear to be the same for the division of vegetative cells and of differentiating cells [1,2], the cells are able to regulate the timing and position of septum formation, and by doing so determine the daughter cells' fate. The atypical division is often the first visible indication that a cell or filament is differentiating, and provides the basis for further development.

Many filamentous cyanobacteria are able to form different cell and filament types with specialised functions relating to the ecology of the organism. It has been proposed that an asymmetric cell division plays a role in the differentiation of heterocysts [3]. However, it is in the formation of hormogonia that cell division is clearly part of the differentiation process.

1.1. Cell Division during Hormogonia Formation

Hormogonia are short motile filaments that occur in *Nostoc* and *Calothrix* and whose formation is triggered by a range of environmental conditions. These include changes in the composition of the media, changes in the spectral quality of the light, or the presence of exogenous compounds from the host plant in a symbiotic association [4]. In *Fremyella diplosiphon* (*Calothrix* PCC 7601) differentiation of hormogonia is triggered by

The Phototrophic Prokaryotes, edited by Peschek *et al.*
Kluwer Academic / Plenum Publishers, New York, 1999.

a change in the light quality from white to red light. Cultures grown under white light have a typical vegetative filament morphology; only a small number of cells are dividing at any one time and these are distributed throughout the filament (Figure 1). Between 4 and 5 hours after the transfer to red light whole filaments or domains of filaments respond by undergoing cell division. This occurs more or less synchronously in all cells, so that in the space of about 2 hours the cells will have divided at least once, compared to a normal cell division cycle of around 20 hours. The division takes place irrespective of the position of the cell in the cell cycle, which is possible because of the high copy number of the genome in these organisms. In *Fremyella diplosiphon* gas vesicles develop shortly after division and by 12 hours after the shift to red light the filaments have fragmented into smaller sections and have become motile. The hormogonia remain motile for around 20 hours, before returning to the vegetative filament morphology.

In addition to forming hormogonia, *Fremyella diplosiphon* is also able to undergo complementary chromatic adaptation (CCA) on transfer to red light. There is now evidence for the role of a phytochrome-like sensor protein and phosphorelay regulating CCA in *Fremyella* [5,6]. However, work by Campbell *et al.* [7] has shown that hormogonia formation may be regulated independently, using a separate mechanism dependent upon the redox status of components of the photosynthetic electron transport chain.

In studying the regulation of hormogonia formation in cyanobacteria we have focused on cell division as the earliest visible event in differentiation. We have started by asking a number of questions about cell division in prokaryotes in general. Do cyanobacteria share a common mechanism of cell division with other prokaryotes? The current model for cell division in prokaryotes has been developed in *Escherichia coli*—is this model true for cyanobacteria?

1.2. The Cell Division Model in *E. coli*

A number of essential cell division genes and their products have been identified in *E. coli*, of which the best characterised is *ftsZ*. FtsZ is a GTP-binding protein which has some homology to eukaryotic tubulin [8,9] and which is able to self-assemble into filaments under conditions favouring tubulin assembly [10,11]. In the non-dividing cell it is present in the cytoplasm, but as the cell approaches division, FtsZ aggregates in the form of a ring at the midpoint of the cell. It is proposed that constriction of the FtsZ ring brings about division, either

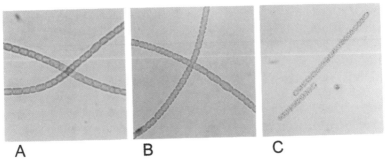

A   **B**   **C**

Figure 1. Differentiation of hormogonia in *Fremyella diplosiphon*. Cultures were grown to mid-logarithmic phase under white light in BG11+17.5 mM NaNO$_3$, then resuspended in fresh medium and transferred to red light. (A) 0 hours, (B) 4 hours (C) 24 hours after transfer to red light.

by pulling in the cell membrane and/or by directing the synthesis of peptidoglycan at the newly-forming septum [12,13]. Two other proteins, FtsA and ZipA, have also been shown to be localised to a ring at division and to interact with FtsZ [14–16]. FtsA is a peripheral membrane protein associated with the inner face of the cytoplasmic membrane [17]. ZipA was recently isolated on the basis of its ability to bind to FtsZ *in vitro*. ZipA is an integral membrane protein with a large cytoplasmic domain and it has been suggested that this protein anchors the FtsZ ring to the membrane [16]. The essential cell division genes *ftsI, ftsK, ftsL, ftsN, ftsQ* and *ftsW* also code for transmembrane proteins [18–23] and may form part of the cell division apparatus, although their interaction with the Z-ring has not been demonstrated. The product of *ftsI* is a penicillin-binding protein PBP3; the function of the other gene products is not yet known.

2. CHARACTERISATION OF CELL DIVISION GENES IN CYANOBACTERIA

We have cloned *ftsZ* from *Anabaena* sp. strain PCC 7120 using a probe derived by PCR with oligonucleotides based on conserved regions of known FtsZ amino acid sequences [24]. We have also cloned and sequenced the same region from *Fremyella diplosiphon* (Doherty and Adams, unpublished results). In both organisms *ftsZ* lies downstream of a small unidentified open reading frame (Orf1) and upstream of a convergently-transcribed cluster including the gene for glutathione synthetase (*gsh-II*). The *Synechocystis* PCC 6803 genome sequence has also revealed a single homologue of *ftsZ*, however, in this organism *ftsZ* is not clustered with known genes [25]. The three cyanobacterial genes are well conserved through most of the sequence, and each codes for a protein of about 420 amino acids from the first in-frame AUG codon. Figure 2 shows an alignment of the amino-terminal region of the deduced amino acid sequences, compared to the corresponding *E. coli* sequence. It can be seen that the cyanobacterial sequences each have a unique N-terminal extension of around 50 amino acids. This is not present in any of the other prokaryotic FtsZ sequences known, including those from the archaea. The only exception to this is an FtsZ homologue from the higher plant *Arabidopsis thaliana*, which is believed to function in chloroplast division. This also has an N-terminal extension, but this has been shown to function as a transit peptide targeting the FtsZ precursor to the chloroplast [26].

For each of the cyanobacterial genes the nucleotide sequence lacks a recognisable ribosome-binding site upstream of the first AUG codon. In the case of *Anabaena* PCC 7120 at least it is possible that an alternative GUG start codon is used, which is marked above the sequence.

In many of the organisms where it has been studied *ftsZ* is present in a gene cluster with other genes involved in cell division, in particular with *ftsQ* and *ftsA* [27]. In *E. coli* it has been proposed that co-transcription of *ftsA* with *ftsZ* is important to maintain the stoichiometry of the two gene products [28]. In both *Anabaena* 7120 and *Fremyella diplosiphon* we found a third open reading frame transcribed in the same direction as *ftsZ*, lying upstream of ORF1. The deduced proteins were homologous to each other and also to a hypothetical protein encoded by ORF slr1632 from *Synechocystis* PCC 6803 (Figure 3). In *Synechocystis 6803*, slr1632 lies directly upstream of *ftsZ* (Figure 4). The proteins are all of a similar size (approximately 280 amino acids) and show an overall homology to the cell division protein FtsQ. All of the proteins have the signal sequence and hydrophobic region expected of a protein with a single membrane spanning domain, which is the structure described for FtsQ.

```
                                                                         ⇓
Anabaena 7120    MTLDNNQELTYRN..SQSLQPGFSLAV.NS....SNPFNHSGLNFGQNNDSKKISVENNRIGEIVPGRVANIKVIG
F. diplosiphon   MTLDNNQGLTYKN..SQSVGQSGFSLAVNNS....TNPFNHSGLNFSQNNDGKKITAESNRIGEIVPGRVANIKVIG
Synechocystis    MTLNNDLPLNNIGFTGSGLNDGTEGLDDLFS....SSIVDNEPLEALVETPTFASPSPNLKRDQIVPSNIAKIKVIG
Chloroplast      MAIIPLAQLNEITISSSSSSFLTKSISSHSLHSSCICASSRISQFRGGFSKRRSDSTRSKSMRLRCSFSPMESARIKVIG
E. coli                                                                         MFEPMELTNDAVIKVIG


Anabaena 7120    VGGGGGNAVNRMIESDVSGVEFWSINTDAQALTLAGAPSRLQIGQKLTRGLGAGGNPAIGQKAAEESRDEIATALEGADL
F. diplosiphon   VGGGGGNAVNRMIESDVSGVEFWSINTDAQALTLAGAPSRLQIGQKLTRGLGAGGNPIAGQKAAEESRDEIATALEGADL
Synechocystis    VGGGGCNAVNRMIASGVTGIDFWAINTDSQALTNTNAPDCIQIGQKLTRGLGAGGNPAIGQKAAEESRDEIARSLEGTDL
Chloroplast      VGGGGNAVNRMISSGLQSVDFYAINTDSQALLQFSAENPLQIGELLTRGLGTGGNPLLGEQAAEESKDAIANALKGSDL
E. coli          VGGGGNAVEHMVRERIEGVEFFAVNTDAQALRKTAVGQTIQIGSGITKGLGAGANPEVGRNAADEDRDALRAALEGADM


Anabaena 7120    VVFITAGMGGGTGTGAAPIVAEVAKEMGALTVGVVTRPFVEGRRRTSQAEQGIEGLKSRVDTLIIIPNNKLLEVIPEQT
F. diplosiphon   VVFITAGMGGGTGTGAAPIVAEVAKEMGALTVGVVTRPFVEGRRRTSQAEQGIEGLKSRVDTLIIIPNNKLLEVIPEQT
Synechocystis    VVFITAGMGGGTGTGAAPIVAEVAKEMGCLTVGIVTRPFTFEGRRRAKQAEEGINALQSRVDTLIVIPNNQLLSVIPAET
Chloroplast      VVFITAGMGGGTGSGAAPVVAQISKDAGYLTVGVVTYPFSFEGRKRSLQALEAIEKLQKNVDTLIVIPNDRLLDIADEQT
E. coli          VVFIAAGMGGGTGTGAAPVVAEVAKDLGILTVAVVTKPFNFEGKKRMAFAEQGITELSKHVNSLITIPNDKLLKVLGRGI
```

Figure 2. Comparison of the amino terminal region of the deduced amino acids sequences of FtsZ from *Anabaena* PCC 7120, *Fremyella diplosiphon*, *Synechocystis* PCC 6803, *Arabidopsis thaliana* and *Escherichia coli*. Shaded residues are those identical in three or more sequences. The arrow marked above the sequence indicates the N-terminus of the deduced protein derived from a second potential start codon in *Anabaena* PCC 7120.

```
E. coli                  MSQAALNTRNSEEEVSSRRNNGTRLAGILFLLTVLTTVLVSG--
Anabaena 7120            MPGIASVSQKELAQRR----KKLRRQRQMKIIQAIWQ-TIAISGLAGGLL
Fremyella diplosiphon   MADLVSLSRADLAQRR----KRLRRKQQMKIIQAIWR-SFAITSLAGCLL
Synechocystis 6803       MTDLVVSDSLKNRR-EQLMWQRRWKRLRSC---WQ-FVCVCGLTGGMV

E.coli                   WV------VLGWMEDAQRLPLSKLVLTGERHYTRNDDIRQSILALGEPGT
Anabaena 7120            WVALQPIWVLKTPEQVLMKSDNQL-LSQEAIKSLLVLSYPQSLWRIIQPA
Fremyella diplosiphon   WLAIQPVWVLNTPKQIVMKSGNEL-LPEETVKSLLVLSYPQSLWRI-QPS
Synechocystis 6803       WVMSWPEWSIRSDRQV-----NKL-VSRETLYEDLDLEYPQAVWQL-STQ

E.coli                   FMTQDVNIIQTQIEQRLPWIKQVSV-RKQWPDELKIIHLVEYVPIA-RWN
Anabaena 7120            AIADSLK--------KQPTIAQATVNRRLFPPGLII-EIEERIPVAVAQR
Fremyella diplosiphon   AIAESLK--------RQPTIAQASVSRRLFPPGLIV-DITERVPVAIVQT
Synechocystis 6803       ALGDELA--------KNPALLRVEVTRQLFPAQVNV-AVQERQPVAIAVA

E.coli                   DQHMVDAEGNTFSVPPERTSKQVLPMLYGPEPSANEV--LQGYREMGQML
Anabaena 7120            RREQSNSTSNKQTHTGLI-ANGVWDSSGKIYTNVNPQFKLPTLKVIGLPE
Fremyella diplosiphon   PRSGNSEAANKQASVGLIDASGVWIPLEKTY---NPTRKLPSLKVIGLPE
Synechocystis 6803       DQGPGYLDGEGNYIPASLYSQAVRKTLPQTPQF-----------LGYGP

E.coli                   LKDRFTLK-------EAAMTARRSWQLTNNDI-KLNLGRGDTMKRLARFE
Anabaena 7120            QYAPYWSKLYPYISQSSIKITEIDYQDPNNLILKTELGTVYLGATSAQSS
Fremyella diplosiphon   QYLPYWNQLYLSINQSLVKIQEINCQDPTNLILKTELGNVYLGPPSPQLS
Synechocystis 6803       QYRSFWQTHQILIQQSPVNIRIINGNNPSNISLTTDLGLVFIGSDLSRFG

E.coli                   LVYPVLQQQAQTDGK----RISYVDLRYDSGAAVGWAPLPPEESTQQQNQ
Anabaena 7120            STTNLLAQLSHINTKLNPSEIDYIDLKNPESPLVHMIQNKETIKNQTP*
Fremyella diplosiphon   EQIKTLAQMRHLSSKLNSSQIEYIDLQNPEAPLVQLNQKNQKINVRNP*
Synechocystis 6803       QQVQVLEKMQNLPSRVPKERLLFIDLTNPDSPSIQLRPQPPKEKAAVNKP
```

Figure 3. Alignment of the deduced amino acid sequences of FtsQ homologues from *Anabaena* PCC 7120, *Fremyella diplosiphon* and *Synechocystis* PCC 6803 (slr1632), compared to FtsQ from *Escherichia coli*. Shaded residues are those identical in three or more of the sequences. The region of hydrophobic residues which may act as a membrane-spanning segment is underlined.

No homologue of *ftsA* has yet been located in association with *ftsZ* in *Anabaena* 7120 or *Fremyella diplosiphon*. Analysis of the *Synechocystis* 6803 genome sequence and comparison with known FtsA sequences has failed to reveal a homologue of FtsA in this organism. It is possible therefore that cyanobacteria lack this gene and that the components of the cell division apparatus in cyanobacteria differ from those proposed in the *E. coli* cell division model.

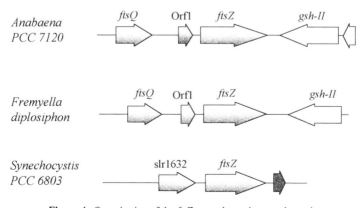

Figure 4. Organisation of the *ftsZ* gene cluster in cyanobacteria.

3. EXPRESSION OF CELL DIVISION GENES IN CYANOBACTERIA

Using *Anabaena* 7120 *ftsZ* as a probe in Northern hybridisations, three major transcripts could be seen in RNA from *Anabaena* 7120 cultures grown in BG11+17.5 mM $NaNO_3$ at the mid-log phase of growth (Figure 5A). The transcripts had sizes of approximately 1.8, 1.6 and 1.4 kb. Further Northern hybridisations with probes from within and upstream of *ftsZ* indicate that the transcripts vary in the position of their 5′ ends (Figure 5B). Preliminary mapping experiments suggest that these lie upstream of Orf1 and *ftsZ*, and within the N-terminal region of the FtsZ coding sequence (Doherty and Adams, unpublished results) .

We have investigated *ftsZ* expression in *Fremyella diplosiphon* during the initial stages of hormogonia formation. As was the case for *Anabaena* 7120, the *ftsZ* probe hybridised to a number of transcripts (Figure 6). All of the transcripts increased in abundance transiently between 1–4 hours after the shift to red light, preceding the peak in cell division which occurred 4–5 hours after the transfer to red light. This implies that the burst of cell division seen during hormogonia formation depends on a substantial increase in the amount of *ftsZ* transcript and on an increase in the level of FtsZ protein in the cell. Cell-cycle dependent transcription of *ftsZ* in *E. coli* suggests that although FtsZ protein is present throughout the cell cycle, the amount of FtsZ could be a rate-limiting factor in division [29].

Northern hybridisations of *Fremyella diplosiphon* RNA using the *Fremyella ftsQ* gene as a probe revealed only very low levels of transcript hybridising to the probe, and we were unable to clearly show full-length transcripts for *ftsQ*. This suggests that *ftsQ* transcripts are less stable than we have shown with *ftsZ*, and that since the two gene probes did not hybridise to a common RNA species on Northern blots, the genes are not likely to be co-transcribed.

It has been shown by Damerval *et al.* [30] that the synthesis of gas vesicles during hormogonia formation in *Calothrix* is preceded by increased transcription of the *gvp* operon. Another early event is the formation of pili or fimbriae on the surface of the hormogonia cells. The role that pili play in hormogonia function is not known; they may have a role in motility,

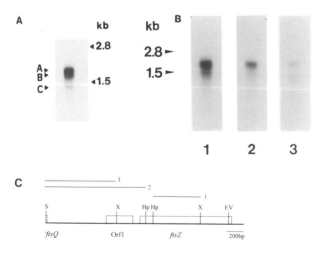

Figure 5. (A) Northern hybridisation of *Anabaena* PCC 7120 RNA probed with an *Xba*I-*Hpa*I fragment internal to *Anabaena ftsZ*. The three major transcripts are marked. (B) *Anabaena* PCC 7120 RNA hybridising to probes 1, 2, 3. (C) Restriction map of the region of pOE1 [24] containing *ftsZ* and Orf1, showing the derivation of probes used in Northern hybridisations. Abbreviations used are: S, *Sau*3AI; X, *Xba*I; Hp, *Hpa*I; Ev, *Eco*RV.

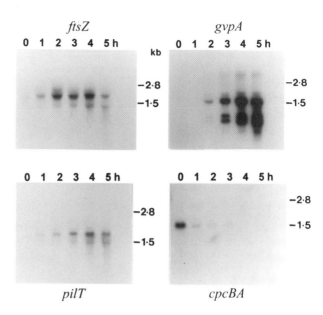

Figure 6. Northern hybridisations of *Fremyella diplosiphon* RNA. Cultures grown in BG11+17.5 mM NaNO₃ under white light were resuspended in fresh culture medium and transferred to red light at the start of the experiment. RNA was extracted from cells harvested 0–5 hours after transfer to red light. Equal amounts (5 µg) of RNA were electrophoresed under denaturing conditions and hybridised with *Anabaena* PCC 7120 *ftsZ* (*Xba*I-*Hpa*I fragment), *Anabaena flos-aquae gvpA* (contained on an *Xba*I fragment of pGVP17, a gift of P. Hayes), *Fremyella diplosiphon pilT* [31], and *Synechococcus* PCC 7002 *cpcBA* (contained on a *Hinc*II fragment of pAQPRI [33]).

cell-cell communication, or affect surface properties such as hydrophobicity. We were interested to know how the expression of *ftsZ* correlates with that of other genes involved in hormogonia formation, for example in the synthesis of gas vesicles and pili. Figure 6 shows the expression of *ftsZ* compared to that of the *gvp* operon, and to *pilT* which codes for a protein believed to be involved in pilus function. The increase in *ftsZ* transcript abundance appears to precede the increase in *gvp* transcription. Although an increase in *gvp* transcripts can be seen as little as 1–2 hours after the transfer to red light, the peak in expression occurs later than for ftsZ. Transcripts hybridising to *pilT* also increase in abundance within the first few hours of hormogonia formation, again with the peak of expression occurring slightly later than for *ftsZ*. The *pilT* probe was derived from a random sequencing project in *Fremyella diplosiphon* [31] and was identified as a homologue of *Pseudomonas aeruginosa pilT*, which codes for a protein involved in the retraction of pili and in 'twitching motility' [32].

Damerval *et al.* [30] have also shown that expression of phycobiliprotein genes is repressed during the early stages of hormogonia differentiation, although once the mature hormogonia are formed transcription of the *cpcBA* genes resumes and a second, red light-inducible phycocyanin is expressed. Figure 6 shows that transcripts hybridising to a heterologous *cpcBA* probe from *Synechococcus* 7002 disappear in the first hour following transfer to red light.

4. SUMMARY

Cyanobacteria contain homologues of the key cell division genes *ftsZ* and *ftsQ*, although genes encoding other components of the *E. coli* cell division apparatus such as FtsA may be absent. An increase in *ftsZ* expression precedes cell division during hormogonia formation and is one of the earliest events in differentiation. *ftsZ* is one of a number of genes that are expressed during the differentiation of hormogonia in *Fremyella diplosiphon* and *ftsZ* expression forms part of a co-ordinated program of gene expression beginning prior to the first visible event which is the actual division itself.

REFERENCES

1. Beall, B., Lowe, M. and Lutkenhaus, J.(1988) Cloning and characterisation *of Bacillus subtilis* homologues of *Escherichia coli* cell division genes *ftsZ* and *ftsA*. J. Bacteriol. 170, 4855–4864.
2. McCormick, J.R., Su, E.P., Driks, A. and Losick, R. (1994) Growth and viability of a *Streptomyces coelicolor* mutant for the cell division gene *ftsZ*. Mol. Microbiol. 14, 243–254.
3. Adams, D.G. (1992) Multicellularity in cyanobacteria. In Mohan, S.,. Dow, C and Cole, J.A. (Eds.) Prokaryotic Structure and Function. Society for General Microbiology Symposium Volume 47, Cambridge University Press, Cambridge UK. pp. 341–384.
4. Herdman, M., and Rippka, R. (1988) Cellular Differentiation: Hormogonia and Baeocytes. In Packer, L. and Glazer, A.N. (Eds.) Methods in Enzymology Vol. 167. Academic Press. pp232–242.
5. Kehoe, D.M. and Grossman, A.R. (1996) Similarity of a chromatic adaptation sensor to phytochrome and ethylene receptors. Science 273, 1409–1412.
6. Kehoe, D.M. and Grossman, A.R. (1997) New classes of mutants in complementary chromatic adaptation provide evidence for a novel four-step phosphorelay system. J. Bacteriol. 179, 3914–3921.
7. Campbell, D., Houmard, J. and Tandeau de Marsac, N. (1993) Electron transport regulates cellular differentiation in the filamentous cyanobacterium *Calothrix*. Plant Cell 5, 451–463.
8. RayChaudhari, D. and Park, J.T. (1992) *Escherichia coli* cell-division gene *ftsZ* encodes a novel GTP-binding protein. Nature 359, 251–254.
9. De Boer, P.A.J., Crossley, R.E. and Rothfield, L.I. (1992) The essential bacterial cell-division protein FtsZ is a GTPase. Nature 359, 254–256.
10. Mukherjee, A. and Lutkenhaus, J. (1994) Guanine nucleotide-dependent assembly of FtsZ into filaments. J. Bacteriol.176, 2754–2758.
11. Bramhill, D. and Thompson, C.C. (1994) GTP-dependent polymerisation of *Escherichia coli* FtsZ protein to form microtubules. Proc. Natl. Acad. Sci. USA. 91, 5813–5817.
12. Bi, E. and Lutkenhaus, J. (1991) FtsZ ring structure associated with division in *Escherichia coli*. Nature 354, 4855–4864.
13. Rothfield, L.I., Justice, S.S. (1997) Bacterial cell division: the cycle of the ring. Cell 88, 581–584.
14. Addinall, S.G. and Lutkenhaus, J. (1996) FtsA is localized to the septum in an FtsZ-dependent manner. J. Bacteriol. 178, 7167 -7172.
15. Wang, X, Huang, J, Mukherjee, A., Cao, C., Lutkenhaus, J. (1997) Analysis of the interaction of FtsZ with itself, GTP, and FtsA. J. Bacteriol. 179, 5551–5559.
16. Hale, C.A. and de Boer, P.A.J. (1997) Direct binding of FtsZ to ZipA, an essential component of the septal ring structure that mediates cell division in *E. coli*. Cell 88, 175–185.
17. Pla, J., Dopazo, A. and Vicente, M. (1990) The native form of FtsA, a septal protein of *Escherichia coli*, is located in the cytoplasmic membrane. J. Bacteriol. 173, 5097–5102.
18. Begg, K.J., Dewar, S.J. and Donachie, W.D. (1995) A new Escherichia coli cell division gene, ftsK. J. Bacteriol. 177, 6211–6222
19. Bowler, L.D. and Spratt, B.G. (1989) Membrane topology of penicillin-binding protein 3 of *Escherichia coli*. Mol. Microbiol. 3, 1277–1286.
20. Carson, M.J., Barondess, J.J., Beckwith, J. (1991) The FtsQ protein *of Escherichia coli*: membrane topology, abundance and cell division phenotypes due to overproduction and insertion mutations. J. Bacteriol. 173, 2187–2195.
21. Guzman, L.-M., Barondess, J.J. and Beckwith, J. (1992) FtsL, an essential cytoplasmic membrane protein involved in cell division in *Escherichia coli*. J. Bacteriol. 174, 7716–7728.
22. Ikeda, M., Sato, T., Wachi, W., Jung, H.K., Ishino, F., Kobayashi, Y. and Matsuhashi, M. (1989) Structural similarity among *Escherichia coli* FtsW and RodA proteins and *Bacillus subtilis* SpoVE protein which function in cell division, cell elongation, and spore formation, respectively. J. Bacteriol. 171, 6375–6378.
23. Dai, K., Xu, Y. and Lutkenhaus, J. (1996) Topological characterisation of the essential *Escherichia coli* cell division protein FtsN. J.Bacteriol. 178, 1328–1334.
24. Doherty H.M. and Adams, D.G. (1995) Cloning and sequence of *ftsZ* and flanking regions from the cyanobacterium *Anabaena* PCC 7120. Gene 163, 93–96.
25. Kaneko, T., Sato, S., Kotani, H., Tanaka, A., Asamizu, E., Nakamura, Y., Miyajima, N., Hirosawa, M., Sugiura, M., Sasamoto, S., Kimura, T., Hosouchi, T., Matsuno, A., Muraki, A., Nakazaki, N., Naruo, K., Okumura, S., Shimpo, S., Takeuchi, C., Wada, T., Watanabe, A., Yamada, M., Yasuda., M., Tabata, S. (1996) Sequence analysis of the the genome of the unicellular cyanobacterium *Synechocystis* sp. strain PCC6803. II. Sequence determinationof the entire genome and assignment of potential protein-coding regions. DNA Res. 3, 109–136.

26. Osteryoung, K.W. and Vierling, E. (1995). Conserved cell and organelle division. Nature, 376, 473–474.

27. Pucci, M.J., Thanassi, J.A., Discotto, L., Kessler, R.E. and Dougherty, T.J. (1997) Identification and characterization of cell wall-cell division gene clusters in pathogenic gram-positive cocci. J. Bacteriol. 179, 5632–5635.

28. Dai, K., and Lutkenhaus, J. (1992) The proper ratio of FtsZ to FtsA is required for cell division to occur in *Escherichia coli*. J. Bacteriol. 174, 6145–6151.

29. Garrido, T. Sanchez, M., Palacios, P., Aldea, M. and Vicente, M. (1993) Transcription of *ftsZ* oscillates during the cell cycle of *Escherichia coli*. EMBO J. 12, 3957–3965.

30. Damerval. T, Gugliemi, G., Houmard, J. and Tandeau de Marsac, N. (1991) Hormogonium differentiation in the cyanobacterium *Calothrix*: a photoregulated developmental process. Plant Cell 3, 191–201.

31. Gupta, A, Robinson, N.J. and Robinson, P.J. (1995) Tabulation of thirty-one putative new genes from cyanobacteria. Plant Mol. Biol. 29, 617–620.

32. Whitchurch, C.B, Hobbs, M., Livingstone, S.P., Krishnapillai, V. and Mattick, J.S. (1990) Characterisation of a *Pseudomonas aeruginosa* twitching motility gene and evidence for a specialised protein export system widespread in eubacteria. Gene 101, 33–44.

33. De Lorimer, R., Bryant, D.A., Porter, R.D., Lui, W.Y., Jay, E.and Stevens, S.E. Jr. (1984) Genes for the α and β genes of phycocyanin. Proc. Natl. Acad. Sci. USA. 81, 7946–7950.

54

CYANOBACTERIAL NITROGEN ASSIMILATION GENES AND NtcA-DEPENDENT CONTROL OF GENE EXPRESSION

Enrique Flores, Alicia M. Muro-Pastor, and Antonia Herrero

Instituto de Bioquímica Vegetal y Fotosíntesis
CSIC-Universidad de Sevilla
Centro de Investigaciones Científicas Isla de la Cartuja
Avda. Américo Vespucio s/n, E-41092 Sevilla, Spain

1. INTRODUCTION

1.1. Nitrogen Sources

The sources of nitrogen more commonly used for the growth of cyanobacteria include ammonium, nitrate, and dinitrogen. Some cyanobacteria are also able to use urea and certain amino acids, but current knowledge on the mechanisms of assimilation of urea and amino acids is rudimentary, especially with regard to the genetic aspects. The presence of ammonium in the culture medium determines repression of the proteins that constitute the assimilatory pathways for alternative nitrogen sources so that cyanobacteria generally assimilate ammonium with preference over other nitrogen sources. In some filamentous cyanobacteria that express the nitrogen fixation machinery in specialized cells called heterocysts, the development of these differentiated cells is also repressed by ammonium. In this article, we shall first summarize our current knowledge, at the genetic level, of the pathways of assimilation of ammonium, nitrate (and nitrite) and dinitrogen in the cyanobacteria and shall then describe what is known about the molecular mechanism underlying repression by ammonium in these organisms. No detailed description of the process of heterocyst development will be presented since this topic is covered in another chapter of this book.

1.2. Cyanobacterial Gene Promoters

The structure of the cyanobacterial RNA polymerase has been reviewed by J. Houmard [1]. The major form of RNA polymerase isolated from vegetative cells of the heterocyst-forming cyanobacterium *Anabaena* sp. PCC 7120 contains a σ factor, SigA, that is homologous to the *Escherichia coli* RNA polymerase σ^{70} factor encoded by the *rpoD* gene [2,3].

The Phototrophic Prokaryotes, edited by Peschek *et al.*
Kluwer Academic / Plenum Publishers, New York, 1999.

Multiple genes encoding RpoD-related polypeptides are present in the genomes of *Anabaena* sp. PCC 7120 [4], *Synechococcus* sp. PCC 7942 [5] and PCC 7002 [6], and *Synechocystis* spp. [7,8]. The cyanobacterial RNA polymerase-SigA holoenzyme can recognize promoters bearing a sequence similar to that of the canonical *E. coli* σ^{70} promoters [9], composed of two hexamers, the −35 (consensus sequence: TTGACA) and −10 (consensus sequence: TA-TAAT) boxes, that are separated by ca. 17 nucleotides. Indeed, some cyanobacterial house-keeping genes are transcribed from promoters that bear this structure [9,10]. Nonetheless, the putative transcription start point(s) (*tsp*) of some other cyanobacterial genes either are not preceded by any sequences that could be recognized as those of a SigA-dependent promoter or are preceded only by a −10 box, normally in the form TAN$_3$T [11,12]. This implies that some other transcription factor(s) should be operative in the cyanobacteria.

2. AMMONIUM UPTAKE

Ammonium transport has been characterized in a number of cyanobacteria using [^{14}C]methylammonium as a probe. The cyanobacteria investigated include the unicellular strains *Synechococcus* sp. PCC 7942 [13] and *Synechocystis* sp. PCC 6803 [14], and the het-erocyst-forming strains *Anabaena variabilis* ATCC 29413 and *Anabaena azollae* [15]. The ammonium/methylammonium permease seems to mediate a membrane potential-driven transport process that is active in spheroplasts [13], suggesting that it is a monocomponent permease rather than a binding protein- and ATP-dependent transporter. The ammo-nium/methylammonium transport activity is repressed by growth of the cells in ammonium-containing medium [16] and, therefore, appears not to be required for assimilation of ammonium when this nutrient is available to the cells at relatively high concentrations (for instance, in the order of 1 mM). Under these conditions, diffusion of ammonia through the cytoplasmic membrane [17] is high enough to support growth.

The recently published complete genomic sequence of *Synechocystis* sp. PCC 6803 contains three ORFs whose protein products would be homologous to the ammonium/methylammonium permeases of *Arabidopsis thaliana*, *Saccharomyces cerevisiae*, and *Co-rynebacterium glutamicum*. Mutagenesis of these ORFs has shown that one of them (*sll0108*, which we shall call *amt1*) is responsible for more than 95% of the [^{14}C]methylam-monium uptake activity of strain PCC 6803 whereas the other two (*sll1017*, *amt2*; and *sll0537*, *amt3*) seem to make a minor contribution to methylammonium uptake [14]. The *Synechocystis* Amt proteins are highly hydrophobic polypeptides that bear 12 putative mem-brane spanning regions and, as stated above, can represent monocomponent permeases.

Expression of the three *amt* genes is higher in nitrogen-starved cells than in cells in-cubated in the presence of a source of nitrogen (ammonium or nitrate), but *amt1* is ex-pressed at much higher levels than the other two *amt* genes [14]. The fact that *amt1* is expressed at highest levels under nitrogen stress would be consistent with a role in uptake of ammonium in natural habitats where this nutrient might be found at very low concen-trations. DNA/DNA hybridization analysis has shown that *amt1* is the only *amt* gene that is highly conserved and widespread among the cyanobacteria [14].

3. NITRATE ASSIMILATION

3.1. Nitrate Transport

Cyanobacteria are able to take up low concentrations of nitrate in a process that re-quires metabolic energy in the form of ATP (see [18]). The K_s of the cells for nitrate is about 1 μM, and operation of a nitrate transporter builds up an intracellular nitrate concentration that allows the nitrate assimilation system to work. The cyanobacterial nitrate transporter is of the ABC type, and some of the *nrt* genes encoding this system, first described in *Synechococcus* sp. PCC 7942 [19], have now been identified in some other cyanobacteria as well [8,20–23]. The Nrt system is able to transport not only nitrate but also nitrite [24,25]. The *nrtA* gene encodes a membrane-anchored periplasmic binding protein [25], the previously known ammonium-repressed, 48-kDa component of the nitrate transporter [26]. The *nrtB* gene would encode the membrane-spanning protein component of the system, that might act as a dimer. The *nrtD* gene would encode an ATPase for the system, the so called "conserved component". Finally, the *nrtC* gene would encode a polypeptide with an N-terminal half homologous to NrtD, that could represent a second ATPase subunit for the system, and a C-terminal half homologous to NrtA.

Short-term inhibition by ammonium of nitrate uptake appears to be a general phenomenon in cyanobacteria, and ammonium has been shown to inhibit the activity of the nitrate/nitrite permease [18]. In *Synechococcus* sp. PCC 7942, it has recently been observed that the inhibition by ammonium of nitrate and nitrite uptake is not operative in a mutant of the *glnB* gene [27]. This gene encodes the P_{II} protein which is phosphorylated in a serine residue in response to ammonium deprivation [28]. It follows that the non-phosphorylated form of the P_{II} protein is required for the ammonium-promoted inhibition of the nitrate/nitrite permease. Expression of the *glnB* gene has recently been found to be depressed by ammonium not only in strain PCC 7942 [29] but also in *Synechocystis* sp. PCC 6803 [30].

3.2. Nitrate Reduction

Once inside the cells, nitrate is reduced to ammonium. This takes place in two sequential steps: the 2-electron reduction of nitrate to nitrite catalyzed by nitrate reductase and the 6-electron reduction of nitrite to ammonium catalyzed by nitrite reductase. Cyanobacterial nitrate reductase uses reduced ferredoxin as an electron donor, bears a molybdenum cofactor and probably also an S-Fe cluster, and shows a K_m for nitrate in the order of 1 mM (see [18,31]). The *narB* gene encoding the nitrate reductase apoenzyme has been cloned from several cyanobacteria including *Synechococcus* sp. PCC 7942 [31,32], *Anabaena* sp. PCC 7120 [22], and *Oscillatoria chalybea* [33]. This gene would encode a polypeptide of about 730 amino acid residues (the exact number of residues varies depending on the cyanobacterial strain) homologous to bacterial oxidoreductases that bear a guanine molybdopterin dinucleotide-type of molybdenum cofactor.

A cluster of genes involved in molybdopterin biosynthesis whose mutation prevents expression of nitrate reductase activity has recently been characterized in *Synechococcus* sp. PCC 7942. This gene cluster corresponds to the previously described *narA* locus [32] and consists of *moeA* and, in the opposite direction, an operon that includes *moaCDEA* [34].

3.3. Nitrite Reduction

Cyanobacterial nitrite reductase, which is homologous to higher-plant nitrite reductases, also uses reduced ferredoxin as an electron donor, bears siroheme and an S-Fe cluster as prosthetic groups, and shows a K_m for nitrite in the order of 0.1 mM (see [18]). The *nir* gene encoding nitrite reductase has been cloned from several cyanobacteria including *Synechococcus* sp. PCC 7942 [35,36], *Anabaena* sp. PCC 7120 [22,23], *Phormidium laminosum* [37], and *Plectonema boryanum* [21]. The *Plectonema* polypeptide is unique in that it contains a C-terminal extension homologous to 2Fe-2S ferredoxins.

3.4. Nitrate Assimilation Genes: Operon Structure and Regulation

In *Synechococcus* sp. PCC 7942, the *nir*, *nrt*, and *narB* genes are clustered together and constitute an operon with the structure: *nir-nrtABCD-narB* [19,36,38]. This operon structure is also found in *Anabaena* sp. PCC 7120 [22,23], but not in other cyanobacteria like *Synechocystis* sp. PCC 6803, where a gene cluster with the structure *nrtABCD-narB* is found in map position 1010 to 1002 kb, approximately, of the chromosome, whereas the *nir* gene is located in map position 2770 kb close to a cluster of molybdopterin biosynthesis genes [8]. In *Phormidium laminosum*, only the *nir*, *nrtA*, *nrtB* and *nrtC* genes are found clustered together [20].

Consistent with classic physiological data concerning the expression of nitrogen assimilation enzymes (see [18]), mRNA levels of the *nir* operon are depressed upon growth of the cells in ammonium-supplemented media [12,20,21,23,36]. Although in most cyanobacterial strains expression of the *nir* operon takes place in media lacking any source of nitrogen (see also [22]), higher levels of the *nir* operon mRNA are usually found in cells incubated in the presence of nitrate [12,23]. In *Synechococcus* sp. PCC 7942, *ntcB*, a gene encoding a protein homologous to LysR, has been implicated in this positive effect of nitrate [39]. The *ntcB* gene is found, along with a gene called *nirB* that appears to be necessary to attain maximum levels of nitrite reductase, in the complementary strand, upstream from the *nir* operon [40]. The expression of both *nirB* and *ntcB* is also repressed by ammonium. On the other hand, in *Phormidium laminosum*, high levels of mRNA of some nitrate assimilation genes have only been observed after nitrogen starvation of the cells [20].

4. NITROGEN FIXATION

4.1. Nitrogenase

Reduction of molecular nitrogen to two molecules of ammonium is catalyzed by the nitrogenase complex which is homologous in the different bacteria that are able to carry out this process. The nitrogenase complex contains nitrogenase reductase (a homodimer of NifH) and nitrogenase (composed of two NifD and two NifK polypeptides), and bears a 4Fe-4S cluster (carried in nitrogenase reductase), an Fe-S cluster known as "P-cluster" [41], and the Fe-Mo-cofactor (carried in nitrogenase) that in some nitrogenases can be substituted by a cofactor containing vanadium instead of molybdenum.

Nitrogenase is rapidly and irreversibly inactivated in the presence of oxygen. Since the cyanobacteria perform oxygen-evolving photosynthesis, this poses a restriction to nitrogen fixation in these organisms that, in many cyanobacteria, has been solved by separating the two processes either in time or in space [18]. Thus, some cyanobacteria perform nitrogen fixation in the dark, when photosynthesis is not operative [42], whereas some filamentous strains are able to develop heterocysts, differentiated cells specialized in nitrogen fixation that do not perform oxygenic photosynthesis [43]. Other cyanobacteria like *Trichodesmium* sp. do not separate the two processes and must have mechanisms for protection of nitrogenase that are yet to be defined [44]. Still some other cyanobacteria perform nitrogen fixation only under anaerobic or "microaerobic" conditions [42].

4.2. Nitrogen Fixation Genes

The cyanobacterial *nifH*, *nifD*, and *nifK* genes encoding a conventional Mo-nitrogenase were first cloned and characterized from *Anabaena* sp. PCC 7120 in the laboratory of

R. Haselkorn [18,43,45]. These genes have more recently been characterized in some other cyanobacteria including heterocyst-forming *Fischerella* sp. [46] and unicellular strains *Cyanothece* sp. ATCC 51142 [47] and *Synechococcus* sp. RF-1 [48]. Production of an active nitrogenase requires the products of a number of other *nif* genes as well. In the chromosome of the heterocyst of strain PCC 7120, a cluster of nitrogen fixation genes is found that shows the following structure: *nifB-fdxN-nifS-nifU-nifH-nifD-nifK-nifE-nifN-nifX*-ORF3-*nifW*-ORF1-ORF2-*fdxH* [43,45]. (In the vegetative cell chromosome, the *fdxN* and *nifD* genes are interrupted by DNA elements of 55 and 11 kb, respectively, that are excised during hetero-cyst development [49].) The *nif* genes are named after their homologues from other diazot-rophs, and *fdxH* and *fdxN* encode a ferredoxin that acts as an electron donor to nitrogenase and a ferredoxin-like protein of unknown function, respectively. To date, four transcriptional units have been reported in this gene cluster: *nifB-fdxN-nifSU* [50], *nifHDK* [51], ORF1-ORF2 [52], and one that includes *fdxH* [53]. The genes in this cluster are expressed in the heterocysts [54]. In some strains of heterocyst-forming *Nostoc* sp., a gene, *glbN*, encoding a myoglobin-like protein ("cyanoglobin"), has been found between *nifU* and *nifH* [55,56].

A gene cluster carrying three additional *nif* genes (*nifVZT*) has been recently identi-fied in *Anabaena* sp. PCC 7120 [57]. This gene cluster, that appears to constitute an op-eron, is, however, not essential for nitrogen fixation in this cyanobacterium [57]. This implies that this cyanobacterium bears another protein that can perform, at least partially, the function of NifV (homocitrate synthase) in the biosynthesis of the Fe-Mo-cofactor. Also a *nifJ* gene has been described in strain PCC 7120 [58]. This gene, which would en-code a protein homologous to pyruvate flavodoxin oxidoreductase from other bacteria, is essential for growth only in media that is both iron- and nitrogen-limited.

Anabaena variabilis ATCC 29413 (PCC 7937) and a few other heterocyst-forming cy-anobacteria are unique among nitrogen-fixing organisms in that they are able to express two functional Mo-dependent nitrogenases [59–61]. The complete DNA sequence of the second *nif* cluster (*nif2*) is now available (acc. no. AVU49859 and Z46890). In contrast to the *nif1* cluster (that is equivalent to the strain PCC 7120 *nif* cluster), the *nif2* cluster is also ex-pressed in vegetative cells and appears to be more similar to the *nif* cluster of non-hetero-cystous cyanobacteria like *Plectonema boryanum*. In addition to the two Mo-dependent nitrogenases, *Anabaena variabilis* ATCC 29413 also expresses a vanadium-dependent nitro-genase. Two genes specific for this alternative nitrogenase, *vnfDG* and *vnfK*, have been characterized [62], and the *nifB* gene from the *nif1* cluster has been shown to be required not only for Mo-nitrogenase but also for V-nitrogenase [63]. Expression of the V-nitrogenase appears to be confined to the heterocyst [61].

4.3. Regulation of Expression of Nitrogen Fixation Genes

Synthesis of nitrogenase and development of heterocysts are suppressed in ammo-nium- or nitrate-supplemented media [18]. The *nifHDK* mRNA is not detected in cells grown in the presence of a source of combined nitrogen. This has been observed both in heterocyst-forming strains (e.g., in *Anabaena* sp. PCC 7120 [64,65]) and in non-hetero-cyst-forming strains like the unicellular *Synechococcus* sp. RF-1 [66] or the filamentous *Plectonema boryanum* [67]. Some other genes in the strain PCC 7120 main *nif* cluster have also been studied and found to be repressed by ammonium. These include *nifB* [50], ORF1 and ORF2 [52], and *fdxH* [53]. In strain PCC 7120, also the *nifVZT* operon is repressed in the presence of combined nitrogen [57].

In *Anabaena variabilis* ATCC 29413, genes in the *nif2* cluster are also repressed by combined nitrogen, but anaerobiosis is required, in addition to nitrogen deficiency, to observe expression of those genes [59,60]. With regard to the *vnf* genes of this strain, they are repressed not only by ammonium but also by a source of molybdenum [62].

As mentioned above, in heterocyst-forming cyanobacteria nitrogenase is generally confined to heterocysts. Because heterocyst development is itself repressed by ammonium, repression of *nifHDK* and of other *nif* (and *fdx*) genes might be indirect. Indeed, expression of heterocyst-specific *nif* genes is thought to be subjected to developmental, rather than environmental, control [54]. Some direct effects of environmental factors on *nif* gene expression seem to exist, however, since ammonium can repress nitrogenase synthesis in *Anabaena* cultures bearing developed heterocysts (discussed in [18]) and iron deficiency, in the presence of combined nitrogen, has recently been shown to induce *nifV* [57], *nifJ* [58], *nifH*, and *fdxH* [68] in *Anabaena* sp. PCC 7120.

Although discussion of the developmental process that leads to the differentiation of heterocysts [43,69] is outside the scope of this review, the important role played in this process by the *hetR* regulatory gene should be mentioned. Mutation of *hetR* prevents heterocyst differentiation, and expression of *hetR* is increased in response to nitrogen deficiency, in a process that involves autoregulation, in those cells that will become heterocysts [70,71].

5. INCORPORATION OF AMMONIUM INTO CARBON SKELETONS

5.1. Glutamine Synthetase/Glutamate Synthase Pathway

Former biochemical and physiological experiments had shown that incorporation of ammonium into carbon skeletons in the cyanobacteria takes place mainly through the glutamine synthetase/glutamate synthase (GS/GOGAT) pathway, with possible minor contributions of other enzymes like alanine dehydrogenase (reviewed in [18]). Glutamine synthetase catalyzes the ATP-dependent synthesis of glutamine from glutamate and ammonium, and glutamate synthase catalyzes the synthesis of two glutamate molecules from glutamine and α-ketoglutarate, a reaction that requires the supply of two electrons. The combined action of the two enzymes results in the production of one glutamate molecule from ammonium and α-ketoglutarate.

5.2. Glutamine Synthetase

The *glnA* gene encoding a conventional eubacterial glutamine synthetase is widely distributed among the cyanobacteria [72–74]. In different cyanobacteria, glutamine synthetase activity and/or protein levels are about 50% in ammonium-grown cells as compared to those of cells grown with nitrate or dinitrogen as nitrogen sources, a difference that roughly reflects levels of *glnA* mRNA [72–75]. In *Synechococcus* sp. PCC 7942, however, *glnA* mRNA levels in ammonium-grown cells are extremely low [12,76] suggesting some sort of post-transcriptional regulation that might help to attain relatively high levels of glutamine synthetase. On the other hand, an ammonium-promoted reversible inactivation of glutamine synthetase is known to occur in *Synechocystis* sp. PCC 6803 [18].

A second glutamine synthetase-encoding gene, *glnN*, has been characterized in *Synechocystis* sp. PCC 6803 that would encode a protein homologous to the peculiar glutamine synthetase of *Bacteroides fragilis* and *Butyrivibrio fibrisolvens* [77,78]. This gene is found in some other cyanobacteria as well, but not in heterocyst-forming strains. Expression of *glnN* takes place at significant levels only under nitrogen starvation [75].

5.3. Glutamate Synthase and the *gltS* and *gltB* Genes

A gene, *gltS*, encoding a ferredoxin-dependent glutamate synthase that is homologous to higher-plant ferredoxin-dependent glutamate synthases is widespread in cyanobacteria, while a second gene for glutamate synthase, *gltB*, is present only in some cyanobacterial strains [79]. In contrast to expression of glutamine synthetase, expression of glutamate synthase in cyanobacteria is constitutive, not being influenced by the nitrogen source used for growth (see [18]).

5.4. Isocitrate Dehydrogenase

Isocitrate dehydrogenase, the enzyme producing α-ketoglutarate from isocitrate with the simultaneous reduction of $NADP^+$ to NADPH, and the *icd* gene encoding isocitrate dehydrogenase have been recently characterized in both *Anabaena* sp. PCC 7120 and *Synechocystis* sp. PCC 6803 [80,81]. Interestingly, expression of *icd* is subjected to regulation by nitrogen, being highest under nitrogen stress in strain PCC 6803 or under dinitrogen-fixing conditions in strain PCC 7120 [82].

6. NtcA-MEDIATED NITROGEN CONTROL

6.1. Nitrogen Control

As summarized above, repression by ammonium of nitrogen assimilation genes is common in cyanobacteria. Thus, ammonium is a preferred nitrogen source in these organisms. Among the genes whose expression is stimulated upon withdrawal of ammonium from the culture medium, two groups can be distinguished: (i) those genes that are expressed at high levels in the presence of nitrate (e.g., *glnA*, the *nir* operon), and (ii) those whose expression takes place at high levels only in the absence of combined nitrogen (e.g., *amt1*, *hetR*, *nifHDK*). Consistent with the suggestion that the nitrate repressive effect involves its metabolism to ammonium [83], repression by nitrate of *nifHDK* expression in *Anabaena* sp. PCC 7120 does not take place in a nitrate reductase-less mutant [65]. Nitrate might be considered as a nutrient that leads to intermediate levels of nitrogen in the cell. At a molecular level, the difference between ammonium and nitrate nutrition may result from the degree of saturation by ammonium of glutamine synthetase, since the intracellular concentration of ammonium might be lower in cells generating ammonium from nitrate reduction than in cells exposed to externally added ammonium. The absence of a source of combined nitrogen, on the other hand, leads to a transient nitrogen deficiency which becomes severe in non-nitrogen-fixing strains and permits expression of the nitrogen fixation machinery in the nitrogen fixers.

In cyanobacteria, repression by nitrate or ammonium of nitrogen assimilation elements generally requires ammonium metabolism through glutamine synthetase to take place [18]. This brings the different repressive phenomena to a unified framework and suggests that a common molecular mechanism might be responsible for all of them.

6.2. The NtcA Transcriptional Regulator

The *ntcA* gene encodes a positive transcriptional regulator of genes that are subjected to repression by ammonium in cyanobacteria. It was first isolated by complementation of a pleiotropic mutant of *Synechococcus* sp. PCC 7942 that was unable to grow using nitrate (or nitrite) as a nitrogen source and was impaired in expression, in the absence of ammonium, of several enzyme activities or proteins that are repressed by ammonium in the wild type strain [84]. The NtcA protein is composed of 222 amino acids and is homologous to transcription factors like Crp from *Escherichia coli* that bear a DNA-binding helix-turn-helix motif close to their carboxyl terminus [85]. DNA sequences homologous to *ntcA* are present in a wide range of cyanobacteria, and the *ntcA* genes from *Synechocystis* sp. PCC 6803 [86] and *Anabaena* sp. PCC 7120 [86,87] have also been cloned and sequenced. More recently, the *ntcA* gene has been cloned and sequenced from three other cyanobacteria including *Anabaena variabilis* ATCC 29413 [88], *Cyanothece* sp. ATCC 51142 (strain BH68K) [89], and *Synechococcus* sp. WH7803 [90]. The predicted NtcA polypeptides are strongly similar proteins ($\geq 63.2\%$ identity), and the putative helix-turn-helix motif is identical in the six proteins, with a single exception, Val[189] (numeration of NtcA from strain PCC 7942) is substituted by Ile in NtcA from strain WH7803.

In *Synechococcus* sp. PCC 7942, a transcript of ca. 1.3 kb is observed for *ntcA* that is much more abundant in cells incubated in nitrate medium or in medium lacking combined nitrogen than in ammonium-grown cells [12]. Since *ntcA* is 666 bp in length and the *ntcA tsp* is only about 110 bp upstream from the start codon of the gene, a long downstream region appears to be present in the *ntcA* mRNA. Transcripts of ca. 1.4 and 1.3 kb are observed for *ntcA* in *Anabaena* sp. PCC 7120 [91] and *Cyanothece* sp. ATCC 51142 [89], respectively. Additional transcripts of 0.8 and 1.0 kb are also evident in strain PCC 7120 [92] and of ca. 1 kb in strain ATCC 51142 [89]. In contrast to what is observed in strain PCC 7942, no strong differences in total *ntcA* mRNA levels are observed in strain PCC 7120 incubated with different sources of nitrogen [92,93]. With regard to strain ATCC 51142, levels of *ntcA* mRNA are higher in cells grown on nitrate than on dinitrogen, whereas the levels in ammonium-grown cells depend on the concentration of ammonium used, being highest with 1 mM [89]. A note of caution is necessary, however, regarding strain ATCC 51142 since interpretation of data on regulation by the nitrogen source can be complicated by the fact that cells subjected to dark-light cycles were used by Bradley and Reddy in some experiments [89].

Mutants of the *ntcA* gene are available for *Synechococcus* sp. PCC 7942 [84] and *Anabaena* sp. PCC 7120 [91,94]. Analysis of the phenotype of these mutants has permitted the identification of a number of physiological functions that require an intact *ntcA* gene to be manifest: (i) derepression of the nitrate assimilation system in strains PCC 7942 and PCC 7120 [84,94]; (ii) increased expression of glutamine synthetase in the absence of ammonium in strains PCC 7942 and PCC 7120 [84,91,94]; (iii) derepression of the ammonium/methylammonium uptake system in strain PCC 7942 [84]; (iv) derepression of the *ntcA* gene itself in strain PCC 7942 [12]; (v) heterocyst development and *nif*HDK expression [91,94] as well as induction of *hetR* in strain PCC 7120 [94]; (vi) derepression of *nirB* and *ntcB* in strain PCC 7942 [40]; and (vii) increased expression of the *glnB* gene in the absence of ammonium in strain PCC 7942 [29]. The *cynS* gene encoding a cyanase whose role in cyanobacterial nitrogen metabolism is yet to be defined is also repressed by ammonium in strain PCC 7942 as well as in *Synechocystis* sp. PCC 6803, and derepression of *cynS* does not take place in a strain PCC 7942 *ntcA* mutant [95].

Consistent with the presence of a putative DNA-binding helix-turn-helix motif in NtcA, the product of the *Synechococcus* sp. PCC 7942 *ntcA* gene cloned in *E. coli* is able to specifically bind to DNA fragments corresponding to DNA sequences located upstream from NtcA-regulated genes of strain PCC 7942 including *nir*, *glnA*, and *ntcA* itself [12]. The NtcA-binding sites in those DNA fragments have been precisely localized by means of "mobility shift" and DNase I "footprinting" assays, and they have been found to contain the characteristic sequence GTAN$_8$TAC that is palindromic in nature [12]. Using *Anabaena* sp. PCC 7120 partially purified cell-free extracts as a source of NtcA, binding of strain PCC 7120 NtcA to some DNA fragments has also been characterized, and the consensus recognition sequence TGT(N$_{9\ or\ 10}$)ACA has been reported [96]. The difference in the consensus DNA recognition sequence of NtcA reported by Luque et al. [12] and by Ramasubramanian et al. [96] might derive from the particular genes studied by each group since, as discussed above, both NtcA proteins, from strains PCC 7942 and PCC 7120, bear identical amino acid sequences in their putative helix-turn-helix motifs. The NtcA-binding sites studied in strain PCC 7942 correspond to genes whose expression is activated by NtcA, whereas those investigated in strain PCC 7120 correspond to one NtcA-activated gene (*glnA*) and two genes whose interaction with NtcA is of unknown function (*rbcL* encoding a subunit of ribulose bisphosphate carboxylase/oxygenase, and *xisA* encoding the site-specific recombinase responsible for excision of the *nifD* 11-kb intervening element). At least one of the two NtcA-binding sites found upstream from *rbcL* could have a repressor role [96].

6.3. NtcA-Activated Promoters

As described above, the *Synechococcus* sp. PCC 7942 *nir*, *glnA*, and *ntcA* genes are expressed at low levels in ammonium-grown cells. In the case of *ntcA*, this transcription takes place from a promoter similar to the canonical *E. coli* σ^{70} promoters [12]. When the cells are incubated in the absence of ammonium, transcription of the *nir*, *glnA*, and *ntcA* genes is increased and takes place from *tsp* that are preceded by a putative -10 promoter box in the form TAN$_3$T, but not by a recognizable −35 box [12]. As described above, NtcA binds upstream from each of these genes. The NtcA-binding site characteristic recognition sequence, GTAN$_8$TAC, is located in the three cases 22 nucleotides upstream from the −10 box (Table 1). This has permitted to suggest that the NtcA-activated *Synechococcus* sp. PCC 7942 promoters bear an NtcA-binding site that would substitute for the −35 box found in promoters similar to the canonical *E. coli* σ^{70} promoters [12]. The requirement of the GTAN$_8$TAC sequence for transcription of *glnA* in *Synechococcus* sp. PCC 7942 is supported by the observation that deletion of nine nucleotides that include the 5′ half of this NtcA-binding site prevents expression of a *glnA::cat* fusion [97].

In *Synechococcus* sp. PCC 7942, in addition to the NtcA-binding site centered at −40.5, two other NtcA-binding sites are present upstream from *nir* centered at −109.5 and −180.5, respectively [12]. Whereas the function of the −109.5 site is unknown, the −180.5 site represents a perfect NtcA-binding site for the promoter of the *nirB* gene (Table 1), whose transcription takes place divergently from that of the *nir* operon and appears to be activated by NtcA [40].

The *Synechococcus* sp. PCC 7942 *glnB* gene has recently been found to be transcribed from at least two *tsp* [29]. One of them is not used in an *ntcA* mutant and is preceded by the sequence shown in Table 1 that compares well to those of other NtcA-activated promoters. A DNA fragment containing this region specifically binds NtcA *in vitro* [29]. It is interesting that *glnB*, as the *nir* operon, is activated by NtcA, what would increase the production of the GlnB (P$_{II}$) protein that is a regulator of the activity of the nitrate uptake system encoded by the *nrt* genes in the *nir* operon.

The *tsp* of the *Anabaena* sp. PCC 7120 *nir* operon is preceded by the sequence shown in Table 1 that shows the characteristics of an NtcA-activated promoter [23]. This *tsp* is not used in ammonium-grown wild type cells or under any nitrogen regime in an

Table 1. NtcA-activated promoters in cyanobacteria. Bases corresponding to the signature sequence for NtcA binding sites and to the most conserved positions in the −10 box are indicated in boldface; transcription start point (*tsp*) are shown underlined

Strain	gene/operon	Promoter sequence
PCC 7942	*nir-nrtABCD-narB*	AAAGTTGTAGTTTCTGTTACCAATTGCGAATCGAGAACTGCC.TAATCTGCCGAG
	glnA	TTTTATGTATCAGCTGTTACAAAAGTGCCGTTTCGGGCTACC.TAGGATGAAAGC
	ntcA	GAAAAAGTAGCAGTTGCTACAAGCAGCAGCTAGGCCG.TACGGTAACGA
	nirB-ntcB	TTTTTAGTAGCAATTGCTACAAGCCTTGACTCTGAAGCCCGC.TTAGGTGGAGCCATTA
	glnB	TTGCTGTGTAGCAGTAACTACAACTGTGGTCTAGTCAGCGGTGTTACCAAAGAGTC
PCC 7120	*nir-nrtABCD-narB*	AATTTTGTAGCTACTTATACTATTTTACCTGAGATCCCGACA.TAACCTTAGAAGT
	glnA (RNA_I)	CGTTCTGTAACAAAGACTACAAAACTGTCTAATGTTTAGAATCTACGATATTTCA
PCC 6803	*amt1*	TGAAAAGTAGTAAATCATACAGAAAACAATCATGTAAAAA...TTGAATACTCTAA
	glnA	AAAATGGTAGCGAAAAATACATTTTCTAACTACTTGACTCTT.TACGATGATAGTCG
	glnB	CAAACGGTACTGATTTTTACAAAAAACTTTTGGAGAACATGTTAAAAGTGTCTGG
	icd	AATTTCGTAACAGCCAATGCAATCAGAGCCTCCAGAAAGGAT.TATGATCGTCTCCG
Consensus		......GTA...N(8).TAC......N(20-23).........TA...T.N(4-9).tsp

ntcA mutant [23]. Additionally, specific binding of NtcA to a DNA fragment containing this region has recently been observed (our unpublished results). Therefore, the strain PCC 7120 *nir* promoter represents an NtcA-activated promoter.

The *glnA* gene of *Anabaena* sp. PCC 7120 shows a complex promoter structure, with at least three putative *tsp* corresponding to different RNA species observed by S1 mapping or by primer extension analysis [2,9,72,94]. The use of the *tsp* corresponding to RNA$_I$ is increased in the absence of ammonium [72], and this activation does not take place in an *ntcA* mutant [94]. Binding of NtcA to a DNA fragment carrying a *glnA* upstream region including this *tsp*, studied by "mobility shift" and DNase I "footprinting" assays [96], indicates that an NtcA-binding site is located upstream from RNA$_I$ as shown in Table 1. The *tsp* corresponding to RNA$_{II}$ was first described to be used in ammonium-containing medium [72] but has been later characterized as constitutive and is not NtcA-dependent [94]. This *tsp* is also used in an *E. coli* strain carrying the *Anabaena glnA* gene cloned in a plasmid [72] and in *in vitro* transcription assays with both *Anabaena* sp. PCC 7120 and *E. coli* RNA polymerases [2], and is preceded by a sequence that resembles the canonical *E. coli* σ^{70} promoters [9]. Therefore, the promoter for RNA$_{II}$ appears to be a typical constitutive RNA polymerase-SigA promoter. The *tsp* corresponding to RNA$_{IV}$ has been observed, in different experiments, to be more abundant in cells incubated in the presence [72] or in the absence of ammonium [72,94] and is not detected in primer extension assays with RNA from an *ntcA* mutant [94]. In contrast to the NtcA-dependent RNA$_I$, RNA$_{IV}$ is detected in *in vitro* transcription assays with RNA polymerase from both strain PCC 7120 and *E. coli* and is preceded by sequences that may be considered similar to those of the canonical *E. coli* σ^{70} promoters [2,9]. The *tsp* corresponding to RNA$_{IV}$ may therefore be only indirectly regulated by NtcA. Two other RNA 5' ends have been detected for strain PCC 7120 *glnA*, RNA$_{III}$ that can represent a degradation product of RNA$_{IV}$ [9,94] and RNA$_V$ that has only been detected in *in vitro* transcription assays [2]. The combined expression of constitutive and inducible promoters would account for the doubling of glutamine synthetase levels that takes place when strain PCC7120 cells are transferred from ammonium-containing medium to media with nitrate or no combined nitrogen.

Sequences with structures similar to those of the well-characterized NtcA-dependent promoters described above are found in the promoter regions of four ammonium-repressed genes from *Synechocystis* sp. PCC 6803 whose putative *tsp* have been determined (*amt1* [14], *glnA* [75], *glnB* [30], and *icd* [82]; see Table 1). In the four cases, specific binding of NtcA to DNA fragments carrying these sequences has been observed *in vitro*. Unfortunately, however, a *Synechocystis* sp. PCC 6803 *ntcA* mutant is not yet available and, therefore, no final proof for the requirement of NtcA for expression of these genes has been obtained. In the promoter of the strain PCC 6803 *glnA* gene, in addition to the putative NtcA-binding site, a putative −35 box is found 15 nucleotides upstream from the −10 box [75]. The -35 and -10 boxes might together constitute a constitutive promoter responsible for the transcription of *glnA* observed in ammonium-grown cells. On the other hand, the *Synechocystis* sp. PCC 6803 *glnN* gene is expressed at highest levels under nitrogen stress, and its promoter shares some similarity to the sequence of NtcA-activated promoters, but a DNA fragment containing the *glnN* promoter has failed to bind NtcA *in vitro* [75].

Calothrix sp. PCC 7601, a strain that contains DNA sequences that hybridize with *ntcA* [86], produces *glnA* transcripts of 1.8 and 1.6 kb, the latter being more abundant in nitrate than in ammonium-grown cells [73]. The *tsp* that probably corresponds to the 1.6-kb transcript is preceded by a sequence strongly similar to that of the RNA$_I$ promoter of the *Anabaena* sp. PCC 7120 *glnA* gene, suggesting that it could also correspond to an NtcA-activated promoter [98].

Comparison of the sequences of the promoters included in Table 1 corroborates the validity of the $GTAN_8TAC$ sequence as a signature sequence for the NtcA-binding sites found in NtcA-activated promoters which also bear a −10 box in the form TAN_3T. Several A or T residues are found both 5' and 3' of these NtcA-binding sites, and a C and an A are common in the second and third position, respectively, downstream from the 5'GTA triplet.

6.4. NtcA in Heterocysts

Expression of the *hetR* regulatory gene is impaired in an *ntcA* mutant [94], suggesting that NtcA represents a regulatory link between nitrogen nutrition and heterocyst development. However, no DNA sequence similar to the consensus sequence of the NtcA-binding site is present at the promoters for *hetR* (W. J. Buikema, unpublished). The identification of gene(s) that may act in heterocyst differentiation between NtcA and *hetR* is currently being pursued.

It can be suspected, however, that NtcA has also a function in the developed heterocyst. The *ntcA* gene is expressed, in *Anabaena* sp. PCC 7120, both during heterocyst differentiation and in developed heterocysts [92,99], and heterocyst extracts contain NtcA protein that can be detected by means of "mobility shift" assays [96]. NtcA binds to a DNA fragment from the upstream region of *nifH* [96], although this fragment does not carry any good putative NtcA-binding site. On the other hand, sequences showing some similarity to those of the NtcA-activated promoters are found upstream of the *tsp* for the strain PCC 7120 *nif* cluster ORF1 [52] and for the *Anabaena variabilis* ATCC 29413 *vnfDG* gene [62]. Further work is necessary to discern whether the role of the NtcA protein present in the heterocysts includes activation of expression of these genes.

7. CONCLUSIONS AND PROSPECTS

Ammonium behaves as a nutritional repressor of proteins involved in the assimilation of alternative nitrogen sources, like nitrate or dinitrogen, in cyanobacteria. Substantial evidence for the involvement in this ammonium effect of the transcription factor NtcA, that is necessary for expression of ammonium-repressible genes, is available. An array of cyanobacterial nitrogen assimilation genes whose expression is activated by NtcA has been established in *Synechococcus* sp. PCC 7942, the strain from which the original *ntcA* mutant was isolated. Mutants of *ntcA* have also been isolated from *Anabaena* sp. PCC 7120 that have shown that NtcA is also required for nitrogen fixation and heterocyst development. The isolation of *ntcA* mutants from some other cyanobacteria should also be tried in order to get a comprehensive picture of the range of genes that belong to the NtcA regulon. A good example is *Cyanothece* sp. ATCC 51142. It has been suggested that in this strain *ntcA* might not be involved in the regulation of nitrogen fixation [89], but analysis of the phenotype of an *ntcA* mutant would be necessary to elucidate the function(s) of NtcA in this cyanobacterium.

The structure of the NtcA-activated promoters is similar to that of the Class II Crp-activated promoters of *E. coli* and is composed of an NtcA-binding site that conforms to the sequence signature $GTAN_8TAC$ and, about 22 nucleotides downstream from it, a -10 promoter box similar to the consensus sequence TAN_3T (see Table 1). The presence of this -10 box suggests that the NtcA-activated genes are transcribed by an RNA polymerase bearing SigA or a SigA-type σ factor, but this is yet to be experimentally tested.

The *ntcA* gene is autoregulated in *Synechococcus* sp. PCC 7942 so that the low level of NtcA protein that appears to be present in ammonium-grown cells would be capable, after withdrawal of ammonium, of activating expression of the *ntcA* gene itself. It is unknown how the nitrogen nutrition of the cell would influence the capability of NtcA to promote transcription. It will be of interest to test whether the NtcA protein is activated somehow, and whether different levels of active protein can be reached as a function of the severity of the nitrogen deficiency as to explain the hierarchy observed in the assimilation of different nitrogen sources in cyanobacteria: ammonium is assimilated with preference over nitrate which in turn is assimilated with preference over dinitrogen.

An aspect only briefly mentioned above is the possible role of NtcA as a repressor. This could happen when the NtcA-binding site is found in a location where binding of NtcA would interfere with the binding or function of RNA polymerase, as is the case for one of the NtcA-binding sites found upstream of the *Anabaena* sp. PCC 7120 *rbcL* gene. Further studies with more NtcA-binding sites of this type could help to define how important a repressor role of NtcA might be in control of gene expression in cyanobacteria.

In this review, we have focused on one aspect of the control of expression of cyanobacterial nitrogen assimilation genes, namely, the NtcA-dependent activation of expression that accounts for repression by ammonium. Some other environmental factors do appear to influence expression of some of those genes as well. Nitrate has a positive effect on the expression of the nitrate assimilation system, especially in heterocyst-forming cyanobacteria, but the molecular mechanism for this nitrate effect is not yet known. Some nitrogenase-encoding genes are only expressed in the absence of oxygen in some cyanobacteria, but the regulatory mechanism behind this effect is completely unexplored. Finally, light may constitute an important regulatory environmental factor (see, e.g., [100]) that is yet to be investigated. Obviously, much work is still required in order to elucidate the fine mechanisms of control of gene expression in cyanobacterial assimilatory nitrogen metabolism.

ACKNOWLEDGMENTS

We thank N. Tandeau de Marsac, T. Omata, and W. J. Buikema for communicating results prior to publication. Work in the authors' laboratory is currently supported by grants no. PB94-0074 and PB95-1267 from Dirección General de Enseñanza Superior and by Plan Andaluz de Investigación (group CVI-129), Spain.

REFERENCES

1. Houmard, J. (1994) Microbiol. 140, 433–441.
2. Schneider, G. J., Tumer, N.E., Richaud, C., Borbely, G. and Haselkorn, R. (1987) J. Biol. Chem. 262, 14633–14639.
3. Brahamsha, B. and Haselkorn, R. (1991) J. Bacteriol. 173, 2442–2450.
4. Brahamsha, B. and Haselkorn, R. (1992) J. Bacteriol. 174, 7273–7282.
5. Tanaka, K., Masuda, S. and Takahashi, H. (1992) Biosci. Biotechnol. Biochem. 56, 1113–1117.
6. Gruber, T. M. and Bryant, D. A. (1997) J. Bacteriol. 179, 1734–1747.
7. Sakamoto, T., Shiral, M., Asayama, M., Aida, T., Sato, A., Tanaka, K.Takahashi, H. and Nakano, M. (1993) Inter. J. System. Bacteriol. 43, 844–847.
8. Kaneko, T., Sato, S., Kotani, H., Tanaka, A., Asamizu, E., Nakamura, Y., Miyajima, N., Hirosawa, M., Sugiura, M., Sasamoto, S., Kimura, T., Hosouchi, T., Matsuno, A., Muraki, A., Nakazaki, N., Naruo, K.,

Okumura, S., Shimpo, S., Takeuchi, C., Wada, T., Watanabe, A., Yamada, M., Yasuda, M., and Tabata, S. (1996) DNA Research 3, 109–136.

9. Schneider, G. J. , Lang, J. D. and Haselkorn, R. (1991) Gene 105, 51–60.
10. Floriano, B., Herrero, A. and Flores, E. (1994) J. Bacteriol. 176, 6397–6401.
11. Curtis, S. E. and Martin, J. A. (1994) in: The Molecular Biology of Cyanobacteria (Bryant, D. A., Ed.), pp. 613–639, Kluwer Academic Publishers, Dordrecht.
12. Luque, I., Flores, E. and Herrero, A. (1994) EMBO J. 13, 2862–2869.
13. Boussiba, S., Dilling, W. and Gibson, J. (1984) J. Bacteriol. 160, 204–210.
14. Montesinos, M. L., Herrero, A. and Flores, E. (1998) Submitted.
15. Rai, A. N., Rowell, P. and Stewart, W. D. P. (1984) Arch. Microbiol. 137, 241–246.
16. Boussiba, S. and Gibson, J. (1987) FEMS Microbiol. Lett. 43, 289–293.
17. Ritchie, R. J. and Gibson, J. (1987) J. Membr. Biol. 95, 131–142.
18. Flores, E. and Herrero, A. (1994) in: The Molecular Biology of Cyanobacteria (Bryant, D. A., Ed.), pp. 487–517, Kluwer Academic Publishers, Dordrecht.
19. Omata, T., Andriesse, X. and Hirano, A. (1993) Mol. Gen. Genet. 236, 193–202.
20. Merchán, F., Kindle, K. L., Llama, M. J., Serra, J. L. and Fernández, E. (1995) Plant Mol. Biol. 28, 759–766.
21. Suzuki, I., Kikuchi, H., Nakanishi, S., Fujita, Y., Sugiyama, T. and Omata, T. (1995) J. Bacteriol. 177, 6137–6143.
22. Cai, Y. and Wolk, C. P. (1997) J. Bacteriol. 179, 258–266.
23. Frías, J. E., Flores, E. and Herrero, A. (1997) J. Bacteriol. 179, 477–486.
24. Luque, I., Flores, E. and Herrero, A. (1994) Biochim. Biophys. Acta 1184, 296–298.
25. Maeda, S. and Omata, T. (1997) J. Biol. Chem. 272, 3036–3041.
26. Madueño, F., Vega-Palas, M. A., Flores, E. and Herrero, A. (1988) FEBS Lett. 239, 289–291.
27. Lee, H.L., Flores, E., Herrero, A., Houmard, J. and Tandeau de Marsac, N. (1998) Submitted.
28. Forchhammer, K. and Tandeau de Marsac, N. (1994) J. Bacteriol. 176, 84–91.
29. Lee, H.L., Vázquez, M. F. and Tandeau de Marsac, N. (1998) Submitted.
30. García-Domínguez, M. and Florencio, F. J. (1997) Plant Mol. Biol. 35, 723–734.
31. Rubio, L. M., Herrero, A. and Flores, E. (1996) Plant Mol. Biol. 30, 845–850.
32. Kuhlemeier, C. J., Logtenberg, T., Stoorvogel, W., van Heugten, H. A. A., Borrias, W. E. and van Arkel, G. A. (1984) J. Bacteriol. 159, 36–41.
33. Unthan, M., Klipp, W. and Schmid, G. H. (1996) Biochim. Biophys. Acta 1305, 19–26
34. Rubio, L. M., Flores, E. and Herrero, A. (1998) Submitted.
35. Luque, I., Flores, E. and Herrero, A. (1993) Plant Mol. Biol. 21, 1201–1205.
36. Suzuki, I., Sugiyama, T. and Omata, T. (1993) Plant Cell Physiol. 34, 1311–1320.
37. Merchán, F., Prieto, R., Kindle, K. L., Llama, M. J., Serra, J. L. and Fernández, E. (1995) Plant Mol. Biol. 28, 759–766.
38. Luque, I., Herrero, A., Flores, E. and Madueño, F. (1992) Mol. Gen. Genet. 232, 7–11.
39. Aichi, M. and Omata, T. (1997) J. Bacteriol. 179, 4671–4675.
40. Suzuki, I., Horie, N., Sugiyama, T. and Omata, T. (1995) J. Bacteriol. 177, 290–296.
41. Dean, D. R., Bolin, J. T. and Zheng, L. (1993) J. Bacteriol. 175, 6737–6744.
42. Fay, P. (1992) Microbiol. Rev. 56, 340–373.
43. Wolk, C. P., Ernst, A. and Elhai, J. (1994) in: The Molecular Biology of Cyanobacteria (Bryant, D. A., Ed.), pp. 769–823, Kluwer Academic Publishers, Dordrecht.
44. Bergman, B., Gallon, J. R., Rai, A. N. and Stal, L. J. (1997) FEMS Microbiol. Rev. 19,139–185.
45. Haselkorn, R. and Buikema, W. J. (1992) in: Biological Nitrogen Fixation (Stacey, G., Burris, R.H. and Evans, H.J., Eds.), pp. 167–190, Chapman & Hall, New York, London.
46. Luo, X.-Z. J. and Stevens, S. E. (1996) Data bank entry FSU49514.
47. Colón-López, M. and Sherman, L. A. (1997) Data bank entries AF003336, AF003337, and AF003338.
48. Chen, H.-M. (1995) Data bank entry SSU22146.
49. Haselkorn, R. (1992) Annu. Rev. Genet. 26, 113–130.
50. Mulligan, M. E. and Haselkorn, R. (1989) J. Biol. Chem. 264, 19200–19207.
51. Haselkorn, R., Golden, J. W., Lammers, P. J. and Mulligan, M. E. (1986) Trends in Genetics 2, 255–259.
52. Borthakur, D., Basche, M., Buikema, W. J., Borthakur, P. B. and Haselkorn, H. (1990) Mol Gen Genet 221, 227–234.
53. Böhme, H. and Haselkorn, R. (1988) Mol. Gen. Genet. 214, 278–285.
54. Elhai, J. and Wolk, C. P. (1990) EMBO J. 9, 3379–3388.
55. Potts, M., Angeloni, S. V., Ebel, R. E. and Bassam, D. (1992) Science 256, 1690–1692.

56. Hill, D. R., Belbin, T. J., Thorsteinsson, M. V., Bassam, D., Brass, S., Ernst, A., Böger, P., Paerl, H., Mulligan, M. E. and Potts, M. (1996) J. Bacteriol. 178, 6587–6598.

57. Stricker, O., Masepohl, B., Klipp, W. and Böhme, H. (1997) J. Bacteriol. 179, 2930–2937.

58. Bauer, C. C, Scappino, L. and Haselkorn, R. (1993) Proc. Natl. Acad. Sci. USA 90, 8812–8816.

59. Schrautemeier, B., Neveling, U. and Schmitz, S. (1995) Mol. Microbiol. 18, 357–369.

60. Thiel, T., Lyons, E. M., Erker, J. C. and Ernst, A. (1995) Proc. Natl. Acad. Sci. USA 92, 9358–9362.

61. Thiel, T., Lyons, E. M. and Erker, J. C. (1997) J. Bacteriol. 179, 5222–5225.

62. Thiel, T. (1993) J. Bacteriol. 175, 6276–6286.

63. Lyons, E. M., and Thiel, T. (1995) J. Bacteriol. 177, 1570–1575.

64. Rice, D., Mazur, B. J. and Haselkorn, R. (1982) J. Biol. Chem. 257, 13157–13163.

65. Martín-Nieto, J., Herrero, A. and Flores, E. (1991) Plant Physiol. 97, 825–828.

66. Huang, T. C. and Chou, W. M. (1991) Plant Physiol. 96, 324–326.

67. Fujita, Y., Takahashi, Y., Shonai, F., Ogura, Y. and Matsubara, H. (1991) Plant Cell Physiol. 32, 1093–1106.

68. Razquín, P., Schmitz, S., Fillat, M. F., Peleato, M. L. and Böhme, H. (1994) J. Bacteriol. 176, 7409–7411.

69. Wolk, C. P. (1996) Annu. Rev. Genet. 30, 59–78.

70. Buikema, W. J. and Haselkorn, R. (1991) Genes Dev. 5, 321–330.

71. Black, T.A., Cai, Y. and Wolk, C. P. (1993) Mol. Microbiol. 9, 77–84.

72. Tumer, N. E., Robinson, S. J., and Haselkorn, R. (1983) Nature 306, 1–6.

73. Elmorjani, K., Liotenberg, S., Houmard, J. and Tandeau de Marsac, N. (1992) Biochem. Biophys. Res. Comm. 189, 1296–1302.

74. S.J. Wagner, S. J., Thomas, S. P., Kaufman, R. I., Nixon, B. T. and Stevens, S. E. (1993) J. Bacteriol. 175, 604–612.

75. Reyes, J. C., Muro-Pastor, M. I. and Florencio, F. J. (1997) J. Bacteriol. 179, 2678–2689.

76. Cohen-Kupiec, R., Gurevitz, M. and Zilberstein, A. (1993) J. Bacteriol. 175, 7727–7731.

77. Reyes, J. C. and Florencio, F. J. (1994) J. Bacteriol. 176, 1260–1267.

78. García-Domínguez, M., Reyes, J. C. and Florencio, F. J. (1997) Eur. J. Biochem. 244, 258–264.

79. Navarro, F., Chávez, S., Candau, P. and Florencio, F. J. (1995) Plant Mol. Biol. 27, 753–767.

80. Muro-Pastor, M. I. and Florencio, F. J. (1992) Eur. J. Biochem. 203, 99–105.

81. Muro-Pastor, M. I. and Florencio, F. J. (1994) J. Bacteriol. 176, 2718–1726.

82. Muro-Pastor, M. I., Reyes, J. C. and Florencio, F. J. (1996) J. Bacteriol. 178, 4070–4076.

83. Ramos, J. L. and Guerrero, M. G. (1983) Arch. Microbiol. 136, 81–83.

84. Vega-Palas, M. A., Madueño, F., Herrero, A. and Flores, E. (1990) J. Bacteriol. 172, 643–647.

85. Vega-Palas, M. A., Flores, E. and Herrero, A. (1992) Mol. Microbiol. 6, 1853–1859.

86. Frías, J. E., Mérida, A., Herrero, A., Martín-Nieto, J., and Flores, E. (1993) J. Bacteriol. 175, 5710–5713.

87. Wei, T. F., Ramasubramanian, T. S., Pu, F. and Golden, J. W. (1993) J. Bacteriol. 175, 4025–4035.

88. Thiel, T. (1997) Data bank entry U89516.

89. Bradley, R. L. and Reddy, K. J. (1997) J. Bacteriol. 179, 4407–4410.

90. Lindell, D., Padan, E. and Post, A. F. (1997) Data bank entry AF017020.

91. Wei, T.-F., Ramasubramanian, T. S. and Golden, J. W. (1994) J. Bacteriol. 176:4473–4482.

92. Ramasubramanian, T. S., Wei, T.-F., Oldham, A. K. and Golden, J. W. (1996) J. Bacteriol. 178, 922–926.

93. Frías, J. E. (1996) Thesis, Universidad de Sevilla, Sevilla.

94. Frías, J. E., Flores, E. and Herrero, A. (1994) Mol. Microbiol. 14, 823–832.

95. Harano, Y., Suzuki, I., Maeda, S.-I., Kaneko, T., Tabata, S. and Omata, T. (1997) J. Bacteriol. 179, 5744–5750.

96. Ramasubramanian, T. S., Wei, T.-F. and Golden, J. W. (1994) J. Bacteriol. 176, 1214–1223.

97. Cohen-Kupiec, R., Zilberstein, A. and Gurevitz, M. (1995) J. Bacteriol. 177, 2222–2226.

98. Liotenberg, S. (1995) Thesis, Institut Pasteur, Paris.

99. Bauer, C. C. and Haselkorn, R. (1995) J. Bacteriol. 177, 3332–3336.

100. Reyes, J. C. and Florencio, F. J. (1995) Plant Mol. Biol. 27, 789–799.

55

ELECTRON TRANSPORT TO NITROGENASE IN *RHODOSPIRILLUM RUBRUM*

Anders Lindblad and Stefan Nordlund

Department of Biochemistry
Arrhenius Laboratories for Natural Sciences
Stockholm University
S-106 91 Stockholm, Sweden

1. INTRODUCTION

Biological nitrogen fixation is a strictly prokaryotic phenomenon which, in all systems hitherto examined, is catalysed by the nitrogenase protein complex. Nitrogenase consists of two proteins, dinitrogenase reductase (Fe-protein) and dinitrogenase (MoFe-protein) [1]. Dinitrogenase reductase is constituted of two identical subunits, encoded by *nifH*, bridged by one [4Fe-4S] cluster. Dinitrogenase is an $\alpha_2\beta_2$ tetramer, encoded by *nifD* and *nifK* respectively, which contains two Fe-Mo-cofactors (FeMoco) and two 2[4Fe-4S] clusters (P-clusters) [2]. Nitrogen fixation is an energetically costly process, at least 16 ATP and 8 electrons are required for each dinitrogen molecule reduced [1]. The expression of genes, encoding polypeptides involved in nitrogen fixation, the *nif*-genes, (and other genes, often encoding proteins involved in nitrogen metabolism) are transcriptionally regulated by the nitrogen status in the cell [3]. Many of the *nif*-genes are transcribed from a promoter recognised by an alternative σ-factor of the RNA-polymerase [3]. In addition the activity of nitrogenase is metabolically regulated in some diazotrophs, *e.g. Rhodospirillum rubrum*, by reversible ADP-ribosylation of one of the subunits of dinitrogenase reductase, a physiological phenomenon referred to as the "switch off"-effect [4]. Effectors, such as ammonium ions, glutamate and also darkness, regulate the extent of ADP-ribosylation. The pathway generating reductant to nitrogenase has been genetically and biochemically well characterised only in the enterobacterium *Klebsiella pneumoniae*. In this organism a pyruvate: flavodoxin oxidoreductase, encoded by *nifJ*, has been shown to transfer electrons from pyruvate to nitrogenase via a nitrogen fixation specific flavodoxin, encoded by *nifF* [5].

In contrast little is known about the electron transport pathway to nitrogenase in photosynthetic diazotrophs. These bacteria represent a totally different metabolic strategy with respect to energy and reductant metabolism. A direct involvement of the photosynthetic machinery in production of reducing equivalents to nitrogenase is not plausible as

The Phototrophic Prokaryotes, edited by Peschek *et al.*
Kluwer Academic / Plenum Publishers, New York, 1999.

the primary photosynthetic acceptor has a midpoint potential (E_m) too positive to function as a reductant to nitrogenase.

2. PYRUVATE OXIDOREDUCTASE IN *R. RUBRUM*

In the photosynthetic bacterium *R. rubrum*, a pyruvate dependent nitrogenase activity has been demonstrated *in vivo* [6] and a pyruvate oxidoreductase from *R. rubrum* has been purified and partially characterised [7]. This enzyme shows a high degree of similarity to the enzyme from *K. pneumoniae* with respect to molecular and substrate properties. The enzyme is a dimer of two identical subunits with an estimated molecular mass of 252 kDa. The iron content was determined to 13±4 atoms per mol of dimer and thiamine pyrophosphate identified as a cofactor. The activity of the enzyme, determined as methyl viologen-reducing activity, was strictly dependent on the addition of pyruvate and coenzyme A. The K_m value for pyruvate and CoA was determined to 179 µM and 9 µM respectively. In contrast to results from previous studies of cell extracts of *R. rubrum* [6], 2-oxoglutarate and oxaloacetate could not substitute for pyruvate as electron donor. The purified protein is very sensitive to oxygen, a 5 minutes exposure to air resulted in a total and irreversible inactivation. The purified enzyme was clearly shown to support pyruvate dependent nitrogenase activity in an *in vitro* assay, in the presence of added ferredoxin suggesting that ferredoxin rather then flavodoxin is the electron acceptor of the pyruvate oxidoreductase in *R. rubrum*.

3. THE *nifJ* GENE IN *R. RUBRUM*

3.1. The *nifJ* Sequence

A *nifJ* homologue was identified in *R. rubrum* by Southern hybridisation [8], using the *nifJ* gene from *K. pneumoniae* as probe. No hybridisation to the *nifF* gene from *K. pneumoniae* was detected. The gene encoding the pyruvate oxidoreductase from *R. rubrum*, including the upstream region, was sequenced. The deduced amino acid sequence of the *nifJ*-like gene in *R. rubrum* was compared to published sequences from *K. pneumoniae* and *Anabaena* 7120. The total percentage of identity of the *R. rubrum* gene was 47% and 56% respectively.

3.2. Expression of *nifJ*

In contrast to *K. pneumoniae*, pyruvate oxidoreductase activity was detected in crude extracts of *R. rubrum* even under non nitrogen fixing growth conditions. When grown on ammonia as nitrogen source (*nif* gene repressing conditions) the specific activity of pyruvate oxidoreductase, measured as methyl viologen-reduction, was 26% of the activity in nitrogen fixing cells (unpublished results). The transcription of *nifJ*, detected as Northern hybridisation, was approximately 50% in the same experiment. Bauer and coworkers have shown [9] that transcription of *nifJ* in *Anabaena* 7120 is dependent on the iron status of the cell. In *Anabaena* there was no transcription of *nifJ* in the presence of added iron. In contrast there is no increase in transcription of *nifJ* in *R. rubrum* under iron limiting growth conditions (unpublished results). Surprisingly, a probable σ^{54} promoter was identified by sequence comparison and primer extension experiments, in the upstream region of the *nifJ* -like gene from *R. rubrum.* [8]. No other promoter that would lead to expression under non nitrogen fixing conditions could be detected (unpublished results).

3.3. Mutation of the *nifJ* Gene

A *nifJ* mutant (SNJ-1) was constructed by insertion of a kanamycin resistance cassette in an internal *SphI* site in the *nifJ* gene[8]. The mutation was confirmed by, Southern and Northern hybridisation and by the lack of pyruvate oxidoreductase activity in the mutant. The mutant strain was able to grow at the same rate as the wild type *R. rubrum* with molecular nitrogen as the sole nitrogen source. The mutant strain showed no differences in nitrogenase activity compared to the wild type with either pyruvate or malate as the carbon source. The growth rates under iron limiting diazotrophic conditions showed no difference between the wild type and the mutant. Taken together, these results indicate that the main function of the pyruvate oxidoreductase in *R. rubrum*, is not in the electron transport to nitrogenase. Further studies are required to determine the main function of pyruvate oxidoreductase in *R. rubrum*, which in fact could be to catalyse the reverse reaction i.e. work as a pyruvate synthase.

4. INVOLVEMENT OF THE MEMBRANE POTENTIAL IN THE ELECTRON TRANSPORT TO NITROGENASE

The results presented above, indicate the operation of an alternate pathway of electron transfer to nitrogenase is present in *R. rubrum*. A direct involvement of the photosynthetic machinery in production of reducing equivalents to nitrogenase in *R. rubrum* would be an elegant solution of the requirement of reducing power to nitrogenase. However this is not electrochemically favourable as the primary photosynthetic acceptor has a midpoint potential (E_m) top positive to reduce a direct electron carrier to nitrogenase. It has previously been suggested that the energization of the chromatophore membranes could be involved in producing reductant to nitrogenase. In the phototroph *Rhodobacter spheroides* whole cell nitrogenase activity could be inhibited by lowering the potential across the cytoplasmic membrane while the ATP/ADP ratio remained constant [10]. By using ionophores on bacteroids of *Rhizobium leguminosarum* it was shown that valinomycin and nigericin led to a loss in nitrogenase activity without significant changes in the ATP/ADP ratio [11]. Similar results were obtained with *Azotobacter vinelandii*, using TTFB as uncoupler [12]. In addition to these findings, more recent studies have identified a new class of nitrogen fixing genes in the phototroph *Rhodobacter capsulatus* [13,14]. Further studies of the *rnf* genes have indicated that some of the products are membrane bound and that they are subunits of a membrane protein complex [15].

4.1. The Effect of Altering the Membrane Potential in *R. rubrum*

We suggest that maintaining a high membrane potential is crucial not only for a sufficient supply of ATP but also for generating reductant to nitrogenase, possibly involving a membrane bound protein complex. To initially investigate the possibility of effecting nitrogenase activity by altering the energization of the chromatophore membranes, we have studied the effect of different light intensities and of uncouplers on *R. rubrum* [16]. Two different strains of *R. rubrum* were used, the wild type (S1) and a mutant (UR212) lacking the ability to modify nitrogenase reductase by ADP-ribosylation.

In the light experiments a decrease in nitrogenase activity was observed when the light intensity was lowered from saturation (4000 lux) down to darkness, but the ATP/ADP ratio remained high. This was observed both for the wild type and strain UR212. At 10% of the original light intensity the activity had dropped to 18% for the wild type but there was no detectable modification of dinitrogenase reductase, by ADP-ribosylation, as detected by immuno-blotting. This indicates that the lowered acetylene reduction activity of nitrogenase is not due to a lack of ATP and that nitrogenase activity is more sensitive to a decrease in light intensity than the ATP/ADP ratio. From this experiment we conclude that a high membrane potential probably is critical for generating a reduced electron carrier to nitrogenase in *R. rubrum*. A different pattern was observed with the use of uncouplers. At a low concentrations (1–15 μM) of CCCP or FCCP the ATP/ADP ratio collapsed. This was the case both for the wild type and strain UR212. The cells were, in spite of this, capable of retaining a high acetylene reducing activity. Modification ("switch-off") of dinitrogenase reductase was observed only at high concentrations of uncoupler in the wild type. Thus nitrogenase activity is less sensitive to uncouplers than is the ATP/ADP ratio.

In conclusion: the observed decrease in nitrogenase activity in response to lowered light intensity or uncouplers does not correlate with the ATP/ADP ratio nor is it due to inactivation of the Fe protein by ADP-ribosylation. These results support a model in which reduction of electron donors to nitrogenase in *R. rubrum* is coupled to the energization of the chromatophore membranes by photosynthesis.

5. NAD(P)H AS A ELECTRON SOURCE TO NITROGENASE IN *R. RUBRUM*

As the cyclic electron flow in the photosynthetic machinery of *R. rubrum* does not give any net inflow of electrons that could be used in the nitrogenase reaction, we have investigated the transfer of electrons from the carbon sources. The presence of the TCA cycle as well as pyruvate dehydrogenase has previously been demonstrated in *R. rubrum*. [17–20]. The role of the TCA cycle in producing reducing equivalents, that could be used as electron donors to nitrogenase, was investigated by inhibiting the TCA cycle with fluoroacetate. Fluoroacetate is metabolised by citrate synthase to fluorocitrate which is a potent inhibitor to aconitase, thereby inhibiting the TCA cycle [17,21]. Whole cells, grown diazotrophically on molecular nitrogen and with pyruvate as carbon source, showed a nearly complete inhibition of nitrogenase activity after addition of 1 mM fluoroacetate (final concentration) [22]. The nitrogenase activity could be restored by adding NADH (1.5–3.0 mM, final concentration). In additions, the consumption of added NADH corresponded well with the amount of electrons used in the nitrogenase reaction (measured as acetylene reduction). The inhibition of nitrogenase activity was less pronounced if malate was used as electron source, which can be due to the reaction of malate dehydrogenase, generating NADH without a complete TCA cycle [23]. It should, however, be pointed out that the difference in midpoint potential between NAD(P)H and ferredoxin, the most likely electron carrier to nitrogenase in *R. rubrum*, suggests that there is a requirement of additional input of energy to produce a potent reductant to nitrogenase.

6. CONCLUDING REMARKS

Summarising the above results make it less plausible that the pyruvate oxidoreductase is the main electron transport pathway for supplying reductants to nitrogenase in *R.*

rubrum. A model involving the energization of the chromathophore membranes is supported by the findings that light and uncouplers can effect the ability of whole cells to fix nitrogen. This is further supported by the experiment showing that a functional TCA cycle is critical for maintaining a high nitrogen fixing ability in whole cells of *R. rubrum*, under the growth conditions we have studied. Studies have been initiated to identify the components that are required to couple the membrane potential to the reduction of the electron donor to nitrogenase.

ACKNOWLEDGMENTS

We are very grateful to Dr. Gary P. Roberts for the gift of strain UR212. These investigations were supported by grants to S.N. from the Swedish Natural Science Research Council and Carl Tryggers Stiftelse för Vetenskaplig Forskning

REFERENCES

1. Burris, R. H. (1991) J. Biol. Chem. 266, 9339–9342.
2. Kim, K. and Rees, D. C. (1994) Biochemistry 33, 389–397.
3. Merrick, M. J. (1992) in: Biological Nitrogen Fixation (Stacey, G., Burris, R. H. and Evans, H. J. eds.) pp. 825–876, Chapman and Hall, London.
4. Ludden, P. W. and Roberts, G. P. (1989) Curr. Top. Cell. Regu. 30, 23–55.
5. Ludden, P. W. (1991) Curr. Top. Bioenerg. 16,369–380.
6. Ludden, P. W. and Burris, R. H. (1981) Arch. Microbiol. 130, 155–158.
7. Brostedt, E. and Nordlund, S. (1991) Biochem. J. 279, 155–158.
8. Lindblad, A., Jansson, J., Brostedt, E., Johansson, M., Hellman, U. and Nordlund, S. (1997) Mol. Microbiol 20, 559–568.
9. Bauer, C. C., Scappino, L. and Haselkorn, R. (1993) Proc. Natl. Acad. Sci. USA 90, 8812–8816.
10. Haaker, H., Laane, C., Hellingwerf, K., Houwer, B., Konings, W. N. and Veeger, C. (1982) Eur. J. Bichem. 127, 639–645.
11. Laane, C., Krone, W., Konings, W. N., Haaker, H. and Veeger, C. (1979) FEBS Lett. 103, 328–332.
12. Haake, H., de Kok, A. and Veeger, C. (1974) Biochim. Biophys. Acta. 357, 344–357.
13. Schmehl, M., Jahn, A., Mayer zu Vilsendorf, A., Hennecke, S., Masepohl, B., Scuppler, M., Marxer, M., Olelze, J. and Klipp, W. (1993) Mol. Gen. Genet. 241, 602–615.
14. Saeki, K., Tokuda, K-i., Fujiwara, T. and Matsubara, H. (1993) Plant Cell Physiol. 34(2), 185–199.
15. Kumagai, H., Fujiwara, T., Matsubara, H. and Saeki, K. (1997) Biochemistry 36, 5509–5521.
16. Lindblad, A. and Nordlund, S. (1997) Photosynt. Res, In press.
17. Elsden, S. R. and Ormerod, J. G. (1956) Biochem. J. 63, 691–701.
18. Eisenberg, M. A. (1953) J. Biol. Chem. 203, 815–835.
19. Gorell, T. E. and Uffen, R. L. (1978) J. Bacteriol. 134, 830–836.
20. Lüderitz, R. and Klemme, J-H. (1977) Z. Naturforsch. 32c, 351–361.
21. Gest, H., Ormerod, J. G.and Ormerod, K. S. (1962) Arch. Biochem. Biophys 97, 21–33.
22. Brostedt, E., Lindblad, A, Jansson, J. and Nordlund, S. (1997) FEMS Microbiol. Lett. (1997) 150, 263–267.
23. Ormerod, J. G.and Gest, H. (1962) Bacteriol. Rev. 26, 51–66.

56

NITROGEN FIXATION IN THE MARINE CYANOBACTERIUM *TRICHODESMIUM*

A Challenging Model for Ecology and Molecular Biology

J. P. Zehr,[1] B. Dominic,[1] Y. -B. Chen,[1] M. Mellon,[1] and J. C. Meeks[2]

[1]Department of Biology
Rensselaer Polytechnic Institute
Troy, New York 12180-3590
[2]Section of Microbiology
University of California
Davis, California 95616-5224

1. INTRODUCTION

Over the past decade, the marine cyanobacteria *Trichodesmium* spp. have received increasing attention [1,2], because of their ability to fix nitrogen in the presence of oxygen, and the recognition of their importance in ocean N budgets [3,4]. *Trichodesmium* spp. are distributed worldwide in tropical and subtropical seas [2]. Dugdale et al. [5] first demonstrated that nitrogen fixation occurred in association with *Trichodesmium* aggregates in the sea, although it was many years before it became clear that the nitrogen fixation observed was indeed due to the cyanobacterium, and not associated heterotrophic bacteria. Evidence based on DNA sequences (*nifH*) amplified from natural populations, and immunological studies eventually confirmed that *Trichodesmium* itself was the primary nitrogen fixer [6–8], even though other diazotrophs are sometimes found associated with *Trichodesmium* aggregates [9,10].

Nitrogenase, which catalyzes the reduction of N_2 to ammonium, is composed of two proteins. The molybdenum iron protein (sometimes called Component I or dinitrogenase) is a tetramer composed of identical subunits of two types (α and β), which bind a molybdenum iron sulfur cluster between the β subunits [11]. The iron protein (Component II or dinitrogenase reductase) is a dimer of identical subunits that coordinate an FeS center. The structural proteins are encoded by the *nifHDK* genes, which have been mapped in several cyanobacterial genera using hybridization techniques [12–14]. The structures of the Fe protein and the MoFe protein have been solved [11] and *nifHDK* genes have been sequenced from representatives of diverse organisms [15]. While not extensive, there are representative sequences from heterocystous filamentous, nonheterocystous filamentous, and unicellular cyanobacte-

The Phototrophic Prokaryotes, edited by Peschek et al.
Kluwer Academic / Plenum Publishers, New York, 1999.

rial species [16]. Less is known about the mechanisms of regulation in cyanobacteria than in enterobacteria (e.g. *Klebsiella*), but there is a growing body of literature on regulatory mechanisms involved in nitrogen fixation by cyanobacteria [17,18]. In general, cyanobacterial nitrogenase structural genes are the "conventional" *nifH,D,* and *K* genes, which encode subunits of nitrogenase requiring a molybdenum metallocenter. Alternative nitrogenases, which substitute either vanadium or iron for molybdenum in the metallocenter, have been well-studied in the proteobacterial genera *Azotobacter* and *Rhodobacter* [19]. Multiple copies of nitrogenase have been reported in the cyanobacterium *Anabaena variabilis*, including a VFe protein [20] and a second MoFe nitrogenase that was expressed only in vegetative cells under anoxic conditions [21]. The *nifHDK* genes in most heterocystous cyanobacteria undergo a rearrangement, with an excision of a circular 11 kilobase piece of DNA, during the development of the heterocyst [22]. The occurrence of different nitrogenase operons, and the structure of the nitrogenase genes, clearly are involved in determining the mechanisms by which cyanobacteria fix nitrogen in different cells and at different times. Until recently, little was known about the biochemistry and molecular biology of nitrogen fixation in *Trichodesmium*.

Since nitrogenase is rapidly inactivated by oxygen, cyanobacteria typically either spatially or temporally separate nitrogen fixation activity from photosynthetic oxygen evolution [23]. In general, spatial mechanisms are characteristic of heterocystous species, and separation of nitrogen fixation activity and photosynthesis into the dark and light phases of a daily cycle in nonheterocystous species. *Trichodesmium* only fixes nitrogen during the light phase of a light-dark cycle [2,24]. It is this restriction of nitrogen fixation to the light phase of a light-dark cycle, that distinguishes *Trichodesmium* from certain other nonheterocystous cyanobacteria that can fix aerobically under certain conditions, such as *Gloeothece* or *Oscillatoria* [25]. This pattern of nitrogen fixation presents a paradox with respect to the classical view of mechanisms of nitrogen fixation in oxygen-evolving phototrophs [1]. *Trichodesmium* provides contrast to both nonheterocystous and heterocystous models of nitrogen fixation in cyanobacteria, and presents unique physiological and biochemical research questions.

2. BASIC ECOLOGY OF TRICHODESMIUM RELEVANT TO NITROGEN FIXATION

From an ecological perspective, *Trichodesmium* is interesting since it plays an important role in the carbon and nitrogen fluxes of the open ocean [2,4]. *Trichodesmium* grows as long unbranching filaments, which cluster together to form aggregates of various morphologies. The aggregate or colony morphology has often been used as one of the characteristics to distinguish *Trichodesmium* species [10,26]. There is some uncertainty over the taxonomic status of the various species of *Trichodesmium*, but several have been described on the basis of colony morphology, cellular dimensions and ultrastructure [10,27]. Molecular evidence suggests that some of the species or strains are very closely related [2,28–30]. For many years, *Trichodesmium* proved difficult to cultivate, but there are now at least two strains that can be maintained in stable culture on completely artificial seawater media [31,32].

Trichodesmium is buoyant due to its gas vacuoles, and can migrate great distances through the water column [2]. Thus, even though the maximum concentration of colonies is often found between 10 and 30 meters depth, *Trichodesmium* may migrate through large gradients of light and nutrients over time scales of hours to days. The ability to vertically migrate may be particularly important in understanding the biochemistry and physiology of nitrogen fixation in *Trichodesmium*, and the potential physiological diversity of different species should be noted prior to the ensuing discussion on biochemistry and molecular biology of *Trichodesmium*. Most of the molecular information has been obtained from the cultivated isolates *Trichodesmium* sp. IMS 101 [32], or *Trichodesmium* sp. NIBB 1067 [32]. Both of these strains may be most closely related to *T. erythraeum* [33].

3. NITROGEN UPTAKE AND ASSIMILATION

Many phototrophic organisms growing in nutrient-depleted oceanic waters depend on rapidly cycled ammonium (i.e. nutrient regeneration), but *Trichodesmium* does not appear to be competitive in the use of regenerated nitrogen sources *in situ* [34]. *Trichodesmium* isolates do grow on nitrate, urea or ammonium as sole nitrogen sources [31,35]. Interestingly, nitrogenase synthesis is repressed when *Trichodesmium* sp. NIBB 1067 is grown on urea, but not by nitrate [31]. Although the nitrogenase protein is still present when the culture is grown on fixed nitrogen sources, it is inactive [31]. Nitrate does repress nitrogenase synthesis in *Trichodesmium* sp. IMS101 [36]. *Trichodesmium* spp. must contain nitrate and nitrite reductases and urease, since the cultures can be cultivated on nitrate and urea, but this has not been demonstrated by biochemical or immunological assays. In fact, it is sometimes reported that *Trichodesmium* is unable to utilize nitrate [1]. In other cyanobacteria, it has been suggested that the regulatory mechanism involved in nitrate or ammonium repression is through the same intermediate, since inhibition of glutamine synthetase (GS) prevented nitrate from inhibiting nitrogenase synthesis [17]. It is curious, therefore, that nitrate does not inhibit nitrogenase synthesis in *Trichodesmium* sp. NIBB1067.

Trichodesmium can take up amino acids such as glutamate (Glu), glutamine (Gln), as well as the glutamate analog and GS inhibitor, methionine sulfoximine (MSX) [37], but it is not known whether these amino acids and/or the transport rate will support growth. Whereas MSX lowered the rate of Glu and Gln uptake [37], neither Glu nor Gln blocked

Table 1. Effect of glutamine or glutamate on *in vivo* MSX inhibition of GS transferase activity

Treatment	nmol gamma-glutamyl hydroxymate formed per min per mg protein	% inhibition
Control	4617 ± 1108	
MSX	225 ± 162	95.1
MSX + 0.05 mM Glu	166 ± 39	96.3
MSX + 0.05 mM Gln	430 ± 108	90.7
MSX + 0.20 mM Glu	250 ± 18	94.6
MSX + 0.20 mM Gln	357 ± 275	92.2
MSX + 1.0 mM Glu	597 ± 488	87.1
MSX + 1.0 mM Gln	228 ± 6	92.6

Values are means ± SD of two replicate samples. Cells were collected at 9:00 AM. Supplements added first, followed by MSX at a final concentration of 0.2 mM. Samples were incubated in supplemented, filtered sea water at ambient temperature for 2 h and the reaction stopped by filtration and freezing in LN. Samples were thawed, lysed by homogenization with glass beads in a break buffer of 20 mM Hepes, pH 7.0, 1 mM PMSF, 0.2 mM $MnCl_2$. Immediately after lysis, the samples were assayed for 5 min for transferase activity and time zero values of gamma-glutamylhydroxamate subtracted. The assay mixture was that of (Lee et al., [43]) in 100 mM Hepes pH 7.7. Protein was determined with a Pierce kit.

MSX transport when at equal or fivefold excess concentrations, as determined by *in vivo* inhibition of GS activity (Table 1). Thus, it is not clear whether the amino acids and analog are transported by separate pathways in *Trichodesmium*, or by a common pathway as occurs in nonmarine cyanobacteria [17].

Ammonium, the product of nitrogen fixation, nitrite reduction and urease, is presumably assimilated via the GS-glutamate synthase (GOGAT) pathway; GS activity and protein has been detected in *Trichodesmium* [37], but the results of neither GOGAT nor glutamate dehydrogenase assays have been reported. A fragment of *glnA*, the gene encoding GS, has been cloned from *Trichodesmium* and sequenced [38]. *glnA* is transcribed on a diel cycle with 2 peaks during the day; the first coinciding with maximum nitrogenase transcript abundance, and the second during the "switch-off" of nitrogen fixation near the end of the light phase [38]. However, a diel pattern is not reflected in GS protein abundance [37] or GS catalytic activity (Table 2). Both nonphysiological transferase and physiological biosynthetic activities of GS show some variability over the diel cycle, but the variation is similar in the two activities (Table 2). In many microorganisms, GS may be modified by adenylation [39] or ADP ribosylation [40], which result in lower biosynthetic but not transferase activities. The consistent coupling between transferase and biosynthetic activities implies a lack of modification of the GS protein in *Trichodesmium*. The measured transferase activities are three- to fourfold higher than activities seen in crude extracts of heterocystous cyanobacteria [41–43], but the biosynthetic activities are not correspondingly higher; consequently there is an exceptionally high fourtyfold ratio between transferase and biosynthetic activities. Typically, transferase activity is about eight to tenfold higher than biosynthetic activity [39,42,43]. Whether this high ratio might reflect an unknown modification of the GS protein affecting biosynthetic activity prior to or during cell lysis is not known. However, the GS of another filamentous marine cyanobacterium, *Anabaena* sp. CA, also had a similarly high transferase to biosynthetic activity ratio [41].

4. NITROGEN FIXATION APPARATUS

Subsequent to the amplification of a portion of *nifH* from natural populations of *Trichodesmium* collected in the Caribbean Sea [8], the nitrogenase structural genes

Table 2. Average GS activities (biosynthetic and transferase) of *Trichodesmium* colonies collected over a diel cycle in eastern Caribbean Sea during January, 1995

Time	Activity as nmol per min per mg protein		Ratio as T/B
	Biosynthetic	Transferase	
4:00 AM	103.7	3731	36.0
7:00 AM	111.6	3812	34.2
10:30 AM	91.9	3132	34.1
1:00 PM	114.2	3849	33.7
4:30 PM	87.9	3371	38.4
6:30 PM	126.4	3950	31.3
10:00 PM	103.4	3245	31.4
12:00 PM	119.3	3834	32.1

Values are of a single experiment and are representative of two diel sampling cycles with slightly different sampling times. Activity was measured during 5 min assays immediately after cell lysis with time zero values of gamma-glutamylhydroxamate subtracted as in Table 1. Assay mixtures (Lee et al., [43]) in an assay buffer of 100 mM Hepes, pH 7.5 (biosynthetic) or pH 7.7 (transferase). All other conditions as in Table 1.

(*nifHDK*) were mapped in *Trichodesmium* sp. NIBB1067. The *nifHDK* genes, identified with heterologous probes from *Anabaena* sp. PCC 7120 (generously provided by R. Haselkorn), were found to be contiguous, an arrangement that is consistent with the arrangement in other nonheterocystous cyanobacteria such as *Plectonema boryanum* [13], *Pseudanabaena* [14] and *Cyanothece* (*Synechococcus* sp. RF-1) sp. (GenBank U22146; [44]). The Southern analysis of *Trichodesmium* sp. NIBB 1067 showed that the map of the *nif* structural operon was identical regardless of whether the culture had been grown under nitrogen fixing conditions or with fixed nitrogen [45]. Thus, there do not appear to be any gene rearrangements associated with nitrogen-fixing and non-nitrogen fixing *Trichodesmium* filaments. Restriction mapping of *nifHDK* using heterologous probes from *Anabaena* sp. PCC 7120 hybridized to single bands for *nifH, D* and *K* [45], indicating that there is a single copy of these genes. Multiple copies of *nif* structural genes have been discovered in a number of cyanobacterial species [18,21]. There is no evidence for multiple copies or alternative non-molybdenum containing *nif* genes in *Trichodesmium*. It is possible that the *nif* probes would not recognize alternative *nifH* genes, particularly the non-vanadium, non-molybdenum-containing second alternative nitrogenase. Very similar *nifH* sequences have been amplified from multiple samples and different species of *Trichodesmium* [33,46]. Interestingly, attempts to grow *Trichodesmium* without molybdenum have yielded inconsistent results. However, this may be due to the ability of *Trichodesmium* to grow on the very low concentrations of molybdenum found as contaminants.

Trichodesmium also contains *nifE, N, W, X, fdxH*, and *hesAB* [47], previously identified in *Anabaena* sp. PCC 7120 [18]. The arrangement and size of these genes corresponds to the arrangement found in *Anabaena* sp. PCC 7120 as well [47]. *nifE* and *nifN* are involved in synthesis of the MoFe protein, probably as scaffolding [48]. *nifX* may encode a regulatory protein [49]. The functions of the proteins encoded by *nifW* and the *hesAB* genes are not known, but they are required for nitrogen fixation in *Anabaena* sp. PCC 7120 [18,50,51]. The *fdxH* gene is expressed only in heterocysts in *Anabaena* sp. PCC 7120 [52], with the *fdxN* gene, located upstream of *nifH*, being expressed in vegetative cells. Unfortunately, it is not yet known which other *nif* genes are present in *Trichodesmium*, but it may be that the ferredoxin encoded by *fdxH* is the primary electron donor for the Fe protein in *Trichodesmium*. The *fdxH* gene has been found in another nonheterocystous cyanobacterium (*Plectonema boryanum*, [53]).

In some cyanobacteria, a flavodoxin is synthesized under Fe limiting conditions [54]. Iron availability was shown to be limiting growth of natural populations of *Trichodesmium* [55]. Furthermore, Fe availability may have regulated nitrogen fixation rates in the oceans over geological time scales [56]. Thus, it would seem advantageous for *Trichodesmium* to have the genetic capability to synthesize a flavodoxin to replace ferredoxin in at least some biochemical reactions. A protein in *Trichodesmium* sp. IMS101 cross-reacted with antisera to flavodoxin from *Anabaena*, but the protein appeared to be synthesized constitutively [57]. The electron donor for nitrogen fixation in *Trichodesmium* is not known, but a constitutively synthesized flavodoxin may be important in the ecological success of *Trichodesmium* in Fe-poor oligotrophic oceans.

4.1. Evolutionary Considerations

The structural genes for nitrogenase have been sequenced from a variety of organisms including representatives of the α, β, γ and δ *proteobacteria*, cyanobacteria and archaea. The structural genes of the conventional Mo-containing nitrogenase (*nifHDK*) are highly conserved, and are very similar among cyanobacteria. Analysis of available full length *nifH*

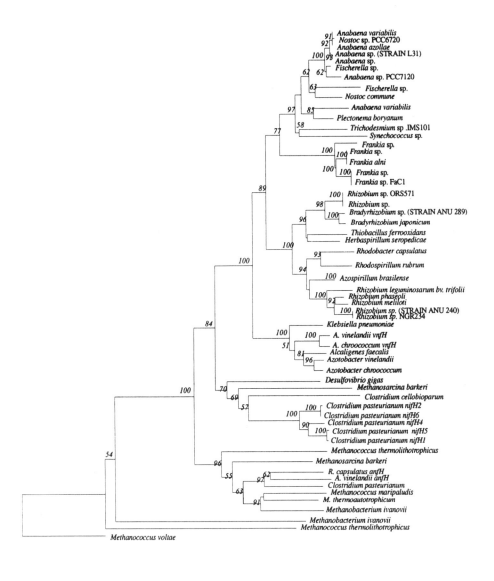

0.1 substitutions/site

Figure 1. Phylogenetic analysis of *nifH* sequences from A) diverse microorganisms based on full-length deduced amino acid sequences, and B) cyanobacteria based on partial deduced amino acid sequences. Sequences were analyzed by distance methods and bootstrapped 100 times as described by Zehr et al. (1997). Bootstrap values are shown at respective nodes. *Trichodesmium*, as well as other nonheterocystous cyanobacteria branch deeply relative to the heterocystous cyanobacterial sequences.

B

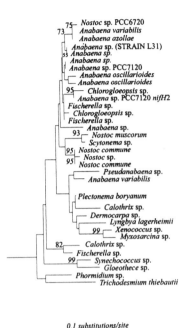

0.1 substitutions/site

Figure 1. (*continued*)

sequences indicates that cyanobacterial *nifH* sequences are more closely related to each other than to *nif* sequences from other nitrogen-fixing organisms, and that *Trichodesmium nifH* is relatively distantly related to *nifH* from other cyanobacteria (Fig. 1A). In general, the nonheterocystous cyanobacterial *nifH* genes are deeply branching, and do not form a coherent cluster (Fig. 1B). The *Trichodesmium nifH* sequence may be relatively closely related to that from *Microcoleus* (*Symploca*) *chthonoplastes*, another filamentous nonheterocystous cyanobacterium [30]. The distance between *Trichodesmium nifH* and the *nifH* of other cyanobacteria suggests that: (i) *Trichodesmium* diverged early in the evolution of the cyanobacteria and/or (ii) that the *Trichodesmium nifH* has unique primary sequence that could affect the secondary and tertiary structure of the Fe protein.

There are few *nifD* and *nifK* sequences available, but *Trichodesmium nifD* and *nifK* are more closely related to cyanobacterial than bacterial *nifDK* (Table 3). Interestingly, the *Trichodesmium nifH* gene is equally similar to unicellular and heterocystous filamentous cyanobacterial *nifH*, but the *Trichodesmium* deduced amino acid sequences of *nifD* and *nifK* are most similar to those of the unicellular *Cyanothece* (*Synechococcus*) spp. *nifD* and *nifK* sequences (Table 3). These data imply different lines of descent of *nifH, D* and *K*, or that there are varying degrees of biochemical constraints on the evolution of the proteins. The higher degree of identity among unicellular and filamentous nonheterocystous (*Plectonema* and *Trichodesmium*) *nifD* and *nifK* sequences, than between these sequences and *Anabaena*

Table 3. Percent identity between deduced amino acid sequences of *nifH*, *D* and *K* (H/D/K) from cyanobacteria and representative bacteria (*Klebsiella pneumoniae* and *Clostridium pasteurianum*)

	Clostridium pasteurianum	*Klebsiella pneumoniae*	*Nostoc (Anabaena)* sp. PCC 7120	*Synechococcus* sp.	*Cyanothece* sp.	*Plectonema boryanum*	*Trichodesmium thiebautii*
Trichodesmium thiebautii	65 / 42 / 41	71 / 67 / 50	79 / 66 / 68	79 / 75 / 70	76 / 75 / 71	78 / 72 / NA	—
Plectonema boryanum	59 / 41 / NA	71 / 66 / NA	84 / 71 / NA	77 / 78 / NA	75 / 78 / NA	—	
Cyanothece sp.	57 / 43 / 40	63 / 68 / 51	69 / 70 / 69	82 / 82 / 76	—		
Synechococcus sp.	62 / 42 / 41	67 / 70 / 54	75 / 71 / 69	—			
Nostoc (Anabaena) sp. PCC 7120	63 / 40 / 42	71 / 63 / 51	—				
Klebsiella pneumoniae	67 / 42 / 36	—					
Clostridium pasteurianum	—						

NA = data not available.

sp. PCC 7120 *nifD* and *K*, indicate that the phylogenetic analysis of *nifD* and *nifK* may provide interesting insights into the evolution of nitrogenase in the cyanobacteria.

The deduced amino acid sequence of the cloned *Trichodesmium glnA* gene fragment is only 61–74% identical to the gene from other cyanobacteria [38], and slightly more similar to the sequences of *Vibrio alginolyticus* and *E. coli* [38]. Sequences amplified from natural populations and *Trichodesmium* sp. IMS101 were 93% identical (at the DNA level), indicating that the amplified fragment was indeed derived from *Trichodesmium*. A less likely explanation is that the *glnA* fragment was amplified from a bacterium associated with natural populations that was also a contaminant in the culture. However, the rather distant relationship of the *glnA* gene to the gene from other cyanobacteria is consistent with the nitrogenase phylogeny, where *Trichodesmium* is a deeply branched lineage with respect to other cyanobacteria (Fig. 1). The database for *gln* genes is not as extensive as the database for *nif*, and conclusions regarding the phylogenetic relationship on the basis of *glnA* will have to wait until more sequence data are available.

Trichodesmium clusters with *Oscillatoria sp.* PCC 7515 with a high bootstrap value on the basis of 16S ribosomal RNA sequences [58]. However, *Oscillatoria* sequences are scattered through the 16S rRNA phylogenetic tree [58]. On both the *nifH* and 16S rRNA trees, the nonheterocystous filamentous genera are deeply branching (Fig. 1, [16]). Unfortunately, direct comparisons of the topology of the 16S rRNA and *nifH* trees cannot be made, since sequences from different genera are used to construct the two trees. The inclusion of the *Microcoleus nifH* sequence [30] creates a cluster of *Microcoleus*, *Phormidium*, *Pseudanabaena* and *Trichodesmium*. It appears that the topology of the *nifH* tree may approach that of the 16S rRNA tree as more sequences become available. In both trees, it is clear that many nonheterocystous filamentous genera are deeply branching and are not closely related to each other. Thus, *Trichodesmium*, and many other filamentous genera are probably ancient lineages.

The phylogeny of *Trichodesmium nif* could be affected by selection for oxygen-resistance. If the nitrogenase (Fe protein) has evolved in *Trichodesmium* as a result of selective pressures, then it implies that the protein has changed to adopt a structure which is beneficial. The structure of the Fe protein of Trichodesmium, which was investigated by computer modeling based on the *Azotobacter vinelandii* Fe protein structure determined by X-ray diffraction [59], shows that there is very little difference between the structures [60]. Thus, it is unlikely that the *Trichodesmium* Fe protein structure is more tolerant to oxygen than any other Fe proteins. Similar analysis of the MoFe protein has not been performed. However, the *Trichodesmium* amino acid sequences of the α and β subunits of the MoFe protein are more divergent from the corresponding sequences of other cyanobacteria than is *nifH* (Table 3). If the nitrogenase of *Trichodesmium* has unusual properties that are involved in simultaneous oxygen evolution and photosynthesis, they are likely to be structural features of the MoFe subunits, and not the Fe protein. The structure of the Fe protein does not appear to be affected by the primary sequence of *Trichodesmium nifH*, indicating that the *nifH* sequence reflects the phylogenetic relationship, rather than selective pressure.

4.2. Regulation of Nitrogen Fixation

The nitrogenase structural operon is generally believed to be transcribed from a single promoter upstream of *nifH* [18]. The polycistronic messenger RNA molecule includes *nifH*, *D* and *K*, although fragments corresponding to *nifHD*, and *nifH* have been reported as stable transcripts [22]. The transcriptional start site of *Trichodesmium* sp. IMS101 *nifH*

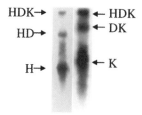

Figure 2. Analysis of *nifHDK* transcripts in *Trichodesmium* sp. IMS101 using cloned *Trichodesmium nifHDK* DNA as probe. Transcripts identified by Dominic and Zehr (submitted) are indicated by arrows. Results indicate multiple transcript start sites.

has been identified by primer extension, and is located 212 base pairs upstream of the *nifH* ATG codon [61]. This leader region is rather long in comparison to those for *nif* genes in *A. variabilis* (*vnfH*, 80 bp [19]) or *Anabaena* sp. PCC 7120 (127 bp, [62]). Transcript analysis, using cloned *nifH*, *nifD* and *nifK* DNA probes, showed that in addition to transcripts initiating upstream of *nifH*, additional transcripts were initiated upstream of *nifD* and *nifK*, resulting in *nifD*, *nifDK* and *nifK* transcripts as well as *nifH*, *nifHD* and *nifHDK* (Fig. 2; [61]). These transcripts can be observed in Northern blots probed with a *nifHDK* probe (Fig. 2). Reasonable terminator sequences were found in the intergenic regions between *nifH* and *nifD*, and between *nifD* and *nifK* [61]. Although there appear to be multiple transcript start points, analysis over a diel cycle indicate that the transcripts may be coordinately regulated [36]. The *nifH* transcript is the most abundant (Fig. 2), which may be a result of a high turnover rate of the Fe protein relative to the MoFe protein, or the requirement of the Fe protein for other activities, such as its role in the synthesis of the FeMo-cofactor for the MoFe protein.

In many prokaryotes, nitrogen metabolism is controlled through the NtrB/NtrC (*ntr*) two-component system [17,63,64]. A uridylyltransferase and the P_{II} protein, encoded by *glnD* and *glnB*, repectively, are involved in the regulatory cascade [63]. When cells are N-limited, a phosphorylation cascade presumably initiated by P_{II} is initiated and the NtrC protein is phosphorylated [63,64]. Phosphorylated NtrC is the transcriptionally active form which binds to the DNA consensus sequence GCAC(N_7)GTGC [63], which is usually located 100–200 bases upstream of the transcription start site of *ntr* regulated promoters including the nitrogen fixation regulatory protein, *nifA* [63,65]. In enterobacteria, the *nifA* gene codes for the DNA binding protein NifA which binds to the consensus sequence TGT(N_{10})ACA, located 100–200 bases upstream of the transcription start site of nitrogen-fixation genes of many diazotrophs [65].

In cyanobacteria, repeated efforts have failed to identify homologs for *nifA*, *ntrB* and *ntrC*, but the recently identified *ntcA* gene was shown to have regulatory effects on N-assimilation genes [17]. Homologs for *ntcA* have been identified in at least nine different cyanobacterial species, and NtcA was shown to belong to the Crp (the cAMP-receptor protein) family of bacterial regulators [66]. The DNA consensus sequence to which NtcA binds has been determined to be GTA(N_8)TAC located approximately 35 bases upstream of transcription start site of nitrogen-regulated genes [67]. It is suggested that NtcA exerts its regulatory action by positively influencing mRNA levels of the nitrogen-regulated genes, although NtcA expression was inversely proportional to *nifHDK* expression in *Cyanothece* sp. [67,68]. So far, the molecular mechanisms that regulate *ntcA* expression or NtcA activity in response to the N-status of the cell have not been established [17].

Trichodesmium contains a protein that crossreacts with antiserum to the P_{II} protein from *Synechococcus* sp. strain PCC 7942, where it functions as the carrier of the cellular nitrogen status by modification/demodification [69]. The P_{II} protein was present in

A

B

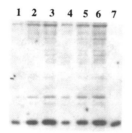

Figure 3. P_{II} protein immunoblots of SDS-PAGE from *Trichodesmium* extracts (A) of a diel cycle or (B) following in vivo ammonium shock. Antibody to the *Synechococcus* sp. strain PCC 7942 P_{II} protein (Forchhammer and Tandeau de Marsac, 1994) was used at a 1:20,000 dilution and detected using horseradish peroxidase conjugated to goat antirabbit antibody and the *p*-iodophenol/luminol protocol. Frozen and thawed samples were homogenized with glass beads in a SDS-β-mercaptoethanol break buffer, clarified by centrifugation and heated before loading onto a SDS-10% polyacrylamide gel. Each lane in panel A represents about 2 *Trichodesmium* colonies. Lanes: 1. 4:00 AM; 2, 5:30 AM, 3, 7:00 AM; 4, 8:30 AM; 5, 11:30 AM; 6, 2:30 PM; 7, 4:30 PM; 8, 6:00 PM; 9, 7:30 PM; 10, 9:00 PM; 11, reference sample of N_2-grown *Nostoc punctiforme*; 12, reference sample of NH_4^+-grown *Nostoc punctiforme*. Panel B is of *Trichodesmium* colonies following a 30 min incubation at ambient temperature in the filtered sea water, without or with 0.2 mM ammonium. Incubation was stopped by filtration and freezing of samples in LN. Samples were processed as above and varying concentrations loaded onto a SDS-10% polyacrylamide gel. Samples incubated in the absence of ammonium are in lanes 1–3, those in the presence of ammonium lanes 4–6. Lanes 1 and 4, 2 and 5, and 3 and 6 contain the protein equivalent of 0.5, 1.0 and 2.5 *Trichodesmium* colonies, in the order stated.

Trichodesmium throughout the diel cycle (Fig. 3A) and there was no evidence for its modification (data not shown). In contrast, *nifH* expression in natural populations follows a diel pattern [70], which has been confirmed in the culture of *Trichodesmium* sp. IMS101. It is possible that P_{II} may play a more general role in coordinating carbon and nitrogen metabolism, rather than only nitrogen source utilization; under these conditions its presence would be necessary throughout the diel cycle. This is consistent with the fact that P_{II} abundance was not affected by ammonium shock (Fig. 3B).

The transcription start site upstream of *Trichodesmium nifHDK* has recently been determined by primer extension [61]. Unfortunately, sequence information upstream of the *nifH* transcription start site, located over 200 bp upstream of the translation start, is not yet available. NtcA consensus binding sites could not be found within the 220 bp leader sequence of the *Trichodesmium* sp. *nifH* coding region [61], although they may exist upstream of the transcription start site. NtcA binding consensus sequences were not found in the intergenic region between *nifH* and *nifD*; *nifD* and *nifK* [61]; or in the entire non-coding region upstream of *nifE* of *Trichodesmium* sp. [71]. As discussed above, it appears that transcription is initiated at *nifD* and *nifK* in coordination with *nifHDK* transcription in *Trichodesmium*. Interestingly, perfect consensus NifA binding site sequences can be identified upstream of the *nifH* and *nifD* coding regions of *Trichodesmium* sp. IMS101 [61]. A less perfect consensus NifA binding site ($TGTN_{11}ACA$), was also identified upstream of *nifE* of *Trichodesmium* sp. IMS101 [71]. Although these potential regulatory binding sites are positioned (*nifH* and *nifD*) downstream of the transcription start, it has been shown that the NifA binding site (also known as the upstream activator sequence, UAS) can positively affect transcription of NifA-dependent promoters independent of orientation, and even when located up to 4 kb from the -24 and -12 promoter sequences [63]. This could mean that the UAS's identified in *Trichodesmium* may not necessarily activate the gene immediately downstream, but could affect the transcriptional activity of a *nif* gene further downstream. The presence of these consensus sequences, however, have not been reported among the *nif* genes in other cyanobacteria. The presence of these consensus sequences suggests the interesting possibility of a NifA homologue in *Trichodesmium*, and potentially different regulatory cascades from other nitrogen-fixing cyanobacteria.

4.3. The Daily Cycle. It is not only curious that *Trichodesmium* fixes nitrogen in the presence of oxygen and in the light, but that nitrogen fixation is constrained to the light phase. In several strains of unicellular and filamentous nonheterocystous cyanobacteria, nitrogen fixation usually occurs in only the dark phase of the light-dark cycle [dark], although fixation can be forced to occur in the light by shifting the culture to continuous light conditions [1,25]. In contrast, *Trichodesmium* will only fix nitrogen during a fraction of a 24-hour period, even if grown under continuous light [35]. It is not clear why the nitrogen fixation cycle is so strongly cued to this pattern, or what factors are involved in controlling or modifying the daily cycle.

The initial finding that natural populations of *Trichodesmium* fixed nitrogen only during the light [24] was eventually shown to correlate with a cycle of synthesis and degradation [72] and later to also be related to a cycle of apparent modification-demodification [73]. The Fe protein is synthesized and increases in abundance at the beginning of the light cycle. In coordination, the protein abundance of the lower molecular mass form of the Fe protein increases, and the abundance of the higher molecular mass form decreases. The Fe protein in *Trichodesmium* switches between two forms that can be identified by different mobilities on SDS-PAGE gels [73]. The rate of nitrogen-fixation activity correlates with the abundance of the upper molecular mass form [33] which is consistent with the results of studies with other cyanobacteria [74–76]. In *Trichodesmium*, the unmodified form appears in the light phase, in contrast to nonheterocystous cyanobacteria, such as *Gloeothece* in which the unmodified active form appears during the dark phase [76]. The modification of the Fe protein may be similar to that of other cyanobacteria, but biochemical mechanisms for modification have not yet been successfully identified [77].

The daily cycle of nitrogen fixation in *Trichodesmium* appears to be driven by a cycle of protein synthesis, and modification-demodification. The total abundance of the Fe protein (in modified and unmodified forms) appears to cycle, at least in natural populations [72], as well as in cultures of *Trichodesmium* sp. IMS101 [35]. The cycle of protein synthesis and degradation is less marked in cultures than in natural populations [31], and

1ay be due to differences in physiological state, rather than strictly part of the normal aily cycle of nitrogen fixation. This pattern of modification-demodification appears to be t least superficially similar to the pattern of Fe protein expression in other cyanobacteria, nd may not be unique to *Trichodesmium*.

The cycle of protein modification-demodification appears to be driven by a circadian rhythm [35], which is probably modulated by the light-dark cycle. The cycle of nitrogen fixation continues for 6 days after switching the culture to continuous light [35], while maintaining a cycle of *nif* transcription and protein modification [36]. Analysis of *nif* transcripts in cultures grown under a light-dark cycle or continuous light show the same cycle of transcript abundance (Fig. 4). The cycle of the clock appears to be slightly onger than 24 hours in *Trichodesmium*, leading to drift in the cycle when the culture is witched to continuous light [35]. The cycle has recently been shown to be entrainable by light cue [36]. Circadian rhythms of nitrogen fixation have previously been identified in nicellular cyanobacteria [78], and were some of the first convincing evidence that circadian rhythms existed in cyanobacteria [79]. The finding of the circadian clock in *Trichodesmium* indicates that circadian clocks are characteristic of filamentous as well as nicellular cyanobacteria.

Perhaps the most interesting aspect of the finding of the circadian-driven *nif* expression is that the cycle is offset from the cycle in most other cyanobacteria. The model for he circadian rhythm in cyanobacteria includes a clock that globally stimulates transcription [79]. The clock is reset by external factors (light/dark) and the output modulated by actors to ultimately regulate expression [79]. Thus, characterization of the circadian hythm in *Trichodesmium* spp. may provide information on the mechanisms involved in nodulation of output of the cyanobacterial clock.

5. NITROGEN FIXATION AND PHOTOSYNTHESIS: THE OXYGEN PROBLEM

Despite an increasing understanding of *Trichodesmium* biology, it is still unclear how *Trichodesmium* fixes nitrogen simultaneous with photosynthetic oxygen evolution. Early 1ypotheses were that the aggregate allowed interior cells to act as "heterocysts" [80]. This loes not appear to be the case [1,25], and thus, *Trichodesmium* continues to be an enigma. Numerous mechanisms have been proposed to be involved in mediating the effect of photosynthetically generated oxygen on the nitrogen fixation apparatus including respiratory activity [81,82], oxygenase activity of Rubisco [83], superoxide dismutase activity [84], Mehler reaction activity [81], and spatial separation of photosynthesis and nitrogen fixation within a filament [7,27,85]. Spatial separation would perhaps provide an intellectually satis-

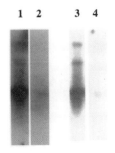

Figure 4. Diel *nifH* transcript abundance showing that the cycle of transcription (lane 1: 12:00 hours, light; lane 2: 24:00 hours, dark) continues even under constant light conditions (lane 3: 12:00 hours, light; lane 4: 24:00 hours, light).

fying explanation, but even reported cellular differences in nitrogenase localized with antisera [7,85] can be explained with alternative explanations to those proffered in the reports. None of the data published thus far provide irrefutable evidence for explaining oxic nitrogen fixation, although they suggest important and intriguing avenues for further research.

Perhaps most illuminating has been the repeated observation of high respiratory rates in *Trichodesmium* [2,81]. This observation could be related to the observation that cytochrome oxidase abundance was correlated with nitrogenase abundance [82]. High Mehler reaction activity also provides a mechanism for reduction of oxygen while generating ATP [81].

The solution to the question of oxic nitrogen fixation in *Trichodesmium* is not likely to be a "magic bullet". The high respiratory rate, potential spatial separation of photosynthesis and nitrogen fixation, and unique molecular features of *Trichodesmium* may all be involved in explaining the phenomenon of nitrogen fixation in a filamentous cyanobacterium, in oxic waters, simultaneous with oxygenic photosynthesis. Much of the work on *Trichodesmium* in the past decade has been based on work with natural populations, with large variability in species, environmental conditions (light, nutrients) and working conditions (sea-state). The uncontrolled experiments thus add to variance of resulting data and, in turn, to the difficulty in data interpretation. One factor that has not been considered extensively, is the effect of physiological stress on cellular structure and variability of cell structure. *Trichodesmium* cells grown under nutrient-limited conditions exhibit different abundance of intra-cellular structures and arrangement of membranes. Electron micrographs of natural populations, as well as some cultivated isolates that are stressed, appear to be variable in cellular ultrastructural characteristics [86]. Thus, the availability of cultures that can be grown under constant conditions and in completely artificial media [31,35] should facilitate the application of state-of-the-art molecular, and biochemical analyses to solve the riddle of nitrogen fixation in *Trichodesmium*.

REFERENCES

1. Gallon, J. R., Jones, D. A. and Page, T. S. (1996) Arch. Hydrobiol. Algol. Stud. 83, 215–243.
2. Capone, D. G., Zehr, J. P., Paerl, H. W., Bergman, B. and Carpenter, E. J. (1997) Science 276, 1221–1229.
3. Lipschultz, F. and Owens, N. J. P. (1996) Biogeochemistry 35, 261–274.
4. Karl, D., Letelier, R., Tupas, L., Dore, J., Christian, J. and Hebel, D. (1997) Nature 388, 533–538.
5. Dugdale, R. C., Menzel, D. W. and Ryther, J. H. (1961) Deep-Sea Res. 7, 298–300.
6. Paerl, H. W., Priscu, J. C. and Brawner, D. L. (1989) Appl. Environ. Microbiol. 55, 2965–2975.
7. Bergman, B. and Carpenter, E. J. (1991) J. Phycol. 27, 158–165.
8. Zehr, J. P. and McReynolds, L. A. (1989) Appl. Environ. Microbiol. 55, 2522–2526.
9. Paerl, H. W., Bebout, B. M. and Prufert, L. E. (1989) J. Phycol. 25, 773–784.
10. Siddiqui, P. J. A., Carpenter E. J. and Bergman B. B. (1992) in: Marine Pelagic Cyanobacteria: *Trichodesmium* and Other Diazotrophs (Carpenter, E. J., Capone, D. G., and Rueter, J. G., eds.), pp. 9–28, Kluwer Academic Publishers, Dordrecht.
11. Howard, J. B. and Rees, D. C. (1996) Chem. Rev. 96, 2965–2982.
12. Saville, B., Straus, N. and Coleman, J. R. (1987) Plant Physiol. 85, 26–29.
13. Apte, S. K. and Thomas, J. (1987) J. Genet. 66, 101–110.
14. Singh, R. K., Stevens, S. E. Jr. and Bryant, D. A. (1987) FEMS Microbiol. Lett. 48, 53–58.
15. Young, J. P. W. (1992) in: Biological Nitrogen Fixation (Stacey, G., Evans, H. J., and Burris, R. H., eds.), pp. 43–86, Chapman and Hall, New York.
16. Zehr, J. P., Mellon, M. T., and Hiorns, W. D. (1997) Microbiology 143, 1443–1450.
17. Flores, E., and Herrero, A.(1994) in: The Molecular Biology of Cyanobacteria. (Advances in Photosynthesis; V. 1) (Bryant, D. A., ed.), pp. 487–517, Kluwer Academic Publishers, Dordrecht.
18. Haselkorn, R. and Buikema, W. J. (1992) in: Biological Nitrogen Fixation (Stacey, S., Burris, R. H., and Evans, H. J., eds.), pp. 166–190, Chapman and Hall, New York.

19. Eady, R. R. (1996) Chem. Rev. 96, 3013–3030.
20. Thiel, T. (1993) J. Bacteriol. 175, 6276–6286.
21. Thiel, T., Lyons, E. M., Erker, J. C. and Ernst, A. (1995) Proc. Natl. Acad. Sci. USA 92, 9358–9362.
22. Golden, J. W., Whorff, L. L. and Wiest, D. R. (1991) J. Bacteriol. 173, 7098–7105.
23. Fay, P. (1992) Microbiol. Rev. 56, 340–373.
24. Saino, T. and Hattori, A. (1978) Deep-Sea Res. 25, 1259–1263.
25. Bergman, B. B., Gallon, J. R., Rai, A. N. and Stal, L. J. (1997) FEMS Microbiol. Rev. 19, 139–185.
26. Carpenter, E. J., O'Neil, J. M., Dawson, R., Capone, D. G., Siddiqui, J. A., Roenneberg, T. and Bergman, B. (1993) Mar. Ecol. Prog. Ser. 95, 295–304.
27. Janson, S., Siddiqui, P. J. A., Walsby, A. E., Romans, K. M., Carpenter, E. J. and Bergman, B. (1995) J. Phycol. 31, 463–477.
28. Ben-Porath, J., Carpenter, E. J. and Zehr, J. P. (1993) J. Phycol. 29, 806–810.
29. Zehr, J. P., Limberger, R. J., Ohki, K. and Fujita, Y. (1990) Appl. Environ. Microbiol. 56, 3527–3531.
30. Janson, S. Personal communication.
31. Ohki, K., Zehr, J. P., and Fujita, Y. (1992) in: Marine Pelagic Cyanobacteria: *Trichodesmium* and Other Diazotrophs (Carpenter, E. J., Capone, D. G., and Rueter, J. G., eds.), pp. 307–318, Kluwer Academic Publishers, Dordrecht.
32. Prufert-Bebout, L., Paerl, H. W. and Lassen, C. (1993) Appl. Environ. Microbiol. 59, 1367–1375.
33. Zehr, J. P., Braun, S, Chen, Y.-B., and Mellon, M. (1996) J. Exp. Mar. Biol. Ecol. 203, 61–73.
34. Carpenter, E. J. and McCarthy, J. J. (1975) Limnol. Oceanogr. 20, 389–401.
35. Chen, Y.-B., Zehr, J. P. and Mellon, M. T. (1996) J. Phycol. 32, 916–923.
36. Chen Y.-B. Manuscript in preparation.
37. Carpenter, E. J., Bergman, B., Dawson, R., Siddiqui, P. J. A., Soderback, E. and Capone, D. G. (1992) Appl. Environ. Microbiol. 58, 3122–3129.
38. Kramer, J. G., Wyman, M., Zehr, J. P. and Capone, D. G. (1996) FEMS Microbiol. Ecol. 21, 187–196.
39. Stadtman, E. R., Mura, U., Chock, P. B. and Ree, S. G. (1980) in: Glutamine: Metabolism, Enzymology and Regulation, (Mora, J., and Palacios, R., eds.), pp. 41–59, Academic Press, New York.
40. Sillman, N. J., Carr, N. G. and Mann, N. H. (1995) J. Bacteriol. 177, 3527–3533.
41. Stacey. G., Tabita, F. R., and van Baalen, C. (1977) J. Bacteriol. 132, 596–603.
42. Tuli, R and Thomas, J. (1981) Arch. Biochem. Biophys. 206, 181–189.
43. Lee, K.Y., Joseph, C. M. and Meeks, J. C. (1988) Antonie van Leeuwenhoek 54, 345–355.
44. Chen, H.-M. Unpublished data.
45. Zehr, J. P., Ohki, K. and Fujita, Y. (1991) J. Bacteriol. 173, 7055–7058.
46. Sroga, G. E., Ladegren, U., Bergman, B. and Lagerström-Fermér, M. (1996) FEMS Microbiol. Lett. 136, 137–145.
47. Dominic, B. Manuscript in preparation.
48. Brigle, K. E., Weiss, M. C., Newton, W. E. and Dean, D. R. (1987) J. Bacteriol. 169, 1547–1553.
49. Gosink, M. M., Franklin, N. M. and Roberts, G. P. (1990) J. Bacteriol. 172: 1441–1447.
50. Dean, R., and Jacobson, M. R. (1992) in: Biological Nitrogen Fixation. (Stacey, S., Burris, R. H., and Evans, H. J., eds.), pp. 763–834, Chapman and Hall, New York.
51. Borthakur, D., Basche, M., Buikema, W. J., Borthakur, P. B. and Haselkorn, R. (1990) Mol. Gen. Genet. 221, 227–234.
52. Böhme, H. and Haselkorn, R. (1988) Mol. Gen. Genet. 214, 278–285.
53. Schrautemeier, B., Cassing, A. and Böhme, H. (1994) J. Bacteriol. 176, 1037–1046.
54. Straus, N. A. (1994) in: The Molecular Biology of Cyanobacteria (Bryant, D. A., ed.), pp. 731–750, Kluwer Academic Publishers, Dordrecht.
55. Rueter, J. G. (1988) J. Phycol. 24, 249–254.
56. Falkowski, P. G. (1997) Nature 387, 272–275.
57. LaRoche, J. Personal communication.
58. Wilmotte, A., Neefs, J.-M. and De Wachter, R. (1994) Microbiology 140, 2159–2164.
59. Georgiadis, M. M., Komiya, H., Chakrabarti, P., Woo, D., Kornuc, J. J. and Rees, D. C. (1992) Science 257, 1653–1659.
60. Zehr, J. P., Harris, D., Dominic, B. and Salerno, J. (1997) FEMS Microbiol. Lett. 153, 303–309.
61. Dominic, B., and Zehr, J. P. (1997) Microbiology (submitted).
62. Haselkorn, R., Rice, D, Curtis, S. E. and Robinson, S. J.(1983) Ann. Microbiol. 134B: 181–193.
63. Merrick, M. J. (1992) in: Biological Nitrogen Fixation (Stacey, G., Evans, H. J., and Burris, R. H., eds.), pp. 835–877, Chapman and Hall, New York.
64. Stock, J. B., Ninfa, A. J., and Stock, A. M. (1989) Microbiol. Rev. 53: 450- 490.
65. Drummond, M. H., Contreras, A., and Mitchenall, L. A. (1990) Mol. Microbiol. 4: 29–37.

66. Frías, J. E., Mérida, A., Herrero, A., Martín-Nieto, J., and Flores, E. (1993) J. Bacteriol. 175: 5710–5713.

67. Luque, I., Flores, E. and Herrero, A. (1994) EMBO J. **13:** 2862–2869.

68. Bradley, R. L. and Reddy, K. J. (1997) J. Bacteriol. 179, 4407–4410.

69. Forchhammer, K. and Tandeau de Marsac, N. (1994) J. Bacteriol. 176: 84–91.

70. Wyman, M., Zehr, J. P. and Capone, D. G. (1996) Appl. Environ. Microbiol. 62, 1073–1075.

71. Dominic, B., and Zehr, J. P. Unpublished data.

72. Capone, D. G., O'Neil, J. M., Zehr, J. and Carpenter, E. J. (1990) Appl. Environ. Microbiol. 56, 3532–3536.

73. Zehr, J. P., Wyman, M., Miller, V., Duguay, L., and Capone, D. G. (1993) Appl. Environ. Microbiol. 59, 669–676.

74. Villbrandt, M., Stal, L. J., Bergman, B. and Krumbein, W. E. (1992) Bot. Acta 105, 90–96.

75. Ernst, A., Reich, S. and Böger, P. (1990) J. Bacteriol. 172, 748–755.

76. Du, C. and Gallon, J. R. (1993) New Phytol. 125, 121–129.

77. Dougherty, L. J., Brown, E. G. and Gallon, J. R. (1996) Biochem. Soc. Trans. 24, 477S.

78. Huang, T.-C., Tu, J., Chow, T.-J. and Chen, T.-H. (1990) Plant Physiol. 92, 531–533.

79. Golden, S. S., Ishiura, M., Johnson, C. H. and Kondo, T. (1997) Ann. Rev. Plant Physiol. and Plant Mol. Biol. 48, 327–354.

80. Carpenter, E. J. and Price, C. C. (1976) Science 191, 1278–1280.

81. Kana, T. M. (1993) Limnol. Oceanogr. 38, 18–24.

82. Bergman, B., Siddiqui, P. J. A., Carpenter, E. J. and Peschek, G. A. (1993) Appl. Environ. Microbiol. 59, 3239–3244.

83. Siddiqui, P. J. A., Carpenter, E. J. and Bergman, B. (1992) J. Phycol. 28, 320–327.

84. Cunningham, K. A. and Capone D. G. (1992) in: Marine Pelagic Cyanobacteria: *Trichodesmium* and Other Diazotrophs (Carpenter, E. J., Capone, D. G., and Rueter, J. G., eds.), pp. 219–238, Kluwer Academic Publishers, Dordrecht.

85. Fredriksson, C. and Bergman, B. (1997) Protoplasma 197, 76–85.

86. Zehr, J. P. Unpublished data

57

MOLECULAR BIOLOGY AND EVOLUTION OF THE Ntr SYSTEM

Comparisons of Phototrophs to Other Prokaryotes

Robert G. Kranz and William C. Bowman

Department of Biology
Washington University
One Brookings Dr., St. Louis, Missouri 63130

1. THE Ntr SYSTEM

1.1. Nitrogen Control in *Rhodobacter capsulatus*: The NtrB/NtrC Two-Component Regulatory System

R. capsulatus is a purple non-sulfur photosynthetic bacterium that fixes nitrogen microaerobically or anaerobically when levels of intracellular fixed nitrogen are limiting. Because the nitrogenase enzyme is inactivated by oxygen and the process of converting atmospheric nitrogen to ammonia is energy intensive, the synthesis of proteins required for nitrogen fixation is highly regulated by oxygen and fixed nitrogen. Approximately 36 genes encoding both structural and regulatory proteins have been identified as nitrogen fixation (*nif*) genes in *R. capsulatus* for a review see [1]. Functions for some of the *nif* genes have been demonstrated directly or inferred from sequence homology with genes from *Klebsiella pneumoniae*, a free-living bacterium whose *nif* genes have been studied extensively [reviewed in 2,3]. A model for the *R. capsulatus nif* transcriptional circuit has been proposed [4,5] and *R. capsulatus nif* gene expression has recently been reviewed with respect to the *K. pneumoniae* model [6,7]. In *R. capsulatus* and *K. pneumoniae*, two levels of control exist: the first involves proteins that sense and relay the nitrogen status of the cell while the second level involves mechanisms of oxygen repression. Figure 1 shows a general overview of the *R. capsulatus* regulatory cascade that will be briefly discussed here.

In *K. pneumoniae* the first level of *nif* regulation is mediated by the *ntr* system which includes the regulatory genes *glnD* (UTase), *glnB*, *ntrB*, *ntrC* and *rpoN* (*ntrA*, *glnF*) [2 and references therein]. The nitrogen sensing network in *R. capsulatus* is similar in many respects. The proteins encoded by *R. capsulatus ntrB*, *ntrC*, and *rpoN* genes (pre-

The Phototrophic Prokaryotes, edited by Peschek *et al.*
Kluwer Academic / Plenum Publishers, New York, 1999.

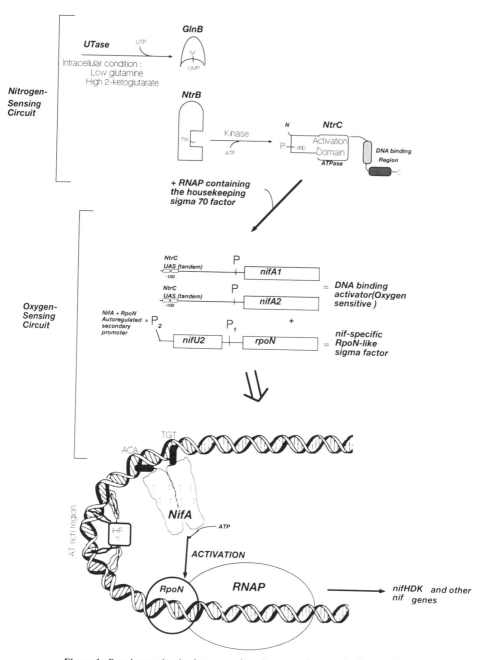

Figure 1. Regulatory circuits that respond to nitrogen and oxygen in *R. capsulatus*.

viously called *nifR2*, *nifR1*, and *nifR4* respectively) act as positive regulators of *nif* transcription; GlnB acts as a negative regulator which when mutated results in the constitutive expression of nitrogenase polypeptides (via constitutive *nifA*) under anaerobic growth with glutamine as nitrogen source, a typically repressing condition [4]. Each of these genes is homologous to the *K. pneumoniae* counterpart at the sequence level [8] and was placed in a regulatory hierarchy similar to the *K. pneumoniae* cascade. It is predicted that in both organisms, the nitrogen status of the cell is sensed as a ratio of glutamine to 2-ke-toglutarate. When this ratio is low, UTase uridylylates GlnB, releasing NtrB to act primar-ily as a kinase on NtrC (see Figure 1 for diagram of *R.capsulatus* circuitry). In *K. pneumoniae* NtrC⁻P is proposed to activate *nifLA* transcription under anaerobic conditions by binding to upstream *cis*-acting regulatory sequences, contacting the RNAP/RpoN by DNA looping, and activating open complex formation via ATP hydrolysis [e.g., 9–11]. The central domains of NifA and NtrC from *K. pneumoniae* are homologous, including the conserved nucleotide binding fold, and NifA is proposed to activate the transcription of the remaining *nif* genes by a similar RpoN-dependent mechanism [e.g., 12–15].

In contrast to *K. pneumoniae*, two functional copies of *nifA*, which appear to be iden-tical in their coding regions but differ in sequences upstream of the translational start, have been identified in *R. capsulatus* [16]. We have shown that the transcription of *nifA1*, *nifA2*, and *glnB* in *R. capsulatus* requires *ntrC* but not *rpoN* [17,18]. Our recent results have demonstrated that this is a direct effect of RcNtrC on these three promoters [18,19]. Importantly, RcNtrC contains the nucleotide binding fold but is missing a domain con-served in all other NtrC-like activators that is implicated as a domain that interacts with RpoN [19–21]. The *R.capsulatus* NifA protein also possesses the nucleotide binding fold and the putative RpoN interaction domain. Thus, subsequent to *nifA1* and *nifA2* activation by RcNtrC, NifA is proposed to activate all other *nif* genes at typical RpoN-type promot-ers [6]. Consistent with this proposal *R. capsulatus* RpoN⁻ strains show *nif*-specific rather than general nitrogen control defects (i.e. *ntr*) [22]. Additionally, expression of the *rpoN* gene in *R. capsulatus* is controlled by oxygen and nitrogen by a complex superoperon via a primary promoter and an autoregulated secondary promoter; the secondary promoter was shown to be necessary under specific environmental stress conditions such as low iron or in the presence of salt [23]. Recently, it has been proposed that a NtrC-independent nitro-gen-sensing mechanism may also be operating at the level of the NifA protein [24]. When *nifA* is placed behind the constitutive Kan^r promoter, a *nifH-lacZ* is constitutively ex-pressed at approximately 100 units activity in media containing 15 mM ammonia. (Wild-type expression is typically less then 1 unit under these growth conditions). However, under 2.5 mM ammonia *nifH-lacZ* expression is increased to approximately 500 units in the *nifA* constitutively-expressed strain [24]. Thus, although the RcNtrB/RcNtrC system is clearly the major nitrogen-sensing mechanism for nitrogen fixation in this organism, an-other control point may be present, indirectly or directly dependent on ammonia concen-trations at the level of the NifA protein or at the *nifHDK* promoter. Our lab concentrates on the NtrB/NtrC-mediated control mechanisms in photosynthetic bacteria and this is the topic of the remainder of this review.

1.2. Paradigms of Transcriptional Activation in Prokaryotes: Two Old Ones and a Newly Described One in *Rhodobacter*

During the last ten years, enormous progress has been made in defining the molecu-lar requirements for transcription in procaryotes [e.g., see 25,26 and references therein]. Reviews of transcription in *E. coli* have suggested two distinct types (or mechanisms) of activation [e.g., 26,27]. Figure 2 depicts the salient ideas behind these two types.

1.2.1. The First Paradigm Involves RNAP Containing the σ⁷⁰ Subunit and Activators That Bind Adjacent to RNAP at the Promoter. Based on many studies of different activa-tion systems in *E. coli*, the first general type of transcriptional activation employs a σ⁷⁰

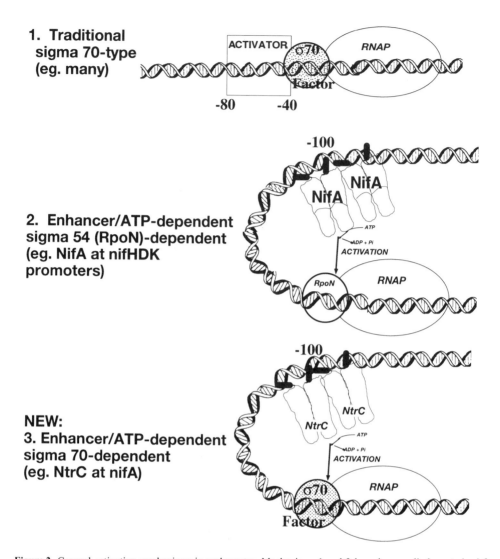

1. Traditional
sigma 70-type
(eg. many)

2. Enhancer/ATP-dependent
sigma 54 (RpoN)-dependent
(eg. NifA at nifHDK
promoters)

NEW:
3. Enhancer/ATP-dependent
sigma 70-dependent
(eg. NtrC at nifA)

Figure 2. General activation mechanisms in prokaryotes. Mechanisms 1 and 2 have been well characterized for many regulatory systems in *E. coli*. The third mechanism has been newly discovered in *Rhodobacter capsulatus* (as described in the text).

factor (or σ^{70}-like, which includes all other sigmas except σ^{54}) that directs RNAP to a region −10 and −35 nucleotides upstream of the transcription start site. Activation of transcription by specific DNA binding proteins is predicted to occur by binding to DNA directly adjacent to the RNAP holoenzyme with subsequent formation of the open complex [25]. The activator proteins in nearly all cases bind to DNA less than 80 bp upstream from the transcriptional start sites. At these types of promoters, it has been shown that RNAP only binds transiently in the closed complex; although alternative hypotheses are possible, one suggests that activator binding may facilitate a more stable RNAP complex and subsequent melting of the DNA by RNAP [see 28 for a review]. Within the last few years remarkable progress has been made in understanding the protein:protein interactions that take place between activators and σ^{70}/RNAP holoenzymes [see 25 for review, 29,30]. The binding sites of class I activators are typically centered near −61 while those of classII are near −41, overlapping the RNAP footprint. In examples of class I activation, the CAP protein interacts with the a subunit of RNAP at the *lac* promoter, as do some other DNA binding proteins [e.g. OmpR at ompC, AraC at ara—see ref. cited in 31]. In class II activation for example, the CAP protein interacts with the a RNAP subunit at two positions (N- and C-termini) as well as apparently the σ^{70} subunit at the *gal* promoter [29,32] and the cI activator of lambda also apparently contacts the σ^{70} polypeptide [33]. At *E. coli rrn* promoters the a subunit of RNAP directly contacts AT-rich sequences of DNA that are located between −40 and −60 bps upstream of the transcription start sites, called UP elements, resulting in increased transcription [34].

1.2.2. The Second Paradigm Involves RNAP Containing the σ^{54} Subunit and Activators That Bind Distant to RNAP. A second general type of activation in *E coli* requires RNAP that is poised in a closed complex at the promoter and is activated (to form an open complex) by proteins bound at distant sites upstream of the RNAP complex. This type of activation requires a specific sigma factor, called σ^{54} (=RpoN=ntrA) that directs RNAP to a DNA region at −24 to −12 bps upstream of the start site (at $GG-N_{10}-GC$ as minimal consensus sequence) for review see [20,35,36]. The activator proteins in this process can sometimes bind to DNA hundreds to thousands of bps upstream of the promoter and, as elucidated using the NtrC activator, open complex formation requires the hydrolysis of ATP by the activator protein [21 see below]. NtrC, bound to DNA, contacts the RNAP holoenzyme and/or promoter by a DNA looping mechanism [37]. In addition to this activation at a distance, a number of other characteristics of the RpoN type of prokaryotic activation distinguish it from the σ^{70} process. These include the presence of a predicted acidic domain [38] and glutamine-rich region in RpoN and the formation of a stable RNAP/RpoN closed complex. During the last three years a number of domains in RpoN have been intensely studied, including those required for binding to DNA [39,40], binding to core RNAP [41]and required for activation [40,42–45].

Since the discovery of NtrC (and the evidence that it acts as an enhancer binding protein by activating RNAP/RpoN was first reported) many prokaryotic NtrC-like proteins have been shown or implicated as activators that function from a distance [e.g., 46,47]. Many of these are confirmed members of the 2-component systems that have been reviewed previously [48,49]. Briefly, a sensor protein, such as NtrB (see below), in response to a specific environmental signal, phosphorylates the response regulator (= activator protein in the case of NtrC). Only phosphorylated NtrC acts as a transcriptional activator, although in most cases both NtrC and NtrC-P apparently bind to the upstream enhancer sites. Concatemerization experiments with two separate plasmids, one with the RpoN-activated *glnA* promoter and one with the NtrC binding site, suggest that the function of the NtrC upstream binding site is to increase the local concentration of NtrC in order to optimize NtrC/RNAP-RpoN interaction [50]. The intervening DNA, between NtrC and RNAP/RpoN is predicted to loop out to allow this interaction. Looping has been confirmed directly by electron microscopy of NtrC/RNAP/RpoN complexes [37].

It is predicted that phosphorylation of NtrC facilitates ATP hydrolysis at a conserved nucleotide binding fold in NtrC and NtrC-like activators [21]. This ATPase activity is somehow required to activate transcription of RNAP/RpoN. Recent studies have indicated that in addition to ATPase activity, oligomerization of NtrC at tandem dimer binding sites is essential for activation [51]. The mechanism of activation, particularly with respect to ATPase and oligomerization properties, have been recently reviewed [20,36]. As already described, the presence of NifA/RpoN in R.capsulatus indicates that this activation system is operating in phototrophs.

1.2.3. A Third Paradigm Has Emerged That Is Repesented by the Rhodobacter NtrC.

In all cases that have been studied from diverse microorganisms, NtrC-like proteins that possess the conserved nucleotide binding fold activate the RNAP-RpoN at consensus RpoN-type promoters (i.e. the −12 to −24 region). The only exception to this is the NtrC system in *R.capsulatus*; although RcNtrC possesses the nucleotide binding fold and binds at a distance it does not require *rpoN* for activation (see below). It is also important to emphasize that activation of RNAP containing the housekeeping σ^{70} subunit by NtrC or an enhancer protein of the NtrC class has never been observed. Attempts have been made to isolate *glnA* promoter mutations that allow NtrC activation of RNAP containing σ^{70} [e.g., 52 and references therein]. Such mutations could have been located in (1) promoters that allow NtrC activation of σ^{70}/RNAP or (2) promoters that facilitate NtrC-independent transcription (ie. creation of a new σ^{70}-dependent promoter). Only the latter class has been discovered [52]. In a related respect, Lee et al. have shown that the *Salmonella typhimurium* NtrC does not require the RNAP a subunit (C-terminus) for activation of RNAP/RpoN [53]. However, the exact mechanism(s) by which NtrC, upon ATP hydrolysis, activates RNAP/RpoN is unknown. Kustu and colleagues prefer a model which involves direct interaction of NtrC with the RpoN subunit to induce melting [21]. In this respect, a clue to the region on the RpoN-dependent NtrC like activators that may be involved in this contact comes from a comparison to the *R.capsulatus* NtrC protein [19,20]. RcNtrC contains a natural deletion of this domain, amino acid residues GAFTGA, that is present in all other NtrC-type activators. This theory is also consistent with the characterization of mutant StNtrCs in this domain which still possess ATPase activity and DNA binding properties but do not activate transcription [42].

Our laboratory has purified all of the proteins involved in the *R. capsulatus* NtrB/NtrC two component system and the RNAP σ^{70} housekeeping holoenzyme to greater then 95% homogeniety. To reconstitute the two component system in vitro, the RcNtrB protein was overexpressed as a maltose binding protein fusion (MBP-RcNtrB). RcNtrC was purified as a dimer by traditional procedures. MBP-RcNtrB autophosphorylates *in vitro* to the same steady state level and with the same stability as the *Salmonella typhimurium* NtrB (StNtrB)protein but at a lower initial rate [StNtr proteins were provided by Dr. Sydney Kustu]. MBP-RcNtrB~P phosphorylates the RcNtrC and StNtrC *in vitro* [54]. The rate of phosphotransfer to RcNtrC and autophosphatase activity of phosphorylated RcNtrC (RcNtrC~P) are comparable to the StNtrC[54]. However, the RcNtrC protein appears to be a specific RcNtrB~P phosphatase since RcNtrC is not phosphorylated by the StNtrB protein or by small molecular weight phosphate compounds. RcNtrC~P binds the upstream tandem binding sites of the *glnB* promoter four-fold better than the unphosphorylated RcNtrC protein, presumably due to oligomerization of RcNtrC~P[54]. We conclude that the *R. capsulatus* NtrB and NtrC proteins form a two-component system similar to other NtrC-like systems, where specific RcNtrB phosphotransfer to the RcNtrC protein results in increased oligomerization at the enhancer but with subsequent activation of a σ^{54}-independent promoter.

The *R. capsulatus* σ^{70} RNAP was purified by traditional procedures and in vitro transcription by the RcRNAP/σ^{70} holoenzyme was initially characterized using two classical σ^{70} promoters, the bacteriophage T7A1 and the plasmid RNAI promoters [55]. RcRNAP/σ^{70} used the same start sites as the EcRNAP/σ^{70} holoenzyme for these promoters, and transcription was sensitive to rifampicin and σ^{70} monoclonal antibody 2G10. Putative *R. capsulatus* σ^{70}-dependent promoters were analyzed by *in vitro* transcription assays from linear and supercoiled templates. Specific transcripts were detected for R.capsulatus cytochrome c2 (*cycA*), a fructose-inducible gene (*fruB*), and genes involved in photosynthesis (*puc*, *bchC*, *puf*). The specific transcripts were of the sizes predicted from *in vivo* primer extension analysis. These results confirm that that true σ^{70}-dependent systems are operating in *R. capsulatus*, as previously proposed based on genetic analysis [56] and promoter alignments [57].

Very recently, we have isolated *R. capsulatus ntrC* genes which are mutated such that the NtrC protein is partially active even when not phosphorylated and hyperactive (responsive) to NtrB [58]. Using these purified NtrC proteins we have demonstrated that the the *R. capsulatus* σ^{70} housekeeping RNAP is activated at the *nifA1* promoter [58]. This reconstituted system requires the two upstream enhancer binding sites of RcNtrC, a supercoiled template, and the ATPase function of RcNtrC appears to be necessary [58]. These results indicate that the *Rhodobacter* NtrC activation mechanism represents a third prototype for transcriptional activation in prokaryotes (see Figure 2). That is, an enhancer binding protein, dependent on ATP for activation, that uses the housekeeping RNAP. It is worth considering that this protype respresents the evolutionary transition from the traditional σ^{70} type to the classic σ^{54} types. Moreover, we speculate that the reason that *Rhodobacter* has retained all three systems is that this provides for more diverse mechanisms of control that parallel metabolic diversity.

REFERENCES

1. Klipp, W. (1990) in: Nitrogen Fixation: Achievements and Objectives, pp. 467–474 (Gresshoff, P.M., Roth, L.E., Stacy, G. and Newton, W.E., Eds.) Chapman and Hall, New York.
2. Gussin, G.N., Ronson, C.W. and Ausubel, F.N. (1986) Annu. Rev. Genet. 20, 567–591.
3. Arnold, W., Rump, A., Klipp, W., Priefer, U.B. and Puhler, A. (1988) J. Mol. Biol. 203, 715–738.
4. Kranz, R.G., Pace, V.M. and Caldicott, I.M. (1990) J Bacteriol 172, 53–62.
5. Hübner, P., Willison, J.C., Vignais, P.M. and Bickle, T. (1991) J.Bacteriol. 173, 2993–2999.
6. Kranz, R.G. and Cullen, P.J. (1995) in: Anoxygenic Photosynthetic Bacteria, pp. 1191–1208 (Blankenship, R.E., Madigan, M. and Bauer, C.E., Eds.) Kluwer Academic Publishing, Norwell, MA.
7. Masepohl, B. and Klipp, W. (1996) Arch. Microbiol. 165, 80–90.
8. Jones, R. and Haselkorn, R. (1989) Mol. Gen. Genet. 215, 507–16.
9. Hunt, T.P. and Magasanik, B. (1985) Proc Natl Acad Sci U S A 82, 8453–7.
10. Keener, J. and Kustu, S. (1988) Proc Natl Acad Sci U S A 85, 4976–80.
11. Popham, D.L., Szeto, D., Keener, J. and Kustu, S. (1989) Science 243, 629–35.
12. Buck, M., Cannon, W. and Woodcock., J. (1987) Mol. Microbiol. 1, 243–249.
13. Huala, E. and Ausubel, F.M. (1989) J Bacteriol 171, 3354–65.
14. Santero, E., Hoover, T., Keener, J. and Kustu, S. (1989) Proceedings of the National Academy of Sciences of the United States of America 86, 7346–50.
15. Drummond, M.H., Contreras, A. and Mitchenall, L.A. (1990) Mol Microbiol 4, 29–37.
16. Masepohl, B., Klipp, W. and Pühler, A. (1988) Mol.Gen.Genet. 212, 27–37.
17. Foster-Hartnett, D. and Kranz, R.G. (1992) Mol. Microbiol. 6, 1049–1060.
18. Foster-Hartnett, D. and Kranz, R.G. (1994) J Bacteriol 176, 5171–6.
19. Foster-Hartnett, D., Cullen, P.J., Monika, E.M. and Kranz, R.G. (1994) J. Bacteriol. 176, 6175–6187.
20. Morett, E. and Segovia, L. (1993) J. Bacteriol. 175, 6067–6074.
21. Weiss, D.S., Batut, J., Klose, K.E., Keener, J. and Kustu, S. (1991) Cell 67, 155–67.
22. Kranz, R.G. and Haselkorn, R. (1985) Gene 40, 203–15.
23. Cullen, P.J., Foster-Hartnett, D., Gabbert, K. and R.G., K. (1994) Mol.Microbiol. 11, 51–65.
24. Hubner, P., Masepohl, B., Klipp, W. and Bickle, T.A. (1993) Mol Microbiol 10, 123–32.
25. Busby, S. and Ebright, R.H. (1994) Cell 79, 743–746.
26. Gralla, J.D. and Collado-Vides, J. (1996) in: *Escherichia coli* and *Salmonella*, Vol. 1, pp. 1232–1245 (Neidhardt, F.C., Ed.) ASM Press, Washington D.C.

27. Gralla, J.D. (1991) Cell 66, 415–418.
28. Reznikoff, W.S. (1992) J. Bacteriol. 147, 655–658.
29. Busby, S. and Ebright, R.H. (1997) Mol. Micro. 23:853–859.
30. Adhya, S., Gottesman, M., Garges, S. and Oppenhein, A. (1993) Gene 132, 1–6.
31. Ishihama, A. (1993) J. Bacteriol. 175, 2483–2489.
32. Niu, W., Kim, Y., Tau, G., Heyduk,T. and Ebright, R.H. (1996) Cell 87, 1123–1134.
33. Li, M., Moyle, H. and Susskind, M. (1994) Science 253, 75–77.
34. Ross, W., Gosink, K.K., Salomon, J., Igarashi, K., Zou, C., Ishihama, A., Severinov, K. and Gourse, R.L. (1993) Science 262, 1407–1413.
35. Kustu, S., Santero, E., Keener, J., Popham, D. and Weiss, D. (1989) Microbiol. Revs. 53, 367–376.
36. North, A.K., Klose, K.E., Stedman, K.M. and l993., K.S. (1993) J.Bacteriol. 175, 4267–4273.
37. Su, W., Porter, S., Kustu, S. and Echols, H. (1990) Proc Natl Acad Sci U S A 87, 5504–8.
38. Sasse-Dwight, S. and Gralla, J. (1990) Cell 62, 945–954.
39. Cannon, W., Missailidis, S., Smith, C., Cottier, A., Austin, S., Moore, M. and Buck, M. (1995) J. Mol. Biol. 248, 781–803.
40. Wang, J.T. and Gralla, J.D. (1996) J. Biol. Chem 271, 32707–32713.
41. Tintut, Y., Wong, C., Jiang, Y., Hsieh, M. and Gralla, J.D. (1994) Proc. Natl. Acad. Sci. USA 91, 2120–2124.
42. North, A.K., Weiss, D.S., Suzuki, H., Flashner, Y. and Kustu, S. (1996) J Mol Biol 260, 317–31.
43. Hsieh, M., Tintut, Y. and Gralla, J. D. (1994) J. Biol. Chem 269, 373–378.
44. Tintut, Y., Wang, J.T. and Gralla, J.D. (1995) Genes and Dev. 9, 2305–2313.
45. Wedel, A. and Kustu, S. (1995) Genes Dev 9, 2042–52.
46. Reitzer, L.J. and Magasanik, B. (1986) Cell 45, 785–92.
47. Reitzer, L.J., Movas, B. and Magasanik, B. (1989) J. Bacteriol. 171, 5512–5522.
48. Parkinson, J.S. (1993) Cell 73, 857–871.
49. Volz, K. (1993) Biochemistry 171, 5512–5522.
50. Wedel, A., Weiss, D.S., Popham, D., Droge, P. and Kustu, S. (1990) Science 248, 486–90.
51. Porter, S.C., North, A.K., Wedel, A.B. and Kustu, S. (1993) Genes Dev 7, 2258–73.
52. Reitzer, L.J., Bueno, R., Cheng, W.D., Abrams, S.A., Rothstein, D.M., Hunt, T.P., Tyler, B. and Magasanik, B. (1987) J. Bacteriol. 169, 4279–84.
53. Lee, H.S., Ishihama, A. and Kustu, S. (1993) J. Bacteriol. 175, 2479–82.
54. Cullen, P.J., Bowman, W.C. and Kranz, R.G. (1996) J. Biol. Chem. 271:6530–6536.
55. Cullen, P.J., Kaufman, C.K., Bowman, W.C. and Kranz, R.G. (1997) J. Biol. Chem..in press
56. Ma, D., Cook, D.N., O'Brien, D.A. and Hearst, J.E. (1993) J. Bacteriol. 175, 2037–2045.
57. Bauer, C.E. (1995) in: Anoxygenic Photosynthetic Bacteria (Blankenship, R.E., Madigan, M. and Bauer, C.E., Eds.) Kluwer Academic Publishing, Norwell, MA.
58. Bowman, W.C. and Kranz, R.G. (1997). Submitted.

GENETIC ANALYSIS OF HETEROCYST FORMATION

C. Peter Wolk, Jinsong Zhu, and Renqui Kong

MSU-DOE Plant Research Laboratory
Michigan State University
East Lansing, Michigan 48824

Fogg's observation (1) that heterocyst differentiation is regulated by the availability of fixed nitrogen has been abundantly confirmed (references in 2; see, however, 3,4). Fig. 1 summarizes what has been learned about transcriptional dependencies during the early stages of heterocyst formation. In *Anabaena* sp. strain PCC 7120, *nirA* and *narB* encode NO_2^- reductase and NO_3^- reductase, respectively, assimilatory genes that are activated within 1 h after nitrogen stepdown and that bracket genes of the NO_2^-/NO_3^- uptake apparatus (8,9). It is known from the work of Frías et al. and of Wei et al. (10,11) that for a vegetative cell growing on ammonium, the DNA binding protein NtcA, encoded by the gene *ntcA*, is required both for heterocyst formation and activation of nitrite reductase, and thus nitrate assimilation. After the NtcA-mediated step, the regulation of NO_3^- assimilation and heterocyst differentiation divides: mutations in, or between, *nirA* and *narB* do not prevent heterocyst formation, which is evidence (see below) that they do not prevent activation of *hetR*, and mutations in *hetR* do not block activation of the genes required for assimilation of NO_2^- and NO_3^- (8,12).

hetR, identified by Buikema and Haselkorn, is essential for heterocyst formation (13). This gene, which is activated within about two h, probably within 1 h, in response to nitrogen stepdown, is autoregulatory (14), that is, the product of *hetR* is required for the induction of *hetR*. After only 3.5 h of nitrogen deprivation (12 h are required for differentiation observable by transmission light microscopy, and 18 to 24 h for aerobic N_2 fixation), *hetR* transcription can be visualized at spaced loci (14). Haselkorn and coworkers (15,16) also showed that *Anabaena*, like other bacteria, has a protein called HU that binds single-strand DNA, and Khudyakov and Wolk (17) observed that a mutation in the *Anabaena* gene, *hanA*, that encodes HU allows nitrate assimilation but prevents heterocyst formation. Moreover, there is a good reason for the latter effect, namely, that such a mutation prevents normal induction of *hetR*. The dependency of *hetR* upon *hanA* is depicted in Fig. 1 by the *hanA* arrow leading to the *hetR* arrow.

The Phototrophic Prokaryotes, edited by Peschek *et al.*
Kluwer Academic / Plenum Publishers, New York, 1999.

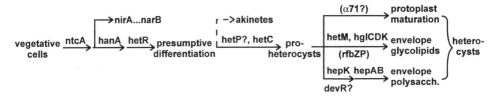

Figure 1. Working model of dependency relationships during early stages of heterocyst differentiation. X→Y denotes that activation of Y depends upon X. See text for details. There is suggestive evidence that *ntcA* acts also at later stages of heterocyst differentiation (5; see also 6) and that *hepA* also affects akinete envelope deposition (7). A dashed line indicates a possible relationship about which there is an especial dearth of information.

Conditions have not been found that permit *Anabaena* sp. strain PCC 7120, with which much of the genetic study of heterocyst formation has been conducted, to make akinetes. Leganés et al. (18) made the interesting observation that in *Nostoc ellipsosporum*, *hetR* is also required for the formation of akinetes, and becomes highly transcribed in maturing akinetes, but a gene *hetP* that is required for heterocyst formation (19) is apparently not required for akinete formation. Unfortunately, some *hetP* mutations are leaky, and eventually allow the formation of heterocysts, occasioning a "?" in Fig. 1. The next evident open reading frame 5' from *hetP* is a gene, *hetC*, that is also required for morphological differentiation of heterocysts in *Anabaena* 7120, and whose *hetC* mutations have not shown the leakiness that sometimes affects *hetP* mutants (6). Activation of *hetC* and *hetP* is observed starting at about 4 or 5 h, much later than activation of *hetR*, and recent evidence shows that a *hetR* mutation blocks the induction of *hetC* and, in the presence of nitrate, reduces its expression (20). When viewed by bright-field microscopy, mutants in *hanA*, *hetR* and *hetC* show no morphological differentiation; however, when viewed by fluorescence microscopy, *hetC* (and *hetP*) mutants show a very clear pattern of cells with low fluorescence (6,21), whereas no such pattern is seen with *hanA* or *hetR* mutants. Therefore, the activations of *hetR* and *hetC* may bracket a branch point that determines the initiation of heterocyst differentiation.

HetR and HetP bear little resemblance to other proteins in the databases. In contrast, HetC is highly similar throughout the C-terminal 2/3 of its 1044 amino acids to ATP-binding-cassette (ABC) exporters of toxic, often pore-forming, proteins through the plasmalemma of bacterial cells (6; see 22). If HetC also transports a pore-forming protein, the phenotype of a *hetC* mutant might be interpreted in at least two different ways. (i) Perhaps an inhibitor of heterocyst maturation must normally move out from a presumptive heterocyst and possibly into adjacent cells, but in the mutant is prevented from so moving. Alternatively, (ii) a substrate whose movement from vegetative cells to heterocysts is normally required for heterocyst differentiation may fail to move in a *hetC* mutant. Whatever is transported by HetC might move only to, and within, the periplasmic cylinder that runs the length of the filament inside its outer membrane (23,24), or may traverse the outer membrane of the filament.

Before the heterocyst can support nitrogen fixation under aerobic conditions, and—in fact—before the heterocyst can prevent inactivation of nitrogenase by ambient oxygen, the cell must form a glycolipid layer to block the penetration of oxygen, a polysaccharide layer to provide physical protection for the glycolipid layer, a capacity for respiration to reduce whatever oxygen penetrates the cell, and must "turn off" its oxygen-producing photosystem II. The heterocyst retains photosystem I as a source of

ATP in the light (25). As discussed below, it appears that the further development of each of the two envelope layers is independently regulated; whether further development of the protoplast is also independently regulated remains unclear.

Black et al. (26) showed that gene *hglK* is required for deposition of heterocyst envelope glycolipids outside of the vegetative wall. (Fox⁻ mutants are defective in nitrogen fixation in the presence of oxygen, and may or may not be capable of nitrogen fixation under other conditions [28]. Most genes known to be required for heterocyst formation have been identified as such by complementation, or transposon-generation, of Fox⁻ mutants (27–29). *hglK* has also been tagged by a transposon, in Fox⁻ mutant M8 [30], and the mutation was found also to result in an aberrant heterocyst envelope polysaccharide layer [20].) Maldener and coworkers (31,32) investigated the genetic region of gene *devA* whose mutation prevents maturation of the heterocyst envelope. Normal expression of *devA* was shown to require wild-type *hetR* (12). Whether normal expression of *devA* requires expression of *hetC* has not been tested. The predicted product of *devA* is again an ATP-binding-cassette transporter. Maldener and Bauer (32,33) have independently proposed that *hetM* (34), whose predicted product is related to fatty acid- or polyketide-biosynthetic enzymes, and nearby genes denoted *hglC* and *hglD* (33) are engaged in the synthesis of the heterocyst-specific glycolipids (35) that comprise the glycolipid layer (36).

Some transposition-derived Fox⁻ mutants of *Anabaena* 7120 proved resistant to one or both of cyanophages A-1(L) and A-4(L), and some spontaneous phage-resistant mutants showed a Fox⁻ phenotype. Following up these observations, Xu et al. (37) found that a galactosyl transferase and a mannosyl transferase whose genes are contiguous, and that are involved in synthesis of the extracellular lipopolysaccharide of vegetative cells, are required for a Fox⁺ phenotype. The connection appears to be that if there is a defect in the vegetative-cell lipopolysaccharide through which the heterocyst glycolipids must pass, deposition of those glycolipids is abnormal (37).

Two genes, *hepA* and *hepB*, that are involved in normal deposition of the heterocyst envelope polysaccharide have been characterized. The morphological effect of *hepA* is subtle, and is seen as a diffluence, or non-compactness, of the polysaccharide layer of the heterocyst envelope (27,38). The predicted product of *hepA*, a gene that is activated nearly exclusively in developing heterocysts starting about 4 to 7 h after nitrogen stepdown (12,39,40) and is *hetR*-dependent (14), also resembles ABC transporters (39). HepA may be involved in transport of a substituent used in the synthesis of a non-stoichiometrically incorporated side-branch of the envelope polysaccharide (see 41). A *hepB* mutant, in contrast, appears to lack the polysaccharide layer completely, except for short spurs at the heterocyst poles. HepB (GenBank accession no. ASU68035) shows similarity to glycosyl transferases (30).

A transposon mutation in a gene that we denote *hepK* blocks transcription of *hepA* completely, and leaves the glycolipid layer with no external polysaccharide layer (42). Sequence analysis shows that *hepK* has great similarity to sensory protein-histidine kinases of two-component regulatory systems: the H, N, D/F, and G boxes that are characteristic of such proteins, and the H-box histidine residue that is presumptively phosphorylated (43), are all observed (30). Interestingly, a similar phenotype has been found by Campbell et al. (44,45) for a *devR* mutant of a *Nostoc* sp. DevR shows great similarity to the effector portion of two-component regulatory systems, but for such an effector portion that lacks a DNA-binding motif. Thus, there is presumptive evidence that synthesis of the heterocyst envelope polysaccharide is dependent upon a two-component regulatory system.

D-glutamate and muramic acid are common constituents of the peptidoglycan component of bacterial walls (46). In a transposon mutant denoted HNL3, a gene is intercepted that shows great similarity to glutamate racemases and that is contiguous to a gene that shows great similarity to muramic acid amidase; the mutation blocks both activation of *hepA* and synthesis of heterocyst envelope polysaccharide. Glutamate racemase and muramic acid amidase are presumably active in cell wall metabolism; and since growth of the mutant is unimpaired, it may be that deposition of envelope polysaccharide is contingent upon some prior catabolism of the peptidoglycan layer of the vegetative cell from which the heterocyst is differentiating (45).

A transcriptional start site of *hepA* was localized 104 bp 5′ from the translational initiation codon of that gene. A *hepA::luxAB* fusion was transferred to *Anabaena* in an RSF1010 replicon, preceded by either the entire intergenic region or that region with "windows" of, on average, about 100 bp missing. It was found that regions from nt −574 to nt −440 and from −344 to −169 are required for abundant transcription of *hepA*. The gene 5′ from *hepA* appears to be involved in regulation of transcription of *hepA*. Whether the product of that gene may bind to the intergenic region(s) remains to be determined (45).

Least information is available about genes that affect differentiation of the heterocyst protoplast. In transposon mutant α71, only the differentiation of the heterocyst protoplast may be affected. Reconstruction of the mutation showed that the phenotype is linked to the transposon (47), but the mutant awaits further analysis.

As the above discussion has illustrated, a transcriptional regulatory cascade underlies heterocyst differentiation. In an attempt to elucidate transcriptional regulation during that differentiation, Brahamsha and Haselkorn identified two sigma factors that were transcriptionally activated upon nitrogen stepdown, inactivated them individually and jointly, and found no significant change in differentiation or nitrogen fixation (48). Thus, how the progress of the differentiation process is organized, and, in particular, the biochemical basis of the regulatory cascade described, remains unknown.

However, there is a reasonable explanation as to why mutational analysis may not yet have uncovered genes whose role is central to regulation of the early stages of heterocyst differentiation. Experimental evidence indicates that the pattern of spaced heterocysts results from an inhibition, by mature and developing heterocysts, of the differentiation of nearby cells into heterocysts, and suggests that the inhibition is mediated by the elaboration of some differentiation-inhibiting substance that moves outward along a filament (review: ref. 2). Because the differentiation of heterocysts is organized by inhibition, it appears appropriate for *Anabaena* to have genes whose products inhibit the initiation or progression of differentiation. Moreover, supernumerary copies of *hetR* or the presence of azatryptophan and certain other amino acid analogs that lead to the formation of clusters of heterocysts in the absence of fixed nitrogen also elicit the formation of spaced heterocysts in the presence of fixed nitrogen (13,49,50). Therefore, the inhibitory system that regulates heterocyst formation (leading to spacing of heterocysts) in the absence of fixed nitrogen appears to be operative also during vegetative growth on fixed nitrogen. A mutation that blocks the expression of, or response to, any of the component steps of that inhibitory system could lead all cells to initiate the process of heterocyst formation. However, since heterocysts are a type of cell that does not grow and divide, growth might thereby be prevented. If such mutations were previously obtained, they may have been scored as lethal, rather than as developmentally highly interesting.

A kind of mutation is needed whose effect can be "turned on" to permit observation of the mutant phenotype but "turned off" to permit growth. Conditional mutants are such mutants: their mutant phenotype is expressed under conditions referred to as restrictive conditions, but not under other conditions referred to as permissive conditions. Conditional mutants are powerful tools for analysis of metabolic systems that are essential to normal replication (e.g., 51). We are seeking, by analysis of certain kinds of conditional mutations, to identify possible differentiation-inhibiting genes in *Anabaena*.

Heterocysts, immature as well as mature, are nutritionally dependent upon neighboring vegetative cells (52,53). We therefore anticipate that immature heterocysts would probably be unable to progress to fully mature heterocysts in the absence of vegetative cells. In consequence, we will rely not on the formation of morphologically differentiated heterocysts, but on reporter constructions that link an easily observable phenotype to the extensive *initiation* of heterocyst formation. I would like to emphasize that results obtained with such conditional mutations may elucidate mechanisms that underlie pattern formation as well as cellular differentiation.

ACKNOWLEDGMENTS

HNL3 was isolated by T.A. Black. Our recent studies have been supported by the U.S. Department of Energy under grant DOE-FG02-90ER20021.

REFERENCES

1. Fogg GE (1949) Growth and heterocyst production in *Anabaena cylindrica* Lemm. II. In relation to carbon and nitrogen metabolism. Ann. Bot., N.S. 13:241–259.
2. Wolk CP (1982) Heterocysts. *In* Carr NG & Whitton BA, eds, The Biology of Cyanobacteria. Blackwell, Oxford, pp 359–386.
3. Wilcox M (1970) One-dimensional pattern found in blue-green algae. Nature (London) 228:686–687.
4. Campbell D, Houmard J & Tandeau de Marsac N (1993) Electron transport regulates cellular differentiation in the filamentous cyanobacterium *Calothrix*. Plant Cell 5:451- 463.
5. Ramasubramanian TS, Wei T-F & Golden JW (1994) Two *Anabaena* sp. strain PCC 7120 DNA-binding factors interact with vegetative cell- and heterocyst-specific genes. J. Bacteriol. 176:1214–1223.
6. Khudyakov I & Wolk CP (1997) *hetC*, a gene coding for a protein similar to bacterial ABC protein exporters, is involved in early regulation of heterocyst differentiation in *Anabaena* sp. strain PCC 7120. Submitted for publication.
7. Leganés F (1994) Genetic evidence that *hepA* gene is involved in the normal deposition of the envelope of both heterocysts and akinetes in *Anabaena variabilis* ATCC 29413. FEMS Microbiol. Lett. 123:63–67.
8. Cai Y & Wolk CP (1997) Nitrogen deprivation of *Anabaena* sp. strain PCC 7120 elicits rapid activation of a gene cluster that is essential for uptake and utilization of nitrate. J. Bacteriol. 179:258–266.
9. Frías JE, Flores E, Herrero A (1997) Nitrate assimilation gene cluster from the heterocyst-forming cyanobacterium *Anabaena* sp. strain PCC 7120. J. Bacteriol. 179:477–486.
10. Frías JE, Flores E & Herrero A (1994) Requirement of the regulatory protein NtcA for the expression of nitrogen assimilation and heterocyst development genes in the cyanobacterium *Anabaena* sp. PCC 7120. Mol. Microbiol. 14:823–832.
11. Wei T-F, Ramasubramanian TS & Golden JW (1994) *Anabaena* sp. strain PCC 7120 *ntcA* gene required for growth on nitrate and heterocyst development. J. Bacteriol. 176:4473–4482.
12. Cai Y & Wolk CP (1997b) *Anabaena* sp. strain PCC 7120 responds to nitrogen deprivation with a cascade-like sequence of transcriptional activations. J. Bacteriol. 179:267–271.
13. Buikema WJ & Haselkorn R (1991b) Characterization of a gene controlling heterocyst differentiation in the cyanobacterium *Anabaena* 7120. Genes Dev. 5:321–330.
14. Black TA, Cai Y & Wolk CP (1993) Spatial expression and autoregulation of *hetR*, a gene involved in the control of heterocyst development in *Anabaena*. Mol. Microbiol. 9:77–84.
15. Haselkorn R & Rouvière-Yaniv J (1976) Cyanobacterial DNA-binding protein related to *Escherichia coli* HU. Proc. Natl. Acad. Sci. USA 73:1917–1920.
16. Nagaraja R & Haselkorn R (1994) Protein HU from the cyanobacterium *Anabaena*. Biochemie 76:1082–1089.

17. Khudyakov I & Wolk CP (1996) Evidence that the *hanA* gene coding for HU protein is essential for hetero-
    cyst differentiation in, and cyanophage A-4(L) sensitivity of, *Anabaena* sp. strain PCC 7120. J. Bacteriol.
    178:3572–3577.
18. Leganés F, Fernández-Piñas F & Wolk CP (1994) Two mutations that block heterocyst differentiation have
    different effects on akinete differentiation in *Nostoc ellipsosporum*. Mol. Microbiol. 12:679–684.
19. Fernández-Piñas F, Leganés F & Wolk CP (1994) A third genetic locus required for the formation of het-
    erocysts in *Anabaena* sp. strain PCC 7120. J. Bacteriol. 176: 5277–5283.
20. Wolk CP & Zarka K. Unpublished observations.
21. Wolk CP, Zarka K (1997) Genetic dissection of heterocyst differentiation. In press.
22. Fath MJ & Kolter R (1993) ABC transporters: bacterial exporters. Microbiol. Rev. 57: 995–1017.
23. Ris H & Singh RN (1961) Electron microscope studies on blue-green algae. J. Biophys. Biochem. Cytol.
    9:63–80.
24. Bergman B, Lindblad P, Pettersson A, Renström E & Tiberg E (1985) Immuno-gold localization of glu-
    tamine synthetase in a nitrogen-fixing cyanobacterium (*Anabaena cylindrica*). Planta 166:329–334.
25. Wolk CP, Ernst A & Elhai J (1994) Heterocyst metabolism and development. *In* Bryant D, ed, Molecular
    Genetics of Cyanobacteria. Kluwer Acad Publ, Dordrecht, The Netherlands, pp 769–823.
26. Black K, Buikema WJ & Haselkorn R (1995) The *hglK* gene is required for localization of heterocyst-spe-
    cific glycolipids in the cyanobacterium *Anabaena* sp. strain PCC 7120. J. Bacteriol. 177:6440–6448.
27. Wolk CP, Cai Y, Cardemil L, Flores E, Hohn B, Murry M, Schmetterer G, Schrautemeier B & Wilson R
    (1988) Isolation and complementation of mutants of *Anabaena* sp. strain PCC 7120 unable to grow aerobi-
    cally on dinitrogen. J. Bacteriol. 170:1239–1244.
28. Ernst A, Black T, Cai Y, Panoff J-M, Tiwari DN & Wolk CP (1992) Synthesis of nitrogenase in mutants of
    the cyanobacterium *Anabaena* sp. strain PCC 7120 affected in heterocyst development or metabolism. J.
    Bacteriol. 174:6025–6032.
29. Buikema WJ & Haselkorn R (1991a) Isolation and complementation of nitrogen fixation mutants of the cy-
    anobacterium *Anabaena* sp. strain PCC 7120. J. Bacteriol. 173:1879–1885.
30. Kong R & Wolk CP. Unpublished observations.
31. Maldener I, Fiedler G, Ernst A, Fernández-Piñas F & Wolk CP (1994) Characterization of *devA*, a gene re-
    quired for the maturation of proheterocysts in the cyanobacterium *Anabaena* sp. strain PCC 7120. J. Bacte-
    riol. 176: 7543–7549.
32. Maldener I. Personal communication.
33. Haselkorn R (1995) Molecular genetics of nitrogen fixation in photosynthetic prokaryotes. *In* (Tikhonovich
    IA, Provorov NA, Romanov VI & Newton WE, eds.) *Nitrogen Fixation: Fundamentals and Applications*,
    Kluwer Academic Publishers, Dordrecht, pp. 29–36.
34. Black TA & Wolk CP (1994) Analysis of a Het⁻ mutation in *Anabaena* sp. strain PCC 7120 implicates a
    secondary metabolite in the regulation of heterocyst spacing. J. Bacteriol. 176:2282–2292.
35. Gambacorta, A., I. Romano, A. Trincone, A. Soriente, M. Giordano, and G. Sodano (1996) Heterocyst gly-
    colipids from five nitrogen-fixing cyanobacteria. Gazz. Chim. Ital. 126:653–656.
36. Winkenbach F, Wolk CP & Jost M (1972) Lipids of membranes and of the cell envelope in heterocysts of a
    blue-green alga. Planta 107:69–80.
37. Xu X, Khudyakov I & Wolk CP (1997) Lipopolysaccharide dependence of cyanophage sensitivity and
    aerobic nitrogen fixation in *Anabaena* sp. strain PCC 7120. J. Bacteriol. 179:2884–2891.
38. Murry MA & Wolk CP (1989) Evidence that the barrier to the penetration of oxygen into heterocysts de-
    pends upon two layers of the cell envelope. Arch. Microbiol. 151:469–474.
39. Holland D & Wolk CP (1990) Identification and characterization of *hetA*, a gene that acts early in the proc-
    ess of morphological differentiation of heterocysts. J. Bacteriol. 172:3131–3137. [*hetA* was later renamed
    hepA: see ref. 28.]
40. Wolk CP, Elhai J, Kuritz T & Holland D (1993) Amplified expression of a transcriptional pattern formed
    during development of *Anabaena*. Mol. Microbiol. 7:441–445.
41. Cardemil L & Wolk CP (1979) The polysaccharides from heterocyst and spore envelopes of a blue-green
    alga. Structure of the basic repeating unit. J. Biol. Chem. 254:736–741.
42. Zhu J, Kong R & Wolk CP. Unpublished observations.
43. Stock JB, Surette MG, Levit M & Park P (1995) Two-component signal transduction systems: structure-
    function relationships and mechanisms of catalysis, *in* (Hoch JA & Silhavy TJ, eds.) *Two-component Sig-
    nal Transduction*, ASM Press, Washington, DC, pp. 25–51.
44. Campbell EL, Hagen KD, Cohen MF, Summers ML & Meeks JC. 1996. The *devR* gene product is charac-
    teristic of receivers of two-component regulatory systems and is essential for heterocyst development in
    the filamentous cyanobacterium *Nostoc* sp. strain ATCC 29133. J. Bacteriol. 178:2037–2043.
45. Zhu J & Wolk CP. Unpublished observations.

46. Stanier RY, Adelberg EA & Ingraham JL (1976) *The Microbial World*, Prentice-Hall, Englewood Cliffs, N.J.
47. Ernst A. Personal communication.
48. Brahamsha B & Haselkorn R. (1992) Identification of multiple RNA polymerase sigma factor homologs in the cyanobacterium *Anabaena* sp. strain PCC 7120: cloning, expression, and inactivation of the *sigB* and *sigC* genes. J. Bacteriol. 174:7273–7282.
49. Bottomley PJ, van Baalen C & Tabita FR (1980) Heterocyst differentiation and tryptophan metabolism in the cyanobacterium *Anabaena* CA. Arch. Biochem. Biophys. 203:204- 213.
50. Rogerson AC (1979) Modifiers of heterocyst repression and spacing and formation of heterocysts without nitrogenase in the cyanobacterium *Anabaena variabilis*. J. Bacteriol. 140:213–219.
51. Hartwell LH (1974) *Saccharomyces cerevisiae* cell cycle. Bacteriol. Rev. 38:164–198.
52. Wolk CP (1968) Movement of carbon from vegetative cells to heterocysts in *Anabaena cylindrica*. J. Bacteriol. 96:2138–2143.
53. Giddings Jr, TH, Wolk CP, Shomer-Ilan A (1981) Metabolism of sulfur compounds by whole filaments and heterocysts of *Anabaena variabilis*. J. Bacteriol. 146:1067–1074.

59

ORGANIZATION AND REGULATION OF TWO CLUSTERS OF *nif* GENES IN THE CYANOBACTERIUM *ANABAENA VARIABILIS*

Teresa Thiel, Eilene M. Lyons, and Jessica Thielemeier

Department of Biology
University of Missouri-St. Louis
8001 Natural Bridge Rd., St. Louis, Missouri 63121

1. NITROGENASE GENES IN CYANOBACTERIA

Many species of cyanobacteria are capable of nitrogen fixation, either aerobically or anaerobically. Many filamentous strains that fix aerobically differentiate heterocysts that are specialized cells for nitrogen fixation [reviewed 16]. Other filamentous strains that do not make heterocysts, such as *Plectonema boryanum*, can also fix, but only under anaerobic conditions [reviewed 3,5]. In addition, several unicellular strains fix nitrogen aerobically in the absence of any specialized cells. *Trichodesmium* sp., which forms bundles of filaments, fixes aerobically, but only in certain cells or regions of the bundle. The mechanisms that allow nitrogenase to function in aerobic cells of cyanobacteria are as yet poorly understood [reviewed 3,5].

We have studied the genetics of nitrogen fixation in the filamentous, heterocystous cyanobacterium, *Anabaena variabilis* ATCC 29413. As is true in other heterocystous strains that have been characterized, nitrogenase genes are expressed only in heterocysts, which are made only in response to nitrogen deprivation [13]. The cluster of genes that controls nitrogen fixation in heterocysts of *A. variabilis* is called *nif1* and comprises a large group of genes including *nifB1*, *fdxN*, *nifU1*, *nifH1*, *nifD1*, *nifK1*, *nifE1*, *nifN1*, *nifX1*, *nifW1* as well as several small open reading frames of unknown function (Fig. 1) [1,7,9]. Like the homologous cluster in *Anabaena* sp. PCC 7120, this cluster is interrupted by an 11-kb excision element in *nifD1* that is removed by the site-specific recombinase, XisA, during heterocyst differentiation [1]. Unlike *Anabaena* sp. PCC 7120, the *nif1* genes of *A. variabilis* lack the 55-kb excision element in *fdxN* [1,7]. For the coding regions that have been sequenced, the *nif* cluster of *Anabaena* sp. PCC 7120 and the *nif1* cluster of *A. variabilis* are about 95% identical [1,9]. *In situ* localization of *nifH* gene expression in these strains has been studied using the reporter gene *luxAB* in *Anabaena* sp. PCC 7120

The Phototrophic Prokaryotes, edited by Peschek *et al.*
Kluwer Academic / Plenum Publishers, New York, 1999.

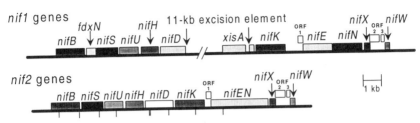

Figure 1. Maps of the *nif1* and *nif2* gene clusters of *A. variabilis*. The *nif2* genes were identified by their similarity to known *nif* genes. Transcription is from left to right. The *nifD1* gene is interrupted by the 11-kb excision element (1); thus, the carboxy-terminal end of the *nifD1* gene is just upstream of *nifK1*.

and *lacZ* in *A. variabilis* [2,13]. In both strains the *nifH* gene is expressed only in heterocysts, even under strictly anaerobic conditions.

In *A. variabilis* another nitrogenase functions in heterocysts. This alternative nitrogenase system, encoded by *vnfH*, *vnfDG*, *vnfK*, *vnfE*, and *vnfN*, contains a vanadium cofactor, in contrast to the molybdenum cofactor encoded by the conventional *nif1* system [11,12]. The *vnf* genes are controlled by the availability of molybdenum: when molybdenum is absent the genes are expressed, but only in heterocysts. Thus these genes are regulated by both environmental factors (metal availability) as well as developmental factors (heterocyst differentiation). NifB1 is essential for activity of the V-nitrogenase but NifS1 and NifU1 are not [9]. Thus, the *nif1* and *vnf* systems appear to share some genes and probably some regulatory elements.

2. *nif1* AND *nif2* GENE ORGANIZATION

A third nitrogenase in *A. variabilis* is distinct from, but homologous to the *nif1* system in that same strain. This system, *nif2*, contains all the genes of the *nif1* system except *fdxN* (Fig. 1). It lacks the 11-kb excision element and has a fusion of the *nifEN* genes. Unlike the *nif1* and *vnf* genes, the *nif2* genes are not expressed under aerobic growth conditions. They function under strictly anaerobic conditions in vegetative cells and in heterocysts after heterocysts form [13,14]. A comparison of the *nif2* genes with the *nif1*-type genes of *Anabaena* sp. PCC 7120 is given in Table 1.

There are three ORFs in the *nif* cluster of *Anabaena* sp. PCC 7120 and the *nif2* cluster that do not show significant similarity to any known *nif* genes. ORF1 of the *nif2* cluster and its homolog in *Anabaena* sp. PCC 7120 are both located just downstream from *nifK*. ORF1 is very similar to ORF4 of *Fischerella* sp., which is also downstream from *nifK* (GenBank U49514). The ORF2 and ORF3 homologs of the *nif2* cluster and the *nif* cluster of *Anabaena* sp. PCC 7120, both downstream from *nifX*, showed significant similarity to two ORFs in the *nifX*, *nifW*, *nifZ*, *nifB* cluster in *Frankia alni* [6] (GenBank L29299) as well as to two ORFs downstream from *nifX* in *A. vinelandii* [8] (GenBank M20568). ORF2 of the two *Anabaena* spp. *nif* clusters has about 55% amino acid similarity to ORF3 of the *F. alni nif* cluster and about 40% similarity to ORF3 of *A. vinelandii*. ORF3 of the two *Anabaena* spp. *nif* clusters has about 36% amino acid similarity to ORF1 of *F. alni* and about 22% similarity to ORF4 of *A. vinelandii*. The conservation of the two ORFs

Table 1. Comparison of *nif* gene clusters

Gene	Percent identity DNA	Percent similarity a. a.	Deduced M.W. 7120	Nif2
nifB	73	78	52,882	53,956
nifS	69	78	43,661	42,953
nifU	64	71	32,135	33,031
nifH	73	79	32,818	32,281
nifD	67	80	55,898	55,180
nifK	66	69	57,581	57,650
ORF1	65	63	12,518	11,136
nifE	69	79	52,811	98,817*
nifN	63	73	48,337	
nifX	64	65	15,044	14,944
ORF2	59	59	17,708	17,792
ORF3	57	61	8,228	9,805
nifW	62	59	12,262	12,284

*Deduced product of *nifEN2* gene of *A. variabilis*.

among these unrelated strains suggests that these ORFs encode proteins that function in nitrogen fixation.

3. PHYLOGENETIC ANALYSIS OF *nif* GENES

Sequence alignment of the deduced proteins for the *nifH*, *nifD* and *nifK* genes from *A. variabilis* with the homologs of *Plectonema boryanum* revealed greater similarity between the *nif2* cluster and the *P. boryanum* genes than between the *nif1* and *nif2* clusters in the same strain. A phylogenetic tree showing the relationships among nitrogenase genes from several bacterial species reveals that the *nif1*-type genes that function in heterocysts cluster together, while the *nif2* genes and the genes of *P. boryanum* form a different taxonomic group (Fig. 2).

4. REGULATION OF *nif1* AND *nif2* GENES

Both the localization of expression of the *nif2* genes to vegetative cell and their similarity to the *nif* genes of *P. boryanum*, which lacks heterocysts, suggests that this gene cluster evolved a functional and possibly regulatory independence from the *nif1* system. In fact, even when the *nif2*-encoded nitrogenase is fully functional and supporting cell growth, heterocysts differentiate, as if the cells were nitrogen starved [13,14]. This suggests a lack of common regulation or even coordinate regulation for these two independent systems.

Little is known concerning regulation of nitrogenase genes in cyanobacteria. Only one global regulatory gene, *ntcA*, that controls nitrogen metabolism in cyanobacteria has been identified [4,10,15]. This gene is present in both filamentous and unicellular cyanobacteria and is required for growth of cells on nitrate, and in filamentous cyanobacteria, is required for heterocysts differentiation. Since expression of the *nif1*-type genes in cyanobacteria depends on heterocyst differentiation, the *nif1*-type nitrogenase is not expressed in a *ntcA* mutant. The exact role of NtcA in regulation of gene expression is not

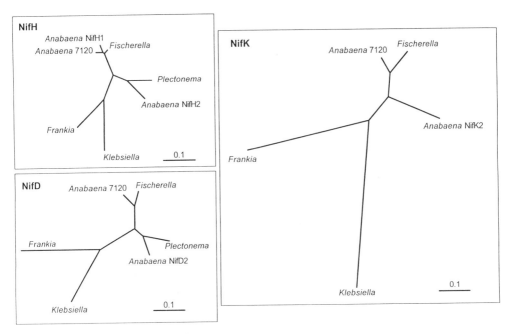

Figure 2. Unrooted phylogenetic trees for deduced proteins for NifH, NifD, and NifK. Bar represents 10% distance as determined by the Dayhoff Pam algorithm using the Protdist program of Phylip (Phylogeny Inference Package) version 3.5c. Distributed by J. Felsenstein, Department of Genetics, University of Washington, Seattle.

known; however, the protein binds to specific sites upstream of several genes, such as *glnA*, that are critical for maintenance of nitrogen balance in the cells.

We have cloned and sequenced the *ntcA* gene of *A. variabilis* (GenBank U89516). The gene is over 99% identical to its homolog in *Anabaena* sp. PCC 7120. We have constructed an interposon mutant in this gene. The mutant was very difficult to obtain and for many months the mutant and wild-type copies would not segregate. The mutant now appears to be segregated and preliminary experiments have been conducted using this strain. Like the mutant in *Anabaena* sp. PCC 7120, the *ntcA* mutant of *A. variabilis* fails to grow with nitrate as nitrogen source and fails to differentiate heterocysts. In preliminary experiments we have not detected any nitrogenase activity in cells that were anaerobically induced, suggesting that the *nif2* system is not functional in this strain. A protein fraction from wild-type cells, but not from mutant cells binds strongly to the *glnA* promoter region, suggesting that this fraction has the NtcA protein. In preliminary experiments we have found that that protein fraction shows no binding to the promoter region of the *nifH2* gene, suggesting that NtcA may not directly control expression of the *nif2* system. These preliminary results suggest that if NtcA regulates the *nif2* system, the effect is indirect.

In summary, the *nif2* system of *A. variabilis*, which functions under strictly anaerobic conditions in vegetative cells and heterocysts, is phylogenetically related to the *nif* genes of *P. boryanum*, a filamentous strain that fixes nitrogen only under anaerobic conditions. The *nif1* and *nif2* systems appear to be independently regulated. Although NtcA is required for heterocyst differentiation and, hence, for expression of *nif1*, there is no evidence as yet that it is directly involved in regulation of *nif1* or *nif2* in *A. variabilis*.

ACKNOWLEDGMENTS

This work was supported by National Science Foundation grant DCB-9106802 and USDA grant 93-37305-9309.

REFERENCES

1. Brusca, J.S., Hale, M.A., Carrasco, C.D. and Golden, J.W. (1989) J. Bacteriol. 171, 4138–4145.
2. Elhai, J. and Wolk, C.P. (1990) EMBO J. 9, 3379–3388.
3. Fay, P. (1992) Microbiol. Rev. 56, 340–373.
4. Frías, J.E., Flores, E. and Herrero, A. (1994) Mol. Microbiol. 14, 823–832.
5. Gallon, J.R. (1992) New Phytol. 122, 571–609.
6. Harriott,O. T., T. J. Hosted and Benson, D. R.. (1995). Gene 161:63–67.
7. Herrero, A., and Wolk, C. P. (1986) J. Biol.Chem. 261, 7748–7754.
8. Jacobson, M. R., K. E. Brigle, L. T. Bennett, R. A. Setterquist, M. S. Wilson, V. L. Cash, J. Beynon, W. E. Newton and Dean, D. R.. (1989) J. Bacteriol. 171,1017–1027.
9. Lyons, E.M. and Thiel, T. (1995) J. Bacteriol. 177, 1570–1575.
10. Ramasubramanian, T.S., Wei, T.F. and Golden, J.W. (1994) J. Bacteriol. 176, 1214–1223.
11. Thiel, T. (1993) J. Bacteriol. 175, 6276–6286.
12. Thiel, T. (1996) J. Bacteriol. 178, 4493–4499.
13. Thiel, T., Lyons, E.M., Erker, J.C. and Ernst, A. (1995) Proc. Natl. Acad. Sci. USA 92, 9358–9362.
14. Thiel, T., Lyons, E.M., and Erker, J.C. (1997) J. Bacteriol. 179, 5222–5225.
15. Wei, T.F., Ramasubramanian, T.S. and Golden, J.W. (1994) 176, 4473–4482.
16. Wolk, C.P., Ernst, A. and Elhai, J. (1994) in: The Molecular Biology of the Cyanobacteria. (Bryant, D.A, ed.) pp. 769–823, Kluwer Academic Press. Dordrecht, The Netherlands.

60

BIOCHEMICAL CHARACTERIZATION OF HetR PROTEIN OF *ANABAENA* PCC 7120

Ruanbao Zhou, Xing C. Wei, Nan Jiang, Yu Q. Dong, and Jindong Zhao

College of Life Sciences
Peking University
Beijing 100871, China

1. INTRODUCTION

The cyanobacteria are a diverse group of prokaryotes which carry out oxygenic photosynthesis. Some cyanobacteria can also fix nitrogen. Because nitrogenase is sensitive to oxygen, the process of nitrogen fixation in cyanobacteria is separated from oxygenic photosynthesis either temporally or spacially. Under nitrogen deprivation conditions, some filamentous cyanobacteria form thick-walled heterocysts. Heterocysts are terminally differentiated cells which are usually regularly spaced along the filaments (pattern formation). The main function of heterocysts is nitrogen fixation. Differentiation from a vegetative cell to a heterocyst is a complex process and many genes are involved in this process (1–3). Buikema and Haselkorn (4) reported cloning and sequencing of *hetR* gene from *Anabaena* PCC 7120. They showed that the *hetR* gene was up-regulated when nitrogen was removed from growth medium. A single nucleotide mutation in *hetR* open reading frame (S179N) resulted in a totally nonfunctional *hetR* gene product. The strain with such a mutant gene could not form heterocysts under nitrogen deprivation conditions. On the other hand, presence of multiple copies of *hetR* led to formation of strings of heterocysts. Black et al. (5) analysed *hetR* expression and found that *hetR* gene expression required the presence of a functional *hetR* gene product, suggesting that *hetR* was positively auto-regulatory. These features of *hetR* gene and its product led to suggestion that *hetR* gene is a master switch controlling heterocyst differentiation (2). The function of *hetR* gene may not be restricted to regulation of heterocyst differentiation. Southern analysis showed that *hetR* gene may be present in non-heterocystous cyanobacteria such as *Plectonema* while it was not detected in unicellular cyanobacteria (4). Leganes et al. (6) showed that *hetR* is required for akinete formation in *Nostoc ellipsosporum*.

In spite of the important roles of *hetR* gene in controlling cell differentiation and pattern formation, little is known about its product, HetR protein. The deduced amino acid

The Phototrophic Prokaryotes, edited by Peschek *et al.*
Kluwer Academic / Plenum Publishers, New York, 1999.

sequence showed no DNA binding motifs and sequence search in databases revealed no apparent homology to any other proteins. Although the importance of Ser179 provided a clue that HetR might be a proteinase (1), no data supporting this suggestion was available. In this presentation, we summarise our results of biochemical characterization of HetR protein. Our results suggest that HetR may function as proteinase *in vivo*. The results also show that HetR *in vivo* is post-translationally modified, possibly by phosphorylation. The possible mechanism of HetR in regulation of heterocyst differentiation is discussed.

2. MATERIALS AND METHODS

Overproduction of recombinant HetR (rHetR) in *Escherichia coli* was as follows: *hetR* gene of *Anabaena* PCC 7120 was PCR amplified and cloned into pET-3a plasmid. The resulted plasmid, pET-3a/hetR was transformed into BL21(DE3). Overproduction of rHetR was achieved with IPTG induction as described (7). Most of rHetR was present in *E. coli* as inclusion bodies. Isolation of inclusion bodies and refolding of rHetR was according to Zhao et al. (8). Crystallization of rHetR was carried out at room temperautre and 4°C with NaCl as precipitant. Immunoblotting was performed with anti-rHetR antibodies raised in rabbits as primary antibodies and enzyme-conjugated anti-IgG as secondary antibodies. Isoelectric points of proteins were determined with horizontal IEF gel with an apparatus from Pharmacia. Labeling of rHetR with PMSF and Dansyl fluoride (DnsF) was performed according to Vaz et al. (9). The DnsF-labeled protein was separated from free DnsF with acetone precipitation followed by SDS-PAGE. The protein was then electroblotted onto a PVDF membrane. For *in situ* digestion, the PVDF membrane containing DnsF-labeled rHetR was cut into small pieces and the protein was digested with trypsin. The tryptic peptides were then washed off the PVDF membrane with 1% trifluoroacetic acid in 50% acetonitrile. The peptides were analysed with ABI 173 microHPLC equipped with a dynamic blotter so that the eluted peptides were blotted directly onto a PVDF membrane. The fluorescent peptide was observed under UV light and sequenced with an ABI 491 protein sequencer.

3. RESULTS AND DISCUSSION

Examination of *hetR* gene sequence shows that none of the Met residues at the beginning of HetR open reading frame has a typical ribosome binding site, indicating that *hetR* gene may not be translated at a high rate. The fact that *hetR* gene plays a regulatory role in heterocyst differentiation also suggests that HetR is likely present in small amounts *in vivo*. Direct isolation of HetR could thus be difficult to achieve. We decided to first study rHetR and try to find some clues of the biochemical properties of HetR *in vivo*.

Overexpression of *hetR* gene of *Anabaena* PCC 7120 in BL21 (DE3) containing pET-3a/*hetR* was induced with IPTG. rHetR was mostly present as inclusion bodies. Solubilization of inclusion bodies and refolding of rHetR were performed. The refolded rHetR was purified with a DEAE-Sephadex column followed by a gel filtration column. N-terminal amino acid sequencing revealed that rHetR had a sequence of SNDIDLIKRLGPSAM, which matched with the amino acid sequence of the deduced HetR from 2nd residue to 16th residue. This result confirmed that the overproduced protein was indeed rHetR and showed that the first Met residue of rHetR was post-translationally removed in *E. coli*.

To characterize HetR *in vivo*, antibodies against rHetR raised in rabbits were used to perform immunoblotting. Total cell extracts were prepared from *Anabaena* PCC 7120 grown under different conditions. The proteins were separated with SDS-PAGE followed by electroblotting onto a PVDF membrane. Immunoblotting results showed that when *Anabaena* PCC 7120 was grown in medium with nitrate, there was a weak band at 33 kDa which cross-reacted with anti-rHetR antibodies. The intensity of the same band increased two to four fold in cell extracts of *Anabaena* grown in the absence of nitrogen. In cell extracts of isolated heterocysts, the amount of the protein cross-reacting with anti-rHetR antibodies was 20 to 30 times higher than that of *Anabaena* grown with nitrogen. In all cell extracts, the molecular mass of the protein cross-reacting with anti-rHetR antibodies was apparently identical to that of rHetR. No such band in immunoblotting was observed in cell extracts from an *Anabaena* PCC 7120 strain lacking functional *hetR* gene. These results led us to conclude that (i) the first ATG in *hetR* ORF was indeed the translation initiation codon *in vivo*, (ii) most of HetR *in vivo* was present in heterocysts or proheterocysts and (iii) the antibodies were specific for HetR.

It has been shown that transcription of *hetR* was up-regulated within one to two hours upon removal of nitrogen and most of the transcripts were located in those cells which were to become heterocysts (4). To study HetR metabolism *in vivo*, *Anabaena* PCC 7120 was transferred from a nitrogen repletion to nitrogen depletion condition and the amount of HetR was determined. The results showed that 3 hours after the shifting, HetR reached its maximum level in total cellular extracts and remained at the level as long as no nitrogen was present in growth medium. This result agreed with the result of Northern analysis, which showed that *hetR* was one of the earlist genes up-regulated in heterocyst differentiation.

Degradation of HetR in vivo upon addition of nitrate to the growth medium was studied with immunoblotting and the results showed that HetR degradation *in vivo* depend upon differentiation states. When nitrate was added to a culture which had been deprived of nitrogen for only 6 hours, HetR degradation was fast. The amount of HetR in total cell extracts decreased to the minimal level within 30 min. However, when nitrate was added to a culture of *Anabaena* with fully developed heterocysts, degradation of HetR was slow. The amount of HetR decreased to the minimal level in approximately 48 hours. Because most of HetR is located in heterocysts, these results suggest that metabolism of HetR in heterocysts is quite different from that in vegetative cells or proheterocysts.

Isoelectric focusing (IEF) of the total cell extracts followed by immunoblotting showed that rHetR had a pI of 6.2, close to the value calculated based on amino acid sequence. The pI of HetR in filaments grown without nitrogen was about 3.5, regardless if heterocysts were present. There were several pI forms of HetR in filaments grown in the presence of nitrate in medium: one was a basic form while the others were acidic. One of the possiblities for an acidic pI in HetR from culture of nitrogen deprivation is that HetR is phosphorylated when cells receive a signal of nitrogen depletion. To investigate this possibility, we added $32P-Na_2HPO_4$ to the culture of *Anabaena* PCC 7120 in the absence of nitrogen. Membrane fractions were removed from total cell extracts and immunoprecipitation was performed. After SDS-PAGE separation and electroblotting onto a PVDF membrane, radioactive bands were determined with x-ray film. One weak band at the position of about 33 kDa and a strong band at position of 12 kDa were detected. They both cross-reacted with anti-rHetR antibodies when immunoblotting was performed with the same PVDF membrane. This result shows that it is possible that HetR is phosphorylated when nitrogen is removed from growth medium. The 12 kDa band might be a degradation product of HetR. Further study of HetR isolated from *Anabaena* PCC 7120 will help to determine the site of modification if it is indeed phosphorylated.

Figure 1. Dansyl fluoride labeling of wild type rHetR and S179Q. rHetR proteins were incubated with DnsF at room temperature for 1 hour before SDS-PAGE/electroblotting. Fluorescent bands were recorded with a digital camera under UV illumination.

During our research we noticed that rHetR is not stable at 4°C. Degradation products were often observed even if rHetR was apparently pure when it was first purified. Incubation of rHetR at 37°C resulted in obvious degradation of rHetR. This degradation could be prevented by PMSF, a serine type proteinase inhibitor. This observation plus the fact that a S179N mutation resulted in a nonfunctional HetR *in vivo* led us to investigate if rHetR is a serine type proteinase. When rHetR was incubated with 100 µM PMSF followed by analysis with native gel electrophoresis, we found that the mobility of rHetR was much slower without incubation with PMSF. This result suggests that PMSF covalently modified rHetR, resulting in a mobility shift in native gel. Since PMSF reacts with active serine of proteinases and a few esterases, our results suggest that there is an active serine in rHetR. To further investigate the possiblity that rHetR contains an active Ser, Dansyl fluoride (DnsF) was used to covalently modify rHetR. The advantage of DnsF is that it is fluorescent when covalently attached to the active serines of proteinases and that this reaction is very specific. The results showed that rHetR was covalently modified by DnsF (lane 1, Figure 1). Mass spectrum of the labeled rHetR confirmed that the labeling was specific: only one Dansyl group was attached to the protein. The labeled protein

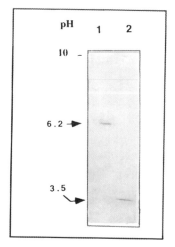

Figure 2. IEF analysis of native HetR. Affinity purified HetR from *Anabaena* PCC 7120 (lane 2) and rHetR (lane 1) were subjected to IEF gel electrophoresis followed by Coomassie staining. The isoelectric points of these proteins were determined and shown on the left.

electroblotted onto a PVDF membrane was digested with trypsin *in situ* and the tryptic peptides were analysed with microHPLC. The eluted tryptic peptides from the HPLC were directly blotted onto a strip of PVDF membrane. Under UV illumination, only one fluorescent spot was observed. The peptide was sequenced and the sequencing result (LPSNLPDAQLVTSFEFLELIEFLHKR) matched with a peptide fragment of HetR from residue 140 to 164. This fragment contained two Ser residues. We are currently determining which Ser is involved in DnsF labeling of rHetR.

The Ser179 residue of HetR is very critical to HetR function since a single mutation which led to S179N resulted in a nonfunctional HetR. To study this Ser position further, we performed site-specific mutagenesis which changed Ser179 to a Gln. S179Q was overproduced in *E. coli*. It was found that refolding of S179Q from inclusion bodies was very difficult: less than 0.1% of S179Q dissolved in 5 M Guanidine HCl remained in solution when the chaotrophic agent was removed. When the soluble S179Q was incubated with DnSF, it was found that the protein can still be labeled (lane 2, Figure 1), supporting the data of peptide sequencing. It is apparent that Ser179 is not involved in DnsF attachment directly.

Affinity purification of HetR from *Anabaena* PCC 7120 grown in the absence of nitrogen was carried out with Sepharose conjugated with anti-rHetR antibodies. A few micrograms of HetR were obtained. SDS-PAGE followed by Coomassie staining showed that it was apparently pure. Isolated native HetR had a molecular mass of 35 kDa. IEF and immunoblotting showed that it had a pI of 3.5, agreeing with the data obtained from whole cell extracts. The degradation pattern of the native HetR was similar to that of rHetR, suggesting that rHetR degradation was not caused by a contaminating protease. Protein sequencing of the native HetR did not reveal any signal, probably because the N-terminus was blocked.

Our results show that HetR is probably an enzyme with an active Ser residue. It is likely that HetR functions as a proteinase *in vivo*. Proteinases have been shown to play very important roles in regulation of cell differentiation and development from bacteria to animals. They are also shown to be critical to heterocyst differentiation. One possibility is that HetR spefically digests repressors so that transcription of some genes involved in cell differentiation could be initiated. We are currently looking for proteins which could interact with HetR.

The Ser179 has been shown to be very critical to HetR's function. Although it is not the DnsF modification site, we can not rule out the possibility that it is involved in HetR's activity such as substrate binding if HetR is a proteinase. Another possibility is that it is the phosphorylation site of HetR upon shifting to nitrogen deprivation conditions. The two component regulatory systems have been characterized in many bacteria and are shown to be present in cyanobacteria. HetR could be one of the downstream proteins of nitrogen depletion signal transduction.

REFERENCES

1. Buikema, W. J. and Haselkorn, R. (1993) Annu. Rev. Plant Physiol. Plant Mol. Biol. 44, 33–52.
2. Wolk, C. P., Ernst, A. and Elhai, J. (1995) in: Molecular Biology of the Cyanobacteria, (Bryant, D. A., ed.) pp. 769–823, Kluwer Academic Publishers, Dordrecht.
3. Wolk, C. P. (1996) Annu. Rev. Genet. 30, 59–78.
4. Buikema, W. J. and Haselkorn, R. (1991) Genes. Devel. 5, 321–330.
5. Black, T. A., Cai, Y. and Wolk, C. P. (1993) Mol. Microbiol. 9, 77–84.
6. Leganes, F., Fernandez-Pinas, F. and Wolk, C. P. (1994) Mol. Microbiol. 12, 679–684.

7. Studier, F. W., Rosenberg, A. H., Dunn, J. J. and Dubendorff, J. W. (1991) Meth. Enzymol. 185, 60–89.
8. Zhao, J., Li, N., Warren, P. V., Golbeck, J. H. and Bryant, D. A. (1992) Biochemistry 31, 5093–5099.
9. Vaz, W. L. C. and Schoellmann, G. (1975) Biochim. Biophys. Acta 439, 194–205.

61

AN ABC EXPORTER IS ESSENTIAL FOR THE LOCALISATION OF ENVELOPE MATERIAL IN HETEROCYSTS OF CYANOBACTERIA

Gabriele Fiedler, Matthias Arnold, Stefan Hannus, and Iris Maldener

Lehrstuhl für Zellbiologie und Pflanzenphysiologie
Universität Regensburg
93040 Regensburg, Germany

1. INTRODUCTION

To perform oxygenic photosynthesis and fix dinitrogen simultaneously the filamentous cyanobacterium *Anabaena* sp. protects the extremely oxygen-sensitive nitrogenase by spatial separation of the two processes in two different cell types, the oxygen evolving vegetative cell and the N_2-fixing heterocyst. A thick envelope, consisting of heterocyst-specific glycolipids and polysaccharides, forms outside the gram-negative cell wall to reduce the diffusion of gases into the heterocyst and to establish a microaerobic environment tolerated by the nitrogenase [1]. Transposon mutagenesis of *Anabaena* 7120 was used to create mutants that are arrested in different stages of heterocyst development [2]. One of these mutants, M7, is able to fix dinitrogen under anaerobic conditions (Fix+), but not under aerobic conditions (Fox−). This defect is due to an aberrant heterocyst envelope (Hen−) and an arrest in protoplast maturation, visible by lack of heterocyst-specific oxidation of diamino benzidine (Dab−) [3]. Maldener et al. (1994) [4] showed that the phenotype of mutant M7 was caused by transposition of Tn1063a into the *devA* gene. Expression studies using *luxAB* as reporter genes showed that about four hours after nitrogen stepdown *devA* expression increases ca. 8 fold in whole filaments [4]. The deduced amino acid sequence of DevA shows striking similarity to the ATP-binding subunit of ABC transporters [5]. These are export and import systems common in bacteria and eukaryotes, catalysing an ATP-dependent transport of a great variety of substrates. Prokaryotic ABC transporters consist of several subunits that are organized in an operon [6]. The components are: one or two ATP-binding subunits functioning as hetero- or homodimer, one or two membrane spanning components, also working as hetero- or homodimer, a periplasmic substrate-binding protein in case of importers or, in case of exporters, a membrane fusion protein working as homodimer and connecting the inner and the outer membrane of

The Phototrophic Prokaryotes, edited by Peschek *et al.*
Kluwer Academic / Plenum Publishers, New York, 1999.

gram-negative bacteria [6,7]. Sequencing of the region flanking *devA* revealed two open reading frames, *devB* and *devC*, that encode proteins, which together with *devA* could function as an ABC exporter. *Anabaena* 7120 strains were constructed bearing a mutated *devA*, *devB* or *devC* gene. The genotypical and phenotypical characterization of these mutants and the ultrastructural analysis of mutant M7 by transmission electron microscopy suggest that the DevBCA proteins function as an ABC exporter of heterocyst-specific glycolipids, which are essential components of the heterocyst envelope.

2. MATERIALS AND METHODS

2.1. Strains and Growth Conditions

Strains of *Anabaena* sp. strain PCC7120 and *Anabaena variabilis* ATCC 29413 FD were grown as described earlier [3]. Mutant strains of *Anabaena* 7120 were grown in the presence of 5 mM NO_3 and 200 μg/ml neomycin sulfate, mutant strains of *Anabaena variabilis* were grown in the presence of 5 mM NO_3 and 50 μg/ml kanamycin. Strains of *E. coli* were grown on LB medium under standard conditions [8]. Transfer of plasmids by conjugation between *Anabaena* and *E. coli* was achieved as described earlier [9]. Selection for recombinants was performed as described by Cai and Wolk (1990) [10].

2.2. DNA Isolation and Analysis

Total DNA of *Anabaena* strains was isolated as described earlier [10]. Plasmids were purified from *E.coli* with the QIAGEN-plasmid kit. Sequencing was done with the T7 Sequencing kit from Pharmacia using oligonucleotides complementary to the DNA fragments. Sequence analysis was performed with the UWGCG package of the University of Wisconsin Genetics Computer Group, Inc., version 7.3 [11]. Sequence comparisons were made by using the blastp program of the Heidelberg Unix Sequence Analysis Resources, version 4.

2.3. Analysis of Glycolipids

From 50 ml-cultures of *Anabaena* 7120 that had been deprived for nitrogen for 48 h total lipids were extracted as described by Ernst et al. (1992) [3].

2.4. Electron Microscopy

Fixation with 2.5% glutaraldehyde and 2% $KMnO_4$ and dehydratation was done as described earlier [12]. After dehydratation the samples were incubated in a 1:1 mixture of Durcupan (Fluka) and propylene oxide overnight, at 37°C, followed by embedding in Durcupan for 24 h at 37°C and 48 h at 60°C. Thin sections of 70 to 90 nm were collected on copper grids and stained with uranyl acetate for 20 min and lead citrate for 5 min. The samples were examined with a Zeiss EM109 electron microscope at 80 kV.

2.5. Mutation of *devA*, *devB*, and *devC*

The genes *devA, devB* and *devC* of *Anabaena* 7120 were disrupted as will be described by Fiedler et al. (1997) [13]. The *devA* gene of *Anabaena variabilis* was subcloned

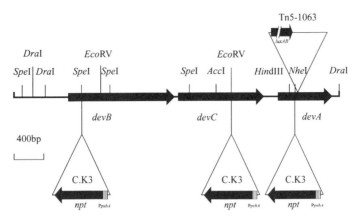

Figure 1. Restriction map of the dev region (4.4 kb). The three open reading frames devA, devB and devC of the putative operon, encoding the subunits of the ABC exporter are shown as filled arrows, which indicate the direction of transcription. Restriction sites were mapped by sequencing, and all sites for each enzyme are shown. Sites of directed insertion of the cassette C.K3 and of transposition by Tn5–1063 are shown. The npt gene encoding neomycin resistance under control of the psbA promoter was used as selection marker. The luxAB genes served as reporter and are located at the left end of the transposon.

as *PstI/Ecl*136I fragment in pRL271[14] digested with *Pst*I and *Ecl*136I resulting in pIM86. PIM86 was digested with *Nhe*I, and the *devA* gene was disrupted by insertion of the C.K3 cassette derived from pRL448 [15] as *Xba*I fragment resulting in pIM87.

3. RESULTS

3.1. The *devBCA* Gene Cluster

Two genes are located upstream of *devA* that have the same orientation as *devA* and may, together with *devA,* form an operon (Figure 1). The nucleotide sequences are available from the GenBank-EMBL Database under the accession number X99672. *DevC*, the open reading frame 180 bp upstream of *devA* encodes a protein of 384 amino acids, having a molecular weight of 43.3 kDa. Upstream of *devC*, spaced by a non-coding strech of 41 bp the first open reading frame of the gene cluster, named *devB*, could be identified. It encodes a protein of 474 amino acids with a molecular weight of 51.6 kDa. A stem-loop structure, which may be part of a rho-independent signal for termination of transcription, could be identified 33 bp downstream from the *devA* stop codon.

3.2. Sequence Comparisons

Sequence comparisons revealed that the amino acid sequence of DevA is very similar to sequences of ATP-binding subunits of ABC importers, e.g. 37% identity with the maltose importer of *E. coli* [4], as well as to ATP-binding subunits of exporters, e.g. 36% identity to the C-terminal half of the hemolysin exporter of *E. coli*. Figure 2A and B show the high degree of conservation of the two nucleotide-binding sites between DevA and the ATP-binding subunits of several exporters. Furthermore, the hydropathy profile of DevA is amphipatic as expected for this type of protein (Figure 3A). The deduced amino acid se-

A

```
DevA (A)     26 LFDINLEIYPGEIVIMTGPSGSGKTTLLSLIGGLRSVQEGNLQFLG  71
PrtD (Sma)  339 LHGVSFRLEAGEVLGVIGASGSGKTLLMRQLVGALTPISGDGGAEQ 384
PrtD (Ech)  347 LQNIHFSLQAGETLVILGASGSGKSSLARLLVGAQSPTQGKVRLDG 392
LktB (Pha)  486 LNNVNLEIRQGEVIGIVGRSGSGKSTLTKLLQRFYIPENGQVLIDG 531
HlyB (Eco)  485 LDNINLSIKQGEVIGIVGRSGSGKSTLTKLIQRFYIPENGQVLIDG 530

                 *  *** *  **        * ***** **   *   *      *    *
```

B

```
DevA (A)    135 GLENRVDYYPENLSGGQKQRVAIARALVNNPPLVLADEPTAALD 178
PrtD (Sma)  437 GYETELGEGGSGLSGGQRQRVALARALYGSPALVVLDEPNANLD 480
PrtD (Ech)  456 GYDTELGDGGGGLSGGQRQRIGLARAMYGDPCLLILDEPNASLD 499
LktB (Pha)  595 GYNTIVGEQGAGLSGGQRQRIAIARALVNNPKILIFDEATSALD 638
HlyB (Eco)  594 GYNTIVGEQGAGLSGGQRQRIAIARALVNNPKILIFDEATSALD 637

                  *       *      ***** ** ********** *   ********
```

C

```
DevB (A)     77 KLSAPVGGMQSASRVKQLFVKEGERVRKGQVIAILDNHDTQ 117
SapE (Lsa)   56 KQNVPIIQGSTNSTIKQNYLKEGQLVKKGQTLLIYKNSRNK  96
ORF  (Hin)   86         QNAGAVSQVLVQNGQNVKKGEVLVELDS          130
PrtE (Sma)   60 AGNRKAVQHPSGGVVSQIQVHEGDRVRAGQVLLLMDTVDSR 100
LktB (Aac)  105               SLVKHIFVKEGERYVKKGELL         124
HlyD (Eco)   83 SGRSKEIKPIENSIVKEIIVKEGESVRKGDVLLKLTALGAE 133
SppE (Lsa)   60       PILQASTNSVLKQNYLKEGKFVKKGQTLLVYQNTKNQ  96
PrtE (Ech)   65 SGNRKVIQHMQGGIVDRIQVKDGDRVAAGQVLLTLNAVDAR 105
LktD (Pha)   83 SGRSKEIKPIENAIVQEIFVKDGQFVEKGQLLVSLTALGSD 133
                 *        * ***  ***********    *** *
```

D

```
DevB (A)    334 VRRAEAELKLSYIQAPSAGEI..LKI...YSKSGEAITADGIAEMGETE 377
PrtE (Sma)  311 LAKAEADLGHTQVKAPVAGTVVGLSVFTEGGVIGAGQQLMEIVPSD..R 357
PrtE (Ech)  286 REKADFNLANVQVRAPVAGTVVDMKIFTEGGVIAPGQVMMDIVPED..Q 332
LktD (Pha)  315 LEKNNQRRQASMIRAPVSGTVQQLKIHTIGGVVTTAETLMIIVPED..D 361
HlyD (Eco)  318 LAKNEERQQASVIRAPVSVKVQQLKVHTEGGVVTTAETLMVIVPED..D 364
CvaA (Eco)  238 LVNTD.VEGEIIIRALSDGKVDSLSV.TVGQMVNTGDSLLQVIPENI.E 283
                 **  *   *  * ** **     ***       *  *      *

consensus       ------------I-A-----V-Q----T-----T---TL----P-----
```

```
DevB (A)    378 QMMIIAEVPEDSIGRVRIGQTVT..ANSDNGAFSGEIRGTVT 417
PrtE (Sma)  358 GLQVEARIPVELIDKVQVGLPVELLFSAFNQSTTPRVEGEVT 399
PrtE (Ech)  333 PLLVDGRIPVEMVDKVWSGLPVELQFTAFSQSTTPRVPGTVT 374
LktD (Pha)  362 VLEATALVPNKDIGFVAAGQEVIIKVETFPYTRYGYLTGRIK 403
HlyD (Eco)  365 TLEVTALVQNKDIGFINVGQNAIIKVEAFPYTRYGYLVGKVK 406
CvaA (Eco)  284 NYYLILWVPNDAVPYISAGDKVNIRYEAFPSEKFGQFSATVK 325

                   *  **   ** * *  ** *            *    *  **

consensus       -L-----V-------V--G--------A----------G-V-
```

Figure 2. Multiple alignments of regions of DevA and DevB with high similarity to other known subunits of ABC exporters. (A) Walker motif A and (B) Walker motif B of DevA and corresponding sequences in PrtD (metalloprotease exporter of S. marcescens, Sma [19], and E. chrysanthemi, Ech [20]), LktB (leukotoxin exporter of P. haemolytica, Pha [21]) and HlyB (hemolysin exporter of E. coli, Eco [22]). The highly conserved nucleotide binding sites are underlined. (C) Aligment of the conserved N-terminal regions of DevB and of the membrane fusion proteins SapE (sakacin exporter of L. sake, Lsa [23]), ORF (of H. influenzae, Hin), PrtE (of S. marcescens, Sma), LktD (of A. actinomycetemcomitans, Aac [24]), HlyD (of E. coli, Eco), SppE (sakacin exporter of L. sake, Lsa), PrtE (of E. chrysanthemi, Ech) and LktD (of P. haemolytica, Pha) and (D) of the conserved C-terminal regions of DevB and PrtE, LktD, HlyD and CvaA (colicinV exporter of E. coli, Eco [25]). Asterisks show identities of DevA or DevB to at least 50% of the aligned sequences, derived from Dinh et al. (1994) [7]. The high percentage of conservative exchanges is not emphasized. Residue numbers of each protein are provided at the beginning and end of each line. The consensus sequence of membrane fusion proteins [7] is shown below.

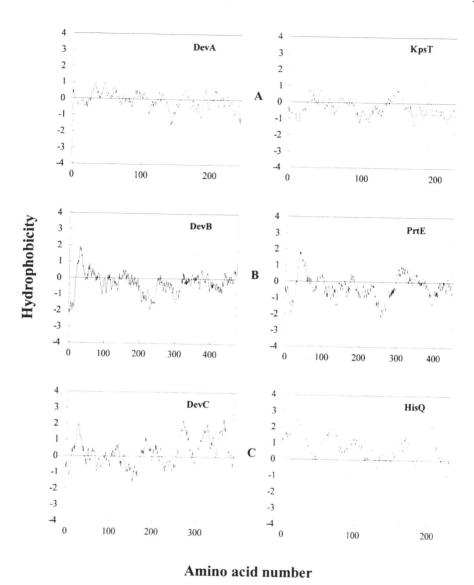

Amino acid number

Figure 3. (A) Comparison of hydropathy profiles of DevA and KpsT (ATP binding subunit of the capsular polysaccharide exporter of *E. coli* [26]), (B) DevB and PrtE (MFP of metalloprotease exporter of *E. chrysanthemi*) and (C) DevC and HisQ (membrane spanning subunit of histidine importer of *E. coli*).

quence of DevB shows similarity to several membrane fusion proteins (MFPs) (Figure 3C and D). Dinh et al. (1994) [7] made a multiple alignment of various MFP sequences, which revealed a consensus sequence in the C-terminal part of these proteins. As shown in Figure 2D, DevB has conserved 8 of 16 residues of this consensus sequence. The residues in the N-terminal region are conserved with up to 52% identity between DevB and the corresponding region of various MFPs (Figure 2C). Besides the sequence similarity of DevB and several MFPs, the hydropathy profile of DevB renders very well the profile of typical MFPs like that of PrtE of *S. marcescens* (Figure 3B). Both proteins contain a hydrophilic region at the N-terminus followed by a highly hydrophobic part, that might be responsible for anchoring the MFP in the cytoplasmic membrane. Adjacent is a region of moderate hydrophobicity followed by a strikingly hydrophilic part that possibly traverses the periplasmic space. The C-terminus consists of a slightly hydrophobic β-strand, typical for outer-membrane associated domains [7]. The hydropathy profile of DevC shows similarity to the profile of the membrane-spanning domain of e.g. a histidine transporter, HisQ, that forms a heterodimer with HisM in that transporter complex [16]. Five significant hydrophobic stretches may cross the membrane as transmembrane helices (Figure 3C).

3.3. Mutation of the *dev* Gene Cluster in *Anabaena* 7120 and of *devA* in *Anabaena variabilis*

To prove the assumption that the three *dev* genes are functionally related, *devB* and *devC* of *Anabaena* 7120 were directly mutagenized by insertion of the C.K3 cassette (Figure 1) as will be described in Fiedler et al. (1997) [13]. The *devB* and *devC* mutants show the same phenotype as the *devA* mutant, M7, i.e. Fox⁻, Het⁺, Hen⁻ and Dab⁻. Southern blot analysis of mutant DNA showed the correct insertion of the C.K3 cassette.

To see whether the *dev* genes are important in other heterocyst forming cyanobacteria like *Anabaena variabilis*, the three genes were fished in an *Anabaena variabilis* genbank and sequenced. The deduced amino acid sequences of DevA, B and C of *Anabaena variabilis* show 98%, 97% and 98% identity to the corresponding sequences of *Anabaena* 7120. The *devA* gene of *Anabaena variabilis* was disrupted by insertion of the C.K3 cassette. First characterization of the mutant revealed that it is not able to grow on N_2 as sole nitrogen source under aerobic conditions and does not make mature heterocysts.

3.4. Analysis of Heterocyst Specific Glycolipids in the *devBCA* Mutants

According to Ernst et al. (1992) [3], mutant M7 lacks the heterocyst specific glycolipids (Hgl⁻) as determined by thin layer chromatography of extracts of filaments deprived for nitrogen. Cells of *devC* and *devB* mutant together with cells of wild type and M7 were analysed the same way (Figure 4). In the wild type and all mutants that had been deprived for nitrogen a spot could be detected in a region that has been attributed to heterocyst specific glycolipids [17], whereas this spot is not present in cells grown on NO_3 or in cells of Hgl⁻ mutant P2 [3]. In conclusion, the phenotypes of M7 and the mutants *devB* and *devC* in general have to be defined as Hgl⁺.

3.5. Ultrastructure of Heterocysts of Mutant M7

Despite the presence of heterocyst-specific glycolipids in extracts of whole filaments, the heterocyst envelope of the *dev* mutants looks thin and less refractile in the light microscope [3, 4]. This prompted us to examine the ultrastructure of filaments of mutant M7 deprived for nitrogen for 48 h and wild-type filaments deprived for nitrogen for 36 h (Figure 5). The heterocysts of wild type and mutant M7 show the thick homogenous layer consisting of polysaccharides. The innermost laminated layer composed of heterocyst-specific glycolipids can be clearly seen near the junction between heterocyst and vegetative cell of the wild type, but is completely absent in mutant M7. The formation of the so-called honeycomb membranes close to the poles is arrested in an early stage in the mutant heterocysts. Furthermore, a polar nodule consisting of cyanophycin is not built up in the

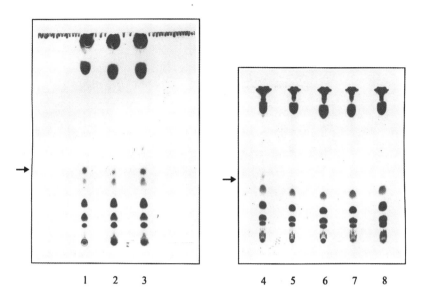

Figure 4. Thin-layer chromatography of glycolipid extracts from cultures containing 50 μg chlorophyll. The position of heterocyst-specific glycolipids is indicated by an arrow. (A) Glycolipids of NO_3^--deprived cells of the wild type (lane 1), mutant M7 (lane 2) and *devC* mutant (lane 3). (B) Glycolipids of NO_3^- grown cells of wild type, in which heterocyst differentiation was not totally repressed (lane 4), mutant M7 (lane 5), the *devC* mutant (lane 6), mutant P2 (lane 7) and N_2-deprived cells of mutant P2 (lane 8); mutant P2 was used as Hgl⁻ control (Ernst et al. 1992) [3].

mutant, whereas a small septum at the junction to the vegetative cell, which is characteristic for heterocysts is clearly visible.

4. DISCUSSION

The sequence data of DevA and DevB and hydropathy profiles of the three proteins encoded by the *devBCA* gene cluster clearly demonstrate that they represent the subunits of an ABC-protein mediated export system. The functional linkage between the three *dev* genes was shown by site-directed mutagenesis of each open reading frame, followed by a detailed analysis of the resulting phenotype, which was identical for all three mutants. Disruption of the *devA* gene in *Anabaena variabilis*, which shows 98% identity to *devA* of *Anabaena* 7120 leads to a Fox⁻ phenotype. This observation suggests that the *dev* gene cluster plays a similar role in different filamentous heterocyst forming cyanobacteria. Chromatographic analysis of the *dev* mutants showed that, in contrast to the original description of Ernst et al. (1992) [3], heterocyst specific glycolipids were present in all *dev*-mutant strains. Nevertheless, ultrastructural analysis showed that the laminated layer consisting of these glycolipids was not present in mutant M7. This layer is a primary barrier to diffusion of oxygen [18] and its lack could explain the Fox⁻ phenotype of the *devBCA* mutants. The polysaccharide layer is exposed in mutant M7; this observation shows that the formation of the two layers is regulated independently. The pleiotropic phenotype of mutant M7 could be explained by assuming that the maturation of the protoplast, e.g. the reorganization of intracellular membranes, can only proceed after an intact oxygen barrier has been formed. We suggest that the formation of the laminated glycolipid

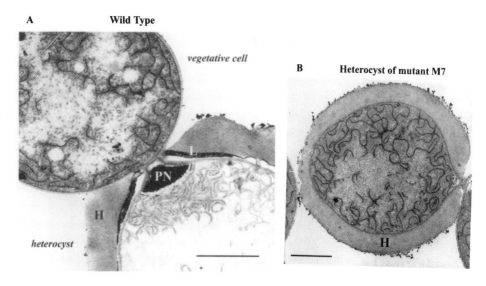

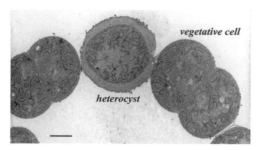

Figure 5. Transmission electron micrograph of ultra-thin sections of a (A) connection between a vegetative cell and a heterocyst of the wild type of *Anabaena* sp. strain PCC 7120, (B) heterocyst of M7; and (C) filament from which B was magnified. H = homogeneous layer (polysaccharides), L = laminated layer (glycolipids), PN = polar nodule (cyanophycin). The bar represents 1 µ n. For details see text.

layer, resulting in an intracellular microaerobic environment, is the prerequisite for further development of the protoplast. To summarize, sequence data and mutational analysis of the *devBCA* gene cluster demonstrate an ABC-protein mediated exporter to be essential for the maturation of heterocysts. The ultrastructural characterization of mutant M7 shows that the heterocyst envelope lacks the laminated layer although heterocyst-specific glycolipids are present inside the cells. The simplest interpretation of these data is that the DevBCA transporter allows the export of glycolipids or as alternative of an enzyme required for assembly of the laminated layer.

REFERENCES

1. Wolk, C.P. (1982) In Carr, N.G. and Whitton, B.A. (ed). The Biology of Cyanobacteria. Blackwell Scientific Publications Ltd.,Oxford.

2. Wolk, C.P., Cai, Y. and Panoff, J.-M. (1991)Proc. Natl Acad.Sci. USA, 88, 535–5359.

3. Ernst, A., Black, T., Cai, Y., Panoff, J.-M., Tiwari, D.N. and Wolk, C.P. (1992) J. Bacteriol., 174, 6025–6032.

4. Maldener, I., Fiedler, G., Ernst, A., Fernández-Piñas, F. and Wolk, C.P. (1994) J. Bacteriol., 176, 7543–7549.

5. Higgins, C.F., Hyde, S.C., Mimmack, M.M., Gileadi, U., Gill, D.R. and Gallagher, M.P. (1990) J. Bioen. Biomem., 22, 571–592.

6. Ames, G. F.-L. (1986) Ann. Rev. Biochem., 55, 397–425.

7. Dinh, T., Paulsen, I.T. and Saier ,M.H., JR (1994) J. Bacteriol., 176, 3825–3831.

8. Maniatis, T., Fritsch, E.F. and Sambrook, J. (1982) Molecular cloning. A laboratory manual. Cold Spring Harbor Laboratory, Cold Spring Harbor, NY.

9. Wolk, C.P., Vonshak, A., Kehoe, P. and Elhai, J. (1984) Proc. Natl. Acad. Sci. USA, 81, 1561–1565.

10. Cai, Y. and Wolk, C.P. (1990) J. Bacteriol., 179, 267–271.

11. Devereux, J., Haeberli, P. and Smithies, O. (1984) Nucleic Acids Res., 12, 387–395.

12. Black, K., Buikema, W.J. and Haselkorn, R. (1995) J. Bacteriol., 177, 6440–6448.

13. Fiedler, G., Arnold, M., Hannus, S. and Maldener, I., in preparation.

14. Black, T.A., Cai, Y. and Wolk, C.P. (1993) Molec. Microbiol., 9,77–84.

15. Elhai J., and Wolk, C.P. (1988) Gene, 68, 119–138.

16. Kerppola, R.E., Shyamala, V. K., Klebba, P. and Ames, G.F.-L. (1991) J. Biol. Chem., 266, 9857–9864.

17. Winkenbach, F., Wolk, C.P. and Jost, M. (1972) Planta, 107, 69–80.

18. Murray, M.A. and Wolk, C.P. (1989) Arch. Microbiol., 151, 469–474.

19. Létoffé, S., Ghigo, J.-M. and Wandersman, C. (1993) J. Bacteriol., 175, 7321–7328.

20. Létoffé, S., Delepelaire, P. and Wandersman, C. (1990) EMBO J., 9, 1375–1382.

21. Strathdee, C.A. and Lo, R.Y.C. (1989) J. Bacteriol., 171, 916–928.

22. Schulein, R., Gentschev, I., Mollenkopf, H.J. and Goebel, W. (1992). Mol. Gen. Genet., 234, 155–163.

23. Axelsson, L.and Hoeck, A. (1995) J. Bacteriol., 177, 2125–2137.

24. Guthmiller, J.M., Kraig, E., Cagle, M.P. and Kolodrubetz, D. (1990) Nucleic Acids Res. 18, 5292.

25. Gilson,L., Mahanty, H.K. and Kolter, R. (1990) EMBO J., 9, 3875–3884.

26. Pavelka, M.S., Wright, L.F. and Silver, R.P. (1991) J. Bacteriol., 173, 4603–4610.

62

REGULATION OF CARBON AND NITROGEN METABOLISM IN THE UNICELLULAR CYANOBACTERIA *SYNECHOCOCCUS* SPP.

Nicole Tandeau de Marsac and Hyun-Mi Lee

Unité de Physiologie Microbienne
Département de Biochimie et Génétique Moléculaire
Institut Pasteur
28 rue du Docteur Roux, 75724 Paris Cedex 15, France

Cyanobacteria are photosynthetic prokaryotes that mainly use CO_2 and bicarbonate as carbon sources, and ammonium and nitrate as nitrogen sources to fulfil their requirement for growth. Some strains can also fix molecular N_2 either under aerobic or anaerobic conditions[16,27]. Under autotrophic conditions, the enzymatic reactions required for the utilization of any form of inorganic nitrogen depend upon the availability of energy (ATP) and reductant generated from photosynthesis, as well as upon CO_2-fixation products, which in part act as amino acceptors to generate amino acids and other organic nitrogenous compounds. There might thus exist tight interactions between nitrogen and carbon metabolism to balance the intracellular N/C ratio.

Given the diversity of the physiological properties of cyanobacteria, we have chosen to focus this chapter on unicellular strains that do not fix N_2, in particular *Synechococcus* sp. strain PCC 7942 and taxonomically very closely related strains. Following a short description of some aspects of both carbon and nitrogen assimilation, required to understand how these two biological processes are co-ordinated, the regulatory pathways will be described sequentially, to end up with a global scheme presenting our current knowledge of the various interactions between elements of these two major metabolic pathways (Fig. 1). For more details, the reader is invited to refer to excellent comprehensive reviews available in the literature.[16,27,33,53,74]

1. CARBON ASSIMILATION

There exist two means for the entrance of inorganic carbon into the cells, the uptake of CO_2 and that of bicarbonate[33,51,74]. The major route for CO_2 fixation requires the ribu-

The Phototrophic Prokaryotes, edited by Peschek *et al.*
Kluwer Academic / Plenum Publishers, New York, 1999.

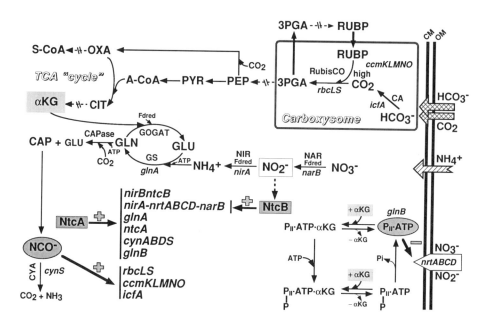

Figure 1. Scheme for the regulation of carbon and nitrogen metabolism in *Synechococcus* spp.. Abbreviations: A-CoA, acetyl CoA; CA, carbonic anhydrase (*icfA* gene); CAP, carbamoyl phosphate; CAPase, carbamoyl synthetase; Ci-concentrating mechanism, inorganic carbon-concentrating mechanism (*ccmKLMNO* genes); CIT, citrate; CM, cytoplasmic membrane; CYA, cyanase (*cynS* gene); *cynABD*, genes encoding a putative transport system for cyanate; Fdred, reduced ferredoxin; GLN, glutamine; GLU, glutamate; αKG, α–ketoglutarate; GOGAT, glutamate synthase; GS, glutamine synthetase (*glnA* gene); NAR, nitrate reductase (*narB* gene); NCO⁻, cyanate; NIR, nitrite reductase (*nirA* gene); *nirB*, gene required for maximum expression of NIR activity; *nrtABCD*, genes coding for the nitrate/nitrite permease complex; NtcA, transcriptional regulator (*ntcA* gene); NtcB, transcriptional regulator (*ntcB* gene); OM, outer membrane; OXA, oxaloacetate; P_{II}/P_{II}-P, unphosphorylated/phosphorylated P_{II} protein (*glnB* gene); PEP, phosphoenolpyruvate; 3PGA, 3-phosphoglycerate; PYR, pyruvate; RubisCO, ribulose-1,5-bisphosphate carboxylase/oxygenase (*rbcLS* operon); RUBP, ribulose-1,5-bisphosphate; S-CoA, succinyl CoA; TCA cycle, tricarboxylic acid cycle.

lose-1,5-bisphosphate carboxylase/oxygenase (RubisCO)[74,75], an enzyme that displays a low affinity for CO_2 and thus necessitates activation of a concentrating mechanism for inorganic carbon (Ci). The two main components of the Ci-concentrating mechanism are a machinery for the uptake of Ci[2–4,31,32] and specific structures, the carboxysomes[8]. Three modes of uptake of Ci have been described[4,11,12,31,78]: i) an active CO_2 transport system which functions under high- and low-Ci conditions; ii) a Na⁺-dependent bicarbonate transport mechanism; iii) a Na⁺-independent system that converts bicarbonate into CO_2 for its coupled and immediate use by the CO_2 transport system; the latter two operate under low-Ci conditions. Regardless of whether bicarbonate or CO_2 is supplied, the former is always the predominant species in the cytoplasm. Bicarbonate then passes through the thylakoid membrane and enters the carboxysomes which are polyhedral bodies[8] that contain: i) the RubisCO enzyme[48,74,75], an heterooctamer L_8S_8, whose large and small subunits, of 52 and 13 kDa, respectively, are encoded by the *rbcLS* operon[67–69]; ii) the carbonic anhydrase[58], a 30 kDa enzyme that converts bicarbonate into CO_2 and is encoded by the *icfA* gene[26]; iii) a few other polypeptides. These structural entities generate locally high concentration of CO_2 required to activate the RubisCO enzyme. Relatively little is known about the genes whose products are involved in the Ci-concentrating mechanism. A gene cluster that consists of five genes, *ccmKLMNO*, and *ccmJ* have been found upstream of the *rbcLS* operon[25,59,65]. Mutants affected in these genes contain defective carboxysomes and use the internal Ci pool less efficiently than the wild-type strain, confirming that carboxysomes are required for efficient CO_2 fixation. The *ccmJ* gene has been shown to be involved in the formation of the carboxysomal shell[65].

The RubisCO enzyme catalyzes the fixation of CO_2 into ribulose-1,5-bisphosphate to yield 3-phosphoglycerate (Fig. 1). This compound is subsequently metabolized by enzymes common to intermediary metabolism to regenerate ribulose-1,5-bisphosphate on the one hand, and to yield phosphoenolpyruvate through steps that involve enzymes from the glycolytic pathway, on the other hand. Phosphoenolpyruvate is then converted into pyruvate, and pyruvate into acetyl CoA that enters the tricarboxylic acid cycle by reacting with oxaloacetate to generate citrate. The latter is then converted into α-ketoglutarate through different steps of the oxidative branch of the tricarboxylic acid cycle. This cycle being incomplete in cyanobacteria, since it lacks the α–ketoglutarate dehydrogenase complex[70], succinyl CoA is produced from oxaloacetate through the reductive branch of the biosynthetic form of the tricarboxylic acid cycle. Depending on environmental conditions, there also exist alternative pathways for CO_2 fixation which involve the phosphoenolpyruvate carboxylase and possibly the pyruvate carboxylase, as well as the carbamoyl phosphate synthetase[74]. The phosphoenolpyruvate carboxylase and the pyruvate carboxylase generate oxaloacetate by fixation of CO_2 on phosphoenolpyruvate and pyruvate, respectively; the carbamoyl phosphate synthetase converts glutamine into glutamate and carbamoyl phosphate in the presence of CO_2 and ATP.

2. NITROGEN ASSIMILATION

In *Synechococcus* spp., the two means for the entrance of inorganic nitrogenous compounds into the cells are the uptake of ammonium and that of nitrate/nitrite ions[16,18,27]. The uptake of ammonium occurs both via an active transport system that depends on the membrane potential and via passive diffusion of unprotonated ammonia[5,6,60]. Both nitrate and nitrite ions utilize the same active transport system[39,42,61]. Moreover, nitrate can enter by passive diffusion at a concentration higher than 1 mM[56], and diffusion of nitrous acid occurs at neutral pH[17]. The nitrate/nitrite active transport is performed by an ATP-binding cassette-type (ABC) system[16,53]. The permease complex consists of four different polypeptides[39,41,43,44,52,55,56]: NrtA is a 48 kDa membrane-bound lipoprotein that binds nitrate and nitrite ions; NrtB is a 30 kDa hydrophobic integral membrane protein with structural similarities to components of the ABC transporters; NrtC and NrtD of 72 and 30 kDa, respectively, share similarities with the ATP binding subunits from ABC transporters. According to recent results from T. Omata and co-workers[35], NrtC consists of two domains, the N-terminus being similar to NrtD and the C-terminus 30% identical to NrtA. As also shown by this group, the C-terminal domain is involved in the inhibition by ammonium of the nitrate/nitrite transporter. The enzymes for nitrogen assimilation are the nitrate and nitrite reductases which reduce nitrate into nitrite and nitrite into ammonium, respectively. The nitrate reductase, encoded by the *narB* gene, consists of a single 75 kDa polypeptide and possesses a Mo cofactor and Fe-S clusters whose number is not yet well defined[36,41,57,66]. The nitrite reductase, encoded by the *nirA* gene, consists of a single 56 kDa polypeptide and possesses a siroheme and a Fe_4-S_4 cluster as prosthetic groups[38,41,72]. Both enzymes depend on reduced ferredoxin as the immediate electron donor[45]. The *nirA* and *narB* genes are clustered with the genes encoding the permease complex and form the so-called *nirA* operon with, in order, the *nirA-nrtABCD-narB* genes[41,72]. The GS/GOGAT cycle is the major pathway for ammonium assimilation. The glutamine synthetase (GS) catalyzes the ATP-dependent ligation of glutamate and ammonium to yield glutamine, and the glutamate synthase (GOGAT) transfers the amido group of glutamine to α-ketoglutarate to produce glutamate. The glutamine synthetase, encoded by the *glnA* gene, is an enzyme, made up of 12 identical polypeptides of 47 kDa each, that requires a divalent ion, generally Mg^{++}, to be active[13,76]. The activity of the enzyme displays a short-term regulation by light-dark transition and depends on the growth phase, but the exact process remains unclear[47,76]. The glutamate synthase is a 156 kDa flavoprotein containing one molecule of flavin mononucleotide and possibly Fe-S clusters as prosthetic groups[46].

3. REGULATORY ELEMENTS THAT CONTROL CARBON AND NITROGEN ASSIMILATION

A few years ago a global nitrogen regulator, NtcA, was characterized by the group of E. Flores and A. Herrero[24,79,80]. This regulator, encoded by the *ntcA* gene, is a 25 kDa protein that shares similarities with bacterial transcriptional activators of the CRP-type family. NtcA has been shown to control the expression of a number of genes and operons regulated by ammonium, such as *nirBntcB*[71] and *nirA-nrtABCD-narB*[40], *glnA*[9,10,40] and *ntcA* itself[40] (Fig. 1). It generally activates transcription of these genes in the absence of ammonium, or under conditions of nitrogen limitation, by binding to the consensus sequence GTA-N$_8$-TAC[40] located in their promoter region.

A putative transcriptional activator, NtcB, has been identified by T. Omata and co-workers[71]. This 35 kDa protein shares similarities with the bacterial transcriptional factors of the LysR-type family. NtcB has been shown to activate the *nirA-nrtABCD-narB* operon in response to nitrite in nitrogen-starved cells[1,34] (Fig. 1). This regulatory system would allow cells to rapidly adapt to changing availability of nitrate and nitrite in their natural environment. No effect of nitrite on the expression of *ntcA, nirBntcB* and *glnA* has been observed. The *ntcB* gene forms with *nirB* an operon located upstream from the *nirA* operon and divergently transcribed from it. The *nirB* gene is required for expression of maximum nitrite reductase activity, but its precise function is currently unknown[71].

As shown by Miller and Espie[50], cyanate arises from spontaneous dissociation of carbamoyl phosphate, which in turn is synthesized from glutamine, CO_2 and ATP as mentioned earlier. Cyanate is decomposed into CO_2 and ammonia by cyanase, a 16 kDa enzyme encoded by the *cynS* gene[28]. This gene, together with three other *cyn* genes (*cynABD*) found in the same gene cluster, have been cloned by the groups of both T. Omata and G.S. Espie. The *cynABD* genes share similarities with the *nrtABD* genes from the ABC-type nitrate/nitrite transporter mentioned earlier and are thus likely to be involved in the transport of cyanate into the cells[28], but this remains to be demonstrated. The *cynABDS* operon has been shown to be positively regulated by NtcA[28]. According to T. Omata's group[54,73], cyanate is involved in the regulation of genes for carbon assimilation (Fig. 1). It activates the expression of the *rbcLS* operon encoding RubisCO[73], as well as the *ccmKLMNO* genes involved in the Ci-concentrating mechanism and the *icfA* gene encoding carbonic anhydrase[54].

Another important partner in this regulatory circuit is the P$_{II}$ protein (Fig. 1). This protein, encoded by the *glnB* gene, is a 12.4 kDa polypeptide whose amino acid sequence is highly similar to its bacterial counterpart known to be involved in the control of nitrogen assimilation[77]. The cyanobacterial P$_{II}$, in its native form, is a homotrimeric protein of

36 kDa which can be posttranslationally phosphorylated at a seryl residue in position 49 (Ser$_{49}$) instead of being uridylylated at a tyrosyl residue in position 51 as in *E. coli*, despite the presence of a tyrosyl residue at the same position in the cyanobacterial sequence[21]. In vivo, the phosphorylation state of P$_{II}$ depends on the nitrogen and carbon status of the cells[22]. In the presence of nitrate, under high CO$_2$ concentration or under nitrogen starvation, P$_{II}$ is phosphorylated by a kinase[23]. Three isoforms of the phosphorylated P$_{II}$ can be resolved by electrophoresis on native gels and revealed by specific antibodies[21]. P$_{II}$1, P$_{II}$2 and P$_{II}$3 correspond to the trimeric protein with one, two and three phosphorylated monomers, respectively. Upon transfer to a medium containing ammonium or under a low CO$_2$ concentration, P$_{II}$ is dephosphorylated by a phosphatase[21,29] and converted into the P$_{II}$0 isoform which migrates more slowly on a native gel. Moreover, it has been shown that in the absence of P$_{II}$, photosynthetic nitrate reduction is uncoupled from CO$_2$ fixation[22]. Taken together these results indicated that the cyanobacterial P$_{II}$ protein is a signal transducer that coordinates nitrogen and carbon assimilation.

To get a better understanding of the role of the P$_{II}$ protein, mutants have been constructed by in vitro mutagenesis[22; H.M. Lee, to be published]. MP2 is a P$_{II}$-null mutant in which the *glnB* gene encoding P$_{II}$ has been inactivated by a kanamycin cartridge. MP2-S is the MP2 mutant in which the wild-type *glnB* gene has been reintroduced at a "neutral site"[7] in the genome; MP2-S behaves like the wild-type strain. MP2-A is a mutant in which the residue Ser$_{49}$ in the amino acid sequence has been changed into an alanine, this mutant synthesizes a protein which mimics an unphosphorylated P$_{II}$ whatever the nutrient conditions. MP2-D and MP2-E are mutants in which the residue Ser$_{49}$ has been replaced by an aspartate and a glutamate, respectively, leading in both cases to a protein which mimics a phosphorylated P$_{II}$ whatever the nutrient conditions. In MP2-S, which synthesizes a wild-type P$_{II}$ protein, nitrate uptake was inactivated in the presence of ammonium but high in its absence. In the mutant MP2 lacking P$_{II}$, nitrate uptake occurred both with or without ammonium, while in MP2-A, nitrate uptake was not observed. These results indicated that P$_{II}$ plays a role in this process. Finally, in the mutants MP2-E and MP2-D, the uptake of nitrate was inactivated in the presence of ammonium, but high in its absence, as in the wild-type cells. Consequently, not only the unphosphorylated but also the phosphorylated form of P$_{II}$ might inactivate the uptake process and an additional factor is required for its recovery in the absence of ammonium.

According to recent results of in vitro experiments[20,23,29], P$_{II}$ binds α-ketoglutarate and ATP in a mutually dependent manner, and both the phosphorylation and dephosphorylation of P$_{II}$ depend on the availability of α-ketoglutarate but not on that of a nitrogenous compound such as glutamate and glutamine, for example. It is thus tempting to postulate that the additional factor required to control the active transport system, which is common for both nitrate and nitrite ions, is α-ketoglutarate. P$_{II}$ would thus exist under four different states, the unphosphorylated and phosphorylated P$_{II}$ forms either liganded or not to α-ketoglutarate, whose equilibrium would vary with environmental conditions. In the presence of a low intracellular concentration of α-ketoglutarate, P$_{II}$ would be mainly under its unphosphorylated state and not liganded to α-ketoglutarate. This P$_{II}$ state would inactivate the nitrate/nitrite active transport system. One of the possible target(s) would be the permease complex with which P$_{II}$ could interact directly or indirectly. In contrast, in the presence of a high intracellular concentration of α-ketoglutarate, P$_{II}$ would be under its phosphorylated form and bound to α-ketoglutarate; this state of P$_{II}$ would have no affinity for the target(s) and thus nitrate/nitrite could freely enter the cells. A negative control of the nitrate/nitrite active transport system is consistent with the phenotype of the P$_{II}$-null mutant. In the mutant MP2-A, in which P$_{II}$ cannot be phosphorylated, the uptake of nitrate/nitrite would be inactivated whatever the intracellular concentration of α-ketoglutarate, and in the mutants MP2-D and MP2-E that synthesize a protein which mimics a phosphorylated form of P_{II}, the ability of the cells to take up nitrate/nitrite would depend on the intracellular concentration of α-ketoglutarate, the entry of nitrate/nitrite ions being possible only in the presence of a high concentration of this metabolite. This interpretation of the phenotype of the different MP2 mutants implies that the unphosphorylated P_{II} liganded to α-ketoglutarate and the phosphorylated P_{II} are intermediary states that also display some affinities for the target(s) and negatively control the nitrate/nitrite active transport system. Moreover, it emphasizes the necessity that P_{II} should be both liganded to α-ketoglutarate and phosphorylated for the uptake process to occur. Such a regulatory system is in good agreement with the results of a series of excellent experiments performed by a Spanish group in the 80's, which led to the conclusion that nitrate utilization was modulated by a short-term process involving a "sensitive regulatory system integrating the photosynthetic metabolism of carbon and nitrogen"[14,15,19,37,62–64].

4. CONCLUDING REMARKS

The complex network of interactions that controls and co-ordinates carbon and nitrogen assimilation in *Synechococcus* spp. involves at least two transcriptional effectors, NtcA and NtcB, which activate the expression of genes whose products mainly participate in nitrogen metabolism, cyanate which activates those for carbon metabolism, as well as the P_{II} protein, a signal transducer that controls the nitrate/nitrite active transport system to appropriately balance the intracellular N/C ratio, by primarily sensing the availability of α-ketoglutarate. This metabolite has recently been found to be a key regulatory element for global nitrogen control via P_{II} not only in cyanobacteria, but also in proteobacteria in which an additional system sensing glutamine through the uridylyltransferase/uridylyl-removing enzyme has evolved[30,49].

ACKNOWLEDGMENTS

We wish to thank E. Flores, A. Herrero, J. Houmard, R. Rippka, M. Herdman and K. Forchhammer for helpful discussions. We are grateful to K. Forchhammer, T. Omata and A. Kaplan for providing results prior to publication.

REFERENCES

1. Aichi, M., and T. Omata. 1997. Involvement of NtcB, a LysR family transcription factor, in nitrite activation of the nitrate assimilation operon in the cyanobacterium *Synechococcus* sp. strain PCC 7942. J. Bacteriol. *179*:4671–4675.
2. Aizawa, K., and S. Miyachi. 1986. Carbonic anhydrase and CO_2 concentrating mechanisms in microalgae and cyanobacteria. FEMS Microbiol. Rev. *39*:215–233.
3. Badger, M. R. 1987. The CO_2-concentration mechanism in aquatic phototrops. Pages 219–274, *in* The Biochemistry of Plants. Academic Press, New York.
4. Badger, M. R. and Price, G. D. 1992. The CO_2-concentrating mechanism in cyanobacteria and microalgae. *Physiol. Plant.* 84: 606–615.
5. Boussiba, S., and J. Gibson. 1991. Ammonia translocation in cyanobacteria. FEMS Microbiol. Rev. *88*:1–14.

6. Boussiba, S., C. M. Resch, and J. Gibson. 1984. Ammonia uptake and retention in some cyanobacteria. Arch. Microbiol. *138*:287–292.

7. Bustos, S. A., and S. S. Golden. 1992. Light-regulated expression of the *psbD* gene family in *Synechococcus* sp. strain PCC 7942: Evidence for the role of duplicated *psbD* genes in cyanobacteria. Mol. Gen. Genet. *232*:221–230.

8. Codd, G.A. 1988. Carboxysomes and ribulose bisphosphate carboxylase/oxygenase. Pages 115–164, *in* A.H. Rose, and D.W. Tempest, ed., Advances in Microbial Physiology. Academic Press, London.

9. Cohen-Kupiec, R., M. Gurevitz, and A. Zilberstein. 1993. Expression of *glnA* in the cyanobacterium *Synechococcus* sp. strain PCC 7942 is initiated from a single *nif*-like promoter under various nitrogen conditions. J. Bacteriol. *175*:7727–7731.

10. Cohen-Kupiec, R., A. Zilberstein, and M. Gurevitz. 1995. Characterization of *cis* elements that regulate the expression of *glnA* in *Synechococcus* sp. strain PCC 7942. J. Bacteriol. *177*:2222–2226.

11. Espie, G. S., and R. A. Kandasamy. 1992. Na^+-independent HCO_3^--transport and accumulation in the cyanobacterium *Synechococcus* UTEX 625. Plant Physiol. *98*:560–568.

12. Espie, G. S., and R. A. Kandasamy. 1994. Monensin inhibition of Na^+-dependent HCO_3^- transport distinguishes it from Na^+-independent HCO_3^- transport and provides evidence for Na^+/HCO_3^- symport in the cyanobacterium *Synechococcus* UTEX 625. Plant Physiol. *104*:1419–1428.

13. Florencio, F. J., and J. L. Ramos. 1985. Purification and characterization of glutamine synthetase from the unicellular cyanobacterium *Anacystis nidulans*. Biochim. Biophys. Acta *838*:39–48.

14. Flores, E., M. G. Guerrero, and M. Losada. 1980. Short-term ammonium inhibition of nitrate utilization by *Anacystis nidulans* and other cyanobacteria. Arch. Microbiol. *128*:137–144.

15. Flores, E., M. G. Guerrero, and M. Losada. 1983. Photosynthetic nature of nitrate uptake and reduction in the cyanobacterium *Anacystis nidulans*. Biochim. Biophys. Acta *722*:408–416.

16. Flores, E., and A. Herrero. 1994. Assimilatory nitrogen metabolism and its regulation. Pages 487–517, *in* D.A. Bryant, ed., The Molecular Biology of Cyanobacteria. Kluwer Academic Publishers, The Netherlands.

17. Flores, E., A. Herrero, and M. G. Guerrero. 1987. Nitrite uptake and its regulation in the cyanobacterium *Anacystis nidulans*. Biochim. Biophys. Acta *896*:103–108.

18. Flores, E., J. L. Ramos, A. Herrero, and M.G. Guerrero. 1983. Nitrate assimilation by cyanobacteria. Pages 363–387, *in* G. C. Papageorgiou, and L. Parker, ed., Photosynthetic Prokaryotes: Cell Differentiation and Function. Elsevier Biomedical, New York.

19. Flores, E., J. M. Romero, M. G. Guerrero, and M. Losada. 1983. Regulatory interaction of photosynthetic nitrate utilization and carbon dioxide fixation in the cyanobacterium *Anacystis nidulans*. Biochim. Biophys. Acta *725*:529–532.

20. Forchhammer, K., and A. Hedler. 1997. Phosphoprotein P_{II} from cyanobacteria. Analysis of functional conservation with the P_{II} signal-transduction protein from *Escherichia coli*. Eur. J. Biochem. *244*:869–875.

21. Forchhammer, K., and N. Tandeau de Marsac. 1994. The P_{II} protein in the cyanobacterium *Synechococcus* sp. strain PCC 7942 is modified by serine phosphorylation and signals the cellular N-status. J. Bacteriol. *176*:84–91.

22. Forchhammer, K., and N. Tandeau de Marsac. 1995. Functional analysis of the phosphoprotein P_{II} (*glnB* gene product) in the cyanobacterium *Synechococcus* sp. strain PCC 7942. J. Bacteriol. *177*:2033–2040.

23. Forchhammer, K., and N. Tandeau de Marsac. 1995. Phosphorylation of the P_{II} protein (*glnB* gene product) in the cyanobacterium *Synechococcus* sp. strain PCC 7942: Analysis of in vitro kinase activity. J. Bacteriol. *177*:5812–5817.

24. Frías, J. E., A. Mérida, A. Herrero, J. Martín-Nieto, and E. Flores. 1993. General distribution of the nitrogen control gene *ntcA* in cyanobacteria. J. Bacteriol. *175*:5710–5713.

25. Friedberg, D., A. Kaplan, R. Ariel, M. Kessel, and J. Seijffers. 1989. The 5'-flanking region of the gene encoding the large subunit of ribulose-1,5-biphosphate carboxylase/oxygenase is crucial for growth of the cyanobacterium *Synechococcus* sp. strain PCC 7942 at the level of C02 in air. J. Bacteriol. *171*:6069–6076.

26. Fukuzawa, H., E. Suzuki, Y. Komukai, and S. Miyachi. 1992. A gene homologous to chloroplast carbonic anhydrase (*icfA*) is essential to photosynthetic carbon dioxide fixation by *Synechococcus* PCC7942. Proc. Natl. Acad. Sci. USA *89*:4437–4441.

27. Guerrero, M. G. and C. Lara. 1987. Assimilation of inorganic nitrogen. Pages 163–186, *in* P. Fay, and C. Van Baalen, ed., The cyanobacteria. Elsevier, Amsterdam.

28. Harano, Y., I. Suzuki, S.-I. Maeda, T. Kaneko, S. Tabata, and T. Omata. 1997. Identification and nitrogen regulation of the cyanase gene from the cyanobacteria *Synechocystis* sp. strain PCC 6803 and *Synechococcus* sp. strain PCC 7942. J. Bacteriol. *179*:5744–5750.

29. Irmler, A., S. Sanner, H. Dierks, and K. Forchhammer. 1997. Dephosphorylation of the phosphoprotein P_{II} in *Synechococcus* PCC 7942: Identification of an ATP and 2-oxoglutarate-regulated phosphatase activity. Mol. Microbiol. *25*::In press.

30. Kamberov, E. S., M. R. Atkinson, and A. J. Ninfa. 1995. The *Escherichia coli* P_{II} signal transduction protein is activated upon binding 2-ketoglutarate and ATP. J. Biol. Chem. *270*: 17797–17807.

31. Kaplan, A., R. Schwarz, R. Ariel, and L. Reinhold. 1990. The "CO_2 concentrating mechanism" of cyanobacteria: Physiological molecular and theoretical studies. Bot. Mag. Tokyo *2*:53–71.

32. Kaplan, A., R. Schwarz, J. Lieman-Hurwitz, and L. Reinhold. 1991. Physiological and molecular aspects of the inorganic carbon-concentrating mechanism in cyanobacteria. Plant Physiol. *97*:851–855.

33. Kaplan, A., R. Schwarz, J. Lieman-Hurwitz, M. Ronen-Tarazi, and L. Reinhold. 1994. Physiological and molecular studies on the response of cyanobacteria to changes in the ambient inorganic carbon concentration. Pages 469–485, *in* D.A. Bryant, ed., The Molecular Biology of Cyanobacteria. Kluwer Academic Publishers, The Netherlands.

34. Kikuchi, H., M. Aichi, I. Suzuki, and T. Omata. 1996. Positive regulation by nitrite of the nitrate assimilation operon in the cyanobacteria *Synechococcus* sp. strain PCC 7942 and *Plectonema boryanum*. J. Bacteriol. *178*:5822–5825.

35. Kobayashi, M., R. Rodríguez, C. Lara, and T. Omata. 1997. Involvement of the C-terminal domain of an ATP-binding subunit in the regulation of the ABC-type nitrate/nitrite transporter of the cyanobacterium *Synechococcus* sp. strain PCC 7942. J. Biol. Chem. :In press.

36. Kuhlemeier, C. J., T. Logtenberg, W. Stoorvogel, H. A. A. van Heugten, W. E. Borrias, and G. A. van Arkel. 1984. Cloning of nitrate reductase genes from the cyanobacterium *Anacystis nidulans*. J. Bacteriol. *159*:36–41.

37. Lara, C., J. M. Romero, and M. G. Guerrero. 1987. Regulated nitrate transport in the cyanobacterium *Anacystis nidulans*. J. Bacteriol. *169*:4376–4378.

38. Luque, I., E. Flores, and A. Herrero. 1993. Nitrite reductase gene from *Synechococcus* sp. PCC 7942: Homology between cyanobacterial and higher-plant nitrite reductases. Plant Mol. Biol. *21*:1201–1205.

39. Luque, I., E. Flores, and A. Herrero. 1994. Nitrate and nitrite transport in the cyanobacterium *Synechococcus* sp. PCC 7942 are mediated by the same permease. Biochim. Biophys. Acta *1184*:296–298.

40. Luque, I., E. Flores, and A. Herrero. 1994. Molecular mechanism for the operation of nitrogen control in cyanobacteria. EMBO J. *13*:2862–2869.

41. Luque, I., A. Herrero, E. Flores, and F. Madueño. 1992. Clustering of genes involved in nitrate assimilation in the cyanobacterium *Synechococcus*. Mol. Gen. Genet. *232*:7–11.

42. Madueño, F., E. Flores, and M. G. Guerrero. 1987. Competition between nitrate and nitrite uptake in the cyanobacterium *Anacystis nidulans*. Biochim. Biophys. Acta *896*:109–112.

43. Madueño, F., M. A. Vega-Palas, E. Flores, and A. Herrero. 1988. A cytoplasmic-membrane protein repressible by ammonium in *Synechococcus* R2: Altered expression in nitrate-assimilation mutants. FEBS Lett. *239*:289–291.

44. Maeda, S.-I., and T. Omata. 1997. Substrate-binding lipoprotein of the cyanobacterium *Synechococcus* sp. strain PCC 7942 involved in the transport of nitrate and nitrite. J. Biol. Chem. *272*:3036–3041.

45. Manzano, C., P. Candau, C. Gomez-Moreno, A. M. Relimpio, and M. Losada. 1976. Ferredoxin-dependent photosynthetic reduction of nitrate and nitrite by particles of *Anacystis nidulans*. Mol. Cell. Biochem. *10*:161–169.

46. Marqués, S., F. J. Florencio, and P. Candau. 1992. Purification and characterization of the ferredoxin-glutamate synthase from the unicellular cyanobacterium *Synechococcus* sp. PCC 6301. Eur. J. Biochem. *206*:69–77.

47. Marqués, S., A. Mérida, P. Candau, and F. J. Florencio. 1992. Light-mediated regulation of glutamine-synthetase activity in the unicellular cyanobacterium *Synechococcus* sp. PCC 6301. Planta *187*:247–253.

48. McKay, R. M. L., S. P. Gibbs, and G. S. Espie. 1993. Effect of dissolved inorganic carbon on the expression of carboxysomes, localization of Rubisco and the mode of inorganic carbon transport in cells of the cyanobacterium *Synechococcus* UTEX 625. Arch. Microbiol. *159*:21–29.

49. Merrick, M. J., and R. A. Edwards. 1995. Nitrogen control in bacteria. Microbiol. Rev. *59*:604–622.

50. Miller, A. G., and G. S. Espie. 1994. Photosynthetic metabolism of cyanate by the cyanobacterium *Synechococcus* UTEX 625. Arch. Microbiol. *162*:151–157.

51. Miller, A. G., G. S. Espie, and D. T. Canvin. 1990. Physiological aspects of CO_2 and HCO_3^- transport by cyanobacteria: A review. Can. J. Bot. *68*:1291–1302.

52. Omata, T. 1991. Cloning and characterization of the *nrtA* gene that encodes a 45-kDa protein involved in nitrate transport in the cyanobacterium *Synechococcus* PCC 7942. Plant Cell Physiol. *32*:151–157.

53. Omata, T. 1995. Structure, function and regulation of the nitrate transport system of the cyanobacterium *Synechococcus* sp. PCC7942. Plant Cell Physiol. *36*:207–213.

54. Omata, T. 1996. Nitrogen regulation of carbon and nitrogen assimilation genes in the cyanobacterium *Synechococcus* sp. strain PCC 7942. *In* US-Japan Workshop on Advances in the Molecular Biology of Photosynthesis - Grand Canyon National Park, USA.

55. Omata, T., X. Andriesse, and A. Hirano. 1993. Identification and characterization of a gene cluster involved in nitrate transport in the cyanobacterium *Synechococcus* sp. PCC7942. Mol. Gen. Genet. *236*:193–202.

56. Omata, T., M. Ohmori, N. Arai, and T. Ogawa. 1989. Genetically engineered mutant of the cyanobacterium *Synechococcus* PCC 7942 defective in nitrate transport. Proc. Natl. Acad. Sci. USA *86*:6612–6616.

57. Peschek, G. A. 1979. Nitrate and nitrite reductase and hydrogenase in *Anacystis nidulans* grown in Fe- and Mo-deficient media. FEMS Microbiol. Lett. *6*:371–374.

58. Price, G. D., J. R. Coleman, and M. R. Badger. 1992. Association of carbonic anhydrase activity with carboxysomes isolated from the cyanobacterium *Synechococcus* PCC7942. Plant Physiol. *100*:784–793.

59. Price, G. D., S. M. Howitt, K. Harrison, and M. R. Badger. 1993. Analysis of a genomic DNA region from the cyanobacterium *Synechococcus* sp. strain PCC7942 involved in carboxysome assembly and function. J. Bacteriol. *175*:2871–2879.

60. Rai, A. N., P. Rowell, and W. D. P. Stewart. 1984. Evidence for an ammonium transport system in free-living and symbiotic cyanobacteria. Arch. Microbiol. *137*:241–246.

61. Rodríguez, R., C. Lara, and M. G. Guerrero. 1992. Nitrate transport in the cyanobacterium *Anacystis nidulans* R2. Kinetic and energetic aspects. Biochem. J. *282*:639–643.

62. Romero, J. M., T. Coronil, C. Lara, and M. G. Guerrero. 1987. Modulation of nitrate uptake in *Anacystis nidulans* by the balance between ammonium assimilation and CO_2 fixation. Arch. Biochem. Biophys. *256*:578–584.

63. Romero, J. M., and C. Lara. 1987. Photosynthetic assimilation of NO_3^- by intact cells of the cyanobacterium *Anacystis nidulans*. Influence of NO_3^- and NH_4^+ assimilation on CO_2 fixation. Plant Physiol. *83*:208–212.

64. Romero, J. M., C. Lara, and M. G. Guerrero. 1985. Dependence of nitrate utilization upon active CO2 fixation in *Anacystis nidulans*: A regulatory aspect of the interaction between photosynthetic carbon and nitrogen metabolism. Arch. Biochem. Biophys. *237*:396–401.

65. Ronen-Tarazi, M., J. Lieman-Hurwitz, C. Gabay, M. I. Orus, and A. Kaplan. 1995. The genomic region of *rbcLS* in *Synechococcus* sp. PCC 7942 contains genes involved in the ability to grow under low CO_2 concentration and in chlorophyll biosynthesis. Plant Physiol. *108*:1461–1469.

66. Rubio, L. M., A. Herrero, and E. Flores. 1996. A cyanobacterial *narB* gene encodes a ferredoxin-dependent nitrate reductase. Plant Mol. Biol. *30*:845–850.

67. Shinozaki, K., and M. Sugiura. 1983. The gene for the small subunit of ribulose-1,5-bisphosphate carboxylase/oxygenase is located close to the gene for the large subunit in the cyanobacterium *Anacystis nidulans* 6301. Nucleic Acids Res. *11*:6957–6964.

68. Shinozaki, K., and M. Sugiura. 1985. Genes for the large and small subunits of ribulose-1,5-biphosphate carboxylase/oxygenase constitute a single operon in a cyanobacterium *Anacystis nidulans* 6301. Mol. Gen. Genet. *200*:27–32.

69. Shinozaki, K., C. Yamada, N. Takahata, and M. Sugiura. 1983. Molecular cloning and sequence analysis of the cyanobacterial gene for the large subunit of ribulose-1,5-biphosphate carboxylase/oxygenase. Proc. Natl. Acad. Sci. USA *80*:4050–4054.

70. Stanier, R. Y., and G. Cohen-Bazire. 1977. Phototrophic prokaryotes: The cyanobacteria. Annu. Rev. Microbiol. *31*:225–274.

71. Suzuki, I., N. Horie, T. Sugiyama, and T. Omata. 1995. Identification and characterization of two nitrogen-regulated genes of the cyanobacterium *Synechococcus* sp. strain PCC7942 required for maximum efficiency of nitrogen assimilation. J. Bacteriol. *177*:290–296.

72. Suzuki, I., T. Sugiyama, and T. Omata. 1993. Primary structure and transcriptional regulation of the gene for nitrite reductase from the cyanobacterium *Synechococcus* PCC 7942. Plant Cell Physiol. *34*:1311–1320.

73. Suzuki, I., T. Sugiyama, and T. Omata. 1996. Regulation by cyanate of the genes involved in carbon and nitrogen assimilation in the cyanobacterium *Synechococcus* sp. PCC 7942. J. Bacteriol. *178*:2688–2694.

74. Tabita, F.R. 1987. Carbon dioxide fixation and its regulation in cyanobacteria. Pages 95–117, *in* P. Fay, and C. Van Baalen, ed., The cyanobacteria. Elsevier, Amsterdam.

75. Tabita, F.R. 1994. The biochemistry and molecular regulation of carbon dioxide metabolism in cyanobacteria. Pages 437–467, *in* D.A. Bryant, ed., The Molecular Biology of Cyanobacteria. Kluwer Academic Publishers, The Netherlands.

76. Tischner, R., and A. Schmidt. 1984. Light mediated regulation of nitrate assimilation in *Synechococcus leopoliensis*. Arch. Microbiol. *137*:151–154.

77. Tsinoremas, N. F., A. M. Castets, M. A. Harrison, J. F. Allen, and N. Tandeau de Marsac. 1991. Photosynthetic electron transport controls nitrogen assimilation in cyanobacteria by means of posttranslational modification of the *glnB* gene product. Proc. Natl. Acad. Sci. USA *88*:4565–4569.

78. Tyrrell, P. N., R. A. Kandasamy, C. M. Crotty, and G. S. Espie. 1996. Ethoxyzolamide differentially inhibits CO_2 uptake and Na^+-independent and Na^+-dependent HCO_3^- uptake in the cyanobacterium *Synechococcus* sp. UTEX 625. Plant Physiol. *112*:79–88.

79. Vega-Palas, M. A., E. Flores, and A. Herrero. 1992. NtcA, a global nitrogen regulator from the cyanobacterium *Synechococcus* that belongs to the Crp family of bacterial regulators. Mol. Microbiol. *6*:1853–1859.

80. Vega-Palas, M. A., F. Madueño, A. Herrero, and E. Flores. 1990. Identification and cloning of a regulatory gene for nitrogen assimilation in the cyanobacterium *Synechococcus* sp. strain PCC 7942. J. Bacteriol. *172*:643–647.

63

THE P_{II} PROTEIN IN *SYNECHOCOCCUS* PCC 7942 SENSES AND SIGNALS 2-OXOGLUTARATE UNDER ATP-REPLETE CONDITIONS

Karl Forchhammer

Lehrstuhl für Mikrobiologie
Universität München
Maria-Ward-Str 1a, D-80638 München, Germany

1. INTRODUCTION TO THE P_{II} SYSTEM

The *glnB* gene product, termed P_{II} protein, is widely distributed among bacteria and functions as a signal transduction protein in the central regulation of nitrogen metabolism (reviewed in 9). In proteobacteria, P_{II} is modified by uridylylation at a conserved tyrosyl residue (Tyr 51); under nitrogen replete conditions, P_{II} is present in its unmodified state whereas P_{II}-UMP signals nitrogen-deficiency. Recently the 3-D structure of P_{II} from *Escherichia coli* has been resolved (1), showing that the site of modification is located at the apex of a large solvent-protruding loop, termed T-loop. In contrast to proteobacteria, P_{II} in cyanobacteria is not modified by uridylylation but is phosphorylated at a seryl residue (Ser 49) separated only by one amino acid from the conserved tyrosyl residue (2,4). This indicates that phosphorylation also occurs at the solvent-exposed T-loop. In *Synechococcus* PCC 7942, the trimeric P_{II} protein was shown to be involved in the coordination of carbon and nitrogen assimilation, in particular mediating the dependence on CO_2 fixation for nitrate utilization (3). In vivo analyses revealed that the phosphorylation state of P_{II} responds to the status of nitrogen and carbon assimilation (2,3). In the presence of ammonium, P_{II} is predominantly present in its unmodified form. In nitrate grown cells, the extent of P_{II} phosphorylation depends on the CO_2 supply to the cells: Under CO_2-limiting conditions, only a low degree of P_{II} phosphorylation is observed whereas under CO_2 sufficiency, P_{II} is efficiently phosphorylated. The highest level of phosphorylation is found in nitrogen-starved cells. Studying the P_{II} modification system, therefore, offers the possibility to investigate a mechanism used by cyanobacteria to sense the environmental changes in the nitrogen and carbon supply.

To approach the mechanism of P_{II} phosphorylation and dephosphorylation in cyanobacteria, it is useful to consider the system of P_{II} modification in *Escherichia coli* as re-

The Phototrophic Prokaryotes, edited by Peschek *et al.*
Kluwer Academic / Plenum Publishers, New York, 1999.

vealed during three decades of investigation (7,10). The sensory mechanism resulting in the uridylylation/deuridylylation of P_{II} comprises two proteins of which one is the P_{II} protein itself, which binds the metabolites ATP and 2-oxoglutarate with high affinity (8). Only the liganded form of P_{II} efficiently interacts with the bifunctional enzyme GlnD, which catalyzes the uridylylation of P_{II}. Glutamine inhibits the uridylyl-transferase activity of GlnD and promotes its uridylyl-removing activity. When consideration is taken of the physiologically relevant concentrations of glutamine, 2-oxoglutarate and ATP, the GlnD-P_{II} monocycle functions essentially as a glutamine-sensing device (7).

2. METABOLITE BINDING BY THE *SYNECHOCOCCUS* PCC 7942 P_{II} PROTEIN

A question of primary importance concerns the metabolite-binding capacity of the *Synechococcus* PCC 7942 P_{II} protein. Therefore, we determined the binding of ^{3}H-labelled ATP and ^{14}C-labelled 2-oxoglutarate to purified *Synechococcus* P_{II} by equilibrium dialysis or ultrafiltration (5). No 2-oxoglutarate binding to P_{II} could be detected in the absence of ATP; however, in its presence high affinity binding of 2-oxoglutarate was detected. The dependence of 2-oxoglutarate-binding on the ATP concentration was analyzed in detail and it was found that the dissociation constant for 2-oxoglutarate-binding decreased with increasing ATP concentration. For example, in the presence of 25 µM ATP, an apparent K_d for 2-oxoglutarate binding of 6 µM was calculated and the stoichiometry of binding was 0.72 2-oxoglutarate/P_{II} trimer, whereas in the presence of 2.5 mM ATP, a K_d of 3.8 µM and a stoichiometry of 1.42 2-oxoglutarate/P_{II} trimer was determined. The ATP concentration required to obtain half maximal 2-oxoglutarate binding affinity was calculated to 17 µM. Conversely, binding of radiolabelled ATP to P_{II} could already be shown in the absence of 2-oxoglutarate at the limit of detection, with a calculated K_d of 37 µM and a stoichiometry of 0.88 ATP/P_{II} trimer. In the presence of 0.5 mM 2-oxoglutarate, the K_d of ATP binding was 2 µM, the stoichiometry was 1.5 ATP/P_{II} trimer and in the presence of 10 mM 2-oxoglutarate, the K_d was lowered to 0.38 µM whereas the stoichiomentry increased to 1.9 ATP/P_{II} trimer, indicating that 2-oxoglutarate stimulated ATP-binding. The 2-oxoglutarate concentration required to obtain half maximal ATP binding affinity was 3.9 mM. These results illustrate a synergism between the binding of ATP and 2-oxoglutarate which may result from cooperativity of the ATP and 2-oxoglutarate binding sites. Fig. 1 presents a model which takes into account the observed mutual dependence on ATP and 2-oxoglutarate for P_{II}-binding. Under experimental conditions, the non-labelled partner ligand is in excess over the radiolabelled ligand for which binding is to be determined. According to this model, P_{II} will be preloaded by the non-labelled partner ligand, which by cooperativity creates a high affinity binding site for the labelled ligand. The concentration of the partner ligand which provides half-maximal binding of the labelled ligand would correspond to the dissociation constant of the first binding event. According to that assumption, the dissociation constant of 2-oxoglutarate binding to P_{II} would be in the millimolar range. Since the analytical method employed in this study only permits the determination of dissociation constants equal to or lower than the concentration of the binding protein, the low affinity 2-oxoglutarate-binding would exceed the limit of detection by two orders of magnitude. Considering that the determination of weak binding by equilibrium dialysis is rather inaccurate, the dissociation constant for ATP binding in the absence of 2-oxoglutarate determined by direct measurement is in the same range as the ATP concentration required for half-maximal stimulation of 2-oxoglutarate binding. Since the physiologically

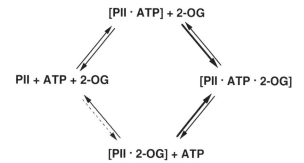

Figure 1. Proposed model of the synergistic binding of ATP and 2-oxoglutarate (2-OG) to P$_{II}$. For details see text.

relevant concentration of ATP is by two orders of magnitudes higher, *Synechococcus* P$_{II}$ will almost always be liganded by ATP. Therefore, its actual ligandation status will depend on the cellular 2-oxoglutarate concentration.

3. THE P$_{II}$ PHOSPHORYLATION AND DEPHOSPHORYLATION ACTIVITIES IN *SYNECHOCOCCUS* PCC 7942

From the analysis presented above, it appeared that the P$_{II}$ protein has conserved its ligand-binding properties in the divergently evolved P$_{II}$ signal transduction pathways in proteobacteria and cyanobacteria. In contrast to proteobacteria, the modification of P$_{II}$ in cyanobacteria resembles a eukaryotic type of signal transduction system involving protein serine phosphorylation and dephosphorylation. What are the modifying enzymes of the *Synechococcus* P$_{II}$ protein and how are these activities regulated? Detailed in vivo analysis of the phosphorylation state of P$_{II}$ employing inhibitors of carbon and nitrogen metabolism showed that all conditions which induced P$_{II}$ phosphorylation were found to inhibit the ammonium-promoted dephosphorylation of phosphorylated P$_{II}$ protein (P$_{II}$-P), indicating that these reactions are inversely regulated (3,6). Secondly, permeabilization of *Synechococcus* cells was found to promote a rapid P$_{II}$-P dephosphorylation (6). Since this treatment leads to the loss of small molecules, P$_{II}$-P dephosphorylation seemed to be regulated by inhibitory metabolites.

In an attempt to understand the molecular basis of the P$_{II}$ modifying system, an in vitro system to analyze P$_{II}$ phosphorylation and dephosphorylation was devised using extracts from a P$_{II}$ deficient mutant of *Synechococcus* PCC 7942, termed MP2. Extracts of MP2 cells were prepared and incubated together with purified non-phosphorylated P$_{II}$ protein in the presence of ATP and stimulatory metabolites to assay P$_{II}$ phosphorylation. Likewise, purified P$_{II}$-P was incubated together with MP2 extracts and analyzed for P$_{II}$-P dephosphorylation. With this approach, we could identify a P$_{II}$ kinase activity in the soluble fraction (S100) of MP2 extracts (4). The kinase activity was specifically stimulated by 2-oxoglutarate and ATP was used as phosphoryl-donor. The requirement for 2-oxoglutarate was absolute as shown by the fact that in the absence of this metabolite there was no trace of P$_{II}$ phosphorylation. In analogy to the inhibition of *E. coli* P$_{II}$ uridylyltransferase inhibition by glutamine, we analyzed a variety of nitrogen-containing metabolites for their ability to inhibit P$_{II}$ kinase activity. No metabolite, however, was found to have any influence on P$_{II}$ phosphorylation in vitro.

In vitro phosphatase assays revealed a strong magnesium- or manganese-dependent P_{II}-P phosphatase activity in crude extracts of MP2 cells. Since such an activity could be unspecific or not related to the in vivo P_{II}-P phosphatase, we attempted to purify that activity and assayed for P_{II} phosphatase as well as for P_{II} kinase activities. After gel filtration through Superdex 200, P_{II} kinase could be physically separated from the P_{II}-P phosphatase activity, indicating that these two enzymes are distinct, in contrast to the bifunctional nature of the P_{II}-modifying GlnD enzyme in proteobacteria. P_{II}-P phosphatase activity eluted as a single peak which was further purified by High-Q anion-exchange chromatography, yielding a preparation which was purified about 200-fold compared to the soluble extract. The biochemical properties of this phosphatase preparation were investigated in more detail (6). The substrate spectrum of the phosphatase activity was assayed. P_{II}-P phosphatase was unable to dephosphorylate p-nitrophenylphosphate (PNPP), a commonly used substrate for a wide range of phosphatases, including acid and alkaline phosphatases and most protein phosphatases except the PP1 type of protein phosphatases. However, P_{II}-P phosphatase was highly active against serine/threonine-phosphorylated casein or mixed histones. Dephosphorylation of these substrates showed the same dependence on magnesium or manganese ions as P_{II}-P dephosphorylation. This dependence, as well as the sensitivity profile against a variety of protein phosphatase inhibitors specific for different protein phosphatase families, indicated a resemblance between P_{II}-P phosphatase with the PP2C type of protein phosphatases (6).

The intriguing question was, how is P_{II}-P phosphatase activity regulated? In the absence of any metabolites, P_{II}-P phosphatase was active with the activity depending only on divalent cations. In the presence of both 2-oxoglutarate and ATP, P_{II}-P phosphatase was inhibited, whereas both metabolites on their own were not able to inhibit the activity. The mutual dependence on ATP and 2-oxoglutarate for phosphatase inhibition resembles the synergism of these metabolites in P_{II}-binding. Indeed, ATP and 2-oxoglutarate were without any effect in phosphatase assays using phosphocasein or phosphohistones as substrate, suggesting that the inhibition of P_{II}-P dephosphorylation does not occurr by direct inhibition of the enzyme but by the ligandation status of P_{II}. According to the model presented in Fig. 2, liganding both ATP and 2-oxoglutarate would induce a conformational change in P_{II}, which would be recognized by the modifying enzymes. In the liganded form, P_{II} would be phosphorylated whereas liganded P_{II}-P would be protected from dephosphorylation; non-liganded P_{II}-P would be the substrate for the P_{II}-P phosphatase. This mechanism would be a genuine 2-oxoglutarate-sensing apparatus under ATP replete conditions, since the 2-oxoglutarate pool would determine the ligandation status of P_{II} which in turn would determine the phosphorylation state of P_{II}.

Why would cyanobacteria only rely on 2-oxoglutarate as an indicator of the state of nitrogen and carbon assimilation? If the structure of the assimilatory pathways is considered, this metabolite seems ideally suited for that purpose, since it is located at the inter-

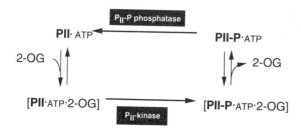

Figure 2. Proposed model of the control of PII phosphorylation and PII-P dephosphorylation by the ligandation status of PII. Note that in vivo PII will almost always be liganded by ATP.

section between carbon and nitrogen metabolism. 2-oxoglutarate is formed directly from carbon assimilation products by glycolytic reactions and the first steps of the TCA cycle. The only 2-oxoglutarate-consuming reaction is its conversion to glutamate by the GOGAT enzyme using the amido-group of glutamine as nitrogen donor. Therefore, changes in the carbon and nitrogen assimilating reactions would immediately influence the pool size of 2-oxoglutarate. In contrast, in organisms with a complete TCA cycle, like most proteobacteria, the 2-oxoglutarate pool depends on the flux through the TCA cycle which would explain the need of these organisms to use a nitrogen-containing metabolite, particulary glutamine, to sense the state of nitrogen assimilation.

ACKNOWLEDGMENTS

I whish to thank my students Angelika Irmler, Henning Dierks and Silvia Sanner for their contributions to the work as well as Andrea Hedler for excellent technical assistence. Gary Sawers is acknowledged for critically reading the manuscript. This work was supported by a grant from the Deutsche Forschungsgemeinschaft.

REFERENCES

1. Cheah, E., Carr, P.D., Suffolk, P.M., Vasudevan, C., Dixon, N.E. and Ollis, D.L. (1994) Structure 2, 981–990
2. Forchhammer, K. and Tandeau de Marsac, N. (1994) J. Bacteriol. 176, 84–91.
3. Forchhammer, K. and Tandeau de Marsac, N. (1995) J. Bacteriol. 177, 2033–2040.
4. Forchhammer, K. and Tandeau de Marsac, N. (1995) J. Bacteriol. 177, 5812–5817
5. Forchhammer, K. and Hedler, A. (1997) Eur. J. Biochem. 244, 869–875
6. Irmler, A., Sanner, S., Dierks, H. and Forchhammer, K. (1997) Mol. Microbiol. (in press)
7. Jiang, P., Zucker, P. and Ninfa, A. (1997) J. Bactriol. 179, 4354–4360
8. Kamberov, E.S., Atkinson, M.A. and Ninfa, A.J. (1995) J. Biol. Chem. 270, 17797–17807
9. Merrik, M.J and Edwards, R.A. (1995) Microbiol. Rev. 59, 604–622
10. Stadtman, E.R., Shapiro, M, Ginsburg, A, Kingdon, H.S. and Denton, M.D. (1968) Brookhaven Symp. Biol. 21, 378–396

64

INVOLVEMENT OF THE *cmpABCD* GENES IN BICARBONATE TRANSPORT OF THE CYANOBACTERIUM *SYNECHOCOCCUS* SP. STRAIN PCC 7942

Tatsuo Omata,[1] G. Dean Price,[3] Murray R. Badger,[3] Masato Okamura,[1] and Teruo Ogawa[2]

[1]Department of Applied Biological Sciences
School of Agricultural Sciences
Nagoya University
Nagoya 464-01, Japan
[2]Bioscience Center
Nagoya University
Nagoya 464-01, Japan
[3]Molecular Plant Physiology Group
Research School of Biological Sciences
Australian National University
P.O. Box 475, Canberra A.C.T. 2601, Australia

1. INTRODUCTION

Cyanobacteria possess a CO_2-concentrating mechanism (CCM) which elevates the CO_2 concentration around the active site of Rubisco and thereby compensates for the low selectivity of Rubisco for CO_2 [1]. Active transport into the cell of inorganic carbon (CO_2 and HCO_3^-; designated hereafter Ci) is an essential function of the CCM, and physiological studies have suggested the occurrence of multiple forms of CO_2 and HCO_3^- transporters. However, no proteins or genes directly involved in the process of Ci transport have been identified.

It is well known that cyanobacterial cells grown under CO_2-limited conditions (L-cells) have a higher affinity for Ci transport than the cells grown under elevated CO_2 concentrations (H-cells). In an attempt to biochemically identify the proteins involved in Ci

The Phototrophic Prokaryotes, edited by Peschek *et al.*
Kluwer Academic / Plenum Publishers, New York, 1999.

transport, we compared the polypeptide compositions of the cytoplasmic membrane preparations from L-cells and H-cells of *Synechococcus* sp. strain PCC 6301 and *Synechococcus* sp. strain PCC 7942, and found a 42-kD protein that is synthesized under the conditions of carbon limitation [2,3]. Although co-induction of the 42-kDa protein and Ci-transporting activity by carbon limitation suggested that the protein may play a role in the Ci-transporting mechanism [2–4], a deletion mutant (M42) of the gene encoding the protein, *cmpA*, could grow under moderately low CO_2 conditions (300 ppm CO_2) and showed appreciable activities of CO_2 and HCO_3^- transport, suggesting that the protein is unlikely to be involved in Ci transport [5]. Subsequent to these studies, however, another cytoplasmic membrane protein of 45 kD apparent molecular mass, encoded by *nrtA*, was found to be involved in active transport of nitrate [6] and shown to be 47% identical in deduced amino acid sequence to the *cmpA*-encoded 42-kD protein [7]. Later studies identified three additional genes *nrtB*, *nrtC*, and *nrtD* essential for nitrate transport, which are located downstream of *nrtA* and tightly clustered with *nrtA* [8]. From the structure of the *nrtBCD* genes, the nitrate transporter was deduced to be an ABC-type transporter [8]. *cmpA* was also found to be clustered with *cmpB*, *cmpC*, and *cmpD* genes that are strongly similar to the *nrtB*, *nrtC* and *nrtD* genes, respectively [9] (Fig. 1). Thus, the *cmp* gene cluster was hypothesized to encode a low CO_2-inducible ABC transporter. In this study, we reinvestigated the role of the *cmp* genes in Ci transport by carefully comparing the HCO_3^- transport activity of the wild-type (WT) and M42 cells. As previously shown, M42 retained inducible activities of CO_2 and HCO_3^- transport, but the HCO_3^- uptake by L-cells of M42 showed lower affinity for external HCO_3^- than that by L-cells of WT, indicating that *cmpABCD* encodes a high-affinity bicarbonate transport system of *Synechococcus* sp. strain PCC 7942.

2. MATERIALS AND METHODS

2.1. Strains and Growth Conditions

Cells of the wild-type *Synechococcus* sp. strain PCC 7942 and a targeted mutant M42 (Δ*cmpAB*::*Kan*, see Fig. 1) [5] were grown photoautotrophically at 30°C under continuous illumination provided by fluorescent lamps (70 µE m^{-2} s^{-1}) with nitrate as the nitrogen source in a modification of BG11 medium as previously described [10]. The cultures were routinely maintained under high CO_2 conditions, i.e., with aeration with 2% [v/v] CO_2 in air. For induction of Ci transport activities, the high CO_2-grown cells (H-cells) were collected by centrifugation, washed twice with the growth medium, inoculated into fresh medium, and aerated for 20–40 h with air containing 20–50 ppm CO_2.

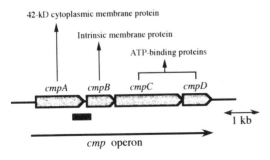

Figure 1. Organization of the genes in the *cmp* gene cluster of *Synechococcus* sp. strain PCC 7942. The thick bar below the map shows the genome region deleted and replaced by a kanamycin resistance gene cartridge in the M42 mutant.

2.2. Measurements of the Activities of Ci Uptake

The initial rate of HCO_3^- uptake after onset of light was measured using a membrane filtration technique, using $H^{14}CO_3^-$ as the substrate. After incubation periods of 5–20 s in the light, the cell suspensions were rapidly filtered onto a glass filter (Whatman, GF/B), washed with the incubation medium (see below), and the ^{14}C on the filter was counted using a scintillation counter. The steady-state rates of CO_2 and HCO_3^- uptake during photosynthesis were measured by a mass spectrometric disequilibrium technique in an aqueous phase-sampling mass spectrometer as previously described [11]. The assay medium used for measurements of Ci uptake was a modification of BG11 medium, in which $NaNO_3$ has been replaced by NaCl (10 mM). The pH of the medium was adjusted to 8.2 and 9.0 with BTP buffer for the measurements of steady-state rates of Ci uptake and initial rate of $H^{14}CO_3^-$ uptake, respectively.

3. RESULTS AND DISCUSSION

3.1. Involvement of the *cmp* Genes in Bicarbonate Transport

In both WT and M42, the H-cells incorporated only negligible amounts of HCO_3^- in 20 s after onset of light, from media containing <100 µM of HCO_3^-. The L-cells of either of the strains took up significant amounts of HCO_3^- from medium, indicating that HCO_3^- uptake activity had been induced in both WT and M42. In repeated experiments using different batches of cultures, the rate of HCO_3^- uptake in M42 ranged from 22 to 70 µmol per mg of Chl per h and that in WT ranged from 70 to 220 µmol per mg of Chl per h. Nevertheless, the activity of M42 was always 2 to 3 fold lower than that of WT grown under the identical growth conditions.

Fig. 2 compares the CO_2 and HCO_3^- uptake rates of H- and L-cells of WT and M42 during steady-state photosynthesis. The H-cells of WT and M42 were essentially the same in their ability to take up CO_2 (Fig. 2A) and HCO_3^- (Fig. 2B) from external medium. Incubation of the cells under low CO_2 conditions increased the affinity for CO_2 uptake activity

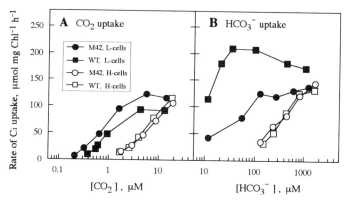

Figure 2. Dependence of the rates of CO_2 (A) and HCO_3^- (B) uptake on external concentrations of CO_2 and HCO_3^-, respectively, during steady-state photosynthesis. H-cells (open symbols) and L-cells (closed symbols) of the wild-type *Synechococcus* sp. strain PCC 7942 (squares) and the M42 mutant (circles) are shown.

in WT and M42 to a similar extent; the $K_{1/2}(CO_2)$ value for CO_2 uptake, the CO_2 concentration required for half-maximal CO_2 uptake, was decreased from 5 µM to 1 µM in both WT and M42 (Fig. 2A). By contrast, the responses of HCO_3^- transport of WT and M42 to carbon limitation were different; While $K_{1/2}(HCO_3^-)$ value for HCO_3^- uptake in WT decreased from 300 µM to 10 µM, that in M42 decreased only to 33 µM. Also, the maximal rate of HCO_3^- uptake was increased by 50% in WT after acclimation to low CO_2, but there was no change in the maximal rate of HCO_3^- uptake in M42 (Fig. 2B).

On the basis of the presence of the inducible Ci transport activities in M42, we previously thought that *cmpA* was not involved in Ci transport [5]. The present results confirm the existence of the inducible CO_2 and HCO_3^--transporting activities in M42 as well as in WT, but demonstrate that M42 is impaired specifically in induction of a high-affinity HCO_3^- uptake mechanism. These results indicate that there are at least two low-CO_2-inducible HCO_3^- transporters and that the *cmp* genes encode one of the inducible HCO_3^- transporters, which has a high affinity for the substrate.

3.2. Deduced Structure of the HCO_3^- Transporter Encoded by the *cmp* Genes

The proteins encoded by the *cmp* genes (Fig. 3A) are structurally very similar to the corresponding proteins encoded by the *nrt* genes [12]; CmpA has a putative signal peptide typical of a lipoprotein; CmpB is an intrinsic membrane protein; CmpD and the N-terminal domain of CmpC have ATP-binding motifs and are similar to each other; the C-terminal domain of CmpC is 29% identical to the CmpA protein. The *nrt* transporter transports nitrite as well as nitrate [13] and the NrtA protein has been shown to be the substrate (nitrate and nitrite)-binding lipoprotein anchored to the cytoplasmic membrane [14]. The strong similarities between the NrtA and CmpA proteins suggest that CmpA is a membrane-anchored lipoprotein and functions as the HCO_3^--binding protein (Fig. 3B). The C-

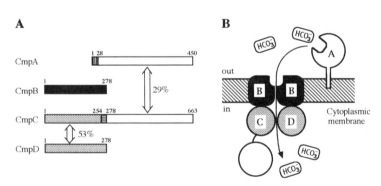

Figure 3. (A) Schematic presentation of the structure of the proteins (CmpA, CmpB, CmpC, and CmpD) constituting the ABC-type bicarbonate transporter of *Synechococcus* sp. strain PCC 7942. The intrinsic protein (CmpB) and the ATP-binding protein/domain (CmpD and the N-terminal domain of CmpC) are filled and dotted, respectively. The vertically shaded bar indicates the presumed signal peptide of CmpA and the horizontally shaded bar indicates the linker peptide connecting the N-terminal and C-terminal domains of CmpC. Homologies between the proteins and domains are also indicated. (B) Diagrammatic representation of the organization of the bicarbonate transporter encoded by the *cmpABCD* gene cluster of *Synechococcus* sp. strain PCC 7942 drawn according to the structure of the nitrate/nitrite transporter encoded by the *nrtABCD* gene cluster. As in (A), the intrinsic protein and the ATP-binding protein/domain are filled and dotted, respectively.

terminal domain of NrtC, on the other hand, has been shown to be involved in ammonium-promoted inhibition of nitrate/nitrite transport [15]. By analogy, it is inferred that the C-terminal domain of CmpC is a regulatory domain of the bicarbonate transporter. Biochemical and molecular biological studies on the Cmp proteins are being performed to elucidate the structure-function relationships of the bicarbonate transporter.

REFERENCES

1. Kaplan, A., Schwarz, R., Lieman-Hurwitz, J., Ronen-Tarazi, M. and Reinhold, L. (1994) in: The Molecular Biology of Cyanobacteria (Bryant, D.A. ed), pp. 469–485, Kluwer, Dordrecht.
2. Omata, T. and Ogawa, T. (1985) Plant Cell Physiol. 26, 1075–1081.
3. Omata, T. and Ogawa, T. (1986) Plant Physiol. 80, 525–530.
4. Omata, T., Ogawa, T., Marcus, Y., Friedberg, D. and Kaplan, A. (1987) Plant Physiol. 83, 892–894.
5. Omata, T., Carlson, T.J., Ogawa, T. and Pierce, J. (1990) Plant Physiol. 93, 305–311.
6. Omata, T., Ohmori, M., Arai, N. and Ogawa, T. (1989) Proc. Natl. Acad. Sci. USA 86, 6612- 6616.
7. Omata, T. (1991) Plant Cell Physiol. 32, 151–157.
8. Omata, T., Andriesse, X. and Hirano, A. (1993) Mol. Gen. Genet. 236, 193–202.
9. Omata, T. (1992) in: Research in Photosynthesis (Murata, N. ed) vol. 3, pp. 807–810, Kluwer, Dordrecht.
10. Suzuki, I., Horie, N., Sugiyama, T. and Omata, T. (1995) J. Bacteriol. 177, 290–296.
11. Badger et al. (1994) Physiol. Plant. 90, 529–536.
12. Omata, T. (1995) Plant Cell Physiol. 36, 207–213.
13. Luque, I., Flores, E. and Herrero, A. (1994) Biochim. Biophys. Acta 1184, 296–298.
14. Maeda, S. and Omata T. (1997) J. Biol. Chem. 272, 3036–3041.
15. Kobayashi, M., Rodríguez, R., Lara, C. and Omata, T. (1997) J. Biol. Chem. 272, 27197–27201.

THE INORGANIC CARBON-CONCENTRATING MECHANISM OF CYANOBACTERIA

Genes and Ecological Significance

Aaron Kaplan, Michal Ronen-Tarazi, Dan Tchernov, David J. Bonfil, Hagit Zer, Daniella Schatz, Assaf Vardi, Miriam Hassidim, and Leonora Reinhold

Department of Plant Sciences
The Moshe Shilo and the Avron-Evenari Minerva Centers
The Hebrew University of Jerusalem
91904 Jerusalem, Israel

1. INTRODUCTION

In this chapter we briefly present and discuss recent progress in the elucidation of certain physiological and molecular aspects of the cyanobacterial inorganic carbon (Ci)-concentrating mechanism (CCM). The reader is referred to earlier chapters and reviews [1–14] for a comprehensive account of other important aspects, including the acclimation of cyanobacteria to changing CO_2 concentration.

1.1. Brief Description of the CCM

Photosynthetic microorganisms are capable of adapting to a wide range of ambient CO_2 concentrations by modulating the expression of certain genes [4,8,9,11]. Some of these genes are involved in the operation of the CCM. It is now widely accepted that the CCM consists of two major components: the energy-dependent mechanisms whereby Ci accumulates within the cytoplasm; and the carboxysomes where HCO_3^- is converted to CO_2 by carbonic anhydrase (CA) confined to these bodies [4,15–18]. The generated CO_2 is fixed by the carboxysomal-located [19] ribulose-1,5-bisphosphate carboxylase/oxygenase (rubisco). The active accumulation of Ci to levels ranging from 2 mM to over 50 mM Ci within the cell, consequent on the activity of the CCM, enables the cells to perform efficient photosynthesis in spite of the relatively low affinity of their rubisco for CO_2. Moreover, the elevated concentration of CO_2 in close proximity to rubisco enables

The Phototrophic Prokaryotes, edited by Peschek *et al.*
Kluwer Academic / Plenum Publishers, New York, 1999.

proper activation of the enzyme and efficient carboxylation [20], and reduces competition by O_2 and hence photorespiration.

Mutants impaired in the ability to grow under low CO_2 conditions are being used as a major tool to study the adaptive responses of cyanobacteria to changing CO_2 concentrations, and to identify and characterize genes involved in the operation of the CCM [8,9,11,20–35].

2. UPTAKE OF INORGANIC CARBON

2.1. Nature of the Ci Species Taken Up from the Medium

Regardless of which Ci species is supplied, bicarbonate is the species which accumulates internally and light energy is used to maintain the cytoplasmic CO_2 concentration below that expected at chemical equilibrium with HCO_3^- [18,36]. Expression of a human CA gene in the cytoplasm of *Synechococcus* PCC 7942 resulted in a high-CO_2-requiring mutant due to excessive leak of CO_2 from the cells [17], confirming that CO_2 and HCO_3^- are not at chemical equilibrium within the cytoplasm. These findings led to the suggestion that utilization of CO_2 from the medium involves a light-energized vectorial CA-like moiety which converts CO_2 to HCO_3^- on transit [8,36]. The system would have to depend on metabolic energy supply since the release of bicarbonate in the cytoplasm would occur against its electrochemical gradient. The discovery of plasma membrane-associated CA activity in *Synechocystis* PCC 6803 [37] lent support to this suggestion. The presence of CA-like activity would also serve the important function of scavenging CO_2 molecules from the cytoplasm and thus minimizing their leak to the medium. Moreover, this postulate obviates the earlier suggestion that uptake of CO_2 in cyanobacteria is mediated by a membrane-located CO_2 transport mechanism. Diffusion of CO_2 across the cytoplasmic membrane, and subsequent energy-dependent conversion and release of bicarbonate, might account for the reported saturation kinetics and apparent uphill transport of CO_2 [18].

Uptake of bicarbonate also displays saturation kinetics with rising external HCO_3^- concentration but the kinetic parameters are different from those observed in the case of CO_2 uptake. Further, the presence of sodium in the mM range is required for HCO_3^- transport [38,39] whereas µM concentrations suffice to saturate sodium-dependent CO_2 uptake [10]; bicarbonate transport is associated with transient hyperpolarization of the cytoplasmic membrane, not observed for CO_2 uptake [7]; and CO_2 and HCO_3^- utilization exhibit different sensitivity to CA inhibitors [40]. Thus, there are numerous physiological indications that uptake of the two Ci species is mediated by at least two distinct mechanisms. Studies of the unidirectional HCO_3^- influx in "zero-trans" experiments indicated a large increase in the Vmax but minor changes in the apparent affinity for Ci during the acclimation of high-CO_2-grown *Anabaena variabilis* and *Synechococcus* PCC 7942 [41 and 42, respectively] from high- to low CO_2 conditions. On the other hand, measurements performed at steady state photosynthesis [43–45] suggested that the affinity for Ci, rather than the Vmax, was the main difference between high- and low-CO_2-adapted cells. It is not known whether the difference between these two sets of results reflects initial rates versus steady state measurements or an intrinsic problem associated with the experimental approach (filtering centrifugation technique or membrane inlet mass spectrometry, respectively), or the interpretation of the data. Finally, there are clear indications that the ratio of CO_2 to HCO_3^- taken up at a given Ci concentration varies in different cyanobacterial species [6,46].

2.2. Genes Involved in Ci Transport

It is widely accepted that acclimation of cyanobacteria from high to low CO_2 conditions involves changes in the expression of various genes and the formation of new proteins [see 8,9]. On the other hand, brief exposure to low CO_2 (4 min. in the presence of CA which facilitated removal of Ci) in the absence of protein synthesis resulted in a complete expression of the high affinity HCO_3^- transporting capability in both the fresh water *Synechococcus* PCC 7942 and the marine *Synechococcus* PCC 7002 [47]. While the mechanisms involved in the "fast induction" are poorly understood, its sensitivity to various kinase inhibitors suggests posttranslational modifications. An important implication of these findings is that the "hunt" for genes encoding components of the high affinity HCO_3^- transporting system should include constitutively expressed genes rather than those solely induced under low CO_2 conditions [22]. Thus, experimental approaches such as differential display may not be appropriate.

It is expected that mutants impaired in the ability to accumulate Ci internally would demand high CO_2 for growth and would include: mutants bearing lesions in the transporting components; those impaired in the energization of Ci transport; mutants showing enhanced efflux of CO_2 from the cells due to a lesion in the structure of the carboxysomes; and mutants defective in genes involved in the acclimation to low CO_2. The latter are easily distinguishable since they are expected to exhibit a photosynthetic performance, with respect to Ci concentration, similar to that of high-CO_2-grown wild type. This was the case in mutants JR12 [48] and D4 of *Synechococcus* PCC 7942 [49] where purine biosynthesis, and hence growth, ceased when the cells were exposed to low CO_2. All other high-CO_2-requiring mutants defective in the ability to accumulate Ci within the cells exhibit an apparent photosynthetic $K_{1/2}$ (Ci) approximately 50-fold higher than the high-CO_2-grown wild type cells.

2.2.1. Genes Encoding Putative Ci Transporters. Physiological studies, some of which are mentioned in section 2.1., above, suggested the presence of several Ci transporters engaged in CO_2 and in HCO_3^- uptake. It is frustrating that while the entire sequence of a cyanobacterial genome is available, genes encoding Ci transport components were not yet recognized. Identification of the genes involved in CO_2 uptake will help to clarify whether CO_2 enters the cell by diffusion or via a cytoplasmic-membrane-located transport mechanism. Mutants apparently defective in CO_2 transport were isolated by Ogawa and colleagues [50] and a gene, *cotA*, which complemented the mutations was identified. Direct inactivation of *cotA*, however, resulted in a mutant which exhibited significant CO_2 transport activity but lost the light- and sodium-dependent proton extrusion [51]. It was therefore concluded that *cotA* (now renamed *pexA*) is probably involved in the regulation of internal pH and hence in anion transport rather than in CO_2 uptake directly [34, 51]. It is interesting to note that *cotA* is highly homologous to *cemA* encoding a chloroplast envelope-located protein [51], the role of which is yet to be established.

Transfer of high-CO_2-grown *Synechococcus* PCC 7942 to low CO_2 conditions resulted in a marked accumulation of the *cmpA*-encoded, 42 kDa polypeptide, in the cytoplasmic membrane [52]. Sequence analysis indicated significant homology between *nrtA-D*, encoding components of an ABC-type nitrate transporter, and *cmpA-D* in *Synechococcus* PCC 7942 [53]. It was therefore proposed that the latter encode a high affinity HCO_3^- transporter in cyanobacteria. Inactivation of *cmpA* did not affect the ability of *Synechococcus* PCC 7942 to grow under 350 ppm CO_2, as in normal air [54] but lowered growth under 20 ppm CO_2, and raised the apparent photosynthetic $K_{1/2}$(Ci) from 10 μM to 30 μM Ci. In *Synechocystis* PCC 6803, on the other hand, inactivation of a *cmpA* homologue did not result in an observable phenotype [34]. It was concluded that while CmpA-D probably constitute components of an ABC-type HCO_3^- transporter, we should seek for other genes encoding putative HCO_3^- transporters. It is likely that an important function such as supply of Ci to photoautotrophic microorganisms is mediated and regulated by several genes.

Several high-CO_2-requiring mutants of *Synechococcus* PCC 7942, defective in HCO_3^- uptake, were isolated with the aid of an inactivation library [22,35,53]. Figure 1A presents a schematic map of the relevant genomic region of the wild type where the insertion of the vector resulted in the observed phenotype of mutant IL-2 [35]. Comparison of the sequences in the wild type and in IL-2 showed that the single cross-over recombination event resulted in two separate deletions of a few base pairs each and the addition of one base pair, at the insertion site. These mutations caused frame shifts in both repeated copies (the relevant genomic region is duplicated due to the single cross-over event) leading to an inactivation of a gene, designated DC14.

Inactivation of DC14 by the insertion of a cassette encoding kanamycin-resistance (Kan^r) in several sites did not produce a high-CO_2-requiring mutant. The Kan^r mutants obtained grew significantly slower than the wild type under air level of CO_2. Southern analysis showed that the cells were merodiploids, containing both the normal and the interrupted genomic region. Analysis of the putative protein sequence encoded by DC14 showed that it contains ten trans-membrane regions and is most likely located in the inner membrane. Alignment of its sequence with that of various proteins in the data bases suggested that it is highly homologous to slr1515 in *Synechocystis* PCC 6803 and is probably involved in membrane transport. However, a firm conclusion as to whether DC14 encodes a high affinity HCO_3^- transporter should await confirmation.

A comprehensive account of the nature and role of all the open reading frames identified here (Fig. 1) is beyond the scope of the present paper. It is important to note, however, that a deletion of a fragment located upstream of DC14, bearing DC13 and DI33 (Fig. 1A), resulted in another high-CO_2-requiring mutant which exhibited photosynthetic characteristics, with respect to CO_2, similar to that observed in an adaptation mutant (see section 2.2., above).

2.2.2. Genes Involved in Energization of Ci Transport. Studies by Ogawa and colleagues [56,57] established that photosynthetic light energy is essential for Ci uptake and implicated cyclic PSI electron transport in its energization. Further, it was demonstrated that the accumulated Ci affects photosynthetic electron transport around PSI [58,59]. Inactivation of *psaE*, encoding a subunit of PSI in *Synechococcus* sp. PCC 7002, resulted in a reduced efficiency of HCO_3^- utilization, and hence lower ability to grow under low CO_2 [44], supporting the suggested role of PSI in energization of Ci uptake. Insertion of an inactivating vector (see section 2.2.1., above) in the genomic region of *psaI-L* (encoding subunits VIII and XI of PSI) in *Synechococcus* sp. PCC 7942 (Fig. 1B), resulted in a high-CO_2-requiring mutant impaired in the high affinity HCO_3^- transport. Since this mutant is also defective in other bioenergetic processes (e.g. the ability to perform efficient photosynthesis at alkaline pH's, presumably due to a defect in the ability to maintain the internal pH [35]), we conclude that the lesion in HCO_3^- uptake stemmed from an energization problem rather than from a defective component of the transporting system itself.

Inactivation of various *ndh* genes in several cyanobacteria [27,28,30,32,60] resulted in a defective ability to utilize HCO_3^- from the medium. These data suggested that NAD(P)H dehydrogenase may be involved in cyclic photosynthetic electron transport, and

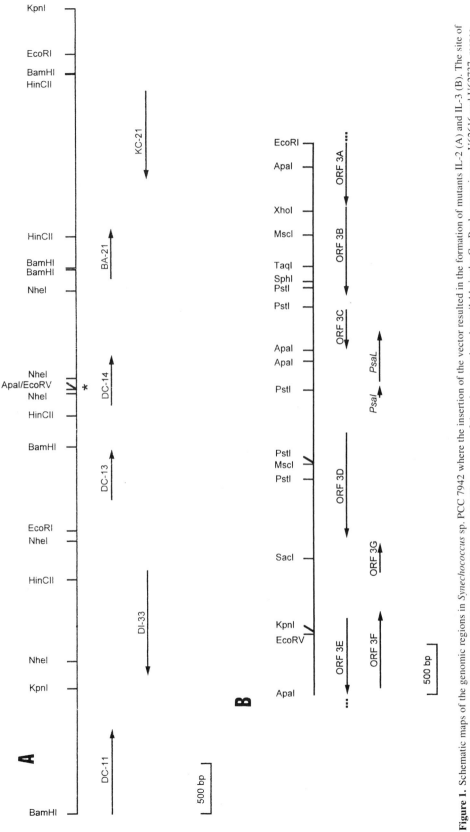

Figure 1. Schematic maps of the genomic regions in *Synechococcus* sp. PCC 7942 where the insertion of the vector resulted in the formation of mutants IL-2 (A) and IL-3 (B). The site of insertion of the inactivating vector in mutant IL-2 is marked with an asterisk. The sequence of the relevant regions is available in the GenBank, accession no. U62616 and U62737, respectively.

hence in Ci uptake [34,61]. It was proposed that NAD(P)H donates electrons to plastoqui-none and that ATP production, coupled to the cyclic electron flow, may serve as the direct energy source for Ci transport [34]. Interestingly, there are 5 different *ndhD* genes in *Synechocystis* PCC 6803. Northern analysis indicated that two of these genes were consti-tutively expressed whereas transcripts originated from the others were only detected in cells exposed to low CO_2 [34]. Inactivation of *ndhD3* resulted in a slow growing mutant when the cells were exposed to 50 ppm CO_2 at pH 6.5 [34], in agreement with results ob-tained using an *ndhD*-defective mutant of *Synechococcus* PCC 7002 [60]. The roles of the various *ndhD* gene products in the regulation of photosynthetic electron transport, in up-take of Ci, and its interrelations with respiratory electron transport are yet to be eluci-dated.

3. CARBOXYSOMES

The carboxysomes are polyhedral bodies characteristic of certain photo- and chemoautotrophic bacteria. They are surrounded by a proteinaceous shell and exhibit most of the cellular rubisco and CA activities [4,19,62]. Studies by Shively and colleagues [63, 64] identified the *cso* operon encoding carboxysomal shell proteins in *Thiobacillus neapolitanus*. Analysis of these genes indicated high homology to the *pud* and *eut* operons encoding components of polyhedral bodies assembled when propanediol or ethanolamine are used as an energy source by *Salmonella* exposed to anaerobic conditions [63]. The *cso* genes are also highly homologous to the *ccm* genes in *Synechococcus* sp. PCC 7942 [8,21,34]; five of them (*ccmK-ccmL-ccmM-ccmN-ccmO*) are clustered immediately up-stream of *rbcLS* [21]. In *Synechocystis* PCC 6803, on the other hand, the *ccm* genes are not located in the vicinity of the *rbc* operon and are clustered in the order of *ccmK-ccmK-ccmL-ccmM-ccmN* [34]. A *ccmO* was not recognized in *Synechocystis* but five different *ccmK* genes (which show high homology to *ccmO*, [63]) were identified in this strain [34]. It is yet to be established whether the different organization of the *ccm* clusters and their location in these cyanobacteria are of functional significance.

Inactivation of the various *ccm* genes resulted in high-CO_2-requiring mutants con-taining aberrant carboxysomes or lacking these bodies. These mutants showed an apparent photosynthetic affinity for Ci approximately 100-fold lower than that of the wild type but, nevertheless, exhibited normal ability to accumulate inorganic carbon (Ci) within the cells. It was therefore suggested, and experimentally verified, that these mutants are defec-tive in the ability to utilize the internal Ci pool in photosynthesis [4,8,20,23,34]. These data are in agreement with the predictions made by the quantitative models which as-signed a critical role to the structural organization of the carboxysomes in the operation of the CCM [16,18]. The models suggest that the packing of rubisco and CA into car-boxysomes enables the formation of a high CO_2 concentration at the fixation site and, fur-ther, minimizes leakage of CO_2 from the cell; and that defective carboxysomes might well be incapable of forming or maintaining this CO_2 pool. A lesion in the carboxysomal struc-ture might well result in a CO_2 level being too low for adequate activation of rubisco. Fur-ther, the concentration of CO_2 as a substrate of the enzyme would be lower in aberrant carboxysomes than in the wild type [16]. Analysis of the activity of rubisco, *in situ*, dem-onstrated that transfer of high-CO_2-requiring mutants which possess defective car-boxysomes to low CO_2 conditions resulted in a low state of activation of the enzyme. This underlines the low apparent photosynthetic affinity for extracellular Ci observed in such mutants and their resulting demand for high CO_2 for growth [20].

Failure to maintain the structural organization of carboxysomes might be expected to render the cytoplasmic HCO_3^- pool readily accessible to the carboxysomal CA. This should have resulted in a lower internal Ci pool in view of data which demonstrated that expression of a foreign CA resulted in dissipation of the internal Ci pool due to an excessive leak of CO_2 ([17], see section 2.1.). However, for an unknown reason, this does not seem to be the case as the characterized carboxysomal-defective mutants were able to accumulate Ci internally like the wild type.

Very little information is as yet available on the biogenesis of the carboxysomes and the processes involved in the organization of their constituents. It is well established that a native, unaltered, rubisco is essential for the organization of the carboxysomes in cyanobacteria [8,20]. Substitution of *rbcLS* in *Synechocystis* sp. PCC 6803 with the *rbc* encoding rubisco in *Rhodospirillum rubrum* resulted in a high-CO_2-requiring mutant which lacked visible carboxysomes [65]. Extension of *rbcS* encoding the small subunit of rubisco in *Synechococcus* sp. PCC 7942 resulted in a high-CO_2-requiring mutant which possesses swollen carboxysomes [20].

Further progress in the identification of carboxysome constituents and their respective genes may enable use of carboxysomes as a model system to study assembly of cellular bodies in prokaryotes, in response to changing environmental conditions.

4. ECOPHYSIOLOGICAL SIGNIFICANCE OF THE CCM

Ever since its discovery [41,66] it was clear that the CCM plays a major role in the ability of aquatic photosynthetic microorganisms to grow under a wide range of ambient CO_2 concentrations. The elevated internal CO_2 concentration in close proximity to rubisco enables substantial photosynthetic rate under low external Ci concentration and inhibits photorespiration.

Recent studies pointed to the significance of the Ci fluxes associated with the CCM to major cellular and ecophysiological processes. These studies [67–69] demonstrated that certain photosynthetic marine microorganisms can serve as a source of CO_2 rather than a sink during photosynthesis. *Synechococcus* WH7803, which belongs to the phycoerythrin-possessing cyanobacteria (significant contributors to oceanic primary productivity), evolved CO_2 at a rate which increased with light intensity and attained a value approximately 5-fold that of photosynthetic O_2 evolution. The external CO_2 concentration reached was substantially higher than that predicted for chemical equilibrium between HCO_3^- and CO_2, as confirmed by its rapid decline on addition of CA (Fig. 2). Measurements of oxygen exchange between water and CO_2, by means of stable isotopes, demonstrated that the CO_2 evolved originated from HCO_3^- taken up and intracellularly converted to CO_2 in a light-dependent process. These findings may have several interesting implications. For instance, the fact that the net CO_2 efflux curve lies well above that for O_2 evolution (i.e. CO_2 fixation, Fig. 3), and continues to rise at light intensities beyond those which saturate photosynthetic O_2 evolution, suggests that Ci cycling may serve as a means for dissipating excess light energy, since significant energy expenditure is required for HCO_3^- uptake against its electrochemical potential (see [67,69]). Massive HCO_3^- uptake and CO_2 efflux may have a substantial impact on the energy demand for the maintenance and regulation of internal pH since the excess OH^- ions must be expelled or neutralized. The eco-physiological significance of this apparently futile, energy-dependent, Ci circulation and its possible role in affording some protection against photoinhibitory damage in natural populations, is yet to be investigated.

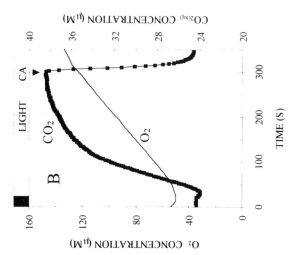

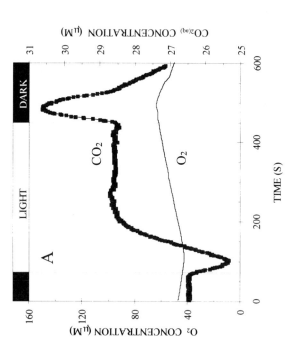

Figure 2. The effect of illumination on $CO_{2(aq)}$ and O_2 concentrations in suspensions of *Synechococcus* WH7803. During measurements (using a membrane inlet quadrupole mass spectrometer, Balzers QMG 421) 3 ml cell suspension, corresponding to 5–8 µg chlorophyll·ml^{-1}, were exposed to light/dark cycles. Light intensities were 50 µmole quanta·m^{-2}·s^{-1} in (A) and 420 µmole quanta·m^{-2}·s^{-1} in (B). Bovine carbonic anhydrase was provided where indicated.

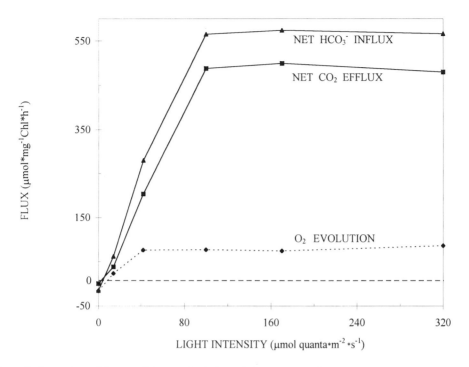

Figure 3. Comparison of the rate of photosynthetic O_2 evolution, net CO_2 efflux and calculated net HCO_3^- influx in *Synechococcus* WH7803 as a function of light intensity. The cells were exposed to the indicated light intensities until a steady state level of $CO_{2(aq)}$ was reached. At this steady state, net HCO_3^- influx is equal to the sum of photosynthetic rate and net CO_2 efflux. The latter is equal to the rate of net CO_2 hydration in the medium, which was assessed from the slope of the progress curve for decline in $[CO_{2(aq)}]$ following injection of a known quantity of $CO_{2(aq)}$ to a cell-free medium in a closed system (see [67]).

Massive Ci cycling may also have a considerable effect on the composition of the stable carbon isotopes ($\delta^{13}C$) in the organic matter produced. It is well established that the carboxylating enzyme, rubisco, discriminates against $^{13}CO_2$ (see [67,70], and references therein) leading to an elevated concentration of ^{13}Ci in the internal Ci pool used as its substrate. As a consequence, in organisms which possess a CCM, the organic matter produced is heavier than in those which lack this mechanism [67,70]. Excessive cycling of Ci between the bulk medium and the internal Ci pool (in CCM-possessing organisms exposed to high light intensity, Fig. 3) remove part of the ^{13}Ci, leading to a lighter composition of the internal Ci pool and hence of the organic matter produced. The large Ci cycling may thus have a considerable impact on the potential use of $\delta^{13}C$ as a paleobarometer probe for CO_2 concentration and for oceanic productivity. Finally, it remains to be seen whether organisms engaged in massive cycling of Ci through their CCM may serve as a CO_2 source for preferential CO_2 users in their immediate vicinity in the phytoplankton consortium.

A CO_2 concentrating mechanism has been identified in *Synechococcus* sp. WH7803 and observed to be severely inhibited by short exposure to elevated photon flux densities [71]. Similar light treatments resulted in a considerable reduction in the efficiency of energy flow from phycocyanin to the phycobilisome terminal acceptor and chlorophyll *a*. Consequently, while the maximal photosynthetic rate, at saturating light and CO_2 concentration, was hardly affected by a high photon flux density treatment, the light intensity required to reach that maximum increased with the duration of the light treatment [71]. The mechanisms whereby the high light treatment resulted in the changing efficiency of energy transfer are poorly understood. Nevertheless, the possibility that they might provide an important means for dissipation of excess light intensity and protection of the photosynthetic reaction centers from photodamage should be examined.

ACKNOWLEDGMENTS

Research in this laboratory was supported by grants from the USA-Israel Binational Science Foundation (BSF), Jerusalem, the Israel Science foundation founded by the Israel Academy of Science and Humanities and from DISUM-BMFT (German-Israeli Ecology Research). We thank Drs. T. Ogawa, T. Omata, J. Shively, D. Sültemeyer, D. Price and M. Badger for making unpublished results available to us.

REFERENCES

1. Aizawa, K. and Miyachi, S. (1986) FEMS Microbiol. Rev. 39, 215–233.
2. Badger, M.R. (1987) in: The Biochemistry of Plants: A Comprehensive Treatise (Hatch, M.D. and Boardman, N.K., Eds.), Vol. 10, pp 219–274, Academic press, New York.
3. Badger, M. R. and Price, G. D. (1992) Physiol. Plant. 84, 606–615.
4. Badger, M.R. and Price, G. D. (1994) Ann. Rev. Plant Physiol. Plant Mol. Biol. 45, 369–399.
5. Coleman, J.R. (1991) Plant Cell Environ. 14, 861–867.
6. Kaplan, A., Schwarz, R., Ariel, R., and Reinhold, L. (1990) in: Regulation of Photosynthetic Processes (Kanay, R., Katoh, R.S. and Miyachi, S. Eds.), Vol. 2. pp 53–71, Special Issue of the Botanical magazine, Tokyo.
7. Kaplan, A., Schwarz, R., Lieman-Hurwitz, J., and Reinhold, L. (1991) Plant Physiol. 97, 851–855.
8. Kaplan, A., Schwarz, R., Lieman-Hurwitz, J., Ronen-Tarazi, M., and Reinhold, L. (1994) in: The Molecular Biology of the Cyanobacteria (Bryant, D. Ed.), pp 469–485 Kluwer Academic Pub., Dordrecht, The Netherlands.
9. Kaplan, A., Ronen-Tarazi, M., Zer, H., Schwarz, R., Tchernov, D., Bonfil, D.J., Schatz, D., Vardi, A., Hassidim, M. and Reinhold, L. (1998) Can. J. Bot. special issue, (in press).
10. Miller, A.G., Espie, G.S. and Canvin, D.T. (1990) Can. J. Bot. 68, 1291–1302.
11. Ogawa, T. (1993) in: Photosynthetic Responses to the Environment. (Yamamoto, H. and Smith, C. Eds.), pp. 113–125, American Society of Plant Physiologists Series, Rockville.
12. Raven, J. A. (1991) Can. J. Bot. 69, 908–924.
13. Raven, J. A. (1996) Exp. Marine Biol. Ecol. 203, 39–47.
14. Raven, J. A. (1997) Plant Cell Environ. 20, 147–154.
15. Reinhold, L., Zviman, M. and Kaplan, A. (1989) Plant Physiol. Biochem. 27, 945–954.
16. Reinhold, L., Koslof, R. and Kaplan, A. (1991) Can. J. Bot. 69, 984–988.
17. Price, G.D. and Badger, M. R. (1989) Plant Physiol. 91, 505–513.
18. Fridlyand, L., Kaplan, A. and Reinhold, L. (1996) Biosystems 37, 229–238.
19. Codd, G. A. (1988) in: Advances in Microbial physiology (Ross, A.H. and Tempest, D.W. Eds.) 29, 115–164, Academic Press London.
20. Schwarz, R., Reinhold, L. and Kaplan, A. (1995) Plant Physiol. 108, 183–190.
21. Ronen-Tarazi, M., Lieman-Hurwitz, J., Gabay, C., Orus, M. and Kaplan, A. (1995) Plant Physiol. 108, 1461–1469.
22. Ronen-Tarazi, M., Schwarz, R., Bouevitch, A., Lieman-Hurwitz, J., Erez, J. and Kaplan. A. (1995) in: Molecular Ecology of Aquatic Microbes (Joint, I. Ed.), Vol G38 pp. 323–334, NATO ASI series, Springer-Verlag, Berlin.
23. Friedberg, D., Kaplan, A., Ariel, R., Kessel, M. and Seijffers, J. (1989) J. Bacteriol. 171, 6069–6076.
24. Bedu, S., Peltier, G., Sarrey, F. and Joset, F. (1990) Plant Physiol. 93, 1312–1315.
25. Bedu, S., Pozuelos, P., Cami, B. and Joset, F. (1995) Mol. Microbiol. 18, 559–568.
26. Fukuzawa, H., Suzuki, E., Komukai, Y. and Miyachi, S. (1992) Proc. Natl. Acad. Sci. USA 89, 4437–4441.

27. Ogawa, T. (1991) Proc. Nat. Acad. Sci. USA 88, 4275–4279.
28. Ogawa, T. (1992) Plant Physiol. 99, 1604–1608.
29. Price, G. D., Howitt, S. M., Harrison, K. and Badger, M. R. (1993) J. Bacteriol. 175, 2871–2879.
30. Marco, E., Ohad, N., Schwarz, R., Lieman-Hurwitz, J., Gabay, C. and Kaplan, A. (1993) Plant Physiol. 101, 1047–1053.
31. Marco, E., Martinez, I., Ronen-Tarazi, M., Orus, M.I. and Kaplan, A. (1994) App. Environ. Microbiol. 60, 1018–1020.
32. Sültemeyer, D., Price, G. D., Bryant, D. A. and Badger, M. R. (1997) Planta 201, 36–42.
33. Sültemeyer, D., Klughammer, B., Ludwig, M., Badger, M. R. and Price, G. D. (1997) Aust. J. Plant Physiol. 24, 317–327.
34. Ohkawa, H., Sonoda, M., Katoh, H. and Ogawa, T. (1998) Can. J. Bot. special issue, (in press).
35. Ronen-Tarazi, M., Bonfil, D. J., Schatz, D. and Kaplan, A. (1998) Can. J. Bot., special issue, (in press).
36. Volokita, M., Zenvirth, D., Kaplan, A. and Reinhold, L. (1984Plant Physiol. 76, 599–602.
37. Bedu, S., Beuf, L. and Joset, F. (1992) in: Research in Photosynthesis (Murata, N. Ed.) Vol. III, 819–822, Dodrecht, The Netherlands.
38. Kaplan, A., Volokita, M., Zenvirth, D. and Reinhold, L. (1984) FEBS Lett. 176, 166–168.
39. Espie, G.S., Miller, A.G., Kandasamy, R. and Canvin, D.T. (1991) Can. J. Bot. 69, 936–944.
40. Tyrrell, P. N., Kandasamy, R. A., Crotty, C. M. and Espie, G. S. (1996). Plant Physiol. 112, 79–88.
41. Kaplan, A., Badger, M.R. and Berry, J.A. (1980) Planta 149, 219–226.
42. Schwarz, R., Friedberg, D., Reinhold, L. and Kaplan, A. (1988) Plant Physiol. 88, 284–288.
43. Badger, M.R., Palmqvist, K. and Yu, J-W. (1994) Physiol. Plant. 90, 529–536.
44. Sültemeyer, D., Price, G.D. and Badger, M. R (1995) Planta 197, 597–607.
45. Yu, J-W., Price, G. D. and Badger, M.R. (1994) Aus. J. Plant Physiol. 21, 185–195.
46. Skleryk, R.S., Tyrrell, P.N. and Espie, G. S. (1997) Physiol. Plant. 99, 81–88.
47. Sültemeyer, D., Klughammer, B., Badger, M. R. and Price, G. D. (1998) Can. J. Bot., special issue, (in press).
48. Lieman-Hurwitz, J., Schwarz, R., Martinez, F., Maor, Z., Reinhold, L. and Kaplan, A. (1990) Can. J. Bot. 69, 945–950.
49. Schwarz, R., Lieman-Hurwitz, J., Hassidim, M. and Kaplan, A.(1992) Plant Physiol. 100, 1987–1993.
50. Katoh, A., Sonoda, M., Katoh, H., and Ogawa, T. (1996) J. Bacteriol. 178, 5452–5455.
51. Katoh, A., Sonoda, M., Katoh, H. and Ogawa, T. (1997) J. Bacteriol. 178, 5452–5455.
52. Omata, T. and Ogawa, T. (1986) Plant Physiol. 80, 525–530.
53. Omata, T. (1992) in: Research in Photosynthesis (Murata, N. Ed.), Vol III, 807–810, Dodrecht, The Netherlands.
54. Omata, T., Carlson, T. J., Ogawa, T. and Pierce, J. (1990) Plant Physiol. 93, 305–311.
55. Dolganov N. and Grossman, A. (1993) J. Bacteriol. 175, 7644–7651.
56. Ogawa, T., Miyano, A. and Inoue, Y. (1985) Biochim. Biophys. Acta 808, 77–84.
57. Kaplan, A., Zenvirth, D., Marcus, Y., Omata, T. and Ogawa, T. (1987) Plant Physiol. 84, 210–213.
58. Badger, M.R. and Schreiber, U. (1993) Photosynth. Res. 37, 177–191.
59. Li Q. and Canvin, D.T. (1997) Plant Physiol. 114, 1273–1281.
60. Price, G. D., Sültemeyer, D., Klughammer, B., Ludwig, M. and Badger, M. R. (1998) Can. J. Bot., special issue, (in press).
61. Mi, H., Endo, T., Ogawa, T. and Asada, K. (1995) Plant Cell Physiol. 36, 661–668.
62. McKay, R. M. L., Gibbs, S. P. and Espie, G. S. (1992) Arch Microbiol 159: 21–29
63. Shively, J. M., Bradburne, C. E., Aldrich, H. C., Bobik, T. A., Mehlman, J. L., Jin, S. and Baker, S. H. (1998) Can J Bot., special issue (in press).
64. English, R. S., Lorbach, S. C., Qin, X. and Shively, J. M. (1994) Mol Microbiol 12: 647–654.
65. Pierce, J., Carlson, T. J. and Williams, J. G. K. (1988) Proc Nat Acad Sci USA 86: 5753–5757.
66. Badger, M. R., Kaplan, A. and Berry, J. A. (1980) Plant Physiol. 66, 407–413.
67. Tchernov, D., Hassidim, M., Luz, B., Sukenik, A., Reinhold, L. and Kaplan, A. (1997) Current Biology (in press).
68. Sukenik, A., Tchernov, D., Huerta, E., Lubian, L.M., Kaplan, A. and Livne, A. (1997) J. Phycol. (in press).
69. Tchernov, D., Hassidim, M., Vardi, A., Luz, B., Sukenik, A., Reinhold, L. and Kaplan, A. (1998) Can. J. Bot., special issue, (in press).
70. Erez, J., Bouevitch, A. and Kaplan, A. (1998) Can. J. Bot. special issue, (in press).
71. Hassidim, M., Keren, N., Ohad, I., Reinhold, L. and Kaplan, A. (1997) J. Phycol. (in press).

66

HIGH-AFFINITY C4-DICARBOXYLATE UPTAKE IN *RHODOBACTER CAPSULATUS* IS MEDIATED BY A 'TRAP' TRANSPORTER, A NEW TYPE OF PERIPLASMIC SECONDARY TRANSPORT SYSTEM WIDESPREAD IN BACTERIA

David J. Kelly, Neil R. Wyborn, Mark Gibson, Jason A. Forward, and Simon C. Andrews

Department of Molecular Biology and Biotechnology
University of Sheffield
Firth Court, Western Bank, Sheffield S10 2TN, United Kingdom

1. INTRODUCTION

Anoxygenic photosynthetic bacteria in the purple non-sulphur group have long been known to grow rapidly and with high yields on certain citric-acid cycle intermediates and their precursors, particularly pyruvate and the C4-dicarboxylates malate and succinate, under a wide variety of conditions. The transport of these molecules across the cytoplasmic membrane of *Rhodobacter sphaeroides* was first studied by Gibson[1], who concluded that transport of D,L-malate, succinate and fumarate was mediated by a common, inducible system of high-affinity. The closely related species *Rhodobacter capsulatus* also possesses a high-affinity transport system for D and L-malate, succinate and fumarate, with a K_t value for L-malate of about 3 μM in intact cells[2]. This system has been studied in some detail in our laboratory and it has now become clear that it represents a distinct type of periplasmic transporter.

Early experiments indicated that malate uptake in aerobically grown cells was extremely sensitive to osmotic shock and that radiolabelled malate and succinate bound specifically to periplasmic protein fractions prepared by this method, indicating the presence of a dicarboxylate binding-protein[2] which was subsequently purified[3]. Mutants which were unable to grow on D,L-malate under aerobic conditions in the dark but which grew normally on pyruvate were isolated by Tn5 mutagenesis[4]. Of five mutants isolated, all were unable to grow aerobically in the dark on succinate and fumarate in addition to D,L-malate and all were found to be unable to transport C4-dicarboxylates from low external

The Phototrophic Prokaryotes, edited by Peschek *et al.*
Kluwer Academic / Plenum Publishers, New York, 1999.

concentrations. Interestingly, these mutants were still able to grow slowly under phototrophic (anaerobic/light) conditions on malate and succinate[4] due to the presence of additional transport systems of low affinity which appear to be absent in aerobically grown cells. Mapping of the Tn*5* insertions indicated clustering within a 4.0 kb *Eco*RI fragment, defining a locus designated *dct*, which was cloned from a wild-type cosmid gene bank[4]. Subsequent complementation analysis[4–6] using subclones in the broad-host range vector pRK415 indicated that five *dct* genes in two transcriptional units were located on two contiguous *Eco*RI fragments of 4.0 and 4.3 kb. This region has now been completely sequenced[3,5,6], which has allowed the primary structure of all of the *dct* gene products to be determined (see Fig. 1). The *dct* locus has also been physically mapped at high resolution on the *Rb. capsulatus* chromosome[7]. Here, we describe the general features of the Dct system in *R. capsulatus*, and show that it is a representative of a completely new type of secondary transporter, to which we have given the general name 'TRAP' (Tripartite ATP-independent Periplasmic) transporter[6].

2. COMPOSITION AND STRUCTURE OF THE Dct SYSTEM

2.1. The DctS and DctR Proteins

The *dct* locus consists of two regulatory and three structural genes, organised in two divergently transcribed operons (*dctSR* and *dctPQM* respectively; Fig. 1), closely linked to the *dra* locus. The *rypA* gene, downstream of *dctM*, encodes a putative protein-tyrosine phosphatase which appears to overlap *draG*, and is not part of the transport system[6]. The *dctS* and *R* genes are translationally coupled and encode proteins with sequence similarity to the so-called "two-component" sensor-regulator systems[5]. DctS is predicted to be a sensor-kinase and DctR is predicted to be a response-regulator. It is thus likely that DctR transcriptionally activates expression of the *dctPQM* operon in response to a signal from DctS. When either *dctS* or *dctR* are insertionally inactivated by interposon mutagenesis, the resulting mutants are unable to grow aerobically on C4-dicarboxylates and do not synthesise the periplasmic C4-dicarboxylate binding-protein[5], demonstrating the essential role of these proteins as positive regulatory components. DctS contains two potential membrane-spanning sequences in the N-terminal region, with the intervening hydrophilic se-

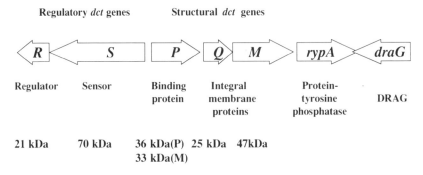

Figure 1. Gene organisation and deduced products encoded at the *dct* locus in *R. capsulatus*. The deduced sizes (in kDa) of the *dct* gene products are shown. P, precursor form of DctP; M, mature form of DctP; DRAG, dinitrogenase reductase activating glycohydrolase.

quence likely to be located in the periplasm[5]. It is tempting to speculate that this periplasmic region forms a C4-dicarboxylate sensing domain. Sequence comparisons have shed some light on relationship of DctS and DctR to other sensor and regulatory proteins[5]. DctS was found to be most similar to three rhizobial sensor-kinases; FixL, a haemoprotein involved in the microaerobic induction of symbiotic nitrogen fixation, NodV, a *Bradyrhizobium* sensor essential for the formation of root nodules, and DctB, a sensor involved in rhizobial C4-dicarboxylate transport. In each case the similarity is restricted to the C-terminal domains of the proteins, which contain the conserved histidine, asparagine and glycine residues characteristic of the sensor-kinase family. No N-terminal sequence similarity is apparent between DctS and DctB. This is of interest since the position of the predicted membrane-spanning regions of DctS and DctB are very similar, suggesting that the overall organisation of the two proteins is analogous. Either the periplasmic domains of these proteins are folded in a similar way, despite the absence of any primary sequence homology, suggesting they sense the same signal or else they are responding to different stimuli.

DctR shares sequence similarity with a number of proteins in the conserved family of prokaryotic response regulators which function as transcriptional activators. The greatest degree of similarity is found with the regulator counterparts of two of the three rhizobial systems discussed above, i.e. FixJ and NodW. DctR is also similar to the rhizobial DctD protein. However, the similarity is confined to the N-terminal region; the central and C-terminal domains in DctD are not present in DctR. The data indicate that DctR is a member of the FixJ-UhpA subfamily of response regulators[8]. These proteins are smaller than those of the NtrC-DctD subfamily and are not dependant on the NtrA sigma factor (σ^{54}) for transcriptional activation. In *Rb. capsulatus*, an NtrA homologue (NifR4) is known, and *nifR4* mutants transport C4-dicarboxylates normally[5]. FixJ has been proposed to have a modular organisation, consisting of a C-terminal DNA binding domain (FixJC) containing a helix-turn-helix motif and an N-terminal domain (FixJN) homologous with the phospho-accepting receiver portion of other response-regulators. It has been proposed that FixJN negatively regulates the activity of FixJC[8]. Secondary structure predictions provide evidence for a similar modular structure in DctR. There are two regions of dyad symmetry in the *dctS-dctP* intergenic region which are potential binding sites for DctR[5]. Although the start of an additional open reading frame appears to be present just downstream of *dctR* in the same transcriptional unit, its product is not involved in transport, as it is possible to complement *dctR* mutants with plasmids containing a wild-type copy of the *dctR* gene alone[5]. We have overexpressed DctR in *E. coli* under the control of an IPTG inducible T7 promoter, and have purified the overexpressed protein to homogenity (M.Gibson and D.J. Kelly, unpublished) in order to carry out DNA binding studies with the *dctP* promoter region.

2.2. DctP

The *dctP* gene encodes the periplasmic binding-protein of the transport system, and binds malate, succinate and fumarate with high affinity[3]. Comparison of the deduced sequence with the N-terminal sequence of the purified protein revealed the presence of a typical signal-sequence which is cleaved to give a mature 307 residue protein of Mr-33,567[3]. Exploitation of the intrinsic tryptophan fluorescence properties of periplasmic binding-proteins has proven a useful way of studying their interaction with ligands. The rates of ligand binding to the galactose, maltose, arabinose and histidine binding-proteins have been determined by stopped-flow fluorescence spectroscopy and this has revealed a simple association of protein and ligand, with a linear concentration dependence of the association rate constant[9,10]. The mechanism of ligand-binding to DctP has been studied in some detail by using both steady-state and stopped-flow fluorescence spectroscopy[11,12]. The fluorescence properties of DctP are characterised by excitation and emission peaks centred around 280 nm and 319 nm respectively. The addition of fumarate to the protein causes a 17% quench in fluorescence without shifting the emission maximum, while succinate causes a 4.5% enhancement in fluorescence. D- or L-malate also cause a small (2%) enhancement[11]. Titration of the fluorescence indicated a binding stoichiometry of 1:1 and ligand dissociation constants (equilibrium k_d values) of 0.05 μM for L-malate, 0.17 μM for succinate, 0.25 μM for fumarate and 6.3 μM for D-malate were determined[11]. Stopped-flow fluorescence spectroscopy revealed that DctP displays kinetics which are inconsistent with a mechanism in which there is a simple association of ligand and protein[11]. The most important observation on which this conclusion is based was that the rate of ligand-induced fluorescence change was found to decrease in a hyperbolic fashion as the concentration of ligand was increased. The same kinetic behaviour was observed with malate, succinate or fumarate as ligands and with protein that had been reversibly denatured with urea, thus removing any endogenously bound ligand which may have been responsible. These data are consistent with a slow isomerisation of the protein, with the ligand binding to only one form of an equilibrium mixture of two pre-existing protein conformations:

$$
\begin{array}{ccccccc}
 & k_1 & & k_2\,[L] & & k_3 & \\
BP1 & \Leftrightarrow & BP2 & \Leftrightarrow & BP2\text{-}L & \Leftrightarrow & BP3\text{-}L \\
 & k_{-1} & & k_{-2} & & k_{-3} &
\end{array}
\qquad (1)
$$

BP1 represents the non-binding (closed) conformation, BP2 represents the binding (open) form, L denotes the ligand and BP3-L denotes the closed-liganded form. The DctP protein is thus kinetically rather distinct from other binding-proteins which have been characterised to date, with regard to the remarkable stability of the closed unliganded (BP1) conformation. It can be calculated from the values obtained[11] for k_1 and k_{-1} that about 96% of the protein molecules adopt the closed conformation in the absence of ligand. The molecular basis for the position of the equilibrium between the open and closed forms of DctP is almost certainly the presence of a salt bridge in the molecule, as indicated by the pH and ionic strength dependence of the rate constants[12]. We have overexpressed DctP in *E. coli* and are currently conducting crystallisation trials with the purified protein, with the eventual aim of establishing its tertiary structure.

2.3. DctQ and DctM

The deduced products of the *dctQ* and *dctM* genes are not related in terms of primary sequence to the integral membrane proteins of other periplasmic transport systems. However, the amino-acid composition of DctQ and DctM reveals that they are very hydrophobic, with 62% and 73% apolar residues (i.e. MIVLPGAFW) respectively. This suggests strongly that they are both integral membrane proteins. We have confirmed this by the heterologous expression of the cognate genes in *E. coli* under the control of a T7 promoter, and localisation of the products to the cytoplasmic membrane after IPTG induction[6]. Although DctQ migrates in SDS-PAGE at about 26 kDa, consistent with its deduced size of 24.6 kDa, DctM runs anomalously (at 30 kDa rather than 47 kDa), presumably due to its more hydrophobic nature. The hydropathy profile of DctQ shows that there are four particularly hydrophobic regions of at least 20 amino-acids which are potential membrane-spanning sequences. In contrast, the profile of DctM implies a total of twelve potential membrane-spanning segments, arranged as two groups of six separated by a markedly hydrophilic region. We are currently determining the actual topology of both proteins using BlaM fusions, and the data thus far indicates a model for DctQ in which both N- and C-terminii are located in the cytoplasm. Using the hydropathy profile and applying von Heijnes rules[13], the predicted orientation of DctM is likely to be similar. DctQ is somewhat unusual in possessing 4 transmembrane helices, and its precise role in the transport system is unknown, but there are examples of small membrane transport proteins in which four transmembrane α-helices are predicted[14]. We believe it may be involved in either (i) mediating interactions with the binding-protein (ii) energy-coupling, or (iii) as a chaperonin for the correct assembly of DctM. The hydropathic profile of DctM is remarkably similar to that of the so-called twelve transmembrane helix transporters exemplified by many families of membrane transport proteins in bacteria and eukaryotes[15]. This is consistent with a central role in solute translocation across the cytoplasmic membrane, even though no significant sequence similarity can be demonstrated with known proteins in any of these families.

3. REGULATION OF THE SYNTHESIS AND ACTIVITY OF THE Dct SYSTEM

Malate transport activity via the Dct system in *Rb. capsulatus* is inducible by C4-dicarboxylates and is repressed by other organic acids (e.g. lactate) and by fructose or glucose. These effects are presumably mediated, at least in part, by the DctS and DctR gene products although, apart from the phenotypes of mutants in the corresponding genes, this has not been studied directly by the use of reporter gene fusions to *dctP* or *dctQ*. Under inducing conditions, DctP becomes one of the most abundant periplasmic proteins and can easily be visualised by SDS-PAGE of crude periplasmic extracts[3]. However, the *dctP* operon organisation suggests that DctQ and M are likely to be much less abundant; although *dctP* mutants are polar on *dctQ/M*, and cannot be complemented unless intact *dctQ and M* genes are also present[3,6], a GC-rich region of dyad symmetry which has the potential to form a stable stem-loop structure ($\Delta G^{\circ\prime}$ −30 kcal mol^{-1}) is located immediately 3′ of *dctP*, and is followed by a repeated CTTT motif[3]. This type of structure has been noted in some other *Rb. capsulatus* operons[16]. It may act as a partial transcriptional terminator, which permits some readthrough into *dctQM*. This has been confirmed by northern blots using either *dctP* or *dctQ* probes hybridised to mRNA derived from fructose or malate grown cells. The *dctP* probe detects an abundant 1kb transcript (from *dctP*) *and* a less abundant 3 kb transcript (derived from *dctP, Q* and *M*) in malate, but not fructose grown cells. The *dctQ* probe only detects the 3 kb transcript (M. Gibson and D.J. Kelly, unpublished).

Many bacterial transport systems can be modulated in their activity by variations in the intracellular pH value. Although, *in vivo*, cytoplasmic buffering and pH homeostasis normally keeps the internal pH value within narrow limits, this form of regulation may be important under some conditions. The activity of periplasmic binding-protein dependent systems in particular has been shown to vary significantly with internal pH, including that for alanine transport in *Rb. sphaeroides*[17]. A study of the effect of external and internal pH on the activity of the *Rb. capsulatus* Dct system in intact cells[2] showed that a marked activation of transport occurred above an internal pH value of 7 and that below neutrality, the rate of transport fell to zero. The data were obtained using the ionophore nigericin (which catalyses an electroneutral exchange of protons and potassium ions) over a range of external pH values. The molecular mechanism of such regulation is unknown, but conceivably may involve changes in the protonation state of key amino-acid residues in DctP or DctQ/M. The effect may be physiologically significant, since utilisation of C4-dicarboxylates as sole carbon source in batch cultures leads to a large rise in the external pH, which to a smaller extent will increase the intracellular pH, thus increasing the activity of the transport system, particularly as the culture approaches stationary-phase. Control of the activity of the transport system by the value of the intracellular pH might also be important in other situations. For example, despite the fact that a significant membrane potential exists under anaerobic-dark conditions in *Rb.capsulatus* cells (thought to be generated by the electrogenic efflux of K^+,[18]), malate transport via the Dct system was not detectable[4]. This is probably due to a lowered internal pH resulting in the formation of a reversed ΔpH across the cytoplasmic membrane under these conditions[19].

4. ENERGY COUPLING

Although clearly binding-protein dependent, the Dct system has no associated AT-Pase or ATP-binding-cassette (ABC) protein subunits(s) and consists of just three structural proteins which alone are able to form a functional transport system[6]. It is also clear that there is no relationship between the integral membrane proteins of the system, and those of classical ABC transporters (see below). Thus, it is unlikely that ATP hydrolysis is the energy source for uptake. At physiological pH, the C4-dicarboxylates are all negatively charged (e.g. for malate pK1 = 3.40, pK2 = 5.11) and thus the anionic forms cannot be accumulated by diffusion or uniport against an opposing membrane potential, inside negative. If ATP is not the energy source, symport with either protons or another cation is the most likely alternative possibility. The overall driving force for the operation of a proton-symport system for a dicarboxylate anion with two negative charges is given by the following equation:

$$\frac{\Delta\mu D^{2-}}{F} = (n-2)\cdot\Delta\Psi - n\cdot Z\Delta pH \tag{2}$$

where $\Delta\mu D^{2-}/F$ is the transmembrane electrochemical gradient of the dicarboxylate, n is the number of protons co-transported, $\Delta\psi$ is the membrane potential, ΔpH is the pH gradient across the cytoplasmic membrane and Z is a term that converts ΔpH into mV. At external pH values at or above neutrality, ΔpH is small, and the proton-motive force is dominated by the membrane potential in neutrophilic bacteria, including *Rhodobacter*[20]. Thus, symport with more than two protons is needed to obtain active transport according to eq. 2. A role for other cations (e.g. Na^+) is not ruled out. In intact cells of *Rb. capsulatus* treated with increasing concentrations of structurally different uncouplers, an excellent correlation has been found between the decrease in membrane potential (measured by carotenoid electrochromism) and the decrease in the rate of succinate transport via the Dct system, in the absence of any change in intracellular pH[6]. Importantly, no correlation was found with the intracellular ATP concentration, which remained above 1 mM at each uncoupler concentration tested[6]. These observations provide evidence for the membrane potential as the driving force for C4-dicarboxylate transport via the Dct system. It must be emphasised that although the transport of C4-dicarboxylates in both Gram-negative[21] and Gram-positive[22] bacteria is well known, such systems appear to be "classical" secondary transporters which are binding-protein *in*dependent. The *Rhodobacter* Dct system thus appears to be an example of a binding-protein dependent secondary transporter. Experiments to confirm the energy coupling mechanism using membrane vesicles have been partially successful using *R. sphaeroides*, but have proven more difficult with *R. capsulatus*, apparently due to inhibition of transport by sucrose and EDTA, which are used in vesicle preparation. In vesicles derived from *R. sphaeroides*, succinate uptake is binding-protein dependent and does not require ATP to be addeed; either photosynthetic or respiratory electron flow can energise transport.

The orthovanadate anion is a potent inhibitor of E_1-E_2-type ATPases in eukaryotic cells. It has also been demonstrated to inhibit transport through ABC-type periplasmic binding-protein dependent transporters[23], indicating the involvement of a phosphorylated intermediate in the transport mechanism. In a study of the effect of vanadate on a number of solute transport systems in intact cells of *Rb. sphaeroides,* Abee[24] found that succinate uptake was insensitive to 1 mM vanadate while binding-protein dependent alanine transport and the transport of a number of other amino-acids was severely inhibited. Succinate transport via the Dct system in *Rb. capsulatus* is also vanadate insensitive but that of the alanine analogue aminoisobutyrate is abolished by this reagent[6]. Although the precise mode of action of vanadate on the ABC-transporters is unknown, these findings would be consistent with the absence of an ABC-protein in the Dct system and suggest that vanadate sensitivity may prove to be a useful tool in the investigation of energy coupling in other binding-protein dependent transporters.

The data available thus far therefore strongly suggest that the energy coupling mechanism of the *Rhodobacter* Dct system is distinct from the ABC-family of periplasmic transporters and may involve proton-symport, although co-transport with another cation cannot be ruled out. This is consistent with what is known about the structure of the system as determined from molecular genetic and protein studies. A model of the operation of the Dct system based on the available data is shown in Fig. 2.

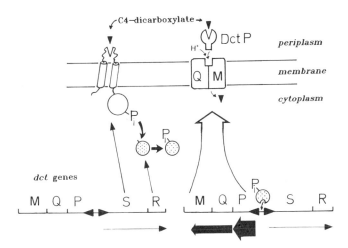

Figure 2. Model for the operation and synthesis of the *Rb. capsulatus* Dct system. DctS is presumed to be a C4-dicarboxylate sensor. Phosphotransfer to DctR results in transcriptional activation from the *dctP* promoter. The presence of a partial transcriptional terminator between *dctP* and *dctQM* results in greater abundance of *dctP* transcripts (thick arrows) compared to those for *dctQM*. It is assumed that *dctS* and *dctR* are expressed constitutively at low levels (thin arrows).

Table 1. The TRAP transporter family: Bacteria
which contain DctPQM homologues

Bacterium	Homologues of *R. capsulatus* Dct proteins and cognate gene order (5' to 3')
Gram-negative eubacteria	
Escherichia coli	QMP
Salmonella typhimurium	M
Vibrio cholerae	QMP
Haemophilus influenzae	PQM
	PQM
	QMP
Bordetella pertussis	PQM
Synechocystis PCC6803	QM(GlnH)
	P (separate locus)
Thiosphaera pantotropha	M
Treponema pallidum	M
Gram-positive bacteria	
Bacillus subtilis	P, M (separate loci, weak homology)
Enterococcus faecalis	M
Deinococcus radiodurans	M
Archaea	
Archaeoglobus fulgidus	M

5. HOMOLOGOUS SYSTEMS IN OTHER BACTERIA: EVIDENCE FOR A NEW TYPE OF TRANSPORTER, THE 'TRAP' TRANSPORTER FAMILY

Database searching has revealed that homologues of the *R. capsulatus* DctPQM proteins are found in a number of Gram-negative bacteria, some Gram-positive bacteria and at least one archeon (Table 1). In *E. coli*, the three ORF's (*orf157a, orf424* and *orf328*, encoding products corresponding to DctQ, DctM and DctP respectively) are located in the 80 min region sequenced by Sofia *et al.*[25] and appear to be part of a large operon containing genes known to be involved in pentose sugar metabolism, and which are required for growth of *E. coli* on L-lyxose[26]. In *B. pertussis*[27,28], the product of an ORF located at the 3' end of the fimbrial gene cluster ('*orfC*') is homologous with DctM from *Rb. capsulatus* and the Orf424 DctM homologue in *E. coli*. However, examination of the region upstream of '*orfC*' revealed two other ORF's, here designated *orf279* and *orf166* which encode, respectively, (i) an incompletely sequenced DctP homologue, also homologous to the product of the *E. coli orf 328* and (ii) a product with significant sequence similarity to DctQ and to the Orf 157a protein in *E. coli*. In *H. influenzae*[29], there are no less than three separate sets of genes encoding DctPQM homologues (Table 1). In *Synechocystis sp.* PCC6803[30], there is a cluster of three genes encoding, in addition to DctQ/M like proteins, a periplasmic binding-protein which is very similar (34.5% amino-acid sequence identity) to the glutamine binding protein (GlnH) from *Bacillus stearothermophilus*, rather than to DctP, suggesting that this may be a glutamine transporter. Interestingly, however, there is also a DctP homologue encoded elsewhere in the *Synechocystis* genome[30]. In all these cases, the homologous transport system appears to be composed of three proteins, as in *R. capsulatus*, although the gene order varies somewhat. In other bacteria, e.g. *Thiosphaera pantotropha* and *Salmonella typhimurium*, only one homologue has so far been identified, but this may be due to incomplete sequencing. The DctQ homologues are the most divergent in terms of sequence similarity and have been difficult to detect in database searches, while the DctM homologues are the most conserved. It is thus clear that these systems represent a distinct family of transporter that is not closely related to known primary or secondary transport systems. As the main features of this new family are (i) the presence of three component proteins, (ii) no ABC-protein and ATP-independent uptake (ii) the presence of a periplasmic binding-protein, we have designated them 'TRAP' (Tripartite ATP-independent Periplasmic) transporters. There is already evidence for additional TRAP transporters in *Rhodobacter*; a binding-protein dependent glutamate transporter which is vanadate insensitive and driven by the proton-motive force has been reported recently[31]. We believe that new members of this family able to transport a range of substrates will come to light as more microbial genomes are sequenced, and that they will all prove to have basic properties in common with the *R. capsulatus* Dct system.

ACKNOWLEDGMENTS

This work has been supported by grants and studentships from the UK Biotechnology and Biological Sciences Research Council to DJK, for NRW and MG.

REFERENCES

1. Gibson, J. (1975) J. Bacteriol. 123, 471–480.
2. Shaw, J.G. and Kelly, D.J. (1991) Arch. Microbiol. 155, 466–472.
3. Shaw, J.G., Hamblin, M.J. and Kelly, D.J. (1991) Mol. Microbiol. 5, 3055–3062.
4. Hamblin, M.J., Shaw, J.G., Curson, J.P. and Kelly, D.J. (1990) Mol. Microbiol. 4, 1567–1574.
5. Hamblin, M.J., Shaw, J.G. and Kelly, D.J. (1993) Mol.Gen.Genet. 237, 215–224.
6. Forward, J.A., Behrendt, M.C., Wyborn, N.R., Cross, R. and Kelly, D.J. (1997) J. Bacteriol. 179, 5482–5493.
7. Fonstein, M. and Haselkorn, R. (1993) Proc. Natl. Acad. Sci. USA 90, 2522–2526.
8. Khan, D. and Ditta, G. (1991) Mol. Microbiol. 5, 987–997.
9. Miller, D.M., Olson, J.S. and Quiocho, F.A. (1980) J. Biol. Chem. 255, 2465–2471.
10. Miller, D.M., Olson, J.S., Pflugrath, J.W. and Quiocho, F.A. (1983) J. Biol. Chem. 258, 13665–13681.
11. Walmsley, A.R., Shaw, J.G. and Kelly, D.J. (1992) J. Biol. Chem. 267, 8064–8072.
12. Walmsley, A.R., Shaw, J.G. and Kelly, D.J. (1992) Biochemistry 31, 11175–11181.
13. von Heijne, G. (1986) EMBO J. 5, 3021–3027.
14. Grinius, L., Dreguniene, G., Goldberg, E.B., Liao, C-H. and Projan, S.J. (1992) Plasmid 27, 119–129.
15. Griffith, J.K., Baker, M.E., Rouch, D.A., Page, M.G.P., Skurray, R.A., Paulsen, I.T., Chater, K.F., Baldwin, S.A. and Henderson, P.J.F. (1992) Curr.Opin.Cell Biol. 4, 684–695.
16. MacGregor, B.J. and Donohue, T.J. (1991) J. Bacteriol. 173, 3949–3957.
17. Abee, T., van der Waal, F.-J., Hellingwerf, K.J. and Konings, W.N. (1989) J. Bacteriol. 171, 5148–5154.
18. Golby, P., Carver, M. and Jackson, J.B. (1990) Eur. J. Biochem. 187, 589–597.
19. Abee, T., Hellingwerf, K.J. and Konings, W.N. (1988) J. Bacteriol. 170, 5647–5653.
20. Jackson, J.B. 1988. Bacterial photosynthesis. p 317–375. *In* Anthony, C. (ed). Bacterial Energy Transduction. Academic Press, London.
21. Finan, T.M., Wood, J.M. and Jordan, D.C. (1981) J. Bacteriol. 148, 192–202.
22. Ghei, O.K., and Kay, W.W. (1973) J. Bacteriol. 114, 65–79.
23. Richarme, G., Elyaagoubi, A. and Kohiyama, M. 1993. J. Biol. Chem. 268: 9473–9477.
24. Abee, T. (1989) Ph.D. thesis, University of Groningen, Netherlands.
25. Sofia, H.J., Burland, V., Daniels, D.D., Plunkett III, G. and Blattner, F.R. (1994) Nucl. Acids Res. 22, 2576–2586.

26. Sanchez, J.C., Gimenez, R., Schneider, A., Fessner, W.D., Baldoma, L., Aguilar, J. and Badia, J. (1994) J. Biol. Chem., 269, 29665–29669.

[27] Willems, R.J.L., van der Heide, H.G.J. and Mooi, F.R. (1992) Mol. Microbiol. 6, 2661–2671.

28. Willems, R.J.L., Geuijen, C., van der Heide, H.G.J., Renauld, G., Bertin, P., van den Akker, W.M.R., Locht, C. and Mooi, F.R. (1994) Mol. Microbiol. 11, 337–347.

29. Fleischmann, R.D., Adams, M.D., White, O., Clayton, R.A., Kirkness, E.F., Kerlavage, A.R., Bult, C.J., Tomb, J-F., Dougherty, B.A., Merrick, J.M., McKenney, K., Sutton, G., FitzHugh, W., Fields, C., Gocayne, J.D., Scott, J., Shirley, R., Liu, L-I., Glodek, A., Kelley, J.M., Weidman, J.F., Phillips, C.A., Spriggs, T., Hedblom, E., Cotton, M.D., Utterback, T.R., Hanna, M.C., Nguyen, D.T., Saudek, D.M., Brandon, R.C., Fine, L.D., Fritchman, J.L., Furhmann, J.L., Geoghagen, N.S.M., Gnehm, C.L., McDonald, L.A., Small, K.V., Fraser, C.M., Smith, H.O. and Venter, J.C. (1995) Science 269, 496–512.

30. Kaneko, T., Sato, S., Kotani, H., Tanaka, A., Asamizu, E., Nakamura, Y., Miyajima, N., Hirosawa, M., Sugiura, M., Sasamoto, S., Kimura, T., Hosouchi, T., Matsuno, A., Muraki, A., Nakazaki, N., Naruo, K., Okumura, S., Shimpo, S., Takeuchi, C., Wada, T., Watanabe, A., Yamada, M., Yasuda, M. and Tabata, S. (1996) DNA Res. 3, 109–136.

31. Jacobs, M.H.J., van der Heide, T., Driessen, A.J.M. and Konings, W.N. (1996) Proc. Natl. Acad. Sci. USA 93, 12786–12790.

67

THE BIOCHEMISTRY AND GENETICS OF THE GLUCOSYLGLYCEROL SYNTHESIS IN *SYNECHOCYSTIS* SP. PCC 6803

Martin Hagemann,[*] Arne Schoor, Uta Effmert, Ellen Zuther, Kay Marin, and Norbert Erdmann

Department of Biology
University Rostock
Doberaner Strasse 143, D-18051 Rostock, Germany

1. SUMMARY

Salt-loaded cells of the cyanobacterium *Synechocystis* sp. PCC 6803 accumulate mainly the osmoprotective compound glucosylglycerol (GG). The enzymes involved in GG biosynthesis were found to be strictly dependent on enhanced salt concentrations. By modulation of the NaCl concentration *in vitro*, it was possible to activate and to inactivate the enzymes. Using salt-sensitive mutants generated by random cartridge mutagenesis, the genes encoding for the GG synthesizing enzymes were cloned. Their expression was about threefold enhanced in salt stressed cells. A model is presented showing the dual role of NaCl as a stressor and a regulator in salt-stressed cells.

2. INTRODUCTION

Cyanobacteria have successfully colonized a wide range of biotopes. They are found in virtually all aquatic ecosystems (marine, brackish and freshwaters) as well as in the soil, on naked rocks, in deserts and even in the air. Their remarkable morphological and physiological properties allow an amazing high capability of adaptation to environmental parameters and, consequently, a world distribution [19]. During their adaptation to different salt concentrations, a balanced water potential is achieved by the accumulation of so called osmoprotective compounds (compatible solutes), while excess inorganic ions are extruded from the cells. Osmoprotective compounds are low molecular mass organic substances showing no net charge. Even in molar concentrations they are compatible with the

* Tel: +49–381–4942076; Fax: +49–381–4942079; Email: mh@boserv.bio4.uni-rostock.de

The Phototrophic Prokaryotes, edited by Peschek *et al.*
Kluwer Academic / Plenum Publishers, New York, 1999.

cellular metabolism and protect macromolecules from harmful influences. The accumulation of osmoprotective compounds represents the central subprocess of the acclimation to high salt concentrations in cyanobacteria. The analyses of a large number of cyanobacteria allowed the assignment of strains to three salt resistance groups, which distincly differ in the principal osmoprotective substance they accumulate. No link has been found between the kind of osmoprotective substance accumulated and the taxonomic group [15].

1. The least halotolerant strains accumulate sucrose or trehalose and can tolerate up to 0.7 M NaCl.
2. Moderately halotolerant cyanobacteria accumulate glucosylglycerol (GG) and their upper tolerance limit is 1.8 M NaCl.
3. The highest halotolerance is exhibited by strains that accumulate glycine betaine and glutamate betaine. These strains can tolerate salt concentrations of up to saturation. They are truly halophilic, since they grow only in salt-enriched media.

While sucrose, trehalose, and betaines were often found also in salt-treated heterotrophic bacteria, yeast, and plants, the occurrence of glucosylglycerol is typical for salt-stressed cyanobacteria with only a few findings in other organisms [6].

3. BIOCHEMISTRY OF GLUCOSYLGLYCEROL SYNTHESIS

The biochemical pathway leading to GG has been elucidated recently [4]. GG is synthesized from ADP-glucose and glycerol-3-phosphate *via* the intermediate glucosylglycerol-phosphate (GGP) by cooperation of GGP-synthase (GGPS) and GGP-phosphatase (GGPP). The GGPS was found to be strictly dependent on ADP-glucose and on Mg^{2+} as well. The salt-dependence of GGPS and GGPP represents one of the most interesting features of this enzyme system. Enzyme activities could be only detected in the presence of enhanced salt concentrations and about 100 mM NaCl in the enzyme assay were found to provide maximal activities in crude protein extracts [6], but, to a lesser extent, other salts were also effective [17]. Protein extracts obtained from salt-adapted cells with a high internal GG concentration showed no GGPS and GGPP activity, when homogenized and assayed under NaCl-free conditions. The activities could be restored after adding NaCl to the assays. Furthermore, in protein extracts obtained from low-salt-grown cells with no internal GG, the activities of GGPS and GGPP emerged, when the buffers used for homogenization and/or assay contained higher concentrations of NaCl or other salts [4,6]. These experiments allowed to conclude that the GG synthesizing enzyme system is present, but inactive in cells growing in basal medium. This is confirmed by *in vivo* measurements of GG accumulation, which starts immediately after the shock without any lag phase and could not be inhibited by the addition of chloramphenicol [5]. The activation of GG synthesis occurs immediately by addition of NaCl or other salts *in vivo* and *in vitro* without the need of *de novo* protein synthesis. Furthermore, the activation and inactivation was found to be reversible ruling out the involvement of proteolytic processes as an possible activation mechanism.

4. GENETICS OF GLUCOSYLGLYCEROL SYNTHESIS

For the cloning of genes involved in GG synthesis, the method of random cartridge mutagenesis [10] was used to generate salt-sensitive mutants [7]. Beside several mutants with unknown defects, nine mutants impaired in GG synthesis were found. The remaining salt resistance limit of the GG minus mutants was lower than that in the salt sensitive mutants, which were able to accumulate this osmoprotective compound, showing the importance of GG accumulation for the establishment of salt tolerance. Externally supplied GG was effective to complement such mutants to salt tolerance underlining the osmoprotective function of GG [12]. In one salt-sensitive mutant, the intermediate GGP was found to be accumulated. After cloning and sequencing of the chromosomal region affected in this mutant and the corresponding region of the wild type, it became obvious that the *stpA* gene encodes the enzyme GGPP [7,8]. The *stpA* (salt tolerance protein A) gene has been isolated before from another salt-sensitive mutant of *Synechocystis*, but it has not been functionally characterized so far [13]. The StpA protein showed only to a very small extent similarity to functionally related proteins like trehalose-6-phosphate, sucrose-6-phosphate, and glucose-1-phosphate phosphatases. This divergence might be responsible for the high specificity of GGPP, since in extracts from *E. coli* no other protein was able to dephosphorylate GGP [8].

In another salt-sensitive mutant, which was completely impaired in GG-synthesis, a deletion of about 10 kb had been occured during the integration of the antibiotic resistance gene cassette. This deletion covers 7 open reading frames [9]. One of them resembles the *otsA* gene (42% identity), which encodes trehalose-phosphate synthase in *Escherichia coli* [18]. After mutation of only this single gene in the wild type, the defect in GG synthesis was reproduced, which led to the identification of the gene encoding GGPS named *ggpS* [unpublished results]. In Northern blot experiments an about threefold increase in transcripts specific for *stpA* and *ggpS* were found in salt-stressed and also salt-adapted cells indicating that, additionally to the biochemical regulation of GGPS and GGPP also their gene expression seems to be controlled by salt [8 and unpublished results].

5. REGULATION OF GLUCOSYLGLYCEROL SYNTHESIS BY SALTS

The results regarding the biosynthesis of GG allow to develop a simple model of the events occurring in salt-shocked cells of *Synechocystis* (Fig. 1). Immediately after applying an hyperosmotic salt shock, high amounts of ions, especially Na^+ and Cl^- are entering the cyanobacterial cell [14,2]. On the one hand, these high internal ion concentrations lead to an immediate osmotic equilibrium. On the other hand, unfavourable conditions for cellular metabolism are created. The high cellular ionic concentrations inhibit many physiological processes like photosynthesis [1] and protein synthesis [5], but they stimulate at the same time processes necessary for salt adaptation like GG synthesis and uptake as well as export of ions [6,12]. Parallel to the GG accumulation, the concentration of inorganic ions is diminished leading to the restoration of cellular metabolism [14,1]. Under these more favourable conditions, further modifications of cellular metabolism are initiated including changes in the expression of several genes, which encode proteins necessary in stressed cells. The concerted action of these basic processes ensure to reach a salt-adapted state as quick as possible. A direct activation of enzymes related to salt adaptation by NaCl and other salts were also found for several enzymes involved in the synthesis of other osmolytes, e.g., for trehalose-6-phosphate synthase in bacteria [3,11] and for mannitol-1-phosphate dehydrogenase in algae [16]. The NaCl-mediated activation of GG-synthesizing and other enzymes involved in salt adaptation suggests that the concentration of NaCl might also play an important direct regulatory role in triggering the salt adaptation

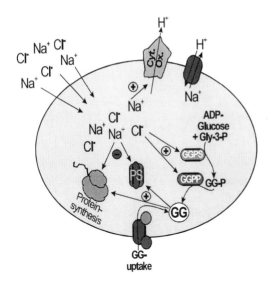

Figure 1. Schematic drawing summarizing events which occur in cells of *Synechocystis* sp. PCC 6803 immediately after a hyperosmotic salt shock. (Cyt. Ox. - cytochrome oxidase, Na^+/H^+ - Na^+/H^+-antiporter, PS - photosynthesis, GG - glucosylglycerol, Gly-3-P - glycerol-3-phosphate, GGPS - glucosylglycerol-phosphate synthase, GGPP - glucosylglycerol-phosphate phosphatase, $\oplus$ - ion-stimulated process, $\ominus$ - ion inhibited process)

[6]. Furthermore, the remarkable high salt tolerance of GGPS and GGPP ensures that almost all fixed carbon is used for GG synthesis in salt-shocked cells, since enzymes which would compete for carbon are inhibited by salt at the same time.

6. FUTURE PROSPECTS

In order to study the biochemical features of GGPS and GGPP in more detail, it is necessary to purify the enzymes in their active state from salt-treated cells as well as in their inactive state from cells grown in basal medium. The cloned genes will enable GGPS and GGPP to be overexpressed in *E. coli* and to be purified. Consequently, the comparative characterization of the GG-synthesizing enzymes obtained after biochemical purification from *Synechocystis* and after overexpression from *E. coli* will allow a further elucidation of the NaCl-mediated activation mechanism of the GG-synthesizing enzyme system. Additionally, the salt-dependent regulation of the expression of these genes will be studied in *Synechocystis*.

ACKNOWLEDGMENTS

Generous financial support by grants of the DFG (Deutsche Forschungsgemeinschaft) is greatly appreciated.

REFERENCES

1. Erdmann, N., Fulda, S. and Hagemann, M. (1992) J. Gen. Microbiol. 138, 363–368.
2. Gabbay-Azaria, R. and Tel-Or, E. (1993) in: Plant Responses to the Environment (Gresshoff, P.P., Ed.) pp. 123–132, CRC Press, Boca Raton.
3. Giæver, H.M., Styrvold, O.B., Kaasen, I. and Strøm, A.R. (1988) J. Bacteriol. 170:2841–2849.
4. Hagemann, M. and Erdmann, N. (1994) Microbiol. 140, 1427–1431.

5. Hagemann, M., Wölfel, L. and Krüger, B. (1990) J. Gen. Microbiol. 136, 1393–1399.
6. Hagemann, M., Schoor, A. and Erdmann, N. (1996a) J. Plant Physiol. 149, 746–752.
7. Hagemann, M., Richter, S. and Zuther, E. (1996b) Arch. Microbiol. 166, 83–91.
8. Hagemann, M., Schoor, A., Jeanjean, R., Zuther, E. and Joset, F. (1997) J. Bacteriol. 179, 1727–1733.
9. Kaneko, T., Sato, S., Kotani, H., Tanaka, A., Asamizu, E., Nakamura, Y., Miyajima, N., Hirosawa, M., Sugiura, M., Sasamoto, S., Kimura, T., Hosouchi, T., Matsuno, A., Muraki, A., Nakazaki, N., Nruo, K., Okumura, S., Shimpo, S., Takeuchi, C., Wada, T., Watanabe, A., Yamada, M., Yasuda, M. and Tabata, S. (1996) DNA Res. 3, 109–136.
10. Labarre, J., Chauvat, F. and Thuriaux, P. (1989) J. Bacteriol. 171, 3449–3497.
11. Lippert, K., Galinski, E.A. and Trüper, H.G. (1993) Antonie van Leeuwenhoek 63:85–91.
12. Mikkat, S., Hagemann, M. and Schoor, A. (1996) Microbiol. 142, 1725–1732.
13. Onana, B., Jeanjean, R. and Joset, F. (1994) Russian Plant Physiol. 41, 1176–1183.
14. Reed, R.H., Warr, S.R.C., Richardson, D.L., Moore, D.J. and Stewart, W.D.P. (1985) FEMS Microbiol. Lett. 28, 225–229.
15. Reed, R.H., Borowitzka, L.J., Mackay, M.A., Chudek, J.A., Foster, R., Warr, S.R.C., Moore, D.J. and Stewart, W.D.P. (1996) FEMS Microbiol. Rev. 39, 51–56.
16. Richter, D.F.E. and Kirst, G.O. (1987) Planta 170, 528–534.
17. Schoor, A., Hagemann, M. and Erdmann, N. (1997) Arch. Microbiol., submitted
18. Strøm, A.R. and Kaasen, I. (1993) Mol. Microbiol. 8, 205–210.
19. Tandeau de Marsac, N. and Houmard, J. (1993) FEMS Microbiol. Rev. 104, 119–190.

68

HYDROGENASES IN CYANOBACTERIA

H. Bothe, G. Boison, and O. Schmitz

Botanisches Institut
Universität zu Köln
Gyrhofstr.15, D-50923 Köln

1. THE NUMBER OF HYDROGENASES IN CYANOBACTERIA

Cyanobacteria can express at least two different hydrogenases catalyzing the consumption of molecular H_2 [1,2]. A so called uptake hydrogenase is component of the thylakoid membrane [3]. It is particularly active in heterocysts of filamentous forms [4], but also occurs in the unicellular *Anacystis nidulans* (*Synechococcus* PCC 6301) [5–7]. Electron acceptor for H_2-utilization catalyzed by the uptake hydrogenase is either cytochrome *b* or plastoquinone. The electrons are then allocated via the cytochrome *bc* complex to either photosystem I or to the respiratory terminal oxidase [1,7]. H_2 uptake of the thylakoid membranes is activated by the thioredoxin system [2]. The enzyme is encoded by the *hupLS* genes. In *Anabaena* PCC 7120 [8], but not in *Nostoc* PCC 73102 [9], *hupL* is subject to gene rearrangement in parallel with heterocysts differentiation. The *hupLS* genes have not been detected in unicellular cyanobacteria and are not present on the completely sequenced chromosome of *Synechocystis* PCC 6803 [10].

The second hydrogenase, the reversible or bidirectional enzyme, catalyzes *in vitro* H_2-uptake with PMS or methylene blue as electron acceptor and the $Na_2S_2O_4$-dependent evolution of the gas. This latter activity differentiate this enzyme from the uptake hydrogenase. The bidirectional hydrogenase occurs both in heterocysts and vegetative cells of filamentous forms [11] and also in unicellular cyanobacteria [12]. It is, however, absent in *Nostoc* PCC 73102 [13]. Its activity levels in the cells are enhanced by cultivation under anaerobic conditions and in the presence of H_2 [1,2]. Immunogold-labeling experiments with *A. nidulans* cells indicated that it resides at the cytoplasmic membrane [12], however, only loosely attached to it [14].

In addition, cyanobacteria contain further enzymes which catalyze partial reactions also performed by hydrogenases. Nitrogenase is known to catalyze the ATP dependent reduction of H^+ in the absence of any other substrate. In addition, the reduction of N_2 to ammonia is accompanied by the production of H_2. Cyanobacteria can contain two different Mo-nitrogenases, a V-enzyme [15,16] and possibly also an Fe-only nitrogenase [17]. Remarkably, the alternative nitrogenases produce more H_2 than the Mo-containing enzymes.

The Phototrophic Prokaryotes, edited by Peschek *et al.*
Kluwer Academic / Plenum Publishers, New York, 1999.

A hydrogenase from *Anabaena cylindrica* was described to consist of a 50 kDa and a 42 kDa protein, of which the smaller subunit alone catalyzed the tritium-exchange reaction [18]. Sequence data indicated that, physiologically, this enzyme more likely is a transaminase than a hydrogenase [19,20]. The tritium exchange reaction is often considered as being indicative for a hydrogenase. In the photosynthetic bacterium *Rhodobacter capsulatus*, the *hupUV* gene products have recently been shown also to catalyze the tritium exchange reaction [21]. It would not be too surprising to us if cyanobacteria contain further enzymes catalyzing the tritium exchange reaction and/or the uptake and/or the evolution of H_2.

2. THE DNA SEQUENCE OF THE BIDIRECTIONAL HYDROGENASE

The first "real" hydrogenase cloned and sequenced was the bidirectional enzyme from *Anabaena variabilis* [22]. This hydrogenase is encoded by the four genes *hoxFUYH* and shows high sequence similarities to the NAD^+-reducing enzyme from *Alcaligenes eutrophus* and to the MV reducing protein from archaebacteria and thus to organisms otherwise unrelated to cyanobacteria. The gene cluster is interspersed with two ORFs and an 0.9 kb apparently non-coding region (Fig. 1). One of the ORFs (no. 3) shows distinct sequence similarities to the gene coding for the CP12 protein involved in glyceraldehyde-3-

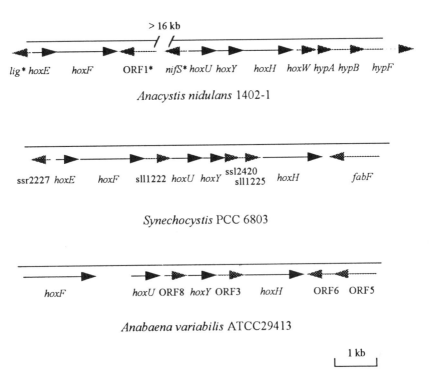

Figure 1. Gene clusters of the bidirectional hydrogenase of three cyanobacteria in comparison. The arrangement of the genes is taken from [6,24] for *A. nidulans*, [10,28] for *S.* PCC 6803 and [22] for *A. variabilis*. Genes only partially sequenced are marked by asterisks.

phosphate dehydrogenase assembly of higher plants chloroplasts [23], whereas ORF8 has no apparent counterpart in the databases. In the meantime the data of the bidirectional hydrogenase genes are available for the unicellular *Synechocystis* 6803 (present on the completely sequenced chromosome [10]) and for *A. nidulans* [6]. The order in the arrangement of the *hoxFUYH* is the same in the three cyanobacteria (Fig. 1). In *Synechocystis* 6803, the hydrogenase gene cluster is interspersed with also three ORFs, however, at different positions and totally unrelated to those in *A. variabilis*. In *A. nidulans,* the *hoxF* gene is separated by at least 16 kb from the residual *hoxUYH* genes [24]. The accessory genes *howW hypA hypB* and *hypF* reside downstream of *hoxH* in the case of *A. nidulans* but apparently not in *A. variabilis* and *Synechocystis* 6803. Thus the bidirectional hydrogenase genes in three cyanobacterial examples show a totally different organization for unknown reasons. Outside of cyanobacteria, hydrogenase genes are contiguous with the exception of the *hydLS* genes of *Thiocapsa roseopersicina* which are separated by a ~2 kb region [25].

The crystallized hydrogenase of *Desulfovibrio gigas* [26] contains the Ni active site on the larger subunit (corresponding to HoxH in cyanobacteria), whereas the small subunit contains three [Fe-S] clusters. In cyanobacteria, only one [Fe-S] binding motif is detectable on the HoxY sequence. Four [Fe-S] clusters probably reside in HoxU [22]. The *hoxFU* genes code for the diaphorase part which catalyzes the transfer of electrons to NAD^+ [27]. The characteristic NAD^+ and FMN binding motifs are clearly discernible on the HoxF sequence. The unicellular *Synechocystis* 6803 and *A. nidulans* possess another hydrogenase gene (*hoxE*) upstream of *hoxF* [24,28]. This gene could not be detected in the heterocystous *A. variabilis* by sequencing upstream of *hoxF* and by heterologous probing with *hoxE* from *A. nidulans* [24].

Recently, defined cyanobacterial hydrogenase mutants have been obtained by insertion of a kanamycin resistance cassette into the structural genes. A mutant in *hoxH* of *A. nidulans* was completely inactive in $Na_2S_2O_4$ and methyl viologen dependent H_2-evolution, but could still perform PMS-dependent H_2-uptake with ~50% of the wild-type activity [6]. This clearly indicates that the unicellular *A. nidulans* possess an uptake hydrogenase in addition to the bidirectional enzyme. It remains to be shown whether this uptake hydrogenase is encoded by the same *hupLS* genes occurring in the filamentous *Anabaena* PCC 7120 or *Nostoc* PCC 73102.

An insertion of a *lacZKm*[R] cassette into *hoxH* and *hoxU* provided some mutants inactive in H_2-evolution [24]. In such mutants, the expression of the hydrogenase, monitored by the b-galactosidase assay, was not stimulated by incubating *A. nidulans* under anaerobic conditions and in the presence of H_2 [29]. The enzyme is, however, expressed in *A. nidulans* cells. The enzyme had previously been purified and the amino acid sequence of digests of subunits has been reported [22]. Sequences of fragments of a 47 kDa and a 60 kDa subunit were identical with the deduced amino acid sequence of *hoxH* and *hoxF*, respectively.

3. THE INTERRELATIONSHIP BETWEEN RESPIRATORY COMPLEX I AND THE DIAPHORASE PART OF THE BIDIRECTIONAL HYDROGENASE

It had already been noted earlier that hydrogenase genes coding for the smaller and larger subunits (corresponding to *hoxH* and *hoxY*) [30] and for the diaphorase subunits of the soluble hydrogenase of *A. eutrophus* [31,32] show sequence similarities with subunits

Complex I of *E. coli*

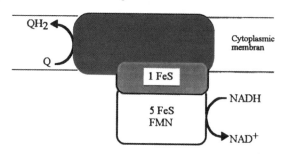

Bidirectional hydrogenase of cyanobacteria

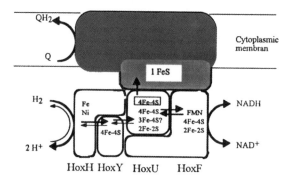

Figure 2. Model of the coupling of bidirectional hydrogenase to respiratory complex I of the cytoplasmic membrane in cyanobacteria [33].

of respiratory complex I. In all organisms, this complex I is composed of at least 14 subunits. Out of these, only 11 have been identified for cyanobacteria as yet. It has been proposed by Oliver Schmitz at the Cost Action 818 „Hydrogenases" meeting in London in December 1995 that the remaining three subunits could be encoded by the *hoxFU* genes of the diaphorase part of the bidirectional hydrogenase (Fig. 2, [33]).

The diaphorase subunit HoxF carries the motifs typical for the binding of NAD^+ and FMN. The electrons from H_2 could be allocated to either NAD^+ or respiratory terminal oxidase (Fig. 2). A special [4Fe-4S] cluster in HoxU (framed in Fig. 2) could serve as a link to respiration. The 24, 51 and 75 kDa proteins of bovine heart mitochondria or the NuoE, NuoF and NuoG proteins of complex I of *E. coli* show distinct sequence similarities to the bidirectional hydrogenase proteins HoxF and HoxU, respectively (see Fig. 3 with the sequence similarities values). Remarkably, NuoE of *E. coli* has distinct similarities both to the first 104 amino acids of HoxF and to HoxE. The *hoxE* gene is present in the unicellular *Synechocystis* [28] and in *A. nidulans* but appparently not in *A. variabilis* [24]. The gene *hoxE* may have arisen by duplication from the 104 amino acids of the N-terminus of *hoxF* or *vice versa*.

The idea of a link between respiratory complex I and bidirectional hydrogenase subunits was readily adopted by others subsequently [28,34]. Some of the already mentioned mutants in *hoxU* and those in *hoxF* [34] showed virtually no $Na_2S_2O_4$-dependent H_2-evolution, but performed respiratory O_2-uptake unimpaired. Thus it could be argued that a common link between respiratory complex I and bidirectional hydrogenase subunits

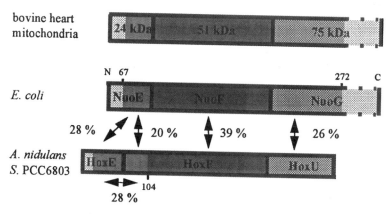

Figure 3. Sequence similarities of the diaphorase subunits of the bidirectional hydrogenase of cyanobacteria with three subunits of respiratory complex I. Corresponding parts showing similarities are shown in the same pattern. The number indicates the amino acid position.

does not exist [34]. However, since the sequence similarities are so tempting to assume such a common usage, we tend to favour another interpretation of the results with the mutants [24]. The autotrophic cyanobacteria investigated thus far respire only with low activity and have multiple respiratory chains [see 35]. Thus any defect in one respiratory chain may be compensated by another, explaining the low but unimpaired respiratory O_2-uptake in the mutants. Thus the attractive hypothesis cannot be proven simply with such defined mutants.

ACKNOWLEDGMENT

The work by the authors cited in the text was kindly supported by the Deutsche Forschungsgemeinschaft.

REFERENCES

1. Houchins, J.P. (1984) Biochim. Biophys. Acta 768, 227–255.
2. Papen, H., Kentemich, T., Schmülling, T. and Bothe, H. (1986) Biochimie 68, 121–132.
3. Eisbrenner, G., Roos, P. and Bothe, H. (1981) J. Gen. Microbiol. 125, 383–390.
4. Peterson, R.B. and Wolk, C.P. (1978) Plant Physiol. 61, 688–691.
5. Peschek, G. (1979) Arch. Microbiol. 123, 81–92.
6. Boison, G., Schmitz, O., Mikheeva, L., Shestakov, S. and Bothe, H. (1996) FEBS Lett. 394, 153–158.
7. Bothe, H., Distler, E. and Eisbrenner, G. (1978) Biochimie 60, 277–289.
8. Carrasco, C.D., Buettner, J.A. and Golden, J.W. (1995) Proc. Natl. Acad. Sci. USA 92, 791–795.
9. Lindblad, P., Hansel, A., Oxelfelt, F., Tamagnini, P. and Troshina, O. (1997) In: Vth International conference on the molecular biology of hydrogenases , Albertville, France, 12–17 July, 1997, Vignais, P.M.; Frey, M. (eds.), 102.
10. Kaneko, T. et al. (1996) DNA Res. 3, 109–136, 185–209.
11. Houchins, J.P. and Burris, R.H. (1981) J. Bacteriol. 146 , 209–214.
12. Kentemich, T., Bahnweg, M., Mayer, F. and Bothe, H. (1989) Z. Naturforsch. 44c, 384–391.
13. Tamagnini, P., Troshina, O., Oxelfelt, F., Salema, R. and Lindblad, P. (1997) Appl. Environm. Microbiol. 63, 1801–1807.

14. Kentemich, T., Casper, M. and Bothe, H. (1991) Naturwiss. 78, 559–560.
15. Kentemich, T., Danneberg, G., Hundeshagen, B. and Bothe, H. (1988) FEMS Microbiol. Lett. 51, 19–24.
16. Thiel, T. (1993) J. Bacteriol. 175, 6276–6286.
17. Kentemich, T., Haverkamp, G. and Bothe, H. (1991) Z. Naturforsch. 46c, 217–222.
18. Ewart, G.D. and Smith, G.D. (1989) Arch. Biochem. Biophys. 268, 327–337.
19. de Zoysa, P.A. and Danpure, C.J. (1993) Mol. Biol. Evol. 10, 704–706.
20. Ouzounis, C. and Sander, C. (1993) FEBS Lett. 322, 159–164.
21. Vignais, P.M., Dimon, B., Zorin, N.A., Colbeau, A. and Elsen, S. (1997) J. Bacteriol. 179, 290–292.
22. Schmitz, O., Boison, G., Hilscher, R., Hundeshagen, B., Zimmer, W., Lottspeich, F. and Bothe, H. (1995) Eur. J. Biochem. 233, 266–276.
23. Pohlmeyer, K., Paap, B.K., Soll, J. and Wedel, N. (1996) Plant Mol. Biol. 32, 969–978.
24. Boison, G., Schmitz, O., Schmitz, B. and Bothe, H. (1997) Curr. Microbiol., *in press*.
25. Kovács, K.L., Rákhely, G., Colbeau, A. and Vignais, P.M. (1997) In: Vth International conference on the molecular biology of hydrogenases , Albertville, France, 12–17 July, 1997, Vignais, P.M.; Frey, M. (eds.), 74.
26. Volbeda, A., Charon, M-H., Piras, C., Hatchikian, E.C., Frey, M. and Fontecilla-Camps, J.C. (1995) Nature 373, 580–587.
27. Schmitz, O. and Bothe, H. (1996) FEMS Microbiol. Lett. 135, 97–101.
28. Appel, J. and Schulz, R. (1996) Biochim. Biophys. Acta 1298, 142–147.
29. Boison, G. (1997) PhD Thesis, Köln.
30. Albracht, S.P.J. (1993) Biochim. Biophys. Acta 1144, 221–224.
31. Patel, S.D., Aebersold, R. and Attardi, G. (1991) Proc. Natl. Acad. Sci. USA 88, 4225–4229.
32. Pilkington, S.J., Skehel, J.M., Gennis, R.B. and Walker, J.E. (1991) Biochemistry 30, 2166–2175.
33. Schmitz, O. and Bothe, H. (1996) Naturwiss. 83, 525–527.
34. Howitt, C.A. and Vermaas, W.F.J. (1997) In: IX International Symposium on Phototrophic Prokaryotes. Vienna, Austria, Sept. 6.-13., 1997, Peschek, G.A., Löffelhardt, W. and Schmetterer, G. (eds.), pp. 595 *ff.*
35. Schmetterer, G., (1994) In: The Molecular Biology of Cyanobacteria (Bryant, D.A., ed.), pp. 409–435, Kluwer, Dordrecht, The Netherlands.

69

SUBUNITS OF THE NAD(P)-REDUCING NICKEL-CONTAINING HYDROGENASE DO NOT ACT AS PART OF THE TYPE-1 NAD(P)H-DEHYDROGENASE IN THE CYANOBACTERIUM *SYNECHOCYSTIS* SP. PCC 6803

Crispin A. Howitt and Wim F. J. Vermaas

Department of Plant Biology and Center for the Study of Early Events in
    Photosynthesis
Arizona State University
Box 871601, Tempe, Arizona 85287-1601

1. INTRODUCTION

Membrane bound bacterial pyridine nucleotide dehydrogenases can be divided into two classes that are commonly called type-1 (NDH-1) and type-2 (NDH-2) (for extensive reviews see [1,2]): NDH-1 is a multisubunit complex that has 14 or more subunits, depending on the organism, in systems where a functional NDH-1 has been isolated [3,4]. An operon encoding 14 subunits of NDH-1 and 6 unidentified reading frames has been identified in *Paracoccus denitrificans* ([5] and references therein). An operon encoding 14 similar subunits in the same order has been cloned and sequenced from *Escherichia coli* [6]; the unidentified open reading frames are missing in this organism. NDH-1 contains 4–6 iron-sulfur clusters, translocates protons across the membrane, has a flavin mononucleotide as the prosthetic group, and is at least partially inhibited by rotenone. NDH-2 complexes consist of a single subunit, do not contain iron-sulfur clusters, do not translocate protons across the membrane, have flavin adenine dinucleotide as the prosthetic group and are inhibited by flavone rather than by rotenone.

Little is know about pyridine nucleotide dehydrogenases in cyanobacteria (see [7] for a review). Evidence for their presence has developed from both biochemical studies and the cloning of *ndh* genes. In *Synechocystis* sp. PCC 6803 antibodies raised against fu-

The Phototrophic Prokaryotes, edited by Peschek *et al.*
Kluwer Academic / Plenum Publishers, New York, 1999.

sion proteins containing the products encoded by *ndhK* and *ndhJ* cross reacted with proteins on both thylakoid and cytoplasmic membranes [8,9]. Both membranes catalyzed rotenone-sensitive NADH oxidation. Antibody affinity columns have been used to purify a complex that contains NdhK, NdhJ, NdhI and NdhH [10]. Initial studies have shown that homologs of eleven of the fourteen *ndh* genes are found in cyanobacteria [11]. Inactivation of *ndh* genes in *Synechocystis* sp. PCC 6803 have shown that NDH-1 plays a role in both photosynthesis and respiration. Strains in which *ndhB, ndhC, ndhK* or *ndhL* have been inactivated are unable to concentrate inorganic carbon and require an environment enriched in CO_2 to grow. While strains lacking *ndhB* or *ndhL* showed severely impaired oxygen consumption rates ([12] and references therein). By following P700 oxidation and reduction kinetics in wild type and the *ndhB⁻* and *ndhK⁻* strains it has been shown that the thylakoid-bound NDH-1 mediates cyclic electron transport around PS I that is dependent on both NADPH and ferredoxin [13]. However, this is not the case in all cyanobacteria, as inactivation of *ndhF* in the marine organism *Synechococcus* sp. PCC 7002 did not lead to a high-CO_2 requiring phenotype or impair cyclic electron transport around PS I [14].

NDH-1 in cyanobacteria appears to differ in some respects from the much better characterized NDH-1s from mitochondria and bacteria. Specifically, no genes for homologs of the three subunits thought to be involved in binding of FMN and the substrate, NAD(P)H, have been identified (NuoE, NuoF, and NuoG in *Escherichia coli* [6]). With the sequencing of the genome of *Synechocystis* sp. PCC 6803 [15] it has become clear that indeed no obvious homologs to the three subunits from bacterial and mitochondrial NDH-1 are present in this cyanobacterium. The absence of these three subunits has lead to a number of theories as to the nature of the NAD(P)H binding/oxidizing subunits: (a) subunits of the NAD(P)-reducing nickel-containing hydrogenase act as the NAD(P)H binding/oxidizing subunits of NDH-1 [16,17]; (b) FNR is the NADPH oxidizing subunit of NDH-1 in cyanobacteria and chloroplasts [11]; (c) ferredoxin may act as the electron donor to NDH-1 [11]; and/or (d) the three missing subunits are present but their sequences are so divergent from their bacterial and mitochondrial homologs that they can not be detected by conventional means.

Here we report the results of studies designed to investigate the first of the theories outlined above, in which respiration was characterized in a strain in which genes encoding subunits of the NAD(P)-reducing nickel-containing hydrogenase were insertionally inactivated.

2. MATERIALS AND METHODS

Synechocystis sp. PCC 6803 was cultivated at 30°C in BG-11 medium containing 10 mM TES-NaOH (pH 8.0) [18] in which $NaNO_3$ was partially substituted with NH_4NO_3 (final concentration of ammonia was 4.5 mM). For photomixotrophic growth the medium was supplemented with 5 mM glucose. On plates, 1.5% (w/v) agar and 0.3% (w/v) sodium thiosulfate were added, and BG-11 was supplemented with antibiotics appropriate for the particular strain (35 μg ml⁻¹ chloramphenicol). Cultures were grown under normal illumination (50 μE m⁻² s⁻¹).

DNA from *Synechocystis* sp. PCC 6803 was prepared essentially as described in [19]. After restriction digest of genomic DNA the fragments were separated by agarose gel electrophoresis and transferred to GeneScreen Plus (Du Pont NEN) according to the manufacturer's instructions. Probes were prepared by hot PCR with ³²P-dATP using wild type genomic DNA as the template. The sequences of the primers used to prepare probes were 5′ GGA AGG AAA

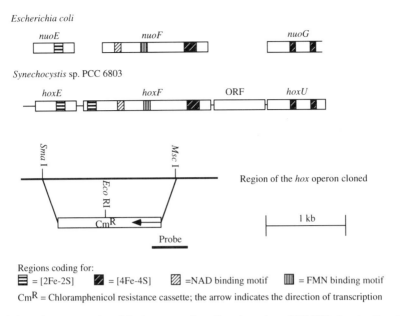

Regions coding for:
▤ = [2Fe-2S]   ◪ = [4Fe-4S]   ▨ =NAD binding motif   ▥ = FMN binding motif
CmR = Chloramphenicol resistance cassette; the arrow indicates the direction of transcription

Figure 1. Schematic representation of the *hox* operon from *Synechocystis* sp. PCC 6803 showing the relative positions of important functional groups and an alignment of these with *nuoE*, *nuoG* and *nuoF* from *Escherichia coli*. Only the first 240 amino acids of *nuoG* are represented as the rest of *nuoG* does not show homology with the hydrogenase subunits downstream from *hoxU*. The region of the operon cloned, the deletion made and the *hoxF* region to which the probe hybridizes is also shown.

ACG GGG AAC GCC CCG 3′ and 5′ CAT CCA TGA CAA TCA TGC CCC CGG 3′, recognizing nucleotides 1,677,071–1,677,048 and 1,676,650–1,676,673 respectively (numbering according to CyanoBase (http://www.kazusa.or.jp/cyano/cyano.html)).

Oxygen consumption measurements were carried out in a manner similar to that described previously [20]. Chlorophyll *a* concentrations were determined according to [21].

3. RESULTS

Part of the putative operon of the NAD(P)-reducing nickel-containing hydrogenase was amplified using the polymerase chain reaction (PCR). The sequences of the primers used were 5′ GCA AAC CAA CTA Gga TTC CAT TCG 3′ (nucleotides 1,678,735–1,678,712) and 5′ TTG GAT AAA agC TTC CCA GGT GGC 3′ (nucleotides 1,676,257–1,676,280); nucleotides in lower case represent changes made to the sequence in order to introduce restriction sites to facilitate cloning of the PCR product into pUC19. A 1.6 kb *Sma* I-*Msc* I (positions 1,678,376 and 1,676,788) fragment was deleted from the putative coding region of *hoxE* and *hoxF* and replaced by a chloramphenicol resistance gene from pACYC184 [22] (Fig. 1). Wild type *Synechocystis* sp. PCC 6803 was transformed with this construct and transformants were selected for on chloramphenicol. Transformants were subcultured in the presence of chloramphenicol in air to allow segregation to occur and thus to obtain a homozygous genotype, resulting in strains that lacked *hoxE* and *hoxF* (designated *hox⁻*).

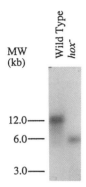

MW
(kb)

12.0—
6.0—

3.0—

Figure 2. Southern analysis of wild type and the *hox⁻* strain. DNA was cut with *Eco* RI and probed with a ³²P-labeled 400 bp PCR product that hybridized to a region towards the 3' end of *hoxF*.

Southern analysis was used to determine if the strain created was homozygous for the *hoxEF* deletion. As shown in Figure 2, indeed homozygosity was obtained for the deletion, with the probe hybridizing to a 12 kb *Eco* RI fragment in wild type (Fig. 2) and a 6 kb *Eco* RI fragment in the *hox⁻* strain (Fig. 2).

As the *hox⁻* strain was segregated at air levels of CO_2 and no increased CO_2 requirement for photoautotrophic growth was obtained (data not shown), HoxE and HoxF do not appear not essential components of NDH-1 in *Synechocystis* sp. PCC 6803. In order to further characterize this strain, its doubling time was determined under photoautotrophic and photomixotrophic conditions. Oxygen consumption rates were also determined for photomixotrophically grown cells. The results are presented in Table 1. No significant difference between the growth rates of wild type and the *hox⁻* strain was observed, and also the respiratory rates of whole cells of wild type and the mutant strains were similar when grown under normal illumination (50 µE m⁻² s⁻¹). Respiration in the *hox⁻* strain was almost completely sensitive to KCN, as is the case in wild type.

4. DISCUSSION

As shown in Figure 1 the order of the three genes in the *Synechocystis* sp. PCC 6803 hydrogenase operon is identical to that of their homologs in the *nuo* operon of *E. coli* [6]. This is also the case in the *nqo* operon of *Paracoccus denitrificans* [5] which also contains an ORF in the same position as the one found in the *hox* operon of *Synechocystis* sp. PCC

Table 1. Doubling times of photoautotrophically (PA) and photomixotrophically (PM) grown cells and oxygen consumption rates of cells grown photomixotrophically

Strain	Doubling time (h)		Oxygen consumption (μmol O_2 mg chl⁻¹ h⁻¹)	
	PA	PM	No additions	+KCN
Wild type	11.4 ± 0.5	8.3 ± 2.1	42 ± 11	2.0 ± 3.5
hox⁻	12.5 ± 0.3	9.3 ± 0.6	46 ± 9.1	1.1 ± 1.9

Values given are averages ± standard deviation from at least 3 determinations. Oxygen consumption of whole cells was measured in 10 mM HEPES-NaOH (pH 7.0); where indicated KCN was added to a final concentration of 1 mM.

Table 2. Amino acid sequence identity of HoxE, HoxF and HoxU from *Synechocystis* sp. PCC 6803 to homologs from other organisms

Species	Identity (%)		
	HoxE	HoxF	HoxU
Escherichia coli	29	40	28
Bos taurus	28	40	30
Alcaligenes eutrophus	28*	34	36
Anabaena variabilis	n.a.	65	63

Sequence identities were determined using the FASTA program from the GCG package using the default settings. Identities calculated are to NuoE, NuoF, and NuoG in *E. coli* and the appropriate homologs in *B. taurus* NDH-1. n.a., not applicable; gene has not been sequenced. * HoxE has not been sequenced from *A. eutrophus* and the identity shown is with the N terminus of HoxF (see text).

6803. The degree of identity of *Synechocystis* sp. PCC 6803 HoxE, HoxF and HoxU to homologs from various organisms are shown in Table 2. HoxF and HoxU have the highest homology to HoxF and HoxU from *Anabaena variabilis*. However there is little difference in the degree of identity between the HoxE, HoxF and HoxU from *Synechocystis* sp. PCC 6803 and the homologs in *E. coli, Bos taurus* and *Alcaligenes eutrophus*. Interestingly, no homolog of HoxE has been found in *A. eutrophus*; however, HoxE from *Synechocystis* sp. PCC 6803 shows significant homology to the N-terminus of HoxF in *A. eutrophus*.

In addition to having high sequence identity with the subunits of NDH-1 that are thought to make up the NADH binding/oxidizing domain from other systems, the HoxE/HoxF/HoxU polypeptides have conserved sequence motifs as shown in Figure 1. HoxE contains a motif that may coordinate a [2Fe-2S]-cluster. However, this motif is not conserved in the region of HoxF from *A. eutrophus* that shows homology to HoxE, and thus this may not be an iron-sulfur protein [23]. HoxF contains motifs for an NADH binding site, a possible FMN binding site as well as a [2Fe-2S]-cluster and a [4Fe-4S]-cluster. The motif for the [2Fe-2S] cluster is in an N-terminal extension of HoxF that is not present in homologs from NDH-1. HoxU and its homolog in NDH-1s contain three conserved blocks of cysteine residues that are thought to coordinate two [4Fe-4S] clusters [23].

Based on this homology and the conserved sequence motifs it has been proposed that NDH-1 arose as a result of modular evolution, from an *en bloc* association of pre-existing smaller complexes that had different functions [23]. According to this theory, complete sections of the electron transfer pathway and the mechanisms of proton translocation would have evolved independently as separate structural modules and then come together to form Complex I [23,24]. This theory is supported by the findings that Complex I, at least in *Neurospora crassa*, assembles as two independent sub-complexes that aggregate to form the functional complex [25,26]. Subunits within each of these two sub-complexes have been shown to have significant homology with other simpler bacterial enzymes, further strengthening the modular theory of evolution for Complex I. As discussed above, the module made up of the three subunits which are missing in the genome of *Synechocystis* sp. PCC 6803 contains the binding sites for NADH, FMN and up to five FeS clusters. These three subunits show significant homology to the diaphorase part of the NAD+-reducing hydrogenase of *A. eutrophus* [27] which contains covalently bound FMN and an NADH. Similarly, subunits of the second sub-complex show significant homology to a different bacterial enzyme, the *E. coli* formate hydrogenlyase [23,24].

The results presented here show that (1) *hoxE* and *hoxF* are easily deleted from *Synechocystis* sp. PCC 6803 under photomixotrophic conditions at ambient CO_2 concentration, suggesting that these gene products are neutral for growth and survival under these conditions, (2) a strain in which the genes encoding HoxE and HoxF have been deleted respired at rates similar to wild type, and (3) were able to grow photoautotrophically at air levels of CO_2. As HoxF contains NADH and FMN binding motifs it would be expected that a strain lacking these would be unable to respire in darkness if they were to constitute the nucleotide-domain. Indeed this phenotype is observed in strains which lack *ndhB* [28] or *ndhL* [29], which are single copy genes like *hoxE*, *hoxF* and *hoxU*. Furthermore, the fact that the *hox⁻* strain created was able to grow at ambient CO_2 levels sets it apart from strains in which single copy *ndh* genes have been deleted [11] and which require an environment enriched in CO_2 for growth. Thus it can be concluded that in *Synechocystis* sp. PCC 6803 HoxE, HoxF and HoxU do not have dual roles and do not also act as the NAD(P)H binding/oxidizing module of NDH-1. The nature of this module in cyanobacterial NDH-1 remains to be elucidated, but HoxE, HoxF and HoxU are not involved.

ACKNOWLEDGMENTS

We thank Dr. Satoshi Tabata (Kazusa DNA Research Institute) for the opportunity to search the genome sequence of *Synechocystis* sp. PCC 6803 prior to its release. This research was supported by a grant from the US Department of Agriculture to W.V.

REFERENCES

1. Yagi, T. (1991) J. Bioenerg. Biomembr. 23, 211–225.
2. Yagi, T. (1993) Biochim. Biophys. Acta 1141, 1–17.
3. Walker, J.E. (1992) Quart. Rev. Biophys. 25, 253–324.
4. Leif, H., Sled, V.D., Ohnishi, T., Weiss, H. and Friedrich, T. (1995) Eur. J. Biochem. 230, 538–548
5. Xu, X., Matsuno-Yagi, A. and Yagi, T. (1993) Biochemistry 32, 968–981.
6. Weidner, U., Geire, S., Ptock, A., Freidrich, T., Leif, H. and Weiss, H. (1993) J. Mol. Biol. 233, 109–122.
7. Schmetterer, G. (1994) in: The Molecular Biology of Cyanobacteria, pp. 409–435 (Bryant, D.A., Ed.) Kluwer Academic Publishers, Dordrecht.
8. Berger, S., Ellersiek, U. and Steinmüller, K. (1991) FEBS Lett. 286, 129–132.
9. Dzelzkalns, V.A., Obinger, C., Regelsberger, G., Niederhauser, H., Kamensek, M., Peschek, G.A. and Bogorad, L. (1994) Plant Physiol. 106, 1435–1442.
10. Berger, S., Ellersiek, U., Kinzelt, D. and Steinmüller, K. (1993) FEBS Lett. 326, 246–250.
11. Friedrich, T., Steinmüller, K. and Weiss, H. (1995) FEBS Lett. 367, 107–111.
12. Ogawa, T. (1992) in: Research in Photosynthesis, Vol. III, pp. 763–770 (Murata, N., Ed.) Kluwer Academic Publishers, Dordrecht.
13. Mi, H., Endo, T., Ogawa, T. and Asada, K. (1995) Plant Cell Physiol. 36, 661–668.
14. Schluchter, W.M., Zhoa, J. and Bryant, D.A. (1993) J. Bacteriol. 175, 3343–3352.
15. Kaneko, T., Sato, S., Kotani, H., Tanaka, A., Asamizu, E., Nakamura, Y., Miyajima, N., Hirosawa, M., Sugiura, M., Sasamoto, S., Kimura, T., Hosochi, T., Matsuno, A., Muraki, A., Nakazaki, N., Naruo, K., Okumura, S., Shimpo, S., Takeuchi, C., Wada, T., Watanabe, A., Yamada, M., Yasuda, M. and Tabata, S. (1996) DNA Res. 3, 109–136.
16. Appel, J. and Schulz, R. (1996) Biochim. Biophys. Acta 1298, 141–147.
17. Schmitz, O. and Bothe, H. (1996) Naturwissenschaften 83, 525–527.
18. Rippka, R., Derulles, J., Waterbury, J.B., Herdmane, M. and Stanier, R.Y. (1979) J. Gen. Microbiol. 111, 1–61.
19. Williams, J.G.K. (1988) Methods Enzymol. 167, 766–778.
20. Vermaas, W.F.J., Shen, G. and Styring, S. (1994) FEBS Lett. 337, 103–108.

21. Porra, R.J., Thompson, W.A. and Kriedemann, P.E. (1989) Biochim. Biophys. Acta 975, 384–394.

22. Chang, A.C.Y. and Cohen, S.W. (1978) J. Bacteriol. 134, 1141–1156.

23. Weiss, H., Freidrich, T., Hofhaus, G. and Preis, D. (1991) Eur. J. Biochem. 197, 563–576.

24. Walker, J.E., Arizmendi, J.M., Dupuis, A., Fearnley, I.M., Finel, M., Medd, S.M., Pilkington, S.J., Runswick, M.J. and Skehel, J.M. (1992) J. Mol. Biol. 226, 1051–1072.

25. Tuschen, G., Sackmann, U., Nehls, U., Haiker, H., Buse, G. and Weiss, H. (1990) J. Mol. Biol. 213, 845–857.

26. Nehls, U., Friedrich, T., Schmiede, A., Ohnishi, T. and Weiss, H. (1992) J. Mol. Biol. 227, 1032–1042.

27. Pilkington, S.J., Skehel, J.M., Gennis, R.B. and Walker, J.E. (1991) Biochemistry 30, 2166–2175.

28. Ogawa, T. (1991) Proc. Natl. Acad. Sci. USA 88, 4275–4279.

29. Ogawa, T. (1992) Plant Physiol. 99, 1604–1608.

70

MASS SPECTROMETRIC ANALYSIS OF HYDROGEN PHOTOEVOLUTION IN THE FILAMENTOUS NON-HETEROCYSTOUS CYANOBACTERIUM *OSCILLATORIA CHALYBEA*

Klaus P. Bader and Refat Abdel-Basset

Lehrstuhl für Zellphysiologie
Fakultät für Biologie
Postfach 10 01 31, D-33501 Bielefeld, Bundesrepublik Deutschland

1. INTRODUCTION

Few unicellular or filamentous heterocystous cyanobacteria have been shown to photoevolve molecular hydrogen under anaerobic conditions. The phenomenon as such can be relevant in the future under the aspects of hydrogen as being an extremely clean and efficient fuel. For this reason light-induced hydrogen evolution has become an increasingly important topic in scientific research [1–6]. Hydrogen evolution is in part based on the function of the nitrogenase system for the fixation of atmospheric nitrogen in the respective organism necessary for the nitrogen supply under nitrogen-limiting conditions. In other cases hydrogenase systems have been described which catalyze the oxidation of hydrogen as an uptake hydrogenase or which can act as a reversible hydrogenase using hydrogen as an electron source under specific conditions and reducing protons depending on the redox situation of the cells.

The so-called Ni- and Fe-hydrogenases can be distinguished not only by their function with the latter being reversible—and the former uptake hydrogenases, but also by their distinct structural properties [7,8]. The uptake hydrogenases contain a Ni-catalytic centre and two or more [4Fe-4S] clusters, they are heterodimers with subunits of about 30 and 60 kD and—in general—they are membrane-bound. These enzymes oxidize molecular hydrogen using hydrogen as a substrate and source for electrons which might play a role for the removal of redox equivalents. The reversible hydrogenase which can operate in both directions thus leading to the production of molecular hydrogen has been described as being a 60 kD monomeric Fe-hydrogenase [5,9–13]. The reversible hydrogenase from *Anabaena variabilis* requires reductive activation before hydrogen can be oxidized and carbon monoxide appears to act as a competitive inhibitor [14]. At present, the reversible

The Phototrophic Prokaryotes, edited by Peschek *et al.*
Kluwer Academic / Plenum Publishers, New York, 1999.

hydrogenase does not seem to be an ubiquitous component in hydrogenase-containing cyanobacteria, as according to recent reports this enzyme seems to be missing in *Nostoc* sp. strain PCC 73102. It can not be excluded, however, that more sensitive and specific physiological and molecular techniques are required to detect the reversible hydrogenase also in this organism [15]. Starting from the first detection of a light-induced hydrogen evolution signal in *Oscillatoria chalybea* [20] we approach a physiological characterization of the hydrogen gas exchange in this filamentous non heterocystous cyanobacterium in the present work.

2. MATERIALS AND METHODS

Oscillatoria chalybea was originally obtained from the Algal Collection in Göttingen (Germany) and has been cultivated since then in a climatized room at about 26°C. The cyanobacteria were grown in medium D described by Kratz and Myers (1955) on clay plates as porous mechanical support in large Petri dishes. During growth the cultures were kept in a 14 h light/10 h dark cycle at a light intensity of approx. 12 $\mu E \cdot m^{-2} \cdot s^{-1}$. Protoplasts from *Oscillatoria chalybea* were isolated following enzymic digestion (glucuronidase, cellulase) essentially as previously described [16].

Mass spectrometry was carried out by means of a specifically modified magnetic sector field 'Stable Isotope Ratio Mass Spectrometer' type 'Delta' from Finnigan MAT (Bremen, Germany). The modifications of the apparatus leading to a substantial increase in the sensitivity and the improvement of the response time for dynamic measurements in particular, have been developed earlier and described in detail [17]. Further specifications and the calibration of the set-up for quantification of the respective signals have also been outlined [18,19]. The hydrogen evolution and uptake signals were directly detected at m/e = 2 in a H/D collector of the mass spectrometric set-up and recorded on a SE 130-03 three-channel recorder from Goertz Metrawatt. The respective assays were illuminated either by short (5 μs) saturating white flashes provided by a stroboslave 1539A from General Radio or by red light from a projector (Leitz, Prado Universal).

3. RESULTS AND DISCUSSION

In previous papers we have shown that *Oscillatoria chalybea* is capable of fixing atmospheric nitrogen under conditions of nitrogen limitation which was detected and analyzed in our mass spectrometric set-up as $^{15}N_2$-uptake from the gas phase over the assay [19]. In addition, we observed a substantial light-induced hydrogen evolution signal under anaerobic conditions [20]. This hydrogen evolution, however, is at least in part independent of the respective nitrogenase system, as nitrogen free-grown and nitrate-grown cultures yielded almost identical hydrogen amplitudes. Nitrogen fixation measured as nitrogen uptake, on the other hand, was restricted to nitrogen free-grown cells. Thus, it appears that continuous light- or flash-induced hydrogen gas exchange must be analyzed in terms of uptake hydrogenase- and/or reversible hydrogenase systems in this organism. Illumination of *Oscillatoria* protoplasts with 10 short (5 μs) saturating light flashes (Fig. 1a) or for 1 minute with red light (Fig. 1b) leads to a hydrogen signal which apparently is composed of several phases. Initially, a steep increase in the hydrogen partial pressure is observed which is followed by a continuously increasing uptake phenomenon. Even at the off-set of light, however, the uptake tracing still contains a significant evolution portion as

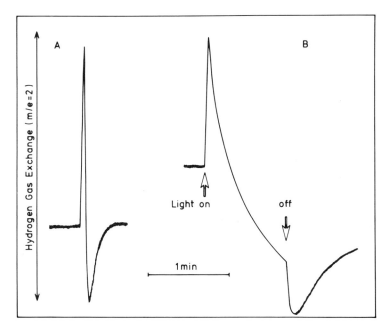

Figure 1. Mass spectrometric recording at m/e = 2 of the hydrogen gas exchange in *Oscillatoria chalybea* induced by a sequence of 10 short (5 μs) saturating light flashes spaced 300 ms apart (A) or by illumination for 1 min with red light (B). The signals were detected in a H/D collector of a 'Delta' mass spectrometer and directly recorded.

can be concluded from the additional down-step of the mass 2 signal. In one of the few earlier reports such a transition has been attributed to an initially non-active Calvin cycle. During illumination the cycle is activated deriving more electrons to CO_2-fixation rather than to proton reduction. Under this assumption, hydrogen evolution would gradually decrease and hydrogen uptake increase [21,22]. This interpretation can possibly be substantiated by the analysis of the concomitant mass spectrometric recording of CO_2 at m/e = 44. At present we suggest that *Oscillatoria chalybea* contains both the uptake- and the reversible hydrogenase. This seems not to be self-evident as recent investigations report on cyanobacteria lacking the bidirectional enzyme [15].

The hydrogen evolution signal requires relatively short dark adaptation periods between sequential illuminations, hence it can virtually not be light-exhausted. Fig. 2 shows that even upon re-illumination after 5 s dark adaptation following a preillumination about two thirds of the maximum signal are observed. Thus, the electron supply which is necessary for proton reduction seems to be very efficient and very fast in particular. This could be brought in line with an interrelationship between the respiratory and photosynthetic electron transport chains (*vide infra*). Not only the dark time between the illumination periods (1 min red light or sequences of short [5 μs] saturating Xenon flashes) but also the flash frequency in the case of flash illumination play an important role for the hydrogen yield. Decrease of the flash frequency from 10 Hz to 1 Hz still further increases the signal amplitudes. In contrast to this parallel recording of the photosynthetic oxygen evolution resulted in a decrease of the oxygen amplitudes due to a substantial decrease of the higher oxidized S-states in the case of 1 Hz [23]. It appears that the electron transfer (most probably from the respiratory chain) to the hydrogen formation systems is biphasic with com-

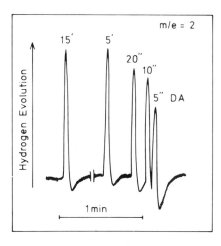

Figure 2. Dependence of the flash-induced evolution of molecular hydrogen on the dark time between flash series in *Oscillatoria chalybea*. The signals represent the cumulated hydrogen yields induced by a train of 10 5μs-flashes.

pletely different time constants for the two parts—a very fast reaction which requires only a few seconds and a slowly (order of minutes) filled pool which maximizes the yield.

The process of light-induced hydrogen evolution is oxygen sensitive and is only pronounced after extended flushing with argon or nitrogen, hence requires anaerobic conditions. At oxygen concentrations of 6.16 μM the hydrogen evolution is inhibited by about 50%. This oxygen sensitivity is in line with earlier reports in the literature. On the other hand, recent papers also report on largely oxygen insensitive hydrogen evolution even at very high oxygen partial pressures [24].

No substantial difference was observed when the protoplasts were flushed to anaerobiosis with either argon or nitrogen. The age of the cultures, however, plays an important role as the capacity, *i.e.* the evolution rate decreases with increasing age. This can be brought into context with the idea that internal substrates *e.g.* glycogen are used up or become more and more limiting. A similar interpretation has been made in [24]. If we assume that in early phases of growth endogenous substrates function as electron donors for proton reduction it could be expected that exogenously added carbohydrates play only a minor role at least much less than has been described for other organisms. In fact, we observed that neither glucose nor fructose nor sucrose nor pyridine nucleotides induce a stimulation of the hydrogen evolution signal. Instead, in most of the cases a slight inhibition was observed what we ascribe to an unspecific chemical stress situation. As *Oscillatoria chalybea* is a prokaryote with any electron transport system being located in the same membrane system and lacking cell organelles, it could be expected that the required electrons for proton reduction are supplied by common redox carriers *e.g.* between the respiratory and the photosynthetic electron transport system. (Similar interactions have been reported on several occasions.) Under this assumption, however, it should be possible to somehow exhaust and deplete this endogenous substrate pool and this in turn should lead to an immediate inhibition of the hydrogen evolution signals. For this purpose we pre-illuminated the reaction assays for two hours with weak red light *prior* to the mass spectrometric measurements and we observed that under such conditions only 50% of the hydrogen evolution remained detectable. Moreover, we were then capable to increase *i.e.* restore the hydrogen evolution by the addition of the above mentioned substrates. At concentrations of about 0.1 mM of the respective compound stimulations of 50–60% (in relation to the depleted samples) were detected. This means that in fact endogenous substrates in *Oscillatoria chalybea* completely suffice to sustain a maximally working hydrogenase system and that only under the conditions of artificially achieved depletion of the endogenous pool exogenously added substrates can exert any stimulatory effect in comparison to the impaired signals.

In further experiments we dealt with the question which electron transport system is responsible for the effective funnelling of electrons to proton reduction. In any of the investigated cases, antimycin A, an inhibitor which is known to block the electron transport of the respiratory chain in the region of the cytochrome b/c_I–complex (III), reduced the hydrogen evolution to 30% of its initial value. Moreover, the classical respiration inhibitor cyanide with its effect on the cytochrome a/a_3–complex decreased the hydrogen evolution to about the same extent. Most interestingly, salicylhydroxamic acid (SHAM) an inhibitor of the alternate respiration also inhibited the proton reduction rate in *Oscillatoria chalybea* with the effects of cyanide and SHAM being additive. In the presence of one of the two inhibitors—applied at concentrations of 1 mM—30–40% of the signals remained, whereas with cyanide and SHAM applied simultaneously less than 10% of the control rates were detected. Our results suggest that the respiratory electron transport chain in *Oscillatoria chalybea* is intimately linked to the photosynthetic electron transport system and supplies the bulk of electrons which are necessary for the reduction of protons in this organism. Moreover, it can be concluded that also the alternate respiration which is present in *Oscillatoria chalybea* [25] must be involved in the light-induced hydrogen evolution.

In future experiments we will try to isolate the hydrogenase(s) from *Oscillatoria chalybea* and to characterize the isolated enzymes. Moreover, it seems to be promising to approach a genetic analysis of the responsible genes in this filamentous non-heterocystous cyanobacterium.

ACKNOWLEDGMENT

The authors wish to thank the 'Alexander von Humboldt-Stiftung' for the generous financial support to R.A.-B.

REFERENCES

1. Boichenko, V.A. and Hoffmann, P. (1994). Photosynthetica 30 (4), 527–552.
2. Boison, G., Schmitz, O., Mikheeva, L., Shestakov, S. and Bothe, H. (1996). FEBS Lett. 394, 153–158.
3. Brass, S., Ernst, A. and Böger, P. (1992). Arch. Microbiol. 158, 422–428.
4. Gallon, J.R. (1992). New Phytol. 122, 571–609.
5. Peschek, G.A. (1979). Arch. Microbiol. 123, 81–92.
6. Villbrandt, M., Stal, L.J. and Krumbein, W.E. (1990). FEMS Microbiol.Ecol. 74(1), 59–72.
7. Adams, M.W.W. (1990). Biochim. Biophys. Acta 1020, 115–145.
8. Przybyla, A.E., Robbins, J., Menon, N. and Peck, H.D. Jr. (1992). FEMS Microbiol. Rev. 88, 109–136.
9. Bothe, H., Yates, M.G. and Cannon, F.C. (1983). Encyclopedia of Plant Physiol. 15A, Läuchli, A. and Bielski, R. (eds.), Springer: Heidelberg, Berlin, New York, pp. 241–282.
10. Böger, P. (1978). Naturwissenschaften 65, 407–412.
11. Kentemich, T., Haverkamp, P. and Bothe, H. (1990). Naturwissenschaften 77, 12–18.
12. Ernst, A., Kerfin, W., Spiller, H. and Böger, P. (1979). Z. Naturforsch. 34c, 820–825.
13. Spiller, H., Ernst, A., Kerfin, W. and Böger, P. (1978). Z. Naturforsch. 33c, 541–547.
14. Serebryakova, L.T., Medina, M., Zorin, N.A., Gogotov, I.N. and Cammach, R., (1996). FEBS Lett. 383(1–2), 79–82.
15. Tamagnini, P., Troshina, O., Oxelfelt, F., Salema, R. and Lindblad, P. (1997). Applied And Environmental Microbiology Vol. 63(5), 1801–1807.

16. Bader, K.P., Thibault, P. and Schmid, G.H. (1983). Z. Naturforsch. 38c, 778–792.
17. Bader, K.P., Thibault, P. and Schmid, G.H. (1987). Biochim. Biophys. Acta 893, 564–571.
18. Bader, K.P., Schmid, G.H., Ruyters, G. and Kowallik, W. (1992). Z. Naturforsch. 47c, 881–888.
19. Bader, K.P. and Röben, A. (1995). Z. Naturforsch. 50c, 199–204.
20. Bader, K.P. (1996). Ber. Bunsenges. Phys. Chem. 100, 2003–2007.
21. Lee, J.W., Tevault, C.V., Owens, T.G. and Greenbaum, E. (1996). Science Vol. 273, 364–367.
22. Greenbaum, E., Lee, J.W., Tevault, C.V., Blankinship, S.L. and Mets, L.J. (1995). Nature Vol. 376, 438–441.
23. Abdel-Basset, R. and Bader, K.P. (1997). submitted
24. Luo, Y.H. and Mitsui, A. (1994). Biotechnology and Bioengineering 44 (10), 1255–1260.
25. Bader, K.P. and Schmid, G.H. (1989). Biochim. Biophys. Acta 974, 303–310.

71

A STUDY OF THE CYTOPLASMIC MEMBRANE OF THE CYANOBACTERIUM *SYNECHOCOCCUS* PCC 7942 CELLS ADAPTED TO DIFFERENT NITROGEN SOURCES BY LASER DOPPLER ELECTROPHORESIS

Chantal Fresneau, Maria Zinovieva, and Bernard Arrio

Laboratoire de Bioénergétique Membranaire
ERS 571 C.N.R.S. Institut de Biochimie, Bât. 433
Université de Paris-Sud
91405 Orsay Cedex, France

1. INTRODUCTION

Plasmalemma surface represents the first sensor of environment changes. Among all its responses, the plasmalemma protein composition reflects the physiological processes involved in adaptation. This is easily perceived on the different polypeptide compositions of the *Synechococcus* PCC 7942 cytoplasmic membrane from cells adapted to nitrate or ammonium. Substitution of ammonium by nitrate in the growth medium stimulates the occurence of 45 kDa [1–3] and 126 kDa [4] proteins. The former has been already described as the nitrate binding protein from experiments carried out on a soluble truncated form [5]. The amount of these proteins is 25% [3] and 2.5% [4] of the plasmalemma proteins, respectively, in low nitrate content medium. They are absent in plasmalemma from ammonium grown cells. The amplitude of the expression-repression processes is large enough to result in surface charge density variations. Laser Doppler electrophoresis measurements are particularly designed to provide informations not only on the surface charge variations related to the adaptation process but also on the interactions between the membrane surface and ions [6–8]. Our approach in this paper is to establish the differentiation of the adapted membranes and to shed light on the nitrate ions interactions at a greater level of organisation than a modified protein. For the purpose of electrokinetic properties comparison, three conditions of adaptation were selected *i.e.* 2 mM or 175 mM nitrate and 5 mM ammonium. The vesicles obtained from these membranes were compared with regard to their interactions with different salts.

The Phototrophic Prokaryotes, edited by Peschek *et al.*
Kluwer Academic / Plenum Publishers, New York, 1999.

2. MATERIALS AND METHODS

2.1. Culture Conditions

The *Synechococcus* PCC 7942 cells were grown at 30°C in sterilized liquid medium, under gentle stirring and bubbling with 2% CO_2 enriched air and illuminated by 50 $\mu E\ m^{-2}\ s^{-1}$ white light. The basal medium was prepared according to Van Allen [9] and Herdmann [10], without nitrogen source. Ammonium-containing medium was obtained by addition of 5 mM NH_4Cl to the basal medium. Nitrate-containing medium was prepared by addition of 2 mM or 175 mM $NaNO_3$ to the basal medium. Prior to harvesting, the non-contamination of each flask of culture was controlled on minimum medium agar plates incubated in the dark.

2.2. Membrane Preparation and Characterization

The preparation and purification of the plasmalemma vesicles, the protein determination of the purified membranes and the polypeptide analysis by SDS-PAGE (sodium dodecylsulphate-polyacrylamide gel electrophoresis) of the plasmalemma vesicles were performed using the techniques already described [4,11]. The lipid and fatty acid compositions of the three types of vesicles were determined by TLC and GLC, respectively [12,13].

2.3. Laser Doppler Electrophoresis Measurements

Purified plasmalemma vesicles from all *Synechococcus* adapted cells were incubated for 1 h 30 at 4°C in a large volume of 5 mM Hepes-NaOH (N-2-hydroxyethylpiperazine-N,-2-ethane-sulphonic acid) buffer pH 7.4. These suspensions were centrifuged for 1 h at 100 000 × g at 4°C and the pellets were resuspended in the same buffer at a final concentration of 2 mg/ml. Samples of 100 µg protein were incubated at 4°C in 1 ml of 5 mM Hepes-NaOH buffer pH 7.4 containing (or not) 0.2, 0.4, 0.8, 1.2, 1.6, 2.0 and 2.5 mM $NaNO_3$, KNO_3, NaCl, KCl or NH_4Cl. After incubation for 45 min, electrophoretic mobility and particle size measurements were performed on a laser Doppler electrophoresis apparatus described previously [6, 7]. Electrophoretic mobility measurements were carried out on 3 different membrane preparations and each measurement was repeated 4 times. Mobility values are always negative, the minus signs are not represented in the figure ordinates.

3. RESULTS

As shown in Fig. 1, the comparison between the three media clearly indicates that the 45 kDa and the 126 kDa proteins are exclusively expressed in the membranes from nitrate grown cells. The amount of these proteins was inversely related to the nitrate concentration.

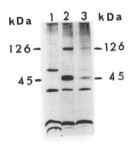

Figure 1. Influence of the nitrogen source on the polypeptide composition of cytoplasmic membranes. Cytoplasmic membranes were prepared from 5 mM ammonium-grown cells (1) and from 2 mM (2) or 175 mM (3) nitrate-grown cells. Samples containing 5 µg proteins were subjected to SDS PAGE on a 8–25% polyacrylamide gradient gel and stained by silver nitrate using a Phast System (Pharmacia). Broad-range molecular mass standards from Bio-Rad were run simultaneously.

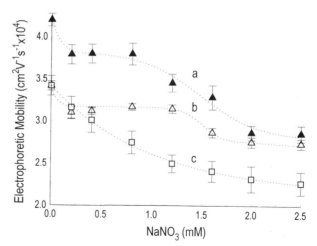

Figure 2. Effect of nitrate and chloride on the electrophoretic mobility of plasmalemma vesicles from different nitrate adapted cells. Effect of $NaNO_3$ on the electrophoretic mobility of plasmalemma vesicles from 175 mM $NaNO_3$ (a) and 2 mM $NaNO_3$ (b) adapted cells. The same curves were obtained when KNO_3 was added. Effect of NaCl on the electrophoretic mobility of plasmalemma vesicles from 2 mM adapted cells (c). The same curve was obtained when KCl was added.

Indeed, the 45 kDa protein was 25% [3] of the membrane proteins from 2 mM nitrate grown cells and 3 times less of those from 175 mM (Fig. 1, lanes 2 and 3). The 126 kDa protein was estimated at about 2.5% and less than 1% of the plasmalemma proteins from 2 mM and 175 mM nitrate grown cells, respectively (Fig. 1, lanes 2 and 3). The repression effect of ammonium ions is shown in Fig. 1, lane 1.

The electrophoretic mobility of the membrane vesicles from nitrate grown cells was dependent on the medium nitrate content. Fig. 2 a and b shows that the mobility of the vesicles from 175 mM adapted cells was greater than that of the 2 mM adapted ones. When $NaNO_3$ was added, a more or less extended zone of constant mobility was observed from 0.2 mM but above 1.2 mM both curves decreased and reached the same level at 2.5 mM. This was reproducible with potassium nitrate. When nitrate was replaced by chloride, asymptotic decay of mobility was observed whatever the cation (Fig. 2, c).

Unlike the membranes from nitrate grown cells, those from ammonium had the same variation of mobility in the presence of $NaNO_3$ (Fig. 3, a) or NaCl (data not shown). How-

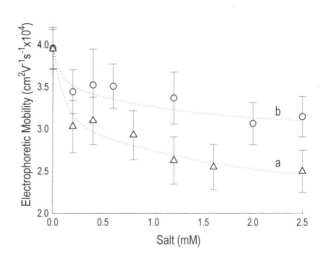

Figure 3. Effect of $NaNO_3$ (a) and NH_4Cl (b) on the electrophoretic mobility of plasmalemma vesicles from 5 mM NH_4Cl adapted cells.

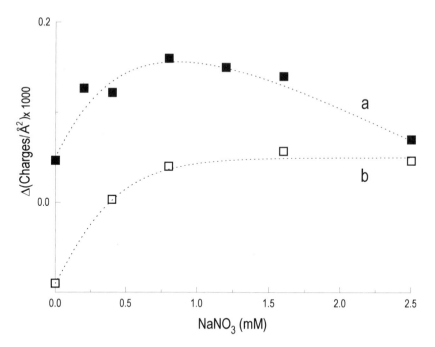

Figure 4. Surface charge density variation of plasmalemma vesicles from 175 mM (a) and 2 mM (b) nitrate adapted cells. The data are the difference between the charge density of each type of membranes and that of membranes from ammonium grown cells.

ever, these membranes were more negatively charged when NH_4Cl (Fig. 3, b) was added instead of $NaNO_3$ or NaCl.

Fig. 4 illustrates the difference of charge density between the reference, *i.e.* vesicles from ammonium grown cells (no nitrate binding protein) and vesicles from nitrate grown cells. The membranes from the low nitrate medium had a lower charge density than those from high nitrate medium. The aspect of saturation curve is more pronounced with the membranes from cells grown in low nitrate (Fig. 4, b). Assuming a saturation of nitrate binding sites, the dissociation constant would be approximately 0.2 mM.

The different vesicles presented the same mean diameter: 250 ± 20 nm and it was independent of the salt concentrations used to measure the mobility. The growth medium did not induce significant variations of the lipids and fatty acids composition.

4. DISCUSSION

The electrophoretic mobility and the charge density of the plasmalemma vesicles from *Synechococcus* PCC 7942 was sensitive not only to the adaptation process to nitrate and ammonium but also to the nature of the interactions between ions and the surface. These interactions are usually distinguished depending on wether the ions are indifferent or involved in specific adsorption or binding [for review see 14].

In the absence of added salt, the plasmalemma vesicle mobility was inversely related to the expression of the nitrate transport proteins. When they were absent or slightly expressed *i.e.* in the plamalemma of cells adapted to ammonium or high nitrate concentration, the mobility values were high and almost identical. Conversely, when they were greatly expressed *i.e.* in the plasmalemma from cells adapted to 2 mM nitrate the mobility value was lower. Normally, a higher mobility and surface charge density are expected when the protein content increases, the inverse effect was observed. A specific adsorption of cations could explain the low level of negative charges observed on vesicles from cells grown in a low nitrate medium. But the addition of NaCl or KCl to these vesicles resulted in an asymptotic decay of the mobility typical of indifferent salts which reduce the mobility as the ionic strength increases. The possible contribution of vesicle size variations and lipid composition modifications were experimentally excluded. Indeed, the 250 ± 20 nm mean diameter remained constant and the lipid composition was not significantly different. To explain the paradox of a low charge density accompanying the full expression of the nitrate binding protein, it was assumed that positively charged segments of the 45 kDa polypeptide, rising upward the membrane surface, might reduce the net charge of the vesicles. This view is supported by the fact that the 45 kDa sequence [15] presents several positively charged segments: 162–188 (8 Lys), 264–288 (5 Lys) and 314–364 (5 Lys, 4 Arg and 1 His). Although the spatial structure of this protein is still unknown, the eventual existence of positive zones, at the membrane surface, brought by a protein which represents 25% of the protein content could account for the charge density differences between the membranes. As expected from this assumption, in the absence of the binding protein *i.e.* membranes from ammonium grown cells, an identical electrophoretic mobility was observed either by NaNO$_3$ addition or NaCl. But, in the presence of ammonium chloride the mobility of these membranes was greater, indicating adsorption or binding on proteins of the ammonium transport system [16,17].

Thus, the nitrate effect observed was related to the presence of specific proteins related to adaptation and the estimate of the dissociation constant was at least hundred times greater than the value obtained on the soluble truncated protein [5]. This comparison is limited by the difference of the materials used and subsequent studies on reconstituted binding systems could provide complementary informations on the nitrate interactions with the plasmalemma.

REFERENCES

1. Madueno, F., Vega-Palas, M.A., Flores, E. and Herrero, A., (1988) FEBS Lett., 239, 289–291.
2. Omata, T., Ohmori, M., Arai, N. and Ogawa, T., (1989) Proc. Natl. Acad. Sci. USA, 86, 6612–6616.
3. Sivak, M.N., Lara, C., Romero, J.M., Rodriguez, R. and Guerero, M.G., (1989) Biochem. Biophys. Res. Commun., 158, 257–262.
4. Zinovieva, M., Fresneau, C. and Arrio, B., (1997) FEBS Letters 416, 179–182.
5. Maeda, S.-I. and Omata, T., (1997) J. Biological Chem. 272, 3036–3041.
6. Arrio, B., Johannin, G., Carrete, A., Chevallier, J. and Brèthes, D., (1984) Arch. Biochem. Biophys., 228, 220–229.
7. Rivière, M.E., Johannin, G., Gamet, D., Molitor, V., Peschek, G.A. and Arrio, B., (1988) Methods Enzymol.,167, 691–700.
8. Rivière, M.E., Missiakine, E., Barot, R., Nobrega, T., Fresneau, C., Tlaskalova-Hogenova, H., Dumitrescu, M. and Arrio, B., (1995) Cell. Mol. Biol., 41, 289–296.
9. Van Allen, R., (1968) J. Physiol., 4, 1–4.
10. Herdmann, M., Delany, S.F. and Carr, N.G., (1973) J. Gen. Microbiol., 79, 233–237.
11. Fresneau, C., Rivière, M.E. and Arrio, B., (1993) Arch. Biochem. Biophys., 306, 254–260.
12. Trémolière, A. and Lepage, M., (1971) Plant Physiol., 47, 329–334.
13. Sato, N. and Murata, N., (1988) Methods Enzymol., 167, 251–259.

14. Hunter, R.J., (1981) in : Zeta potential in colloid science, (Ottewill R.H. and Rowell R.L., Eds) Academic Press, London.
15. Omata, T., (1991) Plant Cell Physiol., 32, 151–157.
16. Boussiba, S. and Gibson, J., (1991) FEMS Microbiol. Rev., 88 ,1–14.
17. Kashyap, A.K., Shaheen, N. and Prasad, P., (1995) J. Plant Physiol. 145, 387–389.

72

PEPTIDE SYNTHETASE GENES OCCUR IN VARIOUS SPECIES OF CYANOBACTERIA

Elke Dittmann,[1] Guntram Christiansen,[1] Brett A. Neilan,[2] Jutta Fastner,[3] Rosmarie Rippka,[4] and Thomas Börner[1‡]

[1]Institut für Biologie (Genetik)
Humboldt Universität
Chausseestr. 117, 10115 Berlin, Germany
[2]School of Microbiology and Immunology
The University of New South Wales
Sydney, 1051 NSW, Australia
[3]Institut für Wasser-Boden und Lufthygiene
Corrensplatz 1, 14195 Berlin, Germany
[4]Institut Pasteur
Physiologie Microbienne
75724 Paris Cedex 15, France

1. INTRODUCTION

Bloom-forming cyanobacteria are known to produce a variety of bioactive peptides. Among these secondary metabolites, the potent hepatotoxin microcystin has been most extensively investigated. More than 50 isoforms of this heptapeptide are known, sharing the structure cyclo(-D-Ala-L-X-D-MeAsp-L-Z-Adda-D-Glu-Mdha), where X and Z are variable L-amino acids, Adda is (2S,3S,8S,9S)-3-Amino-9-methoxy-2,6,8-trimethyl-10-phenyldeca-4,6-dienoic acid, D-MeAsp is D-*erythro*-β-*iso*-aspartic acid, and Mdha is N-methyl-dehydroalanine. Microcystins were shown to be specific inhibitors of the eukaryotic protein phosphatases 1 and 2A, causing liver damage in humans and livestock [1]. Other peptides produced by cyanobacteria include the depsipeptides cyanopeptolin and micropeptin, the thricyclic microviridins, anabaenapeptolins and aeruginosins, with most of them being protease inhibitors [2].

Small size, cyclic structure and the content of unusual amino acids suggest that these peptides are synthesized nonribosomally via peptide synthetases. These multifunctional proteins are composed of homologous modules. Each of these synthetase units contain specific functional domains for recognition, aminoacyl-adenylation, thioesterification, and modification of its amino acid substrate, as well as for elongation of the growing peptide

The Phototrophic Prokaryotes, edited by Peschek *et al.*
Kluwer Academic / Plenum Publishers, New York, 1999.

product [3]. These domains exhibit a high degree of sequence conservation [4]. We have previously reported that both toxic and non-toxic strains of the cyanobacterium *M. aeruginosa* contain DNA sequences homologous to known peptide synthetase genes and that they differ in their complement of peptide synthetase genes [5]. Furthermore a 2982 bp DNA fragment was isolated, which was specific for toxic strains. As the contribution of the gene product to the production of non-microcystin peptides could be not discounted, we had to establish a procedure for genetic manipulation of *Microcystis* [6]. We could show that transformation of a bloom-forming cyanobacterium using natural competence is possible. The obtained mutants were unable to produce microcystin, whilst still able to produce other small peptides (cyanopeptolins) [6,7]. Therefore the *mapep1* containing peptide synthetase gene cluster was named *mcy* gene cluster. Here we describe further investigations aimed to answer the following questions: Do cyanobacteria in general possess peptide synthetase genes or is the presence of these genes limited to a few species? Is the occurrence of the *mcy* gene cluster limited to *Microcystis aeruginosa*?

2. MATERIALS AND METHODS

2.1. Cyanobacterial Strains, Culturing, and DNA Isolations

Cyanobacterial strains were grown at 25°C in BG-11 [8] at a light intensity of 40 μE/m^2/sec.

Genomic DNA was extracted as described by Franche and Damerval [9].

2.2. PCR

PCR was performed using Goldstar thermostable DNA-polymerase (Eurogentec, Belgium) according to manufacturer's instructions. Primer pairs Tox1P (5´-CGATTGTTACTG-ATACTCGCC-3´) and Tox1M (5´-TAAGCGGGCAGTTGCTGC-3´) and Tox2P (5´-GGAACAAGTTGCACAGAATCCGC-3´) and Tox2M (5´-CCAATCCCTATCTTAA-CACAGTAACTCGG-3´) were used for detection of the *mcyB* gene in the reference strains of the Pasteur Culture Collection. Primers MTF2 (5´-GCNGG(C/T)GG(C/T)GCN-TA(C/T)GTNCC-3´) and MTR2 (5´-CCNCG(A/C/T)AT(C/T)TTNAC(C/T)TG-3´) were used to detect peptide synthetase genes in general [10].

2.3. HPLC

Lyophilized cells (up to 20 mg dry weight) were extracted twice with 1.5 ml of methanol for 1 hr at room temperature. After centrifugation (13000 × g, 20 min) the supernatant was removed and lyophilized. Unquantified measurement of microcystin content was performed by HPLC according to the method of Lawton *et al.* [11] using a LiChroCART RP18 column (Serva, Heidelberg, Germany).

3. RESULTS

3.1. Inactivation of *M. aeruginosa* PCC 7806 *mapep1* Gene

Recently, we reported on the first transformation of a bloom-forming cyanobacterium, *M. aeruginosa* PCC 7806 [6]. It was possible to knock out a peptide synthetase gene, that was hypothesised to contribute to microcystin biosynthesis [5]. Insertional inactivation of this gene cluster was successful using a construct that was essentially the pGEM-vector and peptide synthetase coding sequence interrupted by a chloramphenicol resistence cassette (pACYC184) [12]. The derived mutant cells were shown to lack the characteristic molecular ion peaks for microcystin-LR and (D-Asp^3)-microcystin-LR in MALDI analyses and moreover specific enzymatic activities necessary for microcystin biosynthesis [6]. Figure 1 shows the results of HPLC analyses from both wild-type and mutant cell extracts using the method of Lawton et al. [11]. Identification of microcystins was based on specific retention times and Adda spectra (absorption maximum at 239 nm). Wild-type extracts revealed previously described major peaks for MCYST-LR and (D-Asp^3) MCYST-LR with minor peaks detected for MCYST-RR and MCYST-YR [13]. In the mutant lines all microcystins were absent and no Adda spectrum was detectable. These results clearly indicated, that the mutated gene encodes a peptide synthetase involved in microcystin synthesis. This gene was given the designation *mcyB* [6].

3.2. Distribution of Peptide Synthetase Genes in Various Cyanobacterial Species

Based on this strong evidence for the contribution of *mcyB* to microcystin biosynthesis we have been looking for the contribution of the *mcy* gene cluster [6] in different cyanobacterial species. For this purpose reference strains of the Pasteur Culture Collection of species belonging to sections I and III were checked for the presence of peptide synthetase genes in general and for the presence of the *mcy* gene cluster. As adenylate-forming domains of peptide synthetases are well conserved, there was a good chance to detect peptide synthetase genes using PCR amplification with the degenerate primer set MTF2/MTR2. These primers were designed to bind at conserved motifs repeating in each module and encoding part of the adenylate-forming domain. Interestingly, all *Microcystis* strains belonging to section I and most of the *Oscillatoria* strains, belonging to section III were shown to contain peptide synthetase genes, even when they were non-toxic. In addition *Chamaesiphon* PCC 6605, *Microcoleus* PCC 7420 and *Spirulina* PCC 6313 also gave positive signals. Strains from the other species tested, including *Synechocystis* and *Synechococcus,* did not yield a PCR product. Two more primer pairs (Tox1P/Tox1M, Tox2P/Tox2M) were used to specifically detect the *mcyB* gene [6]. These primers were designed to bind outside the conserved motifs within the *mcyB* sequence. The results are listed in Tables 1 and 2. We were able to amplify fragments of the *mcyB* gene from DNA of all known microcystin producers, with one or both of the specific primer pairs and also from strains *M. aeruginosa* PCC 9355 and PCC 9443. The PCR data for the *Microcystis* strains were supported by Southern hybridisation experiments (data not shown).

4. DISCUSSION

The successful transformation of *Microcystis* has allowed us to confirm our former hypothesis, that *mapep1* (now called *mcyB*) is involved in the biosynthesis of microcystin [6]. The mutants obtained can now be applied to determine the function of microcystins in the cyanobacterial cells. Furthermore, it is now possible to distinguish between genes for microcystin and non-microcystin peptide synthetases in the cyanobacteria using DNA techniques.

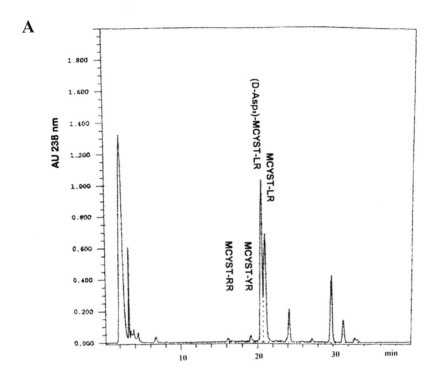

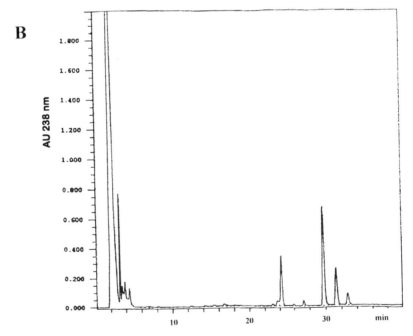

Figure 1. HPLC spectra of wild-type (A) and *mcyB* mutant (B) methanol extracts [11]. Characteristic peaks for microcystin-LR, (D-Asp³)-microcystin-LR , microcystin-YR and microcystin-RR are indicated and lack in the mutant cell extracts.

Table 1. Distribution of peptide synthetase genes in cyanobacterial section I

Genus, PCC no.	Toxin present (catalog data)	Primer[a] MTF2/MTR2	Primer[b] Tox1P/Tox1M	Primer[b] Tox2P/Tox2M
Chamaesiphon 6605	0	+	−	−
Cyanothece 7424	0	−	−	−
Cyanothece 7425	0	−	−	−
Gloeobacter 7421	0	−	−	−
Gloeocapsa 73106	0	−	−	−
Gloeothece 6501	0	−	−	−
Microcystis 7005	0	+	−	−
Microcystis 7806	+	+	+	+
Microcystis 7813	+	+	+	+
Microcystis 7820	+	+	+	+
Microcystis 7941	+	+	+	+
Microcystis 9354	+	+	+	+
Microcystis 9355	0	+	+	+
Microcystis 9432	−	+	−	−
Microcystis 9443	0	+	+	+
Microcystis 9603	0	+	−	−
Synechococcus 6301	0	−	−	−
Synechococcus 6307	0	−	−	−
Synechococcus 7002	0	−	−	−
Synechocystis 6308	0	−	−	−
Synechocystis 6701	0	−	−	−
Synechocystis 6803	0	−	−	−
Synechocystis 7008	0	−	−	−

[a]Primers MTF2/MTR2 were used to detect peptide synthetase genes in general.
[b]Tox-Primers were used to detect the *mcyB* gene.

Table 2. Distribution of peptide synthetase genes in cyanobacterial section III

Genus, PCC no.	Toxin present (catalog data)	Primer[a] MTF2/MTR2	Primer[b] Tox1P/Tox1M	Primer[b] Tox2P/Tox2M
Arthospira 7345	0	−	−	−
Leptolyngbya 6306	0	−	−	−
Microcoleus 7420	0	+	−	−
Oscillatoria 6304	0	−	−	−
Oscillatoria 6412	0	+	−	−
Oscillatoria 6602	0	+	−	−
Oscillatoria 7112	0	+	−	−
Oscillatoria 7515	0	+	−	−
Oscillatoria 7805	+	+	+	+
Oscillatoria 7811	+	+	−	+
Oscillatoria 7821	+	+	−	+
Oscillatoria 9214	−	−	−	−
Pseudoanabaena 6802	0	−	−	−
Pseudoanabaena 7429	0	−	−	−
Spirulina 6313	0	+	−	−

[a]Primers MTF2/MTR2 were used to detect peptide synthetase genes in general.
[b]Tox-Primers were used to detect the *mcyB* gene.

Here we describe the investigation of the peptide synthetase gene content of the well characterized axenic strains of the Pasteur Culture Collection belonging to sections I and III of the cyanobacterial phylum using three different PCR approaches. Approximately half of the strains examined were shown to contain peptide synthetase genes. The distribution of these genes does not reflect proposed phylogenetic relations among the investigated species [14,15]. Interestingly, of section I only the genera *Microcystis* and *Chamaesiphon* were shown to contain peptide synthetase genes but none of the other species. For section III, strains of *Oscillatoria, Spirulina* and *Microcoleus* gave PCR products of the correct length as predicted from other peptide synthetase genes. For most of the *Microcystis* strains and a few of the *Oscillatoria* strains these data were obviously attributed to the presence of microcystin synthetases, as indicated from both the PCC catalogue data and from the PCR data obtained with the Tox primers. In all the other cases it appears that we now have evidence for unknown peptide synthetases. In the last years great strides have been made in detecting peptide metabolites in cyanobacteria [16]. Strains of the Japanese NIES collection of cyanobacteria have been shown to produce a number of peptide metabolites containing unusual amino acids, most of them being cyclic [2,7,17]. It is feasible that these are also products of peptide synthetases. One could expect similar metabolites of strains of the Pasteur Culture Collection, although this has yet to be shown. Many strains, like *Microcystis* PCC 7806, contain more than one non-ribosomally made peptide [6] and only the knock out of the peptide synthetase genes present would allow inference between the peptide and the DNA data. Nevertheless, the ability to produce such peptides is obviously limited to certain cyanobacterial species. The presence of the *mcyB* gene correlated with the permanent presence of gas vesicles, except for PCC 7813. However, *mcyB* could not be shown to be the exclusive peptide synthetase gene in the microcystin-producing species. One could speculate, whether the peptides produced by the nontoxic strains play a role similar to that of the microcystins and whether this function could be related to the gas vacuolation. Moreover *Chamaesiphon, Microcoleus* and *Spirulina* are three genera shown to contain peptide synthetase genes which have not been known to produce any characteristic small bioactive peptide.

ACKNOWLEDGMENTS

This work was supported by the Deutsche Forschungsgemeinschaft and the European Commission (BASIC project) to T.B.

REFERENCES

1. W.W. Carmichael, Sci. Am. 270 (1994) 78–86.
2. K. Ishida, M. Murakami, H. Matsuda, K. Yamaguchi, Tetrahedron Lett. 36 (1995) 3535–3538.
3. H. Kleinkauf and H. von Döhren, Eur. J. Biochem. 236 (1996) 335–351.
4. T. Stachelhaus and M.A. Marahiel, FEMS Microbiol. Lett. 125 (1995) 3–14.
5. K. Meißner, E. Dittmann, and T. Börner, FEMS Microbiol. Lett. 135 (1996) 295–303.
6. E.Dittmann, B.A. Neilan, M. Erhard, H. von Döhren and T.Börner, Mol. Microbiol., in press.
7. C. Martin, L. Oberer, T. Ino, W.A. König, M. Busch and J. Weckesser, J. Antibiot. 46 (1993) 1550–1556.
8. R. Rippka, J. Deruelles, J.B. Waterbury, M. Herdman and R.Y. Stanier, J. Gen. Microbiol. 111 (1979) 1–61.
9. C. Franche and T. Damerval, Methods Enzymol. 167 (1988) 803–808.
10. B. Neilan *et al.*, manuscript in preparation
11. L.A. Lawton, C. Edwards and G.A. Codd, Analyst 119 (1994) 1525–1530.
12. A.C. Chang, S.N. Cohen, J. Bacteriol. 134 (1978) 1141–1156.

13. R. Dierstein, I. Kaiser, J. Weckesser, U. Matern, W.A. König and R. Krebber, System. Appl. Microbiol. 13 (1990) 86–91.

14. Willmotte, A. (1994) in: The Molecular Biology of Cyanobacteria (D. Bryant, Ed.) pp. 1–25, Kluwer Academic, Dordrecht.

15. B.A. Neilan, D. Jacobs, T. Del Dot, L.L. Blackall, P.R. Hawkins, P.T. Cox, A.E. Goodman, Int. J. Syst. Bacteriol. 47 (1997) 693–697.

16. M. Erhard, H. von Döhren and P. Jungblut, Nature Biotechn. 15 (1997) 906–910.

17. T. Okino, H. Matsuda, M. Murakami, K. Yamaguchi, Tetrahedron Lett. 51 (1995) 10679–10686.

73

WIDENING PERCEPTIONS OF THE OCCURRENCE AND SIGNIFICANCE OF CYANOBACTERIAL TOXINS

G. A. Codd, C. J. Ward, K. A. Beattie, and S. G. Bell

Department of Biological Sciences
University of Dundee
Dundee DD1 4HN, United Kingdom

1. INTRODUCTION

Cyanobacterial toxins (CTX) are associated annually with episodes of animal-, bird- or fish-poisonings and, almost annually, with examples of human illness, somewhere or other, in the world. However, recognition of the occurrence and impact of CTX has been, and remains, limited among various relevant groups. These include biologists and limnologists, environmental- and health-care professionals and in the water industry. Constraints have included: inadequate investigation and reporting protocols for suspected cyanobacterial intoxications; a lack of adequate and widely available CTX analysis methods; and a lack of awareness among non-professional water-users of the hazards of cyanobacterial blooms and of signs of cyanobacterial poisoning[1,2]. These shortcomings have been partly addressed by education and information programmes, the introduction of sensitive and quantitative toxin analysis methods and by the performance of state and national CTX surveys. Understanding of the health hazards presented by CTX is gradually being advanced from several sources. These include: fundamental studies of the toxicology of purified CTX, ranging from molecular *in vitro* studies to toxicity trials with animals[3,4] and investigations, albeit limited, of animal poisonings and human health problems associated with—or attributed to—CTX[5-8]. Data on the toxicity of cyanobacterial microcystin hepatotoxins have been used this year in risk assessments for the derivation of a guideline value for microcystin-LR levels in drinking water. Examples of the effects of CTX on human health continue to emerge, a recent example being the deaths of haemodialysis patients exposed to microcystin hepatotoxins in Caruaru, Brazil in 1996[9-11]. Implications and consequences of the Caruaru incident, and their implications are discussed here. We also discuss the CTX exposure routes, the widening understanding of which is benefitting from basic and applied research on the effects and accumulation of CTX in diverse organisms.

The Phototrophic Prokaryotes, edited by Peschek *et al.*
Kluwer Academic / Plenum Publishers, New York, 1999.

2. EARLY AWARENESS OF CYANOBACTERIAL MASS GROWTHS AND OF THE (LIKELY) EFFECTS OF CYANOBACTERIAL TOXINS

Those who investigate cyanobacterial blooms, CTX and associated environmental and health problems, are asked by members of the general public, and by affected water-users, whether cyanobacterial blooms and CTX are "new" and whether they are increasing in distribution, frequency and abundance. It can be argued that blooms merely seem to be more common because researchers and water authorities are surveying and monitoring more often and more widely, and that statistical, data are improving. However, evidence indicates that freshwater and estuarine blooms of cyanobacteria are indeed increasing in distribution, frequency and duration[1,12,13].

There are indications that earlier societies were familiar with mass growths of phytoplankton, likely including cyanobacteria, which caused the discolouration of lakes and rivers and the formation of scums. Such growths were described in Europe's River Dnieper in AD77 by Pliny the Elder and were clearly apparent in 1188 in Llangorse Lake in Wales[14,15]. A monastery, founded in 1175 at Soulseat Loch, Scotland, became known as the *Monasterium Viridis Stagni*, suggested to be "traceable to the green appearance of the water from time to time by the presence of spore-like vegetable growths"[12,16].

Early awareness of the toxic actions of cyanobacterial blooms is apparent from the folk-lore of Canadian Indians who recognised "poison lakes" where dead animals were found, and among Australian Aborigines who used water filtered by bank-side filtration for drinking, rather than directly from the billabong containing cyanobacterial scum[12,17]. "Sick lakes" were recorded in Denmark in 1833 where cattle and fish deaths occurred during periods of surface accumulations of presumed cyanobacteria[18,19]. Awareness and a reporting system existed along stretches of the Lower River Murray in Australia between about 1878 and at least 1888, to warn pastoralists and farmers to protect their livestock from drinking cyanobacterial scums[20]. Livestock deaths and the experimental intoxication of sheep were attributed to the ingestion of *Nodularia spumigena* scum by George Francis at that time[21].

3. PROPERTIES OF CYANOBACTERIAL TOXINS

CTX are conveniently categorised according to the main organs, cells or physiological systems affected in animals and *in vitro* toxicity studies. Among the neurotoxins, anatoxin-a and a methylated form, homoanatoxin-a, are postsynaptic cholinergic nicotine agonists and neuromuscular blocking agents[4,6,22] (Figure 1a). Anatoxin-a(s) is a guanidine methyl phosphate ester (Figure 1b) and potent inhibitor of cholinesterases[23]. As with other organophosphate neurotoxins, anatoxin-a(s) causes hypersalivation in mammals, denoted by the suffix "s" in the toxin's name. A range of sodium channel-blocking neurotoxins of the paralytic shellfish poison (PSP) family is also becoming recognised in cyanobacteria (e.g. Figure 1c). These include saxitoxin, neosaxitoxin, C toxins and GTX toxins[4,22].

Heptatotoxins are found to account for the toxicity of several cyanobacterial blooms, with the cyclic heptapeptide microcystins and pentapeptide nodularins being most often reported. These toxins cause the disruption of liver architecture and function. External signs of poisoning in animals include weakness, recumbency, pallor, vomiting and diarrhoea. More than 50 microcystin variants are known. Variations on the general structure (Figure 1d) are provided by substitutions of L-amino acids at positions 2 and 4

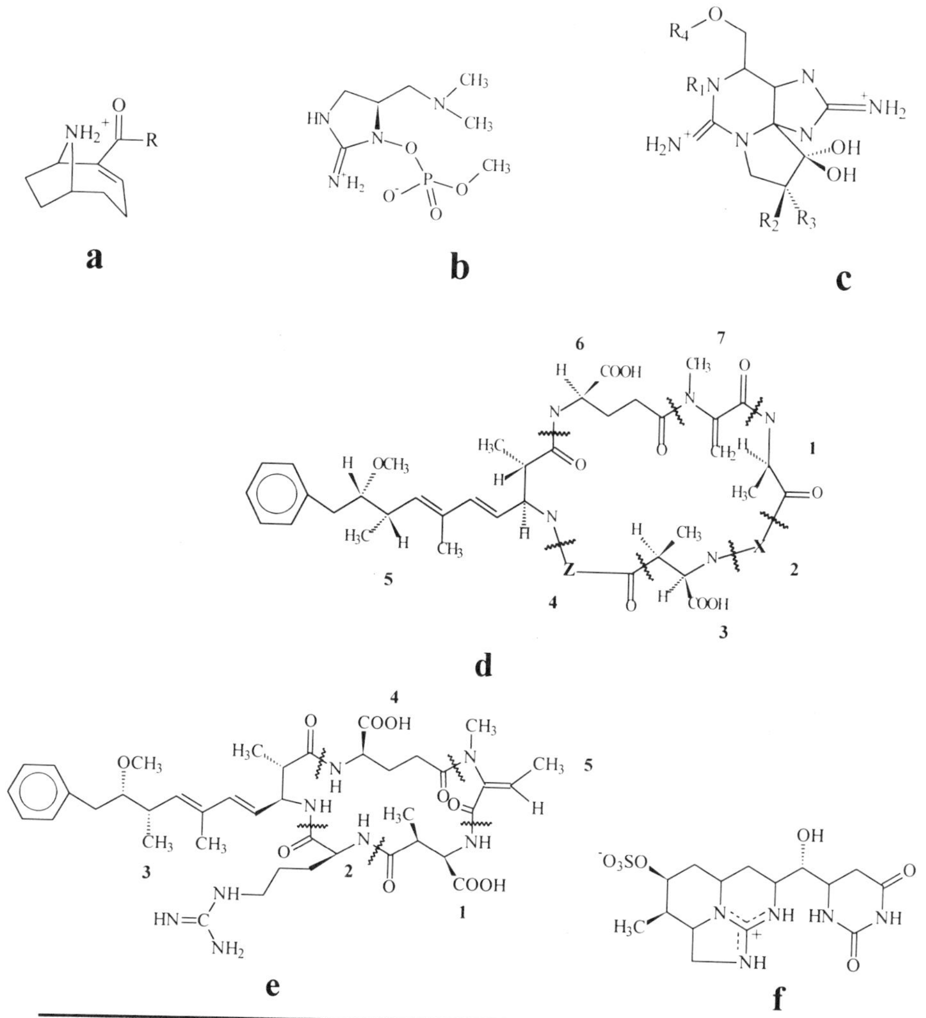

PSP toxins		R_1	R_2	R_3	R_4
C1		H	H	OSO_3^-	$CONHSO_3^-$
C2		H	OSO_3^-	H	$CONHSO_3^-$
C3		OH	H	OSO_3^-	$CONHSO_3^-$
C4		OH	H	H	$CONHSO_3^-$
Gonyautoxin I	(GTX1)	OH	OSO_3^-	OSO_3^-	$CONH_2$
Gonyautoxin II	(GTX2)	H	OSO_3^-	OSO_3^-	$CONH_2$
Gonyautoxin III	(GTX3)	H	H	H	$CONH_2$
Gonyautoxin IV	(GTX4)	OH	H	H	$CONH_2$
Gonyautoxin V	(GTX5)	H	H	H	$CONHSO_3^-$
Gonyautoxin VI	(GTX6)	OH	H	H	$CONHSO_3^-$
Decarbamoyl GTX2	(dc-GTX2)	H	H	OSO_3^-	H
Decarbamoyl GTX3	(dc-GTX3)	H	OSO_3^-	H	H
Saxitoxin	(STX)	H	H	H	$CONH_2$
Neosaxitoxin	(NEO)	OH	H	H	$CONH_2$
Decarbamoyl saxitoxin	(dc-STX)	H	H	H	H

Figure 1. Structures of some cyanobacterial toxins. a, anatoxin (R = CH_3) and homoanatoxin (R = CH_2CH_3); b, anatoxin-a(s); c, PSP toxins, see inset table for explanation; d, general structure of microcystins, X and Y are variable L-amino acids; residue 5 is Adda; e, nodularin; f, cylindrospermopsin.

in the ring, amino acid demethylation at positions 3 and/or 7 and by further modifications or substitutions at the remaining amino acids[4]. The amino acid Adda, (2S,3S,8S,9S)-3-amino-9-methoxy-2,6,8-trimethyl-10-phenyldeca-4,6-dienoic acid (Figure 1d), or O-acetyl-O-demethylAdda, or the stereoisomer (6Z)-Adda, are a consistent feature of microcystins and of the cyclic heptapeptide nodularins (Figure 1e). Fewer variations of nodularin are known and include D-asparate nodularin, O-demethylAdda nodularin and L-valine nodularin[4]. The valine variant of nodularin, also known as motuporin, was isolated from the marine sponge *Theonella swinhoei*[24], raising the possibility that CTX may accumulate in marine invertebrates. Adda and its derivatives are necessary for the toxicity of microcystins and have not been found, apparently, other than in microcystins and nodularins. This raises a further prospect of using Adda and its derivatives as markers for the presence of microcystins, nodularins, and their biosynthetic precursors and breakdown products, in environmental and clinical material.

Microcystins and nodularins irreversibly inhibit the protein phosphatases PP1 and PP2A of animals and plants[25,26]. The implications of this mode of inhibition, which occurs *in vitro* and *in vivo*, are considerable since PP1 and PP2A have widespread functions in the regulation of genetic, developmental, metabolic and physiological processes in mammals and plants[25-27]. Liver cells are particularly susceptible to damage by microcystins *in vivo* and *in vitro* because the toxins enter hepatocytes by the bile-acid carrier, a broad-specificity anion transport system[3]. Examples are accumulating that microcystins can also adversely affect organisms and tissues, without the involvement of a bile-acid transport system. The repeated topical application of microcystin to *Phaseolus vulgaris* (French Bean) primary leaves, at a concentration of the toxin which can occur in eutrophic waterbodies (20 µg per l), causes a progressive, and eventually, irreversible inhibition of whole-leaf CO_2 fixation, with leaf necrosis at higher microcystin exposure levels[28].

Cylindrospermopsin (Figure 1f) is a cytotoxic guanidine alkaloid and inhibitor of protein synthesis[29,30]. This toxin causes necrotic injury to the mammalian liver, adrenals, kidneys, lungs and intestine. It is likely that a range of cylindrospermopsin variants and of cylindrospermopsin-producer organisms will emerge. Other CTX, including skin irritants and gastrointestinal toxins, are reviewed elsewhere[4,8,22].

4. OCCURRENCE OF CYANOBACTERIAL BLOOM TOXICITY AND CYANOBACTERIAL TOXINS

Cyanobacterial bloom toxicity appears to be a global phenomenon, judging from reports of toxicity assessments and CTX assays after poisoning episodes, and of local, state or national monitoring programmes (Table 1). In most cases, findings refer to blooms and scums of planktonic cyanobacteria in freshwaters, although CTX (including anatoxin-a and microcystins) have also been found in mats of benthic cyanobacteria[6,34,35] in freshwaters and in terrestrial habitats[36]. Marine and estuarine blooms of cyanobacteria (e.g. *Nodularia, Trichodesmium*) have commonly been found to be toxic when tested[37,38].

Individual blooms and scums of potentially toxigenic cyanobacteria range widely in their toxicity to mammals according to intraperitoneal mouse bioassay, although in all cases where a regional/national survey has been performed, a high incidence of acute toxicity has been found (Table 2). The high probability of an individual sample being acutely toxic to mice, argues for the use of a precautionary principle in cyanobacterial bloom-management and water-treatment, i.e that a bloom, scum or mat should be assumed to be toxic, unless analysed and found to be otherwise[7,8].

Table 1. Geographical findings of toxic cyanobacterial blooms, scums, or mats

Europe	Belgium, Czech Republic, Denmark, Estonia, Finland, France, Germany, Greece, Hungary, Ireland, Italy, Latvia, Netherlands, Norway, Poland, Portugal, Russia, Slovakia, Slovenia, Spain, Sweden, Switzerland, Ukraine, United Kingdom
Americas	Argentina, Bermuda, Brazil, Canada, Chile, Mexico, USA (at least 27 states), Venezuela
Middle East and Asia	Bangladesh, India, Israel, Japan, Jordan, Malaysia, Peoples' Republic of China, Saudi Arabia, Sri Lanka, South Korea, Thailand
Australasia	Australia (New South Wales, Queensland, South Australia, Tasmania, Western Australia), New Caledonia, New Zealand
Africa	Egypt, Ethiopia, Morocco, South Africa, Zimbabwe
Marine	Atlantic Ocean, Baltic Sea, Caribbean Sea, Indian Ocean

Sources: [31–33].

Table 2. Regional or national surveys of cyanobacterial blooms and scums for toxicity by intraperitoneal mouse bioassay

Country	Years	No. of sites sampled	% sites positive	Reference
UK	1981–89	24	75	
Norway, Sweden and Finland		51	89	see 39, 40
Sweden		27	56	
Finland		103	44	
UK		91	68	
Hungary	pre-1991	35	83	41
Norway	1989–91	36	88	42
Australia	1990–91	130+	42	43
Netherlands	1992	29	48	44
Belgium	pre-1993	17	59	45
UK	1989–93	48	51	46
France	1991–93	12	58	47
Denmark	1994	42	90	48
Slovenia	1994–95	9	89	49

A range of analytical methods is being developed for the quantification of CTX, including physico-chemical-, enzyme-based- and immunoassays (ELISA), with minimum detection limits between 10^{-2} and 10^{-5} times lower than the mouse assay[50,51]. Higher numbers of environmental cyanobacterial samples have been found to be positive according to high-performance liquid chromatography (HPLC) than by mouse bioassay, e.g. [46,48], clearly indicating that mouse bioassays can underestimate the occurrence of CTX. However, the more sensitive analytical methods do not currently include detection of some, only partially-characterised, toxic principles in cyanobacterial blooms e.g. the "protracted toxic factor" found in blooms in Norway[42].

Based on the environmental monitoring of natural blooms, and laboratory studies with monocyanobacterial, but not necessarily axenic strains, CTX production is inferred to occur widely among at least 25 cyanobacterial genera[31,52]. Unambiguous assignations of the origins of CTX currently depend upon the identification of the toxins in monocyanobacterial, axenic strains. Thus, microcystins are produced by axenic strains of *Anabaena*, *Microcystis*, *Nostoc* and *Oscillatoria* and nodularins are produced by *Nodularia* strains[4,31]. Current screening of Pasteur Culture Collection isolates for CTX and the emerging application of DNA probes for enzymes of microcystin synthesis[53,54] will provide necessary information to provide a wider understanding of the origins and distribution of microcystins.

5. HUMAN HEALTH EFFECTS ATTRIBUTED TO CYANOBACTERIAL TOXINS AND ASSOCIATED EXPOSURE ROUTES

Examples of human illness associated with contact with, or the ingestion of CTX are available in several recent reviews[7,8,32,55]. Episodes of illness or poisoning associated with, or attributed to CTX have included gastroenteritis, skin irritation, allergenic responses, muscular pains, liver and kidney damage. That CTX present hazards to human health is established from: *a*, toxicity studies with animals, including dose-response trials; *b*, knowledge of the modes of action of purified CTX from the molecular to the tissue organ level; *c*, assessment of human health incidents according to the Bradford-Hill epidemiological criteria[56]. Most of the investigations into human health incidents have been (unavoidably) incomplete investigations after single- or relatively short-term exposures to cyanobacterial blooms, scums or CTX, e.g. [57].

The need to investigate the consequences of long-term exposure to CTX is apparent since: *a*, CTX can be present in drinking water in the absence of effective water treatment; *b*, long-term or at least multiple exposures to CTX may occur during recreational water activities and bathing or showering; *c*, microcystins are potent tumour-promoters in animals studies[3,8,32]. Studies are progress in the Peoples' Republic of China in regions where a higher incidence of human primary liver cancer occurs among populations using surface water containing microcystins for drinking, than in neighbouring areas where groundwater, apparently free from microcystins is used for drinking[58,59]. Further epidemiological studies have recently been carried out in New South Wales, Victoria and South Australia in an Australian study of cyanobacterial contamination and drinking water quality. Out of a range of adverse health outcomes studied (prematurity, low birth weight, congenital birth anomalies, gastrointestinal cancer mortality and overall mortality) no consistent pattern of exposure to cyanobacterial blooms *via* drinking water was apparent with the exception of the following: a high rate of congenital defects in relation to first trimester exposure[60]. These studies indicate the need for further epidemiological investigations into the possible consequences of human exposure to CTX.

A major human poisoning episode, involving the death of at least 55 people, and attributed to exposure to microcystin(s), occurred in 1996 in Brazil. The deaths occurred after haemodialysis treatment at a clinic in Caruaru, NE Brazil[9-11], with victims presenting liver damage consistent with microcystin poisoning and the presence of microcystins in clinical and post-mortem specimens. Microcystins have been detected and quantified in victims' serum by mass spectrometry, HPLC, ELISA and protein phosphatase inhibition assay[10,11]. The consequences of the Caruaru haemodialysis deaths are many-fold and far-reaching (Table 3). They have also contributed to the recognition of exposure routes by which CTX can affect human health and may provide valuable data on relations between exposure levels and health outcomes as the incident continues to be investigated.

Investigations of human CTX exposure routes and exposure levels have so far largely concentrated on drinking water and recreational water activities (Table 4). The addition of haemodialysis to the list of perceived exposure routes is clearly required after the Caruaru incident and that of inhalation is merited from studies on the exposure of animals

Table 3. Consequences of the human deaths attributed to microcystins
in haemodialysis water, Caruaru, Brazil, 1996

1. More than 55 patients have died so far.
2. More than 50 additional exposed patients survived with effects of exposure uncertain.
3. Closure of haemodialysis clinics in Caruaru.
4. Massive media coverage, public reaction and increase at international level in awareness of hazards presented by CTX.
5. Major involvement of human health-care professionals, clinicians, pathologists and epidemiologists in a case of human deaths attributed to CTX.
6. National (Brazilian) requirement announced for action plans for health protection and quality assurance of water used for haemodialysis with respect to CTX.
7. Increased international recognition of the health significance of CTX.

Table 4. Some cyanobacterial toxin exposure routes in human water-based and other activities

	Activity
1. Drinking water	• Accidental ingestion of scum in raw (untreated) water
	• Ingestion of cyanobacterial cells and/or free CTX in raw, or ineffectively treated water
2. Skin and mucosa	• Direct contact with scums, blooms and/or free CTX during recreation, work practices, bathing or showering
3. Haemodialysis	• Contact with CTX during dialysis treatment
4. Inhalation	• Showering, work practices, water sports
5. Food consumption	• Consumption of shellfish and finfish if containing CTX
	• Consumption of plant products and cyanobacterial cell dietary supplements, if containing CTX

to microcystins and anatoxin-a *via* aerosols and intranasal administration[61]. The accumulation of nodularin and PSP toxins in marine and freshwater mussels respectively[62,63], following exposure to toxin-containing blooms of *Nodularia* and *Anabaena*, requires the addition of shellfish as a possible exposure route. Finfish as a possible CTX exposure route are listed (Table 4) due to the finding of microcystin-type compounds in farmed salmon livers[64]. If edible plants are exposed to CTX during crop spray irrigation[28] then the possibility occurs that the toxins may accumulate in or on plant material. This scenario has apparently not been investigated beyond limited laboratory studies[28]. Analyses for CTX in spray irrigation water, when taken from sources containing toxigenic cyanobacterial blooms, are needed to enable risk assessment studies to be made of this possible exposure route of plant foods/products and of the inhalation route during crop spraying. Dietary supplements containing cyanobacterial cells would also constitute an exposure route if they contained CTX and the need also exists to analyse the cyanobacterial blooms and cells used in the manufacture of these products.

World Health Organization (WHO) Guidelines for human drinking water quality have not hitherto included Guideline Values (GV) for CTX. However, based on exposure assessments of studies of the effects of microcystin-LR in animals and risk assessment to humans *via* drinking water, a GV of 1 microgram per litre has been proposed[65]. This development provides a precedent for risk assessments and the derivation of GV's for other CTX with respect to drinking water and other exposure routes (Table 4). The WHO GV for microcystin-LR in drinking water is also expected to influence the development of policies for the management and monitoring of waterbodies containing cyanobacterial blooms if needed for potable water, and on the development of CTX removal or destruction methods during water treatment[8].

6. CONCLUDING REMARKS

A history of sporadic animal and human poisoning incidents and human health problems attests to the health hazards presented by CTX. Proactive monitoring and toxin analysis programmes show that CTX commonly occur in waters with cyanobacterial blooms worldwide and that a precautionary principle in cyanobacterial bloom management and water treatment is appropriate with respect to CTX.

Awareness of CTX health hazards is becoming established in some developed countries after poisoning incidents *via* education programmes and the establishment of cyanobacterial bloom management action plans. However, a need for further education and awareness programmes in developing countries remains. Risk assessment and CTX exposure studies are contributing to the derivation of guideline values (GV) for maximum acceptable levels of CTX in drinking water. These approaches can be expected to be increasingly applied in the future to derive GV's for recreational waterbodies.

Space does not permit here a review of research in progress on genetic[53,54], biochemical or physiological aspects of CTX production, or on the possible natural functions of CTX[31]. Such research is clearly needed to answer questions on the rationale and regulation of CTX production and will undoubtedly help in formulating rational and effective strategies for the control of CTX production in waterbodies, for CTX removal in water treatment and the protection of health.

REFERENCES

1. Skulberg, O.M., Codd, G.A. and Carmichael, W.W. *Ambio* **13**, 244–247 (1984).
2. Codd, G.A., Brooks, W.P., Lawton, L.A. and Beattie, K.A. In: Watershed '89. The Future for Water Quality in Europe. Vol. II, eds. D. Wheeler, M.L. Richardson and J. Bridges, Pergamon Press, Oxford, pp. 211–220 (1989).
3. Falconer, I.R. In: Algal Toxins in Seafood and Drinking Water. ed. I.R. Falconer Academic Press, London, pp. 177–186 (1993).
4. Sivonen, K. *Phycologia* **35** (no. 6 suppl.) 12–24 (1996).
5. Beasley, V.R., Cook, W.O., Dahlem, A.M., Hooser, S.B., Lovell, R.A. and Valentine, W.M. *Food Animal Practice* **5**, 345–361 (1989).
6. Edwards, C., Beattie, K.A., Scrimgeour, C.M. and Codd, G.A. *Toxicon* **30**, 1165–1176 (1992).
7. Carmichael, W.W. and Falconer, I.R. In: Algal Toxins in Seafood and Drinking Water, ed. I.R. Falconer, Academic Press, London, pp. 187–209 (1993).
8. Bell, S.G. and Codd, G.A. *Rev. Med. Microbiol.* **5**, 256–264 (1994).
9. Dunn, J. *Br. Med. J.* **312**, 1183–1184 (1996).
10. Barreto, V., Lira, V., Figueiredo, J., Fittipaldi, H., Juca, N., Gayotto, L.C., Raposo, F., Barbosa, J., Holmes, C., Cardo, D., Azevedo, S. and Carmichael, W.W. *Hepatol.* **24**, 189A (1996).
11. Ward, C.J., Preiser, W., Poon, G.K., Pouria, S., Neild, G.H. and Codd, G.A. *The Phycologist* No. **46**, 29 (1997).
12. Codd, G.A. *Harmful Algae News* No. **15**, 4–5 (1996).
13. Anderson, N.J. *Freshw. Biol.* **34**, 367–378 (1995).
14. Griffiths, B.M. *Proc. Linn. Soc. Lond.* **151**, 12–19 (1939).
15. Codd, G.A. and Beattie, K.A. *Publ. Health Lab. Serv. Digest* **8**, 82–86 (1991).
16. Dick, C.H. Highways and Byways in Galloway and Carrick, (1916) reprinted by G.C. Book Publishers, Wigtown, Scotland, pp. 315 (1994).
17. Hayman, J. *Med. J. Austr.* **157**, 794–796 (1992).

18. Hald, J.C., Bidrag til Kundskabs om de danske Provindsers naervaerende Tilstand i oekonomisk Henseende. Ottende Stykke, Ringkjøbirg Amt, Landhusholdningsselskabet, Copenhagen, p.56 (1833).
19. Moestrup, Ø. *Phycologia* 35, (no. 6 suppl.) 5 (1996).
20. Codd, G.A., Steffensen, D.A., Burch, M.D. and Baker, P.D. *Austr. J. Mar. Freshw. Res.* 45, 731–736 (1994).
21. Francis, G. *Nature* London. 18, 11–12 (1878).
22. Carmichael, W.W. *J. Appl. Bacteriol.* 72, 445–459 (1992).
23. Rinehart, K.L., Namikoshi, M. and Choi, B.W. *J. Appl. Phycol.* 6, 159–176 (1994).
24. De Silva, E.D., Williams, D.E., Andersen, R.J., Klix, H., Holmes, C.F.B. and Allen, T.M. *Tetrahedron Letts.* 33, 1561–1564.
25. Mackintosh, C., Beattie, K.A., Klumpp, S. Cohen, P. and Codd, G.A. *FEBS Letts.* 264, 187–192 (1990).
26. Matsushima, R., Yoshizawa, S., Watanabe, M.F., Harada, K-I., Furusawa, M., Carmichael, W.W. and Fujiki, H. *Biochem. Biophys. Res. Comm.* 171, 867–874 (1990).
27. Zolnierowicz, S. and Hemmings, B.A. *Trends Cell Biol.* 6, 359–362 (1996).
28. Abe, T., Lawson, T., Weyers, J.D.B. and Codd, G.A. *New Phytol.* 133, 651–658 (1996).
29. Ohtani, I. Moore, R.E. and Runnegar, M.T.C. *J. Amer. Chem. Soc.* 114, 7941–7942 (1992).
30. Hawkins, P.R., Chandrasena, N.R., Jones, G.J., Humpage, A.R. and Falconer, I.R. *Toxicon* 35, 341–346 (1997).
31. Codd, G.A. *Wat. Sci. Technol.* 32, 149–156 (1995).
32. Yoo, R.S., Carmichael, W.W., Hoehn, R.C. and Hrudey, S.E. Cyanobacterial (Blue-Green Algal) Toxins: A Resource Guide, American Waterworks Association Research Foundation, Denver, pp. 229 (1995).
33. Reguera, B., (ed.) Abstracts and Posters, VIII International Conference on Harmful Algae, Instituto Espanol de Oceanografia, Vigo, pp. 269 (1997).
34. Mez, K., Beattie, K.A., Codd, G.A., Hanselmann, K., Hauser, B., Naegeli, H. and Preisig, H.R. *Eur. J. Phycol.* 32, 111–118 (1997).
35. Naegeli, H., Sahin, A., Braun, U., Hauser, B., Mez, K., Hanselmann, K., Preisig, H.R., Bivetti, A. and Eitel, J. *Arch. Tierheilkunde* 139, 201–209 (1997).
36. Patterson, G.M.L., Larsen, L.K. and Moore, R.E. *J. Appl. Phycol.* 6, 151–157 (1994).
37. Sivonen, K., Kononen, K., Carmichael, W.W., Dahlem, A.M., Rinehart, K.L., Kiviranta, J. and Niemela, S.I. *Appl. Env. Microbiol.* 55, 1990–1995 (1989).
38. Hawser, S.P., O'Neil, J.M., Roman, M.R. and Codd, G.A. *J. Appl. Phycol.* 4, 79–86 (1992).
39. Codd, G.A., Bell, S.G. and Brooks, W.P. *Wat. Sci. Technol.* 21, 1–13 (1989).
40. National Rivers Authority. Toxic Blue-Green Algae. Water Quality Series No. 2, National Rivers Authority, London, pp. 125 (1990).
41. Törökné Kosma, A. and Mayer, G. *Hidrölögiai Közlöny* 68, 49–54 (1988).
42. Skulberg, O.M., Underdahl, B. and Utkilen, H. *Arch. Hydrobiol.* (suppl. 105) Algal Studies 75, 279–289.
43. Baker, P.D. and Humpage, A.R. *Austr. J. Mar. Freshw. Res.* 45, 773–786 (1994).
44. Heinis, F. Toxinen in Cyanobacterien. De Situatie in Nederland, Hageman Verpakkers, Zvetemeer, pp. 41 (1994).
45. van Hoof, F., van Es, T., D'hout, D. and de Pauw, N. In: Detection Methods for Cyanobacterial Toxins, eds. G.A. Codd, T.M. Jefferies, C.W. Keevil and E. Potter, The Royal Society of Chemistry, Cambridge, pp. 136–138 (1994).
46. Codd, G.A. and Bell, S.G. The Occurrence and Fate of Blue-Green Algal Toxins in Freshwaters. National Rivers Authority R and D Report 29, HMSO, London, pp. 30 (1996).
47. Vezie, C., Benoufella, F., Sivonen, K., Bertru, G. and Laplanche, A. *Phycologia* 35, (no. 6 suppl.) 198–202 (1996).
48. Henriksen, P. *Phycologia* 35, (no. 6 suppl.) 102–110 (1996).
49. Sedmak, B. and Kosi, G. *Natural Toxins* 5, 64–73 (1997).
50. Codd, G.A., Jefferies, T.M., Keevil, C.W. and Potter, E. (eds). Detection Methods for Cyanobacterial Toxins, The Royal Society of Chemistry, Cambridge, pp. 191 (1994).
51. An, J. and Carmichael, W.W. *Toxicon* 12, 1495–1507 (1994).
52. Skulberg, O.M., Carmichael, W.W., Codd, G.A. and Skulberg, R. In: Algal Toxins in Seafood and Drinking Water, ed. I.R. Falconer, Academic Press, London, pp. 145–164 (1993).
53. Dittman, E., Meissner, K. and Börner, T. *Phycologia* 35 (no. 6 suppl.) 62–67 (1996).
54. Dittman, E., Neilan, B.A., Ehrhardt, M., von Dören, H. and Börner, T. In: Abstracts of the IX International Symposium on Phototrophic Prokaryotes, eds. G.A. Peschek, W. Löffelhardt and G. Schmetterer, University of Vienna, p. 85 (1997).

55. Ressom, R., Soong, F.S., Fitzgerald, J., Turczynowicz, L., El Saadi, O., Roder, D., Maynard, T. and Falconer, I.R. Health Effects of Cyanobacteria (Blue-Green Algae). National Health and Medical Research Council, Canberra, pp. 108 (1994).

56. Hunter, P.R. In: Detection Methods for Cyanobacterial Toxins, eds. G.A. Codd, T.M. Jefferies, C.W. Keevil, and E. Potter, The Royal Society of Chemistry, Cambridge, pp. 11–18 (1994).

57. Turner, P.C, Gammie, A.J., Hollinrake, K. and Codd, G.A. *Br. Med. J.* **300**, 1440–1441 (1990).

58. Yu, S.Z. In: Primary Liver Cancer, eds. Z.Y. Tang, M.C. Wu, S.S. Xia, China Academic Publishers, Beijing, pp. 30–37 (1989).

59. Yu, S.Z. *J. Gastroenterol. Hepatol.* **10**, 674–682 (1995).

60. Pilotto, L.S., Kliewer, E.V., Burch, M.D., Attewell, R.G. and Davies, R.D. Prematurity, Birth Weight, Congenital Anomalies, Overall Mortality and Gastrointestinal Cancer Mortality in Relation to Cyanobacterial Contamination in Drinking Water Sources, CRC for Water Quality and Treatment, Adelaide, Australia, pp. 62 (1997).

61. Fitzgeorge, R.B., Clark, S.A. and Keevil, C.W. In: Detection Methods for Cyanobacterial Toxins, eds. G.A. Codd, T.M. Jefferies, C.W. Keevil and E. Potter, The Royal Society of Chemistry, Cambridge, pp. 69–74 (1994).

62. Falconer, I.R., Choice, A. and Hosja, W. *Env. Toxicol. Wat. Qual.* **7**, 119–124 (1992).

63. Negri, A.P. and Jones, G.J. *Toxicon* **33**, 667–678 (1995).

64. Andersen, R.J., Luu, H.A., Chen, D.Z., Holmes, C.F.B., Kent, M.L., Le Blanc, M., Taylor, F.J.R. and Williams, D.E. *Toxicon* **31**, 1315–1323 (1993).

65. World Health Organization. Report of the Working Group Meeting on Chemical Substances in Drinking Water, Geneva, 22–26 April, 1997. Section 5.2, Microcystin-LR. World Health Organization, Geneva.

74

THE NATURAL SELECTION OF GAS VESICLES

A. E. Walsby, S. J. Beard, and P. K. Hayes

School of Biological Sciences
University of Bristol
Woodland Road, Bristol BS8 1UG, United Kingdom

1. NATURAL SELECTION OF GAS VESICLE WIDTH: THE STORY IN A NUTSHELL

Gas vesicles, which provide cyanobacteria and other prokaryotes with buoyancy, are hollow shells formed from protein. Although they have a simple structure which shows a basic uniformity in various groups of microorganisms, they exhibit some differences which can be explained by natural selection of smaller, stronger structures withstanding higher pressures in deeper lakes.

The gas vesicle is like a nutshell, rigid but brittle. Given a gentle squeeze it shrinks imperceptibly but under sufficient force it cracks and collapses. Just as large nuts are easier to crack than small ones with the same thickness of shell, so large gas vesicles are weaker than small ones.

In lakes, various cyanobacteria regulate their depth by using the buoyancy provided by gas vesicles. The hydrostatic pressure increases with depth and wide gas vesicles will collapse at lesser depths than narrow ones. Once the gas vesicles have collapsed the bacteria lose their buoyancy and the ensuing advantages. They will sink to greater depths and will be unable to accumulate new gas vesicles under the greater pressure there. There is therefore an advantage in making narrower, stronger gas vesicles. If this is so, why does the width of gas vesicles vary in different bacteria?

The buoyancy provided by a gas vesicle increases with its width. Wide gas vesicles have a higher ratio of gas space to wall volume than narrow ones and, put in terms of benefit and cost, they provide buoyancy more cheaply. The variation in width of gas vesicles that occurs in different organisms can therefore be seen as a trade-off between efficiency, which increases with width, and the requirement for strength, which decreases with width. The consequences of this are that in organisms adapted to shallow lakes, where the hydrostatic pressure is low, wider and weaker gas vesicles occur, whereas in those adapted to deeper lakes and seas, narrower and stronger gas vesicles have evolved.

The Phototrophic Prokaryotes, edited by Peschek *et al.*
Kluwer Academic / Plenum Publishers, New York, 1999.

2. NATURAL SELECTION OF GAS VESICLE WIDTH: THE STORY IN MORE DEPTH

The process of natural selection of gas vesicle width can be summarised by the cycle of events illustrated in Fig. 1, which indicates the steps that must be understood to provide a full description of the process.

We start at the top comparing two cyanobacterial cells one of strain W, which has genes that specify the production of wide gas vesicles, like those found in *Anabaena flos-aquae*, and one of strain N, whose genes specify narrow gas vesicles, like those in *Planktothrix (Oscillatoria) rubescens*.

Studies of the relation between critical collapse pressure (p_c) and cylinder diameter (d) of gas vesicles have shown that the cylinder diameter of $d = 84$ nm in strain W, confers a critical pressure of $p_c = 0.6$ MPa (6 bar), whereas the diameter of $d = 62$ nm in N confers a critical pressure of $p_c = 1.0$ MPa. The wider gas vesicles in W have a lower density, only 119 kg m^{-3}; the consequence is that the cell must devote 5% of its protein to the production of gas vesicle wall proteins in order to achieve buoyancy. The narrower gas vesicles in N have a higher density, of 162 kg m^{-3}, and their production requires 7% of the cell protein for buoyancy. If the accumulation of protein is the rate-limiting step for growth then strain N will grow 2% more slowly than strain W. In a temperate lake, over the growing season of about 15 weeks, W would pass through 50 generations while N would pass through only 49: the progeny of W would number twice that of N.

In winter, the lake cools and mixes to the bottom. If the lake were shallow, say only 15 m deep, both the weak gas vesicles in W and the strong gas vesicles in N would survive the 0.15 MPa hydrostatic pressure; on restratification next spring, similar proportions of the overwintering progeny would float up to form the inoculum for the next season's growth, i.e., the starting proportions of W:N would be 2:1. By the next year it would be 4:1 and after only a decade 1024:1. This illustrates how quickly natural selection could produce changes in a population.

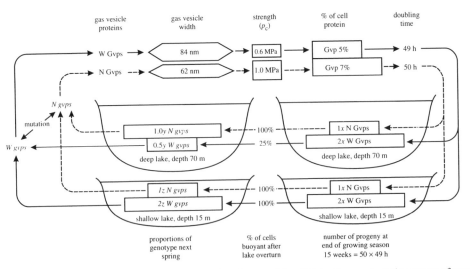

Figure 1. Stages in the annual cycle of natural selection of gas vesicle width by the counteracting agents of pressure and economy (after Walsby, 1994).

In a deeper lake the outcome would be the same up until the end of the first summer because the populations of cyanobacteria grow only in the epilimnion where the depths are insufficient to cause gas vesicle collapse. Differences occur after winter mixing. As the water circulates to greater depths the cyanobacteria are exposed to increasing pressure. The pressure here has two principal components. The first is the hydrostatic pressure, p_h, which rises by about 0.1 MPa for each increment of 10 m. The second is the turgor pressure of the cell, p_t, which usually exceeds 0.2 MPa even in cells kept at low irradiance. The combination of p_h and p_t is therefore just sufficient to cause gas vesicle collapse in strain W when it is mixed to a depth of 40 m; in strain N, however, gas vesicles will remain intact at depths up to 80 m. In such lakes deep circulation would result in significant buoyancy loss in strain W but no loss in strain N. If 75% of W were to lose buoyancy then in spring the situation in the shallow lake would be reversed: with only a quarter of W floating up, the proportions of W:N would be 1:2, and after a decade 1:1024.

3. THE COMPLETE QUANTITATIVE ACCOUNT: A JIGSAW PUZZLE

This description of the selection process provides a quasi-quantitative description of the process of natural selection of gas vesicle width but it is incomplete, like a picture on an unfinished jigsaw puzzle. By describing the incomplete picture, however, perhaps we can identify the missing pieces.

3.1. Gas Vesicle Genes

Natural selection acts on the phenotype, gas vesicle width, but its consequence is to change the ratio in the population of alleles of the genes that control gas vesicle width.

The identity of the gene(s) that controls gas vesicle width is a vital missing piece of the jigsaw. Width may be controlled by one or more of the six different genes, identified in the gas vesicle operon in *Anabaena flos-aquae*, or by some other gene.

3.1.1. gvpA and gvpC. Two structural genes have been characterised in cyanobacteria, *gvpA* (Tandeau de Marsac *et al.*, 1985), which encodes the main structural protein, GvpA (Walker *et al.*, 1984; Hayes *et al.*, 1986), and *gvpC*, a gene containing 4 partially conserved 99-nucleotide repeats (Damerval *et al.*, 1987), which encodes an outer protein, GvpC (Walsby & Hayes, 1988; Hayes *et al.*, 1988). The possible roles of GvpA and GvpC in determining width are discussed below.

Homologues of *gvpA* have been found in all gas-vesicle containing organisms. The sequence is very highly conserved; in many cyanobacteria the GvpA amino acid sequence differs from that in *Calothrix* at only 1 to 3 residues, though in one halophilic cyanobacterium, *Dactylococcopsis salina*, there are 8 differences, mostly conservative changes (Griffiths, Walsby & Hayes, 1992). The *gvpC* genes show much less sequence conservation and, except in closely related species, they cannot be found by looking for homologous sequences.

3.1.2. gvpN, gvpJ, gvpKA, and gvpFL. The cyanobacterial gene sequences were used to discover a homologous *gvpA* in halobacteria, which was located in a cluster of 12 or 13 *gvp* genes (Horne *et al.*, 1991; Halladay *et al.*, 1993). There is a functional equivalent of *gvpC*, which produces a minor component of the structure, but it has little homology to

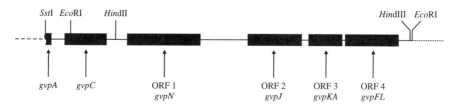

Figure 2. The *gvp* operon in *Anabaena flos-aquae*.

cyanobacterial *gvpC*. Some of the other genes have been shown to be involved in the regulation of gas vesicle expression and morphology, though the precise role of many of them is unknown. (See recent reviews by Pfeifer *et al.*, 1997, and DasSarma & Arora, 1997). These discoveries have led back to the discovery of further *gvp* genes in cyanobacteria: by sequencing downstream from *gvpC* in *Anabaena flos-aquae* four further ORFs were found that encoded proteins that shared homology with the products of *gvp*s in the halobacterial *gvp* cluster (Kinsman & Hayes, 1997): on the basis of this the *Anabaena* genes have been designated *gvpN*, *gvpJ*, *gvpKA*, and *gvpFL* (Fig. 2). There are other examples where the product of one bacterial gene performs the functions of two archaeal genes (Zillig *et al.*, 1991). It is conjectured that GvpN has a role in gas vesicle assembly, because it has sites that may be involved in hydrolysis of ATP or GTP (Pfeifer & Englert, 1992) and because it shows homology to other proteins involved in the post-translational activation or assembly of specific multiprotein complexes (Kinsman & Hayes, 1997). GvpJ and GvpK both show regions of homology with the main structural protein GvpA and they may therefore also be components of the gas vesicle, but present in such small amounts that they have escaped detection; the function of GvpFL is unknown.

Until the genetic control of width has been resolved we need to be open to the possibility that genes outside the characterised *gvp* operon could be involved in the process.

3.2. Gas Vesicle Proteins

3.2.1. The Rib Protein, GvpA. The wall of the cyanobacterial gas vesicle is made entirely of protein. The principal gas vesicle protein, GvpA, is a small molecule of 70 or 71 amino acid residues arranged along ribs that run around the conical ends and cylindrical mid-section, normal to the long axis of the gas vesicle (Fig. 3a). X-ray crystallography showed that there are two layers of β-sheet within the thickness of the wall. The repeating unit cell of GvpA contains two pairs of β-chains, one in each layer, which are oriented at

➤

Figure 3. Structure of the gas vesicle in *Anabaena flos-aquae*. a) Whole gas vesicle showing the conical ends and cylindrical mid-section with 2-nm thick wall and 4.6-nm wide ribs drawn to scale. b) Part of 7 ribs showing the orientation of GvpA molecules, as viewed from outside; the GvpA molecules are depicted as rhomboids 6.6 nm long and 1.15 nm wide formed from β-sheet; the GvpC molecules, each comprising five 33-residue repeats (33RRs), are attached at the outer surface; they are depicted as α-helical rods: the arrangement shown is conjectural, based on the observed GvpA:33RR ratio of 5:1. c) The β-sheet structure of GvpA: 62 of the 70-amino acids may be in the two layers of β-sheet within the thickness of the wall, each layer containing a pair of antiparallel β-chains. d) The 54° angle of the β-chains, indicated by x-ray crystallography, is explained by the stacking arrangement when adjacent chains differ in length by 1 dipeptide.

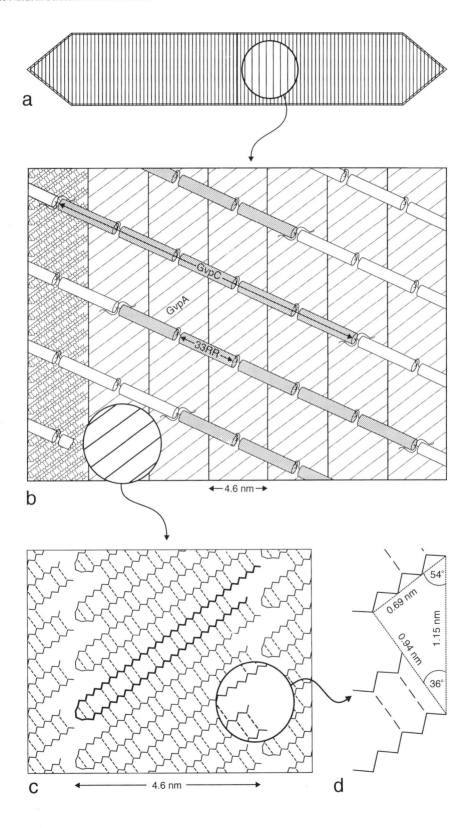

an angle of ~55° to the rib axis (Fig. 3b), i.e. ~35° to cylinder axis (Blaurock & Walsby, 1976). Atomic force microscopy (AFM) reveals the chirality at the outer surface: the β-chains slope from top left to bottom right (Fig. 3c) (McMaster, Miles & Walsby, 1996). The structure of the turn joining adjacent chains is unknown but if the chains were joined by a γ-turn (as shown in Fig. 3c) they would lie in the same plain (Panitch *et al.*, 1997).

An explanation for the angle of the chains is provided by a model in which a β-chain of 7 dipeptides is followed by an antiparallel β-chain of 8 dipeptides. Such antiparallel chain pairs would stack at an angle of $\alpha = \tan^{-1}(2a/2h)$, where a is the distance between β-chains (0.472 nm) and $2h$ is the dipeptide repeat (0.69 nm): α is then 53.8° (Fig. 3d).

An interesting consequence of this stacking angle is that the hydrogen bonds between adjacent chains would lie at the complementary angle of 36.2° to the rib axis, and at 53.8° to the axis of the cylinder. This is within 1° of the angle, 54.7°, at which transverse and longitudinal stresses are equal in the wall of a cylinder; the hydrogen bonds, which are weaker than the covalent bonds in the β-chains in the molecular structure, are therefore protected from twisting when the gas vesicle is placed under pressure. It seems, therefore, that this feature must increase the stability and strength of the gas vesicle. One imagines that primitive gas vesicles had GvpAs in which adjacent β-chains were of equal length and crossed the ribs at 90°, a common arrangement in other fibrous proteins such as silks (Panitch *et al.*, 1997); a chance mutation that moved the position of the *g*-turn then set the β-chains at 54° with the ensuing increase in stability. Such meanderings could be given substance by mutating the *gvpA* but before sensible experiments can be performed further details are required of the crystalline structure of the protein. These might be obtained by cryo-EM (Böttcher *et al.*, 1997).

3.2.2. *The outer Protein, GvpC.*

In *Calothrix* sp. the inferred amino acid sequence of GvpC contains four 33-residue repeats (33RRs) (Damerval *et al.*, 1987); in *Anabaena flos-aquae* there are five such 33RRs (Hayes *et al.*, 1988). GvpC is located on the outer surface of the gas vesicle. It can be stripped off by rinsing isolated gas vesicles with 1% SDS (Walsby & Hayes, 1988) or 6M urea (Hayes, Buchholz & Walsby, 1992); the stripped gas vesicles remain intact but their critical collapse pressure is greatly reduced, suggesting that this outer protein has the function of stabilising the structure. This idea has been proved by reassembling recombinant GvpC on gas vesicles stripped of the native protein with 6M urea (Hayes, Buchholz & Walsby, 1992): the reassembled structures regained their original strength.

From changes in the amino acid composition that occur when GvpC is stripped off intact gas vesicles, it has been shown that the ratio of GvpA:GvpC is 25:1 in gas vesicles of *Anabaena flos-aquae*. Recombinant GvpC binds back onto stripped gas vesicles in the same ratio (Buchholz, Hayes & Walsby, 1993). This indicates that the 33RRs bind in a ratio of 1:5 with GvpA. The amino acid sequence suggests that the secondary structure of GvpC is α-helical (Damerval *et al.*, 1987). If this is correct then one way in which the two proteins might interact is for the α-helical rods of GvpC to cross the ribs, tying them together (Buchholz *et al.*, 1993): Fig. 3b shows a model of a possible interaction between the two proteins that is compatible with the size of the 33RR and GvpA repeats.

The evolution of the repeating structure in GvpC has evidently involved replication of the 99-nucleotide sequence that encodes the 33RR. Kinsman, Walsby & Hayes (1995) produced recombinant forms of the *Anabaena* GvpC that had, respectively, 2, 3, 4 or 5 of the 33RRs and showed that they would bind to gas vesicles that had been stripped of their native GvpC in the expected ratios. On binding, they restored the strength of the gas vesicles but the degree of strengthening increased with the number of repeats. The evolution of longer GvpC molecules would therefore have been supported by their greater efficacy in stabilising the gas vesicle. It is interesting that in *Calothrix* sp., in which the native GvpC molecule contains only four of the 33RR segments (Damerval *et al.*, 1987), the gas vesicles are not quite as strong as the slightly wider gas vesicles fond in *Anabaena flos-aquae* (Walsby & Bleything, 1988).

4. WHAT DETERMINES THE WIDTH OF GAS VESICLES?

The gas vesicle forms by assembly from a small biconical structure (Waaland & Branton, 1969). It may grow by insertion of new GvpA molecules at the base of each cone, i.e. at the centre of the biconical structure. The diameter of each successive rib will be larger than the previous one. Eventually, a maximum rib diameter will be reached and further addition of proteins will result in cylindrical extension at the base of each cone. The point at which this occurs is characteristic of the gas vesicles in each species, though the diameter is not determined with molecular precision: the width of gas vesicles within a cell varies by about 5% (Walsby & Bleything, 1988). What determines the width at which this change occurs is not known. It might be determined by the shape of the GvpA molecule itself, which can be envisaged as a flexible brick whose curvature in its relaxed form influences the cylinder diameter (Walsby, 1978a). It might be determined by GvpC, which in assembling onto the ribs of the growing bicone forms a corset that constricts its expansion.

These possibilities are being investigated by genetic manipulation. In halobacteria gas vesicles produced by expression of the genes in the *gvp* cluster from the plasmid are wider than those produced from the chromosomal cluster (Simon, 1981). Pfeifer and her collaborators transformed *Haloferax volcanii*, which does not have gas vesicles, with the entire p-*vac* cluster from *Halobacterium salinarium*: the transformant produced mainly wide spindle-shaped gas vesicles (Offner, Wanner & Pfeifer, 1996). When *Hf. volcanii* was transformed with the mc-*vac* cluster of *Hf. mediterranei*, the gas vesicles produced were predominantly cylindrical (Englert, Wanner & Pfeifer, 1992). In collaboration with Pfeifer we have shown that when only the p-*gvpA* is replaced by mc-*gvpA*, the gas vesicles produced by the p-*vac* transformant change from spindle to cylindrical morphology and their critical pressure increases. The proteins encoded by p-*gvpA*, c-*gvpA* and by mc-*gvpA* differ in only 5 or 6 amino acid residues; changing individual or combinations of different residues also changes the critical pressure (Beard, Hayes, Pfeifer & Walsby, in preparation). Halobacterial gas vesicles produced in mutants with insertions or deletions in *gvpC* also show large differences in width (DasSarma *et al.*, 1994; Offner *et al.*, 1996).

Further investigations in cyanobacteria are needed, however, because the halobacterial gas vesicles are so pleiomorphic. Of the two known structural proteins, we would guess that GvpA is less likely to determine the width variation in closely related cyanobacteria because its amino acid sequence seems to show insufficient variation to account for it.

5. THE RELATIONSHIP BETWEEN WIDTH AND STRENGTH

There is a causal relationship between width and critical pressure of the gas vesicle that can be explained in terms of mechanics (Hayes & Walsby, 1986, Walsby 1991). This has been reviewed in detail by Walsby (1994) and is therefore briefly summarised here.

When pressure is applied to a hollow cylinder the stress is proportional to the cylinder diameter (d) (because the area of the wall is proportional to d but force on the cross-section of the cylinder increases as d^2). Thin-walled cylinders fail through instability (by going out of round) and the buckling pressure, p_b, varies inversely as d^3. The critical pressures of gas vesicles stripped of GvpC are close to the theoretical p_b values; GvpC stiffens the structure and postpones buckling (Walsby, 1991). The relationship between p_c and d found in 8 genera of cyanobacteria (Fig. 4) can be described by the expression

$$p_c = 875(d/\text{nm})^{-1.67} \text{ MPa.}$$

This formulation is based on empirical measurements from only a few organisms. Walsby & Bleything (1988) concluded that part of the variation in critical pressure is contributed by other factors (such as the degree of stiffening provided by GvpC). In order to make predictions on changes in p_c caused by changes in width, more data is needed on gas vesicles of closely related organisms. Measurements on strains of *Planktothrix (Oscillatoria) rubescens* isolated from Lake Zürich indicate that the expression can be extrapolated to include narrower gas vesicles of greater strength (D. Bright & A. E. Walsby, unpublished).

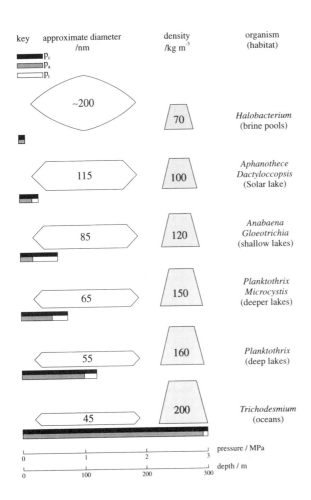

Figure 4. The relation between mean critical pressure and mean diameter of gas vesicles in cyanobacteria. The black bar indicates mean critical collapse pressure, p_c. The white bar indicates turgor pressure, p_t. The shaded bar indicates mean apparent critical pressure, p_a, at which gas vesicle collapse occurs in turgid cells, and also the depth at which 50% of gas vesicles will be collapsed by the hydrostatic pressure, p_h. Data from Hayes & Walsby (1986), Walsby & Bleything (1988), Kinsman (1994), and D. Bright (unpublished).

The narrowest and strongest gas vesicles are found in *Trichodesmium* spp. from the deep tropical ocean; they have a width of 45 nm (Gantt, Okhi, & Fujita, 1984) and a p_c in excess of 3.0 MPa. They will resist collapse at depths exceeding 300 m when the organisms are mixed down by storms, and this will then enable them to float back towards the sea surface when calm conditions return (Walsby, 1978b). The widest and weakest gas vesicles are found in two cyanobacteria from a 5-m deep brine pool in the Sinai peninsular: in *Dactylococcopsis salina d* is 109 nm and p_c is 0.33 MPa (Walsby & Bleything, 1988); in *Aphanothece halophytica d* is 116 nm and p_c is 0.30 MPa (Kinsman, 1994). Between these extremes there is a trend of gas vesicles with increasing width and decreasing critical pressure with waters of decreasing depth (Fig. 4); further details are given by Walsby (1994).

Although these gross differences in gas vesicle width are found in cyanobacteria of different genera, substantial differences also occur in different species of the same genus. In *Anabaena flos-aquae* (Hayes & Walsby, 1986) and *Anabaena minutissima* (Walsby *et al.*, 1989) the p_c is about 0.6 MPa but in *Anabaena lemmermannii* it is 0.93 MPa (Walsby *et al.*, 1991). Isolates of *Nodularia* spp. from the Baltic Sea show a range of p_c from 0.53 to 1.14 MPa (Walsby, Hayes & Boje, 1995; Hayes, Barker & Walsby, 1997). And we have recently found significant variation in p_c of gas vesicles in 12 clonal isolates of *Planktothrix rubescens* from Lake Zürich (Walsby & Schanz, submitted), which we describe in the next section.

6. THE PROTEIN BURDEN

If we hope to produce a quantitative description of the processes of natural selection, the cost of producing gas vesicles must be related to its effect on growth rate. This cost is referred to as the protein burden: the assumption is that when protein synthesis is the rate limiting step, the amount of protein devoted to the production of a structure will affect growth rate. How does this apply to gas vesicles of different width? By what mechanism could a cell make savings by producing wider gas vesicles? Is this another missing piece in the puzzle?

The first step in calculating the cost is to determine the proportion of the cell protein required to construct sufficient gas vesicles for buoyancy. This has been calculated by mechanical considerations, based on the buoyant density of a cell and the dimensions and density of the gas vesicle in different species (see Walsby, 1994). In *Anabaena flos-aquae* it takes 5% of the cell protein; in *Planktothrix (Oscillatoria) agardhii*, which has narrower and stronger gas vesicles, it requires 7%. These are the minimum values required for neutral buoyancy; in *Anabaena* cells grown in low irradiance the gas vesicles may contain over 10% of the cell protein.

These calculations show that a cell could make savings by making wider gas vesicles, but it would do so only if it avoided excess gas vesicle production. It is unlikely that a cell could regulate its Gvp production according to either its gas-vesicle gas content or its specific gravity; it has no mechanism for measuring these quantities. In a lake, however, the consequences of gas vesicle over-production are that the cyanobacterium floats up and receives more light. It then responds to high irradiance by losing buoyancy and sinking. One of the mechanisms involved in buoyancy loss is the repression of gas vesicle synthesis (Utkilen *et al.*, 1985; Damerval *et al.*, 1991). A cyanobacterium with wide gas vesicles will use less Gvp in making gas vesicles and floating up before this repression occurs. A specific sensor is not required to make the saving; regulation is matched to and implemented by the natural light conditions. In order to quantify the relative costs we need to know more about the regulation of expression of the gas vesicle proteins.

7. CHANGES IN THE PROPORTIONS OF DIFFERENT STRAINS: ECOLOGICAL STUDIES

To this point we have dealt with all the parts of the puzzle that can draw on laboratory studies; the remaining parts depend on studies on the planktonic cyanobacteria in their natural environment. We are investigating two ecosystems where spatial or temporal variations in the depth of winter mixing has resulted in the selection of gas vesicle variants with different critical pressures. The first concerns populations of *Nodularia* in the Baltic Sea (Walsby, Hayes & Boje 1995; Hayes & Barker 1997). The second, which we consider here, concerns a population of *Planktothrix* in Lake Zürich. This study is in progress and so there are gaps in it but a number of pieces are fitting together.

7.1. Planktothrix Rubescens in Lake Zürich

The red-coloured cyanobacterium *Planktothrix* (*Oscillatoria*) *rubescens* (Agnostidis & Komàrek, 1988) is one of the dominant phytoplankton in Lake Zürich. Thomas and Märki (1949) showed that it grew in the metalimnion throughout the summer, was entrained in the epilimnion in the autumn and mixed to progressively greater depths during the winter. We have measured the changes in the population over a two-year period by collecting filaments on membrane filters and using epifluorescence microscopy and computer image analysis techniques (Walsby & Avery, 1996) to measure the filament concentration. We have demonstrated a similar annual cycle and found that over 70% of the autumnal population persists throughout the winter and that 85% or more of the filaments remain viable. Thermal stratification begins again in April and the population reforms in the metalimnion (Walsby & Schanz, submitted).

We initially isolated 12 strains of *Planktothrix rubescens* from the lake: the mean p_c of their gas vesicles varied from 0.9 to 1.1 MPa. The turgor pressure in filaments from the lake was about 0.2 MPa and so those with the weakest gas vesicles would lose buoyancy at depths exceeding 70 m whereas those with the strongest would lose buoyancy at 90 m or more. During the warm winter of 1994–5 the lake mixed to a depth of about 100 m; towards the end of the winter period it was found that about 80% of the filaments in the top 80 m retained sufficient gas vesicles to provide buoyancy. In the colder winters of 1993–4 and 1996–7, mixing occurred to the lake bottom at 136 m and less than 30% of the filaments remained buoyant in the top 80 m. Only the buoyant filaments would have been able to float up and form the inoculum for the next season's population. According to the criteria in Fig. 1, there would have been selection for the strains with stronger gas vesicles in colder years that produced deeper mixing (Walsby & Schanz, submitted).

In order to determine if this selection occurs we need a way of comparing the proportions of different strains in the *Planktothrix* population at the end of the growth period and in the population that floats up to form the spring inoculum. We have established over 60 clonal cultures, which we are using to investigate the range of phenotypic and genotypic variation present in Lake Zürich. It is difficult to quantify the proportions of different strains in lakewater samples by culturing, however, because only a proportion (30–85%) of the isolates grow, and there may therefore be a sampling bias that favours particular strains.

An alternative approach is to determine the genotypic variation between different strains and to correlate this with gas vesicle phenotype. Hayes & Barker (1997) have developed a method of amplifying DNA from single filaments of cyanobacteria picked from samples of lake or seawater. The cyanobacterial cells lyse and release sufficient DNA to provide PCR templates, when the filaments are heated to 94°C in a PCR reaction mixture. The alleles at specific loci can then be identified in one of several ways: (1) by sequencing or restriction mapping of the purified PCR products produced using locus-specific primers; (2) from the diagnostic length of amplified fragments generated by using allele-specific primers; (3) by using primers designed to generate groups of amplification fragments that are characteristic of different allelic variants. The first two of these approaches have been used to describe the genetic variation of *Nodularia* population in the Baltic Sea (Hayes & Barker, 1997; Hayes Barker & Walsby, 1997). It is possible to obtain amplification products from at least two loci from a single filament: the filament is divided into two parts from which DNA is amplified separately.

We have used the same approach to investigate the genetic variation of *Planktothrix rubescens* in Lake Zürich. We have made primers that can be used to amplify DNA sequences within genes that encode rRNA, phycocyanin, RUBISCO, and gas vesicle proteins. Specific combinations of sequences serve to characterise specific strains. In characterising the *gvp* genes in the isolates described by Walsby & Schanz (submitted), we have found four different arrangements of *gvpA* and *gvpC*. Individual isolates contain between 2 and 4 identical copies of *gvpA* and at least two copies of *gvpC*. Within each strain there may be *gvpC*s of different length, 417 bp, 516 bp or ~600 bp (the first two of which would encode proteins differing in length by 33 amino acid residues) and additionally some strains possess a 72 bp sequence that corresponds to the 3'-end of *gvpC*. The net result of this variation, in the number and arrangement of the *gvp*-genes, is that individual isolates generate a characteristic set of PCR amplification fragments when oligonucleotides primers that prime DNA synthesis outward from the central portion of *gvpA*, either used alone or in combination with primers complimentary to *gvpC*.

In collaboration with F. Schanz (University of Zürich) we have collected large numbers of individual *Planktothrix* filaments from Lake Zürich, which we will characterise by these techniques. In this way we will be able to make a quantitative description of the changes that occur in the proportions of different genotypes, which results from the process of natural selection.

8. CAN A QUANTITATIVE STUDY BE MADE OF NATURAL SELECTION?

Whilst acknowledging Rutherford's dictum that descriptions lacking numbers are incomplete it has been recognised that "not all scientific concepts are to be found expressed mathematically, and some theories of high science, such as the theory of evolution by natural selection can be expressed without it" (Atkins, 1994). There is no intrinsic reason, however, why the process should not be subjected to quantitative analysis. What discourages the attempt is the complexity of most biological systems.

What are the prospects of a complete quantitative description of the natural selection of gas vesicle width? First, they are favoured by the simplicity of gas vesicle structure, composition and the encoding genetic elements. Secondly, the width, strength and buoyant density of the gas vesicle are related in a simple quantitative and mechanistic way. Thirdly, the selective agent, which in other instances is so difficult to define, is here a simple physical quantity, pressure, which can be precisely measured and has a direct quantitative relationship with the environmental variable, depth. Paradoxically, however, it is here that complexity arises.

When a lake is mixed in winter some filaments, by chance, are swept to the greatest depths others may circulate within the surface layers. In any year it may be difficult to predict the proportions that are mixed to depths where the critical collapse pressure of their gas vesicles is exceeded, though it may be possible to determine the proportions *post facto* from the proportions that remain buoyant (Walsby & Schanz, submitted). For similar reasons it may difficult to predict the outcome of natural selection of different strains but it should be possible now to obtain a quantitative measure of it by determining the changing proportions of different *gvp* genotypes.

ACKNOWLEDGMENTS

We thank Tim Colbourn for preparing the figures. Our investigations are supported by grants from the Natural Environment Research Council (GR9/1201 and GR3/10790) and the European Commission Environment Programme (EV5V-CT94–0404).

REFERENCES

Anagnostidis, K. & Komàrek, J. 1988. Modern approach to the classification system of cyanophytes. 3 - Oscillatoriales. *Arch. Hydrobiol. Suppl.* **80**, 327–472.

Blaurock, A.E. & Walsby, A.E. 1976. Crystalline structure of the gas vesicle wall from *Anabaena flos-aquae. J. Mol. Biol.* **105**, 183–199.

Atkins, P. 1994. Science as a social construct. Against the idea. *The Times Higher Education Supplement* No. 1143, pp 17–19.

Böttcher, B., Wynne, S.A. & Crowther, R.A. 1997. Determination of the fold of the core protein of hepatitis B virus by electron cryomicroscopy. *Nature* **386**, 88–91.

Buchholz, B.E.E., Hayes, P.K. & Walsby, A.E. 1993. The distribution of the outer gas vesicle protein, GvpC, on the *Anabaena* gas vesicle, and its ratio to GvpA. *J. Gen. Microbiol.* **139**, 2353–2363.

Damerval, T., Castets, A-M., Houmard, J. & Tandeau de Marsac, N. 1991. Gas vesicle synthesis in the cyanobacterium *Pseudanabaena* sp.: occurrence of a single photoregulated gene. *Mol. Microbiol.* 5, 657–664.

Damerval, T., Houmard, J., Guglielmi, G., Csiszàr, K. & Tandeau de Marsac, N. 1987. A developmentally regulated *gvpABC* operon is involved in the formation of gas vesicles in the cyanobacterium *Calothrix* 7601. *Gene* **54**, 83–97.

DasSarma, S. & Arora, P. 1997. Genetic analysis of the gas vesicle gene cluster in haloarchaea. *FEMS Microbiol. Lett.* **153**, 1–10.

DasSarma, S., Arora, P., Lin, F., Molinari, E. & Yin, L. R-S. 1994. Wild-type gas vesicle formation requires at least ten genes in the *gvp* gene cluster of *Halobacterium halobium* plasmid pNRC100. *J. Bacteriol.* **176**, 7646–7652.

Englert, C., Wanner, G. & Pfeifer, F. 1992. Functional analysis of the gas vesicle gene cluster of the halophilic archaeon *Haloferax mediterranei* defines the vac-region boundary and suggest a regulatory role of the *gvpD* gene or its product. *Mol. Microbiol.* **6**, 3543–3550.

Gantt, E., Okhi, K. & Fujita, Y. 1984. *Trichodesmium thiebautii*; structure of a nitrogen-fixing marine blue-green alga (Cyanophyta). *Protoplasma* **119**, 188–196.

Griffiths, A.E., Walsby, A.E. & Hayes, P.K. 1992. The homologies of gas vesicle proteins. *J. Gen. Microbiol.* **138**, 1243–1250.

Halladay, J.T., Jones, J.G., Lin, F., MacDonald, A.B. & DasSarma, S. 1993. The rightward gas vesicle operon in *Halobacterium* plasmid pNRC100: identification of the *gvpA* and *gvpC* gene products by use of antibody probes and genetic analysis of the region downstream of *gvpC. J. Bacteriol.* **175**, 684–692.

Hayes, P.K. & Barker, G.L.A. 1997. Genetic diversity within Baltic Sea populations of *Nodularia* (Cyanobacteria). *J. Phycol.* (In press).

Hayes, P.K., Barker, G.L.A. & Walsby, A.E. 1997. The genetic structure of *Nodularia* populations. *Proc. IX Internatl. Symp. Phototrophic Prokaryotes, Vienna: Abstracts.*

Hayes, P.K., Buchholz, B. & Walsby, A.E. 1992. Gas vesicles are strengthened by the outer-surface protein, GvpC. *Arch. Microbiol.* **157**, 229–234.

Hayes, P.K., Lazarus, C. M., Bees, A., Walker, J. E. & Walsby, A. E. 1988. The protein encoded by *gvpC* is a minor component of gas vesicles isolated from the cyanobacteria *Anabaena flos-aquae* and *Microcystis* sp. *Mol. Microbiol.* **2**, 545–552.

Hayes , P.K. & Walsby, A.E. 1986. The inverse correlation between width and strength of gas vesicles in cyanobacteria. *Br. Phycol. J.* **21**, 191–197.

Hayes, P.K., Walsby, A.E. & Walker, J.E. 1986. Complete amino acid sequence of cyanobacterial gas-vesicle protein indicates a 70-residue molecule that corresponds in size to the crystallographic unit cell. *Biochem. J.* **236**, 31–36.

Horne, M., Englert, C., Wimmer, C., & Pfeifer, F. 1991. A DNA region of 9 kbp contains all genes necessary for gas vesicle synthesis in halophilic archaebacteria. *Mol. Microbiol.* **5**, 1159–1174.

Kinsman, R. 1994. Ph.D. thesis. University of Bristol, Bristol, United Kingdom.

Kinsman, R. & Hayes, P.K. 1997. Genes encoding proteins homologous to halobacterial Gvps N, J, K, F & L are located downstream of *gvpC* in the cyanobacterium *Anabaena flos-aquae*. *DNA Sequence* **7**, 97–106.

Kinsman, R., Walsby, A.E. & Hayes, P.K. 1995. GvpCs with reduced numbers of repeating sequence elements bind to and strengthen cyanobacterial gas vesicles. *Mol. Microbiol.* **17**, 147–154.

McMaster, T.J., Miles, M.J. & Walsby, A.E. 1996. Direct observation of protein secondary structure in gas vesicles by atomic force microscopy. *Biophys. J.* **70**, 2432–2436.

Offner, S,. Wanner, G. & Pfeifer, F. 1996. Functional studies of the *gvpACNO* operon of *Halobacterium salinarium* reveal that the GvpC protein shapes gas vesicles. *J. Bacteriol.* **178**, 2071–2078.

Panitch, A., Matsuki, K., Cantor, E.J., Cooper, S.J., Atkins, E.D.T., Fournier, M.J., Mason, T.L. & Tirrell, D.A. 1997. Poly(L-alanylglycine): multigram-scale biosynthesis, crystallization, and structural analysis of chain-folded lamellae. *Macromolecules* **30**, 42–49.

Pfeifer, F. & Englert, C. 1992. Function and biosynthesis of gas vesicles in halophilic Archaea. *J. Bioenerg. Biomembr.* **24**, 577–585.

Pfeifer, F., Krüger, K., Röder, R., Mayr, A., Ziesche, S. & Offner, S. 1997. Gas vesicle formation in halophilic Archaea. *Arch. Microbiol.* **167**, 259–268.

Simon, R. D. 1981. Morphology and protein composition of gas vesicles from wild type and gas vacuole defective strains of *Halobacterium halobium* strain 5. *J. Gen. Microbiol.* **125**, 103–111.

Tandeau de Marsac, N., Mazel, D., Bryant, D.A. & Houmard, J. 1985. Molecular cloning and nucleotide sequence of a developmentally regulated gene from the cyanobacterium *Calothrix* PCC 7601: a gas vesicle protein gene. *Nucl. Acids Res.* **13**, 7223–7236.

Thomas, E. A. and Märki, E. 1949. Der heutige Zustand des Zürichsees. *Mitt. Internat. Verein. Limnol.,* **10**, 476–488.

Utkilen, H.C., Oliver, R.L. & Walsby, A.E. 1985. Buoyancy regulation in a red *Oscillatoria* unable to collapse gas vacuoles by turgor pressure. *Arch. Hydrobiol.* **102**, 319–329.

Waaland, J.R. & Branton, D. 1969. Gas vacuole development in a blue-green alga. *Science* **163**, 1339–1341.

Walker, J.E., Hayes, P.K. & Walsby, A.E. 1984. Homology of gas vesicle proteins in cyanobacteria and halobacteria. *J. Gen. Microbiol.* **130**, 2709–2715.

Walsby, A.E. 1978a. The gas vesicles of aquatic prokaryotes. In "Relations between Structure and Function in the Prokaryotic Cell". *Symp. Soc. Gen. Microbiol* **28**. Edited by R.Y. Stanier, H.J. Rogers and J.B. Ward. Cambridge: University Press, pp. 327–358.

Walsby, A.E. 1978b. The properties and buoyancy-providing role of gas vacuoles in *Trichodesmium* Ehrenberg. *Br. Phycol. J.* **13**, 103–116.

Walsby, A.E. 1991. The mechanical properties of the *Microcystis* gas vesicle. *J. Gen. Microbiol.* **137**, 2401–2408.

Walsby, A.E. 1994. Gas vesicles. *Microbiol. Rev.* **58**, 94–144.

Walsby, A.E. & Avery, A. 1996. Measurement of filamentous cyanobacteria by image analysis. *J. Microbiol. Meth.* **26**, 11–20.

Walsby, A.E. & Bleything, A. 1988. The dimensions of cyanobacterial gas vesicles in relation to their efficiency in providing buoyancy and withstanding pressure. *J. Gen. Microbiol.* **134**, 2635–2645.

Walsby, A.E. & Hayes, P.K. 1988. The minor cyanobacterial gas vesicle protein, GVPc, is attached to the outer surface of the gas vesicle. *J. Gen. Microbiol.* **134**, 2647–2657.

Walsby, A.E., Hayes, P.K. & Boje, R. 1995. The gas vesicles, buoyancy and vertical distribution of cyanobacteria in the Baltic Sea. *Eur. J. Phycol.* **30**, 87–94.

Walsby, A.E., Kinsman, R., Ibelings, B.W. & Reynolds, C.S. 1991. Highly buoyant colonies of the cyanobacterium *Anabaena lemmermannii* form persistent surface waterblooms. *Arch. Hydrobiol.* **121**, 261–280.

Walsby, A.E., Reynolds, C.S., Oliver, R.L. & Kromkamp, J. 1989. The role of gas vacuoles and carbohydrate content in the buoyancy and vertical distribution of *Anabaena minutissima* in Lake Rotongaio, New Zealand. *Arch. Hydrobiol. Ergebn. Limnol.* **32**, 1–25.

Walsby, A.E. & Schanz, F. 1998. The critical pressures of gas vesicles in *Planktothrix rubescens* in relation to the depth of winter mixing in Lake Zürich, Switzerland. *J. Plankton Res.* (submitted).

Zillig, W., Palm, P., Klenk, H-P., Pühler, G., Gropp, F. & Schleper, C. 1991. Phylogeny of DNA-dependent RNA polymerase: testimony for the origin of eukaryotes. In *"General and Applied Aspects of Halophilic Microorganisms"*. Edited by F. Rodriguez-Valera. New York: Plenum Press, pp. 321–332.

75

LIPIDS OF PHOTOSYNTHETIC MEMBRANES IN CYANOBACTERIA

Roles in Protection against Low-Temperature Stress

Z. Gombos,[1,2] Y. Tasaka,[1] O. Zsiros,[1,2] Zs. Várkonyi,[2] and N. Murata[1]

[1]Department of Regulation Biology
National Institute for Basic Biology
Myodaiji, Okazaki 444, Japan
[2]Institute of Plant Biology
Biological Research Center of Hungarian Academy of Sciences
H-6701, P.O. Box 521, Hungary

The photosynthetic machinery of photosynthetic organisms is embedded in membrane composed from pigments, proteins and lipids. The unique lipid composition of thylakoid membrane can determine its functions. The alteration of the fatty-acid composition of glycerolipids in membranes can affect the physical characteristics of the membrane. Consequently it can affect the dependence of photosynthetic activities on the temperature of environment and the tolerance to temperature stresses.

Cyanobacteria provide a simple experimental system as a model for higher-plant chloroplasts. They perform oxygenic photosynthesis and possess lipids similar to chloroplasts of higher plants. Transformable cyanobacterial strains opened the opportunity to manipulate the lipid background of photosynthetic machinery independently of other membrane constituents. Following the isolation of genes encoding all the fatty acid desaturases of *Synechocystis* PCC 6803 (*desA*, *desB*, *desC* and *desD* encoding $\Delta12$, $\omega3$, $\Delta9$ and $\Delta6$, respectively) transformants with disrupted genes and with well-defined fatty acid composition could be obtained. Using a transformant without dienoic fatty acids, *desA⁻/desD⁻*, we could demonstrate the importance of these molecular species in the tolerance to low-temperature stress. The complete replacement of polyunsaturated glycerolipids with monounsaturated ones enhanced the susceptibility of the cells to low-temperature stress in the light. The re-synthesis of D1 protein and its processing was suppressed in the transformed cells at lower temperatures. The introduction of dienoic fatty acids into *Synechococcus* PCC 7942 cells resulted in enhanced chilling tolerance of photosynthesis of the cells and also enhanced tolerance to low-temperature photoinhibition. Generally it could be stated that polyunsaturated fatty acids are important in low-temperature tolerance.

The Phototrophic Prokaryotes, edited by Peschek *et al.*
Kluwer Academic / Plenum Publishers, New York, 1999.

1. INTRODUCTION

1.1. The Relationship between Temperature Tolerance and Membrane Lipids

The photosynthetic machinery is the transformer of light energy to chemical energy in photosynthetic organisms. Since this apparatus is embedded in membranes it can reflect a small alteration in membrane structure by changing the energy production and consequently the photosynthetic organisms are very susceptible to even a minor modification of the constituents of photosynthetic membranes. The molecular mechanism of the tolerance of plants to temperature stresses has been in the focus of research for the last decades. This demand was generated by the chilling and high temperature sensitivity of horticular plants, which can determine the harvest. The correlation between the physical state of biological membranes and the chilling sensitivity of higher plants has been discussed since the early 70s. The most abundant constituents, beside proteins, of photosynthetic membranes are glycerolipids. Thylakoid membranes are constructed from glycerolipids and form bilayers providing the necessary background for the functioning of membrane proteins [1,2]. In thylakoid membranes of chloroplasts there are four abundant glycerolipids; monogalactosyl diacylglycerol (MGDG), digalactosyl diacylglycerol (DGDG), sulfoquinovosyl diacylglycerol (SQDG), and phosphatidylglycerol (PG). These glycerolipids are assumed to play determinative roles in the maintenance of functioning photosynthetic machinery. SQDG is associated with ATP synthase [3], and specific MGDG is bound to the photosystem 2 (PSII) reaction center complex [4]. Recently a study on mutant strain of *Chlamydomonas reinhardtii* lacking SQDG also underlined the involvement of this lipid in maintenance of PSII activity [5].

The physical characteristics of glycerolipids depend on the degree of the unsaturation of fatty acids that are esterified to the glycerol backbone of the lipids [6,7]. Consequently the molecular motions of these glycerolipids can be affected by the alterations in the extent of unsaturation of fatty acids. These changes in unsaturation should alter various functions of membrane bound proteins, such as photochemical and electron-transport reactions in thylakoid membranes. It is of interest whether the unsaturation of fatty acids of glycerolipids in thylakoid membranes can modify the stress tolerance of photosynthetic machinery. The unsaturation of glycerolipids in biological membranes can be modified by various ways. The natural alteration that can affect the level of unsaturation of fatty acids in thylakoid membranes is the change in growth temperature. It was reported that an increase in the growth temperature increased the degree of saturation of fatty acids and enhanced the stability of membranes at high temperature. Such temperature-induced changes are explained in terms of the regulation of membrane fluidity that is necessary to maintain the optimal functioning of biological membrane [8]. This approach failed to clarify the direct relationship between the degree of unsaturation of glycerolipids and the high-temperature stability of photosynthetic machinery, because not only the saturation of fatty acids but also several other metabolic factors are affected by changes in growth temperature [9–11]. For similar reasons one who tries to use this approach to investigate the involvement of fatty-acid unsaturation in chilling tolerance can face the same difficulties to find clear evidence for the relationship between glycerolipid unsaturation and chilling tolerance.

Cyanobacterial cells resemble chloroplasts of eukaryotic plants in terms of membrane structure and glycerolipid composition [12]. The cyanobacterial cells possess a simple membrane system consisting of the plasma membrane, the outer membrane, and the thylakoid membrane [13]. The thylakoid membranes are closed systems and separated from the plasma membrane. Since the techniques for isolation and separation of these membranes are well established [14] the processes bound to various membranes and the involvement of the membrane constituents in these processes can be studied separately. These characteristics make cyanobacterial strains an excellent model system to study photosynthesis-related physiological processes of higher plants without the difficulties of more complex systems. In several cyanobacterial strains the responses to temperature stresses have been studied for a long time. In these photosynthetic organisms the shift in growth temperature results in changes of the unsaturation of glycerolipids as an obligate tool for the manipulation of membrane structure [15,16]. For the manipulation of glycerolipid unsaturation the synthesis of unsaturated fatty acids should be well characterized and the genes encoding enzymes involved in these processes should be known.

1.2. Fatty-Acid Desaturases of Cyanobacteria

Desaturases are the enzymes that introduce double bonds into fatty acids. There are three types of desaturases [17], in cyanobacteria the acyl-lipid desaturases act as the only type of desaturases [18]. These desaturases introduce double bonds into fatty acids that have been esterified to glycerolipids. The acyl-lipid desaturases can be divided into subgroups according to their electron donors. The cyanobacteria use ferredoxin as electron donor [18]. In terms of fatty-acid desaturation, cyanobacterial strains can be classified into four groups [19]. Group 1 is characterized by the presence of only saturated and monounsaturated fatty acids. The strains belonging to the other three groups (2, 3 and 4) can synthesize polyunsaturated fatty acids.

The cells in group 1 contain only one desaturase which enzyme can introduce double bonds at the $\Delta 9$ position of fatty acids. Concerning its thermal tolerance the most studied strain of this group is *Synechococcus* sp. PCC 7942 (*Anacystis nidulans* R2). Strains in group 2 can introduce double bonds at $\Delta 9$, $\Delta 12$ and $\Delta 15$ ($\omega 3$) positions of fatty acids. Strains in group 3 can also introduce three double bonds, but at the $\Delta 6$, $\Delta 9$ and $\Delta 12$ positions [19]. Strains in group 4 can desaturate fatty acids at $\Delta 6$, $\Delta 9$, $\Delta 12$ and $\Delta 15$ ($\omega 3$) positions. *Synechocystis* sp. PCC 6803 the most frequently used strain for genetic manipulations belongs to group 4. Among the four groups, group 2 is the most similar to the chloroplasts of higher plants in terms of desaturation of fatty acids. The specificity of cyanobacterial desaturases to the position in the alkyl chain of fatty acids at which the double bond is introduced was studied in *Synechocystis* sp. PCC 6803 [20]. The double bonds appeared at positions $\Delta 6$, $\Delta 9$ and $\Delta 12$ counted from the carboxyl-terminus, regardless of the chain length of the fatty acids [21]. The double bond at $\Delta 15$ position of C_{18} fatty acids was located at position $\omega 3$, counted from the methyl terminus. From these physiological results, the four desaturases present in *Synechocystis* sp. PCC 6803 were designated $\Delta 6$, $\Delta 9$, $\Delta 12$ and $\omega 3$ acyl-lipid desaturases.

To determine whether the unsaturation of fatty acids contributes to the ability to tolerate temperature stresses, it is necessary to alter the level of unsaturation of glycerolipids exclusively by manipulation of the genes encoding fatty acid desaturases, thereby avoiding effects of any other metabolic factors. We will summarize the recent development on the importance of fatty-acid desaturation in the tolerance and acclimatization of higher plants and cyanobacteria to temperature stresses in relation to modification of photosynthetic membranes by genetic manipulation of desaturases.

2. *SYNECHOCYSTIS* SP. PCC 6803

2.1. Targeted Mutagenesis of Acyl-Lipid Desaturases

An approach to modify the unsaturation level of fatty acid composition of photosynthetic membranes was the sequential elimination of desaturases from *Synechocystis* sp. PCC 6803 [22,23]. Previously chemically induced mutagenesis was used to produce mutants lacking desaturase activities at well-defined steps of the fatty-acid synthesis. Two mutants Fad6 and Fad12 [24] were used for further investigations. The mutation site of Fad6 has not been analyzed. The nucleotide sequence of the *desA* gene from Fad12 mutant was identical with the exception at a nucleotide at 315 position [25]. The mutation converted a codon for leucine, TTG, to a stop codon, TAG. That was the reason why in Fad12 mutant there was no active enzyme. Insertional disruption [26] of the *desA* and *desB* genes provided a unique experimental system. Using this system the number of double bonds in lipid molecules can be manipulated [22,23,27,28]. Wild-type cells and Fad6 mutant, defective in the Δ6 desaturase, were transformed with *desA* gene disrupted by insertion of a Km^r (kanamycin-resistance gene) cartridge. As a result a transformant strain could be obtained without polyunsaturated fatty acids. To avoid the difficulties of using Fad6 with an unidentified mutation site wild type of *Synechocystis* was transformed with a plasmid termed *pdesD*::Cm^r. Wild type and *desA$^-$* [22] were transformed with this plasmid to produce *desD$^-$* and *desD$^-$/desA$^-$* cells [23] having no tetraenoic and polyunsaturated fatty acids, respectively. These results demonstrate that the extent of desaturation of glycerolipid in membranes can be manipulated. Consequently the fatty acid composition of *Synechocystis* sp. PCC 6803 can be modified. These transformants mentioned above were objected to physiological studies to establish relationship between unsaturation of glycerolipids in photosynthetic membranes and thermal tolerance of photosynthetic machinery.

2.2. Photosynthetic Activity and Growth Rate

The photosynthetic activities of the mutant strains was very similar to those of the wild type suggesting that the polyunsaturated fatty acids are not essential for the functioning photosynthetic machinery .The growth rates of wild-type and transformed cells were compared. The results demonstrated that the elimination of 18:3 fatty acids from the membranes of *desD$^-$* mutant does not affect the growth rate at all the measured temperatures [23]. The *desA$^-$/desD$^-$* could grow at 35°C with similar growth rate as the wild type and other mutants but the cell division was suppressed under 30°C compared to the wild-type, *desA$^-$* and *desD$^-$* strains. *desA$^-$/desD$^-$* transformed strain could not grow at 20°C [23]. These results suggest that to maintain the necessary membrane structure for cell division and for the function of photosynthetic machinery cells need 18:2 fatty acids in the lower temperature range.

In general we could conclude that the change in desaturation of lipids in thylakoid membrane does not affect the photosynthesis and PS II activity when the cells possessing different degrees of fatty-acid unsaturation when the cells are gown at isothermal condition.

2.3. Low Temperature Tolerance

It was a general observation in wide range of organisms that decrease in the temperature of the environment resulted in an enhancement of unsaturation of fatty acids in various membranes [29]. The change in the extent of unsaturation of glycerolipids accompanied with the effect on thermal motilities of lipid molecules and as a consequence it alters the functions of proteins embedded in the membranes such as the proteins of the photosynthetic machinery and several membrane-bound enzymes. Clear correlations between the membrane fluidity and the degree of unsaturation of glycerolipids were provided by model experiments [30]. It is noteworthy that the downshift in temperature of the environment manifests not only in the increase of the level of unsaturated molecular species of glycerolipids, but also in changes in some other products of cellular metabolisms in relation to the acclimation process to low temperatures. Thus, to establish the role of *in vivo* unsaturation of glycerolipids in acclimation processes to chilling temperatures, it is essential to manipulate the extent of unsaturation of glycerolipids independently of other metabolic products of the cells. To reach this aim several attempts have been tried. To evaluate the role of fatty-acid desaturation on chilling tolerance of photosynthetic organisms we used genetically manipulated strains.

To clarify the physiological roles of polyunsaturated glycerolipids in chilling tolerance two transformable cyanobacterial strains were used. *Synechocystis* sp. PCC 6803, a strain which possesses polyunsaturated fatty acid up to the extent of four double bounds and it is a chilling-resistant strain. The other strain was *Synechococcus* sp. PCC 7942 that was subjected to study the effect of introduction of polyunsaturated fatty acids on chilling tolerance. This strain originally has only one desaturase enzyme, which can introduce double bond at $\Delta 9$ position, and its thermal characteristics showed that this strain is sensitive to chilling stresses. When photosynthetic activities of transformed and wild type strains grown at various temperatures were compared under isothermal condition, those cells grown at lower temperatures are greater than those of cells grown at higher temperatures [31]. The change in photosynthetic activity due to the growth temperature has been regarded as a result of alterations in unsaturation of glycerolipids. Our recent studies clearly demonstrated that changes in the extent of glycerolipid unsaturation in photosynthetic membranes do not affect photosynthesis and electron transport from H_2O to 1,4-benzoquinone, when all strains were grown at isothermal conditions. Consequently we could conclude that the growth-temperature dependent change in photosynthetic activity is unrelated to the extent of lipid unsaturation.

2.4. Tolerance to Low-Temperature Photoinhibition

The combined effect of low-temperature treatment and the exposure of cells to high light intensity was also studied. It is a well-known phenomenon that the low-temperature sensitivity of photosynthesis is considerably enhanced when the photosynthetic organisms are exposed to light [32,33]. This phenomenon is called low-temperature photoinhibition. The primary target of these processes is the D1 protein of the protein complex known as PS II [34, 35]. Wild type of *Synechocystis* sp. PCC 6803, *desD⁻* and *desA⁻/desD⁻* cells grown at isothermal conditions were exposed to various duration of illumination at low temperatures. The extent of photoinhibition of photosynthesis depended on the strain and the temperature of treatment. At all the measured temperatures *desA⁻/desD⁻* cells containing monounsaturated fatty acids as the only unsaturated fatty-acid molecular species were the most susceptible to low-temperature photoinhibition. *desD⁻* cells lacking $\Delta 6$ desaturase was indistinguishable from wild type concerning the sensitivity to low-temperature photoinhibition. Since the extent of photoinhibition depends on the balance between the photoinduced inactivation process and the recovery from photoinactivated state of photosynthetic machinery the two processes were separated. Using an inhibitor of procaryotic protein synthesis, it was demonstrated that the inactivation site of photoinhibitory processes is not affected by the alteration of the extent of fatty-acid unsaturation. This observation was strengthened by measuring the inactivation profile of photosynthesis on isolated thylakoid membrane where the recovery process is absent. Experiments with intact cells underlined the importance of the effect of altered fatty-acid composition on the recovery process [31,36].

Recently the D1-protein turnover was studied in mutant *desD⁻* and *desA⁻/desD⁻* strains. The proteolytic digestion of the damaged D1 in the wild type, *desD⁻* and *desA⁻/desD⁻* cells of *Synechocystis* sp. PCC 6803 was studied by immunoblotting and it was observed that changes in the extent of unsaturation of glycerolipids in thylakoid membranes does not alter the removal of the damaged D1 protein from the membrane as well as the transcription of the *psbA* (gene for the D1 protein). The results suggested that polyunsaturated fatty acids are involved in protein synthesis, at the translational or posttranslational level [37]. Probably either the re-assembly of the apo-PSII complex with the pre-D1 protein or the processing of the pre-D1 protein to the mature D1 protein needs polyunsaturated glycerolipids, since the pre-D1 was accumulated in the transformed cells at low temperatures.

3. SYNECHOCOCCUS SP. PCC 7942

3.1. Introduction of Novel Fatty-Acid Desaturase

The strategy to modify the extent of desaturation of glycerolipids in the membranes was the introduction of fatty-acid desaturase, which is not present in wild type, into *Synechococcus* sp. PCC 7942 (*Anacystis nidulans* R2). This strain originally has only one desaturase, which can introduce double bond at Δ9 position. This strain is the representative of group 1 according to the unsaturation of glycerolipids classified by Murata et al. [18]. *Synechococcus* cells were transformed with the *desA* gene isolated from *Synechocystis* sp. PCC 6803 [22]. As a result the transformed cells could synthesize polyunsaturated glycerolipids. It appears that the transformation with the *desA* gene rendered the *Synechococcus* cells to be able to introduce the second double bond at the Δ12 position of fatty acids. This transformed cells were able to produce 16:2(9,12) and 18:2(9,12) polyunsaturated fatty acids.

3.2. Low-Temperature Tolerance

Synechococcus sp. PCC 7942 (*Anacystis nidulans* R2) belongs to Group 1 according to the classification of Murata et al. [18]. The chilling susceptibility of *Anacystis nidulans* has been studied for a long time. The observation that the chilling susceptibility of photosynthesis, measured by various parameters, the phase behavior of photosynthetic membranes and the growth temperature are correlated with each other led to the suggestion that the membrane structure has a crucial role in thermal behavior of this photosynthetic organisms. Studies on the unsaturation of glycerolipids of this organism revealed that the modulation of fatty-acid unsaturation could be the most important response in temperature adaptation [38,39]. In these reports the importance of both the cytoplasmic and thylakoid membranes were stressed. To get direct evidences for the relationship between chilling temperature and the extent of unsaturation of glycerolipid a novel desaturase gene was introduced to the wild-type strain of *Synechococcus* sp. PCC 7942. The presence of polyunsaturated glycerolipid enhanced the tolerance of photosynthetic functions of this chilling sensitive strain to cold treatment [28]. The cells transformed with *desA* gene could survive and preserve the photosynthetic activities even following a 1 hour long treatment at 5°C in the dark. The wild-type cells in contrast lost their photosynthetic activity irreversibly and they could not recover after this treatment [40]. Consequently, one can state that the presence of polyunsaturated fatty acid in membranes of photosynthetic organisms is essential to protect the cells against dramatic decrease in temperature of natural environment.

3.3. Tolerance to Low-Temperature Photoinhibition

For the study the involvement of dienoic fatty acids in tolerance to low-temperature photoinhibition *desA+* transformant and wild type strains of *Synechococcus* sp. PCC 7942 were used [41]. The transformed strain growing at 25°C contained 36% dienoic glycerolipids, by contrast, the wild type possesses mono unsaturated lipid species only. When the cells of both strains grown at isothermal conditions were exposed to strong illumination at temperature lower than the growth temperature, photosynthesis was remarkable inhibited. Albeit, the cells with dienoic fatty acids were significantly more tolerant to low-temperature photoinhibition than the wild type cells possessing only monounsaturated fatty acids. Using inhibitor of protein synthesis during the photoinhibitory treatment experiments clearly showed that there was no difference in the kinetics of photoinhibition between transformant and wild type cells. On the other hand recovery experiments emphasized the involvement of dienoic lipids in the recovery processes. Therefore, we can conclude that dienoic fatty acids play determinative role on the recovery side of photoinhibition. The analyses of D1 protein turnover also underlined this idea.

4. CONCLUSION AND FUTURE PERSPECTIVES

To summarize the results obtained by using genetically manipulated cyanobacterial strains we can conclude the following:

1. The genetic manipulation of cyanobacterial fatty-acid synthesis provides a unique experimental system to study the importance of fatty-acid unsaturation in adaptive responses to temperature stresses.
2. Fatty-acid unsaturation does not affect the functions of photosynthetic machinery either in cyanobacteria or in higher plants. However, detailed studies on the kinetics of electron transport have not been done.
3. The polyunsaturated fatty acids, especially the dienoic acids, are essential for the adaptation to and tolerance of low-temperature stress. The phase behavior of lipids in photosynthetic membranes from various transformants with altered lipid compositions can be subjected for biophysical studies of membrane structure and its relationship to photosynthetic characteristics.
4. The role of polyunsaturated fatty acids in the recovery from low-temperature photoinhibition was clearly demonstrated. The relationship between the membrane structure and the temperature of the induction of low-temperature photoinhibition has not been answered. The site of the insertion of proteins (D1) into the thylakoid membranes or the membrane related processing of D1 proteins are the most suspected candidate to be the target of this effect. The intimate specific interaction between proteins and lipids should be studied to understand membrane-related processes.

ACKNOWLEDGMENTS

This work was supported, in part, by Grants-in-Aid for Scientific Research on Priority Areas (Nos. 04273102 and 04273103) to N. Murata and by Hungarian Science Foundation (OTKA; nos. T 020293 and F 023794) to Z. Gombos, supported by the Balaton secretariat of the Prime Minister's Office to Zs. Várkonyi.

REFERENCES

1. Doyle, M.F. and Yu, C.-A. (1985) Biochem. Biophys. Res. Commun. 131, 700–706.
2. Trémolières, A., Dubacq, J.-P., Ambard-Bretteville, F. and Remy, R. (1981) FEBS Lett. 130, 27–31.
3. Pick, U., Weiss, M., Gounaris, K. and Barber, J. (1987) Biochim. Biophys. Acta 891, 28–39.
4. Murata, N., Higashi, S. and Fujimura, Y. (1990) Biochim. Biophys. Acta 1019, 261–268. [5] Sato, N., Tsuzuki, M., Matsuda, Y., Ehara, T., Osafune, T. and Kawaguchi, A. (1995) Eur. J. Biochem. 230, 987–993.
6. Chapman, D. (1975) Quart. Rev. Biophys. 8, 185–235.
7. Quinn, P.J. (1988) in: Physiological regulation of membrane fluidity (Aloia, R.C., Curtain, C.C. and Gordon, L.M. Eds.) vol. 3, pp. 293–321, Alan R Liss, Inc., New York.
8. Cossins, A.R. (1994) in: Temperature Adaptation of Biological Membranes (Cossins, A.R., Ed.) pp. 63–76, Portland Press, London.
9. Cooper, P. and Ort, D.R. (1988) Plant Physiol. 88, 454–461.
10. Guy, C.L., Niemi, K.J. and Brambl, R. (1985) Proc. Natl. Acad. Sci. USA 82, 3673–3677.
11. Mohapatra, S.S., Poole, R.J. and Dhindsa, R.S. (1987) Plant Physiol. 84,1172–1176.
12. Stanier, R.Y. and Cohen-Bazire, G. (1977) Annu. Rev. Microbiol. 31, 225–274.
13. Murata, N. and Nishida, I. (1987) in: The Biochemistry of Plants (Stumpf, P.K. Ed.) vol. 9, pp. 315–347, Academic Press, Orlando/Florida.
14. Murata, N. and Omata, T. (1988) in: Methods in Enzymol. (Packer, L. and Glaser, A.N. Eds.) vol. 167, pp. 245–251, Academic Press, San Diego/California.
15. Murata, N. (1989) J. Bioenerg. Biomembr. 21, 61–75.
16. Nishida, I. and Murata, N. (1996) Annu. Rev. Plant Physiol. Plant Mol. Biol. 47, 541–568.
17. Murata, N. and Wada, H. (1995) Biochem. J. 308, 1–8.
18. Wada, H., Schmidt, H., Heinz, E. and Murata, N. (1993) J. Bacteriol. 175, 544–547.
19. Murata, N., Wada, H. and Gombos, Z. (1992) Plant Cell Physiol. 33, 933–941.
20. Wada, H. and Murata, N. (1990) Plant Physiol. 92, 1062–1069.
21. Higashi, .S and Murata, N. (1993) Plant Physiol. 102, 1275–1278.
22. Wada, H., Gombos, Z., Sakamoto, T. and Murata, N. (1992) Plant Cell Physiol. 33, 535–540.
23. Tasaka, Y., Gombos, Z., Nishiyama, Y., Mohanty, P., Ohba, T., Ohki, K. and Murata, N. (1996) EMBO J. 23, 6416–6425.
24. Wada, H. and Murata, N. (1989) Plant Cell Physiol. 30, 971–978.
25. Gombos, Z., Wada, H., Várkonyi, Zs., Los, D. and Murata, N. (1996) Biochim. Biophys. Acta 1299, 117–123.
26. Williams, J.G.K. (1988) Methods Enzymol. 167, 766–778.
27. Sakamoto, T., Los, D.A., Higashi, S., Wada, H., Nishida, I., Ohmori, M. and Murata, N. (1994) Plant Mol. Biol. 26, 249–264.
28. Wada, H., Gombos, Z. and Murata, N. (1990) Nature 347, 200–203.
29. Thompson, G.A. Jr. (1980) in: The regulation of membrane lipid metabolism. pp. 171–196, CRC Press, Boca Raton, Florida
30. Coolbear, K.P., Berde, C.B. and Keough, K.M.W. (1983) Biochemistry 22, 1466–1473.
31. Gombos, Z., Wada, H. and Murata, N. (1992) Proc. Natl. Acad. Sci. USA 89, 9959–9963.
32. Greer, D.H., Ottander, C. and Öquest, G. (1991) Physiol Plant. 81, 203–210.
33. Öquist, G., Greer, D.H. and Ögren, E. (1987) in: Photoinhibition (Kyle, D.J., Osmond, C.B. and Arntzen, C.J. Eds.) pp. 67–87, Elsevier, Amsterdam, The Nederlands
34. Nanba, O. and Satoh, K. (1987) Proc. Natl. Acad. Sci. USA 84,109–112.
35. Aro, E.-M., Hundal, T., Carlberg, I. and Andersson, B. (1990) Biochim. Biophys. Acta 1019, 269–275.
36. Gombos, Z., Wada, H. and Murata, N. (1994) Proc. Natl. Acad. Sci. USA 91, 8787–8791.
37. Kanervo, E., Tasaka, Y., Murata, N. and Aro, E.-M. (1997) Plant Physiol. 114, 841–849.

38. Ono, T.-A. and Murata, N. (1981) Plant Physiol. 67, 176–181.
39. Ono, T.-A and Murata, N. (1981) Plant Physiol. 67, 182–187.
40. Wada, H., Gombos, Z. and Murata, N. (1994) Proc. Natl. Acad. Sci. USA 91, 4273–4277.
41. Gombos, Z., Kanervo, E., Tsvetkova, N., Sakamoto, T., Aro, E.-M. and Murata, N. (1997) Plant Physiol. In press.

76

POIKILOTROPH GROWTH PATTERNS IN ROCK INHABITING CYANOBACTERIA

A. A. Gorbushina, W. E. Krumbein,* and K. A. Palinska

Department of Geomicrobiology
Carl von Ossietzky University
26111 Oldenburg, PO Box 2503, Germany

INTRODUCTION

Assemblages of immobilised microbial cells embedded in ambient extracellular polymers termed biofilms are well established on water- and air-exposed solid surfaces [1,2]. Subaqueous biofilms consist of microbial cells and often more than 99% water stabilised and immobilised by a matrix of large amounts of extracellular polymeric substances (EPS) formed by the cells. Subaerial biofilms on exposed rocks in contrast can be regarded as accumulations of cell material and EPS maintaining life at the presence of a minimum of water often amounting to less than 1% of the total cell mass [2]. They survive for long periods of time with stress conditions militating against any physiological process [3–5].

Several ways of surviving stress on the subaerial rock surface are possible for microorganisms: (i) quickly spreading in the vegetative form to places with better living conditions; (ii) forming dormant spores or cysts or resting modifications of the vegetative body (e.g. chlamydospores) and (iii) formation of propagative spores. The organisms in subaerial biofilms posses a remarkable capacity of vegetative cell survival and usually do not form any propagative forms. They protect their vegetative cells from desiccation and isolation by a cell wall with EPS layers and photoprotective pigments. The cyanobacteria encountered in such environments usually do not form hormogonia, akinetes or any other spore-like survival or propagative structure.

Growth patterns and survival principles are presented of stress tolerant rock inhabiting cyanobacteria and compared with the behaviour of other poikilotrophic microorganisms living in subaerial biofilm communities.

---

* Corresponding author.

The Phototrophic Prokaryotes, edited by Peschek *et al.*
Kluwer Academic / Plenum Publishers, New York, 1999.

MATERIALS AND METHODS

Subaerial biofilms dominated by representatives of *Stigonema* and *Chroococcidiopsis* covering large areas of quartzite rocks on the Auyan-Tepui in Venezuela were investigated. Light microscopy (Axioplan Zeiss), thin sections and scanning electronmicroscopy (SEM) analyses have been done. For SEM samples were dehydrated by increasing the concentration of ethanol and finally by the critical point method (Balzers Union CPD 010) before gold sputtering (Balzers Union SCD 030). Samples were analyzed using a DSM 940 ZEISS SEM.

RESULTS AND DISCUSSION

Subaqueous biofilms and their development are very well investigated [1,6]. Some processes well-proved for a subaqueous biofilm also take place and even play a more important role in a subaerial biofilm. The important difference between subaqueous and subaerial biofilm is the content of water. Subaqueous biofilms consist of stabilised water—99% and more [7–9]. Subaerial biofilms consist of drought resistant accumulations of cell material and highly specialised extracellular polymeric substances (EPS) serving for water substitution and holding the minimum of water to guarantee survival [5,2].

The occurrence of cyanobacteria on the air-exposed surface of terrestrial substrates like soil, rock or trees is well documented [10,11]. Epilithic as well as endolithic cyanobacteria have been described [12]. A very expressive and characteristic subaerial biofilm of rock inhabiting algae and cyanobacteria was firstly described by Jaag [13] and later elegantly elaborated by Golubic [14] under the name "Tintenstrichflora". These black stretches of microbial growth on rock surfaces are very characteristic world-wide on exposed rocks and monument surfaces. They are usually dominated by drought-resistant but water demanding cyanobacteria.

Many of the rock inhabiting cyanobacteria grow as a mixed community entangled with a dense network of microfibrils of extracellular polymeric substances (EPS) between the cyanobacteria and the surface of the rock [15,16]. This characteristic growth—well described in marine and intertidal environments [17,18] and typical for desiccation stress protection [5]—is common also on many air-exposed rock surfaces. Typical examples are *Microcoleus, Chroococcidiopsis, Stigonema,* and other members of the Pleurocapsales and branched cyanobacteria groups.

Microorganisms in a subaerial biofilm are highly drought resistant and possess a high potential of meristematic tissue-like adaptations and a remarkable capacity of survival of vegetative cells. The organisms dwelling here usually do not form spores, akinetes or any other propagative forms. They, however, protect all of their vegetative cells by a thick cell wall containing several types of photoprotective and resistant pigments and by accumulated EPS layers and slime sheaths. Very often the cell division pattern is causing the formation of smaller cells dividing inside of the protective envelope. It is also observed that the cell wall layers themselves are highly condensed and thickened and, that high accumulations of water containing proteins are stored within the cells.

Cyanobacteria found on a subaerial rock surface possess certain morphological properties enabling their survival on the substrate. They can dedicate all their energy resources to changing their morphology and metabolism according to the environmental conditions without attempts to produce propagules. The rock surface is representing an en-

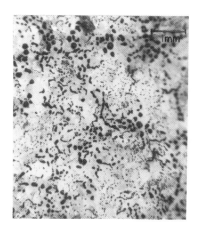

Figure 1. Compact colonies of unicellular cyanobacteria on the subaerial surface of quarzite in nutrient poor and dry environment of Ayan-Tepui (Venezuela).

vironment in which motility of microorganisms and their spreading over the substrate is hindered. Filamentous growth, breakage and hormogonia formation, usually used in nutrient poor environments for reaching new, less extreme and more favourable conditions cannot take place due to the lack of water and nutrients. But growth and cell division is continuing, taking new forms and favouring the formation of colonies, in which each individual cell has a high survival potential. Longevity of individual cells is one of the major principles of the poikilotroph micro-organisms.

On the subaerial rock surface colonial (aggregate) pattern is represented by patchy growth of trichome assemblages and round colonies of unicellular forms (Fig. 1). These forms are usually filamentous with interwined trichomes and coalescent sheaths forming close-knit bundles (Fig. 2). This growth pattern is also found in the strands of *Microcoleus* (soil biofilm), the sheets of *Phormidium* and the bundles of *Oscillatoria*. In others (*Nostoc*, *Rivularia*) and the colonial unicellular forms (e.g. *Gloeocapsa*), the cyanobacteria become aggregated into gelatinous globose colonies. These mucilaginous colonies lose water slowly and have a remarkable capacity to take up water quickly. The sheath in air-exposed cyanobacteria acts as a reservoir of water, where it is bound through strong molecular forces. This thick sheath might allow the cyanobacteria to loose or gain water from the trichomes and appears to be an important factor in desiccation tolerance [5]. These growth patterns have been described previously by Paracelsus [7] and Ehrenberg [3]. The

Figure 2. Subaerial biofilm with filamentous cyanobacteria forming close-knit bundles of interwined trichomes. Patches of colonial growth of unicellular forms are also visible all over the surface.

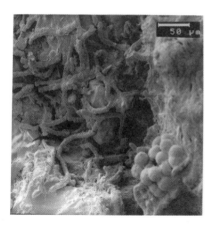

Figure 3. SEM micrograph of subaerial cyanobacterial biofilm clearly demonstrates the tendency to form a covering biofilm with EPS layer, which is jointly used by the whole subaerial rock inhabiting community.

development of an extracellular layer of EPS in subaerial biofilm is not a single cell or colony characteristic, but creates a jointly used layer for the whole community (Fig. 3).

The very stable but hostile conditions on rock surface are allowing an adapted organism to use all the advantages and potentials of the vegetative mode of growth. The genotype able to survive in such conditions could be considered as a favourable one and the organism is going to exploit all present advantages without risking any changes. For instance, for propagation small parts of the vegetative thallus could be used even more successfully than specially produced spores. In rock inhabiting microorganisms it is expressed in diverse protected vegetative structures designed for survival and spreading of the organism.

The formation of compact colonies by rock-inhabiting cyanobacteria (Fig. 4) could be interpreted as a spreading mechanism for the vegetative structure with the same genotype. Growth process itself should not be excluded from the list of spreading mechanisms. Initially cyanobacteria grow as a colony with free, widely spread trichomes, composing a net-like structure. In stressed conditions they change to "compacted" colony growth patterns. This also influences the growth rate: the colony is not expanding and the biomass is not significantly increasing over periods of up to 50 years. This "stationary" phase with extremely long generation times (weeks or months) has been observed by us in natural cy-

Figure 4. The photomicrograph of a well developed community of colonial cyanobacteria biofilm on the surface of a quartz crystall from Tepui. The formation of compact colonies much favoured in subaerial biofilms is shown.

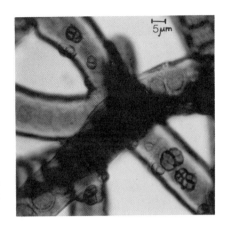

Figure 5. Compact small colonies of *Chroococcidiopsis* on *Stigonema* filaments. Intracellular cell formation of the subaerial colonies of genus *Chroococcidiopsis*.

anobacterial populations of the Ayan-Tepui, an extremely nutrient poor quartzite substrate exposed in addition to high desiccation stress.

A considerable number of native rock inhabitants (*Chroococcidiopsis, Sarcinomyces petricola, Lichenothelia* sp., *Phaeococcus* sp. etc.) have a special form of intracellular cell formation when several genetically identical cells are formed by a founder cell. This process is especially well described for the group of black yeasts [19–21]. Similar division patterns are also observed in the rock inhabiting genus *Chroococcidiopsis* [22]. Almost identical observations were made by us (Figs. 5 and 6) and by Eppard et al. [23] for certain types of actinomycetes of the genera *Geodermatophilus* and *Frankia*.

Other morphological responses of cyanobacteria to unfavourable condition are characteristic changes in pigmentation. The strategies that cyanobacteria use in order to tolerate high sun radiation are varied. In the upper level of the cyanobacterial zone and in subaerial biofilms the cells show obvious signs of the effect of high illumination, such as especially the brown coloration of the cell walls, known as a characteristic response of cyanobacteria to high light intensities [24]. The presence of the photoprotective pigment scytonemin was also demonstrated for cyanobacteria of the Ayan-Tepui. Most commonly the extracellular, UV-adsorbing sheath pigment scytonemin, represents a passive shielding

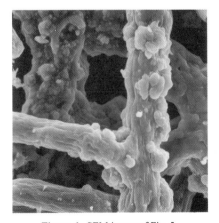

Figure 6. SEM image of Fig. 5.

method which provides protection under metabolic inactivity. The investigated subaerial cyanobacterial biofilm from Venezuela was showing a high level of scytonemin content in the ethanol extract (280–490 nm). From the literature it is known that the scytonemin content of the sheaths increased when cyanobacterial samples were exposed to high light intensity or UV-A [25]. Also all 23 cyanobacterial species from open rock surfaces of the Orinoco lowlands showed intensively coloured sheaths, either brown or yellow [22]. It is also know, that in most of the species without sheaths, cytoplasmic mycosporine-like compounds are elicited.

It was recently found in *Chroococcidiopsis* sp. that nitrogen depletion *in vitro* is causing a change from multicellular aggregates to single viable spore-like cells [26]. Dor et al. [27] observed a similar phenomenon of cell size increasing in response to desiccation. However, during the growth in the natural environment the reduction to propagule-like unicellular growth was not observed by us. This could be explained by the fact that in subaerial biofilms exposed to permanent and sudden changes in water and nutrient availability, colonial growth and the tendency to form a covering biofilm with "joint" EPS layer is stimulated (Fig. 3).

In limnology, marine and soil sciences and in general microbiology it is common to differentiate between zymotrophic or copiotrophic (eutrophic) and oligotrophic microorganisms [28]. In the special and optimal conditions of a subaqueous biofilm the oligotrophic-copiotrophic classification is considered to be unnecessary [1]. But for the subaerial biofilm such a classification needs to be used and can even be expanded.

The oligotrophic organism is adapted to the existence in permanently poor or stressed environments and will be suppressed or eliminated under optimal nutritive and environmental conditions. The copiotrophic (eutrophic) microorganism inhabits nutrient-rich and favourable environments [29]. The rock environment is characterized by generally poor nutrient supply, long periods of dessication, extreme changes of pH, salinity, temperature and irradiation pressure. It is further characterized in desert conditions by sporadic changes of all physiologically important parameters, which in turn make it necessary for an organism to survive periods of "dormancy" lasting as long as 10–25 years. The rock dwelling microflora may, therefore, be dominated by poikilotrophic microorganisms. This new nutritional group consists of genera and species that have a potential of surviving under prolonged conditions of lack of energy sources and/or under lack of water for the metabolic activity. They survive in dormant or close to dormant status which includes minimisation of endogenous metabolism as well as slow and controlled mobilisation of reserve materials [30]. The very moment, however , when water and/or food is available these will immediately be reactivated and run under fast metabolic rates without investing energy in germination of spores. The poikilotroph mode of growth and survival was first described and suggested by Krumbein [31] in a paper on the sociobiology of rock dwelling microbiota.

The behaviour of an organism in a stress situation in the natural environment is practically always limited by resource availability or stress [32]. The conditions on the air-exposed surface of the rock are allowing the growth of only those microorganisms that are highly adapted to the environmental stress and all limitations set by the peculiar environment. Even more: environments such as "hypersaline" "hydrothermal highly alkaline" or "acid" usually described as "extreme" environments as a matter of fact present very narrow, usually constant and optimal (normal or normative) conditions for highly adapted oligotroph or eutroph micro-organisms. The real "extreme" environment can be characterised by extreme and additionally episodic or sporadic changes of the conditions necessary for life. If one wanted to find a micro-organism fit for survival on the surface of Mars one should not screen permanently dry, hypersaline, hot or cold environments. One should screen the environment of the poikilotroph microbiota on rock surfaces exposed to a maximum of environmental changes lasting over irregular periods of time.

The poikilotroph rock inhabiting cyanobacteria are adapted to permanent disturbance which is easily demonstrated by their growth pattern. Among the most typical features of poikilotroph rock inhabiting cyanobacteria the following could be mentioned:

1. Formation of compact colonies consisting of cells with beneficial surface/volume ratio (or trichome bundles).
2. The growth rate is quite low, but any part of the colony or any individual well protected cell is a potential propagative structure.
3. Formation of EPS layers surrounding all the cells in a subaerial biofilm is serving the survival of the whole subaerial biofilm as a community.
4. Protective pigments and cell wall structures form additional barriers against adverse environmental conditions.
5. Internal and external structures and compounds (polysaccharides and proteins) serve as strong binding and incorporation sites of water.
6. The physiological demand for water is reduced to a metabolic minimum.

REFERENCES

1. Characklis, W.G. and Marshall, K.C. (1990) Biofilms: a basis for an interdisciplinary approach. In: Biofilms (Characlis, W. G., Marshall, K.C. Eds.), p 4.
2. Gorbushina, A. A. and Krumbein, W. E. (1997) Microbial effects on subaerial rock surfaces. In: Microbialites (Riding, R. and Awramik, S. Eds.) Spriger, Berlin (in press).
3. Ehrenberg, C.G. (1838) Über das im Jahre 1786 in Curland vom Himmel gefallene Meteorpapier und über dessen Zusammensetzung aus Conferven und Infusorien. Abh. Königl. Akad. d. Wiss. Berlin, Physikal. Kl. 44–66.
4. Krumbein, W.E. and Jens, K. (1981) Biogenic rock varnishes of the Negev Desert (Israel), an ecological study of iron and manganese transformation by cyanobacteria and fungi. Oecologia, vol. 50, pp. 25–38.
5. Potts, M. (1994) Dessication Tolerance in Prokaryotes. Microbiol. Reviews, 58, 755–805.
6. Little, B. J., Wagner, P. A., Lewandowski, Z. (1997) Spatial relationships between bacteria and mineral surfaces. Reviews in Mineralogy, 35, 123–159.
7. Paracelsus, T. B. (1982) Werke. Edited by Peukert. Schwabe and Co., Basel, 426p.
8. Haeckel, E. (1877) *Bathybius* und die Moneren. Kosmos, 1, 293–305.
9. Krumbein, W.E. (1984) Auf den Schultern des Riesen—vom Zeitgeist in der Erforschung geomikrobiologischer Zusammenhänge. Mitt. Geol.-Pal. Institut Hamburg, 56: 435–460.
10. Friedmann, E. I. and Ocampo, R. (1976) Endolithic blue-green Algae in the dry valleys: Primary producers in the Antarctic desert ecosystem. Science, 193,1247–1249
11. Krumbein, W. E. (1983) Damage caused by cyanobacteria and fungi in plasters and casting covering medieval frescos in North German Rural churches. GP News Letter, 4, 27–30.
12. Friedmann, E. I. and Ocampo-Friedmann, R. (1984) Endolithic microorganisms in extreme dry environments : Analysis of a lithobiontic habitat. In: Current Perspectives in Microbial Ecology (Klug, M. J., Reddy, C.A. Eds.), 177–186.
13. Jaag, O. (1945) Untersuchungen über die Vegetation und Biologie der Algen des nackten Gesteins in den Alpen, im Jura und im Schwiezerichen Mittelland. In: Beitr. Kryptogamenfl. Schweiz, 9(3), 560pp. Buchler& Co., Bern.
14. Golubic, S. (1967) Algenvegetation der Felsen, eine ökologische Algenstudie im dinarischen Karstgebiet. Binnengewaesser23, Schweizerbarth, Stuttgart, 183 pp.
15. Le Campion-Alsumard, T. (1975) Etude experimentale de la colonisation d'eclats de cacite par les Cyanophycees endolithes marines. Cah.Biol.Mar, 16: 177–185.
16. Le Campion-Alsumard, T. (1979) Les cyanophycees endolithes marines. Systematique, ultrastructure, ecologie et biodestruction. Oceanologica Acta, 2: 143–156.

17. Schneider, J. and Torunski, H. (1983) Biokarst on Limestone Coasts, Morphogenesis and Sediment Production. Marine ecology, 4 (1): 45–63.
18. Krumbein, W.E. (1994) The year of the slime—instead of an introduction. In: Biostabilization of Sediments (Krumbein W.E., Paterson, D.M., Stal, L.J. Eds.) BIS Oldenburg, pp. 1–7.
19. Gorbushina, A.A.; Krumbein, W.E.; Hamman, C.H.; Panina, L.; Soukharjevski, S. and U. Wollenzien. (1993) Role of Black Fungi in Colour Change and Biodeterioration of Antique Marbles. Geomicrobiol. Journ., 11: 205–22.
20. de Hoog, G.S. (1993) Evolution of black yeasts: possible adaptation to the human host. Anton van Leeuwenhoek, 63, 105–109.
21. Wollenzien, U., de Hoog, G.S., Krumbein, W. E. and Urzi, C. (1995) On the isolation of microcolonial fungi occurring on and in marble and other calcareous rocks. 167, 287–294.
22. Büdel, B., Lüttge, U., Stelzer, R., Huber, O. and Medina E. (1994) Cyanobacteria of rocks and soils of the Orinoco lowlands and the Guayana Uplands, Venezuela. Botanica Acta, 107, 422–431.
23. Eppard, M, Krumbein, WE, Koch, C, Rhiel, E, Staley, J, Stackebrandt, E, (1996) Morphological, physiological and molecular biolocal investigations on new isolates similar to the genus *Geodermatophilus* (Actinomycetes). Archives of Microbiology, 166: 12–22.
24. Friedmann, E. I. (1971) Light and scanning electron microscopy of the endolithic desert algal habitat. Phycologia, 10 (4), 411–428
25. Garcia-Pichel, F. and Castenholz, R. W. (1991) Characterization and biological implications of Scytonemin, a cyanobacterial sheath pigment. Journal of Phycology, 27, 395–409
26. Billi, D. and Grilli Caiola, M. (1997) Effects of nitrogen and phosphorus deprivation on *Chroococcidiopsis* sp. Algological Studies, 83, 93–107
27. Dor, I., Carl, N. and Baldinger, I. (1991) Polymorphism and salinity tolerance as criterion for differentiation of three new species of *Chroococcidiopsis* (Chroococcales). Algological Studies, 64, 411–421.
28. Winogradsky, W. I. (1952) Soil Microbiology, Moscow, 792 pp.
29. Gromov, B.V. and Pavlenko, G.V. (1989) Ecology of Bacteria (in Russian). 248 pp.
30. Hirsch, P., Bernhard, M., Cohen, S.S., Ensign, J.C., Jannasch, H.W., Koch, A.L., Marschall, K.C., Matin, A., Poindexter, J.S., Rittenberg, S.C:, Smith, D.C., Veldkamp, H. (1979) Life Under Conditions of Low Nutrient Concentrations. In: Strategies of Microbial Life in Extreme Environments (M.Shilo Ed.), Berlin: Dahlem Konferenzen, 357–372.
31. Krumbein, W. E. (1988) Biotransformations in monuments - a sociological study, Durability of Building Materials, 5, 359–382.
32. Brock, Th.D. (1966) Principles of Microbial Ecology, Prentice-Hall, Inc., Engelwood Cliffs, New Jersey.

77

DEVELOPMENTAL ALTERNATIVES OF SYMBIOTIC *NOSTOC PUNCTIFORME* IN RESPONSE TO ITS PLANT PARTNER *ANTHOCEROS PUNCTATUS*

Jack C. Meeks, Elsie Campbell, Kari Hagen, Tom Hanson, Nathan Hitzeman, and Francis Wong

Section of Microbiology
University of California
One Shields Avenue, Davis, California 95616

1. INTRODUCTION

In species of certain filamentous cyanobacterial genera of the orders Nostocales and Stigonematales, such as *Nostoc*, vegetative cells can mature in four developmental directions (Table 1). (i) The cells can divide upon reaching a critical size and perpetuate the vegetative growth cycle. (ii) The cells can divide uncoupled from biomass increase and DNA replication to form motile hormogonium filaments. (iii) Some, or all, cells can differentiate into perennating akinetes. (iv) A few cells (typically less than 10%) can differentiate into nitrogen-fixing heterocysts. Entry into a specific developmental state is initiated by environmental conditions that are presumed to be sensed by distinct signal cascade systems resulting in differential gene expression.

An appropriate experimental organism to use in identifying, characterizing and manipulating the genes of these signal cascade systems is the facultatively heterotrophic *Nostoc punctiforme* strain ATCC 29133 (PCC 73102). *N. punctiforme* was originally isolated as a symbiont from a corraloid root of the gymnosperm cycad *Macrozamia* sp. [1]; it has broad symbiotic competence and can establish associations with the angiosperm *Gunnera* spp. [2] and the bryophyte hornwort *Anthoceros punctatus* [3]. The symbiotic associations are of interest because the plant partners produce extracellular signals that stimulate hormogonium formation [4] and heterocyst differentiation is highly derepressed in the symbiotic growth state [3]. Moreover, *N. punctiforme* 29133 is amenable to physiological genetic analysis leading to molecular genetic characterization of genes of interest [5].

The Phototrophic Prokaryotes, edited by Peschek *et al.*
Kluwer Academic / Plenum Publishers, New York, 1999.

Table 1. Developmental alternatives of *Nostoc* species

Developmental alternative	Duration	Environmental signals
Vegetative	Perpetual; hour to day to week	Optimal environmental conditions for the species
Hormogonium	Transient; 24 to 72 hours	Shift to: nutrient excess, red light, solid substratum; autogenic factors
Akinete	Transient; hour, to annual cycle, to > 64 y	Phosphate limitation, energy limitation, autogenic factors
Heterocyst	Terminal; life time of about 2 generations	Combined nitrogen limitation, not specifically ammonium

1.1. Cellular Differentiation Alternatives

Members of the order Nostocales are distinguished by vegetative cell division on a plane that is primarily transverse to the filament, yielding an unbranched filament [6]. The division plane in *Nostoc* spp. tends to drift from a strictly transverse position, thereby yielding filaments that range from a kinky to a coiled morphology of varying degrees. Little is known about what controls the timing or placement of the septum in these cyanobacteria, although an analog of the cell division protein FtsZ has been detected [7].

Hormogonia (Hrm) are associated with vegetative fragmentation of filaments as part of a life cycle. Hormogonia are characterized as gliding filaments lacking heterocysts, with smaller sized and differently shaped cells [8,9]. The cells cease DNA replication and biomass increase upon entry into the hormogonium state [8], and have decreased rates of carbon and nitrogen assimilation [4,10]; as such, they resemble growth precursor cells [11]. Their formation can be induced by a variety of environmental conditions, including light quality, nutritional status and transfer to solid medium [8,9]. When hormogonia differentiate in a heterocyst-containing filament, they arise as the interval between adjacent heterocysts by a detachment at the heterocyst-vegetative cell junction. Thus, hormogonium formation is mutually exclusive to heterocyst differentiation and *Nostoc* spp. cultures in the hormogonium state cannot fix nitrogen. Other than specific initiation of gas vesicle synthesis in *Calothrix* sp. strain PCC 7601, little is known of differential gene expression in hormogonium formation or return to the vegetative state [9]. It is common for *Nostoc* strains maintained by repeated transfer in culture to lose hormogonia motility and experience a muting of their life cycle [6].

The entire population of vegetative cells may differentiate into akinetes (Aki); in some species a few akinetes can form in specific spatial patterns relative to the position of heterocysts. These spore-like cells function in perennation under environmental extremes less harsh than endospores of Gram positive eubacteria [12]. A variety of environmental conditions, including nutrient and energy deprivation [12], and autogenic production of a sporulation factor [13], stimulate their formation. Although surveys of metabolic activities and enzymology of spores have been published [14], little is known of the mechanisms of environmental sensing, or of regulation of differentiation or germination [12].

Heterocysts (Het) are the other type of differentiated cyanobacterial cell. The terminally-differentiated heterocysts are highly specialized for nitrogen fixation in an oxic environment. They form in a semi-regular pattern of spacing in the filaments in response to nitrogen deprivation of free-living populations. Akinetes and heterocysts contain a polysaccharide outer layer that is unique and chemically identical in the two cell types [15].

The Het⁻Aki⁻ phenotype of a *hetR* mutant in *Nostoc ellipsosporum* provides experimental evidence that heterocyst and akinete differentiation is initiated by a common nutrient deprivation signal cascade [16], but continued development into either cell type is determined by specific environmental signals and regulatory genes. It is likely that heterocysts evolved from an akinete precursor [17]. In a physiological context, compared to vegetative cells, heterocysts are sinks of reduced carbon and sources of reduced nitrogen in the filaments [17]. One estimate based on RNA:DNA hybridization suggested that approximately 40% of the cyanobacterial genome is involved in formation and maintenance of functional heterocysts [18]. A number of structural genes of heterocyst formation and function have been identified [17,19,20] and some assigned positions on a physical genomic map of *Anabaena* sp. strain PCC 7120 [21].

Three major states of heterocyst expression in natural populations of cyanobacteria can be identified. (i) Repressed: Growth in the presence of combined nitrogen, especially ammonium, inhibits the differentiation of heterocysts [17]. (ii) Derepressed: During growth in the absence of combined nitrogen, heterocysts are found at relatively regular intervals along the filaments, typically at a frequency of 3–8% of the cells. This state first involves establishing the pattern of heterocyst placement as the combined nitrogen is exhausted and then maintenance of the pattern as the vegetative cells grow and divide utilizing the fixed nitrogen supplied by the heterocysts. Products of proheterocyst and heterocyst metabolism acting by repression are thought to establish and maintain the pattern [17]. (iii) Symbiotically induced: Growth in a physical symbiotic association with a photosynthetic eukaryotic partner results in heterocyst frequencies ranging from 20% to more than 60% of the cells [see 22].

It is generally assumed, but unproved, that states (i) and (ii) operate through a common mechanism. The most prevalent working hypothesis is that this common mechanism involves a type of nitrogen regulatory system functioning through a cellular repressing signal, which is thought to be an organic product of ammonium assimilation. We further hypothesize that the environmental signal stimulating heterocyst differentiation in the symbiotic state (iii) supersedes the nitrogen-dependent system operative in the free-living state.

1.2. A Symbiotic Experimental System

The plant partner in a symbiotic association can be viewed as a unique habitat with specific physicochemical factors that influence growth and development of the cyanobacterial partner. The selective pressure for the formation and maintenance of these associations is presumably provision of N_2-derived ammonium or organic nitrogen by the cyanobacterium for growth of the plant partner. The symbiotic association between *N. punctiforme* and the hornwort *A. punctatus* is an exemplary experimental system because the association can be reconstituted with pure cultures of *A. punctatus* and various isolates and mutant strains of *Nostoc* spp., the reconstituted association grows rapidly in liquid culture and it is amenable to routine experimental manipulations [23]. Based on physiological characterizations, the interactions between the cyanobacterial and plant partners can be categorized into two distinct experimental stages (below) that involve two of the developmental alternatives of *Nostoc*; hormogonium and heterocyst differentiation.

1.2.1. Stage 1. Interactions That Result in Formation of the Symbiotic Association (Infectiveness). All symbiotically competent *Nostoc* spp. strains produce hormogonia and hormogonia are the infective units of *Nostoc* spp.-plant symbiotic associations [10,24]. In the *A. punctatus* association, hormogonia enter a preformed cavity in the gametophyte tissue and establish a colony of macroscopic dimensions [3]. When *A. punctatus* is cultured under nitrogen-limited conditions, it produces an exudate which contains a hormogonium-inducing factor (HIF), that stimulates *Nostoc* spp. to produce hormogonia to 10-fold or higher levels than seen in typical free-living cultures [4]. The identity of HIF is not known, but similar activity is present in the stem gland mucilage of *Gunnera* spp. [25] and in exudate from wheat roots [26]. The consequence of plant-induced hormogonium formation is presumably an increase in the probability of an infection event. *A. punctatus*-induced hormogonium filaments remain in the growth precursor state for the equivalent of two generation times, which effectively defines the initial infection window. Hormogonium filaments return to the vegetative growth state by increasing cell size, replicating DNA and differentiating heterocysts at the ends of the filament [4,8].

Since hormogonia are a nongrowth state, continued entry into this developmental state is presumed to be lethal. Thus, one might predict that *Nostoc* spp. have a mechanism to block the sensing of a HIF, or an equivalent signal, when in its continual presence, or to block the initiation of differentiation. Repression of hormogonium differentiation by autogenic compounds is observed under the following conditions: (i) when carrying across excess spent medium during transfer of cultures [8], and (ii) during coculture with *A. punctatus* following the initial round of hormogonium differentiation [4].

Whereas all symbiotically competent *Nostoc* spp. strains form hormogonia, not all hormogonium-forming strains are symbiotic. An example is *Nostoc* sp. strain ATCC 27896; this strain immediately converts 90% of its vegetative filaments into hormogonia in the presence of HIF [4]. However, its infection of *A. punctatus* tissue is statistically insignificant. Delayed rare infections of *A. punctatus* can be detected about 8 weeks after coculture, in contrast to 10 to 14 d in competent strains [Campbell and Meeks, unpublished results]. The *Nostoc* 27896 colonies fix nitrogen and, as evidenced by greening of the host tissue around the *Nostoc* colony, release the fixed nitrogen. The symbiotic association does not persist due to a lack of reinfection. When reisolated from *A. punctatus*, *Nostoc* 27896 shows the same low frequency, delayed infection phenotype. Thus, *Nostoc* 27896 reflects an infection defective phenotype in a symbiotically effective background. It is possible that *Nostoc* 27896 lacks a chemotactic sensing or responding system to signals produced by *A. punctatus*. Preliminary evidence for the presence of such chemotactic systems in a symbiotic *Nostoc* sp. and a response to exudate of the bryophyte *Blasia pusilla* has recently been presented [27].

1.2.2. Stage 2. Interactions That Yield a Functional Symbiotic Association (Effectiveness). The physiological characteristics of *Nostoc* spp. in association with *A. punctatus* under steady state culture conditions, in contrast to the free-living growth state, are given in Table 2. To achieve a stable association, there is clearly an *A. punctatus* (host)-dependent slowing of the growth rate of associated *Nostoc* sp. In proportion to the slower growth rate of associated *Nostoc* sp. there is an apparent posttranslational inhibition of the specific activity of the initial enzymes of ammonium (glutamine synthetase) [28] and carbon dioxide (ribulose bisphosphate carboxylase) [29] assimilation. Regulation of these enzyme activities appears to vary amongst the various cyanobacterial-host associations, as does the mechanism of regulation [24]. However, there is no obvious causal relationship between the decreased rates of growth and those of assimilation of carbon and nitrogen.

Nitrogen-limited *A. punctatus* appears to produce an environment in the symbiotic cavity that stimulates heterocyst differentiation of associated *Nostoc* sp. to a 5 to 10-fold

Table 2. Summary of growth and *in vivo* or *in situ* carbon and nitrogen assimilatory activities of *Nostoc* sp. in steady state free-living or symbiotic association

Activity	Growth state	
	Free-living	Symbiotic
Growth as doubling time in hours [3]	40	< 240 (17%)[*]
Photosynthesis as nmol $^{14}CO_2$ fixed/min/mg protein [29]	128	15.1 (12%)
	n = 19	n = 12
Ammonium assimilation as cpm of ^{13}N assimilated/min/mg protein [31]	13.9	2.9 (21%)
	n = 11	n = 6
Nitrogen fixation as nmol C_2H_2 reduced/min/mg protein [30]	5.3	23.5 (443%)
	n = 8	n = 42

[*]Percent of free-living values; n = number of replicates; [ ] reference.

higher frequency than the free-living growth state [3]. Relative to the free-living growth state, the rate of symbiotic nitrogen fixation is 4 to 35-fold higher [30], while the rate of assimilation of N_2-derived NH_4^+ is 10-fold lower [31]. The excess N_2-derived NH_4^+ is not utilized as a metabolite by *Nostoc* sp., rather it is made available for growth of *A. punctatus* tissue [31]. Immunocytochemical observations indicate that *Nostoc* sp. in association with *A. punctatus* in natural samples synthesizes nitrogenase only in heterocysts [32]. This observation implies the lack of an alternative Mo-Fe nitrogenase that would be expressed in vegetative cells under anoxic conditions (i.e. the symbiotic cavity, see below) as occurs in free-living *Anabaena variabilis* [33]. Therefore, the increased rate of nitrogen fixation correlates with an increased symbiotic heterocyst frequency. The fact that *Nostoc* sp. in association with *A. punctatus* differentiates a high frequency of heterocysts while growing in the presence of high concentrations of N_2-derived NH_4^+ contributes to a working hypothesis that the environmental sensing component of the regulatory system for heterocyst differentiation in *Nostoc* sp. differs in the symbiotic, compared to the free-living, growth state. Further evidence against negative regulation of heterocyst differentiation through a combined nitrogen sensing system in symbiosis comes from the characterization of a *N. punctiforme* 29133 mutant that is unable to assimilate NO_3^-. In this mutant, NO_3^- does not repress heterocyst differentiation in the free-living growth state, in contrast to the wild-type [34]. However, exogenous NO_3^- did repress nitrogenase expression in both the wild type and mutant strains when they were in association with *A. punctatus* [34]. When reisolated from *A. punctatus* tissue, the mutant retained its NO_3^- defective phenotype; thus, it was unlikely to have directly responded to the exogenous NO_3^- when in the symbiotic cavity. We suggested that *A. punctatus* assimilated the exogenous nitrogen and no longer produced the chemical signal or environment that stimulates heterocyst differentiation.

2. GENETIC ANALYSIS OF INTERACTIONS WITH *A. PUNCTATUS* THAT INFLUENCE DEVELOPMENTAL ALTERNATIVES OF *N. PUNCTIFORME*

A physiological genetic approach to analysis of *N. punctiforme* 29133 was recently developed [5]. This approach is based on mutation by transposition, using derivatives of Tn*5* that were constructed by C. P. Wolk and his colleagues for use in *Anabaena* 7120 [35]. Phenotypic characterization of the resulting mutant in the free-living and symbiotic growth states is followed by recovery of the transposon and flanking genomic DNA from the mutant, sequence analysis of the flanking genomic DNA, or of the complete open reading frame isolated from a cosmid library of wild-type *N. punctiforme* total DNA, and analysis of transcriptional regulation of the presumptive gene. Gene replacement with insertionally inactivated, or site directed mutagenized, genes via homologous recombination is also possible in *N. punctiforme* 29133 [36]. Progress in the application of this genetic approach to identify and characterize genes of *N. punctiforme* that respond to plant signals during symbiotic interactions is described in the remainder of this paper.

2.1. Identification and Characterization of *N. Punctiforme* Genes and Gene Products Involved in the Infection Process

Mutants with defects in genes that are involved in the *A. punctatus*-dependent initiation of hormogonium differentiation have yet to be isolated. However, three mutants with a phenotype of increased initial infection of *A. punctatus* have been isolated and the corresponding genes characterized (Table 3). The mutants and genes were identified in *N. punctiforme* 29133 by three different approaches. The *hrm* operon was identified by characterization of a Tn5-1063-induced mutant that displayed a phenotype identical to the wild-type in the free-living growth state. The involvement of an alternate sigma subunit of RNA polymerase in the infection process was identified by screening the symbiotic phenotype of an insertion mutant after the gene had been cloned from *N. punctiforme* by heterologous hybridization. The gene encoding a tetratricopeptide repeat protein was detected by sequence walking on the 3′ side of *devR*, a gene encoding the receiver domain of a response-regulator protein essential for heterocyst development. Infection phenotypes of *N. punctiforme* strains were scored in all cases after two weeks of coculture with *A. punctatus*, in the absence of combined nitrogen, by counting macroscopically visible *Nostoc* colonies in a given weight of gametophyte tissue.

2.1.1. The hrm Operon. *N. punctiforme* 29133::Tn5-1063 mutant strain UCD 328 forms a high frequency of hormogonia in the presence of *A. punctatus* [37]. The increase in hormogonia correlates with an approximate eightfold increase in the initial infection frequency of *A. punctatus* (Table 3). Two open reading frames (ORF) were initially identified by sequencing genomic DNA flanking the site of transposition. The ORFs were desig-

Table 3. Effect of various insertion mutations on the symbiotic infectiveness and effectiveness of *N. punctiforme* 29133 strains in association with *A. punctatus*

Strain	Colonies per mg dry wt per µg Chl *a*	nmol C_2H_2 reduced per min per		Gene induced by
		g FW	Colony ($\times 10^{-3}$)	
ATCC 29133 (WT)	0.21 ± 0.04	6.3 ± 1.2	12.4 ± 3.3	—
	n = 25	n = 14	n = 14	
UCD 328 (*hrmA*)	1.6 ± 0.1	6.1 ± 1.1	8.6 ± 1.3	HRF
	n = 14	n = 10	n = 10	
UCD 398 (*sigH*)	1.2 ± 0.2	8.0 ± 3.9	10.1 ± 4.1	HIF
	n = 19	n = 4	n = 4	
UCD 400 (*tprN*)	0.49 ± 0.2	10.4 ± 1.9	6.7 ± 0.3	HIF & HRF
	n = 9	n = 9	n = 9	

n = number of replicates. HIF = exudate of *A. punctatus* containing hormogonium inducing factor. HRF = aqueous extract of *A. punctatus* containing hormogonium repressing factor identified as inducing the *hrm* operon.

nated *hrmU* and *hrmA*; *hrmA* has no significant similarity to sequences in major data bases, while *hrmU* has similarity to a family of NAD(P)H-oxidoreductases. Tn5-1063, containing *luxAB* as a reporter gene, had formed a transcriptional fusion with *hrmA* in the initial transposition event. The reporter gene was not induced by abrupt shifts in nutritional status of cultures, or by incubation of the mutant in the presence of plant exudate containing HIF, but it was induced by an aqueous extract of *A. punctatus*. Identical induction kinetics were observed in a *hrmU::luxAB* mutant strain. The aqueous extract also suppressed HIF-dependent hormogonium formation in the wild-type, but not the mutant strain. The results were interpreted to indicate that the *hrmUA* gene products are involved in the inhibition of hormogonium formation by the metabolism of an unknown hormogonium-regulating metabolite, perhaps of autogenic origin. The aqueous extract thus appears to contain what was termed a hormogonium repressing factor (HRF). This factor is not released into the growth medium, but may be released into the symbiotic cavity, the consequences of which would be to shift the developmental alternative away from formation of hormogonia and towards the differentiation of heterocysts.

Single recombinants of insertionally inactivated *hrmU* or *hrmA* genes in which the recombination events in the wild-type strain were shown by restriction mapping to have occurred 5′ of the insertion site and between 2.0 and 3.4 kb 5′ of the translational start of *hrmU*, yielded the mutant phenotype of high infectivity [37]. These results implied that the promoter/operator region of the *hrm* operon was between 2.0 and 3.4 kb 5′ of the *hrmU* translational start site and additional genes are likely to be involved in the metabolic pathway. Sequence of the 3.8 kb interval 5′ of *hrmU* has now been obtained (Fig. 1). The region contains three additional ORFs; the two ORFs immediately 5′ of *hrmU* are transcribed in the same direction as *hrmUA* and the third ORF at the 5′ end of the interval is transcribed in the opposite direction. BLAST analysis indicates the newly identified ORFs of the *hrm* operon have similarity to genes encoding proteins involved in gluconate and glucuronate metabolism. The initial gene of the operon has similarity to gene members of the *lacI/galR* family encoding repressor proteins, such as the gluconate operon repressor; we have labeled this gene *hrmR* in *N. punctiforme*. The second gene has similarity to glu-

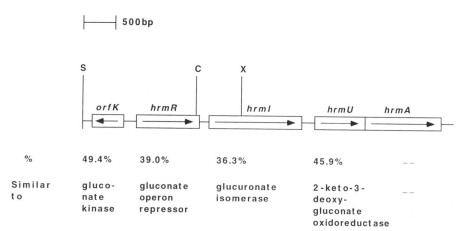

Figure 1. ORF map of the *hrm* operon. Arrows note the direction of transcription. The similarity values are to amino acid sequence from BLAST analysis of double stranded nucleotide sequences of the region. There are no significance sequence similarities to *hrmA*. Restriction enzymes: S = *Spe*I, C = *Cla*I, X = *Xba*I.

curonate isomerase and is designated as *hrmI*. *hrmU* is most similar to several oxidoreductases, including 2-keto-3-deoxygluconate dehydrogenase and 3-ketoacyl-[acyl-carrier protein] reductase. The oppositely transcribed ORF has similarity to gluconate kinase and is provisionally designated *orfK*. Not only do the individual genes have sequence similarity to those of gluconate and glucuronate metabolism, but the organization of the operon is similar to the glucuronate catabolizing *uxu* operon of *Escherichia coli*, with the exception that in *E. coli* the isomerase gene (*hrmI* analog) is in the *exu* regulon [38]. Although it lacks significant sequence similarity, *hrmA* may be equivalent to the dehydratase that forms 2-keto-3-deoxygluconate from mannonate, an intermediate of glucuronate catabolism; none of the known dehydratases in gluconate or glucuronate metabolism show significant sequence similarities to each other. We are currently examining the effects of gluconate and glucuronate on growth, hormogonium formation and infection of *A. punctatus* by wild-type and mutant strains of *N. punctiforme*.

The mutant strain UCD 328 high infection phenotype could not be complemented by a 6.2 kb fragment [37], now known to contain the entire *hrm* operon and *orfK* (Fig. 1). Moreover, this fragment, and a 4.6 kb fragment lacking only *hrmA*, suppressed HRF induction of *hrmA*::Tn*5*-1063-*luxAB* expression. A similar suppression of *hrmA*::Tn*5*-1063-*luxAB* expression occurs when a 1.8 kb fragment containing the presumptive operator region of the *hrm* operon, all of *orfK* and all but 10% of the C terminus of *hrmR* is present in trans (*Spe*I to *Cla*I fragment in Fig. 1). Whether the suppression is due to the presence of the repressor protein or of other trans acting factors that interact in the *hrm* operator region is being investigated.

2.1.2. An Alternate Sigma Subunit and Carboxyl-Terminal Protease. Sigma subunits of RNA polymerase in eubacteria recognize specific promoter sequences and, with or without activator proteins, stimulate formation of open complexes of DNA to initiate transcription. There are two fundamental families of sigma subunits in eubacteria, referred to in *E. coli* as the protein sequences of the sigma-70 family and the sigma-54 family [35]. Based on function and sequence similarity, three classes are recognized within the sigma-70 family: class 1, the primary sigma which is essential for growth; class 2, alternate sigmas with sequence similarity to the primary sigma, but which are dispensable for growth; and class 3, alternate sigmas which differ in sequence similarity from the primary sigma and initiate transcription from distinct promoter regions. Genes encoding two class 2 alternate sigma subunits (*sigB* and *sigC*) were characterized in *Anabaena* 7120 [40]. While transcription of the sigma subunits was responsive to the nitrogen source for growth, neither subunit was obligately involved in cellular differentiation, nitrogen fixation or any other detectable phenotypic trait. Using the *Anabaena* 7120 *sigB* gene as a heterologous probe, at least five restriction fragments in the genome of *N. punctiforme* with similarity to sigma-70 family subunits were identified. One of those fragments, encoding a class 2 alternate sigma subunit designated *sigH* that is involved in infection of *A. punctatus* by *N. punctiforme*, has been characterized.

sigH of *N. punctiforme* shares 72%, 64% and 59% amino acid sequence similarity to *sigB*, *sigC* and *sigA*, in the order stated, of *Anabaena* 7120 (Fig. 2A). The *sigH* gene was interrupted with a *npt*-based antibiotic resistance cassette and *N. punctiforme* mutant strain UCD 398 isolated by gene replacement. Strain UCD 398 has no obvious phenotypic defects under normal culture conditions, in medium with and without combined nitrogen. However, a high infection phenotype, similar to that of *hrmUA⁻* strains, was observed during coculture with *A. punctatus* (Table 3). While *sigH* could be involved in the transcriptional induction of the *hrm* operon, three lines of evidence indicate that *sigH* and *hrmUA*

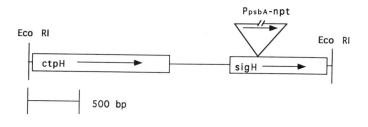

Figure 2. ORF maps of the *ctpH-sigH* (A) and *devR-tprN* (B) regions. The sites of insertion of antibiotic cassettes into *sigH* and *tprN* are shown; neither insertion is drawn to scale.

do not operate in the same developmental or plant interaction pathway. First, transcription of *sigH* is induced from an essentially zero basal level in less than 1.5 h after exposure of cultures to exudate of *A. punctatus* containing HIF and not by the plant extract containing HRF that induces the *hrm* operon. Second, induction of *hrmA*::Tn5-*luxAB* transcription by plant extract containing HRF was not altered in a *sigH⁻* background. Third, mutant strain UCD 398 (*sigH⁻*) does not form an elevated level of hormogonia during coculture with *A. punctatus*, as does mutant strain UCD 328 (*hrmA⁻*). The *hrmUA* high infection phenotype was proposed to be due to an increase in the infection window as a consequence of continued entry into the hormogonium state in the presence of HIF [37]; this is a lethal nongrowth condition. The high infection phenotype of the *sigH* mutant apparently has a quite different basis. One possibility is that it could respond more intensely to other plant signals, such that all hormogonia are infective rather than a few. Preliminary evidence indicates that the exudate from bryophytes also contains chemoattractants of *Nostoc* sp. hormogonia [25]. Synthesis or activation of the chemotactic sensors in *N. punctiforme* could be elevated in a *sigH* mutant background.

A gene (*ctpH*) with 50 to 60% similarity to a three member family of genes encoding carboxyl-terminal proteases of *Synechocystis* sp. strain PCC 6803 is immediately 5' of *sigH* (Fig. 2A). The CtpA of *Synechococystis* 6803 is required for processing of the carboxyl-terminal portion of the photosystem II D1 (32 kDa) protein in the thylakoid lumen and subsequent activation of the oxygen evolving complex [41]. However, since *ctpH* is not transcribed under vegetative growth conditions, it would appear to have a different physiological role in *N. punctiforme*. *ctpH* transcription is induced by plant exudate containing HIF in a pattern similar to *sigH*. Although physically linked in the *N. punctiforme* genome, *ctpH* and *sigH* do not appear to be consistently cotranscribed. *ctpH* is currently being inactivated in *N. punctiforme* 29133 as a step in defining the role of its gene product.

   2.1.3. A Tetratricopeptide Repeat Protein. A gene (*tprN*) with similarity to tetratricopeptide repeat proteins (TPR) was identified 3′ of *devR* in the *N. punctiforme* genome (Fig. 2B). *devR* encodes a receiver domain only response-regulator protein [42]. In most two component sensor-kinase, response-regulator systems involved in cellular response to environmental changes, the two genes are collocated in the genome [43]. Thus, in an attempt to identify the sensor-kinase donor to DevR, sequence was obtained from both sides of *devR*, but a sensor-kinase homolog was not detected. The lack of physical linkage of sensor-kinase and response-regulator genes seems to be characteristics of two component systems involved in developmental processes, such as *Bacillus* sporulation [44]. TPR proteins are relatively wide spread, although they have been studied primarily in eukaryotes, where they may have a variety of functions ranging from cell cycle control, to transcription repression, to protein transport [45].

   tprN was inactivated in the *N. punctiforme* 29133 genome by insertion mutagenesis and gene replacement. Mutant strain UCD 400 has no apparent phenotypic defect in the free-living growth state, but displays an infection phenotype that is about twofold higher than the wild-type strain, but is not as infective as the *hrmUA* or *sigH* mutants (Table 3). The *tprN* gene is transcribed during vegetative growth, but its transcription is elevated following exposure to both *A. punctatus* exudate containing HIF and extract containing HRF. There is currently insufficient information to speculate on the role of *tprN* in the infection process.

2.2. Identification and Characterization of *N. Punctiforme* Genes Involved in Formation of a Functional Symbiotic Association

   Genetic studies on the regulation of growth and assimilatory metabolism of *Nostoc* spp. in the symbiotic growth state have not yet been initiated. The initiation of heterocyst differentiation, heterocyst maturation and nitrogenase expression has been the primary focus. It is of interest to note that the high infection phenotypes of *hrmA*, *sigH* and *tprN* are not accompanied by a statistically significant, correspondingly high rate of nitrogen fixation (Table 3). Various treatments which affect the size of the *Nostoc* spp. colony in *A. punctatus* gametophyte tissue do not markedly alter the rate of nitrogen fixation per unit of *A. punctatus* biomass, within limits. For examples; blocking of reinfection by penicillin treatment leads to a few large colonies with proportionally higher rates of nitrogen fixation per colony [3], and growth of the association under high light and enriched CO_2 leads to fewer and smaller colonies also with high rates of nitrogen fixation per colony and per *Nostoc* biomass [30]. It seems that *A. punctatus* may sense the amount of N_2-derived NH_4^+ that is required and by an unknown mechanism regulate the extent of nitrogenase expression and/or heterocyst differentiation by the symbiotically associated *Nostoc* sp.

2.2.1. A Working Model of the Initiation Cascade for Heterocyst Differentiation and Maturation. The model we are testing is illustrated in Fig. 3, with known genes placed in an epistatic relationship. Major portions of Fig. 3 were adapted from Wolk et al. [17] and then various sources. The model can be considered in three phases of heterocyst formation. (i) Initiation of differentiation in vegetative cells; inactivations of *ntcA* or *hetR* or *hanA* (not shown) are the only genetic consequences known to block initiation of heterocyst differentiation, and *ntcA* is epistatically dominant over *hetR* [46]. (ii) Proheterocyst development; which includes aspects of the pattern of heterocyst placement in the filaments, and activation or synthesis of regulatory genes, such as *devR* or *hetP,* that are involved in the continued synthesis of structural components of the unique wall (*hglE,* *henA*). (iii) Heterocyst maturation; which involves synthesis of nitrogenase components and metabolic pathways for support of nitrogenase activity.

Based in part on the apparent uncoupling of symbiotic heterocyst differentiation from the direct repressive effects of exogenous or endogenous combined nitrogen [34], we suggest that a symbiotic specific sensing/signaling pathway may operate in parallel with a combined nitrogen sensing/signaling pathway in the initiation of differentiation. Where the two sensing/signaling pathways might converge in a putative common pathway of heterocyst differentiation is speculative. Convergence on *ntcA*, which is the first gene currently known to be activated in the initiation of differentiation, is depicted in Fig. 3, but it is not known how *ntcA* is activated in the nitrogen deprivation pathway. If a symbiotic specific sensing/signaling pathway does operate, mutants with a lesion in certain components of the combined nitrogen sensing/signaling pathway should be able to differentiate heterocysts and fix nitrogen in the A. *punctatus* association. Two genetic approaches to test this model in *N. punctiforme* 29133 are being taken: (i) transposon mutagenesis, fol-

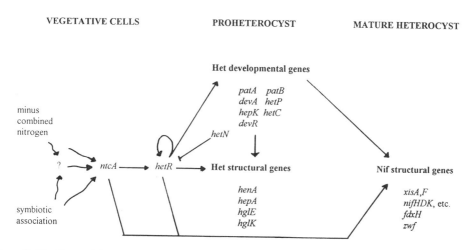

Figure 3. Working model depicting the stages of heterocyst differentiation. Initiation of heterocyst differentiation in vegetative cells of free-living cultures is in response to a signal of combined nitrogen limitation. This signal is proposed to be superseded by a symbiotic specific signal when *Nostoc* spp. are in association with A. *punctatus* and other plants. Proheterocysts are morphologically distinct and reflect a stage where a variety of regulatory systems and metabolic pathways for synthesis of heterocyst cellular structures are active. Under certain conditions, proheterocysts can dedifferentiate to vegetative cells. The transition to a mature heterocyst marks a commitment to terminal differentiation. The metabolic systems characteristic of heterocyst function are active in this stage. The genes have been placed in a relative epistatic order; some relationships are purely speculative and the model is inevitably incomplete. The genes have been identified variously in the literature, but see [17,19,20].

lowed by screening for mutants unable to differentiate heterocysts in the free-living growth state and then screening of Het⁻ mutants for symbiotic competence in the *A. punctatus* association; and (ii) cloning, insertational inactivation and screening the symbiotic phenotypes of mutations in genes known to affect initiation of heterocyst differentiation in other, nonsymbiotic strains. Because vegetative cells become large and distorted in the symbiotic growth state and symbiotic heterocysts may contain phycobiliproteins [10], it is difficult to distinguish vegetative cells and heterocysts by light microscopy. Therefore, nitrogen fixation is used as the provisional indicator of the presence or absence of heterocyst in symbiosis.

2.2.2. Mutagenesis of N. Punctiforme 29133 with Tn5-1063. A collection of 30,100 *N. punctiforme*::Tn5-1063 mutants has been established (Table 4), which represents a mutation frequency of approximately 1×10^{-6}, similar to the frequency observed in the initial Tn5-1063 mutagenesis experiments with *N. punctiforme* 29133 [5]. Mutants with defects in nitrogen fixation in the presence of air (provisionally Fox⁻) were detected by transfer of the mating filters to solidified medium lacking combined nitrogen. Colonies that begin to yellow within 10–14 days due to loss of pigmentation as a consequence of protein turnover during prolonged nitrogen starvation are scored as Fox⁻. Fox⁻ mutants represented 0.2% of the total *N. punctiforme*::Tn5-1063 mutants recovered. Fox⁻ mutants were then transferred to liquid culture conditions with NH_4^+ as the nitrogen source for further analysis. The remaining colonies were collected and are stored as a massed library at −80°C.

Some randomly chosen Fox⁻ mutants were screened for anoxic acetylene reduction capacity (true Fox⁻) and then for symbiotic competence by coculture with *A. punctatus*. The symbiotic cavity occupied by *Nostoc* spp. in *A. punctatus* gametophyte tissue was previously shown to have a microoxic atmosphere and could physiologically protect the nitrogenase from oxygen inactivation in Fox⁻ mutants such as strain UCD 311 (*devR⁻*) [42]. However, only 25% of the Fox⁻ mutants fixed nitrogen in symbiotic association with *A. punctatus*. Thus, it appears that a *N. punctiforme* 29133 Fox⁻Sym⁺ phenotype is more complicated than the singular provision of a low oxygen tension.

All Fox⁻ mutants in liquid culture were subjected to nitrogen step-down and periodically samples were observed microscopically by phase contrast and epifluorescence illumination for the presence or absence of heterocysts and of a pattern of nonfluorescing cells in filaments. Vegetative cells fluoresce orange to red when excited with green (510 to 560 nm) light, while heterocysts lack such fluorescence. Six putative Het⁻ mutants, that also lacked a pattern of nonfluorescing cells in the filaments, were recovered from 69 Fox⁻ mutants examined. This approximate 8% frequency of Het⁻ mutants in a population of Fox⁻ mutants is considerably higher than the 1.5% reported for *Anabaena* 7120 [47]. The transposon and flanking genomic DNA has been recovered from four of the Het⁻ mutants and the flanking DNA sequenced. In three of the mutants, the transposon inserted into the

Table 4. Transposon mutants of *N. punctiforme* 29133 screened for defects in nitrogen fixation and symbiotic competence

Total antibiotic resistant colonies recovered	Number of Fox⁻ mutants	Number of Het⁻ mutants
30,100 (mutation frequency of about 1×10^{-6})	69 (0.2%) (symbiotic frequency about 25% of those tested)	6 (0.02%) (the 4 Het⁻ mutants now tested are Sym⁻)

same ORF, but at different sites and in both orientations (e.g. strain UCD 416). The sequence of this ORF has no significant similarity to sequences in the major data bases. In the fourth mutant (strain UCD 415), the transposon had inserted into a gene whose derived gene product has 92% similarity to HetR of *Anabaena* 7120 [48,49]. The *luxAB* of Tn5-1063 in strains UCD 415 and 416 formed a transcriptional fusion to *hetR* and the *orf416*, respectively, upon transposition. Luciferase expression in strain UCD 415 increased about threefold starting about 3 h after nitrogen step-down, whereas expression in strain UCD 416 remained at a constant low level during the nitrogen starvation period.

2.2.3. Symbiotic Phenotypes of N. Punctiforme Het⁻ Mutants. Strains UCD 415 and UCD 416 both establish symbiotic colonies in *A. punctatus* tissue, but they both failed to support N_2-dependent growth of *A. punctatus* tissue, or to reduce acetylene at a reproducibly detectable rate. These results have significance to two aspects of the symbiotic interactions with *A. punctatus*. First, the infectiveness and effectiveness stages can clearly be distinguished experimentally and infection of the gametophyte tissue is not dependent on the presence of a functional nitrogen-fixing system. While the two stages could be distinguished using different natural isolates, such a separation of infectiveness and effectiveness had not been verified in the same genetic background [10]. Second, if the lack of nitrogen fixation is indicative of a lack of heterocysts, symbiotic heterocyst differentiation requires intact *hetR* and *orf416* genes. Therefore, it would appear that there is a common pathway for symbiotic and free-living heterocyst differentiation, at least beyond the temporal *hetR* control point, and that a symbiotic sensing/signaling system, if present, must enter into this common pathway at a temporal point prior to *hetR* (Fig. 3). The epistatic relationship between *orf416* and *hetR* and *ntcA* is under investigation.

The *ntcA* gene of *N. punctiforme* has also been cloned and sequenced (100% and 82% amino acid and nucleotide identity, respectively, to that of *Anabaena* 7120) [50]. Selection of an insertionally inactivated gene replacement with which to examine the symbiotic phenotype of a *ntcA* mutant is in progress. Results with the *ntcA* mutant strain and those of the currently uncharacterized *N. punctiforme*::Tn5-1065 Het⁻ mutants will contribute to describing the signal cascade system involved in the initiation of heterocyst differentiation in free-living and symbiotic growth states.

ACKNOWLEDGMENTS

Work in the authors' laboratory was supported by grants NSF IBN 95-14787 and NSF MCB 96-04270 from the U.S. National Science Foundation. We gratefully acknowledge Bianca Brahamsha for collaboration in cloning, sequencing and inactivation of the *N. punctiforme sigH* gene.

REFERENCES

1. Rippka, R., Deruelles, J., Waterbury, J.B., Herdman, M. and Stanier, R.Y. (1979) J. Gen. Microbiol. 111, 1–61.
2. Johansson, C. and Bergman, B. (1994) New Phytol. 126, 643–652.
3. Enderlin, C.S. and Meeks, J.C. (1983) Planta 158, 157–165.
4. Campbell, E.L. and Meeks, J.C. (1989) Appl. Environ. Microbiol. 55, 125–131.
5. Cohen, M.F., Wallis, J.G., Campbell, E.L. and Meeks, J.C. (1994) Microbiology 140, 3233–3240.

6. Castenholz, R.W. and Waterbury, J.B. (1989) in: Bergy's Manual of Systematic Bacteriology, vol. 3 (Staley, J.T., Bryant, M.P., Pfenning, N and Holt, J.G., Eds.), pp. 1710–1728, Williams and Wilkins Co., Baltimore.
7. Doherty, H.M. and Adams, D.G. (1995) Gene 163, 93–96.
8. Herdman, M. and Rippka, R. (1988) Methods Enzymol. 167, 232–242.
9. Tandeau de Marsac, N. (1994) in: The Molecular Biology of Cyanobacteria (Bryant, D., Ed.), pp. 825–842, Kluwer Academic Publishers, Dordrecht.
10. Meeks, J.C. (1990) in: Handbook of Symbiotic Cyanobacteria (Rai, A.N., Ed.), pp. 43–63, CRC Press, Boca Raton, FL.
11. Dow, C.S., Whittenbury, R. and Carr, N.G. (1983) in: Microbes in Their Natural Environment (Slater, J.H., Whittenbury, R. and Wimpenney, J.W.T., Eds.), pp. 187–247, Cambridge Univ. Press, Cambridge.
12. Herdman, M. (1987) in: The Cyanobacteria (Fay, P. and van Baalen, C, Eds.), pp. 227–250, Elsevier, Amsterdam.
13. Hirosawa, T. and Wolk, C.P. (1979) J. Gen. Microbiol. 114, 433–441.
14. Theil, T. and Wolk, C.P. (1983) J. Bacteriol. 156, 369–374.
15. Cardemil, L. and Wolk, C.P. (1981) J. Phycol. 17, 234–240.
16. Leganes, F., Fernandez-Pinas, F. and Wolk, C.P. (1994) Mol. Microbiol. 12, 679–684.
17. Wolk, C.P., Ernst, A. and Elhai, J. (1994) in: The Molecular Biology of Cyanobacteria (Bryant, D., Ed.), pp. 769–823, Kluwer Academic Publishers, Dordrecht.
18. Lynn, M.E., Bantle, J.A. and Ownby, J.D. (1986) J. Bacteriol. 167, 940–946.
19. Buikema, W.J. and Haselkorn, R. (1993) Annu. Rev. Plant Physiol. Plant Mol. Biol. 44, 33–52.
20. Wolk, C.P. (1996) Annu. Rev. Genet. 30, 59–78.
21. Bancroft, I., Wolk, C.P. and Oren, E.V. (1989) J. Bacteriol. 171, 5940–5948.
22. Rai, A.N. (1990) Handbook of Symbiotic Cyanobacteria, CRC Press, Boca Raton, FL.
23. Meeks, J.C. (1988) Methods Enzymol. 167, 113–121.
24. Bergman, B., Johansson, C. and Soderback, E. (1992) New Phytol. 122, 379–400.
25. Rasmussen, U., Johansson, C. and Bergman, B. (1994) Mol. Plant-Microbe Interact. 7, 696–702.
26. Gantar, M., Kerby, N.W. and Rowell, P. (1993) New Phytol. 124, 505–513.
27. Knight, C.D. and Adams, D.G. (1996) Physiol. Mol. Plant Path. 49, 73–77.
28. Joseph, C.M. and Meeks, J.C. (1987) J. Bacteriol. 169, 2471–2475.
29. Steinberg, N.A. and Meeks, J.C. (1989) J. Bacteriol. 171, 6227–6233.
30. Steinberg, N.A. and Meeks, J.C. (1991) J. Bacteriol. 173, 7324–7329.
31. Meeks, J.C., Enderlin, C.S. Joseph, C.M., Chapman, J.S. and Lollar, M.W.L. (1985) Planta 164, 406–414.
32. Rai, A.N., Borthakur, M., Singh, S. and Bergman, B. (1989) J. Gen. Microbiol. 135, 385–395.
33. Theil, T., Lyons, E.M., Erker, J.C. and Ernst, A. (1995) Proc. Natl. Acad. Sci. USA 92, 9358–9362.
34. Campbell, E.L. and Meeks, J.C. (1992) J. Gen. Microbiol. 138, 473–480.
35. Wolk, C.P., Cai, Y. and Panoff, J.-M. (1991) Proc. Natl. Acad. Sci. USA 88, 5355–5359.
36. Summers, M.L., Wallis, J.G., Campbell, E.L. and Meeks, J.C. (1995) J. Bacteriol. 177, 6184–6194.
37. Cohen, M.F. and Meeks, J.C. (1997) Mol. Plant-Microbe Interact. 10, 280–289.
38. Lin, E.C.C. (1996) in: Escherichia coli and Salmonella : Cellular And Molecular Biology (Neidhardt, F.C. Ed.), pp. 307–342, Amer. Soc. Microbiol. Washington D.C.
39. Gross, C.A., Lonetto, M. and Losick, R. (1992) in: Transcriptional Regulation (Yamamoto, K. and McNight, S., Eds.), pp. 129–176, Cold Spring Harbor Lab. Press, Cold Spring Harbor, NY.
40. Brahamsha, B. and Haselkorn, R. (1992) J. Bacteriol. 174, 7273–7282.
41. Anbudurai, P.R., Mor, T.S., Ohad, I., Shestakov, S.V. and Pakrasi, H.B. (1994) Proc. Natl. Acad. Sci. USA 91, 8082–8086.
42. Campbell, E.L., Hagen, K.D., Cohen, M.F., Summers, M.L. and Meeks, J.C. (1996) J. Bacteriol. 178, 2037–2043.
43. Wanner, B.L. (1992) J. Bacteriol. 174, 2053–2058.
44. Hoch, J.A. (1993) Annu Rev. Microbiol. 47, 441–465.
45. Lamb, J.R., Tugendreich, S. and Hieter, P. (1995) TIBS 20, 257–259.
46. Frias, J.E., Flores, E. and Herrero, A. (1994) Mol. Microbiol. 14, 823–832.
47. Ernst, A. Black, T., Cai, Y., Panoff, J.-M., Tiwari, D.N. and Wolk, C.P. (1992) J. Bacteriol. 174, 6025–6032.
48. Buikema, W.J. and Haselkorn, R. (1991) Genes Develop. 5, 321–330.
49. Black, T.A. and Wolk, C.P. (1993) Mol. Microbiol. 9, 77–84.
50. Wei, T.-F., Ramasubramanian, T.S. and Golden, J. (1994) J. Bacteriol. 176, 4473–4482.

78

CHARACTERISATION OF PLANT EXUDATES INDUCING CHEMOTAXIS IN NITROGEN-FIXING CYANOBACTERIA

Simon D. Watts,[1] Celia D. Knight,[1] and David G. Adams[2]

[1]Department of Biology
[2]Department of Microbiology
University of Leeds
Leeds LS2 9JT, United Kingdom

1. INTRODUCTION

Cyanobacteria form symbiotic associations with a wide range of plants including cycads, the angiosperm *Gunnera*, the water fern *Azolla,* and bryophytes such as the liverwort *Blasia* and the hornwort *Anthoceros* (Bergman *et al.*, 1992, 1996). The cyanobacterial symbionts in these plant symbioses are almost always members of the genus *Nostoc* that possess two important characteristics: they are capable of nitrogen fixation, in differentiated cells known as heterocysts (Wolk *et al.*, 1994), and they produce specialised filaments known as hormogonia (Tandeau de Marsac, 1994). The latter are motile filaments that serve as the infective agents in most if not all the plant symbioses, and that develop from immotile parent trichomes in response to a variety of environmental stimuli (Tandeau de Marsac, 1994) including signals from potential plant hosts. For example, a hormogonia inducing factor is excreted by the hornwort *Anthoceros* when grown free of its symbiotic cyanobacteria in combined nitrogen-free medium (Campbell and Meeks, 1989). Similarly, the acidic mucilage secreted by *Gunnera* stem glands contains a hormogonia inducing activity thought to be a small, heat-labile protein (Rasmussen *et al.*, 1994; Bergman *et al.*, 1996). Even the roots of wheat, which forms only loose associations with cyanobacteria, release hormogonia inducing factors (Gantar *et al.*, 1993).

Nitrogen-starved plants are therefore capable of improving their chances of becoming infected by nitrogen-fixing cyanobacteria by stimulating potential partners to produce hormogonia. However, many of the plant structures occupied by symbiotic cyanobacteria are not readily accessible, and their rapid infection and colonisation implies that hormogo-

The Phototrophic Prokaryotes, edited by Peschek *et al.*
Kluwer Academic / Plenum Publishers, New York, 1999.

nia are guided by chemotaxis, and the likely source of chemoattractants is the plant host. We have recently developed an assay that permits the reliable estimation of chemotaxis in hormogonia (Knight & Adams, 1996). We report here on the use of this assay to facilitate the preliminary characterisation of a chemoattractant excreted by the liverwort *Blasia*, and to demonstrate the release of a chemoattractant by germinating wheat seeds.

2. METHODS

2.1. Organisms and Growth Conditions

The cyanobacterium *Nostoc* sp. strain LBG1 was originally isolated from symbiosis with the hornwort *Phaeoceros laevis* (Babic, 1996). It was grown at 30°C in 100 ml of BG-11$_0$ medium (BG-11 medium with NaNO$_3$ ommitted; Rippka *et al.*, 1979) supplemented with 30 mM glucose, in 250 ml Erlenmeyer flasks on an orbital shaker (120 rpm), under constant illumination (12–22 Wm^{-2}). The liverwort *Blasia pusilla* was originally purified by S. Babic (Babic, 1996) and was grown axenically (without cyanobacterial symbionts) in 50 ml of five-fold diluted BG-11 in 150 ml Erlenmeyer flasks. The flasks were shaken at 100 rpm in an orbital incubator at 20°C under an 8 h:16 h dark:light (3–4 Wm^{-2}) cycle.

2.2. Preparation of Hormogonia

Cultures of *Nostoc* LBG1, growing exponentially in BG-11$_0$ medium supplemented with 30 mM glucose, were induced to form hormogonia by the removal of glucose from the medium by centrifugation at 3200 rpm for 12 min. The cells were resuspended in fresh BG-11$_0$ and incubated under the same conditions used for growth. Transformation to motile hormogonia was complete after 15–20 h, and these hormogonia were used immediately for the chemotaxis assay described below.

2.3. Nitrogen Starvation of *Blasia Pusilla*

A culture of *Blasia pusilla* growing in BG11 was washed in fresh BG-11$_0$, resuspended in 50 ml of BG-11$_0$ and returned to the original growth conditions. Subsamples (3 ml) of medium (referred to as *Blasia* exudate) were taken weekly and stored at −20°C before testing in the chemotaxis assay described below. Fresh BG-11$_0$ was used as a negative control, and a known active exudate as a positive control.

2.4. Dialysis of Exudate

A 2 ml subsample of *Blasia* exudate, in 3500 M$_r$ cut off dialysis membrane, was dialysed at 4°C for 24 h against 2 l of five-fold diluted BG-11$_0$ medium. The dialysate was removed and tested in the chemotaxis assay described below, using an undialysed sample kept at 4°C for 24 h as a positive control.

2.5. Filtration of Exudate

A 500 µl subsample of *Blasia* exudate was centrifuged at 6500 g for 30 minutes in a 1000 M$_r$ cut off Microsep microconcentrator (Filtron Technology Corp., Northborough,

MA). The filtrate was tested in the chemotaxis assay, similarly filtered medium being used as a negative control.

2.6. Base Treatment of Exudate

A subsample (1 ml) of *Blasia* exudate was freeze-dried and resuspended in 1 ml of a mixture of pyridine, ethyl acetate and water (5:3:3, v:v:v). The mixture was agitated vigorously, centrifuged, and the supernatant was removed and freeze-dried. The resulting solid was resuspended in 1 ml distilled water and assayed as described below. A second sample, freeze-dried and resuspended in distilled water, but otherwise untreated, was used as a control.

2.7. Acid Treatment of Exudate

A subsample (1 ml) of *Blasia* exudate was freeze-dried and resuspended in 1 ml of a mixture of acetic acid, dichloromethane and water (3.5:3:0.5, v:v:v). The remainder of the protocol was the same as the base treatment described above.

2.8. Germination of Wheat Seeds

Twenty wheat seeds (cultivar *Alexandria*) were surface sterilised for 10 min in 10% (v/v) sodium hypochlorite, rinsed three times in sterile distilled water, and transferred to sterile Magenta GA7 vessels (Magenta Corp., Chicago, IL) containing four squares of sterile Whatman 3MM filter paper and 10 ml of sterile distilled water. The seeds were incubated in the dark at 20°C until germination after 48 h, at which point they were placed under an 8 h:16 h dark:light cycle. A subsample (0.5 ml) of the water was taken after 4 d and stored at −20°C.

2.9. Chemotaxis Assay

To estimate chemoattractant activity the capillary based method of Knight and Adams (1996) was employed, all samples being assayed in quadruplicate. All assays included positive (untreated exudate) and appropriate negative controls.

3. RESULTS

In symbiotic association with *Blasia*, cyanobacteria occupy structures known as auricles on the ventral side of the thallus. The addition of hormogonia to *Blasia pusilla* cultures grown in medium containing 17.5 mM $NaNO_3$ (BG-11) did not produce infection of auricles after 48 h, whereas auricles infected with hormogonia could be seen within 30 min of the addition of hormogonia to cultures grown in medium lacking combined nitrogen (data not shown). The concentration of chemoattractant in the medium of *Blasia pusilla* cultured in the presence of nitrate was extremely low, but began to increase following nitrate removal, and was still increasing after 6 weeks (Fig. 1). The active component was able to pass through 3500 M_r cut off dialysis membrane, and a significant proportion also passed through a filter with a 1000 M_r cut off (Table 1). The activity of the chemoattractant in the *Blasia* exudate was little affected by heating, retaining 70% of its initial activity after heating at 95°C for 10 min (Table 1). Activity in-

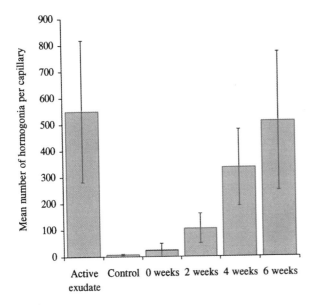

Figure 1. Production of chemoattractant by nitrogen starved *Blasia pusilla.* A symbiont-free culture of *Blasia pusilla*, growing on medium supplemented with sodium nitrate, was transferred to medium lacking combined nitrogen at time zero. Samples of the growth medium were taken at the times indicated and tested for chemoattractant activity. The values given are the mean of quadruplicate subsamples; error bars indicate SE.

creased significantly after freeze-drying and treatment with the base pyridine, but was lost following exposure of the exudate to acetic acid (Table 1). Chemoattractant activity was also produced by germinating wheat seeds (Table 2), but its properties were not determined.

4. DISCUSSION

The availability of a reliable method for measuring the chemotaxis of hormogonia has enabled us to demonstrate that the liverwort *Blasia* excretes a chemoattractant when starved of combined nitrogen. The M_r of the active component(s) is likely to be less than

Table 1. Effect of a variety of treatments on the chemoattractant activity in *Blasia* exudate

	Activity (% of untreated exudate)[a]	
Treatment	Negative control	Treated sample
3500 MWCO dialysis	0.64	4.60[b]
1000 MWCO filtration	4.23	62.22[c]
65°C for 10 min	2.71	66.17
80°C for 10 min	2.71	107.00
95°C for 10 min	2.71	70.11
Freeze drying	2.72	209.66
Acetic acid	2.62	0.15
Pyridine	2.72	227.63

[a]Every assay included the original untreated exudate as a positive control and all other activities are expressed as a percentage of this. All values are the mean of quadruplicate subsamples.
[b]Activity of the dialysate.
[c]Activity of the filtrate.

Table 2. Production of chemoattractant activity by germinating wheat seeds[a]

Sample	Hormogonia per capillary
Blasia exudate	340
Negative control	6
Germinating wheat seed exudate	599

[a]Seeds were germinated on filter paper soaked in water. After 4 days a sample of the water was assayed for chemoattractant activity and compared with a positive *Blasia* exudate. The values are the mean of quadruplicate subsamples.

1000. The reason for the doubling in chemoattractant activity following freeze drying and treatment with pyridine (which involved a freeze-drying step) is not clear. It is not a concentration effect, because the samples were always resuspended in the original volume following freeze-drying. It may result from the loss, during freeze-drying, of a component inhibitory to chemotaxis. The answer may become clear when the chemoattractant has been purified and its chemical identity determined; this work is in progress.

The ability to assay for plant extracellular compounds that induce hormogonia formation should also permit the chemical identification of this second signal. However, there are likely to be additional signals, of both plant and cyanobacterial origin, responsible for the post-infection changes that occur to both partners (Bergman *et al.*, 1996). For example, the host down regulates the activity of both glutamine synthetase and ribulose bisphosphate carboxylase of its cyanobacterial partner. In the hornwort *Anthoceros* this is thought to involve an undetermined post-translational modification of the enzymes (Meeks, 1990). As the symbiotic colony develops it is invaded by finger-like protrusions of the host tissue that presumably increase metabolic exchange between the partners. The heterocyst frequency of cyanobacteria within the symbiotic colony can be as high as 45%, compared with 5% in the same cyanobacterium when free-living (Meeks, 1990). The signals responsible for changes such as these are unknown. Such signalling is not limited to those plants that form intimate symbioses with cyanobacteria, because wheat roots are known to stimulate hormogonia formation in some *Nostoc* strains (Gantar *et al.*, 1993), and we have shown here that germinating wheat seedlings excrete a chemoattractant for hormogonia. With improved understanding of such plant-cyanobacteria signalling may come the ability to enhance the efficiency of nitrogen-fixation and transfer in existing cyanobacterial associations with crop plants such as wheat, and to develop novel associations.

ACKNOWLEDGMENTS

Thanks to Dr. John Arnold for help with purification techniques and to the Natural Environment Research Council for provision of a Ph.D. studentship to SDW.

REFERENCES

Babic, S. (1996) Ph.D. thesis, University of Leeds, England.
Bergman, B., Rai, A. N., Johansson, C. and Söderbäck, E. (1992) Symbiosis 14, 61–81.

Bergman, B., Matveyev, A. and Rasmussen, U. (1996) Trends Plant Sci. 1, 191–197.

Campbell, E. L. and Meeks, J. C. (1989) Appl. Environ. Microbiol. 55, 125–131.

Gantar, M., Kerby, N. W., and Rowell, P. (1993) New Phytol. 124, 505–513.

Knight, C. D. and Adams, D. G. (1996) Physiological and Molecular Plant Pathol. 49, 73–77.

Meeks, J. C. (1990) In: Handbook of Symbiotic Cyanobacteria (Rai, A. N. ed.) CRC Press Inc., Boca Raton, pp. 43–63.

Rasmusssen, U., Johansson, C. and Bergman, B. (1994) Mol. Plant-Microbe Interactions 7, 696–702.

Rippka, R., Deruelles, J., Waterbury, J. B., Herdman, M. and Stanier, R. Y. (1979) J. Gen. Microbiol. 111, 1–61.

Tandeau de Marsac, N. (1994) In: The Molecular Biology of Cyanobacteria (Bryant, D. A. ed.) Kluwer Academic Publishers, Dordrecht, pp. 825–842.

Wolk, C. P., Ernst, A. and Elhai, J. (1994) In: The Molecular Biology of Cyanobacteria (Bryant, D. A. ed.) Kluwer Academic Publishers, Dordrecht, pp. 769–823.

79

POSITIVE PHOTORESPONSES IN *RHODOBACTER SPHAEROIDES*

The Primary Signal and the Signalling Pathway

Judith P. Armitage,[1] Ruslan N. Grishanin,[1] David E. Gauden,[1]
Paul A. Hamblin,[1] Simona Romagnoli,[1] and Thomas P. Pitta[2]

[1]Microbiology Unit
Department of Biochemistry
University of Oxford
Oxford OX1 3QU
[2]Rowland Institute of Science
100 Edwin Land Boulevard
Cambridge, Massachusetts 02142

1. INTRODUCTION

Rhodobacter sphaeroides swims using a single, unidirectional flagellum, changing direction by periodically stopping rotation and reorienting during the stops. It shows changes in motile behaviour in response to a wide range of metabolites, to terminal electron acceptors such as oxygen and DMSO and to light. We have investigated the mechanisms involved in sensing a range of stimuli and compared the mechanisms to those characterised in *Escherichia coli*. During these studies it has become apparent that chemosensory behaviour in *R. sphaeroides* is much more complex than that characterised in *E. coli*, with multiple phosphorelay systems and chemoreceptors expressed under different growth conditions.

Chemotaxis in *E. coli* depends on constitutively expressed membrane spanning chemosensory receptor dimers (methyl-accepting chemotaxis proteins, MCPs) which bind chemoeffectors to their periplasmic surface and transmit the subsequent conformational change across the membrane. The conformational change is passed via a pair of linker proteins, CheW, to a histidine protein kinase dimer, CheA. If the attractant concentration has increased the conformational change results in reduced CheA activity. If, however, the concentration has reduced, CheA becomes autophosphorylated on a conserved histidine. The phosphate group is subsequently transferred to either a small 14 kDa protein, CheY,

The Phototrophic Prokaryotes, edited by Peschek *et al.*
Kluwer Academic / Plenum Publishers, New York, 1999.

which when phosphorylated can bind to the flagellar motor and cause it to switch from counterclockwise (CCW) to clockwise (CW) rotation, or to a methyl esterase, CheB, responsible for resetting the receptors into a non-signalling state and causing adaptation. CheY-P is dephosphorylated and the signal terminated by CheZ which increases the CheY-P rate of dephosphorylation; for recent review see [1].

This single phosphorelay pathway controls the CCW/CW switching of the six or so *E. coli* motors in response not only to chemoeffector concentration changes, but also to changes in oxygen [2,3]. An MCP homologue (Aer), lacking the periplasmic domain but with FAD bound to the cytoplasmic surface, responds to the rate of respiratory electron transfer and signals through CheW to CheA. In addition sugars transported through the PEP-dependent phosphotransferase system (PTS) alter the level of CheA phosphorylation and thus chemotaxis to sugars relies on the same pathway [4].

Chemotaxis in *R. sphaeroides* has been confusing for many years. It is apparent that chemoattractants require transport and at least limited metabolism under anaerobic conditions, and therefore do not use membrane receptor [5]. Mutants in chemotaxis proved highly elusive and no chemoreceptors could be identified by methylation. Recent data have helped explain the past confusions. Rather than having a single chemosensory pathway, *R. sphaeroides* has at least two and possibly three operons controlling the activity of the single flagellum. Mutations in one operon were therefore compensated for by another operon. Mutants, unlike those in the *E. coli* chemosensory pathway, still show normal patterns of swimming and therefore show limited spreading in soft agar; the criterion used to isolate mutants. *R. sphaeroides* has now been found to have membrane spanning MCP homologues, but only under aerobic conditions. From the point of view of chemotaxis *R. sphaeroides* is almost like two separate species, aerobically similar to *E. coli*, but with more chemosensory pathways and anaerobically very different, with cytoplasmic sensory proteins, responding to changes in metabolic activity. In this review the current state of our understanding of photoresponses will be described and the possible interaction with the chemosensory pathways.

2. PHOTORESPONSES

2.1. Responses to Light Gradients

It was apparent from the very earliest studies of behaviour of photosynthetic bacteria that many species simply reversed when swimming over a light/dark boundary or from an photosynthetic to non-photosynthetic wavelength. While it became apparent that bacteria could follow a chemical gradient there has been discussion over many years as to whether free-swimming phototrophic bacteria could also show what may be described as "real" phototactic responses i.e. swim up a light gradient [6].

The photoresponses of two non-sulphur bacteria, *R. sphaeroides*, which transiently stops in response to a reduction in a stimulus and *Rhodospirillum centenum*, which changes its reversal frequency in response to a stimulus, were measured. A light beam was propagated at 90° to the optical axis of a microscope through a population of free-swimming cells. The light beam presented the cells with a steep dark/light gradient perpendicular to the direction of propagation and a shallow light gradient in the direction of propagation. Accumulation of cells within the light beam was measured by changes in light scattering. *Chlamydomonas reinhardtii*, a known phototactic green alga swam directly towards the source of light, i.e. showed true phototaxis. *R. centenum* on the other hand accumulated in the light beam, but there was no difference in the number of cells accumulating close to the source compared to distant from the source. Interestingly, *R. sphaeroides*, which stops rather than reverses when faced with a reduction in stimulus strength, accumulated just outside the light beam. Free-swimming cells responded to a steep gradient of light intensity not to a shallow gradient, indicating that they do not sense the direction of light, only its intensity [7].

2.2. Role of Photosynthetic Electron Transport in Photoresponses

Early data suggested that photosynthetic electron transfer was required for photoresponses as mutants with a full complement of photosynthetic pigments, but no reaction centre proteins, did not respond to a reduction in light intensity. These data has recently been substantiated by a more detailed study using specific inhibitors of electron transfer [8]. Antimycin A, stigmatellin and rotenone all independently inhibited photoresponses at concentrations that inhibited photosynthetic electron transfer (Fig. 1a). *R. sphaeroides* grown under high light, and therefore with fewer light harvesting complexes, showed photoresponses over a much wider range of light intensities than cells grown in low light with a large number of light harvesting complexes. This indicates that low light-grown cells remain above the threshold for a photoresponse, and probably maintain saturating photosynthetic electron transfer, at much lower light intensities than high light-grown cells. Photoresponses therefore correlate with the rate of photosynthetic electron transfer and a signal is probably generated when the rate of electron transfer falls.

2.3. Integration with Respiratory Electron Transfer

Purple non-sulphur photosynthetic bacteria can grow using oxygen dependent respiration and anaerobic respiration with terminal acceptors such as dimethyl sulphoxide (DMSO) as well as photosynthetically. Many of the electron transfer components are thought to be common to the pathways e.g. the quinones, cytochrome bc_1 etc. *R. sphaeroides* not only shows a response to a step-down in light, but also to a step-down in oxygen if grown aerobically and DMSO if grown anaerobically with DMSO as a terminal acceptor. The response to DMSO is lost if oxygen or light are present, conditions under which electron transport to DMSO is also lost, although DMSO reductase is still present [9]. Similarly the step-down response to light is reduced if oxygen is present, condition under which the rate of photosynthetic electron transfer is also reduced while light inhibits both electron transfer to the terminal oxidase and aerotactic responses. This suggests that bacteria are not responding to the specific effector, but rather to a change in the rate of electron transfer which is somehow signalling through to the sensory pathway to the flagellar motor.

2.4. Protometer or Redox Sensor?

If the primary signal comes from a change in the rate of electron transfer, the receptor could be sensing either a change in the electrochemical proton gradient (Δp) or a redox change in one of the electron transfer proteins.

In *R. sphaeroides* the primary response is to a step-down in light or chemoeffector concentration. Most stimuli altering the rate of electron transfer also change the size of Δp in parallel, making it difficult to identify the cause to the response. However, the addition of small concentrations of the proton ionophore FCCP decreases the Δp, but increases the rate of electron transfer. This allowed us to investigate whether *R. sphaeroides* was re-

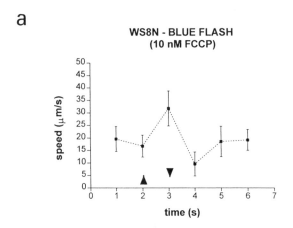

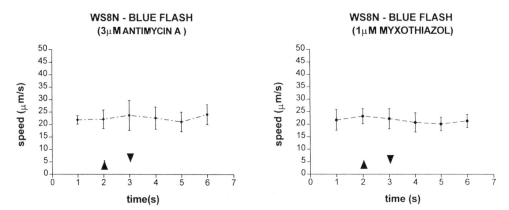

Figure 1. Responses of single cells to a flash of blue light. The responses are the averages of about 20 cells. ▲▼ shows light on and off (a) shows the responses to a range of inhibitors altering electron transport and Δp (b) shows the response of wt WS8 and the two mutants in CheAII to light.

sponding to a reduction in Δp when the light intensity fell, or a reduction in the rate of electron transfer (a redox signal). Using the carotenoid bandshift to monitor the change in Δp we added a small amount of the proton ionophore FCCP, such that the cells continued swimming but the Δp fell. Under these conditions there was no behavioural response, as would be expected if the step-down response was the result of a transient reduction in Δp. A step-down in light causing a reduction in Δp of the same size but accompanied with a reduction in electron transfer caused cells to stop. Removal of FCCP causing an increase in Δp, but a decrease in the rate of electron transfer did cause a stop response. The primary signal is therefore unlikely to be a change in the electrochemical proton gradient and is almost certainly the result of the transient change in the rate of electron transfer. These data were supported by examination of the behaviour of single cells to flashes of light (Fig. 1a). Single cells respond to a 1s flash of light by a transient increase in speed followed by a stop when the flash ends. The response was completely lost when the cell was treated with antimycin A or myxothiazol, but unchanged in the presence of FCCP. These data suggest that the primary signal causing a photoresponse (and probably all electron transport-

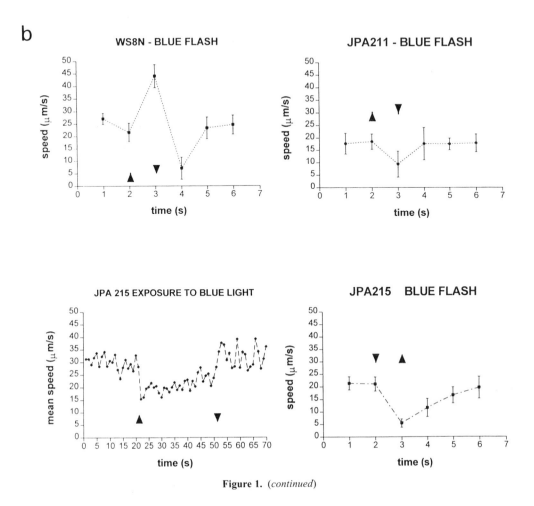

Figure 1. (*continued*)

dependent responses) is a change in redox state of some component of the electron transport chain.

3. GENETICS OF PHOTORESPONSES

3.1. Identification of a Chemotactic Operon

Mutants in chemotaxis have proved impossible to isolate in *R. sphaeroides* until very recently. An operon coding for homologues of the enteric chemosensory proteins was recently identified [10]. It contains homologues of *cheA, cheW, cheR* and two copies of *cheY* in addition to two genes that appear to code for MCP-like proteins. No copies of *cheZ* or *cheB* were identified. However, although mutants in one of the MCPs caused the loss of general aerobic chemotaxis, mutations in the other genes, and even deletion of the complete set of *che* genes had little effect on behaviour (Fig. 2).

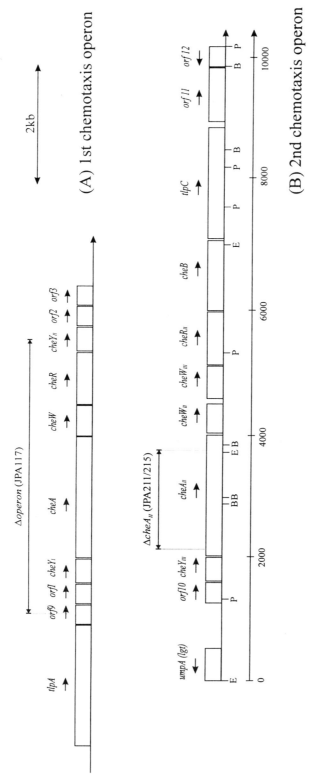

Figure 2. Two operons from *R. sphaeroides* coding for homologues of the enteric *che* genes. *tlp* -transducer like proteins and code for homologues of MCPs. Regions deleted in the different mutants are shown.

Using the mutant in which the *che* operon had been deleted, a second round of Tn5 mutagenesis was used to try to identify a second pathway. The mutants were subjected to multiple passages through a phototaxis screen in which the bacteria had to swim up a darkened tube with a sharp band of light in the middle of the tube. Normally responding cells were slowed at the dark/light/dark boundary while motile but non-photoresponsive cells swam upwards. After up to 12 passages through the screen the mutants were plated onto soft nutrient agar. Mutants that were still chemotactic created their own chemical gradient as a result of metabolism and swam outwards. Mutants that grew but did not swarm were isolated. The mutants identified in this way included metabolic mutants, supporting previous data which showed that metabolism is important for *R. sphaeroides* chemotaxis, and mutants in a second (and possibly third) chemosensory operon. The second operon contains a second copy of *cheA* and *cheR,* a third copy of *cheY*, two more copies of *cheW* and the first copy of *cheB* (Fig. 2). Mutants in this operon show reduced chemotaxis and inverted phototactic responses (Fig. 1b). The mutants, JPA 211, which has a transposon in *cheAII* and also lacks the first operon, and JPA 215, which has the transposon in *cheAII* in a wt background, both stop when light in increased and return to normal swimming when the light is switched off. These data suggest that the photoresponse signal goes through the components coded for by the second pathway and not CheAI, but the observation that there is still a response, albeit inverted, suggests that there may be yet another signalling mechanism.

4. SENSORY RECEPTORS

Oxygen sensing in *E. coli* appears to be through the MCP homologue, Aer, sensing changes in respiratory electron transfer and signalling through the chemosensory phosphorelay system. Recently, using *E. coli* anti-MCP antibody we have identified membrane bound MCPs in *R. sphaeroides*, but primarily under aerobic conditions (J. R. Maddock, D. M. Harrison & J. P. Armitage, unpublished). The number of MCPs cross-reacting with the Ab increased 20X under aerobic as compared to photosynthetic conditions, with the lowest number seen in high light grown cells. High light grown cells had few membrane bound MCPs but a smaller number of MCPs clustered in the cytoplasm. These cytoplasmic MCPs appeared to be constitutive as their numbers changed little under different growth conditions. If all electron transport-dependent signalling uses a common receptor it cannot be the receptor whose expression is controlled by oxygen levels. Either the redox sensor is the constitutive receptor identified under all conditions, or it is present at concentrations too low for the immunoelectron microscopy to detect or it is not recognised by the *E. coli* antibody (the *E. coli* antibody does not recognise either of the two homologues coded for on the first *che* operon).

5. SUMMARY

R. sphaeroides has at least two functional chemosensory pathways, only one of which is involved in photoresponses. Chemotaxis is apparently different under aerobic and anaerobic growth conditions with *E. coli* like receptors being induced or de-repressed by oxygen. Photoresponses are part of a common sensory system responding to any changes in the rate of electron transfer and signalling through CheAII. We are currently trying to identify the primary redox receptor and its mechanism of signalling through to CheAII and thus to the single *R. sphaeroides* motor.

ACKNOWLEDGMENTS

Most of the work on *R. sphaeroides* described here was funded by the UK BBSRC. The light gradient work was part of a project undertaken during the Woods Hole Microbial Diversity course in 1996.

REFERENCES

1. C.D. Amsler, P. Matsumura, in Two-component signal transduction (Hoch, J.A. and Silhavy, T.J. eds.) (1995), pp. 89–103, ASM, Washington, D.C.
2. S.I. Bibikov, R. Biran, K.E. Rudd, J.S. Parkinson, J. Bacteriol. 179 (1997) 4075–4079.
3. A. Rebbapragada, M.S. Johnson, G.P. Harding, A.J. Zuccarelli, H.M. Fletcher, I.B. Zhulin, B.L.Taylor, Proc. Natl. Acad. Sci. USA (1997) in press,
4. R. Lux, K. Jahreis, K. Bettenbrock, J.S. Parkinson, J.W. Lengeler, Proc. Natl. Acad. Sci. USA 92 (1995) 11583–11587.
5. P.S. Poole, M.J. Smith, J.P. Armitage, J. Bacteriol. 175 (1993) 291–294.
6. J.P. Armitage, Archiv. Microbiol. (1997) in press
7. M.J. Sackett, J.P. Armitage, E.E. Sherwood,T.P. Pitta, J. Bacteriol. (1997) 179, in press
8. R.N. Grishanin, D.E. Gauden, J.P. Armitage, J. Bacteriol. 179 (1997) 24–30.
9. D.E. Gauden, J.P. Armitage, J. Bacteriol. 177 (1995) 5853–5859.
10. M.J. Ward, A.W. Bell, P.A. Hamblin, H.L. Packer, J.P. Armitage, Mol. Microbiol. 17 (1995) 357–366.

80

PROPERTIES OF *RHODOBACTER SPHAEROIDES* FLAGELLAR MOTOR PROTEINS

R. Elizabeth Sockett, Ian G. P. Goodfellow, Gabi Günther, Matthew Edge, and Deepan Shah

Department of Life Science
Nottingham University
University Park, Nottingham NG7 2RD, United Kingdom

1. BACKGROUND

The purple non-sulphur bacterium *Rhodobacter sphaeroides* swims due to the rapid rotation of a single unidirectional flagellum [1]. It is well documented that this organism makes tactic swimming responses to a variety of stimuli which include light intensity, pH, partial pressure of oxygen, and concentration gradients of a range of organic chemicals found in the aquatic environments in which they live [1]. *Rhodobacter sphaeroides* swimming behaviour is produced by modulating the stopping and starting of the single, medially located flagellum. Cells reorientate to a new swimming direction when rotation stops and the flagellum coils up. Most work in bacterial motility has been carried out in enteric species *Salmonella typhimurium* and *Escherichia coli* which possess between five and eight flagella [2]. Swimming behaviour is modulated by "switching" the direction of flagellar rotation (counterclockwise rotation results in cells swimming vectorially i.e "running" and clockwise rotation causes the cells to "tumble" until a favourable direction is found). Flagella that work in this manner possess motors that are termed bidirectional.

Bacterial flagella are complex mechanoenzymes consisting of 3 main regions: basal body, hook and a helical filament [1,2]. Bacteria swim due to the proton motive force-driven rotation of flagella emerging from the basal body protein complex which is cytoplasmic-membrane-bound and which spans all layers of the cell (Figure 1). Hook and filament (flagellin) subunits are exported through the hollow core of the basal body during flagellar assembly to their external location where they self-polymerise to form a rigid propeller.

The Phototrophic Prokaryotes, edited by Peschek *et al.*
Kluwer Academic / Plenum Publishers, New York, 1999.

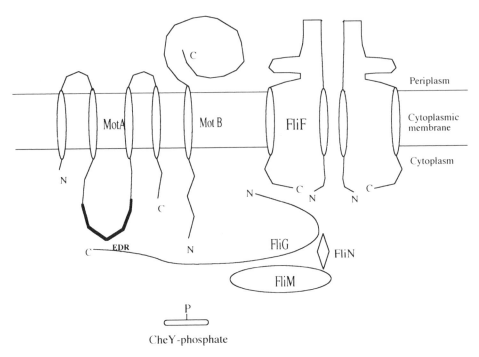

Figure 1. Diagram of a cross-section through the flagellar basal body showing proposed locations of the protein components. Motor and switch proteins are only shown to one side of the FliF MS ring for clarity, in reality some 8–12 copies of these proteins are thought to encircle the MS ring in the membrane. The N terminus of each FliG is thought to be bound to the C-terminus of each FliF molecule leaving the C-terminus of FliG available for potential interaction with the Mot proteins. The flagellar hook and filament are not shown in the diagram but would lie directly above the top of the FliF complex.

We have been characterising the motility proteins of *R. sphaeroides* to try to understand how flagellar rotation and stop-starting of the flagellum are biochemically achieved. We are using mutagenesis and function studies, and looking for similarities and differences in protein sequences from this unidirectional flagellum compared to those from bidirectional flagella.

2. IDENTIFICATION AND CHARACTERISATION OF FLAGELLAR MOTOR PROTEINS

Transposon mutagenesis of *Rhodobacter sphaeroides* WS8 with Tn*phoA* produced a range of mutant strains with impaired motility that showed reduced swarming on soft motility agar plates. The majority of these mutants produced no visible flagellum, but some had paralysed flagella, or in rare cases gave liquid cultures which contained a low percentage of motile swimming cells in mid-log phase . Simple complementation analyses with clones from a genomic library of *R. sphaeroides* WS8 DNA, coupled with subcloning, further complementation studies and DNA sequencing revealed the genes in which the transposon insertions lay [3,4]. Many of the genes sequenced have significant homology to those found previously in *E.coli* and *S. typhimurium* especially where they encode a common function not specific involving the directionality of flagellar rotation. In cases where the function relates to direction switching there are marked differences between *R. sphaeroides* and enteric bacterial proteins, indicative of the altered motor behaviour.

3. DEFECTS IN MOTOR PROTEINS MotA OR MotB CAUSE FLAGELLAR PARALYSIS

MotA contains 253 amino acids, it has 4 predicted membrane spans plus a 100 amino acid charged cytoplasmic loop between helices 2 and 3 [5]. MotB contains 332 amino acids, one predicted membrane span plus a large 270 amino acid periplasmic domain including in *R. sphaeroides* a novel His motif present in place of the usual peptidoglycan-binding domain found in *E. coli* MotB [6]. MotA and B were implicated as the proteins being involved in proton entry. It is not known in any bacteria whether MotA or MotB alone, or a MotAB complex, facilitates proton entry into the flagellar motor. Deletion of the charged cytoplasmic loop (marked in black on Figure 1) from *R. sphaeroides* MotA paralyses the flagellum despite the presence of intact membrane-spanning domains of MotA which could still allow protons to cross the membrane (G. Günther and R. E. Sockett, unpublished). Incorporation of multiple copies of the membrane-spanning domain of *R. sphaeroides* MotB into artificial membrane systems gives high rates of cation-specific conductance at physiological membrane potentials (I. Mellor and R.E. Sockett, in preparation). The amphipathic membrane-spanning domain of MotB is the most highly conserved region between *R.sphaeroides* and other bacterial Mot proteins. It may be that protons enter the cell via a channel composed of MotB (possibly along with MotA) membrane-spanning domains, and alter the charge/conformation of the MotA cytoplasmic loop.

4. THE FliG PROTEIN MAY ACT AS A ROTOR ARM LINKING PROTON ENTRY TO ROTATION OF FliF

Complementation and sequence analysis of a flagellar minus mutant revealed a defect in the gene encoding the 570 amino acid FliF flagellar basal disc protein. 25 subunits of FliF are thought to be arranged like slices in a cake to form the MS ring [8] (Fig. 1). The FliF protein has 2 membrane spanning domains which are poorly conserved between different species. It is these domains which are thought to be rotated within the membrane driving the rest of the flagellum, so it is perhaps surprising that they are not specialised for the task. There is significant sequence conservation in the FliF region predicted to line the hollow core through which flagellin subunits are exported into the growing flagellum. Immediately downstream from *fliF* was found the *fliG* gene predicted to encode a 320 amino acid soluble protein. This had overall 39% homology with *E. coli* FliG, greatest regions of homology lying at the N- and C-termini. *R. sphaeroides* FliF and FliG proteins were overexpressed and found to interact with each other, this confirms reports from *Salmonella* work which shows that FliG is bound at its N-terminus to the C-terminus of FliF [9]. Nine charged amino acids (5D/E, 4 K/R/Q) were absolutely conserved in the last 100 amino acids at the C-terminus of *R.sphaeroides* FliG in comparison to the FliG proteins known from five other bacteria. One model for the generation of flagellar rotation could involve electrostatic interactions between this FliG domain and the charged cytoplasmic loop of

MotA. As MotAs are thought to be arranged radially in the membrane around the FliF basal disc, FliG proteins bound to each FliF subunit could make sequential interactions with each MotA loop, rotating the flagellar base.

5. PROTEIN SEQUENCES IMPLICATED IN FLAGELLAR DIRECTION SWITCHING IN RESPONSE TO TACTIC STIMULI IN *E. COLI* ARE POORLY CONSERVED IN *R. SPHAEROIDES* HOMOLOGUES

The central domain of FliG has a presumed direction-switching role in *E. coli and Salmonella*, it is thought to interact centrally with the centre of another soluble switch protein FliM [2,7,9]. Information indicating the level of tactic stimuli in the environment compared to the recent past is conveyed to FliM in the form of a phosphorylated soluble protein CheY. CheY-phosphate molecules are produced from tactic sensory complexes at the cell membrane when levels of positive stimulus compounds drop, they diffuse to FliM and bind to it. This interaction is thought to alter FliM–FliG interactions and change the conformation of the flagellar motor so that, in enteric bacteria,the rotational direction changes transiently from counterclockwise to clockwise and the bacteria tumble and restart swimming in a new direction. As *R. sphaeroides* flagella stop and start but do not direction- switch, it is unsurprising that the central domain of FliG is not well conserved with that in *E. coli*. Recently analysis of FliM from *R. sphaeroides* shows it to be only 26% identical to *E. coli* FliM, although an IINERF N-terminal motif implicated in binding of CheY-phosphate is conserved. Interestingly, *R. sphaeroides* FliM contains uniquely four cysteine residues, question as to whether these bind metal, have a role in stop-starting the flagellum or even in directly sensing and causing a tactic response to redox changes will have to await further experimentation. Clearly understanding the structure and function of the novel domains from *R. sphaeroides* FliM and FliG is key to discovering the mechanisms which modulate flagellar activity.

6. FLAGELLAR EXPORT PROTEINS

Flagellar export is thought to be identical in all bacterial species, our isolation and analysis of a gene encoding a *R. sphaeroides* flagellar export protein has confirmed this. Western blotting analysis, with anti-flagellin antiserum, of one mutant giving a low percentage motility cultures indicated that it produced a very low level of extracellular (i.e. surface-exposed) filament proteins. This suggested a defect in flagellin export, and this was confirmed when the sequence of the gene which complemented the mutant [10] was found to encode a 49K protein FliI that has homology to the catalytic β subunit of the bacterial F_0F_1 ATPase and a group of proteins responsible for the export of virulence factors in both mammalian and plant pathogens [11]. FliI has the conserved nucleotide binding domains (Walker boxes) noted in ATP hydrolysing enzymes [12] and work in *S. typhimurium* has shown that it supplies energy required for the export of flagellar subunits. As the *R. sphaeroides* mutant is not totally defective in extracellular flagellin, in this bacterium FliI may not be essential for flagellin export, rather it may accelerate the process.

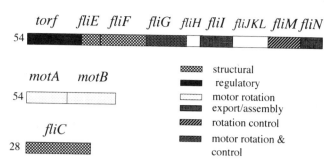

Figure 2. Schematic representation of the organisation of *R. sphaeroides* motility genes. Numbers 28 and 54 indicate the site of consensus sequences in promoter regions for sigma 28 and sigma 54 binding.

7. ORGANISATION OF MOTILITY GENES

Motility genes characterised so far cluster in three loci, all of which have been mapped to approximately 20–200 kb upstream of the photosynthesis cluster on the large *R. sphaeroides* chromosome (Figure 2). The *motAB* genes are part of a two gene operon driven off a promoter with a strong sigma 54 consensus sequence (TGGCACggatcTTGC-38bp-ATG), and preceded 120 bp upstream by a possible enhancer-binding motif containing inverted repeats (CCGCCCGagagccgccgcgaCGGGCGG). The *fliC* flagellar filament gene encodes some 20,000 copies of a 50K protein monomer which polymerise to form the helical flagellar filament. This gene lies immediately downstream of a promoter with a sigma 28 consensus (TAAAAGTTTCTCCGGCCGGCCGTTGAA-54bp-ATG), this promoter motif is not commonly seen in *Rhodobacter*. Complementation analysis of a Tn*phoA*-derived *fliC* mutant indicates the likely presence of other uncharacterised flagellar genes downstream of *fliC* (D. S. Shah, T. Perehinec, S. Stevens and R. E. Sockett, in preparation). All the other flagellar switch, export and motor structural genes map to one large cluster which may be a single operon (Fig 2). At the 5-prime end of this cluster lies an open reading frame which we have named Torf (I. Goodfellow and R. E. Sockett, unpublished EMBL accession X98694). Torf is partly homologous to NtrC from *Rhodobacter capsulatus*, we do not yet have any evidence that it regulates transcription but possibly it could be involved in the regulation of the *motAB* operon which as mentioned above contains a potential enhancer binding site for a protein like NtrC. MotAB proteins would be required immediately after the gene products (such as FliF and FliG) encoded by the Torf gene cluster during sequential assembly of the flagellum. Interestingly, 40bp upstream of the ATG of the *torf* gene is also a sigma 54 consensus site (TTTTAGAGCGCGGTGAAAAATTTT) but this is preceded by different upstream inverted repeat elements to those found in the *motAB* promoter. It is possible that the Torf protein may bind at one of these elements to down-regulate its own expression but this awaits investigation.

ACKNOWLEDGMENTS

This work was supported by a BBSRC grant to RES, a Lawes Trust studentship from IACR Rothamsted for MJE, a BBSRC studentship to IGPG and an E. C. Erasmus studentship to G. G.

REFERENCES

1. Armitage, J.P., Kelly, D.J. & Sockett, R.E. (1995) Flagellate motility, behavioural respnses and active transport in purple non-sulfur bacteria. In Advances in Photosynthesis, volume 2, R.E. Blankenship, M.T. Madigan, & C.E. Bauer (eds.), Anoxygenic Photosynthetic Bacteria, pp 1005–1028. Kluwer Academic Publishers, Dordrecht.

2. Macnab, R.M. Genetics & biogenesis of bacterial flagella. (1992) Annu. Rev. Genet. 26, 131–158.

3. Sockett, R.E., Foster, J.C.A. and Armitage, J.P. (1990) Molecular biology of the *Rhodobacter sphaeroides* flagellum. FEMS Symp. 53, 473–479.

4. Sockett, R.E. and Armitage, J.P. (1991) Isolation, characterization and complementation of a paralysed flagellar mutant of *Rhodobacter sphaeroides* WS8. J. Bacteriol. 173, 2786–2790.

5. Shah, D.S.H. and Sockett, R.E. (1995). Analysis of the motA flagellar motor gene from Rhodobacter sphaeroides, a bacterium with a unidirectional stop-start flagellum. Mol. Microbiol. 17, 961–969.

6. Shah, D. S. H., Armitage, J.P. and Sockett, R.E. (1995). Rhodobacter sphaeroides expresses a polypeptide that is similar to MotB of E. coli. J. Bacteriol. 177, 2929–2932.

7. Lloyd, S.A., Tang, H.T., Wang, X. Billings, S. and Blair, D.F. (1996) Torque generation in the flagellar motor of Escherichia coli : evidence of a direct role for FliG but not FliM or FliN. J. Bacteriol. 178, 223–231.

8. Ueno, T. Oosawa, K. and Aizawa, S.-I. (1994) Domain structures of the MS ring component (FliF) of the flagellar basal body of *Salmonella typhimurium*. J. Mol. Biol. 236, 546–555.

9. Oosawa, K., Ueno, T. and Aizawa, S.-I. (1994) Overproduction of the bacterial flagellar switch proteins and their interactions with the MS ring complex *in vitro*. J. Bacteriol 175. 6041–6045.

10. Goodfellow, I.G., Pollitt, C.E. & Sockett, R.E. (1996) Cloning of the fliI gene from Rhodobacter sphaeroides WS8 by analysis of a transposon mutant with impaired motility. FEMS Microbiol. Lett. 142, 111–116.

11. Vogler, A.P., Homma, M. Irikura, V.M. & Macnab, R.M. (1991) *Salmonella* mutants defective in flagellar filament regrowth and sequence similarity of FliI to F_oF_1 vacuolar and archaebacterial ATPase subunits. J. Bacteriol. 173, 3564–3572.

12. Walker, J.E., Saraste, M., Runswick, M.J. and Gay, N.J. (1982) Distantly related sequences in the a and b subunits of ATP synthase, myosin kinases and other ATP-requiring enzymes and a common nucleotide-binding fold. EMBO J. 1, 945–951.

81

ISOLATION AND CHARACTERIZATION OF PSYCHROPHILIC PURPLE BACTERIA FROM ANTARCTICA

Michael T. Madigan

Department of Microbiology
Southern Illinois University
Carbondale, Illinois 62901-6508

1. INTRODUCTION

Several anoxygenic phototrophic bacteria ("anoxyphototrophs") have been isolated and characterized from extreme environments. These include organisms from thermal [1], hypersaline [2–6], acidic [7,8], and alkaline [2,5,6] environments. In the author's laboratory, thermophilic species of purple and green sulfur bacteria, *Chromatium tepidum* [9,10] and *Chlorobium tepidum* [11], respectively, and of heliobacteria, *Heliobacterium modesticaldum* [12], have been described; all of these organisms are capable of growth above 50°C. In addition, a variety of thermotolerant nonsulfur purple bacteria have been characterized, although none of these show growth above 50°C [13–15].

Conspicuously absent from the known diversity of extremophilic anoxyphototrophs are psychrotolerant or psychrophilic species, organisms capable of growth at low temperature or that actually grow best at low temperature [16]. Ideal habitats for such organisms should be permanently cold (in contrast to temperate) environments, and such exists in polar regions like the Antarctic. Indeed, recent microbiological studies of Antarctic sea-ice microbial communities [17–19] have shown that a variety of microorganisms are present, including phototrophic organisms such as algae and diatoms; chemotrophic bacteria that coexist with these phototrophs have been shown to have growth temperature optima near 4°C [16–18]. In addition, microbial mats exist in various terrestrial and aquatic locales in the Antarctic [20,21], and previous studies of mats in thermal and hypersaline environments have shown them to be ideal habitats for anoxyphototrophs [22].

1.1. Anoxyphototrophs from Antarctica

Anoxyphototrophs have been reported from field studies of a series of meromictic lakes near the Vestfold Hills, Antarctica [23,24] and from near-Antarctic marine sediments

The Phototrophic Prokaryotes, edited by Peschek *et al.*
Kluwer Academic / Plenum Publishers, New York, 1999.

[25,26]. In the meromictic lakes, many of which remain ice-covered for 10–11 months of the year, both purple and green sulfur bacteria as well as purple nonsulfur bacteria were detected [23,24]. Although no laboratory cultures were obtained, good evidence for a *Chlorobium* species (green sulfur bacteria) and a *Thiocapsa* species (purple sulfur bacteria) was obtained from a combination of spectral measurements of bacteriochlorophylls *a* and *c* and from microscopic examinations of water samples [23,24]. Moreover, enrichment cultures from Burton Lake, a well stratified lake containing high levels of biogenic sulfide, yielded a *Chlorobium* sp. that, according to the authors, "grew well at –2°C" [24]. Intense sulfur cycling and a major zone of bacterial photosynthesis was also identified in Ace Lake, a meromictic hypersaline lake nearby Burton Lake [24]. In contrast to these field studies, actual cultures of green and purple sulfur bacteria and nonsulfur purple bacteria were isolated from marine sediments off the South Orkney Islands (62° South) near the Antarctic continent; surprisingly, however, these organisms did not show any adaptive response to cold temperatures and grew best at 25–30°C [25,26]. Finally, anoxyphototrophs have also been reported from algal-bacterial mats in lakes of the Antarctic dry valleys [20,21] but culture studies of these organisms were not pursued.

1.2. The Current Study

From samples of Antarctic algal-bacterial mats kindly supplied by R. W. Castenholz (University of Oregon), enrichment cultures for anoxyphototrophs were established in the spring of 1990. The major selective feature used in these enrichments, besides the usual selective media and growth conditions used for phototrophic bacteria [27,28], was low temperature incubation (5°C). From these enrichments two anoxyphototrophs were obtained, one a purple nonsulfur bacterium and one a purple sulfur bacterium; a brief description of these organisms follows here. In the case of the purple nonsulfur bacterium, a clear preference for low growth temperatures was observed.

2. MATERIALS AND METHODS

2.1. Media and Enrichment Conditions

The mineral salts/malate medium RCVB [29] supplemented with 1% NaCl, 0.01% yeast extract and 20 µg/l vitamin B_{12}, and the sulfide-containing mineral salts medium Pf-7 [10] supplemented with 0.05% ammonium acetate and 1% NaCl, were used as enrichment media; the latter medium contained 2.5 mM sulfide and the final pH of both media was 7. Samples of algal-bacterial mat collected from Cape Royds, Ross Island (77° South), Antarctica, and previously stored for several months at 12°C and 250 lux, were placed in completely filled tubes of both media, the tubes placed at 5°C in darkness for 96 hours, and then illuminated with 400–500 lux. Incubations took place in a refrigerated incubator (Low Temperature Incubator, Fisher Scientific, St. Louis) fitted with incandescent lights; temperature was measured with a calibrated mercury thermometer immersed in a sealed tube of distilled water and placed adjacent to the cultures. For physiological studies of the purple nonsulfur bacterium (designated strain Ant.Br) medium RCVB containing 0.1% NaCl and 0.01% (carbon nutritional studies) or 0.1% (growth rate studies) yeast extract were used. Before transfer of cultures to fresh growth media, all pipettes and culture media were precooled to 5°C.

2.2. Pure Cultures and Measuring Growth

A pure culture of the purple nonsulfur bacterium was obtained using the agar shake method [27] followed by repeated streaking on agar plates incubated phototrophically at 7°C in Gas-Pak® jars (H_2 + CO_2 + N_2 atmosphere). Growth in liquid cultures was measured using a Klett-Summerson photometer (660 nm filter). The purple sulfur bacterium was eventually obtained in pure culture by repeated application of the agar shake method using enrichment medium and low temperature (8°C) incubation.

2.3. Other Methods

Photomicrographs of Antarctic purple bacteria were taken with an Olympus B-Max 60 photomicroscope using the method of Pfennig and Wagener [30]. Electron micrographs were taken following fixation of cells in glutaraldehyde and OsO_4 as previously described [12]. Spectroscopy was done on a Hitachi U2000 spectrophotometer. For absorption spectra of intact cells, 50 μl of dense cell suspensions, previously concentrated by centrifugation, were added to 1 ml of 30% bovine serum albumin (Sigma, St. Louis) and spectra run from 1100 to 400 nm. Assays of nitrogen fixation were run using the acetylene reduction technique as previously described [30]. Phylogenetic analysis on strain Ant.Br was kindly performed by C. R. Woese, University of Illinois, using comparative 16S rRNA sequencing as previously described [11].

3. RESULTS

3.1. Enrichment and Isolation of Antarctic Anoxyphototrophs

Enrichment cultures for anoxyphototrophs using Antarctic microbial mat samples incubated at 5°C yielded, after 4 months incubation, brightly pigmented cultures suggestive of phototrophic purple bacteria. One culture was peach-colored and the other a cherry red. The former, obtained in the malate mineral salts medium, contained small, highly motile rods to curved rods, while the latter, which developed in the sulfide-containing medium, contained fairly large non-motile cocci, as single cells in pairs or occasional tetrads (Fig. 1). Pure cultures of the peach-colored organism (Fig. 1*a*) were fairly easily obtained by a combination of agar shake and conventional plating methods (all incubations done at 5°C). The red coccus was more difficult to purify, but the presence of intracellular S^0 globules in cells grown on sulfide (Fig. 1*b*) suggested that this was likely a purple sulfur bacterium [27].

Pure cultures of the curved rod were given the strain designation Ant.Br. Cells of strain Ant.Br grown phototrophically at 5°C measured 0.7 × 2–3 μm (Fig. 1*a*). Young cultures remained fully suspended in tubes of growth medium but older cultures (> 2 weeks) tended to sediment and form a sticky peach-colored pellet. The red organism consisted of cocci 1.2–1.7 μm in diameter (Fig. 1*b*). In vivo absorption spectra of both organisms clearly indicated the presence of bacteriochlorophyll *a*. For strain Ant.Br, a complex absorption spectrum was obtained showing several peaks in the near infra-red between 800 and 870 nm as well as a series of peaks in the carotenoid region of the spectrum (see Table 1). The red isolate (designated strain Ant.Rd) showed major peaks near 800, 820 and 900 nm as well as a series of carotenoid peaks (see Table 1). Methanol extracts of both organisms showed major peaks at 769–770 and 607–608 nm, confirming the presence of bacteriochlorophyll *a*.

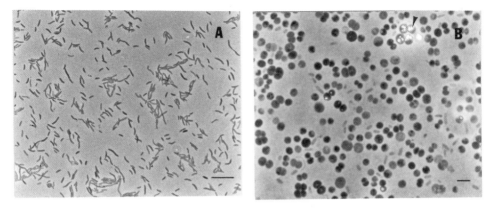

Figure 1. Phase-contrast photomicrographs of cells of Antarctic anoxyphototrophs. (A) Strain Ant.Br, and (B) Strain Ant.Rd. Note sulfur globules (arrow) in some cells of strain Ant.Rd (also note that this photo was taken before the culture was axenic). Marker bar equals 5 μm in A and 2 μm in B.

3.2. Characterization of Strain Ant.Br

Detailed growth and physiological experiments have thus far only been done on strain Ant.Br. Thin sections of cells of this organism showed a gram-negative cell envelope but no obvious intracytoplasmic membrane system (Fig. 2). Growth of strain Ant.Br occurred photohetero-trophically on a variety of carbon sources including organic acids, fatty acids and certain sugars (Table 1). Photoautotrophic growth ($H_2 + CO_2$) was slow but genetic evidence for an operative Calvin Cycle was obtained by strong hybridization of Ant.Br DNA to a *Rhodobacter sphaeroides* ribulose bisphosphate carboxylase (RubisCO) large subunit (*cbbL*) probe (data not shown). As for utilizable nitrogen sources, thus far

Table 1. Basic properties of antarctic anoxyphototrophs Ant.Br and Ant.Rd

	Organism	
Property	Ant.Br	Ant.Rd
Morphology	Curved rods to spirilla (see Fig. 1*a*)	Cocci (see Fig. 1*b*)
Cell size	0.7×2–3 μm	1.2–1.7 μm (diameter)
Color of mass cultures	Peach	Cherry red
Bchl. absorption maxima (in vivo)	866, 819, 799, 582 nm	893, 820, 799, 593 nm
Bchl. absorption maxima (methanol extract)	770, 607 nm	769, 608 nm
Carotenoid absorption maxima (in vivo)	518, 487, 427 nm	549, 512 nm
Intracytoplasmic membrane system	None observed (see Fig. 2)	Not tested
Motility		—[a]
Carbon sources utilized[b]	Malate, succinate, lactate pyruvate, acetate, glucose, fructose, CO_2 (+ H_2)	CO_2 (+ H_2S)
Temperature optimum[c]	15–18°C (0–24°C)	Not tested
Phylogeny	Beta Proteobacteria → *Rhodoferax* relative	Not tested, but probably gamma Proteobacteria

[a]Motility not observed thus far but conditions required for motility may not yet be known.
[b]For photoheterotrophic growth at 7°C. Carbon sources tested but not utilized: butyrate, ethanol, glycerol, ribose.
[c]Temperature range in parentheses.

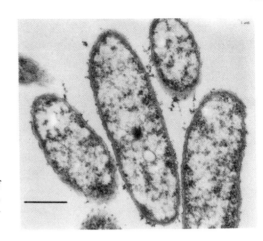

Figure 2. Electron micrograph of thin sections of strain Ant.Br grown at 750 lux incandescent light. Note absence of any obvious intracytoplasmic photosynthetic membranes. Marker bar equals 0.5 μm.

only NH_4^+ and N_2 have been tested and both supported growth of strain Ant.Br. Cells grown on N_2 reduced acetylene at temperatures as low as 0°C (data not shown), thus confirming the existence of a nitrogenase system.

Because of its cold habitat, it was of interest to investigate growth of strain Ant.Br as a function of temperature. In medium RCVB containing 0.1% NaCl and 0.1% yeast extract, growth occurred over a range from 0–24°C, with optimal growth occurring at 15–18°C (data not shown). Growth at 0°C was very slow (generation time > 200 h) but growth at 10°C was much faster and nearly as fast as that at the temperature optimum (generation time approx. 20 h). Although good growth of strain Ant.Br occurred up to 23°C, at 24°C growth was very slow and cells became elongated and pleomorphic; at 25°C or above, no growth occurred.

The phylogenetic status of strain Ant.Br was determined by comparative 16S rRNA sequencing (data not shown). The results showed that strain Ant.Br was a member of the beta purple bacteria (Proteobacteria) [32], with its closest known relative being the anoxyphototroph *Rhodoferax fermentans* [33]. Interestingly, however, but perhaps not surprisingly, strain Ant.Br also showed a close specific relationship to the psychrophilic chemoorganotroph, *Polaromonas vacuolata*, an Antarctic sea-ice bacterium whose growth temperature optimum is 4°C [17].

4. DISCUSSION

Although evidence of anoxyphototrophs in Antarctic environments has existed for over two decades, the isolates described in this paper are the first laboratory cultures of such organisms that show an adaptive response to cold temperatures. At this point one may only guess at the extent of biodiversity of cold-tolerant and cold-loving Antarctic anoxyphototrophs, but the variety of potential habitats that exist in the Antarctic including microbial mats [18,19], meromictic lakes containing biogenic sulfide [23,24], marine inlet waters [23,24], and marine sediments [25,26], suggests that many other types will be found. Indeed, further enrichments in progress in this laboratory bear this out. For example, enrichments established at 10°C have yielded a variety of anoxyphototrophs and developed relatively fast compared to 5°C enrichments; nothing is yet known of the cardinal temperatures of the enriched organisms but isolation and characterization studies are in progress. An interesting but unanswered question at this point is whether extremely psychrophilic anoxyphototrophs (for example, species with growth temperature optima at 5°C or lower) exist. If so, it is likely that only enrichments established at temperatures below 5°C will be successful in isolating them (see below).

Strain Ant.Br is clearly distinct from the nonsulfur purple bacteria isolated by Herbert [25]. None of his isolates resembled strain Ant.Br, either morphologically or in terms of absorption spectra. But more importantly from the standpoint of adaptation to cold temperatures, all of the Herbert isolates grew at 25–30°C while strain Ant.Br cannot. Although it is not clear whether the eventual pure cultures of nonsulfur purple bacteria obtained and characterized by Herbert [25] originated from his 5°C or 18°C enrichment cultures, assuming they originated from the latter (as appears to have been the case), it is possible that psychrophilic species such as Ant.Br were missed because they may only have a competitive growth advantage over mesophilic phototrophs at very low temperatures. To illustrate this, consider the following observations. The anoxyphototroph *Rhodobacter capsulatus*, whose maximum growth temperature is 38.7°C and whose optimum is about 37°C [34], is capable of growing down to about 5°C (data for strain St. Louis, unpublished results of the author); growth at this temperature is very slow. However, at 18°C the generation time of *R. capsulatus* strain St. Louis is about 8 hours; surprisingly, this is less than half that of strain Ant.Br at 18°C even though this temperature is near optimum for the latter organism. Thus, one would predict that in enrichment cultures containing both of these organisms, an enrichment temperature of 18°C would not select for strain Ant.Br even though its temperature optimum is some 20°C lower than that of *R. capsulatus*. Indeed, if it turns out that like strain Ant.Br, other psychrophilic anoxyphototrophs are inherently slow growing species, then temperatures below that in which mesophilic species can develop (and this may be as low as 5°C) will be necessary to successfully enrich them. The comparison between *R. capsulatus* and strain Ant.Br offers an explanation for why the isolates of Herbert [25] were mesophilic—it is likely that his 18°C enrichments were simply too warm to select for psychrophilic species. Comparative enrichment studies of Antarctic environments using low versus moderate enrichment temperatures are needed to shed more light on this problem.

Although not as psychrophilic as some chemotrophic organisms isolated from Antarctic environments [17–19], Ant.Br is the first anoxyphototroph to show a growth temperature optimum below 20°C and a capability for growth at 0°C. Whether or not strain Ant.Br is a "true" psychrophile is a matter of semantics; all definitions of psychrophily include an ability to grow at 0°C and an optimum below 20°C, both criteria of which Ant.Br meets [16]. But some definitions further state that psychrophiles do not grow above 20°C [16], which strain Ant.Br clearly does. However, regardless of the definition used, strain Ant.Br clearly prefers growth at cool temperatures and because of this, and because of its unusual habitat, will be more fully described in a future publication as a new species of the genus *Rhodoferax*, *R. antarcticus*.

Due to its ability to grow in the cold, strain Ant.Br may be useful for biophysical and biochemical studies of photosynthesis at low temperatures, and may be of interest to the rapidly developing field of cold-active enzymes [35]. In particular, the structural features of the pigment-protein complexes and lipids of strain Ant.Br should be of interest for understanding how photosynthetic membrane processes can occur optimally at low temperature. In addition, the fact that strain Ant.Br is a diazotroph, capable of growth on N_2 as sole nitrogen source, is also of general microbiological interest because of the catalytic problems molybdenum nitrogenases experience at low temperature [36]. Thus, further work on the basic properties of strain Ant.Br may yield interesting new results in the field of photosynthesis and related processes.

ACKNOWLEDGMENTS

The author is indebted to Richard W. Castenholz for samples of Antarctic microbial mats, Carl R. Woese for 16S rRNA sequencing, Linda K. Kimble for the RubisCO probe experiment, and John Bozzola, SIU Electron Microscopy Center, for electron microscopy.

REFERENCES

1. Madigan, M.T. (1998) in: Biodiversity, Ecology and Evolution of Thermophiles in Yellowstone National Park (Reysenbach, A-L., Voytek, M., and Mancinelli, R., Eds.), in press. Plenum Press, New York
2. Imhoff, J.F. (1988) in: Halophilic Bacteria, Vol. 1 (Rodriguez-Valera, F., Ed.), pp. 85–108, CRC Press, Boca Raton, Florida
3. Imhoff, J.F. (1992) in: The Prokaryotes, second edition (Balows, A., Trüper, H.G., Dworkin, M., Harder, W., and Schleifer, K-H., Eds.), pp. 3222–3229. Springer-Verlag, New York
4. Imhoff, J.F. (1995) in: Anoxygenic Photosynthetic Bacteria (Blankenship, R.E., Madigan, M.T., and Bauer, C.E., Eds.), pp. 1–15. Kluwer Academic Publishers, Dordrecht, The Netherlands
5. Imhoff, J.F., Hashwa, F., and Trüper, H.G. (1978) Arch. Hydrobiol. 84, 381– 388
6. Imhoff, J.F., and Suling, J. (1997) Arch. Microbiol. 165, 106–113
7. Pfennig, N. (1969) J. Bacteriol. 99, 597–602
8. Pfennig, N. (1974) Arch. Microbiol. 100, 197–206
9. Madigan, M.T. (1984) Science 225, 313–315
10. Madigan, M.T. (1986) Intl. J. Syst. Bacteriol. 36, 222–227
11. Wahlund, T.M., Woese, C.R., Castenholz, R.W., and Madigan, M.T. (1991) Arch. Microbiol. 156, 81–91
12. Kimble, L.K., Mandelco, L., Woese, C.R., and Madigan, M.T. (1995) Arch. Microbiol. 163, 259–267
13. Favinger J., Stadtwald, R., and Gest, H. (1989) Antonie van Leeuwenhoek 55, 291–296
14. Resnick, S.E., and Madigan, M.T. (1989) FEMS Microbiol. Letts. 65, 165–170
15. Stadtwald-Demchick, R., Turner, F.R., and Gest, H. (1990) FEMS Microbiol. Letts. 71, 117–122
16. Baross, J.A., and Morita, R.Y. (1978) in: Microbial Life in Extreme Environments (Kushner, D.J., Ed.), pp. 9–71. Academic Press, New York
17. Irgens, R., Gosink, J.J., and Staley, J.T. (1996) Intl. J. Syst. Bacteriol. 46, 822– 826
18. Bowman, J.P., McCammon, S.A., Brown, M.V., Nichols, D.S., and McMeekin, T.A. (1997) Appl. Environ. Microbiol. 63, 3068–3078
19. Gosink, J.J., and Staley, J.T. (1995) Appl. Environ. Microbiol. 61, 3486–3489
20. Love, F.G., Simmons, G.M. Jr., Parker, B.C., Wharton, R.A. Jr., and Seaburg, K.G. (1983) Geomicrobiol. J. 3, 33–48
21. Wharton, R.A. Jr., Parker, B.C., and Simmons, G.M. (1983) Phycologia 22, 403–405
22. Cohen, Y., and Rosenberg, E., (Eds.) (1989) Microbial Mats: Physiological Ecology of Benthic Microbial Communities. American Society for Microbiology, Washington, D.C.
23. Burke, C.M., and Burton, H.R. (1988) Hydrobiologia 165, 1–11
24. Burke, C.M., and Burton, H.R. (1988) Hydrobiologia 165, 13–23
25. Herbert, R.A. (1976) J. Appl. Bacteriol. 41, 75–80
26. Herbert, R.A., and Tanner, A.C. (1977) J. Appl. Bacteriol. 43, 437–445
27. Pfennig, N., and Trüper, H.G. (1997) in: The Prokaryotes, second edition (Balows, A., Trüper, H.G., Dworkin, M., Harder, W., and Schleifer, K-H., Eds.), pp. 3200–3221. Springer-Verlag, New York
28. Biebl, H., and Pfennig, N. (1981) in: The Prokaryotes (Starr, M.P., Stolp, H., Trüper, H.G., Balows, A., and Schlegel, H.G., Eds.), pp.267–273. Springer-Verlag, New York
29. Tayeh, M.A., and Madigan, M.T. (1987) J. Bacteriol. 196, 4196–4202
30. Pfennig, N., and Wagener, S. (1986) J. Microbiol. Meth. 4, 303–306
31. Kimble, L.K., and Madigan, M.T. (1992) Arch. Microbiol. 155–161
32. Woese, C.R. (1987) Microbiol. Revs. 51, 221–271
33. Hiraishi, A., Hoshino, Y., and Satoh, T. (1991) Arch. Microbiol. 155, 330–336

34. Weaver, P.F., Wall, J.D., and Gest, H. (1975) Arch. Microbiol. 105, 207–216
35. Feller, G., Narinx, E., Arpigny, J.L., Aittaleb, M., Baise, E., Genicot, S., Gerday, C. (1996) FEMS Microbiol. Revs. 18, 189–202
36. Miller, R.W., and Eday, R.R. (1988) Biochem J. 256, 429–432

82

ECOLOGY AND OSMOADAPTATION OF HALOPHILIC *CHROMATIACEAE* IN HYPERSALINE ENVIRONMENTS

P. Caumette,[1] R. Matheron,[2] D. T. Welsh,[1] R. A. Herbert,[3] and R. de Wit[1]

[1]Laboratoire d'Oceanographie
Université Bordeaux I
URA-CNRS 197
2 rue Professeur JolyetF, 33120 Arcachon, France
[2]Laboratoire de Microbiologie
Faculté des Sciences et Techniques de Saint Jerome
F-13397 Marseille Cedex 20, France
[3]Department of Biological Sciences
University of Dundee
Dundee, DD1 4HN, Scotland

1. INTRODUCTION

The *Chromatiaceae* observed to date in hypersaline environments were found in microbial mats which develop in solar salterns. These salterns are man-made structures based on a succession of shallow ponds in which seawater circulates and evaporates until the NaCl concentration reaches saturation, with resulting precipitation and crystalisation. Consequently, a gradient of brines exists between the successive ponds. In the initial ponds, with salinities of 6–8%, most of the isolated phototrophs are marine or slightly halophilic bacteria, according to the classification of Trüper and Galinski [1].

In the intermediary ponds (15–20% salinity), most of the phototrophic bacteria which have been isolated were moderately or *senso stricto* halophilic bacteria (Table 1). The final precipitation ponds are inhabited only by extremely halophilic bacteria, particularly members of the *Halobacteriaceae* and a few phototrophic bacteria of the family *Ectothiorhodospiraceae*. However, these phototrophic bacteria are generally characteristic of hypersaline inland alkaline (soda) lakes rather than marine salterns.

The Phototrophic Prokaryotes, edited by Peschek *et al.*
Kluwer Academic / Plenum Publishers, New York, 1999.

Table 1. Halophilic anoxygenic phototrophic bacteria grouped according to their salt requirements

Bacterial type	Species
Marine to slightly halophilic (1.5 to 6% NaCl)	*Chromatium buderi*
	Chloroherpeton thalassium
	Rhodovulum strictum
	Ectothiorhodospira mobilis
	Rhodovulum sulfidophilum
	Ectothiorhodospira haloalkaliphila
	Pelodictyon phaeum
	Rhodobium marinum
	Rhodospira trueperi
	Ectothiorhodospira vacuolata
	Prosthecochloris phaeoasteroidea
	Ectothiorhodospira shaposhnikovii
	Thiorhodovibrio winogradskyi
	Chlorobium chlorovibrioides
	Chromatium purpuratum
	Rhodovulum euryhalinum
	Rhodovulum adriaticum
	Rhodobium orientis
	Thiorhodococcus minus
	Prosthecochloris aestuarii
	Chromatium vinosum HPC
	Rhabdochromatium marinum
	Ectothiorhodospira marina
	Lamprobacter modestohalophilus
Moderately halophilic (3 to 15% NaCl)	*Chromatium glycolicum*
	Rhodospirillum mediosalinum
	Rhodospirillum salexigens
	Ectothiorhodospira marismortui
	Thiocapsa halophila
	Chromatium salexigens
Halophilic *sensu stricto* ((9 to 24% NaCl)	*Rhodospirillum sodomense*
	Halorhodospira abdelmalekii
	Rhodospirillum salinarum
Extremely halophilic (18 to 30% NaCl)	*Halorhodospira halophila*
	Halorhodospira halochloris

2. *CHROMATIACEAE* IN COASTAL SALTERN MICROBIAL MATS

The halophilic *Chromatiaceae* found in coastal salterns belong to the group of moderately halophilic bacteria [1]. These bacteria develop in microbial mat systems, which grow at the sediment surface, below a gypsum crust of precipitated calcium sulfate, in ponds of intermediate (15–20%) salinity. The gypsum crust is composed of vertically aligned crystals and has a variable thickness, ranging from 5 to 20 mm. The laminated microbial mats which form below this crust are composed of an upper layer of filamentous cyanobacteria mainly of the genera *Phormidium,* or in some cases *Spirulina*. Beneath this cyanobacterial layer is a 1–2 mm thick purple layer, composed of densely packed anoxygenic phototrophic bacteria, from which several strains of halophilic *Chromatiaceae* have been isolated, including three new species, namely, *Chromatium salexigens* [2], *Chromatium glycolicum* [3] and *Thiocapsa halophila* [4]. In synthetic media these species grow optimally between 6–10% NaCl and tolerate up to 18–20% NaCl (Table 1) and thus are well adapted to their environment, where total salinity varies between 15 and 20%. The three species are able to use sulfide, sulfur or thiosulfate as electron donor and CO_2 and some organic compounds, mainly acetate and pyruvate, as carbon source (Table 2).

3. ECOLOGY OF HALOPHILIC *CHROMATIACEAE* IN COASTAL SALTERNS

In these microbial mat ecosystems, the purple layer of *Chromatiaceae* grows in a rather narrow zone where both sulfides and light are present, using the sulfide originating from sulfate reduction in the underlying sediment as electron donor for photosynthesis. Mi-

Table 2. Major properties of halophilic *Chromatiaceae*

	Chromatium salexigens	Thiocapsa halophila	Chromatium glycolicum
Morphology	oviod rod	coccus	rod
Size (μm width × length)	2–2.5 × 4–7.5	2	0.8–1.0 × 2–4
Major caroteniod	spirilloxanthin	okenone	spirilloxanthin
Salinity range (%)	4–20	3–18	2–20
NaCl optimum (%)	10	6	5-6
Vitamin B_{12} requirement	+	+	–
Substrates used:			
Sulfide	+	+	+
Sulfite	+	+	+
Sulfur	+	+	+
Thiosulfate	+	+	+
Fructose	–	+	–
Acetate	+	(+)	(+)
Glycolate	–	–	+
Proprionate	–	(+)	–
Pyruvate	+	+	+
Glycerol	–	–	+
Succinate	–	–	+
Fumerate	–	–	+
Malate	–	–	–
Ethanol	–	–	–

Table 3. Compatible solutes synthesised by representative purple sulfur bacteria[a]

Isolate	Source	Solutes synthesised
Thiocapsa roseopersicina 0P-1	Marine laminated mat	***Sucrose***
Thiocapsa roseopersicina 5811	Brackish lagoon	***Sucrose***
Thiocapsa roseopersicina 5911	Brackish lagoon	***Sucrose***
Thiocapsa halophila SG 3202	Hypersaline saltern	***Glycine betaine***, sucrose, N-acetyl-glutaminyl glutamine amide
Amoebacter roseus 6611	Sewage lagoon	***Sucrose***
Thiocystis violaceae 2311	Estuarine water	***Sucrose***
Chromatium minus 1211	Freshwater pond	***Sucrose***, glycine betaine
Chromatium vinosum D	Ditch water	***Sucrose***
Chromatium purpuratum DSM 1591		***Sucrose***, glycine betaine, N-acetyl-glutaminyl glutamine amide
Chromatium salexigens SG 3201	Hypersaline saltern	***Glycine betaine***, ***sucrose***, N-acetyl-glutaminyl glutamine amide

[a]The dominant compatible solute is shown in bold type. Data from [10,11].

croprofiles of oxygen and sulfide, have revealed the important role of phototrophic bacteria for sulfide reoxidation in these closed systems, where oxygen diffusion is greatly limited by the presence of the gypsum crust [5]. However, this photosynthetic activity is dependent upon the light reaching the purple layer. Within the first mm of sediment, the phototrophic bacteria are able to use the lower wavelenghts of visible light, which are harvested by means of their carotenoids, whereas, in the deeper sediments (up to a few mm) they grow using near infra-red (NIR) radiation which penetrates deeper than visible light in most sediments [6]. NIR radiation is directly harvested by the bacteriochlorophyll antennae. The major taxonomic groups of *Chromatiaceae* have evolved to use different and complementary wavelength regions, for example purple sulfur bacteria with bacteriochlorophylls a and b use NIR radiation of 800–900 and 1000–1050 nm respectively. Thus, different species of *Chromatiaceae* with bacteriochlorophyll a or b can co-exist in the same layer of the microbial mat using different spectral qualities of light. Moreover, specific NIR radiation absorbtion differences are also found amongst the bacteriochlorophyll a containing *Chromatiaceae*, which correspond to differences in the antennae of the light harvesting pigment protein complexes of the photosynthetic membranes. For example, clear absorbtion differences in the NIR region from 800–900 nm exist between *Chromatiaceae* containing okenone as major carotenoid and those containing spirilloxanthin. In microbial mats which develop in the coastal salterns of the Salin-de Giraud, France, the okenone containing *Thiocapsa halophila* (absorbtion maxima, 825–830 nm) coexists with the spirilloxanthin containing *Chromatium salexigens* (absorbtion maxima, 800–805 and 860–870) in a thin purple layer of the microbial mats which form below the gypsum crust. This coexistance is possible due to the their different absorbtion spectra for NIR radiation [7].

4. GLYCOLATE METABOLISM

An important ecological feature of halophilic *Chromatiaceae* found in microbial mats of coastal salterns, is their use of glycolate as a substrate. Glycolate is one of the major photoexcretion products of the cyanobacteria in these mats and can constitute a large proportion of the excreted material when oxygen production is high throughout the day and consequently, CO_2 concentration is low [8]. The excreted glycolate can be utilised as a substrate within the mat by a wide range of microorganisms under both oxic and anoxic conditions. Some cyanobacteria utilise glycolate during photorespiration and under anoxic conditions glycolate is utilised by purple nonsulfur bacteria, fermentative bacteria and some sulfate reducers. However, amongst halophilic bacteria, glycolate utilisation has only been described for the purple sulfur bacterium, *Chromatium glycolicum* [3]. This observation clearly demonstrates the potential importance of carbon source exchange between microorganisms in these mat system, although, these fluxes of carbon are at present unknown. Positive interactions between microorganisms in the mats may not only involve substrates which can be further metabolised, but also the transfer of growth factors such as the osmoregulatory compounds synthesised by the halophilic *Chromatiaceae* in the microbial mats.

5. OSMOADAPTATION OF *CHROMATIACEAE*

Microorganisms which grow in saline or hypersaline environments must maintain a high intracellular osmotic pressure in order to generate cell turgor pressure, which acts as the driving force for cell extension growth and ultimately cell division [9]. In the Halobacteria, this is achieved by the accumulation of intracellular KCl, however, these bacteria often display a relatively narrow range of salt tolerance, since the intracellular machinery is adapted to function at a high, but limited range of KCl concentrations. In contrast, halotolerant and halophilic eubacteria are often capable of growing over a wide range of salt concentrations. In these organisms, osmoadaptation is dependent upon the accumulation of K^+ and a restricted range of organic solutes, termed "compatible solutes". These solutes can be divided into a number of sub-classes, including, polyols, sugars, amino acides, betaines and ectoines.

Among non-halophilic *Chromatiaceae*, osmoadaptation was dependent upon the synthesis of the disaccharide sucrose and intracellular sucrose concentrations were directly dependent upon the osmolarity of the growth medium [10–12]. In contrast, halotolerant and halophilic species synthesised primarily the quaternary amine, glycine betaine, alongwith sucrose and trace quantities of N-acetly-glutaminylglutamine [10–12]. These data are in accord with those reported for cyanobacteria, where nonhalotolerant isolates synthesised sugars (trehalose or sucrose) or the heteroside, glucosylglycerol and halotolerant isolates mainly glycine betaine [13] and thus support the hypothesis of Warr et al. [14], that the degree of halotolerance of a microorganism may be dependent at least in part upon the metabolic effects of the solute(s) accumulated as internal osmotica.

All tested strains of *Chromaticeae* were also capable of accumulating glycine betaine from the growth medium in response to osmotic stress, although the degree of uptake varried widely from isolate to isolate, with halotolerant strains generally being more efficient [11]. We further investigated glycine betaine in a halotolerant *Chromatium* species isolated from the Dead Sea. This isolate possesed a constitutively expressed active transport system, which was saturable with respect to glycine betaine and displayed typical Michaelis-Menten type kinetics: $K_m = 24$ μM, $V_{max} = 306$ nmol·min^{-1}·mg protein^{-1} at an external NaCl concentration of 1 M [15]. The rate of glycine betaine transport declined with increasing growth medium NaCl concentration, decreasing from 192 to 9.7 nmol·min^{-1}·mg ptotein^{-1} in the presence of 1 and 3 M NaCl respectively. This inhibition is possibly due to the direct salt inhibition of a periplasmic glycine betaine binding protein [15]. Thus, at least in this isolate, glycine betaine transport was least efficient during growth at high NaCl concentrations, when the requirement for a high intracellular glycine betaine concentration was greatest. However, the possesion of an active glycine betaine transport system may provide an effective strategy for growth in high osmolarity environments, allowing the cell to scavenge glycine betaine present in the environment and to recycle synthesised glycine betaine which may passively diffuse from the cell. Such a strategy would conserve both cell carbon and energy, since accumulation by transport is energetically more efficient than *de novo* synthesis of glycine betaine.

In addition to their strictly osmotic role in regulating cell turgor pressure, compatible solutes may also perform a dual function as intracellular reserves of carbon and nitrogen. This may be particularily true in organisms which accumulate both sugars and nitrogen containing compatible solutes such as amino acids, betaines or ectoines and indeed, the relative concentrations of these compatible solutes has been shown to be regulated by the nutrient status of the growth medium [16].

6. ROLE OF COMPATIBLE SOLUTES IN SEDIMENT CARBON TURNOVER

In hypersaline environments, the bacteria present accumulate compatible solutes to very high intracellular (molar) concentrations, in order to balance the environmental osmolarity. Thus, particularly in high biomass systems such as microbial mats, these solutes represent a significant pool of organic carbon and nitrogen. These solutes may be released to the environment, through leakage, or upon osmotic downshock or cell death. Thus compatible solutes potentially represent significant sources of carbon, nitrogen and energy for the growth of other bacteria.

The importance of the role played by glycine betaine in sediment carbon turnover can be deduced from the often high rates of methane formation in these ecosystems, since, metabolism of trimethylamine derived from anaerobic degradation of glycine betaine is the most probable source of this methane production [17,18]. Recently, we have carried out a survey of sugar utilisation by both purple sulfur and nonsulfur bacteria. Amongst, the purple nonsulfur bacteria, utilisation of the compatible solutes, sucrose and to a lesser extent trehalose was common and almost all strains were able to utilise their constituent monosaccharides (glucose and fructose). Sugar utilisation was rarer amongst the purple sulfur bacteria, none of the tested cultures was able to use trehalose and only strains of the recently isolated species *Thiorhodococcus minus* [19], were able to utilise sucrose. Fructose was utilised by approximately 50% of the isolates, predominantly *Thiocapsa* species, *Thiorhodococcus minus* strains and one strain of *Chromatium gracile*. However, somewhat surprisingly no strains, including the sucrose utilising strains was able to utilise glucose, even though during growth on sucrose, no glucose accumulated in the growth medium. Sugar utilisation is also common amongst aerobic, fermentative and some sulfate reducing bacteria and thus, sugars synthesised by phototrophic members of the microbial mat community, may be important in the overall turnover of carbon within these microbial mats.

REFERENCES

1. Trüper, H.G.and Galinski, E.A. (1986). Experientia 42, 1181–1187.
2. Caumette, P., Baulaigue, R; and Matheron, R. (1988). System. Appl. Microbiol. 10, 284–292.
3. Caumette, P., Baulaigue, R; and Matheron, R. (1991). Arch. Microbiol. 155, 170–176.
4. Caumette, P., Imhoff, J.F., Süling, J. and Matheron, R. (1997). Arch. Microbiol. 167, 11–18.

5. Caumette, P., Matheron, R., Raymond, N. and Relaxans, J.-C. FEMS Microbiol. Ecol. 13, 273–286.
6. Kulh, M., Larsen, C. and Jørgensen, B.B. (1994). In: Microbial mats, Structure, Development and Environmental Significance. (L.J. Stal and P. Caumette, eds). pp149–168. Springer-Verlag, Berlin.
7. De Wit, R. and Caumette, P. (1994). In: Microbial mats, Structure, Development and Environmental Significance. (L.J. Stal and P. Caumette, eds). pp 377–392. Springer-Verlag, Berlin.
8. Bateson, M.M. and Ward, D.M. (1988). Appl. Environ. Microbiol. 54, 1738–1743.
9. Koch, A.L. (1982). J. Gen. Microbiol. 128: 2527–2540.
10. Severin, J., Wohlfarth, A. and Galinski, E.A. (1992). J. Gen. Microbiol. 138, 1629–1638.
11. Welsh, D.T. and Herbert, R.A. (1993). FEMS Microbiol. Ecol. 13: 145–150.
12. Welsh, D.T. and Herbert, R.A. (1993). FEMS Microbiol. Ecol. 13: 151–158.
13. Reed, R.H., Richardson, D.L., Warr, S.R.C. and Stewart, W.D.P. (1984). J. Gen. Microbiol. 130: 1–4.
14. Warr, S.R.C., Reed, R.H. and Stewart, W.D.P. (1984). J. Gen. Microbiol. 130: 2169–2175.
15. Welsh, D.T. and Herbert, R.A. (1995). FEMS Microbiol. Lett. 128: 27–32.
16. Galinski, E.A. and Herzog, R.M. (1990) Arch. Microbiol. 153: 607–613.
17. Oremland, R.S. and King, G.M. (1989). In: Microbial Mats: Physiological ecology of benthic microbial communities. (Y.Cohen and E.Rosenberg, Eds). American Society of Microbiology, Washington.
18. Oren, A. (1990). Antonie Van Leeuwenhoek 58, 291–298.
19. Guyoneaud, R., Matheron, R., Liesack, W., Imhoff, J.F. and Caumette, P. (1997). Arch. Microbiol. 168, 16–23.

83

CLONING AND MOLECULAR ANALYSIS OF THE GENE *PRQ*R CONTROLLING RESISTANCE TO PARAQUAT IN *SYNECHOCYSTIS* SP. PCC 6803

K. Sidoruk,[1] V. Melnik,[1] M. Babykin,[1] R. Cerff,[2] and S. Shestakov[1]

[1]Department of Genetics
Moscow State University
Moscow 119899, Russia
[2]Institute of Genetics
Technical University of Braunschweig
D-38106 Braunschweig, Germany

1. INTRODUCTION

Oxygen-evolving photosynthetic cells must possess efficient systems for protection against active oxygen species generated by endogenous mechanisms and various environmental factors. Redox-cycling agents, such as paraquat and menadione, can induce the formation of active radicals causing an oxidative stress [1]. Paraquat accepts electrons from Photosystem I and its cation radical reduces oxygen to superoxide which is converted to hydrogen peroxide. The exact site of paraquat action is uncertain. It was suggested that Fe-S center B of Photosystem I is the main site of electron donation to paraquat [2]. The enhancement of resistance to paraquat in plants and bacteria may be associated with alterations in transport of the drug, failure to interact with cell targets, elevation of activities of enzymes involved in detoxification of paraquat, scavenging of active oxygen species or repair of oxidative damage [3,4]. An adaptive response to oxidative stress in some bacteria is mediated by regulatory *sox*RS and *oxy*R gene systems which control at the transcriptional level the induction of antioxidant defensive proteins [1,4]. Bacterial *sox*R-constitutive mutants are characterised by an increased resistance to redox-cycling agents [4,5]. A number of cyanobacterial mutants resistant to paraquat treatment have been isolated [6–8], but the genetic nature of this resistance has not been identified. Cloning and molecular analysis of genes controlling resistance to redox-cycling agents can help to understand the nature of cell targets for these agents and the defence mechanisms against oxidative stress and photoinactivation in cyanobacteria. Here, we report the cloning and sequencing of a novel gene of *Synechocystis* sp. PCC 6803, mutation in which results in elevated cross-resistance to paraquat and menadione.

The Phototrophic Prokaryotes, edited by Peschek *et al.*
Kluwer Academic / Plenum Publishers, New York, 1999.

2. PROCEDURE

Wild type (WT) and mutant cells of *Synechocystis* 6803 were grown in BG11 medium supplemented when necessary with inhibitors at the following concentration: paraquat, 15–20 μg /ml; menadione, 5 μg/ml; kanamycin, 15–30 μg/ml. Transformation of *Synechocystis* 6803 and selection of transformants were carried out as described [9]. Routine molecular biology techniques (plasmid DNA isolation, restriction analysis, Southern blot hybridization etc.) were used as described [10]. DNA sequences were determined by the double stranded dideoxy chain-termination method using Sequenase (USB). The program GeneBee [11] based on PROSIS database was used for amino acid sequence analysis.

3. RESULTS AND DISCUSSION

Strain Prq20 which is cross-resistant to paraquat and menadione (Table 1) was found among spontaneous *Synechocystis* 6803 mutants selected against paraquat [8] and employed for cloning and identification of genes controlling resistance to redox-cycling agents. In contrast to *Synechocystis* 6803 paraquat resistant mutant MR-12 described by Endo and Asada [5] the mutant Prq20 does not differ significantly from the wild type strain in photosynthetic and electron transport characteristics (data not shown).

DNA extracted from Prq20 mutant cells was digested with *Hind*III and fractionated by agarose gel electrophoresis. DNA fragments from fraction (4.3–6.7 kb) that transformed *Synechocystis* 6803 wild type cells to paraquat resistance were cloned into vector pUC20. A recombinant plasmid pPR1 conferring resistance to paraquat was found and used for restriction analysis of chromosomal DNA fragment. A 1.3 kb *Bam*HI-*Sph*I DNA subfragment transforming WT cells to paraquat resistance was sequenced and used as a probe to clone the homologous WT sequence from a cosmid genomic library of the wild type of *Synechocystis* 6803 [12]. Sequence analysis of DNA fragments from WT and Prq20 strains showed that the mutation leading to the elevated resistance to paraquat is located within an ORF coding for a polypeptide of 187 amino acids. In the Cyanobase databank (Kazusa DNA Research Institute, Japan) ORF187 corresponds to *slr*0895 encoding a protein of unidentified function. The ORF187 was named *prq*R1 gene because of its involvement in control of the resistance to paraquat.

Database analysis of the predicted amino acid sequence indicates that only the N-terminal region (about 90 residues) of PrqR shares a partial homology with the N-terminal sequences of some bacterial proteins which act as a transcriptional regulators. The N-terminal

Table 1. Resistance of the wild type and mutant strains of *Synechocystis* 6803 to paraquat and menadione[a]

	Inhibitors (μg/ml)	
Strain	Paraquat	Menadione
WT	5	1
Prq20	20	5
Prin2	15	2

[a]The reported concentration is the threshold amount that allowed growth on solid medium.

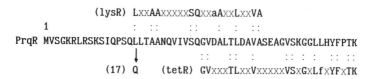

```
        (lysR) LxxAAxxxxxxSQxxaAxxLxxVA
   1            :  ::      ::   :  :  ::
PrqR MVSGKRLRSKSIQPSQLLTAANQVIVSQGVDALTLDAVASEAGVSKGGLLHYFPTK
              ↓          ::  :: :    :: : :: :: :
        (17) Q     (tetR) GVxxxTLxxVxxxxxVSxGxLfxYFxTK
```

Figure 1. Comparison of the deduced amino acid sequences of the N-terminal portion of *Synechocystis* 6803 PrqR protein and DNA-binding domains of bacterial proteins of LysR and TetR families. Arrow shows the position of Prq20 mutation. Symbol "×" indicates presence of any amino acid residue at the corresponding position.

region of PrqR protein shows 50–70% similarity with proteins GusR (repressor of *gus*-RABC operon of *Escherichia coli*), MtrR (repressor of *mtr*C gene of *Neisseria*), TetR (repressor of *tet*A of transposon *Tn*10) and contains a putative DNA-binding domain (Fig. 1) showing a high degree of homology with DNA-binding motifs of proteins which belong to LysR [13,14] and TetR [15] families. Such proteins can serve in regulation of transcription as gene activators and/or repressors.

According to the Cyanobase Databank, *Synechocystis* 6803 genome does not contain regulatory genes which are highly homologous to *oxy*R and *sox*RS genes involved in adaptive response to oxidative stress in enterobacteria. We speculate that PrqR protein serves in regulation of transcription like OxyRS system. On the other hand we can not exclude that photosynthetic cyanobacteria may differ from heterotrophic bacteria in mechanisms of regulation of antioxidant systems.

A comparison of nucleotide sequences from WT and mutant Prq20 revealed a missense mutation (T → A) leading to the replacement of hydrophobic leucine residue at position 17 to hydrophilic glutamine. As shown in Figure 1, mutation Prq20 is localized in the N-terminal region which might represent the functionally important DNA-binding domain of PrqR protein.

To study the involvement of *prq*R1 in control of resistance to paraquat, we performed a targeted insertional mutagenesis of this gene. A kanamycin resistance (Km^R) cartridge from plasmid pUC4K was inserted at *Eco*4711 site located in the middle of *prq*R1 (Fig. 2). A homozygous *Synechocystis* 6803 mutant with disrupted *prq*R1 gene has been obtained after selection by successive cultivation of Km^R transformants in the presence of increasing concentration of antibiotic. The complete segregation of insertional *prq*R mutants was confirmed by Southern blot hybridization analysis (data not shown). Homozygous insertional

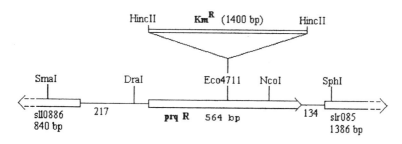

Figure 2. Cartridge mutagenesis of the *prq*R1 gene.

mutants assayed (including strain Prin2) were characterized by an increased resistance to paraquat and menadione (Table 1).

These data indicate that the function of the prqR1 gene is not essential for normal growth of *Synechocystis* 6803 although inactivation of this gene leads to the enhancement of resistance to agents generating superoxide. The phenotype of mutants presumably results from less intracellular concentration of paraquat or superoxide. Preliminary studies on the expression of *prq*R1 and *sod*B permit to suppose that PrqR protein may be involved in the induction of SodB iron superoxide dismutase by paraquat or menadione treatment. Taken together, the sequence information and properties of paraquat resistant mutants tested suggest that the PrqR protein could function in regulation of cellular response to oxidative stress in cyanobacteria. The mode of action of this putative regulatory protein and its transcriptional targets and physiological role remain to be elucidated.

ACKNOWLEDGMENTS

This work was supported by grants from Volkswagen-Foundation (Germany) and the Russian Foundation of Basic Research. S. S. was supported in part by a grant from the International Human Frontier Science Program.

REFERENCES

 1. Farr, S.B. and Kogoma T. (1991) Microbiol. Rev. 55, 561–585.
 2. Fujii, T., Yokoyama, E., Inoue, K. and Sakurai, H. (1990) Bioch. Bioph. Acta 1015, 41–48
 3. Bowler, C., Van Montagu, M. and Inze, D. (1992) Annu. Rev. Plant Physiol.Plant Mol. Biol. 43, 83–116.
 4. Demple, B. (1991) Annu. Rev. Genet. 25, 315–337.
 5. Greenberg, J.T., Monach, P.A., Chou, J.H., Josefy, P. and Demple, B. (1990) Proc. Natl. Acad. Sci. USA 87, 6181–85.
 6. Endo, T. and Asada, K. (1992) in: Research in Photosynthesis, vol. 2 (Murata, N. ed.), pp. 607–610, Kluwer Acad. Publ., Dordrecht.
 7. Campbell, W.S. and Laudenbach, D.E. (1995) J. Bacteriol. 177, 964–972.
 8. Sidoruk, K., Shakhnabatian, L., Belavina, N., Ernst, A., Stal, L., Gallon, J.R. and Shestakov S. (1996) Vestnik Moskovskogo Universiteta, ser.biol.(Russ) 4, 43–49.
 9. Bartsevich, V.V. and Shestakov, S.V. (1995) Microbiology 141, 2915–2920.
10. Sambrook, J., Fritsch, E.F. and Maniatis, T. (1989) Molecular Cloning, Cold Spring Harbor Laboratory Press.
11. Brodsky, L.I., Ivanov, V.V., Kaladzigis, Y.L., Leontovich, R.M., Nikolaev, V.N. and Drachev, A.V. (1995) Biochemistry (Russ) 60, 1221–1230.
12. Shestopalov, V.I., Nashchokina, O.O., Shestakov, S.V. and Yankovskii, N.K. (1994) Rus. J. Genetics, 398–401.
13. Schell, M.A. (1993) Annu. Rev. Microbiol. 47, 527–626.
14. Henikoff, S., Haughn, G.W., Calvo, J.M. and Wallace, J.C. (1988) Proc. Natl. Acad. Sci. USA 85, 6602–6606.
15. Aramaki, H., Yagi, N. and Suzuki, M. (1995) Protein Eng. 12, 1259–1266.

84

HYDROGEN PEROXIDE REMOVAL IN CYANOBACTERIA

Characterization of a Catalase-Peroxidase from *Anacystis nidulans*

Christian Obinger,[1] Günther Regelsberger,[1] Andrea Pircher,[1]
Astrid Sevcik-Klöckler,[1] Georg Strasser,[1] and Günter A. Peschek[2]

[1]Institut für Chemie
Universität für Bodenkultur
Muthgasse 18, A-1190 Wien, Austria
[2]Biophysical Chemistry Group
Institut für Physikalische Chemie
Universität Wien
Althanstrasse 14, A-1090 Wien, Austria

INTRODUCTION

Many lines of geological and biological evidence support the view that most of the atmospheric oxygen has been supplied by photosynthetic organisms [1]. Primordial cyanobacteria have evolved as the most primitive, oxygenic, plant-type photosynthetic organisms appearing on the Earth about 3.5 billion years ago in the Archean [2]. They have left the oldest fossil records currently known (Archean rocks of Western Australia) and it is thought that their photosynthetic activity was responsible that within 1.8 billion years oxygen levels in the Archean of less than 1% of present levels were increased to about 15% in the Paleoproterozoic [3]. Life during this period was bacterial and had to find a way how to balance between death and life through oxygen. The first "pollution crisis" hit the Earth about 2.2 billion years ago. At about this time a fairly rapid increase in levels of oxygen in the atmosphere began and the organisms were challenged to evolve sophisticated biochemical mechanisms for rendering oxygen and activated oxygen species harmless.

Being the first to produce oxygen as a byproduct, cyanobacteria were (among) the first to be affected by it and also to utilize it. Gradual modification of pre-existing photo-

The Phototrophic Prokaryotes, edited by Peschek *et al.*
Kluwer Academic / Plenum Publishers, New York, 1999.

synthetic electron transport and enzyme systems could have changed a photosynthetic into a respiratory chain [4,5]. By elaborating mechanisms for aerobic respiration, cyanobacteria uniquely have accommodated both oxygenic photosynthesis and aerobic respiration within a single prokaryotic cell [6,7].

Metabolism in an oxygen-containing environment results in the generation of reactive oxygen species such as singlet oxygen, superoxide, hydrogen peroxide, and hydroxyl radical. Superoxide and hydrogen peroxide are known to be inevitable byproducts of both photosynthetic and respiratory electron transport chains. Although the production of these species is only below several % that of the total oxygen exchange, their effective scavenging is indispensable to survival of cells under aerobic conditions. Organisms have evolved sophisticated and efficient enzyme systems to neutralize these potentially injurious reactive oxygen species, including superoxide dismutases and hydroperoxidases.

From the three types of superoxide dismutases (distinguished on the basis of the metal in the enzyme), two cyanide insensitive types have been characterized in cyanobacteria, namely a cytosolic hydrogen peroxide sensitive iron-containing superoxide dismutase and a thylakoid-bound hydrogen peroxide insensitive manganese-containing superoxide dismutase. Superoxide dismutases are thought to have a protective role against photooxidative damage [8,9] and oxidative stress [10]. Table 1 summarizes the present knowledge of distribution of superoxide dismutases in cyanobacteria and also demonstrates their occurrence in both vegetative cells and heterocysts of nitrogen-fixing species [8–23].

Table 1. Superoxide dismutase activities in various cyanobacterial species

Species	Fe-SOD	Mn-SOD	Reference
Plectonema boryanum	+	+	11,12
	+	+	13
	+	+	9
	+	+	10
Synechococcus PCC 6301	+	n.d.	8
(*Anacystis nidulans*)			
	+	+	13
	+	+	14
Synechococcus PCC 7942	+	+	15
Synechococcus PCC 7002	+	+	16
Synechocystis PCC 6803	+	+	16
Microcystis aeruginosa	+	n.d.	17
	+	+	18
Anabaena cylindrica			
Vegetative cells	+	n.d.	19
	+	+	12
Heterocysts	+	+	20
	+	n.d.	19
	+	n.d.	21
	+	+	20
	+	n.d.	22
Anabaena variabilis			
Vegetative cells	+	+	12
	+	+	13
Heterocysts	+	n.d.	23
Nostoc muscorum	+	+	12

[a] +, activity detected; −, activity not detected; n.d., not determined.

Two classes of heme proteins designated hydroperoxidases, namely catalase and per-oxidase, are involved in the neutralization of hydrogen peroxide. The former catalyzes the dismutation of hydrogen peroxide to oxygen and water, whereas the latter uses hydrogen peroxide to oxidize a variety of compounds. Research in the past decade has shown, that the distinction between catalases and peroxidases was a matter of degree rather than kind. A novel type of hydroperoxidase, the so-called catalase-peroxidase, has been recently pro-posed to exhibit both catalatic and peroxidatic activities [24]. Bacteria synthesize a wide va-riety of hydroperoxidases, which differ in structure, catalytic properties, physiological function and regulation of their synthesis. Some bacterial strains synthesize only one type of enzyme, whereas others were found to synthesize several types of enzymes in response to environmental factors. Classification of bacterial hydrogen peroxide scavenging enzymes was performed either according to their prostethic groups (heme b, heme d, flavin) or to their electron donor [monofunctional catalase (EC 1.11.1.6), catalase-peroxidase (EC 1.11.1.7), NADH peroxidase (1.11.1.1), cytochrome c peroxidase (EC 1.11.1.5), ascorbate peroxidase (1.11.1.11)]. During the last years as a result of recombinant DNA techniques, as well as techniques for collecting and analyzing X-ray diffraction and NMR data, many and new structural information on heme containing hydroperoxidases became available. On the basis of significant similarities between the core parts of the enzymes and on identification of a number of highly conserved amino acids, now heme containing bacterial hydroperoxi-dases should be classified as either monofunctional catalases (hydrogen peroxide:hydrogen peroxide oxidoreductase, EC 1.11.1.6) with tyrosine as the fifth proximal heme iron ligand, or as members of the so-called class I of the fungal, plant and bacterial peroxidase superfa-mily with histidine as the fifth ligand of the heme iron. Homology-based models have re-vealed the existence of two peroxidase superfamilies, namely one consisting of fungal, plant and bacterial peroxidases [25], whereas the second superfamily was shown to include ani-mal enzymes [26]. On the basis of sequence similarities the fungal, plant and bacterial per-oxidase superfamily has been shown to consist of three major classes [25]: class I, the intracellular peroxidases, including yeast cytochrome c peroxidase [27], a soluble protein found in the mitochondrial electron transport chain, where it probably protects against toxic peroxides; ascorbate peroxidase, the main enzyme responsible for hydrogen peroxide re-moval in chloroplasts and cytosol of higher plants [28]; and bacterial catalase-peroxidases, exhibiting both peroxidase and catalase activities [24]. Class II consists of secretory fungal peroxidases: ligninases, or lignin peroxidases, and manganese-dependent peroxidases, whereas class III has been shown to consist of the secretory plant peroxidases, which have multiple tissue-specific functions.

Concerning the role of hydroperoxidases for detoxification of hydrogen peroxide in cyanobacteria amazingly little is known. Table 2 summarizes the present literature giving evidence of catalatic activity, ascorbate peroxidase activity and glutathione peroxidase ac-tivity in cyanobacteria [16,17,19,21,23,29–36]. There are only two papers reporting a pu-rification procedure down to homogeneity, combined with a comprehensive biochemical and genetic characterization of a cytosolic hydroperoxidase in *Synechococcus* [29,32]. The present contribution tries to focus on the characterization of this type of enzyme as typical class I catalase-peroxidase. The potential role of ascorbate peroxidase in cyanobacteria is discussed shortly, whereas a discussion of glutathione peroxidase has been omitted. Two obvious glutathione peroxidase genes have been found on the genome of *Synechocystis* 6803 [37] but the role of the corresponding proteins as peroxidases has to be clarified. A discussion of monofunctional catalases (EC 1.11.1.6) as potential hydrogen peroxide scav-enging enzymes has been omitted, too, since at the moment there is no evidence for their occurrence in cyanobacteria.

Table 2. Catalatic activities, ascorbate peroxidase, and glutathione peroxidase activities in crude extracts and purified enzyme preparations (*) from various cyanobacterial species[a]

Species	Catalatic activity	Ascorbate peroxidase activity	Glutathione peroxidase activity	Reference
Plectonema boryanum	+	–	–	16
Fremyella diplosiphon	+	–	n.d.	16
Synechococcus PCC 6301	+	–	–	16
(*Anacystis nidulans*)				
	+	–	–	29*
	+	+	–	30,31
Synechococcus PCC 7942	+	–	–	32*
	+	+	n.d.	33
Synechococcus PCC 7002	+	–	–	16
Synechocystis PCC 6803	+	+	–	16
	+	n.d.	+	34
Microcystis aeruginosa	+	n.d.	n.d.	17
Anabaena cylindrica	+	+	–	16
	+	+	n.d.	19
	+	n.d.	n.d.	21
Anabaena variabilis	n.d.	+	n.d.	16
	–	–	+	23
Nostoc muscorum	n.d.	+	–	16
	+	+	–	30,31
	n.d.	n.d.	–	35
Gloethece	–	+	+	36
Aphanothece halophytica	n.d.	+	n.d.	16
Tolypothrix tenuis	n.d.	+	n.d.	16

[a] +, activity detected; –, activity not detected; n.d., not determined.

ASCORBATE PEROXIDASE

Ascorbate peroxidase activity has been demonstrated in higher plants, eukaryotic algae and certain cyanobacteria [16]. Ascorbate peroxidases are typical class I peroxidases, which function as scavengers of hydrogen peroxide in the cytosol and chloroplasts of plants. There are several papers which indicate the presence of ascorbate peroxidase activity in at least some cyanobacterial species, though until now no purification and comprehensive characterization has been done [16,19,30,31,33,36]. No clear correlation between the presence or absence of ascorbate peroxidase and the taxonomy of cyanobacteria can be seen. Ascorbate peroxidase can be found in unicellular and also in filamentous species.

From experiments with a gas permeable membrane-mass-spectrometer for measurements of oxygen isotopes and $H_2^{18}O_2$ [16], it was concluded that cyanobacteria should be divided into two groups, those that have and those that lack ascorbate peroxidase. The first group scavenges hydrogen peroxide with a peroxidase using a photoreductant as electron donor (evolving $^{16}O_2$ but not $^{18}O_2$ on illumination and being dependent on photoreductants generated by photosystem I), whereas the second group only scavenges hydrogen peroxide in a catalatic reaction. In contrast to chloroplasts, which were found to contain 15–25 mM ascorbate, cyanobacteria were found to contain very little intracellular ascorbate (30–100 μM), whereas glutathione was found to be 2–5 mM [30]. Upon peroxidase action ascorbate is oxidized and has to be regenerated. In angiosperm chloroplasts regeneration occurs by dehydroascorbate reductase [38] and monodehydroascorbate reductase [39] using glutathione and NAD(P)H, respectively, as electron donors. If there exists a similar ascorbate-glutathione cycle in cyanobacteria cannot be answered today. Keeping in mind calculated K_m-values from chloroplast ascorbate peroxidases, the corresponding cyanobacterial enzyme would work under suboptimal substrate concentrations [40]. Furthermore in most of the cyanobacterial species examined, dehydroascorbate reductase was not found, with the exception of *Tolypothrix tenuis* [16], *N. muscorum* [30,31] and *Synechococcus* 6311 [30,31]. By contrast, monodehydroascorbate reductase was detected in the ascorbate peroxidase-containing species [16], with the exception of *Synechocystis* 6803 and *Aphanothece halophyta*. Glutathione reductase activity was detected in *Synechococcus lividus* [41], *Nostoc muscorum* [30,42], *Anabaena variabilis* [43,23], *Spirulina maxima* [43] and *Synechococcus* 6301 [30].

CATALASE-PEROXIDASES IN CYANOBACTERIA

Synechococcus PCC 7942 and *Synechococcus* PCC 6301 (*Anacystis nidulans*, an organism closely related to *Synechococcus* PCC 7942) have been shown to possess a typical prokaryotic catalase-peroxidase as the sole cytosolic hydrogen peroxide scavenging enzyme. Protein purification procedures down to homogeneity have been described, as well as both biochemical and genetic characterizations [29,32]. The enzymes exhibit both peroxidase and catalase activities as shown in Figure 1 A and B for the catalase-peroxidase from *Anacystis nidulans* [29]. Native PAGE of both crude cytosolic extracts as well as of the purified enzymes gave single bands of identical migration characteristics detectable with staining procedures either for peroxidase (A) or catalase activity (B). SDS-PAGE of the purified enzyme gave a single band at 80.5 kDa (Figure 1 C), whereas molecular mass analysis by gel filtration gave a molecular mass of about 165 kDa (data not shown) indicating that the enzyme is composed of two subunits of identical size. The isoelectric point of the protein was determined to be at pH 4.7 (Figure 1 E).

The catalase-peroxidase from *Synechococcus* 7942 has been shown to be also homodimeric with subunits of 79.9 kDa in size [32]. A homodimeric or homotetrameric structure is typical for this type of hydroperoxidase. In general the identical subunits of catalase-peroxidases are 78–82 kDa in size as has been demonstrated for the subunits of *E. coli* [44,45], *Streptomyces* sp. IMSNU-1 [46], *Mycobacterium smegmatis* [47], *Mycobac-*

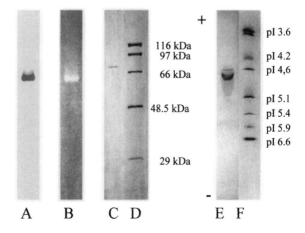

Figure 1. Disc-PAGE. (A) Native disc-PAGE of cytosolic fraction of *Anacystis nidulans* catalase-peroxidase: staining for peroxidase activity with H_2O_2/o-dianisidine; (B) Native disc-PAGE of cytosolic fraction: staining for catalase activity; (C) SDS-PAGE of purified enzyme (4 μg) stained by silver; (D) SDS-PAGE of marker proteins stained by silver; (E) Isoelectric focusing, purified enzyme (4 μg) stained by silver; (F) Isoelectric focusing, marker proteins stained by silver.

116 kDa
97 kDa
66 kDa
48.5 kDa
29 kDa

pI 3.6
pI 4.2
pI 4,6
pI 5.1
pI 5.4
pI 5.9
pI 6.6

A B C D   E F

terium tuberculosis [48], *Rhodobacter capsulatus* [49,50], *B. stearothermophilus* [51], *Halobacterium halobium* [52] and *Klebsiella pneumoniae* [53]. These catalase-peroxidases contain 715–755 amino acid residues per subunit. This is in contrast to other heme-containing peroxidases of plants, fungi, and yeast, which are monomeric, homologous and 290–350 residues in size. The double length of the bacterial peroxidases has been ascribed to gene duplication [25]. Each half is homologous to eukaryotic, monomeric peroxidase and can be modeled into the high-resolution crystal structure of yeast cytochrome c peroxidase [54]. The comparisons and modeling have predicted that (i) the C-terminal half does not bind heme, and bacterial peroxidases should have one heme per subunit and (ii) that the ten dominating helices observed in the yeast enzyme are highly conserved and connected by surface loops which are often longer in the bacterial peroxidases [25].

Spectral properties of the extracted heme from the *Anacystis nidulans* enzyme in the pyridine ferrohemochrome form are shown in Figure 2 A. With α and β bands at 556 nm and 524 nm and the Soret peak at 420 nm a typical iron protoporphyrin IX (heme b) species is revealed. Upon oxidation with ferricyanide the Soret band shifted to 402 nm, and α and β bands disappeared. Absolute spectra of *Anacystis nidulans* catalase-peroxidase showed a strong Soret band at 405 nm (Figure 2 B) and a broad region of low absorbance

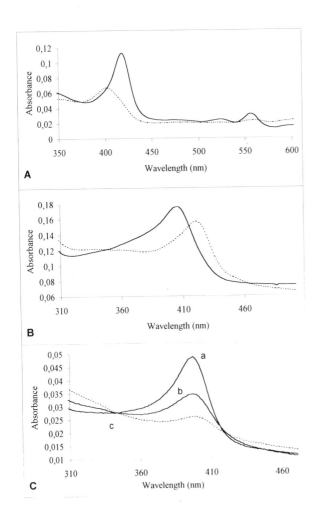

Figure 2. (A) Absolute spectra of Na-dithionite reduced (solid line) and ferricyanide oxidized (dotted line) pyridine adducts of extracted hemes; (B) Absolute spectra of purified enzyme (1.4 μM; solid line) in 67 mM phosphate buffer (pH 7.0) and its cyanide (20 mM KCN) complex (dotted line); (C) Spectral changes upon addition of peroxoacetic acid to the purified enzyme (0.42 μM) in 67 mM phosphate buffer (pH 7.0). Peroxoacetic acid was added discontinuously to the resting enzyme [(a) 0 μM; (b) 11.4 μM; (c) 121.4 μM] and steady-state spectra were recorded immediately after addition using diode-array spectrophotometry.

in the visible range (500–550 nm). These spectra are in accordance with other catalase-peroxidases, which have been shown to contain also heme b at the active site [44–53,55]. Furthermore the absorption bands of the native enzyme are similar to those of high spin heme proteins with histidine as the fifth ligand [56–58]. Upon addition of cyanide to the native catalase-peroxidase from *Anacystis nidulans* a shift of the Soret band from 405 nm to 421 nm (accompanied by 11% hypochromicity) occurred, demonstrating a high to low spin transition of the iron center of the heme (Figure 2 B). The absorption maxima of the CN-adduct of both the cyanobacterial and other catalase-peroxidases are virtually identical to those of low spin, hexacoordinate CN-adducts of other low-spin histidine-ligated peroxidases [56–58]. These conclusions are also supported by both resonance Raman and electron paramagnetic resonance spectroscopy of both the native enzyme and the cyanide-ligated form of catalase-peroxidases from *Streptomyces* sp. [46], *Mycobacterium smegmatis* [47] and *Mycobacterium tuberculosis* [48]. This line of reasoning (resting enzyme in the high-spin out of plane state with histidine as the proximal ligand) has also been considered in the proposal of the potential reaction mechanism of the cyanobacterial enzyme shown in Figure 5.

The A_{405}/A_{280} ratio of the enzyme preparations of *Synechococcus* 6301 varied between 0.41 and 0.48 [29] and of *Synechococcus* 7942 the A_{405}/A_{280} ratio was determined to be 0.54 [32]. This is consistent with one heme per homodimer. This low heme content is not likely due to loss of heme during purification procedure, because for the *Anacystis nidulans* enzyme the mobility on polyacrylamide gel was never modified during purification [29]. It has been reported that catalase-peroxidases found in other bacterial systems have also low heme contents. The A_{405}/A_{280} ratios per subunit of typical catalase-peroxidases are 0.54 in *Streptomyces* sp. [46], 0.56 in *Mycobacterium smegmatis* [47], 0.52 in *E. coli* HPI [44], and 0.3 in *R. capsulatus* [49]. These facts are in contradiction to sequence alignment models proposing one heme binding per subunit [25].

The gene encoding catalase-peroxidase from *Synechococcus* 7942 was cloned from the chromosomal DNA of *Synechococcus* 7942 [32]. A 2160 bp open reading frame (ORF), coding a hydroperoxidase of 720 amino acid residues (approx. 79.9 kDa) was observed [32]. From *Synechocystis* 6803 a potential catalase coding region (754 amino acids) was localized, too [37]. Together with the molecular properties the sequences underline unequivocally that these cyanobacterial hydroperoxidases belong to the class I bacterial catalase-peroxidases [25,55]. The deduced amino acid sequence of the *Synechococcus* enzyme exhibits homologous sequences with those of hydroperoxidases from *E. coli* HPI (56%) [45], *Salmonella typhimurium* (55.5%) [62], *Rhodopseudomonas capsulatus* (58%) [50], and *Bacillus stearothermophilus* (65%) [59]. The similarities have been shown to be higher in the N-terminal half of the proteins and decreased towards the C-terminus. A comparison of the deduced amino acid sequence in the heme iron pocket region of both cyanobacterial enzymes with other class I peroxidases (bacterial catalase-peroxidases, ascorbate peroxidases and cytochrome c peroxidases) is shown in Figure 3. The distal His region of *Synechococcus* 7942 and *Synechocystis* 6803 is RLTWH and RMAWH, respectively. The corresponding sequence in ascorbate peroxidases and cytochrome c peroxidases is RLAWH and in catalase-peroxidases is RMAWH. Thus, in this respect, the cyanobacterial enzymes are similar to all other members of class I peroxidases. The proximal His region of the cyanobacterial enzymes is VALTAGGHT, which is highly conserved in other catalase-peroxidases. Because the conservation of the proximal His regions between catalase-peroxidases and other class I enzymes is less than in the distal region, this is a further evidence that these cyanobacterial enzymes can be described as typical catalase-peroxidases.

Distal haem side

Synechococcus 7942	83	H Y G G L M I	**R** L T	**W**	**H**	A A G T Y R I A G
Synechocystis 6803	112	H Y G G L M I	**R** M A	**W**	**H**	A A G T Y R I A D
Rhodopseudomonas capsulatus	95	H Y G G L M I	**R** M A	**W**	**H**	S A G S Y R A A D
E. coli HPI	95	S Y A G I F I	**R** M A	**W**	**H**	G A G T Y R S I D
Bacillus stearothermophilus	90	S Y A P I F I	**R** M A	**W**	**H**	S A G T Y R S G D
Spinach ascorbate peroxidase	31	Q C A P L M L	**R** L A	**W**	**H**	S A G T F D C T S
Pea ascorbate peroxidase	31	K C A P L I L	**R** L A	**W**	**H**	S A G T F D S K T
Yeast cytochrome c peroxidase	41	G Y G P V L V	**R** L A	**W**	**H**	T S G T W D K H G

Proximal haem side

Synechococcus 7942	247	M A M N D E E T V A L T A G G	**H**	T V G K C	**H**
Synechocystis 6803	275	M A M N D E E T V A L T A G G	**H**	T V G K C	**H**
Rhodopseudomonas capsulatus	250	M G M N D E E T V A L T A G G	**H**	T V G K A	**H**
E. coli HPI	252	M G M N D E E T V A L I A G G	**H**	T L G K T	**H**
Bacillus stearothermophilus	247	M G M N D E E T V A L I A G G	**H**	T F G K A	**H**
Spinach ascorbate peroxidase	131	M G L T D Q D I V A L - S G G	**H**	T L G R C	**H**
Pea ascorbate peroxidase	150	M G L S D Q D I V A L S - G G	**H**	T I G A A	**H**
Yeast cytochrome c peroxidase	161	R L N M D R F V V A L - M G A	**H**	A L G K T	**H**

Figure 3. Comparison of catalase-peroxidase from *Synechococcus* 7942 [32] and *Synechocystis* 6803 [37] with other catalase-peroxidases in regions near the distal and proximal His residues, *Rhodobacter capsulatus* [50], *E. coli* HPI [45], *Bacillus stearothermophilus* [51], Spinach ascorbate peroxidase [59], pea ascorbate peroxidase [60] and yeast cytochrome c peroxidase [61].

The 3-D structures of several peroxidases have been determined, including class I enzymes like yeast cytochrome c peroxidase [54] and pea cytosolic ascorbate peroxidase [63]. All these proteins share the same architecture, consisting of two all-α domains; two antiparallel helices from two domains form a crevice in which the heme group is inserted [27]. One helix, contributed by the C-terminal domain, contains the fifth (proximal) heme iron ligand (His-175 in cytochrome c peroxidase, His-163 in pea cytosolic ascorbate peroxidase and His-262 in *Synechococcus* 7942). The imidazole ring of the proximal His lies approximately perpendicular to the porphyrin plane with $N^{\varepsilon 2}$ bonded to heme iron and $N^{\delta 1}$ hydrogen bonded to the buried carboxylate group of an Asp residue (Asp-235 in cytochrome c peroxidase, Asp-208 in pea cytosolic ascorbate peroxidase). A number of peroxidases (including cytochrome c peroxidase and ascorbate peroxidase) also contain an aromatic residue (Trp or Phe) parallel to and in van der Waals contact with the imidazole ring of the proximal His [27]. The other helix, on the distal side of the heme, contributes three conserved residues (Arg-48, Trp-51 and His-52 in cytochrome c peroxidase; Arg-38, Trp-41 and His-42 in pea cytosolic ascorbate peroxidase; Arg-90, Trp-93 and His-94 in *Synechococcus* 7942), which form a ligand pocket for H_2O_2.

The *Synechococcus* catalase-peroxidases have been shown to possess high catalase activity and as with other catalase-peroxidases a maximum activity around pH 7 [29]. In contrast, the maximum activity of monofunctional catalases plateaus over several pH units [24]. The cyanobacterial enzyme displays saturation kinetics with an apparent K_M between 4.2 mM [32] and 4.3 mM [29] for hydrogen peroxide. Catalatic activity was proportional to enzyme (heme) concentration with an overall rate constant of 3.5×10^6 M^{-1} s^{-1}. For *Anacystis nidulans* catalase-peroxidase a turnover number (k_{cat}) value of about 7200 s^{-1} and a k_{cat}/K_M-ratio of 1.66×10^6 s^{-1} M^{-1} has been calculated [29], which is about one order

Table 3. Catalytic activities of purified bacterial catalase-peroxidases[a,b]

Catalase-peroxidase source	Catalytic activity			Peroxidatic activity					
	K_M (mM)	k_{cat} (s^{-1})	k_{cat}/K_M $(M^{-1}s^{-1})$	AsA	GSH	NAD(P)H	cyt c	o-di	Ref.
E. coli HPI	3.9	16300	4.17×10^6	n.d.	–	+	–	+	44
R. capsulata	4.2	n.d.	n.d.	–	n.d.	+	+	+	49,50
M. smegmatis	1.4	2380	1.7×10^6	n.d.	n.d.	+	n.d.	+	47
M. tuberculosis	5.2	10140	1.95×10^6	n.d.	n.d.	+	n.d.	+	48
Synechococcus 7942	4.2	26000	6.19×10^6	–	n.d.	–	n.d.	+	32
Synechococcus 6301	4.3	7200	1.66×10^6	–	–	–	–	+	29

[a] AsA, ascorbate; GSH, glutathione; cyt c, cytochrome c; o-di, o-dianisidine.
[b] +, electron donor; –, no electron donor; n.d. not determined.

of magnitude below that of beef liver catalase [64]. A comparison of these kinetic parameters with those of other catalase-peroxidases is shown in Table 3.

Table 3 also shows that in addition to their catalatic activity, catalase-peroxidases function as a broad specificity peroxidase, oxidizing various electron donors, including NAD(P)H and cytochrome c [44,47–50]. A significant peroxidase activity with classical artificial electron donors (e.g. o-dianisidine) is shown by all catalase-peroxidases characterized so far. The catalase-peroxidases from Synechococcus 6301 and 7942, respectively, have been shown to oxidize o-dianisidine and pyrogallol; however, no reaction was de-. tected with ascorbate, glutathione, NAD(P)H and cytochrome c [29,32].

Additional characteristics of cyanobacterial catalase-peroxidase shared with other bacterial catalase-peroxidases include inhibition by cyanide and azide and the lack of inhibition by the prototypical catalase inhibitor 3-amino-1,2,4-triazole [65]. A further characteristic property of the cyanobacterial enzyme is its inactivation in the presence of millimolar steady-state hydrogen peroxide concentrations (Figure 4 A). Inactivation was shown to be reversible as long as artificial electron donors (e.g. benzidine or o-dianisidine) are present in substoichiometric concentrations (Figure 4 B). Ascorbate (Figure 4 B), glutathione or cytochrome c (not shown) have been shown not to protect the enzyme from being irreversible inactivated.

A potential reaction mechanism of a cyanobacterial catalase-peroxidase is shown in Figure 5. Upon addition of peroxide, A. nidulans high-spin ferri-catalase-peroxidase forms compound I, as has been demonstrated by the addition of peroxoacetic acid (Figure 2 C). High-spin ferric enzyme reacts with one equivalent of peroxide (e.g. H_2O_2) to low-spin compound I. This is a two-electron oxidation/reduction reaction where H_2O_2 is reduced to water and the enzyme is oxidized. One oxidizing equivalent resides on iron, giving the oxyferryl ($Fe^{IV}=O$) intermediate. The other (designated R•) resides either on an aromatic amino acid (which becomes a transient radical as shown for cytochrome c peroxidase [66,67]) or on the porphyrin ring giving a π-cation radical which has been shown for plant peroxidases [68] or mammalian catalases [69]. Finally, the reaction occurs either via the catalatic or the peroxidatic pathway . In the catalatic pathway oxidation of hydrogen peroxide reduces Compound I to the native resting state. In the peroxidatic pathway compound I oxidizes an organic substrate to give a substrate radical (•AH) with compound I undergoing a one-electron reduction yielding compound II, which contains an oxyferryl center coordinated to a normal (dianionic) porphyrin ligand. Finally, compound II, is reduced back to the native ferric state with concomitant one-electron substrate oxidation.

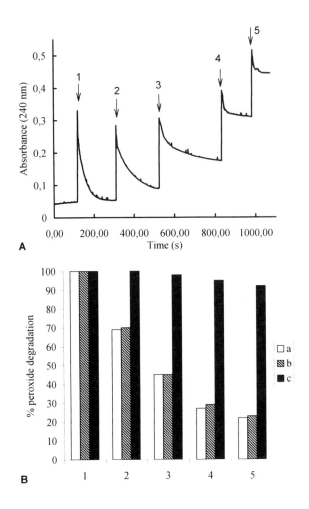

Figure 4. (A), Inhibition of catalatic activity of *Anacystis nidulans* catalase-peroxidase (6.6 nM) by successive addition of 0.8 mM hydrogen peroxide at pH 7.0. Arrows indicate addition of peroxide (1–5). Degradation of hydrogen peroxide was followed spectrophotometrically at 240 nm in the absence of artificial electron donors. (B), Schematic presentation of the influence of ascorbate (b, 33 µM) and benzidine (c, 33 µM) on inhibition of hydrogen peroxide degradation (a, without electron donor). Reaction mixtures as in (A).

During catalatic turnover the cyanobacterial enzyme shows at most 2% hypochromicity of the Soret band, indicating a 0–4% steady-state concentration of compound I [29], compared with 25–30% in bovine liver catalase [70]. These striking differences between typical catalases and the catalase-peroxidases can be explained by postulating an identical rate of formation of compound I (about $1–2 \times 10^7$ M^{-1} s^{-1}) but a significantly higher rate for reduction of compound I by hydrogen peroxide back to the native enzyme. This conclusion is also consistent with the finding that so far no reliable data for the rate of compound I formation of the cyanobacterial enzyme with H_2O_2 by conventional stopped-flow spectroscopy could be obtained. On the contrary, upon addition of peroxoacetic acid it was demonstrated that compound I per se has a Soret absorbance of about 50% that of the ferric enzyme (Figure 2 C) at 405 nm [29], which is in contrast to other class I peroxidases. The Soret peak of thylakoid-bound ascorbate peroxidase compound I has been demonstrated to be 12 nm red-shifted [71], as also observed in the case of cytochrome c peroxidase [72]. These spectra have been proposed to correspond to heme-Fe^{IV}=O with a Trp radical [73]. Thus, the H_2O_2-oxidized intermediate of the cyano-

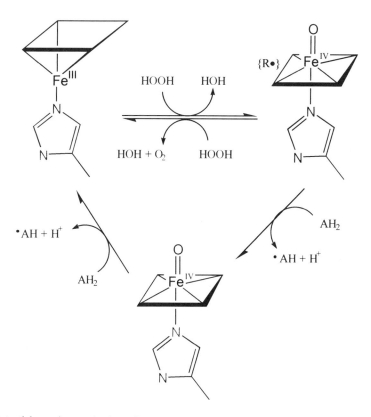

Figure 5. Potential reaction mechanism of cyanobacterial catalase-peroxidase. High-spin ferric enzyme reacts with one equivalent of peroxide (e.g. H_2O_2) to low-spin compound I. This is a two-electron oxidation/reduction reaction where H_2O_2 is reduced to water and the enzyme is oxidized. One oxidizing equivalent resides on iron, giving the oxyferryl ($Fe^{IV}=O$) intermediate. The other resides either on an aromatic amino acid (which becomes a radical) or on the porphyrin ring giving a π-cation radical. Finally reaction occurs either via the catalatic or the peroxidatic pathway. In the catalatic pathway oxidation of hydrogen peroxide returns compound I to the native resting state. In the peroxidatic pathway compound I oxidizes an organic substrate to give a substrate radical ($\cdot AH$) with compound I undergoing a one-electron reduction yielding compound II, which contains an oxyferryl center coordinated to a normal (dianionic) porphyrin ligand. Finally, compound II is reduced back to the native ferric state with concomitant one-electron substrate oxidation.

bacterial catalase-peroxidase is similar to class III peroxidases and therefore seems to correspond to heme-$Fe^{IV}=O$ with a porphyrin π-cation radical.

During peroxidatic turnover and most probably during (reversible) inactivation the catalase-peroxidase compound I is converted into catalatic inactive compound II either by hydrogen peroxide, which has been demonstrated in monofunctional catalases [70], or by various electron donors in the classical peroxidatic pathway. The presence of artificial electron donors diminishes compound II accumulation by formation of the ferric enzyme thus leading back the enzyme to the catalatic cycle (Figure 5). But more detailed experiments are necessary in order to understand the role and the nature of potential endogenous electron donors in protecting the enzyme from inactivation. Furthermore the nature of catalase-peroxidase compound II has to be elucidated, especially because in the cyanobacterial enzyme it does not have distinctive spectral characteristics as known from compound II of class III peroxidases [68] or mammalian catalases [70].

ACKNOWLEDGMENTS

This work was supported by the Austrian Fonds zur Förderung der Wissenschaftlichen Forschung (FWF-Projekt P12371-MOB).

REFERENCES

1. Asada, K. (1993) in Active Oxygens, Lipid Peroxides, and Antioxidants (Yagi, K. , ed.) pp. 289–298, Japan Scientific Societies Press, Tokyo, and CRC Press, New York.
2. Schopf, J. W. (1993) Science 269, 640–646.
3. Berkner, L. V. and Marshall, L. C. (1965 J. Atmos. Sci. 22: 225–261.
4. Broda, E. (1975) The Evolution of the Bioenergetic Processes. - Pergamon Press, Oxford, U.K.
5. Broda, E. and Peschek, G.A. (1979). J. Theor. Biol. 81, 201–212.
6. Peschek, G.A. (1996) Biochem. Soc. Trans. 24, 729–733.
7. Peschek, G.A (1996) Biochim. Biophys. Acta 1275, 27–32.
8. Abeliovich, A., Kellenberg, C. and Shilo, M. (1974) Photochem. Photobiol. 19, 379–382.
9. Steinitz, Y., Mazor, Z. and Shilo, M. (1979) Plant Sci. Lett. 16, 327–335.
10. Campell, W.S. and Laudenbach, D. (1995). J. Bacteriol. 177, 964–972.
11. Asada, K., Yoshikawa, K., Takahashi, M., Maeda, Y. and Enmanji, K. (1975) J. Biol. Chem. 250, 2801–2807.
12. Asada, K., Kanematsu, S. and Uchida, K. (1977) Arch. Biochem. Biophys. 179, 243–256.
13. Okada, S., Kanematsu, S. and Asada, K. (1979) FEBS Lett. 103, 106–110.
14. Cseke, C., Horvath, L.I., Simon, P., Borbely, G., Keszthelyi, L. and Farkas, G.L. (1979) Biochem. 85, 1397–1404.
15. Herbert, S.K., Samson, G., Fork, D.C. and Laudenbach, D.E. (1992) Proc. Natl. Acad. Sci. USA 89, 8716–8720.
16. Miyake, C., Michihata, F. and Asada, K. (1991) Plant Cell Physiol. 32, 33–43.
17. Tytler, E.M., Wong, T. and Codd, G.A. (1984) FEMS Microbiol. Letters 23, 239–242.
18. Canini, A., Donatella, L. and Grilli Caiola, M. (1997) Abstract IX International Symposium on Phototrophic Prokaryotes, pp. 140.
19. Henry, L.E.A., Gogotov, I. N. and Hall, D. O. (1978) Biochem. J. 174, 373–377.
20. Grilli Caiola, M., Canini, A., Galiazzo, F. and Rotilio, G. (1991) FEMS Microbiol. Lett. 80, 161–166.
21. Mackey, E.J. and Smith, G. D. (1983) FEBS Lett. 156, 108–112.
22. Canini, A., Civitareale, P., Marini, S., Grilli Caiola, M.G. and Rotilio, G. (1992) Planta 187, 438–444.
23. Bagchi, S.N., Ernst, A.and Böger, P. (1991) Z. Naturforsch. 46c, 407–415.
24. Nadler, V., Goldberg, I. and Hochman, A. (1986) Biochim. Biophys. Acta 822, 234–241.
25. Welinder, K.G. (1991) Biochim. Biophys. Acta 1080, 215–220.
26. Kimura, S. and Ikeda-Saito, M. (1988) Proteins 3, 113–120.
27. Bosshard, H.R., Anni, H. and Yonetani, T. (1991) in Peroxidases in Chemistry and Biology (Everse, J., Everse, K.E. and Grisham, M.B., eds.) Vol. II, pp. 51–84. CRC Press, Boca Raton.
28. Dalton, D.A. (1991) in Peroxidases in Chemistry and Biology (Everse, J., Everse, K.E. and Grisham, M.B., eds.) Vol. II, pp. 139–153. CRC Press, Boca Raton.
29. Obinger, C., Regelsberger, G., Strasser, G., Burner, U. and Peschek, G. A. (1997) Biochem. Biophys. Res. Commun. 235, 545–552.
30. Tel-Or, E., Huflejt, M.E. and Packer, L. (1985) Biochem. Biophys. Res. Commun. 132, 533–539.
31. Tel-Or, E., Huflejt, M.E. and Packer, L. (1986) Arch. Biochem. Biophys. 246, 396–402.
32. Mutsuda, M., Ishikawa, T., Takeda, T. and Shigeoka, S. (1996) Biochem. J. 316, 251–257.
33. Mittler, R. and Tel-Or, E. (1991) Arch. Microbiol. 155, 125–130.
34. Rady, A.A., El-Sheekh, M.M. and Matkovics, B. (1994) Int. J. Biochem. 26, 433–435.
35. Karni, L., Moss, S. J., Tel-Or, E. (1984) Arch. Microbiol. 140, 215–217.
36. Tözüm, S.R.D. and Gallon, J.R. (1979) J. Gen. Microbiol. 11, 313–326.

37. Kaneko, T., Sato, S., Kotani, H., Tanaka, A., Asamizu, E., Nakamura, Y., Miyajima, N., Hirosawa, M., Sugiura, M., Sasamoto, S., Kimura, T., Hosouchi, T., Matsuno, A., Muraki, A., Nakazaki, N., Naruo, K., Okumura, S., Shimpo, S., Takeuchi, C., Wada, T., Watanabe, A., Yamada, M., Yasuda, M. and Tabata, S. (1996) DNA Res. 3, 109–136.
38. Hossain, M. A. and Asada, K. (1984) Plant Cell Physiol. 25, 85–92.
39. Hossain, M. A. and Asada, K. (1985) J. Biol. Chem. 260, 12920–12926.
40. Asada, K., Miyake, C., Sano, S. and Amako, K. (1993) in Plant Peroxidases: Biochemistry and Physiology (Welinder, K.G., Rasmussen, S.K., Penel, C., Greppin, H., eds.), pp. 242–250. University of Geneva Press, Geneva.
41. Dupouy, D., Conter, A., Croute, F., Murat, M. and Planel,H. (1985) Environmental Experimental Botany 25, 339–347.
42. Bhunia, A.K., Roy, D. and Banerjee, S. K. (1993) Letters Appl. Microbiol. 18, 10–13.
43. Schmidt, A. (1988) Methods Enzymol. 167, 572–583.
44. Clairborne, A. and Fridovich, I. (1979) J. Biol. Chem. 254, 4245–4252.
45. Triggs-Raine, B.L., Doble, B.W., Mulvey, M.R., Sorby, P.A. and Loewen, P. C. (1988) J. Bacteriol. 170, 4415–4419.
46. Youn, H.-D., Yim, Y.-I., Kim, K., Hah, Y.C., and Kang, S.-O. (1995) J. Biol. Chem. 270, 13740–13747.
47. Marcinkevicienne, J.A., Magliozzo, R.S., Blanchard, J.S. (1995) J. Biol. Chem. 270, 22290–22295.
48. Johnsson, K., Froland, W.A., and Schultz, P. G. (1997) J. Biol. Chem. 272, 2834–2840.
49. Hochman, A. and Shemesh, A. (1987) J. Biol. Chem. 262, 6871–6876.
50. Forkl, H., Vandekerckhove, J., Drews, G. and Tadros, M.H. (1993) Eur. J. Biochem. 214, 251–258.
51. Loprasert, S., Negoro, S. and Okada, H. (1988) J. Gen. Microbiol. 134, 1971–1976.
52. Brown-Peterson, N.J. and Salin, M.L. (1993) J. Bacteriol. 175, 4197–4202.
53. Hochman, A. and Goldberg, I. (1993) Biochim. Biophys. Acta 1077, 299–307.
54. Finzel, B.C., Poulos, T.L. and Kraut, J. (1984) J. Biol. Chem. 259, 13027–13036.
55. Hochman, A. (1993) in Plant Peroxidases: Biochemistry and Physiology (Welinder, K. G., Rasmussen, S. K., Penel, C. and Greppin, H., eds.), pp. 242–250. University of Geneva Press, Geneva, Swiss.
56. Rakhit, G. and Spiro, T.G. (1974) Biochemistry 13, 5317–5323.
57. Andersson, L.A., Reganathan, V., Loehr, T.M. and Gold, M.H. (1987) Biochemistry 26, 2258–2263.
58. Mino, Y., Warishi, H., Blackburn, N.J., Loehr, T.M. and Gold, M.H. (1988) J. Biol. Chem. 263, 7029–7036.
59. Loprasert, S., Negro, S. and Okada, H. (1989) J. Bacteriol. 171, 4871–4875.
60. Mittler, R. and Zilinskas, B. A. (1991) Plant Physiol. 97, 962–968.
61. Ishikawa, T., Sakai, K., Takeda, T. and Shigeoka, S. (1995) FEBS Lett. 367, 28–32.
62. Loewen, P.C. and Stauffer, G.V. (1990) Mol. Gen. Genet. 224, 147–151.
63. Patterson, W.R. and Poulos, T.L. (1995) Biochemistry 34, 4331–4341.
64. Ogura, Y. (1955) Arch. Biochem. Biophys. 96, 288–300.
65. Margoliash, E., Novogrodsky, A. and Schejter, A. (1960) Biochem. J. 74, 339–348.
66. Ortiz de Montellano, P.R. (1992) Annu. Rev. Pharmacol. Toxicol. 32, 89–107.
67. Miller, M.A., Bandyopadhyay, D., Mauro, J.M., Traylor, T.G. and Kraut, J. (1992) Biochemistry 31, 2789–2797.
68. Dunford, H.B. (1991) in Peroxidases in Chemistry and Biology (Everse, J., Everse, K.E. and Grisham, M.B., eds.) Vol. II, pp. 1–24. CRC Press, Boca Raton.
69. Dolphin, D., Forman, A., Borg, D.C., Fajer, B. and Felton, R.H. (1971) Proc. Natl. Acad. Sci. USA 68, 614–618.
70. Chance, B., Sies, H. and Boveris, A. (1979) Physiol. Rev. 59, 527–605.
71. Miyake, C. and Asada, K. (1996) Plant Cell Physiol. 37, 423–430.
72. Yonetani, T. and Anni, H. (1987) J. biol. Chem. 262, 9547–9554.
73. Sivaraja, M., Goodin, D.B., Smith, M. and Hoffman, B.M. (1989) Science 245, 738–740.

85

MULTIPLE *clpP* GENES FOR THE PROTEOLYTIC SUBUNIT OF THE ATP-DEPENDENT Clp PROTEASE IN THE CYANOBACTERIUM *SYNECHOCOCCUS* SP. PCC 7942

Importance for Light, Cold, and UV-B Acclimation

Adrian K. Clarke, Jenny Schelin, and Joanna Porankiewicz

Department of Plant Physiology
University of Umeå
901 87 Umeå, Sweden

1. INTRODUCTION

ATP-dependent proteases are important factors for maintaining the protein environment of all cells. As housekeeping enzymes, they remove denatured or otherwise inactive polypeptides that continuously arise from metabolic activities and stress. Degradation of such potentially toxic polypeptides also recycles amino acids necessary for continued protein synthesis. Furthermore, proteases control the turnover of certain unstable regulatory proteins, whose activity thereby relies on continuous synthesis [1]. Many ATP-dependent proteases are induced during stresses that increase the level of protein denaturation and aggregation [2]. Proteins that become irreversibly inactivated are therefore targeted for degradation by these inducible proteases before they can interfere with normal metabolic functions.

The Clp protease best characterized to date is that from *E. coli*. The large protein complex consists of two functionally unrelated subunits: ClpP (21.5 kDa) and ClpA (83 kDa). ClpP possesses the proteolytic active sites, but alone is capable of cleaving only peptides shorter than seven amino acids. For full proteolytic activity, ClpP must be complexed to ClpA. In a process requiring ATP, ClpA selectively binds the target polypeptide and presents it to ClpP in an unfolded state, ready for degradation [3]. Besides regulating the proteolytic activity of ClpP, ClpA also functions independently as a molecular

The Phototrophic Prokaryotes, edited by Peschek *et al.*
Kluwer Academic / Plenum Publishers, New York, 1999.

chaperone [4]. Depending on the conditions, ClpA may either mediate correct protein folding or degrade them via association with ClpP.

ClpA is now recognized as only one member of the Clp/Hsp100 family of molecular chaperones. This family consists of at least six distinct proteins which can be divided into two general groups [5]. The first includes ClpA, B, C and D and, despite variations in size and gene structure, all possess two distinct ATP-binding domains. The second group of ClpX and ClpY contains only one of these two characteristic domains. For most of these proteins, little is yet known of their specific function within different organisms, particularly those in photobionts such as cyanobacteria, or their potential interaction with ClpP in an active Clp protease.

2. RESULTS AND DISCUSSION

2.1. Multiple ClpP Proteins

The ClpP protein functions as the proteolytic subunit of the energy-dependent Clp protease found in eubacteria, mammals and the chloroplasts of plants. Among these organisms, cyanobacteria are unusual in that they possess a multi-gene family coding for up to four distinct ClpP isomers. Using a combined strategy of degenerate-PCR and classical cloning techniques, we have isolated two *clpP* genes from the unicellular cyanobacterium *Synechococcus* sp. PCC 7942. These genes, designated *clpP1* and *clpP2*, are around 80% similar (70% identical) to each other, and 70 to 75% similar to two homologous genes in *Synechocystis* sp. PCC 6803.

The *clpP1* gene from *Synechococcus* sp. PCC 7942 is monocistronic while *clpP2* is the first gene in an operon containing *clpX*, a gene coding for a molecular chaperone that also regulates the proteolytic activity of ClpP in *E. coli* [6]. The gene organization of *clpP2* with *clpX* is similar to that of the *clpP* gene from *E. coli*, and it is likely both encoded ClpP proteins share homologous functions. ClpP1, on the other hand, is likely to be more functionally related to one of the two ClpP proteins in chloroplasts of higher plants. The first of these plant proteins is encoded within the chloroplast genome, as one of three genes in an operon also coding for two ribosomal proteins [7]. The second chloroplast ClpP is encoded by a monocistronic, nuclear gene and is synthesized as a precursor possessing a characteristic chloroplast transit peptide [8]. Little is yet known, however, of the specific structure and function of these two chloroplast ClpP proteins.

2.2. Induction Profiles for ClpP1

ClpP1 is strongly induced during shifts to high light intensity, cold or UV-B irradiation, but not during heat shock in contrast to the ClpP homologue from *E. coli*. In cultures grown to steady-state at four different light intensities (20, 50, 125 and 175 µmol photons $m^{-2} s^{-1}$), the level of ClpP1 protein was four- to five times greater at the two higher irradiances compared to the two lower ones. ClpP1 content in cultures acclimated to 50 µmol photons $m^{-2} s^{-1}$ also increased dramatically during moderate (420 µmol photons $m^{-2} s^{-1}$) or severe (1000 µmol photons $m^{-2} s^{-1}$) photoinhibition for 4 h, rising to 12-times the level prior to the shift. Similar rapid increases in ClpP1 also occurred during shifts from 37 to 25°C for 24 h, or exposures to supplementary UV-B irradiation of 0.5 W m^{-2} (Fig. 1). In contrast, the amount of ClpP1 protein was relatively unaffected during moderate to severe heat shock, when cultures were shifted from 37 to 47 or 50°C for 2 to 4 h.

A.

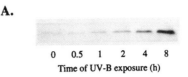

0   0.5   1   2   4   8

Time of UV-B exposure (h)

B.

Figure 1. Induction of ClpP1 by moderate UV-B. *Synechococcus* wild type cultures grown at 50 μmol photons m^{-2} s^{-1} to a Chl concentration of 2.4 to 2.7 μg ml^{-1} were exposed supplementary to 0.5 W m^{-2} of UV-B (280–315 nm) for 8 h. Cellular proteins were extracted from samples taken at the indicated times and separated by polyacrylamide gel electrophoresis on the basis of equal Chl content (0.75 μg ml^{-1}). ClpP1 protein was detected immunologically using a specific polyclonal antibody (A) and later quantified by densitometry (B). The figure shows a representative immunoblot (A) and the average amount of ClpP1 relative to the control value at time zero (B) from three replicates.

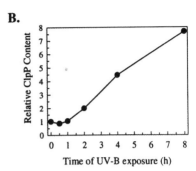

We are now in the process of analyzing the induction characteristics of ClpP2 in order to compare the roles of the two Clp proteins in *Synechococcus* sp. PCC 7942 during normal growth and under different stress conditions.

2.3. Inactivation of clpP1 and clpP2

A *clpP1* inactivation strain of *Synechococcus* (Δ*clpP1*) was obtained using a strategy involving the deletion of the 5′-end of *clpP1* and subsequent insertion of a kanamycin resistance cassette. Observation of Δ*clpP1* cultures acclimated to standard growth conditions revealed that a significant proportion of cells were elongated (ca. 35%), up to 20 times the normal wild type length. All elongated cells were viable as indicated by fluorescence microscopy. Cell wall staining revealed the lack of visible septa within the elongated cells, suggesting each as unicells rather than multicellular filaments. DNA staining revealed the presence of chromosomes evenly distributed throughout the elongated cells.

We have recently used a deletion/insertion strategy similar to that for *clpP1* to inactivate the *clpP2* gene from *Synechococcus*. Analysis of the resulting transformants is currently underway.

2.4. Loss of ClpP1 Affects Light Acclimation

Under standard culture conditions at 50 μmol photons m^{-2} s^{-1}, the Δ*clpP1* strain had a longer average doubling time of 11 ± 0.9 h (*n*=3) compared to the wild type 7.5 ± 0.4 h (*n*=3). This difference in growth rates of the two strains was also observed in cultures acclimated to 20 and 125 μmol photons m^{-2} s^{-1}. At 175 μmol photons m^{-2} s^{-1}, however, the growth rate for the wild type remained unchanged from the level at 125 μmol photons m^{-2} s^{-1}, but that of Δ*clpP1* decreased significantly, indicating inhibition of growth at the higher irradiance. The increasingly stressed phenotype of Δ*clpP1* at high light was confirmed by a corresponding induction of known stress proteins, GroEL and ClpB.

In addition to growth rates, we also examined the ratio of the two primary photosynthetic pigments, chlorophyll *a* (Chl) and phycocyanin (PC), in both strains grown at the

four different irradiances. PC-containing phycobilisomes function as the main light harvesting antennae of PSII in *Synechococcus*, whereas most Chl is located within PSI [9].
The PC/Chl ratio, therefore, is directly affected by the proportion of PSII to PSI, and
changes in PC content within the phycobilisome. From 20 to 175 µmol photons $m^{-2} s^{-1}$,
the PC/Chl ratio for wild type cultures steadily rose from a value of 0.75 to 1.1. This rise
is primarily due to an increase in PSII centers relative to PSI centers at the higher irradiances. In contrast, the $\Delta clpP1$ strain had a significantly lower ratio at 20 µmol photons m^{-2}
s^{-1} (i.e., 0.65), which changed little as the irradiance increased to 175 µmol photons $m^{-2} s^{-1}$
(<0.75); a value still less than the lowest for the wild type. This suggests $\Delta clpP1$ cultures
have a lower proportion of PSII centers compared to PSI centers , and that this remains
low at the higher irradiances.

With the raised ClpP1 content in photoinhibited wild type cells, we also examined
the sensitivity of the $\Delta clpP1$ strain to both moderate and severe high light treatments.
However, no significant difference in the extent of loss or rate of recovery of photosynthetic activity, as measured by both oxygen evolution and Chl fluorescence, was observed
between the wild type and $\Delta clpP1$ strains, indicating ClpP1 is not essential for resistance
to and recovery from photoinhibition.

2.5. ClpP1 Is Essential for UV-B and Cold Acclimation

Due to the high level of ClpP1 induced in wild type cultures during both moderate
UV-B and cold shifts, we examined the ability of both wild type and $\Delta clpP1$ strains to ac-
climate to these new growth regimes. For these experiments, we used exponentially grown
cultures at a Chl concentration of 2.4 to 2.7 µg ml^{-1}. When exposed to 0.5 W m^{-2} UV-B
light, growth of the wild type stopped for the first 1 h but then recovered to a generation
time almost equal to that prior to the UV-B treatment (i.e., 8.2 ± 0.7 h, n=3). In contrast,
growth of the $\Delta clpP1$ strain was severely inhibited by the UV-B dose, and did not recover
during the 8 h treatment (Fig. 2). Moreover, cell survival estimations showed that $\Delta clpP1$
cells steadily lost viability after 2 h of UV-B, dropping to 60% of the control 37°C level
by 8 h. In comparison, over 90% of wild type cells remained viable after the 8 h UV-B
treatment.

When wild type cultures were shifted from 37 to 25°C, growth ceased for 4 h and
then resumed at rate 50% slower than that at 37°C h. The generation time of these cold-ac-
climated cells (15 ± 0.9 h) was consistent with that previously observed for wild type
Synechococcus conditioned to 25°C [10]. Growth of the $\Delta clpP1$ strain also ceased upon
the shift to 25°C but failed to recover afterwards as did the wild type. The mutant did not

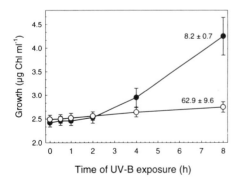

Figure 2. Growth characteristics of wild type (●) and
$\Delta clpP1$ (○) cultures following supplemental exposure of
UV-B at 0.5 W m^{-2}. Values represent averages ± SE for
three independent experiments.

resume growth even after prolonged incubation at 25°C, up to 48 h, at which time cell viability decreased by 50%.

It would appear that ClpP1-mediated protein turnover performs an important function during long-term acclimated growth of *Synechococcus*, especially under sub-optimal conditions such as low temperatures and UV-B exposure. This is clearly shown by the phenotypic changes in wild type cells upon the inactivation of *clpP1*, despite the presence of at least one other *clpP* gene in this cyanobacterium. The importance of ClpP1 in *Synechococcus* is therefore similar to that of the chloroplast ClpP, whose inactivation in algae prevents phototrophic growth [11]. One likely function for ClpP in both cyanobacteria and chloroplasts is as a housekeeping protease, maintaining the correct protein environment during normal growth. If so, then this activity is more critical in photobionts than in *E. coli*, where the role of ClpP is dispensable under all growth conditions [12]. It is equally plausible, however, that in addition to protein housekeeping ClpP proteolysis regulates the turnover of specific short-lived regulatory proteins in cyanobacteria and chloroplasts. During stress acclimation, the degradation of aberrant polypeptides is another likely function of the Clp protease indispensable in photobionts such as *Synechococcus*. We are now in the process of examining the role of ClpP2 from *Synechococcus* during both normal growth and various stress conditions, and comparing it to that of ClpP1. We will also perform pulse/chase experiments with wild type *Synechococcus*, Δ*clpP1* and Δ*clpP2* strains in the hope of identifying those native protein substrates whose turnover is regulated by the different ClpP protease complexes.

REFERENCES

1. Gottesman, S. and Maurizi, M.R. (1992) Microbiol. Rev. 56, 592–621.
2. Parsell, D.A. and Lindquist, S. (1993) Annu. Rev. Genet. 27, 437–496.
3. Maurizi, M.R., Clark, W.P., Katayama, Y., Rudikoff, S., Pumphrey, J., Bowers, B. and Gottesman, S. (1990) J. Biol. Chem. 265, 12536–12545.
4. Wickner, S., Gottesman, S., Skowyra, D., Hoskins, J., McKenney, K. and Maurizi, M.R. (1994) Proc. Natl. Acad. Sci. USA 91, 12218–12222.
5. Clarke, A.K. (1996). J. Biosci. 21, 161–177.
6. Gottesman, S., Clark, W.P., de Crecy-Lagard, V. and Maurizi, M. (1993) J. Biol. Chem. 268, 22618–22626.
7. Clarke, A.K., Gustafsson, P. and Lidholm, J.Å. (1994) Plant Mol. Biol. 26, 851–862.
8. Schaller, A. and Ryan, C.A. (1995) Plant Physiol. 108, 1341.
9. Fujita, Y., Murakami, A., Aizawa, K. and Ohki, K. (1994) in: The Molecular Biology of Cyanobacteria (Bryant, D.A., ed.) Kluwer Academic Publishers, Dordrecht, pp. 677–692.
10. Campbell, D., Zhou, G., Gustafsson, P., Öquist, G. and Clarke, A.K. (1995) EMBO J. 14, 5457–5466.
11. Huang, C., Wang, S., Chen, L., Lemieux, C., Otis, C., Turmel, M. & Liu, X.-Q. (1994) Mol. Gen. Genet. 244, 151–159.
12. Kroh, H.E. and Simon, L.D. (1990) J. Bacteriol. 172, 6026–6034.

86

THE COMPLEX RELATION BETWEEN PHOSPHATE UPTAKE AND PHOTOSYNTHETIC CO_2 FIXATION IN THE CYANOBACTERIUM *ANACYSTIS NIDULANS*

Ferdinand Wagner, Emel Sahan, and Gernot Falkner

Institute of Limnology
Austrian Academy of Sciences
Gaisberg 116, A-5310 Mondsee

1. INTRODUCTION

When algal and cyanobacterial growth in lakes is limited by the supply of available phosphate, the external concentration of this nutrient decreases to nanomolar levels [1], and rapidly approaches a stationary value at which net uptake ceases for energetic reasons [2]. Phosphate uptake and subsequent growth is then only possible if the external phosphate concentration exceeds at least occasionally this stationary or *threshold* value, for example after excretion of this nutrient by zooplankton or fish.

During such pulsewise increases in the external concentration phosphate is rapidly incorporated, and intermediarily stored. This is accomplished by the formation of polyphosphate granules that serve as the proper phosphorus source for the growing organisms. In this special growth situation two requirements must be fulfilled: first, the growth rate has to be adjusted to the size of the polyphosphate granules [3], and second, the uptake system must be regulated such that the balanced adjustment between growth rate and the polyphosphate pool is maintained under conditions of pulsewise phosphate supply. In order to achieve this the uptake system must have some kind of information processing capacity by which information about the pulse pattern "experienced" by the organisms is stored and maintained over the periods in which no uptake occurs.

In previous publications we have shown that during adaptation to phosphate pulses the uptake system attains determinate features that reflect the characteristics of this nutrient pulse which in turn depends on the uptake activity of the prevailing population [4,5]. These determinate features are conserved over a prolonged period of time even if during that time no phosphate uptake occurs. Here we investigate whether adaptation to a phosphate pulse leads to furthergoing metabolic changes. We demonstrate that during a pulse the carbon

The Phototrophic Prokaryotes, edited by Peschek *et al.*
Kluwer Academic / Plenum Publishers, New York, 1999.

flow into glycogen is irreversibly altered and that *via* this mechanism the population as a whole can regulate the metabolism of individual cells.

2. MATERIALS AND METHODS

Anacystis nidulans (*Synechococcus* sp.) (strains PCC 7942, and 1402–1 from Algal Culture Collection, Göttingen) was cultivated photoautotrophically as described previously [4]. Phosphate deficiency was achieved by restricting the total phosphate content of the cultures to 2.5 or 5µM phosphate, the light intensity during growth and experiments was about 80–100 $\mu E \cdot m^{-2} \cdot sec^{-1}$. For determination of net phosphate incorporation the cultures were grown on radioactive phosphate of the same specific activity as used in the experiments. The uptake experiments were performed according to [4].

CO_2 fixation was determined by using ^{14}C-HCO_3^-. In order to be able to follow the time course during the first minutes of CO_2 fixation, the algae had to be concentrated by centrifugation prior to the experiments. The reaction was started by addition of labelled HCO_3^- and stopped by transferring an aliquot of the cyanobacterial suspension into HCl yielding a final concentration of 1 M of this acid. Separation of low molecular weight compounds and high molecular weight compounds was carried out as described in [6].

3. RESULTS AND DISCUSSION

3.1. The Adaptive Phosphate Uptake Behaviour

An analysis of phosphate uptake behaviour of *Anacystis nidulans* in response to fluctuations of this nutrient confronts the experimentalist with the problem that during a phosphate pulse a transition from an adaptive to an adapted state may or may not occur. In consequence, the threshold value itself is determined by the way the experiment is performed, as demonstrated in Fig. 1 A and B. In the experiment of Fig. 1 A, a cyanobacterial suspension containing 20 µg Chl*a*·l^{-1} was exposed to four subsequent phosphate pulses. One can see that after each phosphate addition the external concentration was reduced by the uptake activity of the whole population within a few minutes to a threshold value of about 1 nM phosphate. No obvious change in the uptake kinetics and the threshold value could be observed after subsequent pulses, apparently since these parameters were independent of the amount of stored phosphate under the experimental conditions employed.

Nevertheless the threshold value is not a determinate, objective parameter of cyanobacterial cells in the prevailing physiological state, since it is influenced by the rate of decrease in the external phosphate concentration, as demonstrated in the experiment of Fig. 1 B. Here two differently diluted suspensions of the same batch of algae received different phosphate concentrations, such that the initially available amount of phosphate per average cell was the same. If a suspension with a higher number of cells was given a higher concentration of phosphate, again a low threshold value of a few nanomoles per liter was attained. But if a more diluted suspension was exposed to a lower phosphate concentration, uptake ceased surprisingly at a concentration of about 50 nM. Although here the cells had stored less phosphate than those of the parallel suspension, a persisting adaptation occurred during the uptake process, resulting in a higher threshold value. In a former publication we have shown that the adapted state may be conserved over several hours and that the new kinetic and energetic properties of the uptake system could even be transferred to the daughter cells [4]. It must be emphasized that the threshold value depends on the rate of decrease in the ex-

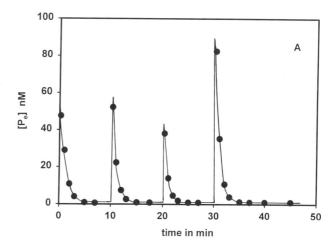

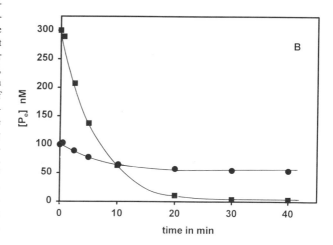

Figure 1. Phosphate uptake in *Anacystis nidulans* after pulsewise addition of phosphate. **A.** Cells were diluted to a final chlorophyll content of 20 µg Chl$a$·l^{-1}, and exposed to four subsequent phosphate pulses at time 0, 10, 20, and 30 min. **B.** Cyanobacteria originating from the same batch of cells were diluted differently and exposed to distinct pulses of phosphate in a way that the amount of initially offered phosphate per cell was identical in the two incubations. The suspension containing 10 µg Chl$a$·l^{-1} received an initial phosphate concentration of 300 nM (■), and the suspension containing 3.3 µg Chl$a$·l^{-1} was incubated with 100 nM phosphate (●).

ternal phosphate concentration which in turn is determined by the uptake activity of the prevailing population. Since this value is decisive for the amount of stored phosphate per cell, the population as a *whole* determines the subsequent growth of the *individual* cells. In this respect the population may be regarded as a superorganism that is more than the sum of the single cells it is composed of.

3.2. Regulation of Carbon flow during Transition from the Non Growing to the Growing State

In order to investigate how the utilization of photosynthetically fixed CO₂ is affected by phosphate pulses—e.g., *via* an alteration of the cytoplasmic phosphate concentration—we studied the flow of carbon into the glycogen pool before and after short term increases in the external phosphate concentration. A key enzyme of glycogen formation is ADP-glucose-pyrophosphorylase which is allosterically inhibited by orthophosphate, and activated by 3-phosphoglyceric acid and fructose-6-phosphate [7]. This enzyme is responsible for the biosynthesis of a precursor of glycogen and plays a regulatory role in the flow of photosynthetically assimilated CO₂ into either low molecular weight compounds (LMWC)

Table 1. CO_2 fixation of the cyanobacterium *Anacystis nidulans* before
and after a pulse of 2.5 µM phosphate

	Total CO_2 fixation nmoles $(10^8 \text{ cells})^{-1}$	Percentage of totally in LMWC	Assimilated carbon in HMWC
Before phosphate pulse	7.96	50.4	49.6
0.5 min after onset of a pulse of 2.5 µM P_i	5.73	71.5	28.5
43 min after onset of a pulse of 2.5 µM P_i	9.35	66.4	33.6

LMWC = low molecular weight compounds. HMWC = high molecular weight compounds.

for immediate utilization, or to glycogen as a carbon storage of high molecular weight (HMWC). Thus, an alteration of the carbon flow into the pool of LMWC and HMWC under different experimental conditions may serve as an indicator of changes in the intracellular phosphate level during and after phosphate pulses.

The result of a representative experiment in which the flow of fixed carbon before and after a phosphate pulse was followed by separating LMWC and HMWC is given in Table 1. Before phosphate pulse the flow of carbon was equally distributed into the pools of LMWC and HMWC. However, when the same cells were exposed to 2.5 µM phosphate, immediately the formation of HMWC was inhibited and about 70% of the assimilated ^{14}C could be detected in the pool of LMWC necessary for the growing cells. 43 min later, when the external phosphate had been taken up by the population and the threshold value was reattained, this mode of carbon utilization was still preserved: the main flow of carbon was directed into LMWC, while synthesis of HMWC was repressed. The observed decrease in glycogen formation can be simply explained by postulating that during the pulse a higher cytoplasmic phosphate concentration had persistingly been established, thus inhibiting ADP-glucose pyrophosphorylase.

The cytoplasmic phosphate concentration is decisive for the energy necessary for phosphate transport across the cell membrane at nanomolar external phosphate concentrations. Moreover, any change in the internal phosphate concentration would change the phosphorylation potential, and thus influence the energetic homeostasis of the cell. The observed maintenance of the higher internal phosphate concentration after a pulse may be explained by three events: First, by an immediate onset of biosynthetic, ATP consuming reactions by which orthophosphate is released. Second, by a futile cycle by which recently formed polyphosphates are hydrolized to orthophosphate by phosphatases. Third, by a change in the H^+/ATP stoichiometry of the ATP synthase. In former studies [8] it was shown that during transition from a phosphate deficient to a non deficient growth state the K_M value for phosphate of the F-ATPase of *Anacystis nidulans* is increased from about 34 µM to 225 µM, while concomitantly the stoichiometry is decreased from about 4 to 3 protons per synthesized ATP (Table 2).

Table 2. Comparison of the kinetic and energetic properties of the thylakoid ATP synthase
of phosphate limited and unlimited *Anacystis nidulans*

	Apparent K_M µM phosphate (at 1 mm ADP)	Maximum pH-gradient	H^+/ATP stoichiometry	Cytoplasmatic phosphate conc. At equilibrium μM
Phosphate deficient algae	34 ± 6	2.3–2.5	4.2 ± 0.30	60–10
Phosphate non-deficient algae	225 ± 35	2.7–2.9	3.4 ± 0.04	800–200

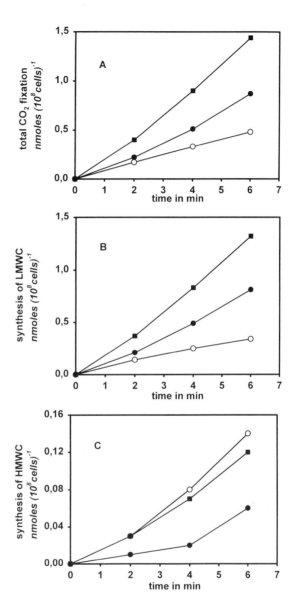

Figure 2. The effect of phosphate pulses on CO₂ fixation in *Anacystis nidulans*. Prior to the CO₂ fixation experiment the cells were diluted to a final chlorophyll content of 40 μg Chl$a$·l^{-1} (●) or to 10 μg Chl$a$·l^{-1} (■), and incubated with four subsequent pulses of 1 μM initial phosphate concentration or one pulse of 1μM phosphate, respectively. The threshold was attained 6–25 min after onset of phosphate preincubation, and the cells were concentrated by centrifugation before the CO₂ fixation experiment was initiated. The open circles (O) refer to cells that did not get a phosphate pulse before the experiment. **A**, rate of total fixation; **B**, rate of LMWC formation; **C**, rate of HMWC formation.

With these parameters one can calculate an intracellular phosphate level at equilibrium of about 10–60 μM for phosphate deficient, and 200–800 μM for non deficient cells, respectively [8]. Comparing these values with the K_M data one can see that the ATP synthase operates under all growth conditions efficiently near equilibrium with minimum entropy production.

It is therefore possible that such an adaptive modification of the ATP synthase occurs during pulse dependent increases in the cytoplasmic phosphate concentration. In this case the ATP synthase of *Anacystis nidulans* should exhibit some type of hysteretic properties, as has been demonstrated with other ATPases [9,10]. A persistingly higher cytoplasmic phosphate concentration would then result from such a hysteretic transition.

Since the cytoplasmic phosphate concentration was affected by phosphate pulses, it could be expected that the pulse pattern would regulate the carbon flow, even under conditions under which the amount of stored phosphate is invariant. To show this, two different dilutions of the same batch of cyanobacteria, yielding a final chlorophyll content of 40 µg Chl$a$·l^{-1} and 10 µg Chl$a$·l^{-1}, respectively, received pulsewise the same amount of phosphate per cell (Fig. 2). The suspension with the four times higher cell density was exposed to four pulses of an initial concentration of 1 µM phosphate, whereas the more diluted suspension was given only one pulse of 1 µM phosphate. After the threshold was attained the two suspensions showed clear differences in carbon utilization, indicating that under these conditions not the amount of stored phosphate but the mode of phosphate supply potentially determines carbon metabolism, and possibly subsequent growth. We conclude that due to distinct adaptive responses exhibited under non stationary conditions, as they exist in a natural environment, a complex relationship between population and individual cells exists which is not accounted for by simple formalistic models.

REFERENCES

1. Rigler, F. (1956) Ecology37, 550–562
2. Falkner, G., Falkner, R. and Schwab, A. (1989) Arch. Microbiol. 152, 353–361
3. Rhee, G.-Y. (1978) Adv. in Aquat. Microbiol. 2, 151–203
4. Wagner, F., Falkner, R. and Falkner, G. (1995) Planta 197, 147–155
5. Falkner, G, Wagner, F., Small J.V. and Falkner, R. (1995) J. Phycol. 31, 745–753
6. Heldt, H.W. et al. (1977) Plant Physiol. 59, 1146–1155
7. Levi, C. and Preiss, J. (1976) Plant Physiol. 58, 753–756
8. Wagner, F. and Falkner, G. (1992) J. Plant Physiol. 140, 163–167
9. Schobert, B. and Lanyi, J. (1989) J. Biol. Chem. 264, 12805–12812
10. Recktenwald, D. and Hess, B. (1977) FEBS Letters 80,187–189

87

DISCOVERY, CHARACTERISTICS, AND DISTRIBUTION OF ZINC-BChl IN AEROBIC ACIDOPHILIC BACTERIA INCLUDING *ACIDIPHILIUM* SPECIES AND OTHER RELATED ACIDOPHILIC BACTERIA

N. Wakao,[1] A. Hiraishi,[2] K. Shimada,[3] M. Kobayashi,[4] S. Takaichi,[5] M. Iwaki,[6] and S. Itoh[6]

[1]Faculty of Agriculture
Iwate University
Morioka 020
[2]Department of Ecological Engineering
Toyohashi University of Technology
Toyohashi 441
[3]Department of Biology
Tokyo Metropolitan University
Tokyo 192–03
[4]Institute of Materials Science
University of Tsukuba
Tsukuba, 305
[5]Biological Laboratory
Nippon Medical School
Kawasaki 211
[6]National Institute for Basic Biology
Okazaki 444, Japan

1. INTRODUCTION

Chlorophylls (Chls or Mg-Chls) and bacteriochlorophylls (BChls or Mg-BChls) are the very important pigments in the natural photosynthesis converting solar energy to chemical energy since its evolution on the Earth. All of Chls and BChls found in photosynthetic organisms (green plants, algae and bacteria) are the porphyrin derivatives

The Phototrophic Prokaryotes, edited by Peschek *et al.*
Kluwer Academic / Plenum Publishers, New York, 1999.

chelated with Mg as the central metal. The other metals than Mg are not known in the natural photosynthesis. Why do plants and photosynthetic bacteria exclusively use Mg as the central metal? Recently, however, we have discovered a new purple photosynthetic pigment, Zinc-chelated BPhe *a* esterified with phytol (Zn-BChl *a*) in the aerobic acido-philic bacteria *Acidiphilium* species [1] and, here, we report the characteristics and distri-bution of the new photosynthetic pigment in acidophilic bacteria.

2. PHOTOSYNTHETIC PIGMENTS OF AEROBIC PHOTOSYNTHETIC BACTERIA

Besides plant and algae with Chls and typical photosynthetic bacteria with BChls, strictly aerobic bacteria containing BChl have been discovered firstly by Japanese micro-biologists and a new category of Aerobic Photosynthetic Bacteria has been proposed [2]. Many kinds of aerobic photosynthetic bacteria have been isolated from different natural environments, such as, *Roseobacter, Erythrobacter, Methylobacterium, Rhizobium, Erythromicrobium* and *Acidiphilium. Acidiphilium (A.)* which belongs to alpha-1 subclass of the *Proteobacteria* is the only acidophilic bacteria among them and contains five spe-cies; *A. rubrum, A. angustum, A. cryptum, A. organovorum* and *A. multivorum*. The con-tents of BChl of the aerobic photosynthetic bacteria were low compared with those of typical purple bacteria. BChls found in the aerobic photosynthetic bacteria were BChl *a* (Mg-BChl) as already reported in typical photosynthetic bacteria. Nevertheless, major BChl in the acidophilic bacteria belonging to *Acidiphilium* was found to be different from that of others.

3. ABSORPTION SPECTRA OF PIGMENTS

Absorption spectra of crude pigment extracts from the cells of *Acidiphilium* sp. with organic solvent showed almost similar absorption patterns to those of purple bacteria. The absorption peaks around 770 nm suggested the occurrence of BChl and the peaks around 500 nm were due to carotenoids. Among the strains, *A. rubrum* contained the largest amount of pigments and was used in this study as the representative of the strains. In gen-eral, BChl *a* has the maximum absorption at 770 nm. However, the peak of *Acidiphilium* sp. was blue-shifted by 6 to 8 nm from that of BChl *a*. The absorption spectra of the mem-brane fractions from *A. rubrum* and *Rhodospirillum (Rs.) rubrum* resembled each other, suggesting that *A. rubrum* had a set of pigment-protein complexes similar to those in *Rs. rubrum* which contains BChl and spirilloxanthin in the light-harvesting complex and the photochemical reaction center complex as well. However, the peak of *A. rubrum* was blue-shifted by 15 nm from that of BChl *a* in *Rs. rubrum*.

4. TLC AND HPLC SEPARATION OF PIGMENTS

Silica-gel TLC of the pigment extracts from *A. rubrum* cells revealed that the main band of unknown purple pigment was preceded by a minor blue band of BChl *a*. BPhe *a* and carotenoids appeared as pink and red bands, respectively. On Silica-gel HPLC elution profiles of the pigment extracts of *A. rubrum*, a large amount of the new pigment was de-tected besides small amounts of BChl *a* and BPhe *a* . The new purple pigment showed an

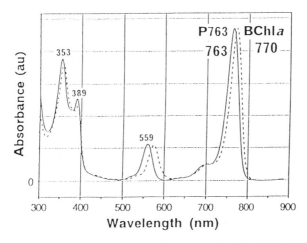

Figure 1. Absorption spectra of the purified P763 and BChl *a* in diethylether.

absorption spectrum which was similar to that of BChl *a* (Figure 1). However, the new purple pigment gave a maximum absorption at 763 nm (diethylether) which was shorter than that of BChl *a* by 7 nm. Therefore, we tentatively designated the pigment as P763. When P763 and BChl *a* were acidified by HCl, both pigments disappeared completely and were converted to the same BPhe *a*, suggesting close similarity between them.

5. MOLECULAR STRUCTURE OF P763

From metal analysis of the purified P763, it contained Zn but not Mg and other metals. The molar ratio of Zn to P763 was equal to 1. The molecular structure of P763 was determined by spectrophotometry, HPLC, metal and elementary analyses, and FAB-mass and ^{1}H-NMR measurements, and found to be Zn-chelated BPhe *a* esterified with a phytol group (Zn-BChl *a*) (Figure 2).

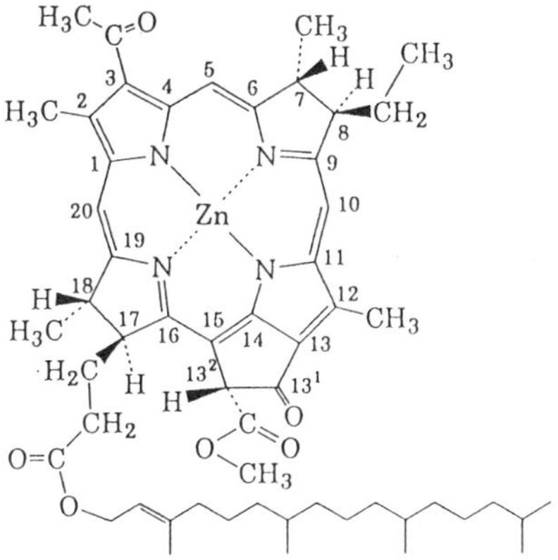

Figure 2. Molecular structure of Zn-BChl *a*.

Table 1. Contents of Zn-BChl, Mg-BChl and BPhe in species of Acidiphilium and other acidophilic bacteria

Species	Contents (nmol/g dry cell)		
	Zn-BChl	Mg-BChl	BPhe
Acidiphilium rubrum	810	120	61
Acidiphilium angustum	180	26	14
Acidiphilium cryptum	56	5	5
Acidiphilium multivorum	87	11	8
Acidiphilium organovorum	7	tr	1
Acidocella facilis	ND	ND	ND**
Acidocella aminolytica	ND	ND	ND
Acidobacterium capsulatum	ND	ND	ND
Acidomonas methanolica	ND	ND	ND
*Rhodopila globiformis**	ND	18,200	1,150
*Rhodopseudomonas acidophila**	ND	16,000	958

*Typical photosynthetic bacteria for reference. **ND, not detected.

6. DISTRIBUTION AND CONTENTS OF PIGMENTS

Table 1 shows the contents of Zn-BChl, Mg-BChl and BPhe in cells of *Acidiphilium* sp., other aerobic acidophilic bacteria and anaerobic acidophilic photosynthetic bacteria. The presence of Zn-BChl was limited to the genus *Acidiphilium*. Zn-BChl was always the dominant in cells. The ratio of three pigments in cells was almost constant and 13:2:1 on the average. Carotenoids pigment of *Acidiphilium* sp. was determined to be spirilloxanthin. In some species, rhodovibrin, an intermediate of spirilloxanthin biosynthesis, was detected in a trace amount.

7. EFFECT OF LIGHT ON Bchls FORMATION

A. rubrum is a strictly aerobic heterotrophic bacterium and therefore can grow aerobically in the dark using glucose as both carbon and energy source. However, the incorporation of CO_2 by whole cells of *A. rubrum* was enhanced under light-aerobic condition [3]. It was confirmed by light-induced absorption change using the membrane fractions that Zn-BChl was photochemically active in living cells. However, continuous illumination was rather harmful to the pigment biosynthesis in cells as in the case of other aerobic photosynthetic bacteria. The growth was not strongly repressed by light illumination as compared to pigment formation.

8. EFFECT OF CARBON SOURCES AND Zn ON Bchls FORMATION

The total contents of Zn-BChl and Mg-BChl in cells of *A. rubrum* were variable depending on different carbon sources of the media. The largest contents were obtained for arabinose and maltose, and the smallest for trehalose and mannitol. However, the ratios of Zn-BChl to Mg-BChl were rather constant. The addition of Zn (0.1–0.5 mM) in the medium did not have a promotive effect on the biosynthesis of Zn-BChl. A trace amount of Zn (1 μM) in the medium was sufficient to biosynthesis of the pigment. These bacteria may take Zn selectively and positively into cells as the central metal of the pigment.

9. EFFECT OF pH ON BChls FORMATION

pH of the medium was found to be one of the important factors influencing the formation of both BChls in living cells. Total contents of both BChls increased along with the decreasing of pHs of the medium. The ratios of Zn-BChl to Mg-BChl were variable at the different pH values; Zn-BChl was dominant at acidic conditions less than pH 3, but Mg-BChl increased at pH 4–5. Supposing the difference of pH-stability between Zn-BChl and Mg-BChl, we measured the acid tolerance of the two pigments. Mg-BChl immediately changed to BPhe at pH 1, but Zn-BChl was extremely stable at the same pH. Pheophytinization rate of Zn-BChl by acid treatment was about 10-fold smaller than that of Mg-BChl. Such tolerance of Zn-BChl to acidity may be one of the effective causes why Zn was chosen by the acidophilic bacteria.

10. PIGMENT-PROTEIN COMPLEXES AND PHOTOCHEMICAL ACTIVITIES

The reaction center and the LH1 type light-harvesting complexes were isolated from *A. rubrum*. Pigment analysis indicated that these complexes contained Zn-BChl but little amount of Mg-BChl. The absorption spectra of the isolated complexes were similar to those of the corresponding complexes of typical purple bacteria, except the 5–15 nm blue-shifted absorption peaks in the near-infrared region. The function of Zn-BChl in the pigment complexes was fundamentally the same as that of Mg-BChl of purple bacteria [4]. *A. rubrum* used Zn-BChl in its LH1-type complex and reaction center complex, which were fully active. Special pair and accessory BChls were Zn-BChl [4].

11. PRIMARY STRUCTURES OF RC PROTEINS

Nucleotide sequences of the genes coding for the photosynthetic proteins (*puf* operon) of *Acidiphilium* sp. and the relatives were determined [5]. The comparison of the deduced amino acid sequences of *puf* genes of *Acidiphilium* sp. showed that His L168, which is highly conserved in the other purple bacteria, was replaced by a glutamic acid in *Acidiphilium* sp. The three dimensional structures of the reaction centers of *Rhodopseudomonas viridis* and *Rhodobacter sphaeroides* suggest that this residue locates closely to the special pair of BChls and may play the important roles for the stability and function of the Zn-BChl in the reaction center of *Acidiphilium* sp.

12. CONCLUSION

It has been believed that the naturally-occurred photosynthetic pigments contain Mg as the central metal without exception. However, we have found the presence of Zn-BChl in the aerobic acidophilic bacteria *Acidiphilium* sp. [1]. This is the first discovery for the natural occurrence of a porphyrin-based photosynthetic pigment including a metal other than Mg. Zn is as functional and efficient as Mg for natural photosynthesis so that some of the oldest photosynthetic organisms might have a chance to choose Zn as well as Mg or other metals in the ancient environments of the Earth. We are just at the entrance of the

World of Zn-BChl Photosynthesis. Many questions remain to be solved over various fields, such as, biosynthesis, genetics, photochemistry, microbiology, evolution, other met-allo-Chls and so on. The finding of the new pigment indicates an unexpectedly variability of biological photosynthesis and will give a new insight into our understanding of photosynthesis.

REFERENCES

1. Wakao, N., Yokoi, N., Isoyama, N., Hiraishi, A., Shimada, K., Kobayashi, M., Kise, H., Iwaki, M., Itoh, S., Takaichi, S. and Sakurai, Y. (1996) Plant Cell Physiol. 37, 889–893
2. Shimada, K. (1995) in: Aerobic Anoxygenic Phototrophs (Blankenship, R.E., Madigan, M.T. and Bauer, C.E. eds) Anoxygenic Photosynthetic Bacteria, pp. 105–122, Kluwer Academic Publishers, Dordrecht.
3. Kishimoto, N. Fukaya, F., Inagaki, K., Sugio, T., Tanaka, H. and Tano, T. (1995) FEMS Microbiol. Ecol., 16, 291–296.
4. Shimada, K. Matsuura, K., Itoh, S., Iwaki, M., Hiraishi, A., Takaichi, S., Kobayashi, M. and Wakao, N. (1997) Abstracts of 9-ISPP (Vienna, Austria), p. 102.
5. Nagashima, K.V.P., Matsuura, K., Hiraishi, A., Wakao, N., and Shimada, K. (1997) Abstracts of 9-ISPP (Vienna, Austria), p. 102.

88

IDENTIFICATION OF A PUTATIVE GAMMA LINKER POLYPEPTIDE GENE IN THE MARINE OXYPHOTOBACTERIUM *PROCHLOROCOCCUS MARINUS*

Implications for the Phylogeny of *Prochlorococcus* Phycoerythrins

Wolfgang R. Hess[1,*] and F. Partensky[2]

[1]Humboldt-Universität Berlin
Institut für Biologie/Genetik
Chausseestr. 117, D-10115 Berlin, Germany
[2]Station Biologique
CNRS, INSU et Université Pierre et Marie Curie
BP 74, F-29682 Roscoff CX, France

1. INTRODUCTION[†]

Prochlorococcus is a marine photosynthetic prokaryote very abundant in most temperate and intertropical oceanic areas [1]. Phylogenetic studies using 16S rRNA or *rpoC1* gene sequence comparisons have suggested that this Chl *b*-containing organism, once called a "prochlorophyte" [2], was closely related to phycobilisome-containing cyanobacteria [3–5]. This was recently corroborated by the characterization of *Prochlorococcus* Chl *a/b* antenna proteins which proved to have originated from CP43', an iron stress-induced cyanobacterial protein binding Chl *a* [6], and by the discovery of one phycobiliprotein (phycoerythrin III or PE III) in the type species *P. marinus* CCMP 1375 [7].

The phylogenetic analysis of this phycobiliprotein is of potential interest for the understanding of oxygenic photosynthesis evolution. Two antagonistic scenarios can be invoked for the origin of *Prochlorococcus* PE III:

---

* Corresponding author.
† abbreviations: chlorophyll, Chl; PE, phycoerythrin; PUB, phycourobilin.

The Phototrophic Prokaryotes, edited by Peschek *et al.*
Kluwer Academic / Plenum Publishers, New York, 1999.

The first one assumes that PE III represents an ancient form of phycoerythrin. In apparent agreement with this hypothesis, the unique pigmentation of *Prochlorococcus marinus* which, besides one (and possibly several) phycobiliprotein(s), also includes (divinyl-) Chl *a*, (divinyl- and monovinyl-) Chl *b* as well as low amounts of a Chl *c* derivative [8,9], resembles the pigment composition previously suggested for an hypothetical ancestor of prokaryotes performing oxygenic photosynthesis [10]. *P. marinus* could derive from such an ancestral form without major pigmentary changes, whereas most other cyanobacterial descendants would have lost the ability to synthesize or use Chls *b* and *c*. Phylogenetic analyses of part of the ribulose-1,5-bisphosphate carboxylase (Rubisco) gene in one *Prochlorococcus* strain (GP2) suggested that this prokaryote could be primitive as they placed it near the root of oxyphototrophs [11]. However, the latter argument is somewhat weakened by the discovery of *Prochlorococcus* strains possessing a form of Rubisco phylogenetically closer to that of phycobilisome-containing cyanobacteria (W. Hess, unpublished data) and the finding that *rbcL* may be transferred laterally between phylogenetically distant organisms [1,12].

The second scenario assumes that *P. marinus* PE III has derived more recently from cyanobacterial PE's. The organization of the genes coding for the two PE subunits (β upstream of α) in the PE operon, as well as the presence of a phycourobilin-like chromophore, the most blue-shifted of all phycobilins, are considered as modern characteristics [13,14]. It can thus be speculated that the development of a constitutive Chl *a/b* antenna [6] allowed the disappearance of phycobilisomes in some lines, such as the high light-adapted *Prochlorococcus* sp. CCMP 1378 [7], whereas in the low light-adapted *P. marinus* CCMP 1375, both light harvesting systems were kept. Hence, *P. marinus* CCMP 1375 would be a relatively modern organism and its phycoerythrin would represent an intermediate stage on a way ultimately leading to the disappearance of any phycobiliprotein. The finding that the "high-light clade" which includes the PE-devoided strain CCMP 1378 is phylogenetically even more recent than the one including CCMP 1375 [4] reinforces this hypothesis [1].

To better understand the questions of the phylogenetic origin, the exact biological function, and the structural arrangement of the particular phycoerythrin found in *Prochlorococcus* (PE-III), analyses of existing *Prochlorococcus* phycoerythrin sequence data as well as further sequencing of upstream and downstream regions of the PE operon were performed. It must be pointed out that PE phylogeny is to some extent complicated by the fact that in cyanobacteria there are two PE types. Phycoerythrin I, which is encoded by the *cpeB* and *cpeA* genes, is widespread in both freshwater and marine cyanobacteria. Phycoerythrin II, which is encoded by the *mpeB* and *mpeA* genes, is only present in some marine cyanobacteria, such as *Synechococcus* WH 8020 and WH 8103 [15,16]. Thus *cpeA* and *mpeA* on one hand, *cpeB* and *mpeB* on the other hand appear to be paralogous genes. With this caveat in mind, analyses reported here support to some degree a closer phylogenetic relatedness between the PE III β subunit of *Prochlorococcus* and those of marine cyanobacteria PE II, whereas the position of the PE III α subunit was less stable in phylogenetic trees. Moreover, a third gene, coding for a protein homologous to several phycoerythrin and phycocyanin linker polypeptides and especially to the phycoerythrin II γ subunit encoded by the *mpeC* gene of the marine cyanobacterium *Synechococcus* sp. WH 8020 [17], has been discovered in the *Prochlorococcus* genome at a distance of only 5 kb to the PE operon. The novel *P. marinus* CCMP 1375 phycoerythrin linker peptide gene therefore is suggested to be called *ppeC* for *Prochlorococcus* phycoerythrin gamma subunit. Its possible role in the structural organization of *P. marinus* PE III is discussed.

2. MATERIALS AND METHODS

Cultures of *P. marinus* were grown as described in [18]. Isolation, analysis, cloning and subcloning of DNA has been described previously [7]. During further analyses of the two selected cosmids containing the complete PE III coding sequences [7], a third gene with homology to phycobiliprotein genes was detected (see results). The 1111 bp of the gene coding region and of adjacent regions were determined using an ABI 373 sequencer (Applied Biosystems, USA). The sequence is available from the DDBJ/EMBL/GenBank under the accession number Z92525. Initial sequence searches were performed by the BLAST-algorithm. The sequences of *P. marinus ppeA* and *B* are available under the accession number Z68890. All other sequences used in this paper were obtained via Internet from the National Center for Biological Information in Washington at http:\\\\www.ncbi.nlm.nih.gov.

The deduced amino-acid sequences were aligned using ClustalW 1.6. [19], followed by an analysis by the neighbor-joining method using the PROTDIST, and NEIGHBOR programs in the PHYLIP 3.572c program package [20]. Bootstrap analysis was obtained by using SEQBOOT, PROTDIST, NEIGHBOR and CONSENSE in PHYLIP 3.572c. All sequences were further analyzed by the maximum likelihood method using the program PUZZLE version 2.5.1. or 3.1 [21]. The sequences of selected species were additionally also analyzed by phylogenetic analysis using parsimony (PAUP).

3. RESULTS AND DISCUSSION

3.1. Analysis of α and β Phycoerythrin

P. marinus α and β PE III can be easily aligned to the homologous PE sequences from other organisms [7], although the α PE III subunit possesses characteristic small in-frame deletions in the central part of the molecule, from amino acids 58–62 and 68–70 (Fig. 1). No

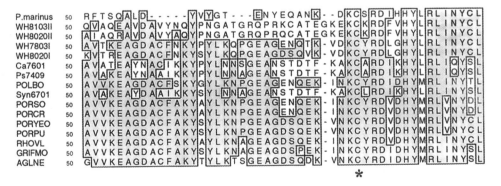

Figure 1. Alignment of the central part (residue 50–88) of the *P. marinus* PE α-subunit to all known α-PE's from pro- and eukaryotic organisms. Identical residues are boxed and shadowed. The conserved chromophore coupling site is indicated by an asterisk. The sequences are designated as follows: WH8103, *Synechococcus* WH 8103; WH8020, *Synechococcus* WH 8020; Syn6701, *Synechocystis* PCC 6701; Ps7409, *Pseudanabaena* PCC 7409; Ca7601, *Calothrix* PCC 7601; WH7803, *Synechococcus* WH 7803; PORSO, *Porphyridium sordidum*; PORCR, *Porphyridium cruentum*; PORYEO, *Porphyra yezoensis*; PORPU, *Porphyra purpurea*; RHOVL, *Rhodella violacea*; GRIFMO, *Griffithsia monolis*; AGLNE, *Aglaothamnion neglectum*; POLBO, *Polysiphonia boldii*, in addition I or II indicates the respective type of PE in case of the marine *Synechococcus*.

equivalent alteration was found so far in any other α PE's. According to existing models on the structure of cyanobacterial phycoerythrins [22,23], the first deletion would shorten the length of the helix a of the α PE III subunit. Interestingly, in the 3-D structure of phycocyanin, deduced from high resolution crystallography [22], this region is found very near from a cysteine, located at position 139/140 and serving for chromophore attachment [23]. At the equivalent position, however, the *Prochlorococcus* α PE III protein has an asparagine and thus lacks the chromophoric group that normally is bound at this site. Both this deletion and the replacement might indicate that, in this *Prochlorococcus* line, a degenerative process has started that might end in the future in the complete loss of phycobilins, as observed nowadays in other lines [7]. Alternatively, these alterations might be the result of some structural constraints. In that case, it is tempting to speculate that the deletion between residues 58–62 as well as the lack of the second chromophore may correspond to a steric adjustment to fit *P. marinus* γ like subunit (see below), if the latter is, as expected, implicated in aggregation of (αβ) PE III complexes. *Synechococcus* PE II and the B-PE's of red algae such as *Porphyridium cruentum,* which include a γ subunit in their rod complex show a common deletion of 8–9 residues in the β subunit [23], an adjustment which is clearly related to the presence of the γ protein. At exactly that position the *P. marinus* PE β subunit has a deletion of only two amino acids [7], possibly too small to fit the γ like subunit, and a complementary adjustment of the α-subunit could be required as well.

The presence of phycobilins in *P. marinus* strongly suggests that the ancestor of all *Prochlorococcus* species was a phycobilisome-containing cyanobacterium. Molecular analyses using the 16S rRNA gene showed that the group of cyanobacteria phylogenetically the most related to *Prochlorococcus* is the *Synechococcus* marine A cluster, represented by strains WH 8103 or WH 8020 [4,5].

Phylogenetic analyses using the α PE sequence did not result in a stable position for the *Prochlorococcus* protein among cyanobacterial phycoerythrins. It was either found as a sister group of the two main branches formed by the type I PE and the type II PE or, with a weak support, it clustered with one or the other branch, depending on the method used (data not shown).

Results from the phylogenetic analysis of *P. marinus* β PE III, made by neighbor joining based on distance values, were more reliable. A representative example of the most frequent trees is shown in Figure 2A. This tree is in agreement with previously published phylogenetic trees using the same molecule [13], although some additional red algal sequences have been incorporated here. The eukaryotic and prokaryotic sequences were well separated in most of the bootstrap replications. The marine cyanobacterial β-PE's (WH 8020 I and II, WH 7803 and WH8103 II), including *P. marinus*', were on different branches than the freshwater cyanobacteria and eukaryotic sequences. However, the PE I

Figure 2. Molecular phylogenetic trees derived from the comparison of the *P. marinus* phycoerythrin β subunit [7] with selected other β-PE amino acid sequences. Full length sequences were aligned using ClustalW [19] (alignment available upon request). **A.** Example of an unrooted neighbor joining tree obtained by the PROTDIST method. Tree topology and evolutionary distance estimations were performed using Kimura distancies (PHYLIP 3.572c release [20]). Bootstrap confidence values [26] were taken from a consensus tree calculated from 100 replicates using the SEQBOOT, PROTDIST, NEIGHBOR and CONSENSE programs of Phylip 3.572 and are indicated at the nodes in per cent. A *Calothrix* PCC 7601 β-C-phycocyanin was selected as outgroup. **B.** The same alignment was analyzed by a maximum likelihood estimation of phylogenies using the mtREV24 model of sequence evolution [24] as part of the PUZZLE 3.1. program package [21]. 2380 quartets were analyzed in 1000 puzzling steps with 52 unresolved quartets (= 2.2%). The sequences are named according to the organism, with I or II indicating the respective type of cyanobacterial PE.

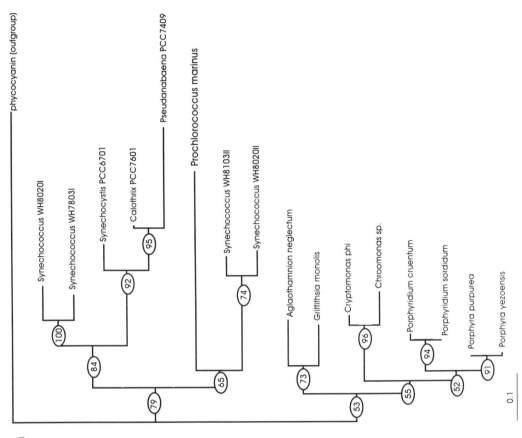

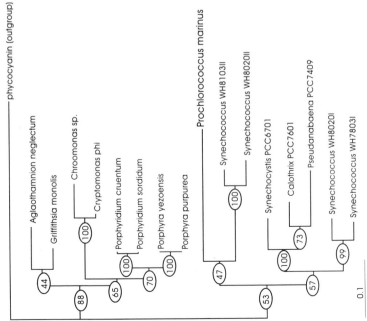

of the marine cyanobacteria WH 8020 was more related to freshwater PE's than to the PE II from the same organism. The only two known marine cyanobacterial β-PE II sequences were gathered on a distinct branch. In most analyses (47% of the bootstrap replications), *P. marinus* β-PE III was placed at the base of this branch, although in some others it was placed either at the base of the cyanobacterial type I PE branch or as a sister group to both type I and type II PE's. Hence this analysis is supporting, although not very strongly, a higher similarity between the *Prochlorococcus* PE III sequence and the β-PE II's. Analysis of the same data set by the maximum likelihood method using the mtREV24 model of sequence evolution [24] resulted in a tree of very similar topology, including a good separation of prokaryote and eukaryote sequences, of cyanobacterial type I and II PE's, and with the marine PE I's located in a distinct clade of the PE I branch (Fig. 2B). In 65% of the analyses *P. marinus* β PE III was again found on the same branch as the marine cyanobacterial PE II sequences (Fig. 2B). For constructing this tree, 2380 quartets were analyzed during 1000 puzzling steps with 52 quartets remaining unresolved. The *Prochlorococcus* sequence was only 14 times involved in one of the unresolved quartets, a comparable score to the other cyanobacterial sequences but lower than some of the eukaryotic sequences (e.g. *Aglaothamnion neglectum*, 20 times). The positioning of the *P. marinus* sequence basal of the *Synechococcus* PE II clade was also obtained using parsimony (PAUP 3.1, not shown). In all programs used, such a position of *P. marinus* β PE III was more strongly supported when the eukaryotic sequences were omitted from the alignments and only cyanobacterial sequences were compared.

3.2. Identification of a Putative Phycoerythrin Gamma Subunit

Sequence analysis of subclones obtained from the same cosmid which was used for cloning and sequencing genes coding for *Prochlorococcus* α and β subunits of PE-III [7] revealed a putative gene of 810 nt (Fig. 3), located at a distance of only 5 kb from the PE operon. The deduced amino acid sequence characterized a protein of 270 residues with a predicted molecular weight of 31.5 kDa. In BLASTP alignments, this protein showed a high sequence similarity to phycocyanin and phycoerythrin linker polypeptides. Furthermore, its physico-chemical characteristics (pI of 10.11) were also similar to those of cyanobacterial linker polypeptides. Alignments actually showed an extended homology domain encompassing approximately 190 residues or two thirds of the amino acid sequence (Fig. 4). With one exception, the protein is in general slightly more similar to phycocyanin-associated linker polypeptides (BLASTP score 271–305) than to phycoerythrin-associated linkers (BLASTP score 272–273). However, in contrast to all cyanobacterial phycoerythrin- and phycocyanin-associated non-bilin-bearing linker polypeptides, the *Prochlorococcus* gene product is characterized by a long N-terminal extension and an approximately similar number of residues lacking at the C terminus (Fig. 4). Only one protein has been described as having such properties, the phycoerythrin γ-subunit [or $\gamma(L_R^{32})$,] of the marine cyanobacterium *Synechococcus* WH 8020 ([17], cf. Fig. 4). Accordingly, alignments on BLASTP basis gave the highest score (355) for this protein. However, this estimation was possibly biased due to the sequence extension/reduction common to these two proteins relative to all other linker polypeptides. Direct sequence comparison showed that the *Prochlorococcus* polypeptide is at the C-terminus even shorter by 20 residues than the *Synechococcus* WH 8020 protein and that this loss is more than offset by an additional N-terminal extension of the coding region. The *Synechococcus* $\gamma(L_R^{32})$-subunit is very different from other cyanobacterial linker polypeptides associated with phycoerythrin in that it bears a single phycourobilin bound through a thioether linkage to γ-Cys-49 [17]. In the *Prochlorococcus* protein,

```
AAGCTTTGTCAAGAGCAATGGGTAAATTAATTGGTGGTGGATTGTTAGACTTAGGAAGAA    60
TTATAGGGGGGAACGATAACTCTCTATTTGCATTTTCTTTTGTATTCTCAATAGAAATAA   120
TCATAATTATAATCTCTATATTTATACTTAATAAAGTGAGTATAAGTAAATTCAAGAATG   180
AAACATCTGCTAAAATGAGCGAGATCCTAATGTCTGACCTAGACAATTGATAATCTTTTA   240
AAATGAAGACTAGCAAAGCAGACTTTATTTCTAGACATTCCCTAGAAGTTAAACTTTCTG   300
     M  K  T  S  K  A  D  F  I  S  R  H  S  L  E  V  K· L  S    19

ATATATCGAAAAGAGAATCATTTAGCTACGGCAAAACACGCGTTTCTGGTAGCAAGCAAA   360
D  I  S  K  R  E  S  F  S  Y  G  K  T  R  V  S  G  S  K  Q    39

CCACTTATATAATGCACTCTAGAAGCGTTTATATGCCCGGCCAGGAAAAACTTTTTACCA   420
T  T  Y  I  M  H  S  R  S  V  Y  M  P  G  Q  E  K  L  F  T    59

GCAATTATCCTGCAAACCCTAATCAAGTAATCGACAAAGAAATGATTCAGCTAAGAAATA   480
S  N  Y  P  A  N  P  N  Q  V  I  D  K  E  M  I  Q  L  R  N    79

TTTATAAGCGAAAGAATTATATAAACTCAAAGCAACCTATTGGCTCCAAAACAATTCATA   540
I  Y  K  R  K  N  Y  I  N  S  K  Q  P  I  G  S  K  T  I  H    99

GATCTCAAGATGCATTTACTTATAAGAGGTTTGCCCCTATCAGTGATGAAGCACTTGAGA   600
R  S  Q  D  A  F  T  Y  K  R  F  A  P  I  S  D  E  A  L  E   119

TTGCTGTTGTAGCAGCATATAAACAAATCTTTGGCAACTTAAATCCAATGGAAAGCGAAC   660
I  A  V  V  A  A  Y  K  Q  I  F  G  N  L  N  P  M  E  S  E   139

GTCCTAAAGAATTAGAAAGAAGATTAAGAAATGGGGACATCCCAATTAGAGAATTTATTC   720
R  P  K  E  L  E  R  R  L  R  N  G  D  I  P  I  R  E  F  I   159

GATCACTTGCTAAGTCAGAATTTTATAGTCGTCATTTTATTGAAAGAGTAAGTCAAATAA   780
R  S  L  A  K  S  E  F  Y  S  R  H  F  I  E  R  V  S  Q  I   179

GATCTGTCGAGCTTAGATTTATGCATCTATTAGGGCGACCACTTAAGGATGAGTCAGAAC   840
R  S  V  E  L  R  F  M  H  L  L  G  R  P  L  K  D  E  S  E   199

TAATCAATAATATAAATTTTATAAGAGAAAAAGGCTTTGAATCTCATATTGATTCCTTAA   900
L  I  N  N  I  N  F  I  R  E  K  G  F  E  S  H  I  D  S  L   219

TGAATTCTTTAGAGTATGAAGAACACTTTGGAGAAGACATAGTGCCATTTCAAAGGCACT   960
M  N  S  L  E  Y  E  E  H  F  G  E  D  I  V  P  F  Q  R  H   239

GGAATTCACCCTGTGGCAGTACAACATCAAGCTTTATAAAGACAGCTTTATTCAGAAAAG  1020
W  N  S  P  C  G  S  T  T  S  S  F  I  K  T  A  L  F  R  K   259

GTTTTGCTTCCAGTGATAACGTAATTTATAAGTAGAGCTTGTCTATTATGACTATAAAAA  1080
G  F  A  S  S  D  N  V  I  Y  K  -                          270

GGCCTCCTTCTAACGAAAGCCCAGGTGATCA                              1111
```

Figure 3. Nucleotide and deduced amino acid sequence of the 1111 bp DNA fragment of *Prochlorococcus marinus* CCMP 1375 containing the *ppeC* gene.

however, there is no cysteine at that position nor at the position corresponding to the non-bilin bearing γ-Cys-64 of the *Synechococcus* sequence. It is therefore most probably colorless. The only cysteine of the *Prochlorococcus* sequence, Cys-244, appears unlikely to bind a chromophore since bilin-binding cysteines are usually close to aspartic acid residues which balance the dipole moment of the chromophore [13]. On another hand, this aspartic

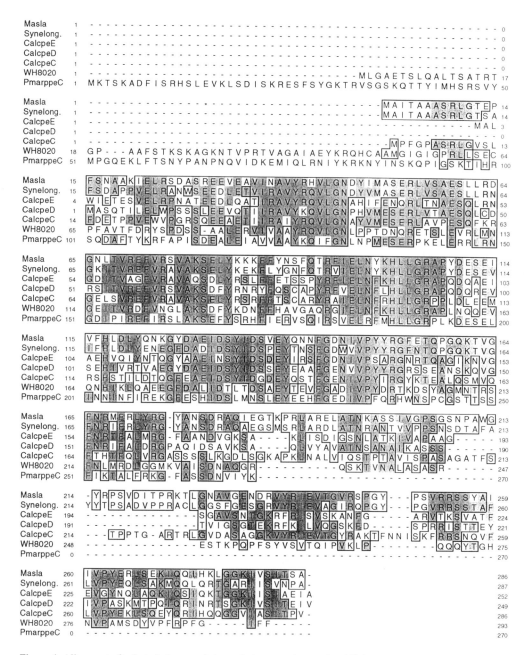

Figure 4. Alignment of selected phycoerythrin- and phycocyanin-associated linker polypeptide sequences from cyanobacteria. The deduced *P. marinus ppeC* sequence (Fig. 3) was compared with the *Synechococcus* WH 8020 PE II γ-subunit (WH8020, accession no. Q02181 [17]), with two phycocyanin-associated linker peptides, *Masticocladus laminosus* (Masla, P11398) and *Synechococcus elongatus* (Synelong, P50034), and with the three phycoerythrin-associated linker polypeptides of *Calothrix* PCC 7601 (CalcpeC-E, P18542, P18543, A43323). Identical and homologous amino acids are displayed by the shadowed and boxed areas, respectively. The core part of the sequences, corresponding to residues 101–270 of *P. marinus ppeC*, was used to perform phylogenetic analysis (Fig. 5).

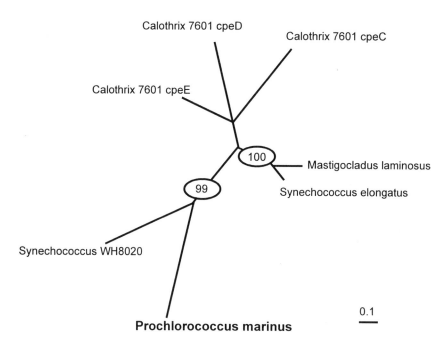

Figure 5. Molecular phylogenetic tree of selected phycoerythrin- and phycocyanin-associated linker proteins based on the alignment shown in Fig. 4. The tree was obtained by a maximum likelihood estimation of phylogenies using the Jones model of evolution [27] and the quartett puzzling search for the best tree as part of the PUZZLE 2.5.1. program package [21]. The *Calothrix* PCC 7601 polypeptide cpeE was set to outgroup. Numbers at the nodes indicate QP support values for the branch distal to it.

acid sometimes is more remote, e.g. in case of the *Aglaothamnion* γ-PE which has a chromophore attachement site at residue 250 [23].

Phylogenetic analyses were performed to clarify the degree of relatedness of the putative *Prochlorococcus* linker protein to those of other cyanobacteria. For this analysis, only the most conserved core sequence from aa 101 to 270 was considered and compared with the corresponding regions from two phycocyanin-associated linker polypeptides and four phycoerythrin-associated linker polypeptides, including the $\gamma(L_{R}^{32})$-PE (Fig. 4). Trees constructed on basis of the maximum likelihood method showed always the topology shown in Figure 5. From 35 quartets analyzed in 1000 puzzling steps, 6 remained unresolved. Neighbor joining trees based on distance values (as well as the consensus tree based upon 1000 replicates) showed the same topology as in Figure 5 with regard to the position of the *Prochlorocccus*, *Synechococcus* WH 8020 and the two phycocyanin linker polypeptides, whereas the positions of the three *Calothrix* PE linker polypeptides were less stable and gave low bootstrap values. Trees of very similar topology were obtained from 100 replicates of parsimony trees (not shown).

Although the conclusions that can be drawn from these trees are obviously limited because of the small data set and the mixture of phycocyanin and phycoerythrin linker polypeptides, it is clear that the *Prochlorococcus* protein was consistently placed on the same branch as the *mpeC* gene product from *Synechococcus* WH 8020, whatever tree-making method was used. This likely indicates a common origin for the two proteins. Therefore it is proposed to designate the corresponding gene *ppeC*, for *Prochlorococcus* phycoerythrin gamma subunit. Interestingly, *Synechococcus* WH 8020 $\gamma(L_{R}^{32})$ was hitherto the sole γ subunit known among cyanobacteria, whereas the γ subunits found in red algal PE complexes are only very distantly related.

3.3. Origin and Putative Organization of *Prochlorococcus* Phycoerythrin

The finding in *Prochlorococcus* of three subunits homologous to the α, β and γ subunits of the marine cyanobacteria *Synechococcus* WH 8020 PE II has important implications both on the phylogenetic and structural viewpoints.

Type II phycoerythrins are considered to be "modern" phycobiliproteins because of the large number and the particular types (PUB and PEB) of bound chromophoric groups. The multiplication of chromophoric groups has enabled an increased light harvesting capacity, whereas their diversification has allowed the use of shorter wavelengths of light than other phycobiliproteins, both tendencies which are assumed to be driving forces for phycobilisome evolution [14]. *Prochlorococcus* PE III is clearly more related to cyanobacteria PE II's and therefore must be considered as a "modern" PE form. However, it seems to make an exception to the chromophore multiplication rule, since the number of chromophoric groups (mainly PUB [7]) that it can bind is low: only 4 per (αβ) -PE instead of 6 in PE II's of marine cyanobacteria such as *Synechococcus* WH 8020 and 5 in freshwater cyanobacteria [15]. This apparent reversion of an evolutionary trend may be related with the development of a constitutive Chl *a/b* antenna in this particular organism, and thus the hypothesis that PE-III is a stage in a degenerative process seems reinforced.

Nevertheless, it remains that *Prochlorococcus* PE-III has all necessary subunits and phycobilins to be functional. Whether it is included in phycobilisomes like in cyanobacteria or red algae is not known yet, but immunological studies suggest that at least allophycocyanin might be present in *P. marinus* as well (G.W.M van der Staay, comm. pers.). In *Acaryochloris marina*, a novel photosynthetic prokaryote containing mainly Chl *d* and therefore probably using a Chl *d*- or a Chl *a/d*-protein antenna as the major light harvesting system, aggregates containing only two types of biliproteins, phycocyanin and allophycocyanin, were recently discovered [25]. This suggests that such incomplete (but apparently functional) phycobilisome structures may occur in *Prochlorococcus* as well and that this may provide an ecological advantage under environmental conditions that this species can experience in the field at the bottom of the euphotic zone.

The exact function of the *Prochlorococcus ppeC* gene product remains to be demonstrated yet, however the presence of this gene in *Prochlorococcus* is intriguing since in *Synechococcus* WH 8020, the *mpeC* gene product is the rod linker polypeptide in (αβ)₆γ PE II hexameric complexes.

Although the phylogenetic origin of *Prochlorococcus* phycoerythrin has now become clearer, further studies aiming at identifying other phycobiliproteins as well as exploring their biophysical characteristics should greatly promote our knowledge on the organization and function of the putative partial or complete phycobilisomes of *Prochlorococcus*.

ACKNOWLEDGMENTS

We are grateful to S. Loiseaux de Goër for helpful comments. This work was supported by grant HE 2544/1–2 from the Deutsche Forschungsgemeinschaft, Bonn, and by the EU MAST III program PROMOLEC (PL96–1031).

REFERENCES

1. Partensky, F., Hess, W.R. and Vaulot, D. (1997) *In:* The Photosynthetic Prokaryotes, Peschek, G.A., Löffelhardt, W., Schmetterer G. (eds.), Plenum Press New York, USA.
2. Chisholm, S.W., Frankel, S.L., Goericke, R., Olson, R.J., Palenik., B, Waterbury, J.B., West-Johnsrud, L. and Zettler, E.R. (1992) Arch Microbiol 157, 297–300.
3. Palenik, B. and Haselkorn, R. (1992) Nature 355, 265–267.
4. Urbach, E., Scanlan, D.J., Distel, D.L., Waterbury, J.B. and Chisholm, S.W. (1998) J. Mol. Evol. 46, 188–201.
5. Urbach, E., Robertson, D.L. and Chisholm, S.W. (1992) Nature 355, 267–270.
6. La Roche, J., van der Staay, G.W.M., Partensky, F., Ducret, A., Aebersold, R., Li, R., Golden, S.S., Hiller, R.G., Wrench, P.M., Larkum, A.W.D. and Green, B.R. (1996) Proc. Natl. Acad. Sci USA 93, 15244–15248.
7. Hess, W.R., Partensky, F., van der Staay, G.W.M., Garcia-Fernandez, J.M., Börner, T. and Vaulot, D. (1996) Proc. Natl. Acad. Sci. USA 93, 11126–11130.
8. Partensky, F., Hoepffner, N., Li, W.K.W., Ulloa, O. and Vaulot, D. (1993) Plant Physiol. 101, 285–296.
9. Moore, L.R., Goericke, R. and Chisholm, S.W. (1995) Mar. Ecol. Prog. Ser. 116, 259–275.
10. Bryant, D.A. (1992) Curr. Biol. 2, 240–242.
11. Shimada, A., Kanai, S. and Maruyama, T. (1995) J. Mol. Evol. 40, 671–677
12. Delwiche, C.F. and Palmer, J.D. (1996) Mol. Biol. Evol. 13, 873–882.
13. Apt, K.E., Collier, J.L. and Grossman, R. (1995) J. Mol. Biol. 248, 79–96.
14. Grossman, A.R., Bhaya, D., Apt, K.E. and Kehoe, D.M. (1995) Annu. Rev. Genet. 29, 231–288.
15. Wilbanks, S.M. and Glazer, A.N. (1993a) J. Biol. Chem. 268, 1226–1235.
16. de Lorimier, R., Chen, C.C. and Glazer, A.N. (1992) Plant Mol Biol 20. 353–356.
17. Wilbanks, S.M. and Glazer, A.N. (1993b) J. Biol. Chem. 268, 1236–1241.
18. Scanlan, D.J., Hess, W.R., Partensky, F., Newman, J. and Vaulot, D. (1996) Eur. J. Phycol. 31, 1–9.
19. Thompson, J.D., Higgins, D.G. and Gibson, T.J. (1994) Nucl. Acids Res. 22, 4673–4680.
20. Felsenstein, J. (1993) PHYLIP (Phylogeny Inference Package) version 3.572c. Department of Genetics, University Washington, Seattle.
21. Strimmer, K. and von Haeseler, A. (1996) Mol. Biol. Evol. 13, 964–969.
22. Wilbanks, S.M., de Lorimer, R. and Glazer, A.N. (1991) J. Biol. Chem. 266, 9535–9539.
23. Sidler, W.A. (1994) In: Bryant, D.A. (ed.) *The Molecular Biology of Cyanobacteria,* Kluwer Academic Publishers, Dordrecht, The Netherlands, pp. 139–216.
24. Adachi, J. and Hasegawa, M. (1996) J. Mol. Evol. 42, 459–468.
25. Marquardt, J., Senger, H., Miyashita, H., Miyachi, S. and Morschel, E. (1997) FEBS Lett. 410, 428–432.
26. Felsenstein, J. (1985) Evolution 39, 783–791.
27. Jones, D.T., Taylor, W.R. and Thornton, J.M. (1992) CABIOS 8, 275–282.

89

A PHYLOGENETICALLY ORIENTED TAXONOMY OF ANOXYGENIC PHOTOTROPHIC BACTERIA

J. F. Imhoff

Institut für Meereskunde
Universität Kiel
Düsternbrooker Weg 20, D-24105 Kiel

INTRODUCTION

Historically, the taxonomy of phototrophic bacteria, like all other bacteria, has been based on easily recognizable phenotypic properties (Pfennig and Trüper, 1974). Although the ability to perform photosynthesis together with the possession of a photosynthetic apparatus and photosynthetic pigments has since Molisch (1907) been considered as primary criterion to distinguish phototrophic from non-phototrophic bacteria and it was implicated that phototrophic bacteria might constitute a separate phylogenetic line, it always had been clearly recognized that the taxonomic system had a practical background and did not constitute a phylogenetic system. Indeed, phylogenetic information became available only with the establishment of sequencing techniques for proteins and nucleic acids on the basis of sequences of cytochromes c (Ambler et al., 1979) and oligonucleotides obtained from 16S rRNA digestion with specific endonucleases (Gibson et al., 1979; Fowler et al., 1984; Stackebrandt et al., 1984), later by complete sequences of the 16S rRNA gene.

It became obvious already from these early investigations on the phylogeny that phototrophic bacteria are found in several major lines of phylogenetic descent (Fox et al., 1980; Woese, 1987), in part isolated from and in part intermixed with non-phototrophic bacteria. Most interesting, major taxonomic groups of phototrophic bacteria, including oxygenic as well as anoxygenic ones, formed different phylogenetic branches. This means that on the basis of phenotypic properties major phylogenetic branches had been recognized. These groups are i) *Chloroflexus* and relatives, ii) Chlorobiaceae, iii) phototrophic purple bacteria and iv) Cyanobacteria. In addition, *Heliobacterium chlorum* and its relatives were found to

The Phototrophic Prokaryotes, edited by Peschek *et al.*
Kluwer Academic / Plenum Publishers, New York, 1999.

represent an additional phylogenetic line located in the branch of the Gram-positive bacteria (Woese et al., 1985). These major groups are distinguished on the basis of a number of well recognized properties such as bacteriochlorophyll structure, structure and function of the photosynthetic apparatus, carotenoid components, and important physiological properties (e.g. biosynthetic pathway of carbon dioxide fixation, see Table 1). Among the phototrophic purple bacteria separate phylogenetic lines are represented by the purple sulfur bacteria (Chromatiaceae and Ectothiorhodospiraceae) in the gamma-Proteobacteria and by the purple nonsulfur bacteria in the alpha-Proteobacteria and the beta-Proteobacteria. Chromatiaceae Bavendamm 1924 and Ectothiorhodospiraceae Imhoff 1984, which can phenotypically easily be differentiated on the basis of elemental sulfur deposition, had already been distinguished by Pelsh (1937) as "Endothiorhodaceae" and "Ectothiorhodaceae". The only "phylogenetic misclassification" between these major groups was the unification of beta-Proteobacteria (*Rhodocyclus purpureus* and relatives) and the diverse assemblage of alpha-Proteobacteria within the Rhodospirillaceae (Pfennig and Trüper, 1971). The recognition of this "phylogenetic misclassification" in regard to group and genus association has led to the transfer of *Rhodospirillum tenue* to *Rhodocyclus tenuis* (Imhoff et al., 1984) and of *Rhodopseudomonas gelatinosa* to *Rhodocyclus gelatinosus* (Imhoff et al., 1984) and later to *Rubrivivax gelatinosus* (Willems et al., 1991). In addition, the use of the family name Rhodospirillaceae was abandoned.

Despite of this high principal correlation of the taxonomic and the phylogenetic system at the higher level, data on the internal phylogenetic structure of these groups became available quite slowly and revealed considerable discrepances between phylogenetic relationship and taxonomic treatment at the genus and species level of purple nonsulfur bacteria (Kawasaki et al. 1993), Chlorobiaceae (Overmann and Tuschack, 1997), and Chromatiaceae (Fowler et al., 1984).

In the following I will review the situation of the Chlorobiaceae (green sulfur bacteria), of the Ectothiorhodospiraceae and Chromatiaceae (purple sulfur bacteria) and of the purple nonsulfur bacteria with special emphasis on the genus *Rhodospirillum* and discuss selected aspects of their current and prospected taxonomic state with the aim of achieving a phylogenetic classification. With the exception of the Ectothiorhodospiraceae (Imhoff and Süling, 1996) none of these groups and families is at present taxonomically in accord with their 16S rDNA defined phylogenetic relationship.

CHLOROBIACEAE

Differentiation of Genera

So far differentiation of the genera of Chlorobiaceae is based on easily recognizable phenotypic properties such as presence of gas vesicles, motility and cell morphology (shape, presence of prosthecae) as principal characters (Pfennig, 1989). However, comparison of the 16S rDNA sequences of representatives of the Chlorobiaceae revealed a phylogenetic relatedness of these bacteria which is not in agreement with their current taxonomic classification (Overmann and Tuschak, 1997; see Figure 1). Differentiation of the genera of the Chlorobiaceae on the basis of phenotypic properties that are in congruence with their phylogenetic relationship on the basis of sequence information at present is quite difficult and more information is necessary regarding determinative properties that are in accord with the phylogenetic data.

Table 1. Diagnostic properties of major groups of phototrophic prokaryotes

	Chlorobiaceae	Chloroflexaceae	Purple bacteria	Heliobacteria	Cyanobacteria
Photosynthesis	anoxygenic	anoxygenic	anoxygenic	anoxygenic	oxygenic
Kind of Chl	bchl c, d, e (+ bchl a)	bchl c, d (+ bchl a)	bchl a, b	bchl g	chl a
Phycobilins	-	-	-	-	+
Type of RC	I	II	II	I	I + II
Location of antenna	chlorosomes	chlorosomes	ICM	CM	ICM phycobilisomes
Autotrophic CO_2-fixation	red. TCC	Hydroxypropionate-pathway	Calvincyclus	-	Calvincyclus
Preferred e-donor	H_2S, H_2	org. compounds (H_2S)	H_2S, H_2 org. compounds	org. compounds	H_2O
Chemotrophic growth	-	-	+/-	-	-/(+)

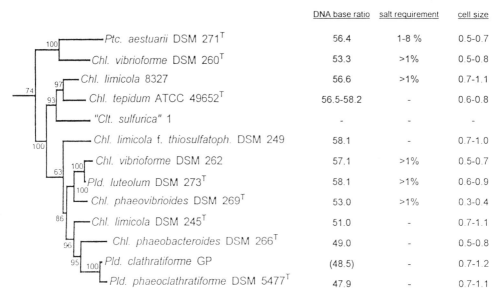

Figure 1. Phylogenetic tree of the green sulfur bacteria (adapted from Overmann and Tuschack, 1997) together with the DNA base ratio (mol% G + C), the salt requirement (% NaCl) and the cell size (μm).

It is now quite clear that the presence of gas vesicles is not a genus determinative property. All *Pelodictyon* species are closely related to *Chlorobium* species and therefore should be regarded as species of this genus. The *Pelodictyon* species *Pld. luteolum*, *Pld. clathratiforme* and *Pld. phaeoclathratiforme* should be transferred to *Chlorobium luteolum* comb. nov., *Chlorobium clathratiforme* comb. nov. and *Chlorobium phaeoclathratiforme* comb. nov., respectively. The last two species are highly similar (more than 99% sequence similarity) and could possibly be considered as a single species. *Pld. luteolum* has a comparably high similarity to *Chl. vibrioforme* DSM 262 and could possibly be combined with this species unless clear and stable phenotypic properties distinguish the two.

Differentiation of Species

From the analysis of all available informations it becomes obvious that besides 16S rDNA sequences the G + C ratio of the DNA and the salt requirement are properties relevant for species recognition. Other properties that might prove to be of significance for a phylogenetic taxonomy are lipid and fatty acid compositions (see Imhoff, 1988). 16S rDNA sequences also have shown that in several cases strains have not been correctly assigned to the species (Overmann and Tuschak, 1997). Major problems arose with *Chl. vibrioforme* and *Chl. limicola f. thiosulfatophilum*.

The type strain of *Chl. vibrioforme* DSM 260 is most similar to *Ptc. aestuarii* and according to its phylogenetic position certainly can not be regarded as a *Chlorobium* species. It is characterized by the low G + C ratio of 53.5 mol%, compared to 57.1 mol% by *Chl. vibrioforme* DSM 262. Taxonomic consequences have to be taken. The phylogenetic position of *Chl. vibrioforme f. thiosulfatophilum* DSM 265 is not yet resolved, though on the basis of the low G + C content (53.5 mol%) it is likely to be related to *Chl. vibrioforme* DSM 260.

Most complicated is the taxonomic situation with *Chl. limicola f. thiosulfatophilum*. The original description of thiosulfate utilizing strains as *Chl. thiosulfatophilum* (Larsen, 1952) was not maintained because of the heterogeneity in the G + C ratios of the various strains that could make use of thiosulfate (Mandel et al., 1971). Instead, a recognition as a subspecies of *Chl. limicola, Chl. limicola f. thiosulfatophilum*, was proposed (Pfennig and Trüper, 1974). Lateron it was found that some strains of *Chl. limicola f. thiosulfatophilum* are not really rod-shaped but curved. This was the reason to consider e.g. strain NCIB 8327 (Lascelles) as *Chl. vibrioforme f. thiosulfatophilum*. Although the catalogue of the DSM and part of the literature have taken this option, to my knowledge this change has never adequately been published. The species designations of this strain are particularly confusing (see below).

Strains recognized as *Chl. limicola* have fairly close values of their G + C ratios (51.0–52.0 mol%). However, strains known as *Chl. limicola f. thiosulfatophilum* belong to two to three different G + C clusters with values of 58.1%, of 56.1–57.1% and of approx. 52.5%. In view of the available data, only strains of the last group could possibly be recognized as strains of *Chl. limicola* or as a subspecies thereof. The type strain of *Chl. limicola f. thiosulfatophilum* DSM 249 is phylogenetically distantly related to *Chl. limicola* and should be regarded as a separate species. Also on the basis of the high G + C ratio of 58.1 mol% it is clearly distinguished from the type strain of *Chl. limicola* DSM 245 with 51.0 mol%. The strain of Lascelles (NCIB 8327) known as *Chl. limicola f. thiosulfatophilum* and recently also as *Chl. vibrioforme f. thiosulfatophilum* is not a *Chl. limicola* and is phylogenetically distinct from the type strain of *Chl. limicola f. thiosulfatophilum* (and also from *Chl. vibrioforme*). It is distinguished from the two other bacteria by an intermediate G + C ratio of 56.6 mol%, the requirement for at least 1% NaCl and the presence of bchl d. It should also be regarded as a separate species.

Among others, the following taxonomic consequences have to be taken: *Chl. vibrioforme* DSM 260 (the type strain of this species) should be transferred to a new genus and species. *Chl. limicola f. thiosulfatophilum* DSM 249 should be redefined and transferred to a new species and also "*Chl. limicola f. thiosulfatophilum*" NCIB 8327 should be classified as a new species.

ECTOTHIORHODOSPIRACEAE

Ectothiorhodospiraceae represent a group of haloalkaliphilic purple sulfur bacteria that form a separate line of phylogenetic descent related to the Chromatiaceae, as was first demonstrated by analysis of their rRNA oligonucleotide catalogues (Stackebrandt et al. 1984). On the basis of physiological and available molecular information the genus *Ectothiorhodospira* had been removed from the Chromatiaceae into the new family of the Ectothiorhodospiraceae (Imhoff 1984a). Both families form distinct groups within the gamma-Proteobacteria (Fowler et al. 1984; Stackebrandt et al. 1984; Imhoff and Süling, 1996).

A number of earlier studies on the molecular systematics of Ectothiorhodospiraceae, which used 16S rRNA cataloguing (Stackebrandt et al. 1984; Oren et al. 1989), DNA-DNA and rRNA-DNA hybridization (Ivanova et al. 1985), analysis of lipopolysaccharides (Zahr et al. 1992), DNA restriction patterns (Ventura et al. 1993), and quinone composition (Imhoff 1984b; Ventura et al. 1993) could not sufficiently resolve the phylogenetic and taxonomic relations within the Ectothiorhodospiraceae because only few strains had been included or because the methods employed did not give the required high resolution.

Therefore, recently sequences of the 16S rDNA have been determined from all type strains of the recognized *Ectothiorhodospira* species and a number of additional strains (Imhoff and Süling, 1996). These data resolved the phylogenetic relations of the whole family, confirmed the established species, and improved the classification of strains of uncertain affiliation. On the basis of sequence similarities and by a number of characteristic signature sequences, two major phylogenetic groups could be recognized and classified as separate genera. The extremely halophilic species were removed from the genus *Ectothiorhodospira* and reassigned to the new genus *Halorhodospira,* including the species *Halorhodospira halophila, Halorhodospira halochloris* and *Halorhodospira abdelmalekii.* Among the slightly halophilic species, the classification of strains belonging to *Ectothiorhodospira mobilis* and *Ectothiorhodospira shaposhnikovii* was improved and the close relationship between *Ect. shaposhnikovii* and *Ect. vacuolata* was demonstrated. Several strains which had been tentatively identified as *Ectothiorhodospira mobilis* formed a separate cluster on the basis of their 16S rDNA sequences and were recognized as two new species: *Ectothiorhodospira haloalkaliphila,* which includes the most alkaliphilic strains originating from strongly alkaline soda lakes and *Ectothiorhodospira marina,* describing isolates from the marine environment (Imhoff and Süling, 1996).

In addition to significant differences of 16S rDNA sequence (similarities of 87.2–89.9%) and the characteristic signature sequences a differentiation of *Halorhodospira* and *Ectothiorhodospira* species is possible on the basis of other molecular and physiological properties. *Ectothiorhodospira* species have MK-7 and either Q-7 or Q-8 as major components, whereas *Halorhodospira* species do not contain significant proportions of homologs with 7 isoprenoid units, but have MK-8 and Q-8 together with a short chain MK component as major components (Imhoff, 1984b; Ventura et al., 1993). Both genera also form separate groups according to their fatty acid composition and are distinguished due to their growth requirements, in particular their salt requirement (Thiemann and Imhoff 1996; Imhoff and Süling, 1996). While *Halorhodospira* species are extremely halophilic bacteria and do not grow below 10% total salts, *Ectothiorhodospira* species have growth optima well below 10% and only strains of *Ect. haloalkaliphila* tolerate and grow up to 15% total salts, although their salt optima are much lower.

The taxonomy of bacteria belonging to the Ectothiorhodospiraceae is in accord with their phylogeny.

CHROMATIACEAE

Sequences of 16S rDNA of Chromatiaceae have been analysed so far to a much lesser extent. First surveys on these bacteria have been made by comparison of oligonucleotide catalogues (Fowler et al., 1984). Conclusions in regard to the phylogenetic relatedness of currently known species and genera within this group have so far not been possible. Sequences of a small number of species have been determined only recently (see Guyoneaud et al., 1997; Caumette et al., 1997). 16S rDNA sequences of most Chromatiaceae species have now been determined and are used to characterize their phylogenetic relationship and to initiate a taxonomic revision (Imhoff et al., in press; Guyoneaud et al., 1998). The results are shortly summarized in the following. Three major phylogenetic branches are resolved within the Chromatiaceae.

The first branch consists of *Chromatium* species which specifically and characteristically have been isolated from marine and hypersaline environments: *Chr. salexigens* (DSM 4395[T]), *Chr. glycolicum* (DSM 11080[T]), *Chromatium* species A (the isolate has been described as *Rhodobacter marinus*, T. Matsunaga, pers. communication), *Chr. gracile* (DSM 203^T) and *Chr. purpuratum* (DSM 1591^T).

There is a specific relation of this group to further species from the marine environment, to *Rhabdochromatium marinum*, *Thiorhodovibrio winogradskyi*, *Thiocapsa halophila*, *Thiorhodococcus minor* and *Chr. buderi* as well as to *Thiocapsa pfennigii*, which are distant, however, at a level that justifies their recognition as separate genera.

The second branch contains species that originate from freshwater habitats in most instances, though individual strains and species also have been found in brackish water and in marine environments. (In some of these cases species identity and salt requirement still have to be proven.) On the basis of the sequence data this branch could be subdivided. *Chr. warmingii* (5810 = DSM 173^T), *Chr. vinosum* (DSM 180^T) and *Chr. minutissimum* (DSM 1376^T) form one group and *Chr. minus* (5710 = DSM 178^T) together with *Thiocystis violacea*, *Thiocystis gelatinosa* and *Chr. violascens* (DSM 168^T) form a second group. *Chr. minus* is closely related to *Thiocystis gelatinosa*. *Chr. tepidum* (DSM 3771^T) and *Chr. okenii* are included in this branch but not specifically related to either of these two groups.

The third branch of the Chromatiaceae contains species assigned to the genera *Thiocapsa* and *Amoebobacter*, with the exception of *Tca. halophila* and *Tca. pfennigii*. Species of this group, *Amb. pedioformis* (DSM 3802^T) and *Amb. purpureus* (DSM 4197^T), *Amb. pendens*, *Amb. roseus* and *Thiocapsa roseopersicina* (DSM 217^T) are characteristically found in freshwater habitats, but a number of isolates from coastal environments do exist (Guyoneaud *et al.* 1998).

Quite significant is the separation of presently known *Chromatium* species into two of these branches: one exclusively contains halophilic species, the other primarily freshwater isolates. *Chr. vinosum* has been isolated from freshwater, brackish water and marine habitats. However, so far identity on the species level between isolates from marine and freshwater habitats is lacking. Further studies have to prove whether differences in the G + C content, which varies considerably in strains designated to this species, and the marine, respectively freshwater nature of these bacteria is in congruence.

In conclusion, relationships among purple sulfur bacteria of the Chromatiaceae on the basis of phylogenetic traits such as 16S rDNA sequences are not in agreement with the current taxonomic classification based on easily recognizable phenotypic properties and taxonomic rearrangements are necessary. The following changes are proposed:

1. *Chromatium minus* is most closely related to *Thiocystis gelatinosa*, but different from this bacterium at the species level. Therefore, I propose to transfer this species to the genus *Thiocystis* with the new name *Thiocystis minor* comb. nov. *Chr. violascens*.

2. *Chromatium okenii* is specifically related to the group with *Thiocystis violacea*, but on the basis of significant phenotypic properties, differences in the DNA base ratio and the 16S rDNA sequences it will constitute a separate genus.

3. *Chr. vinosum*, *Chr. minutissimum* and *Chr. warmingii* are closely related species that should be placed into a new genus. Because *Chr. vinosum* is among the best studied purple sulfur bacteria and the name *Chromatium* is very much associated with this species, we propose as the new genus name *Allochromatium* gen. nov. ("the other Chromatium"), with *Allochromatium vinosum* comb. nov. as the type species and *Allochromatium minutissimum* comb. nov. and *Allochromatium warmingii* comb. nov. as additional species of the genus.

4. Because *Chr. tepidum* quite obviously can not be retained within the genus *Chromatium* and it is neither specifically related to *Thiocystis* and to *Allochro-*

matium, it should be placed in a new genus as well. It is the only thermophilic isolate of the purple sulfur bacteria described so far. Therefore, we propose *Thermochromatium* gen. nov. as the new genus name, although it is so far not well established whether the temperature response will be a usefull criterion for genus differentiation. *Thermochromatium tepidum* comb. nov. will be the new name.

5. *Chr. salexigens* and *Chr. glycolicum* shall be transferred to the new genus *Halochromatium* gen. nov. because of their extended salt dependence and salt tolerance, which is the highest within the Chromatiaceae, as *Halochromatium salexigens* comb. nov. and *Halochromatium glycolicum* comb. nov.

6. *Chr. gracile*, *Chr. purpuratum* and another new isolate (originally described as *Rhodobacter marinus*) are transferred to the new genus *Marichromatium*. The name is proposed according to the marine origin and a moderate salt dependence of the two species as is found in typical marine bacteria. The new names *Marichromatium gracile* comb. nov., *Marichromatium purpuratum* comb. nov. and *Marichromatium marinum* comb. nov. are proposed for these two species.

7. *Thiocapsa pfennigii* certainly is not a species of the genus *Thiocapsa* (with *Thiocapsa roseopersicina* as type species), because it is phylogenetically significantly different from the type species of this genus and belongs to the first branch of Chromatiaceae species. It represents a separate evolutionary line and is not closely related to any of the other known Chromatiaceae species. Therefore, it is proposed to place it into a separate genus. The name *Thiococcus* nom. rev., originally used by Eimhjellen (Eimhjellen et al., 1967) to describe this bacterium (but not validly published) is revived and *Thiocapsa pfennigii* transferred to this genus with *Thiococcus pfennigii* gen. nom. rev., comb. nov. as the new name.

8. In addition, *Thiocapsa halophila* and three of the known *Amoebobacter* species, *Amb. pendens*, *Amb. roseus*, *Amb. pedioformis* need taxonomic reclassification (see Guyoneaud *et al.*, 1998; Imhoff *et al.*, 1998b).

PURPLE NONSULFUR BACTERIA

The purple nonsulfur bacteria classified according to morphological and physiological properties had been assigned to three genera, the spiral-shaped *Rhodospirillum*, the rod-shaped *Rhodopseudomonas* and the stalk-forming *Rhodomicrobium* (Pfennig and Trüper, 1974).

With the availability of sequence information and the recognition of their high phylogenetic diversity numerous taxonomic transfers have been proposed. With the background of similarities of oligonucleotide catalogues obtained from 16S rRNA analyses (Gibson et al., 1979) and additional molecular and phenotypic properties some species were assigned to the new genus *Rhodobacter* and *Rhodopseudomonas globiformis* was transferred to *Rhodopila globiformis* (Imhoff et al., 1984). In addition, *Rhodocyclus tenuis* that originally had been recognized as a *Rhodospirillum* species (*Rhodospirillum tenue*, Pfennig 1969), but belongs to the beta-Proteobacteria was removed from this genus (Imhoff et al., 1984). Later, according to complete sequence comparison *Rhodobacter* species were divided into two branches of marine and of freshwater species. The marine species were transferred to the genus *Rhodovulum* (Hiraishi and Ueda, 1994a). Furthermore, *Rhodopseudomonas marina* was transferred to *Rhodobium marinum* (Hiraishi et al., 1995), *Rhodopseudomonas roseus* to *Rhodoplanes roseus* (Hiraishi and Ueda, 1994b) and *Rho-*

dopseudomonas viridis and *Rps. sulfoviridis* to *Blastochloris viridis* and *Blastochloris sulfoviridis* (Hiraishi, 1997). All of these changes were in accord with the phylogenetic distance of these bacteria and supported by differentiating phenotypic characteristics.

Spiral-Shaped Purple Nonsulfur Bacteria of the Alpha-Proteobacteria, the Genus Rhodospirillum

Like the rod-shaped purple nonsulfur bacteria (*Rhodopseudomonas* Pfennig and Trüper, 1974), spiral-shaped purple nonsulfur bacteria known as *Rhodospirillum* species are phylogenetically extremely heterogeneous and intermixed with chemoheterotrophic non-phototrophic bacteria. At present the genus *Rhodospirillum* comprises seven species, with *Rhodospirillum rubrum* as the type species. In addition, *Rhodocista centenaria*, that was originally described as a species of the genus *Rhodospirillum* (*Rhodospirillum centenum*, Favinger et al., 1989) has later been recognized as a species of the new genus *Rhodocista*, primarily on the basis of significant differences in the rRNA gene sequences (Kawasaki et al., 1992). Another new spiral-shaped species is *Rhodospira trueperi*, which has been described recently as a new species and a new genus on the basis of phenotypic and genotypic properties (Pfennig et al., 1997).

It became obvious already from the work of Kawasaki et al. (1993) that, based on 16S rRNA gene sequences, the recognized species of the genus *Rhodospirillum* of the α-Proteobacteria are phylogenetically quite distantly related to each other and do not warrant classification in one and the same genus. On the basis of 16S rDNA sequences of all recognized bacteria of the genus *Rhodospirillum*, including that of *Rhodospirillum mediosalinum* a reclassification of these bacteria according to their phylogenetic relationship has been proposed (Imhoff et al., 1998).

According to the sequence dissimilarity between the recognized *Rhodospirillum* species different genera are proposed for the *Rhodospirillum fulvum/Rhodospirillum molischianum* group, for *Rhodospirillum rubrum* and *Rhodospirillum photometricum*, for *Rhodospirillum mediosalinum*, for *Rhodospirillum salexigens* and for the group of *Rhodospirillum salinarum* and *Rhodospirillum sodomense*. None of these groups shares sequence similarities higher than 91% with others of these spiral-shaped purple nonsulfur bacteria.

This differentiation is supported by several phenotypic properties, such as major quinone components, fatty acids composition and salt requirement. On the basis of major quinone components and fatty acid composition all proposed new genera of the spiral-shaped purple nonsulfur bacteria can be distinguished. Four of these genera were defined as salt-dependent and three as fresh water bacteria. The salt requirement is considered as a genus specific property of these bacteria, which is in accord with different phylogenetic lines forming fresh water and salt water representatives. Although the structure of the intracytoplasmic membrane system is not considered of primary importance to differentiate the genera, according to their ultrastructure identical internal membrane systems are present in most of the proposed genera (with the exception of *Rhodospirillum*). The DNA base ratio is fairly similar in all genera with values between 63–70 mol% G + C and in this case therefore not of high diagnostic value. Similar, the growth factor requirement varies significantly from species to species and can not be regarded as a suitable tool to differentiate between the genera.

From the available data it is obvious, that phylogenetic relations on the basis of 16S rDNA sequences of these bacteria are in good correlation with differences in major quinone and fatty acid composition and in accord with the requirement for NaCl or sea salt for growth. Therefore, these properties have to be considered of primary importance in defining and differentiating the genera. Accordingly, the majority of these species have to be placed into different genera and only *Rhodospirillum rubrum* and *Rhodospirillum photometricum* belong to the genus *Rhodospirillum*. Following rule 39a of the "International Code of Nomenclature of Bacteria" the genus name *Rhodospirillum* has to be maintained with the type species of this genus, which is *Rhodospirillum rubrum*.

In contrast to the clear separation of the proposed genera and the large phylogenetic distance of them, the species *Rsp. fulvum* and *Rsp. molischianum* are highly similar on the basis of their 16S rDNA sequences and hardly qualify as separate species. The same holds for the couple *Rsp. sodomense* and *Rsp. salinarum*. On the basis of their phenotypic properties, however, they are well recognized species (Trüper and Imhoff, 1989; Nissen and Dundas, 1984; Mack et al., 1993), which should be maintained. The following taxonomic changes have been proposed:

1. Transfer of *Rsp. fulvum* and *Rsp. molischianum* to the genus *Phaeospirillum* nov. gen. as the new combinations *Phaeospirillum fulvum* comb. nov. and *Phaeospirillum molischianum* comb. nov.
2. Transfer of *Rhodospirillum salinarum* and *Rhodospirillum sodomense* to the new genus *Rhodovibrio* as *Rhodovibrio salinarum* comb. nov. and *Rhodovibrio sodomensis* comb. nov.
3. Transfer of *Rhodospirillum salexigens* to the new genus *Rhodothalassium* gen. nov. as *Rhodothalassium salexigens* comb. nov.
4. Transfer of *Rhodospirillum mediosalinum* to the new genus *Roseospira* gen. nov. as *Roseospira mediosalina* comb. nov.

CONCLUSIONS

There was a great excitement among taxonomists due to the large discrepances revealed by the first phylogenetic analysis of phototrophic purple bacteria (Gibson et al., 1979; Fowler et al., 1984) to the existing taxonomic classification. This excitement was in particular due to the apparent lack of phenotypic properties that were in congruence with the phylogenetic relations of these bacteria. Available morphological as well as principal physiological properties allowed a differentiation with respect to phylogeny only to a limited extent or were not suitable for this purpose. In contrast, many examples are now known and have been outlined above, where major quinones, fatty acid and lipid composition are valuable taxonomic tools that support a phylogenetic classification. Most important, the salt requirement proved to be a property of phylogenetic relevance, i.e. marine and halophilic phototrophic bacteria form separate lines of descent on the family level (Ectothiorhodospiraceae), on the genus level (*Halorhodospira*, *Ectothiorhodospira*, *Rhodovulum*, *Prosthecochloris* and a number of new proposed genera as outlined above) and on the subgenus level, as examplified within the genus *Chlorobium*. Finally, also the DNA base ratio expressed as the G + C content, that always had been regarded as a taxonomically important parameter, is of relevance for a phylogenetic classification. In several cases large differences in the G + C content have been ignored or could not be correlated with other parameters, so that taxonomic consideration was difficult. However, with the sequence information as background the G + C content is a valuable additional information. Despite of this first success on defining phenotypic properties that are in accord with phylogenetic relationships, the search for more of such properties has to be continued with emphasis and the phylogenetically relevant properties have to be analysed in the known species with consequence.

REFERENCES

Ambler, R.P., Daniel, M., Hermoso, J., Meyer, T.E., Bartsch, R.G. and Kamen, M.D. 1979. Cytochrome c_2 sequence variation among the recognized species of purple nonsulfur photosynthetic bacteria. Nature (London) **278**: 659–660.

Caumette, P., Imhoff, J.F., Süling. J. and Matheron, R. 1997. *Chromatium glycolicum* sp. nov., a moderately halophilic purple sulfur bacterium that uses glycolate as substrate. Arch. Microbiol. **167**: 11–18.

Eimhjellen, K.E., Steensland, H. and Traetteberg, J. 1967. A *Thiococcus* sp. nov. gen., its pigments and internal membrane system. Arch. Mikrobiol. **59**: 82–92.

Favinger, J., Stadtwald, R. and Gest, H. 1989. *Rhodospirillum centenum*, sp. nov., a thermotolerant cyst-forming anoxygenic photosynthetic bacterium. Antonie van Leeuwenhoek **55**: 291–296.

Fowler, V.J., Pfennig, N., Schubert, W., Stackebrandt, E. 1984. Towards a phylogeny of phototrophic purple sulfur bactria - 16S rRNA oligonucleotide cataloguing of 11 species of Chromatiaceae. Arch. Microbiol. **139**: 382–387.

Fox, G.E., Stackebrandt, E., Hespell, R.B., Gibson, J., Maniloff, I., Dyer, T.A., Wolfe, R.S., Balch, W.E., Tanner, R.S., Magrum, L. J., Zablen, L.B., Blackemore. R., Gupta, R., Bonen, L., Lewis, B.J., Stahl, D.A., Luehrsen, K.R., Chen, K.N. and Woese, C.R. 1980. The phylogeny of prokaryotes. Science **209**: 457–463.

Gibson, J., Stackebrandt, E., Zablen, L.B., Gupta, R. and Woese, C.R. 1979. A phylogenetic analysis of the purple photosynthetic bacteria. Curr. Microbiol. **3**: 59–64.

Guyoneaud, R., Matheron, R., Liesack, W., Imhoff, J.F. and Caumette, P. 1997. *Thiorhodococcus minus*, gen. nov., sp. nov., a new purple sulfur bacterium isolated from coastal lagoon sediments. Arch. Microbiol. **168**: 16–23.

Hiraishi, A. and Ueda, Y. 1994a. Intrageneric structure of the genus *Rhodobacter:* Transfer of *Rhodobacter sulfidophilus* and related marine species to the genus *Rhodovulvum* gen. nov. Int. J. Syst. Bacteriol. **44**: 15–23.

Hiraishi, A. and Ueda, Y. 1994b. *Rhodoplanes* gen. nov., a new genus of phototrophic bacteria including *Rhodopseudomonas rosea* as *Rhodoplanes roseus* comb. nov. and *Rhodoplanes elegans* sp. nov. Int. J. Syst. Bateriol. **44**: 665–673.

Hiraishi, A., Urata, K. and Satoh, T. 1995. A new genus of marine budding phototrophic bacteria, *Rhodobium* gen. nov., which includes *Rhodobium orientis* sp. nov. and *Rhodobium marinum* comb. nov. Int. J. Syst. Bacteriol. **45**: 226–234.

Hiraishi, A. 1997. Transfer of the bacteriochlorophyll *b*-containing phototrophic bacteria *Rhodopseudomonas viridis* and *Rhodopseudomonas sulfoviridis* to the genus *Blastochloris* gen. nov. Int. J. Syst. Bacteriol. **47**: 217–219.

Imhoff J.F. 1984a. Reassignment of the genus *Ectothiorhodospira* Pelsh 1936 to a new family, *Ectothiorhodospiraceae* fam. nov., and emended description of the *Chromatiaceae* Bavendamm 1924. Int. J. Syst. Bacteriol. **34**: 338–339.

Imhoff, J.F. 1984b. Quinones of phototrophic purple bacteria. FEMS Microbiol Lett **256**: 85–89.

Imhoff, J.F. 1988. Lipids, fatty acids and quinones in taxonomy and phylogeny of anoxygenic phototrophic bacteria. In: Green Photosynthetic Bacteria. J.M. Olson (ed.), Plenum Publ. Corp., pp. 223–232.

Imhoff J.F. 1989. Family Ectothiorhodospiraceae. In: Bergey's Manual of Systematic Bacteriology, 1st ed., vol. 3, J.T. Staley, M.P. Bryant, N. Pfennig, J.G. Holt (eds.), Williams & Wilkins, Baltimore, 1654–1658.

Imhoff, J.F. and Süling, J. 1996. The phylogenetic relationship among Ectothiorhodospiraceae. A reevaluation of their taxonomy on the basis of rDNA analyses. Arch. Microbiol. **165**: 106–113.

Imhoff, J.F., Trüper, H.G. and Pfennig, N. 1984. Rearrangement of the species and genera of the phototrophic "purple nonsulfur bacteria". Int. J. Syst. Bacteriol. **34**: 340–343.

Ivanova, T.L., Turova, T.P. and Antonov, A.S. 1985. DNA-DNA and rRNA-DNA hybridization studies in the genus *Ectothiorhodospira* and other purple sulfur bacteria. Arch. Microbiol. **143**: 154–156.

Kawasaki, H., Hoshino, Y., Kuraishi, H. and Yamasato, K. 1992. *Rhodocista centenaria* gen. nov., sp. nov., a cyst-forming anoxygenic photosynthetic bacterium and its phylogenetic position in the *Proteobacteria* alpha group. J. Gen. Appl. Microbiol. **38**: 541–551.

Kawasaki, H., Hoshino, Y. and Yamasato, K. 1993. Phylogenetic diversity of phototrophic purple non-sulfur bacteria in the *Proteobacteria* alpha-group. FEMS: Microbiol. Lett. **112**: 61–66.

Larsen, H. 1952. On the culture and general physiology of green sulphur bacteria. J. Bacteriol. **64**: 187–196.

Mack, E.E., Mandelco, L., Woese, C.R. and Madigan, M.T. 1993. *Rhodospirillum sodomense*, sp. nov., a Dead Sea *Rhodospirillum* species. Arch. Microbiol. **160**: 363–371.

Mandel, M., Leadbetter, E.R., Pfennig, N. and Trüper, H.G. 1971. Deoxyribonucleic acid base compositions of phototrophic bacteria. Int. J. Syst. Bacteriol. **21**: 222–230.

Molisch, H. 1907. Die Purpurbakterien nach neuen Untersuchungen. G. Fischer, Jena.

Nissen, H. and Dundas, I.D. 1984. *Rhodospirillum salinarum* sp. nov., a halophilic photosynthetic bacterium isolated from a Portuguese saltern. Arch. Microbiol. **138**: 251–256.

Overmann J. and Tuschak C. 1997. Phylogeny and molecular fingerprinting of green sulfur bacteria. Arch. Microbiol. **167**: 302–309.

Oren, A., Kessel, M. and Stackebrandt, E. 1989. *Ectothiorhodospira marismortui* sp. nov., an obligatory anaerobic, moderately halophilic purple sulfur bacterium from a hypersaline sulfur spring on the shore of the Dead Sea. Arch. Microbiol. **151**: 524–529.

Pelsh, A.D. 1937. Photosynthetic sulfur bacteria of the eastern reservoir of Lake Sakskoe. Mikrobiologiya **6**: 1090–1100.

Pfennig, N. 1969. *Rhodospirillum tenue* sp. n., a new species of the purple nonsulfur bacteria. J. Bacteriol. **99**: 619–620.

Pfennig, N. 1989. Green sulfur bacteria. In: Bergey's Manual of Systematic Bacteriology, 1ˢᵗ ed., Vol. 3, J.T. Staley, M.P. Bryant, N. Pfennig, J.G. Holt (eds.), Williams & Wilkins, Baltimore, pp. 1682–1697.

Pfennig, N., Lünsdorf, H., Süling, J. and Imhoff, J.F. 1997. *Rhodospira trueperi,* gen. nov. and spec. nov., a new phototrophic Proteobacterium of the alpha-group. Arch. Microbiol. **168**: 39–45.

Pfennig, N. and Trüper, H.G. 1974. The phototrophic bacteria. In: Bergey's manual of determinative bacteriology, 8ᵗʰ ed., R.E. Buchanan and N.E. Gibbons (eds.) Williams & Wilkins, Baltimore, pp. 24–75.

Stackebrandt, E., Fowler V.J., Schubert, W., Imhoff, J.F. 1984. Towards a phylogeny of phototrophic purple sulfur bacteria - the genus *Ectothiorhodospira*. Arch. Microbiol. **137**: 366–370.

Thiemann, B. and Imhoff, J.F. 1996. Differentiation of *Ectothiorhodospiraceae* based on their fatty acid composition. Syst. Appl. Microbiol. **19**: 223–230.

Trüper, H.G. and Imhoff, J.F. 1989. The genus *Rhodospirillum*. In: Bergey's Manual of Systematic Bacteriology, 1ˢᵗ ed., vol. 3, J.T. Staley, M.P. Bryant, N. Pfennig, J.G. Holt (eds.), Williams & Willkins, pp. 1662–1666.

Ventura. S., Giovanetti, L., Gori, A., Viti. C., Materassi, R. 1993. Total DNA restriction pattern and quinone composition of members of the family Ectothiorhodospiraceae. System. Appl. Microbiol. **16**: 405–410.

Willems, A., Gillis, M. and de Ley, J. (1991). Transfer of *Rhodocyclus gelatinosus* to *Rubrivivax gelatinosus* gen. nov., comb. nov., and phylogenetic relationships with *Leptothrix, Sphaerotilus natans, Pseudomonas saccharophila,* and *Alcaligenes latus*. Int. J. Syst. Bacteriol. **41**: 65–73.

Woese, C.R. 1987. Bacterial evolution. Microbiol. Rev. **51**: 221–271.

Woese, C.R., Debrunner-Vossbrinck, B., Oyaizu, H., Stackebrandt, E. and Ludwig, W. 1985. Gram-positive bacteria: possible photosynthetic ancestry. Science **229**: 762–765.

Zahr, M., Fobel, B., Mayer, H., Imhoff, J.F., Campos, V., Weckesser, J. 1992. Chemical composition of the lipopolysaccharides of *Ectothiorhodospira shaposhnikovii, Ectothiorhodospira mobilis,* and *Ectothiorhodospira halophila*. Arch. Microbiol. **157**: 499–504.

90

CRITERIA FOR SPECIES DELINEATION IN THE *ECTOTHIORHODOSPIRACEAE*

Stefano Ventura,[1,2] Alessia Bruschettini,[1,2] Luciana Giovannetti,[1,2] and Carlo Viti[1,2]

[1]Centro di Studio dei Microrganismi Autotrofi, CNR
[2]Dipartimento di Scienze e Tecnologie Alimentari e Microbiologiche
Università di Firenze
p.le delle Cascine 27, Firenze, Italy

1. INTRODUCTION

In a recent taxonomic revision of the family *Ectothiorhodospiraceae* [1], 16S rRNA analysis and coherent physiological properties were used to delineate two distinct genera, *Ectothiorhodospira* and *Halorhodospira*. Less halophilic species have been recognized as belonging to *Ectothiorhodospira*, while highly halophilic species have been assigned to *Halorhodospira*. Key features of the family *Ectothiorhodospiraceae*, and of the two genera belonging to it have been summarized in Table 1. Essentially on the basis of 16S rRNA sequence analysis, two new species of *Ectothiorhodospira* have been described and the existing species have been validated [1]. The family *Ectothiorhodospiraceae*, as presently defined, thus houses the species *Ect. mobilis*, *Ect. shaposhnikovii*, *Ect. marina*, *Ect. marismortui*, *Ect. haloalkaliphila* and *Ect. vacuolata*; *Hlr. halophila*, *Hlr. halochloris* and *Hlr. abdelmalekii*. However, there is still incoherence between this species subdivision and some phenotypic and genotypic traits described for strains of *Ectothiorhodospiraceae*. Moreover, it is a common understanding that taxonomic relationships among highly correlated bacteria cannot be determined on the sole basis of 16S rRNA sequence similarities [2,3]. In the present contribution, the taxonomic status of the *Ectothiorhodospiraceae* have been thus reconsidered integrating phenotypic and genotypic data taken from the literature with new experimental results recently obtained in our laboratory. The following characters have been considered. Salt concentration for optimal growth [4,5]; quinone composition [6,7]; fatty acid composition [5]; lipopolysaccharide composition [8,9]; porins [10]; G+C% content [5,11–16]; DNA reassociation [17]; 16S+23S rDNA RFLP (ribotype) [17]; ARDRA (see the Results section below); 16S rRNA sequence [1].

The Phototrophic Prokaryotes, edited by Peschek *et al.*
Kluwer Academic / Plenum Publishers, New York, 1999.

Table 1. General features of the *Ectothiorhodospiraceae*

Family *Ectothiorhodospiraceae*	
• Anoxyphototrophic purple bacteria • Extracellular sulfur globules produced during sulfide oxidation • Saline and alkaline habitats	• Stacks of intracellular membranes • Polarly flagellated cells • Members of the gamma subdivision of purple bacteria, phylogenetically distinct from *Chromatiaceae*
Genus *Ectothiorhodospira*	Genus *Halorhodospira*
• Medium to low salt requirement • One polar tuft of flagella • Bacteriochlorophyll a • Carotenoids of the spirilloxanthin series • Phylogenetically distinct from *Halorhodospira*	• High salt requirement • Bipolarly flagellated • Bacteriochlorophyll a or b • Phylogenetically distinct from *Ectothiorhodospira*

2. MATERIALS AND METHODS

Amplified 16S rDNA restriction analysis (ARDRA) was performed on all type strains of *Ectothiorhodospira* species, on the type strain of *Hlr. halophila* and on several other strains belonging to *Ectothiorhodospira* species and to *Hlr. halophila*. Culture conditions and DNA extraction were as previously described [7].

In vitro amplification of 16S rDNA of *Ectothiorhodospiraceae* was obtained with primers FD1, 5'-CCGAATTCGTCGACAACAGAGTTTGATCCTGGCTCAG-3' and RD1, 5'-CCCGG-GATCCAAGCTTAAGGAGGTGATCCAGCC-3' [18]. Reaction mixture contained 67 mM Tris HCl, pH 8,8; 16 mM $(NH_4)_2SO_4$; 0,01% Tween-20; 2 mM $MgCl_2$; 0,65 M glycerol; 20 µM each of dNTPs; 0,1 µM of each of the primers; 0.5 U of *Taq* DNA polymerase and 20 ng template DNA. DNA amplifications were carried out in a Perkin-Elmer 9600 thermal cycler using the following program: initial denaturation for 3 min at 95°C; 35 cycles composed by 1 min denaturation at 94°C, 1 min annealing at 59°C, 2 min extension at 72°C each; 3 min final extension at 72°C.

10 µl of PCR products were digested for 3 hours at 37°C with 5 U of restriction endonuclease (Boehringer Mannheim) according to the supplier. Restriction fragments were separated on agarose gel (3% w/v) in TEB buffer (89 mM Tris, 2.5 mM EDTA, 89 mM boric acid, pH 8.3) for 4 hours at 3 V/cm. 1 kb DNA Ladder (Gibco BRL) was used as molecular weight marker. Gel images were captured as TIFF format files with a ccd camera (UVItec Gel Documentation System). Rescaling and normalization of electrophoretic band profiles, band detection, restriction profile comparison and clustering were performed with GelCompar 4.0 (Applied Math). Restriction profiles were compared applying the Dice similarity coefficient (S_D) and clustered with UPGMA. UPGMA clustering was also calculated from composite restriction profiles constructed appending normalized profiles obtained with single restriction enzyme digestions.

3. RESULTS

Results of ARDRA of *Ectothiorhodospira* strains are shown in Figure 1. UPGMA analysis was performed on composite band profiles constructed appending restriction profiles generated by *Cfo*I, *Alu*I, *Rsa*I, *Hae*III, *Msp*I, *Hpa*II and *Hinf*I. An overall high similarity of restriction profiles was detected ($S_D > 85\%$), as expected from strains belonging to the same genus. Nearly complete identity was detected among ARDRA profiles of

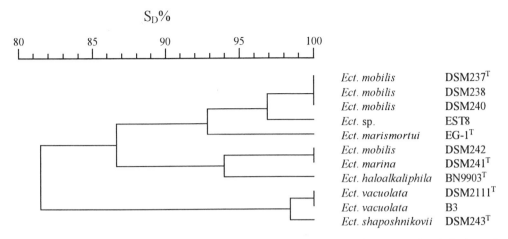

Figure 1. UPGMA dendrogram obtained from ARDRA of strains of *Ectothiorhodospira*. S_D, Dice similarity coefficient. The analysis was performed on composite band profiles constructed appending restriction profiles generated by *Cfo*I, *Alu*I, *Rsa*I, *Hae*III, *Msp*I, *Hpa*II and *Hinf*I.

strains of *Ect. vacuolata* and the type strain of *Ect. shaposhnikovii*. Profiles of *Ect. marina* DSM241[T] and *Ect. mobilis* DSM242 were identical and shared 94% similarity with *Ect. haloalkaliphila* BN9903[T]. The type strain of *Ect. mobilis* and other two strains of this species had identical profiles and were related to *Ectothiorhodospira* sp. EST8 at a 97% similarity level. This latter group of strains showed also 93% similarity with *Ect. marismortui* EG-1[T] (same strain as DSM4180[T]). Results of ARDRA are in good agreement with 16S rRNA sequence analysis [1].

ARDRA of *Halorhodospira halophila* DSM244[T], BN9624 and BN9630 were performed on composite band profiles constructed with only four restriction profiles, generated respectively by *Rsa*I, *Hpa*II, *Hinf*I or *Msp*I. Similarity values among the three strains, ranging from 88.5 to 92.5%, were slightly lower than corresponding values detected among strains of the same species in the genus *Ectothiorhodospira* and also lower than similarities among 16S rRNA sequences of the same strains [1].

4. DISCUSSION

In order to apply a polyphasic approach to the taxonomy of *Ectothiorhodospiraceae*, experimental results presented in this contribution and other data of taxonomic relevance, published in several studies dealing with members of the *Ectothiorhodospiraceae* and summarized in the following paragraphs, have been considered.

Strains of *Ectothiorhodospiraceae* have been subdivided according to salt concentration for optimal growth [4,5]. Salt optima of strains of *Ectothiorhodospira* were always below 10% NaCl. Strains of *Halorhodospira* were subdivided in two groups. Some extremely halophilic strains of *Hlr. halophila* (BN9622, BN9623, BN9624, BN9625, BN9626, BN9627) showed salt optima around 25%; other strains of this species (DSM244[T], BN9620, BN9621, BN9628, BN9630, BN9631) and strains of *Hlr. halochloris* and *Hlr. abdelmalekii* were far less halophilic and had optima around 15%.

Table 2. Groups of species of *Ectothiorhodospiraceae*
based on quinone and fatty acid content

Species/strain	Principal quinone[a]	Fatty acid cluster[b]
Ect. vacuolata	Q7	VI
Ect. shaposhnikovii	Q7	VI
Ect. marina	Q8	VII
Ect. haloalkaliphila	Q8	VII
Ect. mobilis	MK7	V
Ect. marismortui	MK7	V
Ectothiorhodospira sp. EST8	MK7	V
Hlr. halophila (25% salt optima)	MK8	I
Hlr. halophila (15% salt optima)	MK8	II
Hlr. abdelmalekii	MK5	III
Hlr. halochloris	MK5	IV

[a]Data from [6,7].
[b]Data from [5].

Strains could be subdivided into five groups on the basis of their principal quinone and of other quinones present at concentrations higher than 5% [6,7]. Three groups of *Ectothiorhodospira* strains were detected, each one containing strains belonging to two different species. Strains of *Halorhodospira* were subdivided in two groups; one containing strains of *Hlr. abdelmalekii* and *Hlr. halochloris* and another containing only strains of *Hlr. halophila* (Table 2).

Fatty acids, as studied by Thiemann and Imhoff [5], revealed a clear separation between the most halophilic and the less halophilic strains of the *Ectothiorhodospiraceae*. Seven clusters were obtained. Strains belonging to *Hlr. halophila* have been subdivided into two clusters (I and II), one cluster containing extremely halophilic strains, the other containing strains at the borderline between moderately and extremely halophilic bacteria. Among the three strains that have been included in the ARDRA experiments reported in this contribution, strain BN9624 belonged to cluster I, while strains DSM244[T] and BN9630 belonged to cluster II. Strains belonging to the other two *Halorhodospira* species, *Hlr. abdelmalekii* and *Hlr. halochloris* were placed in clusters III and IV, respectively. Species of *Ectothiorhodospira* were grouped in couples exactly in the same way as by quinone content. However, differences between clusters I and II and between clusters III and IV are noticeably smaller (Euclidean distances between 6 and 7) than differences among the other clusters (Euclidean distances > 28). Data on quinone and fatty acid contents have been schematically compared in Table 2.

Lipopolysaccharide composition was determined in *Ect. vacuolata*, *Ect. mobilis*, *Ect. shaposhnikovii* and *Hlr. halophila* by Meißner *et al.* [8] and Zahr *et al.* [9]. Several features of LPS composition differentiated the *Ectothiorhodospiraceae* from the *Chromatiaceae*; the composition of the conservative lipid A region confirmed the separation of *Ectothiorhodospiraceae* into two genera; strong similarities in fatty acid composition were detected between LPSs of *Ect. shaposhnikovii* and *Ect. vacuolata*.

Outer membrane porins were isolated and characterized only from the type strains of *Ect. vacuolata* and *Ect. shaposhnikovii* [10]. The two porins showed remarkable functional and structural similarities. In particular, the sequence of the first eighteen N-terminal amino acids was identical in the two molecules, except for positions 12 and 14. As reported by the authors, these porins were more similar to each other than the porins of two strains belonging to the same species *Rhodobacter capsulatus* that differed in five positions.

G+C% values were determined by several authors [5,11–16] and comparison is indeed difficult, since different experimental approaches were employed. Yet G+C% values of strains of all *Ectothiorhodospira* species were fairly similar, in the range 62–67%, while G+C% values of the three species of *Halorhodospira* were fairly distinctive; *Hlr. halochloris* 50.5–53%, *Hlr. abdelmalekii* 64%, *Hlr. halophila* 66.5–69.5% G+C%.

DNA reassociation among all type strains of *Ectothiorhodospira* species, the type strains of *Hlr. halophila* and *Hlr. abdelmalekii* and other strains have been recently determined in our laboratory [17], extending a previous work of Ivanova et al. [12], restricted to only a few strains. Four genospecies of *Ectothiorhodospira* were delineated at DNA reassociation values between 70 and 100%, that is the range corresponding to infra-specific relationships. A genospecies was composed by *Ect. mobilis* DSM237^T, *Ect. marismortui* EG-1^T and strain EST8, another genospecies by *Ect. vacuolata* DSM2111^T, B3 and *Ect. shaposhnikovii* DSM243^T, and the last two genospecies were represented by *Ect. marina* DSM241^T and *Ect. haloalkaliphila* BN9903^T, respectively. Anyhow, these latter two species were connected at a relatively high value of around 35% DNA similarity, that stayed constant also under stringent hybridization conditions. Under optimal hybridization conditions, inter-group genotypic similarities were in the range of 10–40%. The three examined strains of *Hlr. halophila*, DSM244^T, BN9624 and BN9630, formed a coherent genospecies, well separated from the other *Halorhodospira* species.

Results of 16S+23S rRNA gene RFLP (ribotype) [17], obtained digesting DNA of *Ectothiorhodospira* with restriction endonuclease *Ksp*I, confirmed the genotypic groups revealed by DNA reassociation. Characteristic, species specific banding patterns could be identified for strains of *Hlr. halochloris*, *Hlr. halophila* and *Ect. vacuolata*. Characteristic, nearly identical *Ksp*I ribotypes were also shared by *Ect. mobilis* DSM237^T, *Ect. marismortui* EG-1^T and *Ectothiorhodospira* sp. EST8. In *Ksp*I ribotype, resemblance at the genus level could not be detected either among *Ectothiorhodospira* species or among *Halorhodospira* species.

Carefully examining data of 16S rRNA sequence analysis determined by Imhoff and Süling [1], six clusters or genospecies could be delineated at sequence similarity higher than 97.8%. The sole existing strain of the species *Ect. marismortui* formed the one cluster, another cluster contained *Ect. marina* and *Ect. haloalkaliphila* and a third one *Ect. vacuolata* and *Ect. shaposhnikovii*, while DSM237^T, the only strain of *Ect. mobilis* included in that study, showed the lowest similarity to all the other strains of *Ectothiorhodospira*. Distantly related to *Ectothiorhodospira*, strains of *Halorhodospira* were subdivided in two groups, one corresponding to *Hlr. halophila* and the other containing *Hlr. halochloris* and *Hlr. abdelmalekii*.

The overview of taxonomic data on the *Ectothiorhodospiraceae* revealed the existence of groups of strains, belonging to different described species, having both phenotypic and genotypic coherency. This polyphasic approach delineates an alternative species subdivision of the family *Ectothiorhodospiraceae*, with less species than the nine presently described.

The two species *Ect. vacuolata* and *Ect. shaposhnikovii* have the same quinones, the same fatty acids, very similar porins and lipopolysaccharides. Their genetic relatedness is very high, as it could be deduced from G+C%, DNA reassociation, ARDRA and 16S rRNA sequence analysis. We thus confirm our proposal [17] that strains of *Ect. vacuolata* are renamed as *Ect. shaposhnikovii* and that, consequently, the binomial *Ectothiorhodos-*

pira vacuolata should be considered as a subjective synonym of *Ectothiorhodospira shaposhnikovii*.

Also *Ect. mobilis* DSM237^T, DSM238, DSM240, *Ect. marismortui* EG-1^T and strain EST8 form a tight cluster. In this case, 16S rRNA based approaches tend to emphasize the differences between *Ect. mobilis* DSM237^T and *Ect. marismortui* EG-1^T, in contrast with DNA reassociation and with all phenotypic data. In consideration of the comprehensiveness of the available data, and that only one strain of *Ect. marismortui* is presently available, we propose that *Ect. marismortui* EG-1^T is renamed as *Ect. mobilis*. Strain *Ectothiorhodospira* sp. EST8 should be then assigned to the species *Ect. mobilis*.

The two species *Ect. marina* and *Ect. haloalkaliphila*, although described on the basis of only three strains, show a nearly complete phenotypic identity counterbalanced by an intermediate degree of genotypic relatedness. A description of *Ect. haloalkaliphila* as a subspecies of *Ect. marina* could give a better representation of the reciprocal relationships among the three strains.

Although subdivided into two physiological groups based on salt optima, strains of *Hlr. halophila* showed a fairly good global resemblance belonging to two tightly related fatty acid clusters, having the same quinone content, DNA reassociation values higher than 85%, very similar G+C% and ribotype and high 16S rRNA sequence similarity. The reciprocal relationships among the three strains included in this study are more difficult to be interpreted; fatty acid content, ribotype and total DNA restriction pattern [7] differentiated strain BN9624 from the others; 16S rRNA sequence analysis differentiated strain BN9630; ARDRA, besides giving low similarity among all three strains, differentiated strain DSM244^T. It seems therefore appropriated to maintain strains presently attributed to *Hlr. halophila* in a single species.

REFERENCES

1. Imhoff, J.F. and Süling, J. (1996) Arch. Microbiol. 165, 106–113.
2. Stackebrandt, E. and Goebel, B.M. (1994) Int. J. Syst. Bacteriol. 44, 846–849.
3. Wayne, L.G., Brenner, D.J., Colwell, R.R., Grimont, P.A.D., Kandler, O., Krichevsky, M.I., Moore, L.H., Moore, W.E.C., Murray, R.G.E., Stackebrandt, E., Starr, M.P. and Trüper, H.G. (1987) Int. J. Syst. Bacteriol. 37, 463–464.
4. Ventura, S., De Philippis, R., Materassi, R. and Balloni, W. (1988) Arch. Microbiol. 149, 273–279.
5. Thiemann, B. and Imhoff, J.F. (1996) Syst. Appl. Microbiol. 19, 223–230.
6. Imhoff, J.F. (1984) FEMS Microbiol. Lett. 25, 85–89.
7. Ventura, S., Giovannetti, L., Gori, A., Viti, C. and Materassi, R. (1993) Syst. Appl. Microbiol. 16, 405–410.
8. Meißner, J., Borowiak, D., Fischer, U. and Weckesser, J. (1988) Arch. Microbiol. 149, 245–248.
9. Zahr, M., Fobel, B., Mayer, H., Imhoff, J.F., Campos P.V. and Weckesser, J. (1992) Arch. Microbiol. 157, 499–504.
10. Wolf, E., Zahr, M., Benz, R., Imhoff, J.F., Lustig, A., Schiltz, E., Stahl-Zeng, J. and Weckesser, J. (1996) Arch. Microbiol. 166, 169–175.
11. Imhoff, J.F., Tindall, B.J., Grant, W.D. and Trüper, H.G. (1981) Arch. Microbiol. 130, 238–242.
12. Ivanova, T.L., Turova, T.P. and Antonov, A.S. (1985) Arch. Microbiol. 143, 154–156.
13. Oren, A., Kessel, M. and Stackebrandt, E. (1989) Arch. Microbiol. 151, 524–529.
14. Trüper, H.G. (1968) J. Bacteriol. 95, 1910–1920.
15. Mandel, M., Leadbetter, E.R., Pfennig, N. and Trüper, H.G. (1971) Int. J. Syst. Bacteriol. 21, 220–230.
16. Matheron, R. (1976) PhD Thesis, Marseille, France
17. Ventura, S., Viti, C., Pastorelli, R. and Giovannetti, L. (1998) Int. J. Syst. Bacteriol. (submitted)
18. Weisburg, W.G., Barns, S.M., Pelletier, D.A. and Lane, D.J. (1991) J. Bacteriol. 173, 697–703.

91

THE STRUCTURAL DIVERSITY OF INORGANIC PYROPHOSPHATASES OF PHOTOSYNTHETIC MICROORGANISMS AND THE MOLECULAR PHYLOGENY OF THE HOMOLOGOUS PLASTID ENZYMES

Rosario Gómez,[1] Wolfgang Löffelhardt,[2] Manuel Losada,[1] and
Aurelio Serrano[1]

[1]Instituto de Bioquímica Vegetal y Fotosíntesis
CSIC-Universidad de Sevilla
Centro de Investigaciones Científicas Isla de la Cartuja
E-41092 Seville, Spain
[2]Institut für Biochemie und Molekulare Zellbiologie
Universität Wien, and
Ludwig-Boltzmann-Forschungsstelle für Biochemie
A-1030 Vienna, Austria

1. INTRODUCTION

Inorganic pyrophosphatases (sPPase; pyrophosphate phosphohydrolase, EC 3.6.1.1) are essential enzymes that have been demonstrated in the cytosol and cellular organelles of almost every living cell. sPPases are presumed to perform several important functions [1–3]. They drive anabolism through the hydrolysis of the energy-rich pyrophosphate (PP_i) which is produced in the synthesis of biological polymers, making these processes thermodynamically irreversible. Since this hydrolysis replenishes P_i to the energy-converting systems, sPPases should play an important role in the intracellular "phosphate cycle". In plants and some bacteria, sPPase also participates in sulfur metabolism by coupling PP_i hydrolysis to sulfate activation steps [4,5]. Since PP_i is also found to regulate many enzymes and affects the fidelity of DNA synthesis, sPPase may have an important regulatory role. Besides, some organisms contain enzymes (kinases, membrane-bound H^+-translocating PPases) able to use PP_i as the energy-rich compound in energy-requiring reactions [6,7]. Finally, PP_i (like ATP) can be synthesized by a membrane-bound PP_i synthase in some photosynthetic bacteria as a result of photophosphorylation [8].

The Phototrophic Prokaryotes, edited by Peschek *et al.*
Kluwer Academic / Plenum Publishers, New York, 1999.

The cytoplasmic sPPase of *Saccharomyces cerevisiae* (homodimer, 32 kDa subunit) and the *E. coli* enzyme (homohexamer, 19 kDa subunit), although quite different in amino acid sequence and molecular mass, exhibit a well conserved active site [9] and similar protein core structures [10]. They are the best studied eukaryotic and prokaryotic sPPases, respectively, and have been postulated as examples of divergent structural evolution. Searching for other possible different sPPases should therefore be useful to validate this proposal. However, the eubacterial, archaebacterial and eukaryotic sPPases studied so far can also be classified into these two classes according to their structural features, although some heterogeneity was found among bacterial enzymes [3,11,12]. In this aspect, photosynthetic organisms are worth to be systematically investigated due precisely to their marked diversity. Reports on their sPPases are, however, scarce and somewhat confusing. Thus, whereas virtually no data are available on microalgal enzymes, only one report was published on cyanobacterial sPPases [13], and most of those on anoxygenic phototrophic bacteria are preliminary or contradictory [14,15]. Concerning higher plants, contradictory data have also been reported: whereas the sPPase activity appears to be restricted to the chloroplast [16], two sPPase isoenzymes have been described and claimed to have different cellular localization, namely chloroplastic and cytosolic [17]. The function of the cytosolic enzyme could be performed by a PP_i-dependent phosphofructokinase and the vacuolar H^+-translocating PPase present in plant cells [6,7]. Moreover, primary structures of chloroplast sPPases are not yet known. Two putative plant sPPase cDNA clones have been reported (from potato tuber [18] and *Arabidopsis thaliana* leaf [19]), but the encoded 25–29 kDa proteins lack sequence features (transit peptides) to allow them to be identified clearly as genes for chloroplastic enzymes.

Our aim is to clarify the relationships between the sPPases of a wide range of photosynthetic organisms, from photosynthetic bacteria to higher plants, the first comparative study with phylogenetically diverse microorganisms and organelles being presented here. These sPPases exhibit clear differences in molecular structure and amino acid sequences that could be relevant in the context of the molecular phylogeny of plastids. Moreover, a cyanobacterial *ppa* gene, that from *Synechocystis* sp. strain PCC6803, has been cloned, its real ORF experimentally validated, and the overexpressed protein characterized and used for comparative studies [23].

2. EXPERIMENTAL PROCEDURES

The strains of purple bacteria were kindly provided by A. Verméglio (CEA, St. Paul Lez Durance, France) and M. Baltscheffsky (Arrhenius Laboratory, Univ. of Stockholm, Sweden). The cyanobacterial and algal strains were obtained from the Collections of microorganisms of the Institute Pasteur (PCC, Paris, France) and the Univ. of Göttingen (SAG, RFA), respectively. The endocyanome *Cyanophora paradoxa* UTEX/LB555 was grown at the Institut für Biochemie und Molekulare Zellbiologie of the Univ. of Vienna, Austria. Spinach chloroplasts, isolated as described in [20], and *C. paradoxa* cyanelles, isolated after host-cell breakage by gentle stirring of cell suspensions, were eventually disrupted by sonication. A purification method (cell disruption by sonication, ammonium sulfate fractionation of the soluble proteins, DEAE-cellulose ionic exchange chromatography, Phenyl-Sepharose hydrophobic chromatography and hydroxyapatite chromatography) was optimized and used to purify to electrophoretic homogeneity all sPPases studied in this work. The purified enzymes were analyzed by the following techniques: a) SDS-PAGE (apparent subunit molecular mass); b) FPLC gel filtration (apparent native molecular mass) on Superose12 (Pharmacia Biotech) under native conditions (50 mM Tris-HCl, pH 7.5, 0.15 M NaCl, 2mM $MgCl_2$); c) column chromatofocusing (isoelectric point); and d) matrix assisted laser desorption mass spectroscopy, MALDI-TOF (absolute subunit molecular mass). Their N-terminal sequences were determined by the Edman degradation method using automatic amino acid sequencers. A monospecific polyclonal antibody raised in rabbit against the sPPase of *Synechocystis* sp. strain PCC6803 was used (1:1,000) in Western blot analysis and cross-reaction tests after SDS-PAGE of both purified sPPases and crude protein extracts. The *ppa* gene of *Synechocystis* sp. strain PCC6803 was cloned in the pBS SK(+) vector by a PCR strategy using the *Taq*Plus-Precision System of Stratagene, and the encoded sPPase eventually overexpressed in *E. coli* DH5α. Analysis of biological sequences was performed using programs of the GCG package.

3. RESULTS AND DISCUSSION

Cell-free extracts from all photosynthetic organisms tested, from anoxygenic purple bacteria to higher plants, contain substantial levels (0.1–0.5 U/mg of protein) of an alkaline Mg^{2+}-dependent PP_i hydrolase (optimum pH, 7–9). The sPPases of a variety of strains representative of phylogenetically diverse groups of photosynthetic organisms have been purified to electrophoretic homogeneity (Fig. 1A), and their basic structural parameters have been determined using several experimental techniques (Table 1).

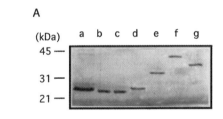

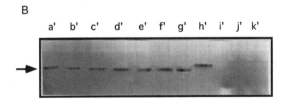

Figure 1. (A) Coomassie-blue-stained SDS-PAGE electrophoretogram of purified preparations of sPPases from the following sources: lane a, *Rhodopseudomonas viridis*; lane b, *Synechocystis* sp. strain PCC6803; lane c, *Anabaena* sp. strain PCC7120; lane d, *Pseudanabaena* sp. strain PCC6903; lane e, *Cyanophora paradoxa* UTEX/LB555 cyanelle; lane f, *Cyanidium caldarium* SAG16.91; lane g, *Chlamydomonas reinhardtii* 6145c. About 5 μg of protein was loaded per lane in all cases. The positions and molecular masses of marker proteins are indicated. (B) Immunoblot analysis of different sPPase preparations with monospecific antibodies raised against the *Synechocystis* sp. strain PCC6803 sPPase. Either 5 μg of purified enzyme or, where indicated, 80 μg of cell-free extract were loaded per lane. Lane a', *Rhodospirillum rubrum* S1 (cell-free extract); lane b', *Rhodopseudomonas palustris*; lane c', *R. viridis*; lane d', *Synechocystis* sp. strain PCC6803; lane e', *Synechocystis* sp. strain PCC6803 (recombinant protein); lane f', *Synechococcus* sp. strain PCC7942 (cell-free extract); lane g', *Anabaena* sp. strain PCC7120; lane h', *Pseudanabaena* sp. strain PCC6903; lane i', *C. paradoxa* UTEX/LB555 cyanelle; lane j', *C. reinhardtii* 6145c; and lane k', spinach chloroplasts. The arrow marks the position of the immunodetected 23-kDa sPPase subunits (apparent molecular mass estimated by SDS-PAGE).

Table 1. Some structural parameters of sPPases purified from phylogenetically
diverse photosynthetic microorganisms and organelles

Source	Subunit molecular mass (kDa)[a]	Native molecular mass (kDa) and oligomeric structure[b]	Isoelectric point[c]
Prokaryotes			
Anoxygenic bacteria			
Rhodopseudomonas viridis	23	240 (dodecamer?)	ND[d]
Rhodopseudomonas palustris	23	85 (tetramer)	ND
Rhodospirillum rubrum	23	ND	ND
Cyanobacteria			
Synechococcus sp. PCC7942	23	120 (hexamer)	ND
Synechocystis sp. PCC6803	23 [19,187[e]]	120 (hexamer)	4.70
Anabaena sp. PCC7120	23	120 (hexamer)	ND
Pseudanabaena sp. PCC6903	24	120 (hexamer)	ND
Eukaryotes			
Cyanidium caldarium	40	160 (tetramer)	ND
Chlamydomonas reinhardtii	36	35 (monomer)	ND
Cyanophora paradoxa (cyanelle)	33	35 (monomer)	5.95
Spinacea oleracea (chloroplast)	37	40 (monomer)	4.90

[a]Except otherwise specified, values were estimated by SDS-PAGE and Western blot analysis (means of at least three independent determinations).
[b]Values estimated by FPLC gel filtration under native conditions in the presence of 2 mM $MgCl_2$ (means of three independent determinations).
[c]Determined by column chromatofocusing.
[d]ND, not determined.
[e]Absolute subunit molecular mass determined by MALDI-TOF.

A marked structural diversity has been found among these enzymes. The sPPases of the photosynthetic prokaryotes studied—purple anoxygenic bacteria and cyanobacteria—showed subunits of virtually identical mass (about 23 kDa) except in the case of *Pseudanabaena* sp. strain PCC6903, which appears to be slightly larger (Fig. 1A). However, whereas all cyanobacterial sPPases are homohexameric (molecular mass, 120 kDa), diverse oligomeric states (tetramer, dodecamer) were exhibited by the enzymes of the purple bacteria examined (Table 1). Thus, these sPPases can in principle be classified as enzymes with a typical prokaryotic structure, although some heterogeneity is observed in its oligomeric native state. In contrast, the sPPases of the eukaryotic photosynthetic organisms investigated—different protists and plants—have clearly larger subunits (35–40 kDa) and are monomers except in the case of *Cyanidium caldarium*, a primitive thermoacidophilic red alga, which seems to have an homotetrameric enzyme (Table 1). The subunit molecular masses of these sPPases are larger than those expected for typical eukaryotic cytosolic sPPases (32 kDa), which are moreover dimeric enzymes. In accordance with what has been reported in several plants [16] and we have confirmed in *Antirrhinum majus* leaves, subcellular localization experiments indicate that the photosynthetic protists lack also cytosolic sPPase, virtually all enzymatic activity being located in their plastids, either cyanelles or chloroplasts (data not shown). Moreover, immunoblot analysis using a monospecific antibody against the sPPase of the unicellular cyanobacterium *Synechocystis* sp. strain PCC6803 indicates cross-reaction with the homologous and similar enzymes of other cyanobacteria and purple bacteria but not with the structurally very different sPPases of eukaryotes, both microalgae and plants (Fig. 1B). The enzymes from other photosynthetic prokaryotes, both oxygenic and anoxygenic, and microalgae (rhodophyceae, chromophyceae, euglenophyceae) are being purified and will be characterized.

The unexpected finding of very different sPPases in cyanobacteria—considered to resemble the ancestral endosymbiotic prokaryote that originated the chloroplasts—and in cyanelles—photosynthetic organelles very similar in structure to free-living unicellular cyanobacteria (still retaining a prokaryotic cell wall) but with a cellular integration level comparable to typical chloroplasts—prompted us to carry out a first detailed comparative study of these enzymes. Cyanelles are considered to represent a different and very early diverging branch of plastid evolution, reminiscent in some aspects to the ancestor of modern chloroplasts [21]. The sPPases from the different cyanobacteria studied (from unicellular to filamentous heterocystous strains) are essentially identical hexameric acidic proteins with subunits of 23–24 kDa (estimated by SDS-PAGE) (Table 1). MALDI mass spectrometry analysis showed however an absolute molecular mass of 19,187 Da for the *Synechocystis* sp. strain PCC6803 sPPase subunit. This discrepancy should be due to an anomalous behaviour of the protein subunits under SDS-PAGE and may explain the strikingly high apparent molecular mass (28 kDa) reported in the only previous study of a cyanobacterial sPPase, that of *Microcystis aeruginosa* [13], claimed to have an unusual trimeric structure. The N-terminal regions of the four cyanobacterial sPPases sequenced so far (three reported in this work, see Table 2, and one in ref. 13) are virtually identical. Although with some peculiar features (rather rich in proline), their marked similarity with the sPPases of other eubacteria and archaebacteria allows to consider them as bacterial-type sPPases (Fig. 2A). Based on the knowledge of the complete genomic sequence of *Synechocystis* sp. strain PCC6803 [22], the *ppa* gene of this cyanobacterium has been confirmed with the N-terminal sequence of the purified sPPase, cloned using a PCR strategy, and experimentally validated by heterologous overexpression [23]. Moreover, our knowledge of the N-terminal end of the protein allowed to define the actual ORF of the gene that uses GTG as the initiation codon and is smaller than the one initially assigned (ORF slr1622) in the genome sequence. The molecular mass predicted for the protein encoded by the smaller ORF (19,088 Da) is in fairly good agreement with that determined by MALDI analysis. The small difference with the MALDI-determined value could be explained either by post-translational modification or by the presence of Mg^{2+} ions (probably four) which are tighly-bound cofactors of the protein [10]. The recombinant cyanobacterial sPPase overexpressed in *E. coli* has been purified and is indistinguishable from the natural one [23] (Fig. 1B). When the complete sequence of the sPPase of *Synechocystis* sp. strain PCC6803 is compared with other sPPases, it appears together with the other bacterial enzymes and the two putative sPPase cDNA clones reported so far in plants (Fig. 2C).

Table 2. N-terminal amino acid sequences of sPPases from phylogenetically diverse photosynthetic microorganisms and organelles[a]

Rhodopseudomonas viridis	MRIDAISICECPPYE
Synechocystis sp. PCC6803	MDLSRIPAQPKAGLINVLIEIPAGSKNKYEFDKDMNNFALDRV
Anabaena sp. PCC7120	MDLSRIPAQPKPGVINILIEIAG
Pseudanabaena sp. PCC6903	MDLSRIPPQPKAGILNVLIEIPAG
Cyanidium caldarium	Not available (blocked N-terminal end)
Chlamydomonas reinhardtii	ATKATTAVTTDKG
Cyanophora paradoxa (cyanelle)	XAITAEPVGTPETLEYRVFIQKDGK
Spinacea oleracea (chloroplast)	KIAKKILIMGGTRFI

[a]Amino acids identified as functional residues of the active site conserved in all sPPases studied so far are marked in bold.

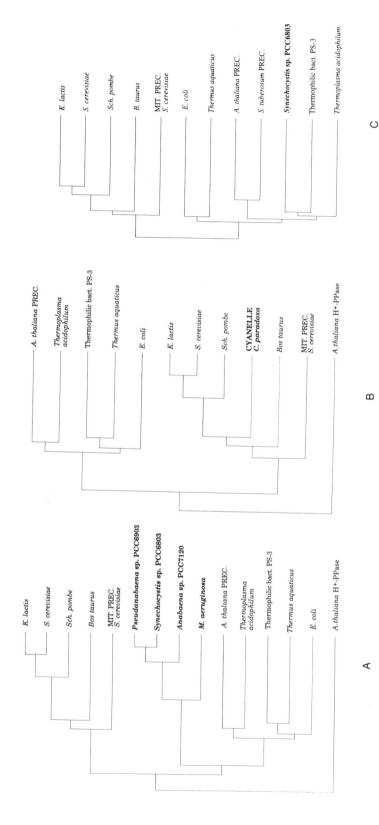

Figure 2. (A) Sequence similarity tree of the sPPases (N-terminal regions) and other sources. (B) Sequence similarity tree of the sPPases from the *C. paradoxa* cyanelle (N-terminal region) and other sPPases. (C) Sequence similarity tree of the *Synechocystis* sp. strain PCC6803 sPPase (experimentally validated complete protein sequence) and other sPPases. In (A) and (B) the amino acid sequence of the *A. thaliana* vacuolar membrane PPase (H⁺-PPase)—very different to those of sPPases—is used as an external reference. In (C) the two putative sPPase cDNAs cloned so far from plants are included. Dendrograms were obtained with the PILEUP program of the GCG package.

In contrast with the cyanobacterial protein, the sPPase of the cyanelle of the endocy-anome *Cyanophora paradoxa* is a monomeric enzyme of 33–35 kDa (Table 1). Column chromatofocusing resolved only one enzyme isoform with the same isoelectric point (5.95) in enzyme preparations from both whole cells and isolated cyanelles. Thus, al-though the presence of a mitochondrial PPase cannot be ruled out, virtually all the sPPase should be located in the photosynthetic organelle. Noteworthy, comparative analysis of the N-terminal amino acid sequence of the *C. paradoxa* sPPase clustered it clearly with the eukaryotic sPPases (Fig. 2B), in agreement with its physico-chemical parameters [24]. The cyanelle sPPase should be therefore an eukaryotic sPPase, encoded by the host cell genome, that during the integrative processs that gave rise to this organelle substituted for its homologous enzyme of the ancestral cyanobacterial-like endosymbiont. It should be noted that the only other organelle sPPase studied so far, namely the mitochondrial en-zyme of *S. cerevisiae* [25], belongs also to the eukaryotic-type sPPases, thus suggesting that loss of the ancestral bacterial endosymbiont sPPase could occur also in this case dur-ing the integrative process. In contrast with this, as stated above, our knowledge on higher plant chloroplast sPPase is still confuse. Although all biochemical studies point to an eu-karyotic-type sPPase (Table 1, cf. [26,27]), the only two reported cDNA sequences known to date [18,19] suggest a relationship with bacterial enzymes. The predicted proteins are however larger (25–29 kDa) than the typical bacterial sPPase subunits because they pos-sess N-terminal and C-terminal extensions with little or no sequence similarity with one another, and are clearly smaller than the sPPase proteins purified from algal or higher plant chloroplasts (35–40 kDa, see Table 1). To clarify this point, the N-terminal region of

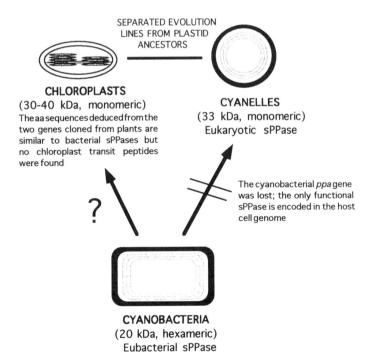

Figure 3. The proposed evolutionary relationships between the sPPases of cyanobacteria, cyanelles and chloro-plasts.

purified chloroplast sPPases of spinach and the green microalga *Chlamydomonas rein-hardtii* has been sequenced (Table 2). However, sequence comparison studies revealed no clear similarity with other eukaryotic or prokaryotic proteins, probably due to the hyper-variable character of the N-terminal regions of the reported plant sPPases [18]. The pro-posed evolutionary relationships between the sPPases of cyanobacteria, cyanelles and chloroplasts are summarized in Figure 3. To clarify further this scenario, the genes encod-ing the actual chloroplastic sPPases should be cloned and sequenced. Oligonucleotide probes have been constructed from the N-terminal sequences of the sPPases purified from the photosynthetic organelles of *C. paradoxa*, *C. reinhardtii* and spinach, and are being currently used for screening of the corresponding cDNA libraries.

The protein sequences referred in this communication have been deposited in the Swiss-Prot and Pir databases under the following accession numbers: P80507 (*Synecho-cystis* sp. strain PCC6803 sPPase), P80562 (*Anabaena* sp. strain PCC7120 sPPase), P80887 (*C. paradoxa* sPPase), P80898 *(Pseudanabaena* sp. strain PCC6903). The other reported sequences were submitted to databases.

ACKNOWLEDGMENTS

Supported by grants DGICYT PB94–0033 of the Spanish Ministry of Education, PAI CVI0198 of the Regional Government of Andalusia and Acción Integrada Hispano-Austriaca 35B. We thank Dr. Günter Allmaier (Univ. of Vienna) for kindly performing the MALDI-TOF analysis of the sPPase of *Synechocystis* sp. strain PCC6803.

REFERENCES

1. Kornberg, A. (1962) *In*: Horizons in Biochemistry (Kasha, H. and Pullman, B., eds.), pp. 251–264, Aca-demic Press, New York.
2. Kukko, E. and Heinonen, J. (1985) Int. J. Biochem. 17, 575–580.
3. Cooperman, B.S., Baykov, A.A. and Lahti, R. (1992) Trends Biochem. Sci.17, 262–266.
4. Schmidt, A. and Jäger, K. (1992) Annu. Rev. Plant Physiol. Plant Mol. Biol. 43, 325–349.
5. Liu, M.Y. and LeGall, J. (1990) Biochem. Biophys. Res. Comm. 171, 313–318.
6. Wood, H.G. (1977) Fed. Proc. 36, 2197–2205.
7. Rea, P.A. and Poole, R.J. (1993) Annu. Rev. Plant Physiol. Plant Mol. Biol. 44, 157–180.
8. Baltscheffsky, M. and Baltscheffsky, H. (1992) *In*: Molecular Mechanisms in Bioenergetics (Ernster, L., ed.), pp. 331–348, Elsevier, Amsterdam.
9. Lahti, R., Kolakowski, L.F., Heinonen, J., Vihinen, M., Pohjanoksa, K. and Cooperman B.S. (1990) Bio-chim. Biophys. Acta 1038, 338–345.
10. Kankare, J., Neal, G.S., Salminen, T., Glumhoff, T., Cooperman, B.S., Lahti, R. and Goldman, A. (1994) Prot. Eng. 7, 823–830.
11. Lahti, R. (1983) Microbiol. Rev. 169–179.
12. Richter, O.H. and Schäfer G. (1992) Eur. J. Biochem. 209, 343–349.
13. Kang, C.B.H. and Ho, K.K. (1991) Arch. Biochem. Biophys. 289, 281–288.
14. Klemme, J.H., Klemme, B. and Gest, H. (1971) J. Bacteriol. 108, 1122–1128.
15. Schwarm, H.M., Vigenschow, H. and Knobloch, K. (1986) Biol. Chem. Hoppe-Seyler 367, 119–126.
16. Weiner, H., Stitt, M. and Heldt, H.W. (1987) Biochim. Biophys. Acta 893, 13–21.
17. Maslowski, P. and Maslowska, H. (1976) Z. Pflanzenphysiol. 79, 23–32
18. Du Jardin, P., Rojas-Beltrán, J., Gebhardt, C. and Brasseur, R. (1995) Plant Physiol. 109, 853–860.
19. Keiber, K.K. and Signer, E.R. (1991) Plant Mol. Biol. 16, 345–348.
20. Kuwavara, T. and Murata, N. (1982) Plant Cell Physiol. 23, 533–539.
21. Löffelhardt, W. and Bohnert, H. (1994) *In*: The Molecular Biology of Cyanobacteria (Bryant, D.A., ed.), pp. 65–89, Kluwer Academic Publishers, Dordrecht.

22. Kaneko, T., Sato, S., Kotani, H., Tanaka, A., Asamizu, E., Nakamura, Y., Miyajima, N., Hirosawa, M., Sugiura, M., Sasamoto, S., Kimura, T., Hosouchi, T., Matsuno, A., Muraki, A., Nakazaki, N., Naruo, K., Okumura, S., Shimpo, S., Takeuchi, C., Wada, T., Watanabe, A., Yamada, M., Yasuda, M. and Tabata, S. (1996) DNA Research 3, 109–136.
23. Gómez, R., Löffelhardt, W., Allmaier, G., Losada, M. and Serrano, A. (in preparation).
24. Serrano, A., Gómez, R., Losada, M., Allmaier, G. and Löffelhardt, W. (in preparation).
25. Lundin, M., Baltscheffsky, H. and Ronne, H. (1991) J. Biol. Chem. 266, 12168–12172.
26. Mukherjee, J.J. and Pal, P.R. (1983) Agric. Biol. Chem. 47, 2973–2975.
27. Pwe, K.H., and Ho, K.K. (1995) Plant Physiol. Biochem. 33, 39–46.

92

AN OVERVIEW OF RNA POLYMERASE SIGMA FACTORS IN PHOTOTROPHS

Tanja M. Gruber and Donald A. Bryant

Department of Biochemistry and Molecular Biology
Pennsylvania State University
University Park, Pennsylvania 16802

1. INTRODUCTION

Sigma factors are dissociable subunits that confer promoter specificity on eubacterial core RNA polymerase and are required for transcription initiation. Two major families of sigma factors occur in eubacteria: the σ^{70} (RpoD) family [1,2]) and the σ^{54} (RpoN) family [3]; which are both named after the originally identified *Escherichia coli* proteins. Based upon sequence comparisons and functional considerations, the σ^{70} family has been divided into three groups [1]. The primary sigma factors, which are essential for cell viability, comprise group 1. Group 2 includes alternative sigma factors that are highly similar in sequence to the respective group 1 members, whereas group 3 sigma factors are alternative sigma factors that vary more significantly in sequence from the other two groups and include functional groupings such as heat shock and sporulation sigma factors. Sigma factors of groups 2 and 3 are not required for cell viability. Group 1 and group 2 members are related sufficiently closely that cross-hybridization is easily detected in genomic Southern analyses.

In the case of *E. coli*, the primary sigma factor, RpoD or σ^{70}, directs transcription from promoters with the consensus DNA sequence motif 5′ TATAAT centered at about −10 with respect to the transcription start at +1, and 5′ TTGACA centered at about −35; these two elements are optimally separated by 17 base pairs of non-conserved sequence. This consensus promoter sequence is suspected to be largely conserved for group 1 sigma factors throughout all eubacteria [2]. As more sigma factors are characterized from various organisms, it becomes apparent that eubacteria employ very different modes of transcriptional activation (concerning sigma factor usage) under various stress conditions. For example, *E. coli* and certain other proteobacteria use a specific heat-shock sigma factor to transcribe a group of genes ("regulon") following certain stresses. However, in a number of other eubacteria, including the gram-positve *Bacillus subtilis* and certain cyanobacteria, the regulatory element CIRCE (*c*ontrolling *i*nverted *r*epeat of *c*haperone *e*xpression) in conjunction with the pri-

The Phototrophic Prokaryotes, edited by Peschek *et al.*
Kluwer Academic / Plenum Publishers, New York, 1999.

Table 1. Sigma factors from five different phototrophic eubacteria

Organism	Phylum	Gene	Amino acids	Group	Accession no.	Function/comments
Synechococcus sp. PCC 7002	cyanobacterium	*sigA*	375	1	U15574	Major vegetative sigma factor
"	"	*sigB*	328	2	U82435	Carbon/nitrogen regulation
"	"	*sigC*	365	2	U82436	Carbon/nitrogen regulation
"	"	*sigD*	319	2	U82484	No function yet identified
"	"	*sigE*	398	2	U82485	Post-exponential phase expression
"	"	*sigF*	233	3	n/a	Most similar to *B. subtilis* SigB
"	"	*sigG*	220	3	n/a	No function yet identified; RpoE1
"	"	*sigH*	189	3	n/a	No function yet identified; RpoE2
"	"	*sigI*	?	3	n/a	Found in *Synechocystis* 6803; RpoE3
Chloroflexus aurantiacus	green gliding bacterium	*sigA*	343	2	U67719	?
"	"	*sigB*	346	2	U67720	Aerobic growth?
"	"	*sigC*	312	1	U67721	Major vegetative sigma factor
"	"	*sigD*	398	2	U67722	?
Chlorobium tepidum	green sulfur bacterium	*sigA*	299	1	U67718	Major vegetative sigma factor
Heliobacillus mobilis	gram-positive bacterium	*sigA*	333	1	U67424	Major vegetative sigma factor
Rhodobacter sphaeroides	α proteobacterium	*sigA*	666	1	U67425	Major vegetative sigma factor
"	"	*rpoN1*	434	σ^{54}	M86823	Meijer and Tabita, 1992
"	"	*rpoN2*	426	σ^{54}	B07848	C. Mackenzie, unpublished
"	"	*rpoH*	298	3	U82397	Heat shock sigma factor, T. Donohue, unpublished

mary sigma factor is responsible for the expression of chaperones under heat shock conditions [4].

In this article we briefly consider the sigma factor family members for representative eubacteria from the five phyla which contain phototrophs: proteobacteria, gram-positive bacteria, green sulfur bacteria, green gliding bacteria, and cyanobacteria (see Table 1). We present initial functional characterizations of a number of sigma factors as well as efforts to use group 1 sigma factors as phylogenetic markers.

2. PROTEOBACTERIA

The best studied proteobacterium, *E. coli,* utilizes at least seven sigma factors: RpoD (σ^{70}), RpoH, RpoS [1], RpoE, FliA, and FecI [5], and RpoN [3]. RpoN is the lone member of the σ^{54} family; RpoD is the group 1 sigma factor of the σ^{70} family, RpoS is a group 2 member, and the remaining four sigma factors belong to group 3. The group 3 members include RpoH, which plays a key role in the heat-shock response of *E. coli*, the extracellular factor (ECF) family (including RpoE and FecI), and a sigma factor involved in flagellar synthesis (FliA).

We have cloned and sequenced the group 1 sigma factor *rpoD* of the photosynthetic α-proteobacterium *Rhodobacter sphaeroides.* In Southern analyses weakly hybridizing bands were detected when chromosomal DNAs were hybridized with probes derived from *rpoD,* suggesting that at least one group 2 sigma factor is present in this organism. The group 3 heat shock sigma factor RpoH (T. Donohue, unpublished) and a σ^{54} member RpoN [6] have also been identified from this organism.

3. GRAM-POSITIVE BACTERIA

Another well studied eubacterium, *B. subtilis,* is believed to contain 10 sigma factors [7], including one member of the σ^{54} family, one group 1 member of the σ^{70} family, and eight members of group 3. In *B. subtilis* sporulation is a starvation response in which the organism converts itself from an actively growing vegetative cell into a dormant endospore. Four sigma factors in this bacterium appear to be solely dedicated to endospore formation, and cascades of sigma factors are partly responsible for the sequential pattern of gene expression that occurs during sporulation.

We cloned and sequenced the group 1 sigma factor of the photosynthetic low-GC gram positive bacterium *Heliobacillus mobilis.* Southern hybridization analyses showed no evidence for the presence of group 2 sigma factors in this bacterium. *H. mobilis* is a endospore-forming bacterium that is closely related to *Clostridium* sp., so it is conceivable that this organism contains a number of sporulation-specific sigma factors like *B. subtilis.*

4. GREEN SULFUR BACTERIA

We cloned and sequenced the presumed group 1 sigma factor of the green sulfur bacterium *Chlorobium tepidum.* Interestingly, a specific amino acid change in region 2.4 has been correlated with an altered preference in the −10 sequence in several promoters that have been characterized from this organism. No further hybridization signals were detected in Southern analyses, indicating that no group 2 sigma factors are present in this organism. No sigma factors of other organisms in the same phylum have been characterized so it is not known how many group 3 members *C. tepidum* might possess.

5. GREEN GLIDING BACTERIA

We identified and characterized four group 1 + 2 sigma factors in the green gliding bacterium (perhaps better referred to as anoxygenic flexibacterium or filamentous, anoxygenic phototrophic bacterium) *Chloroflexus aurantiacus*. This organism can grow aerobically as a light-independent heterotroph, and anaerobically as a photoautotroph or photoheterotroph. We grew *C. aurantiacus* both aerobically and anaerobically and monitored the RNA transcript levels of the four sigma factor genes by northern analysis. One gene, *sigC*, was expressed at approximately equal levels under both conditions. The gene encoding SigB was predominantly expressed under anaerobic conditions and the remaining two genes, *sigA* and *sigD*, were not detectably transcribed under either growth condition. These results imply that SigC is the group 1 sigma factor of this organism, and that SigA, SigB, and SigD belong to group 2. As is the case with *C. tepidum*, no predictions can be made concerning group 3 members.

6. CYANOBACTERIA

Transcription of a large number of genes in several species of cyanobacteria have been shown to be altered by environmental changes [8]. Alignment of sequences believed to encompass the -10 and -35 promoter regions of many cyanobacterial genes has shown that many promoter regions contain sequences similar to the consensus promoter sequence of the *E. coli* group 1 sigma factor RpoD [9]. Although multiple sigma factors have been found in all cyanobacteria studied to date [10–14], the roles of the alternative sigma factors have remained somewhat uncertain. No σ^{54}-type sigma factors have yet been identified in any cyanobacterium.

We identified nine sigma factors of the σ^{70} family in *Synechococcus* sp. PCC 7002 (see Table 1). Sequence analysis and mutational studies of eight of these sigma factors have revealed one group 1 sigma factor [15], four group 2 sigma factors [16,17] and three group 3 sigma factors in this organism. Cloning and characterization of the ninth sigma factor is still in progress. Homologs of each of these nine sigma factors were found in the genome of *Synechocystis* sp. PCC 6803 [13,14], and interestingly, no RpoH (heat-shock sigma factor), RpoS (stationary phase sigma factor), or σ^{54}-type sigma factor homologs are present in *Synechocystis* sp. PCC 6803.

Interposon mutagenesis has been performed on all eight cloned sigma factor genes of *Synechococcus* sp. PCC 7002, and mutants, double mutants, and triple mutants lacking all possible combinations of the group 2 sigma factors have been constructed. Only SigA has proved to be essential for cell viability, and thus it has been designated the group 1 sigma factor of the organism [15]. Two group 2 members (SigB and SigC) appear to be involved in cellular responses to carbon and nitrogen stress [16]. The *sigD* gene, encoding an additional group 2 member, was insertionally inactivated, but no function has yet been identified for this protein. Transcripts of the *sigD* gene have not been detected in cells subjected to a variety of environmental stresses.

It was observed that stationary-phase cells of strains in which the *sigE* gene is insertionally inactivated consistently displayed longer times to re-establish exponential growth than wild-type cells when diluted into fresh medium from a stationary-phase culture. However, once the cells begin to grow again, the growth rate of the *sigE* mutant is identical to the wild type. This behavior was observed both for cells grown in liquid medium or for cells grown on agar plates and was exclusively characteristic of the *sigE* strain and and was not observed for the *sigB*, *sigC*, and *sigD* mutants. These results imply that cells lacking SigE cannot easily re-establish exponential growth after entering into stationary phase. These growth experiments prompted us to track the expression of a stationary-phase specific protein to determine possible differences between the wild-type strain and the *sigE* mutant. The starvation-induced DpsA/PexB protein of *E. coli* is a 19.7 kDa, DNA-binding hemoprotein which forms high-molecular weight complexes that have been implicated in the protection of DNA from oxidative damage during stationary phase [18]. A homolog of the DpsA/PexB protein was identified in nitrate-limited cells of *Synechococcus* sp. strain PCC 7942 [19,20]. Since the expression of this protein is specific to stationary-phase or nutrient-limited cells in *E. coli*, and its transcription in stationary phase was shown to be dependent on RpoS [21], it was reasoned that expression of this protein in *Synechococcus* sp. PCC 7002 might be under the control of SigE. Immunoblot analyses were performed to determine the relative levels of the DpsA/PexB protein in the wild-type and *sigE* mutant strain cells that had been grown into stationary phase for varying times (1, 4, or 10 days). The immunoblots were probed with antibodies raised against the DpsA/PexB protein of *Synechococcus* sp. PCC 7942 [19]. No DpsA/PexB protein was detected in wild type-cells after 1 day of growth, but after 4 or 10 days of growth, a protein of the expected size cross-reacted with the anti-DpsA/PexB antibodies. However, no DpsA/PexB was detected in cells of the *sigE* mutant strain even after 10 days of growth. These results strongly imply that a functional *sigE* gene product is necessary for the expression of the *dpsA/pexB* gene during stationary phase in *Synechococcus* sp. PCC 7002.

7. PHYLOGENETIC ANALYSES

Group 1 sigma factors possess a variety of characteristics that make them excellent candidates as phylogenetic markers. They occur in all members of the eubacterial kingdom and as noted above are essential for cell viability. Functional constraints for these proteins give rise to highly conserved structural features. Even though sigma factor proteins vary in length, four regions of very high sequence conservation have been identified [22]. These regions span approximately 250 contiguous amino acids and include the subregions that are responsible for recognition of the −10 and −35 promoter elements. This 250 amino acid region was aligned for 60 group 1 sigma factor proteins from a variety of eubacterial phyla, and the resulting alignment was used in phylogenetic analyses (see Fig. 1).

Three major groupings become apparent in the sigma factor tree: proteobacteria, gram-positive bacteria, and cyanobacteria. The proteobacterial phylum has been divided into five phylogenetically distinct subdivisions (α, β, γ, δ, ε), based on SS-rRNA sequences [23]. Using sigma factors as the marker, the arrangement of four of the subgroups (α, β, γ, δ) is essentially identical to that for trees derived from SS-rRNA sequences of the same species composition, but the groups are even more strongly supported statistically [2]. The gram-positive phylum is believed to consist of four subdivisions, two of which have been well characterized. These two subgroups, the low-GC and high-GC groups, are well represented in this study, and a member of the third subgroup, namely *H. mobilis* of the photosynthetic subdivision, was also included. The fourth subgroup is comprised of

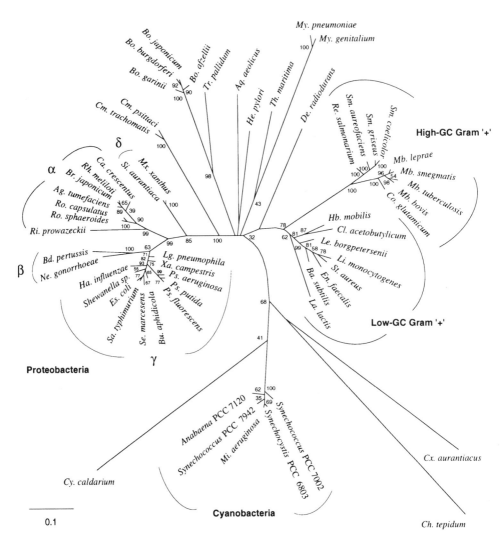

Figure 1. Neighbor-joining tree for group 1 sigma factors. The tree was generated by the method of Kimura from the multiple sequence alignments encompassing regions 2.1 to 4.2 of group 1 sigma factor proteins [2]. The alignment was generated using the *clustalw* multiple sequence alignment program. The distances were calculated using the *protdist* program of the PHYLIP package, using a PAM-matrix based distance correction. The scale bar represents 0.1 substitutions per site. Bootstrap values were obtained after 100 replications and are indicated when over 30. The sequence of SigA of *Aquifex aeolicus* was kindly provided by Ron Swanson of Diversa Corp. prior to publication. (*Ag.* = *Agrobacterium*; *Aq.* = *Aquifex*; *Ba.* = *Bacillus*; *Bd.* = *Bordetella*; *Bo.* = *Borrelia*; *Br.* = *Bradyrhizobium*; *Bu.* = *Buchnera*; *Ca.* = *Caulobacter*; *Ch.* = *Chlorobium*; *Cl.* = *Clostridium*; *Cm.* = *Chlamydia*; *Co.* = *Corynebacterium*; *Cx.* = *Chloroflexus*; *Cy.* = *Cyanidium*; *De.* = *Deinococcus*; *En.* = *Enterococcus*; *Es.* = *Escherichia*; *Ha.* = *Haemophilus*; *Hb.* = *Heliobacillus*; *He.* = *Helicobacter*; *La.* = *Lactococcus*; *Le.* = *Leptospira*; *Li.* = *Listeria*; *Lg.* = *Legionella*; *Mb.* = *Mycobacterium*; *Mi.* = *Microcystis*; *Mx.* = *Myxococcus*; *My.* = *Mycoplasma*; *Ne.* = *Neisseria*; *Ps.* = *Pseudomonas*; *Rh.* = *Rhizobium*; *Ri.* = *Rickettsia*; *Ro.* = *Rhodobacter*; *Sa.* = *Salmonella*; *Se.* = *Serratia*; *Si.* = *Stigmatella*; *St.* = *Staphylococcus*; *Sm.* = *Streptomyces*; *Th.* = *Thermotoga*; *Tr.* = *Treponema*; *Xa.* = *Xanthomonas*)

the genera *Megasphaera, Selenomonas,* and *Sporomusa* but no sigma factor sequences have yet been determined for any of these organisms. There has been much debate over whether the gram-positive bacteria are monophyletic. For example, some studies on rRNA genes [24] and GroEL genes [25] suggested that the gram-positive bacteria are of monophyletic origin, whereas other studies on rRNA genes [26], elongation factor Tu, the β subunit of ATP synthase [27], and RecA sequences [28] suggested that these organisms are not monophyletic. With the species represented in studies using sigma factors as the marker, the gram-positive bacteria were found to be of monophyletic origin, and this conclusion is supported by a very high bootstrap value of 78%. Cyanobacteria, including the closely related prochlorophytes, are the only eubacteria that perform oxygenic photosynthesis. They form a coherent clade in the sigma factor tree that presently contains no non-photosynthetic members. The cyanobacteria are most closely related to several other photosynthetic organisms, namely the green-gliding bacterium *C. aurantiacus* and the green sulfur bacterium *C. tepidum.*

Placement of several other phyla which are not represented by a large number of organisms is well supported in the tree. For example, the chlamydia are positioned on the proteobacterial branch, and *C. aurantiacus* and *C. tepidum* are placed as near relatives of the cyanobacteria. Other less represented phyla are positioned more ambiguously: e.g., *Deinococcus radiodurans (Thermus/Deinococcus* group) and *Aquifex aeolicus* (Aquificales group). The Aquificales have been placed as the most deeply branching eubacteria in the SS-rRNA tree [29]. It is very plausible that *A. aeolicus* is positioned very closely to the hypothetical root of the sigma factor tree, although this organism is clearly more closely related to other proteobacteria than to cyanobacteria or gram-positive bacteria.

REFERENCES

1. Lonetto, M., Gribskov, M. and Gross, C.A. (1992) J. Bacteriol. 174, 3843–3849.
2. Gruber, T.M. and Bryant, D.A. (1997) J. Bacteriol. 179, 1734–1747.
3. Kustu, S., Santero, E., Keener, J., Popham, D. and Weiss, D. (1989) Microbiol. Rev. 53, 367–376.
4. Zuber, U. and Schumann, W. (1994) J. Bacteriol. 176, 1359–1363.
5. Lonetto, M.A., Brown, K.L., Rudd, K.E. and Buttner, M.J. (1994) Proc. Natl. Acad. Sci. USA 91, 7573–7577.
6. Meijer, W.G. and Tabita, F.R. (1992) J. Bacteriol. 174, 3855–3866.
7. Haldenwang, W.G. (1995) Mirobiol. Rev. 59, 1–30.
8. Tandeau de Marsac, N. and Houmard, J. (1993) FEMS Microbiol. Rev. 104, 119–190.
9. Curtis, S.E. and Martin, J.A. (1994) in: The Molecular Biology of Cyanobacteria, pp. 613–639 (Bryant, D.A., Ed.) Kluwer Academic Publishers, Dordrecht.
10. Tanaka, K., Masuda, S. and Takahashi, H. (1992) Biochim. Biophys. Acta 1132, 94–96.
11. Brahamsha, B. and Haselkorn, R. (1992) J. Bacteriol. 174, 7273–7282.
12. Tsinoremas, N.F., Ishiura, M., Kondo, T., Andersson, C.R., Tanaka, K., Takahashi, H., Johnson, C.H. and Golden, S.S. (1996) EMBO J. 15, 2488–2495.
13. Kaneko, T. et al. (1996) DNA Res. 3, 109–136.
14. Kaneko, T. et al. (1996) DNA Res. 3, 185–209.
15. Caslake, L.F. and Bryant, D.A. (1996) Microbiology 142, 347–357.
16. Caslake, L.F., Gruber, T. and Bryant, D.A. (in press) Microbiology.
17. Gruber, T.M. and Bryant, D.A. (in press) Arch. Microbiol.
18. Almiron, M., Link, A., Furlong, D. and Kolter, R. (1992) Genes Dev. 6, 2646–2654.
19. Pena, M.M., Burkhart, W. and Bullerjahn, G.S. (1995) Arch. Microbiol. 163, 337–344.
20. Pena, M.M. and Bullerjahn, G.S. (1995) J. Biol. Chem. 270, 22478–22482.
21. Altuvia, S., Almiron, M., Huisman, G., Kolter, R. and Storz, G. (1994) Mol. Microbiol. 13, 265–272.
22. Helmann, J.D. and Chamberlin, M.J. (1988) Annu. Rev. Biochem. 57, 839–872.
23. Olsen, G.J., Woese, C.R. and Overbeek, R. (1994) J. Bacteriol. 176, 1–6.

24. Woese, C.R. (1987) Microbiol. Rev. 51, 221–271.
25. Viale, A.M., Arakaki, A.K., Soncini, F.C. and Ferreyra, R.G. (1994) Int. J. Syst. Bacteriol. 44, 527–533.
26. van der Peer, Y., Neefs, J.M., De Rijk, P., De Vos, P. and De Wachter, R. (1994) Syst. Appl. Microbiol. 17, 32–38.
27. Ludwig, W. et al. (1994) Antonie Van Leeuwenhoek 64, 285–305.
28. Eisen, J.A. (1995) J. Mol. Evol. 41, 1105–1123.
29. Pace, N.R. (1997) Science 276, 734–740.

93

PHYLOGENY OF CAROTENOGENIC GENES AND ENZYMES IN CYANOBACTERIA

Gerhard Sandmann[1] and Agustin Vioque[2]

[1]Biosynthesis Group
Botanical Institute
J. W. Goethe Universität
P.O. Box 11932, D-60054 Frankfurt, Germany
[2]Instituto de Bioquímica Vegetal y Fotosíntesis
Universidad de Sevilla-CSIC
Americo Vespucio s/n, E-41092 Sevilla, Spain

Carotenoids are essential for photosynthetic organisms as accessory pigments and as protectants against excess light. In contrast to anoxygenic photosynthetic bacteria, cyanobacteria possess carotenoids with cyclic end groups. The presence of β-carotene is universal to all cyanobacteria but the potential to synthesize oxygen-containing derivatives is different among different genera. Myxoxanthophyll is typically found in *Synechocystis* species together with zeaxanthin and echinenone [1] whereas the tri- and tetrahydroxy β-carotene derivatives accumulate in *Synechococcus* as by-products of zeaxanthin [2].

All cyanobacterial carotenoids are formed in a biosynthetic pathway (outlined in Fig. 1) involving the same intermediates as in other bacteria, fungi, algae or higher plants. However, many cyanobacterial enzymes involved in catalyzing individual reaction steps are diverse in their mechanism or originate from non-homologous genes. Especially among the cyanobacteria, several genes are newly acquired or functionally modified. A great number of carotenogenic genes from cyanobacteria have been cloned [3,4]. Furthermore, the whole genome of *Synechocystis* PCC6803 has been sequenced [5] and the identification of carotenoid biosynthetic genes by comparison with others and by functional assignment is in progress. From all genes of the pathway (Table 1), only those encoding enzymes leading to phytoene synthesis are conserved among carotenogenic organisms [4]. For the subsequent 4-step desaturation of 15-*cis* phytoene to lycopene the situation in cyanobacteria is totally different to other prokaryotes. In other bacteria and fungi the whole desaturation sequence is catalyzed by one single *crtI*-type desaturase. However, among the cyanobacteria and also among plants two different genes/enzymes are responsible for the whole desaturation sequence. Phytoene desaturase catalyzes the first two steps and ζ-carotene desaturase the last two desaturations. In plants, both genes

The Phototrophic Prokaryotes, edited by Peschek *et al.*
Kluwer Academic / Plenum Publishers, New York, 1999.

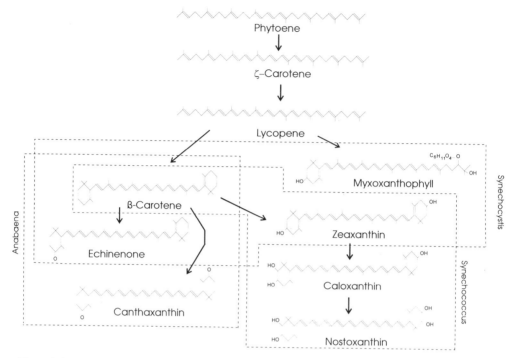

Figure 1. Pathway of carotenoid biosynthesis in cyanobacteria. Carotenoids in individual boxes represent typical end product combinations found in different genera.

(*pds* and *zds*) are unrelated to *crtI* and homologous between themselves. In cyanobacteria the situation is more complex. However, it is not possible to tell how widespread the occurence of either *crtQ-2* or *crtQ-1* are. The latter gene was found in *Anabaena* which is a heterocyst-forming cyanobacterial species. Therefore, it was speculated that the function of the *crtQ-1* related enzyme may be restricted to this organelle which is specialized in N₂ fixation [10].

Regardless of the enzyme type, either kind of phytoene desaturases or ζ-carotene desaturase must have a common feature in recognizing phytoene and ζ-carotene, respectively, as a substrate. For the binding of the substrate phytoene to phytoene desaturase the central trienoic region of the substrate and the *cis* configuration of the central double bond

Table 1. Cyanobacterial carotenogenic genes

Gene (old name[a])	Expressed enzyme	Relationship to genes of other function	Reference
crtB[1,2]	Phytoene synthase		
crtP[1,2] (*pds*)	Phytoene desaturase, 2 steps		
crtQ-1[3] (*zds*)	ζ-Carotene desaturase	*crtI* (Phytoene desaturase, 4 steps)	7
crtQ-2[2]	ζ-Carotene desaturase	*crtP* (Phytoene desaturase, 2 steps)	8
crtL[1] (*lcy*)	Lycopene cyclase		11
crtL-2[2]	Lycopene cyclase	unknown	
crtO[2]	β-Carotene ketolase	*crtI* (Phytoene desaturase)	13
crtR[2]	β–Carotene hydroxylase	*crtW* (β-Carotene ketolase)	14

Source: [1]*Synechococcus*; [2]*Synechocystis*; [3]*Anabaena*. [a]New names as suggested in reference [3].

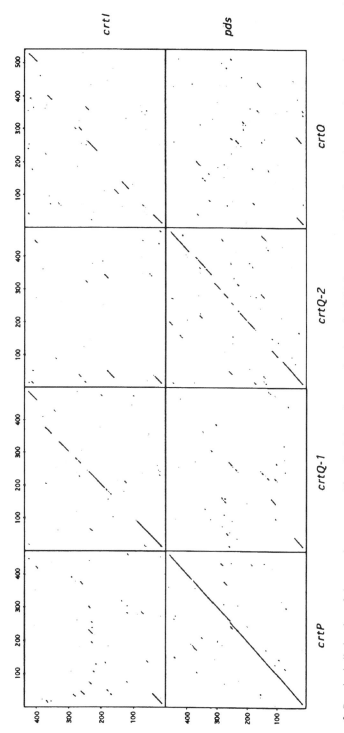

Figure 2. Protein similarity plots of phytoene desaturases *crtI* from *Erwinia uredovora* and a plant *pds* cDNA from tomato without the transit sequence in comparison to cyanobacterial desaturases and a related enzyme (see their characteristics in Table 1).

may be important. From a mechanistic point of view the *crtP* and *crtI*-encoded desaturases differ by their cofactor requirement using NAD(P) in the first case and FAD in the latter and also by their inhibitor sensitivities [6]. The specific binding of ζ-carotene to ζ-carotene desaturase can be explained by the assumption that this enzyme recognizes an even longer conjugated system e.g. a heptaenoic structure adjacent to the position where the double bond is inserted.

The next step after desaturation in the carotenoid biosynthetic pathway is cyclization of lycopene to β-carotene. The only available cyanobacterial lycopene cyclase gene is from *Synechococcus*. This gene, *crtL* (Table 1), exhibits poor sequence conservation to other bacterial cyclase genes (*crtY*) and resembles much more the higher plant enzyme with identities around 23% [11]. However, a gene homologous to *crtL* or *crtY* cannot be found in the *Synechocystis* genome. As *Synechocystis* synthesizes cyclic carotenoids as endproducts of its pathway (Table 1), a structurally different lycopene cyclase gene must exist in this species.

Rather homologous ketolase genes involved in the synthesis of canthaxanthin are available from various bacteria and green algae [12]. However, the only ketolase gene *crtO* (Table 1) identified from *Synechocystis* [13] is completely unrelated to this type. The *crtO*-encoded enzyme shows also a mechanistical difference to the other *crtW*-type ketolases yielding echinenone as the major reaction product. Figure 2 demonstrates that *crtO* is closely related to the *crtI* phytoene desaturase gene. The change of function from a desaturase to a ketolase is difficult to understand especially as another prokaryotic ketolase exists. Therefore, it will be interesting to clone the ketolase genes from other cyanobacteria which synthesize the diketo carotenoid canthaxanthin instead of echinenone and compare their sequence with *crtO*. Very recently, the gene for β-carotene hydroxylase has been identified in *Synechocystis* by functional complementation [14]. It showed clear sequence similarity to the bacterial β-carotene ketolase gene *crtW* but is unrelated to other bacterial or plant β-carotene hydroxylase genes.

Phylogenetic studies revealed the cyanobacterial origin of chloroplasts and the lineage via chlorophytes to higher plants. Looking at the molecular phylogeny of carotenoid biosynthesis, a common evolutionary back-ground between plants and cyanobacteria is evident. Most of the structural basis for carotenogenesis in higher plant chloroplasts was already established within the cyanobacteria with two exceptions. The potential to synthesize ε-ionone end groups in addition to cyanobacterial carotenoids with β-ionone end groups as well as the epoxidation of zeaxanthin evolved in the chlorophytes. This happened in correlation with the evolution of oxygenic photosynthesis. It is tempting to speculate that *pds*-type enzymes are better adapted to work in organisms with oxygenic photosynthesis than the *crtI*-type. One major difference of the catalytic mechanism is the use of the cofactor FAD in *crtI* desaturases which is replaced in *pds* related desaturases by NAD(P) [4]. Especially NADP is very abundant in chloroplasts, the compartment where also the carotenoids are synthesized.

ACKNOWLEDGMENTS

This work was supported in part by a grant from the European Communities RTD Programme and DGICYT (Spain).

REFERENCES

1. Bramley, P.M. and Sandmann, G. (1985) Phytochemistry 24, 2919–2922.
2. Stransky, H., and Hager, A. (1979) Arch. Microbiol. 72, 84–96.
3. Hirschberg, J. and Chamovitz, D. (1994) in: The Molecular Biology of Cyanobacteria (Bryant, D.A. ed.) pp. 559–579, Kluwer Academic Publishers, Dordrecht
4. Sandmann, G. (1994) Eur. J. Biochem. 223, 7–24.
5. Kaneko, T., Sato, S., Kotani, H., Tanaka, A., Asamizu, E., Nakamura, Y., Miyajima, N., Hirosawa, M., Sugiura, M., Sasamoto, S., Kimura, T., Hosouchi, T., Matsuno, A., Muraki, A., Nakazaki, N., Naruo, K., Okumura, S., Shimpo, S., Takeuchi, C., Wada, T., Watanabe, A., Yamada, M., Yasuda, M. and Tabata, S. (1996). DNA Res. 3, 109–136.
6. Sandmann, G. (1994) J. Plant Physiol. 143, 444–447.
7. Linden, H., Misawa, N., Saito, T., and Sandmann, G. (1994) Plant Mol. Biol. 24, 369–379.
8. Breitenbach, J., Fernández-González, B., Vioque, A., and Sandmann, G. (1997) manuscript in preparation.
9. Pecker, I., Chamovitz, D., Linden, H., Sandmann, G., and Hirschberg, J. (1992) Proc. Natl. Acad. Sci. USA 89, 4962–4966.
10. Albrecht, M., Klein, A., Hugueney, P., Sandmann, G. and Kuntz, M. (1995) FEBS Lett. 372, 199–202.
11. Cunningham, F.X., Sun, Z., Chamovitz, D., Hirschberg, J., and Gantt, E. (1994) Plant Cell 6, 1107–1121.
12. Misawa, N., Satomi, Y., Kondo, K., Yokoyama, A., Kajiwara, S., Saito, T., Ohtani, T., and Miki, W. (1995) J. Bacteriol. 177, 6575–6584.
13. Fernández-González, B., Sandmann, G. and Vioque, A. (1997) J. Biol. Chem. 272, 9728–9733.
14. Masamoto, K., Misawa, N., Kaneko, T., Kikuno, R., and Toh, H. (1997) Plant Mol. Biol., submitted.

94

PRIMORDIAL UV-PROTECTORS AS ANCESTORS OF THE PHOTOSYNTHETIC PIGMENT-PROTEINS

Armen Y. Mulkidjanian[1,2] and Wolfgang Junge[1]

[1]Division of Biophysics
Faculty of Biology/Chemistry
University of Osnabrück
D-49069 Osnabrück, Germany
[2]A. N. Belozersky Institute of Physico-Chemical Biology
Moscow State University
119899, Moscow, Russia

1. INTRODUCTION

Photosynthetic reaction centers (RC) drive the charge separation between a special dimeric chlorophyll moiety, the primary electron donor, and a series of electron acceptors, thereby creating oxidizing and reducing power at opposite sides of the protein (and of the membrane). The most abundant reaction centers are photosystem I (PSI) and photosystem II (PSII) which operate in series both in green plants and cyanobacteria. Other photosynthetic prokaryotes possess only a single RC which resembles either PSI or PSII [1]. PSI and its relatives are distinguished by iron-sulfur clusters at the acceptor side and by large heterodimeric (PSI) or homodimeric (green sulfur bacteria and Heliobacteriaceae) cores (RC1, reviewed in [2]). The low-resolution crystal structure of PSI suggests 11 transmembrane α-helices for center subunit PsaA and PsaB [3]. In PSII as well as in the reaction centers of purple bacteria (bacterio)pheophytin and quinones serve as electron acceptors (RC2, see [4] for a review). PSII is distinguished by an oxidising power which is sufficiently high for the production of oxygen from water. This reaction is catalyzed by a Mn-cluster and a redox-active tyrosine residue Y_Z. (Henceforth we denote the reaction centers of bacteria as BRC2 to distinguish them from oxygenic PSII). The crystal structure of the BRC2 of purple bacteria shows two innermost membrane subunits, L and M, each forming five transmembrane α-helices A-E (reviewed in [5]). Based on sequence similarities between PSII and BRC2, the D1 and D2 subunits in the very center of PSII were modeled according to this template [6–9].

The Phototrophic Prokaryotes, edited by Peschek *et al.*
Kluwer Academic / Plenum Publishers, New York, 1999.

Structural and functional differences between the two classes of RCs are obvious. Nevertheless it has been speculated about their cognate origin [1,3,10–16]. Based on a particular sequence alignment between the membrane-spanning segments of RC1, RC2, core chlorophyll-proteins of PSII CP43 and CP47, chlorophyll a/b-containing antenna (CAB) proteins of plants and light-harvesting complexes of purple bacteria we have suggested that all these various pigment-proteins could be traced back to one common ancestor, a large protein with more than 10 transmembrane spans [17]. We speculated that this common ancestor originally served as UV-protector of the ancient cell, and that the productive photochemistry originated from a dissipative, i.e. a UV-protecting one. In this work we (i) elucidate the alignment between RC1 and RC2 in the light of recent identification of the transmembrane segments in PsaB of PSI [18] and (ii) consider the phenomenon of the UV-protection in some more detail.

2. SEQUENCE ALIGNMENTS BETWEEN TWO CLASSES OF REACTION CENTERS

In Figures 1 and 2 six sequential α-helical segments are aligned, those established as such from the crystal structure of RC2 with those predicted for RC1 from hydropathy plots (the identified transmembrane spans formed by PsaB of PSI [18] were used as reference; for the alignment series including also the light-harvesting proteins see [17]). In addition to the transmembrane helices A–E of RC2, we included in the consideration also the CD helix of RC2 which connects the transmembrane helices C and D from the periplasmic side of the membrane. Some similarity between the CD helices of both subunits of BRC2 and the predicted VIIth hydrophobic segment of PsaA and PsaB of PSI was reported by Kuhn and co-workers [16]. We took this motif as a reference and, as shown in Figures 1 and 2, looked for similarities between the helices A–C and D–E of RC2 and the predicted hydrophobic segments IV–VI and VIII–IX of RC1, respectively.

In the alignments in Figures 1 and 2 those conserved amino acids that may serve as ligands to pigments and cofactors are underlined; the histidine residues serve as the key markers. Unlike the ligands, all other amino acid residues in the pigment-proteins serve more or less as fillers and their conservation is very poor even inside the same class of proteins. Still these blocks must meet certain structural requirements, so the conservation of some amino acid stretches just by chance might be expected. This type of conservation is illustrated by pairwise alignments of membrane segments belonging to proteins of different classes depicted in the lower part of Figures 1 and 2. We are aware, on the one hand, that the statistic significance of revealed similarities is rather low. On the other hand, we stress that the similarities were revealed by comparing given segments from one RC type not with a whole sequence database but only with 3 or 4 corresponding segments of the subtending RC type (all of which are depicted in the upper part of Figures 1 and 2). Based on these pairwise alignments we speculate on the common origin of RC1 and RC2 (see [17] for more details).

One can see in Figure 2 that although helix D of BRC2 shows similarity with helix VIII of RC1, the conserved histidine residue in helix D which binds the bacteriochlorophyll(s) of the primary donor (indicated in Figure 2 as P) founds a possible counterpart only in the helix VIII of Hb. mobilis RC but not in the PSI. This corroborates the recent data that the chlorophylls which form the primary donor in RC1 are, to all likelihood, attached to the conserved histidine residues in the helix X [19]. In [17] we speculated that the formation of RC2 took place at the interface between two truncated RC1-type subunits

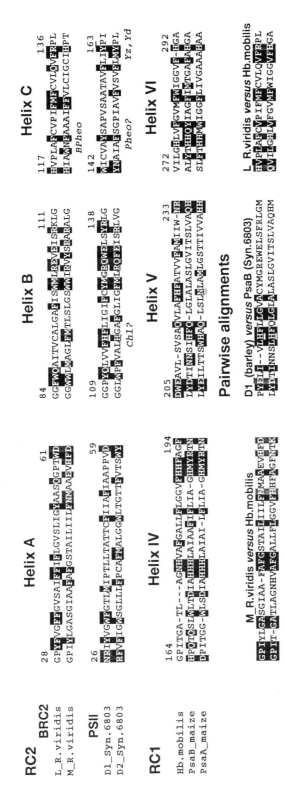

Figure 1. Alignment of α-helices A–C of RC2 with the membrane spans IV–VI of RC1. L and M subunits of *Rps. viridis* and D1 and D2 subunits of *Synechocystis* sp. PCC 6803 were compared with PsaA, PsaB subunits of maize PSI and the single subunit of homodimeric RC1 from *Heliobacterium mobilis*. See Section 3 of the text for the colour scheme and [17] for further details. Pairwise alignments: the identical residues are highlighted by black, whereas those with conserved aromaticity, electrical charge and/or capacity to serve as ligands are highlighted by gray. Other cases of conservative exchange were neglected for simplicity. Note that the alignment of helix A of RC2 with its counterparts in RC1 is rather arbitrary because of absence of any specific markers in the former.

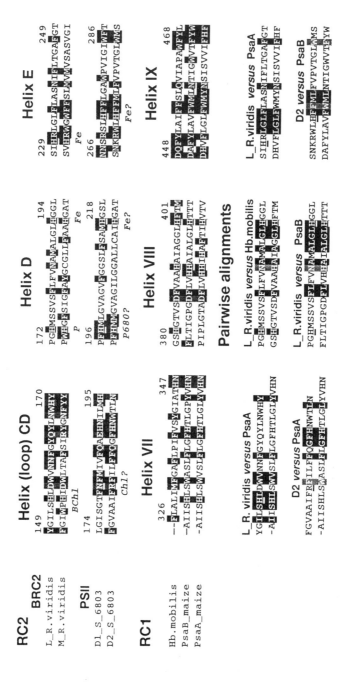

Figure 2. Alignment of α-helices CD, D and E of RC2 with the membrane spans VII–IX of RC1. See text and the caption to Figure 1 for further details.

after two helices from the C-terminus of the latter, X and XI, were lost. (By the way, the principal component analysis of the physico-chemical properties of amino acid residues revealed that the two C-terminal helices of PsaA and PsaB do not find any counterparts in the RC2 sequences [13]).

Another event that, as follows from Figure 2, has accompanied the transition from the large RC1-type reaction center to the small, RC2-type one was the shortening of the RC1 segment which included the helix VII with two adjusting connecting loops to an extent not enabling to form a whole transmembrane span. The rest of this segment was however retained, even with the bound pigment. It forms the CD helix which lies in the membrane plane and carries the accessory chlorophylls of RC2. As one can check in Figure 2 the amino acid sequences of the CD helices show almost no similarity when BRC2 and PSII are compared. The tertiary structure, however, is well conserved, as it was possible to predict correctly that in PSII the redox-active tyrosines Yz (D1-Tyr161) and Y_D (D2-Tyr 161) form hydrogen bonds with D1-His190 and D2-His190, respectively, just basing on the known structure of BRC2 [20,21]. The pairwise alignments in Figure 2 indicate that the CD helices of BRC2 and PSII show structural similarity with *different* parts of helix VII of PSI. We consider this as another reflection of the common origin of all these RC. It is plausible that the CD helix of the ancient RC2 was more similar to the helix VII of RC1 than those of the modern BRC2 and PSII.

Note that the low resolution crystal structure of PSI clearly shows all three pairs of core pigments that form the charge separation chain(s) being bound by the transmembrane α-helices. In BRC2, however, the pigments of the middle pair, the accessory bacteriochlorophylls, are not bound by transmembrane helices but by the laterally lying CD-helices. Hence the structures of the charge separation relays are analogous, but not homologous in RC1 and RC2.

3. ORIGIN OF THE PHOTOSYNTHETIC FUNCTION FROM THE UV-PROTECTION

The alignment series in the upper rows of Figures 1 and 2 show that the UV-absorbing aromatic amino acid residues (highlighted in black) are not randomly distributed along the amino acid chain but rather clustered with each other. Several other residues might participate in the UV-protection and/or serve as ligands to pigments, namely histidine, methionine, arginine, glutamine, asparagine, aspartate and glutamate. They are also found in these clusters (highlighted in gray). In [17] we suggested (i) that these clusters act as sinks for UV-quanta and (ii) that the modern pigment-proteins could have originated from less sophisticated ancient structures that protected the cell interior and DNA from the hazards of UV-light before the ozone layer was formed from the photosynthetically produced oxygen. As one can see from Fig. 1 and 2, the UV-trapping amino acid residues tend to occupy either end of α-helices or the middle. Because of this, three UV-trapping layers are formed after packing of α-helices in a membrane protein: two close to the membrane-water interface and one, less pronounced, in the middle (one can see them in the available crystal structures of BRC2 [22,23]). Such a structure operates as a UV-protector in the following way:

Capturing of the UV-light by *clustered* aromatic amino acid residues decreases the probability of photo-cleavage by decreasing the life time of the excited state due to the energy transfer between excitonically coupled and densely packed residues. This property is well established and is widely used to increase the photostability of technical polymers

[24]. In the next step, excitation is transferred to nearby pigments. It is followed by a rapid internal conversion to the lowest excited singlet state of the pigment. The lower the energy of this singlet state, the better it is for UV-protection. Chlorophyll is particularly favorable in this respect, it has two main excitation levels: one (the Soret band) in the near UV, close to the emission maxima of aromatic amino acids and the low-energy one, (the Q_y band) in the red. The fluorescence spectrum of thylakoid membranes excited at 260 nm shows only a small spectral feature at about 345 nm where the fluorescence of aromatic amino acid residues (dominated by tryptophane) could be expected but a strong fluorescence of chlorophyll at about 685 nm (personal observation). This indicates that the quantum yield of the excitation transfer from aromatic amino acid residues to pigments is close to unity.

Summarizing, *each* chlorin chromophore may be considered as a small photoreactive center (trap) operating in the UV region and surrounded by an antenna formed by interacting UV-absorbing amino acid residues. (The actual arrangement of the aromatic residues around pigments in the crystal structures of BRC2 [22,23] is in line with this view). Participation of the modern pigment-proteins in photosynthesis masks their ability to perform the UV-protection. Still this ability is not lost: comparison of absolute action spectra for pyrimidine dimer formation in alfalfa seedlings versus T7 bacteriophage shows over a 100 fold decrease in the rate of damage induced by UV-irradiation of 260–280 nm in the former case [25].

The three layered arrangement of pigments and their clustering paved the way for the *gradual* invention of the machinery which enabled the separation of electrical charges across the membrane. It is conceivable that the original advantage of a primitive charge separation was in its ability to quench the fluorescence (by chlorophyll cations) and/or to prevent the trapping of excitation energy by the long-living triplet states of chlorophyll. Only later the stability of a charge separation across the membrane may have gradually led to the "secondary" function which is to gain useful work (as discussed in detail in [17]).

After an ancient RC was invented, the necessity to recycle the reduced electron acceptors of RC arose. This task was not trivial in the reducing primordial atmosphere. In modern organisms it is performed by electron transfer chains with NAD(P)H:quinone oxidoreductase, cytochrome *bc*-complex and various cytochrome oxidases as the main participants. The membrane parts of all these enzyme complexes are formed from α-helices, the redox cofactors are coordinated predominantly by histidines. This enables speculation on their common origin with the photosynthetic pigment-proteins from primordial UV-protectors.

Obviously only the organisms that resided close to the water/air interface needed a UV-protecting apparatus. Their chance of survival in the absence of the ozone layer was through formation of microbial communities where the organisms in upper layers absorbed the UV quanta and protected those below. In line with this view, the oldest biogenic fossils, early Archean stromatolites, resemble modern microbial communities [26]. Invention of UV-protecting structures by some inhabitants of the community was highly advantageous because it sharply decreased the thickness of the layer of victimized cells. Using reasonable estimates it is possible to show that a single membrane containing UV-protecting pigment-proteins is able to absorb 1–5% of the UV-irradiation. Hence already 10–50 layers of most primitive cells were enough to protect the whole underlying population (community) from UV. Further improvement could be achieved by using membrane stacks and chlorosomes. The latter suggestion is supported by the high resistance of the chlorosome-containing *Chloroflexus aurantiacus* to UV-radiation (see [27] and references therein).

4. CONCLUSIONS

RC1 and RC2 seem to originate from one common ancestor which may have served to the primordial cell as a UV-protector before the creation of the ozone shield. The photosynthetic function may have evolved *gradually* as follows:

1. The common ancestral pigment-protein carried UV-absorbing residues and pigments in three layers clustered at either side and in the middle of the membrane. The layering could have helped to promote energy transfer towards chlorins/porphyrins which, by rapid internal conversion, partially degraded the potentially harmful UV-quanta. The relics of this mechanism may be still operative in modern plants.
2. A purely dissipative photochemistry (reversible charge separation) started still in the context of UV-protection.
3. The charge separation machinery was invented at least twice in the course of evolution, separately for RC1 and RC2.
4. The loss of certain pigments caused by mutations in their binding sites and the acquisition of redox cofactors, which filled the gaps and re-stabilized the protein, may have paved the way to proceed from dissipative to productive RC and, perhaps, to non-photosynthetic membrane electron transfer complexes.

ACKNOWLEDGMENTS

This study started in the framework of collaborative project supported by the Alexander von Humboldt Stiftung and by INTAS (INTAS-93-2852). The research was financially supported by grants from the Deutsche Forschungsgemeinschaft to A.M. (Mu-1285/1-1, Mu-1285/1-2) and to W.J. (SFB 171-A2).

REFERENCES

1. Olson, J.M. and Pierson, B.K. (1987) International Reviews of Cytology 108, 209–248.
2. Nitschke, W., Mattioli, T., and Rutherford, A.W. (1996) Origin and Evolution of Biological Energy Conservation. (Baltscheffsky, H. ed.), VCH, 177–203.
3. Fromme, P., Witt, H.T., Schubert, W.-D., Klukas, O., Saenger, W., and Krauss, N. (1996) Biochim. Biophys. Acta 1275, 76–83.
4. Rutherford, A.W. and Nitschke, W. (1996) Origin and Evolution of Biological Energy Conversion. (Baltscheffsky, H. ed.), VCH, 143–74.
5. Lancaster, C.R.D., Ermler, U., and Michel, H. (1995) Anoxygenic Photosynthetic Bacteria. (Blankenship, R.E., Madigan, M.T., Bauer, C.E. eds.), Kluver Academic Publishers, 503–26.
6. Trebst, A. (1986) Z. Naturforsch. 41c, 240–245.
7. Michel, H. and Deisenhofer, J. (1988) Biochemistry 27, no. 1, 1–7.
8. Svensson, B., Vass, I., Cedergren, E., and Styring, S. (1990) EMBO J. 9(7), 2051–2059.
9. Ruffle, S.V., Donnelly, D., Blundell, T.L., and Nugent, J.H.A. (1992) Photosynth. Res. 34, 287–300.
10. Mathis, P. (1990) Biochim. Biophys. Acta 1018, 163–167.
11. Nitschke, W. and Rutherford, A.W. (1991) Trends Biochem. Sci. 16, 241–245.
12. Blankenship, R.E. (1992) Photosynth. Res. 33, 91–111.
13. Otsuka, J., Miyachi, H., and Horimoto, K. (1992) Biochim. Biophys. Acta 1118, 194–210.
14. Vermaas, W.F.J. (1994) Photosynth. Res. 41, 285–294.
15. Meyer, T.E. (1994) BioSystems 33, 167–175.
16. Kuhn, M., Fromme, P., and Krabben, L. (1994) TIBS 19, 401–402.
17. Mulkidjanian, A.Y. and Junge, W. (1997) Photosynth. Res. 51, 27–42.

18. Sun, J., Xu, Q., Chitnis, V.P., Jin, P., and Chitnis, P.R. (1997) J. Biol. Chem. 272, 21793–21802.

19. Webber, A.N., Su, H., Bingham, S.E., Kass, H., Krabben, L., Kuhn, M., Jordan, R., Schlodder, E., and Lubitz, W. (1996) Biochemistry 35, 12857–12863.

20. Campbell, K.A., Peloquin, J.M., Diner, B.A., Tang, X.S., Chisholm, D.A., and Britt, R.D. (1997) Journal. of. the. American. Chemical. Society. 119, 4787–4788.

21. Force, D.A., Randall, D.W., and Britt, R.D. (1997) Biochemistry 36, 12062–12070.

22. Deisenhofer, J., Epp, O., Sinning, I., and Michel, H. (1995) J. Mol. Biol. 246, 429–457.

23. Allen, J.P., Feher, G., Yeates, T.O., Komiya, H., and Rees, D.C. (1987) Proc. Natl. Acad. Sci. U. S. A. 84, 6162–6166.

24. Shlyapintokh, V.Y. (1984) Photochemical Conversion and Stabilization of Polymers. Hanser Publishers, Munich.

25. Quaite, F.E., Sutherland, B.M., and Sutherland, J.C. (1992) Nature 358, 576–578.

26. Awramik, S.M. (1992) Photosynthesis Research 33, 75–89.

27. Pierson, B.K. and Castenholz, R.W. (1995) Anoxygenic Photosynthetic Bacteria. (Blankenship, R.E., Madigan, M.T., Bauer, C.E. eds.), Kluwer Academic Publishers. Printed in The Netherlands. 31–47.

AUTHOR INDEX

SUBJECT INDEX

ABC transporter, 464, 511, 514, 523, 529, 535, 536, 540–542, 556, 578, 579, 581
Acetate, 72, 253, 332, 359, 530, 681, 700, 702, 709
Acetate kinase, 69
Acetylene, 482, 676, 677, 701, 703
Acetyl phosphate, 69, 127, 128
Activation energy, 43–45, 383
Active transport, 541, 543, 544, 555, 556, 578, 698, 711
Acyl homoserine lactones, 8
Adaptation, 4, 254, 384, 583, 609, 658, 664
 to anaerobiosis, 203, 218
 to carbon sources, 564
 dark adaptation, 605
 to nitrogen sources, 609, 612, 613
 to phosphate concentration, 739, 740
 salt adaptation, 253, 256, 403, 583, 585, 711
 to sulfidic conditions, 278
 temperature adaptation, 653, 654, 705
Adda ((2S,3S,8S,9S)-3-amino-9-methoxy-2,6,8-trimethyl-10-phenyldeca-4,6-dienoic acid), 615, 617, 625, 626
Adenosine 5′-diphosphate (ADP), 13, 17, 18, 207, 220, 223, 227, 342, 346, 351, 353, 380–382, 481, 482, 488, 584, 742
Adenosine 5′-monophosphate (AMP), 13, 17, 18, 346
Adenosine 5′-triphosphate (ATP), 2, 13, 14, 16–18, 72, 149, 153, 197, 203, 205, 207, 208, 225, 241, 251, 253–255, 257, 281, 293, 337, 342, 346, 351, 353, 365, 375–385, 387, 464, 468, 479, 481, 482, 498, 503, 505–507, 510, 511, 529, 531, 533, 539, 541–543, 546, 549–552, 558, 566, 574, 578, 579, 581, 589, 636, 648, 733, 734, 742, 743, 781, 797
Adenylate energy charge (EC), 13, 14, 16, 18, 150, 155, 342, 359
Adenylate regulation, 342
ADP: see Adenosine 5′-diphosphate
ADP-glucose, 584, 741, 742
ADP-ribosylation, 479, 482
Aeruginosins, 615

Akinetes, 510, 513, 514, 523, 657, 658, 665–667
Alanine dehydrogenase, 396, 397, 468
Alanine transport, 577, 579
Algal-bacterial mats, 226, 626, 627, 699, 700, 703, 705, 707, 709–713
Allophycocyanin (AP), 57, 61, 62, 72, 74, 77, 83, 88, 89, 92, 93, 365, 449, 760
Amicyanin, 46
Aminoisobutyrate, 579
δ-Aminolevulinic acid, 84
Ammonia/ammonium
 as metabolic regulator, 16, 463ff, 479, 487, 495, 541, 543, 549, 551, 559, 609, 611–613, 666–668
 as nitrogen source, 1, 16, 52, 202, 276, 358–361, 368, 369, 463ff, 480, 485, 487, 488, 495, 503, 509, 539, 542, 543, 596, 609–613, 667, 676, 700, 703
 uptake, 463ff, 541, 669
AMP: see Adenosin 5′-monophosphate
Ampicillin, 151, 446
Amplified 16S rDNA restriction analysis (ARDRA), 776–780
Amt proteins, 464, 631
Anabaenapeptolins, 615
Anatoxin-a, 624–626, 629
Anchor protein, 62, 226, 242, 260, 283, 295, 316
Antenna pigments, 22–24, 39, 71, 77, 83, 97, 99, 126, 166, 167, 295, 751, 752, 760, 806, 810
Antimycin A, 208, 225, 230, 232, 235, 236, 242–244, 246, 252, 254–257, 607, 687
Apicomplexa, 202
AppA, 134–138
Arabinose, 748
Arabinose binding protein, 575
AraC, 505
Ascorbate, 26, 84, 87, 244, 245, 259, 287, 288, 308, 332, 400, 401, 404, 406, 721–723, 725–728
 peroxidase, 721–729
Asymmetric cell division, 9, 453
Atomic force microscopy, 638, 645

INDEX OF ORGANISMS

INDEX OF GENES